D0151518

Engineering Geology and Construction

Engineering Geology and Construction

Fred G. Bell
British Geological Survey

LONDON AND NEW YORK

First Published 2004 by Spon Press
11 New Fetter Lane, London EC4P 4EE

Simultaneously published in the USA and Canada
by Spon Press
29 West 35th Street, New York, NY 10001

Spon Press is an imprint of the Taylor & Francis Group

Typeset in Times by Integra Software Services Pvt. Ltd,
Pondicherry, India
Printed and bound in Great Britain by
TJ International Ltd, Padstow, Cornwall

British Library Cataloguing in Publication Data
A catalogue record for this book is available from the British Library

Library of Congress Cataloging in Publication Data

Bell, F.G. (Frederic Gladstone)
Engineering geology and construction / Fred G. Bell.
p. cm.
ISBN 0–415–25939–8 (alk. paper)
1. Engineering geology. I. Title.

TA705. B329 2004
624.1′51—dc22

2003015438

ISBN 0–415–25939–8

Contents

Preface

This book has been written primarily for the professional although that does not mean that those post-graduate students and senior undergraduates following courses in this subject area would not benefit from its pages. In other words, the book should be of help to those who work, or intend to work, on or in the ground, that is, engineering geologists, geotechnical engineers, foundation engineers and mining engineers, and perhaps to a lesser extent water engineers, builders, architects, and planners and developers. Geology and construction are intimately related in that construction takes place on or within soil and/or rock masses. Hence, the character of the ground conditions can have a profound influence on the location and type of building or structure that is to be erected, on infrastructure and on development or redevelopment. Not only that, but geological materials play a vital role in construction.

The author is conscious of the fact that this book is somewhat larger than most of his previous tomes and that the potential purchaser may think that it therefore has a high price tag. However, in terms of cost per page it probably is no more expensive than other books sitting, hopefully comfortably, next to it on the booksellers shelves. What is more, in relative terms it is no more expensive than some of the books the author was expected to buy when he was a student and, of course, existed in abject poverty – the natural lot of students. And a book of this sort should be regarded as an investment, as something that one can turn to as a source of encapsulated knowledge again and again in the future, the proviso being that its contents are worthwhile – and here the author, as far as this book is concerned, is not allowed to comment. Nevertheless, in this context, the author would like to think that he is following in the footsteps of Krynine and Judd, Attewell and Farmer, and Legget and Karrow, who produced texts that are worthy of gracing the shelves of all professionals within this field. Indeed, one would tentatively suggest that the potential purchaser should feel some debt of gratitude towards the publisher for allowing a book of this size to come into existence. Most others would have requested that the axe should have been wielded to the manuscript with murderous blood-dripping abandon (perhaps some future readers may come to share this view and, of course, they are entitled to their opinions, at least that is what we are led to believe in so-called democracies!). Obviously, a book of over 800 pages has a longer than normal gestation period. In fact, the manuscript was completed in October 2002, and so the reader should be able to appreciate the subsequent efforts that the publisher has had to put in for the book to rest in his or her, hopefully well-washed caring, hands.

Even so, no book, however large, can cover all the subject matter concerned in the detail or manner required by every reader, and every book has its bias reflecting the views and experience of the author. Therefore, a host of references have been provided for the dedicated professionals, drawn from all the leading journals in this subject area, to allow them to pursue their interests further. In particular, as every problem created by ground conditions is to some extent unique, then a study of case histories can prove valuable. Case histories accordingly are provided within the text but obviously their coverage is limited. Again, the copious references should help compensate for this.

F.G. Bell

Blyth, Nottinghamshire.

1 Land evaluation and site investigation

Land evaluation and site investigation are undertaken to help determine the most suitable use of land in terms of planning and development or for construction purposes. In the process the impact on the environment of a particular project may have to be assessed, this is especially the case for large projects. Obviously, there has to be a geological input into such assessments. This may be with regard to earth processes and geological hazards, mineral resources and the impact of mining, water supply and hydrogeological conditions, soil resources, ground conditions or disposal of waste. The impact of the development of land is most acute in urban areas where the human pressures on land are greatest. In addition, the redevelopment of brownfield sites can present difficult problems such as derelict heavy industrial sites with extensive foundations, severely contaminated ground and abandoned mineral workings at shallow depth (Bell, 1975; Bell *et al*., 2000).

Investigations in relation to land-use planning and development obviously can take place at various scales, from regional to site investigations. Regional investigations generally are undertaken on behalf of government authorities at, for example, state or county level, and may be involved with the location and use of mineral resources, with the identification of hazards, with problems due to the past types of land use or with land capability studies and zoning for future land use. In this context, it is necessary to recognize those geological factors that represent a resource or constitute a constraint. A constraint imposes a limitation on the use to which land can be put, so that a particular locality that is so affected is less suited to a specific activity than another. Site investigations tend to be undertaken for specific reasons, for instance, to obtain the necessary information for the location of a suitable site for a landfill, for the site of a routeway, a tunnel, or of a reservoir and dam, or for the development of a large brick pit and works.

A site investigation may form part of a feasibility study or is undertaken to assess the suitability of a site and surroundings for a proposed engineering structure. As such it involves exploring the ground conditions at and below the surface (Anon., 1999). The data that is obtained from a site investigation for a feasibility study is used to help determine whether a project is feasible. A site investigation carried out prior to the construction of an engineering project is a pre-requisite for the successful and economic design of engineering structures and earthworks. Insufficient or inadequate geotechnical data may lead to unsatisfactory design, which subsequently may result in serious damage to, or even failure of an engineering structure. Any attempt to save costs by allocating a low budget to a site investigation may be the cause of additional expenditure later if unfavourable ground conditions are encountered during construction, which were not indicated by the original site investigation. Accordingly, a site investigation also should attempt to foresee and provide against difficulties that may arise during construction due to ground and/or other local conditions. Indeed, investigation should not cease once construction begins. It is essential that the prediction of ground conditions that constitute the basic design assumption is checked as construction proceeds, and designs should be modified accordingly if conditions are revealed that differ from those predicted. An investigation of a site for an engineering structure requires not only ground exploration but also sampling of all rock or soil types that may be affected significantly by the engineering project. Data relating to the groundwater conditions, the extent of weathering and the discontinuity pattern in rock masses also are very important. In some areas there are special problems that need investigating, for example, the degree of contamination of the ground or the potential for subsidence in areas of shallow abandoned mine workings. It may be necessary in some construction projects to carry out an investigation to determine whether suitable sources of sufficient material for construction purposes are available. A site investigation also may be required to help explain the failure of an engineering structure when it is suspected that the ground conditions may have been a contributary cause. In this instance, the data obtained is used in the design of any remedial works.

The complexity of a site investigation depends upon the nature of the ground conditions and the type of engineering structure concerned. More complicated ground conditions and more sensitive large engineering structures require more rigorous investigation of the ground conditions. Although a site investigation usually consists of three stages, namely, a desk study, a preliminary reconnaissance and a site exploration, there must be a degree of flexibility in the procedure, since no two sites are the same.

Any investigation begins with the formulation of aims, what it needs to achieve and which type of information is relevant to the particular project in question. Once the pertinent questions have been posed, the nature of the investigation can be defined and the process of data collection can begin. The amount of detail required depends largely on the purpose of the investigation. For instance, less detail is required for a feasibility study for a project than is required by engineers for the design and construction of that project. Various methodologies are employed in data collection. These may include the use of remote sensing imagery, aerial photography, existing literature and maps, fieldwork and mapping, subsurface exploration by boring and drilling, sample collection, geophysical surveying and *in situ* testing. In some instances, geochemical data, notably when water or ground is polluted or contaminated, may need to be gathered, or monitoring programmes carried out. Once the relevant data has been obtained it must be interpreted and evaluated, and then, along with the conclusions, embodied in a report, which will contain maps and/or plans. Geographical information systems (GIS) may be used to help process the data.

1.1 Desk study and preliminary reconnaissance

A desk study is undertaken as the first stage of a site investigation in order to make an initial assessment of the ground conditions and to identify, if possible, any potential geotechnical problems (Herbert *et al.*, 1987). The objective of a desk study is to examine available archival records, literature, maps, imagery and photographs relevant to the area or site concerned to ascertain a general picture of the existing geological conditions prior to a field investigation, that is, to begin the process of constructing what Fookes (1997) referred to as the geological model. Sources of such information in Britain are outlined in Appendix B of Anon. (1999), Appendix 1 of Weltman and Head (1983) and in Perry and West (1996). In addition, a desk study can be undertaken in order to determine the factors that affect a proposed development for feasibility assessment and project planning purposes. The terms of reference for a desk study in both cases need to be defined clearly in advance to the commencement of the study. The effort expended in a desk study depends upon the type of project, the geotechnical complexity of the area or site and the availability of relevant information. Over and above these factors, the budget allocated for the study affects the time that can be spent on it.

Therefore, a desk study for the planning stage of a project can encompass a range of appraisals from the preliminary rapid response to the comprehensive. Nonetheless, there are a number of common factors throughout this spectrum that need to be taken into account. These are summarized in Table 1.1 from which it can be concluded that an appraisal report typically includes a factual and interpretative description of the surface and geological conditions, information on previous site usage, a preliminary assessment of the suitability of the site for the planned development, an identification of potential constraints and provisional recommendations with

Table 1.1 Summary contents of engineering geological desk study appraisals

Item	*Content and main points of relevance*
Introduction	Statement of terms of reference and objectives, with indication of any limitations. Brief description of nature of project and specific ground-orientated proposals. Statement of sources of information on which appraisal is based.
Ground conditions	Description of relevant factual information. Identification of any major features which might influence scheme layout, planning or feasibility.
Site description and topography	Descriptions of existing surface conditions from study of topographic maps and actual photographs, and also from site walkover inspection (if possible).
Engineering history	Review of information on previous surface conditions and usage (if different from present) based on study of old maps, photographs, archival records and related to any present features observed during site walkover. Identification of features such as landfill zones, mineworkings, pits and quarries, sources of contamination, old water courses, etc.
Engineering geology	Description of subsurface conditions, including any information on groundwater, from study of geological maps and memoirs, previous site investigation reports and any features or outcrops observed during site walkover. Identification of possible geological hazards, e.g. buried channels in alluvium, solution holes in chalk and limestone, swelling/shrinkable clays.
Provisional assessment of site suitability	Summary of main engineering elements of proposed scheme, as understood. Comments on suitability of site for proposed development, based on existing knowledge.
Provisional land classification	Where there is significant variation in ground conditions or assessed level of risk, subdivision of the site into zones of high and low risk, and any intermediate zones. Comparison of various risk zones with regard to the likely order of cost and scope of subsequent site investigation requirements, engineering implications, etc.
Provisional engineering comments	Statement of provisional engineering comments on such aspects as: foundation conditions and which method(s) appears most appropriate for structural foundations and ground slabs, road pavement subgrade conditions, drainage, excavatability of soils, and rocks suitability of local borrow materials for use in construction slope stability considerations, nature and extent of any remedial works, temporary problems during construction.
Recommendations for further work	Proposals for phased ground investigation, with objectives, requirements and estimated budget costs.

Source: Herbert *et al.*, 1987. Reproduced by kind permission of The Geological Society.

regard to ground engineering aspects. However, a desk study should not be regarded as an alternative to a ground exploration for a construction project. The information revealed by a desk study also can reduce the risk of encountering unexpected ground conditions that could adversely affect the financial viability of a project.

The effort expended in any desk study depends on the complexity and size of the proposed project and on the nature of the ground conditions. Detailed searches for information can be time consuming and may not be justified for small schemes at sites where the ground conditions are relatively simple or well known. In such cases a study of the relevant topographical and geological maps and memoirs, and possibly aerial photographs may suffice. On large projects literature and map surveys may save time and thereby reduce the cost of the associated site investigations. Unfortunately, however, in some parts of the world little or no literature, or maps are available.

Topographical, geological and soil maps, along with remote sensing imagery and/or aerial photographs, can provide valuable information that can be used during the planning stage of a construction project. Topographic maps offer a general view of the relief of the land surface and drainage, as well as communications and built-up areas. Such maps are available at various scales. Topographic maps are particularly valuable when planning routeways. In some instances a study of past and present topographic maps is necessary as, for example, to assess the industrial or mining development in an area.

Geological maps afford a generalized picture of the geology of an area. Usually, the formation of boundaries and positions of the structural features, especially faults, are interpolated. As a consequence, their accuracy cannot always be taken for granted. Normally, a map memoir accompanies an individual map and provides a detailed description of the geology of the area in question. Nevertheless, from the planner's and engineer's point of view one of the shortcomings of conventional geological maps is that the boundaries are stratigraphical and more than one type of rock may be included in a single mappable unit. What is more, the geological map is lacking in quantitative information that the engineer requires, for instance, relating to the physical properties of the rocks and soils, the amount and degree of weathering, the hydrogeological conditions, etc. However, information such as that relating to the distribution of superficial deposits, landslipped areas and potential sources of construction materials frequently can be obtained from geological maps. A geological map also can be used to indicate those rocks or soils that could be potential sources of groundwater. Now many national geological surveys are producing hazard maps, environmental geology maps and engineering geology maps that provide data more relevant to engineers and planners. Such maps represent an attempt to make geological information more understandable to the non-geologist. Frequently, because it is impossible to represent all environmental or engineering geological data on one map, a series of thematic maps, each of a different topic, are incorporated into a memoir on a given area. Smith and Ellison (1999) have provided an excellent review of applied geological maps for planning and development purposes. Their review describes various ways by which geological data can be represented on maps, with illustrations from 35 mapping surveys undertaken in Britain.

However, these thematic geological maps, derived from the traditionally produced two dimensional map, probably represent the maximum extent of what can be extracted from a map format that has served geologists and their clients for nearly 200 years. The rapid development of information technology in the last 20 years is beginning to revolutionize geological mapping. Many national geological surveys have established digital databases of geological information. Geographical information systems (GIS, see Section 1.5) are being used increasingly to prepare geological map information in digital form. The use of such systems has many advantages.

Individual layers of map information can be held separately. For example, all the landslides mapped in an area can be held as a separate layer in the GIS. This allows more rapid production of thematic maps. In addition, each polygon (mapped area) can have information 'attached' to it. For example, each landslide polygon could have 'attached' some of the information relating to it that might be held in the digital databases.

Further, developments in digital modelling are beginning to allow geologists to produce a whole new type of map, the 3D geological model. Such models are built up from a series of cross sections but the software allows the geologist to move the positions of subsurface geological boundaries between the cross sections in an intuitive way, replicating how geologists might develop these models using pen and paper! These new 3D models, however, require large amounts of subsurface data, which is only likely to be found for older cities in more developed countries. Inevitably, the creation of such models is time consuming, requiring the digitization of borehole logs and their input into the model. However, as borehole logs become increasingly available in digital form, this problem (and the associated cost) should decrease.

The new digital models will create one further problem. Geologists are familiar with the, often undocumented, uncertainties associated with traditional geological map linework. The new 3D maps (models) include surfaces and are presented in a way that is very convincing to the non-specialist. It is likely that geologists will have to find new ways to explain the uncertainties associated with these interpreted surfaces.

Soil is a major natural resource and soil maps are a spatial representation of these resources. Hence, soil maps should be consulted whenever greenfield sites are to be developed for construction purposes. If land is to be put to its optimum use, then good agricultural land should not be used, for example, for house building. Soil maps show the distribution of soil types according to their pedological classification.

The preliminary reconnaissance involves a walk over the site noting, where possible, the distribution of the soil and rock types present, the relief of the ground, the surface drainage and associated features, any actual or likely landslip areas, ground cover and obstructions, earlier uses of the site such as tipping or evidence of underground workings etc. The inspection should not be restricted to the site but should examine adjacent areas to see how they affect or will be affected by construction on the site in question.

The importance of the preliminary investigation is that it should assess the suitability of the site for the proposed works. If the site appears suitable, the data from the desk study and the preliminary reconnaissance will form the basis upon which the site exploration is planned. The preliminary reconnaissance also allows a check to be made on some of the conclusions reached in the desk study.

1.2 Remote sensing imagery and aerial photographs

Remote sensing imagery and aerial photographs have proved valuable aids in land evaluation, particularly in those developing regions of the world where good topographic maps do not exist (Avery and Berlin, 1992). They commonly represent one of the first stages in the process of land assessment. However, the amount of useful information obtainable from imagery and aerial photographs depends upon their characteristics, as well as the nature of the terrain they portray. Remote imagery and aerial photographs usually prove to be of most value during the planning and reconnaissance stages of a project. The information such imagery provides can be transposed to a base map and this is checked during fieldwork. This information not only allows the fieldwork programme to be planned much more effectively, but also should help to shorten the period spent in the field. The data obtained also can be used in geographical information systems.

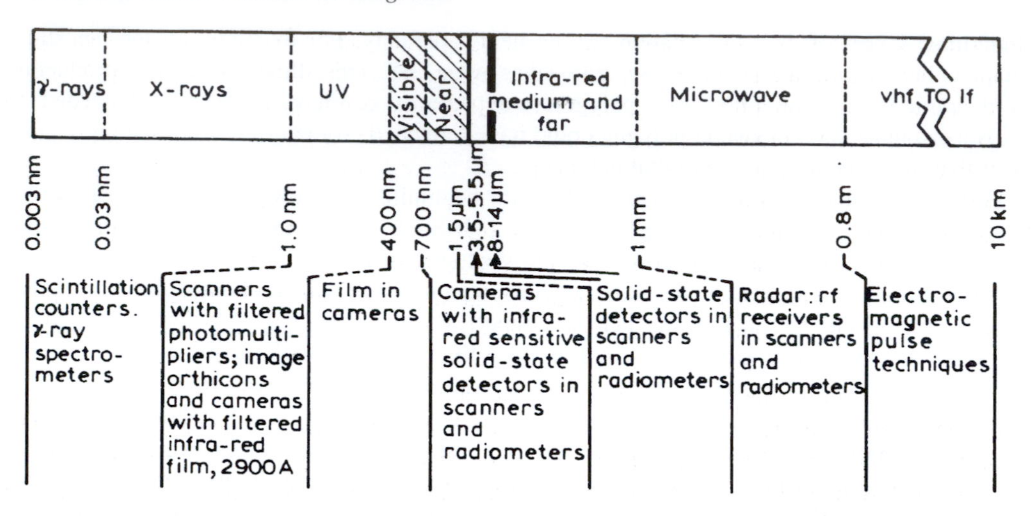

Figure 1.1 Electromagnetic spectrum showing wavelengths of thermal infrared radiation in relation to other wavelengths.

Remote sensing involves the identification and analysis of phenomena on the earth's surface by using devices borne by aircraft or spacecraft. Most techniques used in remote sensing depend upon recording energy from part of the electromagnetic spectrum, ranging from gamma rays through the visible spectrum to radar (Fig. 1.1). The scanning equipment used measures both emitted and reflected radiation, and the employment of suitable detectors and filters permits the measurement of certain spectral bands. Signals from several bands of the spectrum can be recorded simultaneously by multispectral scanners. Two of the principal systems of remote sensing are infrared linescan (IRLS) and side-looking airborne radar (SLAR).

1.2.1 Infrared methods

Infrared linescanning is dependent upon the fact that all objects emit electromagnetic radiation generated due to the thermal activity of their component atoms, that is, infrared images record the radiant heat of materials. Radiant temperature is determined by the kinetic temperature of a material and its emissivity, which is a measure of its ability to radiate and to absorb thermal energy. The emissivities of most materials occur within a very narrow range, namely, 0.81–0.96 μm. Those materials with high emissivities absorb large amounts of incident energy and radiate large quantities of kinetic energy whereas materials with low emissivities absorb and radiate lower quantities of energy. Emission is greatest in the infrared region of the electromagnetic spectrum for most materials at ambient temperature. The reflected infrared region ranges in wavelength from 0.7 to 3.0 μm and includes the photographic infrared band. This can be detected by certain infrared sensitive film. The thermal infrared region ranges in wavelength from 3.0 to 14.0 μm. The most effective waveband used for thermal infrared linescanning for geological purposes is 8–14 μm (Salisbury and D'Aria, 1992).

Infrared linescanning involves scanning a succession of parallel lines across the track of an aircraft with a scanning spot. The spot travels forwards and backwards in such a manner that nothing is missed between consecutive passes. Since only an average radiation is recorded, the limits of resolution depend on the size of the spot. The diameter of the spot usually is around 2–3 milliradians, which means that if an aircraft is flying at a height of 1000 m, then the spot measures 2–3 m across. The radiation is picked up by a detector that converts it to electrical signals that, in turn, are transformed into visible light via a cathode ray tube, thereby enabling a record to be made on a film or a magnetic tape. The data can be processed in colour, as well as black and white, colours substituting for grey tones. Unfortunately, prints are increasingly distorted with increasing distance from the line of flight, which limits the total useful angle of scan to about 60° on either side. In order to reduce the distortion along the edges of the imagery, flight lines have a 50–60 per cent overlap. According to Warwick *et al.* (1979) a temperature difference of 0.15 °C between objects of 500 mm diameter can be detected by an aircraft at an altitude of 300 m. The spatial resolution is, however, much lower than that of aerial photographs, in which the resolution at this height would be 80 mm. At higher altitudes the difference becomes more marked.

The use of infrared linescan depends on clear calm weather. What is more, some thought must be given to the fact that thermal emissions vary significantly throughout the day. The time of the flight therefore is important. From the geological point of view, pre-dawn flying proves most suitable for thermal infrared linescan. This is because radiant temperatures are fairly constant and reflected energy is not important, whereas during a sunny day radiant and reflected energy are roughly equal so that the latter may obscure the former. Also, because sun-facing slopes are warm and shade slopes cool, rough topography tends to obliterate geology in post-dawn imagery.

Although temperature differences of 0.1 °C can be recorded by infrared linescan, these do not represent differences in the absolute temperature of the ground but in emission of radiation. Careful calibration therefore is needed in order to obtain absolute values. Emitted radiation is determined by the temperature of the object and its emissivity, which can vary with surface roughness, soil type, moisture content and vegetative cover.

A grey scale can be used to interpret the imagery, it being produced by computer methods from linescan data that have been digitized. This enables maps of isoradiation contours to be produced. Colour enhancement also has been used to produce isotherm contour maps with colours depicting each contour interval. This method has been used in the preparation of maps of engineering soils.

Identification of grey tones is the most important aspect as far as the interpretation of thermal imagery is concerned, since these provide an indication of the radiant temperatures of a surface. Warm areas give rise to light, and cool areas to dark tones. Relatively cold areas are depicted as purple and relatively hot areas as red on a colour print. Thermal inertia is important in this respect since rocks with high thermal inertia, such as dolostone or quartzite, are relatively cool during the day and warm at night. Rocks and soils with low thermal inertia, for example, shale, gravel or sand, are warm during the day and cool at night. In other words, the variation in temperature of materials with high thermal inertia during the daily cycle is much less than those with low thermal inertia. Because clay soils possess relatively high thermal inertia they appear warm in pre-dawn imagery whereas sandy soils, because of their relatively low thermal inertia, appear cool (Table 1.2). The moisture content of a soil influences the image produced, that is, soils that possess high moisture content may mask differences in soil types. On the other hand, soils like peat that contain large amounts of water may be recognizable. Fault zones often are picked out because of their higher moisture content. Similarly, the presence of old landslides frequently can be discerned due to their moisture content differing from that of their surroundings.

Texture also can help interpretation. For instance, outcrops of rock may have a rough texture due to the presence of bedding or jointing, whereas soils usually give rise to a relatively smooth texture. However, where soil cover is less than 0.5 m, the rock structure usually is observable on the imagery since deeper, more moist soil occupying discontinuities gives a darker signature.

Table 1.2 Thermal properties of geological materials and water at 20 °C

Geological materials	*Thermal conductivity k.* (cal cm^{-1} s^{-1} °C^{-1})	*Density ρ* (gm cm^{-3})	*Thermal capacity c* (cal gm^{-1} °C^{-1})	*Thermal Diffusivity k* (cm^2 s^{-1})	*Thermal inertia P* (cal cm^{-2} $s^{-\frac{1}{2}}$ °C^{-1})	*1/P (often used as thermal inertia value)*
Basalt	0.0050	2.8	0.20	0.009	0.053	19
Clay soil (moist)	0.0030	1.7	0.35	0.005	0.042	24
Dolomite	0.0120	2.6	0.18	0.026	0.075	13
Gabbro	0.0060	3.0	0.17	0.012	0.055	18
Granite	0.0075	2.6	0.16	0.016	0.052	19
	0.0065					
Gravel	0.0048	2.5	0.17	0.011	0.045	22
Limestone	0.0048	2.5	0.17	0.011	0.045	22
Marble	0.0055	2.7	0.21	0.010	0.056	18
Obsidian	0.0030	2.4	0.17	0.007	0.035	29
Peridotite	0.0110	3.2	0.20	0.017	0.084	12
Pumice, loose	0.0006	1.0	0.16	0.004	0.009	111
Quartzite	0.0120	2.7	0.17	0.026	0.074	14
Rhyolite	0.0055	2.5	0.16	0.014	0.047	21
Sandy gravel	0.0060	2.1	0.20	0.014	0.050	20
Sandy soil	0.0014	1.8	0.24	0.003	0.024	42
Sandstone quartz	0.0120	2.5	0.19	0.013	0.054	19
	0.0062					
Serpentine	0.0063	2.4	0.23	0.013	0.063	16
	0.0072					
Shale	0.0042	2.3	0.17	0.008	0.034	29
	0.0030					
Slate	0.0050	2.8	0.17	0.011	0.049	20
Syenite	0.0077	2.2	0.23	0.009	0.047	21
	0.0044					
Tuff, welded	0.0028	1.8	0.20	0.008	0.032	31
Water	0.0013	1.0	1.01	0.001	0.037	27

Source: Warwick *et al.*, 1979. Reproduced by kind permission of The Geological Society.

Free-standing bodies of water usually are readily visible on thermal imagery; however, the high thermal inertia of highly saturated organic deposits may approach that of water masses, the two therefore may prove difficult to distinguish at times.

1.2.2 Radar methods

Radar differs from other remote sensing systems by recording electromagnetic energy as a function of time rather than angular distance. Time is more precisely measured than angular distance and so radar images can be obtained with higher resolution and from longer ranges than images from other remote sensing systems (Sabins, 1996). In side looking airborne radar, SLAR, short pulses of energy, in a selected part of the radar waveband, are transmitted sideways to the ground from antennae on both sides of an aircraft. The pulses of energy strike the ground along successive range lines and are reflected back at time intervals related to the height of the aircraft above the ground. The swath covered by normal SLAR imagery varies from 2 to 50 km. In SLAR the images are obtained as slant-range displays that are geometrically distorted according to the depression angle, that is, the angle between the beam of the antenna to the target on the ground and the horizon. Modern radar systems automatically transform the images into ground-range displays, that is, the images are produced as black and white photographs that are more or less planimetric. Returning pulses cannot be accepted from any point within 45° from the vertical so that there is a blank space under the aircraft along its line of flight.

The roughness of the reflecting surface, that is, ground, vegetation cover or water, influences the strength of the radar return. For instance, a rough surface scatters the incident radar energy whereas a smooth surface reflects all the incident energy at an angle of reflection equal and opposite to the angle of incidence. The dielectric constant is an electrical property that influences the interaction between matter and electromagnetic energy, particularly at radar wavelengths. Dry soils and rocks have dielectric constants between 3 and 8 at radar wavelengths whereas water has a value of 80. Hence, the moisture content of a rock or soil mass increases its dielectric constant and it can increase to above 20 for saturated sands and clays. The brightness of the image increases more or less linearly with increasing moisture content.

Topographic features interact with a radar beam to produce highlights and shadows. In low-relief terrain, it is better to obtain images with a small depression angle so that maximum highlights and shadows are produced. Conversely, in terrain with high relief, an intermediate depression angle is required since a low depression angle would give extensive shadows that may obscure much of the imagery. Natural linear features such as faults in radar images may be enhanced by highlights and shadows if the features run normal or at an acute angle to the look direction. On the other hand, those features that trend more or less parallel to the look direction yield no highlights or shadows and therefore are subdued and difficult to recognize in an image.

There are some notable differences between SLAR images and aerial photographs. For instance, although variations in vegetation produce slightly different radar responses, a SLAR image depicts the ground more or less as it would appear on aerial photographs devoid of vegetation. Displacements of relief are to the side towards the imaging aircraft and not radial about the centre as in aerial photographs. Such displacement in a radar image is referred to as foreshortening or image layover and cannot be corrected easily. It can be minimized by obtaining images at low depression angles but, as mentioned, in regions of high relief, radar shadows may be excessive. Furthermore, radar shadows fall away from the flight line and are normal to it. The shadows on SLAR images form black areas that yield no information, whereas most areas of shadow on aerial photographs are partially illuminated by diffused lighting. The subtle changes of tone and texture that occur on aerial photographs are not observable on SLAR images.

Because the wavelengths used in SLAR are not affected by cloud cover, imagery can be obtained at any time. This is particularly important in equatorial regions, which are rarely free of cloud. Consequently, this technique provides an ideal means of reconnaissance survey in such areas.

Typical scales for radar imagery available commercially are 1:50 000–1:250 000, with a resolution of between 5 and 30 m. Smaller objects than this can appear on the image if they are strong reflectors, and the original material can be enlarged. Mosaics are suitable for the identification of regional geological features and for preliminary identification of terrain units.

Airborne radar surveys normally are flown with a flight line spacing that gives 60 per cent sidelap between adjacent swaths of imagery. Lateral overlapping of images means that the overlapping parts are obtained at different depression angles and so relief features are foreshortened by different amounts. The difference in foreshortening may be used to determine the height of a feature, thereby offering a stereoscopic image. Elevation data are used to produce contour maps. Furthermore, imagery recorded by radar systems can provide appreciable detail of landforms as they are revealed due to the low angle of incident illumination.

Radar interferograms are obtained by two antennae being mounted in an aircraft or spacecraft in positions that are slightly offset. One antenna transmits the energy pulse but the return waves are picked up by both antennae. However, because of the offset of the antennae, there is a phase difference between the two return waves. The two waves are superimposed by digital processing to produce a resultant wave from which an interferogram is produced with colour spectra representing each interference colour. The number of colour cycles or colour fringes affords a measure of topographic relief.

1.2.3 Satellite imagery

Small-scale space imagery provides a means of initial reconnaissance that allows areas to be selected for further, more detailed investigation, either by aerial and/or ground survey methods. Indeed, in many parts of the world a LANDSAT image may provide the only form of base map available. The large areas of the ground surface that satellite images cover give a regional physiographic setting and permit the distinction of various landforms according to their characteristic photo-patterns. Accordingly, such imagery can provide a geomorphological framework from which a study of the component landforms is possible. The character of the landforms may afford some indication of the type of material of which they are composed and geomorphological data aid the selection of favourable sites for field investigation on larger-scale aerial surveys. Small-scale imagery may enable regional geological relationships and structures to be identified that are not noticeable on larger-scale imagery or mosaics.

The capacity to detect surface features and landforms from imagery obtained by multi-spectral scanners on satellites is facilitated by energy reflected from the ground surface being recorded within four specific wavelength bands. These are visible green (0.5–0.6 μm), visible red (0.6–0.7 μm) and two invisible infrared bands (0.7–0.8 μm and 0.9–1.0 μm). The images are reproduced on photographic paper and are available for the four spectral bands plus two false colour composites. The infrared band between 0.7 and 0.8 μm is probably the best for geological purposes. Because separate images within different wavelengths are recorded at the same time, the likelihood of recognizing different phenomena is enhanced significantly. Since the energy emitted and reflected from objects commonly varies according to wavelength, its characteristic spectral pattern or signature in an image is determined by the amount of energy transmitted to the sensor within the wavelength range in which that sensor operates. As a consequence, a unique tonal signature may be identified frequently for a feature if the energy that is being emitted and/or reflected from it is broken into specially selected wavelength bands.

In Fig. 1.2 the reflectance curves for four different rock types illustrate the higher reflectance of brown sandstone at longer (orange-red) wavelengths and the lower reflectance of the siltstone in the shorter (blue) wavelengths of the visible spectrum (Beaumont, 1979). This indicates that if reflected energy from the shorter and longer ends of the visible spectrum is recorded separately, differentiation between rock types can be achieved. The ability to distinguish between different materials increases when imagery is recorded by different sensors outside the visible spectrum, the spectral characteristics then being influenced by the atomic composition and molecular structure of the materials concerned.

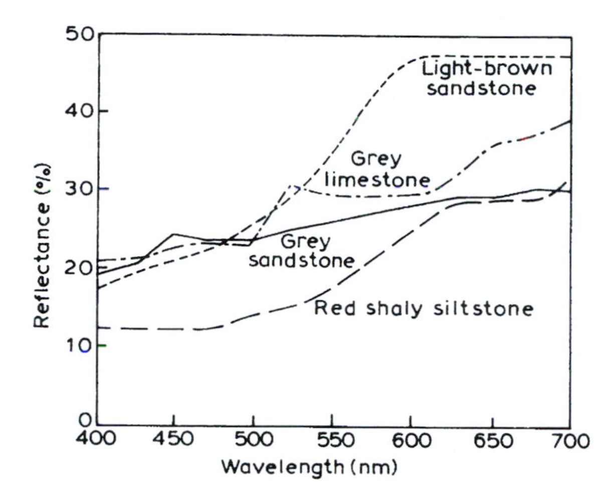

Figure 1.2 Spectral reflectance curves for four different rock types.

The minimal distortion and uniform scale of LANDSAT images mean that the compilation of mosaics is relatively easy. Mosaics that are compiled manually from individual images are referred to as analog mosaics. Those mosaics that are compiled from digitally recorded image data are termed digital mosaics.

Satellite images may be interpreted in a similar manner to aerial photographs, although the images do not come in stereopairs. Nevertheless, a pseudostereoscopic effect may be obtained by viewing two different spectral bands (band-lap stereo) of the same image or by examining images of the same view taken at different times (time-lap stereo). There is also a certain amount of side-lap, which improves with latitude. This provides a true stereographic image across a restricted strip of a print, however, significant effects are produced only by large relief features. Interpretation of satellite data may also be accomplished by automated methods using digital data directly or by using interactive computer facilities with visual display devices.

The value of space imagery is important where existing map coverage is inadequate. For example, it can be of use for the preparation of maps of terrain classification, for regional engineering soil maps, for maps used for route selection, for regional inventories of construction materials, and for inventories of drainage networks and catchment areas (Lillesand and Kiefer, 1994). A major construction project is governed by the terrain, optimum location requiring minimum disturbance of the environment. In order to assess the ground conditions it is necessary to make a detailed study of all the photo-pattern elements that comprise the landforms on the satellite imagery. Important evidence relating to soil types, or surface or subsurface conditions may be provided by erosion patterns, drainage characteristics or vegetative cover. Engineering soil maps frequently are prepared on a regional basis for both planning and location purposes in order to minimize construction costs, the soils being delineated for the landforms within the regional physiographic setting. Satellite imagery is also used for interpretation of geological structure, for geomorphological studies, for compilation of regional inventories of construction materials, for groundwater studies and for site and corridor location (Sabins, 1996).

Later generation LANDSAT satellites carry an improved imaging system called thematic mapper (TM), as well as a multispectral scanner (MSS). The TM is a cross-track scanner with an oscillating scan mirror and arrays of 16 detectors for each of the visible and reflected infrared bands. Thermal mapper images have a spatial resolution of 30 m and excellent spectral resolution. Generally, TM bands are processed as normal and infrared colour images. Data gathered by LANDSAT TM are available as computer-compatible tapes or as CD-ROMS, which can be read and processed by computers. The weakest point in the system is the lack of adequate stereovision capability, however, a stereomate of a TM image can be produced with the help of a good digital elevation model. The French SPOT satellite is equipped with two sensor systems that cover adjacent paths, each with a swath width of 60 m. Potentially higher temporal resolution is provided by the sideways viewing option since the satellite can observe a location not directly under the orbital path. SPOT senses the terrain in a single wide panchromatic band and in three narrower spectral bands corresponding to the green, red and near infrared parts of the spectrum. The spatial resolution in the panchromatic mode is 10 m and the three spectral bands have a spatial resolution of 20 m. Images can be produced for stereoscopic purposes. Better resolution can be produced by the newest satellites. Radar satellite images are available from the European ERS-1 and the Japanese JERS.

1.2.4 Aerial photographs and photogeology

Aerial photographs are generally taken from an aeroplane that is flying at an altitude of between 800 and 9000 m, the height being governed by the amount of detail that is required. Photographs may be taken at different angles ranging from vertical to low oblique (excluding horizon) to high oblique (including horizon). Vertical photographs, however, are the most relevant for photogeological purposes. Oblique photographs occasionally have been used for survey purposes but, because their scale of distortion from foreground to background is appreciable, they are not really suitable. Nevertheless, because they offer a graphic visual image of the ground they constitute a good illustrative material.

Normally, vertical aerial photographs have 60 per cent overlap on consecutive prints on the same run, and adjacent runs have a 20 per cent overlap or sidelap. As a result of tilt (the angular divergence of the aircraft from a horizontal flight path) no photograph is ever exactly vertical but the deviation is almost invariably less than 1°. Scale distortion away from the centre of the photograph represents another source of error.

Aerial photographs are being digitized and distributed on CD-ROMs that are compatible with desktop computers and image processing software. Orthophotographs are aerial photographs that have been scanned into digital format and computer processed so that radial distortion is removed. These photographs have a consistent scale and therefore may be used in the same ways as maps.

Not only does a study of aerial photographs allow the area concerned to be divided into topographical and geological units, but it also enables the geologist to plan fieldwork and to select locations for sampling. This should result in a shorter, more profitable period in the field. When a detailed interpretation of aerial photographs is required, the photographs can be enlarged up to approximately twice the scale of the final map to be produced (Rengers and Soeters, 1980).

Examination of consecutive pairs of aerial photographs with a stereoscope allows observation of a three-dimensional image of the ground surface. This is due to parallax differences brought about by photographing the same object from two different positions. The three-dimensional image means that heights can be determined and contours can be drawn, thereby producing a topographic map. However, the relief presented in this image is exaggerated, therefore slopes appear steeper than they actually are. Nonetheless, this helps the detection of minor changes in slope and elevation. Unfortunately, exaggeration proves a definite disadvantage in mountainous areas, as it becomes difficult to distinguish between steep and very steep slopes. A camera with a longer focal lens reduces the amount of exaggeration and therefore its use may prove preferable in such areas. Digital photogrammetric methods use digital images and a computer instead of a photogrammetric plotter to derive digital elevation models (DEMs), with the advantage that various aspects of the measurement process can be automated (Chandler, 2001).

Aerial photographs may be combined in order to cover larger regions. The simplest type of combination is the uncontrolled print laydown that consists of photographs, laid alongside each other, which have not been accurately fitted into a surveyed grid. Photomosaics represent a more elaborate type of print laydown, requiring more care in their production. Controlled photomosaics are based on a number of geodetically surveyed points. They can be regarded as having the same accuracy as topographic maps.

There are four main types of film used in normal aerial photography, namely, black and white, infrared monochrome, true colour and false colour. Black and white film is used for topographic survey work and for normal interpretation purposes. The other types of film are used for special purposes. For example, infrared monochrome film makes use of the fact that near-infrared radiation is strongly absorbed by water. Accordingly, it is of particular value when mapping shorelines, the depth of shallow underwater features and the presence of water on land, as for instance, in channels, at shallow depths underground or beneath vegetation. Furthermore, it is more able to penetrate haze than conventional photography. True colour photography displays variation of hue, value and chroma, rather than tone only and generally offers much more refined imagery. As a consequence, colour photographs have an advantage over black and white ones as far as photogeological interpretation is concerned, in that there are more subtle changes in colour in the former than in the grey tones in the latter, hence they record more geological information. However, colour photographs are more expensive and it is difficult to reproduce slight variations in shade consistently in processing. Another disadvantage is the attenuation of colour in the atmosphere, with the blue end of the spectrum suffering a greater loss than the red end. Even so at the altitudes at which photographs normally are taken, the colour differentiation is reduced significantly. Obviously, true colour is only of value if it is closely related to the geology of the area shown on the photograph. False colour is the term frequently used for infrared colour photography since on reversed positive film, green, red and infrared light are recorded respectively as blue, green and red. False colour provides a more sensitive means of identifying exposures of bare grey rocks than any other type of film. Lineaments, variations in water content in soil and rock masses, and changes in vegetation that may not be readily apparent on black and white photographs often are depicted clearly by false colour. The choice of the type of photographs for a project is governed by the uses they will have to serve during the project. A summary of the types of geological information that can be obtained from aerial photographs is given in Table 1.3.

Allum (1966) pointed out that when stereopairs of aerial photographs are observed the image perceived represents a combination of variations in both relief and tone. However, relief and tone on aerial photographs are not absolute quantities for particular rock types. For instance, relief represents the relative resistance of rocks to erosion, as well as the amount of erosion that has occurred. Tone is important since small variations may be indicative of different types of rock. Unfortunately, tone is affected by light conditions, which vary with weather, time of day, season and processing. Nevertheless, basic intrusions normally produce darker tones than acid intrusions. Quartzite, quartz schist, limestone, chalk and sandstone tend to give light tones, whilst slates, micaceous schists, mudstones and shales give medium tones, and basalts, dolerites and amphibolites give dark tones.

The factors that affect the photographic appearance of a rock mass include climate, vegetative cover, absolute rate of erosion, relative rate of erosion of a particular rock mass compared with that of the country rock, colour and reflectivity, composition, texture, structure, depth of

Table 1.3 Types of photogeological investigation

Structural geology	Mapping and analysis of folding. Mapping of regional fault systems and recording any evidence of recent fault movements. Determination of the number and geometry of joint systems
Rock types	Recognition of the main lithological types (crystalline and sedimentary rocks, unconsolidated deposits)
Soil surveys	Determining main soil-type boundaries, relative permeabilities and cohesiveness, periglacial studies
Topography	Determination of relief and landforms. Assessment of stability of slopes, detection of old landslides
Stability	Slope instability (especially useful in detecting old failures which are difficult to appreciate on the ground) and rock fall areas, quick clays, loess, peat, mobile sand, soft ground, features associated with old mine workings
Drainage	Outlining of catchment areas, steam divides, surface run-off characteristics, areas of subsurface drainage such as karstic areas, especially of cavernous limestone as illustrated by surface solution features; areas liable to flooding. Tracing swampy ground, perennial or intermittent streams and dry valleys. Levees and meander migration. Flood control studies. Forecasting effect of proposed obstructions. Run-off characteristics. Shoals, shallow water, stream gradients and widths
Erosion	Areas of wind, sheet and gully erosion, excessive deforestation, stripping for opencast work, coastal erosion
Groundwater	Outcrops and structure of aquifers. Water bearing sands and gravels. Seepages and springs, possible productive fracture zones. Sources of pollution. Possible recharge sites
Reservoirs and dam sites	Geology of reservoir site, including surface permeability classification. Likely seepage problems. Limit of flooding and rough relative values of land to be submerged. Bedrock gulleys, faults and local fracture pattern. Abutment characteristics. Possible diversion routes. Ground needing clearing. Suitable areas for irrigation
Materials	Location of sand and gravel, clay, rip-rap, borrow and quarry sites with access routes
Routes	Avoidance of major obstacles and expensive land. Best graded alternatives and ground conditions. Sites for bridges. Pipe and power line reconnaissance. Best routes through urban areas
Old mine workings	Detection of shafts and shallow abandoned workings, subsidence features

weathering, physical characteristics and factors inherent in the type of photography, and the conditions under which the photograph was obtained. Many of these factors are interrelated. Regional geological structures frequently are easier to recognize on aerial photographs, which provide a broad synoptic view, than they are in the field.

Lineaments are any alignment of features on an aerial photograph (Norman, 1968). The various types recognized include topographic, drainage, vegetative and colour alignments. Bedding is portrayed by lineaments that usually are few in number and occur in parallel groups. If a certain bed is more resistant than those flanking it, then it forms a clear topographic lineament. Even if bedding lineaments are interrupted by streams, they usually are persistent and can be traced across the disruptive feature. Foliation may be indicated by lineaments. It often can be distinguished from bedding since parallel lineaments that represent foliation tend to be both numerous and impersistent.

Care must be exercised in the interpretation of the dip of strata from stereopairs of aerial photographs. For example, dips of 50 or 60° may appear almost vertical, and dips between 15 and 20° may look more like 45° because of vertical exaggeration. However, with practice, dips can be estimated reliably in the ranges, less than 10°, 10–25°, 25–45°, and over 45°. Furthermore, displacement of relief makes all vertical structures appear to dip towards the central or principal point of a photograph. Because relief displacement is much less in the central areas of photographs than at their edges, it is obviously wiser to use the central areas when estimating dips. It also must be borne in mind that the topographic slope need bear no relation to the dip of the strata composing the slope. However, scarp slopes do reflect the dip of rocks. Also, as dipping rocks cross interfluves and river valleys, they produce crescent and V-shaped traces respectively. The pointed end of the V always indicates the direction of dip, and the sharper the angle of the V, the shallower the dip. If there are no dip slopes, it may be possible to estimate the dip from bedding traces. Vertical beds are independent of relief.

The axial trace of a fold can be plotted, and the direction and amount of its plunge can be assessed when the direction and amount of dip of the strata concerned can be estimated from aerial photographs. Steeply plunging folds have well-rounded noses and the bedding can be traced in a continuous curve. On the other hand, gently plunging folds occur as two bedding lineaments meeting at an acute angle (the nose) to form a single lineament. Also, the presence of repeated folding may sometimes be recognized by plotting bedding plane traces on aerial photographs.

Straight lineaments that appear as slight negative features on aerial photographs usually represent faults or master joints. In order to identify the presence of a fault there should be some evidence of movement. Usually, this evidence consists of the termination or displacement of other structures. In areas of thick soil or vegetation cover, faults may be less obvious. Faults running parallel to the strike of strata also may be difficult to recognize. Joints, of course, show no evidence of displacement. Jointing patterns may assist the recognition of certain rock types, as for example, in limestone or granite terrains.

Dykes and veins also give rise to straight lineaments, which are at times indistinguishable from those produced by faults or joints. If, however, dykes or veins are wide enough they may give a relief or tonal contrast with the country rock. They then are distinctive. Acid dykes and quartz veins often are responsible for light-coloured lineaments and basic dykes for dark lineaments. Even so, because relative tone depends very much on the nature of the country rock, positive identification cannot be made from aerial photographs alone.

If the area portrayed by the aerial photographs is subject to active erosion, then it frequently is possible to differentiate between different rock masses, although it generally is not possible to identify the rock types. Normally, only general rather than specific rock types are recognizable from aerial photographs, for example, superficial deposits, sedimentary rocks, metamorphic rocks, intrusive rocks and extrusive rocks. However, certain rock types with particular characteristics may be identifiable such as limestone terrains with karstic features. Superficial deposits can be grouped into transported and residual categories. Transported superficial deposits can be recognized by their blanketing effect on the geology beneath, by their association with their mode of transport and with diagnostic landforms such as meander belts, river terraces, drumlins, eskers, sand dunes, etc. and their relatively sharp boundaries. Residual deposits generally do not blanket the underlying geology completely and in places there are gradational boundaries with rock outcrops. Obviously, no mode of transport can be recognized.

It usually is possible to distinguish between metamorphosed and unmetamorphosed sediments as metamorphism tends to make rocks more similar as far as resistance to erosion is concerned. Metamorphism also should be suspected when rocks are tightly folded and associated with multiple intrusions. By contrast, rock masses that are horizontally bedded or gently folded, and are unaffected by igneous intrusions are unlikely to be metamorphic. As noted above, acid igneous rock masses give rise to light tones on aerial photographs and they may display evidence of jointing. The recognition of volcanic cones indicates the presence of extrusive rocks. Extensive areas of lava flows may be identifiable from some of the surface features.

Other features that may be recognized on aerial photographs include landforms and drainage features such as catchment areas, watersheds and flood planes. Aerial photographs prove particularly useful in the detection of old landslides and may indicate where slopes are potentially unstable (Soeters and Van Westen, 1996). Areas of coastal erosion, including landslip areas, and deposition can be determined from aerial photographs, as can solution features in carbonate or evaporitic rocks, and potential areas of settlement due to the consolidation of peat deposits (Norman and Watson, 1975). Soil surveys have been produced from aerial photographs including engineering soil surveys (Garner and Heptinstall, 1974). The sequential study of aerial photographs proves of value when the land surface is subject to rapid change such as where sand dunes are migrating, land is subject to periodic flooding or coastal areas are undergoing continuing erosion or deposition. Old mine workings may be detected on aerial photographs (Anon., 1976a). Again, a sequential study of past aerial photographs can prove useful as certain indicative features may not be present on the latest photographs. In addition, false colour infrared photography has been used to detect hot spots that might relate to concealed mineshafts and for the identification of stressed vegetation, which might indicate problematical ground conditions. Similarly, the sequential study of aerial photographs is important in detecting the presence of old quarries or pits that have been filled with unconsolidated material, or in derelict urban areas where extensive buried foundations, basements or cellars are likely to be present or abandoned chemical works are likely to have contaminated the ground. Aerial photographs also can be used to help locate sites for dams and reservoirs, for waste disposal, for power stations and for bridges, and to help locate routeways and construction materials.

1.2.5 Some recent developments

Airborne and satellite imagery has gradually improved in its resolution over time so that its use has extended from regional geological mapping to larger scale geomorphological mapping and geohazard identification. However, new techniques now are becoming available. High resolution airborne geophysical surveys involving magnetic, gamma spectrometry and very low frequency electromagnetic sensors are improving the ability, for example, to locate unmapped buried pipelines and to identify landfill with high ferrous contents. In addition, these techniques help to identify contaminated sites, and naturally high concentrations of radio-active elements such as uranium, thorium and radon. They also help to map high conductivities that might be related to abandoned mine sites or groundwater that has been affected by salts leached from various types of waste deposit such as spoil heaps or landfills. Such surveys offer relative ease of access to 'difficult' sites, are non-invasive and provide rapid comprehensive data coverage, which permits focused confirmatory ground follow-up. This is especially advantageous when investigating hazardous sites.

Radar and laser sensors on airborne platforms are being used to produce high resolution (centimetre to metre) digital terrain models. These are finding particular application in flood plain studies. The Light Detecting And Ranging (LIDAR) system sends a laser pulse from an airborne platform to the ground and measures the speed and intensity of the returning signal.

From this changes in ground elevation can be mapped. Radar systems use radar rather than lasers to achieve the same end. Synthetic Aperture Radar (SAR) interferometry provides a means of mapping from a satellite continuous displacements, over large areas (100 × 100 km), with a spatial resolution of 10 m and an accuracy of a few centimetres. The technique has been used to produce large-scale displacement maps of co- and post-seismic movements, of post-eruptive deformation of volcanic eruptions, and of displacements associated with landslides (Zebker *et al.*, 1994; Fruneau and Achache, 1996). The displacement maps are produced by differentiating radar images taken by satellite during two successive passes over the same area.

A similar satellite technique known as Permanent Scatterer Interferometry (PSInSAR) uses radar data collected by satellites 800 km out in space. The PSInSAR method exploits a dense network of 'natural' reflectors that can be any hard surface such as a rock outcrop, a building wall or roof, or a road kerb. These reflectors are visible to the radar sensor over many years, typically in urban regions but also in mixed urban/rural areas. They are known as permanent scatterers. These features are derived from the analysis of a stack of 30 or more different radar scenes relating to repeated satellite passes, spanning from 5 up to 10 years in time, over a specific region of interest. In urban areas, the density of permanent scatterers detected is of the order of 100 per square km. Using this dense natural network of points common to all 30-plus images, precise correction filters for the atmospheric conditions at the time of each acquisition are calculated during the processing, as well as the exact elevation values at every permanent scatterer position. Permanent Scatterer Interferometry produces maps showing rates of displacement, accurate to a few millimetres per year, over extensive time periods, currently up to a decade long. The process provides the millimetric displacement histories for each reflector point across the entire time period analysed, as calculated at every individual radar scene acquisition. Small incremental ground movements therefore can be detected that might be caused, for example, by mining subsidence, groundwater withdrawal, slow foundation settlement or subsidence due to tunnelling.

1.3 Terrain evaluation

Terrain evaluation only is concerned with the uppermost part of the land surface of the earth, that is, with that which lies at a depth of less than 6 m, excluding permanent masses of water. Mitchell (1991) described terrain evaluation as involving analysis (the simplification of the complex phenomena that makes up the natural environment), classification (the organization of data in order to distinguish and characterize individual areas) and appraisal (the manipulation, interpretation and assessment of data for practical ends) of an area of the earth's surface. There are two different approaches to this in terrain evaluation, namely, parametric evaluation and landscape classification. Parametric land evaluation refers to the classification of land on a basis of selected attribute values appropriate to the particular study, such as class of slope or the extent of a certain kind of rock. The simplest form of parametric map is one that divides a single factor into classes. Landscape classification is based on the principal geomorphological features of the terrain.

In terrain evaluation the initial interpretation of landscape can be made from large-scale maps and aerial photographs (Webster and Beckett, 1970). Observation of relief should give particular attention to direction (aspect) and angle of maximum gradient, maximum relief amplitude and the proportion of the total area occupied by bare rock or slopes. In addition, an attempt should be made to interpret the basic geology and the evolution of the landscape. An assessment of the risk of erosion (especially the location of slopes that appear potentially unstable) and the risk of excess deposition of water-borne or wind-blown debris also should be made.

Terrain evaluation provides a method whereby the efficiency and accuracy of preliminary surveys can be improved. In other words, it allows a subsequent investigation to be directed towards the relevant problems. It also offers a rational means of correlating known and unknown areas, that is, of applying information and experience gained on one project to a subsequent project. This is based on the fact that landscape systems of terrain evaluation have indicated that landscapes in different parts of the world are sufficiently alike to make predictions from the known to the unknown. The most appropriate use of terrain evaluation is for feasibility studies, and in civil engineering, this is especially related to routeway selection.

The following units of classification of land have been recognized for purposes of terrain evaluation, in order of decreasing size, namely, land zone, land division, land province, land region, land system, land facet and land element (Brink *et al.*, 1966; Anon., 1982). The land system, land facet and land element are the principal units used in terrain evaluation (Anon., 1978; Lawrance, 1978; Lawrance *et al.*, 1993).

A land systems map shows the subdivision of a region into areas with common physical attributes that differ from those of adjacent areas. Land systems usually are recognized from aerial photographs, the boundaries between different land systems being drawn where there are distinctive differences between landform assemblages. For example, the character of land units can be largely determined from good stereopairs of photographs with an optimum scale of about 1:20 000, depending on the complexity of the terrain. Field work is necessary to confirm the landforms and to identify soils and bedrock.

In order to establish the pattern identified on the aerial photographs as a land system, it is necessary to define the geology and range of small topographic units referred to as land facets. A land system extends to the limits of a geological formation over which it is developed or until the prevailing land-forming process gives way and another land system is developed. Land systems maps usually are prepared at scales of 1:250 000 or 1:1 000 000. However, Waller and Phipps (1996) showed that mapping at a scale of 1:50 000 was most suitable for construction projects when integrating satellite imagery and geological mapping. More detailed maps may be required in complex terrain. These provide background information that can be used in a preliminary assessment of the ground conditions in an area and permit locations to be identified where detailed investigations may prove necessary.

A land system comprises a number of land facets. Each land facet possesses a simple form, generally being developed on a single rock type or superficial deposit. The soils, if not the same throughout the facet, at least vary in a consistent manner. An alluvial fan, a levee, a group of sand dunes or a cliff are examples of a land facet. Indeed, geomorphology frequently provides the basis for the identification of land facets. Land facets occur in a given pattern within a land system. They may be mapped from aerial photographs at scales between 1:10 000 and 1:60 000 although Waller and Phipps (1996) showed their applicability at scales of 1:2500–1:5000.

A land facet, in turn, may be composed of a small number of land elements, some of which deviate somewhat in a particular property, such as soils, from the general character. They represent the smallest unit of landscape that normally is significant. For example, a hill slope may consist of two land elements, an upper steep slope and a gentle lower slope. Other examples of land elements include small river terraces, gully slopes and small outcrops of rock.

Although nearly all terrain evaluation mapping is carried out at the land system level, the land region may be used in a large feasibility study for some project. A land region consists of land systems that possess the same basic geological composition and have an overall similarity of landforms. Land regions are usually mapped at a scale between 1:1 000 000 and 1:5 000 000.

Most land systems maps are accompanied by a report that gives the basic information used to establish the classification of landforms within the area surveyed. The occurrence of land facets

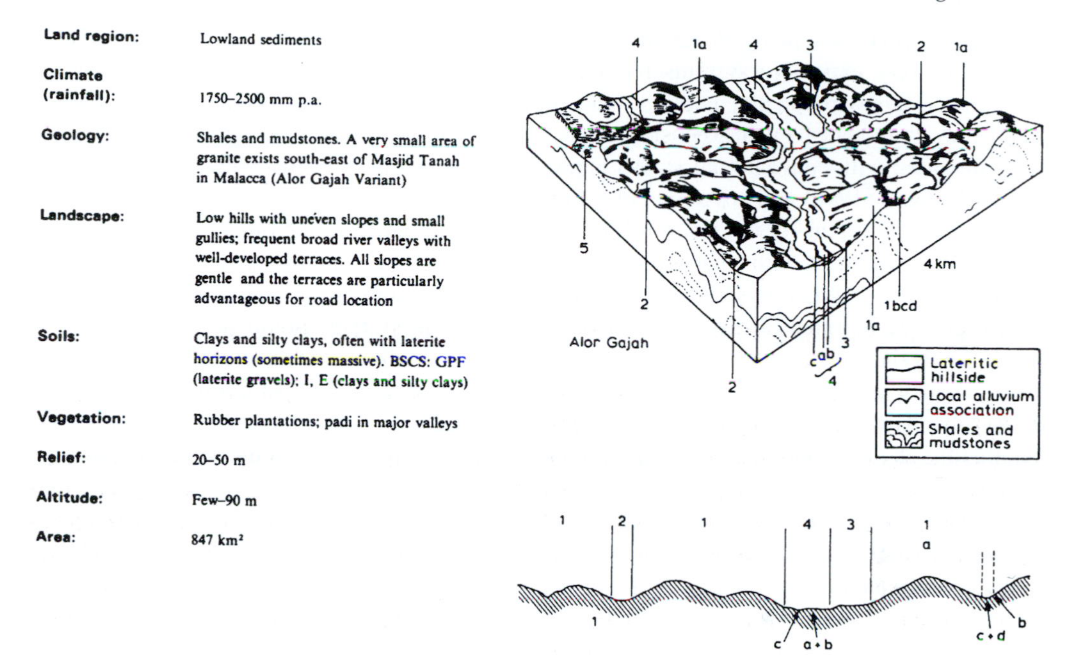

Figure 1.3 Alor Gajah land system.
Source: Lawrance, 1978. Crown copyright 1978. Reproduced by kind permission of the controller of HM Stationery Office.

normally is shown on a block diagram (Fig. 1.3), cross section or a map; maps are more often used in areas where the relative relief is very small such as alluvial plains. The descriptions of land facets include data on slope and soil profile, with vegetation and water regime referred to where appropriate.

1.4 Geographical information systems (GIS)

One means by which the power, potential and flexibility of mapping may be increased is by developing a GIS. Geographical information systems represent a form of technology that is capable of capturing, storing, retrieving, editing, analysing, comparing and displaying spatial environmental information. For instance, Star and Estes (1990) indicated that a GIS consists of four fundamental components, namely, data acquisition and verification, data storage and manipulation, data transformation and analysis, and data output and presentation. The GIS software is designed to manipulate spatial data in order to produce maps, tabular reports or data files for interfacing with numerical models. An important feature of a GIS is the ability to generate new information by the integration of existing diverse data sets sharing a compatible referencing system (Goodchild, 1993). Data can be obtained from remote sensing imagery, aerial photographs, aero-magnetometry, gravimetry and various types of maps. This data is recorded in a systematic manner in a computer database. Each type of data input refers to the characteristics of recognizable point, or linear or areal geographical features. Details of the features usually are stored in either vector (points, lines and polygons) or raster (grid cell) formats. The manipulation and analysis of data allows it to be combined in various ways to evaluate what will happen in certain situations.

Currently, there are many different GISs available, ranging from public domain software for PCs to very expensive systems for mainframe computers. Since most data sets required in engineering/environmental geology data processing are still relatively small they can be readily accommodated by inexpensive PC-based GIS applications. The advantages of using GIS compared with conventional spatial analysis techniques have been reviewed by Burrough and McDonnell (1998) and are summarized in Table 1.4.

An ideal GIS for many engineering/environmental geological situations combines conventional GIS procedures with image processing capabilities and a relational database. Because frequent map overlaying, modelling, and integration with scanned aerial photographs and satellite images are required, a raster system is preferred. The system should be able to perform spatial analysis on multiple-input maps and connected attribute data tables. Necessary GIS functions include map overlay, reclassification, and a variety of other spatial functions incorporating logical, arithmetic, conditional and neighbourhood operations. In many cases modelling requires the iterative application of similar analyses using different parameters. Consequently, the GIS should allow for the use of batch files and macros to assist in performing these iterations. Hellawell *et al.* (2001) outlined a number of case histories where GIS methodology has been advantageous in geotechnical engineering including surveys of contaminated land and construction-planning projects.

Mejía-Navarro and Garcia (1996) described a decision-support system for planning purposes that evaluates a number of variables by use of GIS. This integrated computer-support system was designed to assist urban planning by organizing, analysing and evaluating existing or needed spatial data for land-use planning. The system incorporates GIS software that allows comprehensive modelling capabilities for geological hazards, vulnerability and risk assessment by using data on topography, aspect, solid and superficial geology, structural geology, geomorphology, soil, land cover and use, hydrology and floods, and historical data on hazards. As a consequence, it has been able to delineate areas of high risk from those where future urban development could take place safely and is capable of producing hazard-susceptibility maps. Dai *et al.* (2001) provide a further example of the use of GIS in geoenvironmental evaluation

Table 1.4 Advantages and disadvantages of GIS

Advantages	*Disadvantages*
A much larger variety, analysis, techniques are available. Because of the speed of calculation, complex techniques requiring a large number of map overlays and table calculations become feasible	A large amount of time is needed for data entry. Digitizing is especially time consuming
It is possible to improve models by evaluating their results and adjusting the input variables. Users can achieve the optimum results by a process of trial and error, running the models several times, whereas it is difficult to use these models even once in the conventional manner. Therefore, more accurate results can be expected	There is a danger in placing too much emphasis on data analysis as such at the expense of data collection and manipulation based on professional experience. A large number of different techniques of analysis are theoretically possible, but often the necessary data are missing. In other words, the tools are available but cannot be used because of the lack of uncertainty of input data
In the course of a hazard assessment project, the input maps derived from field observations can be updated rapidly when new data are collected. Also, after completion of the project, the data can be used by others in an effective manner	

for land-use planning, their study being undertaken in the urban area in and around Lanzhou city in northwest China.

1.5 Mapping

One of the important ways in which the geologist can be of service is by producing maps to aid those who are concerned with the development of land. As mentioned above, a variety of maps can be produced, some of the most useful being engineering geomorphological maps, environmental geological maps and engineering geological maps. As Varnes (1974) pointed out, maps represent a means of storing and transmitting information, in particular, of conveying specific information about the spatial distribution of given factors or conditions.

1.5.1 Morphological mapping

The classical method of landform mapping is through surveyed contours. Waters (1958), however, devised a technique, which was further refined by Savigear (1965), that defined the geometry of the ground surface in greater detail than normally is found on contour maps. They proposed that the ground surface consisted of planes that intersected in convex and concave, angular or curved 'discontinuities'. An angular discontinuity was defined as a break of slope and a curved discontinuity as a change of slope. A morphological map therefore is divided into slope units that are delineated by breaks of slope, thereby defining the pattern of the ground (distinction between breaks and changes of slope provides a more precise appreciation of landform than is possible from reading contours).

When available, aerial photographs should be used for preliminary morphological mapping since they furnish an idea of the terrain and may be used to locate boundaries between the morphological units. These are recorded on the photographs and then transferred, by plotter, to the base map. The best scale for the field map depends on the objectives of the survey. Whatever the scale, some units will be recognized as having boundaries that are too close together to be represented separately. If small features, which are regarded as important, cannot be incorporated on the base map, then they should be mapped on a larger scale. For example, Savigear (1965) maintained that certain features, such as cliff units, should always be represented in morphological mapping. Most standard geomorphological features can be represented on a base map that has a scale of 1:10 000. However, not only does clear representation of all morphological information on one map provide difficult cartographic problems, but it also gives rise to difficulties in interpretation and use, thus limiting the value of the map. This problem can, to some extent, be overcome by using overlays to show some special aspect of the land surface.

Convex and concave boundaries are distinguished in morphological mapping, and measurements can be made of slope steepness and, if present, slope curvature. Knowledge of slope angles is needed for the study of present day processes and to understand the development of relief. Steepness can be shown by an arrow lying normal to the slope, pointing downhill, with the angle of the slope being marked in degrees on the arrow (Fig. 1.4a). Special symbols are used for very steep slopes such as cliffs. Differences in slope steepness can be emphasized by shading or colours according to defined slope classes (Fig. 1.4b). Slope-category maps depict average inclination over an area and make it easier to perceive the distribution of steep slopes, planation surfaces and valley asymmetry, than is possible from contour maps. Slope steepness is of considerable importance in land management, for example, it frequently poses a restricting factor on route selection and urban development (Table 1.5). Morphological mapping may prove

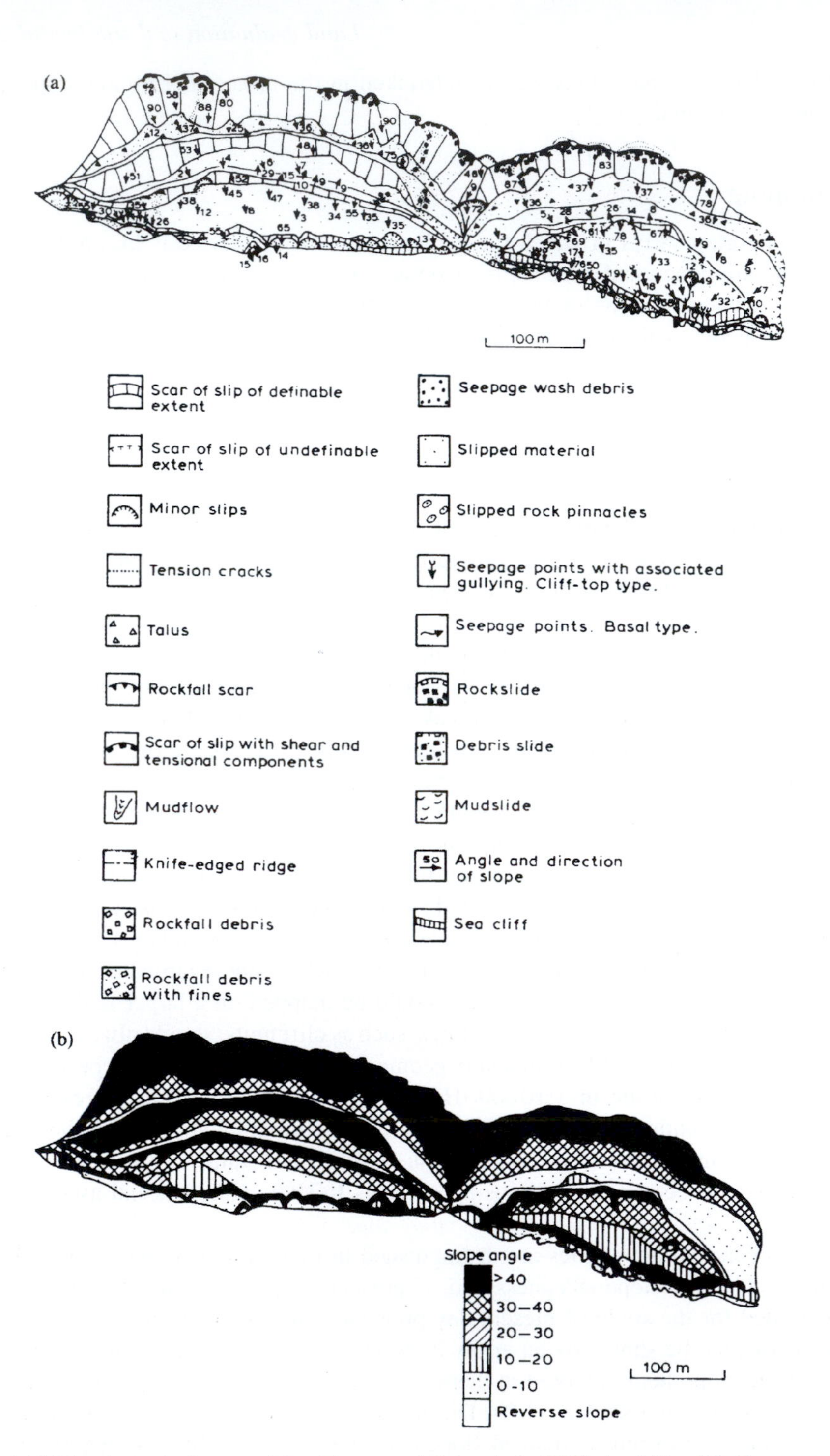

Figure 1.4 (a) Morphological map of the Haven and Culverhole Cliffs landship. (b) Slope categories of the Haven and Culverhole Cliffs.
Source: After Pitts, 1979. Reproduced by kind permission of The Geological Society.

Table 1.5 Critical slope steepness for certain activities

Steepness (%)	*Critical for*
1	International airport runways
2	Main-line passenger and freight rail transport
	Maximum for loaded commercial vehicles without speed reduction
	Local aerodrome runways
	Free ploughing and cultivation
	Below 2% – flooding and drainage problems in site development
4	Major roads
5	Agricultural machinery for weeding, seeding
	Soil erosion begins to become a problem
	Land development (constructional) difficult above 5%
8	Housing, roads
	Excessive slope for general development
	Intensive camp and picnic areas
9	Absolute maximum for railways
10	Heavy agricultural machinery
	Large-scale industrial site development
15	Site development
	Standard wheeled tractor
20	Two-way ploughing
	Combine harvesting
	Housing-site development
25	Crop rotations
	Loading trailers
	Recreational paths and trails

useful as a quick reconnaissance exercise prior to a site investigation or as a more extensive undertaking where difficult or inaccessible terrain is concerned and therefore restricts the use of some site investigation techniques.

1.5.2 Engineering geomorphological mapping

Mapping the surface form is the first step in geomorphological mapping. The next is to make interpretations regarding the forms and to ascribe an origin to them. This must be done in relation to the geological materials that compose each feature and in relation to the past and present processes operating in the area concerned. As such, geomorphological maps provide a comprehensive, integrated statement of landform and drainage. Consequently, they contain much information of potential value as far as land-use planning and construction projects are concerned (Doornkamp *et al.*, 1979).

Engineering geomorphological maps are of value for planning and engineering purposes since surface form and aerial pattern of geomorphological processes often influence the choice of a site. Such maps provide a rapid appreciation of the nature of the ground and thereby help the design of more detailed investigations, as well as focusing attention on problem areas (Demek, 1972). Engineering geomorphological mapping involves the recognition of landforms along with their delimitation in terms of size and shape. The initial phase of an engineering geomorphological investigation is carried out prior to the fieldwork and involves familiarization with the project and the landscape. The amount of information that can be obtained from a literature survey varies with location. In some developing countries little or nothing may be available, even worthwhile topographical maps, which are normally a pre-requisite of a geomorphological mapping programme, may not exist. Base maps then can be made from aerial photographs that can be specially commissioned. A study of aerial photographs enables many of the significant landforms and their boundaries to be defined prior to the commencement of fieldwork. The scale of the photographs is usually 1:10000. Field mapping permits the correct identification of landforms, recognized on aerial photographs, as well as geomorphological processes, and indicates how they will affect the project. Mapping of a site can provide data on the nature of the surface materials. The recognition of both the interrelationships between landforms on site and their individual or combined relationships to landforms beyond the site is fundamental. This is necessary in order to appreciate not only how the site conditions will affect the project but, just as importantly, how the project will affect the site and the surrounding environment.

The scale of an engineering geomorphological map is influenced by the project requirement and the map should focus attention on the information relevant to the particular project. Maps produced for extended areas, such as needed for route selection, are drawn on a small scale. Small-scale maps also have been used for planning purposes, land-use evaluation, land reclamation, flood plain management, coastal conservation, etc. These general engineering geomorphological maps concentrate on portraying the form, origin, age and distribution of landforms, along with their formative processes, rock type and surface materials. In addition, if information is available, details of the actual frequency and magnitude of the processes can be shown by symbols, annotation, accompanying notes or successive maps of temporal change. On the other hand, large-scale maps and plans of local surveys provide an accurate portrayal of surface form, drainage characteristics and properties of surface materials, as well as an evaluation of currently active processes. If the maximum advantage is to be obtained from a geomorphological survey, then derivative maps should be compiled from the geomorphological sheets. Such derivative maps generally are concerned with some aspect of ground conditions, for example, landslip areas, areas prone to flooding or over which sand dunes migrate.

A field survey provides additional data that enable the preliminary views on causative processes to be revised, if necessary. Further precision can be afforded geomorphological interpretations by obtaining details from climatic, hydrological or other records and by analysis of the stability of landforms. What is more, an understanding of the past and present development of an area is likely to aid prediction of its behaviour during and after any construction operations. Engineering geomorphological maps therefore should show how surface expression will influence a project and should provide an indication of the general environmental relationship of the area concerned. The information shown on engineering geomorphological maps should help the planning of the subsequent investigation. For instance, it should aid the location of boreholes, and these hopefully will confirm what has been discovered by the geomorphological survey. Engineering geomorphological mapping therefore may help to reduce the cost of an investigation.

1.5.3 Engineering and environmental geological mapping

Topics which are included on engineering and environmental geology maps vary but may include solid geology, unconsolidated deposits, geotechnical properties of soils and rocks, depth to rockhead, hydrogeology, mineral resources, shallow undermining and opencast workings, landslides, floodplain hazards, etc. Each aspect of geology can be presented as a separate theme on a basic or element map. Derivative maps display two or more elements combined to show, for example, foundation conditions, ease of excavation, aggregate potential, landslide susceptibility, subsidence potential, groundwater resources or capability for solid waste disposal. Environmental potential maps present, in general terms, the constraints on development such as areas where land is susceptible to landslip or subsidence, or where land is likely to be subjected to flooding. They also can present those resources with respect to mineral, groundwater, or agricultural potential that might be used in development or that should not be sterilized by building over, or contamination from landfill sites.

An engineering geology map is produced from the information collected from various sources (literature survey, aerial photographs and imagery, and fieldwork). The preparation of engineering geological maps of urban areas frequently involves systematic searches of archives (Dearman, 1991). Information from site investigation reports, records of past and present mining activity, successive editions of topographical maps, etc. may prove extremely useful. Once the data have been gathered, follow the problems of how they should be represented on the map and at which scale should the map be drawn. The latter is very much influenced by the requirement. A map represents a simplified model of the facts and the complexity of various geological factors can never be portrayed in its entirety. The amount of simplification required is governed principally by the purpose and scale of the map, the relative importance of particular engineering geological factors or relationships, the accuracy of the data and on the techniques of representation employed (Anon., 1976b). The major differences between maps of different scales are, first, the amount of data they show (the more detailed a map needs to be, the larger its scale) and, second, the manner in which it is presented. Engineering geological maps usually are produced on the scale of 1:10 000 or smaller whereas engineering geological plans, being produced for particular engineering purposes, have a larger scale (Anon., 1972). The scale of the mapping will depend on the engineering requirement, the complexity of the geology, and the staff and time available. For example, on a large project geological mapping frequently is required on a large and detailed scale. In other words thin, but suspect horizons such as clay bands should be recorded. As far as presentation is concerned, this may involve not only the choice of colours and symbols, but also the use of overprinting. Overprinting frequently takes the form of striped or stippled shading, both of which can be varied, for instance, according to frequency, pattern, dimension or colour. More than one map of an area may be required to record all the information that has been collected during a survey. Preparation of a series of engineering geological maps can reduce the amount of effort involved in the preliminary stages of a site investigation, and indeed may allow site investigations to be designed for the most economical confirmation of the ground conditions. Engineering geological maps should be accompanied by cross-sections, and an explanatory text and legend.

An engineering geological map provides an impression of the geological environment, surveying the range and type of engineering geological conditions, their individual components and their interrelationships. These maps also provide planners and engineers with information that will assist them in land-use planning and the location, construction and maintenance of engineering structures of all types. Engineering geological maps may be simply geological maps to which engineering geological data have been added. If the engineering information is extensive, then it can be represented in tabular form (Fig. 1.5; Tables 1.6a and 1.6b) perhaps accompanying the map on the reverse side. Where possible rock and soil masses should be mapped according to their lithology and, if possible, their presumed engineering behaviour, that is, in terms of their engineering classification. Alternatively, the rocks and soils may be presented as mapped units defined in terms of engineering properties. Particular attention should be given to the nature of the superficial deposits and, where present, made-over ground. Details of geological structures should be recorded, especially fault and shear zones, as should the nature of the discontinuities and grade of weathering where appropriate. Geomorphological

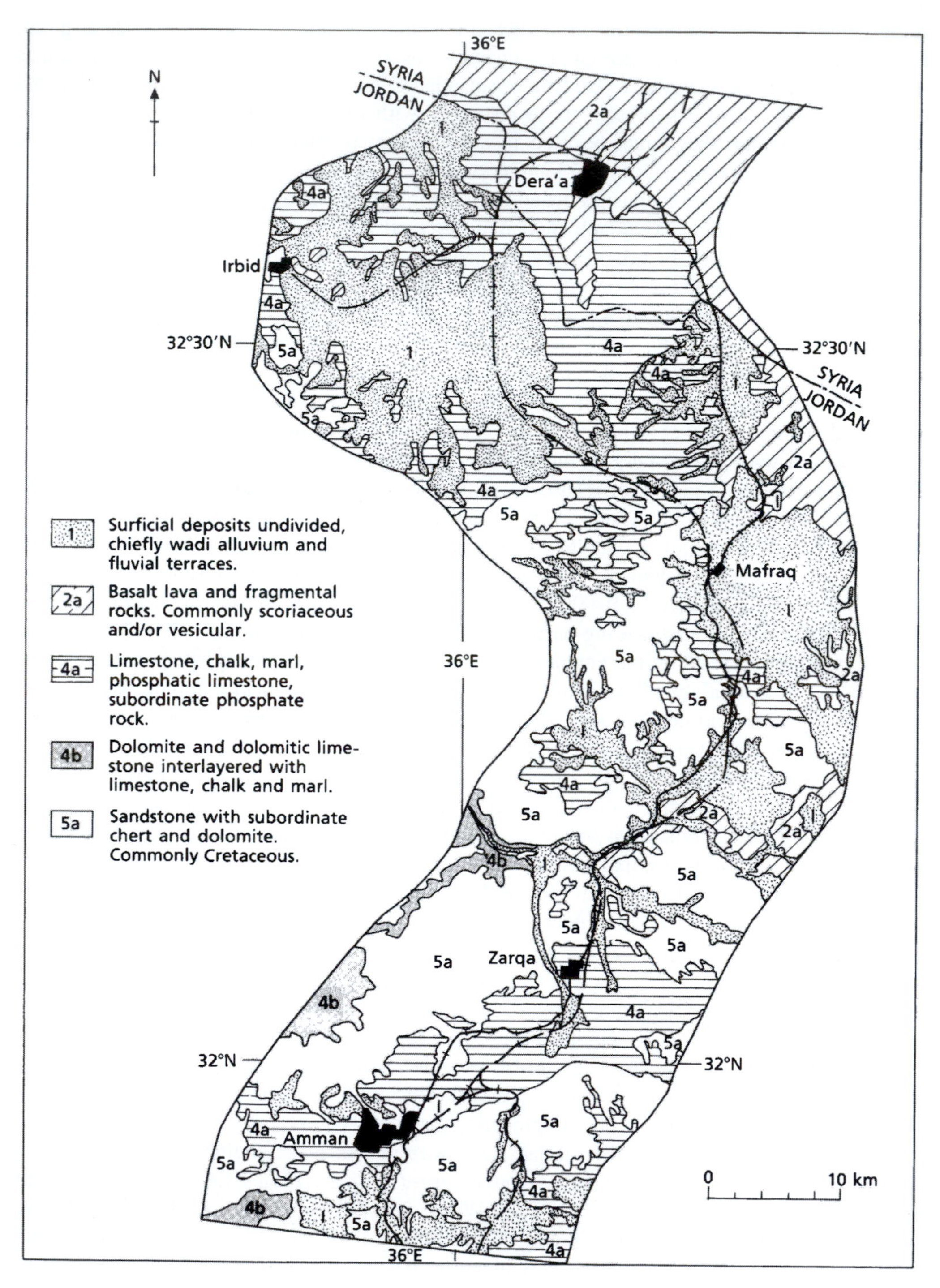

Figure 1.5 Segment of the engineering geological map for the Hijaz railway in Jordan.
Source: Briggs, 1987. Reproduced by kind permission of Association of Engineering Geologists.

Table 1.6a Excerpts from the engineering geology table illustrating the variety of materials in the study area for Hijaz railway. Symbols 1, 2a, 4a, 4b and 5a occur in the area shown in Fig. 1.5

Map symbol	*Geological description*	*Distribution*	*Map segments*	*Engineering characteristics*	*Suitability as source of material for*	*Moderate water-supply favourability in shallow aquifers*	*Topographic expression*
1	Surficial deposits undivided, chiefly wadi alluvium and fluvial and marine terraces	Most common in Saudi Arabia and southern Jordan	Present on most map segments	Excavation: easy Stability: poor Strength: fair Tunnel support: maximum	Ballast – 0 Coarse aggregate – + Sand – +++ Embankments – 0 Riprap – 0	Fair to good with seasonal fluctuations. Coastal areas poor	Generally flat, locally steeply dissected
2a	Basalt lava and fragmental rocks. Commonly scoriaceous and/or vesicular	Widespread in southern Syria. Locally elsewhere	01, 02, 13, 19, 20 and 28	Excavation: difficult Stability: good Strength: good Tunnel support: moderate	Ballast – + Coarse aggregate – + Sand – + Embankments – ++ Riprap – +	Generally poor. Locally fair to good, depending on interlayering	Flat to mountainous. Surfaces commonly bouldery
3c	Sandstone and conglomerate with limestone and marl. Loosely cemented. Locally hard	Along coastal plain between Al Wajh and Yanbu, Saudi Arabia	26, 27, 28 and 30	Excavation: intermediate Stability: fair Strength: fair Tunnel support: moderate to maximum	Ballast – 0 Coarse aggregate – 0 Sand – + Embankments – ++ Riprap – 0	Poor	Flat to rolling, locally hilly and dissected

Table 1.6a (Continued)

Map symbol	*Geological description*	*Distribution*	*Map segments*	*Engineering characteristics*	*Suitability as source of material for*	*Moderate water-supply favourability in shallow aquifers*	*Topographic expression*
4a	Limestone, chalk, marl, phosphatic limestone, subordinate phosphate rock	Widespread in Jordan	02–05 and 29	Excavation: difficult Stability: fair to good Strength: fair to good Tunnel support: moderate to minimum	Ballast – 0 Coarse aggregate – + Sand – 0 Embankments – + Riprap – +	Generally poor	Hilly, locally rolling or mountainous
4b	Dolomite and dolomitic limestone interlayered with limestone, chalk, and marl	Central Jordan	02 and 03	Excavation: moderately difficult Stability: fair to good Strength: fair to good Tunnel support: moderate to minimum	Ballast – + Coarse aggregate – + Sand – 0 Embankments – ++ Riprap – +	Generally poor	Hilly, locally mountainous
5a	Sandstone with subordinate chert and dolomite. Commonly calcareous	Widespread in Jordan	02–05 and 29	Excavation: moderately difficult Stability: fair to good Strength: good Tunnel support: moderate to minimum	Ballast – 0 Coarse aggregate – 0 Sand – + Embankments – ++ Riprap – 0	Poor to fair	Hilly to mountainous

6b	Chiefly andesite lava and fragmental rocks. Common medium-grade metamorphism, greenstone	Widespread in Hijaz Mountains	10–13, 18–25, 27, 30 and 31	Excavation: difficult Stability: good Strength: good Tunnel support: minimum	Ballast – +++ Coarse aggregate – +++ Sand – 0 Embankments – ++ Riprap – +++	Poor	Core of Hijaz Mountains. Relief locally greater than 2000 m
7b	Early and altered granites, granodiorite, quartz monzonite. Includes some gneiss	Common in the Hijaz Mountains and southern Jordan	10–13 and 19–31	Excavation: difficult Stability: good Strength: good Tunnel support: minimum to moderate	Ballast – + Coarse aggregate – + Sand – + Embankments – ++ Riprap – ++	Poor	Chiefly mountainous. Mostly more resistant than other instrusive rocks

Source: Briggs, 1987. Reproduced by kind permission of Association of Engineering Geologists.

Table 1.6b Key to the engineering characteristics column of the engineering geology table (Table 1.6a)

Excavation facility	*Stability of cut slopes*	*Foundation strength*	*Tunnel support requirements*
Easy – can be excavated by hand tools or light power equipment. Some large boulders may require drilling and blasting for their removal. Dewatering and bracing of deep excavation walls may be required	Good – these rocks have been observed to stand on essentially vertical cuts where jointing and fracturing are at a minimum. However, moderately close jointing or fracturing is common, so slopes not steeper than 4:1 (vertical:horizontal) are recommended. In deep cuts debris-catching benches are recommended	Good – bearing capacity is sufficient for the heaviest classes of construction, except where located on intensely fractured or jointed zones striking parallel to and near moderate to steep slopes	Minimum – support probably required for less than 10% of length of bore, except where extensively fractured
Moderately easy – probably rippable by heavy power equipment at least to weathered rock – fresh rock interface and locally to greater depth	Fair – cut slopes ranging from 2:1 to 1:1 are recommended; flatter where rocks are intensely jointed or fractured. Rockfall may be frequent if steeper cuts are made. Locally, lenses of harder rock may permit steeper cuts	Fair – choice of foundation styles is largely dependent on packing of fragments, clay content and relation to the water table. If content of saturated clay is high, appreciable lateral movement of clay may be expected under heavy loads. If packing is poor, settling may occur	Moderate – support may be required for as much as 50% of length of bore, more where extensively fractured
Intermediate – probably rippable by heavy power equipment to depths chiefly limited by the manoeuvrability of the equipment. Hard rock layers or zones of hard rock may require drilling and blasting	Poor – flatter slopes are recommended. Some deposits commonly exhibit a deceptive temporary stability, sometimes standing on vertical or near-vertical cuts for periods ranging from hours to more than a year	Poor – foundations set in underlying bedrock are recommended for heavy construction, with precautions taken to guard against failure due to lateral stress	Maximum – support probably required for entire length of bore
Moderately difficult – probably require drilling and blasting for most deep excavations, but locally may be ripped to depths of several metres			
Difficult – probably require drilling and blasting in most excavations except where extensively fractured or altered			

Source: Briggs, 1987. Reproduced by kind permission of Association of Engineering Geologists.

conditions, hydrogeological conditions, subsidences, borehole and field test information all can be recorded on engineering geological maps. The unit boundaries are drawn for changes in particular properties. Frequently, the boundaries of such units coincide with stratigraphical boundaries. In other instances, as for example, where rock masses are deeply weathered, they may bear no relation to geological boundaries. Unfortunately, one of the fundamental difficulties in preparing such maps arises from the fact that frequent changes in physical properties of rock and soil masses are gradational. As a consequence, regular checking of visual observations by *in situ* testing or sampling is essential to produce a map based on engineering properties. This type of map often is referred to as a geotechnical map (Dearman, 1991).

There are two basic types of engineering geological or geotechnical plans, namely, the site investigation plan, and the construction or foundation stage plan (Dearman and Fookes, 1974). The site investigation plan is prepared during the early stages of an investigation, thereby aiding the ground exploration to be planned and engineering problems to be anticipated. The scale of such plans varies from 1:5000 to as large as 1:500 or even 1:100, depending on the size and nature of the site and the engineering requirement (Fig. 1.6). The foundation plan records the ground conditions exposed during construction operations. It may be drawn to the same scale as the site investigation plan or the construction drawings. The foundation plan provides a record of the ground conditions and so allows the ground model to be refined (Griffiths, 2001). Such plans may be based on large-scale topographic maps or large-scale base maps produced by surveying or photogrammetric methods, with scales as large as 1:100.

1.6 Site exploration: direct methods

The aim of a site exploration is to try to determine, and thereby understand, the nature of the ground conditions on site and those of its surroundings (Clayton *et al.*, 1996). It expands, and hopefully verifies, the information gathered by the desk study and preliminary reconnaissance, and should identify possible construction problems. The data should be continually reassessed as the exploration proceeds and the programme altered if the data so indicates. The extent to which the site exploration is taken depends, to some extent, upon the size and importance of the construction operation on the one hand and the complexity of the ground conditions on the other. A report embodying the findings and conclusions of the exploration programme must be produced at the end of the site investigation, which can be used for design purposes. This should contain geological plans of the site with accompanying sections, and possibly fence diagrams, thereby conveying a three-dimensional model of the subsurface strata.

There are no given rules regarding the location of boreholes or drillholes, or the depth to which they should be sunk. This depends upon two principal factors, the geological conditions and the type of project concerned. The information provided by the desk study, preliminary reconnaissance and from any trial trenches should provide a basis for the initial planning and layout of the borehole or drillhole programme. Holes should be located so as to detect the geological sequence and structure. Obviously, the more complex this is, the greater the number of holes needed. In some instances, it may be as well to start with a widely spaced network of holes. As information is obtained, further holes can be put down if and where necessary.

Exploration should be carried out to a depth that includes all strata likely to be significantly affected by the structural loading. Experience has shown that damaging settlement usually does not take place when the added stress in the soil due to the weight of a structure is less than 10 per cent of the effective overburden stress. It therefore would seem logical to sink boreholes on compact sites to depths where the additional stress does not exceed 10 per cent of the stress due to the weight of the overlying strata. It must be borne in mind that if a number of loaded

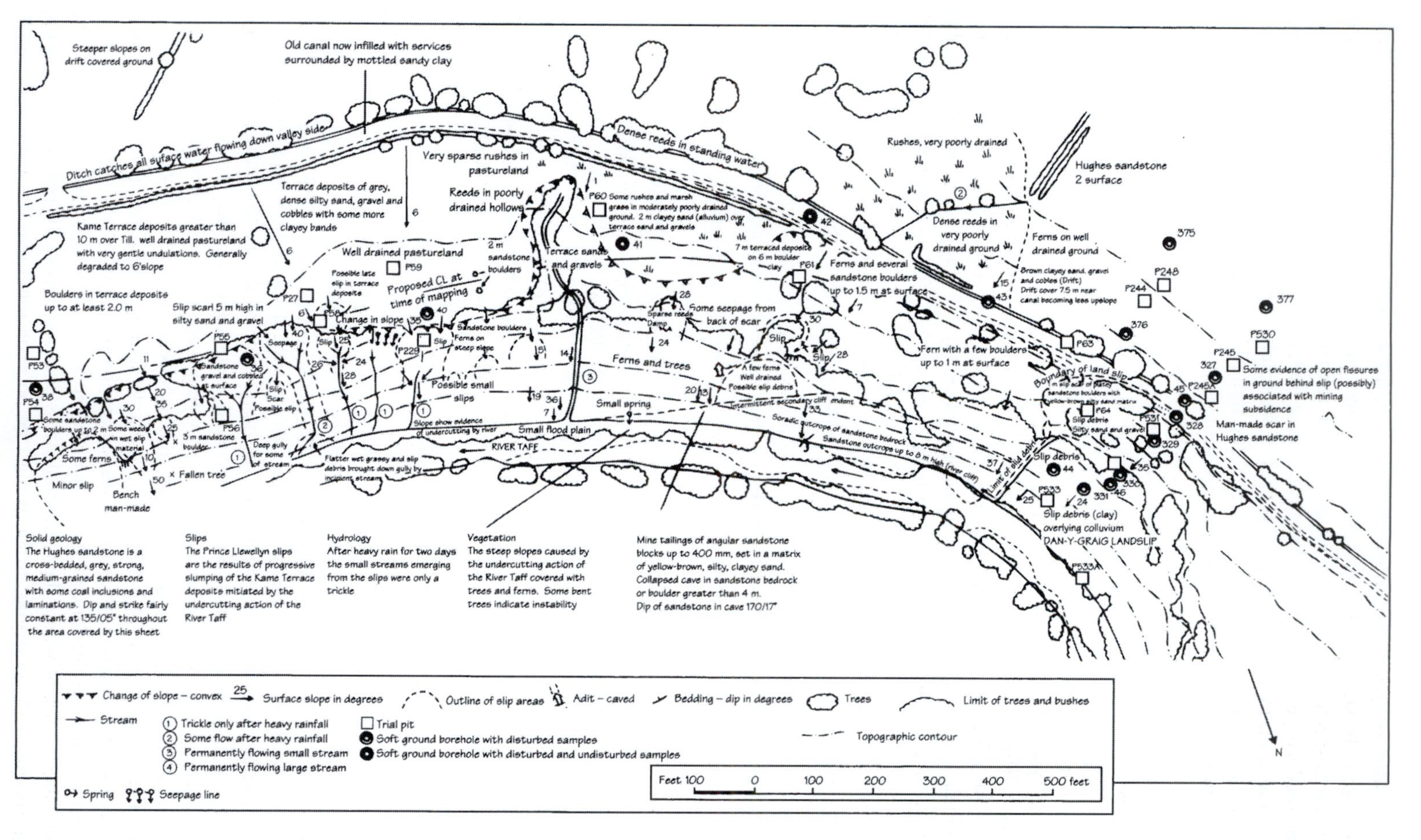

Figure 1.6 Engineering geological plan produced at the site investigation stage, Prince Llewellyn area, Stage IV, Taff Vale trunk road, South Wales. The contours are in feet above ordnance datum.
Source: Dearman and Fookes, 1974. Reproduced by kind permission of The Geological Society.

areas are in close proximity the effect of each is additive. Under certain special conditions holes may have to be sunk more deeply as, for example, when voids due to abandoned mining operations are suspected, or when it is suspected that there are highly compressible layers, such as interbedded peats, at depth. If possible, boreholes should be taken through superficial deposits to rockhead. In such instances, adequate penetration of the rock should be specified to ensure that isolated boulders are not mistaken for the solid formation.

The results from a borehole or drillhole should be documented on a log (Fig. 1.7). Apart from the basic information such as number, location, elevation, date, client, contractor and engineer responsible, the fundamental requirement of a borehole log is to show how the sequence of strata changes with depth. Individual soil or rock types are presented in symbolic form on a borehole log. The material recovered must be adequately described, and in the case of rocks this frequently includes an assessment of the degree of weathering, fracture index, and relative strength. The type of boring or drilling equipment should be recorded, the rate of drilling progress made being a significant factor. The water level in the hole and any water loss, when it is used as a flush during rotary drilling, should be noted, as these reflect the mass permeability of the ground. If any *in situ* testing is done during boring or drilling operations, then the type(s) of test and the depth at which it/they were carried out must be recorded. The depths from which samples were taken must be noted. A detailed account of the logging of cores for engineering purposes is provided by Anon. (1970). Description and classification of soils, and of rocks and rock masses, can be found in Anon. (1999) and in Norbury and Gosling (1996), whilst a description and classification of weathered rocks is given in Anon. (1995), and of discontinuities in Barton (1978). Bell (2000) also has provided extensive summaries of descriptions and classifications of soils and rocks, and of weathered rocks and discontinuities.

1.6.1 Subsurface exploration in soils

The simplest method whereby data relating to subsurface conditions in soils can be obtained is by hand augering. However, hand augering is not suitable in unstable soils or where groundwater flows into the hole. The two most frequently used types of augers are the post-hole and screw augers (Fig. 1.8). These are used principally in fine-grained soils. Soil samples that are obtained by augering are badly disturbed and invariably some amount of mixing, and even loss, of soil types occurs. Critical changes in the ground conditions therefore are unlikely to be located accurately. Even in very soft soils it may be very difficult to penetrate more than 7 m with hand augers. The Mackintosh probe and Lang prospector are more specialized forms of hand tools.

Power augers are available as solid stem or hollow-stem, both having an external continuous helical flight (Fig. 1.9). The hollow stem can be sealed at the lower end with a combined plug and cutting bit, which is removed when a sample is required. Hollow-stem augers are useful for investigations where the requirement is to locate bedrock beneath overburden. Solid-stem augers are used in stiff clays that do not need casing, however, if an undisturbed sample is required they have to be removed and replaced by a sampling device. Disturbed samples taken from auger holes often are unreliable. In favourable ground conditions, such as firm and stiff homogeneous clays, auger rigs are capable of high output rates.

The development of large earth augers and patent piling systems have made it possible to sink 1 m diameter boreholes in soils more economically than previously. The ground conditions can be inspected directly from such holes. Depending on the ground conditions, the boreholes may be either unlined, lined with steel mesh or cased with steel pipe. In the latter case, windows are provided at certain levels for inspection and sampling.

Pits and trenches allow the ground conditions in soils and highly weathered rocks to be examined directly, although they are limited to a few metres in terms of their depth. Both

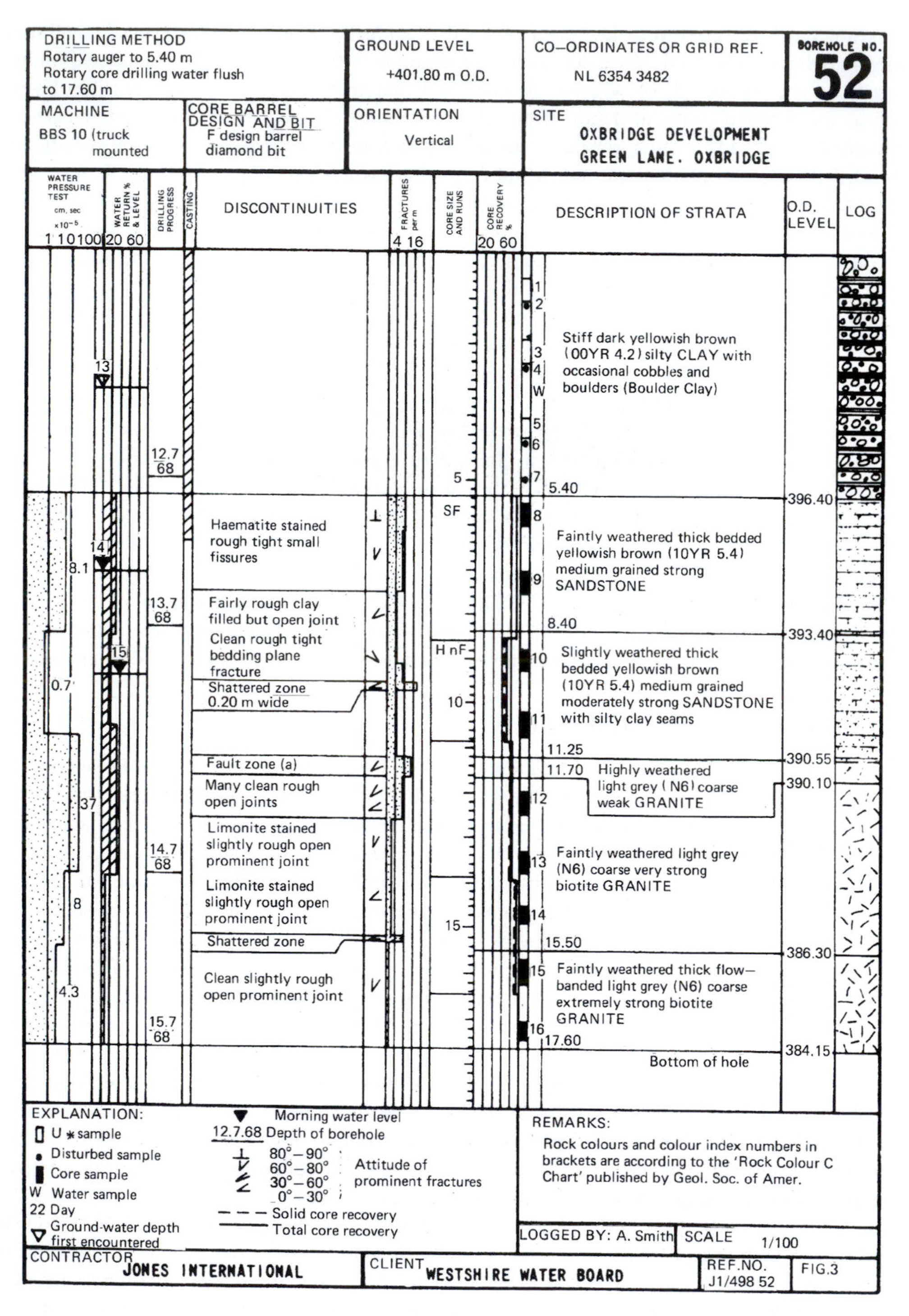

Figure 1.7 A drillhole log.
Source: Reproduced by kind permission of The Geological Society.

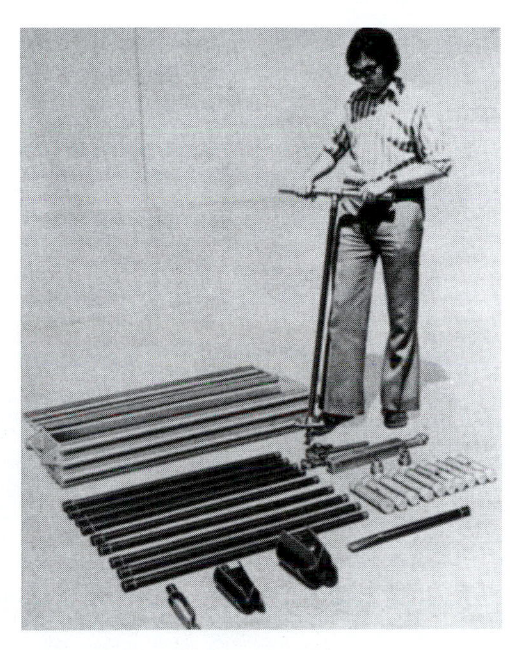

Figure 1.8 Hand augering equipment.
Source: ELE International. Reproduced by kind permission of ELE International.

can be excavated quickly in suitable ground conditions and can provide flexible and economic methods of obtaining information (Hatheway and Leighton, 1979). Groundwater conditions and stability of the sides obviously influence whether or not they can be excavated, and safety must at all times be observed, this at times necessitates shoring the sides. The soil conditions in pits and trenches can be mapped, sketched and photographed (Fig. 1.10). Undisturbed, as well as disturbed, samples can be collected as required, their location being located on the map or sketch plan. Such excavations are used to locate slip planes in landslides.

The light cable and tool boring rig is used for investigating soils (Fig. 1.11). The hole is sunk by repeatedly dropping one of the tools into the ground. A power winch is used to lift the tool, suspended on a wire cable, and by releasing the clutch of the winch the tool drops and cuts into the soil. Once a hole is established it is lined with casing, the drop tool operating within the casing. This type of rig usually is capable of penetrating about 60 m of soil, in doing so the size of the casing in the lower end of the borehole is reduced. The basic tools are the shell and the clay-cutter, which are essentially open-ended steel tubes to which cutting shoes are attached. The shell, which is used in coarse-grained soils, carries a flap valve at its lower end that prevents the material from falling out on withdrawal from the borehole. The material is retained in the cutter by the adhesion of the clay.

When boring in stiff clay soils the weight of the claycutter may be increased by adding a sinker bar. A little water frequently is added to assist boring progress in very stiff clay soils. This must be done with caution so as to avoid possible changes in the properties of soil about to be sampled. Furthermore, in such clay soils the borehole can often be advanced without lining, except for a short length at the top to keep the hole stable. If cobbles or small boulders are encountered in clay soils, particularly tills, then these can be broken by using heavy chisels.

When boring in soft clay soils, although the hole may not collapse it tends to squeeze inwards and prevent the cutter operating, hence the hole must be lined. The casing is driven in and

Figure 1.9 Multi-purpose auger attached to mobile drill.

winched out, however, in difficult ground conditions it may have to be jacked out. Casing tubes have internal diameters of 150, 200, 250 and 300 mm, the most commonly used sizes being 150 and 200 mm (the large sizes are used in coarse gravels).

The usual practice is to bore ahead of the casing for about 1.5 m (the standard length of a casing section) before adding a new section of casing and surging it down. The reason for surging the casing is to keep it 'free' in the borehole so that it can be extracted easily on completion. Smaller diameter casing is introduced when the casing can no longer be advanced by surging. However, if the hole is near its allotted depth, then the casing may be driven into

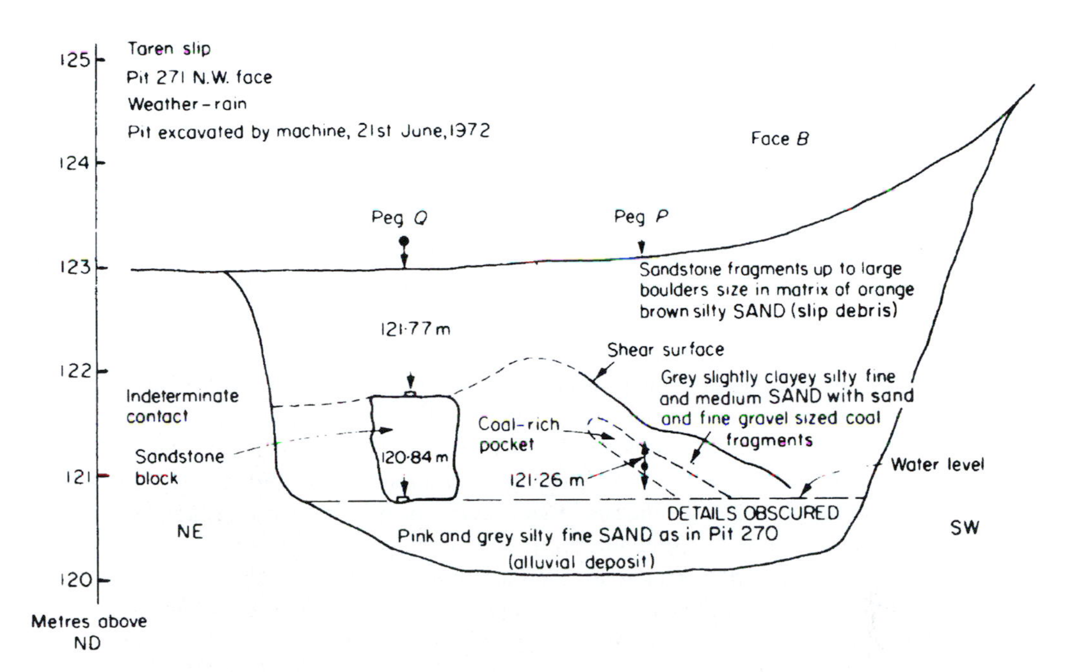

Figure 1.10 Log of trial pit.

Figure 1.11 Cable and tool rig.
Source: English Drilling Equipment Co. Reproduced by kind permission of English Drilling Equipment Co.

the ground for quickness. Where clay soil occurs below a coarse-grained deposit, the casing used as a support in the latter soil is driven a short distance into the clay soil to create a seal and the shell is used to remove any water that might enter the borehole.

Boreholes in sands or gravels almost invariably require lining. The casing should be advanced with the hole or overshelling is likely to occur, that is, the sides collapse and prevent further progress. Because of the mode of operation of the shell, the borehole should be kept full of water so that the shell may operate efficiently. Where coarse-grained soils are water bearing, all that is necessary is for the water in the borehole to be kept topped up. If flow of water occurs, then it should be from the borehole to the surrounding soil. Conversely, if water is allowed to flow into the borehole, piping may occur. Piping usually can be avoided by keeping the head of water in the borehole above the natural head. If artesian conditions are encountered, then the casing should be extended above ground and kept filled with water. The shell generally cannot be used in highly permeable, coarse gravels since it usually is impossible to maintain a head of water in the borehole. Fortunately, these conditions often occur at or near ground level and the problem can sometimes be overcome by using an excavator to open a pit either to the water table or to a depth of 3 or 4 m. Casing then can be inserted and the pit backfilled, after which boring can proceed. Another method of penetrating gravels and cobbles above the water table is to employ a special grab with a heavy tripod and winch, and casing of 400 mm diameter or greater.

Rotary attachments are available that can be used with light cable and tool rigs. However, they are much less powerful than normal rotary rigs and tend to be used only for short runs as, for example, to prove rockhead at the base of a borehole.

Wash boring is a quick method of sinking a hole in soils that often is used in the United States, the hole being advanced by a combination of chopping and jetting the soil, the cuttings thereby produced being washed from the hole by the water used for jetting (Fig. 1.12). The method has been used rarely in western Europe because it is not suitable for mixed soils, neither

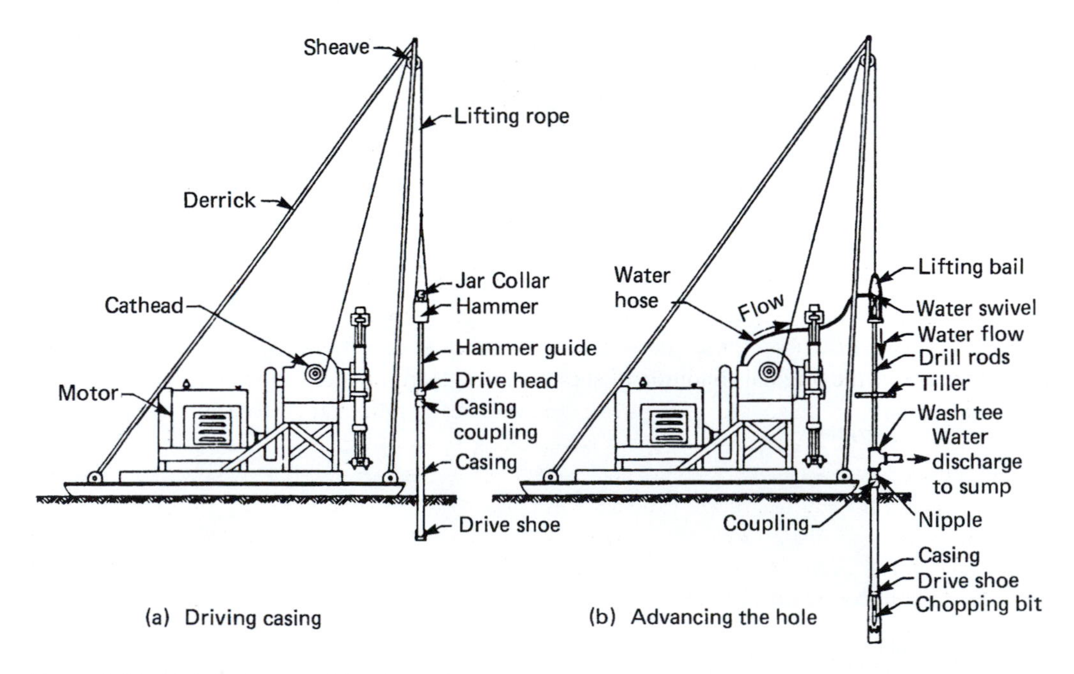

Figure 1.12 Wash boring rig: (a) driving the casing; (b) advancing the hole.

does it penetrate coarser gravels. The wash boring method may be used in both cased and uncased holes. Casing obviously has to be used in coarse-grained soils to avoid collapse of the sides of the hole. Several types of chopping bits are used. Straight and chisel bits are used in sands, silts, clays and very soft rocks whilst cross bits are used in gravels and soft rocks. Bits are available with either the jetting points facing upwards or downwards. The former type of bits are better at cleaning the base of the hole than are the latter. Some indication of the type of ground penetrated may be obtained from the cuttings carried to the surface by the wash water, from the rate of progress made by the bit or from the colour of the wash water. Wash boring cannot be used for direct sampling, which has to be done separately by removing the bit and replacing it with a sampler, frequently a split-spoon sampler. In fact, standard penetration tests in sands often are carried out, in the United States, in a hole sunk by wash boring.

Cable and tool or churn drilling is a percussion drilling method that can be used in more or less all types of ground conditions but the rate of progress tends to be slow. The rig can drill holes up to 0.6 m in diameter to depths of about 1000 m. The hole is advanced by raising and dropping heavy drilling tools that break the soil or rock. Different bits are used for drilling in different formations and an individual bit can weigh anything up to 1500 kg, so that the total weight of the drill string may amount to several thousand kilograms. A slurry is formed from the broken material and the water in the hole. The amount of water introduced into the hole is kept to the minimum required to form the slurry. The slurry is periodically removed from the hole by means of bailers or sand pumps. In unconsolidated materials, casing should be kept near the bottom of the hole in order to avoid caving. Again, changes in the type of strata penetrated can be inferred from the cuttings brought to the surface, the rate of drilling progress or the colour of the slurry. Sampling, however, has to be done separately. Unfortunately, the ground that has to be sampled may be disturbed by the heavy blows of the drill tools.

1.6.2 Sampling in soils

As far as soils are concerned samples may be divided into two types, disturbed and undisturbed. Disturbed samples can be obtained by hand, by auger or from the clay-cutter or shell of a boring rig. Samples of fine-grained soil should be approximately 1 kg in weight, this providing a sufficient size for index testing. They are placed in screw-top jars and should be sealed with tape or wax after the top has been screwed on to maintain the natural moisture content. A larger sample (e.g. 5 kg) is necessary if the particle size distribution of coarse-grained soil is required and this may be retained in a tough plastic sack. Care must be exercised when obtaining such samples to avoid loss of fines.

An undisturbed sample can be regarded as one that is removed from its natural condition without disturbing its structure, density, porosity, moisture content and stress condition. Although it must be admitted that no sample is ever totally undisturbed, every attempt must be made to preserve the original condition of such samples. Unfortunately, mechanical disturbances produced when a sampler is driven into the ground distort the soil structure. Furthermore, a change of stress condition occurs when a borehole is excavated.

Undisturbed samples may be obtained by hand from surface exposures, pits and trenches. Careful hand trimming is used to produce a regular block, normally a cube of about 250 mm dimensions. Block samples are covered with muslin and sealed with wax. When sampling more friable material it may be necessary to cover it with muslin to stop the sample from breaking before cutting it free. Once removed, the sample should be carefully packed in a sample box. The box should be marked to show its orientation and other details. Block samples are particularly useful when it is necessary to test specific horizons, such as shear zones. They also can be used to assess fissure classification and fabric in fissured soils (Fookes and Denness, 1969).

The fundamental requirement of any undisturbed sampling tool is that on being forced into the ground it should cause as little remoulding and displacement of the soil as possible. The amount of displacement is influenced by a number of factors. First, there is the cutting edge of the sampler. A thin cutting edge and sampling tube minimizes displacement but it is easily damaged, and it cannot be used in gravels and hard soils. Second, the internal diameter of the cutting edge (D_i) should be slightly less than that of the sample tube, thus providing inside clearance that reduces drag effects due to friction. Third, the outside diameter of the cutting edge (D_o) should be from 1 to 3 per cent larger than that of the sampler, again to allow for clearance. The relative displacement of a sampler can be expressed by the area ratio (A_r):

$$A_r = \frac{D_i^2 - D_o^2}{D_o^2} \times 100 \qquad (1.1)$$

This ratio should be kept as low as possible, for example, according to Hvorslev (1949) displacement is minimized by keeping the area ratio below 15 per cent. It should not exceed 25 per cent. Lastly, friction also can be reduced if the tube has a smooth inner wall. A coating of light oil may also prove useful in this respect.

Open-tube samplers can be either thick-walled or thin-walled. The standard open-tube sampler for obtaining samples from fine-grained soils with a shear strength exceeding 50 kPa is thick walled and is referred to as the U100 sampling tube. It has a diameter of 100 mm, a length of approximately 450 mm and its walls are 1.2 mm thick (Fig. 1.13). A cutting shoe is screwed to the bottom of the sample tube, which should meet the requirements noted above. The upper end of the tube is fastened to a head with a check valve that allows air or water to escape during driving and helps to hold the sample in place when it is being withdrawn. On withdrawal from the borehole the head and shoe are removed from the tube, and the sample is sealed within it with paraffin wax and the end caps screwed on. The tube can incorporate a detachable liner to facilitate sample retrieval. The influence of the increased area ratio is counteracted by using a thin-angle cutting shoe that is longer than that used in the unlined tube. A core-catcher is used when sampling silts and silty sands, hopefully to retain the sample within the tube. In soft materials two or three tubes may be screwed together to reduce disturbance in the sample.

In thin-walled open-tube samplers the lower edge of the tube is shaped and turned slightly inwards to provide a small inside clearance, the area ratio being around 10 per cent. A thin-walled piston sampler should be used for obtaining clay soils with a shear strength less than

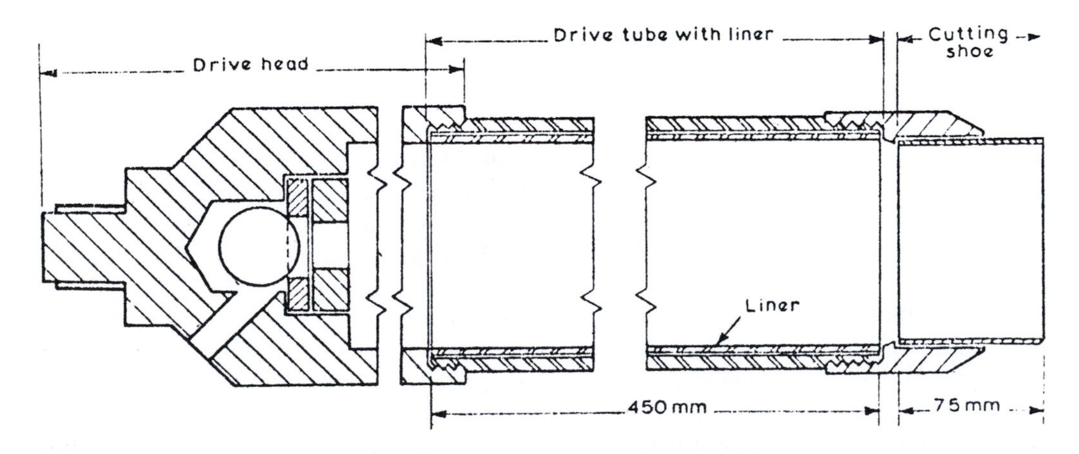

Figure 1.13 U100 open drive sampler.

50 kPa, since soft clays tend to expand into the sample tube. Expansion is reduced by a piston that remains stationary in the sampler as it is jacked into the soil (Fig. 1.14). Piston samplers can be pushed below any disturbed soil at the bottom of a borehole prior to taking a sample. The piston is attached to a rod inside hollow drill rods, the lowermost of which is attached to the head of the sampler. When locked into position, the tube is sealed by the piston as they are jacked to the chosen level, boring having ended some 500 mm above this position. Next, the inner rods are unlocked and kept stationary at the surface so that the piston remains at the chosen level as the sampling tube is pushed into the soil. After the sample has been taken the piston rod is clamped at the upper end of the tube and the sampler is extracted. Piston samplers

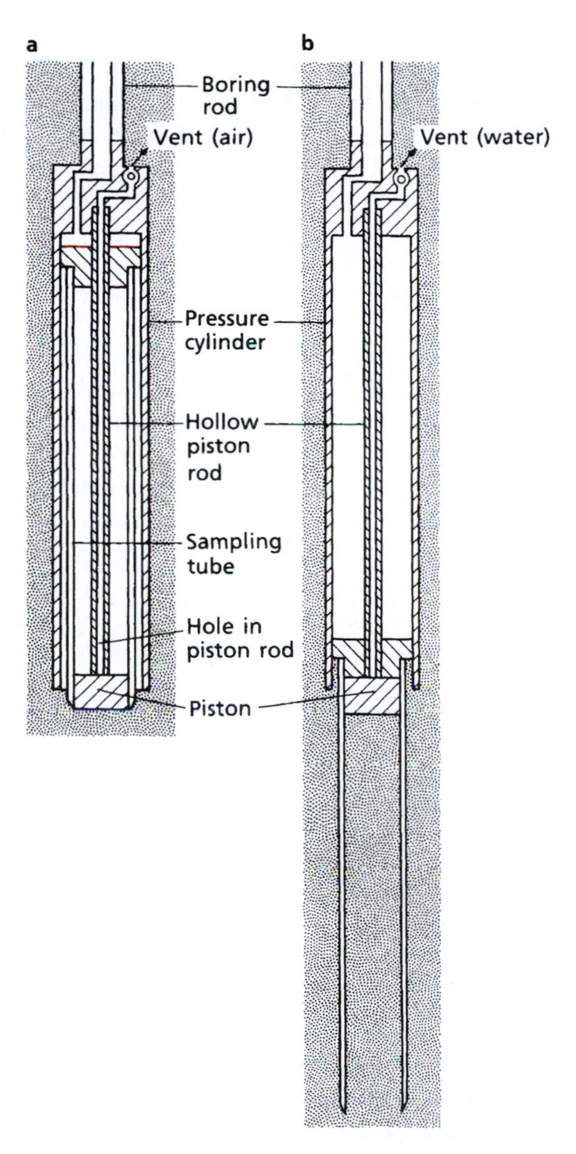

Figure 1.14 Piston sampler: (a) lowered to the bottom of a borehole with boring rod clamped in position; (b) sampling tube forced into soil by water supplied through boring rod.

range in diameter from 75 to 250 mm, although the larger sizes are not in frequent use except where samples are required for testing in a Rowe consolidation cell. A vacuum tends to be created between the piston and the soil sample, and thereby helps to hold it in place.

Special samplers have to be used when continuous samples are required as the excessive side adhesion, which is likely to develop in long sampling tubes, has to be overcome. Continuous sampling may be necessary in rapidly varying or sensitive soils. In the Swedish soil sampler, thin strips of steel foil are fed into the tube as it is advanced into the soil, so forming an inner lining that wraps the soil and prevents it touching the walls of the tube. Hence, friction is reduced. Samples up to 25 m in length have been obtained in this way. However, the foil sampler is complicated to use and so tends not to be used in Britain. Alternatively, a Delft sampler can be used to obtain a continuous sample (Fig. 1.15). There are two sizes of Delft sampler, providing samples of 29 and 66 mm diameters. The larger size sampler is jacked into the soil by a conventional 10 Mg deep sounding machine to a depth of up to 18 m. However, in a suitable soil it can be modified to extend to depths of 30 m. During sampling a self-vulcanizing nylon sleeve feeds automatically from a magazine above the cutting shoe, surrounding the sample as it passes into an inner plastic tube. The latter contains a bentonite-barytes supporting fluid. Extension tubes, 1 m in length, are added as sampling proceeds. A 2 Mg machine is used with the 29 mm sampler. Samples from the latter frequently are split down the middle, one half being used for index testing whilst the other is used for visual examination, both in the natural and dry states. The soil fabric is more readily seen in the dry state, in which condition it may be photographed.

The most difficult undisturbed sample to obtain is that from saturated sand, particularly when it is loosely packed, the problem being to retain the sample in the sampler. In some cases a long sample tube can be made by coupling three U100 tubes together, and fitting a core-catcher and cutting shoe. The sand in the central tube is less disturbed than that in the end tubes. Disturbance can be reduced by using a thin-walled piston sampler but it is still possible to lose the sample on extraction. Because it is necessary to break the vacuum at the head of the sample before removing the piston, some samplers incorporate a vacuum-release device. The Bishop sand sampler makes use of compressed air and incorporates a thin-walled sampling tube housed in an outer tube. The inner tube is driven into the soil and compressed air introduced into the outer tube expels the water. Then the sampling tube is retracted into the outer tube, the air pressure creating capillary zones that retain the soil. A modified Bishop sand sampler incorporates a stationary piston to determine the exact length to which the sample tube is driven into the sand. This makes a correction for volume change during sampling.

1.6.3 Subsurface exploration in rocks

Rotary drills are either skid mounted, trailer mounted or, in the case of larger types, mounted on lorries (Fig. 1.16). They are used for drilling through rock, although they can, of course, penetrate and take samples from soil. A heavy rig is required to drill through hard rock. Compressed air, water or mud may be used as the flush.

Rotary-percussion drills are designed for rapid drilling in rock. The rock is subjected to rapid high-speed impacts whilst the bit rotates, which bring about compression and shear in the rock. Full-face bits, which produce an open hole, are used. These are usually of the studded, cruciform or tricone roller bit type (Fig. 1.17). Full-face diamond bits rarely are used in site investigation. The technique is most effective in brittle materials since it relies on chipping the rock. The rate at which drilling proceeds depends upon the type of rock, particularly on its strength, hardness and fracture index, the type of drill and drill bit, the flushing medium and

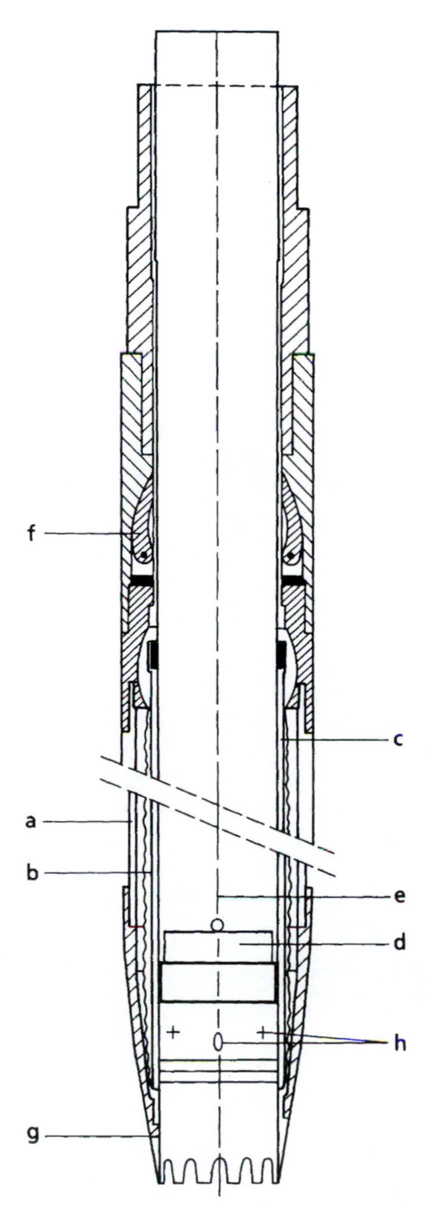

Figure 1.15 Delft continuous sampling apparatus: (a) outer tube; (b) stocking tube over which precoated nylon stocking is slid; (c) plastic inner tube; (d) cap at top of sample; (e) steel wire to fixed point at ground surface tension; (f) sample retaining clamps; (g) cutting shoe; (h) holes for entry of lubricating fluid.

the pressures used, as well as the experience of the drilling crew. If the drilling operation is standardized, then differences in the rate of penetration reflect differences in rock types. Drill flushings should be sampled at regular intervals, at changes in the physical appearance of the flushings and at significant changes in penetration rates. Interpretation of material penetrated in rotary-percussion drillholes should be related to a cored drillhole near by. The rotary-percussion method is sometimes used as a means of advancing a hole at low cost and high speed between

Figure 1.16 Medium size, skid-mounted rotary drill.

intervals where core drilling or *in situ* testing is required. Open holes also can be used for inspection by drillhole camera or closed circuit television. Because holes can be sunk rapidly, rotary-percussion drilling is frequently used when searching for abandoned mine workings at shallow depth (Fig. 1.18).

For many engineering purposes a solid, and as near as possible continuous rock core is required for examination. The core is cut with a bit and housed in a core barrel. The bit is

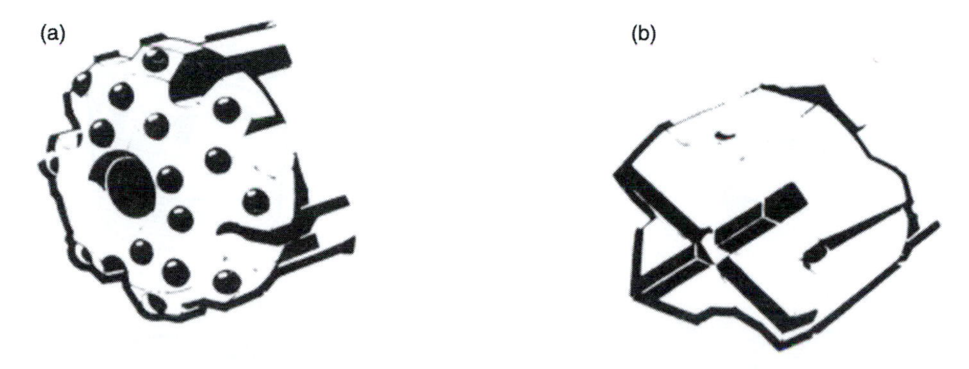

Figure 1.17 Full-face bits for rotary-percussion drilling: (a) studded bit; (b) cross-chisel or cruciform bit.

Figure 1.18 Mobile rotary-percussion rig.

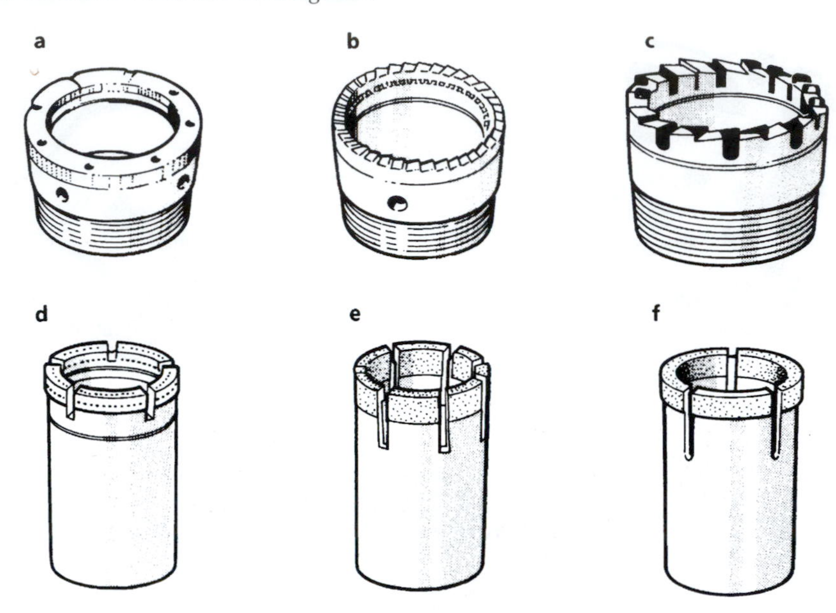

Figure 1.19 Some common types of coring bits: (a) surface set diamond bit (bottom discharge); (b) stepped sawtooth bit; (c) tungsten carbide bit; (d) impregnated diamond bit; (e) 'Diadril' corebit impregnated; (f) 'Diadrill' corebit impregnated.

set with diamonds or tungsten carbide inserts (Fig. 1.19). The coarser surface set diamond and tungsten carbide-tipped bits are used in softer formations. These bits generally are used with air rather than with water flush. Impregnated bits possess a matrix impregnated with diamond dust and their grinding action is suitable for hard and broken formations. In fact, most core drilling is carried out using diamond bits, the type of bit used being governed by the rock type to be drilled. In other words, the harder the rock, the smaller the size and the higher the quality of the diamonds that are required in the bit. Tungsten bits are not suitable for drilling in very hard rocks. Thick-walled bits are more robust but penetrate more slowly than thin-walled bits. The latter produce a larger core for a given hole size. This is important where several reductions in size have to be made. Core bits vary in size and accordingly core sticks range between 17.5 and 165 mm diameter (Fig. 1.20). Other factors apart, generally the larger the bit, the better is the core recovery.

A variety of core barrels are available for rock sampling. The simplest type of core barrel is the single-tube but because it is suitable only for hard, massive rock types, it is rarely used. In the single-tube barrel, the barrel rotates the bit and the flush washes over the core. In double-tube barrels the flush passes between the inner and outer tubes. Double-tubes may be of the rigid or swivel type. The disadvantage of the rigid barrel is that both the inner and outer tubes rotate together and in soft rock can break the core as it enters the inner tube. It therefore is only suitable for hard rock formations.

In the double-tube swivel core barrel the outer tube rotates whilst the inner tube remains stationary (Fig. 1.21). It is suitable for use in medium and hard rocks and gives improved core recovery in soft friable rocks. Both the bit and core barrel are attached by rods to the drill, by which they are rotated. Either water or air is used as a flush. This is pumped through the drill rods and discharged at the bit. The flushing agent serves to cool the bit and to remove the cuttings from the drillhole. Bentonite is sometimes added to water. It eases the running

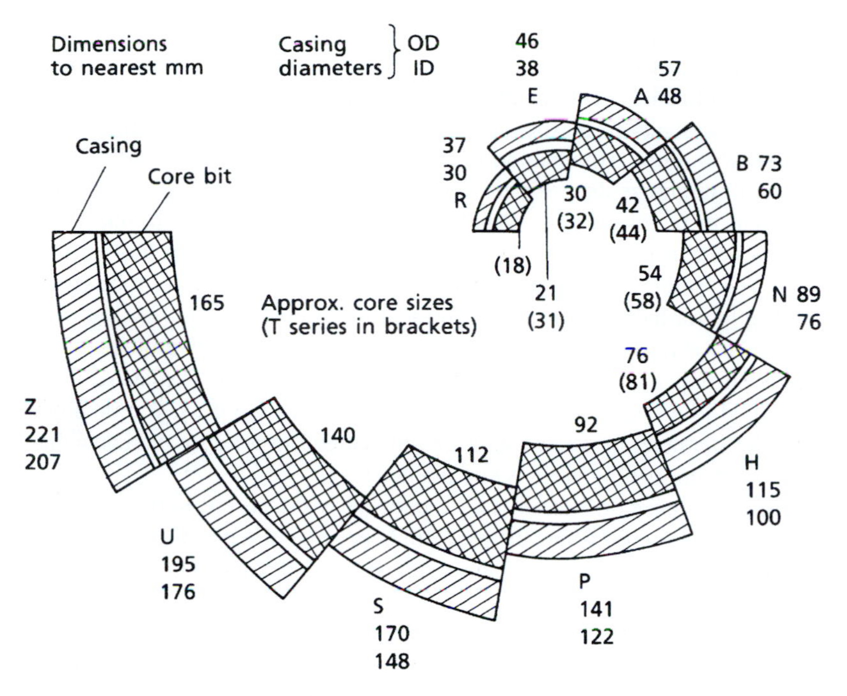

Figure 1.20 Core barrel and casing diameters.

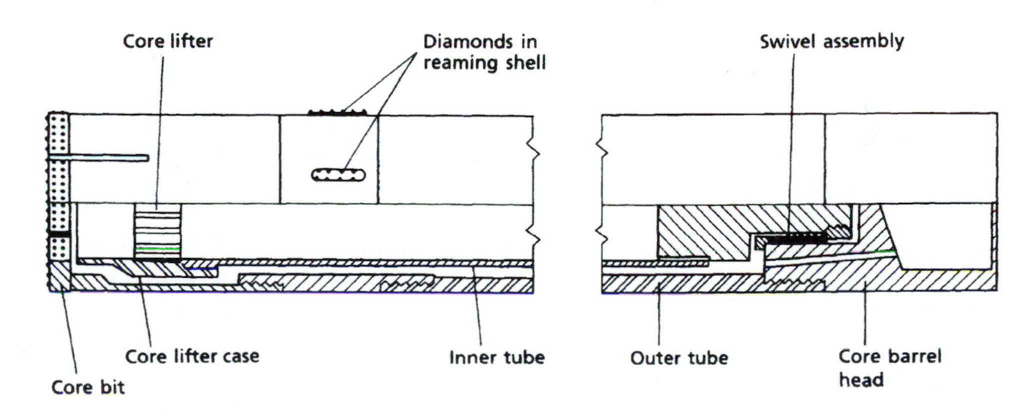

Figure 1.21 Double tube swivel type core barrel.

and pulling of casing by lubrication, it holds chippings in suspension, and it promotes drillhole stability by increasing flush returns through the formation of a filter skin on the walls of the hole. Disturbance of the core is likely to occur when it is removed from the core barrel. Most rock cores should be removed by hydraulic extruders whilst the tube is held horizontally. In order to reduce disturbance during drilling and extrusion, the inner tube of double core barrels can be lined with a split plastic sleeve before drilling commences. On completion of the core-run the plastic sleeve containing the core is withdrawn from the barrel.

The face-ejection barrel is a variety of the double-tube swivel type in which the flushing fluid does not affect the end of the core. This type of barrel is a minimum requirement for coring badly shattered, weathered and soft rock formations. Triple-tube barrels have been developed for obtaining cores from very soft rocks and from highly jointed and cleaved rock. The triple-tube barrel is a modification of the double-tube swivel barrel with a liner incorporated in the inner tube. Liners are made of plastic, steel or aluminium and the metal liners sometimes are split longitudinally to allow examination of the core. The core is extracted from the barrel in the liner in order to minimize disturbance. The retractable triple-tube barrel has a spring-loaded inner tube adapted to receive a liner. When drilling in very weak strata, the leading edge of the inner tube is held in position by the spring, and the core entering the barrel is protected from the flush. The spring allows the inner tube to retract into the outer tube if harder rock is encountered. The barrel then acts as a conventional triple-core barrel.

If casing is used for drilling operations, then it is drilled into the ground using a tungsten carbide or diamond-tipped casing shoe with air, water or mud flush. The casing may be inserted down a hole drilled to a larger diameter to act as conductor casing when reducing and drilling ahead in a smaller diameter or it may be drilled or reamed in a larger diameter than the initial hole to allow continued drilling at the same diameter.

Many machines will core drill at any angle between vertical and horizontal. Unfortunately, inclined drillholes tend to go off-line, the problem being magnified in highly jointed formations. In deeper drilling, the sag of the rods causes the hole to deviate. Drillhole deviation can be measured by an inclinometer.

The weakest strata are generally of greatest interest but these are the very materials that are most difficult to obtain and if recovered may deteriorate after extraction. Hawkins (1986) introduced the concept of lithology quality designation (LQD), which he defined as the percentage of solid core present greater than 100 mm in length within any lithological unit. He, however, did not attempt to devise grades of LQD, as has been done for the rock quality designation. Hawkins also recommended that the total core recovery (TCR) and the maximum intact core length (MICL) should be recorded. Zones of core loss or no recovery must be recorded as these could represent problem zones. Shales and mudstones are particularly prone to deterioration and some may disintegrate completely if allowed to dry. If samples are not properly preserved they will dry out. Deterioration of suspect material may be reduced by wrapping the cores with aluminium foil or plastic sheeting. Core sticks may be photographed before they are removed from the site.

A simple but nonetheless important factor is labelling. This must record the site, the drillhole number and the position in the drillhole from which material was obtained. The labels themselves must be durable and properly secured. When rock samples are stored in a core box the depth of the top and bottom of the core contained, as well as the separate core runs, should be noted both outside and inside the box. Zones of core loss also should be identified.

Direct observation of strata, discontinuities and cavities can be undertaken by cameras or closed-circuit television equipment, and drillholes can be viewed either radially or axially. Remote focusing for all heads and rotation of the radial head through 360° are controlled from the surface. The television heads have their own light source. The camera can be used in boreholes/drillholes down to a minimum diameter of 100 mm. Focusing, light intensity, rotation and digital depth control on the image are made by means of a surface control unit and the image is recorded on standard VHS format video-tape. Colour changes in rocks can be detected as a result of the varying amount of light reflected from the drillhole walls. Discontinuities appear as dark areas because of the non-reflection of light. However, if the drillhole is deflected from the vertical, variations in the distribution of light may result in some lack of picture definition. In open-hole sections, the inclination azimuth (from a compass attachment), frequency and aperture of discontinuities can be determined, as well as any ingress of water from above water level.

A televiewer emits ultrasonic energy from a piezoelectric transducer to the wall of a drillhole via the fluid in the hole. Some of the energy is reflected and picked up by the transducer, which also acts as a receiver. The transducer is rotated in the hole at some $3\,\text{rev}\,\text{s}^{-1}$ and is orientated relative to the earth's magnetic field by a downhole magnetometer in the sonde. The amplitude of the reflected signal is proportional to the reflected energy, which is a function of acoustic impedance of the wall of the hole. Ultrasonic scanners, lowered down drillholes, have been used to survey the interior of abandoned mine workings that are flooded (Braithewaite and Cole, 1986). Horizontal scans are made over the height of the old workings and then tilting scans are taken from the vertical to the horizontal. The echoes received during the horizontal scans are processed by computer to give a plan view of the mine and vertical sections are produced from the vertical scans.

Adits are at times used on major construction projects such as dams, tunnels and underground chambers. They are driven into sloping ground or as headings from shafts. Support is needed in soft ground and some rock conditions. Adits provide an opportunity for detailed examination of the ground conditions and frequently are used for *in situ* testing purposes.

1.7 Recording discontinuity data

1.7.1 Direct discontinuity surveys

Before a discontinuity survey commences the area in question must be mapped geologically to determine rock types and delineate major structures. It is only after becoming familiar with the geology that the most efficient and accurate way of conducting a discontinuity survey can be devised. Comprehensive reviews of the procedure to be followed in discontinuity surveys have been provided by Barton (1978) and by Priest (1993).

One of the most widely used methods of collecting discontinuity data is simply by direct measurement on the ground. A direct survey can be carried out subjectively in that only those structures that appear to be important are measured and recorded. In a subjective survey the effort can be concentrated on the apparently significant joint sets. Nevertheless, there is a risk of overlooking sets that may be important. Conversely, in an objective survey, all structures intersecting a fixed line or area of a rock face are measured and recorded.

Several methods have been used for carrying out direct discontinuity surveys. Halstead *et al.* (1968) used the fracture set mapping technique by which all discontinuities occurring in 6 by 2 m zones, spaced at 30 m intervals along a face, were recorded. On the other hand, Piteau (1971) maintained that using a series of line scans provides a satisfactory method of joint surveying. The technique involves extending a metric tape across an exposure, levelling the tape and then securing it to the face. Two other scanlines are set out as near as possible at right angles to the first, one more or less vertical, the other horizontal. Four or five scanlines, running parallel to each other, are used in multiple scanline mapping. The distance along a tape at which each discontinuity intersects is noted, as is the direction of the pole to each discontinuity (this provides an indication of the dip direction). The dip of the pole from the vertical is recorded as this is equivalent to the dip of the plane from the horizontal. The strike and dip directions of discontinuities in the field can be measured with a compass and the amount of dip with a clinometer. Hadjigeorgiou *et al.* (1995) referred to the use of an automated structural logger to record the dip and dip direction of joint surfaces. It is linked to a portable data acquisition system. Measurement of the length of a discontinuity provides information on its continuity. It has been suggested that measurements should be taken over distances of about 30 m, and to ensure that the survey is representative, the measurements should be continuous over that distance. The line scanning technique yields more detail on the incidence of discontinuities and their attitude than other methods (Priest and Hudson, 1981). At least 200 readings per locality are recommended to ensure statistical reliability. A summary of the details that should be recorded regarding discontinuities is given in Fig. 1.22.

Hudson and Harrison (1997) pointed out that where discontinuities occur in sets, the discontinuity frequency along a scanline is a function of scanline orientation. They showed that the spacing distributions of discontinuities often are negative exponential distributions with the mean spacing of discontinuities being the reciprocal of the average number of discontinuities per metre. This value can be calculated by dividing the number of scanline intersections by the total scanline length. Furthermore, Hudson and Harrison maintained that the relationship between RQD and the average number of discontinuities per metre, λ, can be represented as follows:

$$\mathrm{RQD} = 100(0.1\lambda + 1)e^{-0.1\lambda} \tag{1.2}$$

They also showed how the distributions of block areas, for most locations, can be predicted adequately from discontinuity frequency measurements made along scanlines, and how to derive cumulative frequency curves for block volumes from scanline data.

1.7.2 Drillholes and discontinuity surveys

The information gathered by any of the above methods can be supplemented with data from orientated cores from drillholes. Drill core (and drillholes) represent line samples of rock mass. The value of the data depends in part on the quality of the drilling and of the rock concerned, in that poor quality rock is likely to be lost during drilling. Discontinuity orientation, frequency and number of sets usually cannot be adequately recorded from one hole without prior knowledge of the orientation and number of sets. However, if the core recovery is good and it has been orientated, then discontinuity spacing (length between adjacent discontinuities of one set along the core axis and the acute angle they subtend with the core axis) and orientation can be obtained. Core orientation can be achieved by a core orientator (Fig. 1.23). Barton (1987) suggested that it usually is possible to determine the general degree of planarity (planar, curved, irregular) and some degree of smoothness (slickensided, smooth, rough). The thickness of weathered material along the walls of discontinuities may be observed from the core and wall strength may be assessed by a Schmidt hammer. Unfortunately, it is impossible to assess the persistence of discontinuites from cores. What is more, infill material, especially if it is soft, is not recovered by the drilling operation unless triple-tube core barrels and controlled flushing, or better still, integral sampling is used. Similarly, the aperture of discontinuities cannot be determined unless integral sampling is used. Integral sampling is used to preserve the positions of discontinuities in a length of core stick (Rocha, 1971; Fig. 1.24).

Drillhole inspection techniques include the use of drillhole periscopes, drillhole cameras or closed-circuit television. The drillhole periscope affords direct inspection and can be orientated from outside the hole. However, its effective use is limited to about 30 m. The drillhole camera also can be orientated prior to photographing a section of the wall of a drillhole. The television camera provides a direct view of the drillhole and a recording can be made on videotape. These three systems are limited in that they require relatively clear conditions and so may be of little use below the water table, particularly if the water in the drillhole is murky. The televiewer produces an acoustic picture of the drillhole wall. One of its advantages is that drillholes need not be flushed prior to its use.

Discontinuity survey data sheet

General information

Seq. no. [ ] Site [ ] Date (Day Month Year) [ ] Operator [ ] Discontinuity data sheet No. [ ] of [ ]

Nature and orientation of discontinuity

Chainage or No.	Type	Dip	Dip direction	Persistence	Aperture	Nature of infilling	Consistency of infilling	Surface roughness	Trend of lineation	Waviness wavelength	Waviness amplitude	Water/flow	Remarks

Type
0 Fault zone
1 Fault
2 Joint
3 Cleavage
4 Schistosity
5 Shear
6 Fissure
7 Tension crack
8 Foliation
9 Bedding

Dip, Dip direction and trend of lineation
(expressed in degrees)

Persistence
(expressed in metres)

Aperture
1 Wide (>200 mm)
2 Mod. wide (60–200 mm)
3 Mod. narrow (20–60 mm)
4 Narrow (6–20 mm)
5 Very narrow (2–6 mm)
6 Ext. narrow (<2 mm)
7 Tight

Nature of infilling
1 Clean
2 Surface staining
3 Non-cohesive
4 Inactive clay or clay matrix
5 Swelling clay or clay matrix
6 Cemented
7 Chlorite, talc or gypsum
8 Others – specify

Compressive strength of infilling
1 Very soft (<40 kPa)
2 Soft (40–80 kPa)
3 Firm (80–150 kPa)
4 Stiff (150–300 kPa)
5 Very stiff (300–500 kPa)
6 Hard/v. weak (600–1250 kPa)
7 Weak (1.25–5 MPa)
8 Mod. weak (5–12.5 MPa)
9 Mod. strong (12.5–50 MPa)
10 Strong (50–100 MPa)
11 Very strong (100–200 MPa)
12 Ext. strong (>200 MPa)

Roughness
1 Polished
2 Slickensided
3 Smooth
4 Rough
5 Defined ridges
6 Small steps
7 Very rough

Waviness
(express wavelength and amplitude in metres)

Water
1 Dry
2 Seepage

Flow
3 <10 ml/s
4 10–100 ml/s
5 0.1–1 l/s
6 1–10 l/s
7 10–100 l/s
8 >100 l/s

Figure 1.22 Discontinuity survey data sheet.
Source: Anon., 1977. Reproduced by kind permission of The Geological Society.

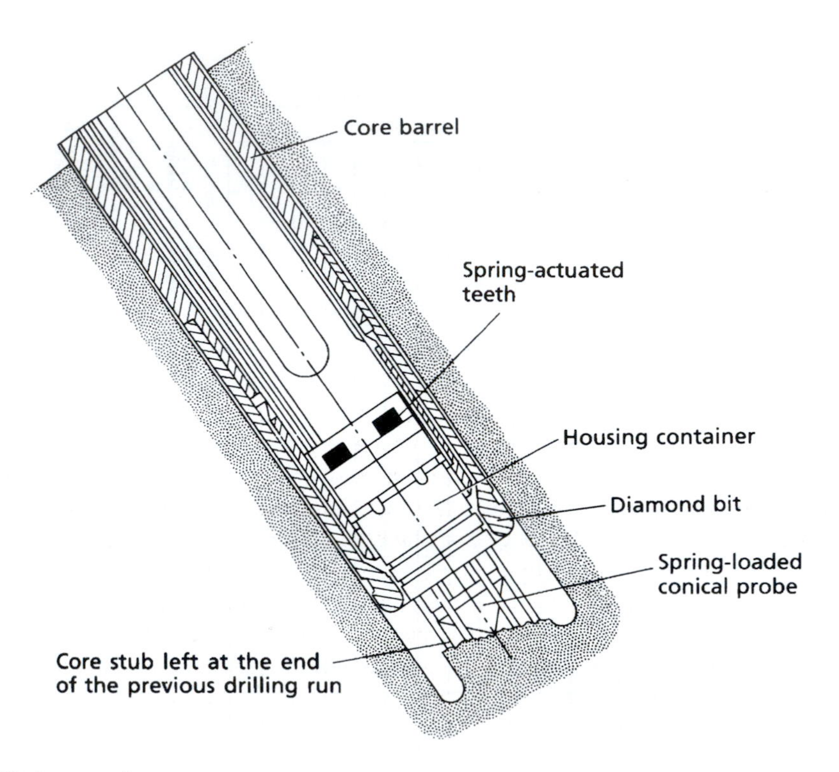

Figure 1.23 A core orientator.

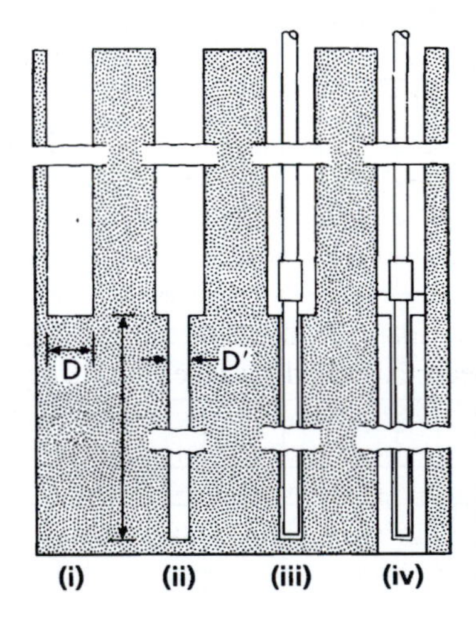

Figure 1.24 Stages of integral sampling: (i) start position; (ii) drilling hole for connecting rod; (iii) fixing connecting rod; (iv) overcoring for integral sample.

1.7.3 Photographs and discontinuity surveys

Many data relating to discontinuities can be obtained from photographs of exposures (Franklin *et al.*, 1988). Photographs may be taken looking horizontally at the rock mass from the ground or they may be taken from the air looking vertically, or occasionally obliquely, down at the outcrop. These photographs may or may not have survey control. Uncontrolled photographs are taken using hand-held cameras, stereo-pairs being obtained by taking two photographs of the same face from positions about 5 per cent of the distance of the face apart, along a line parallel to the face. Delineation of major discontinuity patterns and preliminary subdivision of the face into structural zones can be made from these photographs. Unfortunately, data cannot be transferred with accuracy from them onto maps and plans. On the other hand, discontinuity data can be accurately located on maps and plans by using controlled photographs. Controlled photographs are obtained by aerial photography with complementary ground control or by ground-based phototheodolite surveys. Aerial and ground-based photography usually are done with panchromatic film but the use of colour and infrared techniques is becoming more popular. Aerial photographs, with a suitable scale, have proved useful in the investigation of discontinuities. Photographs taken with a phototheodolite also can be used with a stereo-comparator that produces a stereoscopic model. Measurements of the locations or points in the model can be made with an accuracy of approximately 1 in 5000 of the mean object distance. As a consequence, a point on a face photographed from 50 m can be located to an accuracy of 10 mm. In this way the frequency, orientation and continuity of discontinuities can be assessed. Such techniques prove particularly useful when faces that are inaccessible or unsafe have to be investigated.

1.7.4 Recording discontinuity data

The simplest method of recording discontinuity data is by using a histogram on which the frequency is plotted along one axis and the strike direction along the other. Directional information, however, is represented more effectively on a rose diagram. This provides a graphical illustration of the angular relationships between joint sets. The strikes of the joints and their frequencies are represented by the directions on each rose diagram, the lengths of the vectors being plotted either on a half or full circle. Directions usually are plotted for data contained in 5° arcs, while magnitudes are plotted to scale.

Today, however, data from a discontinuity survey usually are plotted on a stereographic projection. The use of spherical projections, commonly the Schmidt or Wulf net, means that traces of the planes on the surface of the 'reference sphere' can be used to define the dips and dip directions of discontinuity planes. In other words, the inclination and orientation of a particular plane are represented by a great circle or a pole, normal to the plane, which are traced on an overlay placed over the stereonet. The method whereby great circles or poles are plotted on a stereogram has been explained by Hoek and Bray (1981). When recording field observations of the amount and direction of dip of discontinuities it is convenient to plot the poles rather than the great circles. The poles then can be contoured (e.g. by using the computer program DIPS) in order to provide an expression of orientation concentration. This affords a qualitative appraisal of the influence of the discontinuities on the engineering behaviour of the rock mass concerned (see Section 2.6).

1.7.5 Discontinuities and rock quality indices

Several attempts have been made to relate the numerical intensity of fractures to the quality of unweathered rock masses and to quantify their effect on deformability. For example, the concept of rock quality designation (RQD) was introduced by Deere (1964). It is based on the percentage core recovery when drilling rock with NX (57.2 mm) or larger diameter diamond core drills. Assuming that a consistent standard of drilling can be maintained, the percentage of solid core obtained depends on the strength and frequency of discontinuities in the rock mass concerned. The RQD is the sum of the core sticks in excess of 100 mm (measured along the core axis) expressed as a percentage of the total length drilled. Hawkins (1986) suggested that it would be better if the length measured when assessing RQD was the distance between discontinuities in solid core rather than simply along the centreline. Barton (1987) proposed that RQD values should be determined for variable rather than fixed lengths of core run. For instance, values for individual beds or for weak zones would indicate variability in a sequence and would afford a more reliable assessment of the location and width of zones with low RQD. The RQD does not take account of the joint opening and condition, a further disadvantage being that with fracture spacings greater than 100 mm, the quality is excellent irrespective of the actual spacing (Table 1.7). This particular difficulty can be overcome by using the fracture spacing index as suggested by Franklin *et al.* (1971). This simply refers to the frequency, per metre run, with which fractures occur within a rock core (Table 1.7).

Hobbs (1975) introduced the concept of the rock mass factor, j, which he defined as the ratio of the deformability of a rock mass within any readily identifiable lithological and structural component to that of the deformability of the intact rock comprising the component. Consequently, it reflects the effect of discontinuities on the expected performance of the intact rock (Table 1.7). The value of j depends upon the method of assessing the deformability of the rock mass, and the value beneath an actual foundation is not necessarily the same as that determined even from a large-scale field test. According to Hobbs, the greatest difficulties that occur in a jointed rock mass in relation to foundation design are experienced when the fracture spacing falls within a range of about 100–500 mm, in as much as small variations in fracture spacing and condition result in exceptionally large changes in j value.

The effect of discontinuities in a rock mass can be estimated by comparing the *in situ* compressional wave velocity with the laboratory sonic velocity of an intact core sample obtained from the rock mass. The difference in these two velocities is caused by the discontinuities that exist in the field. The velocity ratio, V_{pf}/V_{pl}, where V_{pf} and V_{pl} are the compressional wave velocities of the rock mass *in situ* and of the intact specimen respectively, was first proposed by Onodera (1963). For a high-quality massive rock with only a few tight joints, the velocity ratio approaches unity. As the degree of jointing and fracturing becomes more severe, the velocity ratio is reduced (Table 1.7). The sonic velocity is determined for the core sample in the laboratory under an axial stress equal to the computed overburden stress at the depth from which the rock material was taken, and at a moisture content equivalent to that assumed for the

Table 1.7 Classification of rock quality in relation to the incidence of discontinuities

Quality classification	*RQD (%)*	*Fracture frequency per meter*	*Mass factor (j)*	*Velocity ratio (V_{cf}/V_{cl})*
Very poor	0–25	Over 15		0.0–0.2
Poor	25–50	15–8	Less than 0.2	0.2–0.4
Fair	50–75	8–5	0.2–0.5	0.4–0.6
Good	75–90	5–1	0.5–0.8	0.6–0.8
Excellent	90–100	Less than 1	0.8–1.0	0.8–1.0

in situ rock. The field seismic velocity preferably is determined by uphole or crosshole seismic measurements in drillholes or test adits, since by using these measurements, it is possible to explore individual homogeneous zones more precisely than by surface refraction surveys.

1.8 *In situ* testing

There are two categories of penetrometer tests, the dynamic and the static. Both methods measure the resistance to penetration of a conical point offered by the soil at any particular depth. Penetration of the cone creates a complex shear failure and thus provides an indirect measure of the *in situ* shear strength of the soil.

The most widely used dynamic method is the standard penetration test. This empirical test was developed initially for testing sand. It uses a split-spoon sampler, with an outside diameter of 50 mm, which is driven into the soil at the base of a borehole. Drivage is accomplished by a trip hammer, weighing 63 kg, falling freely through a distance of 750 mm onto a drive head, which is fitted at the top of the rod assembly (Fig. 1.25). First, the split-spoon sampler is driven 150 mm into the soil at the bottom of the borehole. Then it is driven a further 300 mm and the number of blows required to drive this distance is recorded. The blow count is referred to as the N value, from which the relative density, degree of compaction and angle of friction (ϕ) of sandy soil can be assessed (Table 1.8). Refusal is regarded as 100 blows. The results obtained from the standard penetration test also can be used to evaluate the allowable bearing capacity (Fig. 1.26), details of the procedures being given in texts on soils mechanics or foundation

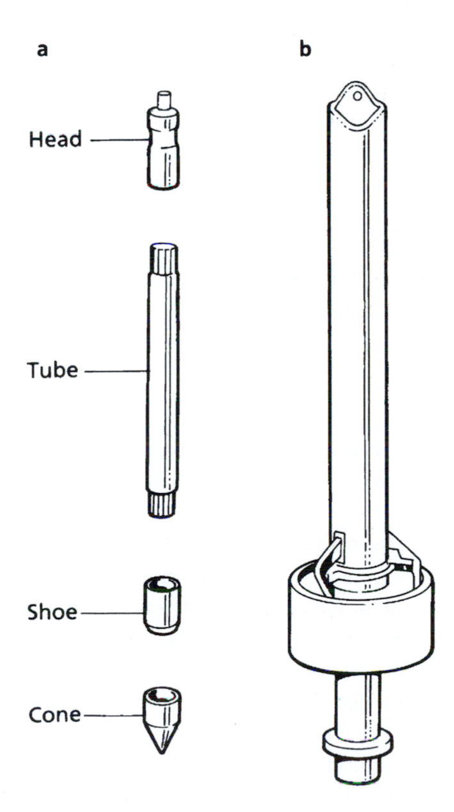

Figure 1.25 Standard penetration test equipment.

Table 1.8 Relative density and consistency of soil

(a) Relative density of sand and SPT values, and relationship to angle of friction

SPT (N)	*Relative density* (D_r)	*Description of compactness*	*Angle of internal friction* (φ)
4	0.2	Very loose	Under 30°
4–10	0.2–0.4	Loose	30–35°
10–30	0.4–0.6	Medium dense	35–40°
30–50	0.6–0.8	Dense	40–45°
Over 50	0.8–1.0	Very dense	Over 45°

(b) N-values, consistency and unconfined compressive strength of cohesive soils

N	*Consistency*	*Unconfined compressive strength* (kPa)
Under 2	Very soft	Under 20
2–4	Soft	20–40
5–8	Firm	40–75
9–15	Stiff	75–150
16–30	Very stiff	150–300
Over 30	Hard	Over 300

engineering (e.g. Simons and Menzies, 2000). Burland and Burbridge (1985) provided a method for deriving settlement from SPT results. In deep boreholes the standard penetration test suffers the disadvantage that the load is applied at the top of the rods so that some of the energy from the blow is dissipated in the rods. Hence, with increasing depth the test results become more suspect.

Terzaghi and Peck (1967) suggested that for very fine or silty submerged sand with a standard penetration value N' greater than 15, the relative density would be nearly equal to that of a dry sand with a standard penetration value N where:

$$N = 15 + \frac{1}{2}(N' - 15) \tag{1.3}$$

If this correction was not made, Terzaghi and Peck suggested that the relative density of even moderately dense very fine or silty submerged sand might be overestimated by the results of standard penetration tests.

In gravel deposits care must be taken to determine whether a large gravel size may have influenced the results. Usually, in the case of gravel, only the lowest values of N are taken into account. The lowest values of the angle of internal friction given in Table 1.8 are conservative estimates for uniform, clean sand and they should be reduced by at least 5° for clayey sand. The upper values apply to well-graded sand and may be increased by 5° for gravelly sand. The standard penetration test can also be employed in clays (Table 1.8), weak rocks and in the weathered zones of harder rocks.

The most widely used static method employs the Dutch cone penetrometer (Fig. 1.27). It is particularly useful in soft clays and loose sands where boring operations tend to disturb *in situ* values (Meigh, 1987). In this technique a tube and inner rod with a conical point at the base are

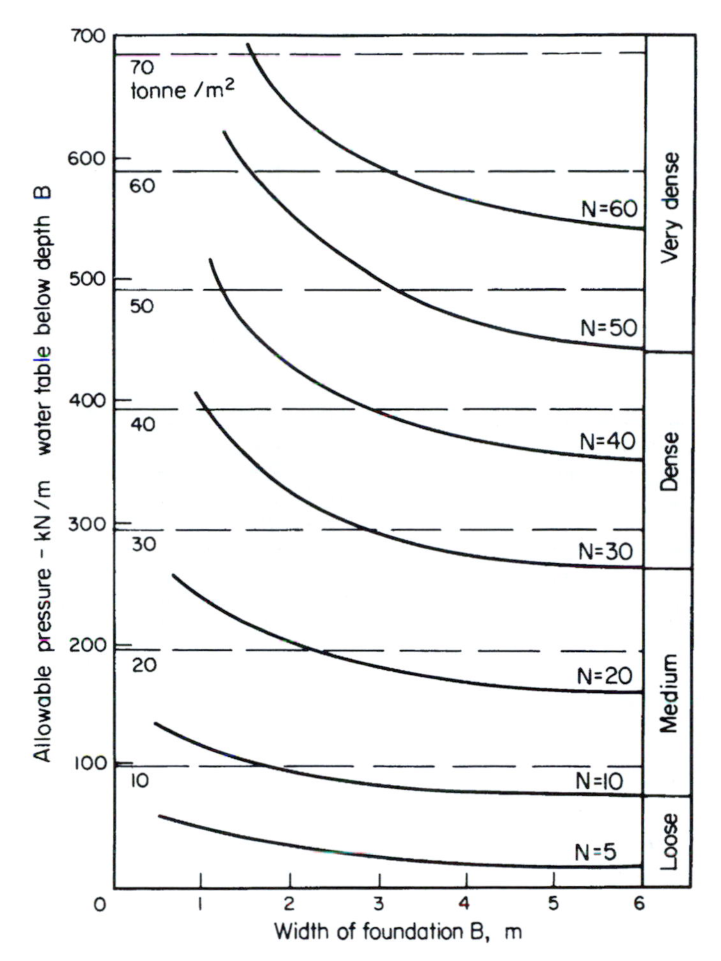

Figure 1.26 Chart for estimating the allowable bearing pressure for 25 mm of settlement in sand from the results of the standard penetration test.
Source: Terzaghi and Peck, 1967. Reproduced by kind permission of John Wiley & Sons, Inc.

hydraulically advanced into the ground, the reaction being obtained from pickets screwed in place. The cone has a cross-sectional area of 1000 mm^2 with an angle of 60° (Sanglerat, 1972). At approximately every 300 mm depth, the cone is advanced ahead of the tube a distance of 50 mm and the maximum resistance is noted. Then, after measurement, the tube is advanced to join the cone and the process repeated. The resistances are plotted against their corresponding depths so as to give a profile of the variation in consistency (Fig. 1.28). One type of Dutch cone penetrometer has a sleeve behind the cone that can measure side friction. The ratio of sleeve resistance to that of cone resistance is higher in fine than coarse-grained soils thus affording some estimate of the type of soil involved.

Because soft clay soils may suffer disturbance when sampled and therefore give unreliable results when tested for strength in the laboratory, a vane test is often used to measure the *in situ* undrained shear strength. Vane tests can be used in clay soils that have a consistency varying from very soft to firm. In its simplest form the shear vane apparatus consists of four blades arranged in cruciform fashion and attached to the end of a rod (Fig. 1.29). To eliminate the

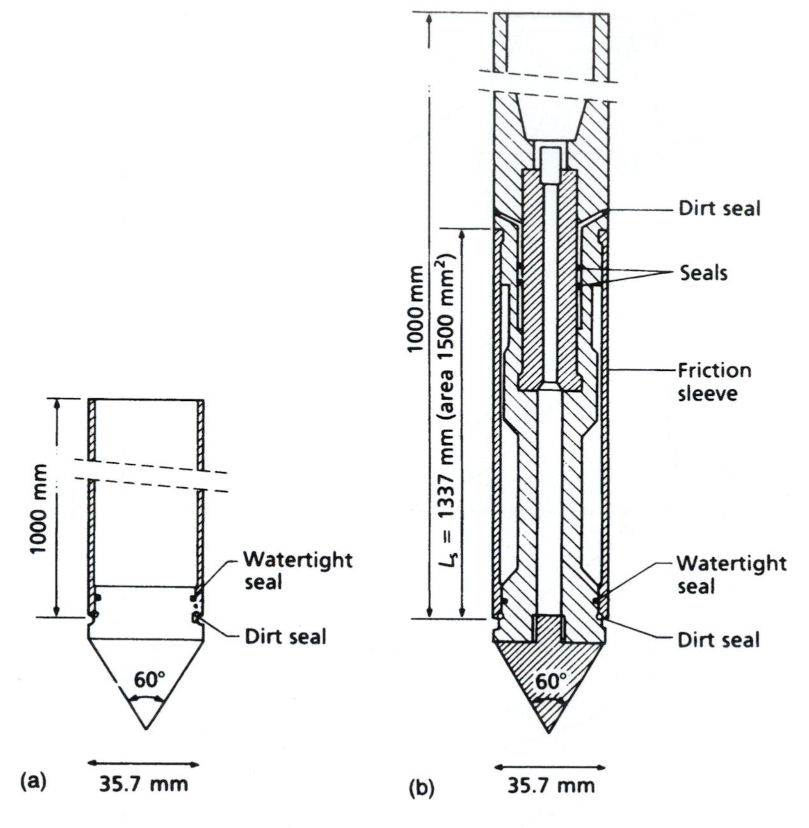

Figure 1.27 Cone penetrometer test equipment: (a) without friction sleeve; (b) with friction sleeve.

effects of friction of the soil on the vane rods during the test, all rotating parts, other than the vane, are enclosed in guide tubes. The vane normally is housed in a protective shoe. The vane and rods are pushed into the soil from the surface or the base of a borehole to a point 0.5 m above the required depth. Then the vane is pushed out of the protective shoe and advanced to the test position, where it is rotated at a rate of 6–12° per minute. The torque is applied to the vane rods by means of a torque measuring instrument mounted at ground level and clamped to the borehole casing or rigidly fixed to the ground. The maximum torque required for rotation is recorded. When the vane is rotated the soil fails along a cylindrical surface defined by the edges of the vane, as well as along the horizontal surfaces at the top and bottom of the blades. The shearing resistance is obtained from the following expression:

$$\tau = \frac{M}{\pi\left(\dfrac{D^2 H}{2} + \dfrac{D^3}{6}\right)} \tag{1.4}$$

where τ is the shearing resistance, D and H are the diameter and height of the vane respectively, and M is the torque. Tests in clay soils with a high organic content or with pockets of sand or silt are likely to produce erratic results. The results should therefore be related to borehole evidence.

Loading tests can be carried out on loading plates. The plate load test provides valuable information by which the bearing capacity and settlement characteristics of a foundation can be

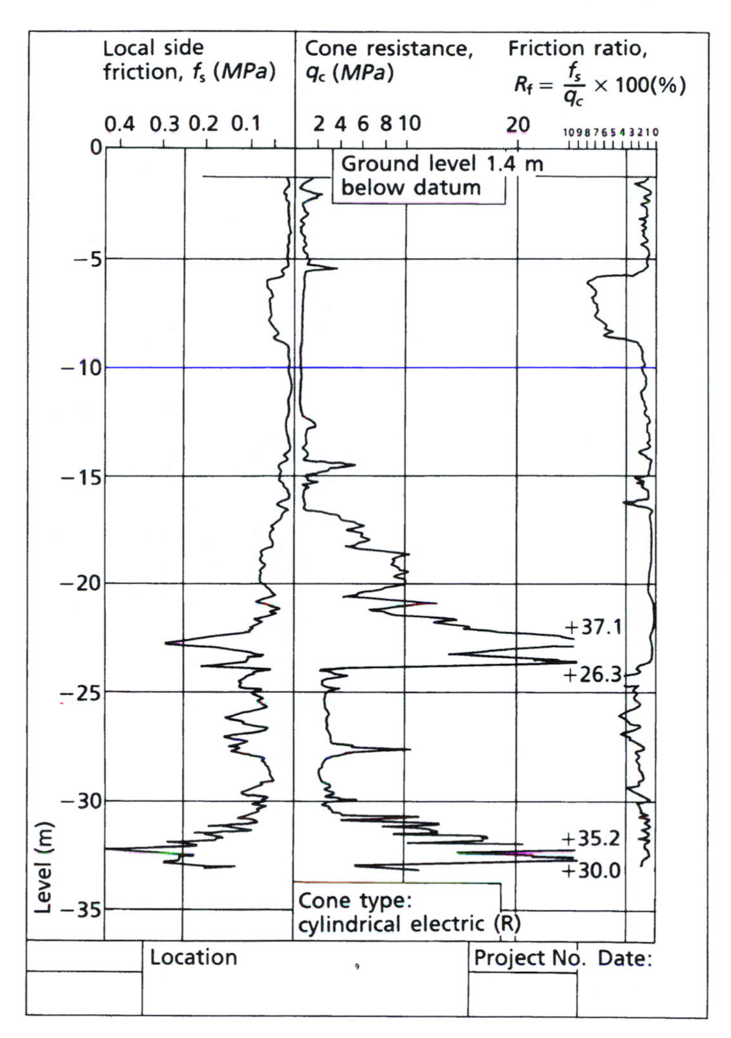

Figure 1.28 Typical record of cone penetrometer test.

assessed. However, just because the ground immediately beneath a plate is capable of carrying a heavy load without excessive settlement, this does not necessarily mean that the ground will carry the proposed structural load. This is especially the case where a weaker horizon occurs at depth but is still within the influence of the bulb of pressure that will be generated by the structure (Fig. 1.30). A plate load test is carried out in a trial pit, usually at excavation base level (Fig. 1.31). Plates vary in size from 0.15 to 0.61 m in diameter, the size of plate used being determined by the spacing of discontinuities. The plate should be properly bedded and the test carried out on undisturbed material so that reliable results can be obtained. The load is applied by a jack, in increments, either of one-fifth of the proposed bearing pressure or in steps of 25–50 kPa (these are smaller in soft soils, that is, where the settlement under the first increment of 25 kPa is greater than $0.002D$, D being the diameter of the plate). Successive increments should be made after settlement has ceased. The test generally is continued up to two or three times the proposed loading, or in clay soils until settlement equal to 10–20 per cent of the plate

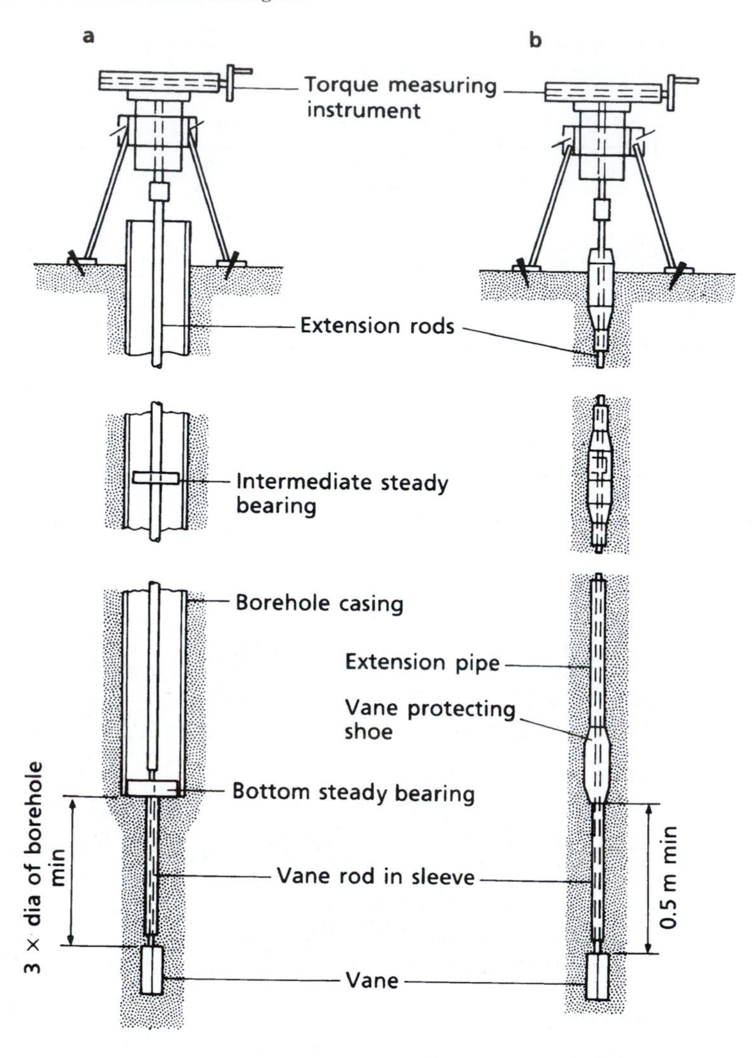

Figure 1.29 Shear vane tests: (a) borehole vane test; (b) penetration vane test.

dimension is reached or the rate of increase of settlement becomes excessive. Consequently, the ultimate bearing capacity at which settlement continues without increasing the load rarely is reached. When the final increment is applied in clay soils the load should be maintained until the rate of settlement becomes less than 0.1 mm in 2 hours. This can be regarded as being the completion of the primary consolidation stage. Settlement curves can be drawn with this information from which the ultimate loading can be determined and an evaluation of Young's modulus made. At the end of the consolidation stage the plate can be unloaded in the same incremental steps in order to obtain an unloading curve.

The screw plate is a variant of the plate load test in which a helical screw is rotated into the ground to the depths at which the test is to be conducted (Kay and Parry, 1982). The test has the advantage that no excavation or drilling is needed and it can be performed beneath the water table. Unfortunately, however, screwing the plate into the soil may cause disturbance around the plate.

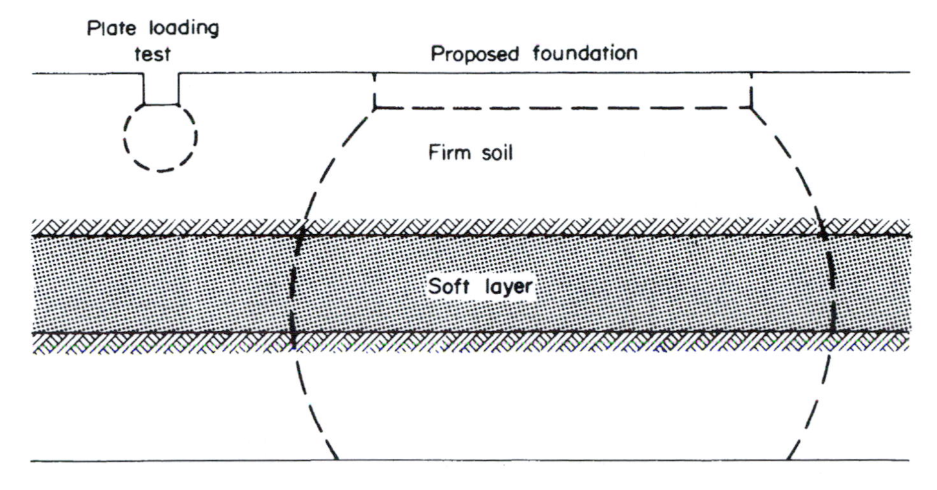

Figure 1.30 Bulb of pressure developed beneath a foundation compared with one developed beneath a plate load test.

Figure 1.31 Plate-bearing test using rock anchors to provide the reaction.

Large plate-bearing tests are frequently used to determine the value of Young's modulus of the foundation rock mass at large civil engineering sites, such as dam sites. Loading of the order of several meganewtons is required to obtain measurable deformation of representative areas. The area of rock load is usually $1\,m^2$. Tests generally are carried out in specially excavated galleries in order to provide a sufficiently strong reaction point for the loading jacks to bear against. The test programme normally includes cycles of loading and unloading. Such tests show that during loading, a noticeable increase in rigidity occurs in the rock mass and that during unloading, a very small deformation occurs for the high stresses applied, with very large recuperation of deformations being observed for stresses near zero. This is due to joint closure. Once the joints are closed, the adhesion between the faces prevents their opening until a certain unloading is reached. However, when brittle rocks like granite, basalt and some limestones have been tested, they generally have given linear stress–strain curves and have not exhibited hysteresis.

Variations of this type of test include the freyssinet jack. This is placed in a narrow slit in the rock mass and then grouted into position so that each face is in uniform contact with the rock (Wareham and Skipp, 1974). Pressure then is applied to the jack. Unless careful excavation, particularly blasting, takes place in the testing area, the results of a flatjack test may be worthless. All loose material must be removed before cutting the slot. Flatjacks can be used to measure residual stress as well as Young's modulus.

The Menard pressuremeter is used to determine the *in situ* strength of the ground (Fig. 1.32). It is particularly useful in those soils from which undisturbed samples cannot be obtained readily (Mair and Wood, 1987). This pressuremeter consists essentially of a probe that is placed in a

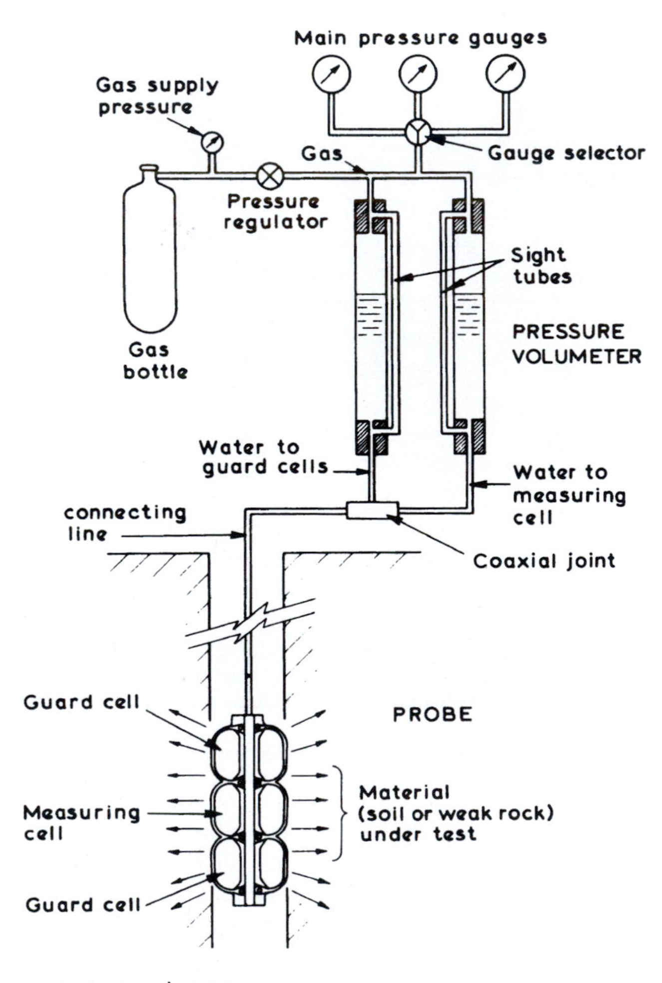

Figure 1.32 Pressuremeter test equipment.

borehole at the appropriate depth and then expanded. Where possible the test is carried out in an unlined hole, but if necessary a special slotted casing is used to provide support. The probe consists of a cylindrical metal body over which are fitted three cylinders. A rubber membrane covers the cylinders and is clamped between them to give three independent cells. The cells are inflated with water and a common gas pressure is applied by a volumeter located at the surface; thus a radial stress is applied to the soil. The deformations produced by the central cell are indicated on the volumeter. A simple pressuremeter test consists of ten or more equal pressure increments with corresponding volume change readings, taken to the ultimate failure strength of the soil concerned. Four volume readings are made at each pressure step at time intervals of 15, 30, 60 and 120 s after the pressure has stabilized. It is customary to unload the soil at the end of the elastic phase of expansion and to repeat the test before proceeding to the ultimate failure pressure. The results of each test are presented in the form of two graphs, that is, the pressure-volume change curve (PV) and the creep curve (Fig. 1.33). The PV curve can be divided into three parts. First, the initial part of the curve corresponds to the phase in which the earth stress is restored. Second, that part of the curve that is more or less straight is regarded as the elastic phase, and Young's modulus is derived from its slope. Third, the part of the curve in which the rate of volume change accelerates represents the phase of plastic deformation. This test provides the ultimate bearing capacity of soils, as well as their deformation modulus (Anderson *et al*., 1990). The test can be applied to any type of soil, and takes into account the influence of discontinuities. It also can be used in weathered zones of rock masses and in weak rocks such as shales. It provides an almost continuous method of *in situ* testing.

The major advantage of a self-boring pressuremeter is that a borehole is unnecessary. Accordingly, the interaction between the probe and the soil is improved. Self-boring is brought about either by jetting or using a chopping tool (Fig. 1.34). For example, the camkometer has a special cutting head so that it can be drilled into soft ground to form a cylindrical cavity of its exact dimensions and thereby create a minimum of disturbance. The camkometer measures the lateral stress, undrained stress–strain properties and the peak stress of soft clays and sands *in situ*. Clarke (1990) described the use of the self-boring pressuremeter test to determine the *in situ* consolidation characteristics of clay soils. An account of how a self-boring pressuremeter was used to assess the *in situ* lateral stress in the London Clay has been provided by Corke (1990).

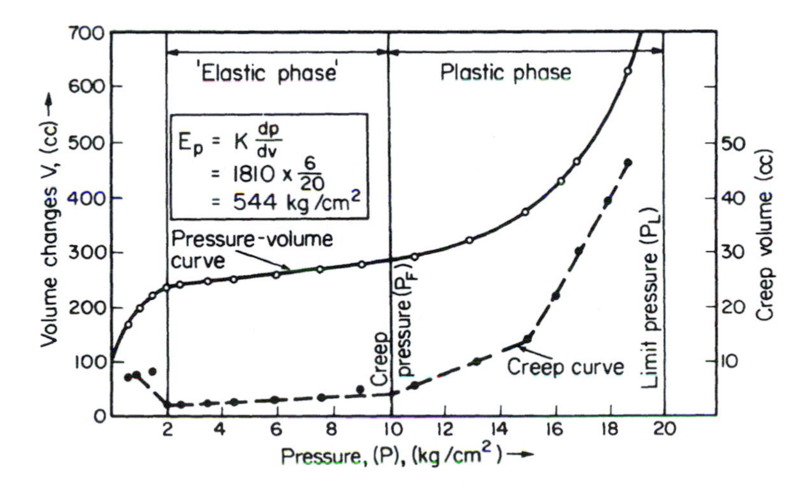

Figure 1.33 Typical pressuremeter test curve.

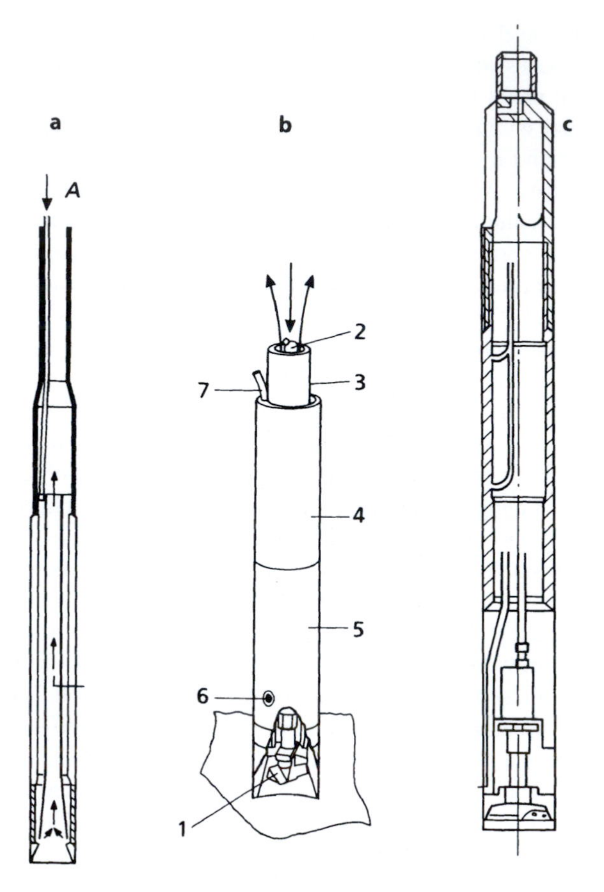

Figure 1.34 (a) Self-boring pressuremeter available in 65, 100 and 132 mm diameters. The injection fluid flow that can be applied is limited by the section of flexible pipe (A). Consequently, this pressuremeter only can be used in fine sandy or soft clayey soils. (b) The camkometer: the soil is broken by a chopping tool rotated by a drill string from the surface. The chopping tool (1) is driven by a hollow middle rod (2) through which water is injected under pressure. This rod turns freely inside a tube (3) used for removing the soil to the surface. There is also a tube (4) that carries the pressiometric cell (5) that may be equipped with a pore pressure tap (6). The pressiometric cell supply and measurement lines (7) run through the annulus between the two tubes. (c) Self-boring pressuremeter probe with built-in motor.

A dilatometer can be used in a drillhole to obtain data relating to the deformability of a rock mass (Fig. 1.35). These instruments range up to about 300 mm in diameter and over 1 m in length, and can exert pressures of up to 20 MPa on the drillhole walls (Rocha *et al.*, 1966). Diametral strains can be measured either directly along two perpendicular diameters or by measuring the amount of liquid pumped into the instrument. The last method is less accurate and is only used when the rock is very deformable.

In an *in situ* shear test a block of rock is sheared from the rock surface whilst a horizontal jack exerts a vertical load. It is advantageous to make the tests inside galleries, where reactions for the jacks are readily available (Fig. 1.36). The tests are performed at various normal loads and give an estimate of the angle of shearing resistance and cohesion of the rock. The value of this test in very jointed and heterogenous rocks is severely limited both because of the difficulty in isolating undisturbed test blocks and because the results cannot be translated to the scale of

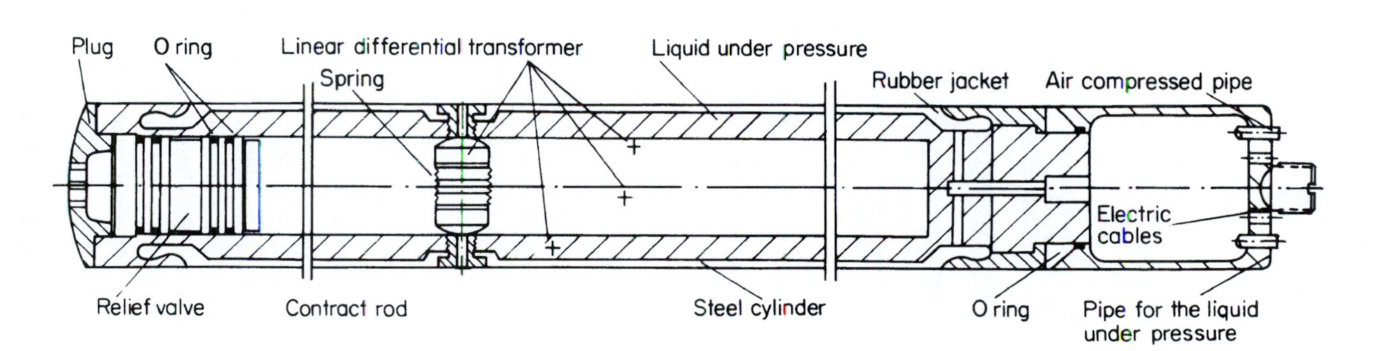

Figure 1.35 The dilatometer.

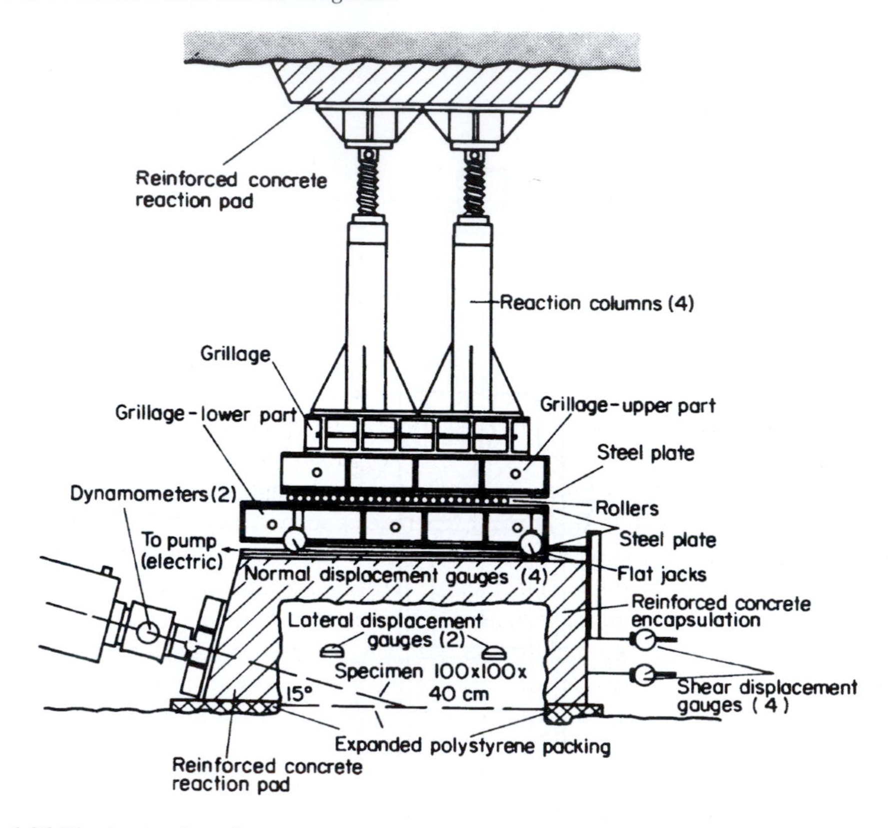

Figure 1.36 The *in situ* shear box test.

conditions of the actual structure and its foundation. *In situ* shear tests usually are performed on blocks, 700×700 mm, cut in a rock mass. These tests can be made on the same rock where it shows different degrees of alteration and can be used to derive the shear strength parameters along notable discontinuities.

1.9 Indirect methods of exploration: geophysical techniques

A geophysical exploration may be included in a site investigation for an important engineering project in order to provide subsurface information over a large area at reasonable cost. The information obtained may help eliminate less favourable alternative sites, may aid the location of test holes in critical areas and may prevent unnecessary repetitive drilling in fairly uniform ground. A geophysical survey helps to locate the position of boreholes and also detects variations in subsurface conditions between boreholes. Boreholes provide information about the strata where they are sunk but tell nothing about the ground in between. Nonetheless, boreholes to aid interpretation and correlation of the geophysical measurements are an essential part of any geophysical survey. Therefore, an appropriate combination of direct and indirect methods often can yield a high standard of results.

Geophysical methods are used to determine the geological sequence and structure of subsurface rocks by the measurement of certain physical properties or forces. The properties that are made most use of in geophysical exploration are density, elasticity, electrical conductivity, magnetic susceptibility and gravitational attraction. For example, seismic and resistivity methods record artificial fields of force applied to the area under investigation whilst magnetic and gravitational methods measure natural fields of force. The former techniques have the advantage over the latter in that the depth to which the forces are applied can be controlled. By contrast, the natural fields of force are fixed and can only be observed, not controlled. Seismic and resistivity methods are more applicable to the determination of horizontal or near horizontal changes or contacts, whereas magnetic and gravimetric methods generally are used to delineate lateral changes or vertical structures.

In a geophysical survey, measurements of the variations in certain physical properties usually are taken in a traverse across the surface thereby providing more or less continuous data along the traverse. Geophysical measurements also may be used to log a borehole. Geophysical surveys can be carried out on land, over water or from the air, depending on the properties being measured. Anomalies in the physical properties measured generally reflect anomalies in the geological conditions. The ease of recognizing and interpreting these anomalies depends on the contrast in physical properties, depth of the target, and the nature and thickness of the overburden that, in turn, influence the choice of the method employed.

The actual choice of method to be used for a particular survey may not be difficult to make. The character and situation of the site have to be taken into account, especially in built-up areas, which may be unsuitable for one or other of the geophysical methods either because of the presence of old buildings or services on the site, interference from some source, or lack of space for carrying out the survey. When dealing with layered rocks, provided their geological structure is not too complex, seismic methods have a distinct advantage in that they give more detailed, precise and unambiguous information than any other method. On the other hand, electrical methods may be preferred for small-scale work where the structures are simple. However, as McCann *et al.* (1997) pointed out, if there is any doubt about the feasibility of a geophysical survey, then a trial survey should be undertaken to determine the most suitable method. The results of the trial survey are used to refine the specification for the main geophysical survey or may indicate that additional geophysical work will not provide the data required. Furthermore, there may be occasions when a single geophysical method may not provide sufficient data about subsurface conditions so that more than one method must be employed. McCann *et al.* also noted that modern geophysical equipment allows downloading of the results to a suitable portable PC so that a preliminary interpretation can be carried out daily. In this way data can be checked, plotted and evaluated, and then can be compared with data gathered by other means. As a result, any errors can be recognized and, if necessary, traverses re-run to gather better information. Indeed, the survey programme can be modified in the light of the new information obtained.

Generally speaking, observations should be close enough for correlation between them to be obvious, so that enabling interpolation could be carried out without ambiguity. Nevertheless, it must be admitted that an accurate and unambiguous interpretation of geophysical data is only possible where the subsurface structure is simple and even then there is no guarantee that this will be achieved.

1.9.1 Seismic methods

The sudden release of energy from the detonation of an explosive charge in the ground or the mechanical pounding of the ground surface generates shock waves that radiate out in a hemispherical wave front from the point of release. Mechanical sources of energy include electromechanically vibrating the ground, striking the ground with a gas propelled piston and striking a plate on the ground with a sledgehammer. In marine surveys, a sparker (an electric arc discharge device) generally is used where shallow depth penetration is required. Alternatively, a boomer (an electromechanical sound source) can be used. When seismic waves pass from one layer to another some energy is reflected back towards the surface whilst the remainder is refracted. Thus, two methods of seismic surveying can be distinguished, that is, seismic reflection and seismic refraction. Measurement of the time taken from the generation of the shock waves until they are recorded by detector arrays of geophones forms the basis of the two methods.

The waves generated are compressional (P), dilational shear (S) and surface waves. The velocities of the P and S waves are derived as follows:

$$V_p = \sqrt{\frac{\left(\frac{K+4}{3G}\right)}{\rho}} \tag{1.5a}$$

$$V_p = \sqrt{\frac{E(1-\nu)}{\rho(1+\nu)(1-2\nu)}} \tag{1.5b}$$

$$V_s = \sqrt{\frac{G}{\rho}} \tag{1.5c}$$

$$V_s = \sqrt{\frac{E}{2\rho(1+\nu)}} \tag{1.5d}$$

where K is the bulk modulus or compressibility, G is the shear modulus or rigidity, ρ is the density, E is Young's modulus and ν is Poisson's ratio. The compressional waves travel faster and are more easily generated and recorded than shear waves. Therefore, they are used most in seismic exploration. The velocity of shock waves depends mainly upon the elastic modulus of the media, which is influenced by density, fabric, mineralogy and pore water. Velocity generally increases with depth below the surface since the elastic moduli increase with depth. In general, velocities in crystalline rocks are high to very high (Table 1.9). Velocities in sedimentary rocks increase concomitantly with consolidation and decrease in pore fluids, and with increase in the degree of cementation and diagenesis. Unconsolidated sedimentary accumulations have maximum velocities varying as a function of the volume of voids, either air filled or water filled, mineralogy and grain size. Poorly consolidated dry materials have very low P wave velocities and absorb S waves, and do not respond elastically. Saturated, poorly consolidated materials

Table 1.9 Velocities of compressional waves of some common rocks

	V_p (kms^{-1})		V_p (kms^{-1})
Igneous rocks		*Sedimentary rocks*	
Basalt	5.2–6.4	Gypsum	2.0–3.5
Dolerite	5.8–6.6	Limestone	2.8–7.0
Gabbro	6.5–6.7	Sandstone	1.4–4.4
Granite	5.5–6.1	Shale	2.1–4.4
Metamorphic rocks		*Unconsolidated deposits*	
Gneiss	3.7–7.0	Alluvium	0.3–0.6
Marble	3.7–6.9	Sands and gravels	0.3–1.8
Quartzite	5.6–6.1	Clay (wet)	1.5–2.0
Schist	3.5–5.7	Clay (sandy)	2.0–2.4

tend to have velocities a little higher than water, and the water table may be a notable seismic interface in such materials. Normally, velocities travelling in the directions parallel to planar structures in anisotropic rocks are greater than in directions perpendicular to these structures.

In the seismic reflection method the travel times of shock waves reflected from subsurface interfaces are recorded at the surface to determine information regarding the depths and shapes of the interfaces. The amount of wave energy reflected from an interface separating two media depends upon the contrast in acoustic impedance (density × seismic velocity) of the media on the one hand and the angle of incidence of the shock wave at the interface on the other (Sharma, 1997). Reflection surveying formerly was used almost exclusively in the oil industry where the depth of investigation is large compared with the distance from the shotpoint to detector array, thereby excluding refraction waves. Indeed, the method is able to record information from a large number of horizons down to depths of several thousands of metres. However, advances in instrumentation and data collection in the 1980s have meant that shallow reflection can be used to target depths at tens of metres so that the method can be used for civil engineering purposes. The number of linear shotpoints in shallow reflection surveys is often an order of magnitude greater than in conventional refraction surveys. Signal enhancement in the field is brought about by stacking records from multiple inputs of the same energy source at the same shotpoint and so energy input should be from a repetitive source.

A walkaway noise test affords an initial appreciation of the seismic signals and noise (extraneous data) in the area being surveyed. It is conducted with a closely spaced geophone array that is shot from both ends at increasing intervals, with the geophones remaining in the same positions. The distance between geophones is small (a metre or so) and depends on the target depth. Shooting from both ends allows detection of dipping reflectors.

Arrays for reflection surveys are always shorter than for refraction surveys if probing to the same sort of depth. The distance from the origin of the shock waves to the nearest geophone depends upon the strength of the shock waves at the origin. For example, it may be as small as 2 m if the shock waves are generated by a sledgehammer striking a plate, but even when explosive charges are used, it normally does not exceed 10 m for shallow depth surveys. The most satisfactory array must be determined by trial, as the most important factors are the arrival times of the direct wave and any strong refracted waves. Reflected waves ideally should arrive after refracted waves if the depth of investigation is very small. Where a horizontal interface separates two formations in which the shock wave velocities are V_1 and V_2 respectively (Fig. 1.37), then the travel time, T, for the reflection path is given by:

$$T = \frac{2\sqrt{\left(Z^2 + \frac{x^2}{4}\right)}}{V_1} \tag{1.6a}$$

or

$$T^2 = T_o^2 + \frac{x^2}{V_1^2} \tag{1.6b}$$

where the normal incidence time $T_o = 2Z/V_1$, and the depth, Z, can be obtained from:

$$Z = \frac{1}{2}(V_1^2T^2 - x^2)^{1/2} \tag{1.7}$$

The increase in travel time $T - T_o$ due to offset x is referred to as the normal moveout, NMO, or the dynamic correction. The NMO helps determination of whether a seismic record is a

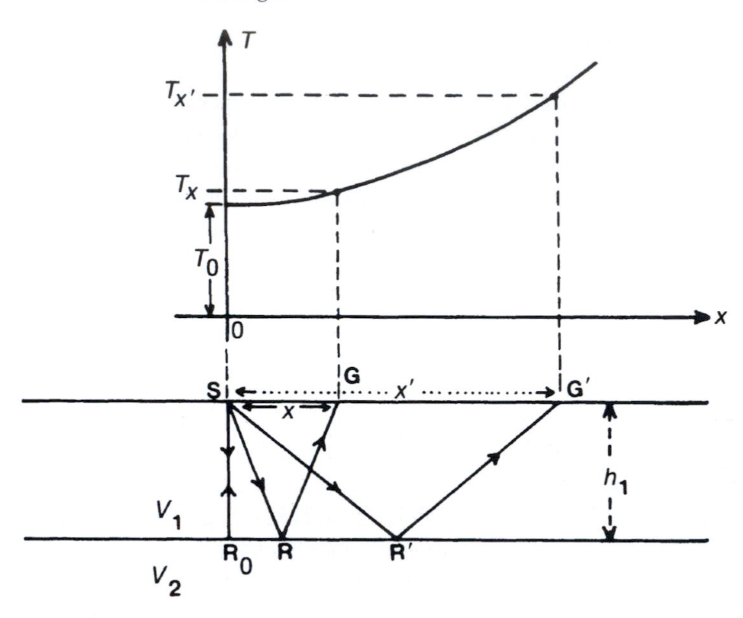

Figure 1.37 Principle of the reflection seismic method. The extent of the reflecting boundary mapped by geophones G, G′ is R, R′. The travel–time curve is a hyperbola and is shown in the upper part of the Figure. T_0 is the two-way vertical travel time, $2SR_0/V_1$.

reflection or not and so is used in processing reflection data. For small offsets, the expression for NMO ($T - T_o = x^2/2T_oV_1^2$) can be used to determine the approximate travel time. Where an interface has a uniform dip, the reduction in time on the up-dip side of the source of the shock waves compensates, up to a point, for the offset, and travel times that are shorter than the normal incidence time may be recorded. However, where the depth of investigation is shallow, these differences only are detectable when the arrays are long and the dips are large.

Refraction is governed by Snell's Law, which relates the angles of incidence, i, and refraction, r, to the seismic velocities of the two layers as follows:

$$\frac{\sin i}{\sin r} = \frac{V_1}{V_2} \tag{1.8}$$

If V_2 is greater than V_1, then refraction is towards the interface but if $\sin i$ is equal to V_1/V_2, then the refracted ray travels parallel to the interface. In the latter case, some energy returns to the surface after leaving the interface at the original angle of incidence. This provides the basis for the refraction method.

In the seismic refraction method one ray approaches the interface between two rock types at a critical angle that means that if the ray is passing from a low, V_1, to a high velocity, V_2, layer it will be refracted along the upper boundary of the latter layer (Fig. 1.39). After refraction, the pulse travels along the interface with velocity V_2. The material at the boundary is subjected to oscillating stress from below. This generates new disturbances along the boundary that travel upwards through the low-velocity rock and eventually reach the surface.

At short distances from the point where the shock waves are generated the geophones record direct waves whilst at a critical distance both the direct and refracted waves arrive at the same time. Beyond this, because the rays refracted along the high-velocity layer travel faster than those through the low-velocity layer above, they reach the geophones first. In refraction

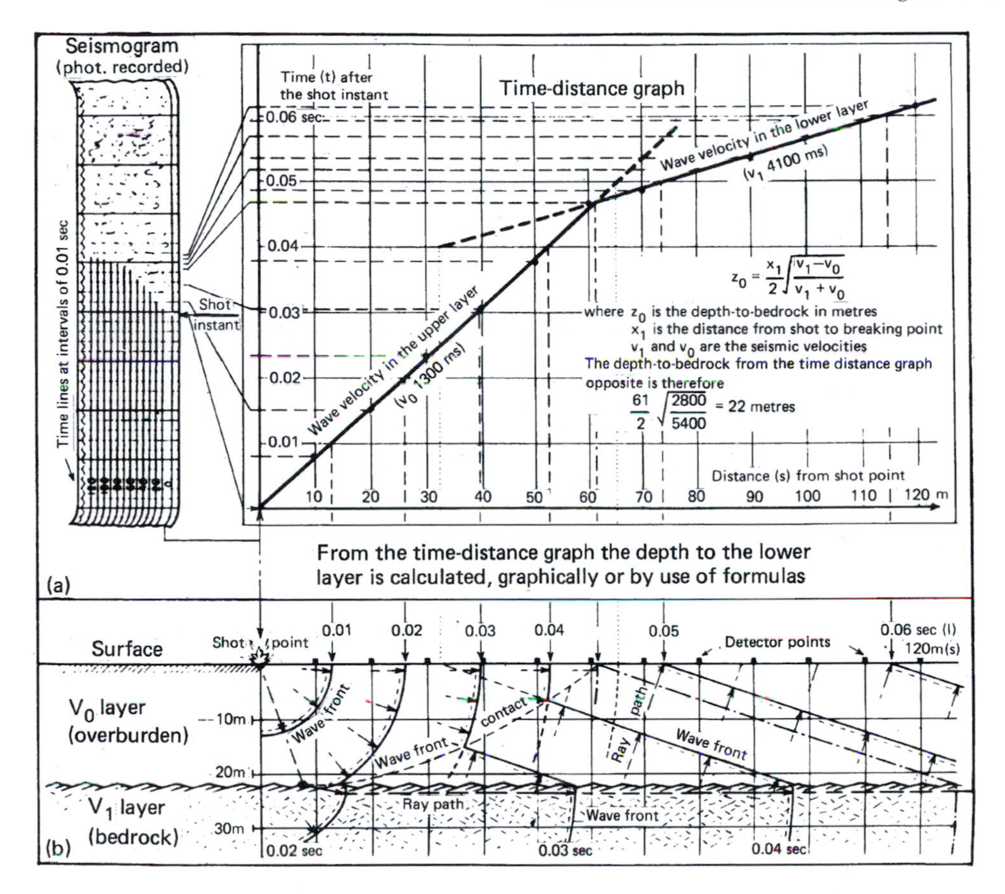

Figure 1.38 Time–distance graph of seismic refraction for a simple two-layered system with parallel interfaces.

work the object is to develop a time–distance graph that involves plotting arrival times against geophone spacing (Fig. 1.38). Thus, the distance between geophones, together with the total length and arrangement of the array, has to be carefully chosen to suit each particular problem.

The most common arrangement in refraction work is profile shooting. Here the shot points and geophones are laid out in long lines, with a row of geophones receiving refracted waves from the shots fired. The process is repeated at uniform intervals down the line. For many surveys for civil engineering purposes where it is required to determine depth to bedrock it may be sufficient to record from two shotpoint distances at each end of the receiving spread. By traversing in both directions the angle of dip can be determined.

In the simple case of refraction by a single high-velocity layer at depth, the travel times for the seismic wave that proceeds directly from the shot point to the detectors and the travel times for the critical refracted wave to arrive at the geophones, are plotted graphically against geophone spacing (Fig. 1.39). The depth (Z) to the high-velocity layer then can be obtained from the graph by using the expression:

$$Z = \frac{x}{2}\sqrt{\left(\frac{V_2 - V_1}{V_2 + V_1}\right)} \tag{1.9}$$

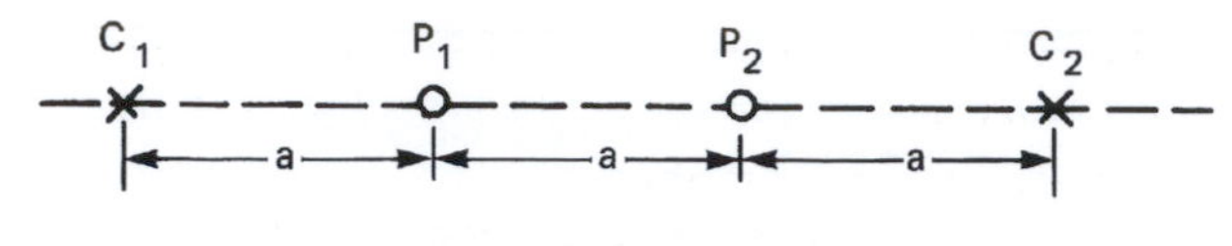

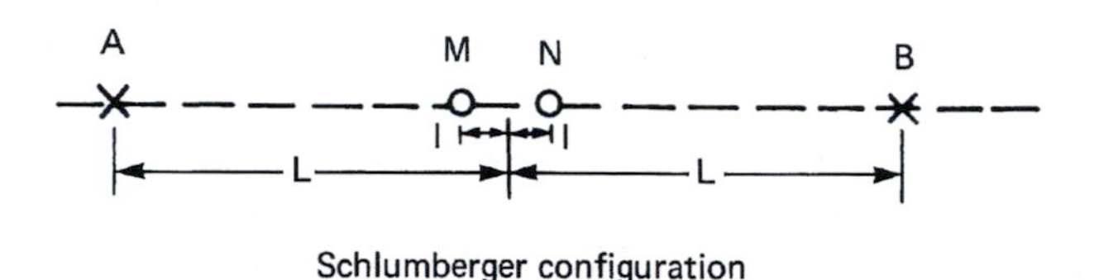

Figure 1.39 Wenner and Schlumberger configurations.

where V_1 is the speed in the low-velocity layer, V_2 is the speed in the high-velocity layer and x is the critical distance. The method also works for multilayered rock sequences if each layer is sufficiently thick and transmits seismic waves at higher speeds than the one above it. However, in the refraction method a low-velocity layer underlying a high-velocity layer usually cannot be detected, as in such an inversion the pulse is refracted into the low-velocity layer. Also, a layer of intermediate velocity between an underlying refractor and overlying layers can be masked as a first arrival on the travel–time curve. The latter is known as a blind zone. The position of faults also can be estimated from time–distance graphs.

As noted above, the velocity of shock waves is closely related to the elastic moduli and can therefore provide data relating to the engineering performance of the ground. Young's modulus, E, and Poisson's ratio, υ, can be derived if the density, ρ, and compressional, V_p, and shear, V_s, wave velocities are known by using the following expressions:

$$E = \rho V_p^2 \frac{(1+\upsilon)(1-2\upsilon)}{(1-\upsilon)} \tag{1.10}$$

or

$$E = 2V_s^2 \rho(1+\upsilon) \tag{1.11}$$

or

$$E = V_s^2 \rho \left[\frac{3(V_p/V_s)^2 - 4}{g(V_p/V_s)^2 - 1} \right] \tag{1.12}$$

$$\upsilon = \frac{1/2(V_p/V_s)^2 - 1}{(V_p/V_s)^2 - 1} \tag{1.13}$$

These dynamic moduli correspond to the initial tangent moduli of the stress–strain curve for an instantaneously applied load and are usually higher than those obtained in static tests. Because the seismic pulse is of very short duration and, more importantly, the stress level associated with a seismic pulse is so small, the motions are entirely elastic. The frequency and nature of discontinuities within a rock mass affect its deformability. Accordingly, a highly discontinuous rock mass exhibits a lower compressional wave velocity than a massive rock mass of the same type. The influence of discontinuities on the deformability of a rock mass can be estimated from a comparison of its *in situ* compressional velocity, V_{pf}, and the laboratory sonic velocity, V_{pl}, determined from an intact specimen taken from the rock mass. The velocity ratio, V_{pf}/V_{pl}, reflects the deformability and so can be used as a quality index. A comparison of the velocity ratio with other rock quality indices is given in Table 1.7.

Site investigations for large projects, notably tunnels, involve the geomechanical assessment of rock masses in terms of their engineering classification such as the RMR system developed by Bieniawski (1989). McCann *et al.* (1990) showed that the seismic properties of a rock mass can help make such an assessment, noting that rocks can be classified according to whether they are very weak, weak, strong or very strong. Previously, Grainger *et al.* (1973) had conducted a seismic refraction survey in the Middle Chalk at Mundford, Norfolk, England, and had been able to correlate their results with the classification of weathered chalk developed by Ward *et al.* (1968) at the same site.

The porosity tends to lower the velocity of a shock wave through a material. In fact, the compressional wave velocity, V_p, is related to the porosity, n, of a normally consolidated sediment as follows:

$$\frac{1}{V_p} = \frac{n}{V_{pfl}} + \frac{1-n}{V_{pl}} \tag{1.14}$$

where V_{pfl} is the velocity in the pore fluid and V_{pl} is the compressional wave velocity for the intact material as determined in the laboratory. The compressional wave velocities may be raised appreciably by the presence of water. Because of the relationship between seismic velocity and porosity, seismic velocity is broadly related to the intergranular permeability of sandstone formations. However, in most sandstones, fissure flow makes a more important contribution to groundwater movement than does intergranular or primary flow.

1.9.2 Resistivity methods

The resistivity of rocks and soils varies within a wide range. Since most of the principal rock forming minerals are practically insulators, the resistivity of rocks and soils is determined by the amount of conducting mineral constituents and the content of mineralized water in their pores. The latter condition is by far the dominant factor and in fact most rocks and soils conduct an electric current by virtue of the water they contain. The widely differing resistivity values of the various types of impregnating water can cause variations in the resistivity of rocks ranging from a few tenths of an ohm-metre (Ω-m) to hundreds of ohm-metres, as can be seen from Table 1.10.

In the resistivity method an electric current is introduced into the ground by means of two current electrodes and the potential difference between two potential electrodes is measured. It is preferable to measure the potential drop or apparent resistance directly in ohms rather than observe both current and voltage. The ohms value is converted to apparent resistivity by use of a factor that depends on the particular electrode configuration in use. The electrodes are normally arranged along a straight line, the potential electrodes being placed inside the current electrodes, and all four are symmetrically disposed with respect to the centre of the configuration. Configurations of the symmetric type are used most frequently, namely, those introduced by Wenner and by Schlumberger. In the Wenner configuration the distances between all four electrodes are equal (Fig. 1.39). The spacings can be progressively increased about the centre of the array in depth sounding or the whole array, with fixed spacings, can be shifted

Table 1.10 Resistivity of some types of natural water

Type of water	*Resistivity* (Ω-m)
Meteoric water, derived from precipitation	30–1000
Surface waters, in districts of igneous rocks	30–500
Surface waters, in districts of sedimentary rocks	10–100
Groundwater, in areas of igneous rocks	30–150
Groundwater, in areas of sedimentary rocks	larger than 1
Sea water	about 0.2

along a given line when profiling. In the Schlumberger arrangement, the potential electrodes maintain a constant separation about the centre of the station whilst if changes with depth are being investigated the current electrodes are moved outwards after each reading (Fig. 1.39). The expressions used to compute the apparent resistivity (ρ_a) for the Wenner and Schlumberger configurations are as follows:

Wenner:

$$\rho_a = 2\pi a R \tag{1.15}$$

Schlumberger:

$$\rho_a = \frac{\pi(L^2 - l^2)}{2l} \times R \tag{1.16}$$

where a, L and l are shown in Fig. 1.39 and R is the resistance reading. Other configurations include the dipole–dipole, the pole–dipole and the square arrays. The Lee and two electrode arrays, and the gradient array are variants of the Wenner configuration and Schlumberger configuration respectively (Reynolds, 1997).

The resistivity method is based on the fact that any subsurface variation in conductivity alters the pattern of current flow in the ground and therefore changes the distribution of electric potential at the surface. Since the electrical resistivity of such features as superficial deposits and bedrock differ from each other (Table 1.11), the resistivity method may be used in their detection, and to give their approximate thicknesses, relative positions and depths. The first step in any resistivity survey should be to conduct a resistivity depth sounding at the site of a borehole in order to establish a correlation between resistivity and lithological layers. If a correlation cannot be established, then an alternative method is required.

Table 1.11 Resistivity values of some common rock types

Rock type	*Resistivity* (Ω-m)
Topsoil	5–50
Peat and clay	8–50
Clay, sand and gravel mixtures	40–250
Saturated sand and gravel	40–100
Moist to dry sand and gravel	100–3000
Mudstones, marls and shales	8–100
Sandstones and limestones	100–1000
Crystalline rocks	200–10 000

Horizontal profiling is used to determine variations in apparent resistivity in a horizontal direction at a pre-selected depth. For this purpose an electrode configuration, with fixed inter-electrode distances, is moved along a straight traverse, with resistivity determinations being made at stations at regular intervals. For example, when the position of a steeply dipping interface such as a fault or igneous contact has to be located, especially if there is a notable difference across the interface, its position can be determined by making a series of traverses across the presumed location of the interface. The length of the electrode configuration must be carefully chosen because it is the dominating factor regarding depth penetration.

The data of a constant separation survey consisting of a series of traverses arranged in a grid pattern may be used to construct a contour map of lines of equal resistivity (Fig. 1.40). In other words, the apparent resistivity recorded at each station is plotted on a base map and isoresistivity contours are interpolated between these values. Such maps often are extremely useful in locating areas of anomalous resistivity such as gravel pockets in clay soils or the trend of buried channels. Even so interpretation of resistivity maps as far as the delineation of lateral variations is concerned is mainly qualitative.

The relationship between the depth of penetration and the electrode spacing is given in Fig. 1.41 from which it can be seen that 50 per cent of the total current passes above a depth equal to about half the electrode separation and 70 per cent flows within a depth equal to the electrode separation. Analysis of the variation in the value of apparent resistivity with respect to electrode separation enables inferences to be drawn about the subsurface formations.

Resistivity sounding provides information concerning the vertical succession of different conducting zones, and their individual thicknesses and resistivities (Fig. 1.42). The method is

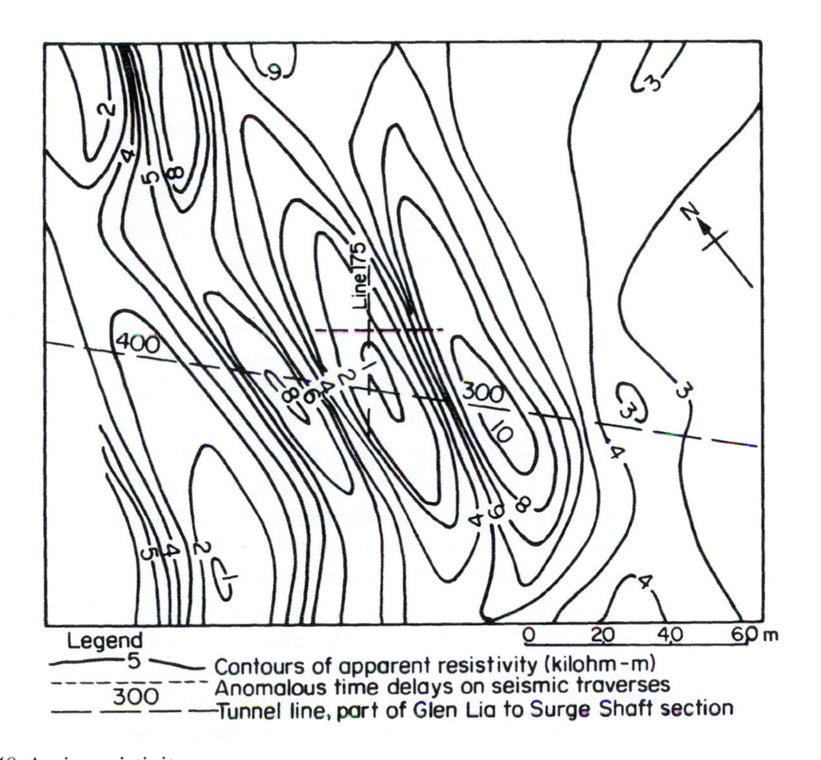

Figure 1.40 An isoresistivity map.

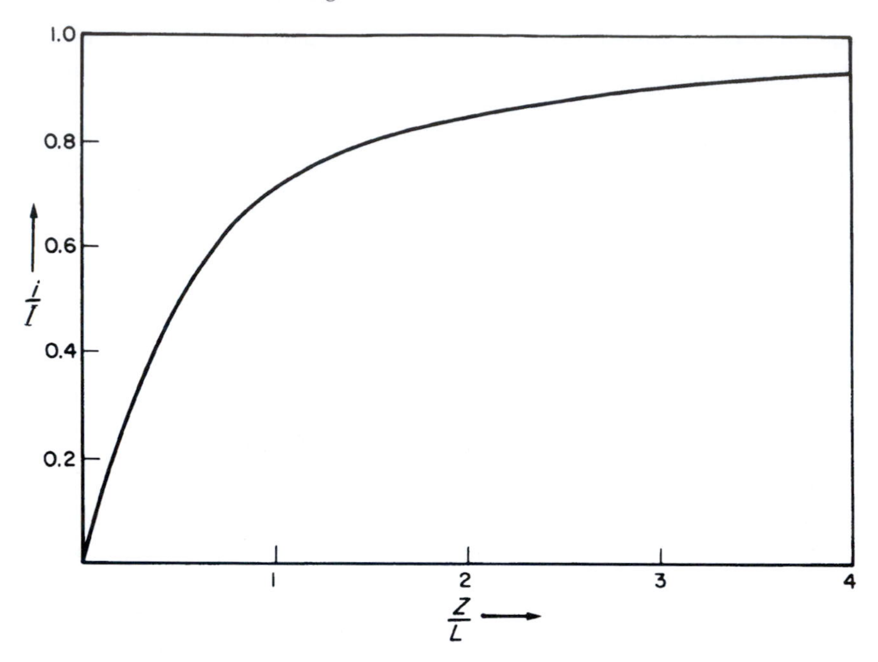

Figure 1.41 Fraction of total current, I, that passes above a horizontal plane at depth, Z, as a function of the distance, L, between two current electrodes.

particularly valuable for investigations on horizontally stratified ground. In electrical sounding the midpoint of the electrode configuration is fixed at the observation station while the length of the configuration is increased gradually. As a result the current penetrates deeper and deeper, the apparent resistivity being measured each time the current electrodes are moved outwards. The readings therefore become increasingly affected by the resistivity conditions at greater depths. The Schlumberger configuration is preferable to the Wenner configuration for depth sounding. The data obtained usually is plotted as a graph of apparent resistivity against electrode separation in the case of the Wenner array, or half the current electrode separation for the Schlumberger array. The electrode separation at which inflection points occur in the graph provides an idea of the depth of interfaces. The apparent resistivities of the different parts of the curve provide some idea of the relative resistivities of the layers concerned.

If the ground approximates to an ideal condition, then a quantitative solution, involving a curve fitting exercise, should be possible. The technique requires a comparison of the observed curve with a series of master curves prepared for various theoretical models.

Generally, it is not possible to determine the depths to more than three or four layers. If a second layer is relatively thin and its resistivity much larger or smaller than that of the first layer, the interpretation of its lower contact will be inaccurate. For all depth determinations from resistivity soundings it is assumed that there is no change in resistivity laterally. This is not the case in practice. Indeed, sometimes the lateral change is greater than that occurring with increasing depth and so corrections have to be applied for the lateral effects when depth determinations are made. Resistivity surveys can give rather large errors in depth sounding at times (up to ± 20 per cent) and they give the best results when used in relatively simple geological conditions. Furthermore, in arid regions precipitation of salts in the soil can give rise to near-surface layers of high conductivity that effectively short-circuit current flow and so permit little penetration to greater depths.

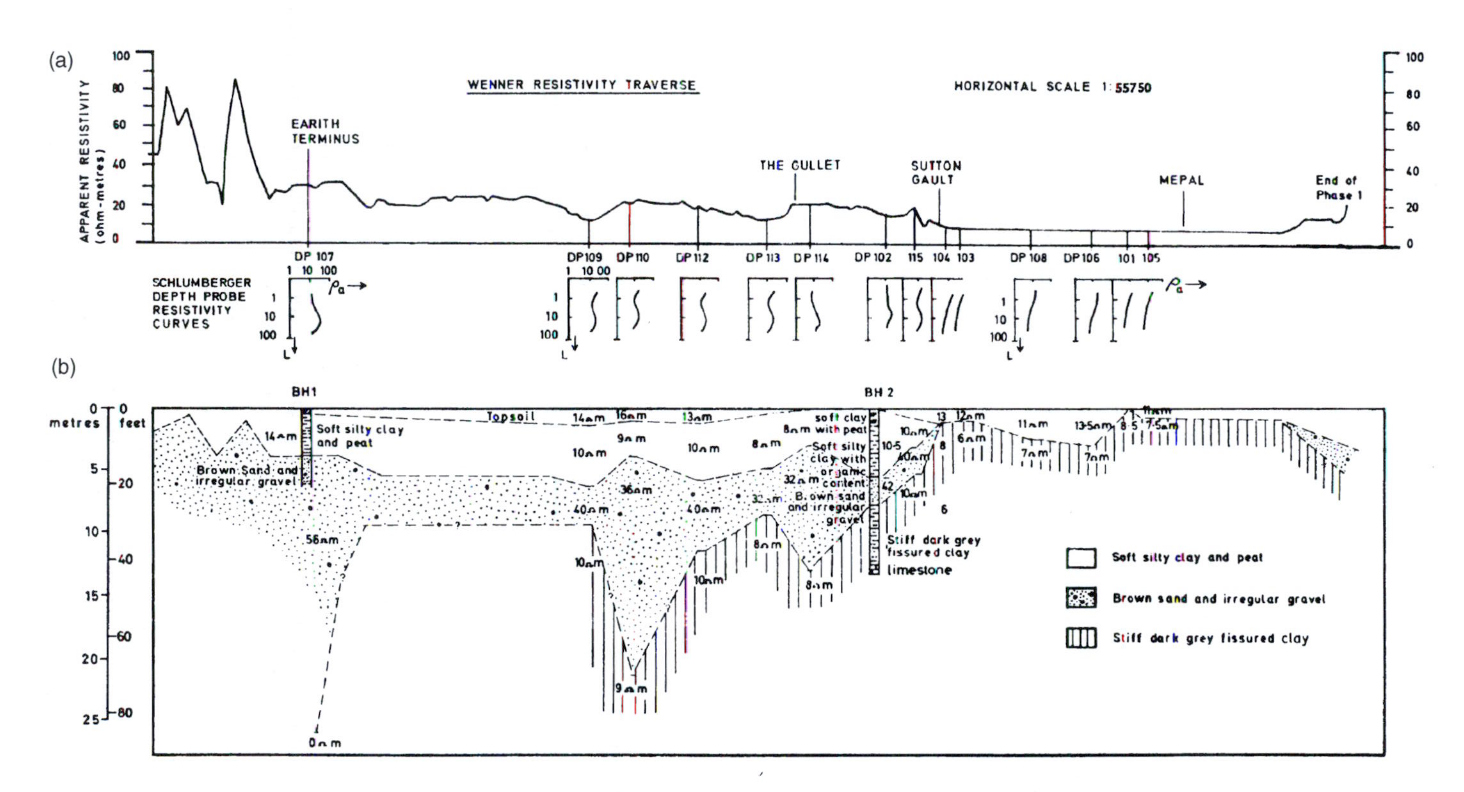

Figure 1.42 (a) Resistivity and depth probes. (b) Provisional section based upon resistivity survey results and boreholes BH1 and BH2. *Source*: McDowell, 1970. Reproduced by kind permission of The Geological Society.

Resistivity is the most frequently used geophysical method in groundwater exploration. However, the resistivity method does not provide satisfactory quantitative results if the potential aquifer(s) being surveyed are thin, that is, 6 m or less in thickness, especially if they are separated by thick argillaceous horizons. In such situations either cumulative effects are obtained or anomalous resistivities are measured, the interpretation of which is extremely difficult, if not impossible. Moreover, the method is more successful when used to investigate a formation that is thicker than the one above it. As noted, most rocks and soils conduct an electric current only because they contain water. However, the widely differing resistivity of the various types of pore water can cause variations in the resistivity of soil and rock formations ranging from a few tenths of an ohm-metre to hundreds of ohm-metres. In addition, the resistivity of water changes markedly with temperature and temperature increases with depth. Hence, for each bed under investigation the temperature of both rock and water must be determined or closely estimated, and the calculated resistivity of the pore water at that temperature converted to its value at a standard temperature (i.e. 25 °C).

As the amount of water present is influenced by the porosity of a rock, the resistivity provides a measure of its porosity. For example, in granular materials in which there are no clay minerals, the relationship between the resistivity, ρ, on the one hand and the density of the pore water, ρ_w, the porosity, n, and the degree of saturation, S_r, on the other is as follows:

$$\rho = a\rho_w n^{-x} S_r^{-y} \tag{1.17}$$

where a, x and y are variables (x ranges from 1.0 for sand to 2.5 for sandstone and y is approximately 2.0 when the degree of saturation is greater than 30 per cent). For those formations that occur below the water table and are therefore saturated, the above expression becomes

$$\rho = a\rho_w n^{-x} \tag{1.18}$$

since $S_r = 1$ (i.e. 100 per cent).

In a fully saturated sandstone, a fundamental empirical relationship exists between the electrical and hydrogeological properties that involves the concept of the formation resistivity factor, F_a, defined as:

$$F_a = \frac{\rho_o}{\rho_w} \tag{1.19}$$

where ρ_o is the resistivity of the saturated sandstone and ρ_w is the resistivity of the saturating solution. In a clean sandstone, that is, one in which the electrical current passes through the interstitial electrolyte during testing with the rock mass acting as an insulator, the formation resistivity factor is closely related to the porosity. The formation resistivity factor is related to the true formation factor, F, by the expression.

$$F = \frac{\rho_A F_a}{\rho_A - F_a \rho_w} \tag{1.20}$$

in which ρ_A is a measure of the effective resistivity of the rock matrix. In clean sandstone ρ_A is infinitely large, consequently $F = F_a$. Generally, F is related to the porosity, n, by the equation:

$$F = \frac{a}{n^m} \tag{1.21}$$

where a and m are constants for a given formation.

The true formation factor in certain formations also has been shown to be broadly related to intergranular permeability, k_g, by the expression:

$$F = \frac{b}{k_g^n} \qquad (1.22)$$

where b and n are constants for a given formation. In sandstones in which intergranular flow is important, the above expressions can be used to estimate hydraulic conductivity and hence, if the thickness of the aquifer is known, transmissivity. The techniques are not useful in highly indurated sandstones, where the intergranular permeabilities are less than $1.0 \times 10^{-7}\,\mathrm{m\,s^{-1}}$, and the flow is controlled by fissures. Similarly, these methods tend to be of little value in multilayered aquifers.

The resistivity method also has been used in investigations to determine the extent of groundwater pollution since pollution increases the salinity of water. For example, the resistivity method was used by Finch (1979) to delineate the contamination of groundwater in the Sherwood Sandstone aquifer, by run-off from colliery spoil, in an area of Nottinghamshire, England, where there is little variation in lithology. Similarly, Oteri (1983) used the results of a resistivity survey to determine the extent of saline intrusion in the shingle aquifer at Dungeness, Kent, England. In both instances, the level of chloride ions in the pore water was related to the resistivity data. However, if the depth to water is too great, then the thickness of the overlying unsaturated sediments can mask any constrasts between polluted and unpolluted water. In addition, the geological conditions in the area have to be relatively uniform so that the resistivity values and profiles can be compared.

1.9.3 Electromagnetic methods

A wide variety of electromagnetic survey methods are available, each involving the measurement of one or more electric or magnetic field components induced in the ground by a primary field. A primary field is produced by a natural (transient) current source or an alternating current artificial source, and spreads out in space above and below the ground inducing currents in subsurface conductors. Secondary electromagnetic fields are produced by these currents that distort the primary field. The resultant field differs from the primary field in intensity, phase and direction, and so can be detected by a suitable receiving coil. The secondary field induced in the subsurface conductor fades gradually when a transient primary field is switched off, fading being slower in media of higher conductivity. Hence, measurement of the rate at which the secondary currents fade and their field offers a means of detecting anomalously conducting bodies. Electromagnetic methods can be classified as either frequency domain systems, FEM, or time domain systems, TEM. Frequency domain equipment uses either one or more frequencies whereas time domain equipment takes measurements as a function of time.

The terrain conductivity meter represents a means of measuring the conductivity of the ground. Electromagnetic energy is introduced into the ground by inductive coupling produced by passing an alternating current through a coil. The receiver also detects its signal by induction. The conductivity meter is carried along traverse lines across a site and can provide a direct continuous readout. Hence, surveys can be carried out rapidly. A popular ground conductivity meter is the Geonics EM34, which is operated by two persons and can probe to depths of 15, 30 and 60 m. The Geonics EM31 is operated by a single person and can be used to survey 5–10 km of profile daily, with readings taken at 5 or 10 m intervals. It has a depth of investigation of less than 6 m. Conductivity values are taken at positions set out on a grid pattern. Corrected values of conductivity can be plotted as profiles or as contoured maps of conductivity. The most useful application is in two-layer cases where the layers have laterally consistent conductivities (e.g. clay, high conductivity, overlying bedrock, low conductivity; where the conductivity contours are highest, the clay would be thickest). Where the thickness of overburden varies within fairly narrow limits and the conductivities of the overburden and bedrock do not change appreciably, the depth to bedrock can be estimated from standard curves. As these depth values are approximate, they need to be checked against borehole evidence or data obtained from more quantitative geophysical methods (Anon., 1988).

The very low-frequency (VLF) method is the most widely used fixed source method operating on a single frequency, making use of powerful radio transmitters in the 15–25 kHz range. An advantage of the method is that measurements can be made quickly by a single operator. However, a drawback is that the method is dependent on there being an appropriate transmitter operational, as such transmitters can be switched off. Another disadvantage is that wave penetration is limited. The method also is adversely affected by topography. The interpretation of VLF data generally is qualitative and it frequently is used for reconnaissance work. The method is well suited to detecting near vertical contacts and fracture zones. As a consequence, the method has found particular application in site investigation for the delineation of faults, especially those occupied by water.

The ground probing radar method is based upon the transmission of pulsed electromagnetic waves in the frequency range 1–1000 MHz, the transmitting and receiving antenna being mounted on a mobile trolley (Leggo, 1982). In this method the travel times of the waves reflected from subsurface interfaces are recorded as they arrive at the surface, and the depth (Z) to an interface is derived from:

$$Z = VT/2 \quad (1.23)$$

where V is the velocity of the radar pulse and T is its travel time. The conductivity of the ground imposes the greatest limitation on the use of radar probing, that is, the depth to which radar energy can penetrate depends upon the effective conductivity of the strata being probed. This, in turn, is governed chiefly by the water content and its salinity. The nature of the pore water exerts the most influence on the dielectric constant. Furthermore, the value of effective conductivity is also a function of temperature and density, as well as the frequency of the electromagnetic waves being propagated. The least penetration occurs in saturated clayey materials or where the moisture content is saline. For example, attenuation of electromagnetic energy in wet clay and silt mean that depth of penetration frequently is less than 1 m. The technique appears to be reasonably successful in sandy soils and rocks in which the moisture content is non-saline. Rocks like limestone and granite can be penetrated for distances of tens of metres and in dry conditions the penetration may reach 100 m. Dry rock salt is radar-translucent, permitting penetration distances of hundreds of metres. Ground-probing radar has been used for a variety of purposes in geotechnical engineering, for example, the determination of the thickness of permafrost, the detection of fractures and faults in rock masses, the location of subsurface voids, groundwater investigations and the delineation of contaminated plumes. Grasmück and Green (1996) described a three-dimensional method of ground probing radar. According to them this system is capable of producing vivid images of the subsurface up to depths of 50 m.

1.9.4 Magnetic methods

All rocks, minerals and ore deposits are magnetized to a lesser or greater extent by the earth's magnetic field. As a consequence, in magnetic surveying, accurate measurements are made of the anomalies produced in the local geomagnetic field by this magnetization. The strength of a magnetic field is measured in nanoteslas, nT, and the average strength of the earth's magnetic field is some 50 000 nT. Obviously, the variations associated with magnetized rock formations are very much smaller than this. The intensity of magnetization and hence the amount by which the earth's magnetic field is changed locally, depends on the magnetic susceptibility of the material concerned. In addition to the magnetism induced by the earth's field, rocks possess a permanent magnetism that depends upon their history.

Rocks have different magnetic susceptibilities related to their mineral content. Some minerals, for example, quartz and calcite are magnetized reversely to the field direction and therefore have negative susceptibility. They are described as diamagnetic. Paramagnetic minerals, which are the majority, are magnetized along the direction of magnetic field so that their susceptibility is positive. The susceptibility of the ferromagnetic minerals, such as magnetite, ilmentie, pyrrhotite and hematite, is a very complicated function of the field intensity. However, since the magnitudes of their susceptibility amount to 10–105 times the order of susceptibility of the paramagnetic and diamagnetic minerals, the ferromagnetic minerals can be found by magnetic field measurements.

If the magnetic field ceases to act on a rock, then the magnetization of paramagnetic and diamagnetic minerals disappears. However, in ferromagnetic minerals the induced magnetization is diminished only to a certain value. This residuum is called remanent magnetization and is of great importance in rocks. All igneous rocks have a very high remanent magnetization acquired as they cooled down in the earth's magnetic field. In the geological past, grains of magnetic materials were orientated by ancient geomagnetic fields during sedimentation in water so that some sedimentary rocks show stable remanent magnetization.

Aeromagnetic surveying has almost completely supplanted ground surveys for regional reconnaissance purposes. Accurate identification of the plan position of the aircraft for the whole duration of the magnetometer record is essential. The object is to produce an aeromagnetic map, the base map with transcribed magnetic values being contoured.

The aim of most ground surveys is to produce isomagnetic contour maps of anomalies to enable the form of the causative magnetized body to be estimated (Fig. 1.43). Profiles are surveyed across the trend of linear anomalies with stations, if necessary, at intervals of as little as 1 m. A base station is set up beyond the anomaly where the geomagnetic field is uniform. The reading at the base station is taken as zero and all subsequent readings are expressed as plus-or-minus differences. Corrections need to be made for the temperature of the instrument as the magnets lose their effectiveness with increasing temperature. A planetary correction is also required which eliminates the normal variation of the earth's magnetic field with latitude. Large metallic objects like pylons are a serious handicap to magnetic exploration and must be kept at a sufficient distance, as it is difficult to correct for them. In addition, Anon. (1988) pointed out that blanket surveys in urban and industrial areas often are of little practical value as sites may be littered with debris made of iron or steel.

A magnetometer may also be used for mapping geological structures. For example, in some thick sedimentary sequences it is sometimes possible to delineate the major structural features because the succession includes magnetic horizons. These may be ferruginous sandstones or shales, tuffs or basic lava flows. In such circumstances anticlines produce positive and synclines negative anomalies. Faults and dykes are indicated on isomagnetic maps by linear belts of somewhat sharp gradient or by sudden swings in the trend of the contours. However, in many areas the igneous and metamorphic basement rocks, which underlie the sedimentary sequence, are the predominant influence controlling the pattern of anomalies since they are usually far more magnetic than the sediments above. Where the basement rocks are brought near the surface in structural highs the magnetic anomalies are large and characterized by strong relief.

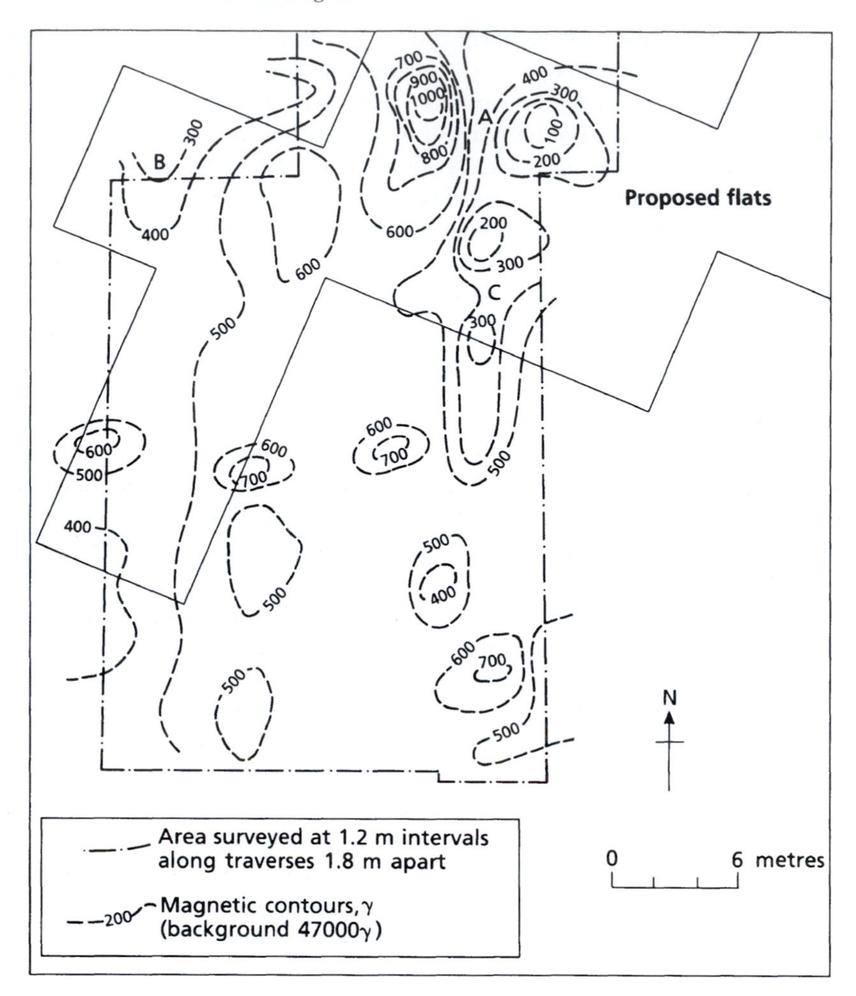

Figure 1.43 Isomagnetic map of a site proposed for development in which mine shafts occurred at A, B and C.
Source: Cripps *et al*., 1988.

Conversely, deep sedimentary basins usually produce contours with low values and gentle gradients on isomagnetic maps.

Magnetic surveying has been used to detect abandoned mine shafts, a proton precession magnetometer normally being used (Bell, 1988). A good subsurface magnetic contrast may be obtained if the shaft is lined with iron tubbing or with bricks, or if the shaft is filled and the filling consists of burnt shale or ash, or contains scrap iron. On the other hand, if a shaft is unfilled and unlined or lined with timber, then it may not give rise to a measurable anomaly.

1.9.5 Gravity methods

The earth's gravity field varies according to the density of the subsurface rocks but at any particular locality its magnitude is also influenced by latitude, elevation, neighbouring topographical features and the tidal deformation of the earth's crust. The effects of these latter factors have to be eliminated in any gravity survey where the object is to measure the variations in acceleration due to gravity precisely. This information can then be used to construct a contoured gravity map. In survey work, modern practice is to measure anomalies in gravity units (g.u. $= 10^{-6}\,\mathrm{ms}^{-2}$). Formerly, the unit of measurement was the milligal (mGal) which was 0.001 Gal, 980 Gal being the approximate acceleration at the earth's crust due to gravity. Hence, 10 g.u. is equal to 1 mGal. Modern gravity meters used in exploration do not measure the absolute value of acceleration due to gravity but measure the small differences in this value between one place and the next.

Gravity methods are mainly used in regional reconnaissance surveys to reveal anomalies that may be subsequently investigated by other methods. Since the gravitational effects of geological bodies are proportional to the contrast in density between them and their surroundings, gravity methods are particularly suitable for the location of structures in stratified formations. Gravity effects due to local structures in near surface strata may be partly obscured or distorted by regional gravity effects caused by large-scale basement structures. However, regional deep-seated gravity effects can be removed or minimized in order to produce a residual gravity map showing the effects of shallow structures that may be of interest.

A gravity survey is conducted from a local base station at which the value of the acceleration due to gravity is known with reference to a fundamental base where the acceleration due to gravity has been measured accurately. The way in which a gravity survey is carried out largely depends on the objective in view. Large-scale surveys covering hundreds of square kilometres, carried out in order to reveal major geological structures, are done by vehicle or helicopter with a density of only a few stations per square kilometre. For more detailed work such as the delineation of ore bodies or basic minor intrusions or the location of faults, spacing between stations may be as small as 20 m. Because gravity differences large enough to be of geological significance are produced by changes in elevation of several millimetres and of only 30 m in north–south distance, the location and elevation of stations must be established with very high precision.

Micro-gravity meters have been used to detect subsurface voids such as caverns in limestone, or abandoned mine shafts or shallow workings (Cripps *et al.*, 1988). Gravity 'lows' are recorded over voids and they are more notable over air-filled than water- or sediment-filled voids.

1.9.6 Drillhole logging techniques

Considerable information is needed regarding the engineering and hydrogeological properties of rock masses for large construction projects, especially for underground structures. Drillholes provide core material that can be tested in the laboratory but core recovery usually is less than 100 per cent and rock material may be disturbed to a varying degree. Furthermore, laboratory tests on core samples provide only an approximate representation of the actual *in situ* engineering properties and likely behaviour of the rock mass from which they were obtained. As geophysical properties are related to lithological and engineering properties, geophysical measurements taken in and between drillholes can be used to provide an enhanced picture of the ground conditions.

The electrical resistivity method makes use of various electrode configurations down-the-hole. As the instrument is raised from the bottom to the top of the hole it provides a continuous record of the variations in resistivity of the wall rock. In the normal or standard resistivity configuration there are one current and two potential electrodes in the sonde. The depth of penetration of the electric current from the drillhole is influenced by the electrode spacing. In a short normal resistivity survey, spacing is about 400 mm, whereas in a long normal resistivity survey spacing is generally between 1.5 and 1.75 m. Unfortunately, in such a survey, because of the influence of thicker adjacent beds, thin resistive beds yield resistivity values that are much too low, whilst thin conductive beds produce values that are too high. The microlog technique may be used in such situations. In this technique the electrodes are very closely spaced (25–50 mm) and are in contact with the wall of the drillhole. This allows the detection of small lithological changes so that much finer detail is obtained than with the normal electric log (Fig. 1.44). A microlog is particularly useful in recording the position of permeable beds.

If, for some reason, the current tends to flow between the electrodes on the sonde instead of into the rocks, then the laterolog or guard electrode is used. The 'laterolog 7' has seven electrodes in an array that focuses the current into the strata of the drillhole wall. The microlaterolog, a focused microdevice, is used in such a situation instead of the microlog.

A dipmeter generally is a four-arm side-wall micro-resistivity device. It measures small variations in the resistivity of a formation that allows the relative vertical shift of characteristic

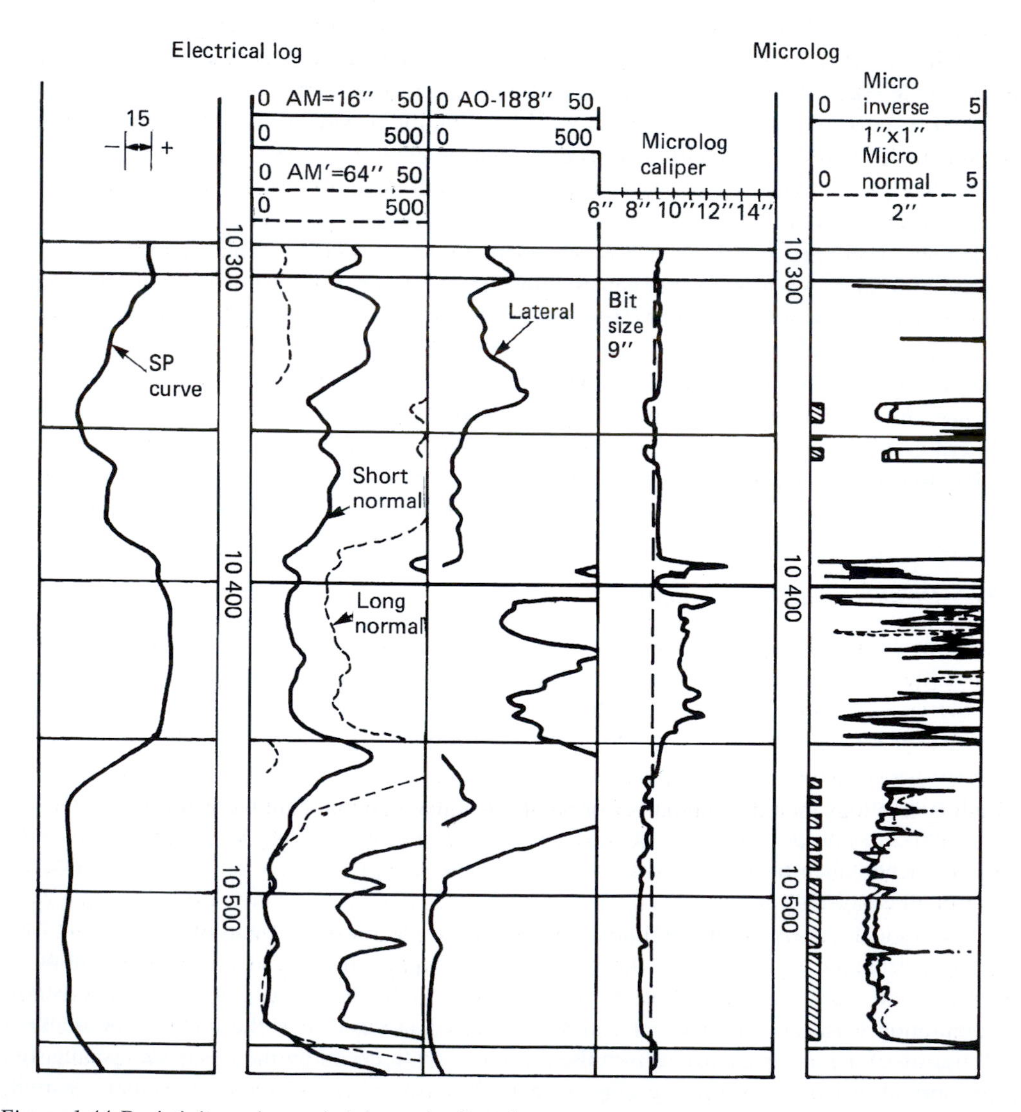

Figure 1.44 Resistivity, microresistivity and caliper log curves.

pattern variation produced by bedding planes, discontinuities or lithological changes to be used to determine, by aid of computer analysis, the attitude of a plane intersecting a drillhole.

Induction logging may be used when an electrical log cannot be obtained. In this technique the sonde sends electrical energy into the strata horizontally and therefore only measures the resistivity immediately opposite the sonde, unlike in normal electrical logging where the current flows between electrodes. As a consequence, the resistivity is measured directly in an induction log whereas in a normal electrical log, since the current flows across the stratal boundaries, it is measured indirectly from the electrical log curves. A gamma-ray log is usually run with an induction log in order to reveal the boundaries of stratal units.

A spontaneous potential (SP) log is obtained by lowering a sonde down a drillhole that generates a small electric voltage at the boundaries of permeable rock units, especially between such strata and less permeable beds. For example, permeable sandstones show large SPs, whereas shales are typically represented by low values. If sandstone and shale are interbedded, then the SP curve has numerous troughs separated by sharp or rounded peaks, the widths of which vary in proportion to the thicknesses of the sandstones. Variations of SP in low-permeability rock such as granite may be related to fracture zones. SP logs are frequently recorded at the same time as resistivity logs. Interpretation of both sets of curves yields precise data on the depth, thickness and position in the sequence of the beds penetrated by the drillhole. The curves also enable a semi-quantitative assessment of lithological and hydrogeological characteristics to be made.

The sonic logging device consists of a transmitter–receiver system, transmitter(s) and receiver(s) being located at given positions on the sonde. The transmitters emit short high-frequency pulses several times a second, and differences in travel times between receivers are recorded in order to obtain the velocities of the refracted waves. The velocity of sonic waves propagated in sedimentary rocks is largely a function of the character of the matrix. Normally, beds with high porosities have low velocities, and dense rocks are typified by high velocities. Hence, the porosity of strata can be assessed. For instance, Wyllie *et al.* (1956) suggested that the porosity, n, of a saturated formation could be derived from:

$$n = \frac{(V_{\text{log}} - V_{\text{m}})}{(V_{\text{f}} - V_{\text{m}})} \tag{1.24}$$

where V_{log}, V_{f} and V_{m} are the compressional wave velocities of the rock mass, the fluid and the rock matrix respectively. In the three-dimensional sonic log one transmitter and one receiver are used at a time. This allows both compressional and shear waves to be recorded, from which, if density values are available, the dynamic elastic moduli of the beds concerned can be determined. As velocity values vary independently of resistivity or radioactivity, the sonic log permits differentiation amongst strata that may be less evident on the other types of log.

Radioactive logs include gamma-ray or natural gamma, gamma–gamma or formation density, and neutron logs. They have the advantage of being obtainable through the casing in a drillhole. On the other hand, the various electric and sonic logs can only be used in uncased holes. The natural gamma log provides a record of the natural radioactivity or gamma radiation from elements such as potassium 40, and uranium and thorium isotopes, in the rocks concerned. This radioactivity varies widely among sedimentary rocks, being generally high for clays and shales and lower for sandstones and limestones. Evaporites give very low readings. The gamma–gamma log uses a source of gamma-rays that are sent into the wall of the drillhole. There they collide with electrons in the rocks and thereby lose energy. The returning gamma-ray intensity is recorded, a high value indicating low electron density and hence low formation density. The gamma–gamma log responds to vertical and horizontal fractures. It tends to show higher porosities than those obtained from a sonic log over a comparable length of drillhole intersected by a high-angled discontinuity. A cross plot of porosity derived from the two logs may help indicate the presence of significant fracture zones within a rock mass (McCann *et al.*, 1997). The curve derived by a neutron–neutron sonde is a recording of the effects caused by bombardment of the strata with neutrons. As the neutrons are absorbed by atoms of hydrogen, which then emit gamma-rays, the log provides an indication of the quantity of hydrogen in the strata around the sonde. The amount of hydrogen is related to the water (or hydrocarbon) content and therefore provides another method of estimating porosity. The neutron sonde responds to the total water content, which includes water adsorded by clay minerals. Hence, the porosity determined, for example, of a shale would be greater than its actual porosity. In low-porosity crystalline rocks the log is very subdued except where fracture zones increase the porosity. Since carbon is a good moderator of neutrons, carbonaceous rocks are liable to yield spurious indications as far as porosity is concerned.

The caliper log measures the diameter of a drillhole. Different sedimentary rocks show a greater or lesser ability to stand without collapsing from the walls of the drillhole. For instance, limestones may present a relatively smooth face slightly larger than the drilling bit whereas soft shale may cave to produce a much larger diameter. A caliper log is obtained along with other logs to help interpret the characteristics of the rocks in the drillhole.

1.9.7 Cross-hole methods

The cross-hole seismic method is based on the transmission of seismic energy between drillholes. In its simplest form cross-hole seismic measurements are made between a seismic source in one drillhole (i.e. a small explosive charge, an air gun, a drillhole hammer, or an electrical sparker) and a receiver at the same depth in an adjacent drillhole (Fig. 1.45a). The receiver can

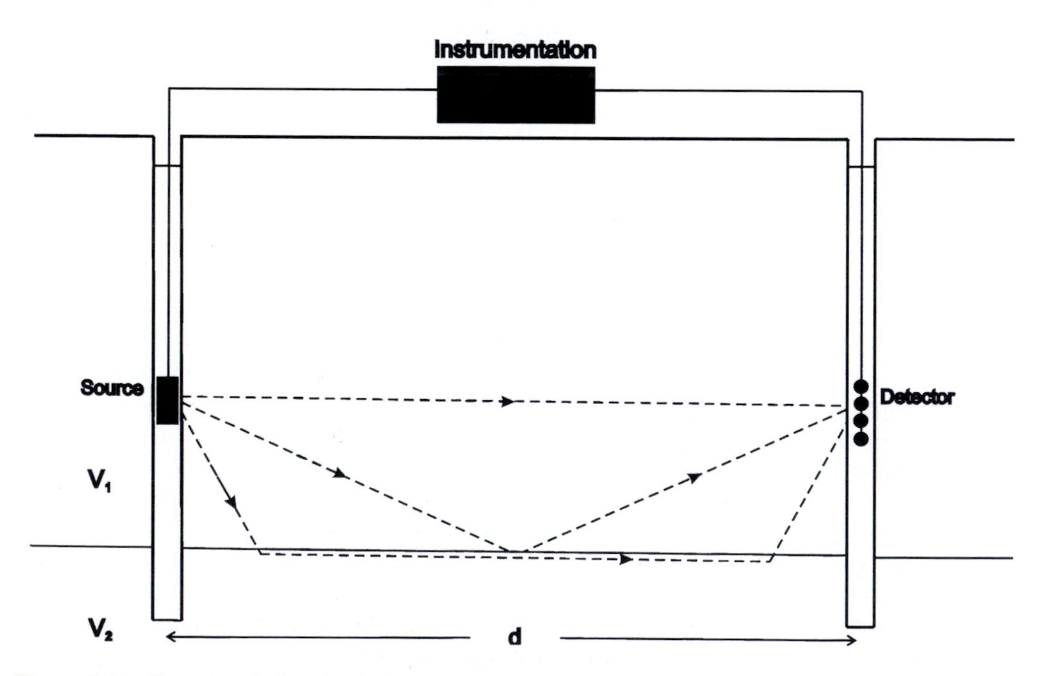

Figure 1.45a Cross-borehole seismic measurements.

either be a three-component geophone array clamped to the drillhole wall or a hydrophone in a liquid-filled drillhole to receive signals from an electric sparker in another drillhole similarly filled with liquid. The choice of source and receiver is a function of the distance between the drillholes, the required resolution and the properties of the rock mass. The best results are obtained with a high-frequency repetitive source (McCann *et al.*, 1986). Generally, the source and receiver in the two drillholes are moved up and down together. Drillholes must be spaced closely enough to achieve the required resolution of detail and be within the range of the equipment. This is up to 400 m in Oxford Clay, 160 m in the Chalk, and 80 m in sands and gravels. By contrast, because soft organic clay is highly attenuating, transmission is only possible over a few metres. These distances are for saturated material and the effective transmission is reduced considerably in dry superficial layers.

Such simple cross-hole seismic surveys are limited in the amount of data they produce. Hence, a system has been developed that uses a multitude of wave paths, thereby enabling the location, shape and velocity contrast relating to an anomaly or target in the rock mass between drillholes to be delineated in an unambiguous fashion (Fig. 1.45b). This is referred to as seismic tomography (tomography means a technique used to obtain an image of a selected plane section of a solid object). The method utilizes two or more drillholes, and possibly the ground surface, for the location of sources and detectors, the object being to derive one or more two-dimensional images of seismic properties within the rock mass (Jackson and McCann, 1997). Cross-hole seismic measurements provide a means by which the engineering properties of the rock mass between drillholes can be assessed. For example, the dynamic elastic properties can be obtained from the values of the compressional and shear wave velocities, and the formation density. Other applications include assessment of the continuity of lithological units between drillholes, identification of fault zones and assessment of the degree of fracturing, and the detection of subsurface voids. Cartmell *et al.* (1997) described the use of cross-hole seismic tomography to

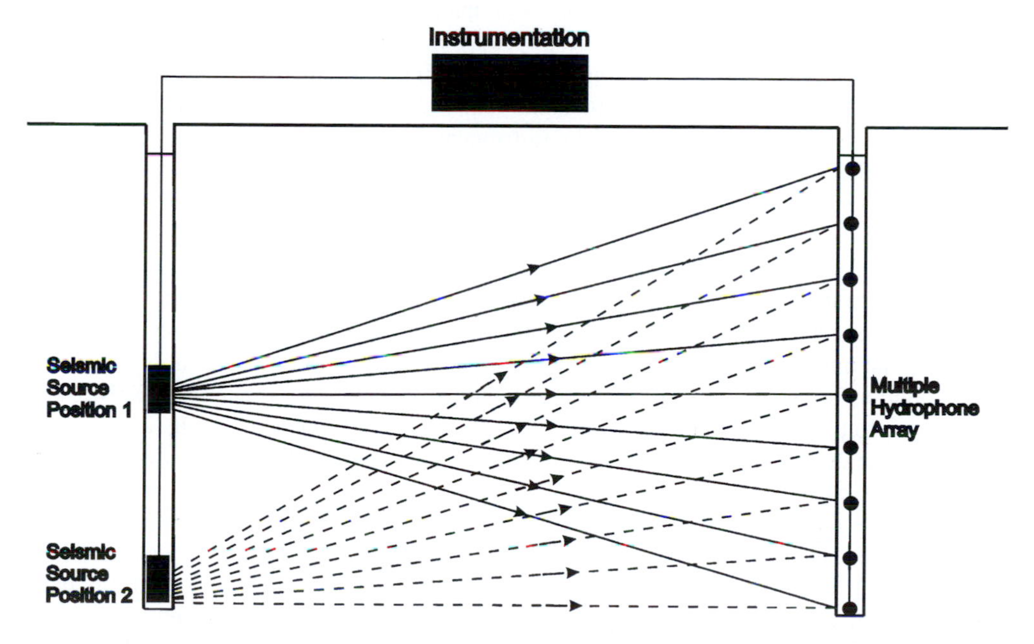

Figure 1.45b Experimental set-up for cross-borehole seismic tomographic survey.
Source: Jackson and McCann, 1997. Reproduced by kind permission of The Geological Society.

assess the dynamic elastic properties of the foundation area at Muela Dam site in Lesotho by using 10 drillholes and the uphole seismic technique.

Electromagnetic and electrical resistivity techniques also have been used to produce tomographic imagery. For example, Corin *et al.* (1997) used drillhole radar tomography to assess the foundation conditions for a long viaduct to be constructed in limestone that was regarded as highly karstified. They concluded that cross-hole methods are probably the best tools available at present to provide the required detailed information, particularly in regions of karstic limestone, for good foundation design. Electrical resistance tomography is a relatively new geophysical imaging technique that uses a number of electrodes in drillholes, and sometimes the ground surface, to image the resistivity of the subsurface.

1.10 Field instrumentation

Instrumentation and monitoring of soil and rock masses not only provide a check on the assumptions made in design, but also can be used where detailed analytical design is not justified, but the stability of the soil or rock masses is still questionable. In other words, instrumentation may allow less costly construction, the instrumentation providing a continual check on the stability and safety of a structure during its life-time. In such situations monitoring can provide advance warning of adverse or unpredicted behaviour thereby affording time for remedial action. Generally, monitoring may be carried out in order to record values of, or variations in geotechnical parameters such as ground stress, ground movements, groundwater level and pore water pressures, etc. (Hanna, 1985; Dunnicliff, 1988). However, an instrumentation programme does not usually constitute part of a site investigation.

In order that a field instrumentation and monitoring system should fulfil its intended function economically and reliably, it should be easy to install and read, possess adequate sensitivity to provide accurate and reproducible results, and be robust enough to endure the period of installation. Even where instrumentation is as simple, reliable and robust as possible, some proportion of it may fail to work or to be destroyed by construction plant or vandals. Hence, it is necessary that additional instrumentation should be available to allow for losses. Furthermore, each instrument should be able to determine the required parameter without inducing a change in that parameter due to the presence of the instrument.

Because of their importance, measurement of groundwater level and pore water pressure represent some of the most frequent types of *in situ* measurements recorded. At the least the water table and its seasonal fluctuations should be determined, since such data help to assess the geotechnical information gathered by an investigation and, of course, groundwater conditions, including pore water pressure, influence the type of foundation structure chosen.

A standpipe or a piezometer can be used to assess pore water pressure. Piezometers should be able to respond quickly to changes in groundwater conditions, should cause as little interference to the ground as possible and should be rugged enough to function over long periods of time. The response to piezometers in rock masses can be very much influenced by the incidence and geometry of the discontinuities so that the values of water pressure obtained may be misleading if due regard is not given to these structures.

The standpipe represents the simplest method of monitoring pore water pressure. It consists of an open-ended plastic tube up to 50 mm in diameter, which is perforated over about 1 m near the base. It is placed in a borehole or drillhole. A filter of fine gravel or sand is placed around the tube and the top of the hole is sealed with puddle clay or concrete to keep out surface water. The water level in the standpipe is measured by lowering an electrical 'dipmeter' down the open standpipe. The dipmeter produces a visual or audible signal when it comes in contact with the water level. Unfortunately, the standpipe suffers a number of disadvantages. For example, as no attempt is made to measure pore water pressure at a particular level, it therefore is assumed that a simple groundwater regime exists. However, if flow occurs between adjacent formations with differing permeabilities, then the water level in the standpipe will be meaningless. Another disadvantage is the length of time required for the level of water in the standpipe to stabilize with that in the ground, unless the latter is highly permeable.

Because changing soil or rock types at different depths mean varying permeabilities and pore water pressures, pore water pressures should be measured over a limited zone by sealing off a section of the hole. In this case, the standpipe piezometer consists of a porous tip surrounded by sand or gravel at the level where the pore water pressure is to be measured and is connected to a plastic tube that extends to the ground surface (Fig. 1.46). The filter is sealed above and below with grout, which typically consists of cement and bentonite. The installation of a piezometer must be carried out carefully otherwise its performance and measurements may be affected adversely.

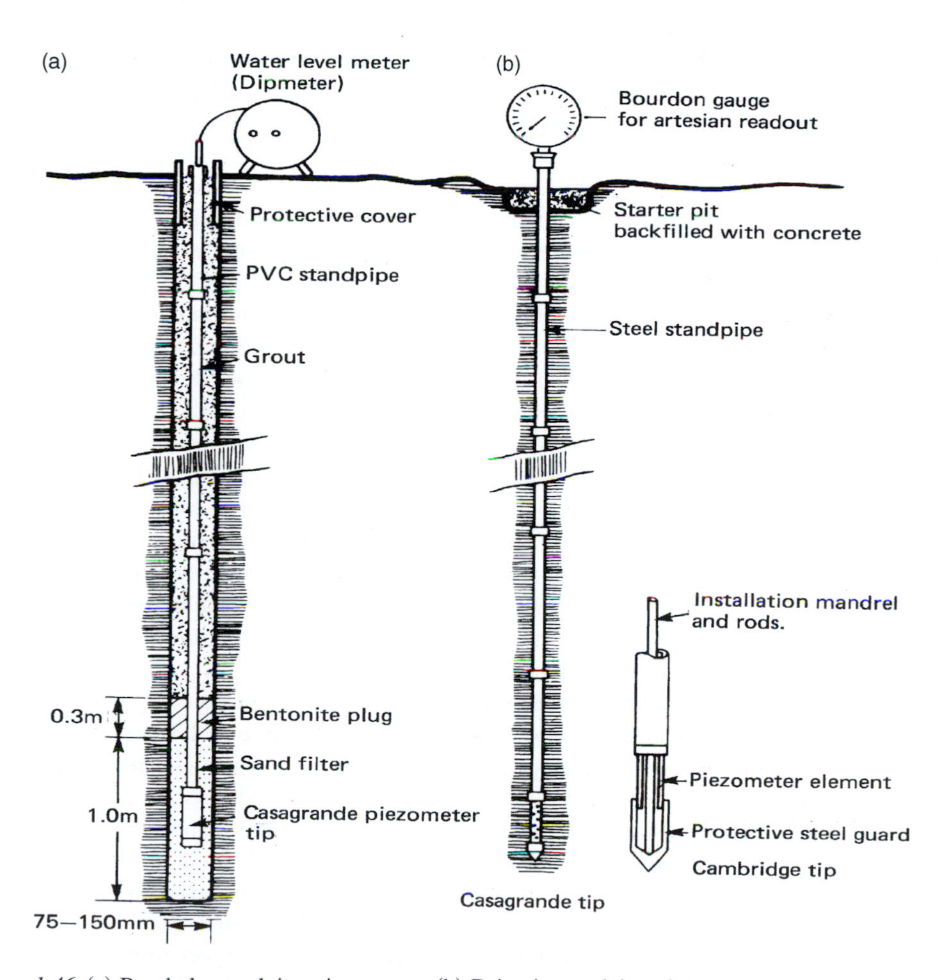

Figure 1.46 (a) Borehole standpipe piezometer. (b) Drive-in standpipe piezometers.

The pore water pressure in low-permeability ground can be measured with a hydraulic piezometer or a pneumatic piezometer. Various methods have been developed to measure the distribution of piezometric head along drillholes in rock masses. The continuous piezometer uses a single continuous membrane as a packer and a probe able to take readings at any position along the drillhole. The modular piezometer works with a chain of packers and with a portable probe, permitting readings in the measurement zones defined by neighbouring packers. The piezodex system also employs a packer chain for defining mutually independent measurement intervals and a portable probe for taking measurements (Kovari and Koeppel, 1987).

Surface deformation either in the form of settlements or horizontal movements can be assessed by precise surveying methods, the use of EDM or laser equipment provides particularly accurate results. Settlement, for example, can be recorded by positioning reference marks on the structures concerned, readings being taken by precise surveying methods. The observations are related to nearby bench marks. Vertical movements also can be determined by settlement tubes or by water-level or mercury-filled gauges.

Borehole extensometers are used to measure the vertical displacement of the ground at different depths (Fig. 1.47). Burland *et al.* (1972) described a precise borehole extensometer that consists of circular magnets embedded in the ground, which act as markers, and reed switch sensors move in a central access tube to locate the positions of the magnets (Fig. 1.48). Subsequently, Londe (1982) referred to a multiple-point extensometer that operates on a self-inductance principle. It consists of a central rod fixed at one end of a drillhole and carries a set of inductance displacement sensors passing through coaxial rings fixed by springs to the rock at selected points.

An inclinometer is used to measure horizontal movements below ground. It consists of a probe that is fitted with wheels and contains a gravity-operated tilt sensor (Fig. 1.48). This produces a signal from which the angle between the axis of the probe and the vertical can be determined. A grooved guide tube is grouted into a drillhole of 100–150 mm in diameter. Readings are made by lowering the probe down the guide tube, and making readings about every 0.5 mm. The probe transmits signals of tilt and depth to a recorder and a vertical profile thereby is obtained. Sets of readings over a period of time enable both the magnitude and rate of any horizontal movement to be determined.

The state of stress that exists in the earth's crust is disturbed by engineering operations that produce a new distribution of induced stress within the surrounding ground. The readjustment of the state of stress in the ground can give rise to stress concentrations that can have a major influence on the stability of surface excavations and underground openings. The original or virgin stresses include gravitational stress attributable to the weight of overburden, tectonic stress due to earth movements, as well as stresses due to factors such as weathering, consolidation, dehydration, hydration and pore water pressure. As a consequence, the *in situ* state of stress in the ground has proved difficult to assess with accuracy.

Stress measurements in rock involve determination of the absolute state of stress on the one hand and measurement of relative stress or change of stress on the other. The measurement of stress, contact pressures and change of stress can be made in two ways, namely, strain may be measured and then converted to stress or stress may be measured directly. Measurement of absolute stress in rock masses that behave more or less in an elastic manner may be achieved by a stress-relief technique, in which rock containing the instrumentation is relieved from the stress produced by the confinement of the surrounding rock. The strain resulting from this stress-relief is measured. Change of stress may be determined by measuring absolute stress over a given interval of time. Generally, the instruments used in both types of measurement are similar.

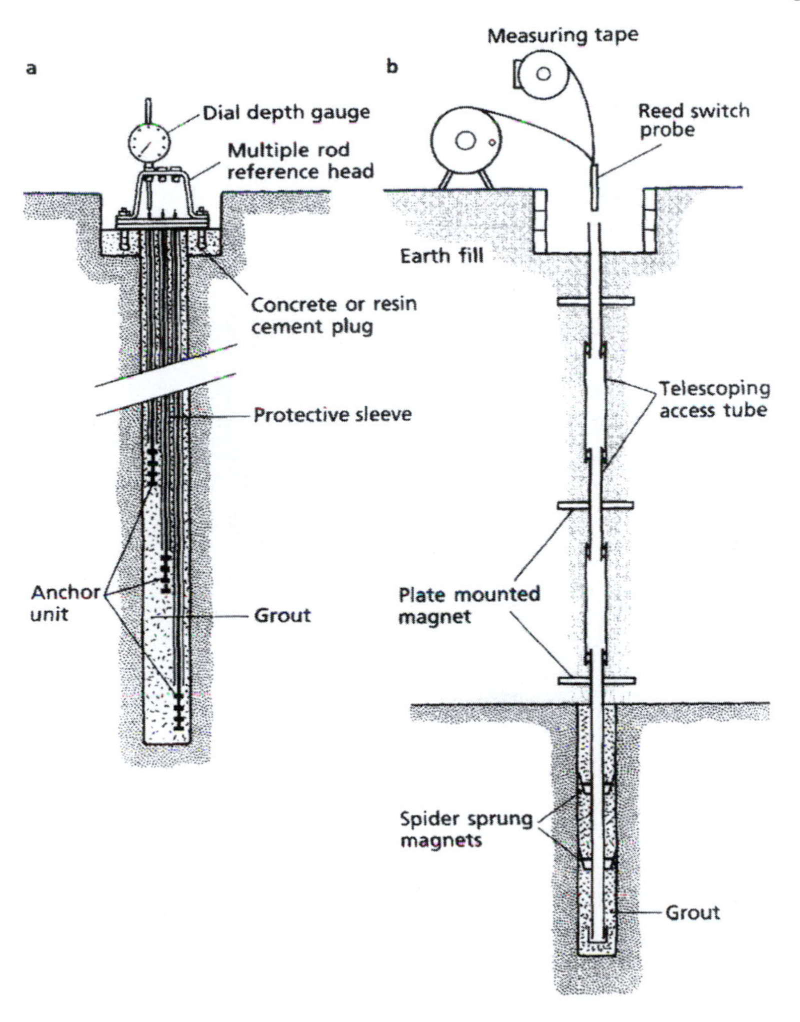

Figure 1.47 (a) Multiple rod extensometer. (b) Magnetic probe extensometer.

In the determination of absolute stress, drillhole deformation meters and drillhole strain cell devices depend upon assumptions of elasticity, as they measure elastic rebound that occurs on overcoring (Bell, 1992). Therefore stress-relief strain measurement techniques are only valid if applied to isotropic homogeneous strong rocks that display near-elastic deformation characteristics. They are of little value in soft sediments, rocks with marked anisotropy or those that undergo time-dependent deformation when loaded. Even in rocks that behave in a more or less elastic manner, there is the problem of determining the appropriate deformation modulus. If measurement extends over lengthy periods, then the time-dependent characteristics of the rock must be taken into account in order to interpret the results correctly.

Strain may be measured and then converted to stress or stress may be measured directly by an earth pressure cell such as the Glotzl cell. This is a hydraulic (flat diaphragm) cell that has a high stiffness at constant temperature and is used for measuring contact pressures. An earth pressure cell must be placed in position in such a way as to minimize disturbance of the stress and strain distribution.

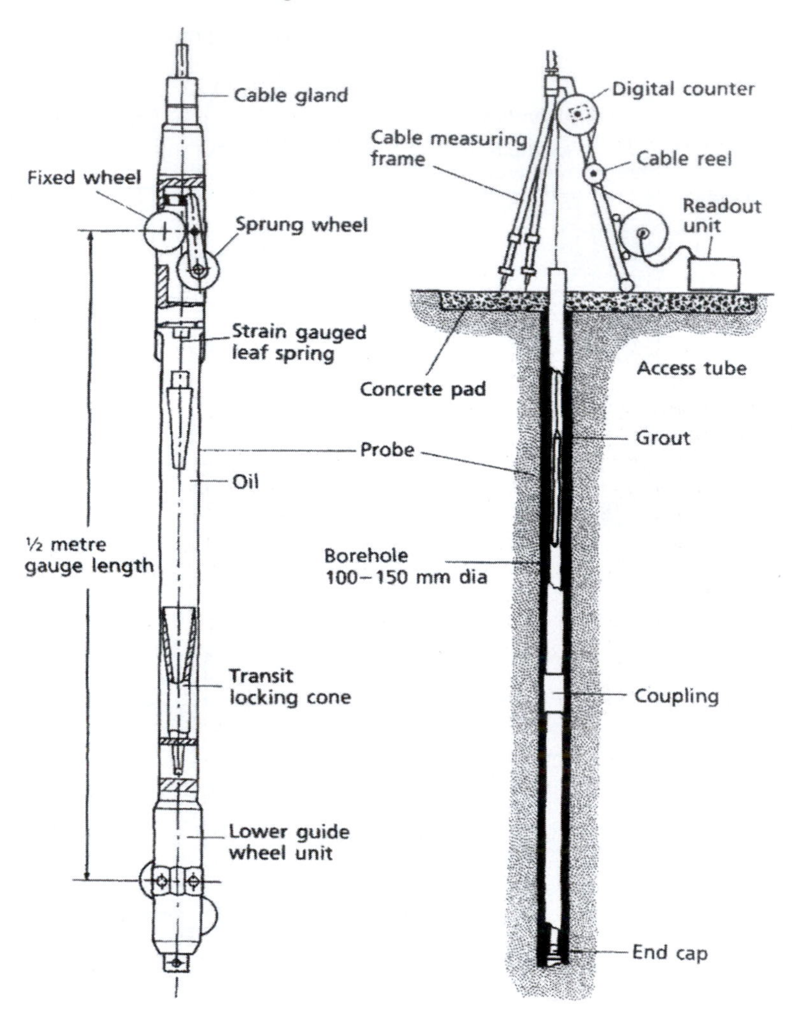

Figure 1.48 Borehole inclinometer and set-up in borehole.

Ideally, an earth pressure cell should have the same elastic properties as the surrounding soil. This, of course, cannot be attained, and in order to minimize the magnitude of error (i.e. the cell action factor) the ratio of the thickness to the diameter of the cell should not exceed 0.2 and the ratio of the diameter to the deflection of the diaphragm must be 2000 or greater. Most stress measurements in soils and soft rocks are deformation measurements that are interpreted in terms of strain. Strain is sometimes measured directly in hard rocks. This can be done by mounting strain gauges onto the end of a probe in a drillhole, thereby monitoring deformation of the drillhole.

A strain cell is required to move with the soil without causing it to be reinforced. In order to record strain it is necessary to monitor the relative movements of two fixed points at either end of a gauge length. Strain cells with a positive connection between their end plates have difficulty in measuring small strains. This can be overcome by substituting separated strain cells at either end of the gauge length.

References

Allum, J.A.E. 1966. *Photogeology and Regional Mapping*. Pergamon, Oxford.

Anderson, W.F., Pyrah, I.C. and Pang, L.S. 1990. Strength and deformation parameters from pressuremeter tests in clays. In: *Field Testing in Engineering Geology,* Engineering Geology Special Publication No. 6, Bell, F.G., Culshaw, M.G., Cripps, J.C. and Coffey, J.R. (eds), Geological Society, London, 23–31.

Anon. 1970. Logging of cores for engineering purposes. Working Party Report. *Quarterly Journal Engineering Geology*, **3**, 1–24.

Anon. 1972. The preparation of maps and plans in terms of engineering geology. Working Party Report. *Quarterly Journal of Engineering Geology*, **5**, 293–381.

Anon. 1976a. *Reclamation of Derelict Land: Procedure for Locating Abandoned Mine Shafts.* Department of the Environment, London.

Anon. 1976b. *Engineering Geology Maps. A Guide to their Preparation.* UNESCO Press, Paris.

Anon. 1977. The description of rock masses for engineering purposes. Engineering Group Working Party Report. *Quarterly Journal Engineering Geology*, **10**, 43–52.

Anon. 1978. *Terrain Evaluation for Highway Engineering and Transport Planning.* Transport Road Research Laboratory, Report SR448, DOE, Crowthorne.

Anon. 1982. Land surface evaluation for engineering practice. Engineering Group Working Party Report. *Quarterly Journal Engineering Geology*, **15**, 265–316.

Anon. 1988. Engineering geophysics. Working Party Report. *Quarterly Journal Engineering Geology*, **21**, 207–273.

Anon. 1995. The description and classification of weathered rocks for engineering purposes. Working Party Report. *Quarterly Journal Engineering Geology*, **28**, 207–242.

Anon. 1999. *Code of Practice on Site Investigations, BS 5930.* British Standards Institution, London.

Avery, T.E. and Berlin, G.L. 1992. *Fundamentals of Remote Sensing and Airphoto Interpretation.* Fifth Edition. Macmillan, New York.

Barton, N. 1978. Suggested methods for the quantitative description of discontinuities in rock masses. International Society of Rock Mechanics Commission on Standardization of Laboratory and Field Tests. *International Journal of Rock Mechanics and Mining Sciences and Geomechanical Abstracts*, **15**, 319–368.

Barton, N. 1987. Discontinuities. In: *Ground Engineer's Reference Book,* Bell, F.G. (ed.), Butterworths, London, 5/1–5/15.

Beaumont, T.E. 1979. Remote sensing for location and mapping of engineering construction materials. *Quarterly Journal Engineering Geology*, **12**, 147–158.

Bell, F.G. 1975. *Site Investigations in Areas of Mining Subsidence*. Newnes-Butterworths, London.

Bell, F.G. 1988. Land development: state-of-the-art in the search for old mine shafts. *Bulletin International Association of Engineering Geology*, **37**, 91–98.

Bell, F.G. 1992. Instrumentation and monitoring in rock masses. In: *Engineering in Rock Masses*, Bell, F.G. (ed.), Butterworth-Heinemann, Oxford, 190–208.

Bell, F.G. 2000. *Engineering Properties of Soils and Rocks.* Fourth Edition. Blackwell Scientific Publishers, Oxford.

Bell, F.G., Genske, D.D. and Bell, A.W. 2000. Rehabilitation of industrial areas: case histories from England and Germany. *Environmental Geology*, **40**, 121–134.

Bieniawski, Z.T. 1989. *Engineering Rock Mass Classifications*. Wiley-Interscience, New York.

Braithewaite, P.A. and Cole, K.W. 1986. Subsurface investigations of abandoned workings in the West Midlands of England by use of remote sensors. *Transactions Institution Mining and Metallurgy*, **95**, Section A, Mining Industry, A181–A190.

Briggs, R.P. 1987. Engineering geology and seismic and volcanic hazards in the Hijaz railway region – Syria, Jordan and Saudi Arabia. *Bulletin Association Engineering Geologists*, **24**, 403–423.

Brink, A.B.A., Mabbutt, J.A., Webster, R. and Beckett, P.H.T. 1966. *Report on the Working Group on Land Classification and Data Storage.* Military Engineering Experimental Establishment, Report No. 940, Christchurch.

Burland, J.B. and Burbridge, M.C. 1985. Settlement of foundations on sand and gravel. *Proceedings Institution Civil Engineers*, **78**, 1325–1381.

Burland, J.B., Moore, J.E.A. and Smith, P.D.K. 1972. A simple and precise borehole extensometer. *Geotechnique*, **22**, 174–177.

Burrough, P.A. and McDonnell, R.A. 1998. *Principles of GIS*. Oxford University Press.

Cartmell, S.J., Conn, P.J. and Pugh, T.D. 1997. An example of the use of crosshole tomography in dam wall foundation studies. In: *Modern Geophysics in Engineering Geology*, Engineering Geology Special Publication No. 12, McCann, D.M., Eddleston, M. Fenning, P.J. and Reeves, G.M. (eds), Geological Society, London, 141–151.

Chandler, J.H. 2001. Terrain measurement using automated digital photogrammetry. In: *Land Surface Evaluation for Engineering Practice,* Engineering Geology Special Publication No. 18, Geological Society, London, 13–18.

Clarke, B.G. 1990. Consolidation characteristics of clays from self-boring pressuremeter tests. In: *Field Testing in Engineering Geology*, Engineering Geology Special Publication No. 6, Bell, F.G., Culshaw, M.G., Cripps, J.C. and Coffey, J.R. (eds), Geological Society, London, 33–37.

Clayton, C.R.I., Matthews, M.C. and Simons, N.E. 1996. *Site Investigation*. Second Edition. Blackwell Scientific Publishers, Oxford.

Corin, L., Couchard, B., Dethy, B., Halleux, L., Mojoie, A., Richter, T. and Wauters, J.P. 1997. Radar tomography applied to foundation design in a karstic environment. In: *Modern Geophysics in Engineering Geology*, Engineering Geology Special Publication No. 12, McCann, D.M., Eddleston, M., Fenning, P.J. and Reeves, G.M. (eds), Geological Society, London, 167–173.

Corke, D.J. 1990. Self-boring pressuremeter *in situ* lateral stress assessment in the London Clay. In: *Field Testing in Engineering Geology*, Engineering Geology Special Publication No. 6, Bell, F.G., Culshaw, M.G., Cripps, J.C. and Coffey, J.R. (eds), Geological Society, London.

Cripps, J.C., McCann, D.M., Culshaw, M.G. and Bell, F.G. 1988. The use of geophysical methods as an aid to the detection of abandoned shallow mine workings. In: *Minescape '88, Proceedings Symposium on Mineral Extraction, Utilisation and Surface Environment*, Harrogate, Institution of Mining Engineers, Doncaster, 281–289.

Dai, F.C., Lee, C.F. and Zhang, X.H. 2001. GIS-based geo-environmental evaluation for urban land-use planning: a case study. *Engineering Geology*, **61**, 257–271.

Dearman, W.R. 1991. *Engineering Geological Maps*. Butterworths-Heinemann, Oxford.

Dearman, W.R. and Fookes, P.G. 1974. Engineering geological mapping for civil engineering practice in the United Kingdom. *Quarterly Journal Engineering Geology*, **7**, 223–256.

Deere, D.U. 1964. Technical description of cores for engineering purposes. *Rock Mechanics and Engineering Geology*, **1**, 18–22.

Demek, J. 1972. *Manual of Detailed Geomorphological Mapping*. Academia, Prague.

Doornkamp, J.C., Brunsden, D., Jones, D.K.C., Cooke, R.U. and Bush, P.R. 1979. Rapid geomorphological assessment for engineers. *Quarterly Journal Engineering Geology*, **12**, 189–204.

Dunnicliff, J. 1988. *Geotechnical Instrumentation for Monitoring Field Performance*. Wiley, New York.

Finch, J.W. 1979. An application of surface electrical resistivity methods to the delineation of spoil tip leachate. *Proceedings Symposium on Engineering Behaviour of Industrial and Urban Fill*, Birmingham, Midlands Geotechnical Society, C35–C44.

Fookes, P.G. 1997. Geology for engineers: the geological model, prediction and performance. *Quarterly Journal Engineering Geology*, **30**, 293–424.

Fookes, P.G. and Denness, D. 1969. Observational studies on fissure patterns in the Cretaceous sediments of south east England. *Geotechnique*, **19**, 453–477.

Franklin, J.A., Maerz, Z.H. and Bennet, C.P. 1988. Rock mass characterization using photoanalysis. *International Journal Rock Mechanics Mining Science and Geomechanical Abstracts*, **25,** 97–112.

Franklin, J.L., Broch, E. and Walton, G. 1971. Logging the mechanical character of rock. *Transactions Institution of Mining and Metallurgy*, **81**, Mining Section, A1–A9.

Fruneau, B. and Achache, J. 1996. Satellite monitoring of landslides using SAR Interferometry. *News Journal, International Society Rock Mechanics*, **3**, No. 3, 10–13.

Garner, J.B. and Heptinstall, S.M. 1974. Aerial photo interpretation of engineering and soil surveys. *The Highway Engineer*, August/September issue, 24–29.

Goodchild, M.P. 1993. The state of GIS for environmental problem solving. In: *Environmental Modelling with GIS*, Goodchild, M.H., Parks, B.O. and Steyaert, L.T. (eds), Oxford University Press, Oxford, 8–15.

Grainger, P., McCann, D.M. and Gallois, R.W. 1973. The application of the seismic refraction technique to the study of fracturing of the Middle Chalk at Mundford, Norfolk. *Geotechnique*, **28**, 219–232.

Grasmück, M. and Green, A.G. 1996. 3-D georadar mapping: looking into the subsurface. *Environmental and Engineering Geosciences*, **2**, 195–200.

Griffiths, J.S. 2001. Engineering geological mapping. In: *Land Surface Evaluation for Engineering Practice*, Engineering Geology Special Publication No. 18. Griffiths, J.S. (ed.), Geological Society, London, 39–42.

Hadjigeorgiou, J., Lessard, J.F., Villaescusa, E. and Germain, P. 1995. An appraisal of structural mapping techniques. *Proceedings 2nd International Conference on Mechanics of Jointed and Faulted Rock*, Vienna, Rossmanith, H. (ed.), A.A. Balkema, Rotterdam, 193–199.

Halstead, P.N., Call, P.D. and Rippere, K.H. 1968. Geological structural analysis for open pit slope design, Kimberley pit, Ely, Nevada. *Annual Conference American Institute Mining Engineers*, New York, Reprint.

Hanna, T.H. 1985. *Field Instrumentation in Geotechnical Engineering*. Trans Tech Publications, Clausthal-Zellerfeld.

Hatheway, A.W. and Leighton, F.B. 1979. Trenching as an exploration method. In: *Geology in the Siting of Nuclear Power Plants, Reviews in Engineering Geology*, Geological Society America, Boulder, Colorado, **4**, 169–195.

Hawkins, A.B. 1986. Rock descriptions. *In Site Investigation Practice: Assessing BS 5930*, Engineering Geology Special Publication No. 2, Hawkins, A.B. (ed.), The Geological Society, London, 59–66.

Hellawell, E.E., Lamont-Black, J., Kemp, A.C. and Hughes, S.J. 2001. GIS as a tool in geotechnical engineering. *Proceedings Institution Civil Engineering, Geotechnical Engineering*, **149**, 85–93.

Herbert, S.M., Roche, D.P. and Card, G.B. 1987. The value of engineering geological desk study appraisals in scheme planning. In: *Planning in Engineering Geology*, Engineering Geology Special Publication No. 4, Culshaw, M.G., Bell, F.G., Cripps, J.C. and O'Hara, M. (eds), The Geological Society, London, 151–154.

Hobbs, N.B. 1975. Factors affecting the prediction of settlement of structures on rocks with particular reference to the Chalk and Trias. In: *Settlement of Structures*, British Geotechnical Society, Pentech Press, London, 579–610.

Hoek, E. and Bray, J.W. 1981. *Rock Slope Engineering*. Third Edition. Institution of Mining and Metallurgy, London.

Hudson, J.A. and Harrison, J.P. 1997. *Engineering Rock Mechanics: An Introduction to the Principles*. Pergamon, Oxford.

Hvorslev, M.J. 1949. *Subsurface Exploration and Sampling for Civil Engineering Purposes*, American Society Civil Engineers Report, Waterways Experimental Station, Vicksburg.

Jackson, P.D. and McCann, D.M. 1997. Cross-hole seismic tomography for engineering site investigation. In: *Modern Geophysics in Engineering Geology*, Engineering Geology Special Publication No. 12, McCann, D.M., Eddleston, M., Fenning, P.J. and Reeves, G.M. (eds), Geological Society, London, 247–264.

Kay, J.N. and Parry, R.H.G. 1982. Screw plate tests in stiff clay. *Ground Engineering*, **15**, No. 6, 22–30.

Kovari, K. and Koeppel, K. 1987. Head distribution monitoring with a sliding piezometer system "Piezodex". *Proceedings Second International Symposium Field Measurements in Geomechanics*, Kobi, Sakurai, S. (ed.), Balkema, Rotterdam, 255–267.

Lawrance, C.J. 1978. *Terrain Evaluation in West Malaysia, Part 2 – Land Systems of South West Malaysia*. Transport Road Research Laboratory, Report SR 378, DOE, Crowthorne.

Lawrance, C.J., Byard, R.J. and Beaven, P.J. 1993. *Terrain Evaluation Manual. Transportation Research Laboratory, State-of-the-Art Review 7*. Her Majesty's Stationery Office. London.

Leggo, P.J. 1982. Geological application of ground impulse radar. *Transactions Institution Mining and Metallurgy*, **91**, Section B, Applied Earth Sciences, B1–B5.

Lillesand, T.M. and Kiefer, R.W. 1994. *Remote Sensing and Image Interpretation*. Third Edition. Wiley, New York.

Londe, P. 1982. Concepts and instruments for improved monitoring. *Proceedings American Society Civil Engineers, Journal Geotechnical Engineering Division*, **108**, 820–834.

Mair, R.J. and Wood, D.M. 1987. *Pressuremeter Testing: Methods and Interpretation*. Construction Industry Research and Information Association Report, Butterworths, London.

McCann, D.M., Baria, R., Jackson, P.D. and Green, A.S.P. 1986. Application of crosshole seismic measurements to site investigation. *Geophysics*, **51**, 914–925.

McCann, D.M., Culshaw, M.G. and Northmore, K. 1990. Rock mass assessments from seismic measurements. In: *Field Testing in Engineering Geology*, Engineering Geology Special Publication No. 6, Bell, F.G., Culshaw, M.G., Cripps, J.C. and Coffey, J.R. (eds), The Geological Society, London, 257–266.

McCann, D.M., Culshaw, M.G. and Fenning, P.J. 1997. Setting the standard for geophysical surveys in site investigation. In: *Modern Geophysics in Engineering Geology*, Engineering Geology Special Publication No. 12, McCann, D.M., Eddleston, M., Fenning, P.J. and Reeves, G.M. (eds), The Geological Society, London, 3–34.

McDowell, P.W. 1970. The advantages and limitations of geophysical methods in the foundation engineering of the track hovercraft experimental site in Cambridgeshire. *Quarterly Journal Engineering Geology*, **3**, 119–126.

Meigh, A.C. 1987. *Cone Penetration Testing: Methods and Interpretation*. Construction Industry Research and Information Association Report, Butterworths, London.

Mejía-Navarro, M. and Garcia, L.A. 1996. Natural hazard and risk assessment using decision support systems, application: Glenwood Springs, Colorado. *Environmental and Engineering Geoscience*, **2**, 299–324.

Mitchell, C.W. 1991. *Terrain Evaluation*. Second Edition. Longmans, London.

Norbury, D.R. and Gosling, R.C. 1996. Rock and soil description and classification – a view from industry. In: *Advances in Site Investigation Practice*, Craig, C. (ed.), Thomas Telford Press, London, 293–305.

Norman, J.W. 1968. Photogeology of linear features in areas covered by superficial deposits. *Transactions Institution Mining and Metallurgy*, **78**, Section B, Applied Earth Science, B60–B77.

Norman, J.W. and Watson, I. 1975. Detection of subsidence conditions by photogeology. *Engineering Geology*, **9**, 359–381.

Onodera, T.F. 1963. Dynamic investigation of foundation rocks. *Proceedings Fifth Symposium on Rock Mechanics*, Minnesota, Pergamon Press, New York, 517–533.

Oteri, A.U.E. 1983. Delineation of saline intrusion in the Dungeness shingle aquifer using surface geophysics. *Quarterly Journal Engineering Geology*, **16**, 43–51.

Perry, J. and West, G. 1996. *Sources of Information for Site Investigations in Britain.* TRL Report No. 192, Transportation Research Laboratory, Crowthorne, Berkshire.

Piteau, D.R. 1971. Geological factors significant to the stability of slopes in cut rock. *Proceedings Symposium on Planning Open Pit Mines*, Johannesburg, A.A. Balkema, Rotterdam, 43–53.

Pitts, J. 1979. Morphological mapping in the Axmouth-Lyme Regis undercliffs, Devon. *Quarterly Journal Engineering Geology*, **12**, 205–218.

Priest, S.D. 1993. *Discontinuity Analysis for Rock Engineering.* Chapman & Hall, London.

Priest, S.D. and Hudson, J.A. 1981. Estimation of discontinuity spacing and trace length using scanline surveys. *International Journal Rock Mechanics Mining Science and Geomechanical Abstracts*, **18**, 183–197.

Rengers, N. and Soeters, R. 1980. Regional engineering geological mapping from aerial photographs, *Bulletin International Association Engineering Geology*, **21**, 103–111.

Reynolds, J.M. 1997. *An Introduction to Applied and Environmental Geophysics.* Wiley, Chichester.

Rocha, M. 1971. Method of integral sampling. *Rock Mechanics*, **3,** 1–12.

Rocha, M., Da Silveira, A., Grossman, N. and De Oliveira, E. 1966. Determination of the deformability of rock masses along boreholes. *Proceedings First International Congress on Rock Mechanics* (ISRM), Lisbon, **3**, 697–704.

Sabins, F.F. 1996. *Remote Sensing – Principles and Interpretation*, W. Freeman and Co., San Francisco.

Salisbury, J.W. and D'Aria, D.M. 1992. Emissivity of terrestrial materials in the 8 to 14 μm atmospheric window. *Remote Sensing of the Environment*, **42**, 83–106.

Sanglerat, G. 1972. *The Penetrometer and Soil Exploration.* Elsevier, Amsterdam.

Savigear, R.G.A. 1965. A technique of morphological mapping. *Annals Association American Geographers*, **53**, 514–538.

Sharma, P.V. 1997. *Environmental and Engineering Geophysics.* Cambridge University Press, Cambridge.

Simons, N.E. and Menzies, B.K. 2000. *A Short Course in Foundation Engineering.* Second Edition. Thomas Telford Press, London.

Smith, A. and Ellison, R.A. 1999. Applied geological maps for planning and development: a review of examples from England and Wales, 1983 to 1996. *Quarterly Journal Engineering Geology*, **32** (Supplement), S1–S44.

Soeters, R. and Van Westen, C.J. 1996. Slope instability recognition, analysis and zonation. In: *Landslides: Investigation and Mitigation*, Turner, A.K. and Schuster, R.L. (eds), Transportation Research Board Special Report 247, National Academy Press, Washington, DC, 129–177.

Star, J. and Estes, J. 1990. *Geographical Information Systems.* Prentice-Hall, Englewood Cliffs, New Jersey.

Terzaghi, K. and Peck, R.B. 1967. *Soil Mechanics in Engineering Practice.* Second Edition. Wiley, New York.

Varnes, D.J. 1974. The logic of geological maps with reference to their interpretation and use for engineering purposes. *USGS Professional Paper 873.* United States Geological Survey, Washington, DC.

Waller, A.W. and Phipps, P. 1996. Terrain systems mapping and geomorphological studies for the Channel Tunnel rail link. In: *Advances in Site Investigation Practice*, Craig, C. (ed.), Thomas Telford Press, London, 25–38.

Ward, W.H., Burland, J.B. and Gallois, R.W. 1968. Geotechnical assessment of a site at Mundford, Norfolk, for a large proton accelerator. *Geotechnique*, **18**, 399–431.

Wareham, B.F. and Skipp, B.O. 1974. The use of the flatjack installed in a sawcut slot in the measurement of in situ stress. *Proceedings Third International Congress International Society Rock Mechanics*, Denver, **2**, 481–488.

Warwick, D., Hartopp, P.G. and Viljoen, R.P. 1979. Application of thermal infrared linescanning technique to engineering geological mapping in South Africa. *Quarterly Journal Engineering Geology*, **12**, 159–180.

Waters, R.S. 1958. Morphological mapping. *Geography*, **43**, 10–17.

Webster, R. and Beckett, P.H.T. 1970. Terrain classification and evaluation using air photography. *Photogrammetric*, **26**, 51–75.

Weltman, A.J. and Head, J.M. 1983. *Site Investigation Manual.* CIRIA Special Publication 25, Construction Industry Research and Information Association, London.

Wyllie, M.R.J., Gregory, A.R. and Gardner, L.W. 1956. Elastic wave velocities in heterogeneous porous media. *Geophysics*, **21**, 41–70.

Zebker, H.A., Rosen, P., Goldstein, R.M., Gabriel, A. and Werner, C.L. 1994. On the derivation of co-seismic displacement fields using differential radar interferometry: the Landers earthquake. *Journal Geophysical Research*, **99**, 617–634.

2 Open excavation and slopes

Open excavation refers to the removal of material, within certain specified limits, for mineral exploitation and construction purposes. For this to be accomplished economically and without hazard, the character of the rock and soil masses involved and their geological setting must be investigated. Indeed, the method of excavation and the rate of progress are very much influenced by the geology of the site (Kummerle and Benvie, 1988). Obviously, the stability of a rock or soil mass is important in excavation, as are the position of the water table in relation to the base level of the excavation, and any possible effects of excavation on the surrounding ground and structures. Open joints in rock masses facilitate weathering and generally aid slope failure. Fissure zones usually represent zones of weakness along which rock masses may have been altered to appreciable depth by weathering. Faults that traverse the area in which excavation is to be made may cause serious trouble. This is principally because of the greater freedom afforded rock masses to move along fault planes. In particular, if a fault intersects a prominent joint or bedding plane in such a way that it produces a wedge that daylights into the excavation, then this is likely to slide.

The cross-section of an open excavation is influenced by the dimensions of the base, its depth and the profile of its slopes. In terms of construction, in particular, slopes should be as steep as possible, consistent with safety, in order to minimize the volume of material to be removed. Allowance must be made for any drainage works as far as the dimensions of excavations are concerned.

2.1 Methods of excavation: drilling and blasting

The method of excavation is very much determined by the geology of the site; however, consideration must also be given to the surroundings. For instance, drilling and blasting, although generally the most effective and economical method of excavating hard rock, are not desirable in built-up areas since damage to property or inconvenience may be caused. Neither is it wise to blast where landslides or rock falls might result.

2.1.1 Drilling

If drilling is not carried out properly, blasts are unable to provide muckpiles having the characteristics required for subsequent operations. Optimum drilling therefore is a pre-requisite of optimum blasting.

According to McGregor (1967) the properties of a rock mass that influence drillability include strength, hardness, toughness, abrasiveness, grain size and discontinuities. The strength of a rock has an appreciable influence on the drilling force required. The hardness of a rock depends not only upon the hardness of the individual minerals involved but also upon their texture. The harder the rock, the stronger the bit that is required for drilling since higher pressures need to be exerted. Toughness is related to hardness and represents the work required to bring about fracture. Abrasiveness with respect to drilling may be regarded as the ability of a rock to wear away drill bits. This property is closely related to hardness and in addition is influenced by particle shape and texture. The size of the chippings produced during drilling operations also influences abrasiveness. For example, large chippings may cause scratching but comparatively little wear of a bit whereas the production of dust in tougher but less-abrasive rock causes polishing. This may lead to the development of high skin hardness on tungsten carbide bits that, in turn, may cause them to spall. Even diamonds lose their cutting ability upon polishing. Rock fabric determines the characteristics of the chippings produced. Generally, coarse-grained rocks can be drilled more quickly than can fine-grained varieties or those in which the grain size is variable.

The ease of drilling in rocks in which there are many discontinuities is influenced by their orientation in relation to the drillhole. Usually, the rate of drilling is less difficult and therefore quicker when a hole runs at a high angle to the discontinuities. Drilling over an open discontinuity means that part of the energy controlling drill penetration is lost. Where a drillhole crosses discontinuities at a low angle this may cause the bit to stick. It also may lead to excessive wear and to a hole going off line. A drillhole may require casing in badly broken ground. If discontinuities are filled with clay, this may penetrate the flush holes of the bit, causing it to bind or deviate from alignment.

In addition to the properties of the rock mass, the rate of penetration is influenced by the power of the rock drill, the shape of the cutting edge of the bit, the air pressure at the rock drill and the diameter of the drillhole (Kahraman *et al.*, 2000). In this context, the relative proportions of transmitted and reflected energy are important, and are governed by the bit-rock contact and the reaction of the rock to impact. There is a certain critical level of stress for any particular rock below which penetration cannot be brought about by impact. The type of flush used and the experience of the drilling crew also influence the rate of penetration.

Penetration is effected by the drill bit causing the rock to be pulverized and chipped. The latter action is the more effective as far as drilling is concerned and if chipping action is to be successful, then the pulverized rock material must be removed as this has a cushioning effect on drilling. Hence, an important function of the flushing medium is to clear the cutting edge of the bit of rock fragments.

A bit with a wide wedge angle only is able to achieve a small depth of penetration. Unfortunately, however, although a bit with a sharp angle is more effective at chipping, it suffers rapid wear (Furby, 1964). Therefore, the shape of the bit represents a compromise between a high initial rate of penetration and the rate of wear.

In softer rocks, rotary drills give faster penetration than the percussive types, but in hard to medium rocks the rotary drill is not economical because of the severity of bit wear. Holes of up to 50 mm in diameter can be sunk in soft to medium hard rocks (for example, some sandstones, shale, coal, gypsum and rock salt) with rotary drills. Larger holes up to 100 mm diameter can be made in soft rocks by reaming the original hole.

Rotary percussion drills are designed for rapid drilling in rock. They provide a continuous rotary cutting action upon which axial percussive blows are superimposed. In this way, the rock is subjected to rapid high-speed impacts whilst the bit rotates, which causes fracture at a lower value of rotary thrust and torque. Consequently, faster rates of penetration are achieved at lower thrust than with rotary drilling (the latter can still produce a faster rate of penetration but with much higher thrust). The technique is most effective in brittle materials since it relies upon chipping the rock. The impact mechanism is coupled to the bit in a down-the-hole hammer drill and accompanies the bit down the hole. Generally, compressed air is used with a down-the-hole hammer, the air pressure being around 1.7 MPa, but in some cases pressures of up to 2.8 MPa have been used. High operating pressures, however, lead to heavy abrasive wear and impact damage to the bit.

Holes with a diameter of 100 mm cater for most normal requirements. As the diameter of blastholes is increased, so the charge is increased. Ground vibrations may present a problem when charges are fired. However, vibrations generally can be reduced to an acceptable level by, for example, using short-delay blasting. Also, as the diameter of a blasthole increases, the burdens and spacings are increased, so that the discontinuities in the rock mass become more significant in terms of fragmentation.

Depending on the pattern of holes and firing technique, subgrade drilling, that is, drilling blastholes some 300–900 mm below excavation grade, usually is necessary to ensure that the area being blasted is broken down to grade. Instantaneous blasts usually require greater subgrade drilling than short-delay blasts. It is only when there is a notable parting, such as a prominent bedding plane that is coincident with the grade of the excavation, that subgrade drilling is unnecessary.

The dip of a drillhole can be measured with an inclinometer. If the deviation is excessive, then this entails re-drilling. Generally speaking, the possibility of deviation tends to increase with distance from the top of the hole and its incidence tends to be greater in angled than in vertical holes.

2.1.2 Explosives for blasting operations

The essential characteristic of an explosive material is that, on initiation, it reacts suddenly to form large volumes of gases at high temperatures, the almost instantaneous release of these gases giving rise to very high pressures (Anon., 1972). Initiating explosives are used to detonate high explosives and, as such, they are extremely sensitive and relatively easy to explode. High explosives detonate at velocities from about 1500 to 7500 $\mathrm{m\,s^{-1}}$, depending upon the composition of the explosive. The performance of a high explosive depends on the velocity of detonation (i.e. the rate at which the detonation wave travels through a column of explosive) and on the volume and temperature of the gases that form. Its power is governed by the amount of energy released when it is detonated. The shock wave developed by the explosive is transmitted through, and produces fractures in the rock mass concerned. The movement of the high-pressure gases through the rock mass completes its breakage.

The density of an explosive is dependent upon its composition. If it has a high density, then the energy of the explosion is concentrated. On the other hand, a low-density explosive can be used when excessive fragmentation has to be avoided, since a low explosive distributes the energy more evenly along the blasthole. Slurry explosives have a similar range of densities to conventional explosives. However, as they are designed to fill a blasthole, they afford a higher effective concentration of energy than normal cartridge explosives used under the same conditions. On the other hand, ammonium nitrate explosives possess a lower density than many of the conventional explosives. Nonetheless, as they also can be used to fill a blasthole completely, they therefore can produce a similar concentration of energy to cartridge explosives when employed in the same situations.

If water is present in a drillhole but the time involved in loading and firing is limited, then an explosive that possesses a good water resistance proves satisfactory. However, a very good to excellent water resistance is required of an explosive if the exposure to water is prolonged. Usually, the best water resistance is provided by blasting gelatines. The higher-density dynamites possess fair to good resistance, whereas low-density dynamites offer little or none.

An extensive range of explosive types and grades is available to satisfy all blasting requirements (Russell, 1997). The gelatine group of explosives is unrivalled as far as use in the hardest rocks such as dolerite, basalt and granite is concerned. They can be used in wet conditions because of their high resistance to water. In addition, because they have a relatively high density they can be used where a powerful concentration of explosive energy is required. Opencast gelignite is a powerful high-density explosive with good water resistance. It was developed for blasting operations in opencast coal mining. Special gelatine (80 per cent strength) and opencast gelignite are recommended for most types of excavation (Anon., 1972). They are medium-strength nitroglycerine gelatines. Because they possess good water resistance, they offer reliable performance in wet conditions. In addition, they have a high velocity of detonation and accordingly produce good fragmentation. When exceptionally good fragmentation is required the high-strength nitroglycerine gelatine, special gelatine 90 per cent, should be used. This explosive has excellent water resistance.

The ammon dynamites have lower densities than the gelatines of the same strength grade. Because they have good spreading action they are suitable for blasting conditions that are not too severe. They have satisfactory water-resistant properties.

Belex explosives are characterized by high weight strength but have lower density. Hence, they represent a range of explosives with varying bulk strengths from which a particular type may be chosen for certain given conditions. The lower density of these explosives can provide advantages as far as fragmentation is concerned. This is because the explosive charge is spread over a greater length of the blasthole and, when compared with an equal weight of gelatine explosive, a larger area of the rock mass comes into direct contact with the explosive charge. Belex explosives possess good water resistance.

The powder explosives have low densities and relatively high weight strengths. They are especially suitable for use in soft to medium strength rocks that are moderately dry. Trimonite powders are medium-density explosives that must be used in dry conditions since they are not moisture resistant. However, if they are packed in sealed containers they may be used in wet conditions.

Blasting agents can be used in opencast mining and quarrying where large diameter drilling is employed for overburden blasting. Nobelite and Anobel are common types. Because blasting agents are insensitive, they are very safe to handle. They are not sensitive to detonators or detonating fuse and are therefore initiated by primer cartridges of a conventional high explosive.

Although ammonium nitrate fuel oil (ANFO) mixtures have proved popular in quarrying and opencast mining, they suffer several disadvantages (Mather, 1997). For example, ANFO has a low bulk strength and a low density. It also has a low velocity of detonation and is not water resistant. These limitations are to some extent compensated for by its low cost per unit weight and high order of safety. ANFO can be pre-packed or mixed on site. It can be fed directly from a bulk explosives truck into dry holes. Where ANFO mixtures have to be packed in sleeves in order to provide water resistance, the addition of metallized powder gives extra density, as well as energy. A blast can be improved by increasing the bulk strength of ANFO through the addition of aluminium fines to the toe charge of the explosive column. The use of aluminized ANFO (ALANFO) in the main charge allows the blasthole pattern to be enlarged (Sandy, 1989). Heavy ANFO has a higher bulk strength and therefore blastholes can be positioned further apart. In addition, it is water resistant. Heavy ANFO has two main components, an oxidizing agent (porous ammonium nitrate prill) and a dense high-energy fuel (HEF). The latter is an emulsion made from a hot solution of ammonium nitrate and calcium nitrate, which then is mixed with diesel oil and patented emulsifier. As the sensitivity of all these products is extremely low, they require a heavy boost to initiate detonation.

Like ANFO, slurry explosives can be pre-packed or mixed on site. However, they involve a higher initial explosive cost than does ANFO. Water is an essential ingredient in slurry explosives, since it makes a solution that turns into a gel and holds solids in suspension. The gel provides the water resistance.

Charges usually are decked by separating zones of the explosive column by using inert materials, that is, stemming. The peak blasthole pressure within each charge deck is not reduced but the rate of decay of pressure can be increased appreciably (Hagan, 1979). Because of availability, drillhole chippings are used most frequently as stemming. In dry blastholes, however, angular crushed rock (around 15–25 mm in size for holes with diameters between 225 and 380 mm) is better than chippings since it possesses a higher effective (air) void ratio, thereby increasing the rate of decay of pressure within the adjacent charge deck. But even crushed rock is not as good as air decks in this respect (Jhanwar *et al.*, 2000). These are constructed by locating closed rigid empty cylinders between the charges in the blasthole.

Plain detonators and safety fuse, electric detonators (instantaneous and short delay), Cordtex detonating fuse and detonating relays may be used for initiation. Because remote control is easy, electric shotfiring is used for most operations. Short-delay firing is recommended for most excavation work. This is achieved by using either short-delay electric detonators or Cordtex and detonating relays. The former have a nominal delay interval between consecutive numbers of 22 ms in the early numbers with a slight increase in the later numbers. As far as detonating relays are concerned, a range of five relays with delay times of 10–45 ms is available. The introduction of short delays yields the highest efficiency from the explosive, giving maximum rock fragmentation and minimum ground vibration.

2.1.3 Blasting operations

If a rock is to be blasted efficiently it must be capable of transmitting the explosive energy some distance from the blastholes. When this does not happen, then the rock immediately surrounding the hole is pulverized and the area between holes is not fractured. The early movement of the rock face is most important for an efficient blast, that is, one that achieves good fragmentation per drillhole with minimum quantity of explosive (Fourney, 1993). Rock breakage in blasting, apart from the character of the rock itself, especially the fracture index, depends largely on the relation between the burden and the hole spacing, as well as the time of ignition between the holes (Muller, 1997).

Obviously, efficient blasting should produce rock fragments sufficiently reduced in size so that they can be easily loaded without resort to secondary breakage (Latham *et al.*, 1999). Accordingly, blastholes should be drilled accurately to the requisite pattern and proper depth to ensure satisfactory fragmentation. The greater the capacity of the buckets of the loading machines, the larger the fragment size that is acceptable. Good fragmentation also reduces wear and tear on loading machinery.

The quantity of a particular explosive required to blast a certain volume of rock is difficult to estimate since it depends upon the strength, toughness and incidence of discontinuities within the rock mass. Consequently, no rigid sequence for progressive resistance to blasting offered by different rock types can be formulated. Nevertheless, it can be said that an explosive with the greatest energy and concentration is required for removing very hard rock; in medium hard rocks a high-velocity detonation produces a shattering effect; a medium to high explosive can be used in medium to hard laminated rocks; and the greatest efficiency is obtained with fairly bulky explosive in soft to medium rocks. Table 2.1 gives an idea of the amount of charge in relation to burdens for primary blasting.

Table 2.1 Typical charges and burdens for primary blasting by shot-hole methods

Minimum finishing diameter of hole (mm)	*Cartridge diameter* (mm)	*Depth of hole* (m)	*Burden* (m)	*Spacing* (m)	*Explosive charge* (kg)	*Rock yield* (tonnes)	*Blasting ratio*	*Tonnes of rock per metre of drillhole*
25	22	1.5	0.9	0.9	0.3	3	10.0	2
35	32	3.0	1.5	1.5	1.8	19	10.5	6.3
57	50	6.1	2.4	2.4	9.5	97	10.2	15.9
75	64	9.1	2.7	2.7	18.0	180	10.0	19.8
75	64	12.2	2.7	2.7	25.0	245	9.8	20.1
75	64	18.3	2.7	2.7	36.3	365	10.1	20.0
100	83	12.2	3.7	3.7	43.0	430	10.0	35.3
100	83	30.5	3.7	3.7	104.3	1120	10.7	36.7
170	150	18.3	6.7	6.7	216.5	2235	10.3	122.1
170	150	30.5	6.7	6.7	363.0	3660	10.1	120.0
230	200	21.3	7.6	7.6	329.0	3350	10.2	157.3
230	200	30.5	7.6	7.6	476.3	4670	9.8	153.1

Source: Sinclair, 1969. Reproduced by kind permission of Kluwer Academic Publishers.

Roberts (1981) suggested the following expression for obtaining the thickness of the burden (B), in metres:

$$B = 0.024d + 0.85 \tag{2.1}$$

where d is the diameter of the blasthole. Vutukuri and Bhandari (1961) suggested that the spacing (S), in metres, between blastholes could be derived from the expression

$$S = 0.9B + 0.91 \tag{2.2}$$

Afrouz and Rostami (1997) examined a number of empirical expressions relating spacing between drillholes and thickness of burden to the dynamic properties of rock. They then used their findings to determine the spacing in relation to variations in rock mass rating. However, careful trials provide the only means of determining the burden and blasting pattern in any rock (Tatiya and Al Ajmi, 2000). As a rule, the spacing of blastholes varies between 0.75 and 1.25 times the burden.

Calculation of the burden and charge also must take account of the toe section since this is confined to the back and the base. In other words, there needs to be a higher concentration of charge at the base than in the column, whilst at the top of the hole no charging is required. Hence, the concentration of explosive in the bottom of the hole generally is approximately 2.5 times greater than in the column. The basal charge may be regarded as extending from the bottom of a hole to a point above the floor level equal to the burden. The column charge extends from there to a point below the crest equal to the burden.

Changes in blasthole diameter usually necessitate changes in burden distance, blasthole spacing and stemming length. The blasthole diameter is governed by the properties of the rock mass being blasted, the degree of fragmentation required and the bench height. If the blasthole diameter is too small, then the costs of drilling, priming and initiation are too high, and charging, stemming and connecting-up operations are too time consuming. On the other hand, if the blasthole diameter is too large, then excessively large blasthole patterns give inadequate fragmentation, especially in rock masses that contain widely spaced open discontinuities. An increase in blasthole diameter normally has to be accompanied by an increase in energy factor for the degree of fragmentation to remain unchanged, the increase being greatest for blocky and least for highly fissured rock masses.

Blasting of two or more dissimilar rock masses frequently occurs in open excavation. Where it is impractical to select a different blasthole diameter for each rock mass a compromise value of blasthole diameter has to be chosen. Inevitably, this will be suboptimum for at least one of the rock types. According to Hagan (1986), the diameter of the blasthole in blocky areas may be so unsuitable that it requires the use of pocket charges (Fig. 2.1).

With vertical front-row blastholes, the burden distance generally increases from the top to the bottom of the face, particularly when the face is high and/or when the dip is shallow. In such instances, front-row blastholes often are collared as close to the crest as safety permits, in order to provide toe burden resembling the design toe burden. However, the top of a charge of normal length is then unburdened, which frequently means that explosive gases burst from the upper face, with resultant high levels of airblast and/or flyrock. This, in turn, reduces blasthole pressure at bench/floor level and makes it more difficult to break and move the toe.

Vertical hole blasting generally is preferred when blasting to a free face, but in special circumstances horizontal holes may be used. For example, horizontal blastholes (or angled holes drilled from the bottom of the face) frequently are used when the ground surface makes

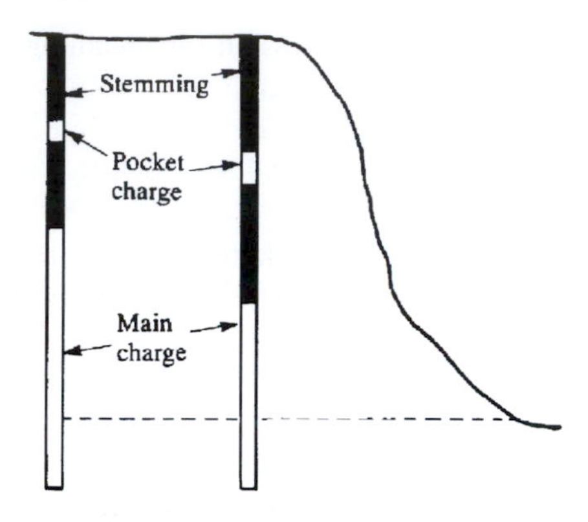

Figure 2.1 Use of pocket charges to assist in breaking collar rock.

access for drilling equipment difficult. In such cases horizontal holes can be used to prepare the surface so that access is provided for drilling vertical holes in the main excavation. Horizontal drilling sometimes is used when excavating steeply dipping rocks but generally it is undesirable since it can lead to dangerous overhangs and less stable faces.

Inclined blastholes are more difficult to drill, but are very effective in eliminating excessive front-row toe burdens. Indeed, the success of multirow blasting depends largely on the ability of front-row charges to heave their burden forwards. If front-row charges fail to displace their burden, progressive relief is not achieved and the blast never recovers, irrespective of the number of rows of blastholes. Inclined blastholes also yield good fragmentation, displacement and muckpile looseness. Fragmentation usually improves due to the reduction in pre-mature venting of explosive gases to atmosphere from alongside the tops of front-row charges, and to faster movement of the front-row toe. This gives rise to improved progressive relief of burden. Inclined blastholes also reduce disruption of, and drilling difficulties in, a bench beneath, and produce smoother and more stable faces. They reduce the tendency for stumps being left at the base of an excavation after blasting. Stumps generally occur in hard rock when the bottom charge generates insufficient energy at the toe of the hole. For example, if vertical holes are drilled to grade only, then a large proportion of the blast is dissipated as compression pulses that are absorbed by the bedrock. Hence, the remaining tensile energy may be insufficient to break out to a level base, so stumps are left at the toe. Accordingly, blastholes should be taken below grade level. If improved fragmentation is not required, then inclined blastholes permit slightly larger blasthole patterns and lower energy factors to be used. Moreover, inclined blastholes cause less surface overbreak (through cratering behind the collars of blastholes) and therefore allow the use of longer inter-row delays, which give greater progressive relief of burden, and hence better fragmentation and muckpile looseness.

When a blast causes considerable overbreak, then the mean inclination of the newly created face is often so small that the toe burden for front-row vertical blastholes in the next blast is excessive (Hagan, 1986). These front-row charges then fail to perform properly, and blasting results are suboptimal. In any blast, the front-row blastholes are the most important. This is because they are most prone to errors in burden distance (especially at bench floor level) and, to a lesser extent, blasthole spacing. Therefore, every possible step should be taken to ensure that overbreak does not cause collaring errors, blocked blastholes or the redrilling of caved blastholes.

In order to obtain good fragmentation and thereby ease loading operations, the explosive consumption in excavation is somewhat greater than in quarrying (Adhikari, 2000). When firing is confined to a single row of blastholes in soft laminated strata, the charging ratios may be as low as 0.15–0.25 kg m^{-3}. In harder sedimentary strata the charging ratios generally are around 0.45 kg m^{-3} while they may be about 0.6 kg m^{-3} in jointed igneous rocks. Even higher charging ratios have been necessary in order to obtain satisfactory results in some metamorphic rocks such as mica schists, which absorb much of the energy of blasts. Generally, 1 kg of high explosive will bring down about 8–12 tonnes of rock.

Staggered blasthole patterns are more effective than square or rectangular patterns. Hagan (1986) maintained that in hard rocks blastholes should be drilled on equilateral triangular grids, since these provide the optimum two-dimensional distribution of energy within a rock mass, whilst allowing a high degree of flexibility in initiation sequence, and consequently direction of firing. In other words, for a given cost of drilling and blasting, equilateral triangular patterns produce the best fragmentation. However, when blasting weak rocks, good fragmentation is readily achieved by using square or rectangular patterns.

Initiation should commence at that point in a blast that gives the best possible progressive relief for the maximum number of blastholes. If there is no free end, initiation should commence near, but not at, one end of the blast block (Fig. 2.2a). This produces less overbreak and lower ground vibrations. If there is a free end, initiation should commence at that end (Fig. 2.2b). If this end is choked by muck broken by a previous blast, initiation should commence near that end of the blast that is remote from the buffer (Fig. 2.2c).

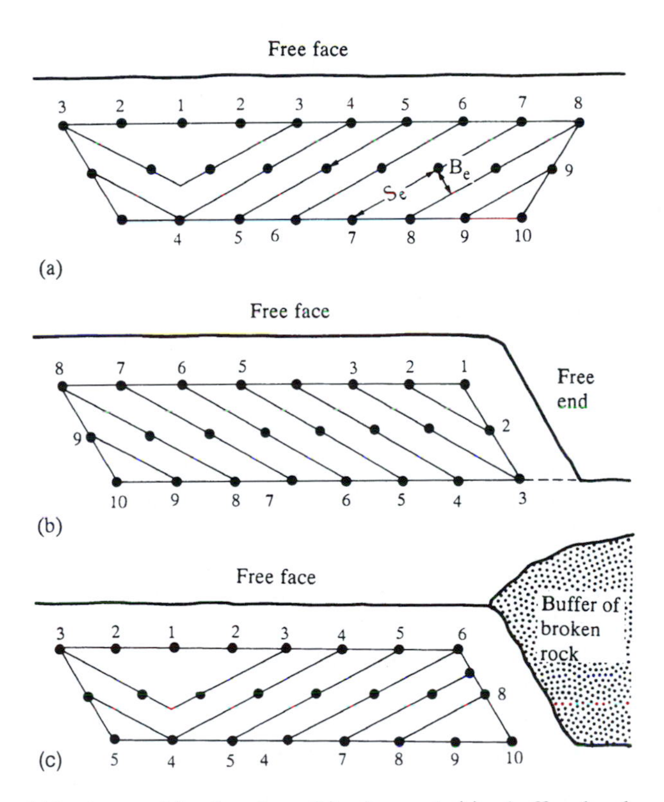

Figure 2.2 Staggered blasting to: (a) a free face; (b) a free end; (c) a buffered end.
Source: Hagan, 1986.

The interaction between adjoining blastholes when short-delay firing is used is an important factor in the design of the blast pattern. Hence, the delay interval between blastholes should allow sufficient time between shots for the fracture process to be completed. In other words, the time factor must provide for the travel of compressive and tensile waves through the burden, for the adjustment of the ambient stress field and for the movement of rock fragments. A cratering type of blast is produced if the delay time is too short. Roberts (1981) therefore suggested that the minimum delay should not be less than 10–20 ms or 1.5–30 ms m^{-1} of burden. The maximum delay time is that required for optimum movement of rock fragments and is around 30–50 ms or 4.5–7.5 ms m^{-1} of burden.

One of the fundamental requirements of blasting is that a free face should exist in front of the blastholes when they are fired. When an excavation has to be made into solid rock from the surface, then the rock is broken so as to provide free faces at which subsequent blasting can take place. Because there is no free face, the blast must occur upwards from the ground surface. For efficient operation, the width of the cut should be at least twice the depth, and the area broken by each charge should be in the form of a cone with its apex below the bottom of the cut. Hence, there are a large number of holes drilled to overlap below grade level. The spacing of blastholes is often about half the depth to grade, and nearly double the quantity of explosive is needed as is used in blasting a free face in the same rock.

The blasting in a sinking cut can be done by instantaneous shots or by short-delay firing. The former method can be used where excessive ground vibration, air-blast, fly-rock and overbreak are not likely to constitute problems. Either instantaneous electric detonators or Cordtex detonating fuse can be used for initiation. A comparatively large number of holes are fired simultaneously and the spacing of blastholes, subgrade drilling and width of burden should ensure that breakage occurs to the required depth. Moreover, the blastholes must be well stemmed. There may be inadequate fragmentation if hard bands of rock occur in the upper part of the excavation. Line drilling is required around the perimeter of the excavation in order to prevent excessive overbreak if such a blast is fired over the entire excavation area.

Figure 2.3 illustrates a typical pattern of blastholes for a short-delay sinking cut. With such an arrangement the zero-delay shots fracture the central area producing an upward movement of rock. This then provides sufficient relief of burden to permit succeeding shots to blast towards the central area.

In large excavations, additional blasting is necessary to expand the preliminary excavation area. These succeeding blasts can be regarded as firing to a free face and can be fired either

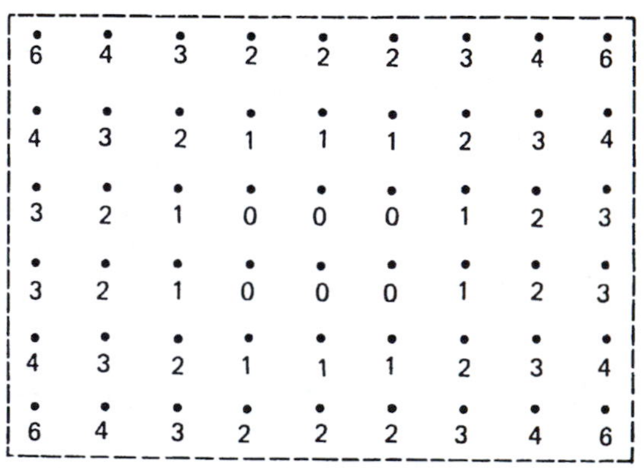

Figure 2.3 Typical pattern of blastholes for a short delay sinking cut.

before or after the rock from a sinking cut has been removed. In the former case, although broken rock remains in the area initially blasted, the burden has been sufficiently relieved to allow satisfactory performance of succeeding blasts. The sinking cut need not necessarily be located at the centre of the excavation and especially where buildings are in close proximity, the initial cut should be located as far away from them as possible.

Two of the principal methods of drilling and blasting used in large excavations are benching and well-hole blasting. As the name suggests, in benching the face is worked in a series of levels or benches, the maximum height of each bench generally not exceeding 10 m. Individual benches can be worked simultaneously, the rows of blastholes being drilled parallel to the free face. When fired, a row of holes forms a new free face for the succeeding row. Single or multiple rows of holes may be used.

Single face with well-hole blasting is commonly used for working large excavations, indeed it is the standard method for working quarries. Here a line of holes is drilled parallel to the face and they usually are inclined, often between 10 and 15°, for safety reasons. Sometimes multiple rows of holes may be used. The height of the face brought down is greater than that in benching and tends to vary between 15 and 35 m.

Under certain conditions heading blasting may be used. Small tunnels are driven into the base of the face and cross headings are driven from them. Chambers are excavated at intervals and filled with explosive. The headings then are backfilled. This is time consuming and there is little control over fragmentation.

The drilling pattern adopted for excavating trenches in rock depends upon their width. In wide trenches the ground is opened up initially with a pattern of angled holes drilled to successively greater depths, the object being to develop a free face. Conversely, in narrow trenches the rock may be blasted free with holes drilled along the edges of the excavation. The simultaneous detonation of charges produces a linear crater.

When the explosive in a blasthole is fired it is transformed into a gas, the pressure of which may sometimes exceed 100 000 atmospheres (1 atmosphere = 100 kPa). The tremendous energy liberated shatters a zone around the blasthole and exposes the rock beyond to enormous tensile stresses (Langefors and Kihlstrom, 1962). This takes place under the influence of shock waves that radiate from the explosion at between 3000 and 5000 m s^{-1}. A zone of intense deformation occurs about the blasthole, its thickness frequently approximating to the diameter of the hole, and radial cracking extends appreciably further (Persson *et al.*, 1970).

Under the influence of the pressure of gases from the explosive, the primary radial cracks expand, and the free face yields and is moved forwards. When this occurs the pressure is unloaded and tension increases in the primary cracks that incline obliquely forwards. If the burden is not too great several of these cracks expand to the exposed surface and the rock is loosened completely. In this way the burden is torn from the rock mass. The lateral pressure in the shock wave is initially positive but falls rapidly to negative values implying a change from compressive to tensile conditions. Hence, the area beyond the hole is exposed to vast tangential stresses. The shock wave itself is not responsible for breakage of rock, but it does provide the basic conditions for the process. Shock wave energy when using high explosive may only account for 5–15 per cent of the total energy liberated.

The discontinuities within a rock mass act as free surfaces that reflect shock waves generated by an explosion. They also provide paths of escape along which energy is dissipated. The geometry of the discontinuity pattern is very important since the greatest loss of energy occurs where most discontinuities intersect. At such locations the explosive energy opens existing breaks in the rock mass but generates few new ones. Secondary blasting therefore is required to break large masses of rock that have only been loosened.

2.1.4 Controlled blasting

In many excavations it is important to keep overbreak to a minimum. Apart from the cost of removing extra material that then has to be replaced, damage to rock forming the walls or floor may lower the bearing capacity and necessitate further excavation. Also, smooth faces allow excavation closer to the payline and are more stable.

The properties of the rock mass have a very great influence on the amount of overbreak resulting from a blast (Singh, 2001). The most important of these properties are the *in situ* dynamic tensile breaking strain and, more particularly, the nature, frequency, orientation and continuity of structural features, notably discontinuities.

The type of explosive and its density also affect the amount of overbreak. A reduction in the density of an explosive produces a significant reduction in the velocity of detonation. For instance, when the density of ANFO is reduced, typically by the addition of polystyrene beads, the peak blasthole pressure amounts to a small fraction of its initial value. Accordingly, when multirow blasting is undertaken and overbreak has to be minimized, such low-density mixtures can be used in back-row blastholes.

Overbreak also can be reduced by decoupling and/or decking back-row charges (Hagan, 1979). In both cases the charge weight in the back-row blastholes is lowered so as to reduce the strain wave energy, density and volume of explosion gases. However, less overbreak only can be guaranteed when the effective overburden on the back row decreases in proportion to the lower energy yield per blasthole. Charges are decoupled where the charge diameter is less than that of the blasthole. In particular, decoupling provides a means by which the amount of overbreak can be reduced in densely fissured rocks. Furthermore, the use of highly decoupled charges in the upper sections of back-row blastholes brings about a large reduction in the amount of crest damage.

Starfield (1966) indicated that the replacement of end-initiated by central-initiated charges produces an increase of around 37 per cent in peak strain in a rock mass. Consequently, minimization of overbreak, and of ground vibrations, is achieved by individual charges being initiated at one end (preferably the bottom) rather than at the centre. Both overbreak and ground vibrations are increased by multiple-primed charges.

Ouchterlony *et al.* (2000) indicated that the use of notched holes for perimeter blasting led to a decrease in blast damage, the faces of the an excavation so formed being relatively smooth. Most of the holes have radial bottom slots, which are cut with high-pressure abrasive water jets. These slots are about 75 mm deep and decoupled charges with a primer are used instead of a bottom charge. The technique accordingly involves no subdrilling, and the burden and spacing of holes is 1 and 0.8 m respectively. A continuous fracture forms along the toe line, which helps the development of an undamaged nose between the holes. Sanchidrian *et al.* (2000) also referred to the use of notched-hole blasting. They found that the technique either allowed a significant increase in drillhole spacing or smaller explosive loads in blastholes that were not spaced further apart, thereby decreasing damage to the rock mass. They noted that the notches must be drilled carefully, the notching being done by a specially designed machine with hammers.

Line drilling is most commonly used to improve the peripheral shaping of excavations. It consists of accurately drilling alternate small-diameter holes between the pattern blastholes forming the edge of the excavation. The quantity of explosives placed in each line hole is significantly smaller and, indeed, if these holes are closely spaced, from 150 to 250 mm, then explosive may be placed only in every second or third hole. The closeness of the holes depends upon the type of rock being excavated and on the payline. These holes are timed to fire ahead, with or after the nearest normally charged holes of the blasting pattern. The time of firing depends largely on the character of the rock involved. Line drilling is not always successful in preventing overbreak although it helps to reduce it. Generally, line holes in sedimentary and some metamorphic rocks are not as effective as in igneous rocks.

Fissile rocks such as slates, phyllites and schists tend to split along the planes of cleavage or schistosity if these run at a low angle to the required face. Paine *et al.* (1961) suggested that the raggedness of faces in sedimentary and metamorphic rocks may occur more readily when blastholes are fired from 1 to 2 m away from the line holes. Compressive strain pulses moving out from the blastholes are reflected from joint, cleavage, schistosity and bedding planes, and the reflected tensile strain pulses cause rupturing of the rock beyond the holes. On the other hand, homogeneous igneous rocks sometimes split quite readily from hole to hole when line drilled, because the major free surfaces are the holes and not a series of natural shear planes.

Cushion blasting is a type of line drilling where large holes, about 170 mm diameter, are located between the line holes to act as guides to the crack direction. Frequently one large hole to three small holes are used with either all three small holes or only the large central one being charged. The holes are loaded with light charges at intervals separated by stemming. The trimming holes are fired after the main blast. Cushion blasting often gives better results than line drilling.

Pre-splitting can be defined as the establishment of a free surface or a shear plane in rock by the controlled usage of explosives in approximately aligned and spaced drillholes. A line of trimming holes around the perimeter of the excavation is charged and fired to produce a shear plane. This acts as a limiting plane for the blast proper and is carried out prior to blasting the main round inside the proposed break lines. The spacing of the trimming holes is governed by the type of rock and the diameter of the holes. The diameter of the trimming holes is smaller than that of the main blastholes in order to avoid the extra radial cracking of the walls that would be produced by large charges. Continuously loaded, moderately coupled charges are recommended. Pre-splitting also requires accurate drilling. Air decking is sometimes used, that is, the charges are separated by spacers instead of stemming. The air space provides effective decoupling and damps the shock wave transmitted to the rock mass. Nevertheless, this is time-consuming and can prove troublesome. It is not recommended when ANFO mixtures are used. In weak rock, alternate advance pre-split holes may be left uncharged. The uncharged holes form relief holes that guide the shear fracture along the required plane of separation. Maximum effectiveness is achieved when the pre-split holes are fired simultaneously. A trial should be made to determine whether the site conditions require modifications in the hole spacing, or charging prior to extensive pre-splitting taking place.

In most rocks a shear plane can be induced to the bottom of the trimming holes, that is, to base level, but in massive rocks difficulty may be experienced in breaking out the main blast to base level. In such instances, the spacing between the outer blastholes and the shear plane may need to be reduced by 50–75 per cent. Once pre-split the rock can be blasted with a normal pattern of holes. If a rock mass is heavily fractured the trimming and primary blastholes can be fired together, delays causing the latter to fire immediately after the former.

2.1.5 Blasting and vibrations

When excavation in rock has to be carried out near to existing structures, especially in urban areas, then their presence may determine whether or not blasting is used, since noise, air concussion and ground movement may cause inconvenience or the latter may even damage structures. Experience has shown that the great majority of complaints and even law suits against blasting operations are due to irritation and that subjective response to vibrations normally leads a person to react strongly long before there is any likelihood of damage occurring to his property (Fig. 2.4). The duration of the operation and the frequency of occurrence of blasts are almost as important as the level of the physical effects.

The three most commonly derived quantities relating to vibration are amplitude, particle velocity and acceleration. Of the three, particle velocity appears to be the one most closely related to damage in the frequency range of typical blasting vibrations. Edwards and Northfield (1960) defined three categories of damage attributable to vibrations:

1 *Threshold damage*: widening of old cracks and the formation of new ones in plaster, dislodgement of loose objects.
2 *Minor damage*: damage does not affect the strength of the structure, it includes broken windows, loosened or fallen plaster, hairline cracks in masonry.
3 *Major damage*: damage seriously weakens the structure, it includes large cracks, shifting of foundations and bearing walls, distortion of the superstructure caused by settlement and walls out of plumb.

They proposed that a vibration level with a peak particle velocity of 50 mm s^{-1} could be regarded as safe as far as structural damage was concerned, 50–100 mm s^{-1} would require caution, and above 100 mm s^{-1} would present a high probability of damage occurring (Fig. 2.5). The United States Bureau of Mines subsequently lent support to the idea that 50 mm s^{-1} provides a reasonable safety from the possibility of damage (Nicholls *et al.*, 1971). However, Oriard (1972) maintained that no single value of velocity, amplitude or acceleration could be used

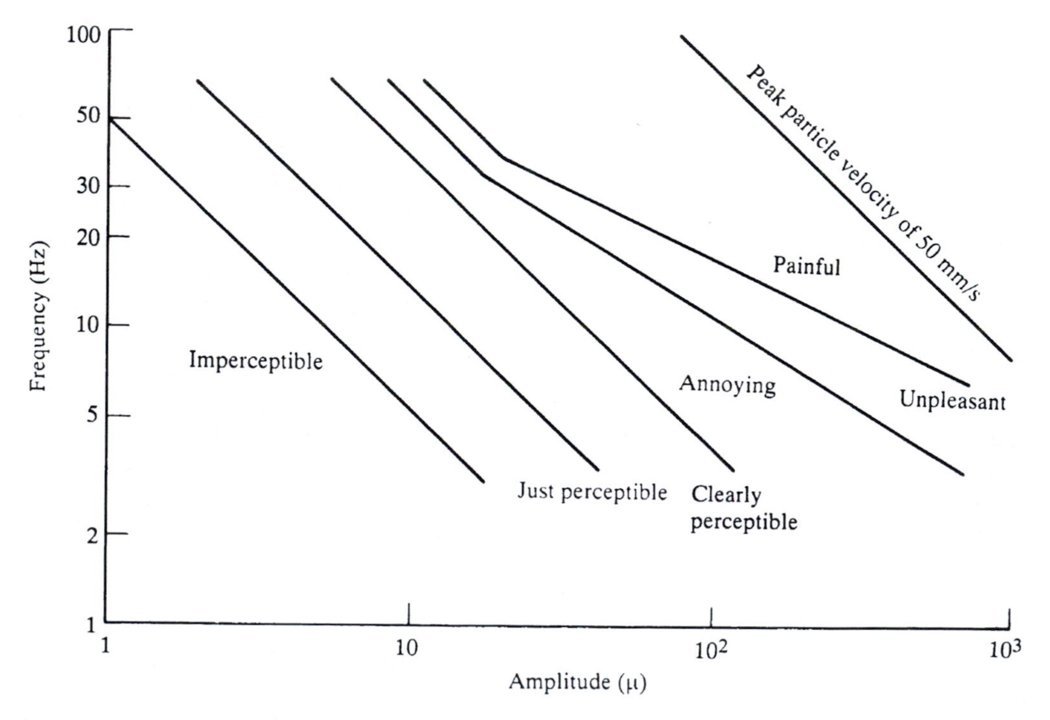

Figure 2.4 Human sensitivity to vibration. If peak particle velocity of 50 mm s^{-1} is taken as the threshold of damage, then the Figure indicates that humans react strongly long before there is any reason to apprehend damage to their property.

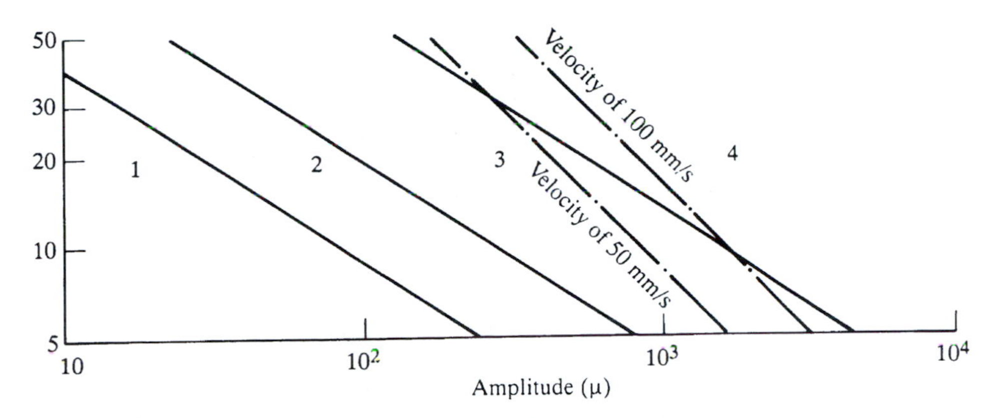

Figure 2.5 Possible damage to buildings for frequencies between 5 and 50 Hz, the range most frequently encountered in buildings: 1 – no damage; 2 – possibility of damage; 3 – possible damage to load bearing structural units; 4 – damage to load bearing structural units.

indiscriminately as a criterion for limiting blasting vibrations. In particular, low vibration levels may disturb sensitive machinery and in this case it is impossible to specify a limit of ground velocity; hence each instance should be separately assessed. Moreover, blasting vibrations at 50 mm s^{-1} particle velocity, in terms of human response, are regarded as highly unpleasant or intolerable. Indeed, Roberts (1971) indicated that the threshold of subjective perception had been variously placed from as low as 0.5 to 10 mm s^{-1}. People react more unfavourably to large-amplitude vibrations of long duration than to low-amplitude short-duration vibrations of the same intensity. It is likely that this sensitivity is increased in the low frequency range 3–10 Hz.

When dealing with high levels of shock and vibration, the time history of the motion and the characteristic response of the structure concerned to the type of motion imposed become increasingly important. For instance, a structure with a slow response, such as a tall chimney, when subjected to vibrations of large amplitude, low frequency and long duration, would come closer to the resonant response of the structure and therefore this would be more dangerous than vibrations with small amplitude, high frequency and short duration, even though both may have the same acceleration or velocity. Because of the dependence of response on frequency, conservative limits should be accepted when applying single values of velocity or acceleration as criteria for different types of structures subjected to different kinds of motion.

Vibrations associated with blasting generally fall within the frequency range 5–60 Hz. The types of vibration depend on the size of the explosive charge, the delay time sequence, the spatial pattern of the blastholes, the volume of the ground set into vibration, the attenuation characteristics of the ground and the distance from the blast (Blair, 1999). A small explosive charge generates a low vibration with relatively high frequency and relatively low amplitude. By contrast, a large explosive charge produces a vibration with relatively low frequency and relatively high amplitude. The shock waves are attenuated with distance from the blast, the higher frequencies being maintained more readily in dense rock masses. In other words, these pulses are rapidly attenuated in unconsolidated deposits that are characterized by lower frequencies.

Vibrographs can be placed in locations considered susceptible to blast damage in order to monitor ground velocity. A record of the blasting effects compared with the size of the charge and the distance from the point of detonation normally is sufficient to reduce the possibility of

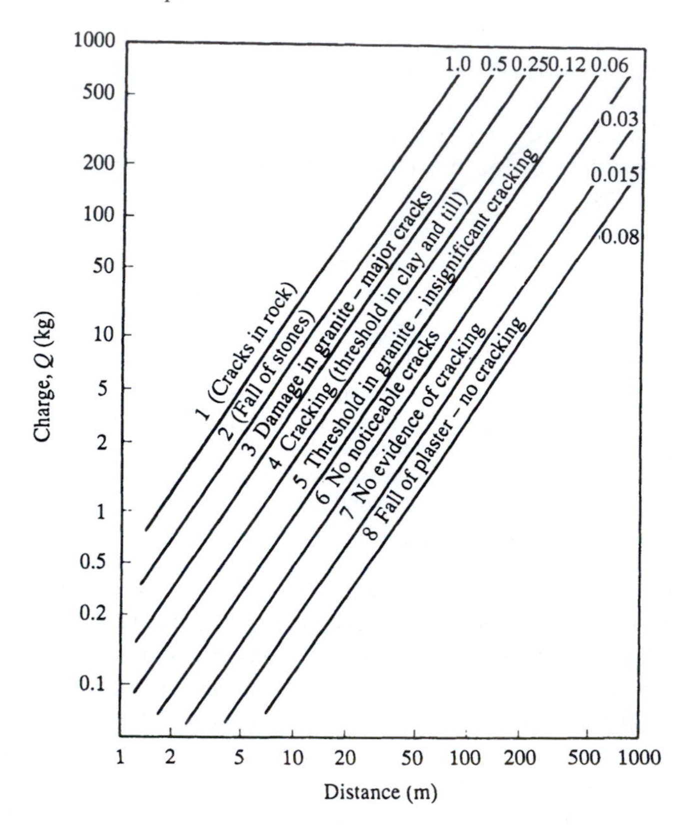

Figure 2.6 Charge as a function of distance for various charge levels. Numbers 3 to 8 inclusive describe damage expected in normal houses.
Source: Langefors and Kihlstrom, 1962.

damage to a minimum (Fig. 2.6). A pre-blast survey informs the owners as to the condition of their buildings before the commencement of blasting and offers some protection to the contractor against unwarranted claims for damage.

The effects of vibration due to blasting operations can be reduced by:

1 Time dispersion, which includes the use of delay intervals in ignition. Delay intervals mean that shock waves generated by individual blasts are mutually interfering, thereby reducing vibration. For instance, Duvall (1964) noted that a delay interval of around 9 ms reduced ground vibrations and Oriard (1972) suggested that even shorter delay intervals were important as far as the control of vibration and potential damage was concerned. Blair and Armstrong (1999) maintained that the use of electronic delay detonators can control the blasting sequence and so the vibrations in resonant structures such as houses. However, where the geological conditions are variable or complex, it may not be possible to control the vibration frequency adequately.
2 Spatial dispersion, which involves the pattern and orientation of the blastholes.
3 The way in which the charge is distributed in the blasthole, the diameter of the hole and the depth of lift.
4 The confinement of the charge, which involves the type of decking and the powder factor as well as the width of burden and the spacing of the blastholes.

If the use of conventional explosives in urban areas is prohibited, then alternative methods of rock breakage must be used. These include hand-held pneumatic breakers or lorry-mounted diesel-powered breakers. Unfortunately, compressed-air or diesel breakers are very noisy and cause vibrations. Consequently, in some situations the use of these tools also is excluded. In such instances, the employment of hydraulic bursters has been found to be the most suitable method for producing large excavations. A burster is essentially a multiple hydraulic jack with several circular rams operating from one side. A series of holes are drilled parallel to the rock face and one or more bursters inserted. Pressure then is exerted by the rams causing the rock to split along the line of holes. With holes about 150 mm diameter, going to a depth of 2–3 m, the burden and spacing of holes should be about 1 m. Obviously, use is made of discontinuities, especially bedding planes, when this method is employed. The holes for the hydraulic burster can be drilled without much noise or vibration by rotary diamond core drills or by oxygen lance if no noise is permitted. Splitting also can be accomplished by freezing water in drillholes with the aid of liquid nitrogen.

2.2 Methods of excavation: ripping

Ripping is an inexpensive method of breaking discontinuous ground or soft rock masses so that the fragmented material can be removed economically. Rock breakage is accomplished by a steel tyne, attached to a bulldozer, being held down into the rock mass and drawn through it (Fig. 2.7). The width of the cut in which fragmentation occurs decreases along the depth of the tyne, the maximum breakage occurring at the surface. Rippability depends on the type and size of excavating machine used, and the method of working. More specifically, ripping is influenced by the weight of the bulldozer, which determines whether it can develop enough tractive force to advance the tyne, the engine power, which also determines the ability to advance the tyne,

Figure 2.7 A ripper at work at the surge pond, Dinorwic Pumped Storage Scheme, Llanberis, North Wales.

the down-acting pressure on the tyne, which helps determine the degree of penetration of the tyne, and the penetration angle of the tip of the tyne. The tyne may be protected from wear by a detachable steel tip, the rate of wear due to abrasive rock being a factor that has to be considered in the prediction of whether it is economic to rip certain rock masses. If a rock mass cannot be broken satisfactorily by ripping or if ripper production has to be increased, then the rock mass can be loosened by blasting. Such a pre-blasting process uses light charges to open the discontinuities. Hydraulic fracturing also has been used to break surface rock prior to ripping. In fact, Pettifer and Fookes (1994) noted that cost considerations have led to the increased use of impact rippers and hydraulic rock breakers as alternatives to light blasting. Impact rippers subject a rock mass to high-frequency impacts via the tyne. Usually, different machines are used for breaking and ripping.

The geological factors that influence rippability in rock masses include rock type and fabric (e.g. coarse-grained rocks are easier to rip than fine-grained rocks), intact strength and degree of weathering, rock hardness and abrasiveness, and the nature, incidence and geometry of discontinuities (Table 2.2). Pettifer and Fookes (1994) also considered block shape and orientation to be significant factors in determination of rippability (Table 2.3). Well-interlocked or irregular blocks prove difficult to rip but ripping such material is easier when benches are used in excavation, as blocks then have greater freedom of movement. Discontinuities reduce the overall strength of a rock mass, their spacing governing the amount of such reduction. Obviously, the continuity of discontinuities within a rock mass has a significant influence on its engineering behaviour. The greater the amount of gouge or soft fill along discontinuities, the easier it is to penetrate the rock mass and so to rip it. In other words, strong massive rocks and hard abrasive rocks do not lend themselves to ripping. On the other hand, if sedimentary rocks such as sandstone and limestone are well-bedded and jointed or if strong and weak rocks are thinly interbedded, then they can be excavated by ripping rather than by blasting. Indeed, some of the weaker sedimentary rocks (less than 15 MPa compressive strength or 1 MPa tensile strength) such as mudstones are not as easily removed by blasting as their low strength would suggest. The reason for this is that they are pulverized in the immediate vicinity of the blasthole. What is more, blasted mudstones may lift along bedding planes to fall back when the gas pressure has been dissipated. Such rocks, especially if well jointed, are more suited to ripping.

According to Atkinson (1970) the most common method used to determine whether a rock mass is capable of being ripped is seismic refraction. The seismic velocity of the rock mass concerned is compared with a chart of ripper performance, supplied by the manufacturer of the machine, and based on ripping operations in different types of rocks (Fig. 2.8). In fact, the limit of ripper operations frequently is regarded as a seismic velocity (V_p) of $2\,\mathrm{km\,s^{-1}}$. Although such charts, according to McCann and Fenning (1995), can be used as a basic guide to the assessment of rippability, they do not necessarily take into account the influence of weathering. Indeed, Minty and Kearns (1983) maintained that the use of seismic velocity for the assessment of rippability was an oversimplified approach since it does not take account of certain geological conditions that enable the assessment to be made. For example, large unweathered corestones can occur within a weathered granite mass that could cause significant problems during ripping, yet standard seismic refraction surveys are not likely to detect their presence. Kirsten (1982) also argued that seismic velocity could only provide a provisional indication of the way in which a rock mass could be excavated, pointing out that in terms of overall assessment seismic velocity cannot be determined to an accuracy better than about 20 per cent. Furthermore, seismic velocity may vary by as much as $1\,\mathrm{km\,s^{-1}}$ in apparently identical materials. After an in-depth study of rock rippability, MacGregor *et al.* (1994) concluded that seismic charts tend to be over-optimistic in that they indicate that rock masses, which when

Table 2.2 Excavation characteristics

(a) In relation to rock hardness and strength

Rock hardness description	*Identification criteria*	*Unconfined compressive strength* (MPa)	*Seismic wave velocity* (m/s)	*Excavation characteristics*
Very soft rock	Material crumbles under firm blows with sharp end of geological pick; can be peeled with a knife; too hard to cut a triaxial sample by hand. SPT will refuse. Pieces up to 3 cm thick can be broken by finger pressure	1.7–3.0	450–1200	Easy ripping
Soft rock	Can just be scraped with a knife; indentations 1–3 mm show in the specimen with firm blows of the pick point; has dull sound under hammer	3.0–10.0	1200–1500	Hard ripping
Hard rock	Cannot be scraped with a knife; hand specimen can be broken with pick with a single firm blow; rock rings under hammer	10.0–20.0	1500–1850	Very hard ripping
Very hard rock	Hand specimen breaks with pick after more than one blow; rock rings under hammer	20.0–70.0	1850–2150	Extremely hard ripping or blasting
Extremely hard rock	Specimen requires many blows with geological pick to break through intact material; rock rings under hammer	>70.0	>2150	Blasting

(b) In relation to joint spacing

Joint spacing description	*Spacing of joints* (mm)	*Rock mass grading*	*Excavation characteristics*
Very close	<50	Crushed/shattered	Easy ripping
Close	50–300	Fractured	Hard ripping
Moderately close	300–1000	Block/seamy	Very hard ripping
Wide	1000–3000	Massive	Extremely hard ripping and blasting
Very wide	>3000	Solid/sound	Blasting

Table 2.3 Effects of block shape and orientation on rippability

Shape class (ISRM 1981)	*Characteristics*	*Significance for rippability*	*Suggested adjustment of data for revised excavatability graph*
Massive	Few or very widely spaced discontinuities	Very unfavourable	None: accounted for by very high I_f value
Blocky	Approximately equidimensional. Typically occurs in sedimentary rocks with subhorizontal bedding and orthogonal joints; also some granites. Where two of the discontinuity sets are not orthogonal, blocks are rhombic	Ripping becomes progressively more difficult as discontinuity spacing and the strength of the intact rock increase. Ripping may be easier where the run direction is normal to the strike of any vertical discontinuities	None: accounted for by I_f and I_s measurements
Tabular	One dimension is considerably smaller than the other two.* Typically occurs in thinly bedded sedimentary rocks with widely spaced joints; also foliated metamorphic rocks. Flaggy and slaty rock masses are extreme examples of this class	Favourable where the run direction is down-dip in inclined strata, particularly in very thinly bedded or cleaved rocks. In less thinly bedded strata, large slabs may jam under the tractor. It may be possible to overcome this by ripping in a direction at 45° to the dip direction, so as to tip the slabs, which may then be broken by driving over them. Optimum conditions for ripping where the dominant discontinuities dip at 45°	Accounted for by I_f value where there is no restriction on the run direction. Where the run direction is parallel to the strike of the bedding or foliation, increase I_f value by 20% (dip 10–50°) or 40% (dip 50–90°)
		Unfavourable where the closest spaced discontinuity set is subhorizontal (dip 0–10°)	Increase I_f value by 40%
Columnar	One dimension is considerably larger than the other two.† Occurs predominantly in fine-grained igneous rocks	Very unfavourable where dominant discontinuities are subvertical Blocks are generally well interlocked and difficult to dislodge	Increase I_f value by 50% where columns are subvertical Difficult conditions emphasized by high I_s values

Table 2.3 (Continued)

Shape class (ISRM 1981)	*Characteristics*	*Significance for rippability*	*Suggested adjustment of data for revised excavatability graph*
Irregular	Wide variations of block size and shape due to random joint pattern. Occurs in some recrystallized limestones and quartzites, and in some medium- to coarse-grained igneous rocks	Fair to unfavourable due to very good block interlock and uncertain behaviour in relation to run direction	Increase I_f value by 30%
Crushed	Heavily jointed to 'sugar cube' structure	Very favourable	None: accounted for by very low I_f value

Source: Pettifer and Fookes, 1994. Reproduced by kind permission of The Geological Society.

Notes
* Thickness much less than length or width.
† Height much greater than cross section.

ripped are marginal, frequently are predicted as economically rippable. In fact, they related rippability to productivity. For example, they suggested that if the estimated productivity is less than $250\,m^3\,h^{-1}$, then very difficult ripping conditions are likely to be experienced, and if it is less than $750\,m^3\,h^{-1}$, then difficult ripping can be expected.

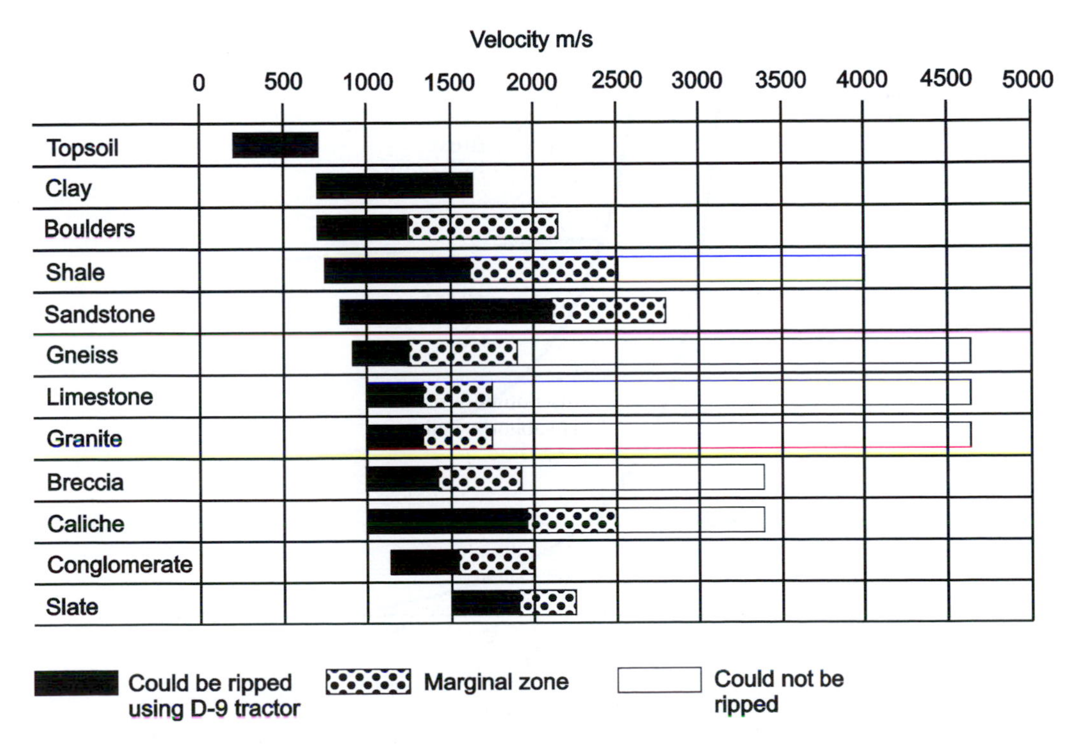

Figure 2.8 Rippability chart.

A number of classification systems related to excavation have been devised and are based on the properties of rock mass(es). A basic requirement of any classification system is that it should be applicable both to drillhole core and field investigation alike. For example, Franklin *et al.* (1971) proposed a simple rock quality classification based on fracture spacing and point load strength (Fig. 2.9). However, Pettifer and Fookes pointed out that, although this system may be useful in terms of initial assessment made during a walkover survey, it now is out of date due to advances in technology. Hence, they revised the system (Fig. 2.10).

Weaver (1975) proposed the use of a modified form of the geomechanics classification of rock masses, proposed by Bieniawski (1973), as a rating system for the assessment of rippability. In addition to the geological factors used by Bieniawski, Weaver included seismic velocity, rock hardness, and strike and dip orientation. Weathering was included in Weaver's system, although it initially was included in the Geomechanics classification and subsequently omitted, but groundwater conditions were not. His rippability rating chart is given in Table 2.4. Minty and Kearns (1983) adapted Weaver's system to include groundwater conditions and surface roughness of discontinuities in order to obtain a geological factors rating (GFR). The latter is multiplied by seismic velocity and compared with tractor size to determine rippability. However,

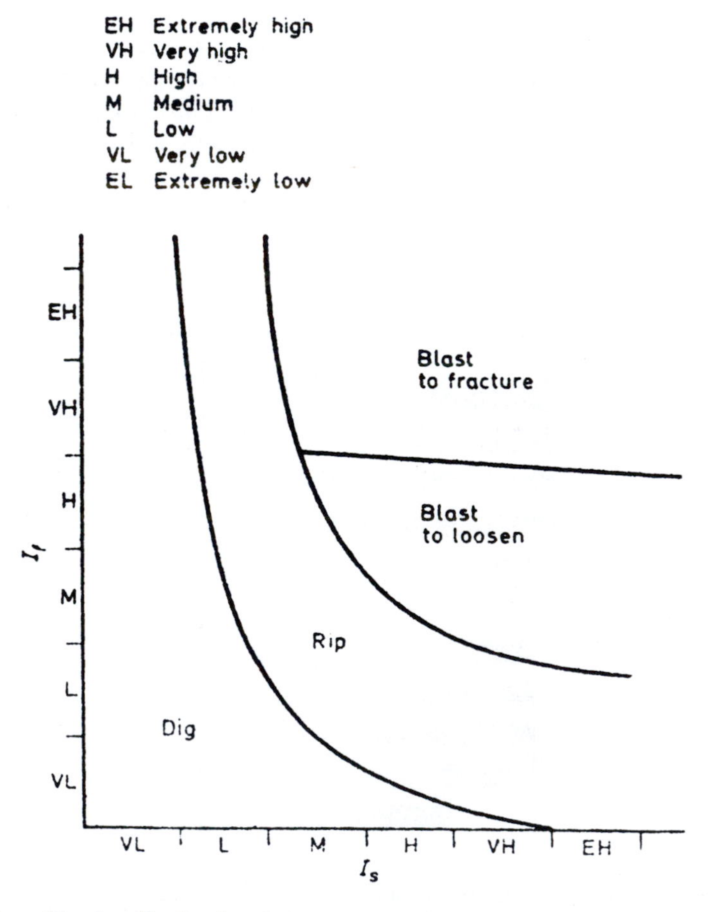

Figure 2.9 Rock quality classification in relation to excavation.
Source: Franklin *et al.*, 1971. Reproduced by kind permission of Institution of Materials, Materials and Mining.

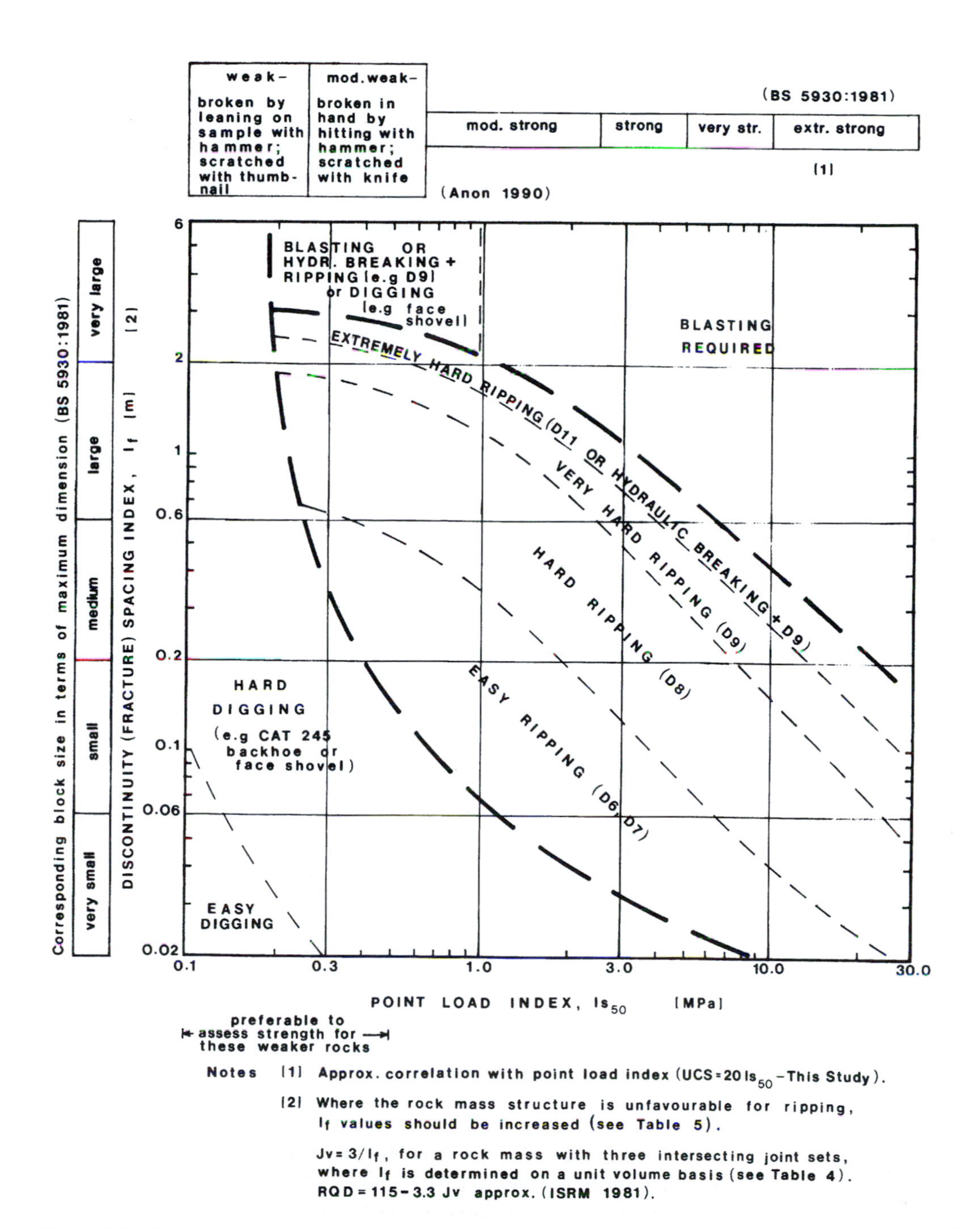

Figure 2.10 Exacavatability chart.
Source: Pettifer and Fookes, 1994. Reproduced by kind permission of The Geological Society.

Table 2.4 Rippability rating chart

Rock class	*I*	*II*	*III*	*IV*	*V*
Description	Very good rock	Good rock	Fair rock	Poor rock	Very poor rock
Seismic velocity (m/s)	>2150	2150–1850	1850–1500	1500–1200	1200–450
Rating	26	24	20	12	5
Rock hardness	Extremely hard rock	Very hard rock	Hard rock	Soft rock	Very soft rock
Rating	10	5	2	1	0
Rock weathering	Unweathered	Slightly weathered	Weathered	Highly weathered	Completely weathered
Rating	9	7	5	3	1
Joint spacing (mm)	>3000	3000–1000	1000–300	300–50	<50
Rating	30	25	20	10	5
Joint continuity	Non-continuous	Slightly continuous	Continuous – no gouge	Continuous – some gouge	Continuous – with gouge
Rating	5	5	3	0	0
Joint gouge	No separation	Slight separation	Separation <1 mm	Gouge < 5 mm	Gouge < 5 mm
Rating	5	5	4	3	1
* Strike and dip orientation	Very unfavourable	Unfavourable	Slightly unfavourable	Favourable	Very favourable
Rating	15	13	10	5	3
Total rating	100–90	90–70 †	70–50	50–25	<25
Rippability assessment	Blasting	Extremely hard ripping and blasting	Very hard ripping	Hard ripping	Easy ripping
Tractor selection	–	DD9G/D9G	D9/D8	D8/D7	D7
Horsepower	–	770/385	385/270	270/180	180
Kilowatts	–	575/290	290/200	200/135	135

Source: Weaver, 1975. Reproduced by kind permission of the South African Institution of Civil Engineers.

Notes

* Original strike and dip orientation now revised for rippability assessment.

† Ratings in excess of 75 should be regarded as unrippable without pre-blasting.

according to the experience of MacGregor *et al.* (1994) both these methods tend to overestimate the ease of ripping, frequently indicating relatively easy ripping when it is very difficult. They also pointed out that as the Weaver system is based upon the largest size of bulldozer being the Caterpillar D9G (the operating weight and engine power have more or less doubled since 1974), the system boundaries are likely to be conservative for larger bulldozers.

Kirsten (1982) adapted the Q system of classification devised by Barton *et al.* (1974) to develop an excavatability index (N) as follows:

$$N = M_s \left(\frac{RQD}{J_n} \right) \times J_s \times \left(\frac{J_r}{J_a} \right) \tag{2.3}$$

The mass strength number, M_s, indicates the amount of effort needed to excavate dry, homogeneous material that contains no discontinuities. The term (RQD/J_n) brings together the rock quality designation *(RQD)*, which affords an equivalent estimate of joint density and the number of joint sets (J_n). This provides an indication of the degree of freedom of rock mass. Hence, together they offer a crude measure of block size, and the smaller the size of the block the less is the effort required to excavate dry, homogeneous, continuous rock. The term J_s represents the reducing effect that the block shape and orientation relative to the excavating force has on the effort to excavate the perfect material. The roughness of the most unfavourable joint set (J_r) and the degree of alteration or filling of the most unfavourable joint set are combined in the last term and as such provide an approximation of the shear strength of the rock mass. In this way it represents the reducing effect that the deformability and weakness of the joints have on the effort to excavate the perfect rock. Once the type of rock mass has been identified, the ratings for the individual terms constituting the excavatability index are determined. Each parameter is divided into a number of categories and given a rating. It is admissible to interpolate intermediate values.

Five categories of mass strength number (M_s) are recognized and their ratings are given in Table 2.5a. As far as the block size number is concerned, only joints that affect the excavation process should be taken into account. The *RQD* can be determined in the conventional way from core stick or from the joint count (J_c), that is, the number of joints per cubic metre, the two being related by the following expression:

$$RQD = 115 - 3.3J_c \tag{2.4}$$

Table 2.5b gives values of J_c in relation to *RQD*. The joint set number (J_n) is given in Table 2.5c in terms of the same categories identified by Barton *et al.* (1974).

The spacing and orientation of discontinuities affect the effort required to penetrate the ground, as well as that needed to dislodge individual blocks. It is easier to rip the ground in the direction in which rocks dip than in the opposite direction. The possibility of ripper penetration along a discontinuity is directly related to its inclination. As more than one set of discontinuities usually are present in a rock mass, the overall kinematic possibility of penetration may be regarded as the average discontinuity inclination weighted by the number of discontinuities per unit length. The values of J_s for various ranges of angle of dip and dip direction of discontinuities are given in Table 2.5d. The angle of dip and direction in Table 2.5d refer to the closer-spaced joint set.

The joint roughness number for various joint conditions is given in Table 2.5e, and the joint alteration number (J_a) is provided in Table 2.5f. No rigorous test can be proposed for the determination of these parameters on their own. As a last resort, Kirsten (1982) suggested that the joint strength number could be determined from:

$$J_r/J_a = \arctan(\tau_p/\sigma_n) \tag{2.5}$$

that represents the total friction angle as defined by Bandis *et al.* (1981). If several joint sets of different strengths are present in the rock mass, the values of J_r/J_a should be determined for each set. The equivalent joint strength for the mass may be determined as the average of the different values of J_r/J_a weighted according to the number of joints in each set per cubic metre.

Table 2.5 Parameters for determination of excavatability index

(a) Mass strength number for rocks (M_s)

Hardness	*Identification in profile*	*Unconfined compressive strength* (MPa)	*Mass strength number* (M_s)
Very soft rock	Material crumbles under firm (moderate) blows with sharp end of geological pick and can be peeled off with a knife. It is too hard to cut a triaxial sample by hand.	1.7	0.87
		1.7–3.3	1.86
Soft rock	Can just be scraped and peeled with a knife; indentations 1–3 mm show in the specimen with firm (moderate) blows of the pick point.	3.3–6.6	3.95
		6.6–13.2	8.39
Hard rock	Cannot be scraped or peeled with a knife; hand-held specimen can be broken with hammer end of a geological pick with a single firm (moderate) blow.	13.2–26.4	17.70
Very hard rock	Hand-held specimen breaks with hammer end of pick under more than one blow.	26.4–53.0	35.0
		53.0–106.0	70.0
Extremely hard rock (very, very hard rock)	Specimen requires many blows with geological pick to break through intact material.	106.0–212.0	140.0
		212.0	280.0

Source: Kirsten, 1982. Reproduced by kind permission of the South African Institution of Civil Engineers.

Table 2.5 (b) Joint count number (J_c)

Number of joints per cubic metre (J_c)	*Rock quality designation (RQD)*	*Number of joints per cubic metre* (J_c)	*Rock quality designation (RQD)*
33	5	18	55
32	10	17	60
30	15	15	65
29	20	14	70
27	25	12	75
26	30	11	80
24	35	9	85
23	40	8	90
21	45	6	95
20	50	5	100

Source: Kirsten, 1982. Reproduced by kind permission of the South African Institution of Civil Engineers.

Table 2.5 (c) Joint set number (J_n)

Number of joints sets	*Joint set number* (J_n)
Intact, no or few joint/fissures	1.00
One joint/fissure set	1.22
One joint/fissure set plus random	1.50
Two joint/fissure sets	1.83
Two joint/fissure sets plus random	2.24
Three joint/fissure sets	2.73
Three joint/fissure sets plus random	3.34
Four joint/fissure sets	4.09
Multiple joint/fissure sets	5.00

Source: Kirsten, 1982. Reproduced by kind permission of the South African Institution of Civil Engineers.

Note
For intact granular materials take $J_n = 5.00$.

Table 2.5 (d) Relative ground structure number (J_s)

Dip direction of closer-spaced joint set* (degrees)	*Dip angle† of closer-spaced joint set* (degrees)	*Ratio of joint spacing, r*			
		1:1	1:2	1:4	1:8
180/0	90	1.00	1.00	1.00	1.00
0	85	0.72	0.67	0.62	0.56
0	80	0.63	0.57	0.50	0.45
0	70	0.52	0.45	0.41	0.38
0	60	0.49	0.44	0.41	0.37
0	50	0.49	0.46	0.43	0.40
0	40	0.53	0.49	0.46	0.44
0	30	0.63	0.59	0.55	0.53
0	20	0.84	0.77	0.71	0.68
0	10	1.22	1.10	0.99	0.93
0	5	1.33	1.20	1.09	1.03
0/180	0	1.00	1.00	1.00	1.00
180	5	0.72	0.81	0.86	0.90
180	10	0.63	0.70	0.76	0.81
180	20	0.52	0.57	0.63	0.67
180	30	0.49	0.53	0.57	0.59
180	40	0.49	0.52	0.54	0.56
180	50	0.53	0.56	0.58	0.60
180	60	0.63	0.67	0.71	0.73
180	70	0.84	0.91	0.97	1.01
180	80	1.22	1.32	1.40	1.46
180	85	1.33	1.39	1.45	1.50
180/0	90	1.00	1.00	1.00	1.00

Source: Kirsten, 1982. Reproduced by kind permission of the South African Institution of Civil Engineers.

Notes
* Dip-direction of closer-spaced joint set relative to direction of rip.
† Apparent dip angle of closer-spaced joint set in vertical plane containing direction of ripping.
For intact material take $J_s = 1.0$.
For values of r less than 0.125 take J_s as for $r = 0.125$.

Table 2.5 (e) Joint roughness number (J_r)

Joint separation	*Condition of joint*	*Joint roughness number* (J_r)
Joints tight or closing during excavation	Discontinuous joint	4.0
	Rough or irregular, undulating	3.0
	Smooth undulating	2.0
	Slicken-sided undulating	1.5
	Rough or irregular, planar	1.5
	Smooth planar	1.0
	Slicken-sided planar	0.5
Joints open and remain open during excavation	Joints either open or containing relatively soft gouge of sufficient thickness to prevent joint wall contact upon excavation	1.0

Source: Kirsten, 1982. Reproduced by kind permission of the South African Institution of Civil Engineers.

Table 2.5 (f) Joint alteration number (J_a)

Description of gouge	*Joint alteration number* (J_a) *for joint separation* (mm)		
	<1.0†	1.0–5.0†	>5.0‡
Tightly healed, hard, non-softening impermeable filling	0.75	–	–
Unaltered joint walls, surface staining only	1.0	–	–
Slightly altered, non-softening, non-cohesive rock mineral or crushed rock filling	2.0	4.0	6.0
Non-softening, slightly clayey non-cohesive filling	3.0	6.0	10.0
Non-softening strongly over-consolidated clay mineral filling, with or without crushed rock	3.0§	6.0§	10.0§
Softening or low-friction clay mineral coatings and small quantities of swelling clays	4.0	8.0	13.0
Softening moderately over-consolidated clay mineral filling, with or without crushed rock	4.0§	8.0§	13.0§
Shattered or micro-shattered (swelling) clay gouge, with or without crushed rock	5.0	10.0	18.0

Source: Kirsten, 1982. Reproduced by kind permission of the South African Institution of Civil Engineers.

Notes

* Joint walls effectively in contact.

† Joint walls come into contact after approximately 100 mm shear.

‡ Joint walls do not come into contact at all upon shear.

§ Values added to Barton's data.

The class limits for the various categories of excavatability are given in Table 2.6. As the limits differ by an order of magnitude any minor inaccuracies in the determination of magnitudes of any of the basic parameters generally should not result in a change of the class of excavation, unless the index is near a class boundary. Kirsten (1982) did not propose that the machines quoted in Table 2.6 should be specified with respect to the different classes of excavation in any conditions of contract but rather that the excavation classes be specified in terms of class intervals for the excavatability index. Although this is one of the more complicated systems of predicting excavatability, MacGregor *et al.* (1994) maintained that it was very conservative, predicting very hard and extremely hard conditions when ripping actually was medium to hard.

Pettifer and Fookes (1994) reported that conditions usually are unfavourable for ripping when the dominant discontinuities are either subhorizontal or subvertical. Therefore, as far as ripping is concerned the run direction should be normal to any vertical joint planes, down-dip to any inclined strata, that is, normal to the strike, and on sloping ground it should be downhill. Atkinson (1970) suggested that long ripping runs of 70–90 m usually gave the best results. He further suggested that, where possible, the ripping depth should be adjusted so that a forward speed of 3 km h^{-1} could be maintained, since this is generally found to be the most productive, reduces track wear significantly and avoids impact shocks. Adequate breakage in rock depends on the spacing between ripper runs that, in turn, is governed by the fracture pattern in the rock mass. The output of a ripper also depends upon the capacity of the bulldozer. Output generally falls within the range of 40–230 m^3 h^{-1}.

2.3 Diggability

Some of the softer rocks such as shales, as well as soils, can be excavated by digging machines. The diggability of ground is of major importance in the selection of excavating equipment and depends primarily upon its intact strength, bulk density, bulking factor and natural moisture content. At present there is no generally accepted quantitative measure of diggability, assessment usually being made according to the experience of the operators. However, a fairly reliable indication can be obtained from similar excavations in the area concerned and the behaviour of the ground excavated in trial pits.

Attempts have been made to evaluate the performance of excavating equipment in terms of seismic velocity (Fig. 2.11). It would appear that most earthmoving equipment operates most effectively when the seismic velocity of the ground is less than 1 km s^{-1} and will not function above 1.8 km s^{-1}, but in areas of complex geology seismic evaluation may prove difficult if not impossible. Scoble and Muftuoglu (1984) proposed a rating system to determine a diggability index. The parameters that were used in this rating system included the degree of weathering, the unconfined compressive strength, the joint spacing and the bed separation. Use of the joint spacing and bedding separation provides a measure of block size. The index was designed to provide a guide for the selection of equipment for use in surface mines.

2.3.1 The bulking factor

When material is excavated it increases in bulk, this being brought about by the decrease that occurs in density per unit volume. The amount of bulking that takes place when a given rock or soil is worked can be ascertained by filling large boxes of known volume with the excavated

Table 2.6 Excavation classification system for rock masses

Material type	*Class*	*Excavation class boundaries*	*Description of excavatability*	*Bulldozer characteristics*					*Backhoe characteristics*			
				Type	*Operating mass* (kg)	*Flywheel power* (kW)	*Drawbar pull*† (kN)		*Type*	*Operating mass* (kg)	*Flywheel power*	*Man-draw bar pull* (kN)
							Stalling speed	1.6 km/h				
Rock	1	1.0–9.99	Easy ripping	D7G	20 230	149	376	220	Cat 235	38 297	145	263
	2	10.0–99.9	Hard ripping	D8K	31 980	224	500	323	Cat 245	59 330	242	472
	3	100.0–999	Very hard ripping	D9H	42 780	306	667	445	RH 40	83 200	360	–
	4	1000.0–9999	Extremely hard ripping/ blasting	D10	77 870	522	1230	778	–	–	–	–
	5	Larger than 10 000	Blasting	–	–	–	–	–	–	–	–	–

Notes
Quoted in Caterpillar Performance Handbook †, or equivalent.
All machines referred to are track mounted.

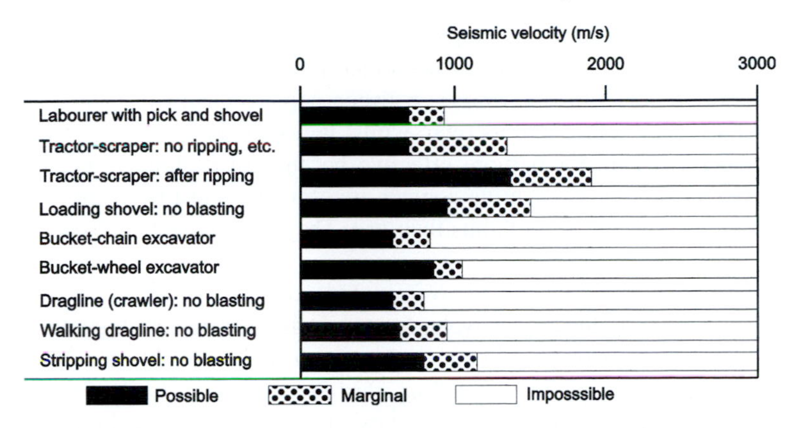

Figure 2.11 Diggability chart.

Table 2.7 Density, bulking factor and diggability of some common soils

Soil type	*Density* (Mg/m^3)	*Bulking factor*	*Diggability*
Gravel, dry	1.8	1.25	E
Sand, dry	1.7	1.15	E
Sand and gravel, dry	1.95	1.15	E
Clay, light	1.65	1.3	M
Clay, heavy	2.1	1.35	M-H
Clay, gravel, and sand, dry	1.6	1.3	M

E, easy digging, loose, free-running material such as sand and small gravel; M, medium digging, partially consolidated materials such as clayey gravel and clay; M-H, medium-hard digging, materials such as heavy wet clay, gravels and large boulders.

material and averaging the results of several tests. This then can be compared with the *in situ* density to give the bulking factor:

$$\text{Bulking factor} = \frac{\text{Intact density per unit volume}}{\text{Disturbed density per unit volume}}$$

Some examples of typical bulking in rocks and soils are given in Table 2.7. The bulking factor obviously is important in relation to loading and removal of material from a working face.

2.3.2 Choice of plant for digging

When choosing the type of plant required for digging, the quantity of the soil, the means by which it is to be transported and the distance must be considered (Church, 1981). For example, if the tip area is within 100 m of a working face, then earth can be moved by a bulldozer. The bulldozer, the blade of which is fixed at right angles to its line of advance, can only push its load in front of it. With an angle-dozer the blade can be angled to cast the spoil from the trailing edge of the blade. A dozer can produce high outputs on shallow excavation work provided, as noted, the haul is kept low, thereby minimizing spillage from the sides.

Scrapers are most effective when employed for large earthmoving works such as road construction. Each type of scraper has an economic range of hauling distance. Under 100 m a bulldozer is more efficient than a scraper, between approximately 100 and 300 m a scraper towed by crawler tractor comes into its own, whilst between about 300 and 1000 m a scraper towed by a rubber tyred tractor is more effective because of its higher speed. Scrapers are unsuitable for deep excavations covering a small area for they must have ample room to turn, and access to such sites generally prohibits their use. They can excavate all types of soil except soft clays and silts.

Where the spoil haul exceeds 1 km, excavators, for example, which load onto dump trucks are required. These include loading shovels, face shovels, backacters, draglines, grabs, drag-scrapers and continuous excavators. Loading shovels only are suitable for cuts up to a few metres in depth. They are capable of handling hard dense soils or badly fragmented ground but they have a poor sub-grade digging capability. The face shovel is the most powerful of the front-end excavators, with the highest output, and can operate in reasonably hard soils. It is designed primarily for bulk excavation above track level, working from the bottom of the excavation upwards. It is most efficient when operating against a high face in firm to hard clay soil, densely packed sand or gravel with clay binder, or soft chalk. The height of the working face that gives the most efficient output is that which allows the bucket to dig a full load in a single cut and is referred to as the optimum depth of cut (Table 2.8). Because of difficulties of access the face shovel is not suited to deep confined excavations.

Although the depth to which backacters can excavate is limited to the length of the jib and bucket arm, they can be used for deep excavation in small areas such as for trenches and column bases. Backacters can dig to close limits and make well-controlled even cuts. They operate satisfactorily in any type of soil that will stand up temporarily and are therefore unsuitable for use in loose sand or soft clay.

With their long reach and dumping radius draglines are especially suited to bulk excavation below track level. Indeed, they lose efficiency when employed on excavation above the level of their own tracks. Draglines can be used in loose soils or for soft ground and inundated areas, for they are designed to stand above the working area on firm ground and move backwards as the excavation proceeds. They also can be used to excavate badly fractured, weathered, and soft rocks such as shale, mudstone and chalk. Draglines can work to appreciable depths, although they cannot cut vertical faces. The most efficient digging area is immediately under the tip of the boom and this decreases as the bucket is drawn towards the base of the machine. One of

Table 2.8 Optimum depths of cut for face shovels

Size of bucket (m^3)	*Optimum depth of cut* (m)		
	Light soil e.g. sand, gravel sandy clay	*Medium soils e.g. dense sand, firm clay*	*Hard soils, e.g. stiff to hard clays, loess chalk*
0.3	1.2	1.4	1.8
0.4	1.4	1.7	2.1
0.6	1.6	2.1	2.4
0.8	1.8	2.4	2.7
1.0	2.0	2.6	3.0
1.2	2.2	2.8	3.3

the advantages of large draglines is that they can pile earth adjacent to the excavation when earth needs to be returned subsequently as backfill.

Grabs or clam-shell buckets are the most suitable type of excavator for deep excavation in confined areas. For example, they are commonly used to excavate ground from within caissons.

Drag-scrapers are infrequently used in earthworks although they sometimes have been employed to excavate ground that is too soft to allow the operation of other types of excavators. They consist of a scraper bucket that travels along a wire cable suspended between two towers.

The two most important types of continuous excavators are the bucket-chain excavator and the bucket-wheel excavator. These are generally massive machines developed for use in open pit mining (Atkinson, 1971). However, smaller bucket-chain excavators can be used to dig trenches. These are capable of very high outputs when digging soils free from large stones.

2.4 Displacement in soils

Displacement in soil, usually along a well-defined plane of failure, occurs when shear stress rises to its value of shear strength. The shear strength of the material along the slip surface is reduced to its residual value so that subsequent movement can take place at a lower level of stress. The residual strength of a soil is of fundamental importance as far as the behaviour of slides is concerned, and in progressive failure (Skempton, 1964).

A slope of 1:1.5 generally is used when excavating dry sand, this more or less corresponding to the angle of repose, that is, 30–40°. This means that a cutting in a coarse-grained soil will be stable, irrespective of its height, as long as the slope is equal to the lower limit of the angle of internal friction, provided that the slope is suitably drained. In other words, the factor of safety (F) with respect to sliding may be obtained from:

$$F = \frac{\tan\phi}{\tan\beta} \tag{2.6}$$

where ϕ is the angle of internal friction and β is the slope angle. As far as sands are concerned, their packing density is important. For example, densely packed sands that are very slightly cemented may have excavated faces with high angles that are stable. Failure on a slope composed of coarse-grained soil involves the translational movement of a shallow surface layer. The slip is often appreciably longer than it is in depth. This is because the strength of coarse-grained soils increases rapidly with depth. If, as is generally the case, there is a reduction in the density of the soil along the slip surface, then the peak strength is reduced ultimately to the residual strength. The soil will continue shearing without further change in volume once it has reached its residual strength. Although shallow slips are common, deep-seated shear slides can occur in coarse-grained soils. They usually are due to the placement of heavy loads at the top of the slope. The influence of water is an important factor as far as slope failure is concerned. For instance, seepage of groundwater through a deposit of sand in which slopes exist can cause them to fail.

The most frequently used gradients in many clay soils vary between 30 and 45°. In some clays, however, in order to achieve stability the slope angle may have to be less than 20°. The stability of slopes in clay soil depends not only on its strength and the angle of the slope but also on the depth to which the excavation is taken and on the depth of a firm stratum, if one exists, not far below the base level of the excavation. In stiff fissured clays, the fissures appreciably reduce the strength below that of intact material. Thus, reliable estimation of slope stability in stiff fissured clays is difficult. Generally, steep slopes can be excavated in such clay soils initially, but their excavation means that fissures open due to the relief of residual stress and there is a change from negative to positive pore water pressures along the fissures, the former tending to hold the fissures together. This change can occur within a matter of days or hours. Not only does this weaken the clay but it also permits more significant ingress of water, which means that the clay is softened. Irregular-shaped blocks may begin to fall from the face, and slippage may occur along well-defined fissure surfaces that are by no means circular. If there are no risks to property above the crests of slopes in stiff fissured clays, then they can be excavated at about 35°. Although this will not prevent slips, those that occur are likely to be small.

In fine-grained soils, especially clays, slope and height are interdependent and can be determined when the shear characteristics of the material are known. Because of their water-retaining capacity, due to their low permeability, pore water pressures are developed in fine-grained soils. These pore water pressures reduce the strength of an element of the failure surface within a slope in the soil, and the pore water pressure at that point needs to be determined to obtain the total and effective stress. This effective stress then is used as the normal stress in a shear box or triaxial test to assess the shear strength of the soil concerned. Skempton (1964) showed that on a stable slope in clay the resistance offered along a slip surface, that is, its shear strength (s), is given by

$$s = c' + (\sigma - u)\tan\phi' \tag{2.7}$$

where c' is the cohesion, ϕ' is the angle of shearing resistance (these are average values obtained around the slip surface and are expressed in terms of effective stress), σ is the total overburden pressure, and u is the pore water pressure. In a stable slope only part of the total available shear resistance along a potential slip surface is mobilized to balance the total shear force (τ), hence:

$$\sum\tau = \sum c'/F + \sum(\sigma - u)\tan\phi'/F \tag{2.8}$$

If the total shear force exceeds the total shear strength, then a slip is likely to occur (i.e. $F > 1.0$).

Clay soils, especially in the short-term conditions, may exhibit relatively uniform strength with increasing depth. As a result, slope failures, particularly short-term failures, may be comparatively deep-seated, with roughly circular slip surfaces. This type of failure is typical of relatively small slopes. Slides on larger slopes are often non-circular failure surfaces following bedding planes or other weak horizons.

When shear failure occurs for the first time in an unfissured clay soil the undisturbed material lies very nearly at its peak strength and it is the effective peak angle of friction (ϕ'_p) and the effective peak cohesion (c'_p) that is used in analysis. In slopes excavated in fissured overconsolidated clay, although stable initially, there is a steady decrease in the strength of the clay towards a long-term residual condition. During the intermediate stages, swelling and softening, due to the dissipation of residual stress and the ingress of water along fissures that open on exposure, take place. Large strains can occur locally due to the presence of the fissures. Considerable non-uniformity of shear stress along a potential failure surface and local overstressing leads to progressive slope failure. Skempton (1964) showed that if clay is fissured, then initial sliding occurs at a value below peak strength. Residual strength, however, is reached only after considerable slip movement has taken place so that the strength relevant to first time slips lies between the peak and residual values. Skempton accordingly introduced the term residual factor *(R)*, which he defined as:

$$R = \frac{\text{Peak shear strength} - \text{mean shear stress at failure}}{\text{Peak shear strength} - \text{residual shear strength}} \tag{2.9}$$

The residual factor represents the proportion of the slip surface over which the strength has deteriorated to the residual value. The residual factor therefore is used in analysis where the residual strength is not developed over the entire slip surface. It allows the mean shear stress at failure to be used in the calculation of the factor of safety.

There are many classifications of slides, which to some extent is due to the complexity of slope movements. However, the most widely used classification is that of Varnes (1978). Varnes classified slides according to the type of movement undergone on the one hand and the type of materials involved on the other (Fig. 2.12). Types of movement were grouped into falls, slides and flows. The materials concerned were simply grouped as rocks and soils. Obviously, one type of slope failure may grade into another, for example, slides often turn into flows. Complex slope movements are those in which there is a combination of two or more principal types of movement. Multiple movements are those in which repeated failures of the same type occur in succession and compound movements are those in which the failure surface is formed of a combination of curved and planar sections.

2.5 Displacements in rock masses

Sliding on steep slopes in hard unweathered rock (defined as rock with an unconfined compressive strength of 35 MPa and over) depends primarily on the incidence, orientation and nature of the discontinuities present. It is only on very high slopes or weak rock masses that failure in intact material becomes significant. Data relating to the spatial relationships between discontinuities affords some indication of the modes of failure that may occur, and information relating to the shear strength along discontinuities is required for use in stability analysis.

Other factors apart, the maximum height that can be developed safely in a rock slope is roughly proportional to its shearing strength, that is, the stronger the rock, the steeper the slopes that may be cut into it. For instance, excavations in fresh massive plutonic igneous rocks such as granite can be left more or less vertical after the removal of loose fragments. Excavation usually is straightforward in stratified rocks that are horizontally bedded and slopes can be determined with some degree of certainty. However, slopes may have to be modified in accordance with how the dip and strike directions are related to an excavation when it occurs in inclined strata. The most stable excavation in dipping strata is one in which the strata dip into the face. Conversely, if the strata daylight into the face (i.e. the dip of the bedding and therefore potential failure planes are into and less than that of the slope), then there is the potential for a slide to occur. This is most critical where the strata dip at angles between 30 and 70°. If the dip exceeds 70° and there is no alternative to working against the dip, then the face should be developed parallel to the bedding planes for safety reasons. Sedimentary sequences in which thin layers of shale, mudstone or clay are present may have to be treated with caution, especially if the bedding planes dip at a critical angle. Indeed, wherever weaker strata underlying stronger rocks are exposed on excavation, then undermining of the latter is likely to occur as the former are removed by agents of denudation. Ultimately, this action will produce a rock fall or slide.

In a bedded and jointed rock mass, if the bedding planes are inclined, the critical slope angle depends upon their orientation in relation to the slope and the orientation of the joints. The relation between the angle of shearing resistance, ϕ, along a discontinuity, at which sliding will occur under gravity, and the inclination of the discontinuity, α, is important. If $\alpha < \phi$ the slope is stable at any angle, whilst if $\phi < \alpha$, then gravity will induce movement along the discontinuity surface and the slope would not exceed the critical angle, which will have a maximum value equal to the inclination of the discontinuities. It must be borne in mind, however, that rock masses generally are interrupted by more than one set of discontinuities.

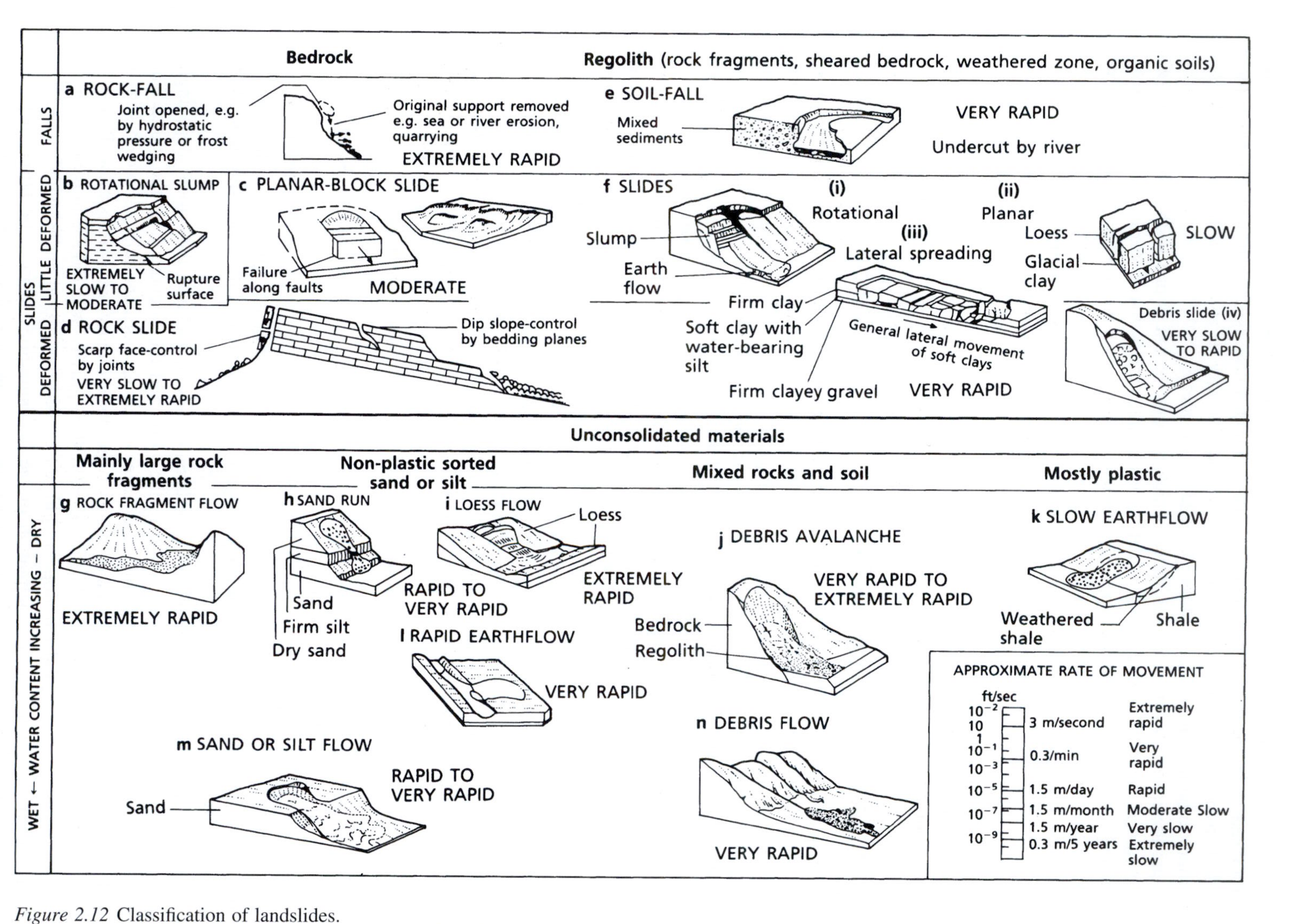

Figure 2.12 Classification of landslides.
Source: *Varnes*, 1978. Reproduced by permission of the Transportation Research Board.

Hard rock masses are liable to sudden and violent failure if their peak strength is exceeded in an excessively steep or high slope. On the other hand, soft materials, which exhibit small differences between peak and residual strengths, tend to fail by gradual sliding. The relative sensitivity of the factor of safety to the variation in importance of each parameter that influences the stability of slopes depends initially on the height of the slope. For example, Richards *et al.* (1978) graded each parameter concerned, in order of importance with respect to their effect on the factor of safety, in relation to slopes with heights of 10, 100 and 1000 m. The results are shown in Table 2.9 and Figure 2.13. The heights and angles of slopes in hard rocks can be estimated roughly from Figure 2.13. The joint inclination is always the most important parameter as far as slope stability is concerned. Friction is the next most important parameter for slopes of medium and large height, whereas unit weight is more important for small slopes than friction. Cohesion becomes less significant with increasing slope height whilst the converse is true as far as the effects of water pressure are concerned.

The shear strength along a joint is attributable mainly to the basic frictional resistance that can be mobilized on opposing joint surfaces. Normally, the basic friction angle (ϕ_b) approximates to the residual strength along the discontinuity. An additional resistance is consequent upon the roughness of the joint surface. Shearing at low normal stresses occurs when the asperities are overriden; at higher confining conditions and stresses, they are sheared through. Barton (1974) proposed that the shear strength (τ) of a joint surface could be represented by the following expression:

$$\tau = \sigma_n \tan\left(JRC \log_{10}\left(\frac{JCS}{\sigma_n}\right) + \phi_b\right) \tag{2.10}$$

where σ_n is the effective normal stress, *JRC* is the joint roughness coefficient and *JCS* is the joint compressive strength. The shear strength along a discontinuity also is influenced by the presence and type of any fill material, and by the degree of weathering undergone along the discontinuity.

Table 2.9 Sensitivity of factor of safety to various parameters. Functions affecting the factor of safety (for a given value of slope height and angle)

Function	*Probable range of magnitude*
Density, ρ	0–300 kPa
Cohesion, c	15–30 kN/m^3
Water pressure, hw/H	0–H m
Friction angle, θ	0–60°
Joint inclination, ϕ	10–50°

Order of importance of functions

Rank	*Slope height* 10 m	100 m	1000 m
1	Joint inclination	Joint inclination	Joint inclination
2	Cohesion	Friction angle	Friction angle
3	Density	Cohesion	Water pressure
4	Friction angle	Water pressure	Cohesion
5	Water pressure	Density	Density

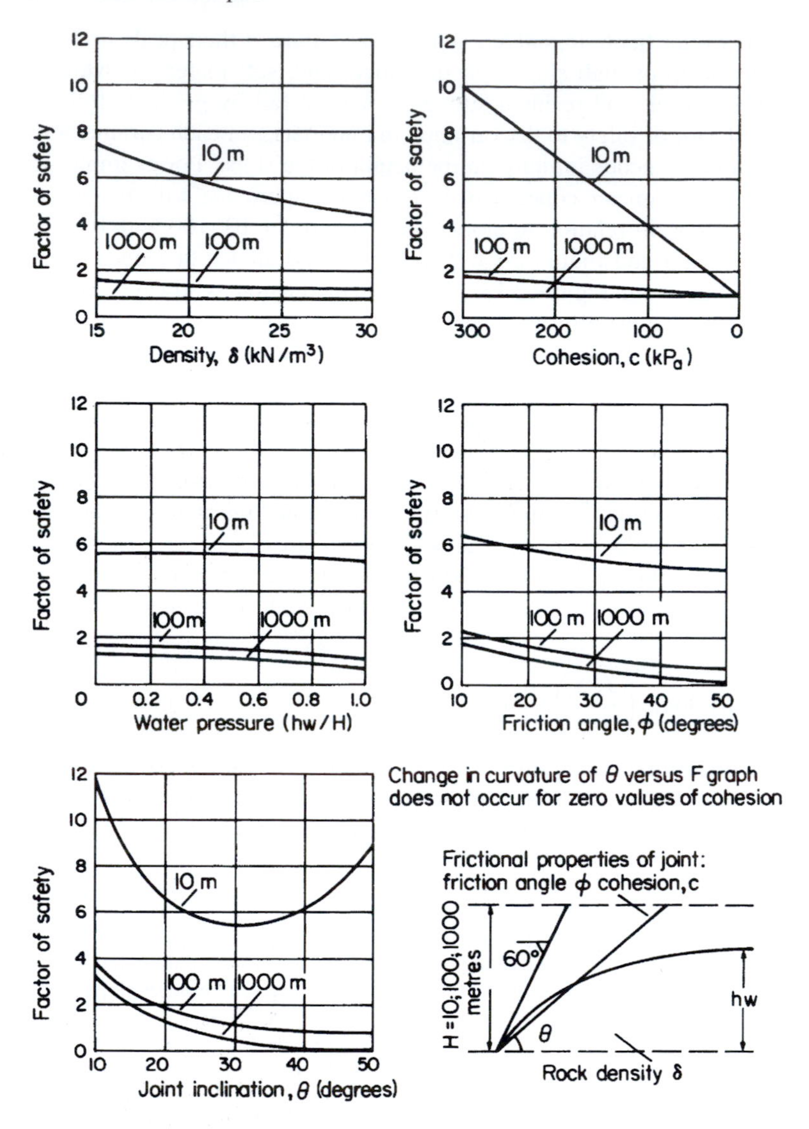

Figure 2.13 Sensitivity analysis for slope stability calculations.
Source: Richards *et al*., 1978.

According to Hoek and Bray (1981), in most hard rock masses neither the angle of friction nor the cohesion are dependent upon moisture content to a significant degree. Consequently, any reduction in shear strength is attributable almost solely to a reduction in normal stress across the failure plane and it is water pressure, rather than moisture content, which influences the strength characteristics of a rock mass.

The principal types of failure that are generated in rock slopes are rotational, translational and toppling modes (Fig. 2.14). Rotational failures normally only occur in structureless overburden, highly weathered material or very high slopes in closely jointed rock. They may develop either circular or non-circular failure surfaces. Circular failures take place where rock masses are intensely fractured, or where the stresses involved over-ride the influence of the discontinuities

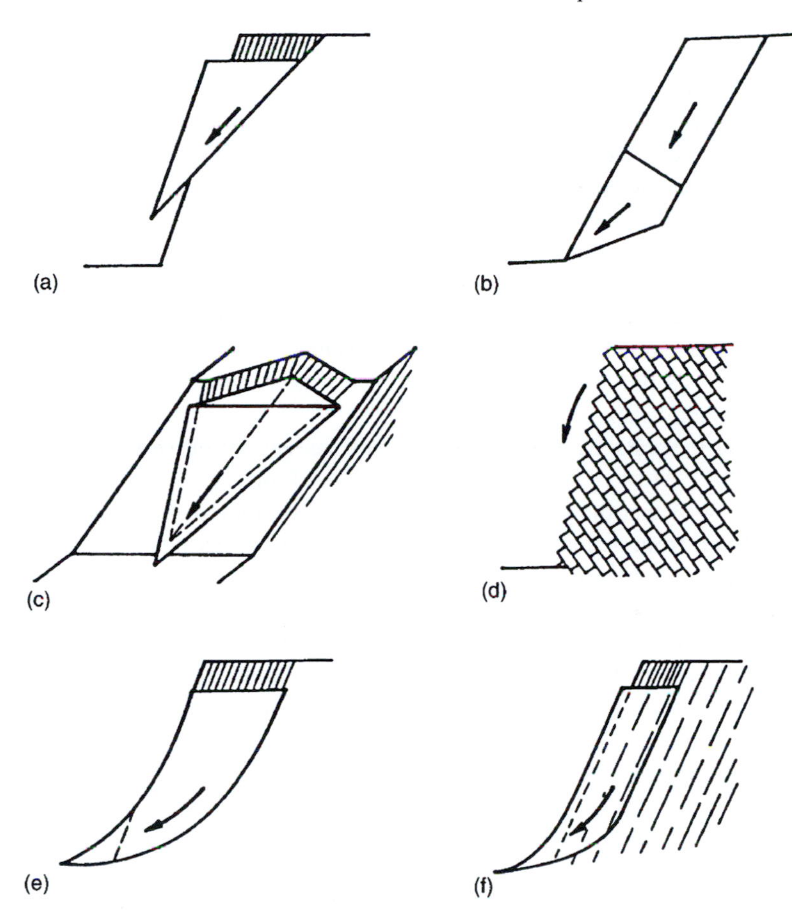

Figure 2.14 Idealized failure mechanisms in rock slopes: (a) plane; (b) active and passive blocks; (c) wedge; (d) toppling; (e) circular; (f) non-circular.

in the rock mass. Relict jointing may persist in highly weathered materials, along which sliding may take place. These failure surfaces are often intermediate in geometry between planar and circular slides.

There are three kinds of translational failures, namely, plane failure, active and passive block failure, and wedge failure. Plane failures are a common type of translational failure and occur by sliding along a single plane that daylights into a slope face. When considered in isolation, a single block may be stable. Forces imposed by unstable adjacent blocks may give rise to active and passive block failures (Fig. 2.14). In wedge failure two planar discontinuities intersect, the wedge so formed daylighting into the face (Fig. 2.14). In other words, failure may occur if the line of intersection of both planes dips into the slope at an angle less than that of the slope.

Toppling failure generally is associated with steep slopes in which the jointing is near vertical. It involves the overturning of individual blocks and therefore is governed by discontinuity spacing as well as orientation. The likelihood of toppling increases with increasing inclination of the discontinuities (Fig. 2.14). Water pressure within discontinuities helps promote the development of toppling.

2.6 A brief note on slope stability analysis

An analysis of stability should determine under what conditions a slope will remain stable, the stability being expressed in terms of the factor of safety (Duncan, 1996). The stability improves as the value of the factor of safety increases above unity. A soil or rock mass under given loading should have an adequate factor of safety with respect to shear failure, and deformation under given loads should not exceed certain tolerable limits.

Analysis provides an evaluation of slope stability by means of quantitative assessment of slope stability behaviour and thereby offers an important input for the design of slopes. Morgenstern (1992) distinguished between analyses used to determine pre-failure conditions and post-failure conditions. The purpose of pre-failure analysis is to evaluate the safety of a slope whereas post-failure or back-analysis provides an explanation of a slide event. The back-analysis provides the data for the design of any remedial measures.

There are several methods available for analysis of the stability of slopes (Bromhead, 1992; Simons *et al.*, 2001). Most of these may be classed as limit equilibrium methods in which the basic assumption is that the failure criterion is satisfied along the assumed path of failure. A free mass is taken from a slope, and starting from known or assumed values of the forces acting upon the mass, calculation is made of the shear resistance required for equilibrium. This shearing resistance then is compared with the estimated or available shear strength to give an indication of the factor of safety.

The two-dimensional methods of analysis have been summarized by Fredlund (1984) and Nash (1987). Computer programs are widely available that allow the use of the most rigorous methods. As Morgenstern (1992) pointed out, two-dimensional methods involve analytical solutions for simplified homogeneous profiles, and numerical methods, primarily based on the method of slices, are used for more complex sections. However, more recently three-dimensional methods of limit equilibrium analysis have been developed (Gens *et al.*, 1988). The use of three-dimensional methods reduces the degree of error in the results of analysis. The three-dimensional equivalent of the method of slices has been termed the method of columns and was developed by Hovland (1977). According to Morgenstern (1992), one of the most useful methods has been proposed by Hungr (1987), which is an extension to the three dimensions of Bishop's (1955) simplified method of analysis.

The drawbacks relating to the reliability of limit equilibrium analysis are attributable to difficulties inherent in site investigation, and material sampling and testing. The two most important parameters in limit equilibrium analysis are pore water pressure distribution and the shear strength of the material. A site investigation and associated laboratory testing programme cannot truly convey the operational conditions that govern stability. Hence, the pore water pressure distributions and shear strength used in an analysis must be modified in some representative manner. Pore water pressures may be affected by rainfall and their equilibration in materials of low permeability is of long duration. This, together with the long-term reduction in strength of many soils, poses problems in terms of the values to be used for stability analyses. In fact, the peak strength as obtained by laboratory testing is not operational over a complete slope when it fails, part of the slope having been stressed beyond the peak. As such, Morgenstern (1992) indicated that deformational analyses are required for this type of problem, coupled with limit equilibrium analyses to obtain an overall assessment of stability. However, in most instances where slope stability has to be determined, if the factor of safety is chosen correctly, then the deformation will be acceptable and the slope will behave as predicted. Progressive failure resulting from strain weakening can be handled by finite element methods (Chan and Morgenstern, 1987; Potts *et al.*, 1990).

Translational or plane failures frequently occur in rock masses. The forces acting on a block in a translational slide (Fig. 2.15a) are the gravitational weight (W), the disturbing force acting down the slide plane ($W \sin \beta$), the normal force acting across the sliding plane ($W \cos \beta$) and the shearing resistance of the surface between the block and the plane (R). According to Hoek (1970) the shearing resistance is given by:

$$R = cA + W \cos \beta \tan \phi \tag{2.11}$$

where C is the cohesion and A is the base area of the block. The condition of limiting equilibrium occurs when

$$W \sin \beta = cA + W \cos \beta \tan \phi \tag{2.12}$$

However, the above expression does not take into account the presence of water within a slope. If, for instance, water is trapped behind the upper face of the block, water pressure distributions are set up along the face and base of the block. The resultant forces, U and V, act in the directions shown in Fig. 2.15b and the limiting equation becomes:

$$W \sin \beta + V = cA + (W \cos \beta - U) \tan \phi \tag{2.13}$$

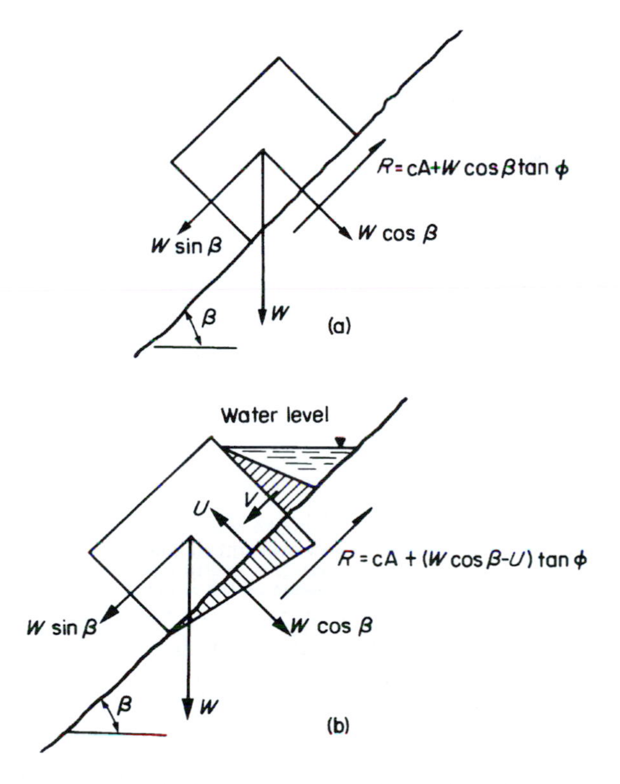

Figure 2.15 (a) Forces acting on a block resting on an inclined plane. W = gravitational weight of the block. $W \sin \beta$ = disturbing force acting down the plane. $W \cos \beta$ = normal force acting across the sliding plane. R = shear resistance of the surface between the block and the plane. (b) Forces acting on a block resting on an inclined plane with water trapped behind the block.

The limiting equilibrium occurs at slope angle α, which is less than the angle of friction, as both water forces (U, V) act in directions that tend to cause instability. The factor of safety (F) then becomes:

$$F = \frac{cA + (W\cos\beta - U)\tan\phi}{W\sin\beta + V} \tag{2.14}$$

There are an infinite number of possibilities for producing limiting equilibrium when sliding occurs on two planes with different frictional values. The factor of safety for wedges often is taken as the value that yields the same factor of safety on both planes. According to Richards *et al.* (1978), this is very arbitrary and probably unrealistic. They suggested that it was more useful to consider the sensitivity of the wedge to possible changes in shear strength characteristics and water pressure conditions.

Hoek (1971) maintained that the condition for toppling is defined by the position of the weight vector in relation to the base of the block. If the weight vector, which passes through the centre of gravity of the block, falls outside the base of the block, toppling will occur (Fig. 2.16).

One of the most important aspects of rock slope analysis is the systematic collection and presentation of geological data so that it can be readily evaluated and incorporated into stability analyses (Bye and Bell, 2001). Spherical projections provide a convenient method for the

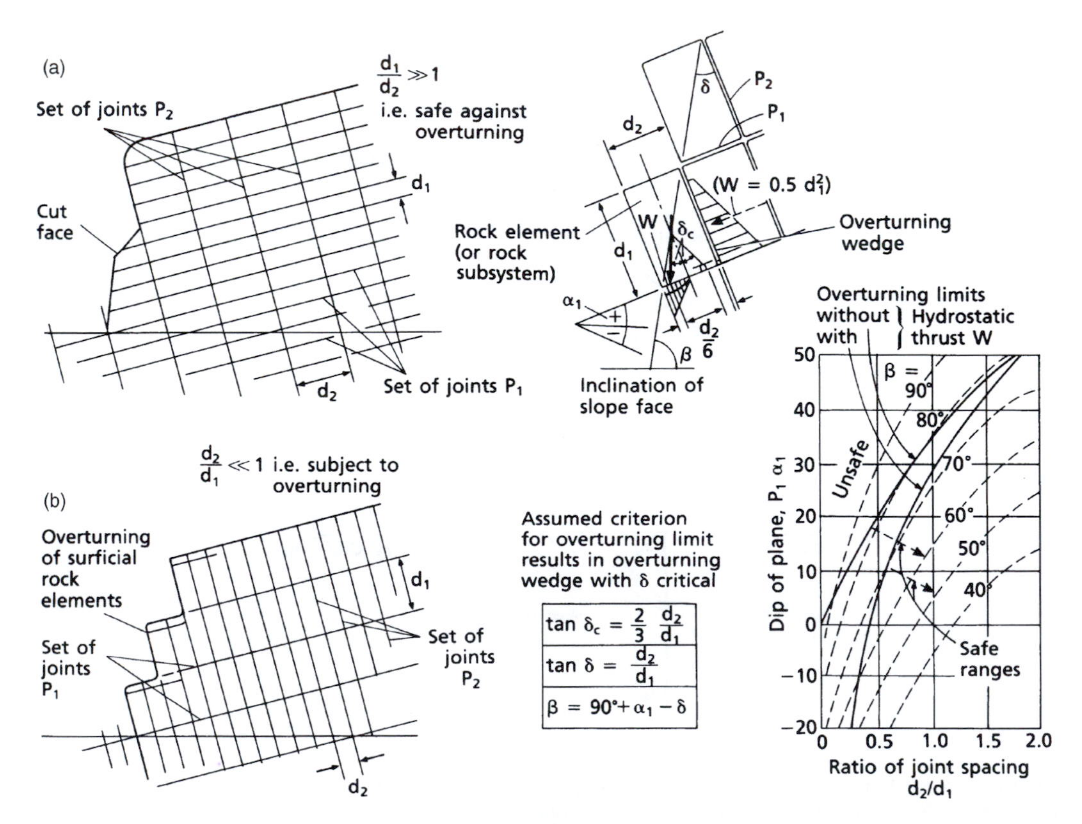

Figure 2.16 Criteria for toppling failure in two dimensions.
Source: Richards *et al.*, 1978.

presentation of geological data. The use of spherical projections, commonly the Schmidt net, means that the traces of planes on the surface of the 'reference sphere' can be used to define the dips and dip directions of the planes. In other words, the inclination and orientation of a particular plane is represented by a great circle or a pole, normal to the plane, which are drawn on an overlay and placed over the stereonet. Hoek and Bray (1981) illustrated how to plot great circles and poles using the stereonet technique and helped pioneer the use of this technique for analysis of the stability of rock slopes. Different types of slope failures are associated with different geological structures and these give different general patterns when analysed by stereonet methods (Fig. 2.17). A summary of analysis of stability of rock slopes has been provided by Norrish and Wyllie (1996).

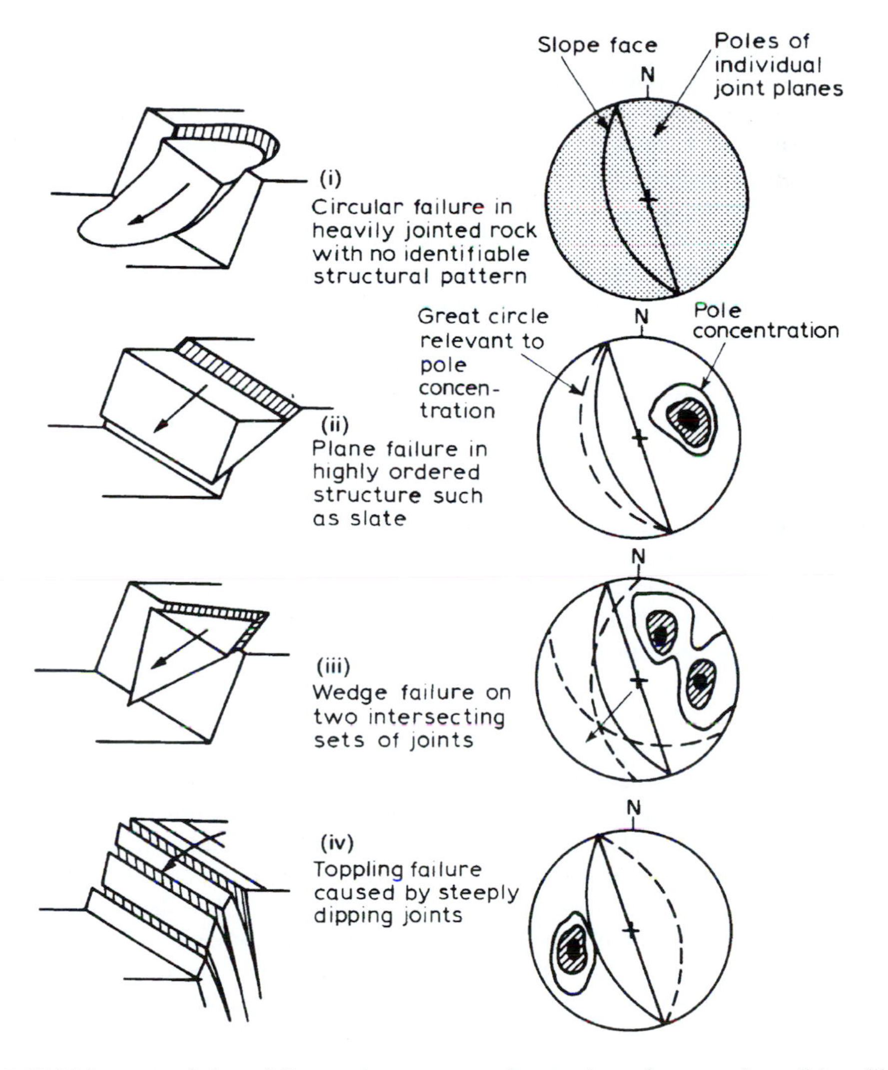

Figure 2.17 Main types of slope failure and appearance of stereoplots of structural conditions likely to give rise to these failures.
Source: Hoek and Bray, 1981. Reproduced by kind permission of Institution of Materials, Minerals and Mining.

2.7 Ground movements and excavation

Excavation causes a reduction in the vertical and horizontal pressure in the ground. It is responsible for an upward movement in the base of the excavation, and inward and vertical movements, both up and down, in the surrounding ground (Burland *et al.*, 1979). Significant movements can take place to an appreciable distance from an excavation, and horizontal movements can be notably larger than vertical movements. The most important factor that governs the magnitude of movement is the type of ground involved. Other factors of importance include the dimensions of the excavation, the method of excavation and bracing employed, and the quality of the workmanship.

2.7.1 *Uplift of floor of excavation*

The stability of the floor of a large excavation may be affected by ground heave. The amount of heave and the rate at which it occurs depend on the degree of reduction in vertical stress during construction operations and on the type and succession of the underlying strata. In other words, heave occurs as a result of elastic rebound. It generally is greater in the centre of a level excavation in relatively homogeneous ground.

Uplift also may occur in an excavation due to long-term swelling involving the absorption of water from the ground surface or of water migrating from below. Swelling is associated particularly with mudrocks. The rate of swelling generally is slow but the amount can be large (many tens of millimetres). If mudrocks contain sulphur compounds, as for example pyrite, then the amount of swelling may be enhanced notably. For instance, Fasiska *et al.* (1974) mentioned that on breakdown, pyrite may be responsible for an increase in volume of almost fourfold. If calcium carbonate is present in mudrock, then the reaction of sulphuric acid, derived from the breakdown of pyrite, with the carbonate will give rise to gypsum, the resultant increase in volume being sevenfold. Penner *et al.* (1973) quoted a case of heave in an excavation in black shale, due to such reactions, where the maximum total heave was 107 mm. The hydration of anhydrite to gypsum also leads to an increase in volume.

The soil around an excavation acts as a surcharge on that remaining below excavation level. A bearing capacity failure may occur if this surcharge is high enough. The possibility of this type of failure occurring is especially likely when the soil beneath excavation level behaves as a frictionless material under undrained conditions.

Peck (1969) pointed out that the extent to which a state of failure below the base of an excavation is approached may be estimated from values of the dimensionless number N_b($N_b = \gamma H \tau_u$ where γ is the unit weight of the soil, H is the height of the excavation and τ_u is the undrained shear strength of the soil below base level). He went on to state that if the strength of the soil representing the surcharge was ignored and if the excavation was regarded as infinitely long, then a plastic zone develops at the lower corners of the excavation when N_b reaches 3.14. The zone spreads with increasing values of N_b until base failure occurs. At this point N_b equals the critical value, that is, $N_{cb} = 5.14$. Accordingly, for values of N_b less than 3.14, heave of the base of the excavation is largely elastic and of relatively small amount. When values exceed 3.14, the heave associated with a certain increase in depth tends to increase significantly until at $N_b = N_{cb} = 5.14$, it occurs continuously and base failure or failure by heave takes place. However, excavations are not of infinite length and the strength of the soil constituting the surcharge is not negligible. Hence, the values of N_{cb} for excavations with normal shapes usually are in the range 6.5–7.5, and the value at which the plastic zone begins to develop therefore exceeds 3.14. Bjerrum and Eide (1956) proposed procedures for estimating the factor of safety against bottom heave in excavations of various rectangular shapes and depths. Excavations carried out after dewatering has taken place usually do not heave. This is because the relief of stress occasioned by excavation is offset, at least in part, by the increase in effective stress due to the water table being lowered (Terzaghi and Peck, 1967).

2.7.2 Movements about an excavation

If an excavation is made in dense sand above the water table or if the groundwater has been lowered and brought under complete control, adjacent settlement usually is inconsequential. In the same situation in loosely packed sands or gravels, settlement may be of the order of 0.5 per cent of the depth of the excavation (Terzaghi and Peck, 1967). On the other hand, if groundwater has not been brought under complete control sands may flow into an excavation and give rise to large erratic damaging settlements in the immediate surrounding area. The location and magnitude of such settlements unfortunately cannot be predicted. Generally, movements resulting from excavations in cohesive sandy soils are small. For instance, Peck (1969) quoted an example of a deep excavation in such material in Oakland, California, where the settlement did not exceed 12.5 mm and usually was less than 6 mm. Furthermore, the cohesion in such soils appreciably reduces their susceptibility to seepage pressures. Although excavation generally is straightforward, lateral support usually is required since the materials tend to spall if left unsupported.

Peck (1969) also reported that saturated plastic clays undergo a consistent pattern of deformation as material is removed from within a temporary bracing system. Excavation reduces the load on the soil beneath the cut, whereupon the underlying soil tends to move upwards. The soil next to the sheeting or other support tends to move inwards, even at levels below the point to which excavation has advanced, before cross-bracing or other types of support are put in place. As a result, the ground surface surrounding the excavation settles. Ou *et al.* (1993) reviewed the characteristics of ground settlement about an excavation. Settlements associated with plastic clays are likely to be significantly greater than those associated with most other types of soils. What is more, settlement may take place over a considerable period of time. Both the amount and the distribution are related to the distance from the excavation. For instance, Peck mentioned settlements in soft clays of up to 0.2 per cent of the depth of an excavation being recorded at distances of three or four times the depth. He concluded that settlement around an excavation can be reduced only if floor heave and inward movement of sheeting can be reduced substantially. A summary of settlements due to excavations of basements in Chicago in relation to the distance from the excavations is given in Fig. 2.18.

Subsequently, Cole and Burland (1972) made measurements, using precise surveying methods, of the inward movement of a retaining diaphragm wall about an excavation some 20 m in depth in clay. They, like Peck (1969), found that these movements were notably time-dependent and necessitated the use of support at an early stage to minimize movements outside the excavation. Cole and Burland suggested that the time-dependent nature of the movements was due to the clay behind the retaining wall becoming softened. The horizontal movements of the ground surface outside the excavation were two to three times larger than the corresponding vertical movements. Cole and Burland went on to point out that structures may be more sensitive to horizontal than to vertical movements and that horizontal ground movements attributable to a wide excavation may be sufficiently large to cause damage or loss of serviceability to neighbouring structures. Burland *et al.* (1977) carried out an investigation at a brickpit, 29 m in depth, which was progressively excavated in Oxford Clay. They showed that the surrounding ground within a distance of 60 m was affected by movement. The release of *in situ* stress consequent upon excavation gave rise to inward displacement of the sides amounting to as

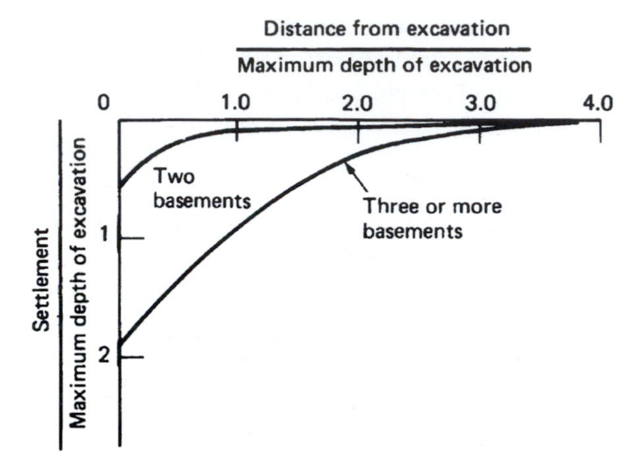

Figure 2.18 Summary of settlements due to excavation of basements in downtown Chicago.
Source: Peck, 1969.

much as 200 mm and to settlement of around 100 mm. Some 100 mm of heave occurred at the base of the pit. Burland *et al.* found that the clay moved inwards as a block, displacement at the base being almost the same as at ground level.

Ou *et al.* (2000) observed ground movements as excavation was deepened. They found that most of the soil behind a diaphragm wall about an excavation underwent horizontal extension and that maximum vertical movement for a given distance from the wall occurred at a certain depth below the ground surface. Volume change occurred in most of the soil behind the wall and Ou *et al.* attributed this to consolidation and/or creep behaviour. They concluded that the performance of neighbouring structures during construction of an excavation was influenced by their size, especially their length, and the type and size of the foundations concerned, as well as the shape of the settlement profile.

The nature of ground movements associated with the excavation of trenches was examined by Irvine and Smith (1983). Like other excavations, potential ground movement depends on soil type, groundwater conditions and the existing load carried by the ground. The actual movement also is influenced by the geometry of the trench, the type of support provided and the method of construction. Irvine and Smith indicated that ground movement occurs as excavation proceeds and before support is installed, as the ground moves into contact with the support system and as the support system deflects under load, as the sheeting for the trench is withdrawn and the trench backfilled, and, after backfilling, depending on the character of the backfill and the compaction achieved. Movements usually decrease rapidly away from a trench wall to a distance of 1.5–2.5 times its depth. Movements ranging from more or less zero to 3 per cent of trench depth have been reported from within these zones, even when trenches are supported. If there is a delay in installing support for the face, or if it is not fully effective, then movement can be significantly greater than this.

2.7.3 Methods of reducing ground movement and controlling the sides of excavations

Peck (1969) provided a review of the construction methods that have been used in deep excavations in order to reduce settlement of the adjacent ground surface and to minimize heave. In order to minimize heave, excavation to subgrade level was frequently carried out in small sections. The floor slab then was cast in the section concerned and it was temporarily backfilled until the other sections were completed. In this way the full weight of the soil was not removed until the lowermost basement floor slab was capable of resisting uplift. Cole and Burland (1972) described a large excavation for a raft where, in order to prevent undue swelling of the clay soil involved, a layer of unreinforced concrete was placed immediately after excavation was completed and prior to the construction of the raft. They also described how the slopes about a large excavation were protected by placing a thin covering of lightly reinforced concrete over them immediately after they were formed. This slowed down the rate of deterioration of the clay soil beneath. Water run-off was conveyed to sumps and then pumped into sewers.

In order to prevent the problem of hydraulic uplift occurring, the pore water pressure must be less than the overburden pressure at any given level. Hence, to keep pore water pressures under control, drainage can be allowed through the basement slab. Huder (1969) described the use of 24 filter wells, sunk to 15 m below the base level of a large excavation, to reduce hydraulic uplift pressure. A suspended basement floor, with a drained void beneath it, was used to deal with large swelling pressures and movements at the base of the excavation for the car park for the House of Commons, London (Burland and Hancock, 1977). The use of a barrier structure such as a diaphragm wall, taken deep enough to cut off any horizontal seepage from suspect layers, also helps control pore water pressures.

The principal factors that influence the design of support systems for the sides of an excavation are the strength and stability of the ground, which involves the relationship between the ground conditions and the depth of the excavation, as well as the length of time the excavation remains open (Anon., 1981). If the soil conditions vary, then the type of support used can vary. However, special attention should be given to the soil around the bottom of the faces since this is where it is most heavily stressed.

Shallow excavations can be made in soil without providing support if there is adequate space to establish slopes at which the soil can stand safely. However, sites in urban areas frequently extend up to the boundaries of adjacent properties and, in such situations, the sides of the excavation, because they will of necessity be excessively steep, require support. Sheeting should be installed at the onset where the ground is likely to collapse. In other words, the placement of sheeting, excavation and insertion of walings and struts should proceed by stages until the full depth of excavation is reached. Installation of the sheeting to the full depth of a cut prior to the commencement of excavation is advantageous where the soil and sheeting sections allow this to occur. For example, if steel sheet piling is used, then this allows the piling to be driven to full depth in most types of ground before excavation begins. As far as steel sheet piling is concerned, the web types are used for shallower excavations whilst the sides of deep excavations are protected by Z piling. In most soils it is advisable to drive sheet piles 1 m or more below the base level of the excavation to avoid local heaves. In fact, in some cases the embedment may eliminate the need for installing struts at the base of the excavation.

Continuous support has to be afforded to faces in water-bearing sands or silts, this being accomplished by using steel sheet piling or trench sheeting. Under such conditions the face support is driven in advance of the excavation. Continuous support also must be provided for faces excavated in loose sands or gravel, or soft clay soils. Sheet piles again can be driven ahead of excavation. Trench sheeting may be used to support the sides of an excavation in which the ground will stand vertically for a metre or so, for a long enough time to allow their installation.

Horizontal struts are positioned between walings or soldier piles to provide ground support in narrower excavations. Hydraulic struts allow adjustment to be made in the support they afford. Walings run horizontally and support sheeting. Soldier piles (vertical beams) can be used to support pre-fabricated panels, as for example, trenchboxes. They are installed at given intervals by driving or pre-drilling a hole and fixing the foot of the pile below formation level with concrete or granular fill. Raking struts can be used in wider excavations to support bracing, which is used to support the sheeting. If bracing is used to support sheeting, then this also runs horizontally and may be adjustable hydraulically. Bracing also can be tied back by anchors. At some sites it is possible to excavate the central area to its maximum depth and to place part of the permanent structure so that this then offers support for raking struts, required when the rest of the site is excavated.

Figure 2.19 illustrates the way in which ground settlements and inward movements of sheet pile walls develop in relation to the insertion of struts as an excavation deepens. Inward movement at and below base level can be kept to the minimum compatible with the size of an excavation and the type of soil involved only if struts are inserted promptly, at closely spaced vertical intervals, after each stage of excavation. As each strut is inserted it restricts the movement of the support system while the next stage of excavation is carried out. When the next lower strut has been placed, the one above is no longer as effective in controlling deep-seated movements. Indeed, it often can be removed without altering to any great extent the nature of ground deformation, providing that the soldier piles, waling or bracing, and remaining struts are able to carry the added load. Hence, after excavation has been completed and all the struts are in place, one or more of the intermediate struts frequently can be removed without causing inward deflections of the sheeting, which would give rise to significant settlements. Their removal may facilitate construction operations.

The effectiveness of a support system may be enhanced by the use of pre-stressed tiebacks or anchors. However, the zone of influence of pre-stressed tiebacks is very limited compared with that involving the relief of *in situ* stresses (Creed *et al.*, 1981). Moreover, they may cause a retaining wall to rotate about its toe, resulting in heave of the ground surface. Nonetheless, pre-stressed tiebacks are effective in reducing wall and surface movements, but they cannot prevent deep-seated movements occurring in the soil. Normally, displacements around an anchor are very small.

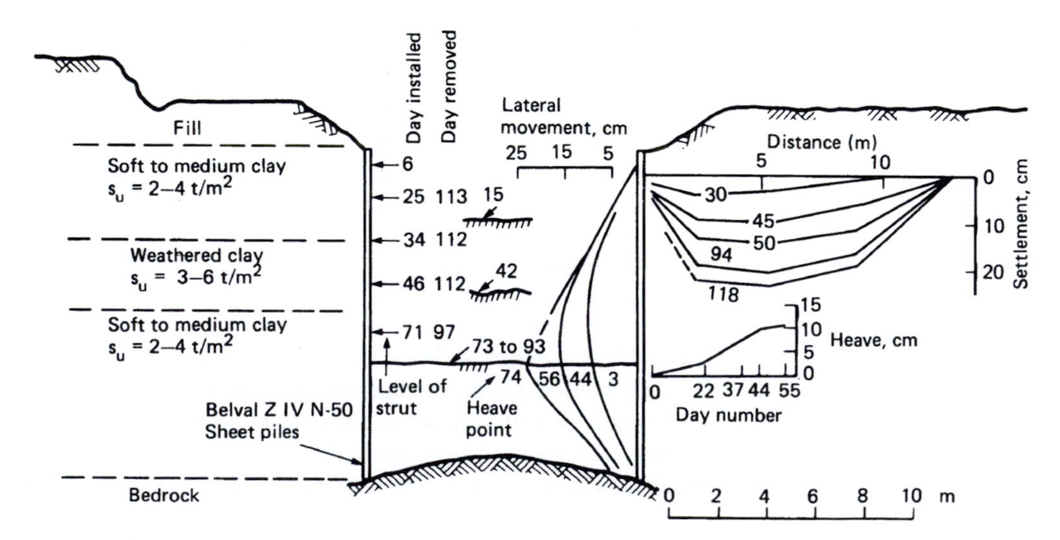

Figure 2.19 Illustration of the way in which ground settlement and inward movement of sheet pile walls developed in relation to the insertion of struts as an excavation in soft clay in Oslo deepened.
Source: Peck, 1969.

An outline of the design methods for support systems in trenches has been provided by Irvine and Smith (1983). In addition to some of the methods mentioned above, they described hydraulic frames, boxes, panel/plate lining systems and shields. Hydraulic frames consist of sheeting or structural sections that are permanently attached to hydraulic struts. They are installed in and jacked against the sides of a trench without anyone having to enter it. Proprietary waling frames are used with trench sheeting. Boxes are modular hydraulically strutted support walls that are installed by lowering them into a pre-dug trench or by digging them in progressively. Lining systems generally consist of hydraulically strutted vertical soldier piles with wall panels/plates spanning between them to support the trench face. Such systems may be used to form a continuous wall. Shields do not provide much support to the sides of a trench and therefore are used principally to afford protection to workers in a trench. Generally, shields consist of vertical plates with bracing apart. They are not suited to ground where the water table is high or where services cross the line of a trench. In trenches, as well as narrow excavations, it is necessary to control the withdrawal of support whilst the excavation is being backfilled to a high standard of compaction.

According to Tomlinson (1986) it is unnecessary to calculate lateral pressures on supports to excavations less than 6 m in depth unless hydrostatic pressure has to be taken into account. Lateral pressures at such depths can be highly variable, for instance, clay soils shrink from behind timbering in dry weather and swell when wetted. Indeed, the swelling pressures may be high enough to deform or rupture the struts. On the other hand, Tomlinson pointed out that economies in material can be achieved by calculating earth pressures for excavations deeper than 6 m, particularly when deep excavations have to be made in unfamiliar ground. Local concentrations of earth pressure can cause high loads on individual bracing members and if one strut fails, then it throws increased loads on to the surrounding members. This can sometimes initiate a general collapse of the system. Calculations of earth pressure should allow an economical spacing of walings and struts to be worked out and permit the development of the full flexural strength of the various components involved.

Diaphragm walls frequently are used to support the sides of large excavations and reinforced diaphragm walls may serve as part of the finished structure. Unless lateral supports such as struts, tiebacks or anchors are placed either before the main excavation or in stages as it proceeds, movements of the ground and associated settlement will occur. Burland and Hancock (1977) described how a diaphragm wall around an excavation for a large car park was strutted by the staged construction of permanent reinforced concrete floors as excavation proceeded, thereby minimizing ground movements. Such continuous strutting was achieved by constructing the floors successively from the top downwards and mining the soil beneath.

Secant piles can be used as perimeter walls for deep excavations and are relatively watertight. They therefore have proved useful in areas where the external water level should not be lowered. Also, they can be used to control running sands and silts.

2.8 Groundwater and excavation

Groundwater frequently provides one of the most difficult problems during excavation and its removal can prove costly. Not only does groundwater make working conditions difficult, but piping, uplift pressures and flow of water into an excavation can lead to erosion and failure of the sides. Collapsed material has to be removed and the damage has to be made good. Subsurface water normally is under pressure, which increases with increasing depth below the water table. Under high pressure, gradients soils and weakly cemented rock can disintegrate. Hence, data relating to the groundwater conditions should be obtained prior to the commencement of operations.

Artesian conditions can cause serious trouble in excavations and therefore if such conditions are suspected it is essential that both the position of the water table and the piezometric pressures should be determined before work commences. Otherwise excavations that extend close to strata under artesian pressure may be severely damaged due to blowouts taking place in their floors. Such action may cause slopes to fail and could lead to the abandonment of the site in question. Sites where such problems are likely to be encountered should be dealt with prior to and during excavation by employing either dewatering techniques or impermeable barriers. It must be remembered that the structures to be placed in these excavations will be acted upon by uplift pressures. If a particular structure is weak then uplift pressure may cause it to fail, for instance, a blowout may occur in a basement floor. On the other hand, if the structure is strong but light in weight, it may be subjected to heave. Uplift pressures can be taken care of by adequate drainage or by resisting the upward seepage force. Continuous drainage blankets are effective but should be designed with filters that will function without clogging. The entire weight of a structure can be used to resist uplift pressures if a raft foundation is employed. Anchors grouted into bedrock also provide resistance to uplift.

2.8.1 Dewatering

Some of the worst conditions are met in excavations that have to be taken below the water table. In such cases the water level may be lowered by dewatering (Bell, 1993). The method adopted for dewatering an excavation depends upon the permeability of the ground and its variation within the stratal sequence, the depth of base level of the excavation below the water table, the piezometric conditions in underlying horizons, the method(s) of providing support to the sides of the excavation and of safeguarding neighbouring structures.

Pumping from a sump within the excavation generally can be achieved where the rate of inflow does not lead to instability of the sides or base of an excavation. Ditches are dug in the floor of the excavation, then lined or filled with gravel, in order to lead water to a sump or sumps, which must be deep enough to ensure that the excavation is drained. Each sump requires its own pump and the method is only capable of lowering the water table by up to approximately 8 m. Sump pumping cannot dewater confined aquifers and may not be able to cope with flow from aquifers with large storage capacities or that transmit copious quantities of groundwater.

In waterlogged silts and sands, inflow may be high enough to cause the sides of an excavation to slump or the floor to boil. Therefore pre-drainage is called for. Pre-drainage of a site can be accomplished by installing wellpoints (Fig. 2.20) or bored wells about the perimeter. Such groundwater lowering techniques depend on excessive pumping that lowers the water table and thereby develops a cone of exhaustion (Fig. 2.21). The radius of the cone of exhaustion at the withdrawal point depends upon the rate of pumping and recharge. The amount of discharge (Q) that is necessary to lower the water table through a given depth can be estimated by using the Dupuit equation, which for the gravity well condition is as follows:

$$Q = \frac{k(H^2 - h_o^2)}{\log_e(R/r_o)} \tag{2.15}$$

where H is the elevation of the original water table above an impermeable horizon, h_o is the elevation of the operating level of the pumping well above this horizon, R is the radius of the area of influence, r_o is the radius of the well, and k is the coefficient of permeability (Fig. 2.22a).

Figure 2.20 A wellpoint system dewatering an excavation.

The equation for a confined aquifer under artesian pressure is:

$$Q = \frac{2\pi kb(H - h_o)}{\log_e(R/r_o)} \tag{2.16}$$

where b is the thickness of the confined layer (Fig. 2.22b).

The installation of wellpoints and surrounding filter is rapid, individual wellpoints can be placed in a matter of minutes in some soils, and the flexibility of the system allows for rearranging their spacing according to the rate of inflow (Cashman and Preene, 2001). The most commonly used wellpoint is the self-jetting type. Such wellpoints are suitable for installation in gravels and sands, in particular for soils that do not yield more than 40–100 l min^{-1}. High-capacity wellpoints are needed for yields exceeding 140 l min^{-1} but these are not jetted into place. Wellpoints using vacuum-assisted drainage are effective in silty sands and silts. The radii of influence of the individual wellpoints overlap and they are laid out so as to lower the water table by approximately a metre below the base level of the excavation. Normally, wellpoints are spaced between 1 and 4 m apart. Some idea of the spacing between individual wellpoints can be gained from nomograms such as shown in Fig. 2.23. If thin layers of clay occur in sands, then the wellpoints can either be closely spaced (i.e. 1–2 m apart) or holes can be positioned outside the line of wellpoints and filled with coarse sand in order to allow water to move more readily to the wellpoints. Layers of highly permeable material can mean that water tends to bypass the wellpoints. In this situation the soil can be mixed by jetting at close intervals. Wellpoint pumping is most useful where the required lowering is not more than about 4–5.5 m. In order to achieve greater lowering another tier of wellpoints, header main and pumps must be installed. The added excavation that this involves may prove a limiting factor as far as cost is concerned. In coarse-grained silty soils the achievable lowering averages about 30 per cent less than in sands. Horizontal wellpoint drainage is particularly suited to groundwater lowering for pipe trenching (Somerville, 1986).

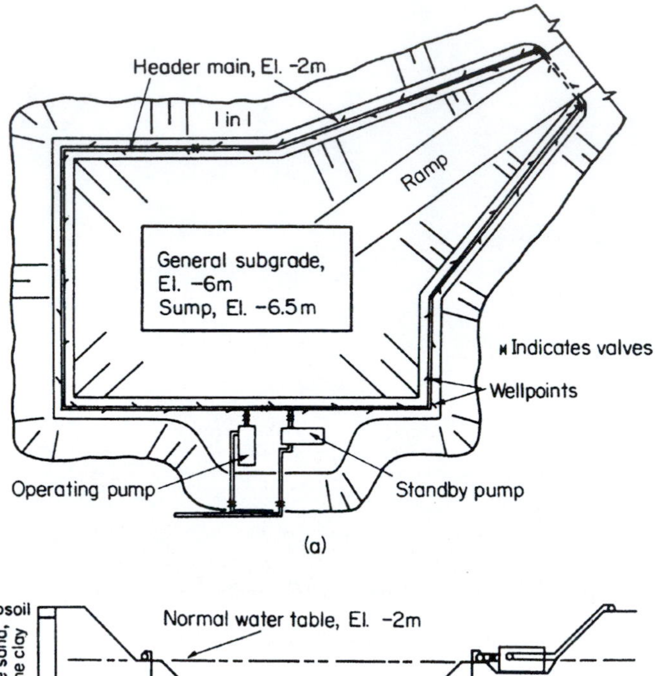

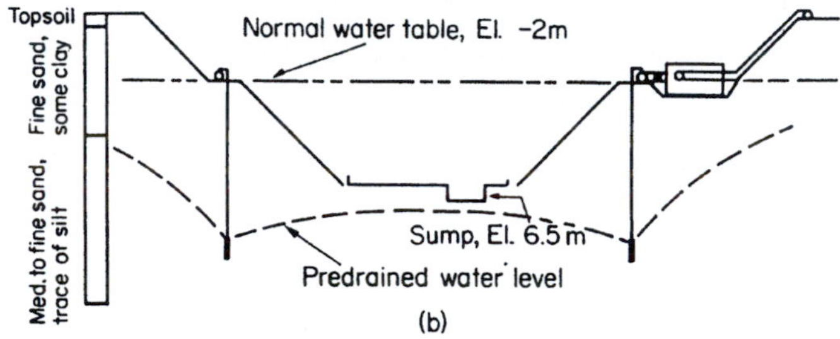

Figure 2.21 (a) Typical layout of a wellpoint system and (b) section through layout.

A bored filter well consists of a perforated tube surrounded by an annulus of filter media and its operational depth, in theory, may be unlimited (Bell and Cashman, 1986). Generally, wells are placed in a 600 mm diameter hole. Having bored the hole to a sufficient depth within an aquifer (in the case of a thin aquifer the well bore is often taken some 1–2 m into the impervious stratum beneath the aquifer), the perforated screen (well screen) is lowered into place and the appropriate filter media is placed. An electric submersible pump is lowered into each well on its own riser pipe and connected to a common discharge main. Bored wells are preferable to wellpointing for deep excavations where the area of the excavation is small in relation to the depth. They also are preferable in ground containing cobbles and boulders where wellpoint installation is difficult. Deep wells are suited particularly to variable soils and multilayer aquifers, as well as to the control of groundwater under artesian or subartesian conditions. For example, they are the appropriate method of dewatering an aquifer at depth beneath an impermeable stratum, in order to prevent a blowout in an excavation terminating within the impermeable material. A bored shallow well system can be used on a highly permeable site where pumping is required for several months, rather than wellpointing where risers at close centres could hinder construction operations.

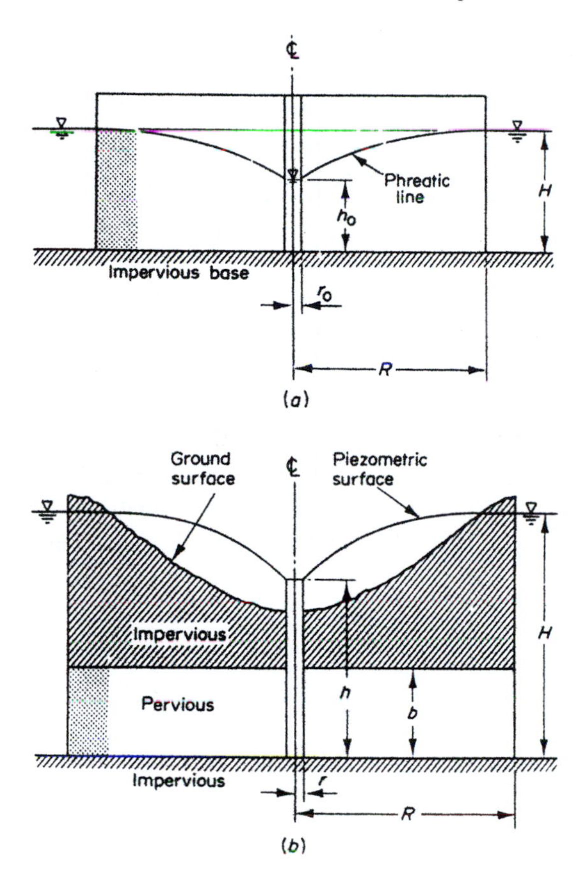

Figure 2.22 (a) Water table or gravity well condition (b) Confined aquifer or artesian well condition.

An ejector is a jet pump used for raising water from a borehole. Ejector systems tend to be used in fine sands and silts and are most effective when used to dewater deep excavations in stratified soils where close spacing is necessary. Such systems are not limited in terms of suction lift, as are wellpoints, and have much lower unit costs than wells. Consequently, they are used where it is inconvenient or impossible to install a multistage wellpoint system, the ejector system being capable of operating at depths in excess of 20 m (Powrie and Roberts, 1990). Spacings of 3–15 m are typical and ejectors are installed by jetting within a casing or by forming a hole by rotary drilling. They are surrounded by a sand filter.

Occasionally, electro-osmosis has been used as a dewatering technique to stabilize soils of low permeability such as silts and clays that cannot be dewatered by the above mentioned methods. Basically electro-osmosis consists of placing electrodes into the ground to a depth of about 2 m below excavation base level, and passing a direct electric current between them. The electric current induces a flow of water from the anodes to the cathodes, the latter acting as wellpoints from which the water can be removed by pumping. Cathodes may be installed in one line, being spaced approximately 8–12 m apart, with anodes mid-way between them. If more than one line of cathodes is employed, then the spacing between cathodes may be increased to 15 m. However, electro-osmosis suffers a fundamental drawback when used solely for dewatering in that as the soil dries out around the anode, its electrical resistance increases. It therefore becomes progressively less efficient as the soil water content is reduced. This means

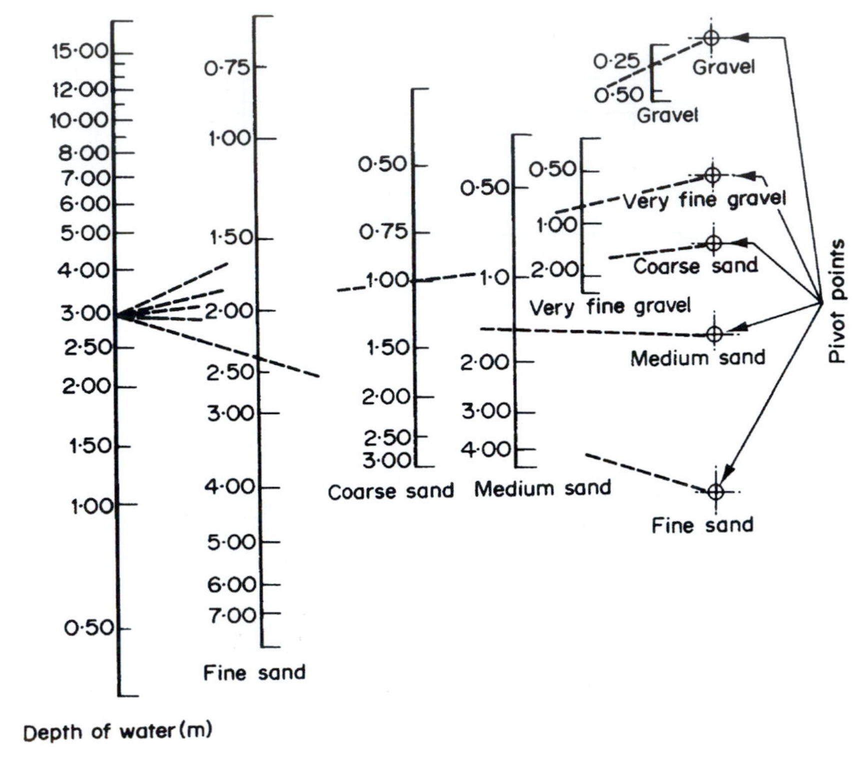

Figure 2.23 Nomogram for spacing, in metres, of wellpoints in sand or gravel.

that the water content in soils is rarely reduced sufficiently to achieve complete stabilization and even where it is, subsequent rehydration often may reverse the process.

A dewatering system not only aids construction operations by lowering the water table and intercepting seepage flow but its employment also means that the stability of the excavated slopes and floor is enhanced. This, in turn, means that the load on any sheeting or bracing is reduced. What is more, dewatering sandy soils improves their digging and hauling characteristics.

However, whenever the phreatic or piezometric surface is lowered, the effective load on the soil is increased, causing compression and consequent settlement. Usually, settlement due to the abstraction of water from clean sands is insignificant unless the sand was initially very loosely packed. On the other hand, pumping from an aquifer containing layers of soft clay, peat or other compressible soils or from a confined aquifer overlain by compressible soils may cause significant settlements. The amount of settlement undergone depends on the thickness of the compressible soils and their compressibility, and on the amount of groundwater lowering (Preene, 2000). The permeability of the soil and the length of the pumping period influence the rate of settlement. As a result it may be necessary to limit the radius of influence of the cones of depression, and so reduce the potential for settlement, by the use of groundwater recharge methods. In such instances, the excavation generally is surrounded by sheet piling with the recharge wells located outside the piling. Relief wells within an excavation can be used to reduce piezometric pressures and thereby overcome the problem of heave. Settlement of adjacent buildings would have occurred as a result of induced consolidation in clay layers in sand that was being dewatered below the excavation for the Latino Americana Building in Mexico City (Zeevaert, 1957). Consequently, water was injected into the sand beneath the adjacent buildings so that seepage gradients were not developed in the clays.

Groundwater flow in a rock mass is controlled primarily by discontinuities and since they generally fall into sets having certain orientations, the permeability of a rock mass varies with direction. Since measurements of permeability represent spot values in the field, an adequate investigation of the rock masses involved and their discontinuity patterns is essential before an assessment of groundwater conditions can be made or means of controlling groundwater designed. The geological structure and topography will indicate the potential for recharge of the rock masses that are to be exposed in an excavation (Fig. 2.24). Analytical methods, using permeability and piezometric data, to determine the groundwater pressure distribution within a

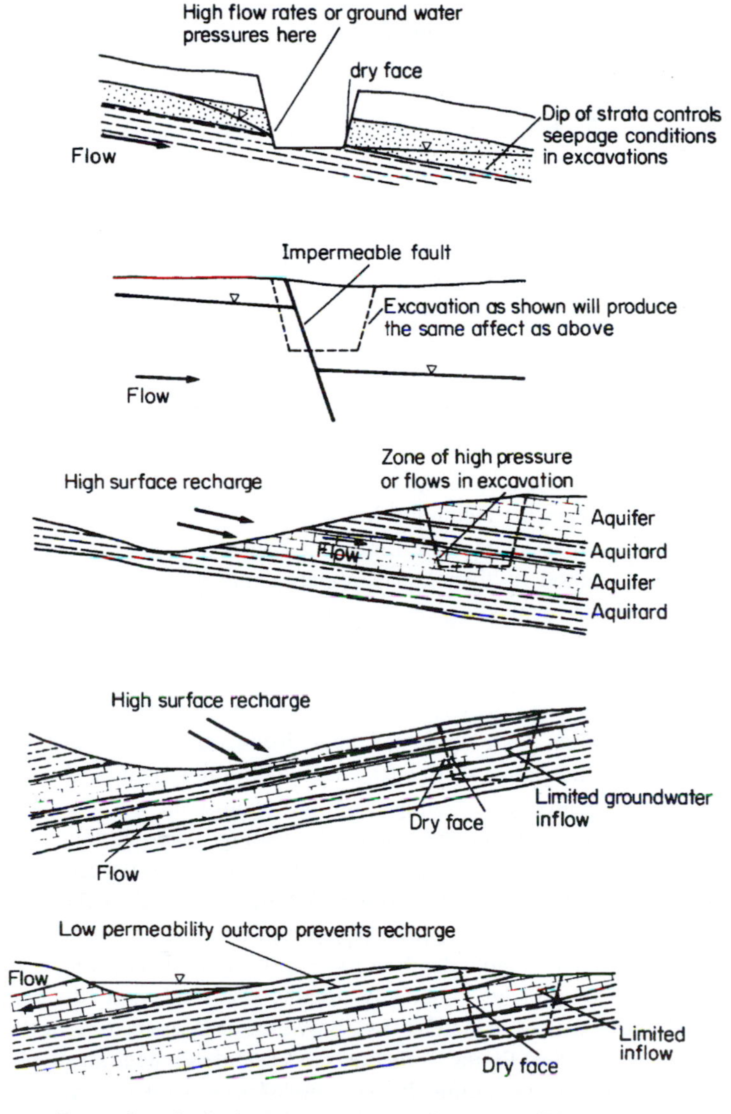

Figure 2.24 Some effects of geological structure on groundwater conditions.

slope formed by excavation include graphical flow net sketching, electrical resistance analogues and computer analyses. Groundwater pressures may adversely affect the stability of rock masses by reducing the shear strength along discontinuities. Drainage of rock slopes includes such measures as drainage trenches constructed down and/or along the slope face, horizontal or near horizontal drain holes drilled into the slope face, vertical pumped wells drilled behind the slope crest or on the slope face, and drainage galleries excavated in the rock mass behind the slope, with or without supplementary holes drilled from the gallery.

2.8.2 Impermeable barriers

In order to help select the type of cut-off required for groundwater control, about an excavation, a number of basic questions have to be answered. First, is a cut-off the best method of groundwater control at the site concerned? Second, if so, then which cut-off technique is best suited to the site conditions? Third, can the cut-off be temporary, that is, remain in existence for the construction period only or has it to be permanent? Fourth, can the cut-off serve any other functions?

Methods of forming such a barrier include sheet piling, contiguous bored pile walls, bentonite cut-off walls, geomembrane barriers, concrete diaphragm walls, grout curtains and panels, and ice-walls (Bell and Mitchell, 1986). The economy of providing a barrier to exclude groundwater depends on the existence of an impermeable stratum beneath the excavation to form an effective cut-off for the barrier. If such a stratum does not exist or if it lies at too great a depth to be practical to use as a cut-off, then upward seepage may occur that, in turn, may give rise to instability at the base of the excavation. The only methods of forming a horizontal barrier are by grouting or freezing.

In urban areas where surrounding property has to be safeguarded it will usually be more appropriate to provide a cut-off about an excavation so as to prevent the inflow of water whilst maintaining the surrounding water table at its normal level, rather than adopt a dewatering technique. By keeping the water table at its normal level, settlement problems are not likely to arise as they could if a dewatering technique was used.

Sheet piling may be composed of steel, timber or concrete piles. Steel sheet piling is the simplest and cheapest method of forming an impermeable barrier, the effectiveness of the integrity of sheet pile walls depending on the interlock between individual piles. However, sheet piling cannot be used in ground containing numerous boulders such as some tills, because the piles become difficult or impossible to drive and because of the risk of piles tearing. One technique that has proved to be effective in such conditions involves excavating the boulders in a slurry trench, backfilling with sand, and then driving the piles. Sheet piling about an excavation frequently requires support. In such cases sheet piling can be driven in advance of excavation. As excavation proceeds, waling is placed against the sheet piling and struts are placed across the excavation and wedged against the waling. Alternatively, soldier beams are driven at intervals along the line of excavation and as the excavation proceeds wooden sheeting planks are inserted horizontally against the ground and are supported by the soldier beams. Tie backs or anchors also may be used as support. Because of its structural strength, watertightness and ability to be driven to appreciable depth in most soils, steel sheet piling is widely used for cofferdams (Fig. 2.25). However, in some soils, notably fine sands and silts beneath the water table, if the critical gradient is exceeded as the excavation is deepened, then there is a danger of quick conditions developing. Piping sometimes occurs along steel sheet piling, especially when it is driven with the aid of jetting. In such situations consideration should be given to lowering the water table below the base of the excavation.

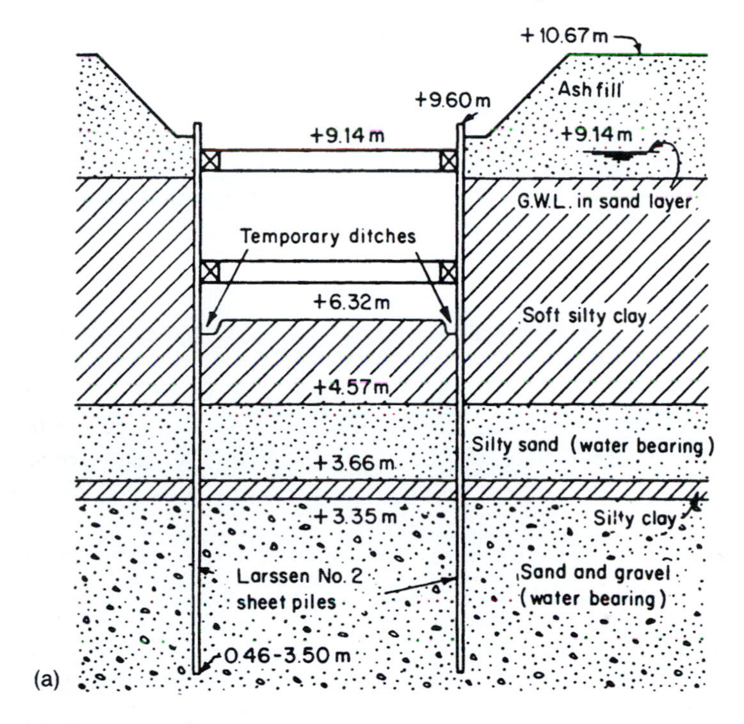

Figure 2.25a Steel sheet piling with bearing in variable soil conditions.

Contiguous bored piles frequently are associated with both shallow and deep excavations, and their length can be varied to suit the ground conditions. Normally, temporary casings have to be used during construction of contiguous bored piles although in clay soils only a short length of casing generally is necessary to provide a seal at the top of the clay. Contiguous bored piles normally are best suited to clay soils. Indeed, they may be impossible to install in saturated loosely packed silts or sands due to the soil undergoing liquefaction. Moreover, a loss of ground may occur when sinking piles in groups in sandy soils. In soils where boring is slow or difficult, for example, due to the presence of boulders, the amount of overbreak resulting from driving and boring out casings at very close centres in line, is likely to be greater than that arising from single piles at wide spacing. Difficulties can arise when constructing bored piles in water bearing or in squeezing ground. For instance, if the bottom of the casing is lifted above the base of the concrete pile while concreting operations are still proceeding, the casing being withdrawn in a number of lifts, then water may enter the borehole and thereby lower the strength of the concrete. Squeezing ground may give rise to 'necking' in a pile. Where contiguous bored piles pass through water bearing strata various grouting techniques can be employed to seal the joints.

Watertightness also can be achieved by using interlocking secant piles. First, alternate piles are installed, then intermediate holes are installed by boring out the soil between alternate pairs of piles, followed by chiselling a groove down the sides of these first formed pile shafts. Last, concrete is placed in the second set of holes, including the grooves, thereby forming a fully interlocking watertight wall. A more recent method of installing secant pile walls is to form alternate bentonite–cement piles and reinforced concrete piles that interlock to provide a continuous wall. The bentonite–cement piles are constructed first, then the reinforced concrete

Figure 2.25b Steel sheet piles used to form a caisson in alluvial soils, near King's Lynn, Norfolk, England.

piles are constructed mid-way between them, parts of the bentonite–cement piles being bored out to allow the reinforced concrete piles to form an interlock.

Slurry trenches are used extensively as a means of groundwater cut-off. There are two types of slurry cut-off walls, namely, soil–bentonite and cement–bentonite. In the soil–bentonite cut-off process, the bentonite slurry, used to support the trench as it is excavated, is displaced by a soil–bentonite. The soil backfill forms a low permeability highly plastic cut-off wall. In the cement–bentonite cut-off process cement is added to a fully hydrated bentonite slurry. The addition of the cement causes the slurry to harden, giving it a strength comparable to that of stiff clay. Because it must have a low permeability, the continuity of slurry trench excavation is important. The depth of a slurry trench is controlled in many cases by the ground conditions, notably the depth to an impermeable formation, and the type of barrier to be constructed. The width is governed by the required permeability of the cut-off wall, the head of water across the wall, the size of the excavation equipment available and the materials that form the wall. However, because the filter cake developed at the walls of the trench provides the principal barrier to groundwater movement, the width of the trench is not a major factor. At sites where slurry trenches have been used in basement construction, pre-cast concrete panels have been lowered into the trenches and exposed by subsequent excavation to form the permanent basement walls.

Obviously, impermeability is the most important property of a cut-off wall. There are three factors that account for the impermeability of a slurry trench, namely, the ‘grouted’ zone, the filter cake and the backfill. Bentonite initially may permeate the pores of the adjacent soil, depending upon the particle size distribution of the pores. As bentonite is thixotropic, when left undisturbed, it gels and thereby seals the pores. The distance to which a soil is affected can range from more or less zero in dense clay to a metre or so in loosely packed sand and gravel. A filter cake of loosely packed bentonite particles, a few millimetres in thickness and with a permeability as low as $10^{-11}\,\mathrm{m\,s^{-1}}$, forms after a relatively short period at the slurry-soil interface and acts as a watertight membrane. The backfill consists of excavated soil, with the addition of material from a borrow pit, mixed with bentonite mud.

Diaphragm walls compare favourably in terms of watertightness, stiffness and mechanical strength, with cut-offs formed of steel sheet piling, pre-cast piles or cast-*in situ* piles (Millet and Perez, 1981). They may be rigid (concrete) or plastic (concrete and bentonite mixture) when load-carrying capacity is not required. Alternatively, diaphragm walls may be used as load bearing and retaining walls. In such cases they are reinforced by incorporating a steel cage. A diaphragm wall is constructed as a series of panels, the trench being excavated, between shallow guide walls, by a grab or hydrofraise (Fig. 2.26). During excavation the trench is kept filled with bentonite slurry to support the sides. If the wall is to be reinforced, then the steel cage is positioned when the excavation has been completed. Then concrete is tremied into the trench, displacing the bentonite. The depths of diaphragm walls have ranged up to 100 m. It may prove impossible to construct diaphragm walls by normal methods in soils with very low strengths. This is because the internal pressure of the bentonite occupying the trench may be less than the active pressure of the adjacent soil. For example, instability commonly occurs in soft marine clays. In fact, any soil in which $\phi = 0$ and the cohesion is less than 10 kPa should be treated with caution. Another disadvantage as far as the construction of a diaphragm wall is concerned is a high water table. The latter hinders the casting of guide walls. It also reduces the differential between the pressure inside the trench and the active pressure of the adjacent soil. Not only does this have a detrimental effect on the stability conditions but it also adversely influences the formation of filter cake. In ground where artesian pressure exists, even a slight tilt of the panels relative to each other or inadequate concreting could cause disastrous flooding in deep excavations.

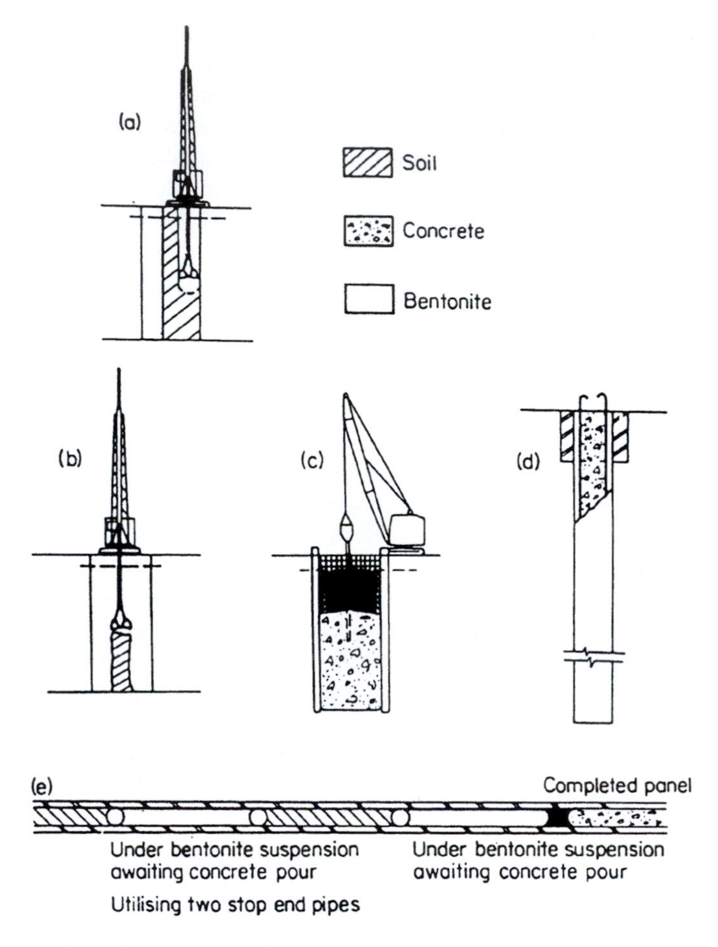

Figure 2.26 Construction of a reinforced diaphragm wall. General procedure for excavation of the panels during which time the excavation is kept filled with bentonite suspension (a) first one end, then the opposite end of the panel is excavated to full depth (b) third and last stage is the excavation of the centre panel. General procedure for concreting panels when the bentonite is displaced by concrete (c) steel stop-end pipes and reinforcement cage positioned and concrete placed through tremie pipe (d) section through completed panel showing guide walls and cage (e) illustrating variety of uses of steel stop-end pipes as required.

Artificial freezing can be employed as an impermeable barrier to stop the flow of groundwater into excavations. The technique is less sensitive to geological conditions than other methods of ground treatment and may be used in any moist soil irrespective of its structure, grain size or porosity (Jessberger, 1985). Weak, running or bouldery ground, and ground in which the pore water pressures are high, lend themselves to freezing. Soil profiles, too heterogeneous to grout predictably, also are amenable to freezing. However, in heterogeneous soils the frozen zone tends to be irregular in shape. For a certain refrigerant temperature the frozen ground is usually thinner in silts, clays and organic soils than in sands and gravels. Artificial freezing of ground usually is carried out in two stages, which are referred to as the active and passive stages of freezing. Active freezing involves freezing the ground to form the ice wall whilst passive freezing is that required to maintain the established thickness of the ice wall against thawing. The refrigeration plant has to operate at a much higher capacity during the active, than the passive stage of freezing.

Jet grouting offers a means of forming an impermeable barrier, as well as providing support (Coomber, 1986). It can be used in all types of soils, and poor soils in relatively inaccessible layers can be replaced. In its simplest form, jet grouting involves inserting an injection pipe into the soil to the required depth (Fig. 2.27). The soil then is subjected to a horizontally rotating jet of water and at the same time mixed with grout to form plastic soil-cement. The injection pipe is raised gradually. Replacement jet grouting involves removal of soil from the zone requiring treatment by a high-energy erosive jet of water and air. The grout is placed simultaneously. Construction of horizontal grouted diaphragms, which are connected to form impermeable cut-offs can be used when excavations extend beneath the water table and an impermeable formation does not exist at suitable depth. Such diaphragms should be 1–2 m thick and, because of uplift pressure, must be constructed at a significant distance below the

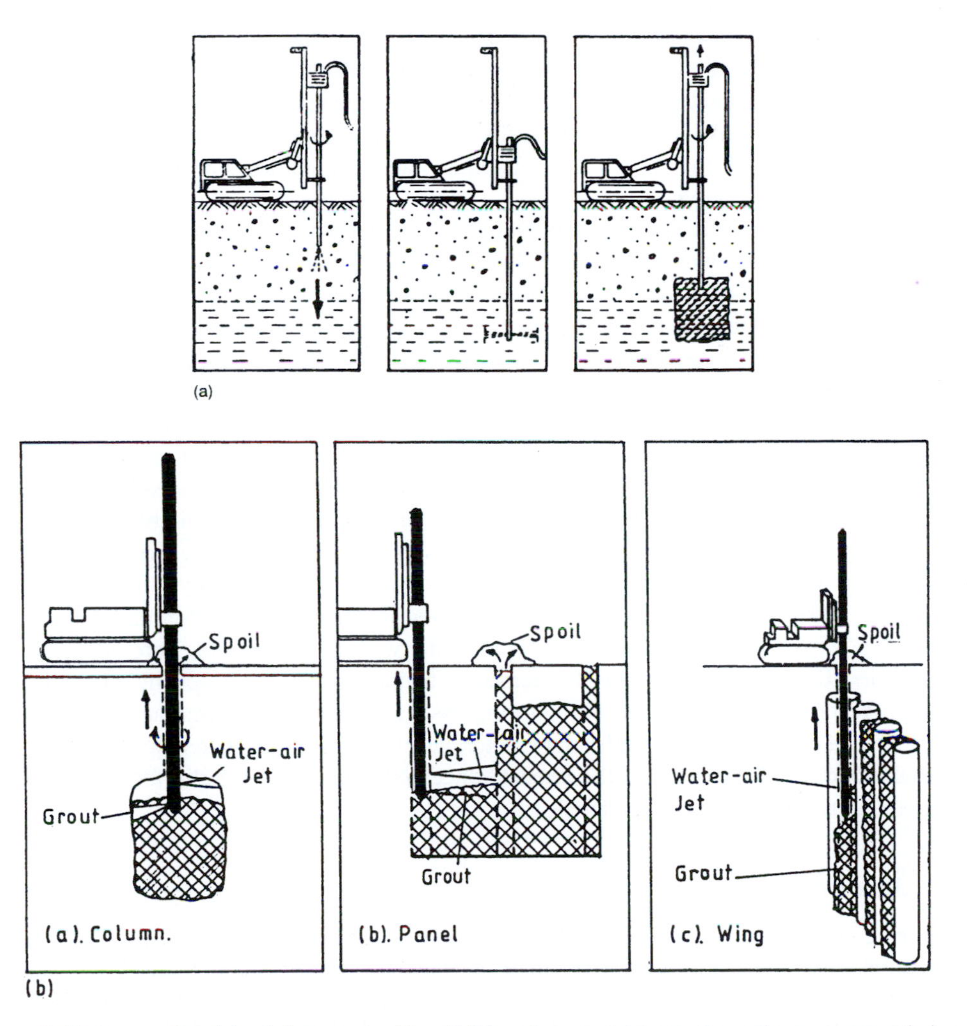

Figure 2.27 Jet grouting (a) mixing grout with soil (b) replacement jet grouting using compressed air and water to remove soil for simultaneous replacement by grout using column, panel and wing methods.

proposed base of the excavation in order to avoid blowout. The grout is placed at the required depth by a *tube-a-manchette*, the groutholes usually being spaced at 1–1.5 m and set out in a triangular pattern. The grout injected from adjacent holes merges to form the diaphragm. Thick diaphragms can be formed by grouting in more than one layer.

In rock masses cut-offs can be formed by the construction of a grout curtain, or in soft rocks by excavating a trench and backfilling with concrete. Alternatively, a row of pumped vertical wells can be sunk to form a well curtain that intercepts flow towards the excavation and provides 'dry' conditions on its downstream side.

2.9 Monitoring slopes

Dunnicliff (1992) emphasized that a monitoring programme for slopes should begin with defining the objective and end with planning how the results of the programme are to be implemented. He noted a number of steps that should be followed in the planning and execution of such a programme. One of the primary requirements is to assess the existing conditions and to determine possible causes of movement if satisfactory remedial treatment is to be undertaken. If the assessment indicates that sliding is a possibility, then three choices are available. First, is to do nothing and accept the consequences of failure. Second, a monitoring programme can be put into effect to warn of instability so that remedial treatment can be undertaken prior to failure occurring. Third, the slope can be stabilized and a monitoring programme installed to verify that the slope is stable.

Small movements usually precede slope failure, particularly a catastrophic failure, and accelerating displacement frequently precedes collapse. If these initial small movements are detected in sufficient time, remedial action can be taken to prevent or control further movement. A slope-monitoring system provides a means of early warning. Other adverse conditions that give rise to instability, notably excess pore water pressures, also require recording.

When there is a lack of adequate data, uncertainties are likely to arise in design. Under such circumstances if the stability of a slope is in doubt, then the expense of a monitoring programme may be justified, provided that remedial measures following the detection of incipient failure are feasible, and that the cost of monitoring and remedial action is less than the cost if a slope failure was to occur. Even if a complete picture of the ground conditions is available, the analytical methods may not be able to deal with the complexity of a real situation. Consequently, data must be simplified into an idealized model with resulting loss of accuracy. Such uncertainties normally are taken account of in the selection of the factor of safety. Monitoring can justify the use of a lower factor of safety than would otherwise be permissible, provided that it is accompanied by contingency plans for remedial action should the slope in question prove unstable. Accordingly, the cost of a monitoring system has to be measured against the cost of operating at an uneconomically high safety factor, necessitating either flatter slopes or expensive remedial work. The total value of the project concerned, and the cost and effect on the project if slopes fail also have to be considered.

2.9.1 *Monitoring movement*

Monitoring of movement provides a direct check on the stability of a slope. Instruments indicate the location, direction and maximum depth of movements, and the data obtained help determine the extent and depth of treatment that is necessary. What is more, the same instruments then can be used to determine the effect of this treatment. Burland *et al.* (1977) indicated that stable reference points must be established at a distance of at least three times the depth of an excavation away from its perimeter in order to monitor movements. Monitoring of surface movement can be done by conventional surveying techniques, the use of electronic distance measurement or laser equipment providing accurate results. Surveys should be designed to suit the topography and the anticipated directions of movements. Surveying should extend beyond the limits of possible movement into the surrounding stable area. In this way any development of surface strain in advance of the appearance of tension cracks can be detected, as can any toe heave. Automated slope-monitoring procedures, using total station surveying instruments can be programmed to take various measurements across a slope (Tran-Duc *et al.*, 1992). Precise results also can be obtained by using close-up photographs taken from ground stations that are then measured in a stereocomparator. Movements may be revealed by examination of a sequence of photographs taken at suitable intervals of time. Photographs can be used to evaluate pre-existing ground topography, for back-analysis of previous landslips and as a basis for engineering geological mapping.

The appearance of tension cracks at the crest of a slope may provide the first indication of instability. Crack measurements, that is, their width and vertical offsets, should be taken since they may provide an indication of slope behaviour. Dunnicliff (1993) described a number of ways in which cracks can be measured. These included using a survey tape, a tensional wire crack gauge or a surface extensometer. The latter consists of a sleeved rod that spans between anchor points on each side of the tension crack, and a mechanical or electrical transducer.

Single-point tiltmeters also can be used to monitor surface movements, when the surface of a slide has a rotational component. Portable tiltmeters can be used with a series of reference places, or tiltmeters can be left in place and connected to a datalogger. Dunnicliff (1992) mentioned multipoint liquid level gauges that have been employed to detect slope movements.

Measurements of subsurface horizontal movements are more important than measurements of subsurface vertical movements. Subsurface movements can be recorded by using settlement gauges, extensometers, inclinometers and deflectometers. Settlement gauges record vertical displacement. Borehole extensometers are used to measure vertical displacement of the ground at different depths. Fixed borehole extensometers include the single and multirod types. A single-rod extensometer is anchored in a borehole, and movement between the rod and the reference sleeve is monitored. However, the rod can bind in the sleeve if significant deformation occurs perpendicular to the length of the extensometer. Their use therefore is limited to where the borehole is almost parallel to the expected direction of movement. Hence, it rarely is practical to install them from the ground surface. Multirod installations monitor displacements at various depths using rods of varying lengths. Each rod is isolated by a close-fitting sleeve and the complete assembly is grouted into place, fixing the anchors to the ground while allowing free movement of each rod within its sleeve (Fig. 1.48a). A precise borehole extensometer has been described by Burland *et al.* (1977) and essentially consists of circular magnets embedded in the ground, which act as markers, and reed switch sensors move in a central access tube to locate the positions of the magnets (Fig. 1.48b). A slope extensometer is a multipoint borehole extensometer that uses tensioned wires instead of rods to monitor deformation perpendicular to the axis of the borehole. Up to ten anchors and wires can be installed in a borehole. At the head of the extensometer each wire passes over a pulley and is attached to a weight. When shear deformation occurs between two anchors, there are no vertical movements of the weights for those anchors above the shear zone. On the other hand, weights attached to anchors below the shear zone move downwards as the wires are displaced by shear. However, the precision with which the shear zone can be located is much less than for inclinometers.

An inclinometer is used to measure horizontal movements below ground (Mikkelsen, 1996). High-accuracy inclinometer measurements frequently represent the initial data relating to subsurface movement. Inclinometers can detect differential movements of 0.17–3.4 mm per 10 m run of hole. Inclinometers designed for permanent installation in a hole usually comprise a chain of pivoted rods. Angular movements between rods may be measured at the pivot points. Another type of permanently installed inclinometer uses a flexible metal strip onto which resistance strain gauges are bonded, which record any bending in the strip induced by ground movements. Fixed-position inclinometers monitor differential lateral movement between the borehole collar and a datum at depth, and are used most frequently in slope stability work. Large diameter inclinometer casing should be used for monitoring slides so as to provide a longer period over which readings can be taken as shear deformation occurs. Probe inclinometers are inserted into a special casing in a borehole each time a set of readings is required (Fig. 1.49). They incorporate a pendulum, the deflection of which indicates movement. Automatic remote winching systems have been developed that permit the inclinometer probe to be lowered and raised at pre-determined intervals and the data transferred to truck-mounted recorders (Lollino, 1992). An in-place inclinometer consists of a series of gravity-sensing transducers (tiltmeters), positioned at intervals along the casing, joined by articulated rods. Borehole deformation data are determined from the distances between the transducers and the measured changes in inclination.

Multiple deflectometers operate on a similar principle to in-place inclinometers but rotation is measured by angle transducers instead of tilt transducers. They usually are installed within the inclinometer casing. These deflectometers can detect shear deformation across a borehole at any inclination.

2.9.2 Monitoring load

Anchors, rock bolts and retaining walls, although designed to a prescribed working load, only develop this load as the material they support starts to move against them. Monitoring of loads and pressures indicates whether the support system has been adequately designed and also can show whether a slope is progressing towards a more stable or an unstable condition. Loads on rock anchors and rock bolts can be monitored by load cells.

Contact pressures on retaining walls can be recorded by pressure cells. A flat jack directly records changes in pressure. Stress may be measured directly by a hydraulically operated diaphragm earth pressure cell such as the Glotzl cell.

2.9.3 Monitoring groundwater

Groundwater is one of the most influential factors governing the stability of slopes. Instability problems may be associated with either excessive discharge or excessive pore water pressure. Pore water pressures are recorded by a piezometer, of which there are several types, the simplest type comprising a standpipe installed in a borehole (Fig. 1.47). If the minimum head recorded is less than 8 m below ground level, ‘closed system’ piezometers connected to mercury manometers normally are used. Pressure transducers are necessary where greater heads have to be measured, especially in low permeability ground. These instruments respond to changes of pressures acting on a flexible diaphragm by recording diaphragm deflections, which are converted to pressure values. Only very small quantities of water are required to produce full-scale deflection and such piezometers therefore are especially helpful when an almost instantaneous response is required.

There are important differences between monitoring water pressures in rock and in soil. Usually, in rock the majority of flow takes place via discontinuities rather than through intergranular pore space. The predominance of fissure flow means that piezometer heads in rock slopes often vary considerably from point to point and therefore a sufficient number of piezometers must be installed to define the overall conditions. They should be located with reference to the geology, especially with regard to the intersection of major discontinuities in rock masses. This can be facilitated by examining the fracture index, by inspecting the drillhole with a television camera, by packer testing or by logging the velocity of flow in the drillhole by using micro-propeller or dilution methods. The piezometer test section in rock, that is, the permeable filter material between sections of grouted hole, may need to be as long as 4 m in order to incorporate a representative number of water-bearing discontinuities.

2.9.4 Monitoring acoustic emissions or noise

Movements in rock or soil masses are accompanied by generation of acoustic emissions or noise. The detection of acoustic emissions is most effective when the amplitude of the signals is high. Hence, detection is more likely in rock masses or coarse-grained soils than fine-grained soils. Obviously, when a slope collapses, noise is audible, but sub-audible noises are produced at earlier stages in the development of instability. Normally, the rate of these microseismic occurrences increases rapidly with the development of instability. Such noises can be picked up by an array of geophones located in the vicinity of a slope or in shallow boreholes. Most movements generating noise originate near or along the plane of failure, so that seismic detection helps locate the depth and extent of the surface of sliding.

2.10 Methods of slope control and stabilization

If slides are to be prevented, then areas of potential sliding must first be identified, as must their type and possible amount of movement. Then, if the hazard is sufficiently real the engineer can devise a method of preventative treatment. Economic considerations, however, cannot be disregarded. In this respect, it is seldom economical to design cut slopes sufficiently flat to preclude the possibility of slides, and indeed many roads in upland terrain could not be constructed with the finance available without accepting some risk of slides. All the same this is no justification for lack of thorough investigation and adoption of all economical means of slide prevention.

Slide prevention may be brought about by reducing the activating forces or by increasing the forces resisting movement. The most frequently used methods of slope stabilization include retention systems, buttresses, slope modification and drainage. Properly designed retention systems can be used to stabilize most types of slopes where large volumes of earth materials are not involved and where lack of space excludes slope modification. The control of subsurface water frequently represents a major component in slope stabilization works (Walker and Mohen, 1987).

2.10.1 Rock fall treatment

It is rarely economical to design a rock slope so that no subsequent rock falls occur. Therefore except where absolute security is essential, slopes should be designed to allow small falls of rock under controlled conditions. The design of systems to prevent rock fall requires data concerning trajectory (height of bounce), velocity, impact energy and total volume of accumulation.

Single-mesh fencing supported by rigid posts will contain small rock falls. Larger heavy duty catch fences or nets are required for larger rock falls. Rolling rocks up to 0.6 m in diameter can be restrained by a chain-link fence but this can suffer severe damage when hit by rocks of this size and is not able to stop larger rocks (Ritchie, 1963). Fences have been developed so that when rocks collide with them, the nets engage energy absorbing friction brakes that extend the time of collision and in this way increase the capacity of the nets to restrain the falling rocks (Fig. 2.28). The braking device is incorporated into tie-backs and cables in the catch fence.

Wire meshing of rock slopes is one of the most effective methods of preventing rock falls from steep slopes (Fig. 2.29). The wire panels are laced together with binding wire. If a more robust method of linking panels is required, then horizontal and vertical steel cables can be shackled to the mesh and fixed to hooks and dowels. The latter provide a strong cable grid at 2–3 m centres, offering a greater resistance to large-scale block movement. The use of cable lashing and cable nets to restrain loose rock blocks was referred to by Piteau and Peckover (1978). Areas of potential instability can be covered with mesh fixed with light cables. High-capacity horizontal cables then are strung across the rock face using anchoring and tensioning methods. Wire mesh suspended from the top of the face provides another method for controlling rock fall.

A rock trap in the form of a ditch and/or barrier can be installed at the foot of a slope. Ritchie (1963) provided a guide for the dimensions of such ditches (Fig. 2.30). These dimensions can be reduced if the bottom of the ditch is filled with gravel to reduce bounce, if a barrier also is used, if the face is netted or if it is excavated in soft rocks. Benches also may act as traps to retain rock fall, especially if a barrier is placed at their edge. Where a road or railway passes along the foot of a steep slope, then protection from rock fall can be provided by the construction of a rigid canopy from the face (Fig. 2.31).

2.10.2 Alteration of slope geometry

Often altering the geometry of a slope is the most efficient way of increasing the factor of safety (Leventhal and Mostyn, 1987). However, such an approach may not be easy to adopt where the geometry is determined by engineering constraints or where a potentially unstable

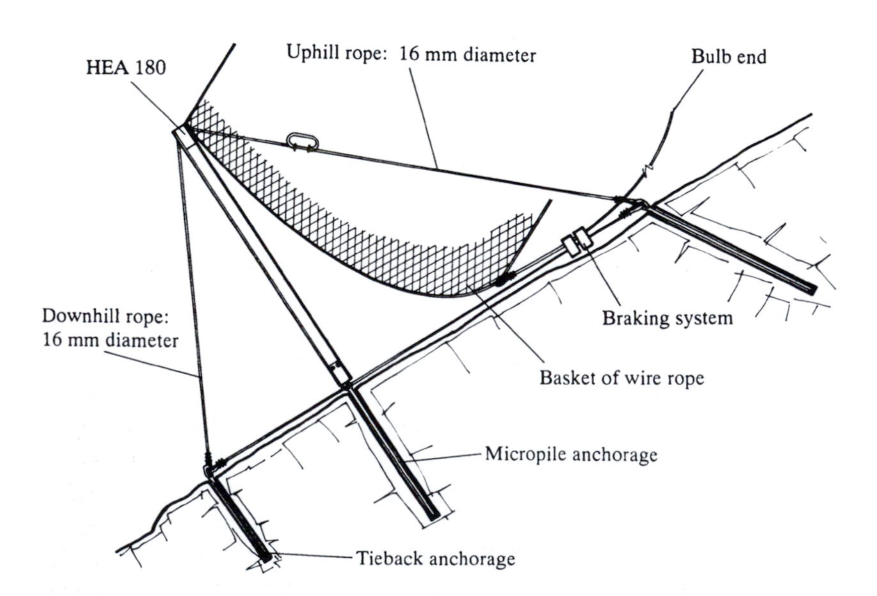

Figure 2.28 A yielding catch-fence.

Figure 2.29 Netting used to prevent rock fall, east of Stockton, California.

area is complex so that a change in topography that improves the stability of one area adversely affects the stability of another. Unstable material may be removed and, if necessary, replaced by stronger material. Alternatively, unstable material can be removed from near the crest of the slope and material added to the toe. In fact, it usually is more practical to load the toe. The material used for the construction of the toe fill should be free-draining or adequate internal drainage should be incorporated (Bell and Maud, 1996).

Benching brings about stability by dividing a slope into segments. Benches ideally should be over 5 m wide to allow access for inspection and therefore should be kept clear. If rock faces are to be scaled efficiently, then benches should not be higher than 12 m. Drainage systems can be installed on benches.

2.10.3 Reinforcement of slopes

If some form of reinforcement is required to provide support for a slope, then it is advisable to install it as quickly as possible after excavation. Dentition refers to masonry or concrete infill

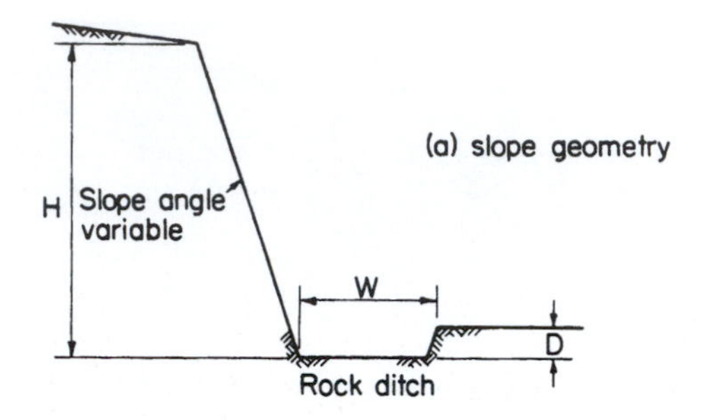

Variables in ditch design for rockfall areas

Rock slope (g) *angle*	*Height* (m) *H*	*Ditch width* (m) *W*	*Ditch depth* (m) *D*	*Notes*
90	5–10	3	1	
	10–20	5	1.5	1
	>20	7	1.5	1
75	5–10	3	1	
	10–20	5	1.5	1
	20–30	7	2	2
	>30	8	2	2
65	5–10	3	1.5	1
	10–20	5	2	2
	20–30	7	2	2
	>30	8	2.5	2
55	0–10	3	1	
	10–20	5	1.5	1
	>20			
45	0–10	3	1	
	10–20	3	2	1
	>20	5	2	2

Notes:
1. **If dimension *D* is greater than 1, rock retaining fence should be used if ditch adjacent to highway.**
2. **Ditch dimension may be reduced to 1.5 if rock fence used.**

Figure 2.30 Design of ditches for rock fall protection.
Source: Ritchie, A.M. 1963. Reproduced by permission of the Transportation Research Board.

placed in fissures or cavities in a rock slope (Fig. 2.32). The use of the same rock material for the masonry to form the slope provides a more attractive finish than otherwise. It often is necessary to remove soft material from fissures and pack the void with permeable material prior to constructing the dentition. Drainage should be provided through the latter.

Thin- to medium-bedded rocks dipping parallel to a slope can be held in place by steel dowels, which are up to 2 m in length. Holes are drilled to beneath the potential slip surface and are normal to the bedding. The dowels are grouted into place and they are not stressed. They are used where low loads are needed to increase stability and where the joint surfaces are at least moderately rough. Deformation in the rock mass stretches the untensioned dowels until sufficient stress is developed to prevent further strain.

Rock bolts may be used as reinforcement to enhance the stability of slopes in jointed rock masses (Fig. 2.32). They provide additional strength on critical planes of weakness within the rock mass. Rock bolts inclined to the potential plane of failure provide greater resistance than those installed normal to the plane. Hence, the design of rock bolt systems depends on prior knowledge of the potential failure mode of a slope (Windsor, 1997). Design charts can be used to estimate the amount of support that has to be provided by the installation of rock bolts. Rock

Figure 2.31 Canopy constructed over the Coquihalla Highway, British Columbia.

bolts may be up to 8 m in length with a tensile working load of up to 100 kN. Bearing plates, light steel sections or steel mesh may be used between bolts to support the rock face. Rock bolts are put into tension so that the compression induced in the rock mass improves shearing resistance on potential failure planes. They are anchored in stable rock beneath potential failure surfaces, the anchorage being provided by mechanical or grouted anchors. When using rock bolts consideration should be given to the influence of uplift and pore water pressures due to water in fractures, also to the effect of this water freezing. In order to counteract these factors the fractures should be grouted and the slope drained by inclined drillholes.

Rock anchors are used for major stabilization works, especially in conjunction with retaining structures (Fig. 2.33). They may exceed 30 m in length. As far as rock anchors are concerned, Littlejohn (1990) noted that there has been a trend towards higher load capacities for individual and concentrated groups of anchors. For example, pre-stressing of the order of 200 t m^{-1} has been used at dam sites, which means that the capacity of an individual anchor is well in excess of 1000 t. Because the stress levels are far greater than those involved in rock bolting, anchor loads are more dependent upon rock type and structure. Because of the risk of laminar failure or of excessive anchor movement, the lengths of closely spaced anchors can be staggered in order to reduce the intensity of stress across discontinuities (Weerasinghe and Adams, 1997). Anchors also can be installed at different inclinations in order to dissipate the load within a rock mass.

Gunite or shotcrete is used frequently to preserve the integrity of a rock face by sealing the surface and inhibiting the action of weathering (Fig. 2.32). The former is pneumatically applied mortar, the latter pneumatically applied concrete. Compressive strengths of up to 40 MPa can be developed within 28 days, and flexural strengths varying between 550 and 700 kPa are normal within that period of time. The modulus of elasticity exceeds 35 GPa and shrinkage varies between 0.03 and 0.106 per cent. Gunite/shotcrete adapts to the surface configuration and can be coloured to match the colour of the surrounding rocks. Coatings may be reinforced

Figure 2.32 Rock bolts, shotcrete and dentition used to stabilize a cutting in limestone along the A55 road in North Wales.

with glass or steel fibres and used in combination with wire mesh and/or rock bolts, or rock anchors. Groundwater must be allowed to drain through the protective cover otherwise it may be affected by frost action and groundwater pressures within the rock mass. It generally is considered that such surface treatment offers negligible support to the overall slope structure. Heavily fractured rocks may be grouted in order to stabilize them.

Figure 2.33 Anchors used with concrete panels to stabilize a cutting at the entrance to a tunnel on the A55 road, North Wales.

2.10.4 Restraining structures

Restraining structures control sliding by increasing the resistance to movement. They include retaining walls, cribs, gabions, buttresses and piling. The following minimum information is required to determine the type and size of a restraining structure:

1. The boundaries and depth of the unstable area, its moisture content and its relative stability, for example, excessive pore water pressures are likely to give difficulties in designing retaining walls.
2. The type of slide that is likely to develop.
3. The foundation conditions since restraining structures require a satisfactory anchorage.

Retaining walls often are used where there is a lack of space for the full development of a slope, such as along many roads and railways. As retaining walls are subjected to unfavourable loading, a large wall width is necessary to increase slope stability, which means that they are expensive. Retaining structures should be designed for a pre-determined load that they have to transmit to a foundation of known bearing capacity. They are located at the foot of a slope and should include adequate provision for drainage, for example, weep holes through the wall and pipe drainage in any backfill. Drainage not only prevents the build-up of pore water pressures but also reduces the effects of frost. Nonetheless, there are certain limitations that must be considered before retaining walls are used as a method of slide control. These involve the ability of the structure to resist shearing action, overturning and sliding on or below the base of the structure.

The use of gravity walls to stabilize a slope generally is restricted in terms of height in that free-standing gravity walls have an upper limit of about 10 m, and slides of only modest proportions can be prevented or stabilized using this type of structure (Morgenstern, 1992). Free-standing gravity walls usually require their bedrock foundation to be located at shallow depth. Hence, stabilization of a slope where the failure surface is deep or where the forces are larger than can be carried by a gravity wall may be brought about by the installation of anchors founded below the potential failure surface. Pre-stressed anchor walls actively oppose the movement of the soil or rock mass. Walls also can be formed of contiguous piles.

Reinforced earth can be used for retaining earth slopes (Fig. 2.34). Reinforced soil structures have advantages over traditional retaining walls in that they are flexible and so can tolerate large deformations, are resistant to seismic loadings and are often less costly to construct. Thus, reinforced earth can be used on poor ground where conventional alternatives would require expensive foundations. Reinforced earth structures are constructed by erecting facing panels at the face of the wall at the same time the earth is placed. Strips of galvanized steel or geosynthetic materials, notably geogrids, are fixed to the panels. The system relies on the transfer of shear forces to mobilize the tensile capacity of the closely spaced reinforcing strips. However, galvanized steel is subject to corrosion with time and so its inclusion is restricted to granular backfills that are free-draining. Epoxy-coated steel reinforcements offer higher resistance to corrosion. Geogrids possess low stiffness compared with steel so that the amount of deformation required to develop maximum shear strength may exceed the allowable deformation of the soil structure. Grid reinforcement systems consist of polymer or metallic elements arranged in rectangular grids, metallic bar mats and wire mesh (Mitchell and Christopher, 1990). The passive resistance developed on the cross members means that grids are more resistant to pull-out than strips. However, full passive resistance only develops after large deformations (50–100 mm). Sheet reinforcement commonly consists of geotextiles that are placed horizontally between layers of granular soil. In this case, facing elements are formed by wrapping the geotextile

Figure 2.34 Reinforced earth used to stabilize a cut-and-fill embankment along the road to Katse Reservoir, northern Lesotho.

around the face of the soil and covering the exposed geofabric with gunite, shotcrete or asphalt emulsion. In anchored earth, passive resistance is developed against anchors at the ends of the reinforcing bars, the other end of the bars being attached to the concrete facing panels (Jones *et al.*, 1985).

Soil nailing has been used to retain slopes, the nails consisting of steel bars, metal rods or metal tubes that are driven into *in situ* soil or soft rock or grouted into bored holes (Fig. 2.35). The nails are passive elements that are not post-tensioned. Normally, one nail is used for each 1 to 6 m^2 of ground surface. The ground surface between the nails is covered with a layer of shotcrete reinforced with wire mesh. Soil nailing is most effective in dense granular soils and low plasticity stiff silty clays. It generally is not cost effective, according to Mitchell and Christopher (1990), in loose sandy soils, poorly graded sandy soils, soft fine-grained soils or highly plastic clays.

Root piles are cast-in-place reinforced concrete piles that vary in diameter from 75 to 300 mm. The root pile system utilizes micropiles to form a monolithic block of reinforced soil that extends beneath a potential failure surface (Lizzi, 1977). The reinforcement provided by the micropiles is strongly influenced by their three-dimensional root-like geometrical arrangement (Fig. 2.36).

Cribs may be constructed of pre-cast reinforced concrete or steel units set up in cells that are filled with gravel or stone (Fig. 2.37). Crib walls offer rapid and easy construction even in difficult terrain. The system is reasonably flexible due to the segmental nature of the elements that comprise the walls and therefore is not particularly sensitive to differential settlements. Plant growth can occur on the faces of crib walls, which masks their presence.

Gabions consist of strong wire mesh surrounding placed stones (Fig. 2.38). Like crib walls, gabions also can be constructed readily, especially in difficult terrain. Gabions also are flexible.

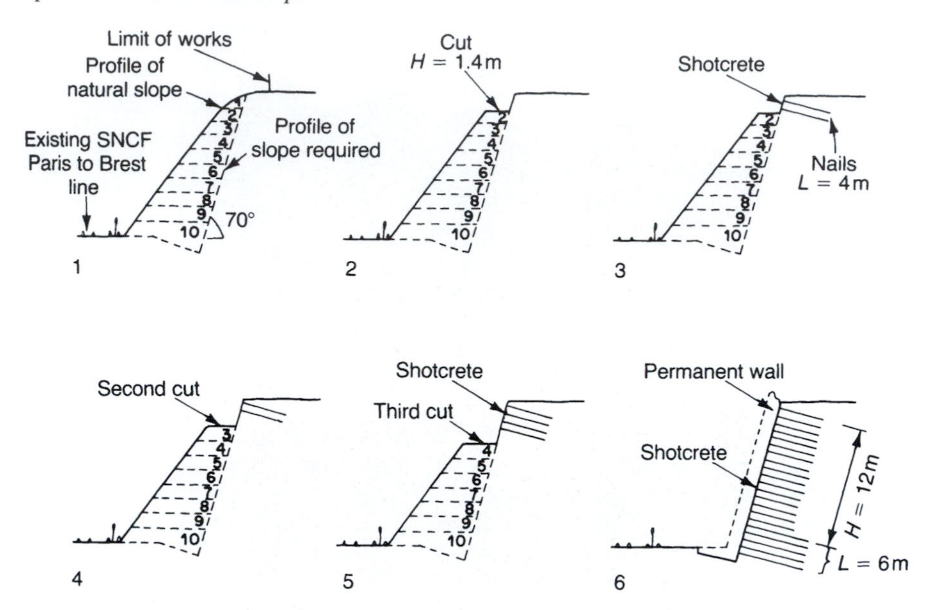

Figure 2.35 Soil nailing sequence in a cutting.

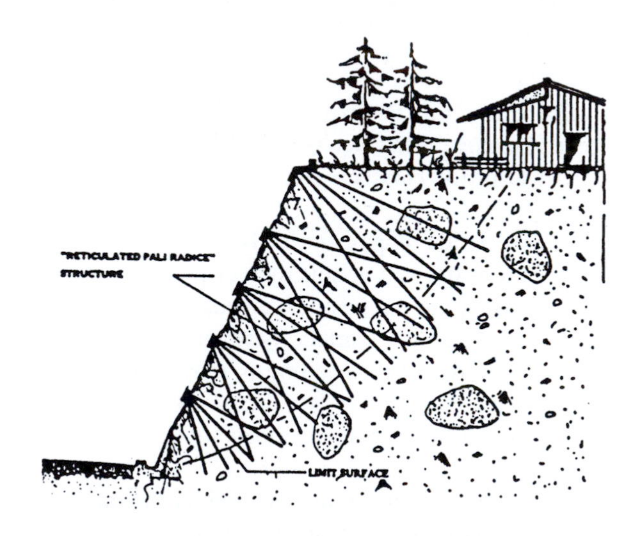

Figure 2.36 Reticulated micropiles used for slope stabilization.
Source: Lizzi, 1977.

The gabion filling provides for good subsurface drainage conditions in the vicinity of the wall, and filtration protection between the gabions and the wall backfill or soil can be afforded by geotextiles.

Concrete buttresses occasionally have been used to support large blocks of rock, usually where they overhang. Piles have been used as a method of controlling slides but they have not always been successful.

Figure 2.37 Crib wall in Hamilton, North Island, New Zealand.

2.10.5 Drainage

Drainage is the most generally applicable method for improving the stability of slopes since it reduces the effectiveness of one of the principal causes of instability, namely, excess pore water pressure. In rock masses groundwater also tends to reduce the shear strength along discontinuities. Surface run-off should not be allowed to flow unrestrained over a slope. This usually is prevented by the installation of a drainage ditch at the top of an excavated slope to collect water drainage from above. The ditch, especially in soils, should be lined to prevent erosion, otherwise it will act as a tension crack. It may be filled with cobble aggregate. Herringbone ditch drainage generally is employed to convey water from the surfaces of slopes. These drainage ditches lead into an interceptor drain at the foot of the slope (Fig. 2.39). Infiltration can be lowered by sealing the cracks in a slope by regrading or by filling with cement, bitumen or clay. A surface covering has a similar purpose and function.

Water may be prevented from reaching a zone of potential instability by a cut-off. Cut-offs may take the form of a trench backfilled with asphalt or concrete, of sheet piling, of a grout curtain or a well curtain whereby water is pumped from a row of vertical wells. Such barriers may be considered where there is a likelihood of internal erosion of soft material taking place due to increased flow of water attributable to drainage measures.

Figure 2.38 Gabions used to stabilize a slope near Port St John's, South Africa.

Trench drains are filled with free draining materials and may be lined with geotextiles. They are used for shallow subsurface drainage. Lew and Graham (1988) described the use of sand drains, 0.85 m in diameter, extending from the base of a trench drain, to stabilize a slope.

Deep wells are used to drain slopes where the depths involved are too deep for the construction of trench drains. For example, Bianco and Bruce (1991) described the use of large diameter

Figure 2.39 Drainage of a slope by gravel-filled drains leading into a toe trench.

vertical wells (up to 2 m diameter), spaced at 5–20 m centres and extending to depths of 50 m. The wells were connected to each other near their bases by means of a horizontal drillhole. The latter was lined with PVC pipe, 100 mm in diameter, and served as a gravity-collector drain. Two out of every three wells were filled with free draining material, the third remaining open for monitoring flow rates. Usually, the collected water flows from the base of the drainage wells under gravity but occasionally pumps may be installed in the bottom of wells to remove water.

Successful use of subsurface drainage depends on tapping the source of water, the presence of permeable material that aids free drainage, the location of the drain on relatively unyielding material to ensure continuous operation (flexible PVC drains are now frequently used) and the installation of a filter to minimize silting in the drainage channel. Drainage galleries are costly to construct, generally being constructed by tunnelling techniques. Most galleries are designed to drain by gravity (Fig. 2.40). Drain holes may be made about the perimeter of a gallery to enhance drainage.

Subhorizontal drainage holes are much cheaper than galleries and are satisfactory over short lengths but it is more difficult to intercept water-bearing layers with them. Subhorizontal drains can be inserted from the ground surface or by drilling from drainage galleries, large diameter wells or caissons. The drain hole is typically 120–150 mm in diameter (Sembenelli, 1988). The drain holes are lined with slotted PVC pipe. The pipes are lined on the outside with a geotextile covering that acts as a filter. When individual benches are drained by horizontal holes, the latter should lead into a properly graded interceptor trench, which is lined with impermeable material.

Schuster (1992) mentioned electro-osmotic dewatering, vacuum dewatering and siphon drainage as techniques that have been employed for slope stabilization. However, their use is very infrequent.

Figure 2.40 Internal drainage gallery in a restored slope, Aberfan area, South Wales.

2.10.6 The use of vegetation

Vegetation influences slope stability through its effect on the soil moisture regime, vegetation intercepting rainfall, as well as drawing moisture from soil by way of evapotranspiration and so enhancing soil strength. The most satisfactory trees in this regard are those that consume most water and have high transpiration rates, this means that deciduous trees are better than conifers. Plant residues reduce the impact of rainfall. In addition, the roots of bushes and trees also enhance the strength of soil (Wu, 1995). Afforestation of slopes may help prevent shallow slides but it cannot prevent the occurrence of deep-seated movements. However, as Rickson (1995) pointed out, some slopes may suffer erosion and instability if they are not protected immediately. In such cases, simulated vegetation in the form of mulches and geotextiles can be employed. Mulch materials are diverse, ranging from crop residues to wood shavings to paper. They can be applied to slopes hydraulically in the form of slurries that also contain seeds, fertilizers and soil binders. Natural geotextiles made from vegetative material may take the form of mats filled with mulch and seeds, which are laid over a slope and held in place by wooden pegs or steel staples.

2.11 A note on cofferdams and caissons

Cofferdams and caissons are used for excavations either in waterlogged or unstable ground that requires lateral support, or when excavations are made below water level (Schroeder, 1987). In other words, they enclose an area down to foundation level from which water is excluded sufficiently to permit excavation and placement of a foundation structure to take place. A cofferdam is a temporary feature whereas a caisson normally is incorporated subsequently into the permanent structure. The principal factors influencing the choice between cofferdams and caissons are, first, the ground conditions and, second, the proposed depth of the foundation level. The construction of caissons today is not as common as it was in the first half of the twentieth century.

Generally, cofferdams are used up to depths of 18 m below water level, depending on the character of the soil. Because they need not exclude all water, indeed it is usually uneconomic to do so, pumping may be necessary. In fact, the water level may be lowered below excavation level by abstracting from wells inside the cofferdam. This is particularly necessary if there is a possibility of a quick condition developing. However, greater passive resistance will be offered at the cofferdam walls if the groundwater level is lowered appreciably below excavation level. This accordingly affects the depth of the cut-off. In permeable soils the groundwater level may be reduced by wells outside the cofferdam.

Open caissons may be used for foundations that extend to and even exceed 45 m below surface water or ground level. However, they are difficult to pass through rock, or soil containing large boulders. What is more, they are not usually employed in hard clays owing to the enormous weight needed to overcome skin friction. A box caisson is one that is closed at the bottom and open at the top, and a pneumatic caisson contains a working chamber in which compressed air excludes the entry of water and loose or soft ground. Caissons may be constructed of reinforced concrete or steel segments.

In dense sand it may be possible to pump water from a caisson without destroying the stability of the floor since the deformation attributable to the seepage pressure does not bring about an increase in the pore water pressure. However, in loose sand, pumping causes uplift pressure that, in turn, causes liquefaction. In order to prevent loss of ground or serious settlement immediately outside an open caisson in wet loose sand, due to sand moving into the bottom of the caisson, the water level inside the caisson can be maintained a metre or more above the water level outside, the sand being excavated by grab. This excess head means that water flows from the caisson and counteracts the tendency of the sand to rise. If pumping from sumps within a caisson is not possible, then abstraction using deep wells or wellpoints should be considered or excavation can be carried out in a slurry-filled hole or by using compressed air in a pneumatic caisson.

As a caisson is advanced into the ground, the downward movement is resisted by skin friction. Adding dead weights to the top of a caisson is a tedious procedure and so the weight of concrete caissons usually exceeds the resistance of skin friction during each stage of construction. Values of skin friction for different soil types were provided by Terzaghi and Peck (1967) and are given in Table 2.10. The skin friction in clay soils can be reduced by painting the outside walls of caissons with a lubricating coat of oil. Clay grout also can be used as a lubricant.

Table 2.10 Values of skin friction

Type of soil	*Skin friction* (kPa)
Silt and soft clay	7–30
Very stiff clay	50–200
Loose sand	12–36
Dense sand	33–67
Dense gravel	50–100

Source: Terzaghi and Peck, 1967. Reproduced by kind permission of John Wiley & Sons, Inc.

References

Adhikari, G.R. 2000. Empirical methods for the calculation of the specific charge for surface blast design. *Fragblast*, **4**, 19–33.

Afrouz, A.A. and Rostami, J. 1997. Semi-theoretical design of blasthole spacing for variable ground conditions. *Mining Technology*, **79**, 17–19.

Anon. 1972. *Blasting Practice*. Nobel's Explosives Co. Ltd, Stevenson, Ayrshire.

Anon. 1981. *Code of Practice on Earthworks, BS 6031*. British Standards Institution, London.

Atkinson, T. 1970. Ground preparation by ripping in open pit mining. *Mining Magazine*, **122**, 458–469.

Atkinson, T. 1971. Selection of open pit excavation and loading equipment. *Transactions Institution Mining Metallurgy*, **80**, Section A, Mining Industry, A101–A129.

Bandis, S., Lumsden, A.C. and Barton, N. 1981. Experimental studies of scale effects on the shear behaviour of rock joints. *International Journal Rock Mechanics and Mining Science & Geomechanical Abstacts*, **18**, 1–20.

Barton, N. 1974. Estimating the shear strength of rock joints. *Proceedings Third International Congress Rock Mechanics* (ISRM), Denver, **2**, 219–220.

Barton, N., Lien, E. and Lunde, J. 1974. Classification of rock masses for the design of tunnel support. *Rock Mechanics*, **6**,189–236.

Bell, F.G. 1993. *Engineering Treatment of Soils*. E & FN Spon, London.

Bell, F.G. and Cashman, P.M. 1986. Groundwater control by groundwater lowering. In: *Groundwater in Engineering Geology*, Engineering Geology Special Publication No. 3, Cripps, J.C., Bell, F.G. and Culshaw, M.G. (eds). The Geological Society, London, 471–486.

Bell, F.G. and Maud, R.R. 1996. Landslides associated with the Pietermaritzburg formation in the greater Durban area, South Africa. *Environmental and Engineering Geoscience*, **2**, 557–573.

Bell, F.G. and Mitchell, J.K. 1986. Control of groundwater by exclusion. In: *Groundwater in Engineering Geology*, Engineering Geology Special Publication No. 3, Cripps, J.C., Bell, F.G. and Culshaw, M.G. (eds). The Geological Society, London, 429–443.

Bianco, B. and Bruce, D.A. 1991. Large landslide stabilization by deep drainage wells. *Proceedings International Conference on Slope Stability Engineering: Applications and Development*, Isle of Wight, Thomas Telford Press, London, 319–326.

Bieniawski, Z.T. 1973. Engineering classification of jointed rock masses. *Transactions South African Institution Civil Engineers*, **15**, 335–344.

Bishop, A.W. 1955. The use of the slip circle in the stability analysis of earth slopes. *Geotechnique*, **5**, 7–17.

Bjerrum, L. and Eide, O. 1956. Stability of strutted excavations in clay. *Geotechnique*, **6**, 32–47.

Blair, D.P. 1999. Statistical models for ground vibration and airblast. *Fragblast*, **3**, 335–364.

Blair, D.P. and Armstrong, I.W. 1999. Spectral control of ground vibration using electronic delay detonators. *Fragblast*, **3**, 303–334.

Bromhead, E.N. 1992. *The Stability of Slopes*. Second Edition. Surrey University Press, London.

Burland, J.B. and Hancock, R.J.R. 1977. Underground car park at the House of Commons, London: geotechnical aspects. *The Structural Engineer*, **55**, 87–100.

Burland, J.B., Moore, J.F.A. and Longworth, T.I. 1977. A study of ground movement and progressive failure caused by a deep excavation in Oxford Clay. *Geotechnique*, **27**, 557–591.

Burland, J.B., Simpson, B. and St John, H.D. 1979. Movements around excavations in London Clay. In: *Design Parameters in Geotechnical Engineering*, British Geotechnical Society, London, 13–29.

Bye, A.R. and Bell, F.G. 2001. Stability assessment and slope design at Sandsloot open pit, South Africa. *International Journal Rock Mechanics and Mining Science*, **38**, 449–466.

Cashman, P. and Preene, M. 2001. *Groundwater Lowering in Construction: A Practical Guide*. Spon Press, London.

Chan, D.H. and Morgenstern, N.R. 1987. Analysis of progressive deformation of the Edmonton Convention Centre excavation. *Canadian Geotechnical Journal*, **24**, 430–440.

Church, H.K. 1981. *Excavation Handbook*. McGraw-Hill, New York.

Cole, K.W. and Burland, J.B. 1972. Observation of retaining wall movements associated with a large excavation. *Proceedings Fifth European Conference Soil Mechanics and Foundation Engineering*, Madrid, **1**, 445–453.

Coomber, D.B. 1986. Groundwater control by jet grouting. In: *Groundwater in Engineering Geology*, Engineering Geology Special Publication No. 3, Cripps, J.C., Bell, F.G. and Culshaw, M.G. (eds), The Geological Society, London, 445–454.

Creed, M.J., Simons, N.E. and Sills, G.C. 1981. Back analysis of the behaviour of a diaphragm wall supported excavation in London Clay. *Proceedings Second International Conference on Ground Movements and Structures*, Cardiff, Geddes, J.D. (ed.), Pentech Press, London, 743–759.

Duncan, J.M. 1996. Soil slope stability analysis. In: *Landslides: Investigation and Mitigation*, Special Report 247, Turner, A.K. and Schuster, R.L. (eds), Transportation Research Board, National Research Council, National Academy Press, Washington, DC, 337–371.

Dunnicliff, J. 1992. Monitoring and instrumentation of landslides. *Proceedings Sixth International Symposium on Landslides*, Christchurch, Bell, D.H. (ed.), Balkema, Rotterdam, **3**, 1881–1886.

Dunnicliff, J. 1993. *Geotechnical Instrumentation for Monitoring Field Performance*. Second Edition. Wiley, New York.

Duvall, W.I. 1964. Design requirements for instrumentation to record vibrations produced by blasting. *Report Investigation No. 6487*, US Bureau of Mines, Washington, DC.

Edwards, A.T. and Northfield, R.D. 1960. Experimental studies of the effects of blasting on structures. *The Engineer*, **210**, 539–546.

Fasiska, A.E., Wagenblast, H. and Dougherty, M.T. 1974. The oxidation mechanism of sulphide minerals. *Bulletin Association Engineering Geologists*, **11**, 75–82.

Fourney, W.L. 1993. Mechanisms of rock fragmentation by blasting. In: *Comprehensive Rock Engineering*, Brown, E.T., Fairchurst, C. and Hock, E. (eds), Pergamon Press, Oxford, **4**, 39–69.

Franklin, J.A., Broch, E. and Walton, G. 1971. Logging the mechanical character of rock. *Transactions Institution Mining Metallurgy*, Section A, Mining Industry, **80**, A1–A9.

Fredlund, D.G. 1984. Analytical methods for slope analysis. *Proceedings Fourth International Symposium on Landslides*, Toronto, **1**, 229–250.

Furby, J. 1964. Tests for rock drillability. *Mine and Quarry Engineering*, **30**, 292–298.

Gens, A, Hutchinson, J.N. and Cavounidis, S. 1988. Three dimensional analysis of slides in cohesive soils. *Geotechnique*, **38**, 1–23.

Hagan, T.N. 1979. Designing primary blasts for increased slope stability. *Proceedings Fourth International Congress Rock Mechanics* (ISRM), Montreux, **1**, 657–664.

Hagan, T.N. 1986. The influence of some controllable blast parameters upon muckpile characteristics and open pit mining costs. *Proceedings Conference on Large Open Pit Mining*, Australia Institute Mining Metallurgy/Institute Engineers, 123–132.

Hoek, E. 1970. Estimating the stability of excavated slopes in opencast mines. *Transactions Institution Mining and Metallurgy*, Section A, Mining Industry, **79**, A109–A132.

Hoek, E. 1971. The influence of structure on the stability of rock slopes. *Proceedings First Symposium on Stability in Open Pit Mining*, Vancouver, American Institute Mining Engineers, 49–63.

Hoek, E. and Bray, J.W. 1981. *Rock Slope Engineering*. Institution Mining and Metallurgy, London.

Hovland, J. 1977. Three dimensional slope stability analysis method. *Proceedings American Society Civil Engineers, Journal Geotechnical Engineering Division*, **103**, 971–986.

Huder, J. 1969. Deep braced excavation with high water level. *Proceedings Seventh International Conference Soil Mechanics and Foundation Engineering*, Mexico City, **2**, 443–448.

Hungr, O. 1987. An extension of Bishop's simplified method of slope stability analysis to three dimensions. *Geotechnique*, **37**, 113–117.

Irvine, D.F. and Smith, R.J.H. 1983. *Trenching Practice*. Construction Industry Research and Information Association, Report 97, London.

Jessberger, H.L. 1985. The application of ground freezing to soil improvement in engineering practice. In: *Recent Developments in Ground Improvement Techniques*. Balasubramaniam, A.S., Chandra, S., Bergado, D.T., Younger, J.S. and Prinzl, F. (eds), Balkema, Rotterdam, 469–482.

Jhanwar, J.C., Jethwa, J.L. and Reddy, A.H. 2000. Influence of air deck blasting on fragmentation in jointed rocks in an open-pit manganese mine. *Engineering Geology*, **57**, 13–29.

Jones, C.J.F.P., Murray, R.T., Temporal, J. and Mair, R.J. 1985. First application of anchored earth. *Proceedings Eleventh International Conference Soil Mechanics and Foundation Engineering*, *San Francisco*, **3**, 1709–1712.

Kahraman, S., Balci, C., Yazici, S. and Bilgin, N. 2000. Prediction of the penetration rate of rotary blasthole drills using a new drillability index. *International Journal Rock Mechanics and Mining Science*, **37**, 729–743.

Kirsten, H.A.D. 1982. A classification system for excavation in natural materials. *The Civil Engineer in South Africa*, **24**, 293–306.

Kummerle, R.P. and Benvie, D.A. 1988. Geologic considerations in rock excavations. *Bulletin Association Engineering Geology*, **25**, 105–120.

Langefors, U. and Kihlstrom, B. 1962. *The Modern Technique of Rock Blasting*. Wiley, New York.

Latham, J.-P., Munjiza, A. and Lu, P. 1999. Rock fragmentation by blasting – a literature study of research in the 1980's and 1990's. *Fragblast*, **3**, 193–212.

Leventhal, A.R. and Mostyn, G.R. 1987. Slope stabilization techniques and their application. *Proceedings Extension Course on Soil Slope Stability and Stabilization*, Walker, B.F. and Fell, R. (eds), Sydney, Balkema, Rotterdam, 121–181.

Lew, K.V. and Graham, J. 1988. Riverbank stabilization by drains in plastic clay. *Proceedings Fifth International Symposium on Landslides*, Lausanne, Bonnard, C. (ed.), **2**, 939–944.

Littlejohn, G.S. 1990. Ground anchorage practice. In: *Design and Performance of Earth Retaining Structures*, Lambe, P.C. and Hansen, L.A. (eds), Geotechnical Special Publication 25, American Society Civil Engineers, New York, 692–733.

Lizzi, F. 1977. Practical engineering in structurally complex formations. The *in situ* reinforced earth. *Proceeding International Symposium on the Geotechnics of Structurally Complex Formations*, Capri, 327–333.

Lollino, G. 1992. Automated inclinometric system. *Proceedings Sixth International Symposium on Landslides*, Christchurch, Bell, D.H. (ed.), **2**, 1147–1150.

MacGregor, F., Fell, R., Mostyn, G.R., Hocking, G. and McNally, G. 1994. The estimation of rock rippability. *Quarterly Journal Engineering Geology*, **27**, 123–144.

Mather, W. 1997. Bulk explosives. *Mining Technology*, **79**, 251–254.

McCann, D.M. and Fenning, P.J. 1995. Estimation of rippability and excavation conditions from seismic velocity measurements. In: *Engineering Geology and Construction*, Engineering Geology Special Publication No. 10, Eddleston, M., Walthall, S., Cripps, J.C. and Culshaw, M.G. (eds), Geological Society, London, 335–343.

McGregor, K. 1967. *The Drilling of Rock*, C.R. Books Ltd (A. McClaren and Co.), London.

Mikkelsen, P.E. 1996. Field instrumentation. In: *Landslides Investigation and Mitigation*, Transportation Research Board, Special Report 247, National Research Council, 278–318.

Millet, R.A. and Perez, J.Y. 1981. Current USA practice: slurry wall specifications. *Proceedings American Society Civil Engineers, Journal Geotechnical Engineering Division*, **107**, 1041–1056.

Minty, E.J. and Kearns, G.K. 1983. Rock mass workability. In: *Collected Case Histories in Engineering Geology, Hydrogeology, Environmental Geology*, Special Publication, Geological Society Australia, Sydney, 59–81.

Mitchell, J.K. and Christopher, B.R. 1990. North American practice in reinforced soil systems. In: *Design and Performance of Earth Retaining Structures*, Geotechnical Special Publication 25, American Society Civil Engineers, New York, 322–346.

Morgenstern, N.R. 1992. The role of analysis in the evaluation of slope stability. *Proceedings Sixth International Symposium on Landslides*, Christchurch, Bell, D.H. (ed.), Balkema.

Muller, B. 1997. Adapting blasting technologies to the characteristics of rock masses in order to improve blasting results and reduce blasting vibrations. *Fragblast*, **1**, 361–378.

Nash, D.F.T. 1987. A comparative review of limit equilibrium methods of stability analysis. In: *Slope Stability*, Anderson, M.G. and Richards, K.S. (eds), Wiley, New York.

Nicholls, H.R., Johnson, C.F. and Duvall, W.I. 1971. Blasting vibrations and their effects on structures. *United States Bureau Mines*, Bulletin 656, Washington, DC.

Norrish, N.I. and Wyllie, D.C. 1996. Rock slope stability analysis. In: *Landslides: Investigation and Mitigation*, Special Report 247, Turner, A.K. and Schuster, R.L. (eds), Transportation Research Board, National Research Council, National Academy Press, Washington DC, 391–425.

Oriard, L.L. 1972. Blasting operations in the urban environment. *Bulletin Association Engineering Geologists*, **9**, 27–46.

Ou, C.Y., Hseih, P.G. and Chiou, D.C. 1993. Characteristics of ground surface settlement during excavation. *Canadian Geotechnical Journal*, **30**, 759–767.

Ou, C.Y., Liao, J.T. and Cheng, W.L. 2000. Building response and ground movements induced by a deep excavation. *Geotechnique*, **50**, 209–220.

Ouchterlony, P., Olsson, M. and Bavik, S.O. 2000. Perimeter blasting in granite with holes with axial notches and radial bottom slots. *Fragblast*, **4**, 55–82.

Paine, R., Holmes, D. and Clarke, H. 1961. Controlling overbreak by pre-splitting. *Proceedings International Symposium on Mining Research, Rolla, Missouri*, Pergamon, New York, **1**, 179–209.

Peck, R.B. 1969. Deep excavation and tunnelling in soft ground. *Proceedings Seventh International Conference Soil Mechanics and Foundation Engineering*, Mexico City, State-of-the-Art Volume, 225–290.

Penner, E., Eden, J.W. and Gillott, J.E. 1973. Floor heave due to biochemical weathering of shale. *Proceeding Eighth International Conference Soil Mechanics and Foundation Engineering*, Moscow, **2**, 151–158.

Persson, R., Lundborg, N. and Johansson, C.H. 1970. The basic mechanism in rock blasting. *Proceedings Second International Congress* (ISRM), Belgrade, Paper 5–3, 19–33.

Pettifer, G.S. and Fookes, P.G. 1994. A revision of the graphical method for assessing the excavatability of rock. *Quarterly Journal Engineering Geology*, **27**, 145–164.

Piteau, D.R. and Peckover, F.L. 1978. Rock slope engineering. In: *Landslides: Analysis and Control.* Schuster, R.L. and Krizek, R.J. (eds), Transportation Research Board, Special Report 176, National Academy of Sciences, Washington, DC, 192–228.

Potts, D.M., Dounias, G.T. and Vaughan, P.R. 1990. Finite element analysis of Carsington embankment. *Geotechnique*, **40**, 79–102.

Powrie, W. and Roberts, T.O.L. 1990. Field trials of an ejector well dewatering system at Conwy, North Wales. *Quarterly Journal Engineering Geology*, **23**, 169–185.

Preene, M. 2000. Assessments of settlements caused by groundwater control. *Proceedings Institution Civil Engineers, Geotechnical Engineering*, **143**, 177–190.

Richards, R.L., Leg, O.M.M. and Whittle, R.A. 1978. Appraisal of stability in rock slopes. In: *Foundation Engineering in Difficult Ground*, Bell, F.G. (ed.), Butterworths, London, 449–512.

Rickson, R.J. 1995. Simulated vegetation and geotextiles. In: *Slope Stabilization and Erosion Control: A Bioengineering Approach*, Morgan, R.P.C. and Rickson, R.J. (eds), E & FN Spon, London, 95–131.

Ritchie, A.M. 1963. The evaluation of rockfall and its control. *Highway Research Board*, Record No. 17, Washington, DC, 13–28.

Roberts, A. 1971. Ground vibrations due to quarry blasting and other sources – an environmental factor in rock mechanics. *Proceedings Twelfth Symposium on Rock Mechanics*, Rolla, Missouri, American Institute Mining Engineers, New York, 427–456.

Roberts, A. 1981. *Applied Geotechnology*. Pergamon, Oxford.

Russell, E. 1997. More bang for your buck with today's explosives. *Tunnels and Tunnelling*, **29**, No. 1, 42–43.

Sanchidrian, J.A., Garcia-Bermundez, P. and Jimeco, C.L. 2000. Optimization of granite splitting by blasting using notched holes. *Fragblast*, **4**, 1–11.

Sandy, D.A. 1989. Drill, blast, load and haul practices at Rosing Mine, Namibia. *Transactions Institution Mining Metallurgy*, **98**, Section A, Mining Industry, A98–A104.

Schroeder, W.L. 1987. Caissons and cofferdams. In: *Ground Engineer's Reference Book*, Bell, F.G. (ed.), Butterworths, London, 40/1–40/16.

Schuster, R.L. 1992. Recent advances in slope stabilization. *Proceedings Sixth International Symposium on Landslides*, Christchurch, Bell, D.H. (ed.), Balkema, Rotterdam, **3**, 1715–1745.

Scoble, M.J. and Muftuoglu, Y.V. 1984. Derivation of a diggability index for surface mine equipment selection. *Mining Science and Technology*, **1**, 305–322.

Sembenelli, P. 1988. Stabilization and drainage. *Proceedings Fifth International Symposium on Landslides*, Lausanne, Bonnard, C. (ed.), Balkema, Rotterdam, **2**, 813–819.

Simons, N.E., Menzies, B. and Matthews, M.C. 2001. *A Short Course in Soil and Rock Slope Engineering*. Thomas Telford Press, London.

Sinclair, J. 1969. *Quarrying, Opencast and Alluvial Mining*. Elsevier, London.

Singh, S.P. 2001. The influence of geology on blast damage. *CIM Bulletin*, **94**, 121–127.

Skempton, A.W. 1964. Long-term stability of clay slopes. *Geotechnique*, **14**, 77–101.

Somerville, S.H. 1986. *Control of Groundwater for Temporary Works*. Construction Industry Research and Information Association, Report 113, London.

Starfield, A.M. 1966. Strain wave theory in rock blasting. *Proceedings Eighth Symposium Rock Mechanics*, Rolla, Pergamon, New York, Supplementary Paper No. 4.

Tatiya, R.R. and Al Ajmi, A. 2000. Estimation of the Atlas Copco relation between burden and blasthole diameter and rock strength at bench blasting – a case study. *International Journal Surface Mining, Reclamation and Environment*, **14**, 151–160.

Terzaghi, K. and Peck, R.B. 1967. *Soil Mechanics in Engineering Practice*. Wiley, New York.

Tomlinson, M.J. 1986. *Foundation Design and Construction*. Fifth Edition, Longman Scientific & Technical, Harlow, Essex.

Tran-Duc, P.O., Ohno, M. and Mawatari, Y. 1992. An automated landslide monitoring system. *Proceedings Sixth International Symposium on Landslides*, Christchurch, Bell, D.H. (ed.), Balkema, Rotterdam, **2**, 1163–1166.

Varnes, D.J. 1978. Slope movement types and processes. In: *Landslides, Analysis and Control*, Schuster, R.L. and Krizek, R.J. (eds), Transportation Research Board, National Academy of Sciences, Special Report 176, Washington, DC, 11–33.

Vutukuri, V.S. and Bhandari, S. 1961. Some aspects of design of open pits. *Colorado School Mines Quarterly*, **56**, 51–61.

Walker, B.F. and Mohen, F.J. 1987. Groundwater prediction and control, and negative pore pressure effects. *Proceedings Extension Course on Soil Slope Stability and Stabilization*, Walker, B.F. and Fell, R. (eds), Sydney, Balkema, Rotterdam, 121–181.

Weaver, J.M. 1975. Geological factors significant in the assessment of rippability. *The Civil Engineer in South Africa*, **17**, 313–316.

Weerasinghe, R.B. and Adams, D. 1997. A technical review of rock anchorage practice 1976–1996. In: *Ground Anchorages and Anchored Structures*, Littlejohn, G.S. (ed.), Thomas Telford Press, London, 481–491.

Windsor, C.R. 1997. Rock reinforcement systems. *International Journal Rock Mechanics and Mining Science*, **34**, 919–951.

Wu, T.H. 1995. Slope stabilization. In: *Slope Stabilization and Erosion Control: A Bioengineering Approach*, Morgan, R.P.C. and Rickson, R.J. (eds), E & FN Spon, London, 221–264.

Zeevaert, L. 1957. Foundation design and behaviour of the Tower Latino Americano in Mexico City. *Geotechnique*, **7**, 115–133.

3 Subsurface excavations

3.1 Introduction

Geology is the most important factor that determines the nature, form and cost of a tunnel (Taylor and Conwell, 1981). For example, the route, design and construction of a tunnel are largely dependent upon geological considerations. Accordingly, tunnelling is an uncertain and sometimes hazardous undertaking because information on ground conditions along the alignment is never complete, no matter how good the site investigation. Estimating the cost of tunnel construction, particularly in areas of geological complexity, therefore is uncertain.

Generally, geological investigations for tunnel sites are conducted in three stages. In the initial stage, a desk study is undertaken using available maps and aerial photographs to obtain an overall impression of the geological conditions and to plan subsequent investigations. The second stage requires a more detailed investigation and is geared to the determination of the feasibility of a particular location. At this stage consideration is given to alternative tunnel alignments. Once a tunnel site is selected, then investigation enters the third phase when special additional work is conducted to assist the final design and estimation of tunnel costs. The investigation should produce a geological map of the area and a cross section along the centre line of the tunnel. Wherever possible, the position of the water table should be shown on the section. The complexity of the surface geology determines the accuracy with which it can be projected to tunnel level. The subsurface geology is explored by means of pits, drifts, drilling and pilot tunnels. Exploration drifts driven before tunnelling proper commences are not usually resorted to unless a particular section appears to be especially dangerous or a great deal of uncertainty exists. Core drilling aids the interpretation of geological features already identified at the surface.

Geophysical investigations can give valuable assistance in determination of subsurface conditions, especially in areas in which the solid geology is poorly exposed. Seismic refraction has been used in measuring depths of overburden in the portal areas of tunnels, in locating faults, weathered zones or buried channels, and in estimating rock quality. Seismic testing also can be used to investigate the topography of a river bed and the interface between the alluvium and bedrock, as well as in sub-seabed investigation. For example, Arthur *et al.* (1997) described a seismic survey carried out to help assess the geological conditions along the route of the Channel Tunnel. This survey also involved geophysical logging of deep drillholes. Seismic logging of drillholes can, under favourable circumstances, provide data relating to the engineering properties of rock (McCann *et al.*, 1990). Resistivity techniques have proved useful in locating water tables and buried faults, particularly those that are saturated. Resistivity logs of boreholes are used in lateral correlation of layered materials of different resistivities and in the detection of permeable rocks. Ground probing radar offers the possibility of exploring large volumes of rock for anomalies in a short time and at low cost, in advance of major subsurface excavations.

A pilot tunnel is probably the best method of exploring tunnel locations and should be used if a major-sized tunnel is to be constructed in ground that is known to have critical geological conditions. It also drains the rock ahead of the main excavation. If the inflow of water is excessive, then rock can be grouted from the pilot tunnel before the main excavation reaches the water-bearing zone. What is more, a pilot tunnel also allows the detection of squeezing pressures in time to determine the required tunnel support and to revise the design for the permanent tunnel support.

Reliable information relating to the ground conditions ahead of the advancing face is obviously desirable during tunnel construction. This can be achieved with a varying degree of success by drilling long horizontal holes between shafts, or by direct drilling from the tunnel face at regular intervals. For example, drilling equipment for drilling in a forward direction can be incorporated into a tunnelling machine. In fact, probing ahead some 300 m, that is, several days ahead of a tunnel boring machine is possible (Williamson and Schmidt, 1972). The penetration rate of a probe drill must exceed that of the tunnel boring machine, ideally it should be about three times faster. In extremely poor ground conditions, tunnelling has to proceed behind an array of probe holes that fan outwards ahead of the tunnel face. This provides advance information on potentially hazardous conditions (notably fault zones, buried channels, weak seams or solution cavities), as well as warning of less drastic changes that, for instance, may entail an alteration in the type of support system. Maintaining the position of an individual probe hole, however, presents a problem when horizontal drilling is undertaken. In particular, variations in hardness of the ground oblique to the direction of drilling can cause radical deviations. Even in uniform ground if the drilling rods are significantly smaller than the bit (the usual practice), gravity combined with axial thrust in the rods leads to the hole going off line. Direction can be improved by employing larger diameter, rigid drilling rods or by rotating and advancing casing behind the bit. The inclination of a hole must be surveyed.

Because probing ahead of a tunnel face needs to be rapid, sampling is reduced to a minimum, the character of the ground being derived from the bit chippings (or auger parings if a continuous flight auger is used in soft ground). Drilling characteristics such as drill torque, speed of rotation and advance, rod feed thrust, change in drilling fluid pressure and loss of drilling water must be recorded. These indicate changes in the hardness of the ground, the presence of large discontinuities or cavities and the presence of water. A borehole or television camera can be used to explore a horizontal hole or it can be examined by means of geophysical logging techniques.

Inspection and mapping of strata should continue during tunnel construction. This information helps to complete the picture of the geological setting as revealed by the site investigation and may help the geologist predict any changing conditions in advance of the tunnel heading. Geological maps of tunnels may be made on scales as low as 1:200 or 1:500. Photographs, preferably in colour, can be taken at frequent intervals, especially where a change in lithology occurs. Tunnelling with a machine makes mapping difficult and it must be done as soon as exposures are available.

3.2 Geological conditions and tunnelling

3.2.1 Stress conditions and rock failure

Rock masses, especially those at depth, are affected by the weight of overburden and the stresses so developed cause them to be strained. In certain areas, particularly orogenic belts, the state of stress also is affected by tectonic factors. However, because the rocks at depth are confined they suffer partial strain. The stress that does not give rise to strain, that is, that which is not dissipated, remains in the rocks and is referred to as residual stress. Rocks encountered in tunnelling operations therefore have been stressed by the weight of overburden, past and present, and by any earth movements to which they have been or are being subjected. While the rocks remain in a confined condition the stresses will accumulate and may reach high values, sometimes in excess of their yield point. If the confining condition is removed, as in tunnelling, then the residual stress can cause displacement. The amount of movement depends upon the magnitude of the residual stress. The pressure relief, which represents a decrease in residual stress, may be instantaneous or slow in character, and is accompanied by movement of the rock mass with varying degrees of violence.

Accurate prediction of the residual stresses likely to be met with may be obtained with the aid of field loading tests, for example, using pairs of flat jacks in the horizontal and vertical directions. One of the early instances of the assessment of the natural state of stress and its significance was made by Moye (1964) at the Eucumberne-Tumat tunnel in the Snowy Mountains project, Australia. There the use of flat jacks in pairs in horizontal and vertical directions showed that the stress in the horizontal direction (18.6 MPa) was some 2.6 times that in the vertical direction. Moye maintained that because there were no major topographic irregularities in the area, the cause of the high horizontal stress must be related to the tectonic history.

Even if there are no residual stresses in a rock mass, the excavation of an opening allows the adjacent rock to move into it. In effect the rock is pulled away by tensile stresses from the rest of the mass that remains in place. Such action will frequently necessitate the provision of adequate support. Underground excavation therefore destroys the existing state of equilibrium in the material around an opening, and a new state of equilibrium is established. This is done by developing self-balanced systems of shearing stresses that give rise to arching around a tunnel. An appraisal of the arching capacity of the rocks around a proposed tunnel must be assessed by the geological investigation preceding construction. Obviously, in badly fractured rocks, arching patterns tend to be poorly developed whereas massive igneous rocks generally offer favourable arching possibilities. This is also the case in horizontal or gently dipping sedimentary rocks where the strike is parallel to the tunnel axis and in steeply dipping formations where the strike is normal to the tunnel axis.

Around the walls of a tunnel the radial stress is zero whilst the circumferential stress is twice the usual stress. With increasing distance from the walls the former increases whilst the latter decreases until at a distance approximately equal to the tunnel diameter the state of stress in the rock practically is unaltered.

Hoek and Brown (1980) discussed the initiation and propagation of fracture in rock under non-uniform stress conditions as occur about underground excavations and suggested that several stages of failure could occur in homogeneous rock when subjected to vertical loading. Figure 3.1 shows the theoretical fracture contours that may develop in such rock surrounding square and elliptical tunnels. They maintained that the critical crack trajectories define the most dangerous crack orientation at any point in rock surrounding a tunnel. Figure 3.1 indicates that the lowest fracture contour occurs in the roof and floor of an excavation so that failure begins in these positions. As the critical crack trajectories in the roof and floor are parallel to the vertical axis of the excavation, a vertical crack develops. However, the formation of vertical cracks in the roof and floor, though giving rise to a redistribution of the stress about the crack, does not lead to instability. Moreover, the development of such cracks is reduced by lateral stress in the surrounding rock.

The initiation of fracture in the sidewall depends upon the shape of the tunnel (Fig. 3.2a). For instance, there is a high concentration of stress at the corners of a square tunnel. In such an instance, Hoek and Brown (1980) noted that the critical crack trajectories indicate that fracture propagation leads to slabs being dislodged from the sidewall. On the other hand, in

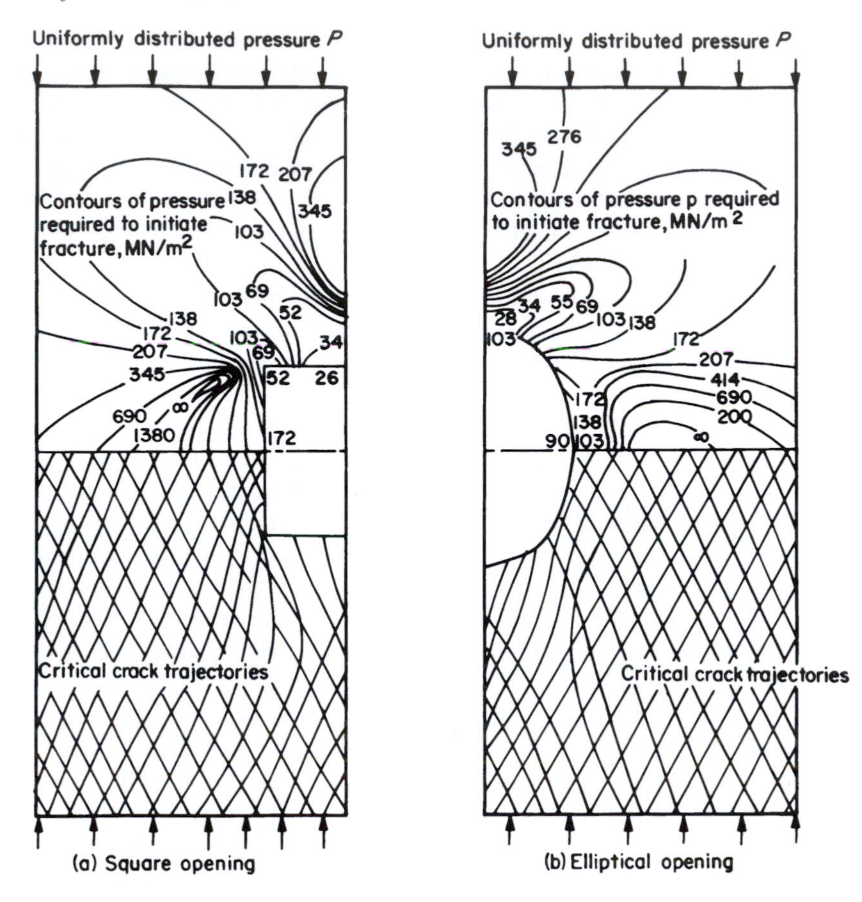

Figure 3.1 Fracture contours and critical crack trajectories in quartzite surrounding square and elliptical tunnels.
Source: Hoek, 1966. Reproduced by kind permission of Institution of Materials, Minerals and Mining.

an elliptical-shaped tunnel as sidewall failure occurs at a higher vertical pressure it is likely to take the form of sidewall scaling (Fig. 3.2b). Indeed, experience in deep mines that use both square and elliptical tunnels confirms that an elliptical tunnel, at the same depth, has a lower tendency for sidewall failure and therefore requires less support. The next stage in failure may be associated with the redistribution of stress in the roof and floor and may follow the pattern shown in Fig. 3.2c, the final fracture configuration being shown in Fig. 3.2d. This applies to tunnels in strong homogeneous rock and not notably discontinuous rock masses.

In tunnels driven at great depths below the surface, rock may suddenly break from the sides of the excavation, such a phenomenon being termed rock bursting. Hundreds of tonnes of rock may be released with explosive force during rock bursts. Rock bursts are due to the development of residual stresses that exceed the strength of the ground around an excavation, and their frequency and severity tend to increase with depth. Indeed, most rock bursts occur at depths in excess of 600 m and the stronger the rock the more likely it is to burst. The most explosive failures occur in rocks that have unconfined compressive strengths and values of Young's modulus greater than 140 MPa and 34.5 GPa respectively. Hardness, grain size and rock structure also are important factors. Controlled de-stressing of the ground can lower the incidence of rock bursts (Obert and Duval, 1967).

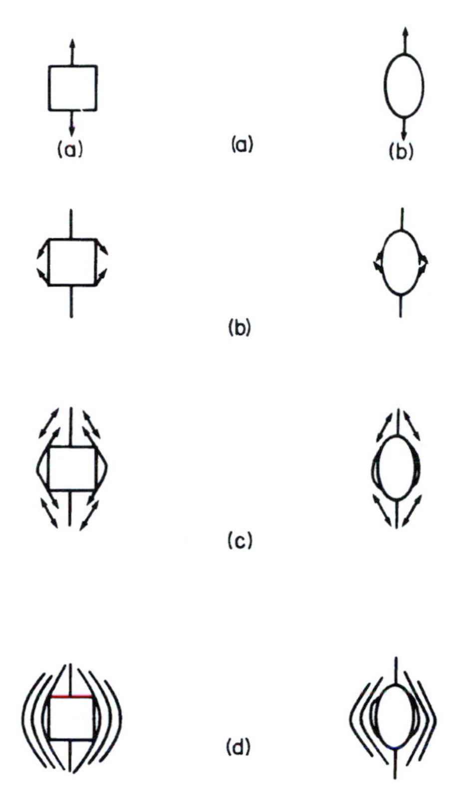

Figure 3.2 Possible fracture sequence for square and elliptical tunnels in quartzite subjected to vertical pressure only.
Source: Hoek, 1966. Reproduced by kind permission of Institution of Materials, Minerals and Mining.

Popping is a similar but less violent form of failure. In this case the sides of an excavation bulge before exfoliating and detached slabs rarely can be fitted back into position on the surface of the excavation, indicating that they have been subjected to considerable stress and subsequent plastic deformation. Spalling tends to occur in jointed or cleaved rocks. To a certain extent such a rock mass can bulge as a sheet, collapse occurring when a key block either fails or is detached from the mass. Another pressure relief phenomenon is bumping ground. Bumps are sudden and somewhat violent earth tremors that at times dislodge rock from the sides of a tunnel. They probably are due to rock displacements consequent upon the newly created stress conditions.

In fissile rock such as shale, the beds may slowly bend into a tunnel. The rock is not necessarily detached from the main mass but the deformation may cause fissures and hollows in the rock surrounding the tunnel. Rocks, particularly those that suffer plastic deformation, can undergo varying degrees of transient and steady-state creep. For instance, stress measurements on the surfaces of excavations in gypsum and salt have shown that the level of stress decreases with time because of creep. Hence, failure in such rocks can occur after a period of time, ranging from days to many years.

3.2.2 Influence of joints

Large planar surfaces form most of the roof in formations that are not inclined at a high angle and strike more or less parallel to the axis of a tunnel. In tunnels where jointed strata dip into the side at 30° or more, the up-dip side may be unstable. Indeed, joints that are parallel to the axis of a tunnel and which dip at more than 45° may prove especially treacherous, leading to slabbing of the walls and fallouts from the roof (Fig. 3.3). The effect of joint orientation in relation to the axis of a tunnel is given in Table 3.1.

The presence of flat-lying joints also may lead to blocks becoming dislodged from the roof. When the tunnel alignment is normal to the strike of jointed rocks and the dips are less than 15° large blocks are again likely to fall from the roof. The sides, however, tend to be reasonably stable. When a tunnel is driven perpendicular to the strike in steeply dipping (Table 3.1) or vertical strata each stratum acts as a beam with a span equal to the width of the cross section. However, such a situation in conventional tunnelling generally means that blasting operations are less efficient. If the axis of a tunnel runs parallel to the strike of vertically dipping rocks,

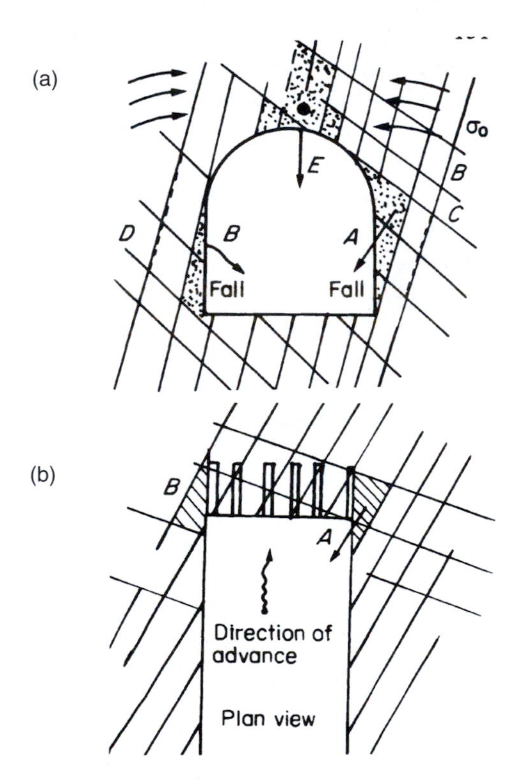

Figure 3.3 Tunnel in rock with steeply dipping joints. (a) steeply dipping joints (45–90°) that are parallel to the tunnel axis, lead to slabbing of the wall and fallouts from the roof. At point A the slab daylights at the feather edge bottom and would probably fall with the force of the blast during tunnel advance. Slab B may not fall, however, it could be loosened by the original blast and would be susceptible to additional loosening by the shocks of later blasts and by working the rock under peak tangential stresses around the tunnel periphery. Unless restructured slab B may fall eventually. Joints at depth such as C and D may tend to open. Joints blocks at E may be extremely dangerous, appearing stable after the blast but becoming unstable as the tunnel advances and the rock adjusts to the new stress field (b) Block A may be loosened and possibly forcefully ejected by gas pressures during blasting. Block B may be loosened but not necessarily removed. Had the tunnel advanced in the opposite direction the relative positions of A and B would be interchanged.

Source: Robertson, 1974. Reproduced by kind permission of the South African Institution of Civil Engineers.

Table 3.1 The effect of joint strike and dip orientations in tunnelling

Strike perpendicular to tunnel axis					
Drive with dip		*Drive against dip*		*Strike parallel to tunnel axis*	
Dip 45–90°	Dip 20–45°	Dip 45–90°	Dip 20–45°	Dip 45–90°	Dip 20-45°
Very favourable	Favourable	Fair	Unfavourable	Very unfavourable	Fair
Dip 0–20°: unfavourable, irrespective of strike					

then the mass of rock above the roof is held by the friction along the bedding planes (Fig. 3.4a). In such a situation the upper boundary of loosened rock, according to Terzaghi (1946), does not extend beyond a distance of 0.25 times the tunnel width above the crown.

When the joint spacing in horizontally layered rocks is greater than the width of a tunnel, then the beds bridge the tunnel as a solid slab and are only subject to bending under their own weight. Thus, if the bending forces are less than the tensile strength of the rock, the roof need not be supported. In conventional tunnelling where horizontally lying rocks are thickly bedded and contain few joints the roof of the tunnel will be flat (Fig. 3.4b). Conversely, if the rocks are thinly bedded and are intersected by many joints, then a peaked roof tends to form. Nonetheless, breakage rarely, if ever, continues beyond a vertical distance equal to half the width of the tunnel above the top of a semi-circular payline (Fig. 3.4c). This type of stratification is more dangerous where the beds dip at 5–10° since this may lead to the roof spalling as the tunnel is driven forward. In fresh rocks where joints have a random orientation the blocks between the joints have little freedom of movement. Tunnels through such rocks require little or no support for the sides, but if the roof is not supported, then a vault-shaped roof ultimately develops.

The presence along joints of substantial thicknesses of gouge may mean that it is squeezed into a tunnel, which can at times result in excessive deformation. In addition, clay gouge may

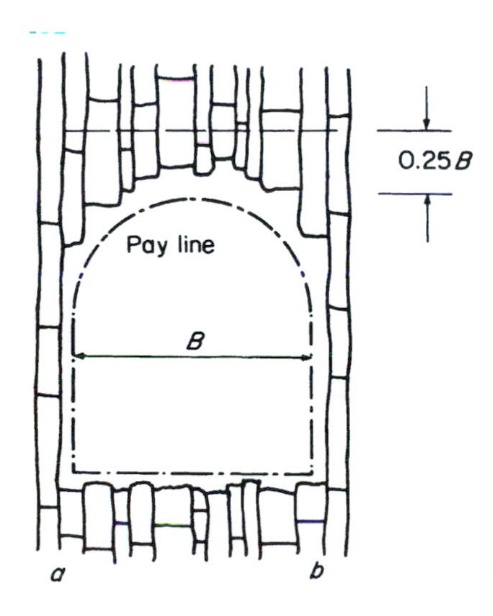

Figure 3.4a Axis of tunnel running parallel with strike in vertically dipping rocks tunnelled by blasting.

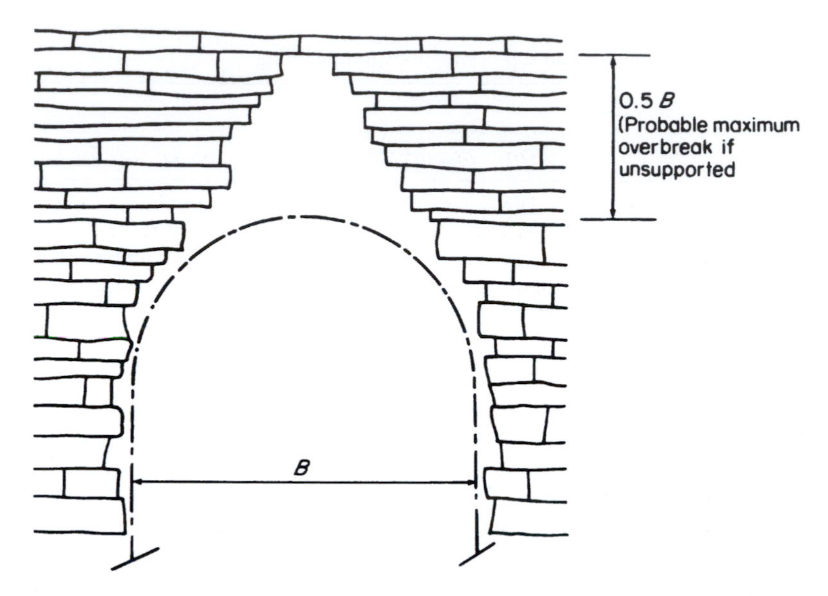

Figure 3.4b Overbreak in thinly bedded horizontal strata with vertical joints during tunnelling by blasting. Ultimate overbreak occurs if no support is installed.

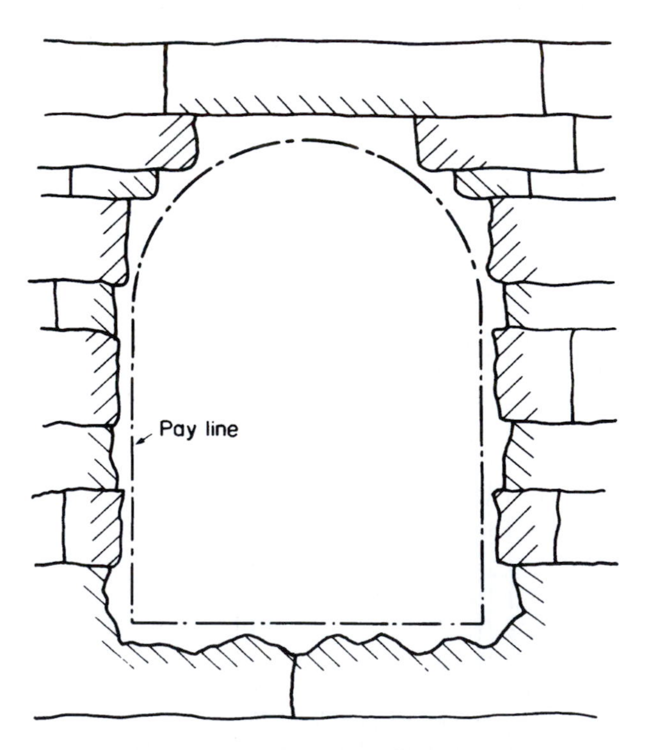

Figure 3.4c Bridge action in strong rocks with few joints during tunnelling by blasting.

absorb water and thereby swell, this again producing deformation. When joints are occupied by sand or silt, this may be washed into a tunnel by water flowing along the joints.

3.2.3 Weathered rock and tunnelling

Weathering of rock is brought about primarily by physical disintegration and chemical decomposition. Ultimately, weathering of rocks leads to formation of soil. In the process the strength of the material is dramatically reduced. The porosity is increased with accompanying reduction in bulk density. However, weathering processes are rarely sufficiently uniform to give gradual and predictable changes in the engineering properties of a weathered profile. In fact, such profiles usually consist of heterogenous materials at various stages of decomposition and/or disintegration. The depth of weathering varies and is governed largely by the type of climatic regime in which weathering took or is taking place.

Some minerals are prone to rapid breakdown on exposure and their reaction products can give rise to further problems. For example, sulphur compounds, notably pyrite, on breakdown give rise to ferrous sulphate and sulphuric acid, which are injurious to concrete. Gypsum, especially when in particulate form, can be dissolved rapidly and generates sulphates and sulphuric acid. Hydration of anhydrite produces an increase in volume of approximately 6 per cent. In certain circumstances carbonate material can be dissolved rapidly (Bell, 2000).

Mudrocks are more susceptible to weathering and breakdown than most other rock types. The lithological factors that govern the durability of mudrocks include the degree of induration, the degree of fracturing, the grain size distribution and the mineralogical composition, especially the amount and nature of the clay mineral fraction (Bell *et al.*, 1997). The breakdown of mudrocks begins on exposure, which leads to the opening and development of fissures as residual stress is dissipated, and to an increase in moisture content and softening. The development of fissures and opening of discontinuities means that access for water becomes more easy, which assists the breakdown process. The two principal controls on the breakdown of mudrocks are slaking and the expansion of mixed-layer clay minerals. Slaking refers to the breakdown of rocks by alternate wetting and drying, and some mudrocks can disintegrate within a few such cycles.

Some basalts and dolerites also are susceptible to rapid weathering (Bell and Jermy, 2000). The factors that are responsible for causing rapid disintegration of basalts and dolerites include swelling and shrinking of smectitic clay minerals on hydration and dehydration respectively, and swelling and shrinking of certain zeolites. Deuteric alteration of primary minerals by hot gases and fluids from a magmatic source migrating through rock gives rise to the formation of secondary clay minerals. The primary rock minerals that tend to undergo the most deuteric alteration are olivine, plagioclase, pyroxene and biotite, and natural glass when present in basalt. The texture of the rock influences the degree of disintegration and how rapidly it occurs in that it governs the access of water to suspect minerals. Bell and Haskins (1997) noted that degradation of the basalts on exposure in the Transfer Tunnel from Katse Dam, Lesotho, initially took the form of crazing, that is, extensive microfracturing (Fig. 3.5). These microfractures expand with time causing the basalt to disintegrate into gravel-sided fragments. They tend to exploit mineralogical and structural weaknesses in the rock such as grain boundaries and altered minerals, and may form a radial pattern around some amygdales. Such deterioration of basalt meant that there was an appreciable delay in the completion of the Transfer Tunnel and that it had to be lined throughout its length with concrete instead of partly lined as initially planned.

Numerous attempts have been made to devise engineering classifications of weathered rock and rock masses. Some schemes have involved quantification of the amount of mineralogical alteration and structural defects with the aid of the petrological microscope. Others have resorted

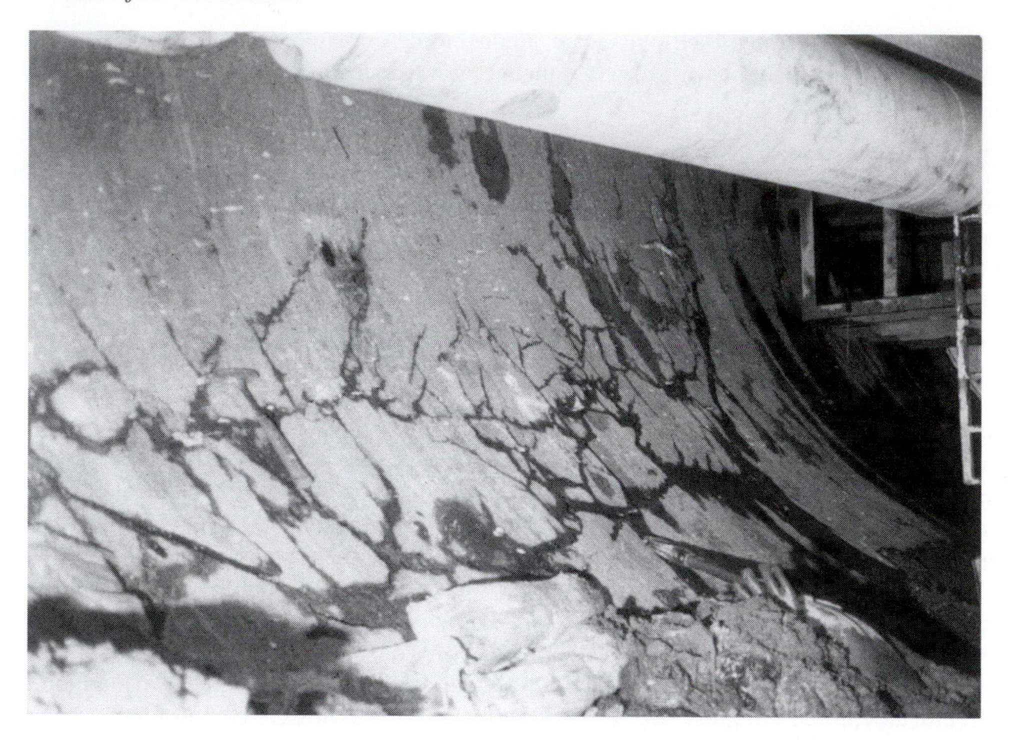

Figure 3.5 Crazing in basalt of the Transfer Tunnel, Lesotho Highlands Water Project.

to some combination of simple index tests to provide a quantifiable grade of weathering. Some of the earliest methods of assessing the degree of weathering were based on a description of the character of the rock mass concerned as seen in the field. Such descriptions recognized different grades of weathering and attempted to relate them to engineering performance. However, grading based on description of the degree of weathering is subjective and accordingly such grading systems now are being coupled with assessments made by index tests to provide better precision and quantification. Anon. (1995) provided one of the latest reviews of engineering classifications of weathered rocks. It considered that five approaches were required in order to cover different situations and scales.

3.2.4 Tunnelling in soft ground

All soft ground moves in the course of tunnelling operations (Peck, 1969). As well as time dependent movements that take place in cohesive ground, some strata change their characteristics on exposure to air. For instance, some volcanic deposits may disintegrate on exposure to air during tunnel construction. Both factors put a premium on speed of advance, and successful tunnelling requires matching the work methods to the stand-up time of the ground.

The difficulties and costs of construction in soft ground tunnelling depend almost exclusively on the stand-up time of the ground and this, in turn, is greatly influenced by the position of the water table in relation to the tunnel (Hansmire, 1981). Above the water table the stand-up time principally depends on the shearing and tensile strength of the ground whereas below the water table, it also is influenced by the permeability of the material involved.

An important feature of tunnelling in non-cohesive ground such as sand and even lightly cemented sandstone concerns the initial state of packing. If it is loosely packed, then thought must be given to the degree of consolidation that will occur during the course of construction with possible consequent distortion of the lining and surface settlement.

Excavation may be difficult if several materials of differing hardness occur in the face of a tunnel. Hence, the contacts between different types of ground, the nature of these contacts and the extent of different ground types, both along the tunnel axis and transverse to it have a significant influence on tunnelling. The contact between two waterlogged soils of different permeability is particularly important. For example, a discontinuous and isolated small pocket of sand in clay may drain a small amount of water into a tunnel and then rapidly dry up. Conversely, if the sand represents the edge of a large sand mass below the water table, then large volumes of water and sand may flow rapidly into the tunnel leading to loss of the heading and to notable subsidence at the ground surface. A sharp and well-defined soil/rock contact that cuts a tunnel at a high angle, meaning only a short length of mixed face, is quite a different problem from an irregular, nearly horizontal contact that wanders in and out of the tunnel face. A mixed face of soft ground over rock generally means slow difficult tunnelling.

Boulders in soft ground may prove difficult to remove, whilst if boulders are embedded in hard clayey soil, they may greatly impede the progress of even a hand-mined shield and may render a mechanical excavator of almost any type impotent (Peck, 1969). Large boulders may be difficult to handle unless they are broken apart by jackhammer or blasting.

Terzaghi (1950) distinguished the following six types of soft ground:

1 *Firm ground*. Firm ground has sufficient shearing and tensile strength to allow a tunnel heading to be advanced without support. Typical representatives are stiff clays with low plasticity and loess above the water table.

2 *Ravelling ground*. In ravelling ground, blocks fall from the roof and sides of the tunnel some time after the ground has been exposed. Several factors may contribute to this delayed failure. The strength of the ground usually decreases with increasing duration of load because of progressive failure mechanisms related to stress concentration around flaws. The strength of the ground also may decrease due to dissipation of excess pore water pressures induced by ground movements in clay or due to evaporation of water with subsequent loss of apparent cohesion in silt and fine sand. Stresses in the ground about a tunnel may increase with time. For instance, three-dimensional arching of loads around the face may initially reduce the loads at the heading, then cause the loads to increase at that point as the face advances beyond. If ravelling begins within a few minutes of exposure it is described as fast ravelling, otherwise it is referred to as slow ravelling. Fast ravelling may take place in residual soils and sands with a clay binder below the water table. These materials above the water table are slow ravelling.

3 *Running ground*. In this type of ground the removal of support from a surface inclined at more than 34° gives rise to a run, the latter occurring until the angle of rest of the material involved is attained. Runs take place in clean loosely packed gravel, and clean coarse to medium-grained sand, both above the water table. In clean fine-grained moist sand a run is usually preceded by ravelling, such behaviour being termed cohesive running.

4 *Flowing ground*. This type of ground moves like a viscous liquid. It can invade a tunnel from any angle and, if not stopped, ultimately will fill the excavation. Flowing conditions occur in any ground below the water table where the effective grain size exceeds 0.005 mm. Such ground above the water table exhibits either ravelling or running behaviour.

5 *Squeezing ground*. Squeezing ground advances slowly and imperceptibly into a tunnel. There are no signs of fracturing of the sides and the ground may not appear to increase in water content. Ultimately the roof may give and this can produce a subsidence trough at the surface. The two most common reasons why ground squeezes on subsurface excavation

are, first, excessive overburden pressure and, second, the dissipation of residual stress, both eventually leading to failure. Soft and firm clays display squeezing behaviour. Rock types in which squeezing conditions may obtain include shales and highly weathered granites, gneisses and schists.

The rate at which squeezing takes place depends upon the degree of overstressing. Peck (1969) showed that the squeezing behaviour of clay in tunnel excavation is related to a stability factor (N_t) given by

$$N_t = \frac{P_z - P_a}{\tau_u} \tag{3.1}$$

where P_z is the overburden pressure at depth Z at the centre line of the tunnel, P_a is the air pressure above atmospheric in the tunnel and τ_u is the undrained shear strength of the clay. It appears that, although squeeze loads on tunnel support systems must be considered for values of N_t greater than unity, the rate of squeeze does not present a problem during excavation if the stability factor is 4 or less. When the value of N_t exceeds about 5, the clay may squeeze rapidly enough to invade the annular void created by the tailskin of a shield before this void can be filled. A value of N_t above about 6 leads to shear failure ahead of the tunnel causing ground movements into the face, even in shield tunnelling. If the value of N_t is greater than 7, clay is overstressed to the extent that general shear failures and ground movements around the tunnel heading mean that control of a shield becomes difficult.

The pressures that squeezing ground develops on supports normally increase with time, but at a decreasing rate as the ground adjusts itself to new conditions. These pressures can be large, particularly in overconsolidated clays, and supports that were sufficient when installed can fail as the ground pressure on them increases. This applies to any support system that restricts the relief of residual stresses, whether in hard or soft materials.

6 *Swelling ground.* Like the former type of ground, swelling ground expands into the excavation but the movement is associated with a considerable volume increase in the ground immediately surrounding the tunnel. Swelling occurs as a result of water migrating into the material of a tunnel perimeter from the surrounding ground. These conditions develop in overconsolidated clay with a plasticity index in excess of about 30 and in certain shales and mudstones, especially those containing montmorillonite. Swelling also has occurred in evaporite formations as a consequence of anhydrite being hydrated to form gypsum. Yuzer (1982) recorded swelling pressures up to 12 MPa when tunnelling through an evaporitic formation in Turkey. Swelling pressures are of unpredictable magnitude and may be extremely large. The development period may take a few weeks or several months. Immediately after excavation the pressure is insignificant but then it increases at a higher rate. In the final stages the increase slows down. This condition usually is dealt with by imposing no restriction on swelling until it has attained a certain limit and by constructing the permanent lining at a later date.

3.2.5 *Effects of folding*

Folds are wave-like in shape and vary enormously in size. As folding movements become intensified, overfolds are formed in which both limbs are inclined in the same direction at different angles. Yet further folding gives rise to recumbent folds in which the beds have been completely overturned so that one limb is inverted and the limbs dip at a low angle. Most folding is disharmonic in that the shape of the individual folds within the structure is not uniform, the fold geometry varying from bed to bed. Disharmonic folding occurs in interbedded competent and incompetent strata. Its essential feature is that incompetent horizons display more numerous and smaller folds than the more competent beds enclosing them. It is developed because competent and incompetent beds yield differently to stress. Localized concentrations of stress may be associated with certain folded rock masses, notably strong brittle rocks. Where these stresses are of sufficient magnitude, they can cause rock failure on excavation.

Cleavage is one of the most notable structures associated with folding and imparts to rocks the ability to split into thin slabs along parallel or slightly sub-parallel planes of secondary origin. The distance between cleavage planes varies according to the lithology of the host rock, that is, the coarser the texture, the further the cleavage planes are apart. Flow cleavage occurs as a result of plastic deformation in which internal readjustments involving gliding, granulation and the parallel reorientation of minerals of flaky habit, such as micas and chlorite, together with the elongation of quartz and calcite, take place. The cleavage planes are commonly only a fraction of a millimetre apart and run parallel to the axial planes of the folds. It is characteristically developed in slates. Fracture cleavage can be regarded as closely spaced jointing, the distance between the planes being measured in millimetres or even in centimetres. Unlike flow cleavage there is no parallel alignment of minerals, fracture cleavage having been caused by shearing forces, and it commonly develops at an angle of approximately 30° to the axis of maximum principal stress. Fracture cleavage frequently is found in folded incompetent strata that lie between competent beds. Where it is developed in competent rocks it forms a larger angle with the bedding planes than it does in the incompetent strata.

When brittle rocks are distorted, tension gashes may develop as a result of stretching over the crest of a fold or by local extension caused by drag exerted when beds slip over each other. Those tension gashes that are the result of bending of competent rocks usually appear as radial fractures concentrated at the crests of anticlines that are sharply folded. Tension gashes formed by differential slip appear on the limbs of folds and are aligned approximately perpendicular to the local direction of extension. Tension gashes are distinguished from fracture cleavage and other types of fractures by the fact that their sides tend to gape. As a result they often contain lenticular bodies of vein quartz or calcite.

Tectonic shear zones lie parallel to the bedding in sedimentary rocks and appear to be due to displacements caused by concentric folding. Such shear zones generally occur in clay beds with high clay mineral contents. The shear zones range up to approximately 0.5 m in thickness and can extend over hundreds of metres.

3.2.6 Problems due to faults

Faults generally mean non-uniform rock pressures on a tunnel and so, at times, may necessitate special treatment such as the construction of box sections with invert arches. Generally, problems increase as the strike of a fault becomes more parallel to a tunnel opening. However, even if the strike is across the tunnel, faults with low dips can represent a hazard. If the tunnel is driven from the hanging-wall, the fault first appears at the invert and generally it is possible to provide adequate support by reinforcement when driving through the rest of the zone. By contrast, when a tunnel is driven from the foot-wall side, the fault first appears in the crown, and there is a possibility that a wedge-shaped block, formed by the fault and the tunnel, will fall from the roof without warning.

Major faults usually are associated with a number of minor faults and the dislocation zone may occur over many metres. If the movement along a fault has been severe, then the rocks involved may have been crushed, sheared or pulverized. Where shales or clays have been faulted the fault zone may be occupied by clay gouge. Fault breccias, which consist of a jumbled mass of angular fragments containing a high proportion of voids, occur when more competent rocks are faulted. Crush breccias and conglomerates develop when rocks are sheared by a regular pattern of fractures. Movements of greater intensity are responsible for the occurrence of mylonite along a fault zone. The ultimate stage in the intensity of movements is reached with the formation of pseudo-tachylite which looks like glass. Problems tend to increase with increasing width of the fault zone. Sometimes sand-sized crushed rock in a fault zone has a tendency to flow into the tunnel. If, in addition, the tunnel is located beneath the water table, a sandy suspension may rush into the tunnel. When a fault zone is occupied by clay gouge and a section of a tunnel follows the gouge zone, swelling of this material may occur and cause displacement or breakage of tunnel support during construction.

Large quantities of water in a permeable rock mass are impounded by a fault zone occupied by impervious gouge and are released when tunnelling operations penetrate through the fault zone. For example, a number of faults were intersected by the San Jacinto tunnel near Balting, California. Unfortunately, the hanging-wall sides of the faults were heavily jointed and so highly permeable. Driving the tunnel towards the hanging-wall sides of the faults did not present serious problems as the rock mass was drained over a period of time prior to reaching the faults. However, when the faults were approached from their foot-wall sides, no pre-drainage could take place due to the impermeable gouge in the faults, and sudden inflows of water occurred. The peak flow from all headings was about $3030 \, \text{l s}^{-2}$, with water pressures commonly ranging between 1 and 2 MPa. Up to $2300 \, \text{m}^3$ of wet fault gouge surged into the tunnel (Thompson, 1966).

Intraformational shears, that is, zones of shearing parallel to bedding are associated with faulting. They often occur in clays, mudstones and shales at the contact with sandstones. Their presence means that the strength of the rock along the shear zone has been reduced to its residual value. Such shear zones tend to die out when traced away from the faults concerned and are probably formed as a result of flexuring of strata adjacent to faults. A shear zone may consist of a single polished or slickensided shear plane, a more complex shear zone may be up to 300 mm in thickness.

3.2.7 Earthquakes

The geological investigation associated with a tunnel project should determine whether the faults in the area can be considered active. Movements along major active faults in certain parts of the world can disrupt the lining of a tunnel and even lead to a tunnel being offset. As a consequence, it is best to shift the tunnel alignment to avoid an active fault or, if this is not possible, to use open cut within the active fault zone.

A deep tunnel in solid rock will be subjected to displacements that are considerably less than those that occur at the surface. Indeed, Howells (1972) suggested that at depths of approximately 250 m the intensity of ground motion is about one tenth of that which is recorded at the surface. The earthquake risk to an underground structure is influenced by the material in which it occurs as well as the depth at which it is located. For instance, a tunnel at shallow depth in alluvial deposits will be seriously affected by the large relative displacements of the ground surrounding it. Unlike structures at the surface, underground structures are relatively rigid and are completely surrounded by the medium through which the earthquake waves travel. As a consequence, they are unlikely to respond to the ground motion with any dynamic modification and the inertia forces acting on them perhaps may be determined as a first approximation by the ground acceleration. According to Howells, the main causes of the stresses in shallow underground structures arise not from inertia forces but from the interaction between the structure and the displacements of the ground. If the structure is sufficiently flexible it will follow the displacements and deformation to which the ground is subjected. He maintained that because earthquake waves closely follow the postulates of the theory of elasticity, a very simplified model of an earthquake wave often is sufficient to enable the strength of a structure to be designed so as to resist ground deformations. Nevertheless, in soft ground, earthquake wave-induced stresses can sometimes cause irreversible displacements, liquefaction providing the most dramatic example.

3.2.8 Water and tunnels

Construction of a tunnel may alter the groundwater regime of a locality as a tunnel usually acts as a drain. The amount of water held in a soil or rock mass depends on its reservoir storage properties that, in turn, influence the amount of water that can drain into a tunnel. Isolated heavy flows of water may occur in association with faults, solution pipes and cavities, abandoned mine workings, or from pockets of gravel. Inflows of water from soft ground are likely to give rise to stability problems. Resulting loss of ground, especially in tunnels located at shallow depth, can lead to subsidence at the ground surface. Tunnels driven under lakes, rivers and other surface bodies of water may tap a considerable volume of flow. Flow is also likely from a perched water table to a tunnel beneath. At one extreme water may drip into a tunnel, this dripping being of variable intensity, while at the other when it is under heavy pressure it may break into the tunnel as a gusher (Fig. 3.6).

Generally, the amount of water flowing into a tunnel decreases as construction progresses. This is due to the gradual exhaustion of water at source and to the decrease in hydraulic gradient, and hence in flow velocity. On the other hand, there may be an increase in flow as construction progresses, if construction operations cause fissuring. For instance, blasting may open new water conduits around a tunnel, shift the direction of flow and in some cases may even cause partial flooding.

Correct estimation of the water inflow into a projected tunnel is of vital importance, as inflow influences the construction programme (Cripps *et al.*, 1993). One of the principal problems created by water entering a tunnel is that of face stability. Secondary problems include removal of excessively wet muck and difficulty of placement of a precision-fitted primary lining or of ribs.

Not only is the value of the maximum inflow required but so is the distribution of inflow along the tunnel section and the changes of flow with time. A series of packer tests may facilitate the selection of the best tunnelling horizon but they are unlikely to detect random flows from fissures, faults or cavities. The greatest groundwater hazard in underground work is the presence of unexpected water bearing zones, and therefore whenever possible the position of hydrogeological boundaries should be located. Indeed, sometimes an impermeable boundary such as a dyke can form an underground dam and so can be used to advantage. Obviously, the location of the water table, and its possible fluctuations, is of major consequence. In coastal areas changes in the level of the water table may be influenced by the daily changes in tidal level (Edmunds and Graham, 1977).

Water pressures are more predictable than water flows as they can be related to the head of water above the tunnel location. They can be very large, especially in confined aquifers. Hydraulic pressures should be taken into account when considering the thickness of rock that will separate an aquifer from the tunnel. Unfortunately, however, the hydrogeological situation is rarely so easily interpreted as to make accurate quantitative estimates possible.

Sulphate bearing solutions attack concrete; thus water quality should be investigated (Anon., 1991a). For example, particular attention should be given to water flowing from formations containing gypsum and anhydrite. In addition, rocks containing iron pyrite may give rise to water carrying sulphates.

Figure 3.6 Major water inflow ($91\,l\,s^{-1}$) from a fissure in the Beachley to Aust drive in the tunnel beneath the River Severn estuary, England.

Most of the serious difficulties encountered during tunnelling operations are directly or indirectly caused by the percolation of water towards the tunnel. As a consequence, most of the techniques for improving ground conditions are directed towards its control. This may be achieved by using drainage, compressed air, grouting or freezing techniques.

Because of the limitation imposed on a wellpointing system by suction lift, the technique can only be used to dewater tunnels in soil at shallow depth (Bell, 1993). According to Powers (1972) deep wells are the most widely used method of pre-draining tunnels, but the unit cost per well tends to be high. Individual ejector systems have lower unit costs than deep wells and Powers noted that they have proved successful in dewatering some very difficult projects, especially in soils sensitive to low flow rates and seepage pressures. The reason for this is that ejectors produce a vacuum in the surrounding soil thereby allowing poorly drained fine-grained soils to be effectively dewatered (Miller, 1988). In order to ensure safety in such soils little or no water should be allowed at the tunnel face, from which it follows that the pre-drainage system must approach total effectiveness. However, ejector systems may be difficult to maintain where the groundwater is hard or contains iron.

Horizontal drainage can be provided in the form of bored wells drilled radially from a deep shaft but this system is costly and depends for its effectiveness on favourable inclination of the strata. The chief problem is in intercepting the more pervious seams. Wellpoints can be installed horizontally or at an angle in order to reduce the head of water some distance beyond a tunnel face at shallow depth (Somerville, 1986).

Where the tunnel invert level is not more than about 15 m below groundwater level, shield driving in compressed air may be an effective and economical method of achieving stable conditions at the tunnel face when driving through some water bearing soils or soft rocks. In soft clay the pressure of the air provides some support and in silts and fine sands the air drives the water back from the exposed area so that capillary attraction between the moist grains helps to stabilize the ground. Where tunnels are driven at depths in excess of 15 m the physiological effects on the men working in compressed air represent a health hazard that worsens with increasing depth so that the method becomes unsafe to use. When compressed air is used in tunnelling operations the heading must be sealed off. Since the air pressure on the heading is greater than the water pressure the compressed air not only stops the flow of water into the tunnel but tends to drive it away. The rate at which water is displaced depends on the air pressure and the effective size (D_{10}) of the grains. In soil, if the effective size of the particles is smaller than 0.01 mm, then the rate of displacement is likely to be zero, whereas if it exceeds 0.2 mm considerable leakage of air is likely to occur. In large subaqueous tunnels such as vehicular tunnels constructed beneath rivers, it is difficult to select an air pressure that will minimize water inflow at the invert, where water pressure is naturally higher, without causing significant air loss at the soffit. In particular, if the heading runs into a highly permeable zone, then the air pressure may drop to zero, whereupon the tunnel may be flooded. This is known as a blowout. For example, a blowout occurred in the soft sediments on the north bank of the River Clyde in Scotland during the construction of a tunnel beneath it. The blowout produced a crater that had to be filled with clay and at the same time the water table was lowered by using bored wells. Blowouts generally occur where the depth of cover is shallow and they are more likely to happen in unconsolidated deposits that lack bedding than in those that are well stratified. The risk of blowouts can be reduced by grouting or freezing. Chemical- or clay-cement grouting in the form of a blanket on the tunnel alignment also have been used as a means of reducing the quantity of compressed air.

Grouting in water-bearing ground often is undertaken ahead of the face of a tunnel to reduce the quantity of water entering the tunnel to readily manageable amounts (Tan and Clough, 1980). In particular, difficulties frequently arise when tunnels are excavated beneath rivers that contain buried channels, especially those occupied by sands and gravels. Such grouting has been referred to as aureole or umbrella grouting and involves drilling groutholes in advance of the tunnel heading, which fan out to form a series of concentric grouted cones (Fig. 3.7a). The tunnel then is excavated through the grouted zone. The length of the individual holes depends upon the type of ground on the one hand, and the quantity and pressure of the water on the other. In most cases, 9 m should prove sufficient. The type of grout used to form the aureole depends on the ground conditions. Sub-horizontal jet grouting has been used for the pre-treatment of the crowns of tunnels ahead of excavation in soils (Fig. 3.7b). However, a disadvantage is that stress relief and loosening of the ground can occur since a void is created before the grout is injected. This can be overcome by direct displacement grouting in which the water erosion jet is removed, the grout being injected on its own or with a stream of air (Anon., 1991b). The columns so formed contain more soil than water-jetted grout columns and therefore possess a lower strength. Nevertheless, the friction around the columns is higher and the surrounding soil is more consolidated. In this way a protective zone is formed ahead of the face of the tunnel in which the stability of the ground is increased and ground movements reduced to a minimum (Dugnani *et al.*, 1989).

Tunnel driving through frozen ground is not normally resorted to except for short lengths. This is because of the difficulties in access for installing the freeze probes and the high cost

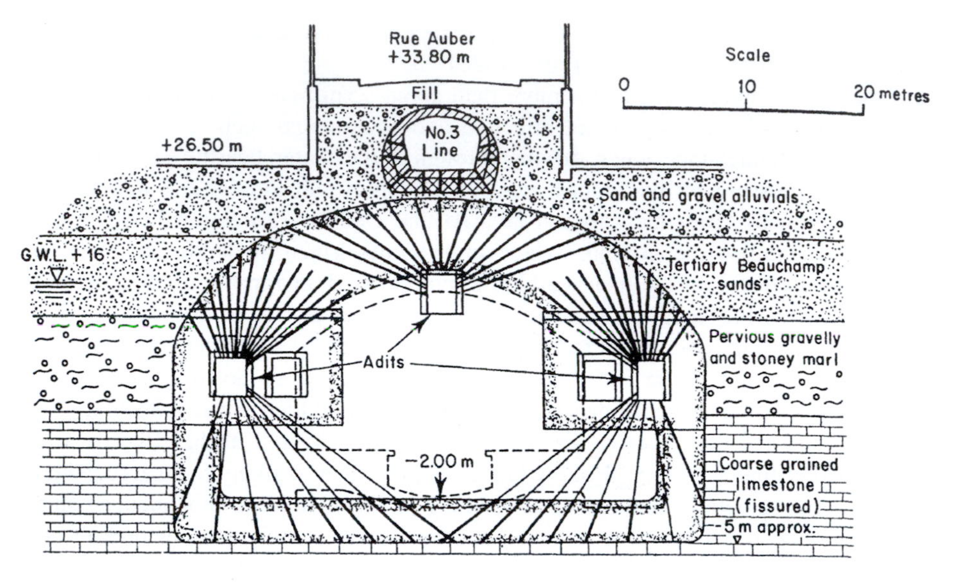

Figure 3.7a Second phase of grout treatment (aureole grouting) at Auber station, Paris, France.

for the relatively short time that the ground is frozen. Freezing with liquid nitrogen via probes driven into the tunnel face is an economical method of dealing with an occasional pocket of water bearing silt or fine sand in otherwise stable ground but the method is again too costly for use throughout the entire construction of a tunnel (Fig. 3.8). Lake and Norie (1982) described the use of ground freezing during two tunnelling operations. They first considered the use of supercooled brine to treat fluvioglacial and post-glacial water bearing soils, which occurred in a deep fault controlled buried gorge, during the construction of the metro system for Helsinski, Finland. The second example was provided by tunnels running beneath the River Mersey and Manchester Ship Canal, England, where the ground conditions consist of alluvium overlying till. Again, super cooled brine was used as the refrigerant.

3.2.9 Gases in tunnels

Naturally occurring gas can occupy the pore spaces and voids in rock. Gas migrates through rock masses via intergranular permeability or, more particularly, along discontinuities. Where strata have been disturbed by subsurface mining gas permeability is enhanced, as is that of groundwater in which gas is dissolved. Gases may be dissolved in groundwater depending on the pressure, temperature or concentration of other gases or minerals in water. Dissolved gases may be advected by groundwater and only when the pressure is reduced and the solubility limit of the gas in water exceeded, do they come out of solution and form a separate gaseous phase. Such pressure release occurs, for example, when coal is removed during mining, tunnelling or shaft sinking operations in strata so affected. An outline of the survey for methane that was carried out for the Channel Tunnel has been provided by Warren *et al.* (1991). It was suspected that gas from underlying coal-bearing strata could have migrated into the Chalk in which the tunnel was to be excavated. Fortunately, methane was not present other than in insignificant amounts.

Gas may be under pressure and there have been occasions where gas under pressure has burst into underground workings, especially coal mines, causing the rock to fail with explosive force

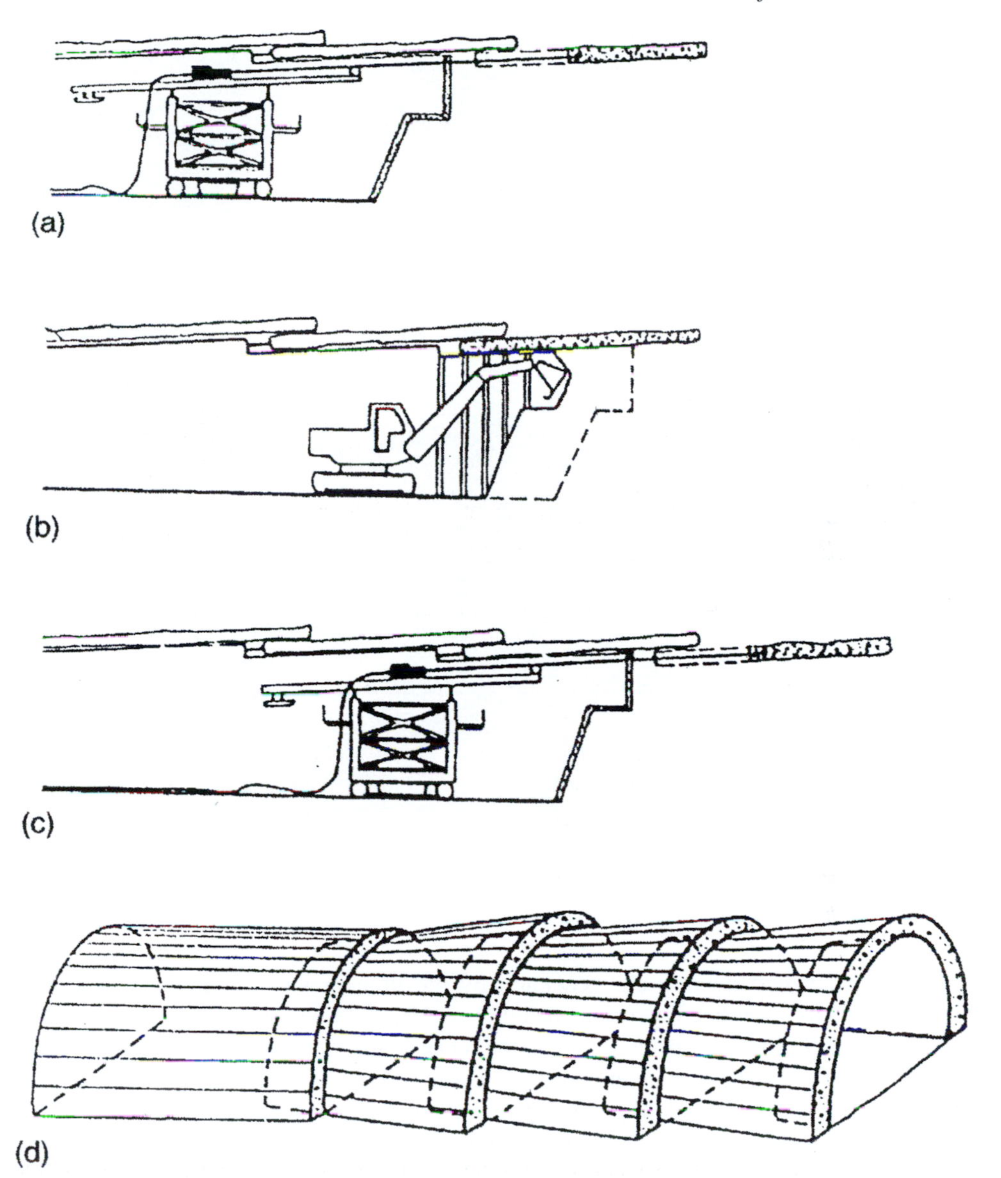

Figure 3.7b Subhorizontal jet grouting execution sequence (a) jet grouting ahead of the face (b) excavation (c) jet grouting at next stage (d) geometrical scheme of treatment. Jet grouted columns are formed ahead of the tunnel face, the overlapping subhorizontal columns forming a protective shell around the tunnel. The length of the treatment is normally between 10 and 15 m but excavation stops 2 or 3 m before the end of the treated zone so as to leave sufficient support for roof stability.
Source: Dugnani *et al*., 1989. Reproduced by kind permission of A.A. Balkema.

(Bell and Jermy, 2002). Wherever possible the likelihood of gas hazards should be noted during the geological investigation, but this is one of the most difficult tunnel hazards to predict. If the flow of gas appears to be fairly continuous, the entrance of the flow may be sealed with concrete. Often the supply of gas is quickly exhausted, but cases have been reported where it continued for up to three weeks.

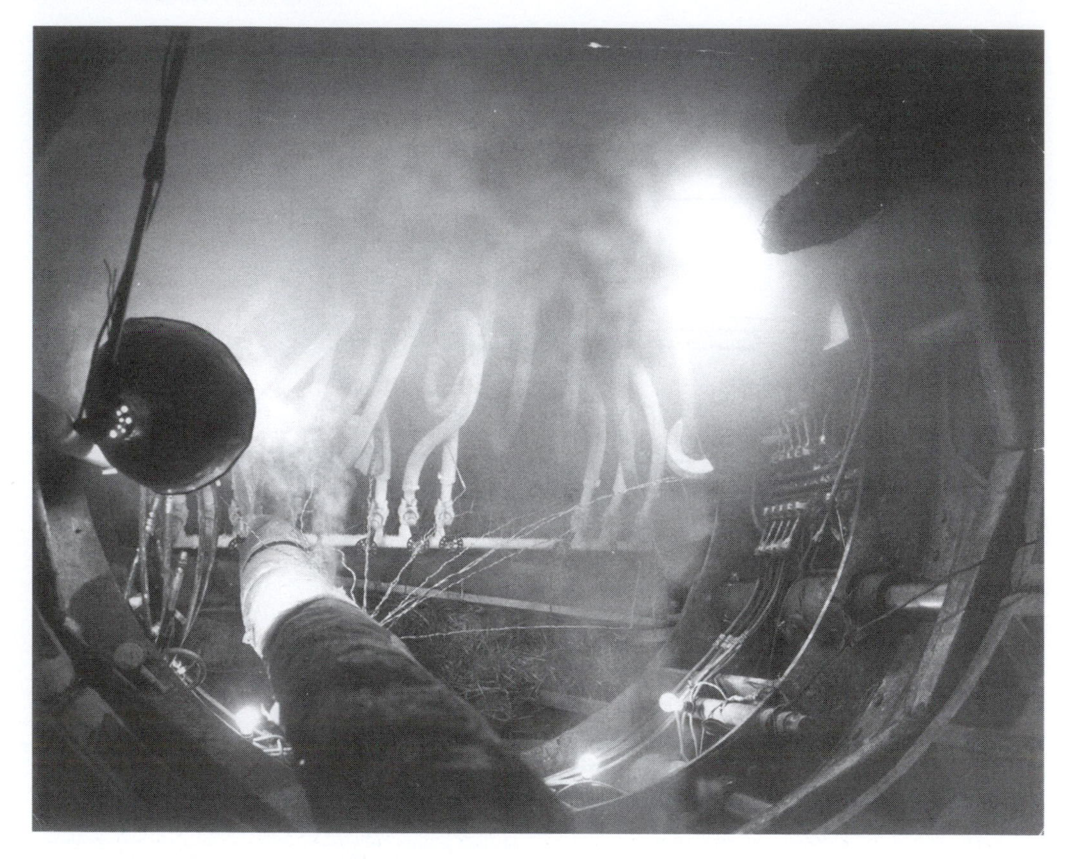

Figure 3.8 Face of tunnel (2.6 m diam) for sewer scheme for Edinburgh, Scotland, showing installation of freeze pipes carrying liquid nitrogen.

Methane and carbon dioxide are generated by the breakdown of organic matter and are associated with the coal-bearing strata. In other words, methane and carbon dioxide are by-products formed during coalification, that is, the process that changes peat into coal. Moreover, methane can be oxidized during migration to form carbon dioxide. Carbon dioxide can be generated both microbially and inorganically in a number of ways that do not involve methane. Methane is lighter than air and can readily migrate from its point of origin. Not only is it toxic but is also combustible and highly explosive when mixed with air (5–15 per cent methane mixed with air forms an explosive mixture). Carbon dioxide and carbon monoxide are both toxic (Table 3.2). The former is heavier than air and hangs about the floor of an excavation. Carbon monoxide is slightly lighter than air, and like carbon dioxide and methane, it is found in coal-bearing strata. Carbon dioxide also may be associated with volcanic deposits and limestone rock masses. Hydrogen sulphide is heavier than air and is highly toxic. It is also explosive when mixed with air. The gas may be generated by the decay of organic substances or by volcanic activity. Hydrogen sulphide may be absorbed by water that then becomes injurious as far as concrete is concerned. Sulphur dioxide is a colourless, pungent, asphyxiating gas that dissolves readily in water to form sulphuric acid. It is usually associated with volcanic emanations or it may be formed by the breakdown of pyrite. Stythe gas is deoxygenated air that occurs in abandoned coal mines and, as such, is a suffocating gas. The oxygen is removed to a varying degree by the oxidation process that takes place between oxygen and coal, rotting

Table 3.2 Effects of noxious cases

Gas	*Concentration by volume in air* p.p.m.	*Effect*
Carbon monoxide	100	Threshold limit value under which it is believed nearly all workers may be repeatedly exposed day after day without adverse effect (TLV)
	200	Headache after about 7 hours if resting or after 2 hours if working
	400	Headache and discomfort, with possibility of collapse, after 2 hours at rest or 45 min exertion
	1 200	Palpitation after 30 min at rest or 10 min exertion
	2 000	Unconsciousness after 30 min at rest or 10 min exertion
Carbon dioxide	5 000	TLV. Lung ventilation slightly increased
	50 000	Breathing is laboured
	90 000	Depression of breathing commences
Hydrogen sulphide	10	TLV
	100	Irritation to eyes and throat: headache
	200	Maximum concentration tolerable for 1 hour
	1 000	Immediate unconsciousness
Sulphur dioxide	1–5	Can be detected by taste at the lower level and by smell at the upper level
	5	TLV. Onset or irritation to the nose and throat
	20	Irritation to the eyes
	400	Immediately dangerous to life

Source: Anon., 1973. Reproduced by kind permission of British Coal Corporation.

Notes

1 Some gases have a synergic effect, that is, they augment the effects of others and cause a lowering of the concentration at which the symptoms shown in the above table occur. Further, a gas which is not itself toxic may increase the toxicity of one of the toxic gases, for example, by increasing the rate of respiration; strenuous work will have a similar effect.

2 Of the gases listed carbon monoxide is the only one likely to prove a danger to life, as it is the commonest. The others become intolerably unpleasant at concentrations far below the danger level.

timber supports and minerals such as pyrite. Hence, stythe gas consists mainly of nitrogen and carbon dioxide, and is heavier than air.

An interesting example of the occurrence of gas was noted by Newbury and Davenport (1975). They reported a concentration of carbon dioxide of up to 6 per cent in shaft B of the Allington Sewer Tunnel at Maidstone, Kent, England. The shaft ran into made-over ground in which ashes from the local gas works had been deposited. These gave rise to acidic groundwater, the pH value of which at times was as low as 3.0. This acidic groundwater percolated downwards to react with calcareous quarry spoil, which produced large volumes of carbon dioxide. The latter accumulated in the voids in the quarry spoil. The gas was bled-off by constructing a series of draught holes linked to an extraction fan.

3.2.10 Temperatures in tunnels

Temperatures in tunnels, except in the case of water tunnels, are not usually of concern unless the tunnel is more than 170 m below the surface. When rock is exposed by excavation the amount of heat liberated depends on the virgin rock temperature (VRT), the thermal properties of the rock, the wetness of rock, the air flow rate, the dry bulb temperature and the humidity of the air.

In deep tunnels high temperatures can make work more difficult. Indeed, high temperatures and rock pressures place limits on the depth of tunnelling. The application of modern rock mechanics techniques, however, has reduced the incidence of rock bursts so that high temperatures are now the more important limit. The moisture content of the air in tunnels is always high and in saturated air the efficiency of labour declines when the temperature exceeds 25 °C, dropping to almost zero when the temperature reaches 35 °C. Conditions can be improved by increased ventilation, by water spraying or by using refrigerated air (McPherson, 1972). Air refrigeration is essential when the virgin rock temperature exceeds 40 °C.

Ventilation obviously is a necessity for any underground workings, not just to help regulate temperature but to supply fresh air to operatives and to help clear any gases. For example, the ventilation system for the Transfer Tunnel from Katse Dam in Lesotho was designed to produce a minimum of 6.75 m^3 of fresh air to every tunnel face and included five ventilation shafts, each of 2.1 m diameter (Bell and Haskins, 1997). These were raise-bore drilled.

The rate of increase in rock temperature with depth depends on the geothermal gradient, which is inversely proportional to the thermal conductivity of the material involved:

$$\text{Geothermal gradient} = \frac{0.05}{k} \text{(approximately)} \ ^{\circ}\text{C m}^{-1} \tag{3.2}$$

where k is the thermal conductivity. Although the geothermal gradient varies with locality, according to rock type and structure, on average it increases at a rate of 1 °C per 30–35 m depth. In geologically stable areas the mean gradient is 1 °C for every 60–80 m whereas in volcanic districts it may be as much as 1 °C for every 10–15 m depth. The geothermal gradient under mountains is larger than under plains; in the case of valleys, the situation is reversed. Consequently, the geothermal gradient found to exist in one tunnel cannot be assumed to exist in another. For instance, if the geothermal gradient found in the St Gottard Tunnel, Switzerland, was applied to the Simplon Tunnel, also in Switzerland, then the temperature would have been around 42 °C when, in fact, it was 55 °C. Szechy (1966) noted that during the construction of the Great Apennine Tunnel, Italy, the ground temperature increased in a clay shale from 27 to 45 °C, and exceptionally to 63 °C after an inrush of methane. He also gave examples of the average geothermal gradients for some long European tunnels (Table 3.3).

Downward percolating meteoric water influences the geothermal temperature, which is also influenced by gases. Fissure water that flows into workings acts as an efficient carrier of heat. This may be more significant locally than the heat conducted through the rock itself. For example, for every litre s^{-1} of fissure water that enters workings at a virgin rock temperature of 40 °C, if the water cools to 25 °C before it reaches the pumps, then the heat added to the

Table 3.3 Gradients for some European tunnels

Tunnel	*Length* (m)	*Depth* (m)	*Average geothermal gradient* (m/°C)	*Maximum temperature* (°C)
Simplon	19 720	2135	37	55
St Gotthard	14 998	1752	47	40
Mont Cenis	12 236	1610	58	30
Tauern	8 551	1567	24	49

Source: Szechy, 1966.

ventilating air stream will be 62.8 kW (McPherson, 1972). On the other hand, the temperature of the rocks influences the temperature of any water or gases they may contain. Water temperatures of up to 60 °C have been recorded in some deep mines.

Earth temperatures can be measured by placing thermometers in drillholes, measurement being taken when a constant temperature is attained. The results, in the form of geoisotherms, can be plotted on the longitudinal section of the tunnel. The possible occurrence of hot water flows during construction also should be indicated.

3.2.11 Abandoned mine workings and tunnels

Tunnels constructed in urban areas may have to contend with old mine workings. In Britain, for example, most of the large urban industrial areas developed on coalfields. Abandoned mine workings may be flooded or partially flooded and old coal mines, in particular, may contain pockets of gas. The conditions of the mine workings may be suspect in that collapse may have occurred in places and zones of future potential collapse may exist. Fractured ground and bedding plane separation probably are present in the roof rocks, which not only weakens the ground but also enhances its permeability. Subsidence of old mine workings beneath a tunnel could have a serious affect upon a tunnel. Subsidence of pillared workings can occur at any time and ground disturbance due to tunnelling operations may help trigger collapse of workings close to a tunnel.

In such circumstances the site investigation for the tunnel must attempt to determine the location and nature of the abandoned mine workings. An obvious first step is to examine old mine records and plans. However, the older the abandoned mine workings are, the less likelihood there is of mine records and plans being in existence. Even if these are in existence, they may be inaccurate. An outline of the type of site investigation required to locate abandoned mine workings has been provided by Healy and Head (1984) and by Bell (1986). Cripps *et al.* (1988) reviewed the various geophysical methods that have been used to detect abandoned mines workings and, more recently, Bishop *et al.* (1997) have described the use of the microgravity technique to determine the location of voids produced by mining. The latter concluded that when the microgravity technique is used with a comprehensive desk study and knowledge of the site geology and topography, then it can be a very effective tool in locating subsurface voids. They also emphasized that a microgravity survey should be followed up by a specifically targeted drillhole programme.

To avoid any problems that might arise during tunnel construction due to the presence of abandoned mine workings, the workings need to be treated. This commonly takes the form of filling the voids with bulk grout. Bulk grouts normally consist of a mixture of gravel, sand, cement and pulverized fly ash (PFA). Alternatively, foam grouts have been used to fill the cavities in old mine workings. Deaves and Cripps (1995) described the treatment of abandoned workings in relation to the construction of a tunnel in Sheffield, England. First, a perimeter curtain wall was formed of grout and then the area within was grouted.

3.3 Excavation of tunnels

The choice of tunnelling method is influenced by a number of factors among which ground conditions are the most important. Tunnelling in soft ground usually employs a shield to provide safe working conditions for operatives, as well as for achieving effective tunnel excavation. The strength of rock masses varies enormously, from less than 10 MPa to over 300 MPa in terms of unconfined compressive strength, and their behaviour is influenced by the presence of discontinuities. Strong massive rocks may preclude machine excavation but will require minimal temporary support. Mixed face conditions can provide problems for machine tunnelling, as well as with the provision of temporary support. The diameter of a tunnel may influence the choice of tunnelling method, as may location. For example, drill and blast excavation may be unacceptable in urban areas.

3.3.1 Machine tunnelling in soft rock

In soft ground, support is vital and so tunnelling is carried out by using a shield (Mayo, 1982). A shield is a cylindrical drum with a cutting edge around the circumference, the cut material being delivered onto a conveyor for removal (Fig. 3.9). The limit of these machines usually is regarded as an unconfined compressive strength of around 20 MPa. Shield tunnelling means that construction can be carried out in one stage at the full tunnel dimension, that constant

Figure 3.9 Double-shielded tunnel boring machine used to excavate the Delivery Tunnel North, Lesotho Highlands Water Scheme.

support is provided to the advancing tunnel and that temporary support is unnecessary because of the immediate installation of the permanent lining.

Robbins (1976) maintained that one of the main drawbacks of rotary cutting heads on tunnelling machines in soft ground was their vulnerability to overexcavation at the face in running ground. This often has led to serious subsidence at the surface. The danger of overexcavation at the face in soils below the water table can be reduced by improving ground stability by employing compressed air in a tunnel if dewatering or grouting is not possible. Subsidence associated with shield-driven tunnels also may be due to ineffective filling of the tail void or to poor ground control at the shield because of the late application of support.

Tunnelling machines are expensive to operate and any delays in drivage due to unforeseen instability of the ground can prove very costly. The consequences are much more serious if the machine is damaged or engulfed by a collapse of the tunnel roof. For instance, during the construction of the Victoria Line for the London underground transport system, the shield broke through the London Clay into water-bearing sand and gravel. The resulting run-in buried the shield completely and the machine had to be extricated by constructing a cofferdam from the surface. Properly located probe holes raked upward from the face of the tunnel would have detected this low in the buried valley of the River Tyburn.

The use of a bentonite slurry to support the face in soft ground in a pressure bulkhead machine was introduced in the late 1960s. It represented a major innovation in mechanized tunnelling, particularly in granular sediments not suited to compressed air. The bentonite slurry counterbalances the hydrostatic head of groundwater in the soil and stability is further increased as the bentonite is forced into the pores of the soil, gelling once penetration occurs. The bentonite also forms a seal on the surface. However, boulders in soils, such as till, create an almost impossible problem for slurry face machines. A mixed face of hard rock and granular soil below the water table presents a similar dilemma.

The earth pressure balance shield was designed for tunnelling through soft ground below the water table without using slurry (Stack, 1985). A rotating cutter head, with drag picks, forms the front of the machine. The excavated debris is collected and compacted in a special chamber immediately behind the cutterhead. The compacted debris forms a plug that provides support for the tunnel face, as well as controlling the effects of ingress of any groundwater on the stability of the face. The compacted material eventually passes through the bulkhead to the disposal system, disposal being a relatively clean operation as no water or slurry is used. Saito and Kobayashi (1979) described the use of an earth pressure balance shield to construct a sewer, with a diameter of 8.48 m, through low-lying soft ground in eastern Tokyo, which caused no subsidence along the route of the tunnel. The slime shield is a type of earth pressure balance shield in which mud is injected into the pressure chamber, thereby increasing the range of ground conditions in which the shield can work. For instance, debris flow may experience difficulties in sandy soils with low fines content in an earth pressure balance shield. Also, poorly controlled running sand and water through the disposal system (usually a screw conveyor) can give rise to collapse at the face where the groundwater pressure is high. Both can be avoided by using a slime shield (Whittacker and Frith, 1990).

3.3.2 Machine tunnelling in hard rock

Machine tunnelling in rock uses either a roadheader machine or a tunnel boring machine (TBM). A roadheader generally moves on a tracked base and has a cutting head, usually equipped with drag picks, mounted on a boom (Fig. 3.10). Twin-boom machines have been developed in order to increase the rate of excavation. Roadheaders can cut a range of tunnel shapes and

Figure 3.10 A roadheader used at the Dinorwic Pumped Storage Scheme, Llanberis, North Wales.

are particularly suited to stratified formations. Some of the heavier roadheaders can excavate massive rocks with unconfined compressive strengths in excess of 100 MPa and even up to 200 MPa. Obviously, the cutting performance is influenced by the presence and character of discontinuities. A high-pressure water jet can assist the cutting operation significantly. One of the principal limiting factors of roadheaders in stronger rocks is the cost of pick replacement. Richardson and Gollick (1988) referred to the use of a roadheader equipped with a shield. Such roadheaders are particularly suited to cutting through wide zones of fault gouge that occur at appreciable depth.

Simultaneous bench operation allows large diameter tunnels to be excavated by roadheaders. Also, roadheaders have proved successful when used in NATM (New Austrian Tunnelling Method) tunnelling. This system has been used in a wide range of geological conditions, and quality of excavation is an important factor. For instance, a two-level operation was used to construct the Karawanken Tunnel between Austria and Yugoslavia (Sandtner and Gehring, 1988). This tunnel had an excavated cross-sectional area of 101 m^2, and is 8.5 km long. The roof section, which was 53 m^2, was driven in advance, the subsequent bench section being excavated whilst rock support, in the form of rock bolts, reinforced shotcrete and steel arches, was installed in the roof section. The tunnel was constructed mainly through limestone, dolostone and sandstone, with difficult ground conditions being encountered in water-bearing brecciated talus and friable black graphitic shale.

Basically excavation by a TBM is accomplished by a cutter head, equipped with an array of suitable cutters, which usually is rotated at a constant speed and thrust into the tunnel face by a hydraulic pushing system (Fig. 3.11). The latter is secured against the sides of the tunnel by hydraulic rams. The stresses imposed on the surrounding rock by a TBM are much less than those produced during blasting and therefore damage to the perimeter is minimized and a sensibly smooth base usually is achieved. What is more, overbreak normally is less during TBM excavation than during drilling and blasting, on average 5 per cent as compared with up

Figure 3.11 One of the tunnel-boring machines used for the construction of the Transfer Tunnel, Lesotho Highlands Water Scheme.

to 25 per cent for conventional methods. This means that less support is required. As a result, machine tunnelling generally is less expensive than the conventional method.

A TBM can be equipped with a rock drill on a boom for probing ahead of the face. The assembly can be mounted on a sliding deck behind the cutter head and probe drilling can take place at the same time as the TBM is advancing. The probe holes are angled to the tunnel alignment and probed several metres ahead of the face. If groundwater is encountered, then the assembly can drill a complete circle of holes in which to inject cover grout.

The rate of tunnel drivage obviously is an important economic factor in tunnelling, especially in hard rock. Whittacker and Frith (1990) referred to the progress made in tunnel drivage due to the developments that have occurred in tunnel boring machines. In other words, these machines have provided increased rates of advance and thereby shortened the time taken to complete tunnelling projects. They have achieved faster rates of drivage than conventional tunnelling methods in rocks with unconfined compressive strengths of up to 150 MPa. For example, Castro and Bell (1995) mentioned an average penetration rate of between 3.4 and 3.8 $m\,h^{-1}$ in sandstones of the Delivery Tunnel South, Lesotho, the rate depending on the strength of the sandstones (Fig. 3.12). In fact, the tunnel was completed 20 months ahead of schedule and in the process an African record of 1344.3 m in one month was achieved. Consequently, tunnels now are excavated much more frequently by TBMs than by conventional drill and blast methods. However, the rate of penetration of TBMs normally is slower in very soft rock because of time spent in ground control and in very hard rock. Moreover, the performance of TBMs is more sensitive to changes in rock properties than conventional drilling and blasting methods. Consequently, their use in rock masses that have not been thoroughly investigated involves a high risk.

According to Parkes (1988), the rock material properties exert a major influence on all aspects of TBM performance. In particular, the unconfined compressive strength commonly is one

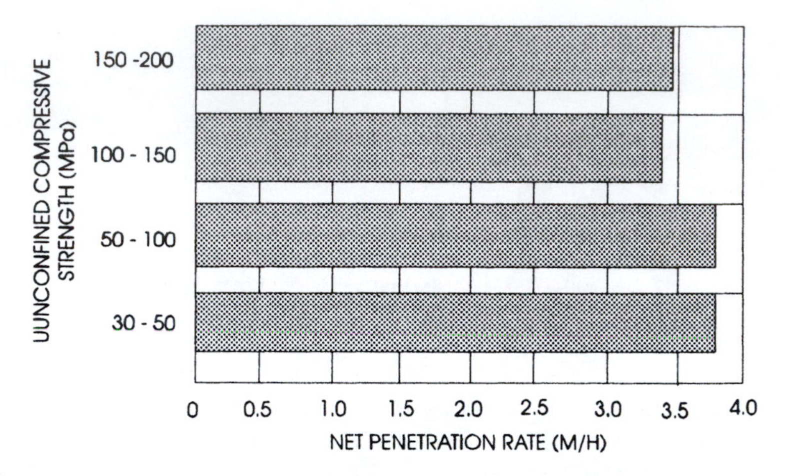

Figure 3.12 Unconfirmed compressive strength and average rate of penetration of TBM through Clarens Sandstone, Ngoajane to Hololo drive, Delivery Tunnel South, Lesotho Highlands Water Scheme. *Source*: Castro and Bell, 1995.

of the most important properties determining the rate of penetration of a TBM. The rate of penetration in low-strength rocks is affected by problems of roof support and instability, as well as gripper problems. The problems associated with rocks of high strength are the increased cutter wear and larger thrust, and hence cost, required to induce rock fracture. The rate of penetration also is influenced by the necessity to replace cutters on the head of a TBM, which involves downtime, the TBM not being in use. Cutter wear depends, in part, on the abrasive properties of the rock mass being bored. Whether a rock mass is massive, jointed, fractured, water-bearing, weathered or folded, also affect cutter life. For instance, in hard blocky ground some cutters are broken by the tremendous impact loads generated during boring. In addition, the rate of penetration is influenced by machine thrust, the rate of rotation of the cutter head, cutter geometry and the excavated diameter of the tunnel.

Estimation of likely rate of advance of a TBM is of considerable importance in relation to establishing the feasibility of a major tunnelling project, especially one with a long drive in hard massive rock masses. It also is required to make a realistic assessment of costs and to help plan the construction programme, and is of use to contractors in preparing their bids (Blindheim *et al*., 1991).

There are two ways of assessing the boreability of rock masses, that is, by using direct and indirect methods. The former involves monitoring the performance of full-size cutters either in testing adits or on large rock samples in the laboratory. One such test is the linear cutting test. Indirect methods involve the interpretation of the results of laboratory tests such as the unconfined compressive strength test, the point load test, the tensile strength test, the Schmidt hammer hardness test, the Taber abrasion test, the Norwegian Institute of Technology (NTH) tests and total hardness tests.

The linear cutting apparatus consists of a large structural frame that houses a TBM cutter. A rock sample is cast so that it fits into a moveable base beneath the cutter. Testing is carried out by drawing the base in a straight path below the cutting disc so that the disc cuts a straight kerf across the rock material. This is done for a preset penetration and the resultant forces generated in the cutter mounting are recorded.

The concept of total hardness (H_t) was developed by Tarkoy (1973) and is derived from the results of the Shore scleroscope (H_s) test and the Taber abrasion (H_a) test as follows:

$$H_t = H_s(H_a)^{0.5} \tag{3.3}$$

The Shore scleroscope test is a rebound hardness test. In the Taber abrasion test a core disc is abraded by a carborundum wheel and after a given number of revolutions the loss in weight of the disc is obtained. The Taber abrasion value is the inverse weight loss. The NTH tests include an impact test, a miniature drill test and an abrasion test, the results from which are used to derive three indices (Chen and Vogler, 1992). These are the drilling rate index (DRI), the bit wear index (BWI) and the cutter life index (CLI). The impact test provides an indication of rock brittleness and is an indirect expression of the amount of energy needed to crush rock. Crushed rock selected from a fraction sieved through the 16 mm and retained on 11.2 mm mesh is subjected to 20 drops in the impact apparatus. The brittleness (S20) value is the percentage of material that passes the 11.2 mm sieve after the test. The miniature drill test provides the Siever's J-value, which is an indirect expression of the surface hardness of rock. It is defined as the depth of hole measured in 0.1 mm after 200 revolutions of the miniature drill with a thrust of 20 kg. Lastly, the abrasion test gives a measure of the abrasion on a piece of cutter steel by rock powder. In other words, crushed rock, passing a 1 mm sieve, is fed onto a steel plate and the abrasion value (AVS) is equal to the mass loss in milligrams of the cutter steel piece after 20 revolutions of the steel plate.

The drilling rate index is derived from an empirical relationship between the brittleness (S20) and the Siever J values. The lower the DRI value, the more difficult it is to bore the material. For instance, values around 20 are not uncommon for hard rocks, while typical values for weak rocks are 80 or above. The bit wear index is based on an empirical relationship between the DRI and AVS. In this case, the lower the BWI, the lower is the wear on a tungsten carbide bit. A high DRI value corresponds with a low BWI value in most rock types and vice versa. The cutter life index is obtained from another empirical relationship between the Siever's J value and the AVS. Low CLI values indicate lower cutter ring life, values ranging from 20 for abrasive rocks to above 60 for quartz free rocks. Some examples of drilling rate index, bit wear index and cutter life index are given in Table 3.4. In the case of the Transfer Tunnel from Katse Dam, Lesotho, the final evaluation using an NTH predictive model suggested that in moderately jointed amygdaloidal basalts (average spacing of about 400 mm), penetration rates would exceed 4 m h^{-1}. If dolerite intrusions were encountered, then the rate of penetration would be reduced to about 1 m h^{-1}. Predictions of cutter life gave estimates ranging from about 300 to more than 1500 m^3 of solid rock per disc ring.

Table 3.4 Examples of drilling rate, bit wear and cutter life indices for tunnels from the Lesotho Highlands Water Project

Index	*Transfer Tunnel*			*Delivery Tunnel North*		
	Basalt 1	*Basalt 2*	*Basalt 3*	*Siltstone*	*Sandstone*	*Dolerite*
DRI	40	44	53	62	91	21
BWI	28	24	16	21	13	43
CLI	31	28	57	60	23	16

Notes
1, olivine basalt; 2, non-amygdaloidal basalt; 3, highly amygdaloidal basalt.

In poor rock conditions a TBM can be equipped to provide rapid erection of medium and heavy rock support. For example, in the Delivery Tunnel North, Lesotho-South Africa, the mudstones were particularly weak and exhibited rapid weathering characteristics (De Graaf and Bell, 1997). They also gave rise to squeezing ground. Moreover, the residual stresses in the weak siltstones and mudstones could cause distress shortly after excavation. For instance, during the earlier phases of investigation, exfoliation and convergence were observed in exploratory excavations. Hence, a double-shielded telescopic TBM was chosen to bore through the varied ground conditions and to provide protection against rock fall from incompetent roof strata, as well as to simultaneously install a structural lining of pre-cast concrete segments. A smooth bore segmental lining was used to provide both initial excavation support behind the cutterhead and the final tunnel lining. In addition, it had to provide strain-convergence compatibility with the rock mass to give acceptable working stresses within the lining system while offering adequate rock support. The segmental lining was developed to be compatible with a range of differing ground conditions, in particular, to provide both initial and long-term support for relatively weak mudstones. The worst conditions were encountered in weak mudstones at depths of 300 m or more. Particular attention also had to be paid to shear failure in bedded strata that could give rise to differential loadings on the lining if not adequately constrained. Each lining ring was 1.4 m long, corresponding to the length of one boring stroke of the cutterhead. A complete ring consisted of five, 250 mm thick, reinforced pre-cast concrete segments and a pre-cast key. They were erected in the tail-shield of the TBM using an erector arm, which lifted and positioned each 2.7 tonne segment by vacuum pad (Fig. 3.13). The segments were bolted together and to the preceding ring. The tail-shield of the TBM overlapped a section of the previously erected ring, providing complete protection during ring building. The segments were manufactured on site. As the TBM advanced, the tail-shield pulled out from under the newly assembled ring. The annulus behind the segmental ring was injected with about 3.2 m^3 of grout as soon as the

Figure 3.13 Installation of segmental lining, Delivery Tunnel North, Lesotho Highlands Water Scheme.

pre-formed grout-holes in the segments were clear of the shield. The grout mix consisted of 100 kg ordinary Portland cement, 400 kg pulverized fly ash, 1130 kg sand and sufficient water to achieve a spread of 175 mm m^{-3}. It was pumped into the annulus at pressures between 1 and 3 bars. The grout had to be thick but pumpable, had to set rapidly enough so as not to travel too far, and had to provide adequate support for the segments and surrounding rock. The annulus represented an overcut of 145 mm. It was designed to cope with potential convergence (squeezing) in the weaker rocks under high overburden pressure. However, this generally was of little consequence. Stress relief under the highest cover (350 m) resulted in some rock slabbing at the face and spalling from the crown. Generally, the fallen rock remained outside the TBM shield and the voids were backfilled with annular grout. Zones of weak rock in the crown were stabilized by proof grouting. This took place from a grouting station on the TBM backup, about 200 m from the face. Consolidation grouting was carried out after tunnelling was completed.

3.3.3 Drilling and blasting

The conventional method of advancing a tunnel in hard rock masses is by full-face driving in which the complete face is drilled and blasted as a unit (Wilbur, 1982). The amount of explosive used varies from about 0.9 kg m^{-3} in large diameter tunnels to around 3.6 kg m^{-3} in small diameter drives. Tunnels of less than 3 m equivalent diameter usually are worked in this way whilst in large tunnels, up to 12 m equivalent diameter, full-face drivage may be used if the ground is good and the tunnel length is sufficient to warrant the employment of major capital equipment. However, full-face driving should be used with caution where the rocks are variable. The usual alternatives are the top heading and bench method and the top heading method whereby the tunnel is worked on an upper and lower section or heading. The sequence of operations in these three methods is illustrated in Fig. 3.14.

Bergh-Christensen and Selmer-Olsen (1970) recognized three general types of rocks that caused difficulties in blasting. First, there were those characterized by high specific gravity and high intergranular cohesion with no preferred orientation of mineral grains. These rocks possessed high tensile strength and very low brittleness, examples being gabbros, greenstones and breccias. The high tensile strength resisted crack initiation and propagation on blasting. The second group included those rocks like certain brittle granites, gneisses and marbles, which were relatively brittle with little resistance to dynamic stresses. Blasting in such rocks gives rise to extensive pulverization immediately around the blastholes, leaving the zone between almost unfractured. The third category of rocks are those possessing marked preferred orientation, mica schist being a typical example. Such rocks split easily along their schistosity but crack propagation parallel to it is limited.

It is a basic principle of tunnel blasting that a cut should be opened up approximately in the centre of the face in order to provide a cavity into which subsequent shots can blast. Delay detonators allow a full-face to be charged, stemmed and fired, the shots being detonated in a pre-determined sequence (Martin, 1988). The first shots in the round blast out the cut and subsequent shots blast in sequence to the free face so formed.

Drilling and blasting can damage the rock structure, depending on the properties of the rock mass and the blasting technique. As far as blasting technique is concerned, attention should be given to the need to maintain adequate depths of pull, to minimize overbreak and to maintain blasting vibrations below acceptable levels. The stability of a tunnel roof in fissured rocks depends upon the formation of a natural arch and this is influenced by the extent of disturbance, the irregularities of the profile, and the relationship between tunnel size and fracture pattern. The amount of overbreak tends to increase with increased depths of pull since drilling inaccuracies

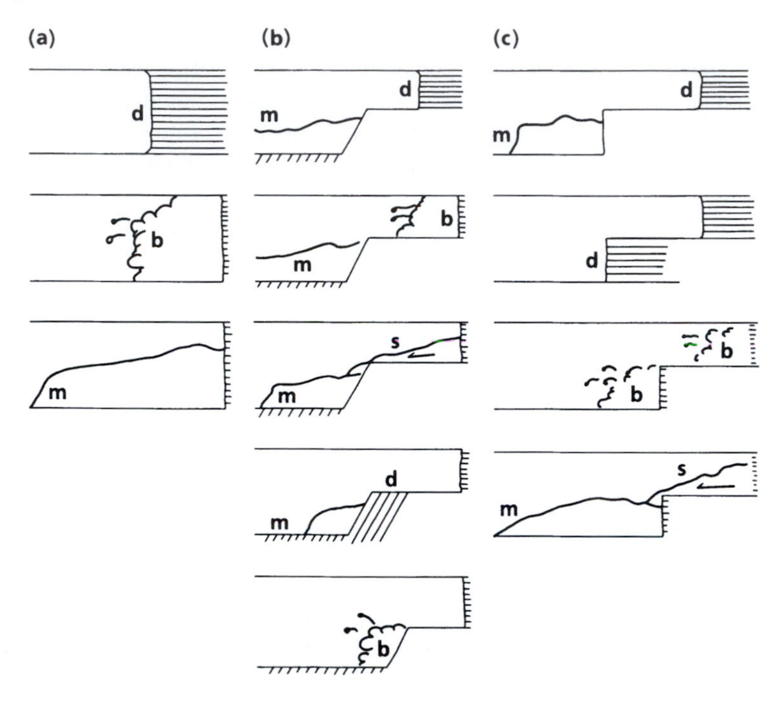

Figure 3.14 Conventional tunnelling methods (i.e. drilling and blasting): (a) full-face; (b) top heading and bench; (c) top heading bench drilled horizontally. Phases: d, drilling; b, blasting; m, mucking; s, scraping.

are magnified. In such situations not only does the degree of overbreak become very expensive in terms of grout and concrete backfill but it may give rise to support problems and subsidence over the crown of a tunnel. However, overbreak can be reduced by accurate drilling and a carefully controlled scale of blasting. Controlled blasting may be achieved either by pre-splitting the face to the desired contour or by smooth blasting.

In the pre-splitting method a series of holes are drilled around the perimeter of the tunnel, loaded with explosives that have a low charging density and are detonated before the main blast. The initial blast develops a fracture that spreads between the holes. Hence, the main blast leaves an accurate profile. The technique is not particularly suited to slates and schists because of their respective cleavage and schistosity. Indeed, slates tend to split along, rather than across their cleavage. Although it is possible to pre-split jointed rock masses adequately, the tunnel profile is still influenced by the pattern of the jointing.

Smooth blasting has proved a more successful technique than pre-splitting. Here again explosives with a low charging density are used in closely spaced perimeter holes. For example, the ratio between burden and hole spacing is usually 1:0.8, which means that crack formation is controlled between drillholes and hence is concentrated within the final contour. The holes are fired after the main blast, their purpose being to break away the last fillet of rock between the main blast and the perimeter. Smooth blasting cannot be carried out without good drilling precision. Normally, smooth blasting is restricted to the roof and walls of a tunnel but occasionally it is used in the excavation of the floor (Gustafsson, 1976). Fewer cracks are produced in the surrounding rock, which means that it is stronger and that there is less penetration of water. The greater rock strength also is closely associated with the fact that the desired roof curvature can be maintained to the greatest possible extent so that the load carrying capacity of the rock is properly utilized. This, in turn, means that less reinforcement is necessary.

Tunnelling in urban areas in close proximity to buildings can mean that vibrations from blasting cause damage to property or nuisance to occupants. In such situations a shallow depth of pull may be employed, thereby allowing smaller charges with delay detonators to be used. Hence, the amount of explosive detonated at any one time is kept to a minimum. A case of local residents alleging damage and nuisance occurred during the construction of the Dunns Bank Tunnel at Stourbridge, England. Explosives were used when sandstone was encountered (Newbury and Davenport, 1975). Initially, these explosives produced peak particle velocities that ranged between 10 and 47 $mm\,s^{-1}$ at 24 m (a peak particle velocity of 50 $mm\,s^{-1}$ normally is accepted as providing an adequate factor of safety against structural damage). These velocities were monitored by vibrographs. Unfortunately, however, the charges could produce peak particle velocities that exceeded 50 $mm\,s^{-1}$ at a distance of 13 m, the minimum distance between houses and tunnel. Although less powerful explosives were then used, so that the peak particle velocity was reduced to 20 $mm\,s^{-1}$, complaints were still forthcoming. Therefore, resort was made to a hydraulic burster to excavate the hard rock. This reduced the rate of progress by some 50 per cent.

3.4 Tunnel support

The time a rock mass may remain unsupported in a tunnel is called its bridging capacity or stand-up time. This mainly depends on the magnitude of the stresses within the unsupported rock mass, which in their turn depend on its span, its strength and its discontinuity pattern. If the bridging capacity of a rock is high, the rock material next to the heading will stay in place for a considerable time. By contrast, if the bridging capacity is low, the rock will immediately start to fail at the heading so that supports have to be erected as soon as possible.

Arch action refers to the capacity of the rock located above the roof of a tunnel to transfer the major part of the total weight of the overburden to the rock on both sides of the tunnel. In the immediate vicinity of the working face, the roof is supported on three sides. Thus, the overburden is carried by a half dome that can support a greater load than an arch of the same span.

From a stability point of view a tunnel cross-section in massive intact rock may be of any desired shape. However, in shattered and unstable rocks a circular shape is most suitable. Similarly, a tunnel in soft ground usually will be circular or elliptical on account of its loading condition.

In some rock masses it is inevitable that some material is removed from outside the perimeter planned as the excavation or pay-line. This material is referred to as the overbreak and its cost has to be met by the contractor. Obviously, every attempt must be made to reduce overbreak to a minimum. The amount of overbreak is influenced by the character of the rock type and its discontinuity pattern, as well as the method of excavation, the distance between the working face and the roof support, the length of time taken to install the support also being important.

The uniaxial compressive strength of intact rock is important if the discontinuities are widely spaced and the rock is weak. The intact strength also is important if the joints are not continuous or if the use of tunnelling machines is contemplated.

The spacing and orientation of discontinuities are of paramount importance as far as the stability of structures within a rock mass is concerned. The presence of discontinuities reduces the strength of a rock mass, and the spacing, as well as their dip and strike, governs the degree of such reduction. The effect of joint orientation in relation to the tunnel axis is given in Table 3.1. The condition of discontinuities includes the amount of separation between adjacent faces, their continuity and roughness, as well as any infill material. Tight discontinuities with rough surfaces and no infill have a high strength. On the other hand, open continuous discontinuities facilitate block movement and unrestricted flow of groundwater. The continuity influences the extent to which the rock material and the discontinuities separately affect the behaviour of the rock mass.

The correct evaluation of the effect of the dip and strike of formations, their joint pattern and the direction of tunnel drive is of paramount importance as regards determination of the type of support system to be used in a tunnel. Steeply dipping rocks (60–90°) lying parallel to the tunnel axis have a more adverse effect on support requirements than rocks dipping at 30°, regardless of direction of drive in the latter case. On the other hand, rocks dipping at 45° and lying perpendicular to the axis vary with respect to direction of drive, either against or with the dip. In the first instance, the rock would have a tendency to fall into the tunnel opening, in the latter the face would confine the rock to some extent.

The joint pattern often proves one of the most difficult and crucial factors to appraise when determining the type of support system to employ. In addition to defining the dimensions and orientation of the joint pattern, it is necessary to evaluate jointing with regard to the conditions of the joint surfaces, tunnel size, direction of drive and method of excavation.

The effect of groundwater on support requirements varies with respect to weathering, joint filler or condition of the joint surfaces and depth of cover. Probably the most difficult support situation experienced in tunnel driving occurs where heavy inflows of water under high pressures are encountered in conjunction with adverse rock properties. Many tunnels, however, have penetrated heavy inflow formations with little difficulty with respect to ground support.

The primary support for a tunnel in rock may be provided by rock bolts, shotcrete or steel arches. Rock bolts maintain the stability of an opening by suspending the dead weight of a slab from the rock above, by providing a normal stress on the rock surface to clamp discontinuities together and develop beam action, by providing a confining pressure to increase shearing resistance and develop arch action, and by preventing key blocks becoming loosened so that the strength and integrity of the rock mass is maintained (Stephansson, 1984). They may be used in conjunction with wire mesh. Shotcrete can be used for lining tunnels. For example, a layer 150 mm thick around a tunnel of 10 m diameter can carry a load of 500 kPa safely, corresponding to a burden of approximately 23 m of rock, more than has ever been observed with rock falls. Shotcrete may be unreinforced or reinforced, depending on the design requirements. Reinforcement may be provided by incorporating glass or steel fibre in the shotcrete or by spraying shotcrete over wire mesh. When combined with rock bolting, shotcrete has proved an excellent primary support system for all qualities of rock (Farmer, 1992). Such a support system also can be used as a permanent lining; in such cases the shotcrete lining usually is thicker. The New Austrian Tunnelling Method (NATM) commonly uses shotcrete for ground support. However, as Beveridge and Rankin (1995) pointed out, the NATM method seeks to share the load between the tunnel support and the ground by allowing some deformation to take place during construction. Accordingly, shotcrete is not a pre-requisite of NATM but is used because it provides relatively safe initial support and allows deformation with the ground, as well as gaining strength with time. In very bad cases steel arches can be used for reinforcement of weaker tunnel sections.

Permanent support may take the form of a concrete lining. This can be a concrete segmental lining, as mentioned above, or can be cast in place (Fig. 3.15). One of the advantages of a cast *in situ* concrete lining is that it can be designed to accommodate any shape of cross-section.

Rock pressures on the lining of a tunnel are influenced by the size and shape of the tunnel with respect to the intact strength of the rock mass(es) concerned and the nature of the discontinuities, the pre-excavation geostatic stress, the groundwater pressures, the method of excavation, the degree of overbreak, the length of time before placing the permanent lining, the stiffness of the

Figure 3.15 Placing concrete lining in an area where rock spalling had occurred in the Transfer Tunnel, Lesotho Highlands Water Scheme.

grout injected behind the lining and the rigidity of the lining itself. Location of a tunnel in an antiform tends to relieve the vertical pressure on the lining; conversely in a synform there is an increase in pressure on the lining. Also in an antiform, lateral pressure on the tunnel is greater close to the portals than at the middle of the tunnel, whereas in a synform the converse again applies. The redistributed ground stresses around circular tunnels in very competent rock are accepted by the rock as low radial and high tangential stresses. This gives a low loading on the tunnel lining. On the other hand, if a rock mass is weathered or highly jointed redistribution of stress may weaken it and eventually lead to partial collapse of overlying strata. Partial collapse results in high stresses on the lining at the crown and redistributes invert and sidewall resisting radial stresses.

A classification of rock masses is of primary importance in relation to the design of tunnel support. Lauffer's (1958) classification represented an appreciable advance in the art of tunnelling, since it introduced the concept of an active unsupported rock span and the corresponding stand-up time, both of which are very relevant parameters for determination of the type and amount of primary support in tunnels. The active span is the width of the tunnel or the distance from support to the face in cases where this is less than the width of the tunnel. The relationships found by Lauffer are given in Fig. 3.16.

3.4.1 The rating concept

Wickham *et al.* (1972) advanced the concept of rock structure rating (RSR) into rock mass classification, which refers to the quality of rock structure in relation to ground support in tunnels. The method rates the relative effect on ground support requirements of three parameters, rock structure, joint pattern and groundwater inflow with respect to each other. Each parameter is given a rating and the RSR value of a particular section of a tunnel is given by the numerical

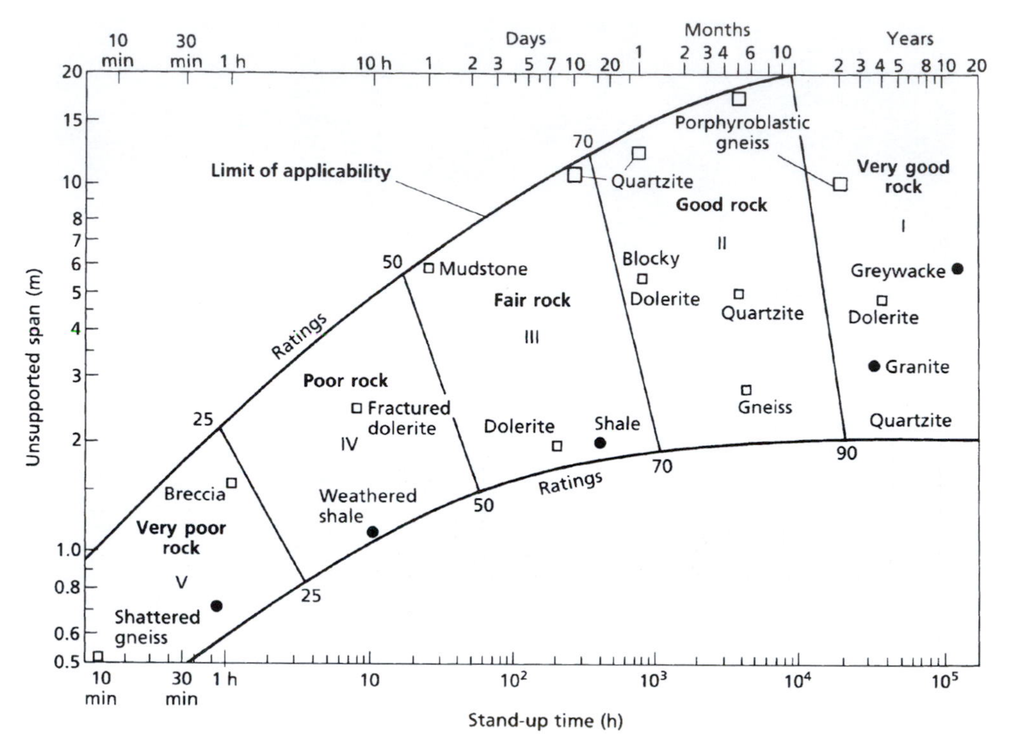

Figure 3.16 Geomechanics classification of rock masses for tunnelling. South African case studies are indicated by squares while those from Alpine countries are shown by dots.
Source: Lauffer, 1958.

sum of the ratings of the parameters. Hence, the RSR reflects the quality of the rock mass irrespective of the size of tunnel opening and the method of excavation. Wickham *et al.* concluded that rock masses with RSR values less than 27 would require heavy support, whilst those with ratings over 77 would probably stand unsupported. Subsequently, Bieniawski (1974) and Barton *et al.* (1975) introduced more sophisticated rating systems to assess rock quality in relation to tunnel support.

3.4.2 The geomechanics or rock mass rating (RMR) classification

Bieniawski (1974, 1983, 1989) maintained that the uniaxial compressive strength of rock material, the rock quality designation, the spacing, orientation and condition of the discontinuities, and groundwater inflow were the factors that should be considered in any engineering classification of rock masses. His classification of rock masses based on these parameters is given in Table 3.5. Each parameter is grouped into five categories and the categories are given a rating. Once determined, the ratings of the individual parameters are summed to give the total rating or class of rock mass. The higher the total rating is, the better the rock mass conditions. Assessment of the rock masses prior to the construction of the Istanbul Metro, Turkey, according to Dalgic (2002), was undertaken by using the RMR system. It was found that the predictions compared well with the actual observations. However, the accuracy of the rock mass rating in certain situations may be open to question; for example, it does not take into account the effects of blasting on rock masses. Neither does it consider the influence of *in situ* stress on stand-up

Table 3.5 The rock mass rating system (geomechanics classification of rock masses)

(a) Classification parameters and their ratings

Parameter		*Ranges of values*				
1 Strength of intact rock material	Point-load strength index (MPa)	>10	4–10	2–4	1–2	For this low range, uniaxial compressive test is preferred
	Uniaxial compressive strength (MPa)	>250	100–250	50–100	25–50	5–25 1–5 <1
Rating		15	12	7	4	2 1 0
2 Drill core quality RQD (%)		90–100	75–90	50–75	25–50	<25
Rating		20	17	13	8	3
3 Spacing of discontinuities		>2 m	0.6–2 m	200–600 mm	60–200 mm	<60 mm
Rating		20	15	10	8	5
4 Condition of discontinuities		Very rough surfaces Not continuous No separation Unweathered wallrock	Slightly rough surfaces Separation < 1 mm Slightly weathered walls	Slightly rough surfaces Separation < 1 mm Highly weathered walls	Slickensided surfaces or Gouge < 5 mm thick or Separation 1–5 mm Continuous	Soft gouge > 5 mm thick or Separation > 5 mm Continuous
Rating		30	25	20	10	0
5 Groundwater	Inflow per 10 m tunnel length (l/min)	None	<10	10–25	25–125	>125
		or	or	or	or	or
	Ratio $= \dfrac{\text{Joint water pressure}}{\text{Major principal stress}}$	0	<0.1	0.1–0.2	0.2–0.5	>0.5
		or	or	or	or	or
	General conditions	Completely dry	Damp	Wet	Dripping	Flowing
Rating		15	10	7	4	0

Table 3.5 (Continued)

(b) Rating adjustment for discontinuity orientations

Parameter		*Ranges of values*				
Strike and dip orientations of discontinuities		Very favourable	Favourable	Fair	Unfavourabole	Very unfavourabole
Ratings	Tunnels and mines	0	−2	−5	−10	−12
	Foundations	0	−2	−7	−15	−25
	Slopes	0	−5	−25	−50	−60

(c) Rock mass classes determined from total ratings

Rating	100←81	80←61	60←41	40←21	<20
Class no.	I	II	III	IV	V
Description	Very good rock	Good rock	Fair rock	Poor rock	Very poor rock

(d) Meaning of rock mass classes

Class no.	I	II	III	IV	V
Average stand-up time	20 year for 15 m span	1 year for 10 m span	1 week for 5 m span	10 hours for 2.5 m span	30 min for 1 m span
Cohesion of the rock mass (kPa)	>400	300–400	200–300	100–200	<100
Friction angle of the rock mass (deg.)	>45	35–45	25–35	15–25	<15

Source: Bianiawski, 1989. Reproduced by kind permission of John Wiley & Sons.

time nor the durability of the rock, notably mudrocks (Varley, 1990). Mudrock durability can be assessed in terms of the geodurability classification, which was introduced by Olivier (1979) during the construction of the Fish River Tunnel, South Africa (Fig. 3.17).

Suitable support measures at times must be adopted to attain a stand-up time longer than that indicated by the class of the rock mass. These measures constitute the primary or temporary support. Their purpose is to ensure tunnel stability until the secondary or permanent support system, for example, a concrete lining, is installed. The form of primary support depends on depth below the surface, tunnel size and shape, and method of excavation. Table 3.6 indicates the primary support measures for shallow tunnels 5–12 m in diameter, driven by drilling and blasting. These primary support measures probably will be able to carry all the load that is ever likely to act on them in a tunnel. Indeed, Bieniawski (1974) suggested that the traditional concept of temporary and permanent support was losing its meaning.

3.4.3 The 'Q' system

Barton *et al.* (1975) pointed out that Bieniawski (1974) in his analysis of tunnel support more or less ignored the roughness of joints, the frictional strength of the joint fillings, and the rock load. They therefore proposed the concept of rock mass quality (Q), which could be used as a

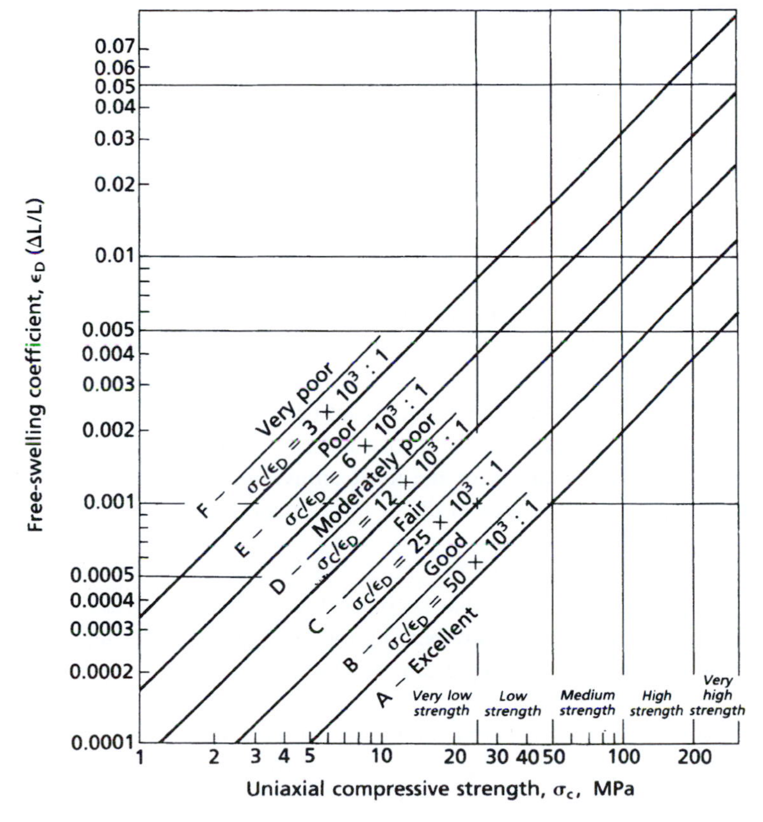

Figure 3.17 Geodurability classification of mudrocks.
Source: Olivier, H.G., 1979. Reproduced with kind permission of Elsevier.

Table 3.6 Guide for selection of primary support in tunnels at shallow depth; size: 5–15 m; construction by drilling and blasting

	Alternative support systems		
Rock mass class	*Mainly rockbolts (20 mm dia., length half tunnel width, resin bonded)*	*Mainly shotcrete*	*Mainly steel ribs*
I	Generally no support is required		
II	Rockbolts spaced 1.5–2.0 m plus occasional wire mesh in crown	Shotcrete 50 mm in crown	Uneconomic
III	Rockbolts spaced 1.0–1.5 m plus wire mesh and 30 mm shotcrete in crown where required	Shotcrete 100 mm in crown and 50 mm in sides plus occasional wire mesh and rockbolts where required	Light sets spaced 1.5–2 m
IV	Rockbolts spaced 0.5–1.0 m plus wire mesh and 30–50 mm shotcrete in crown and sides	Shotcrete 150 mm in crown and 100 mm in sides plus wire mesh and rockbolts, 3 m long spaced 1.5 m	Medium sets spaced 0.7–1.5 m plus 50 mm shotcrete in crown and sides
V	Not recommended	Shotcrete 200 mm in crown and 150 mm in sides plus wire mesh, rockbolts and light steel sets. Seal face. Close invert	Heavy sets spaced 0.7 m with lagging. Shotcrete 80 mm thick to be applied immediately after blasting

Source: Bieniawski, 1974.

means of rock classification for tunnel support. They defined the rock mass quality in terms of six parameters:

1 The RQD or an equivalent estimate of joint density.
2 The number of joint sets (J_n), which is an important indication of the degree of freedom of a rock mass. The RQD and the number of joint sets provide a crude measure of relative block size.
3 The roughness of the most unfavourable joint set (J_r). The joint roughness and the number of joint sets determine the dilatancy of a rock mass.
4 The degree of alteration or filling of the most unfavourable joint set (J_a). The roughness and degree of alteration of the joint walls or filling materials provides an approximation of the shear strength of the rock mass.
5 The degree of water seepage (J_w).
6 The stress reduction factor (SRF) that accounts for the loading on a tunnel caused either by loosening loads in the case of clay-bearing rock masses, or unfavourable stress–strength ratios in the case of massive rock. Squeezing and swelling also is taken account of in the SRF.

They provided a rock mass description and ratings for each of the six parameters (Table 3.7) which enabled the rock mass quality (Q) to be derived from:

$$Q = \frac{\text{RQD}}{J_n} \times \frac{J_r}{J_a} \times \frac{J_w}{\text{SRF}} \tag{3.4}$$

Table 3.7 Classification of individual parameters used in the Norwegian Geotechnical Institute (NGI) tunnelling quality index or 'Q' system.

Description	Value	Notes
1 *Rock quality designation*	RQD	1 Where RQD is reported or measured as ≤ 10 (including 0), a nominal value of 10 is used to evaluate Q
A Very poor	0–25	2 RQD intervals of 5, i.e. 100, 95, 90, etc. are sufficiently accurate
B Poor	25–50	
C Fair	50–75	
D Good	75–90	
E Excellent	90–100	
2 *Joint set number*	J_n	1 For intersections use $(3.0 \times J_n)$
A Massive, no or few joints	0.5–1.0	2 For portals use $(2.0 \times J_n)$
B One joint set	2	
C One joint set plus random	3	
D Two joint sets	4	
E Two joint sets plus random	6	
F Three joint sets	9	
G Three joint sets plus random	12	
H Four or more joint sets, random, heavily jointed 'sugar cube', etc.	15	
J Crushed rock, earthlike	20	
3 *Joint roughness number*	J_r	1 Add 1.0 if the mean spacing of the relevant joint set is greater than 3 m
(a) Rock wall contact and		
(b) Rock wall contact before 10 cm shear		2 $J_r = 0.5$ can be used for planar, slickensided joints having lineations, provided the lineations are orientated for minimum strength
A Discontinuous joints	4	
B Rough or irregular, undulating	3	
C Smooth, undulating	2	
D Slickensided, undulating	1.5	
E Rough or irregular, planar	1.5	
F Smooth, planar	1.0	
G Slickensided, planar	0.5	
(c) No rock wall contact when sheared		
H Zone containing clay minerals thick enough to prevent rock wall contact	1.0	
J Sandy, gravelly, or crushed zone thick enough to prevent rock wall contact	1.0	

Table 3.7 (Continued)

	Description	*Value*		*Notes*
4	*Joint alteration number*	J_a	ϕ_r (approx.)	1 Values of ϕ_r, the residual friction angle, are intended as an approximate guide to the mineralogical properties of the alteration products, if present
	(a) Rock wall contact			
A	Tightly healed, hard, non-softening, impermeable filling	0.75	—	
B	Unaltered joint walls, surface staining only	1.0	(25–35°)	
C	Slightly altered joint walls, non-softening mineral coatings, sandy particles, clay-free disintegrated rock, etc.	2.0	(25–30°)	
D	Silty, or sandy clay coatings, small clay-fraction (non-softening)	3.0	(20–25°)	
E	Softening or low-friction clay mineral coatings, i.e. kaolinite, mica. Also chlorite, talc, gypsum and graphite etc., and small quantities of swelling clays. (Discontinuous coatings, 1–2 mm or less in thickness)	4.0	(8–16°)	
	(b) Rock wall contact before 10 cm shear			
F	Sandy particles, clay-free disintegrated rock, etc.	4.0	(25–30°)	
G	Strongly overconsolidated, non-softening clay mineral fillings (continuous, <5 mm thick)	6.0	(16–24°)	
H	Medium or low overconsolidation, softening, clay mineral fillings (continuous, <5 mm thick)	8.0	(12–16°)	
J	Swelling clay fillings, i.e. montmorillonite (continuous, <5 mm thick). Values of J_a depend on percentage of swelling clay-size particles and access to water	8.0–12.0	(6–12°)	
	(c) No rock wall contact when sheared			
K	Zones or bands of disintegrated or crushed rock and clay	6.0		
L	(see G, H and J for clay conditions)	8.0		
M		8.0–12.0	(6–24°)	
N	Zones or bands of silty or sandy clay, small clay fraction (non-softening)	5.0		
Q	Thick, continuous zones or bands of clay (see G, H and J	10.0–13.0		
P	for clay conditions)			
R		13.0–20.0	(6–24°)	

5 *Joint water reduction factor*	J_w	Approx. water pressure (kgf/cm^2)	1 Factors C to F are crude estimates. Increase J_w if drainage measures are installed 2 Special problems caused by ice formation are not considered
A Dry excavations or minor inflow, i.e. <5 l/min locally	1.0	<1.0	
B Medium inflow or pressure, occasional outwash of joint fillings	0.66	1.0–2.5	
C Large inflow or high pressure in competent rock with unfilled joints	0.5	2.5–10.0	
D Large inflow or high pressure, considerable outwash of fillings	0.33	2.5–10.0	
E Exceptionally high inflow or pressure at blasting, decaying with time	0.2–0.1	>10	
F Exceptionally high inflow or pressure continuing without decay	0.1–0.05	>10	
6 *Stress reduction factor* (a) Weakness zones intersecting excavation, which may cause loosening of rock mass when tunnel is excavated	SRF		1 Reduce these values of SRF by 25–50% if the relevant shear zones only influence but do not intersect the excavation 2 For strongly anisotropic virgin stress field (if measured): when $5 \leq \sigma_1/\sigma_3 \leq 10$, reduce σ_c to $0.8\sigma_c$ and σ_t to $0.8\sigma_t$. When $\sigma_1/\sigma_3 > 10$, reduce σ_c and σ_t to $0.6\sigma_c$ and $0.6\sigma_t$, where σ_c = unconfined compressive strength, and σ_t = tensile strength (point load) and σ_1 and σ_3 are the major and minor principal stresses 3 Few case records available where depth of crown below surface is less than span width. Suggest SRF increase from 2.5 to 5 for such cases (see H).
A Multiple occurrences of weakness zones containing clay or chemically disintegrated rock, very loose surrounding rock (any depth)	10.0		
B Single weakness zones containing clay, or chemically disintegrated rock (excavation depth <50 m)	5.0		
C Single weakness zones containing clay or chemically disintegrated rock (excavation depth >50 m)	2.5		
D Multiple shear zones in competent rock (clay-free), loose surrounding rock (any depth)	7.5		
E Single shear zones in competent rock (clay-free) (depth of excavation <50 m)	5.0		
F Single shear zones in competent rock (clay-free) (depth of excavation <50 m)	2.5		
G Loose open joints, heavily jointed or 'sugar cube' (any depth)	5.0		
(b) Competent rock, rock-stress problems	σ_c/σ_1	σ_t/σ_1	SRF

Table 3.7 (Continued)

	Description	*Value*		
H	Low stress, near surface	>200	>13	2.5
J	Medium stress	200–10	13–0.66	1.0
K	High stress, very tight structure (usually favourable to stability, may be unfavourable for wall stability)	10–5	0.66–0.33	0.5–2
L	Mild rock burst (massive rock)	5–2.5	0.33–0.16	5–10
M	Heavy rock burst (massive rock)	<2.5	<0.16	10–20
	(c) Squeezing rock, plastic flow of incompetent rock under the influence of high rock pressure			
N	Mild squeezing-rock pressure			5–10
O	Heavy squeezing-rock pressure			10–20
	(d) Swelling rock, chemical swelling activity depending upon presence of water			
P	Mild swelling rock pressure			5–10
R	Heavy swelling rock pressure			10–20

Source: Barton *et al.*, 1975. Reproduced by kind permission of Norwegian Geotechnical Institute.

Additional notes on the use of these Tables

When making estimates of the rock mass quality (Q) the following guidelines should be followed, in addition to the notes listed in the Tables:

1 When drillhole core is unavailable, RQD can be estimated from the number of joints per unit volume, in which the number of joints per metre for each joint set are added. A simple relation can be used to convert this number to RQD for the case of clay-free rock masses: RQD = $115 - 3.3J_v$ (approx.) where J_v = total number of joints per m^3 (RQD = 100 for J_v <4.5).

2 The parameter J_n representing the number of joint sets will often be affected by foliation, schistosity, slaty cleavage, or bedding, etc. If strongly developed these parallel 'joints' should obviously be counted as a complete joint set. However, if there are few 'joints' visible, or only occasional breaks in the core due to these features, then it will be more appropriate to count them as 'random' when evaluating J_n.

3 The parameters J_r and J_a (representing shear strength) should be relevant to the weakest significant joint set or clay-filled discontinuity in the given zone. However, if the joint set or discontinuity with the minimum value of (J_r/J_a) is favourably oriented for stability, then a second, less favourably oriented joint set or discontinuity may sometimes be more significant, and its higher value of J_r/J_a should be used when evaluating Q. The value of J_r/J_a should in fact relate to the surface most likely to allow failure to initiate.

4 When a rock mass contains clay, the factor SRF appropriate to loosening loads should be evaluated. In such cases the strength of the intact rock is of little interest. However, when jointing is minimal and clay is completely absent the strength of the intact rock may become the weakest link, and the stability will then depend on the ratio rock stress/rock strength. A strongly anisotropic stress field is unfavourable for stability and is roughly accounted for as in note 2 in the Table for stress reduction factor evaluation.

5 The compressive and tensile strengths (σ_c and σ_t) of the intact rock should be evaluated in the saturated condition if this is appropriate to present or future *in situ* conditions. A very conservative estimate of strength should be made for those rocks that deteriorate when exposed to moist or saturated conditions.

The numerical value of Q ranges from 0.001 for exceptionally poor quality squeezing ground, to 1000 for an exceptionally good quality rock mass that is practically unjointed. Rock mass quality, together with the support pressure, and the dimensions and purpose of the underground excavation are used to estimate the type of suitable permanent support. A fourfold change in Q value indicates the need for a different support system. Zones of different Q value are mapped and classified separately within an underground excavation. However, in variable conditions where different zones occur, each for only a few metres, it is more economic to map the overall quality and to estimate an average value of Q, from which a design of a compromise support system can be made (Barton, 1988).

The Q value is related to the type and amount of support by deriving the equivalent dimensions of the excavation. The latter is related to the size and purpose of the excavation and is obtained from:

$$\text{Equivalent dimension} = \frac{\text{span or height of wall}}{\text{ESR}} \tag{3.5}$$

where ESR is the excavation support ratio related to the use of the excavation and the degree of safety required. Some values of ESR are given in Table 3.8.

Stacey and Page (1986) made use of the Q system to develop design charts to determine the factor of safety for unsupported excavations, the spacing of rock bolts over the face of an excavation, and the thickness of shotcrete on an excavation (Fig. 3.18a, b and c, respectively). For civil engineering applications a factor of safety exceeding 1.2 is required if the omission of support is to be considered. The support values suggested in the charts are for primary support. The values should be doubled for long-term support.

Hoek (1994) introduced the concept of the Geological Strength Index (GSI), which provides a system of estimating the reduction in the strength of a rock mass for different geological conditions. Each category of rock mass can be identified from field observations and then can be used to estimate the value of GSI from Fig. 3.19. Once the GSI has been estimated, then the

Table 3.8 Equivalent support ratio for different excavations

Excavation category	*ESR*
Temporary mine openings	3–5
Vertical shafts:	
Circular section	2.5
Rectangular/square section	2.0
Permanentmine openings, water tunnels for hydropower (excluding high-pressure penstocks), pilot tunnels, drifts, and headings for large excavations	1.6
Storage caverns, water-treatment plants, minor highway and railroad tunnels, surge chambers,access tunnels	1.4
Power stations,major highway or railroad tunnels, civil defence chambers, portals, intersections	1.0
Underground nuclearpower stations, railroad stations, factories	0.8

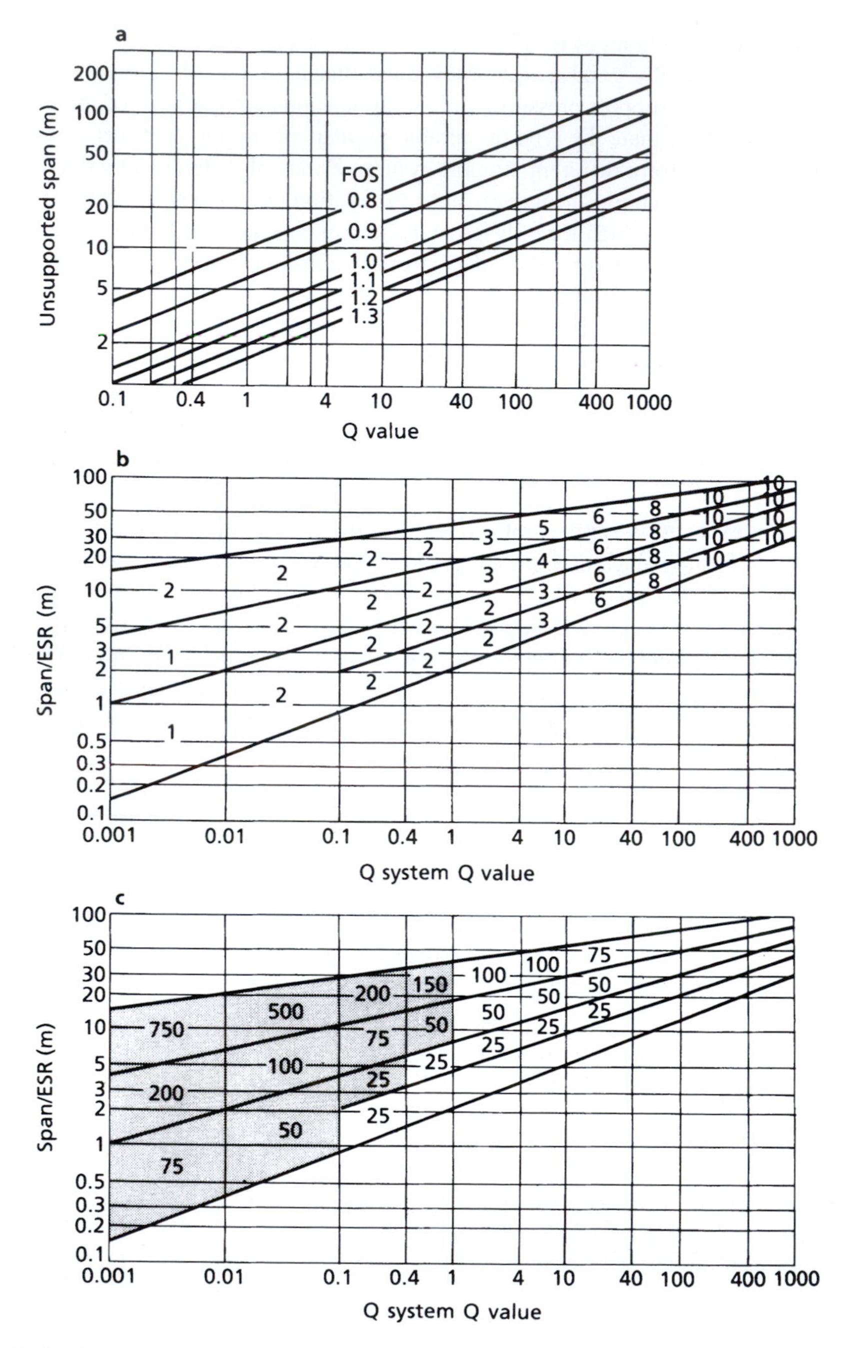

Figure 3.18 (a) Relationship between unsupported span and Q value (b) rock bolt support estimation using the Q system, bolt spacing – m^2 of excavation per bolt, where the area per bolt is greater than $6\,m^2$, spot bolting is implied (c) shotcrete and wire mesh support estimation using the Q system. Thickness of shotcrete in millimetres (mesh reinforcement in shaded areas), Note that very thick applications of shotcrete are not practical but the values are included for completeness. The support intensity given in design charts b and c is appropriate for primary support; where long-term support is required, the design chart values should be modified as follows (i) divide area per bolt by 2 (ii) multiply shotcrete thickness by 2. FOS, factor of safety.

Source: Stacey and Page, 1986. Reproduced by kind permission of Trans-Tech Publication.

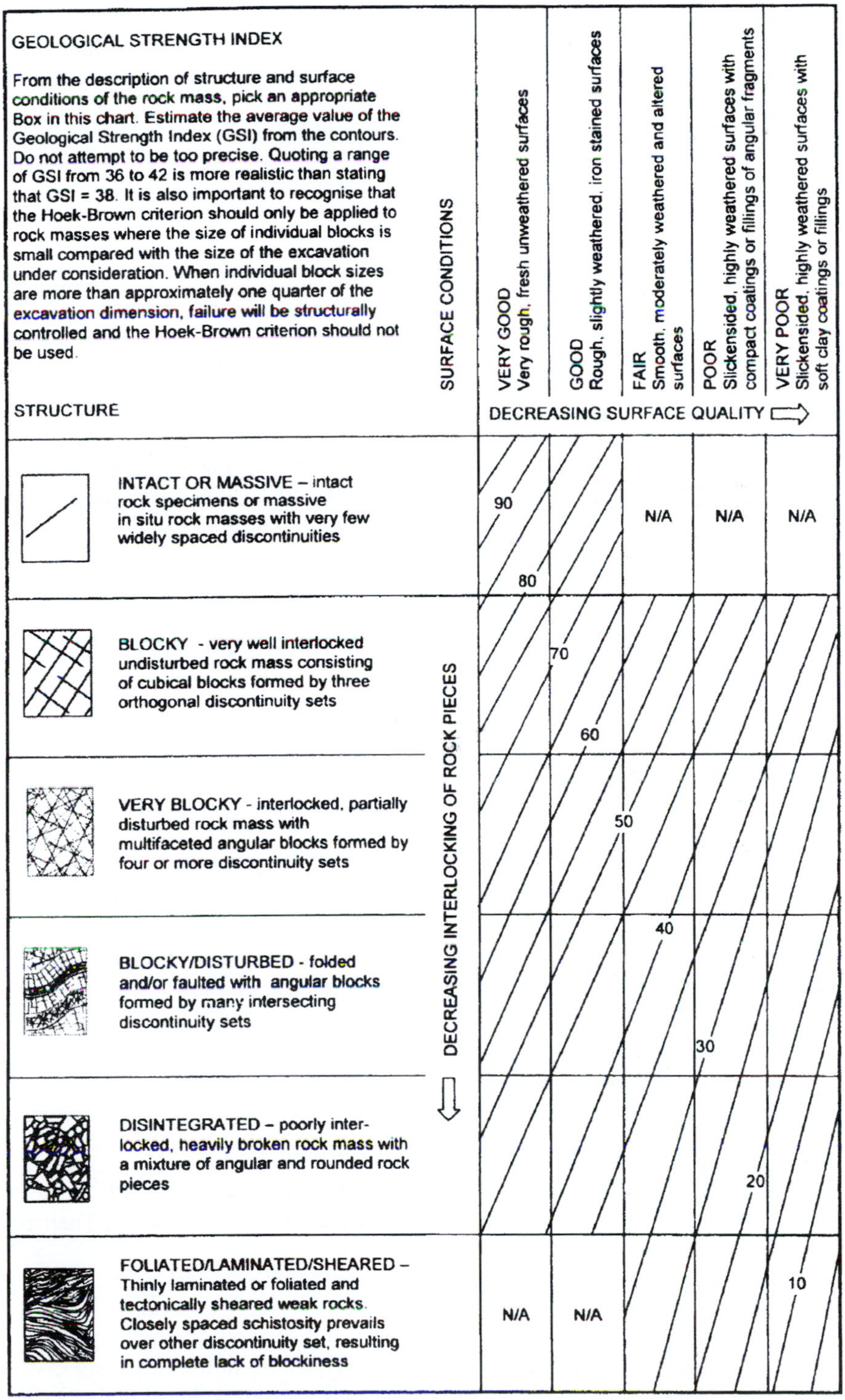

GEOLOGICAL STRENGTH INDEX

From the description of structure and surface conditions of the rock mass, pick an appropriate Box in this chart. Estimate the average value of the Geological Strength Index (GSI) from the contours. Do not attempt to be too precise. Quoting a range of GSI from 36 to 42 is more realistic than stating that GSI = 38. It is also important to recognise that the Hoek-Brown criterion should only be applied to rock masses where the size of individual blocks is small compared with the size of the excavation under consideration. When individual block sizes are more than approximately one quarter of the excavation dimension, failure will be structurally controlled and the Hoek-Brown criterion should not be used.

SURFACE CONDITIONS — DECREASING SURFACE QUALITY ⇨

STRUCTURE — DECREASING INTERLOCKING OF ROCK PIECES ⇩

STRUCTURE	VERY GOOD Very rough, fresh unweathered surfaces	GOOD Rough, slightly weathered, iron stained surfaces	FAIR Smooth, moderately weathered and altered surfaces	POOR Slickensided, highly weathered surfaces with compact coatings or fillings of angular fragments	VERY POOR Slickensided, highly weathered surfaces with soft clay coatings or fillings
INTACT OR MASSIVE – intact rock specimens or massive in situ rock masses with very few widely spaced discontinuities	90 80		N/A	N/A	N/A
BLOCKY - very well interlocked undisturbed rock mass consisting of cubical blocks formed by three orthogonal discontinuity sets		70 60			
VERY BLOCKY - interlocked, partially disturbed rock mass with multifaceted angular blocks formed by four or more discontinuity sets			50		
BLOCKY/DISTURBED - folded and/or faulted with angular blocks formed by many intersecting discontinuity sets			40	30	
DISINTEGRATED – poorly interlocked, heavily broken rock mass with a mixture of angular and rounded rock pieces				20	
FOLIATED/LAMINATED/SHEARED – Thinly laminated or foliated and tectonically sheared weak rocks. Closely spaced schistosity prevails over other discontinuity set, resulting in complete lack of blockiness	N/A	N/A			10

Figure 3.19 Determination of geological strength index.
Source: Hoek *et al*., 1998. Reproduced by kind permission of Springer-Verlag.

parameters relating to the strength characteristics of a rock mass can be derived. Hoek *et al.* (1995) subsequently showed how the RMR system of Bieniawski (1989) and the Q system of Barton *et al.* (1975) could be used to estimate the GSI. In the case of the former, the GSI can be estimated as follows:

$$GSI = RMR - 5 \tag{3.6}$$

Using the Q system, it can be estimated from:

$$GSI = 9\log_e Q' + 44 \tag{3.7}$$

where

$$Q' = \frac{RQD}{J_n} \times \frac{J_r}{J_a} \tag{3.8}$$

and J_n is the number of joint sets, J_r is the joint roughness and J_a is the degree of joint alteration. Note that the minimum value that can be obtained from Bieniawski's 1989 classification is 23. The minimum value of Q' is 0.0208 that results in a value of GSI of approximately 9 for a thick clay-filled fault or shear zone.

The *in situ* value of Young's modulus of a rock mass is an important parameter in large-scale construction operations but may be difficult and expensive to determine in the field. Accordingly, Bieniawski (1978) suggested a means of deriving the intact value of Young's modulus (E_m) from the RMR as follows:

$$E_m = 2RMR - 100 \tag{3.9}$$

Subsequently, Serafim and Pereira (1983) proposed a further relationship involving the RMR, it being:

$$E_m = \frac{10(RMR - 10)}{40} \tag{3.10}$$

Unfortunately, however, this relationship tends to overpredict values of Young's modulus for poor-quality rock masses. Hence, Hoek and Brown (1997) suggested the following modification of the Serafim and Pereira relationship for rock masses where the intact unconfined compressive (σ_{ci}) strength was less than 100 MPa.

$$E_m = \sqrt{\frac{\sigma_{ci}}{100}}\, 100^{(GSI-10/40)} \tag{3.11}$$

In this case the GSI has been substituted for the RMR. Barton *et al.* (1992) suggested that the Q system could be used to estimate the *in situ* value of Young's modulus. Their proposed relationship was as follows:

$$E_m = 25\log_{10} Q \tag{3.12}$$

Hudson and Harrison (1997) pointed out some of the limitations of the RMR and Q rock mass classification systems. For example, they indicated that stress is not included in the RMR system nor is intact strength of rock included in the Q system yet either of these parameters could be a cause of failure in certain circumstances. Furthermore, the measured values of discontinuity frequency and RQD depend on the direction of measurement but this is not taken into account in either of the classification systems.

The RMR and Q systems have been used principally in relation to the design of primary support for tunnels excavated in hard rock by drill and blast methods. Consequently, if a tunnel is constructed in moderately weak rock by TBM, then these two systems may not be wholly suited to the conditions. For example, De Graaf and Bell (1997) referred to a site-specific rock mass classification that was developed for the Delivery Tunnel North in Lesotho-South Africa. The rock types encountered along the tunnel route were mainly mudstones, siltstones and sandstones, and the principal objective of the classification was to characterize the stability conditions. Furthermore, the use of a full-face TBM, which placed a one-pass segmented lining, meant that the data obtainable was restricted to the limited exposures at the tunnel face. This meant that all the data required for the application of the RMR or Q systems was not readily available, so the site-specific classification was based primarily upon rock type, cover and discontinuity spacing.

3.5 Tunnelling and subsidence

O'Reilly and New (1982) stated that one of the basic requirements for the design of a satisfactory tunnel, especially in soft ground in urban areas, was that its construction should cause as little damage as possible to overlying or adjacent structures and services. Accordingly, the risk of damage due to the occurrence of subsidence above a tunnel must be minimized, which means that reliable predictions of its amount and extent are required. The amount of subsidence that occurs as a result of tunnel construction normally depends on the loss of ground that occurs as the ground restabilizes around the tunnel opening. This loss of ground is governed by the depth of cover, the strength and deformation characteristics of the ground, the groundwater conditions and the care with which construction is carried out. Part of the subsidence that may occur during tunnelling in soft ground is due to relaxation of the ground in front of the face, especially when a tunnel is below the water table. Hence, the soil ahead of the face needs to be stabilized by using compressed air, by freezing, or by grouting, or by using a bentonite shield. Most subsidence, however, occurs above the excavation so that any method of supporting a tunnel concurrently with excavation is advantageous. Placement of the permanent lining more or less immediately behind a tunnel face excavated by a tunnelling machine with a shield in soft ground conditions should minimize ground movements. Kettle and Gandais (1995) described the use of jet grouting, in silty sands with clay bands, in order to develop a roof support system for the Brovello Tunnel in northern Italy, which would minimize subsidence.

Other grouting techniques have been used in soft ground to compensate subsidence movements and so restore structures to their original elevation. These techniques include compaction grouting, compensation grouting, claquage or hydrofracture grouting, and squeeze grouting (Kimmance *et al.*, 1995). Compaction grouting, which commonly is referred to as mud-jacking in the United States, involves the injection of low slump grout of moderate to high viscosity and high shear strength under high pressure to form a bulb, which as it develops compacts and raises the soil above (Baker *et al.*, 1983). In this way, a structure that has subsided can be lifted. Compensation grouting can be distinguished from compaction grouting in that it involves injection of grout to balance against ground loss at a tunnel heading so that subsidence does not occur at the surface. In compensation grouting the grout is injected above a tunnel through *tubes-a-manchettes*. The latter consists essentially of a steel tube with rings of small holes, each ring being enclosed by a tightly fitting rubber sleeve that acts as a one-way valve. The *tubes-a-manchettes* may be installed in the ground from specially constructed shafts in which they are located at different levels. Generally, multiple phases of grouting are used incrementally to control subsidence. Claquage grouting uses low to moderate viscosity grouts of low strength to fissure the ground, the grout occupying the fissures. Considerable pressure may be required to start the flow of grout but once it begins to move the pressure can be reduced. In this way an intermeshing network of grout-filled fractures is formed. However, grout occupying horizontal fractures, in particular, may cause the ground to heave. Accordingly, if damaging heave is to be avoided, then ground movement must be monitored. Grouts of moderate viscosity and shear strength are used in squeeze grouting in order to penetrate fissures in the ground. However, these types of grouting can impose loads on a completed tunnel lining that can lead to its deformation and worse. For example, the report of the Health and Safety Executive (Anon., 2000) into the collapse of the Heathrow Express Tunnel at Heathrow Airport, London, sited compaction grouting as a contributary cause of the collapse, in that it had damaged part of the tunnel lining that, in turn, had been inadequately repaired.

Attewell and Farmer (1975) demonstrated a strong similarity between the form of subsidence developed above an advancing shield-driven tunnel in clay soil and subsidence due to longwall mining. In an examination of tunnels excavated in clay soils, they noted that in the vertical plane initial movement was into the tunnel face. This was followed by radial, or near radial, movement of clay into an annular gap created by over-excavation at the shield. Substantial radial movement took place in the plane normal to the centre line of the tunnels, the movement elsewhere being pre-dominantly vertical. Consequently, they concluded that deformation, especially in the immediate vicinity of tunnels, is much greater than predicted by elastic theory. Magata *et al.* (1982) reviewed the factors that influenced subsidence associated with shield tunnelling in soft ground conditions at the Shinozaki Tunnel, Japan. These included lowering of the water table as a result of dewatering, ground collapse at the tunnel face largely due to inadequate improvement in soil strength after injection of chemical grout and disturbance of the soil. Magata *et al.* attributed the soil disturbance to poor-quality excavation. Indeed, over-excavation of ground and deviation in driving the tunnel caused the formation and then the, extension of tail voids.

According to Butler and Hampton (1975) tunnelling took place in gravelly sand, silty sand and silty clay of Pleistocene age in the Washington Metropolitan Area Transit System. Observations at the ground surface indicated that large disturbances occurred as the shield passed beneath a given surface point, with a substantial amount of subsidence occurring after the tail of the shield had passed. Figure 3.20a illustrates how subsidence developed in relation to shield advance.

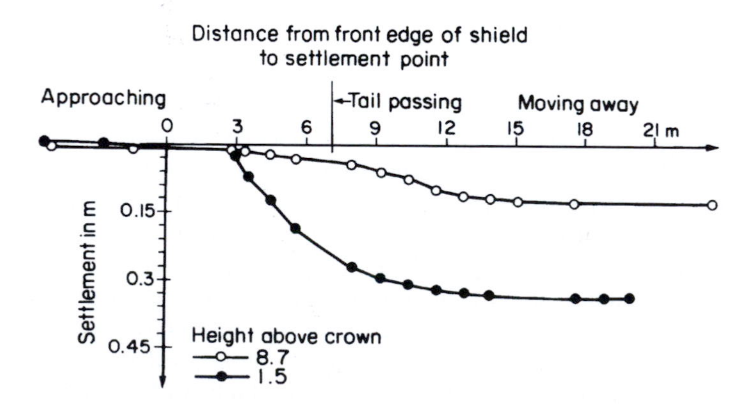

Figure 3.20a Subsidence versus shield advance Washington Metropolitan Area Transit system. *Source*: Butler and Hampton, 1975.

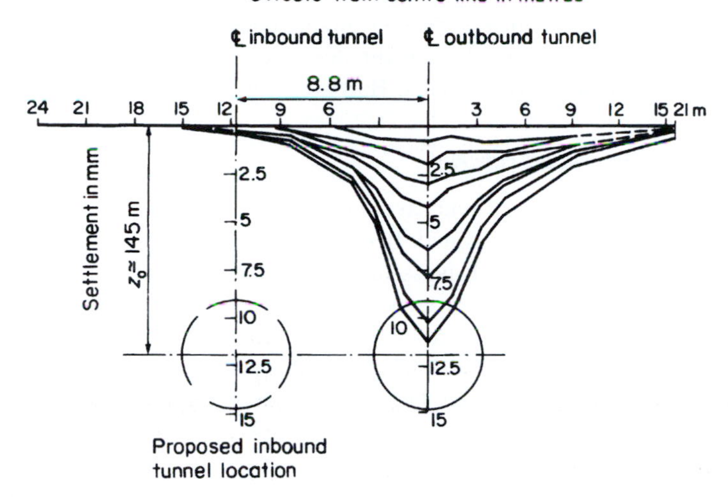

Figure 3.20b Surface soil responses at offsets to tunnel centreline, Washington Metropolitan Area Transit system.
Source: Butler and Hampton, 1975. Reproduced by kind permission of the American Society of Civil Engineers.

Vertical subsidence of up to 410 mm occurred and the subsidence profile took the form shown in Fig. 3.20b. Horizontal movements were recorded up to 16 m from the centre line.

Subsidence associated with tunnel construction is not just associated with soft ground conditions. For example, Priest (1976) described subsidence that occurred above a trial tunnel at Chinnor, England, excavated in chalk. Subsidence at the ground surface ranged from 4.5 to 8.0 mm. However, maximum subsidence of 18 mm was measured in drillholes 0.7 m above the crown of the tunnel.

The seepage of water into tunnels with subsequent lowering of the water table may cause considerable damage. For example, Morfeldt (1970) quoted subsidences that occurred in Oslo after the Holmenkollen Subway was excavated. Comparatively small leakage into the tunnel gave rise to considerable subsidence, a great number of buildings being destroyed or damaged. He also commented on the subsidence that occurred in a clay-filled depression at Karlaplan in Sweden. This again was due to leakage into tunnels that caused the water table to be lowered by as much as 6 m. Subsidence occurred up to 300 m from the tunnels. The damage to many foundations proved too costly to repair so that it was necessary to demolish the dwellings concerned.

3.6 Underground caverns

An understanding of the geology of an area as complete as possible in which an underground cavern is to be constructed is a pre-requisite for its safe design. The data gathered during the desk study should allow alternative locations to be examined, the optimum location being based upon the geological conditions, rock engineering considerations, the excavation technique and the purpose of the cavern, as well as economy (Stephansson, 1992). Once the most likely site for a cavern has been decided, then more detailed investigations can begin in order to develop a three-dimensional model of the site. The drillhole programme should in addition to providing a record of the lithology, provide test data to be used in rock mass classification schemes, such as the RSM and Q systems. Cross-hole seismic tomography and radar tomography can be used to determine rock mass quality, including detection of major discontinuities. The state of stress in the rock mass concerned is critical to the design and construction of an underground cavern. Stress measurements usually are made in drillholes at selected positions using overcoring or hydraulic fracturing techniques (Hudson and Harrison, 1997).

3.6.1 Stability of underground caverns

The site investigation for an underground cavern has to locate a sufficiently large mass of sound rock in which the cavern can be excavated. Because caverns usually are located at appreciable depth below ground surface, the rock mass often is beneath the influence of weathering and consequently the chief considerations are rock quality, geological structure and groundwater conditions. The orientation of an underground cavern generally is based on an analysis of the joint pattern, including the character of the different joint systems in the area and, where relevant, also on the basis of the stress distribution. For example, Gercek and Genis (1999) suggested that when the vertical stress is the minimum component, the orientation of an underground opening parallel to the larger of the two horizontal stresses is the more favourable. Conversely, when the vertical stress is the maximum component, then the more favourable orientation runs parallel to the smaller of the horizontal stresses. In addition, it usually is considered necessary to avoid an orientation whereby the long axis of a chamber is parallel to steeply inclined major joint sets. Wherever possible caverns should be orientated so that fault zones are avoided. For example, the underground chamber at Cruachan hydroelectric power station, Scotland, is crossed by a fault zone some 30 m in thickness, which consists of a series of subparallel shear belts in which closely jointed rock with chloritic seams occur. Problems of instability and associated overbreak that occurred in the granite host rock during construction were attributable to this fault zone, especially where it became partially destressed or developed a tensile condition in excess of the rock mass.

The walls of a cavern may be greatly influenced by the prevailing state of stress, especially if the tangential stresses concentrated around the chamber approach the intact compressive strength of the rock (Gercek and Genis, 1999). In such cases, extension fractures develop near the surface of the cavern as it is excavated, and cracks produced by blast damage become more pronounced. The problem is accentuated if any lineation structures or discontinuities run parallel with the walls of the cavern. Indeed, popping of slabs of rock may take place from the cavern walls.

The depth of excavation influences the mode of rock failure, that is, the rock structure, notably the discontinuities, generally dominates failure near the surface through block movement whereas high concentrations of stress at depth induce failure. Rock is displaced when excavation takes place because excavation removes resistance, and the normal and shear stress that run perpendicular and parallel to the boundary of the excavation respectively are reduced to zero. The excavation of a chamber in blocky rock masses may loosen blocks, leading to the collapse of those blocks that open into the roof and sidewalls. The block theory of Goodman and Shi (1985) is concerned with the three-dimensional configuration of blocks of rock produced by the geometry of discontinuities and how their stability is affected by excavation. It established procedures for locating key blocks, a key block being a block that will move into the excavation unless support is provided. The number of key blocks that occur in the walls or roof of an excavation is limited so that if they are identified and then supported, the excavation is stabilized. Block movement means that the gape of some discontinuities is widened thereby providing an indication of potential collapse. As the size of an excavation increases, the opportunity for blocks to move increases.

The normal stress during displacement of a block-shaped wedge is reduced from an initial value related to the stresses concentrated around the opening to a minimum value that, if not zero, is a function of the weight of additional rock loosened around the wedge and therefore is proportional to the width of the opening (B). If the cavern is located at a depth in excess of 3 times the width of the opening below the surface, then the normal pressure is large in relation to the weight of a wedge-shaped block, especially for small values of θ (one half the included angle), and the wedge does not move into the opening if θ is less than the angle of friction (Robertson, 1974). At very shallow depth the lack of horizontal stresses in the arch about a cavern may lead to blocks falling from the roof irrespective of the value of the angle of friction.

The angle of friction for tight irregular joint surfaces commonly is greater than 45° and as a consequence, the included angle of any wedge would have to be 90° or more, if the wedge is to move into the opening. A tight, rough joint system therefore only presents a problem when it intersects the surface of a cavern at relatively small angles or is parallel to the surface of the cavern. However, if the material occupying a thick shear zone has been reduced to its residual strength, then the angle of friction could be as low as 15° and in such an instance the included angle of the wedge would be 30°. Such a situation would give rise to a very deep wedge that could move into a cavern. Displacement of wedges into a cavern is enhanced if the ratio of the intact unconfined compressive strength to the natural stresses concentrated around the cavern is low. Values of less than 5 are indicative of stress conditions in which new extension fractures develop about a cavern during its excavation. Wedge failures are facilitated by shearing and crushing of the asperities along discontinuities as wedges are displaced.

As can be inferred from above, the effects of stress can aid stability when an excavation is made at intermediate depth. The resulting redistribution of stress gives rise to circumferential stresses around the perimeter of an excavation and the increase in normal stress across discontinuities enhances their frictional resistance to shearing. On the other hand, blocks can be squeezed out at greater depth, and high stresses can be responsible for spalling, popping and rock bursts. For instance, Cruachan underground chamber was constructed at a depth of around 300 m in granitic rocks. Knill (1972) reported that the vertical stress in the underground chamber was some 3.5 times the overburden load and that the horizontal component of stress was about four times the overburden load. He suggested that these relatively high *in situ* stresses partly were due to valley notch concentration of stress and partly to residual stresses associated with former earth movements. The relief of these stresses on excavation led to loosening and spalling of blocks from several locations. Popping occurred in some porphyrite dykes. Rock bursts have occurred in underground caverns at rather shallow depths, particularly where these were excavated in the sides of valleys or fiords (Brekke, 1970). Intense rock bursting has occurred on the inside of a fault when it passed through a cavern and dipped towards an adjacent valley. Bursting can take place at depths of 200–300 m when the tensile strength of the rocks concerned varies between 3 and 4 MPa. A depth in excess of 1000 m may be required to induce bursting when the tensile strength is around 15 MPa.

In a creep-sensitive material, such as may occur in a major shear zone or zone of soft-altered rock, the natural stresses concentrated around an opening cause time-dependent displacements that, if restrained by support, result in a build-up of stress on the support. Conversely, if a rock mass is not sensitive to creep, stresses around an opening normally are relieved as blocks displace towards the opening. However, initial movements may be very much influenced by the natural stresses concentrated around an opening and under certain boundary conditions may continue to act even after large displacements have occurred.

Displacement data, obtained by extensometers, provide a direct means of evaluating cavern stability. In other words, opening of discontinuities and loosening of blocks of rock usually are indicated by displacements several times greater than those predicted, by high rates of displacement that are unrelated to excavation or that do not tend to reduce after excavation, by concentrations of displacements along discontinuities at specific depths, and by displacements that exceed the displacement capacity of the support system. Most of the large displacements in underground caverns have been of the order of 10–75 mm. Displacements that have exceeded the predicted elastic displacements by a factor of 5 or 10 usually have resulted in decisions to modify support and excavation methods.

3.6.2 Construction and support of underground caverns

Three methods of blasting normally are used to excavate underground caverns (Fig. 3.21). First, in the overhead tunnel the entire profile is drilled and blasted together or in parts by horizontal holes. Second, benching with horizontal drilling commonly is used to excavate the central parts of a large cavern. Because there are two faces the amount of linear metre drilling and charge are smaller than in tunnelling operations. Third, the bottom of a cavern may be excavated by benching, the blastholes being drilled vertically. The central part of a cavern also can be excavated by vertical benching providing the upper part has been excavated to a sufficient height or the walls of a cavern are inclined. Smooth blasting is used to minimize fragmentation in the surrounding rock. Once the crown of a cavern has been excavated, economical excavation of the walls requires large deep bench cuts exposing substantial areas of wall in a single blast. Under these conditions an unstable wedge can be exposed and fail before it is supported.

Rock bolts and shotcrete, sometimes in combination with wire mesh, are used as temporary support measures during the excavation stage of an underground chamber and subsequently form part of the permanent support system (Choquet and Hadjigeorgiou, 1993). Tie rods may be used between rock bolts. Grouted cables also may be used with rock bolts and shotcrete. In good quality rock masses in which the stability is controlled by discontinuities, the support has to be designed to reinforce key blocks that may fall into the excavation. This type of support, which is often termed spot bolting, involves the installation of a few rock bolts at certain locations. Pattern bolting is used in weak rock masses and involves the installation of rock bolts in a regular pattern designed to reinforce the entire rock mass. Typically, rock bolts, 5 m in length and of 20 tonne capacity, are installed at 2 m centres. The support pressures required to maintain the stability of a cavern increase as its span increases so that for the larger caverns standard-sized bolts arranged in normal patterns may not be sufficient to hold

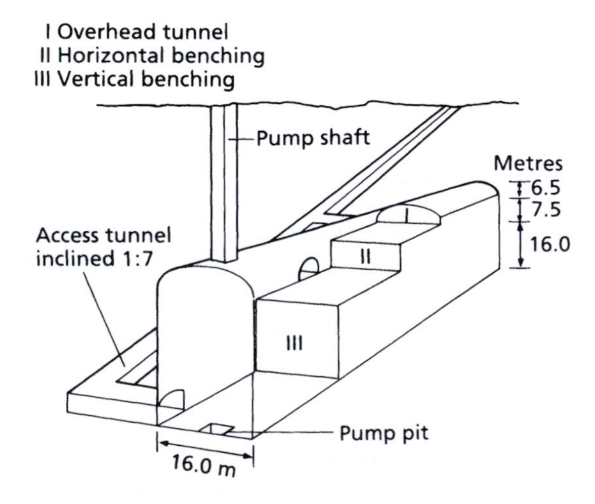

Figure 3.21 Main stages in the excavation of underground chambers.

the rock in place. Frequently, the upper parts of the walls are more heavily bolted in order to help support the haunches and the roof arch, while the lower walls may be either slightly bolted or even unbolted. However, the orientation of the joint systems in relation to the walls frequently means that the large planar surfaces may be subjected to slabbing and buckling, and that unstable wedges may be developed. Most caverns have arched crowns with span to rise ratios (B/R) of 2.5–5.0. In general, higher support pressures are required for flatter roofs.

Shotcrete can be applied quickly to newly exposed rock during excavation and develops strength steadily after application. Hence, it gains strength as load is being applied to it and so can accommodate the redistribution of stress. It can be used for either temporary or permanent support, however, the required thickness of shotcrete differs between the two forms of support (Sem, 1974). High early strength of shotcrete can be achieved by the addition of accelerators, that is, rapid setting agents (Blank, 1974). The inclusion of steel fibres in shotcrete affords considerable tensile and flexural strength, the fibres acting as crack arrestors. About 1–2 per cent by volume of steel fibre is added to the mix, which assists the build-up process and helps to reduce rebound. The addition of up to 10 per cent micro-silica to the mix also reduces rebound and increases the thickness that can build-up in one application. It increases the density and early development of strength in the shotcrete, and is beneficial for the application to damp or wet surfaces. The thickness of single layer of shotcrete typically is between 40 and 80 mm, with the total thickness varying between 100 and 200 mm. A thicker finished layer can be placed when required by the stress conditions.

The roof of a chamber frequently is supported by a reinforced concrete arch (Fig. 3.22). The arch is excavated to its full width and inclined haunches are provided to carry the reaction of the concrete arch. The reinforced concrete is cast in place when the floor of the excavation is in level with the base of the inclined haunches. The rest of the chamber then is excavated. However, Hoek and Moy (1993) pointed out that a concrete arch is very rigid compared with the surrounding rock mass so that deformation due to excavation of the lower part of the chamber may be responsible for excessive bending in the arch. They further maintained that depending on the magnitudes of the displacements in the rock mass, and the curvature and thickness of the concrete arch, stresses in the arch can exceed safe working loads. Repair of a damaged arch is both difficult and expensive. Hoek and Moy suggested that generally the use of concrete arches for roof support should be avoided if underground chambers are constructed in weak rock masses. In such cases, a more flexible system should be used such as that provided by grouted cables and shotcrete. Cheng and Liu (1993) referred to the use of pre-stressed cable anchors and steel fibre micro-silica shotcrete for roof treatment of the power chamber at the Mingtan pumped storage project, Taiwan. There the rocks consisted mainly of sandstones and siltstones. Cheng and Liu also described the treatment of fault zones in the roof of the chamber. Short cross-cuts were excavated along each fault zone by hand mining beyond the final boundary of the excavation. The cross-cuts were used for access from which to remove the clay gouge, to a depth of 4 m, by high-pressure water jets, then the resulting voids were backfilled with cement mortar.

3.7 Shafts and raises

Shafts provide access for people, materials and equipment to subsurface excavations, as well as providing a route for ventilation and services. A shaft usually will be sunk through a series of different rock types and the two principal problems likely to be encountered are varying stability of the walls and ingress of water. These two problems frequently occur together and they are likely to be met within rock masses with high fracture indices, weak zones being particularly hazardous. They are most serious, however, in unconsolidated material, especially loosely packed gravels, sands and silts.

Figure 3.22 Excavation of the cavern for Cruachan Power Station, Scotland.

The geological investigation prior to shaft sinking should provide detailed information relating to the character of the ground conditions. The hydrogeological conditions, that is, position of the water table, hydrostatic pressures, especially artesian pressures, location of inflow and its quantification, as well as chemical composition of the water, are obviously of paramount importance. Indeed, groundwater inflow from deep aquifers presents a major hazard to shaft sinking operations since hydrostatic pressure increases with depth. The data obtained should enable the best method of shaft construction and the design of the lining to be made. In addition, the data should indicate where ground stability and groundwater control measures are needed. Hence, a drillhole should be cored along the full length of the centre-line of a shaft and appropriate down-hole tests conducted. Daw (1984) described various aquifer tests that have been used to determine groundwater flow during site investigations for shaft sinking. He maintained that simple rising and falling head tests, packer tests and slug tests are best suited to near surface investigation either above the water table or where there is an unconfined water table lying within the aquifer. Generally, they have proved unsatisfactory for testing deep multiple-confined aquifers. Daw suggested that the type of test required in these conditions should be based on the measurement of the recovery of aquifer pressure following a period of controlled drillhole flow. The drill-stem test is one such test. Alternatively, the probe hole pressure recovery test can be used. Pressure recovery testing during excavation of a shaft also is required to confirm earlier hydrogeological data obtained from tests carried out in the drillhole sunk down the centre-line of a shaft and to detect any unsuspected localized groundwater conditions.

According to Auld (1992), the best shape of a shaft lining to resist loads imposed by rock deformation and hydrostatic pressure, particularly one that is deep and has a large cross-sectional area, is circular since the induced stress in the lining are all compressional. The circular shape minimizes the effect of inward radial closure exhibited by rock masses under high overburden stresses. In fact, no pressure is applied to the lining of a shaft in strong competent rock masses as inward deformation is minimal, being compatible with the elastic properties of the rock. Consequently, the design of shaft linings normally does not allow for rock loading when sunk in strong competent rock masses. However, the hydrostatic pressures must be taken into account in shaft design. The depth of a shaft is not just controlled by the level to which underground access is required but also by the limits on winding equipment. For instance, at Western Deep Levels gold mine, South Africa, the No. 2 shaft was sunk 3580 m below the ground surface with three separate shaft stages being constructed.

3.7.1 Shaft construction

A shaft is driven vertically or near vertically downwards whereas a raise is driven either vertically or at steep angle upwards. Exacavation is either by drilling and blasting or by a shaft boring machine. In shaft sinking, drilling is usually easier than in tunnelling but blasting is against gravity, and mucking is slow and therefore expensive. Removal of muck is by skip winding to the surface. The first operation in shaft sinking is to construct the shaft collar of reinforced concrete and the foreshaft. The foreshaft is excavated to a depth of some 20–50 m, generally by backactor and may be assisted by a hydraulic impact breaker or by limited blasting. Multiboom sinking jumbos can be used in shaft sump drilling or hand-held pneumatic drills can be used. Full-sump drilling and blasting is used in reasonably dry conditions whilst benching is used in very wet conditions. Benching provides a large sump for water collection below the level of the drilling bench. Rock bolts and wire mesh or rock bolts, wire mesh and shotcrete may be used for temporary support. Liner plates, pre-cast concrete segments or fabricated steel tubbing may be used for temporary support in weak ground conditions. The temporary lining is installed as shaft sinking proceeds. The permanent lining is constructed from the top down, however, if temporary support has been used, then the permanent lining may be installed from the bottom up. Auld (1992) outlined the types of shaft lining that are used.

Shaft drilling machines can either excavate a shaft downwards to its required diameter in a single pass or a small-diameter shaft can be excavated that subsequently is reamed, from the top or bottom, to the required size. These machines either may be purpose built or modified oilfield-drilling rigs. Auld (1992) indicated that muck retrieval normally is by reverse circulation mud flush with air-lift assistance, the shaft excavation being kept full of drilling mud to maintain its stability prior to the installation of the lining (Fig. 3.23). An advantage of this system is that the mud flush prevents ingress of water into the excavation so that no pre-treatment of the ground is required. A shaft boring machine excavates a shaft downwards to the necessary diameter; however, a pilot hole may be drilled first that allows free-fall muck retrieval to a subsurface removal system (Fig. 3.24). Unlike a shaft drilling machine, a shaft boring machine needs to operate in relatively dry competent rock conditions, otherwise pre-treatment of the ground is necessary.

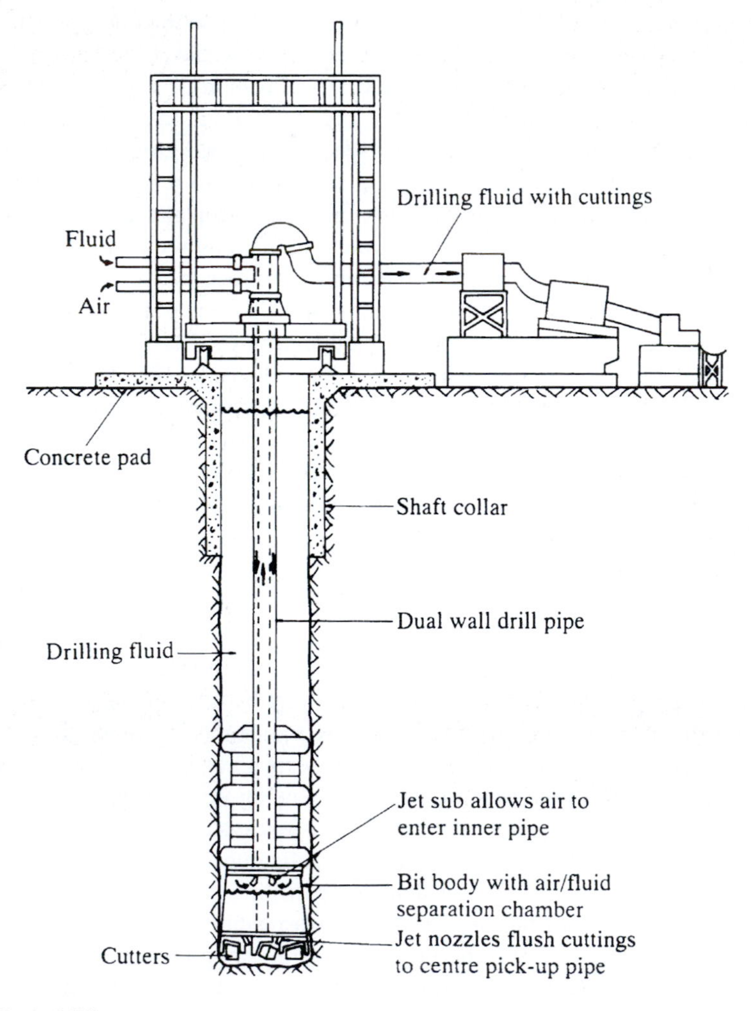

Figure 3.23 Shaft drilling.
Source: Auld; 1992.

Figure 3.24 Shaft boring machine.

Mucking and blasting are simpler in raising operations but drilling is difficult. As the cost of drilling generally is exceeded by that of blasting and mucking, raising is more economic than shaft sinking where, of course, both are practical. Raises also are constructed with raise boring machines (Fig. 3.25). A raise is of small cross-sectional area, if the excavation is to have a large diameter, then enlargement is done from above, the primary raise excavation being used

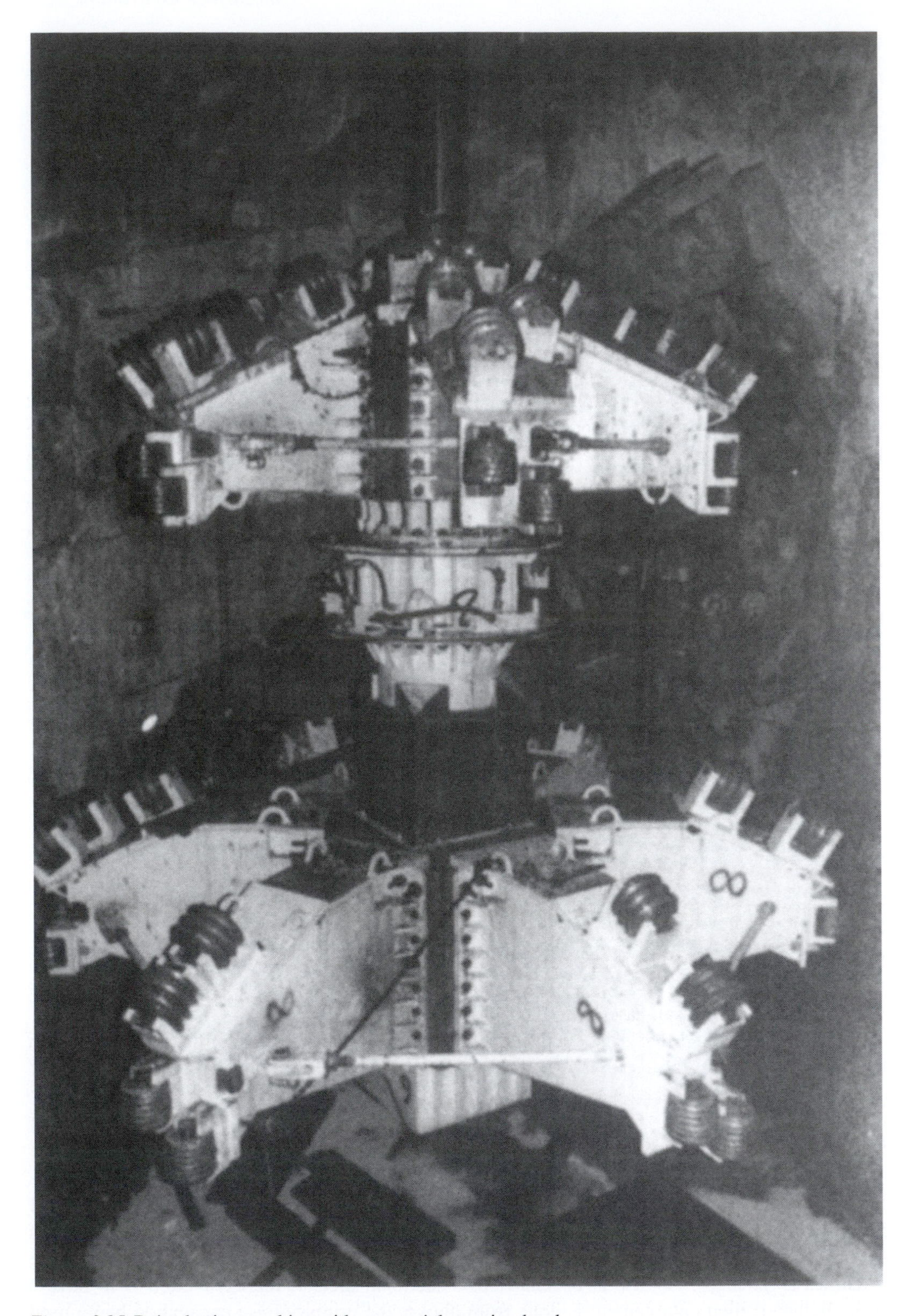

Figure 3.25 Raise-boring machine with sequential reaming head.

as a muck shute. Where a raise emerges at the surface through unconsolidated material, this section is excavated from the surface.

3.7.2 Dealing with groundwater

A simple effective method of dealing with groundwater in shaft sinking is to pump from a sump within the shaft. However, problems arise when the quantity pumped is so large that the rate of inflow under high head causes instability in the sides of the shaft or prevents the fixing and back grouting of the shaft lining. Although the idea of surrounding a shaft by a ring of bored wells is at first sight attractive, there are practical difficulties in achieving effective lowering of the water table. In fissured rocks or variable water bearing soils, there is a tendency for the water flow to by-pass the wells and take preferential paths directly into the excavation.

Where the stability of the wall and/or the ingress of groundwater are likely to present problems in shaft sinking one of the most frequently used techniques is ground freezing. Freezing transforms weak, waterlogged materials into ones that are self-supporting and impervious. It therefore affords temporary support to an excavation, as well as being a means of excluding groundwater. Normally, the freeze probes are installed in linear fashion around a shaft so that an adequate boundary wall encloses the future excavation when the radial development of ice about each probe unites to form a continuous section of frozen ground. The classic coolant used in ground freezing is super-cooled brine although liquid nitrogen also is used. This reduces the freezing time, one of the usually quoted disadvantages of the technique, to about a third.

Ground freezing means heat transfer and, in the absence of moving groundwater, this is brought about by conduction (Bell, 1993). Thus, the thermal conductivities of the materials involved govern the rate at which freezing proceeds. However, these values fall within quite narrow limits for all types of frozen ground. This is why ground freezing is such a versatile technique and can deal with a variety of soil and rock types in a stratal sequence. But a limitation is placed upon the freezing process by unidirectional flow of groundwater. For a brine freezing project, a velocity exceeding 2 m per day seriously affects and distorts the growth of an ice wall. The tolerance is much wider when liquid nitrogen is used. Normally, a natural moisture content of at least 11 per cent is required to achieve impermeability and a significant improvement in strength (Harris and Pollard, 1986). Dry zones can be irrigated to raise the moisture content to an appropriate level.

The success of the freezing process depends upon a precise knowledge of the geological conditions. In particular, groundwater pressures should be investigated. Furthermore, the salinity of water governs its freezing point. Therefore an analysis of groundwater should be made before freezing begins. Sea water presents no special problems in freezing since it solidifies at about 3 °C below fresh water; however, in some situations temperatures as low as −21 °C have had to be achieved before salt-saturated water has frozen.

Because of the bond that exists between water and clay particles a significant proportion of the natural moisture content in a clay soil remains unfrozen even at temperatures as low as −25 °C. However, this does not mean that clay soils cannot be effectively frozen for purposes of ground support. Indeed, the increase in the unconfined compressive strength of clay soil as the temperature is lowered further below freezing point increases exponentially (Fig. 3.26). By contrast, the natural water content of sands and gravels is almost wholly converted to ice at 0 °C so that they exhibit a reasonably high compressive strength only a few degrees below freezing. Consequently, this parameter can be used as a design index of their performance, provided a suitable factor of safety is used.

Even when drilling for the probe holes is carried out with care and skill, the holes may deviate. Deviation, in turn, may mean that areas of ground are left unfrozen. These are referred to as

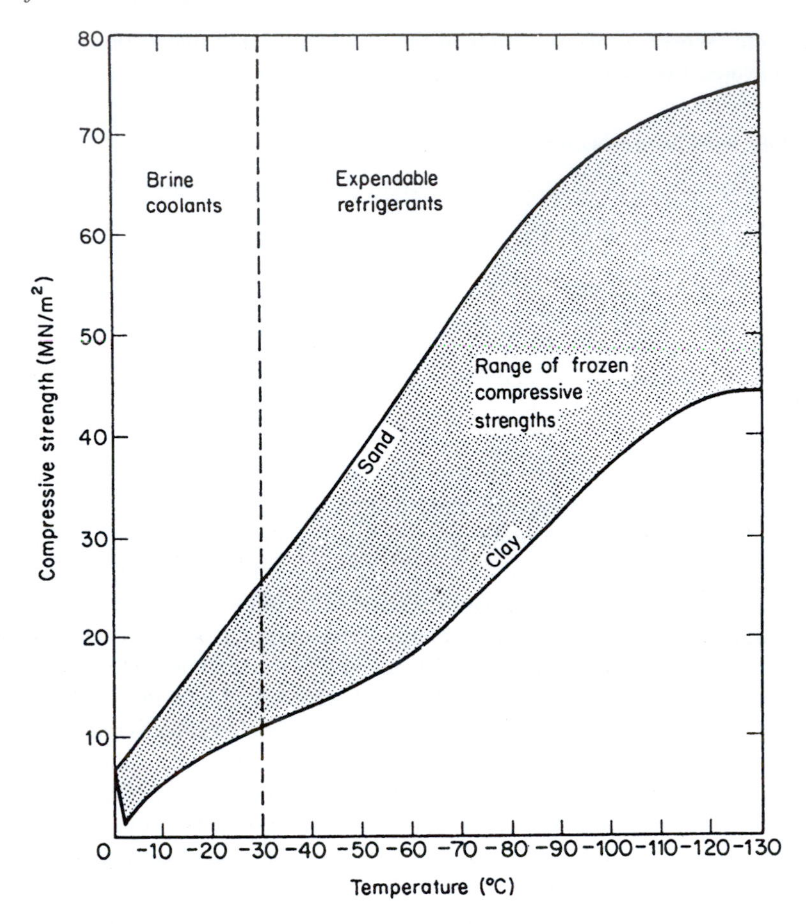

Figure 3.26 Compressive strength of frozen sand and clay.

windows and they can allow the penetration of water into a shaft. Freeze holes must therefore be surveyed to see whether any extra holes need to be drilled so as to avoid the development of windows. Large fissures in rock also may give rise to windows. On rare occasions a frozen wall, although 99 per cent impervious, may still have small windows that may not close, no matter what refrigeration power is used. Pumping must then be resorted to. It generally is agreed that a metre per day is about the maximum acceptable flow.

Ice creeps under uniform loading, as do saturated soils in a frozen state. Hence, the resistance to failure of such materials depends not only upon their temperature but also upon the duration of loading. At ground freezing temperatures, soils rich in silt-clay fractions may creep to a significant extent and if excavations in them are to be open for an extended period of time their long-term creep strength and plastic deformation should be investigated. However, the resistance to failure increases as the temperature below freezing is lowered.

Although frost heave frequently has been mentioned as a problem in silty ground, this rarely occurs when the ground is frozen artificially. Rises in surface level of more than 75 mm are very exceptional. What is more, the likelihood of such movement is reduced if freezing is rapid. After thawing the ground level slowly sinks towards its original level, complete recovery, however, is rarely attained.

Grouting can be an economical method of eliminating or reducing the flow of water into shafts if the soil or rock conditions are suitable for accepting cement or chemical grouts. This is because the perimeter of the grouted zone is relatively small in relation to the depth of the excavation. The depth and thickness of each aquifer zone, together with its permeability and water pressure, should be determined during the site investigation in order to allow the choice of grout to be made. Discontinuities in excess of 200 μm in width can be treated with cement grouts whereas chemical grouts are used to treat those of lesser width. Occasionally, grouting has been carried out from holes drilled from the ground surface but usually grouting operations are undertaken from the advancing face of the excavation, the grout being injected by sealed-in-place steel standpipes. The normal procedure is to advance the grout cover in a number of stages. In certain situations, grout treatment can be combined with pressure relief wells to make groundwater control more effective (Harris and Pollard, 1986).

References

Anon. 1973. *Spoil Heaps and Lagoons*, Technical Handbook, National Coal Board, London.

Anon. 1991a. *Sulphate and Acid Resistance of Concrete in the Ground*. Digest No. 363, Building Research Establishment, Watford, Hertfordshire.

Anon. 1991b. Jet set style. *Ground Engineering*, 28, No. 8, 22–23.

Anon. 1995. The description and classification of weathered rocks for engineering purposes. Working Party Report. *Quarterly Journal Engineering Geology*, **28**, 207–242.

Anon. 2000. *The Collapse of NATM Tunnel at Heathrow Airport*. Sudbury and Health and Safety Executive Books, Health and Safety Executive, London.

Arthur, J.R.C., Philips, G. and McCormick, 1997. High definition seismic for Channel Tunnel marine route. In: *Modern Geophysics in Engineering Geology*, Engineering Geology Special Publication No. 12, McCann, D.M., Eddleston, M., Fenning, P.J. and Reeves, G.M., Geological Society, London, 327–334.

Attewell, P.B. and Farmer, I.W. 1975. Ground deformation resulting from shield tunnelling in the London Clay. *Canadian Geotechnical Journal*, **2**, 380–395.

Auld, A. 1992. Shafts and raises in rock masses. In: *Engineering in Rock Masses*, Bell, F.G. (ed.), Butterworth-Heinemann, Oxford, 465–508.

Baker, W.H., Cording, E.J. and McPherson, H. 1983. Compaction grouting to control ground movements during tunnelling. *Underground Space*, **7**, 205–212.

Barton, N. 1988. Rock mass classification and tunnel reinforcement selection using the Q system. *Proceedings Symposium on Rock Classification for Engineering Purposes*, American Society for Testing Materials, Special Technical Publication 984, Philadelphia, 59–88.

Barton, N., Lien, R. and Lunde, J. 1975. *Engineering Classification of Rock Masses for the Design of Tunnel Support*. Publication 106, Norwegian Geotechnical Institute, Oslo.

Barton, N.R., By, T.-L., Chrysanthakis, L., Tunbridge, L., Kristiansen, J., Loset, F., Bhasin, R.K., Westerdahl, H. and Vik, G. 1992. Predicted and measured performance of the 62 m span Norwegian Olympic Ice Hockey Cavern at Gjovik. *International Journal Rock Mechanics Mining Science and Geomechanical Abstracts,* **31,** 617–641.

Bell, F.G. 1986. Location of abandoned workings in coal seams. *Bulletin International Association of Engineering Geology*, **33**, 123–132.

Bell, F.G. 1993. *Engineering Treatment of Soils*. Spon, London.

Bell, F.G. 2000. *Engineering Properties of Soils and Rocks*. Fourth Edition. Blackwell Scientific Publishers, Oxford.

Bell, F.G. and Haskins, D.R. 1997. A geotechnical overview of the Katse Dam and Transfer Tunnel, Lesotho, with a note on basalt durability. *Engineering Geology*, **46**, 175–198.

Bell, F.G. and Jermy, C.A. 2000. The geotechnical character of some South African dolerites, especially their strength and durability. *Quarterly Journal Engineering Geology*, **33**, 59–76.

Bell, F.G. and Jermy, C.A. 2002. Permeability and fluid flow in stressed strata and mine stability, Eastern Transvaal Coalfield, South Africa. *Quarterly Journal Engineering Geology and Hydrogeology*, **35**,

Bell, F.G., Entwistle, D.C. and Culshaw, M.G. 1997. A geotechnical survey of some British Coal Measures mudstones, with particular emphasis on durability. *Engineering Geology*, **46**, 115–129.

Bergh-Christensen, J. and Selmer-Olsen, R. 1970. On the resistance to blasting in tunnelling. *Proceeding Second International Congress on Rock Mechanics* (ISRM), Belgrade, Paper 5–7, 59–63.

Beveridge, J.P. and Rankin, W.J. 1995. Role of engineering geology in NATM construction. In: *Engineering Geology of Construction*. Engineering Geology Special Publication No. 10, Eddleston, M., Walthall, S., Cripps, J.C. and Culshaw, M.G. (eds), Geological Society, London, 255–268.

Bieniawski, Z.T. 1974. Geomechanics classification of rock masses and its application to tunnelling. *Proceedings Third Congress International Society of Rock Mechanics*, Denver, **1**, 27–32.

Bieniawski, Z.T. 1978. Determining rock mass deformability – experience from case histories. *International Journal Rock Mechanics Mining Science and Geomechanical Abstracts*, **15**, 237–247.

Bieniawski, Z.T. 1983. The Geomechanics classification (RMR System) in design applications to underground excavations. *Proceedings International Symposium on Engineering Geology and Underground Construction*, Lisbon, **2**, 11.33–11.47.

Bieniawski, Z.T. 1989. *Engineering Rock Mass Classifications*. Wiley-Interscience, New York.

Bishop, I., Styles, P. Emsley, S.J. and Ferguson, N.S. 1997. The detection of cavities using the microgravity technique: case histories from mining and karstic environments. In: *Modern Geophysics in Engineering Geology*, Engineering Geology Special Publication No. 12, McCann, D.M., Eddleston, M., Fenning, P.J. and Reeves, G.M.(eds), Geological Society, London, 153–166.

Blank, J.A. 1974. Shotcrete durability and strength – a practical viewpoint. *Proceedings Engineering Foundation Conference on the Use of Shotcrete for Underground Structural Support*, South Berwick, Maine, American Society Civil Engineers/American Concrete Institute, ACI Publication SP-45, 277–296.

Blindheim, O.T., Boniface, A. and Richards, L.A. 1991. Borability assessments for the Lesotho Highlands Water Project. *Tunnels and Tunnelling*, June issue, 55–58.

Brekke, T.L. 1970. A survey of large permanent underground openings in Norway. *Proceedings Symposium on Large Permanent Underground Openings*, Brekke, T.L. and Jorstad, F.A. (eds), Scandinavian University Books, Oslo, 15–38.

Butler, R.A. and Hampton, D. 1975. Subsidence over soft ground tunnel. *Proceedings American Society Civil Engineers, Journal Geotechnical Engineering Division*, **101**, 35–49.

Castro, D. and Bell, F.G. 1995. An engineering geological appraisal of sandstones of the Clarens Formation, Lesotho, in relation to tunnelling. *Geotechnical and Geological Engineering*, **13**, 117–142.

Chen, J.F. and Vogler, U.W. 1992. Rock cuttability/boreability, assessment of research at CSIR. *Proceedings Tuncon'92*, Maseru, South African National Council on Tunnelling, Pretoria, 91–97.

Cheng, Y. and Liu, S.C. 1993. Power caverns of Mingtan pumped storage project, Taiwan. In: *Comprehensive Rock Engineering: Principles, Practice and Projects*, Surface and Underground Case Histories, Hoek E. (ed.), Pergamon Press, Oxford, **5**, 111–131.

Choquet, P. and Hadjigeorgiou, J. 1993. The design of support for underground excavation. In: *Comprehensive Rock Engineering: Principles, Practice and Projects*, Hudson J.A. (ed.), Pergamon Press, Oxford, **4**, 313–348.

Cripps, J.C., Culshaw, M.G., Bell, F.G. and McCann, D.M. 1988. The detection and investigation of abandoned mines by the use of geophysical methods. *Proceedings Symposium on Mineral Extraction, Utilization and the Surface Environment*, Harrogate, Institution of Mining Engineers, 281–289.

Cripps, J.C., Deaves, A., Bell, F.G. and Culshaw, M.G. 1993. The Don Valley intercepting sewer scheme: an investigation of flow into underground workings. *Bulletin Association of Engineering Geologists*, **30**, 409–425.

Dalgic, S. 2002. A comparison of predicted and actual tunnel behaviour in the Istanbul Metro, Turkey. *Engineering Geology*, **63**, 69–82.

Daw, G.P. 1984. Application of aquifer testing to deep shaft investigations. *Quarterly Journal Engineering Geology*, **17**, 367–379.

Deaves, A.P. and Cripps, J.C. 1995. Investigation and treatment of old mine workings for underground excavations: an example from the Don Valley Intercepting Sewer Scheme, Sheffield, England. In: *Engineering Geology of Construction*, Engineering Geology Special Publication No. 10, Eddleston, M., Walthall, S., Cripps, J.C. and Culshaw, M.G. (eds), The Geological Society, London, 269–277.

De Graaf, P.J.H. and Bell, F.G. 1997. The Delivery Tunnel North, Lesotho Highlands Water Project. *Geotechnical and Geological Engineering*, **15**, 95–120.

Dugnani, G., Guatteri, G., Roberti, P. and Mosiici, P. 1989. Subhorizontal jet grouting applied to a large urban twin tunnel in Campinas, Brazil. *Proceedings Twelfth International Conference on Soil Mechanics and Foundation Engineering*, Rio de Janiero, **2**, 1351–1354.

Edmunds, J.M. and Graham, J.D. 1977. Peterhead power station cooling water intake tunnel: an engineering case study. *Quarterly Journal Engineering Geology*, **14**, 281–308.

Farmer, I.W. 1992. Reinforcement and support of rock masses. In: *Engineering in Rock Masses*, Bell, F.G. (ed.), Butterworth-Heinemann, Oxford, 351–369.

Gercek, H. and Genis, M. 1999. Effect of anisotropic *in situ* stresses on the stability of underground openings. *Proceedings Ninth International Congress on Rock Mechanics*, Paris, **1**, 367–370.

Goodman, R.E. and Shi, G. 1985. *Block Theory and its Application to Rock Engineering*. Prentice Hall, Englewood Cliffs, New Jersey.

Gustafsson, R. 1976. Smooth blasting. In: *Tunnelling'76*, Jones, M.E. (ed.), Institution Mining and Metallurgy, London, 141–146.

Hansmire, W.H. 1981. Tunneling and excavation in soft rock and soil. *Bulletin Association of Engineering Geologists*, **18**, 77–89.

Harris, J.S. and Pollard, C.A. 1986. Some aspects of groundwater control by the ground freezing and grouting methods. In: *Groundwater in Engineering Geology*, Engineering Geology Special Publication No. 3, Cripps, J.C., Bell, F.G. and Culshaw, M.G. (eds), The Geological Society, London, 455–466.

Healy, P.R. and Head, J.M. 1984. *Construction Over Abandoned Mine Workings*. Construction Industry Research and Information Association, Special Publication 32, London.

Hoek, E. 1994. Strength of rock and rock masses. *News Journal,* International Society Rock Mechanics, **2**, 4–16.

Hoek, E. and Brown, E.T. 1980. *Underground Excavations in Rock*. Institution of Mining and Metallurgy, London.

Hoek, E. and Brown, E.T. 1997. Practical estimates of rock mass strength. *International Journal Rock Mechanics Mining Science*, **34**, 1165–1186.

Hoek, E. and Moy, D. 1993. Design of large powerhouse caverns in weak rock. In: *Comprehensive Rock Engineering: Principles, Practice and Projects*, Surface and Underground Case Histories, Hoek, E. (ed.), Pergamon Press, Oxford, **5**, 85–110.

Hoek, E., Kaiser, P.K. and Bawden, W.F. 1995. *Support of Underground Excavations in Hard Rock.* A.A. Balkema, Rotterdam.

Hoek, E., Marinos, P. and Benissi, M. 1998. Applicability of the geological strength index (GSI) classification for very weak rock and sheared rock masses. The case of the Athens Schist Formation. *Bulletin Engineering Geology and the Environment*, **57**, 151–160.

Howells, D.A. 1972. Tunnels in earthquake areas. *Tunnels and Tunnelling*, **4**, 437–440.

Hudson, J.A. and Harrison, J.P. 1997. *Engineering Rock Mechanics: An Introduction to the Principles*. Pergamon, Oxford.

Kettle, C.T. and Gandais, M. 1995. A new tunnel roof support system with specific reference to the Brovello Tunnel. In: *Engineering Geology of Construction*, Engineering Geology Special Publication No. 10, Eddleston, M., Walthall, S., Cripps, J.C. and Culshaw, M.G. (eds), The Geological Society, London, 279–288.

Kimmance, J.P., Linney, L.F. and Stapleton, M.J. 1995. Potential of grouting methods to prevent and compensate for tunnelling induced settlement of London Clay. In: *Engineering Geology of Construction*, Engineering Geology Special Publication No. 10, Eddleston, M., Walthall, S., Cripps, J.C. and Culshaw, M.G. (eds), The Geological Society, London, 287–297.

Knill, J.L. 1972. The engineering geology of the Cruachan underground power station. *Engineering Geology*, **6**, 289–312.

Lake, L.M. and Norie, E.H. 1982. Application of horizontal ground freezing in tunnel construction – two case histories. In: *Tunnelling'82*, Jones, M.J. (ed.), Institution Mining and Metallurgy, London, 283–289.

Lauffer, M. 1958. Gebirgsklassifizierung fur den stollenhua. *Geologie und Bauwesen*, **24**, 46–51.

Magata, H., Nakauchi, S. and Sogabe, H. 1982. Subsidence measurements associated with shield tunneling in soft ground. In: *Tunnelling'82*, Jones, M.J. (ed.), Institution of Mining and Metallurgy, London, 231–240.

Martin, D. 1988. Getting the right 'bang' in the right place at the right time. *Tunnels and Tunnelling*, **20**, September issue, 32–37.

Mayo, R.S. 1982. Shield tunnels. In: *Tunnel Engineering Handbook*, Bickel, J.O. and Kuesel, T.R. (eds), Van Nostrand, New York, 93–122.

McCann, D.M., Culshaw, M.G. and Northmore, K.J. 1990. Rock mass assessment from seismic measurement. In: *Field Testing in Engineering Geology*, Engineering Geology Special Publication No. 6, Bell, F.G., Culshaw, M.G., Cripps, J.C. and Coffey, J.R. (eds), The Geological Society, London, 257–266.

McPherson, M.J. 1972. The heat problem underground, with particular reference to South African gold mines. *Transactions Institution Mining and Metallurgy*, **85**, Section A, Mining Industry, A63–A74.

Miller, E. 1988. The eductor dewatering system. *Ground Engineering*, **21**, No. 6, 29–34.

Morfeldt, D.O. 1970. Significance of groundwater at rock constructions of different types. *Proceedings Symposium on Large Permanent Underground Openings*, Brekke, T.L. and Jorstad, F.A. (eds), Scandinavian University Books, Oslo, 285–318.

Moye, D.G. 1964. Rock mechanics in the investigation of the T.I. underground power station, Snowy Mountains, Australia. *Engineering Geology Case Histories*, 1–5, The Geological Society, America, 123–154.

Newbury, J. and Davenport, C.A. 1975. Geotechnical aspects of shallow sewer tunnels in urban areas. *Quarterly Journal Engineering Geology*, **8**, 271–290.

Obert, L. and Duval, W. 1967. *Rock Mechanics and the Design of Structures in Rock*. Wiley, New York.

Olivier, H.G. 1979. A new engineering-geological rock durability classification. *Engineering Geology*, **14**, 255–279.

O'Reilly, M.P. and New, B.M. 1982. Settlements above tunnels in the United Kingdom – their magnitude and prediction. In: *Tunnelling'82*, Jones, M.J. (ed.), Institution Mining and Metallurgy, London, 173–181.

Parkes, D.B. 1988. *The Performance of Tunnel Boring Machines in Rock.* Construction Industry Research and Information Association, Special Publication 62, London.

Peck, R.B. 1969. Deep excavation and tunnelling in soft ground: state-of-the-art-volume. *Proceedings Seventh International Conference on Soil Mechanics and Foundation Engineering*, Mexico City, **3**, 225–290.

Powers, J.P. 1972. Groundwater control in tunnel construction. *Proceedings First North American Tunneling Conference*, American Institute Mining Engineers, New York, 331–369.

Priest, S.D. 1976. Ground movements caused by tunnelling in chalk. *Proceedings Institution Civil Engineers*, **61**, Pt. 8, 23–39.

Richardson, G. and Gollick, M.J. 1988. The Dosco CTM 5 tunnel boring machine. In: *Tunnelling '88*, Jones, M.J. (ed.), Institution Mining and Metallurgy, London, 265–274.

Robbins, R.J. 1976. Mechanized tunnelling – progress and expectations. *Transactions Institution Mining and Metallurgy*, **85**, Section A, Mining Industry, A41–A50.

Robertson, A.M.G. 1974. Joints and gouge materials – their importance and testing. In: *Tunnelling in Rock*, Bieniawski, Z.T. (ed.), South African Institution Civil Engineers/South African National Group Rock Mechanics/C.S.I.R., Pretoria, 125–138.

Saito, T. and Kobayashi, T. 1979. Driving an 8.48 m diameter sewer tunnel by use of an earth pressure balancing shield. In: *Tunnelling '79*, Jones, M.J. (ed.), Institution Mining and Metallurgy, London, 295–304.

Sandtner, A. and Gehring, K.H. 1988. Development of roadheading equipment for tunnelling by NATM. In: *Tunnelling '88*, Jones, M.J. (ed.), Institution Mining and Metallurgy, London, 275–288.

Sem, B. 1974. How to achieve effective roof reinforcement with sprayed concrete. *Tunnels and Tunnelling*, **12**, No. 6, 61–62.

Serafim, J.L. and Pereira, J.P. 1983. Consideration of the Geomechanical Classification of Bieniawski. *Proceedings International Symposium on Engineering Geology and Underground Construction*, Lisbon, **1**, 33–44.

Somerville, S.H. 1986. *Control of Groundwater for Temporary Works*. Construction Industry Research and Information Association, Report 113, London.

Stacey, T.R. and Page, C.H. 1986. *Practical Handbook for Underground Rock Mechanics*. Trans Tech Publications, Clansthal-Zellerfeld.

Stack, B. 1985. Update in trends in soft ground tunnelling machines. *Tunnels and Tunnelling*, **17**, No. 1, 21–23.

Stephansson, O. (ed.) 1984. *Rock Bolting*. A.A. Balkema, Rotterdam.

Stephansson, O. 1992. Underground chambers in hard rock masses. In: *Engineering in Rock Masses*, Bell, F.G. (ed.), Butterworth-Heinemann, Oxford, 440–464.

Szechy, K. 1966. *The Art of Tunnelling*. Academiai Kaido, Budapest.

Tan, D.Y. and Clough, G.W. 1980. Ground control for shallow tunnels by soil grouting. *Proceedings American Society Civil Engineers, Journal Geotechnical Engineering Division*, **106**, 1037–1057.

Tarkoy, P.J. 1973. Predicting tunnel boring machine penetration rates and cutter costs in selected rock types. *Proceedings Ninth Canadian Symposium on Rock Mechanics*, Montreal, 263–274.

Taylor, C.L. and Conwell, F.R. 1981. BART – influence of geology on the construction conditions and costs. *Bulletin Association of Engineering Geologists*, **18**, 195–200.

Terzaghi, K. 1946. Introduction to tunnel geology. In: *Rock Tunneling with Steel Supports*, Proctor, R. and White, T. (eds), Commercial Stamping and Shearing Company, Youngstown, Ohio, 17–99.

Terzaghi, K. 1950. Geological aspects of soft ground tunneling. In: *Applied Sedimentation*, Trask, P.D. (ed.), Wiley, New York, 193–209.

Thompson, F.T. 1966. San Jacinto Tunnel. In: *Engineering Geology of Southern California*, Lang, R. and Proctor, R. (eds), Special Publication Association Engineering Geologists, 56–74.

Varley, P.M. 1990. Susceptibility of Coal Measures mudstone to slurrying during tunnelling. *Quarterly Journal Engineering Geology*, **23**, 147–160.

Warren, C.D., Birch, G.P., Bennett, A. and Varley, P.M. 1991. Methane studies for the Channel Tunnel. *Quarterly Journal Engineering Geology*, **24**, 291–309.

Whittacker, B.N. and Frith, R.C. 1990. *Tunnelling: Design, Stability and Construction*. Institution Mining and Metallurgy, London,

Wickham, G.E., Tiddeman, H.R. and Skinner, E.P. 1972. Support determination based upon geologic predictions. *Proceedings First North American Tunneling Conference*, American Institute Mining Engineers, New York, 43–64.

Wilbur, L.D. 1982. Rock tunnels. In: *Tunnel Engineering Handbook*, Bickel, J.O. and Kuesel, T.R. (eds), Van Nostrand, New York, 123–207.

Williamson, T.H. and Schmidt, R.L. 1972. Probe drilling for rapid tunnelling. *Proceedings First North American Tunneling Conference*, American Institute Mining Engineers, New York, 65–87.

Yuzer, E. 1982. Engineering properties and evaporitic formations of Turkey. *Bulletin International Association Engineering Geology*, **25**, 107–110.

4 Foundation conditions and buildings

Foundation design is concerned primarily with ensuring that movements of a foundation are within the limits that can be tolerated by the proposed structure without adversely affecting its functional requirements. Structures vary widely in their capacity to accommodate movements of their foundations. Hence, it is usual to consider the design of a structure and the foundation as interrelated. The design of a foundation structure requires an understanding of the local geological and groundwater conditions, as well as an appreciation of the various types of ground movement that can occur. Foundation movements can occur when the ground is excavated, as seen in Chapter 2, when the ground is loaded, or independently of construction operations.

4.1 Total and effective pressures

Soil is composed of solid particles between which are voids. These voids may be partially or wholly filled with water. The deformability of a soil as a result of either loading or unloading primarily is attributable to the deformation that takes place in the voids, which usually involves the displacement of pore water. The strength of a soil is its ultimate resistance to loading.

Subsurface water is normally under pressure that increases with increasing depth below the water table to very high values. Such water pressures have a significant influence on the engineering behaviour of most rock and soil masses, and their variations are responsible for changes in the stresses in these masses, which affect their deformation characteristics and failure.

The efficiency of a soil in supporting a structure is influenced by the effective or intergranular pressure, that is, the pressure between the particles of the soil that develops resistance to applied load. In other words, shear stresses are carried by the soil particles, water having no shear strength. Because the pore water offers no resistance to shear, it is neutral and therefore pore water pressure has been referred to as neutral pressure. Since the pore water or neutral pressure (u) plus the effective pressure (σ') equals the total pressure (σ), that is,

$$\sigma = \sigma' + u \qquad (4.1)$$

then a reduction in pore water pressure increases the effective pressure. Reduction of the pore water pressure by drainage consequently affords better conditions for carrying a proposed structure.

The effective pressure at a particular depth is simply obtained by multiplying the unit weight of the soil by the depth in question and subtracting the pore pressure for that depth. In a layered sequence the individual layers may have different unit weights. The unit weight of each layer then should be multiplied by its thickness and the pore water pressure for that layer subtracted.

The effective pressure for the total thickness involved is obtained by summing the effective pressures of the individual layers (Fig. 4.1). Water held in the capillary fringe by soil suction does not affect the values of pore water pressure below the water table. However, the weight of water held in the capillary fringe does increase the weight of overburden and so the total and effective pressures.

The position of the water table governs the pore water pressure and its initial value is referred to as the static pore water pressure. However, when the total vertical stress is increased by loading and the soil is saturated and laterally confined, the increased loading is taken up by the pore water so that the pore water pressure is increased above the static value. This increase in pore water pressure causes water to flow from the area of increased loading until the pore water pressure is reduced to a value again governed by a steady position of the water table, the final value being termed the steady-state pore water pressure. The increase in pore water pressure above the steady-state value is called the excess pore water pressure. The rate at which excess pore water pressure dissipates depends upon the permeability of the soil and when it is completed the soil is described as being in the drained condition. By contrast, the soil is referred to as being in the undrained condition before the dissipation of excess pore water begins. The increased vertical stress is transferred to the soil particles as the excess pore water pressure is dissipated so that when the excess pore pressure is dissipated completely, then the increased

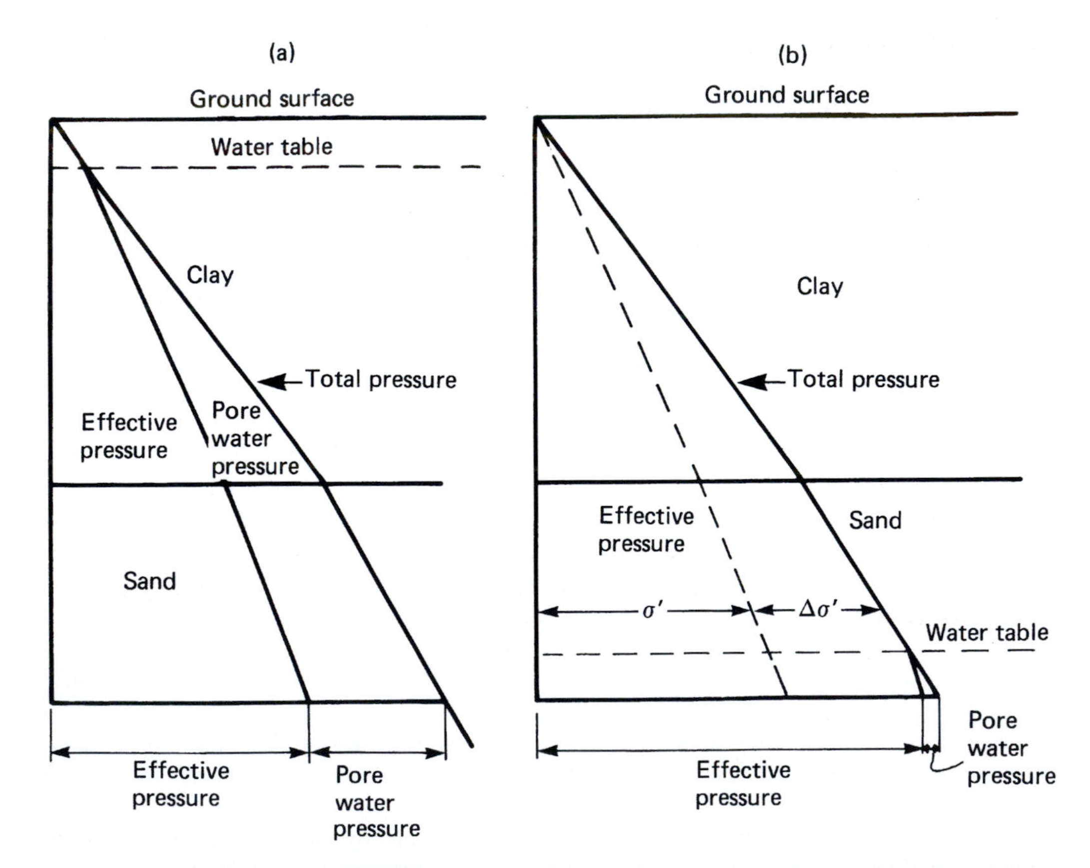

Figure 4.1 Pressure diagram illustrating total, effective and pore water pressures: (a) water table just below the surface; (b) water table has been lowered into the sand and effective pressure is increased with a reduction in pore water pressure.

vertical stress is carried wholly by the soil skeleton. As the excess pore water pressure is dissipated, the effective pressure increases and the volume of the soil, especially compressible soil, is reduced. Volume changes brought about by loading compressive soils depend upon the level of effective stress and are not affected by the area of contact.

In the case of partially saturated soils, the void space is occupied by water and air. The pore water pressure (u) is less than the pore air pressure (u_a) due to surface tension. Unless the degree of saturation is close to unity the pore air forms continuous channels through the soil and the pore water is concentrated around the interparticle contacts. The boundaries between pore water and pore air are in the form of menisci, the radii of which depend on the size of the pore spaces within the soil. The following effective stress equation is used for partially saturated soils:

$$\sigma = \sigma' + u_a - \chi(u_a - u_w) \tag{4.2}$$

where χ is a parameter related to the degree of saturation of the soil. The term $(u_a - u_w)$ is a measure of the suction in the soil. The limiting values are for fully saturated soil ($S = 1$, $\chi = 1$) and for completely dry soil ($S_r = 0$, $\chi = 0$). The value of χ also is influenced, to a lesser extent, by the soil structure and the way the particular degree of saturation was brought about.

Evidence suggests that the law of effective stress holds true for some rocks. Those with low porosity may, at times, prove the exception. However, Serafim (1968) suggested it appeared that pore pressures have no influence on brittle rocks. This is probably because the strength of such rocks mainly is attributable to the strength of the bonds between the component crystals or grains. The changes in stresses and the corresponding displacements due, for example, to construction work influence the permeability of a rock mass. For instance, with increasing effective shear stress the permeability increases along discontinuities orientated parallel to the direction of shear stress, whilst it is lowered along those running normal to the shear stress. Consequently, the imposition of shear stresses and the corresponding strains lead to an anisotropic permeability within joints.

Piezometers are installed in the ground in order to monitor and obtain accurate measurements of pore water pressures (see Chapter 1). Observations should be made regularly so that changes due to external factors such as excessive precipitation, tides, the seasons, etc. are noted, it being most important to record the maximum pressures that have occurred.

4.2 Stress distribution in soil

A reasonable approximation of how stress is distributed in soil upon loading can be obtained by assuming that the soil behaves in an elastic manner as if it was a homogeneous material. In such an instance the vertical pressure (σ_v) produced at any point (N) in the soil by a load (Q) on the surface (Fig. 4.2a) may be derived from the following expressions:

$$\sigma_v = \frac{3Q}{2\pi z^2}\left[\frac{1}{1+(r/z)^2}\right]^{5/2} \tag{4.3}$$

Expression 4.3 has been simplified as follows:

$$\sigma_v = \frac{I_f Q}{z^2} \tag{4.4}$$

where I_f is an influence factor depending on the depth (z) and position at which the stress is required in relation to the point load (Fig. 4.2b).

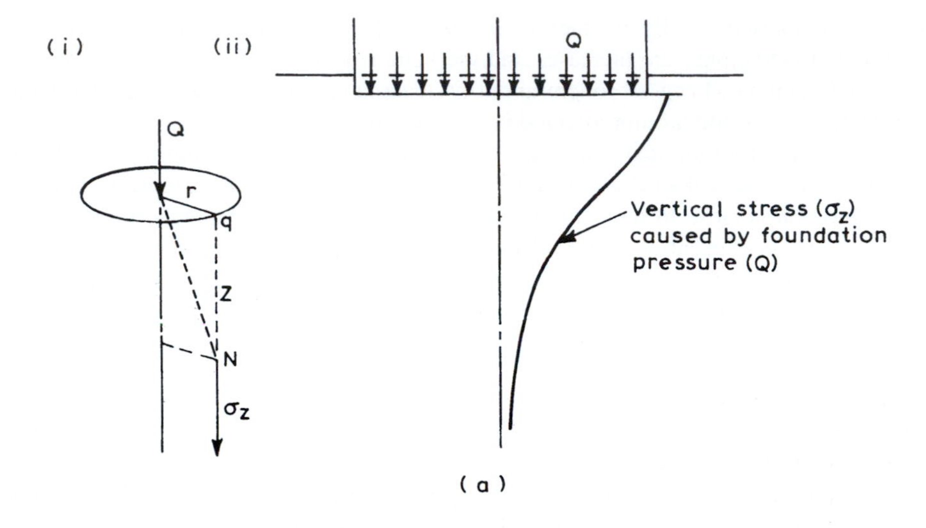

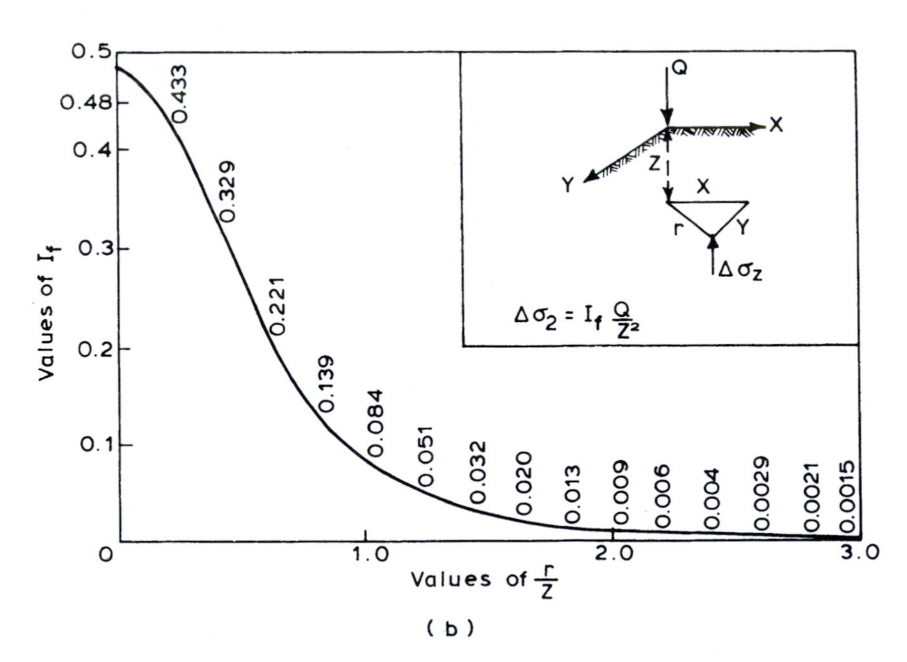

Figure 4.2 Vertical stress distribution beneath a surface point load. (a): (i) Intensity of vertical pressure at point N in the interior of a semi-infinite solid acted on by load Q; (ii) vertical stress (σ_z) with depth generated by foundation pressure (Q). (b) Influence coefficients for vertical stress for a point load.

Various design charts have been proposed for estimating stresses beneath a foundation structure. For example, Fadum (1948) provided a set of curves from which the increment of vertical stress (σ_v) beneath the corner of a uniformly loaded rectangular area could be obtained by substituting values in the following expression:

$$\sigma_v = QI_f \tag{4.5}$$

where Q is the load and I_f is an influence factor depending on the dimensions of the rectangle and the depth concerned (Fig. 4.3). The method also can be used to determine the increment of vertical stress beneath a point anywhere beneath a structure as long as the area can be subdivided into rectangles and the point occurs beneath the coincident corners of the rectangles. The vertical stress then is simply the sum of the stresses produced by the rectangles, if the point is within the foundation area. If the point lies outside the foundation area, then rectangles are extended to include the point. In this case the areas outside the foundation are subtracted in the computation of vertical stress.

A useful design chart for deriving stresses beneath a foundation is that provided by Janbu *et al.* (1956). This chart indicates the increase in vertical stress beneath the centre of a uniformly loaded strip, rectangular or circular foundation structure (Fig. 4.4).

The Newmark chart was introduced by Newmark (1935) in order to provide a graphical method of estimating the vertical stress produced at a given point in a soil mass by a load uniformly

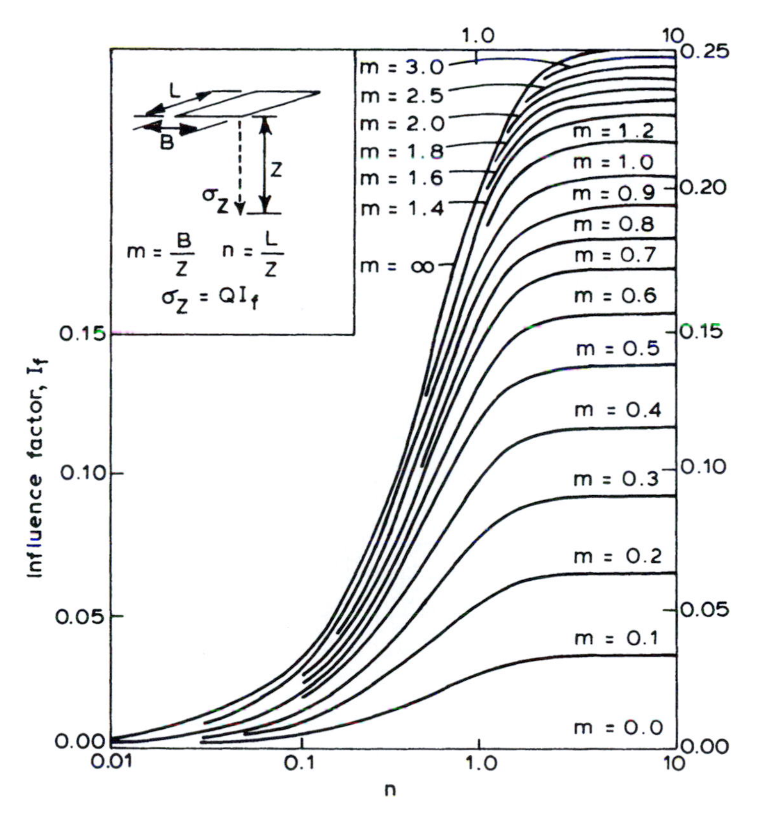

Figure 4.3 Influence factors for vertical stress beneath the corner of a rectangular foundation.
Source: Fadum, 1948.

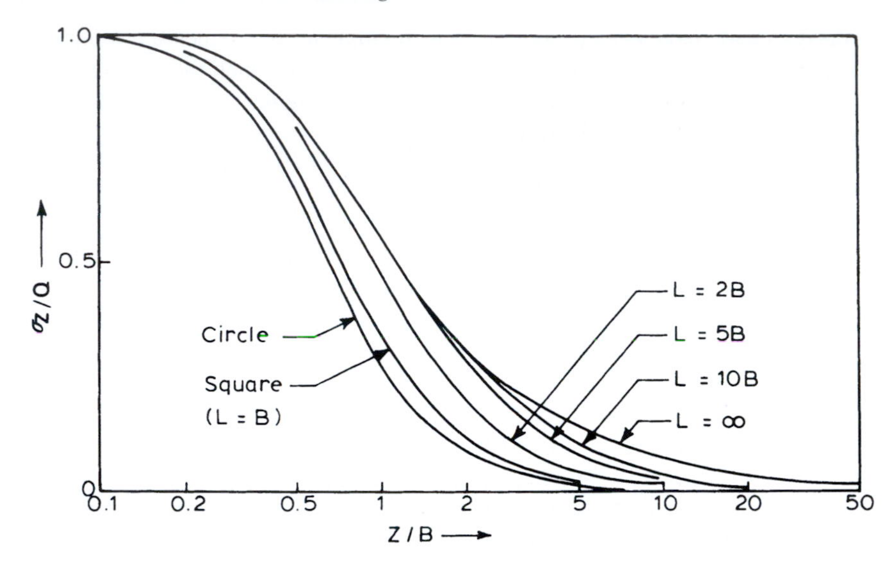

Figure 4.4 Determination of increase in vertical stress under the centre of uniformly loaded flexible footings.
Source: Janbu *et al*., 1956. Reproduced by kind permission of Norwegian Geotechnical Institute.

distributed over an irregularly shaped foundation (Fig. 4.5). A tracing of the foundation is drawn to the same scale as the depth of the position at which the stress is to be evaluated (the line AB on the chart represents the depth and so is used for the scale of the drawing). The scale drawing then is laid over the influence chart with the point, vertically below which the stress is required, being placed at the centre of the chart. Then the number of segments on the chart covered by the drawing is counted and multiplied by the influence factor shown on the chart. This value is multiplied by the loading to give the increase in pressure at the required depth.

A bulb of pressure is a graphical illustration of the manner in which loads imposed by a foundation structure are dissipated by the transfer of stress to increasing volumes of soil as the

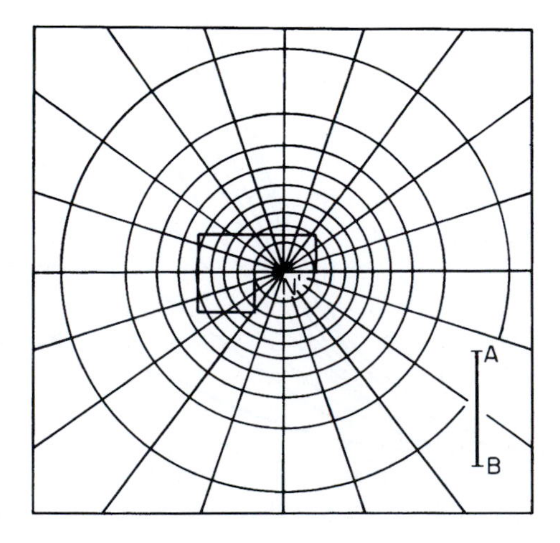

Figure 4.5 The Newmark chart for determination of vertical stress beneath a foundation.
Source: Newmark, 1935.

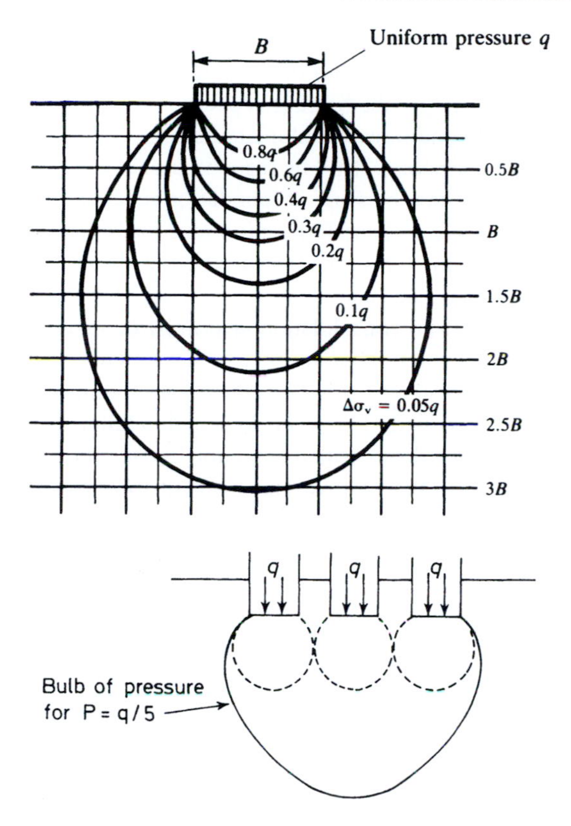

Figure 4.6 (a) Bulbs of pressure. (b) Cumulative effect of adjacent loads on the bulb of pressure.

distance from the foundation structure increases (Fig. 4.6a). Closely spaced loads develop a cumulative pressure bulb as shown in Fig. 4.6b.

4.3 Bearing capacity

4.3.1 Deformation and failure of soil under loading

Movement of foundations under the influence of loading may occur as a result of overstressing of the ground, which gives rise to plastic deformation in the ground beneath the foundation structure. In extreme cases shear failure may occur. In order to avoid shear failure or substantial shear deformation the foundation pressures used in design should have an adequate factor of safety when compared with the ultimate bearing capacity of the foundation. The ultimate bearing capacity is the value of the net loading intensity that causes the ground to fail suddenly in shear. If this is to be avoided, then a factor of safety must be applied to the ultimate bearing capacity, the value obtained being referred to as the maximum safe bearing capacity. In other words, this is the maximum net loading intensity that may be safely carried without the risk of shear failure. However, even this value still may mean that there is a risk of excessive or differential settlement. Thus, the allowable bearing capacity is the value that is used in design, this taking into account all possibilities of failure, and so its value frequently is less than that of the safe bearing capacity. The value of ultimate bearing capacity depends on the type of foundation structure as well as the properties of the ground. For example, for footings the dimensions, shape, depth and inclination of load and base at which a footing is placed all influence the bearing capacity. More specifically the width of the foundation is important in coarse-grained soils, the greater the width, the larger the bearing capacity whilst in saturated clay soils it is of little effect. With uniform soil conditions the ultimate bearing capacity increases with depth of installation of the foundation structure. This increase is associated with the confining effects of the soil, the decreased overburden pressure at foundation level and with the shear forces that can be mobilized between the sides of the foundation and the ground. Where foundations are deep, the contribution to the ultimate bearing capacity by the shear forces on the sides may be large. The density, frictional resistance and cohesion influence the shear strength of a soil, and its permeability, compressibility and consolidation also must be taken into consideration when the allowable bearing pressure, which takes account of settlement, is determined.

When a load is applied to a soil in gradually increasing amounts, the soil deforms and a load–settlement curve can be plotted (Fig. 4.7). When the failure load is reached, the rate of deformation increases and the load–settlement curve goes through a point of maximum curvature that indicates that the soil has failed. The shape of the curve is influenced by the type of soil involved, for example, dense sand and insensitive clay show a more gradual transition, associated with progressive failure.

There are usually three stages in the development of a foundation failure. First, the soil beneath the foundation is forced downwards in a wedge-shaped zone (Fig. 4.8). Consequently, the soil beneath the wedge is forced downwards and outwards, elastic bulging and distortion taking place within the soil mass. Second, the soil around the foundation perimeter pulls away from the foundation and shear forces propagate outward from the apex of the wedge. This is the zone of radial shear in which plastic failure by shear occurs. Thirdly, if a soil is very compressible the failure is confined to fan-shaped zones of local shear. The foundation will displace downwards with little load increase. On the other hand, if the soil is more rigid, the shear zone propagates outwards until a continuous surface of failure extends to the ground surface and the surface heaves. This is termed general shear failure. The weight of the material in the passive zone resists the lifting force and provides the reaction through the other two zones that counteracts downwards motion of the foundation structure. Thus, the bearing capacity is a

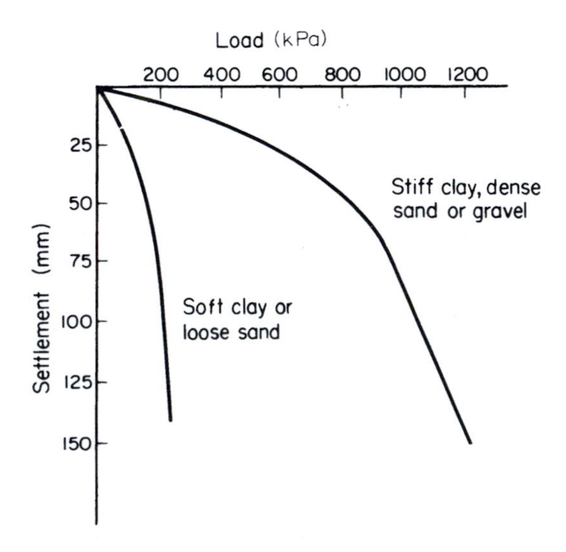

Figure 4.7 Load–settlement curve.

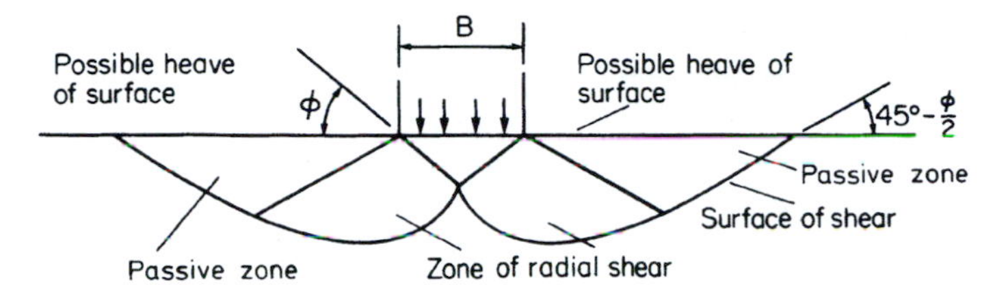

Figure 4.8 Bearing capacity or foundation failure in general shear.

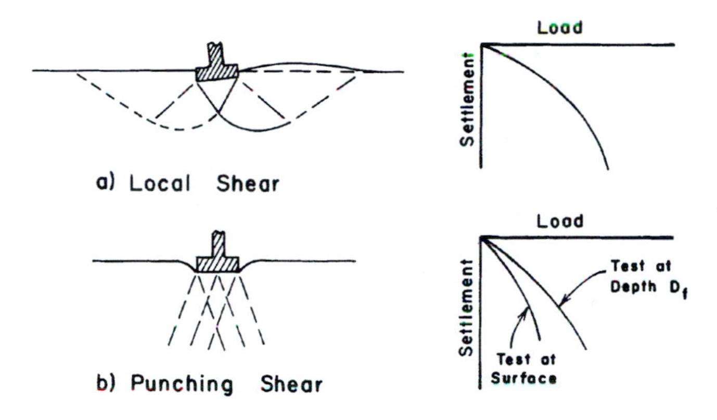

Figure 4.9 Foundation failure: (a) in local shear; (b) in punching shear.

function of the resistance to uplift of the passive zone. This, in turn, varies with the size of the zone (which is a function of the angle of shearing resistance), with the unit weight of the soil and with the sliding resistance along the lower surface of the zone (which is a function of the cohesion, angle of shearing resistance and unit weight of the soil). A surcharge placed on the passive zone, or increasing the depth of the foundation, therefore increases the bearing capacity. Vesic (1973) recognized two further modes of failure that are associated with weaker, more compressible soils and wider or deeper footings (Fig. 4.9). A local shear failure is characterized by well-defined shear surfaces immediately below a footing but they extend for only short distances into the soil and do not reach the surface. There is significant compression of the soil beneath the footing and only slight heaving occurs. A punching mode of failure can occur in highly compressible soils. Compression again occurs beneath a footing and is accompanied by vertical shearing around the footing.

Where a weak horizon overlies a strong one the shear will be confined to the weaker material and the stronger will not be involved in the failure. The bearing capacity therefore should be determined from the strength of the weaker material. Because the shear zone is restricted, the true bearing capacity will exceed that calculated. In the converse situation where a strong layer overlies a weak one, the former spreads the load thus reducing the pressure on the weaker horizon. Failure occurs by shear in the weaker material as the stronger bends under the load.

4.3.2 Determination of bearing capacity

A number of expressions have been proposed to determine the ultimate bearing capacity and make use of bearing capacity factors that depend on various characteristics of the soil and foundation structure. The basic expression for ultimate bearing capacity (q) for general shear for shallow footings (i.e. at a depth not exceeding the width of the footing) was proposed by Terzaghi (1943), and involves cohesion (c), unit weight (γ) and overburden pressure (γz), and is as follows:

$$q = cN_c + \gamma z N_q + 0.5\gamma B N_\gamma \qquad (4.6)$$

where N_c, N_q and N_γ are the bearing capacity factors that depend upon the angle of shearing resistance (ϕ), z is the depth of foundation emplacement and B is the width of the footing. The values of the bearing capacity factors can be obtained from Fig. 4.10. Equation 4.6 should only be used when the water table is deep (i.e. at a depth exceeding the width of the foundation below the base of the foundation). When the water table is located at the base of the foundation the submerged unit weight is used in the bearing capacity expressions. Moreover, when there is little dissipation of pore water pressure during the construction period, as in clay soils because of their low permeability, then because conditions are undrained, determination of the ultimate bearing capacity should be in terms of total stress (i.e. the shear strength values used should be c_u and ϕ_u). However, determination of the ultimate bearing capacity in permeable soils in which excess pore pressures are dissipated on loading means that consolidation and gain in strength are complete at the end of the construction period. This situation applies in free draining gravelly and sandy soils, and so the determination of the bearing capacity should be in terms of effective stress. For example, the ultimate bearing capacity for free draining soils, that is, where the excess pore water pressures beneath a footing are zero at the end of the application of load, and where the water table is high can be expressed as:

$$q = c'N_c + \gamma z'(N_q - 1) + 0.5B\gamma' N_\gamma + \gamma z \qquad (4.7)$$

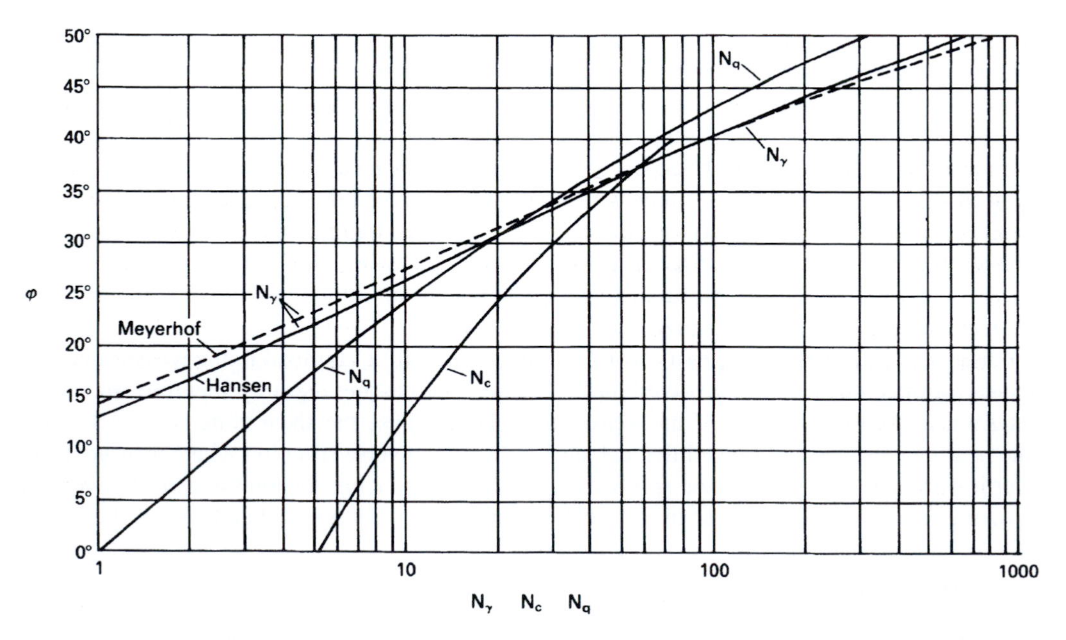

Figure 4.10 Bearing capacity factors for shallow foundations.
Source: Meyerhof, 1963; Brinch Hansen, 1970. Reproduced by kind permission of Danish Geotechnical Institute.

the values c', $\gamma z'$ and γ' being in terms of effective stress. Conversely, if the water table is low, the following expression should be used:

$$q = c'N_c + \gamma z(N_q - 1) + 0.5B\gamma N_\gamma + \gamma z \tag{4.8}$$

Terzaghi suggested the following expression for local shear failure:

$$q = \frac{2}{3}cN_c + \gamma zN_q + 0.5\gamma BN_\gamma \tag{4.9}$$

He also suggested the following expressions for square and circular footings

$$\text{Square } q = 1.3cN_c + \gamma zN_q + 0.4\gamma BN_\gamma \tag{4.10}$$

$$\text{Circular } q = 1.3cN_c + \gamma zN_q + 0.3\gamma BN_\gamma \tag{4.11}$$

When a footing is rectangular or square, the support given by the soil at the ends of the footing must be taken into account and the foundation has a higher ultimate bearing capacity. As seen above, when the soil under a strip footing fails in shear the movement is laterally outward. If a square, rectangular or circular footing fails, then the movement of the soil particles forms a radial pattern. According to Skempton (1951) the ultimate bearing capacity for a rectangular footing is equal to the ultimate bearing capacity for a strip footing multiplied by $1 + 0.2B/L$ where B and L are breadth and length respectively.

In granular soils because the cohesion is negligible the general bearing capacity expression is reduced to:

$$q = \gamma zN_q + 0.5\gamma BN_\gamma \tag{4.12}$$

and the net bearing capacity is derived from:

$$q = \gamma z(N_q - 1) + 0.5\gamma BN_\gamma \tag{4.13}$$

However, obtaining a value of ϕ by laboratory methods, in order to determine the bearing capacity factors, is often virtually impossible. This is due to difficulties in obtaining undisturbed samples of granular soils. Consequently, the allowable bearing capacity of granular soils is commonly determined from the results of *in situ* tests such as the standard penetration test and the cone penetration test (Tomlinson and Boorman, 1996).

If the angle of shearing resistance is zero, as in saturated clay in undrained shear, only the cohesion contributes materially to the bearing capacity. Consequently, Skempton (1951) concluded that the ultimate bearing capacity of a footing on saturated clays in undrained conditions could be obtained from the following expression:

$$q = cN_c + \gamma z \tag{4.14}$$

The net ultimate bearing capacity being simply

$$q = cN_c \tag{4.15}$$

In this case, the bearing capacity factor, N_c, also depends upon the shape of the footing (strip, circular or square) and its depth/breadth ratio (Fig. 4.11). In order to obtain the bearing capacity

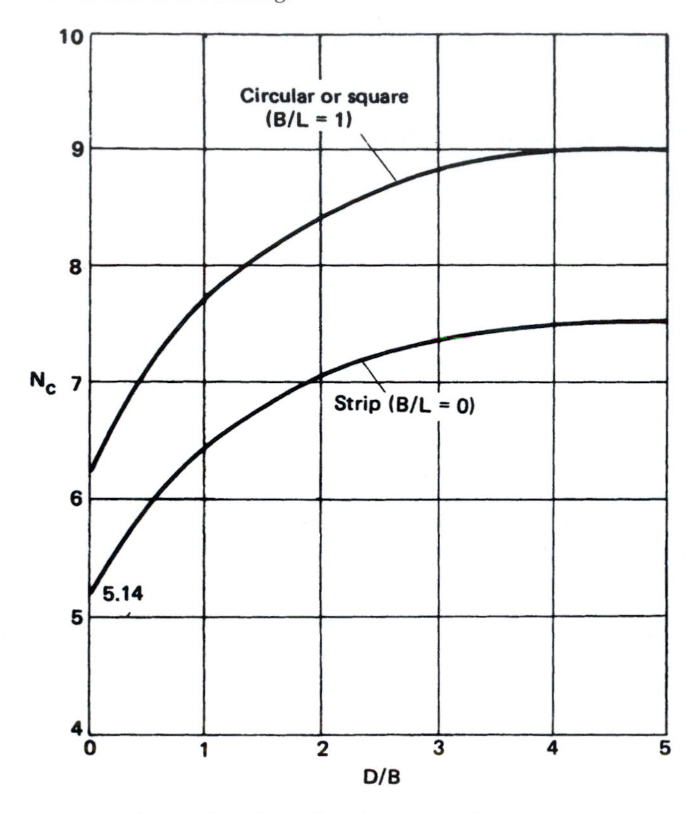

Figure 4.11 Bearing capacity factors for clay soils where $\phi_u = 0$.
Source: Skempton, 1951. Reproduced by kind permission of BRE.

for a rectangular foundation the value of N_c for a strip foundation is multiplied by a shape factor (Fig. 4.12).

Meyerhof (1963) showed that Terzaghi's equations for shallow foundations are conservative since they do not take into account the shearing resistance of the soil along the surface of failure above the level of the base of the foundation. He further pointed out that as the Terzaghi general equation applies to shallow foundations, its application to deep foundations is inappropriate in that when the surface of failure does not extend to ground level, the height over which the shearing resistance of the soil is mobilized is very uncertain. As a consequence, Meyerhof developed his own bearing capacity factors that depend on the depth and shape of the foundation, and the roughness of its base, as well as the angle of shearing resistance. However, he continued to use Terzaghi's general bearing capacity equations but with different curves (Fig. 4.10). He also recommended that in the case of rectangular and circular footings the values of the bearing capacity factors should be multiplied by a shape factor. Footings may be subjected to eccentric and inclined loadings, and in such instances the bearing capacity is reduced. Meyerhof suggested that the effective width of a footing (B'), with an eccentricity (e) of the resultant load on its base of width B, could be determined from:

$$B' = B - 2e \tag{4.16}$$

He also proposed inclination factors to take into account the influence of inclined loading on the bearing capacity. If the angle of inclination of the resultant load to the vertical is α, then

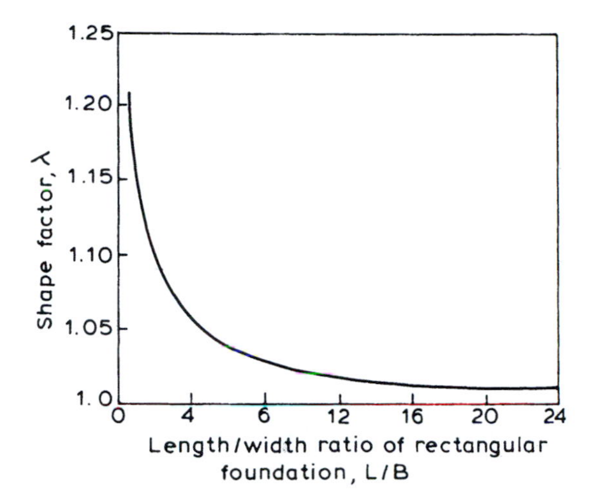

Figure 4.12 Shape factor for rectangular foundations in day.
Source: Skempton, 1951. Reproduced by kind permission of BRE.

the bearing capacity factors N_c, N_q and N_γ should be multiplied by the inclination factors i_c, i_q and i_γ, which are as follows:

$$i_c = i_q = (1 - \alpha/90°)^2 \tag{4.17}$$

$$i_\gamma = (1 - \alpha/\phi)^2 \tag{4.18}$$

Brinch Hansen (1970) also concluded that the Terzaghi expressions required revision and so produced his own bearing capacity curves. He also introduced expressions and curves for shape factors, depth factors, load inclination factors, base inclination factors and ground surface inclination factors for each of the bearing capacity factors. A recent comparison of the methods used to evaluate the bearing capacity of shallow foundations in Europe has been provided by Sieffert and Bay-Gress (2000).

4.3.3 Bearing capacity of rock masses

The bearing capacity of a rock mass must be determined in order to assess the safety of a foundation, although settlement frequently has a controlling influence on design. Generally, compressive loads applied to an embedded foundation are transmitted to the surrounding rock mass through base and side resistance. The relative importance of these two resistances as far as the bearing capacity is concerned depends on the relative stiffnesses of the rock mass and the foundation structure, as well as the geometry of the foundation structure. In addition, there may be some degree of interdependence between the base and side resistances, particularly in jointed rock.

Kulhawy and Carter (1992a) suggested a number of modes of possible bearing capacity failure and that, depending on the geological conditions, one of these modes could be selected as being representative of likely field behaviour. Failure can be by flexure in the case of a thick rigid layer overlying a weaker one. The flexural strength is approximately twice the uniaxial tensile strength of the rock material. Punching failure can occur if a thin rigid layer overlies a weaker one, which is effectively a tensile failure of the rock material. Failure can occur by uniaxial compression of rock columns when a vertical jointed rock mass is loaded. In this instance, the ultimate bearing capacity, q, is given by the Mohr–Coulomb failure criterion:

$$q = 2c\tan(45° + \phi/2) \tag{4.19}$$

in which c is the cohesion and ϕ is the angle of friction of the rock mass. If a rock mass contains closely spaced closed joints, then a general wedge shear zone can develop. The ultimate bearing capacity in this case is given by:

$$q = cN_c + 0.5B\gamma N_\gamma + \gamma z N_q \tag{4.20}$$

in which B is the width of the foundation γ is the unit weight of the rock mass, and N_c, N_γ and N_q are the bearing capacity factors. Equation 4.20 should be modified to consider shape when the foundation is square or cylindrical it becoming:

$$q = i_{cs}cN_c + i_{\gamma s}0.5B\gamma N_\gamma + i_{qs}\gamma z N_q \tag{4.21}$$

where the shape factors are $i_{cs} = 1 + N_q/N_c$, $i_{\gamma s} = 0.6$, and $i_{qs} = 1 + \tan\phi$. Failure occurs by splitting beneath a foundation when joints have a wide spacing, which eventually leads to general shear failure. Assuming that little stress is transmitted across the vertical joints, then according to Kulhawy and Goodman (1980):

$$q = C_f c N_{cr} \tag{4.22}$$

in which C_f is a correction factor (Fig. 4.13a) and N_{cr} is the bearing capacity factor (Fig. 4.13b).

Alternatively, an empirical approach to the assessment of the strength of discontinuous rock masses has been provided by Hoek and Brown (1997), their failure criterion for jointed rock masses being given by:

$$\sigma_1' = \sigma_3' + \sigma_{ci}\left(m_b\frac{\sigma_3'}{\sigma_{ci}} + s\right)^a \tag{4.23}$$

where σ_1' and σ_3' are the respective maximum and minimum effective stresses at failure, m_b is the value of the constant m for the rock mass, s and a are constants that depend on the characteristics of the rock mass, and σ_{ci} is the unconfined compressive strength of the intact rock. In the case of poor quality rock masses where the strength has been partially destroyed by shearing or weathering so that the rock mass has no tensile strength or 'cohesion', then $s = 0$. Hoek and Brown maintained that the failure criterion should be expressed in terms of effective stress when applied to practical design problems and that it therefore was necessary to determine the distribution of pore water pressure in the rock mass concerned. The Hoek–Brown failure criterion is applicable only to intact rock or rock masses in which there are a sufficient number of closely spaced discontinuities, with similar surface characteristics so that the material may be regarded as homogeneous and isotropic. As such, it should not be applied to fissile rock masses such as, slates or to rock masses in which the properties are governed by a single set of discontinuities such as bedding planes.

The Geological Strength Index, GSI, which was introduced by Hoek (1994), provides a means of estimating the reduction in the strength of a rock mass for different geological conditions. Each category of rock mass can be identified from field observations and its value of GSI then

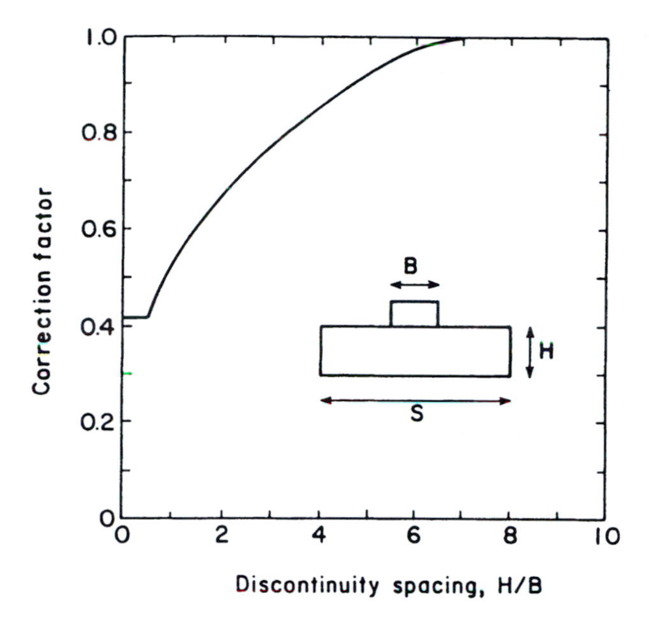

Figure 4.13a Correction factor for discontinuity spacing.
Source: Kulhawy and Carter, 1992.

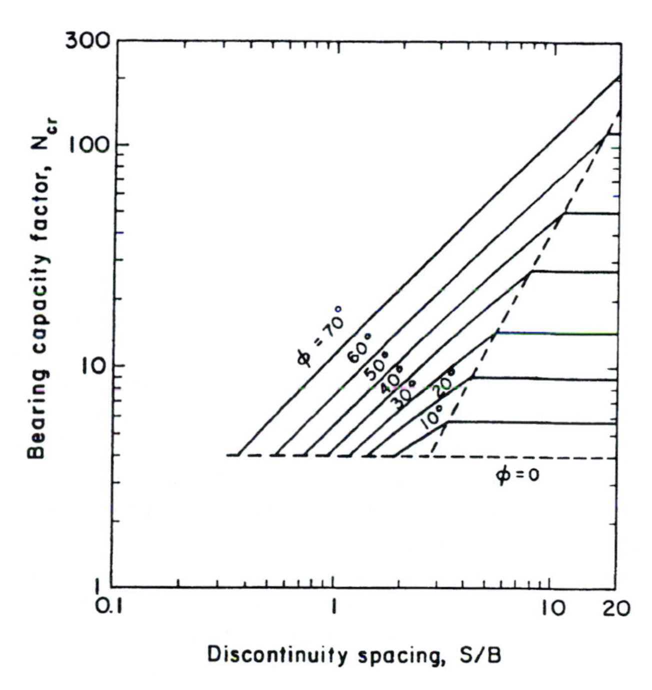

Figure 4.13b Bearing capacity factor for open joints.
Source: Kulhawy and Goodman, 1980.

can be estimated from Fig. 3.19. Once the GSI has been estimated, then the parameters relating to the strength characteristics of a rock mass can be derived, that is:

$$m_b = m_i \exp\left(\frac{\text{GSI} - 100}{28}\right) \tag{4.24}$$

When the GSI > 25, that is, for rock masses of good to reasonable quality:

$$s = exp\left(\frac{\text{GSI} - 100}{9}\right) \tag{4.25}$$

and

$$a = 0.5 \tag{4.26}$$

On the other hand, for rock masses of very poor quality, GSI < 25:

$$s = 0 \tag{4.27}$$

and

$$a = 0.65 - \frac{\text{GSI}}{200} \tag{4.28}$$

4.4 Contact pressure

The pressure acting between the bottom of a foundation structure and the soil is the contact pressure. The assumption that a uniformly loaded foundation structure transmits the load uniformly so that the ground is uniformly stressed is by no means valid. For example, the intensity of the stresses at the edges of a rigid foundation structure on hard clay soil is theoretically infinite (Fig. 4.14a). In fact, of course, the clay soil yields slightly and so reduces the stress at the edges. As the load is increased, more and more local yielding of the ground material takes place until, when the loading is close to that which would cause failure, the distribution is probably very nearly uniform. Therefore, at working loads a uniformly loaded foundation structure on clay soil imposes a widely varying contact pressure. On the other hand, a rigid footing on the surface of dry sand imposes a parabolic distribution of pressure (Fig. 4.14b). Since there is no cohesion in such material no stress can develop at the edges of the footing. If the footing is below the surface of the sand, then the pressure at the edges is no longer zero but increases with depth. The pressure distribution therefore tends to become more nearly uniform as the depth increases. If a footing is perfectly flexible, then it will distribute a uniform load over any type of foundation material.

If a rock mass contains few defects the contact pressure at the surface may be taken conservatively as the unconfined compressive strength of the intact rock. Most rock masses, however, are affected by joints or weathering that may affect their strength and engineering behaviour significantly. Table 4.1 gives values of allowable contact pressure for jointed rocks based on their RQD. If design is based on these values the settlement of foundations should not exceed 12.7 mm even for large loaded areas. The RQD should be the average within a depth below foundation level equal to the width of the foundation, provided the RQD is fairly uniform within that depth. If the upper part of the rock, within a depth equal to about quarter the width

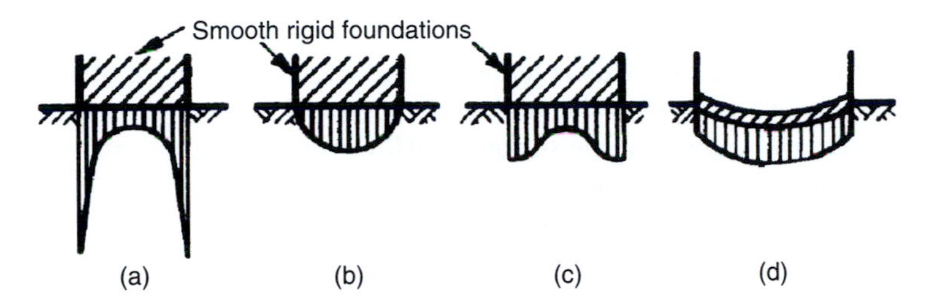

Figure 4.14 Contact pressure distribution beneath a rigid foundation on: (a) clay soil; (b) sand and gravel; (c) mixed soil; (d) a flexible foundation on clay soil.

Table 4.1 Allowable contact pressure for jointed rock*

RQD	*Allowable contact pressure (MPa)*
100	32.2
90	21.5
75	12.9
50	7.0
25	3.2
0	1.1

Source: Peck *et al*., 1974. Reproduced by kind permission of John Wiley & Sons, Inc.

Note
* If the value of the allowable contact pressure exceeds the unconfined compressive strength of intact samples then it should be taken as the unconfined compressive strength.

of the foundation, is of low quality the value for this part should be used or the inferior rock should be removed.

Inclined strata introduce complications to both design and construction of most foundation structures. The presence of faults and shear zones does not necessarily compromise the suitability of the foundations for they may be occupied by material with satisfactory physical properties. On the other hand, some fault zones are occupied by fault gouge that has the characteristics of soft plastic clay. The properties of such materials must be ascertained if heavy structures are to be erected.

The great variation in the physical properties of weathered rock and the non-uniformity of the extent of weathering even at a single site permit few generalizations concerning the design and construction of foundation structures. The depth to bedrock and the degree of weathering must be determined. If the weathered residium plays the major role in the regolith, rock fragments being of minor consequence, then design of rafts or footings should be according to the matrix material. Piles or piers can provide support at depth.

4.5 Consolidation and settlement

The theory of consolidation has enabled engineers to determine the amount and rate of settlement that is likely to occur when structures are erected on clayey soils. When a layer of soil is loaded, some of the pore water is expelled from its voids, moving slowly away from the region of high stress as a result of the hydrostatic gradient created by the load. The void ratio accordingly decreases and settlement occurs. This is termed primary consolidation. Further consolidation, usually of minor amount, may occur due to the rearrangement of the soil particles under stress, this being referred to as secondary consolidations. However, in reality primary and secondary consolidations are not distinguishable.

A clay soil that has been formed recently and achieved equilibrium under its own weight, but not undergone any significant secondary consolidation, is called a normally consolidated clay. Such clay soils only are capable of carrying the overburden weight of soil and any additional load results in relatively large settlements. By contrast, an overconsolidated clay soil is one that has been subjected to a pressure in excess of its present overburden pressure, that is, at some previous time it was buried to a greater depth but subsequently overburden has been removed by erosion. Some heavily overconsolidated clay soils may have been subjected to several cycles of loading and unloading. If a clay soil is overconsolidated, then the estimated amount of settlement is likely to be in excess of that which actually occurs after loading.

As settlement of soils when loaded, particularly many clay soils, frequently represents a problem for construction, the amount and rate at which it is likely to take place needs to be determined. Settlement of clay soils continues after the construction period, indeed settlement of clays may continue for several years. By contrast, settlement in sands normally is complete by the end of the construction period. The average value of settlement undergone by a structure together with the individual settlements experienced by its various parts influences the degree to which the structure serves its purpose. The damage attributable to settlement can range from slight disfigurement to complete failure of the structure. What is more, the zone of settlement on a compressible soil extends beyond the limits of the structure. Thus, nearby buildings may be affected.

4.5.1 *Determination of compressibility*

The relationship between unit load and the void ratio for a clay soil can be represented by plotting the void ratio (e) against the logarithm of the unit load (p) as shown in Fig. 4.15. In the laboratory the relationship between e and p is investigated by using an oedometer (Head, 1994). In the standard oedometer test a disc-shaped specimen is contained in a metal ring, usually 76 mm in diameter, and lies between two porous stones. The apparatus sits in an open cell of water, to which the pore water of the specimen has free access. The specimen is subjected to sequence of loadings, each increment usually being imposed for 24 hours, and is applied when compression due to the previous load has ceased, that is, when excess pore pressure is dissipated. The compression and time are recorded for each load increment. The final load should be greater than any likely to be imposed on the soil concerned. At the end of the test the moisture content of the specimen is determined, its initial moisture content having been measured at the start of the test.

The void ratio e_2 at the end of the test is obtained from the moisture content (m_2)

$$e_2 = m_2 Gs \tag{4.29}$$

where Gs is the specific gravity of the soil. The other values of e are obtained by working backwards from this value of e_2 by finding the change in void ratio for each load increment:

$$\Delta e = \Delta H \frac{(1+e_1)}{H_1} \tag{4.30}$$

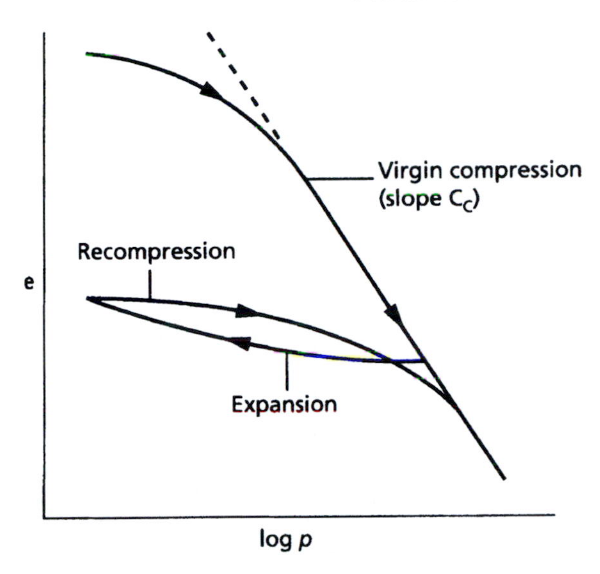

Figure 4.15 Results of a compression test plotted as an *e*–log *P* curve.

where H is the thickness of the specimen and the suffix $_1$ indicates initial, $e_1 = e_2 + \Delta e$. The pressure/voids ratio curve (*e*–log *p* curve) then is plotted. A more detailed explanation is provided by Craig (1987).

The shape of the *e*–log *p* curve is related to the stress history of a clay soil (Fig. 4.15). In other words, the *e*–log *p* curve for a normally consolidated clay soil is linear and is referred to as the virgin compression curve. On the other hand, if a clay soil is overconsolidated, then the *e*–log *p* curve is not straight and the preconsolidation pressure can be derived from the curve (Fig. 4.16). The preconsolidation pressure refers to the maximum overburden pressure to which a deposit has been subjected. Overconsolidated clay is appreciably less compressible than normally consolidated clay.

The compressibility of a clay soil can be expressed in terms of the compression index (C_c) or the coefficient of volume compressibility (m_v). The compression index is the slope of the linear section of the *e*–log *p* curve and is dimensionless. It can be determined from any two points on this part of the curve as follows:

$$C_c = \frac{e_1 - e_2}{\log p_2/p_1} \tag{4.31}$$

The value of C_c for clayey soils ranges between about 0.075 for sandy clays of low compressibility to more than 0.3 for highly compressible soft clay soils. Hence, the compressibility index increases with increasing clay content and so with increasing liquid limit. Indeed, Skempton (1944) found that C_c for normally consolidated clays is related closely to their liquid limit, the relationship between the two being expressed as:

$$C_c = 0.009(LL - 10) \tag{4.32}$$

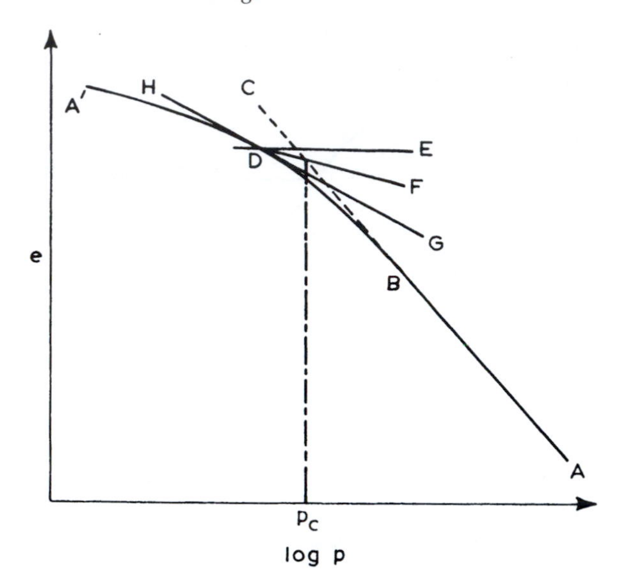

Figure 4.16 Determination of preconsolidation pressure. Extend the straight line part of the curve AB in C. Determine the point of maximum curvature, D, and draw a tangent to the curve at that point. Bisect the angle EDG between the tangent and the horizontal drawn through D to E. Drop a perpendicular from the intersection of AC with DF to the abscissa. This gives an approximate value of the preconsolidation pressure.

The coefficient of volume compressibility is defined as the volume change per unit volume per unit increase in load (p), its units being the inverse of pressure ($m^2\ MN^{-1}$). The volume change can be expressed according to specimen thickness (H) or void ratio (e)

$$m_v = \frac{1}{H_1}\left(\frac{H_1 - H_2}{p_2 - p_1}\right) \tag{4.33}$$

or

$$m_v = \frac{1}{1+e_1}\left(\frac{e_1 - e_2}{p_2 - p_1}\right) \tag{4.34}$$

The value of m_v for a given soil depends upon the stress range over which it is determined. Anon. (1990a) recommended that it should be calculated for a pressure increment of 100 kPa in excess of the effective overburden pressure on the soil at the depth in question. Values of m_v range from below 0.05 for heavily overconsolidated tills to over 1.5 for very highly compressible organic clay soils.

The ratio between the decrease of the void ratio, Δe, at time, t, and the ultimate decrease, Δe_1, represents the degree of consolidation, U, at time, t_1:

$$U = 100\frac{\Delta e}{\Delta e_1} \tag{4.35}$$

With a given thickness, H, of a layer of clay the degree of consolidation at time, t, depends on the coefficient of consolidation, c_v:

$$c_v = \frac{k}{\gamma_w m_v} \tag{4.36}$$

where k is the coefficient of permeability and γ_w is the unit weight of water. The coefficient of consolidation determines the rate at which settlement takes place and is calculated for each load increment. Either a mean value or that value appropriate to the pressure range in question is used (c_v is expressed in m^2 $year^{-1}$). With increasing increments of load (p) both k and m_v decrease, therefore c_v is fairly independent of p. The coefficient of consolidation decreases for normally consolidated clays from about 31.5 m $year^{-1}$ for very lean clays to about 0.03 m $year^{-1}$ for highly colloidal clays. At any value of c_v the time (t) at which a given degree of consolidation (U) is reached, increases in simple proportion to the square of the thickness (H) of the layer.

4.5.2 *Initial settlement*

Initial or immediate settlement is that which occurs under constant volume (undrained) conditions when clay soil undergoes elastic deformation to accommodate the imposed shear stresses. The immediate settlement below the corner of a uniformly loaded rectangular foundation structure can be obtained from the Steinbrenner expression (Terzaghi, 1943):

$$S_i = 0.75Q\frac{B}{E}I_f \tag{4.37}$$

where Q is the load, B is the width of the foundation, E is Young's modulus of the clay soil, and I_f is an influence factor related to the length and width of the foundation and the thickness of the compressible layer. Values of I_f for a value of Poisson's ratio of 0.5 are given in Fig. 4.17. Settlements beneath any point within a rectangular foundation can be calculated by splitting the area into a number of subrectangles, with one of their corners centred on the point concerned, and then summing the individual settlements for each subrectangle. Young's modulus may be determined in the laboratory by compression tests or in the field by plate load tests. A pressuremeter also may be used to determine the undrained deformation modulus of clay.

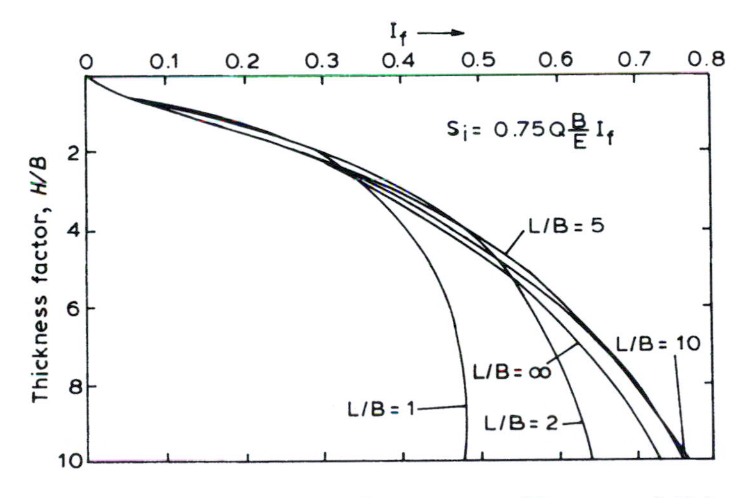

Figure 4.17 Influence factors for loaded area $L \times B$ on a compressible stratum of thickness H.

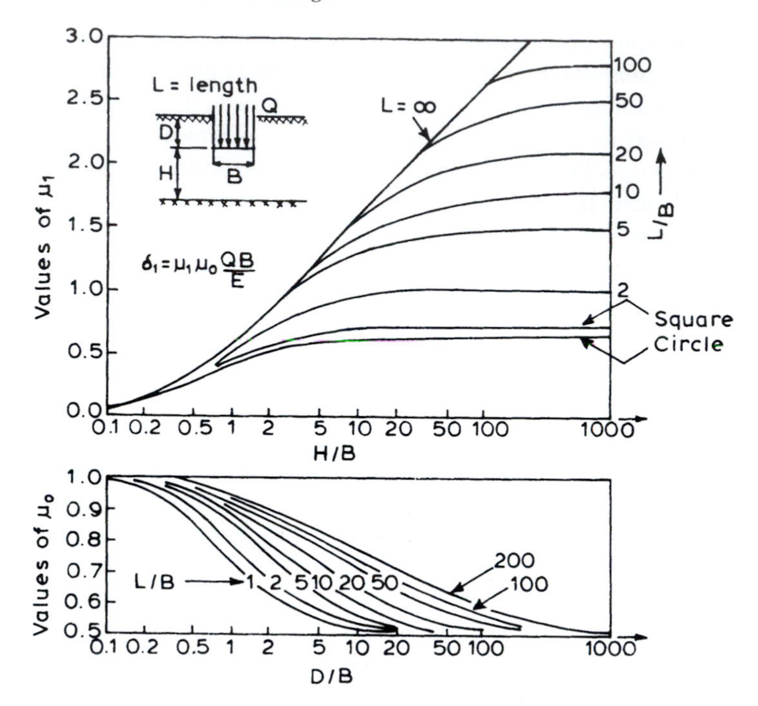

Figure 4.18 Determination of initial settlement.
Source: Janbu *et al.*, 1956. Reproduced by kind permission of Norwegian Geotechnical Institute.

The average immediate settlement can be found by using the method proposed by Janbu *et al.* (1956) whereby:

$$S_i = \mu_1 \mu_0 \frac{B}{E} \tag{4.38}$$

The factors μ_0 and μ_1 are related to the depth of the foundation, the thickness of the compressible layer and the length (L)/breadth (B) ratio of the foundation (Fig. 4.18).

4.5.3 Primary settlement

Primary settlement in a clay soil takes place due to the void space being gradually reduced as the pore water is expelled on loading. The rate at which it occurs depends on the rate at which the excess pore pressure induced by a structural load is dissipated, thereby allowing the structure to be supported entirely by the soil skeleton. Consequently, the permeability of a clay soil is important. The increase in effective stress in the soil gives rise to a decrease in volume, which again is controlled by the rate at which the pore water can escape from the voids. The value of initial settlement is added to that of primary settlement.

Skempton and Bjerrum (1957) showed that primary consolidation settlement, indicated from the results of oedometer tests, may be somewhat greater than actually occurs. This is because the amount of settlement (S_c) is influenced by the type of clay soil concerned, so that

$$S_c = \mu S_o \tag{4.39}$$

where μ is a coefficient depending on the type of clay and S_o is settlement determined from oedometer tests. Skempton and Bjerrum suggested that the following values of μ can be used in most cases:

Very sensitive clays	1.0–1.2
Normally consolidated clays	0.7–1.0
Overconsolidated clays	0.5–0.7
Heavily overconsolidated clays	0.2–0.5

The amount of oedometer settlement (S_o) of a layer of clay soil can be obtained from the expression

$$S_o = m_v \sigma_z H \tag{4.40}$$

where m_v is the average coefficient of volume compressibility for the effective pressure increment in the layer of soil in question, σ_z is the average imposed vertical stress on the layer due to the net foundation pressure, and H is the thickness of the layer. The settlement for each soil stratum, which is likely to be significantly affected by the structural load, is determined. Similarly, thick deposits of clay soil should be subdivided into sublayers and the settlement for each sublayer determined, thereby taking account of variations with depth. The total settlement is equal to the sum of that of the individual layers concerned.

Alternatively, oedometer settlement may be determined from:

$$S_o = \frac{H}{1+e_1}(e_1 - e_2) \tag{4.41}$$

or

$$S_o = \frac{H}{1+e_2} C_c \log_{10} \frac{p_1 + \sigma_z}{p_1} \tag{4.42}$$

where H is the thickness of the layer, e_1 and e_2 are the initial and final void ratios, p_1 is the initial effective overburden pressure and σ_z is the vertical stress at the centre of the layer due to the net foundation pressure. Individual layers of soil again are considered separately and then summed to give total settlement.

The rate at which a foundation settles generally is determined at the time required for 50 or 90 per cent of final settlement to be completed. The time (t) required for a given amount of settlement to take place is obtained from:

$$t = T_v \frac{d^2}{c_v} \tag{4.43}$$

where T_v is a dimensionless number termed the time factor, and d is the length of the drainage path ($d = H$, i.e. the thickness of the layer of soil concerned for drainage in one direction only and $d = H/2$ for two-way drainage). In the field, drainage above and below a stratum depends on the relative permeability of the beds immediately adjacent. Two-way drainage takes place in the oedometer test. The values of the time factor for various degrees of consolidation (U) can be determined from Fig. 4.19. The coefficient of consolidation (c_v) for 50 and 90 per cent final settlement usually is determined by the log time method or the root time method respectively. These two methods are outlined in Figs 4.20a and 4.20b respectively.

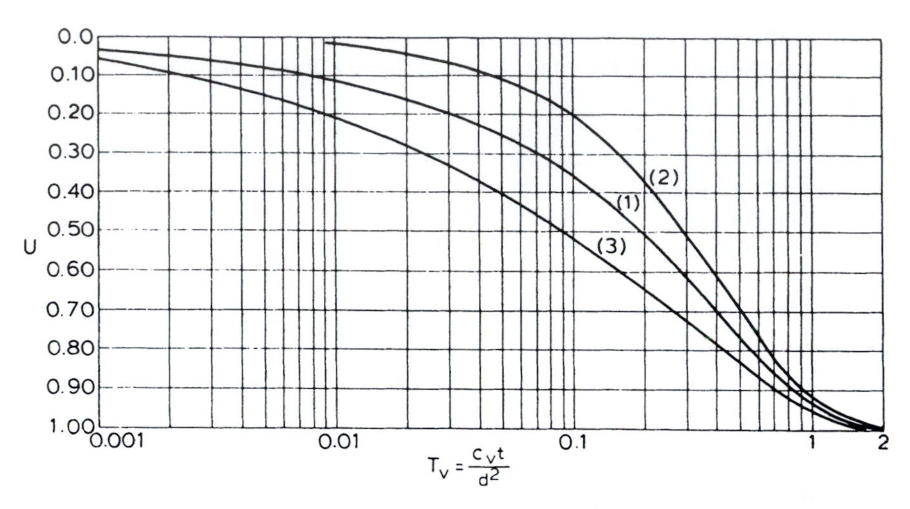

Figure 4.19 Relationship between the average degree of consolidation, U and the time factor, T_v. Curve 1 = two-way drainage and pore water pressure equal throughout. Curve 2 = one-way drainage, pore water pressure increasing downwards. Curve 3 = one-way drainage, pore water pressure increasing upwards.

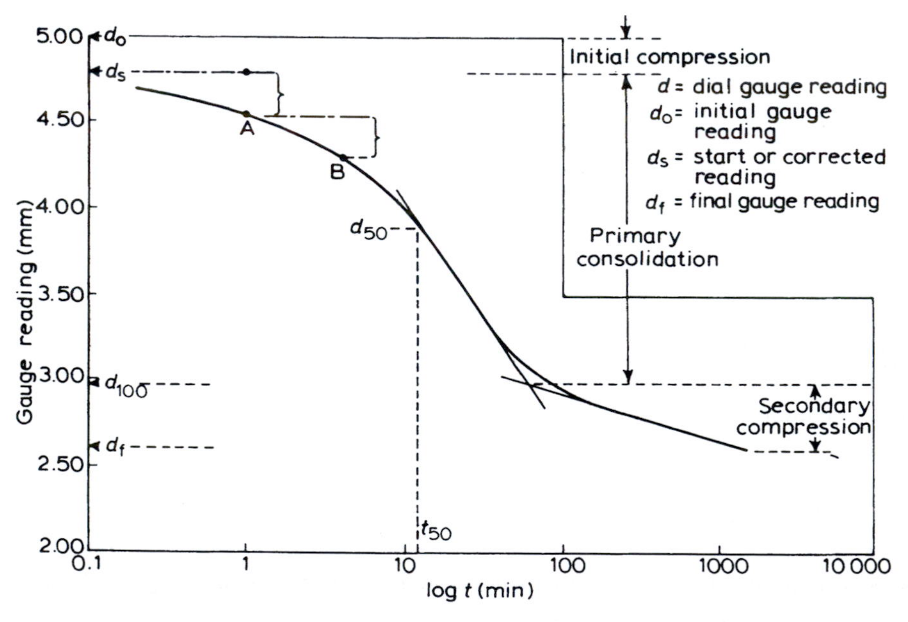

Figure 4.20a Log time method of determination of the coefficient of consolidation, c_v. To obtain the correct position for zero, two values of log t are chosen on the initial part of the curve, for which the values of t are in the ratio 4:1 (see A and B). The vertical distance between these two points is obtained and this distance is stepped off vertically above the upper point, A. The new point is the corrected zero (i.e. $U = 0$). The latter point, d_s, does not correspond with the point representing the initial dial gauge reading, d_o, the difference between them is referred to as initial compression. The point $U = 100\%$ is located at the intersection, d_{100}, of the extension from the straight line parts of the curve. Primary consolidation takes place between d_s and d_{100}, beyond which secondary consolidation occurs. The point corresponding to $U = 50\%$ is midway between d_s and d_{100}. The value of T_v corresponding to $U = 50\%$ is 0.196, hence the coefficient of consolidation at this point is $c_v = 0.196d^2/t_{50}$, where d = half thickness of specimen for a given pressure increment.

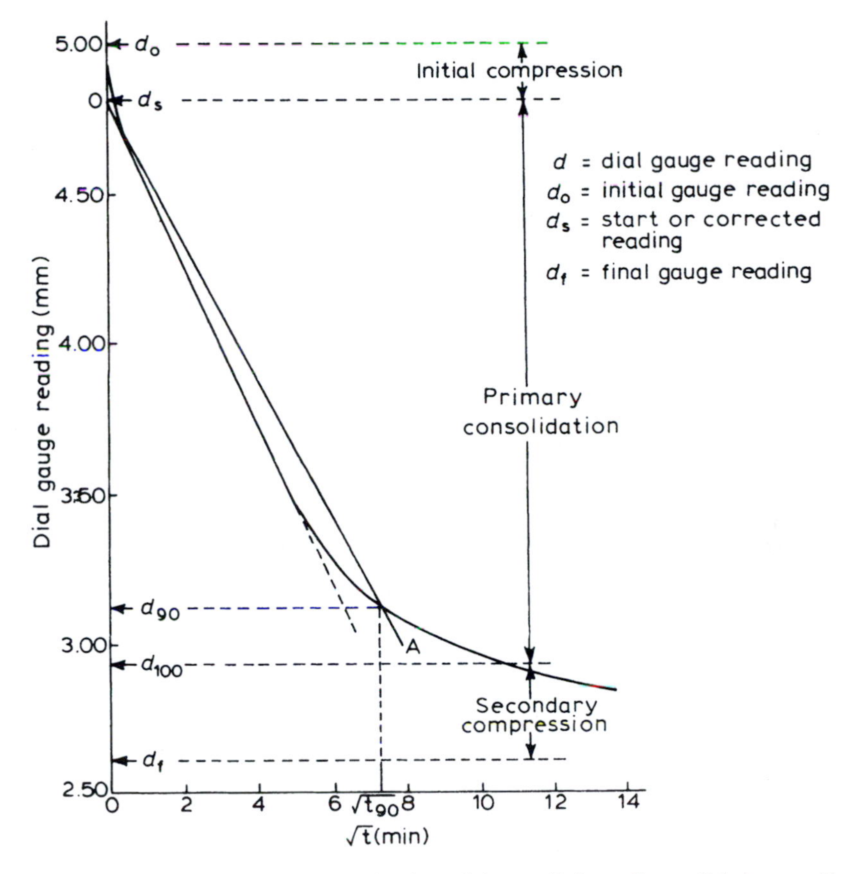

Figure 4.20b The root time method of determination of the coefficient of consolidation, c_v. Extend back the initial straight line part of the curve to intersect the ordinate to give zero time, 0. Multiply the values of $\sqrt{t}$ on the straight-line part of the curve by 1.15 and redraw another curve through these points (i.e. AO). The intersection of AO with the original curve corresponds to $U = 90\%$ and therefore gives the value $\sqrt{t_{90}}$. The value of T_v corresponding to $U = 90\%$ is 0.848 and so the coefficient of consolidation is obtained from $c_v = 0.848d^2/t_{90}$, where d = half thickness of specimen for a given pressure increment. The point on the original curve corresponding to $U = 100\%$ can be obtained by proportioning as shown.

The load imposed on a soil changes during the construction period. Initially, it is reduced due to removal of material on excavation, then as the structure is erected and the load gradually increases. The instantaneous time-settlement curve therefore needs adjusting to take account of the construction period. An empirical method of correction was proposed by Terzaghi (1943) and is outlined in Fig. 4.21.

4.5.4 Secondary settlement

After a sufficient time excess pore pressures approach zero but a clay soil may continue to undergo a decrease in volume. This is referred to as secondary consolidation and involves compression of the soil fabric. Unfortunately, no reliable method is available for determining the amount and rate of secondary consolidation. Consequently, if estimates of secondary consolidation are required in practice, then they generally are based on empirical procedures (Simons,

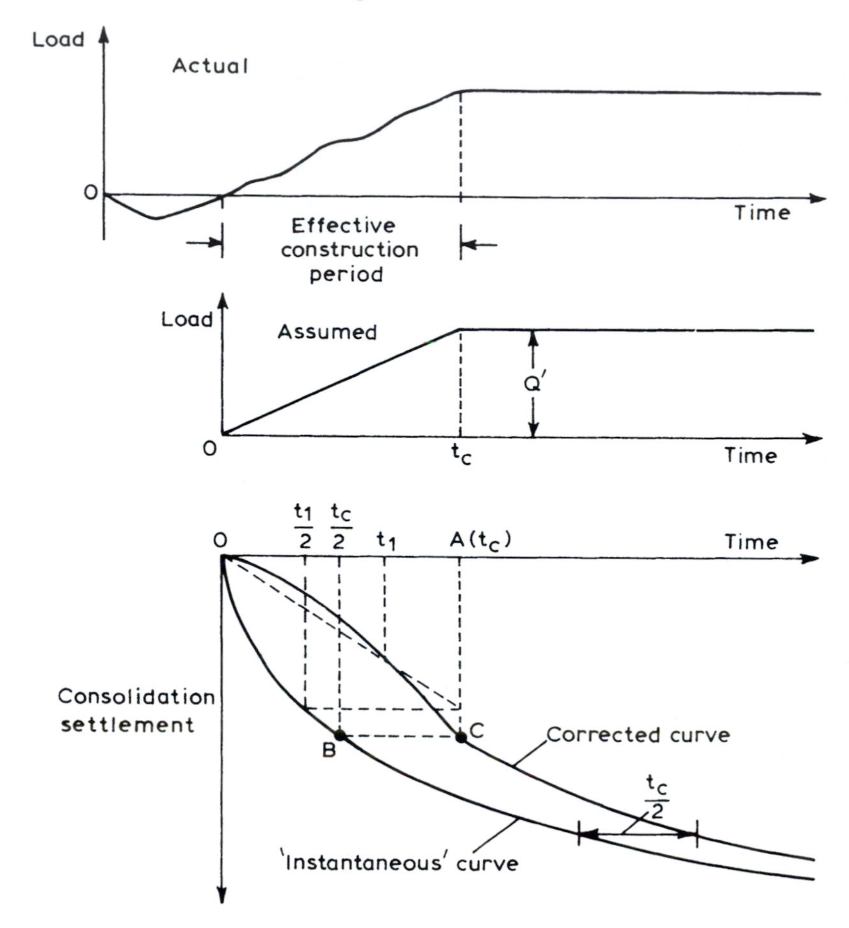

Figure 4.21 Correction of the instantaneous time–settlement curve for the construction period. It is assumed that the net load (given load less mass of soil removed on excavation) is applied uniformly over the construction period, t_c, but that the degree of consolidation at the end of this period is the same as if the load had been acting for half the period, $t_c/2$. Hence, the settlement at any time during the construction period is that which takes place for an instantaneous loading at half the time. Construction: drop perpendicular from A (t_c) and another from $t_c/2$ to intersect the instantaneous curve in B. Draw BC horizontal to cut AC in C, the first point on the corrected curve. Intermediate points are obtained similarly. Beyond C the settlement curve is assumed to be the instantaneous curve offset to the right by half the loading period, that is, offset by distance BC. Corrected total settlement can be derived by adding immediate settlement.

1975). It has been suggested frequently that secondary settlement is of little practical consequence. Nevertheless, this does not mean that the possibility of its occurrence should not be investigated, for it has been noted that in certain circumstances a large part of the observed settlement that has occurred beneath structures has developed after the dissipation of excess pore pressure (Foss, 1969). When considering secondary settlement it should be noted that two different factors may influence the process. The first is reduction in volume at constant effective stress and the second is the vertical strain resulting from lateral movements in the ground beneath the structure. These two factors may result in different types of settlement and their relative importance will vary depending on the level of stress, the type of clay soil and the shape of the structure.

Generally, because overconsolidated clays have a more stable fabric than normally consolidated clays they undergo relatively small secondary settlements compared with the latter. Nevertheless, exceptions do occur, for example, Bjerrum (1966) maintained that even in overconsolidated soils significant secondary settlements can occur in association with structures with large variations in live load.

4.5.5 *Settlement of coarse-grained soils*

A sudden application of a load to a layer of coarse-grained soil composed of sound equidimensional particles produces an instantaneous compression followed by a slight additional compression at a decreasing rate. At low pressure both instantaneous and gradual compression are almost exclusively due to slippage at points of grain contact. As the load increases an increasing proportion of compression is due to grain crushing.

In coarse-grained soils the allowable settlement is exceeded usually before soil rupture considerations become significant. Generally, total settlements of footings on these soils are small, recorded settlements being of the order of 25 mm or less and rarely exceed 50 mm. The commonly accepted basis of design is that the total settlement of a footing should be restricted to about 25 mm as by so doing the differential settlement between adjacent footings can be tolerated by a structure.

The difficulty in assessing the performance of coarse-grained soils is due to the fact that it more or less is impossible to obtain an undisturbed sample from them. Hence, methods of settlement calculation must be based on the data gained from field tests such as the standard penetration test, the cone penetration test and the plate load test (Burland and Burbridge, 1985; Simons and Menzies, 2000).

4.5.6 *Settlement in rocks*

Settlement is rarely a limiting condition in foundations on fresh rocks. In terms of the determination of settlement, such rocks can be regarded as homogeneous and isotropic and so any settlement will take place as the load is applied and there is no time-dependent effect. Settlement (S) beneath a point at the ground surface, in a relatively deep stratum, can be calculated by using elastic theory and appropriate values of Young's modulus (E) and Poisson's ratio (υ):

$$S = \frac{C_s QB(1 - \upsilon^2)}{E} \tag{4.44}$$

where C_s is a shape and rigidity factor (values have been provided by Holtz, 1991), Q is the magnitude of the uniformly distributed load, and B is the characteristic dimension of the loaded area (e.g. width or diameter). However, rock masses tend to be characterized by the presence of discontinuities and sedimentary sequences may comprise rocks that possess notably different geomechanical properties. Discontinuities may be open or filled and the fillings may consist of compressible material. In addition, layered rocks may be inclined and weathering can affect all rocks. Load transmitted through a rock mass is carried either by the transfer of shear across joints or by the spanning effect of blocks of rock. Consequently, the discontinuity pattern and block orientation have an important bearing on the way in which a rock mass behaves as a foundation. Wyllie (1999) indicated how settlement can be calculated for a number of simplified situations such as a compressible layer on a rigid base, a compressible bed within a stiffer formation, a stiff bed overlying a compressible formation, a transversely isotropic rock and an inclined non-uniform compressible rock. Furthermore, if weak or moderately weak rocks are loaded beyond their yield stress they undergo creep, that is, time-dependent strain.

Chalk can be taken as an example of moderately weak to moderately strong rock. Carter and Mallard (1974) found that chalk compressed elastically up to a critical pressure, which they termed the apparent preconsolidation pressure (the latter is the yield stress). They quoted values of preconsolidation pressure of 3.4 MPa for the Upper Chalk of Norfolk, England, and 13.8 MPa for the Lower Chalk of that county. Carter and Mallard obtained coefficients of consolidation (c_v) and volume compressibility (m_v), for loading pressures of 215 and 322 kPa, of 11.25 m^2 $year^{-1}$ and 0.019 m^2 MN^{-1} respectively. Previously, Meigh and Early (1957) had carried out consolidation tests at very high pressures on samples of Upper Chalk from Coulsdon, Surrey, England, and had found that within a range of pressures up to 42.9 MPa the coefficient of volume compressibility was sensibly constant. They quoted an average value of 0.035 m^2 MN^{-1}, with a range of 0.0031–0.038 m^2 MN^{-1}. Therefore, fresh chalk has a low volume compressibility. However, if the yield stress is exceeded, then much more significant consolidation occurs. The yield stress of intact chalk may be associated with pore collapse in high porosity chalks (porosity exceeding 30 per cent), the cement bonds breaking down. Some interparticle slip and local grain crushing may occur. According to Clayton (1990) the yield stress of chalk varies between 200 and 600 kPa for weathered chalks from grades IV and V. The yield values of intact chalk are appreciably higher, Clayton quoted values ranging from 1.7 to 80 MPa. Clayton went on to assert that with initial values of Young's modulus of 100 MPa and over, settlements in chalk normally would not be a problem provided that the yield stress was not exceeded.

Alternatively, settlement can be determined from field tests, as for example, the plate load test or by using penetrometers/dilatometers. One of the largest plate load tests undertaken in Britain was carried out at a site at Mundford, Norfolk, on the Middle Chalk. As noted, the Chalk tends to vary from moderately weak to moderately strong rock and its deformation properties depend on its strength, and the spacing, tightness and orientation of its discontinuities. Weathering also affects these properties. Ward *et al.* (1968) recognized five grades of weathering in the Middle Chalk at Mundford, which extended from completely unweathered to a structureless melange of unweathered and partially weathered chalk fragments in a matrix of deeply weathered chalk. They showed that the values of Young's modulus varied with grade of weathering. The plate load test showed that creep in the two weakest grades of chalk was appreciable. Subsequently, Burland *et al.* (1974) found that settlements of a five-storey building founded in soft low-grade chalk at Reading were very small. Their findings agreed favourably with those obtained at Mundford.

However, large-scale field tests such as the plate load test are expensive. As a consequence, rating classifications have been used to determine both the quality and deformability of rock masses. The determination of the intact value of Young's modulus (E_m) has been referred to in Chapter 3.

4.5.7 Settlement and structures

Settlement that adversely affects the safety or function of a structure is unacceptable. However, with many buildings their appearance is also of concern and therefore significant cracking of architectural features also is unacceptable. Hence, an estimation of the amount of settlement that will adversely affect structural members and/or architectural features is required. This is influenced by many factors, including the type and size of the structure, and the properties of the materials of which it is constructed, as well as the rate and nature of the settlement.

Almost all criteria for tolerable settlement have been established empirically on the basis of observations of ground movement and damage in existing buildings.

Generally, uniform settlements can be tolerated without much difficulty, for example, large storage tanks have been known to settle over a metre. Nevertheless, large settlements are inconvenient and may cause serious disturbance to services even where there is no evident damage to the structure. However, differential settlement is of greater significance than maximum settlement since the former is likely to distort or even shear a structure. Buildings that suffer large maximum settlement are also likely to experience large differential settlement. Both therefore should be avoided.

Two parameters commonly have been used for developing correlations between damage and differential settlement, namely, angular distortion and deflection ratio (Fig. 4.22). Angular distortion (δ/l) describes the rotation of a straight line joining two reference points relative to tilt (ω). The deflection ratio (Δ/L) is defined as the maximum displacement (Δ) relative to a straight line between two points divided by the distance (L) separating the points.

Skempton and MacDonald (1956) selected angular distortion as the critical index of ground movement (Table 4.2a). They concluded that cracking of load-bearing walls or panel walls in frame structures is likely when δ/l exceeds 1/150. Skempton and MacDonald also suggested that a value of $\delta/l = 1/500$ could be used as a design criterion that provides some factor of safety against cracking. Subsequently, Bjerrum (1963) also suggested limits for damage criteria based on angular distortion (Table 4.2b). After an examination of differential settlement in buildings, Grant *et al.* (1974) maintained that a building that experiences a maximum value of angular distortion greater than 1:300 will probably undergo some damage. They further

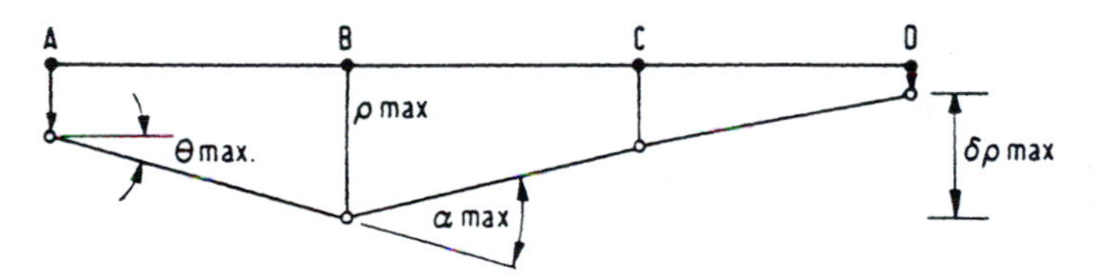

(a) Definitions of settlement ρ, relative settlement $\delta\rho$, rotation θ, and angular strain α

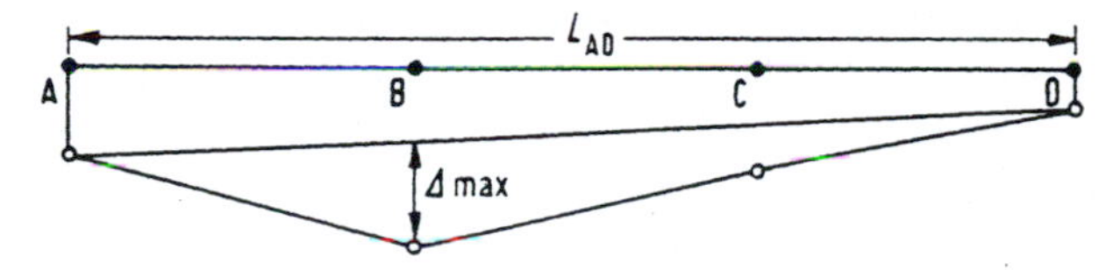

(b) Definitions of relative deflection Δ and deflection ratio Δ/L

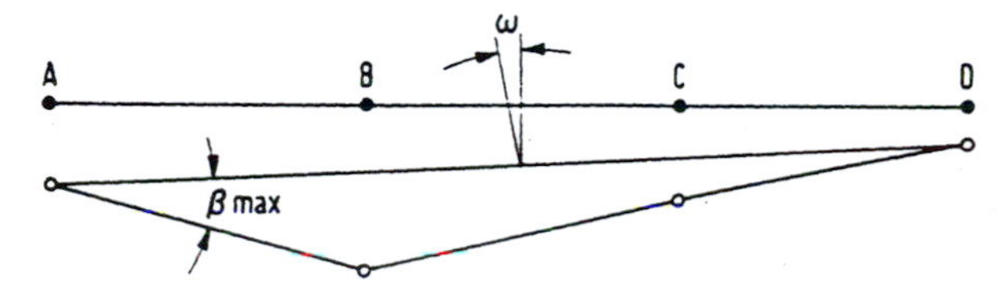

(c) Definitions of tilt ω and relative rotation (angular distortion) β

Figure 4.22 Definitions of ground movement.
Source: Burland *et al.*, 1977. Reproduced by kind permission of Japanese Geotechnical Society.

Table 4.2a Limitations of ground movement

Criterion	*Independent footings*		*Rafts*
Angular distortion (δ/L)	1/300		1/300
Greatest differential movement	Sands	30 mm	30 mm
	Clays	45 mm	45 mm
Maximum movement	Sands	50 mm	75 mm
	Clays	75 mm	75–125 mm

Source: Skempton and MacDonald, 1956. Reproduced by kind permission of Thomas Telford.

Table 4.2b Limiting angular distortion

Category of potential damage	δ/L
Danger to frames with diagonals	1/600
Safe limit for no cracking of buildings*	1/500
First cracking of panel walls	1/300
Tilting of high rigid buildings becomes visible	1/250
Considerable cracking of panel and brick walls	1/150
Danger of structural damage to general buildings	1/150
Safe limit for flexible brick walls, L/H>4*	

Source: Bjerrum, 1963.

Note
* Safe limits include a factor of safety.

suggested that in the case of buildings with raft foundations the estimated maximum allowable differential settlements corresponding to a maximum angular distortion 1:300 were 30 mm for foundations on sand and 56 mm on clay.

Polshin and Tokar (1957) defined allowable displacement in terms of deflection ratio (Δ/L). There are a number of differences between these criteria and those presented by Skempton and MacDonald (1956). For instance, frame structures and load-bearing walls are treated separately. The allowable displacement for frames is expressed in terms of the slope or the differential displacement between adjacent columns. The limiting values quoted by Polshin and Tokar vary between 1/500 for steel and concrete frame infilled structures and 1/200 where there is no infill or danger of damage to cladding. The maximum allowable deflection ratio was assumed to be related to the development of a critical level of tensile strain in a wall. For brick walls, the critical tensile strain was taken as 0.05 per cent. Polshin and Tokar adopted more stringent limits for differential movement of load-bearing brick walls than did Skempton and MacDonald. The deflection ratio at which cracking occurs in brick walls was related to the length to height ratio (L/H) of the wall. For L/H ratios of less than 3, the maximum deflection ratios quoted varied from 0.3×10^{-3} to 0.4×10^{-3}, whilst for L/H ratios greater than 5 the values ranged from 0.5×10^{-3} to 0.7×10^{-3}.

Burland and Wroth (1975), and Burland *et al.* (1977) also used the deflection ratio (Δ/L) at which the critical tensile strain (0.075 per cent) is reached as a criterion for allowable ground movement. They proposed that limiting deflection criteria should be developed for at least three different cases. Diagonal strain is critical in the case of frame structures (which are relatively flexible in shear) and for reinforced load-bearing walls (which are relatively stiff in direct tension). Bending strain is critical for unreinforced masonry walls and structures that have relatively low tensile resistance. Hence, unreinforced load-bearing walls, particularly when subjected to hogging, are more susceptible to damage than frame buildings. For unreinforced load-bearing walls in the sagging mode Burland and Wroth gave values of 0.4×10^{-3} (1/2500) for an L/H ratio of 1, and 0.8×10^{-3} (1/1250) for a ratio of 5. However, they pointed out that cracking in the hogging mode occurs at half these values of deflection ratio. Burland and Wroth accepted the safe limit for angular distortion of 1:500 previously proposed by Skempton and MacDonald (1956) as satisfactory for framed buildings but stated that it was unsatisfactory for buildings with load-bearing walls. Damage in the latter has occurred at very much smaller angular distortions. The rate at which settlement occurs also influences the amount of damage suffered.

Settlements may be reduced by the correct design of the foundation structure. This may include larger or deeper foundations. Most settlement can be reduced if the site is preloaded or surcharged prior to construction or if the soil is subjected to dynamic compaction or vibrocompaction. It is advantageous if the maximum settlement of large structures is reached earlier than later. The installation of sand drains, sandwicks or band drains, which provide shorter drainage paths for the escape of water to strata of higher permeability, is one means by which this can be achieved. Such drains may effect up to 80 per cent of the total settlement in clay soils during the construction stage. Differential settlement also can be accommodated by methods similar to those used to accommodate subsidence.

4.6 Subsidence

Subsidence of the ground surface takes place when mineral deposits, be they gas, fluid or solid, are removed from within the ground. It reflects the movements that occur in the area undergoing extraction or abstraction. Unfortunately, subsidence can have serious effects on surface structures and can require special constructional design in site development or extensive remedial measures in developed areas.

In some parts of the world mining has gone on for centuries and methods of mining have changed with time. In addition, mining methods differ according to the type of mineral deposit exploited. One of the problems associated with old abandoned mineral workings is that there may be no record of their existence. Even if old records exist, they frequently may be inaccurate. Old abandoned workings occur at shallow depth beneath the surface of many urban areas of Western Europe and North America. Such old workings can represent a hazard, especially during any subsequent redevelopment.

4.6.1 Pillared workings

Stratified deposits such as coal, limestone, gypsum, salt, sedimentary iron ore, etc. have been and are worked by partial extraction methods whereby pillars of the mineral deposit are left in place to support the roof of the workings. In pillared workings the pillars sustain the redistributed weight of the overburden, which means that they and the rocks immediately above and below are subjected to added compression. Stress concentrations tend to be located at the edges of pillars and the intervening roof rocks tend to sag. The effects on ground level are normally insignificant. Although the intrinsic strength of a stratified deposit varies, the important factor in the case of pillars is that their ultimate behaviour is a function of bed thickness to pillar width, the depth below ground and the size of the extraction area. Pillars in the centre of the mined-out area are subjected to greater stress than those at the periphery. Individual pillars in dipping seams tend to be less stable than those in horizontal seams since the overburden produces a shear force on the pillars.

Collapse in one pillar can bring about collapse in others in a sort of chain reaction because increasing loads are placed on those remaining. Slow deterioration and failure of pillars may take place after mining operations have long since ceased. The effective pillar width can be reduced by blast damage or local geological weaknesses. Furthermore, the collapse of roof strata around pillars alters their geometry and so can affect their stability. Old pillars at shallow depth have occasionally failed near faults and they may fail if they are subjected to the effects of subsequent mining. The yielding of a large number of pillars can bring about a shallow broad subsidence over a large surface area, which Marino and Gamble (1986) referred to as a sag (Fig. 4.23). The ground surface in a sag displaces radially inwards towards the area of maximum subsidence. This inward radial movement generates tangential compressive strain and circumferential tension fractures frequently are developed. Sag movements depend on the mine layout, in particular the extraction ratio and geology, as well as the topographic conditions at the surface. They tend to develop rather suddenly, the major initial movements lasting, in some instances, for about a week, with subsequent displacements occurring over varying periods of time. The initial movements can produce a relatively steep-sided bowl-shaped area. Nonetheless, the shape of a sag profile can vary appreciably and because it varies with mine layout and geological conditions, it can be difficult to predict accurately. Normally, the greater the maximum subsidence, the greater is the likelihood of variation in the profile. Maximum profile slopes and curvatures frequently increase with increasing subsidence. The magnitude of surface tensile and compressive strains can range from slight to severe.

In the past, especially in coal mining, pillars frequently were robbed on retreat. Extraction of pillars during the retreat phase can simulate longwall conditions, although it can never be assumed that all pillars have been removed. At moderate depths pillars, especially pillar remnants, probably are crushed and the goaf (i.e., the worked out area) compacted, but at shallow depths lower crushing pressures may mean the closure is variable. This causes foundation

Figure 4.23 Subsidence due to pillar collapse, Witbank Coalfield, South Africa.

problems when large or sensitive structures are erected above. The potential for pillar failure should not be ignored, particular warning signs being strong roofs and floors allied with high extraction ratios and moderately to steeply dipping seams. Prediction of subsidence as a result of pillar failure requires accurate data regarding the layout of a mine. Such information frequently does not exist in the case of abandoned mines. On the other hand, when accurate mine plans are available or in the case of a working mine, the method outlined by Goodman *et al.* (1980) may be used to evaluate collapse potential. Basically the method involves calculating vertical stress based upon the tributary area load concept, which assumes that each pillar supports a column of rock with an area bounded by room centres and a height equal to the depth from the surface, and comparing this with the strength of the pillar. When a structure is to be built over an area of old pillared workings the additional load on the pillars can be estimated simply by adding the weight of the appropriate part of the structure to the weight of the column of strata supported by a given pillar. This method is very conservative except when used for large concentrated loads where old workings are located at shallow depth. Bell (1988a) provided a number of methods for calculating the strength of or stresses on pillars.

Squeezes and crushes sometimes occur in a coal mine as a result of pillars being punched into the floor or roof, which might have been weakened by the action of weathering or groundwater. As a result, subsidence at the ground surface adopts a trough-like form with minor strain and tilt problems occurring around the periphery of the basin.

Even if pillars are relatively stable, the surface may be affected by void migration. This can take place within a few months of or a very long period of years after mining. Void migration develops when roof rock falls into the worked out areas or rooms. When this occurs the material involved in the fall bulks, which means that migration eventually is arrested, although the bulked material never completely fills the voids. Nevertheless, the process can, at shallow depth, continue upwards to the ground surface leading to the sudden appearance of a crown hole (Fig. 4.24). The factors that influence whether or not void migration will take place include the width of the unsupported span, the height of the workings, the nature of the cover rocks, particularly their shear strength and the incidence and geometry of discontinuities, the thickness and dip of the seam, the depth of overburden, and the groundwater regime. The maximum height of migration in exceptional cases might extend to 10 times the height of the original room in coal mines, however, it generally is 3–5 times the room height. Exceptionally, void migrations in excess of 20 times the worked height of a seam have been recorded. The self-choking process may not be fulfilled in dipping seams, especially if they are affected by copious quantities of groundwater that can redistribute the fallen material. The redistribution of collapsed material can lead to the formation of supervoids and their migration to rockhead then produces large-scale subsidence at ground level. Supervoids also can form in old pillared coal mines if one seam has been worked above another and void migration from the seam below moves into the workings in the seam above. A thick bed of sandstone above a coal seam usually will arrest a void, especially if it is located some distance above the immediate roof of the workings. However, most voids are bridged when the span decreases through corbelling to an acceptable width, rather than when a more competent bed is encountered. Chimney-type collapses can occur to abnormally high levels of migration in massive strata in which the joints diverge downwards. Weak superficial deposits may flow into voids that have reached rockhead, thereby forming features that may vary from a gentle dishing of the surface to inverted cone-like depressions of large diameter.

Garrard and Taylor (1988) summarized four methods that have been used to predict the collapse of roof strata above rooms. These are clamped beam analysis that considers the tensile strength of the immediate roof rocks (Wardell and Eynon, 1968; Hoek and Brown, 1980);

Figure 4.24 Appearance of a crown hole at the surface due to void migration, Witbank Coalfield, South Africa.

bulking equations that consider the maximum height of collapse before a void is choked (Tincelin, 1958; Price *et al.*, 1969; Piggott and Eynon, 1978; Fig. 4.25); arching theories that estimate the height to which a collapse will occur before a stable arch develops (Terzaghi, 1946; Szechy, 1966); and coefficients based on experience and field observations that act as multipliers of either seam thickness or span width (Walton and Cobb, 1984).

4.6.2 Investigations in areas of abandoned mine workings

A site investigation in an area of abandoned workings should attempt to determine the number and depth of mined horizons, the extraction ratio and the pattern of the layout. Of particular importance is the state of the old workings, for instance, are they open, partially collapsed or collapsed; and the degree of fracturing and bed separation in the roof rocks.

During the desk study all topographic and geological maps of the area in question, going back to the first editions, should be examined for evidence of past workings such as old shafts, adits and spoil heaps. Abandoned mines record offices, when they exist, represent primary sources of information relating to past mining activity. Other sources include public record offices, museums, libraries, specialist contractors and consultants, private collections and geological surveys. The use of aerial photographs for the detection of surface features caused by subsidence is more or less restricted to rural areas and scale is a critical factor. The resolution necessary for the detection of relatively small subsidence features (1.5–3 m across) is provided by aerial photographs with scales between 1:25 000 and 1:10 000. Colour photographs may be more useful than black and white ones in the detection of past workings since they can reveal subtle changes in vegetation related to subsidence and, if there are differences in thermal emission, then infra-red (false colour) photographs should show these differences.

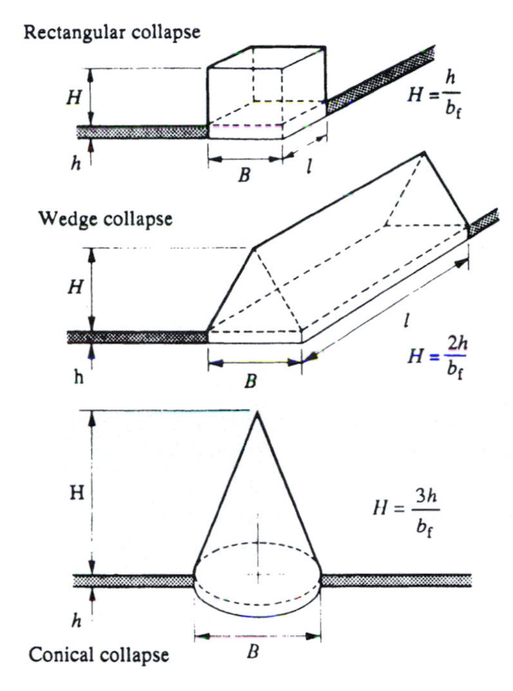

(a) Diagram showing notation relating to maximum height of collapse and geometry

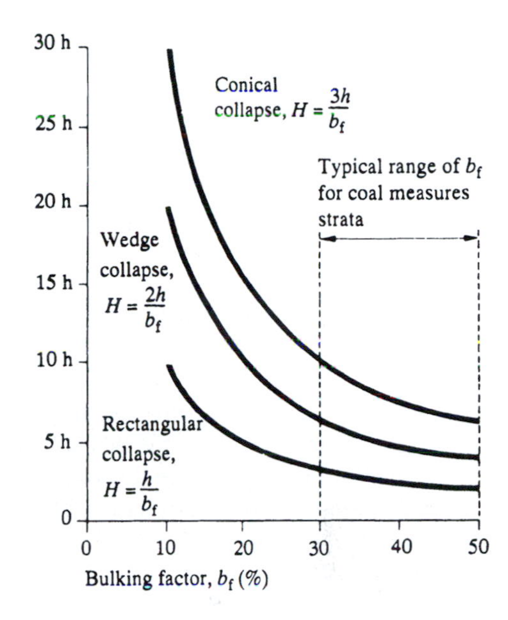

(b) Postulated variation in maximum height of collapse for different modes of failure and bulking factors

Figure 4.25 Type and amount of void migration.
Source: Piggott and Eynon, 1978.

Considerable care should be exercised at the planning stage of a geophysical survey for the location of subsurface voids because of the variable nature of the target. The selection of the most appropriate technique necessitates consideration of four parameters, namely, penetration, resolution, signal to noise ratio and contrast in physical properties. The size and depth of the workings, and the character of any infill control the likelihood of the workings being detected as an anomaly. With the information obtained from the desk study and the reconnaissance survey, many of the available geophysical methods can be assessed at the selection stage, using a model study, and accepted, or rejected, without any requirement for field trials. Of particular importance is that there should be sufficient physical property contrast between the void and the surrounding rock mass so that an anomaly can be detected. However, since the presence of a void is likely to affect the physical properties and drainage pattern of the surrounding rock mass, due mainly to induced fracturing, this can give rise to a larger anomalous zone than that produced by the void alone. In addition, mine workings are connecting and this usually has a cavity enhancement effect.

Seismic refraction has not been used particularly often in searching for voids at shallow depth created by previous mining since such voids are often too small to be detected because of attenuation of seismic waves in the rock mass. Furthermore, if the workings are dry, then high attenuation of the energy from the seismic source occurs so that penetration into the rock mass is poor. For instance, voids less than 20 m in width probably will not be recognized (Anon., 1988).

Electrical resistivity depth sounding can be applied to the location of voids where the width to depth ratio is large. Mine workings that produce an air-filled layer often can be identified on the sounding curve as an increase in apparent resistivity.

Terrain conductivity surveys, in the depth range down to 30 m, are more effective than resistivity traversing but penetration into the ground achieved by electromagnetic radiation can be limited by excessive attenuation in ground of high conductivity. However, problems with detecting an extremely small secondary field in the presence of a strong primary signal can be overcome by the use of pulsed radiation.

Ground-probing radar is capable of detecting small subsurface cavities directly. The method is based upon the transmission of pulsed electromagnetic waves, the travel time of the waves reflected from subsurface interfaces being recorded as they arrive at the surface so allowing the depth to an interface to be obtained (White, 1992). The high frequency of the system provides high resolution and characteristic arcuate traces are produced by air-filled voids. Depths to voids can be determined from the two-way travel times of reflected events if velocity values can be assigned to the strata above the void. The conductivity of the ground imposes the greatest limitation on the use of radar probing in site investigation. In other words, the depth to which radar energy can penetrate depends upon the effective conductivity of the strata being probed. This, in turn, is governed chiefly by the water content and its salinity. Furthermore, the value of effective conductivity is also a function of temperature and density as well as the frequency of the electromagnetic waves being propagated. The least penetration occurs in saturated clayey materials or when the pore water is saline.

Generally speaking, voids in shallow abandoned mine workings are too small and located at depths too great to be detected by normal magnetic or gravity surveys. However, the fluxgate magnetic field gradiometer permits surveys of shallow depths to be carried out. It provides a continuous recording of lateral variations in the vertical gradient of the earth's magnetic field rather than giving the total field strength. The gradiometer tends to give better definition of shallow anomalies by automatically removing the regional magnetic gradient. Quantitative analysis of the depth, size and shape of an anomaly generally can be made more readily for near-surface features, using the gradiometer, than from total field measurements, obtained with a proton-magnetometer. The sensor of gradiometers is small (0.5 or 1 m) in order to record features within 2–3 m of the ground surface. On the other hand, a proton-magnetometer can more easily detect larger and deeper features, and yields results that are more suitable for contouring.

Micro-gravity surveys may be successful when the voids have a significant lateral extent, as in some room and pillar workings (Bishop *et al.*, 1997). The technique locates areas of contrasting subsurface density. A series of traverses, with closely spaced stations, can be used to map the lateral extent of such features but there are still size/depth constraints and an inherent ambiguity in the interpretation of results.

Crosshole techniques can be used when the depth of burial of the void is more than two or three times the diameter of the void. Drillholes must be spaced closely enough to achieve the required resolution of detail. The method can be used to detect subsurface cavities, if the cavity is directly in line between two drillholes and has at least one-tenth of the drillhole separation as its smallest dimension. Air-filled cavities are detectable more readily than those filled with water. Acoustic tomography techniques have been developed to map voids between adjacent drillholes.

The location of old workings generally has been done by exploratory drilling, the locations of drillholes being influenced by data obtained from the desk study or from a geophysical survey. However, it must be admitted that although frequently successful in locating the presence of old mine workings, exploratory drilling is not necessarily able to establish their layout. Nevertheless, when the results from drilling are combined with a study of old mine plans, if they exist, then it should be possible to obtain a better understanding of methods of working, sizes of voids, and directions of roadways and galleries.

Drilling to prove the existence of old mine workings usually is done by open holes, which allow relatively quick probe drilling (Bell, 1986). The drillholes should be taken to a depth where any voids present are not likely to influence the performance of the structure(s) to be erected. If a grid pattern of drillholes is used, some irregularity should be introduced to avoid holes coinciding with pillar positions. The stratigraphic sequence should be established by taking cores in at least three drillholes. The presence of old voids is indicated by the free-fall of the drill string and the loss of flush.

One of the principal objectives of investigations of abandoned mine workings is to determine their extent and condition. Accordingly, core material may need to be obtained. Double-barrel sampling tubes with inner plastic liners can be used to obtain core, which then can be photographed and logged, and the rock quality designation or fracture-spacing index recorded. Drilling penetration rates, water flush returns and *in situ* permeability tests may be used to assess the degree of fracturing. The degree of fracturing is important in that it tends to increase as old workings are approached. Determination of groundwater conditions is necessary, especially where a grouting programme is required to treat old workings.

Below surface workings may be examined by using drillhole cameras or closed circuit television, information being recorded photographically, or on videotape, and used to assess the geometry of voids and, possibly, the percentage extraction. However, their use in flooded old workings has not proved very satisfactory. Occasionally, smoke tests or dyes have been used to aid the exploration of subsurface cavities. It is possible to study the interior of large abandoned mine workings that are flooded by using a rotating ultrasonic scanner (Braithwaite and Cole, 1986). An ultrasonic survey can be carried out within a void, the probe being lowered down a drillhole to mine level. Horizontal scans are made at 1 m intervals over the height of the old workings and then tilting scans at 15° intervals are taken from vertical to horizontal. The echoes received during the horizontal scan are processed by computer to provide a plan view of the mine. Vertical sections are produced from the vertical scan. Hence, it is possible to determine the positions of pillars and the extent of the workings.

Detailed mapping of galleries is best made by driving a heading from an outcrop if this is close at hand or by sinking a shaft to the level of the mineral deposit to obtain access to the workings. Sometimes access can be gained via old shafts. Radial holes may be drilled from a shaft to establish the dimensions of rooms and pillars. As old workings may prove dangerous, it is advisable that any exploration should be undertaken with the advice and aid of experts.

4.6.3 Measures to reduce or avoid subsidence effects due to old mine workings

Where a site that is proposed for development is underlain by shallow old mine workings there are a number of ways in which the problem can be dealt with (Healy and Head, 1984). The first and most obvious method is to locate the proposed structure on sound ground away from old workings or over workings proved to be stable. It is not generally sufficient to locate immediately outside the area undermined as the area of influence should be considered. In such cases the area of influence usually is defined by projecting an angle of 25° to the vertical from the periphery and depth of the workings to the ground surface. Such location, of course, is not always possible.

Where the allowable bearing capacity of the foundation materials has been reduced by mining, it may be possible to use a raft. Rafts can consist of massive concrete slabs or stiff slab and beam cellular rafts. The latter are suitable for the provision of jacking sockets in the upstand beams to permit the columns or walls to be relevelled if subsidence distorts the raft. A raft can span weaker and more deformable zones in the foundation, thus spreading the weight of the structure well outside the limits of the building.

Reinforced bored pile foundations also have been resorted to in areas of abandoned mine workings. In such instances the piles bear on a competent stratum beneath the workings. They also should be sleeved so that concrete is not lost into voids, and to avoid the development of negative skin friction if overlying strata collapse. Some authorities, however, have suggested that piling through old mine workings seems inadvisable because, first, their emplacement may precipitate collapse and, second, subsequent collapse at seam level could possibly lead to piles being either buckled or sheared (Price *et al.*, 1969). There also may be a problem with lateral stability of piles passing through collapsed zones above mine workings or through large remnant voids.

Alternatively, the ground can be treated. Such treatment involves filling the voids in order to prevent void migration or pillar collapse. For example, hydraulic stowing may take place from the surface via drillholes of sufficient diameter. Pneumatic or gravity stowing often is considered where large subsurface voids have to be filled. Barriers can be constructed around the treated area to prevent the fill from being removed by flowing groundwater.

Grouting can be undertaken via drillholes from the surface into the mine workings, the holes usually being set out on a grid pattern, and the voids are filled with an appropriate grout mix (Fig. 4.26). If it has been impossible to obtain accurate details of the layout and extent of the workings, then the zone beneath the intended structure can be subjected to consolidation grouting. The grouts used in these operations commonly consist of cement, fly ash and sand mixes. Pea gravel may be used as a bulk filler where a large amount of grout is required for treatment. Alternatively foam grouts can be used. If the workings are still more or less

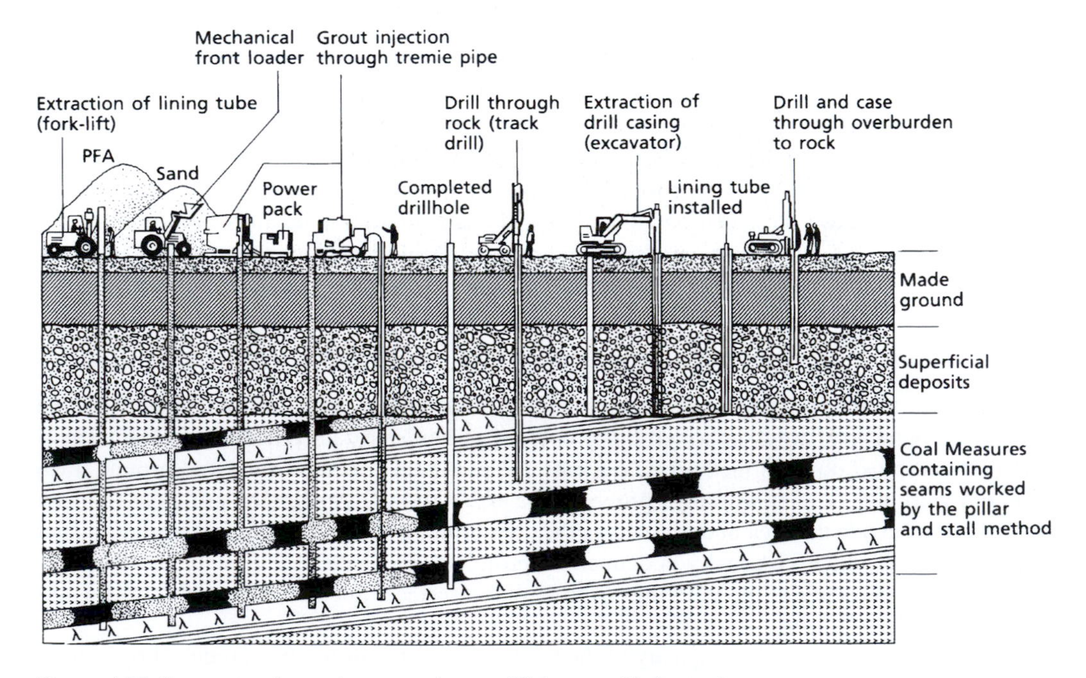

Figure 4.26 Sequence of grouting operations to fill large voids in coal seams.

continuous, there is a risk that grout will penetrate the bounds of the zone requiring treatment. In such instances barriers can be constructed by placing pea gravel down large diameter drillholes around the periphery of the site. When the gravel mound has been formed it is grouted and then the area within is grouted. If the old workings contain groundwater, then a gap should be left in the barrier through which it can drain as the grout is emplaced.

4.6.4 Old mine shafts

Centuries of mining in many countries has left behind a legacy of old shafts. Unfortunately many, if not most, are unrecorded or are recorded inaccurately. In addition, there can be no guarantee of the effectiveness of their treatment unless it has been carried out in recent years. The location of a shaft is of great importance as far as the safety of a potential structure is concerned for although shaft collapse is fortunately an infrequent event, its occurrence can prove disastrous. The ground about a shaft may subside or, worse still, collapse suddenly (Fig. 4.27). Shaft collapse may manifest itself as a hole roughly equal to the diameter of the shaft if the lining remains intact or if the ground around the shaft consists of solid rock. More frequently, however, shaft linings deteriorate with age to a point at which they are no longer capable of retaining the surrounding material. If superficial deposits surround a shaft that is open at the

Figure 4.27 Collapse of a shaft, trapping a car, in Cardiff, South Wales.

top, the deposits eventually collapse into the shaft to form a crown hole at the surface. The thicker these superficial deposits are, the greater will be the dimensions of the crown hole. Such collapses may affect adjacent shafts if they are interconnected.

An investigation to locate an old shaft should include a survey of maps, literature and aerial photographs (Anon., 1976). The principal sources of information include plans of abandoned mines, geological records of shaft sinking, geological maps, all available editions of topographic maps, aerial photographs, and archival and other official records.

The success of geophysical methods in locating old shafts depends on the existence of a sufficient contrast between the physical properties of the shaft and its surroundings to produce an anomaly. The size, especially the diameter, of a shaft influences whether or not it is likely to be detected, as do the physical properties of the lining or any material it contains. Moreover, a shaft often will remain undetected when it is covered by more than 3 m of fill. A resistivity survey may be used where there is a significant contrast in electrical resistance between a shaft and its surroundings. Magnetic surveys, especially those using the proton-magnetometer and, more recently, the fluxgate magnetic field gradiometer, have had some success in locating old mine shafts. An isomagnetic contour map is produced from the results. The microgravity technique also has been used to attempt to detect mine shafts (Emsley and Bishop, 1997). Donnelly and McCann (2000) have described the use of thermal imagery as a method of helping to locate mine shafts.

Geochemical exploration depends on identifying chemical changes such as changes in mineral content or the chemical character of the moisture content in the soil associated with old mine workings. In addition, gases such as carbon dioxide, carbon monoxide, methane and nitrogen may accumulate in open or partially filled shafts. Indeed, the most effective geochemical method yet used in the location of abandoned coal mine shafts is that of methane detection. Methane, being a light gas, may escape from old shafts and methane detectors can record concentrations as low as 1 ppm. Anomalies associated with old coal mine shafts generally have ranged between 10 and 100 ppm. The detector is carried over the site near the ground and should not be used on a windy day. A contour map showing methane concentration is produced.

The confirmation of the existence of a shaft is accomplished by excavation. As this may be a hazardous task, necessary safety precautions should be taken. For example, when a possible position of a shaft has been determined, a mechanical boom-type digger can be used to reveal its presence. The excavator is anchored outside the search area, and harness and lifelines should be used by the operatives. A series of parallel trenches are dug at intervals reflecting the possible diameter of the shaft. If excavation to greater depth than can be achieved by the excavator is required, resort may be made to a dragline. Again the dragline is safely anchored outside the search area.

Where a site is considered potentially dangerous, in that shaft collapse may occur, or where obstructions prevent the use of earthmoving equipment, a light mobile rig can be used to drill exploratory holes. The rig should be placed on a platform or girders long enough to give protection against shaft collapse. Since many old shafts have diameters around 2 m or even less, drilling should be undertaken on a closely spaced grid. Rotary percussion drilling can be done quickly and the holes can be angled. Changes in the rate of penetration may indicate the presence of a shaft or the flush may be lost. Significant differences in the depth of unconsolidated material may indicate the presence of a filled shaft or the fill may differ from the surrounding superficial material.

Once a shaft has been located, the character of any fill occurring within it needs to be determined. A drillhole alongside the shaft to determine the thickness of the overburden and the stratal succession, especially if the latter is not available from mine plans, proves very valuable. In particular, in old coal mines the positions of the seams may mean that mouthings open into the shaft at those levels. It is also important to record the position of the water table and, if possible, the condition of the shaft lining.

If the depth is not excessive and the shaft is open, it can be filled with suitable granular material, the top of the fill being compacted. If, as is more usual, the shaft is filled with debris in which there are voids, these should be filled with pea gravel and grouted. If the exact positions of the mouthings in an abandoned open shaft are known, these areas should be filled with gravel, the rest of the shaft being filled with mine waste. However, the latter will tend to consolidate much more than will gravel. Anon. (1982) supplied details concerning the concrete cappings needed to seal mine shafts. The concrete capping should take the form of an inverted cone. The zone immediately beneath the capping is grouted.

4.6.5 Longwall mining of coal and subsidence

In longwall mining the coal is worked at a face up to 220 m between two parallel roadways (Whittacker and Reddish, 1989). The roof is supported only in and near the roadways, and at the working face. After the coal has been won and loaded, the face supports are advanced – leaving the rocks, in the areas where coal has been removed, to collapse. Subsidence at the surface more or less follows the advance of the working face and may be regarded as immediate (Brauner, 1973a). Trough shaped subsidence profiles associated with longwall mining develop tilt between adjacent points that have subsided different amounts and curvature results from adjacent sections that are tilted by differing amounts. Maximum ground tilts are developed above the limits of the area of extraction and may be cumulative if more than one seam is worked up to a common boundary. Where movements occur, points at the surface subside downwards and are displaced horizontally inwards towards the axis of the excavation (Fig. 4.28). Differential horizontal displacements result in a zone of apparent extension on the convex part of the subsidence profile (over the edges of the excavation) whilst a zone of compression develops on the concave section over the excavation itself. Differential subsidence can cause substantial damage, the tensile strains thereby generated usually being the most effective in this respect. The surface area affected by ground movement is greater than the area worked in the seam. The boundary of the surface area affected is defined by the limit angle of draw or angle of influence, which varies from 8 to 45° depending on the coalfield. It would seem that the angle of draw may be influenced by depth, seam thickness and local geology, especially the location of the self-supporting strata above the coal seam.

One of the most important factors influencing the amount of subsidence is the width–depth relationship of the panel removed. In fact, in Britain it usually has been assumed that maximum subsidence generally begins at a width–depth ratio of 1.4:1 (this assumes an angle of draw of 35°; Anon., 1975). This is the critical condition above and below which maximum subsidence is and is not achieved respectively. However, the width–depth ratio necessary to cause 90 per cent subsidence usually can be achieved only in shallow workings because with deeper workings the critical area of extraction is made up of a number of panels often with narrow pillars of coal left *in situ* to protect one or other of the roadways. For a given subsidence the curvature of the ground surface is more marked over shallow workings than over deep workings and horizontal strains are proportional to subsidence and inversely proportional to the depth of workings.

Ground movements induced at the surface by mining activities are influenced by variations in the ground conditions, especially by the near-surface rocks and superficial deposits. The occurrence of abnormally thick beds of sandstone at or near the surface may resist deflection, in which case stratal separation occurs and the effective movements at the surface are appreciably

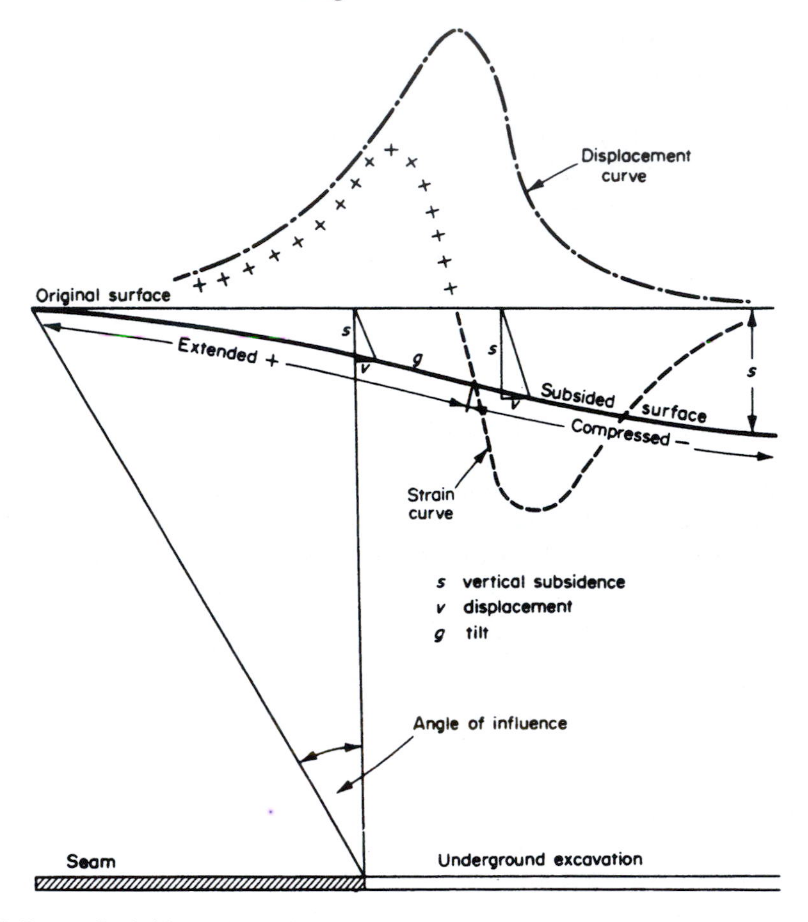

Figure 4.28 Curve of subsidence due to longwall mining showing tensile and compressive strains, vertical subsidence and tilt, together with angle of draw (not to scale).

less than would otherwise be expected. The necessary readjustment in weak strata to subsidence usually can be accommodated by small movements along joints. However, as the strength of the surface rock and the joint spacing increases so the movements tend to become concentrated at fewer points so that in thick limestones and sandstones movements may be restricted to master joints. Well-developed joints or fissures in such rocks concentrate differential displacement. Tensile and compressive strains many times the basic values have been observed at such discontinuities (Bell and Fox, 1991). For example, joints may gape anything up to a metre in width at the surface. According to Shadbolt (1978) it is quite common for the total lateral movement caused by a given working to concentrate in such a manner. In such instances, no strain is measurable on either side of the discontinuity concerned.

Drift deposits are often sufficiently flexible to obscure the effects of movements at rockhead. In particular, thick deposits of till tend to obscure tensile effects. On the other hand superficial deposits may allow movements to affect larger areas than otherwise.

Faults also tend to be locations where subsidence movement is concentrated thereby causing abnormal deformation of the surface. Whilst subsidence damage to structures located close to or on the surface outcrop of a fault can be very severe, in any particular instance the areal extent of such damage is limited, often being confined to within a few metres of the outcrop.

In fact, a subsidence step may occur at the outcrop of the fault. Where a fault passes through block-jointed sandstones or limestones at the surface severe fissures or fractures can occur. The fissures generally are parallel to the line of the fault but can occur up to 300 m away from it. However, many faults have not reacted adversely when subjected to subsidence (Hellewell, 1988). When a coal mine is abandoned, the water table that was lowered to enable coal to be mined begins to rise. This has led to reactivation of faults in certain areas of Britain and has given rise to some concern.

An important feature of subsidence due to longwall mining is its high degree of predictability. Usually, movements parallel and perpendicular to the direction of face advance are predicted. Empirical methods of subsidence prediction such as those developed by the former National Coal Board were developed by continuous study and analysis of survey data from British coalfields (Anon., 1975). Since they do not take into account the topography, the nature of the strata and geological structure involved or how the rock masses are likely to deform, such empirical relationships can be applied only under conditions similar to those in which the original observations were made. Many viable methods of subsidence prediction fall into the category of semi-empirical methods since fitting field data to theory often results in good correlations between predicted and actual subsidence. There are two principal methods of semi-empirical prediction, namely, the profile function and the influence function methods (Brauner, 1973b). The profile function method basically consists of deriving a function that describes a subsidence trough. The equation produced is normally for one half of the subsidence profile, and is expressed in terms of maximum subsidence and the location of the points of the profile. The influence function approach to prediction of subsidence is based upon the principle of superposition. The subsidence trough is regarded as a combination of many infinitesimal troughs formed by a number of infinitesimal extraction elements. Theoretical methods and numerical models also have been developed to predict the amount of subsidence due to longwall mining.

The effects of the magnitude of and rate at which subsidence occurs on a structure are governed by its location and orientation in relation to the underground workings and the depth at which mining is taking place, as well as panel width and extraction. The different types of ground movement associated with mining subsidence affect different structures in different ways. For instance, vertical subsidence may affect services and tilt may cause concern in relation to tall buildings. However, damage to buildings generally is caused by differential horizontal movements, that is, the concavity or convexity of the subsidence profile gives rise to compression or extension respectively in a building, the latter usually being the more serious. It is quite common for a building to undergo compressive strains in one direction and tensile strains in another direction. A building also may be subjected to alternate phases of tensile and compressive ground movements so there is a dynamic effect to consider. Thus, any acceptable design for a structure situated in an active coal-mining area must have regard for the nature, degree and periodicity of the ground movements likely to be caused by mining. The structural factors that influence the amount of damage caused include the size and shape of the structure, the type of foundation, the method of construction and the type of materials used, along with the existing state of repair of the structure.

Investigations carried out by the former National Coal Board (Anon., 1975) in Britain revealed that typical mining damage starts to appear in conventional structures when they are subjected to effective strains of 0.5–1.0 mm m^{-1} and damage can be classified as negligible, slight, appreciable, severe or very severe (Fig. 4.29 and Table 4.3). However, this relationship between damage and change in length of a structure is only valid when the average ground strain produced by mining subsidence is equalled by the average strain in the structure. In fact, this commonly is not the case, strain in the structure being less than it is in the ground.

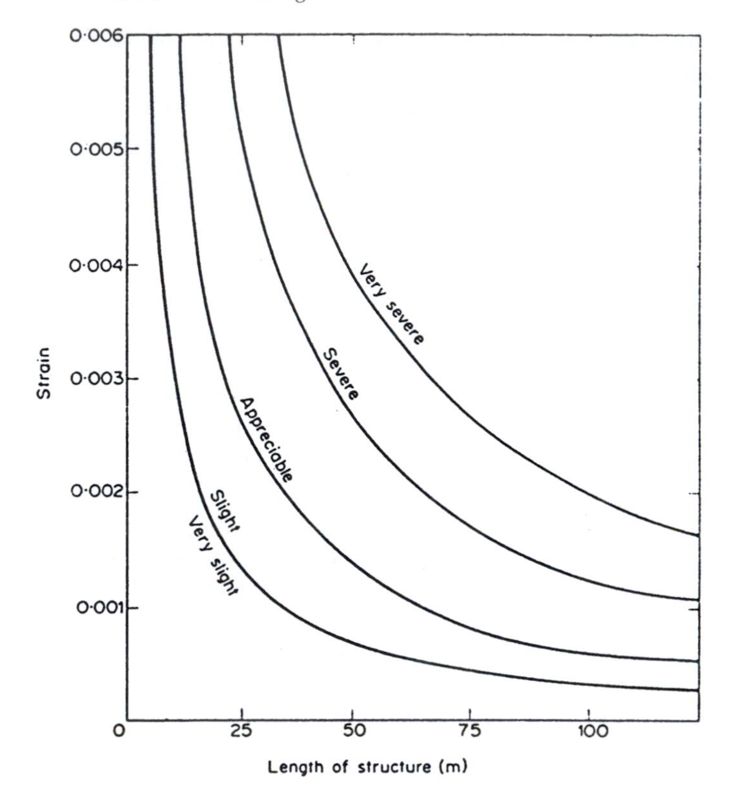

Figure 4.29 Relationship of damage due to longwall mining to length of structure and horizontal strain. *Source*: Anon., 1975. Reproduced by kind permission of British Coal Corporation.

Hence, Fig. 4.29 only serves as a crude guide to the likely damaging effects of ground strains induced by mining subsidence. In addition, it takes no account of the design of the structure or of construction materials. Nevertheless, it indicates that the larger a building, the more susceptible it is to differential vertical and horizontal ground movement.

More recent criteria for subsidence damage to buildings have been proposed by Bhattacharya and Singh (1985) who recognized three classes of damage, namely, architectural that was characterized by small-scale cracking of plaster and doors, and windows sticking; functional damage that was characterized by instability of some structural elements, jammed doors, and windows, broken window panes, and restricted building services; and structural damage in which primary structural members were impaired, there was a possibility of collapse of members, and complete or large-scale rebuilding was necessary. Their conclusions are summarized in Table 4.4. They observed that basements were the most sensitive parts of buildings with regard to subsidence damage and therefore usually suffered more damage than the rest of a building.

The contemporaneous nature of subsidence associated with longwall mining sometimes affords the opportunity to planners to phase long-term surface development in relation to the cessation of subsidence (Bell, 1987). However, the relationship between future programming of surface development and that of subsurface working may be difficult to coordinate. In addition, because subsidence is predictable, damage attributable to subsidence can be reduced or controlled by precautionary measures incorporated into new structures in mining areas, and preventative works applied to existing structures. Several factors have to be considered when

Table 4.3 National Coal Board Classification of subsidence damage

Change in length of structure (mm)	*Class of damage*	*Description of typical damage*
Up to 30	Very slight or negligible	Hair cracks in plaster. Perhaps isolated slight fracture in the building not visible on outside
30–60	Slight	Several slight fractures showing inside the building. Doors and windows may stick slightly. Repairs to decoration probably necessary
60–120	Appreciable	Slight fracture showing on outside of building (or main fracture). Doors and windows sticking, service pipes may fracture
120–180	Severe	Service pipes disrupted. Open fractures requiring rebonding and allowing weather into the structure. Window and door frames distorted, floors sloping noticeably. Some loss of bearing in beams. If compressive damage, overlapping of roof joints and lifting of brickwork with open horizontal fractures
>180	Very severe	As above, but worse and requiring partial or complete rebuilding. Roof and floor beams lose bearing and need shoring up. Windows broken with distortion. Severe slopes on floors. If compressive damage, severe buckling and bulging of the roof and walls

Source: Anon., 1975. Reproduced by kind permission of British Coal Corporation.

Table 4.4 Recommended damage criteria for buildings

Category	*Damage level*	*Angular distortion* ($mm\,m^{-1}$)	*Horizontal strain* ($mm\,m^{-1}$)	*Radius of curvature* (km)
1	Architectural	1.0	0.5	–
	Functional	2.5–3.0	1.5–2.0	20
	Structural	7.0	3.0	–
2	Architectural	1.3	–	–
	Functional	3.3	–	–
	Structural	–	–	–
3	Architectural	1.5	1.0	–
	Functional	3.3–5.0	–	–
	Structural	–	–	–

Source: Bhattacharya and Singh, 1985.

designing buildings for areas of active mining. First, where high ground strains are anticipated, the cost of providing effective rigid foundations may be prohibitive. Second, experience suggests that buildings with deep foundations, on which thrust can be exerted, suffer more damage than those in which the foundations are more or less isolated from the ground. Third, because of the relationship between ground strain and size of structure, very long buildings should be avoided unless their long axes can be orientated normal to the direction of principal ground strain. Last, although tall buildings may be more susceptible to tilt rather than to the effects of horizontal ground strain, tilt can be corrected by using jacking devices.

The most common method of mitigating subsidence damage is by the introduction of flexibility into a structure (Bell, 1988a). In flexible design, structural elements deflect according to the subsidence profile. Flexibility can be achieved by using specially designed rafts. Raft foundations should be as shallow as possible, preferably above ground, so that compressive strains can take place beneath them instead of transmitting direct compressive forces to their edges. They should be constructed on a membrane so that they will slide as ground movements occur beneath them. For instance, reinforced concrete rafts, laid on granular material reduce friction between the ground and the structure. Where relatively small buildings (up to about 30 m in length) are concerned, they can be erected on a 'sandwich raft' foundation. Cellular rafts have been used for multi-storey buildings.

The use of piled foundations in areas of mining subsidence presents its own problems. The lateral and vertical components of ground movement as mining progresses mean that the pile caps tend to move in a spiral fashion and that each cap moves at a different rate and in a different direction according to its position relative to the mining subsidence. Such differential movements and rotations would normally be transmitted to the structure with a corresponding readjustment of the loadings on the pile cap. In order to minimize the disturbing influence of these rotational and differential movements it often is necessary to allow the structure to move independently of the piles by the provision of a pin joint or roller bearing at the top of each pile cap. It may be necessary to include some provision for jacking the superstructure where severe dislevelment is likely to occur.

Preventative techniques frequently can be used to reduce the effects of movements on existing structures. The type of technique depends upon the type of structure and its ability to withstand ground movements, as well as the amount of movement likely to occur. Again the principal objective is to introduce greater flexibility or reduce the amount of ground movement that is transmitted to the structure. In the case of the buildings longer than 18 m, damage can be reduced by cutting them into smaller, structurally independent units. The space produced should be large enough to accommodate deflection. In particular, such items as lift shafts can be planned independently of the other parts of a building and separated from the main structure by joints through foundations, walls, roof and floor that allow freedom of movement. Structures that are weak in tensile strength can be afforded support by strapping or tie bolting together where they are likely to undergo extension.

Damage to surface structures also can be reduced by adopting specially planned layouts of underground workings that take account of the fact that surface damage to structures is caused primarily by ground strains. Harmonic mining reduces the effects of strain but not necessarily the amount of subsidence at the surface. The system involves two or more faces being worked simultaneously so that the resultant strains from each panel tend to cancel each other out. Pillars of coal can be left in place to protect surface structures above them. Maximum subsidence also can be reduced by packing the goaf. Pneumatic stowing can reduce subsidence by up to 50 per cent (Anon., 1975).

4.6.6 Metalliferous mining and subsidence

There are several underground mining methods used for the extraction of metalliferous deposits. In partial extraction methods, solid pillars are left unmined to provide support to the underground workings. In old mines pillars often were extracted or reduced in size towards the end of the life of the mine. In such situations, in particular, subsidence of the surface is a possibility.

The collapse of the roof or hangingwall of the workings between pillars can result in localized surface subsidence. The subsidence profile may be very severe, with large differential subsidence, tilts and horizontal strains. Pillar collapse can give rise to localized, or substantial areas of subsidence. If variable-sized pillars or if larger barrier pillars were provided at regular spacings, the area of subsidence is likely to be restricted. The resulting subsidence profile usually is irregular, with large differential settlements, tilts and horizontal strains at the perimeter of the subsidence area. The presence of faults or dykes can give rise to steps. When mining inclined ore bodies, the steeper the dip, the more localized the subsidence or potential subsidence is likely to be.

Mining tabular stopes with substantial spans, has been and is practised in the gold and platinum mines of southern Africa. When this type of mining occurs at shallow depth, detrimental subsidence has frequently taken place. The surface profile is commonly very irregular and tension cracks have been observed frequently. The effects are dependent on the dip of the reefs, which vary from as little as 10° to vertical. Closure of stopes is more likely for flatter dips and hence surface movements are more likely to occur under such conditions. Steeply dipping stopes tend to remain open and present a longer-term hazard. Adverse subsidence is not usually observed when the mining depth exceeds 250 m. Most of the detrimental subsidence that has occurred has been associated with geological features such as dyke contacts and faults.

Many metalliferous deposits occur in large disseminated ore bodies and often the only way in which such ore bodies can be mined economically is by means of very high-volume production. When the rock mass is sufficiently competent, large open stopes, with spans of between 50 and 100 m, can be excavated and some form of pillar usually separates adjacent open stopes. In caving, material from above is allowed to cave into the opening created by the mined extraction. Since caving occurs above the ore body, it usually progresses through to the surface to form a subsidence crater. This crater usually has several scarps around its perimeter, and contains material that has subsided *en block* overlying caved and rotated material, and loosely consolidated ravelled material. Owing to the large volumes that are extracted during mining, the extent and depth of the subsidence crater can be very large. Obviously, caving cannot take place where any surface development is likely to be affected.

Stacey and Bell (1999) described the influence that subsidence due to past gold mining has had on development in Johannesburg. Gold mining commenced in the late nineteenth century and the early mines were at shallow depth. Their presence has given rise to subsidence, which has imposed limitations on development in the old reef area that runs east to west through the city. Indeed, the erection of buildings on the undermined land has been controlled by the Government Mining Engineer. These controls determined whether or not building took place, as well as the permissible heights of buildings in relation to the depth at which mining occurred. More recently, however, due to the pressure on available space in the central business district, the controls have been relaxed when development proposals have been accompanied by sound methods of mine stabilization. Several methods of mine stabilization have been employed from dynamic compaction to concrete dip pillars to provide rigid in-stope support, to plugging stopes and grouting above the plug.

4.6.7 Subsidence associated with the abstraction of fluids

Subsidence of the ground surface can occur in areas where there is intensive abstraction of groundwater, that is, where abstraction exceeds natural recharge and the water table is lowered, the subsidence being attributed to the consolidation of the sedimentary deposits as a result of increasing effective stress (Bell, 1988b). The total overburden pressure in partially saturated or unsaturated deposits is borne by their granular structure and the pore water. When groundwater abstraction leads to a reduction in pore water pressure, this means that there is a gradual transfer of stress from the pore water to the granular structure. For instance, if the groundwater level is lowered by 1 m, then this gives rise to a corresponding increase in average effective overburden pressure of almost 10 kPa. As a result of having to carry this increased load, the void ratio of the deposits concerned undergoes a reduction in volume, the surface manifestation of which is subsidence. Surface subsidence occurs over a longer period of time than that taken for abstraction.

The amount of subsidence that occurs is governed by the increase in effective pressure (that is, the magnitude of the decline in the water table), the thickness and compressibility of the deposits involved, the depth at which they occur, and the length of time over which the increased loading is applied (Lofgren, 1968). The rate at which consolidation occurs depends on the thickness of the beds concerned as well as the rate at which pore water can drain from the system, which is governed by its permeability. Thick slow-draining fine-grained beds may take years or decades to adjust to an increase in applied stress, whereas coarse-grained deposits adjust rapidly. However, the rate of consolidation of slow-draining aquitards reduces with time and is usually small after a few years of loading.

Abstraction of groundwater from the Chalk between 1820 and 1936 caused subsidence in some areas of London in excess of 0.3 m (Wilson and Grace, 1942). In 1820 the artesian head in the Chalk was approximately +9.1 m AOD but by 1936 this had declined in some places to −90 m AOD. Subsidence in Mexico City due to the abstraction of water from a sand, gravel aquifer surrounded by clay meant that by 1959 most of the old city had suffered at least 4 m of subsidence, and in the northeast part as much as 7 m had been recorded.

Methods that can be used to arrest or control subsidence caused by groundwater abstraction include reduction of pumping draft, artificial recharge of aquifers from the ground surface and repressurizing the aquifer(s) involved via wells, or any combination thereof. The aim is to manage the rate and quantity of groundwater withdrawal so that its level in wells is either stabilized or raised somewhat so that the effective stress is not further increased. Reduction of pumping draft may be brought about by importing water from outside the area concerned, by conserving or reducing the use of water, by treating and re-using water or by decreasing the demand for water. The geological conditions determine whether or not artificial recharge of aquifers from the ground surface is feasible. If confining aquitards or aquicludes inhibit the downward percolation of water, then such treatment is impractical. Repressurizing confined aquifer systems from wells may prove the only viable means of slowing down and eventually halting subsidence.

Subsidence due to the abstraction of oil occurs for the same reason as does subsidence associated with the abstraction of groundwater. In other words, pore pressures are lowered by the removal of not only oil but gas and water also, the increased effective load causing consolidation in compressible beds. Such an explanation first was advanced to account for the spectacular and costly subsidence at Wilmington oil field, California (Gilluly and Grant, 1949). By 1947 subsidence was occurring at a rate of 0.3 m per year and had reached 0.7 m annually by 1951 when the maximum rate of withdrawal was attained. By 1966 an elliptical area of over 75 km^2 had subsided more than 8.8 m. Maximum horizontal displacement at Wilmington oilfield exceeded 35 per cent of the maximum subsidence. Because of the seriousness of the subsidence remedial action was taken in 1957 by injecting water into and thereby repressurizing the abstraction zones. This had brought subsidence to a halt in most of the field by 1962. More than 40 known examples of differential subsidence, horizontal displacement and surface fissuring/faulting, associated with 27 oil and gas fields in California and Texas were reported by Yerkes and Castle (1970). The maximum subsidences recorded in these generally are less than 1 m.

Those deposits that readily go into solution such as rock salt can be extracted by solution mining. For instance, salt has been obtained by pumping natural brine in a number of areas in the United Kingdom, Cheshire being by far the most important. Such pumping accelerated the formation of solution channels. Active subsidence normally was concentrated at the head and sides of a brine run where the fresh water first entered the system (Fig. 4.30). Hence, serious subsidence used to occur at considerable distances, anything up to 8 km, from pumping centres. The maximum strains developed by wild brine pumping could ruin buildings. What was worse as far as structural damage was concerned, was the formation of tension scars (small faults) on the convex flanks of subsidence hollows. The vertical displacement along a scar usually is less than a metre. Consequently, subsidence was an inhibiting factor, especially in parts of Cheshire, as far as major developments were concerned (Bell, 1992). This was chiefly because of unpredictable nature of such subsidence. Today in Cheshire 99 per cent of the brine is obtained by controlled solution mining from cavities formed in the salt, which has not given rise to any subsidence. Wassman (1980) quoted some 1.6 m of subsidence as having taken place in the centre of one subsidence trough that resulted from the solution mining of salt in the area around Hengelo in the Netherlands.

Deere (1961) described differential subsidence (resulting from the development of a concavo–convex subsidence profile), simultaneous horizontal displacements and surface faulting, peripheral to the subsidence basin, associated with the abstraction of sulphur from numerous wells in the Gulf region of Texas. During the first 31 months of operation, differential subsidence amounting to 1.75 m occurred over an elliptical area exceeding 5 km^2. The subsidence basin

Figure 4.30 Subsidence caused by wild brine pumping. Note the tension scar to the right of the fence and the flash in the top right-hand corner.

was centred directly above the narrow linear producing zone. Mining of sulphur and potash in Louisiana has given rise to vertical subsidence ranging up to 9 and 6 m respectively.

Ground failure often is associated with subsidence due to the abstraction of groundwater, oil, natural gas or brines (Fig. 4.31). Some of the best examples of ground failure attributable to the withdrawal of groundwater in the United States are found in southern Arizona, the Houston–Galveston region of Texas and the Fremont valley in California. Two types of ground failure, namely, fissures and faults are recognized. Fissures may appear suddenly and the appearance of some may be preceded by the occurrence of minor depressions at the surface. The longest fissure zone in the United States is 3.5 km and lengths of hundreds of metres are typical (Holzer, 1980). Individual fissures commonly are not continuous but consist of a series of segments with the same trend. Occasionally a zone, defined by several closely spaced parallel fissures, occurs, the width of which may be of the order of 30 m. Secondary cracks frequently are associated with fissures. These cracks develop subparallel to the main fissure and usually are only a few metres long. They occur at distances up to 15 m from the parent fissure. According to Larson (1986) fissures form at points of maximum horizontal tensile stress associated with maximum

Figure 4.31 Severe cracking in brickwork due to subsidence and the development of fissures resulting from groundwater abstraction, Xian City, China.

convex upward curvature of the subsidence profile. The association of fissure occurrence with variable aquifer thickness also suggests that differential consolidation takes place near these fissures as groundwater levels fall. Tensile strains at fissures at times of their formation have ranged from 0.1 to 0.4 per cent. Faults that are suspected of being related to groundwater withdrawal are much less common than fissures. They frequently have scarps more than 1 km in length and more than 0.2 m high. Fault scarps have been found to increase in height by dip-slip creep along the normal fault planes. Measured rates of vertical offset range from 4 to 60 mm annually, however, movement tends to vary with time. The land surface near a scarp may tilt, tilting being greatest near a scarp but has been observed to extend as far as 500 m from a scarp. It also has been maintained that fluid abstraction has brought about the reactivation of faults.

4.6.8 Sinkholes and subsidence

Many sinkholes in carbonate rocks are induced by man's activities, that is, they result from declines in groundwater level, especially those, due to excessive abstraction. For example, Jammal (1986) recorded that 70 sinkholes appeared in Orange and Seminole counties in central Florida in the 20 years up to 1985. Most developed in those months of the year when rainfall was least (i.e. April and May) and withdrawal of groundwater was high. Frequently, collapses forming sinkholes result from roof failures of cavities in unconsolidated deposits. These cavities are created when the unconsolidated deposits move or are eroded downward into openings in the top of bedrock. Collapse of bedrock roofs, as compared with the migration of unconsolidated deposits into openings in the top of bedrock, is rare.

As there is preferential development of solution voids along zones of high secondary permeability because these concentrate groundwater flow, data on fracture orientation and density, fracture intersection density and the total length of fractures have been used to model the presence of solution cavities in limestone. Therefore, the location of areas of high risk of cavity collapse has been estimated by using the intersection of lineaments formed by fracture traces and lineated depressions. For example, Brook and Alison (1986) used such a technique to produce subsidence susceptibility maps of a covered karst terrain in Dougherty County, southwest Georgia.

Dewatering associated with mining in the gold-bearing reefs of South Africa, which underlie dolomite and unconsolidated deposits, has led to the formation of sinkholes and subsidence depressions (De Bruyn and Bell, 2001). Hence, certain areas became unsafe for occupation. Although sinkholes were initially noticed in the 1950s, the seriousness of the situation was highlighted in December 1962 when a sinkhole engulfed a three-storey crusher plant at West Driefontein Mine (Fig. 4.32). Consequently, it became a matter of urgency that the areas at risk of subsidence and the occurrence of sinkholes be delineated. Sinkholes formed concurrently with the lowering of the water table in areas that formerly had been relatively free of sinkholes. Subsidence depressions have occurred as a result of consolidation taking place in unconsolidated material as the water table has been lowered, the amount of subsidence reflecting the thickness, proportion, and original density of unconsolidated deposits that have consolidated. These deposits vary laterally, thereby giving rise to differential subsidence. Furthermore, the risk of sinkhole and subsidence occurrence is increased by development, where interrupted natural surface drainage, increased run-off, and leakage from water bearing utilities can result in the concentrated ingress of water into the subsurface. Where overlying material is less permeable, the risk of instability is lower. In the area underlain by dolomite that extends around Johannesburg and Pretoria the problem has been more notable in recent years because of housing development and the growth of informal settlements. Residential densities may be very high, especially for low-cost housing,

Figure 4.32 Collapse of a three-storey crusher plant into a sinkhole due to dewatering at West Driefontein, South Africa.

the development of which in places has proceeded without recognition of the risk posed by karst-related ground instability. The appearance of significant numbers of small sinkholes has been associated with dolomite at shallow depth, that is, occurring at less than 15 m beneath the ground surface. When dolomite is located at depths greater than 15 m, the sinkholes that appear at the surface usually are larger in diameter. The risk of sinkhole occurrence in areas of shallow dolomite may be greater, although the hazard itself is less severe. A classification system for the evaluation of dolomitic land based on the risk of formation of certain sized sinkholes has enabled such land to be zoned for appropriate development. Ongoing monitoring and maintenance of water-bearing services, and the implementation of precautionary measures relating to drainage and infiltration of surface water are regarded as essential in developed areas underlain by dolomite. Prior to development taking place on a dolomitic area a site investigation must be undertaken to evaluate the area in terms of the risk of sinkhole or subsidence depression occurrence. The evaluation has to identify any restrictions on land use and the type of development that is suitable. Various types of foundation structure have been used in dolomitic areas such as engineered soil mattresses, rafts and piles. Soil stabilization, dynamic compaction and grouting have been used to treat the ground.

4.7 Earthquakes and ground movements

The most serious direct effect of an earthquake in terms of buildings and structures is ground shaking. The types of ground conditions are important in this regard, for instance, buildings on firm bedrock suffer less than those on saturated alluvium. Nonetheless, buildings standing on firm rock still can be affected, so that susceptible buildings should not be located near to a fault trace. Poorly constructed buildings and those that are not reinforced undergo the worst damage.

Ground displacement during an earthquake may be horizontal, vertical or oblique. Displacement is associated with rupture along faults due to the accumulation of strain. Once the elastic limit of the rocks involved is exceeded, then sudden movements occur. However, ground rupture as a cause of damage to buildings can be avoided by not building on or near major active faults.

Regional crustal deformation manifested in uplift or subsidence may be associated with earthquakes. Furthermore, earthquakes may fissure the ground, this being most prevalent next to the fault trace. Fissures may gape up to 1 m wide at times. Fault creep can weaken and ultimately cause structures to fail. Landslides and other types of mass movements frequently accompany earthquakes, most occurring within 40 km of the earthquake concerned.

Earthquake damage to buildings depends upon how well they have been constructed and maintained; the seismic energy generated by the earthquake, as represented by the magnitude, duration and acceleration of strong motions in the ground; the ground response that depends upon the type and character of the rocks and soils involved; and the distance from the epicentre of the earthquake (Ambraseys, 1988). Furthermore, an earthquake does not impart simple motion to a building but generally there are elastic impacts followed by forced and free vibrations of the structure. The damage potential of ground shaking is governed by the amplitude, duration and frequency of the waves, and these are unique to each earthquake. Few buildings are capable of resisting the displacement developed by acceleration exceeding 0.6–0.7 g.

4.7.1 Ground conditions and seismicity

The response of structures on different foundation materials has proven surprisingly varied. In general, structures not specifically designed for earthquake loadings have fared far worse on soft saturated alluvium than on hard rock. In other words, amplitude and acceleration are much greater on deep alluvium than on rock, although rigid buildings may suffer less on alluvium than on rock. This is because the alluvium seems to have a cushioning effect and the motion may be changed to a gentle rocking, which is easier on such a building than the direct effect of earthquake motions experienced on harder ground. Conversely, alluvial ground beneath a structure with any kind of poor construction facilitates destruction. Intensity attenuation on rock is very rapid whereas it is extremely slow on soft formations and speeds up only in the fringe area of the shock. Hence, the character of intensity attenuation in any shock will depend largely on the surface geology of the shaken area. It therefore is important to try to relate the dynamic characteristics of a building to those of the subsoil in which it is founded. The vulnerability of a structure to damage is enhanced considerably if the natural frequencies of vibration of the structure and the subsoil are the same.

Ground vibrations caused by earthquakes often lead to compaction of coarse-grained soil and associated settlement of the ground surface. Loosely packed saturated sands and silts tend to lose all strength and behave like fluids during strong earthquakes. When such materials are subjected to shock, densification occurs. During the relatively short time of an earthquake, drainage cannot be achieved and this densification therefore leads to the development of excessive pore water pressures that cause the soil mass to act as a heavy fluid with practically no shear strength, that is, a quick condition develops (Peck, 1979). Water moves upward from the voids to the ground surface where it emerges to form sand boils. An approximately linear relationship exists between the relative density of sands and the stress required to cause initial liquefaction in a given number of stress cycles.

If liquefaction occurs in a sloping soil mass, the entire mass will begin to move as a flow slide. Such slides develop in loose saturated sandy/silty materials during earthquakes. In addition,

Figure 4.33 Formation of a graben-type structure due to the earthquake that affected Anchorage, Alaska, in 1964, which destroyed the school shown.

loose saturated silts and sands often occur as thin layers underlying firmer materials. In such instances, liquefaction of the silt or sand during an earthquake may cause the overlying material to slide over the liquefied layer. Structures on the main slide frequently are moved without suffering significant damage. However, a graben-like feature often forms at the head of the slide and buildings located in this area are subjected to large differential settlements and often are destroyed (Fig. 4.33). Buildings near the toe of the slide frequently are heaved upward or even pushed over by the lateral thrust.

Clay soils, with the exception of quick clays, do not undergo liquefaction when subjected to earthquake activity, but under repeated cycles of loading large deformations can develop, although the peak strength remains about the same. Nonetheless, these deformations can reach the point where, for all practical purposes, the soil has failed. The damage caused by the 1985 earthquake that affected Mexico City was restricted almost exclusively to saturated deposits of clay characterized by low natural vibration frequencies. The clay amplified the shock waves between 8 and 50 times compared with the motions on solid rock in adjacent areas. Those buildings that sustained the most severe damage were located mainly where the thickness of the clay was greater than 37 m thick. In fact, the severity of ground motion increased in relation to the thickness of the clay.

4.7.2 Seismic zoning

Seismic zoning and micro-zoning provide a means of regional and local planning in relation to the reduction of seismic hazard, as well as being used in terms of earthquake resistant design. While seismic zoning takes into account the distribution of earthquake hazard within a region, seismic micro-zoning defines the distribution of earthquake risk in each seismic zone. A seismic zoning map shows the zones of different seismic hazard in a particular area. Detailed seismic zoning maps should take account of local engineering and geological characteristics, as well as the differences in the spectrum of seismic vibrations and, most important of all, the probability of the occurrence of earthquakes of various intensities.

Usually, seismic zoning maps are linked to building codes and commonly are zoned in terms of macroseismic intensity increments or related to seismic coefficients incorporated in a particular code. The building code may specify the variation of the coefficient in terms of ground conditions and type of structure. Other maps may distinguish zones of destruction. Karnik and Algermissen (1979), however, argued that there is a need for quantities related to earthquake resistant design to be taken into account, such as maximum acceleration or peak particle velocity, predominant period of shaking or probability of occurrence.

Most studies of the distribution of damage attributable to earthquakes indicate that areas of severe damage are highly localized and that the degree of damage can change abruptly over short distances. These differences frequently are due to changes in soil conditions or local geology. Such behaviour has an important bearing on seismic micro-zoning. Seismic micro-zoning maps are most detailed and more accurate where earthquakes have occurred quite frequently and the local variations in intensity have been recorded.

4.7.3 Earthquake analysis and design

The pseudostatic method of earthquake analysis involves the seismic factor concept in building design. In other words, the stability of a structure is determined as for static loading conditions and the effects of an earthquake are taken into account by including an equivalent horizontal force acting on the structure, in the computation. This means that a building is made strong enough to resist a horizontal force equal to a certain proportion of the weight of the building. Since weight is expressed in terms of the acceleration due to gravity, the horizontal force is expressed in the same way, and it is possible to compare this force with measured or computed accelerations of earthquakes. The horizontal force representing earthquake effects is expressed as the product of the weight of the structure and a seismic coefficient, k. In the United States the value of the seismic coefficient normally is selected on the basis of the seismicity of the region, for example, in California the values range from 0.05 to 0.15. However, such values do not necessarily correspond with accelerations measured by strong motion accelerometers. Although experience has shown this to be a practical approach it can underestimate seismic loading.

The dynamic behaviour of a structure depends on the intensity of the loads acting upon it. For small amplitude vibrations, structures behave linearly, but as the amplitude increases non-linear behaviour becomes increasingly more important. In some types of structures, under increasing dynamic loads, cracks or local ruptures can occur that completely change the behaviour of the structure. Accordingly, knowledge of the true dynamic properties of actual structures is essential for earthquake design so that measurements should be taken when structures undergo strong vibration due to earthquakes. The resonance developed in multi-storey buildings shows that they are more susceptible to earthquake damage than single- or two-storey buildings. Buildings, towers and bridges must be designed so that their vibration frequencies are different from those of earthquakes.

The purpose of a design earthquake is to enable an estimate to be made of the severest earthquake that is likely to occur within a 50- or a 100-year period. More simply, it is an indication of the amount of motion, with its spectrum or frequency distribution. This is the critical factor required by the designer for dynamic analysis of a structure and for determining its response to earthquake motion. The initial step in formulating a design earthquake at a given site is the estimation of the magnitudes of earthquakes that can be expected to affect a site during its lifetime. For example, the extrapolated frequency–magnitude relations for southern California indicate that the largest expected earthquake per year has a magnitude of 6.2 and that the largest expected earthquake per 100 years has a magnitude of 8.2.

Earthquake design criteria must be based on the probablility of occurrence of strong ground shaking, the characteristics of the ground motion, the nature of the structural deformations, the behaviour of building materials when subjected to transient oscillatory strains, the nature of the building damage that might be sustained, and the cost of repairing the damage as compared to the cost of providing additional earthquake resistance (Cherry, 1974). Furthermore, design criteria should specify the desired strength of structures in order to achieve a reasonably uniform factor of safety. This usually is done by means of a design spectrum. A design spectrum is not a specification of a particular earthquake ground motion, rather it specifies the relative strengths of structures of different periods (Housner, 1970). When establishing the design spectrum it is necessary to have some idea of the ground shaking that might be experienced at the location under consideration, so that the earthquake history of the region should be examined, together with evidence of recent faulting or other indications of tectonic activity. Once the damping and allowable stresses are prescribed, the earthquake forces for which the structure is to be designed can be determined from the spectrum. In fact, the actual earthquake forces used in the design of a structure largely depend on the damping that the structure is assumed to have.

The behaviour of an actual structure during an earthquake depends on its natural frequency and damping forces. In order to establish the earthquake design forces, models are vibrated at different vibration periods and the corresponding accelerations determined. Thus, an idealized spectrum may be obtained applicable to any structure and any earthquake. Buildings also have been forced into vibration by means of shaking machines. The systematic analysis of the earthquake response of buildings also can be investigated by computer simulation.

4.7.4 Building performance

All buildings have a certain natural period at which they tend to vibrate. If earthquake waves have vibrations with the same period, the amplitude is increased and they will shake a building severely. If the period is very different, the effect will be much less severe. Adjoining buildings may have different fundamental periods and may vibrate out of phase with each other. If the buildings are less than 1 m apart, the resulting interference can cause mutual destruction. Progressive collapse may occur in this way if buildings are closely spaced down a hill. Taller buildings tend to have longer fundamental periods and to be more seriously damaged when located on soft ground that also has a long fundamental period. Low buildings with shorter fundamental periods are most vulnerable on rocks since they have higher frequencies. It is possible to calculate the period a building will have when it is completed. Indeed, measurement of the period can be made by shaking a building. Alternatively, the fundamental periods of vibration of buildings can be determined by recording the small vibrations induced by wind and microseisms. The period itself depends upon the manner in which the weight is distributed in a building and upon the stiffness of the materials involved. The addition or removal of interior partitions or the storage of heavy goods may change the period considerably. The periods of earthquake waves can be determined by using strong motion accelerometers.

The seismic performance of a building is influenced by its mass and stiffness, its ability to absorb energy (i.e. its damping capacity), its stability and its structural geometry and continuity. As such, the earthquake resistance of a building depends on an optimum combination of strength and flexibility since it has to absorb and resist the impact of earthquake waves. Shallow foundations are sensitive to the vertical displacement component, especially if the structure is light and the water table high. Deep foundations, however, generally are not hazardous in an earthquake. Buildings that have been constructed of several types of materials may be vulnerable to earthquake shocks. Generally, well-built simple structures often can come through an earthquake unscathed. If properly constructed, wood-frame buildings can withstand strong shaking. Buildings consisting of brick, stone or concrete bricks are not as resistant but may prove satisfactory if the mortar is good and they are reinforced with steel, especially at weak points. The primary reasons for the collapse of masonry buildings during earthquakes are their low tensile and shear strength, their high rigidity, low ductility and low capacity for bearing reversed loads and the redistribution of stresses. Framed structures are flexible, and capable of resisting large deformations and the redistribution of stresses. The joints between columns and girders represent their weakest points. Joints in pre-fabricated frame structural systems also are weak points. The concentrations of stress are therefore of major importance in determining behaviour under dynamic loading during an earthquake. Damage can occur as a result of buckling of columns and, in high-rise buildings, by lateral deformation causing an eccentricity of the vertical load and thereby generating further bending moments. A reinforced concrete or steel-framed building possesses both strength and flexibility and, although there may be superficial damage to partitions and curtain walls, it usually remains substantially sound even after a large earthquake.

4.8 Problem soils

Unfortunately, many soils can prove problematic in geotechnical engineering, because they expand, collapse, disperse, undergo excessive settlement, have a distinct lack of strength or are corrosive. Such characteristics may be attributable to their composition, the nature of their pore fluids, their mineralogy or their fabric. Soil is formed by the breakdown of rock masses by either weathering or erosion. It may accumulate in place or undergo a certain amount of transport, either of which influence its character and behaviour. However, the type of breakdown suffered by rock masses is influenced profoundly by the climatic regime in which they occur, as well as the stage of maturity that has been reached. For example, laterites are developed in hot wet tropical regions and sabkhas are characteristic of arid regions. The effects of these problems can result in considerable financial loss.

4.8.1 Expansive clays

Some clay soils undergo slow volume changes that occur independently of loading and are attributable to swelling or shrinkage. These volume changes can give rise to ground movements that can cause damage to buildings. Low-rise buildings are particularly vulnerable to such ground movements since they generally do not have sufficient weight or strength to resist.

The principal cause of expansion in clay soils is the presence of swelling clay minerals such as montmorillonite. Differences in the period and amount of precipitation and evapotranspiration are the principal factors influencing the swell–shrink response of a clay soil beneath a building. Poor surface drainage or leakage from underground pipes also can produce concentrations of moisture in clay soils. Trees with high water demand and uninsulated hot process foundations may dry out clay soil causing shrinkage. Cold stores also may cause desiccation of clay soil. The depth of the active zone in expansive clays (i.e. the zone in which swelling and shrinkage occurs in wet and dry seasons respectively) varies. It may extend to over 6 m depth in some semi-arid regions of South Africa, Australia and Israel. Many soils in temperate regions such as Britain, especially in southeast England, possess the potential for significant volume change due to changes in moisture content. However, owing to the damp climate in most years volume changes are restricted to the upper 1.0–1.5 m in clay soils.

The potential for volume change in clay soil is governed by its initial moisture content, initial density or void ratio, its microstructure and the vertical stress, as well as the type and amount of clay minerals present. The type of clay minerals is responsible primarily for the intrinsic expansiveness whilst the change in moisture content or suction (where the pore water pressure in the soil is negative, that is, there is a water deficit in the soil) controls the actual amount of volume change that a soil undergoes at a given applied pressure. The rate of expansion depends upon the rate of accumulation of moisture in the soil.

Grim (1962) distinguished two modes of swelling in clay soils, namely, intercrystalline and intracrystalline swelling. Interparticle swelling takes place in any type of clay deposit irrespective of its mineralogical composition, and the process is reversible. In relatively dry clay soils the particles are held together by relict water under tension from capillary forces. On wetting the capillary force is relaxed and the clay soil expands. In other words, intercrystalline swelling takes place when the uptake of moisture is restricted to the external crystal surfaces and the void spaces between the crystals. Intracrystalline swelling, on the other hand, is characteristic of the smectite family of clay minerals, of montmorillonite in particular. The individual molecular layers that make up a crystal of montmorillonite are weakly bonded so that on wetting water enters not only between the crystals but also between the unit layers that comprise the crystals. Generally, kaolinite has the smallest swelling capacity of the clay minerals and nearly all of its swelling is of the interparticle type. Illite may swell by up to 15 per cent but intermixed illite and montmorillonite may swell some 60–100 per cent. Swelling in Ca montmorillonite is very much less than in the Na variety and ranges from about 50 to 100 per cent. Swelling in Na montmorillonite can amount to 2000 per cent of the original volume, the clay then having formed a gel.

Cemented and undisturbed expansive clay soils often have a high resistance to deformation and may be able to absorb significant amounts of swelling pressure. Remoulded expansive clays therefore tend to swell more than their undisturbed counterparts. In less dense soils expansion initially takes place into zones of looser soil before volume increase occurs. However, in densely packed soil with low void space, the soil mass has to swell more or less immediately to accommodate the volume change. Therefore, clay soils with a flocculated fabric swell more than those that possess a preferred orientation. In the latter, the maximum swelling occurs normal to the direction of clay particle orientation. Because expansive clays normally possess extremely low permeabilities, moisture movement is slow and a significant period of time may be involved in the swelling–shrinking process. Accordingly, moderately expansive clays with a smaller potential to swell but with higher permeabilities than clay soils having a greater swell potential, may swell more during a single wet season than more expansive clays.

The swell–shrink behaviour of a clay soil under a given state of applied stress in the ground is controlled by changes in soil suction. The relationship between soil suction and water content depends on the proportion and type of clay minerals present, their microstructural arrangement and the chemistry of the pore water. Changes in soil suction are brought about by moisture movement through the soil due to evaporation from its surface in dry weather, by transpiration from plants, or alternatively by recharge consequent upon precipitation. The climate governs the amount of moisture available to counteract that which is removed by evapotranspiration (i.e. the soil moisture deficit). The volume changes that occur due to evapotranspiration from clay soils can be conservatively predicted by assuming the lower limit of the soil moisture content to be the shrinkage limit. Desiccation beyond this value cannot bring about further volume change.

Transpiration from vegetative cover can represent an important means of water loss from soils in dry summers. Indeed, the distribution of soil suction in soil may be controlled by transpiration from vegetation and represents one of the significant changes made in loading (i.e. to the state of stress in a soil). The maximum soil suction that can be developed is governed by the ability of vegetation to extract moisture from the soil. The level at which moisture is no longer available to plants is termed the permanent wilting point and this corresponds to a pF value of about 4.2. The moisture characteristic (moisture content v. soil suction) of a soil provides valuable data concerning the moisture contents corresponding to the field capacity (defined in terms of soil suction this has a pF value of about 2.0) and the permanent wilting point, as well as the rate at which changes in soil suction take place with variations in moisture content. This enables an assessment to be made of the range of soil suction and moisture content that is likely to occur in the zone affected by seasonal changes in climate. The suction pressure associated with the onset of cracking is approximately pF 4.6.

The extent to which vegetation is able to increase the suction to the level associated with the shrinkage limit obviously is important. In fact, the moisture content at the wilting point exceeds that of the shrinkage limit in soils with high contents of expansive clay minerals and is less in those possessing low contents. This explains why settlement resulting from the desiccating effects of trees is more notable in low to moderately expansive soils than in expansive ones.

As mentioned, many clay soils in Britain possess the potential for volume change but the climate means that any significant deficits in soil moisture developed during the summer are confined to the upper 1.0–1.5 m of the soil and the field capacity is re-established during the winter. Nonetheless, deeper permanent deficits can be brought about by large trees. With this in mind, Driscoll (1983) suggested that desiccation could be regarded as commencing when the rate of change in moisture content (and therefore volume) with increasing soil suction, increases significantly. He proposed that this point approximates to a suction of pF about 2 (10 kPa). Similarly, notable suction could be assumed to have taken place if, on its disappearance, a low-rise building was uplifted due to the soil swelling. Driscoll maintained that this suction would have a pF value of 3 (100 kPa).

As far as *in situ* testing is concerned, initial effective stresses can be estimated with the aid of a psychrometer used to measure soil suction. Gourley *et al.* (1994) referred to the use of a suction probe for measuring soil suction in the field. In addition, settlement points can be installed at different depths in the ground using sleeved rods to measure the seasonal movements of the soil in conjunction with moisture content. These measurements provide direct evidence of potential shrinkage and swelling movements. However, such measurements are time consuming. In addition, Williams and Pidgeon (1983), after investigations in South Africa, pointed out that the measurement of seasonal ground movements under natural conditions may give appreciable underestimates of potential total or differential movements under buildings, particularly when desiccation extends to some depth.

Methods of predicting volume changes in soils can be grouped into empirical methods, soil suction methods and oedometer methods (Bell and Maud, 1995). Empirical methods make use of the swelling potential as determined from void ratio, natural moisture content, liquid and plastic limits, and activity. For example, Driscoll (1983) proposed that the moisture content (m) at the onset of desiccation (pF = 2) and when it becomes significant (pF = 3) could be approximately related to the liquid limit (LL), in the first instance $m = 0.5$ LL and in the second $m = 0.4$ LL. Anon. (1981) suggested that the plasticity index provided an indication of volume change potential as shown in Table 4.5. A degree of overlap was allowed. The activity chart,

Table 4.5 Volume change potential

Plasticity index (%)	*Potential*
Over 35	Very high
22–48	High
12–32	Medium
Less than 18	Low

proposed by Van der Merwe (1964), frequently has been used to assess the expansiveness of clay soils (Fig. 4.34). However, because the determination of plasticity is carried out on remoulded soil, it does not consider the influence of soil texture, moisture content, soil suction or pore water chemistry, which are important factors in relation to volume change potential. Over reliance on the results of such tests therefore must be avoided. Consequently, empirical methods should be regarded as simple swelling indicator methods and nothing more. As such, it is wise to carry out another type of test and to compare the results before drawing any conclusions.

Soil suction methods use the change in suction from initial to final conditions to obtain the degree of volume change. Soil suction is the stress which, when removed allows the soil to swell, so that the value of soil suction in a saturated fully swollen soil is zero. O'Neill and Poormoayed (1980) quoted the United States Army Engineers Waterways Experimental Station (USAEWES) classification of potential swell (Table 4.6) that is based on the liquid limit,

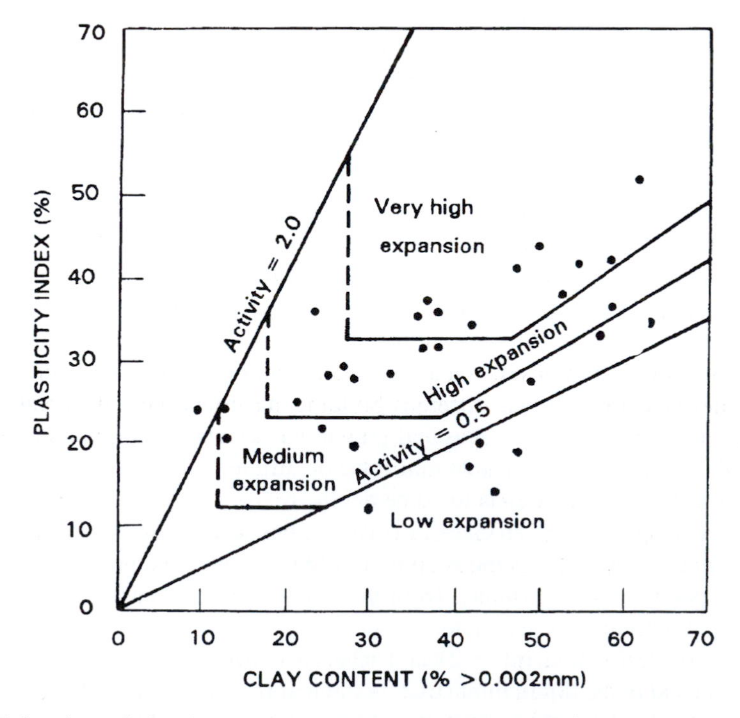

Figure 4.34 Estimation of the degree of expansiveness of clay soil (Van der Merwe, 1964). Some expansive clay soils from Natal, South Africa, are shown (from Bell and Maud, 1995).

Table 4.6 USAEWES classification of swell potential

Liquid limit (%)	*Plastic limit* (%)	*Initial* in situ *suction* (kPa)	*Potential swell* (%)	*Classification*
Less than 50	Less than 25	Less than 145	Less than 0.5	Low
50–60	25–35	145–385	0.5–1.5	Marginal
Over 60	Over 35	Over 385	Over 1.5	High

Source: O'Neil and Poormoayed, 1980. Reproduced by kind permission of American Society of Civil Engineers.

plasticity index and initial (*in situ*) suction. The latter is measured in the field by a psychrometer. Soil suction, however, is not easy to measure accurately. Filter paper has been used for this purpose (McQueen and Miller, 1968). According to Chandler *et al.* (1992) measurements of soil suction obtained by the filter paper method compare favourably with measurements obtained using psychrometers or pressure plates.

The oedometer methods of determining the potential expansiveness of clay soils represent more direct methods (Jennings and Knight, 1975). In the oedometer methods undisturbed samples are placed in the oedometer and a wide range of testing procedures are used to estimate the likely vertical strain due to wetting under vertical applied pressures. The latter may be equated to overburden pressure plus that of the structure that is to be erected. In reality most expansive clays are fissured that means that lateral and vertical strains develop locally within the ground. Even when the soil is intact, swelling or shrinkage is not truly one-dimensional. The effect of imposing zero lateral strain in the oedometer is likely to give rise to overpredictions of heave and the greater the degree of fissuring, the greater the overprediction. The values of heave predicted using oedometer methods correspond to specific values of natural moisture content and void ratio of the sample. Therefore, any change in these affects the amount of heave predicted.

Gourley *et al.* (1994) mentioned the use of a stress path oedometer to determine volume change characteristics of expansive soils. Such a method provides data on vertical and radial total stresses, suction and void ratio.

Effective and economic foundations for low-rise buildings on swelling and shrinking soils have proved difficult to achieve. This is partly because the cost margins on individual buildings are low. Obviously, detailed site investigation and soil testing are out of the question for such individual dwellings. Similarly, many foundation solutions that are appropriate for major structures are too costly for small buildings. Nonetheless, the choice of foundation is influenced by the subsoil and site conditions, estimates of the amount of ground movement and the cost of alternative designs. In addition, different building materials have different tolerances to deflections (Burland and Wroth, 1975). Hence, materials that are more flexible can be used to reduce potential damage due to differential movement of the structure.

Three methods can be adopted when choosing a design solution for building on expansive soils, namely, provide a foundation and structure that can tolerate movements without unacceptable damage, isolate the foundation and structure from the effects of the soil, and alter or control the ground conditions. In addition, moisture control measures should be adopted as far as possible.

The isolation of foundation and structure has been widely adopted for 'severe' and 'very severe' ground conditions. Straight-shafted bored piles can be used in conjunction with suspended floors for severe conditions. The piles are sleeved over the upper part and provided with reinforcement. For severe conditions it may be necessary to place piles at appreciable depth (i.e. below the level of fluctuation of natural moisture content) and/or use underreams to resist the pull-out forces. The use of stiffened rafts is fairly commonplace (Bell and Maud, 1995). The design of the slab is dependent on assumed allowable relative deflections and it usually is necessary to incorporate certain anti-cracking features in the superstructure such as flexible joints.

Moisture control is perhaps the most important single factor in the success of foundations on shrinking and swelling clays. The aim is to maintain stable moisture conditions with minimum moisture content or suction gradients. The loss of moisture around the edges of a building, which leads to the moisture content of the soil under the centre of the building being higher, gives rise to differential heave. In order to control this an attempt should be made to maintain the same moisture content beneath a building. This can be achieved by the use of horizontal and vertical moisture barriers around the perimeter of the building, drainage systems and control of vegetation coverage.

A simple method of reducing or eliminating ground movements due to expansive soil is to replace or partially replace them with non-expansive soils. There is no requirement for the thickness of the replacement material but a minimum of 1 m has been suggested by Chen (1988). The material should be granular but it should not allow surface water to travel freely through the soil so that it wets any swelling soils in lower horizons. Therefore, the presence of a fine fraction is required to reduce permeability or, better still, a geomembrane can surround the granular material.

If expansive soil is allowed to swell by wetting prior to construction and if the soil moisture content then is maintained, the soil volume should remain relatively constant and no heave take place. Ponding is the most common method of wetting. This may take several months to increase the water content to the required depth, notably in areas with deep groundwater surfaces. Vertical wells can be installed to facilitate flooding and thus decrease the time necessary to adjust the moisture content of the soil.

The amount of heave of expansive soils is reduced significantly when compacted to low densities at high moisture contents. Expansive soils compacted above optimum moisture content undergo negligible swell for any degree of compaction. On the other hand, compaction below optimum results in excessive swell.

Many attempts have been made to reduce the expansiveness of clay soil by chemical stabilization. For example, lime stabilization of expansive soils, prior to construction, can minimize the amount of shrinkage and swelling they undergo. In the case of light structures, lime stabilization may be applied immediately below strip footings. However, significant SO_4 content (i.e. in excess of $5000\,mg\,kg^{-1}$) in clay soils can mean that this reacts with CaO to form ettringite with resultant expansion. The treatment is better applied as a layer beneath a raft so as to overcome differential movement. The lime-stabilized layer is formed by mixing 4–6 per cent lime with the soil. A compacted layer, 150 mm in thickness, usually gives satisfactory performance. Furthermore, the lime stabilized layer redistributes unequal moisture stresses in the subsoil so minimizing the risk of cracking in the structure above, as well as reducing water penetration beneath the raft. Alternatively, lime treatment can be used to form a vertical cut-off wall at or near the footings in order to minimize movement of moisture.

Cement stabilization has much the same effect on expansive soils as lime treatment, although the dosage of cement needs to be greater for heavy expansive clays. Alternatively, they can be pretreated with lime, thereby reducing the amount of cement that needs to be used.

4.8.2 Collapsible soils

Soils such as loess, brickearth and certain wind-blown silts may possess the potential to collapse. These soils generally consist of 50–90 per cent silt particles and sandy, silty and clayey types have been recognized by Clevenger (1958), with most falling into the silty category. Collapsible soils possess porous textures with high void ratios and relatively low densities. They often have sufficient void space in their natural state to hold their liquid limit moisture content at saturation. At their natural low moisture content these soils possess high apparent strength but they are susceptible to large reductions in void ratio upon wetting. In other words, the metastable texture collapses as the bonds between the grains break down when the soil is wetted. Hence, the collapse process represents a rearrangement of soil particles into a denser state of packing. Collapse on saturation normally only takes a short period of time, although the more clay such a soil contains, the longer the period tends to be.

The fabric of collapsible soils generally takes the form of a loose skeleton of grains (generally quartz) and microaggregates (assemblages of clay or clay and silty clay particles). These tend to be separate from each other, being connected by bonds and bridges, with uniformly distributed pores. The bridges are formed of clay-sized minerals, consisting of clay minerals, fine quartz, feldspar or calcite. Surface coatings of clay minerals may be present on coarser grains. Silica and iron oxide may be concentrated as cement at grain contacts and amorphous overgrowths of silica occur on some grains. As grains are not in contact, mechanical behaviour is governed by the structure and quality of bonds and bridges. The structural stability of collapsible soils is not only related to the origin of the material, to its mode of transport and depositional environment but also to the amount of weathering undergone.

Popescu (1986) maintained that there is a limiting value of pressure, defined as the collapse pressure, beyond which deformation of these soils increases appreciably. The collapse pressure varies with the degree of saturation. He defined truly collapsible soils as those in which the collapse pressure is less than the overburden pressure. In other words, such soils collapse when saturated since the soil fabric cannot support the weight of the saturated overburden. When the saturation collapse pressure exceeds the overburden pressure, soils are capable of supporting a certain level of stress on saturation and Popescu defined these soils as conditionally collapsible soils. Conditionally collapsible soils therefore can be described as pseudo-stable. The maximum load that such soils can support is the difference between the saturation collapse and overburden pressures. Phien-wej *et al.* (1992) concluded that the critical pressure at which collapse of the soil fabric begins was greater in soils with smaller moisture content. Under the lowest natural moisture content the soils they investigated posed a severe problem on wetting (the collapse-potential was as high as 12.5 per cent at 5 per cent natural moisture content). During the wet season when the natural moisture content could rise to 12 per cent, there was a reduction in the collapse potential to around 4 per cent.

In the Britain brickearth, which occurs mainly in southeast England, is similar to loess. Brickearth is composed largely of silt-sized particles and has a very open, low density structure. The lower parts of a deposit tend to be more rigid and better consolidated than the upper parts. Brickearth generally is calcareous at depth, although the upper parts, and often the full thickness, may be leached of carbonate material. It has a natural moisture content that usually varies from 12 to 25 per cent and is of variable plasticity, ranging from low to high plasticity but most are of low plasticity. Like loess, clayey, silty and sandy brickearth can be recognized. Brickearth has a high degree of compressibility and undergoes rapid consolidation. Northmore *et al.* (1996) investigated the effects of flooding on the undrained and drained shear strength parameters of brickearth. They found that there was a dramatic decrease in undrained shear strength for a small increase in deviator stress.

Several collapse criteria have been proposed for predicting whether a soil is liable to collapse upon saturation. For instance, Clevenger (1958) suggested a criterion for collapsibility based on dry density, that is, if the dry density is less than $1.28\,\mathrm{mg\,m^{-3}}$, then the soil is liable to undergo significant settlement. On the other hand, if the dry density is greater than $1.44\,\mathrm{mg\,m^{-3}}$, then the amount of collapse should be small, while at intermediate densities the settlements are transitional. Gibbs and Bara (1962) suggested the use of dry unit weight and liquid limit as criteria to distinguish between collapsible and non-collapsible soil types. Their method is based on the premise that a soil that has enough void space to hold its liquid limit moisture content at saturation is susceptible to collapse on wetting. This criterion only applies if the soil is uncemented and the liquid limit is above 20 per cent. When the liquidity index in such soils approaches or exceeds 1, then collapse may be imminent. As the clay content of a collapsible soil increases, the saturation moisture content becomes less than the liquid limit so that such deposits are relatively stable. More simply, Handy (1973) suggested that collapsibility could be determined either by the percentage clay content or from the ratio of liquid limit to saturation moisture content. He maintained that soils with a high probability for collapse possessed a clay content of less than 16 per cent, those with a clay content of between 16 and 24 per cent were probably collapsible, those with between 25 and 32 per cent had a probability of collapse of less than 50 per cent, whereas those with a clay content that exceeded 32 per cent were non-collapsible. Soils in which the ratio of liquid limit to saturation moisture content was less than 1 were collapsible, while if it was greater than 1 they were safe. After an investigation of loess in Poland, Grabowska-Olszewska (1988) suggested that loess with a natural moisture content less than 6 per cent was potentially unstable, that in which the natural moisture content exceeds 19 per cent could be regarded as stable, while that with values between these two figures exhibited intermediate behaviour.

Collapse criteria have been proposed that depend upon the void ratio at the liquid limit (e_l), the plastic limit (e_p), and the natural void ratio (e_o). According to Audric and Bouquier (1976) collapse is probable when the natural void ratio (e_o) is higher than a critical void ratio (e_c), which depends on e_l and e_p. They quoted the Denisov and Feda collapse criteria as providing fairly good estimates of the likelihood of collapse:

$$e_c = e_l \qquad \text{(Denisov, 1963)} \tag{4.45}$$

$$e_c = 0.85e_l + 15e_p \qquad \text{(Feda, 1966)} \tag{4.46}$$

Fookes and Best (1969) proposed a collapse index (i_c) that also involved these void ratios as follows:

$$i_c = \frac{e_o - e_p}{e_l - e_p} \tag{4.47}$$

Previously, Feda (1966) had proposed the following collapse index:

$$i_c = \frac{m/S_r - PL}{PI} \tag{4.48}$$

in which m is the natural moisture content, S_r is the degree of saturation, PL is the plastic limit and PI is the plasticity index. Feda also proposed that the soil must have a critical porosity of 40 per cent or above and that an imposed load must be sufficiently high to cause structural collapse when the soil is wetted. He suggested that if the collapse index was greater than 0.85, then this was indicative of metastable soils. However, Northmore *et al.* (1996) suggested that a lower critical value of collapse index, that is 0.22, was more appropriate for the brickearth of south Essex, England. Derbyshire and Mellors (1988) also referred to a lower collapse index for the brickearth of Kent, England.

The absolute collapse index (i_{ac}) also can be used to predict collapse, it being

$$i_{ac} = \frac{m}{S_r} - PL \tag{4.49}$$

Northmore *et al.* (1996) showed that values of absolute collapse index above 6 indicated collapsible brickearth beneath construction loads exceeding 200 kPa and that values above 5 predicted metastability beneath loads greater than 400 kPa.

Although such empirical indices may provide a relatively rapid and inexpensive means of determining collapsibility, it must be borne in mind that some of the parameters that are used for assessment are derived from remoulded soil samples. Such samples do not take account of the initial soil fabric. Consequently, the results should be regarded as only general indicators of collapsibility and indeed some indices need to be used with caution (Bell *et al.*, 2004).

The oedometer test can be used to assess the degree of collapsibility. For example, Jennings and Knight (1975) developed the double oedometer test for assessing the response of a soil to wetting and loading at different stress levels (i.e. two oedometer tests are carried out on identical samples, one being tested at its natural moisture content, whilst the other is tested under saturated conditions, the same loading sequence being used in both cases). They subsequently modified the test so that it involved loading an undisturbed specimen at natural moisture content in the oedometer up to a given load. At this point the specimen is flooded and the resulting collapse strain, if any, is recorded. Then the specimen is subjected to further loading. The total consolidation upon flooding can be described in terms of the coefficient of collapsibility (C_{col}) given by Feda (1988) as:

$$\begin{aligned} C_{col} &= \frac{\Delta h}{h} \\ &= \frac{\Delta e}{1+e} \end{aligned} \tag{4.50}$$

in which Δh is the change in height of the specimen after flooding, h is the height of the specimen before flooding, Δe is the change in void ratio of the specimen upon flooding and e is the void ratio of the specimen prior to flooding. Table 4.7 provides an indication of the potential severity of collapse and shows that those soils that undergo more than 1 per cent collapse can be regarded as metastable. However, in China a figure of 1.5 per cent is taken (Lin and Wang, 1988) and in the United States values exceeding 2 per cent are regarded as indicative of soils susceptible to collapse (Lutenegger and Hallberg, 1988).

In some parts of the world significant settlements beneath structures on collapsible soils have led to foundation failure, especially after the soils have been wetted. Clemence and Finbarr (1981) recorded a number of techniques that could be used to stabilize collapsible soils. These are summarized in Table 4.8. Various methods of compaction have been used to densify collapsible soils such as dynamic compaction and the use of compaction piles. However, if these soils contain relatively high carbonate contents, then it may be difficult to achieve the desired result with dynamic compaction. Concrete compaction piles may be driven into the ground, alternatively compacted soil-cement piles have been employed. Gibbs and Bara (1967) referred to the use of clay grout, it being injected under pressure to compact loess. Compaction also has been brought about by vibration, either by deep vibroflotation or deep explosion. In both cases

Table 4.7 Collapse percentage as an indication of potential severity

Collapse (%)	*Severity of problem*
0–1	No problem
1–5	Moderate trouble
5–10	Trouble
10–20	Severe trouble
Above 20	Very severe trouble

Source: Jennings and Knight, 1975. Reproduced by kind permission of A.A. Balkema.

Table 4.8 Methods of treating collapsible foundations

Depth of subsoil treatment	*Foundation treatment*
	A. Current and past methods
0–1.4 m	Moistening and compaction (conventional extra heavy impact or vibratory rollers)
1.5–10 m	Over-excavation and recompaction (earth pads with or without stabilization by additives such as cement or lime). Vibro-flotation (free draining soils). Vibroreplacement (stone columns). Dynamic compaction. Compaction piles. Injection of lime. Lime piles and columns. Jet grouting. Ponding or flooding (if no impervious layer exists). Heat treatment to solidify the soils in place
Over 10 m	Any of the aforementioned or combinations of the aforementioned, where applicable. Ponding and infiltration wells, or ponding and infiltration wells with the use of explosive
	B. Possible future methods
	Ultrasonics to produce vibrations that will destroy the bonding mechanics of the soil. Electrochemical treatment. Grouting to fill pores

Source: Clemence and Finbarr, 1981. Reproduced by kind permission of American Society of Civil Engineers.

the soil has been wetted beforehand (Litvinov, 1973). Moreover, compaction has been achieved by inundation. Cement, lime and bitumen emulsion have been used to stabilize loess soils. Evstatiev (1988) referred to the use of soil or soil-cement cushions to replace loess, usually to a depth of 1.5 m and rarely reaching 3 m. These cushions are properly compacted in layers. Silicate grouts have been employed to treat loess soils in Russia and more recently jet grouting has proved an effective method of stabilization. Prevention of saturation around the foundations of small buildings, by using flexible drains and pipes, and by ensuring that the run-off is kept away by the use of concrete aprons, may be sufficient to avoid collapse taking place.

4.8.3 Quicksands

As water flows through silts and sands and loses head, its energy is transferred to the particles past which it is moving, which in turn creates a drag effect on the particles. If the drag effect is in the same direction as the force of gravity, then the effective pressure is increased and the soil is stable. Conversely, if water flows towards the surface, then the drag effect is counter to gravity thereby reducing the effective pressure between particles. If the velocity of upward flow is sufficient it can buoy up the particles so that the effective pressure is reduced to zero. This represents a critical condition where the weight of the submerged soils is balanced by the upward-acting seepage force. If the upward velocity of flow increases beyond the critical hydraulic gradient a quick condition develops.

Quicksands can undergo a spontaneous loss of strength if subjected to cyclic or shock loading. For example, liquefaction of quicksands may be brought about by the action of heavy machinery (notably pile driving), blasting and earthquakes. The basic cause responsible for the liquefaction of saturated sands under such conditions, according to Chaney (1987), is the build-up of excess pore water pressure. As a result, the grains of sand are compacted, with a consequent transfer of stress to the pore water and a reduction of stress on the sand grains. If drainage cannot take place, then the decrease in volume of the grains causes an increase in pore water pressure. If the pore water pressure builds up to the point where it is the same as the overburden pressure, then the effective stress is reduced to zero and the sand loses strength with a liquefied state developing. The build-up of pore water pressure is related to cyclic shear strain and the number of cycles involved. In loose sands the pore water pressure can increase rapidly to the value of the overburden or confining pressure. If the sand undergoes more or less unlimited deformation without mobilizing any notable resistance to deformation, then it can be described as having liquefied. However, Norris *et al.* (1998) maintained that loose sand does not lose all strength during liquefaction. Loose sands at low confining pressure, and medium and dense sands undergo only limited deformation due to dilation once initial liquefaction has occurred. Such response is referred to as ranging from limited liquefaction in the case of loose and medium dense sands at low confining pressure to dilative behaviour in dense sands. Liquefaction may develop in any zone of a deposit of sand where the required conditions exist. It also may occur in the upper layers of a sand if liquefaction develops in an underlying zone and dissipation of excess pore water pressure from the liquefied zone moves upwards, developing a sufficient hydraulic gradient to induce a quick condition. This depends upon the density of the sand, its permeability, the boundary drainage conditions, the duration of cyclic loading and the nature of the deformation. There is also a possibility of a quick condition developing in a layered soil sequence where the individual beds have different permeabilities. Hydraulic conditions are particularly unfavourable where water initially flows through a very permeable horizon with little loss of head, which means that flow takes place under a great hydraulic gradient.

One of the most dramatic ways in which buildings and structures suffer damage is by the liquefaction of saturated deposits of sand brought about by earthquakes (Fig. 4.35). Liquefaction is evidenced at the ground surface by the development of boils and mud spouts, by seepage of water through cracks in the ground, and especially by the development of quick conditions. Where quick conditions occur buildings may sink into the ground or, by contrast, lightweight buried structures may rise to the surface.

Maps showing the degree of liquefaction hazard have been produced, for example, for the San Francisco Bay area, California. The zones are based on the likely response of the surface materials to seismic loading. Restriction on the location of certain types of buildings within particular zones can mean that severe damage due to liquefaction can be avoided.

There are several methods that may be employed to avoid the development of quick conditions. In the case of an excavation in sands, such as in a cofferdam, where the upward seepage force becomes more effective as the base of the excavation is lowered, one of the most effective techniques is to prolong the length of the seepage path thereby increasing the frictional losses and so reducing the seepage force. This can be accomplished by placing a clay blanket at the base of an excavation where seepage lines converge. If sheet piling is used in excavation of

Figure 4.35 Collapsed building in the Marina District of San Fransisco as a result of the Loma Prieta earthquake of October 1989. The first storey failed and the second collapsed when the ground was liquefied, leaving only the third storey above ground level.

critical soils, then the depth to which it is sunk determines whether or not quick conditions will develop. Consequently, it should be sunk deep enough to avoid a potential critical condition occurring at the base level of the excavation. The hydrostatic head also can be reduced by means of relief wells and seepage can be intercepted by a wellpoint system placed about the excavation. Furthermore, a quick condition may be prevented by increasing the downward acting force. This may be brought about by laying a load on the surface of the soil where seepage is discharging. Gravel filter beds may be used for this purpose. Suspect soils also can be densified by vibroflotation or vibroreplacement, treated with stabilizing grouts, or frozen. Priebe (1998) maintained that vibroreplacement generally was a more effective method of treatment for prevention of liquefaction due to a seismic event than other methods. The stone columns increase the shearing resistance of the soils concerned and may act as drainage channels. Their flexibility allows them to absorb seismic shock without losing their bearing capacity.

4.8.4 Soils of arid regions

A number of engineering problems arise with arid deposits. Some of these are common features of the materials themselves and others occur due to the climatic conditions. The shortage of surface water, together with the harsh conditions, inhibits biological and chemical activity so that in most arid regions there is only a sparse growth, or perhaps absence, of vegetation. This makes the surface regolith of upland areas and slopes highly vulnerable to intense denudation and redistribution by gravitational or aeolian processes or the action of ephemeral water. Weathering activity tends to be dominated by the physical breakdown of rock masses into poorly sorted assemblages of fragments ranging in size down to silts. Much of the gravel-sized material often consists of relatively weak low durability materials. Many arid areas are dominated by the presence of large masses of sand. For the most part, aeolian sands are poorly graded. In the absence of downward leaching, surface deposits become contaminated with precipitated salts, particularly sulphates and chlorides. Alluvial plain deposits often contain gypsum particles and gypsum cement, and also of fragments of weak weathered rock and clay. Many of the deposits within alluvial plains and covering hillsides are poorly consolidated. As such, they may undergo large settlements, especially if subjected to vibration due to earthquakes or cyclic loading.

Low-lying coastal zones and inland plains with shallow water tables are areas in which sabkha conditions commonly develop. These are extensive saline flats that are underlain by sand, silt or clay and often are encrusted with salt. Highly developed sabkhas tend to retain a greater proportion of soil moisture than moderately developed sabkhas. In addition, the higher the salinity of the groundwater, the greater the amount of water retained by a sabkha at a particular drying temperature (Sabtan *et al.*, 1995). The groundwater surface is held at a particular level by capillary soil moisture. The height to which water can rise from the water table is a function of the size and continuity of the pore spaces in the soil. Capillary rises of up to 3 m can occur in some fine-grained deposits. Soil above this level is subject to aeolian erosion.

Within coastal sabkhas the dominant minerals are calcite [$CaCO_3$], dolomite [$CaMg(CO_3)_2$] and gypsum [$CaSO_4.nH_2O$] with lesser amounts of anhydrite [$CaSO_4$], magnesite [$MgCO_3$], halite [NaCl] and carnalite [$KCl, MgCl_2.6H_2O$], together with various other sulphates and chlorides. For example, James and Little (1994) described the fine-grained soils of the sabkhas on the Gulf Coast of Saudi Arabia as being partially cemented with sodium chloride and calcium sulphate (i.e. gypsum), with little carbonate present. They also referred to the highly saline nature of the groundwater that, at times, contained up to 23 per cent sodium chloride, and occurred close to ground level. In such aggressively saline conditions precautions have to be taken when using concrete. James and Little recommended the use of high-density, low-water/cement-ratio concrete made from sulphate-resisting cement. The sodium chloride content of the water also is high enough to represent a corrosion hazard to steel reinforcement in concrete and to steel piles. Indeed, James and Little suggested that the thickness of steel may be reduced by half within 15 years. In such instances, reinforcement requires protective sheathing.

Minerals that are precipitated from groundwater in arid deposits also have high solution rates so that flowing groundwater may lead quickly to the development of solution features. Problems such as increased permeability, reduced density and settlement are liable to be associated with engineering works or natural processes that result in a decrease in the salt concentration of groundwaters. Changes in the state of hydration of minerals, such as swelling clays and calcium sulphate, also causes significant volume changes in soils. In particular, low density sands that are cemented with soluble salts such as sodium chloride are vulnerable to salt removal by dissolution by freshwater, leading to settlement. Hence, rainstorms and burst water mains present a hazard, as does watering of grassed areas and flower beds. The latter should be controlled and major structures should be protected by drainage measures to reduce the risks associated with rainstorms or burst water pipes. In the case of inland sabkhas, the minerals precipitated within the soil are much more variable than those of coastal sabkhas since they depend on the composition of the local groundwater. The same applies to inland drainage basins and salt playas that are subject to periodic desiccation.

Sabkha soils frequently are characterized by low strength. Furthermore, some surface clay soils that are normally consolidated or lightly overconsolidated may be sensitive to highly sensitive. The low strength is attributable to the concentrated salt solutions in sabkha brines; the severe climatic conditions under which sabkha deposits are formed (e.g. large variations in temperature and excessive wetting–drying cycles) that can give rise to instability in sabkha soils; and the ready solubility of some of minerals that act as cements in these soils. As a consequence, the bearing capacity of sabkha soils and their compressibility frequently do not meet routine design requirements. Various ground improvement techniques therefore have been used in relation to large construction projects such as vibro-replacement, dynamic compaction, compaction piles and underdrainage.

Silts that occur within alluvial plains may be interbedded with evaporite deposits and been affected by periodic desiccation. The latter process leads to the development of a stiffened crust or, where this has occurred successively, to a series of hardened layers within the formation. Many silty deposits formed under arid conditions are liable to undergo considerable volume reduction or collapse when wetted. Thus, infiltration of surface water, including that applied in the course of irrigation, leakage from pipes and rise of water table may cause large settlements to occur.

Various types of calcareous silty clay soils form in arid and sub-arid regions when clayey material is deposited in saline or lime-rich waters. These are characterized by the presence of a desiccated surface layer typically up to 2 m thick, which may be capable of supporting lightly loaded structures, although care needs to be exercised to ensure that bearing capacity failure does not occur in the underlying softer soils (Hossain and Ali, 1988).

A common feature of arid deposits is the cementation of sediments by the precipitation of mineral matter from the groundwater. The species of salt held in solution, and also those precipitated, depends on the source of the water, as well as the prevailing temperature and humidity conditions. The process may lead to the development of various crusts or cretes in which unconsolidated deposits are cemented by gypsum, calcite, silica, iron oxides or other compounds. Cretes and crusts may form continuous sheet or isolated patch-like masses at the ground surface where the groundwater table is at, or near, this level, or at some other position within the ground profile. Well-cemented crusts and cretes may provide adequate bearing capacity for structures. However, care must be taken that the underlying uncemented material is not overloaded. Also, possible changes in the engineering behaviour of the material with any changes in the water conditions must be borne in mind.

4.8.5 Tropical soils

Engineers from temperate climate countries often have experienced difficulties when dealing with certain tropical soils because it has been assumed that they will behave in a similar manner to soils in the temperate zone. Consequently, methods of testing and engineering classification schemes have been used that were not designed to cope with the different conditions found when dealing with soils formed in tropical environments (Gidigasu, 1988). Of course, it would be wrong to assume that all tropical soils are different from those found in other climatic zones. Vargas (1985), for example, pointed out that many soils, such as alluvial clays and sands or organic clays, behave in the same manner and have similar geotechnical properties regardless of the climatic conditions existing in the area of deposition.

The fabric of *in situ* residual soils, particularly of the lateritic type, involves a wide range of void ratios and pore sizes. However, the variability of void ratio does not vary systematically with soil type, parent rock, type of weathering and state of stress. It commonly is due to differential leaching removing varying quantities of material from the soil. The void ratio at a particular state of stress may be in a metastable, stable-contractive or stable-dilatant state. The strains that a residual soil with a bonded structure experiences when it yields depend on its void ratio and degree of bonding. When the yield stress is exceeded the strains undergone are determined by the soil state. A soil is subjected to large contraction if it is metastable whereas it will undergo only small strains if it is stable-dilatant. The void ratio and pore size do not change significantly with stress if the yield stress is not exceeded. Vaughan *et al.* (1988) introduced the concept of relative void ratio (e_R) for residual soils, defining it as

$$e_R = \frac{e - e_{opt}}{e_l - e_{opt}} \tag{4.51}$$

where e is the natural void ratio, e_l is the void at the liquid limit and e_{opt} is the void ratio at the optimum moisture content. They suggested relating engineering properties to relative void ratio rather than *in situ* void ratio.

Water of crystallization may be present within some minerals in many tropical residual soils, as well as free water. Some of the former type of water may be lost during conventional testing for moisture content. In order to avoid this, Anon. (1990b) recommended that comparative tests should be carried out on duplicate samples, taking the measurement of moisture content by drying to constant mass between 105 and 110°C on one sample, and at a temperature not exceeding 50°C on the other. An appreciable difference in the two values indicates the presence of structured water.

Drying brings about changes in the properties of tropical residual soils not only during sampling and testing, but also *in situ*. The latter occurs as a result of local climatic conditions, drainage and position within the soil profile. Drying initiates two important effects, namely, cementation by sesquioxides and aggregate formation on the one hand, and loss of water from hydrated clay minerals on the other. In the case of halloysite, the latter causes an irreversible transformation to metahalloysite. Some consequences of these changes are illustrated in Table 4.9. Drying can cause almost total aggregation of clay size particles into silt and sand size ranges, and a reduction or loss of plasticity. Unit weight, shrinkage, compressibility and shear strength also can be affected. Hence, classification tests should be applied to the soil with as little drying as possible, at least until it can be established from comparative tests that drying has no effect on the results.

It is conventional to classify soils on the basis of their engineering properties. For non-tropical soils this has resulted in a number of fairly well-established classification systems that are based on geotechnical index properties, principally particle size distribution and plasticity. Once a soil has been classified by means of one of these systems it then may be possible to infer some of the ways in which the soil will behave from an engineering point of view. With many tropical soils this is not the case and the use of such systems for tropical soils has been criticized by, for example, De Graft-Johnson and Bhatia (1969). For some soils the tests used are inadequate to determine the property being measured. For example, the liquid and plastic limit values obtained from 'standard' tests for some tropical red clay soils are dependent upon the precise test method employed and, in particular, the amount of energy used in mixing the soil prior to carrying out the test. Consequently, two quite different values of liquid limit can be obtained by two operators testing the same soil. Different results also will be obtained depending upon whether the soil was pre-dried prior to testing or kept close to its natural moisture content. As far as the plasticity of tropical residual soils is concerned disaggregation should be carried out with care so that individual particles are separated but not fragmented. In fact, disaggregation may have to be brought about in some cases by soaking in distilled water (Anon., 1990b). Moreover the sensitivity of the soil to working with water should be checked by using a range of mixing times before testing, in order to determine the shortest time for thorough mixing.

One of the most recent classifications of tropical residual soils was proposed by Fookes (1997). The classification is pedologically based and the main groupings are given in Table 4.10.

Table 4.9 Influence of testing procedure on the results of plastic and liquid limits of tropical residual soils from Indonesia

(a) Hand remoulding for 10 and 60 min

Soil type	*10 min mixing*				*60 min mixing*			
	Natural		*Air-dried*		*Natural*		*Air-dried*	
	PL	*LL*	*PL*	*LL*	*PL*	*LL*	*PL*	*LL*
Latosols	55	88	56	82	55	100	55	100
	46	81	47	82	45	95	48	95
Red latosols	44	81			47	85		
	44	68			44	77		
Andosols	112	152	113	130	111	140	105	120
	43	62	48	53	46	65	47	62
	171	216		149	201			

(b) Prolonged hand remoulding and remoulding by greaseworker

Soil type	*Prolonged hand remoulding*				*Remoulding by greaseworker*			
	Natural		*Air-dried*		*Natural*		*Air-dried*	
	PL	*LL*	*PL*	*LL*	*PL*	*LL*	*PL*	*LL*
Latosols	50	92	52	80	56	99	50	96
	66	89	33	87	68	133	60	132
Andosols	66	89	NP	82	62	100	68	99
	96	128	64	90	98	126	75	94
	143	220	NP	71	145	196	40	183

Source: Northmore *et al*., 1992. Reproduced by permission of The British Geological Society.

Table 4.10 Classification of tropical residual soils

Mature soils	*Duricrusts*
Vertisols	Silcrete
Fersiallitic andosols	Calcrete
Fersiallitic (sensu stricto)	Gypcrete
Ferruginous (sensu stricto)	Ferricrete
Ferrallitic	Alcrete (Alicrete)

Although duricrusts are characteristically developed in arid climates (see Section 4.8.4), ferricretes (includes laterites) and alcretes (bauxites) are common in tropical regions. The term 'laterite' has been applied to soft, clay-rich horizons showing marked iron segregation or mottling and also to gravelly materials comprised mainly of iron oxide concretions or pisoliths. McFarlane (1976) considered that these non-indurated materials formed part of a sequence of lateritic weathering ultimately resulting in the formation of an indurated surface or near-surface sheet of duricrust. Because of the presence of a hardened crust near the surface, the strength of laterite may decrease with increasing depth. The variation of shear strength with depth is influenced by the mode of formation, type of parent rock, depth of water table and its movement, degree of laterization and mineral content, as well as the amount of cement precipitated. Leaching of laterites means that the finer particles that form larger aggregates break down, which is shown by the increase in liquid limit after leaching. Moreover, removal of cement by leaching gives rise to an increase in compressibility, sometimes of more than 50 per cent.

Vertisols are characterized by the presence of clay minerals of the smectite group that typically have a high swell and shrink potential, and possess contraction cracks and slickensides. Anon. (1990b) indicated that these soils are prone to erosion and dispersion as well as swell–shrink problems. Generally, the clay fraction in these soils exceeds 50 per cent, silty material varying between 20 and 40 per cent, and the remainder being sand. Black cotton soils are probably the most common type and are highly plastic silty clays. Shrinkage and swelling of these soils is a problem in many regions that experience alternating wet and dry seasons. Ola (1980) noted an average linear shrinkage of 8 per cent in some Nigerian black cotton soils, with swelling pressure averaging around 120 kPa with a maximum of about 240 kPa. However, the volume changes frequently are confined to an upper critical zone in the soil, which frequently is less than 1.5 m thick. Below this, the moisture content remains more or less constant so that, wherever possible, foundations should extend into this region of the soil.

The 'red' soils of tropical regions include the fersiallitic, ferruginous and ferrallitic soils. Each type relates to a broad set of climatic conditions. Fersiallitic soils are found in subtropical or Mediterranean climates. Smectite is the main clay mineral but on older surfaces, well-drained sites or silica-poor parent rocks, kaolinite may be present. Young volcanic ashes produce fersiallitic andosols characterized by allophane, which alters to imogolite or halloysite on weathering. Ferruginous soils occur in either more humid (without dry season) or slightly hotter regions than Mediterranean and are more strongly weathered than fersiallitic soils. Kaolinite is the dominant clay mineral and smectite is subordinate. Ferrallitic soils develop in the hot humid tropics. All the primary minerals except quartz are weathered with much silica and bases being removed in solution. Any residual feldspar is converted to kaolinite. Gibbsite may be present. Soils can be divided into ferrites and allites depending upon whether iron oxides or aluminium oxides dominate.

With this form of classification the different groupings should be seen as part of a weathering continuum from fersiallitic soils through to ferrallitic soils. Nonetheless, this can lead to practical problems in that, for example, fersiallitic soils can be found in a climatic environment conducive to the formation of ferrallitic soils when the parent material has not been exposed to the climatic conditions for a long enough period of time. In the same way, different soils may occur within the soil profile. Further practical problems may arise when using this classification because of the difficulty in identifying the variation of climate, at a particular location, with time. In other words, the present environmental conditions are not necessarily a good indicator of the soils to be expected if climate has changed since the formation of the soil.

Obviously, the cementation between particles of residual soils has a significant influence on their shear strength, as does the widely variable nature of the void ratio and partial saturation which, as noted, can occur to appreciable depth. A bonded soil structure exhibits a peak shear strength, unrelated to density and dilation, which is destroyed by yield as large strains develop. Many tropical residual soils behave as if they are overconsolidated in that they exhibit a yield stress at which there is a discontinuity in the stress–strain behaviour and a decrease in stiffness. This is the apparent preconsolidation pressure. The degree of overconsolidation depends on the amount of weathering. The cementation of soils formed in regions with a distinct dry season can be weakened by saturation and this leads to a collapse of the soil structure. As a consequence, the apparent preconsolidation pressure decreases and the compressibility of the soil increases.

Collapsible soils have an open-textured fabric that can withstand reasonably large stresses when partly saturated, but undergo a decrease in volume due to collapse of the soil structure on wetting, even under low stresses. Many partially saturated tropical residual soils are of this nature (Vargas, 1990). Collapse and associated settlement normally is due to loss or reduction of bonding between soil particles because of the presence of water. Intensively leached residual soils formed from quartz-rich rocks tend to be prone to collapse. For example, Haskins *et al.* (1998) referred to collapse in saprolites derived from granite in Mpumalanga Province, South Africa. Other ways in which collapse is brought about include loss of the stabilizing effect of surface tension in water menisci at particle contacts in partially saturated soil and loss of strength of 'dry' clay bridges between particles.

4.8.6 Dispersive soils

Dispersion occurs in soils when the repulsive forces between clay particles exceed the attractive forces thus bringing about deflocculation so that in the presence of relatively pure water the particles repel each other (Bell and Maud, 1994). In non-dispersive soil there is a definite threshold velocity below which flowing water causes no erosion. The individual particles cling to each other and are only removed by water flowing with a certain erosive energy. By contrast, there is no threshold velocity for dispersive soil, the colloidal clay particles go into suspension even in quiet water and therefore these soils are highly susceptible to erosion and piping. Dispersive soils occur in semi-arid regions that generally have less than 850 mm of rain annually. They contain a higher content of dissolved sodium (up to 12 per cent) in their pore water than ordinary soils. A fuller account of dispersive soils is given in Chapter 6.

4.8.7 Quick clays

The material of which quick clays are composed is predominantly smaller than 0.002 mm but many deposits seem to be very poor in clay minerals, containing a high proportion of fine quartz. Cabrera and Smalley (1973) suggested that such deposits owe their distinctive properties to the predominance of short-range interparticle bonding forces, which they maintained were characteristic of deposits in which there was an abundance of glacially produced fine non-clay minerals. Certainly, quick clays have a restricted geographical distribution, occurring in certain parts of the northern hemisphere that were glaciated during Pleistocene times.

Gillot (1979) showed that quick clays possess an open fabric and high moisture content Granular particles, whether aggregations or primary minerals, are rarely in direct contact, generally being linked by bridges of particles. Clay minerals usually are non-oriented and clay coatings on primary minerals tend to be uncommon, as are cemented junctions. Networks of platelets occur in some soils. Primary minerals, particularly quartz and feldspar, form a higher than normal proportion of the clay size fraction and illite and chlorite are the dominant phyllosilicate minerals. The presence of swelling clay minerals varies from almost zero to significant amounts.

Quick clays often exhibit little plasticity, their plasticity indices at times varying between 8 and 12 per cent. Their liquidity index normally exceeds 1, and their liquid limit is often less than 40 per cent. Quick clays are usually inactive, their activity frequently being less than 0.5. The most extraordinary property possessed by quick clays is their very high sensitivity (Fig. 4.36). In other words, a large proportion of their undisturbed strength is permanently lost following shear. The small fraction of the original strength regained after remoulding quick clays may be attributable to the development of some different form of interparticle bonding.

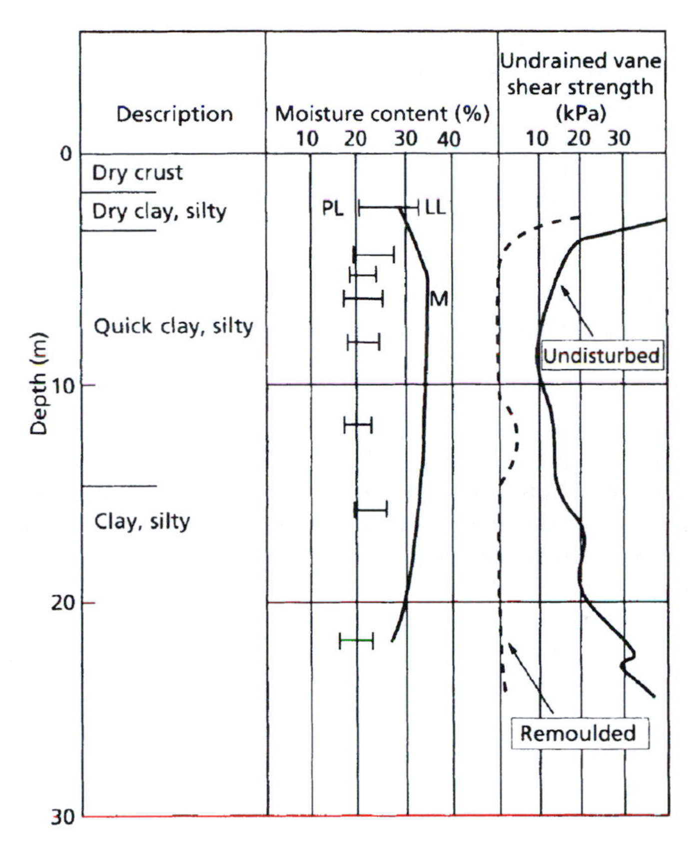

Figure 4.36 Moisture content, plastic and liquid limits, and undisturbed shear strength (indicating sensitivity) of quick clay near Trondhein, Norway.

In fact, the reason why only a small fraction of the original strength can ever be recovered is because the rate at which it develops is so slow. As an example, the Leda Clay is characterized by exceptionally high sensitivity, commonly between 20 and 50, and high natural moisture content and void ratio, the latter is commonly about 2. The natural moisture contents of quick clays always exceed that of the plastic limits and commonly exceed that of the liquid limits. In such cases the liquidity indices are greater than 1. Strength decreases and sensitivity increases dramatically as liquidity index increases. Quick clays can liquefy on sudden shock. This has been explained by the fact that if quartz particles are small enough and if the soil has a high water content, then the solid–liquid transition can be achieved. In other words, when a quick clay is subjected to rapid undrained loading beyond a certain critical shear stress, the volume decrease because of the collapse of the metastable fabric leads to an increase in pore water pressure. This can occur very rapidly resulting in an almost instantaneous loss in strength (Aas, 1965). Consequently, quick clays are associated with several serious engineering problems. Their bearing capacity is low, settlement is high and prediction of consolidation of quick clays by the standard methods is unsatisfactory.

Quick clays can be stabilized by adding salts or lime, which increase the remoulded shear strength. For instance, Locat *et al.* (1996) found that the addition of lime to quick clay in Ontario improved its index properties significantly. Previously, Bryhn and Loken (1987) had described the use of polymeric hydroxy-aluminium to stabilize quick clay (i.e. the shear strength was at least doubled and the remoulded shear strength increased from 0.1 to 12 kPa, the sensitivity being reduced from 200 to 8). Bjerrum *et al.* (1967) referred to the use of electro-osmosis to help dewater and hence stabilize a foundation in quick clay. However, they noted that excessive current density meant that the soil around the anodes dried out, thereby increasing anode-soil resistance so that the process became less effective. Subsequently, Lo *et al.* (1991) carried out tests on Leda Clay to examine its improvement when subjected to electro-osmosis. They found that the preconsolidation pressure was increased by 51–88 per cent, the undrained shear strength by up to a maximum of 172 per cent and the moisture content decreased by 30 per cent. Quick clay in the liquid state can be converted to the solid state by the addition of certain electrolytes such as sodium chloride. For example, Moum *et al.* (1968) referred to the use of sodium chloride introduced via wells to stabilize quick clays in Norway.

4.8.8 Frozen soil

Frozen ground phenomena are found in regions that experience a tundra climate, that is, in those regions where the winter temperatures rarely rise above freezing point and the summer temperatures are only warm enough to cause thawing in the upper metre or so of the soil. Beneath the upper or active zone, which is frozen in winter and thaws out in summer, the subsoil is permanently frozen and so is known as the permafrost layer. Cold, long winters with little snow; short, dry and relatively cool summers; with low precipitation in all seasons, favour the development of permafrost. Other factors which influence the distribution of permafrost include the topography, vegetation and ground cover, hydrology and snowfall, and the type of ground conditions (Harris, 1986). Permafrost is thinner or may be absent beneath rivers and lakes, which are not frozen completely in winter. This is especially the case in areas of discontinuous permafrost. Because of the permafrost summer meltwater cannot seep into the ground, so the active zone then becomes waterlogged. Layers or lenses of unfrozen ground termed taliks may occur, often temporarily, in the permafrost. Permafrost is an important characteristic, although it is not essential to the definition of periglacial conditions. It covers 20 per cent of the earth's land surface and during Pleistocene times it was developed over an even larger area. The temperature of perennially frozen ground below the depth of seasonal change ranges from slightly less than 0 °C to –12 °C. Generally, the depth of thaw is less, the higher the latitude. It is at a minimum in peat or highly organic sediments and increases in clay, silt and sand to a maximum in gravel where it may extend to 2 m in depth.

Frost action in a soil, of course, is not restricted to tundra regions. Its occurrence is influenced by the initial temperature of the soil, as well as the air temperature, the intensity and duration of the freeze period, the depth of frost penetration, the depth of the water table, and the type of ground and exposure cover. If frost penetrates down to the capillary fringe in fine-grained soils, especially silts, then, under certain conditions, lenses of ice may be developed. The formation of such ice lenses may, in turn, cause frost heave and frost boil (Casagrande, 1932). In addition, fossil frozen ground phenomena occur in regions that during Pleistocene times experienced tundra conditions but now have warmer climatic regimes.

Ice may occur in frozen soil as small disseminated crystals whose total mass exceeds that of the mineral grains, as large tabular masses that range up to several metres thick, or as ice wedges. The latter may be several metres wide and may extend to 10 m or so in depth. As a consequence, frozen soils need to be described and classified for engineering purposes. A method of classifying frozen soils involves the identification of the soil type and the character of the ice (Andersland and Anderson, 1978). First, the character of the actual soil is classified according to the Unified Soil Classification System. Second, the soil characteristics consequent upon freezing are added to the description. Frozen soil characteristics are divided into two basic groups based on whether or not segregated ice can be seen with the naked eye (Table 4.11). Third, the ice present in the frozen soil is classified, this refers to inclusions of ice that exceed 25 mm in thickness.

The mechanical properties of frozen soil are very much influenced by the grain size distribution, the mineral content, the density, the frozen and unfrozen water contents, and the presence of ice lenses and layering. The strength of frozen ground develops from cohesion, interparticle friction and particle interlocking, much the same as in unfrozen soils. However, cohesive forces include the adhesion between soil particles and ice in the voids, as well as the surface forces between particles. The relative density influences the behaviour of frozen coarse-grained soils, especially their shearing resistance, in a manner similar to that when they are unfrozen. The cohesive effects of the ice matrix are superimposed on the latter behaviour and the initial deformation of frozen sand is dominated by the ice matrix. Sand in which all the water is more or less frozen exhibits a brittle type of failure at low strains, for example, at around 2 per cent strain. However, the presence of unfrozen films of water around particles of soil not only means that the ice content is reduced, but leads to a more plastic behaviour of the soil during deformation. For instance, frozen clay soil, as well as often containing a lower content of ice than sand, has layers of unfrozen water (of molecular proportions) around the clay particles. Under very rapid loading the ice behaves as a brittle material, with strengths in excess of those characteristic of

Table 4.11 Description and classification of frozen soils

		Classify soil phase by the Unified Soil Classification System			
		Major group		*Subgroup*	
		Description	*Designation*	*Description*	*Designation*
1	Description of soil phase (independent of frozen state)	Segregated ice not visible by eye	N	Poorly bonded or friable	Nf
				Well bonded No excess ice	Nb n
				Well bonded Excess ice	Nb e
2	Description of frozen soil	Segregated ice visible by eye (ice 25 mm or less thick)	V	Individual ice crystals or inclusions	Vx
				Ice coatings on particles	Ve
				Random or irregularly oriented ice formations	Vr
				Stratified or distinctly oriented ice formations	Vs
3	Description of substantial ice strata	Ice greater than 25 mm thick	ICE	Ice with soil inclusions	ICE + soil type
				Ice without soil inclusions	ICE

Source: Andersland and Anderson, 1978, Reproduced by kind permission of the McGraw-Hill Companies.

fine-grained frozen soils. By contrast, the ice matrix deforms continuously when subjected to long-term loading, with no limiting long-term strength.

When loaded, stresses at the point of contact between soil particles and ice bring about pressure melting of the ice. Because of differences in the surface tension of the melt-water, it tends to move into regions of lower stress, where it refreezes. The process of ice melting and the movement of unfrozen water are accompanied by a breakdown of the ice and the bonding with the grains of soil. This leads to plastic deformation of the ice in the voids and to a rearrangement of particle fabric. The net result is time-dependent deformation of the frozen soil, namely, creep. Frozen soil undergoes appreciable deformation under sustained loading, the magnitude and rate of creep being governed by the composition of the soil, especially the amount of ice present, the temperature, the stress and the stress history. The creep strength of frozen soils is defined as the stress level, after a given time, at which rupture, instability leading to rupture or extremely large deformations without rupture occur. Frozen fine-grained soils can suffer extremely large deformations without rupturing at temperatures near to freezing point.

Because frozen ground is more or less impermeable this increases the problems due to thaw by impeding the removal of surface water. What is more, when thaw occurs the amount of water liberated may greatly exceed that originally present in the melted out layer of the soil. As the soil thaws downwards the upper layers become saturated and, since water cannot drain through the frozen soil beneath, it may suffer a complete loss of strength. Indeed, under some circumstances excess water may act as a transporting agent, thereby giving rise to soil flows that can move on gentle slopes. This movement downslope as a viscous flow of saturated debris is referred to as solifluction. It is probably the most significant process of mass wastage in tundra regions. Solifluction deposits commonly consist of gravels, which are characteristically poorly sorted, sometimes gap-graded, and poorly bedded. These gravels consist of fresh, poorly worn, locally derived material. Individual deposits are rarely more than 3 m thick and frequently display flow structures. Sheets, lobes and solifluction debris, transported by mudflow activity, are commonly found at the foot of slopes. These materials may be reactivated by changes in drainage, by stream erosion, by sediment overloading or during construction operations. Solifluction sheets may be underlain by slip surfaces, the residual strength of which controls their stability.

Settlement is associated with thawing of frozen ground. As ice melts, settlement occurs, water being squeezed from the ground by overburden pressure or by any applied loads. Excess pore pressures develop when the rate of ice melt is greater than the discharge capacity of the soil. Since excess pore water pressures can lead to the failure of slopes and foundations, both the rate and amount of thaw settlement should be determined. Pore water pressures also should be monitored.

Further consolidation, due to drainage, may occur on thawing. If the soil was previously in a relatively dense state, then the amount of consolidation is small. This situation only occurs in coarse-grained frozen soils containing very little segregated ice. On the other hand, some degree of segregation of ice is always present in fine-grained frozen soils. For example, lenses and veins of ice may be formed when silts have access to capillary water. Under such conditions the moisture content of the frozen silts significantly exceeds the moisture content present in their unfrozen state. As a result, when such ice-rich soils thaw under drained conditions they undergo large settlements under their own weight.

Shrinkage, which gives rise to polygonal cracking in the ground, presents another problem when soil is subjected to freezing. The formation of these cracks is attributable to thermal contraction and desiccation. Water that accumulates in the cracks is frozen and consequently helps increase their size. This water also may aid the development of lenses of ice. Individual cracks may be 1.2 m wide at their top, may penetrate to depths of 10 m and may be up to 12 m apart. They form when, because of exceptionally low temperatures, shrinkage of the ground occurs. When the ice in an ice wedge disappears an ice wedge pseudomorph is formed by sediment, frequently sand, filling the crack. In addition, frozen soils may undergo notable disturbance as a result of mutual interference of growing bodies of ice or from excess pore water pressures developed in confined water bearing lenses. Involutions are plugs, pockets or tongues of highly disturbed material, generally possessing inferior geotechnical properties, which have been intruded into overlying layers. They are formed as a result of hydrostatic uplift in water trapped under a refreezing surface layer and usually are confined to the active layer. Fossil ice wedges and involutions usually mean that one material suddenly replaces another. This can cause problems in shallow excavations.

There are two methods of construction in permafrost, namely, passive and active. In the former, the frozen ground is not disturbed, and heat from a structure is prevented from thawing the ground below, thereby reducing its stability. The prevention of heat flow to permafrost can be accomplished by providing an air space beneath a structure. Placing an insulating layer, such as a gravel blanket, between a structure and the frozen soil delays, but does not stop, thawing. By contrast, the ground is thawed prior to construction in the active method. It is either kept thawed or removed and replaced by materials not affected by frost action. The latter method is used where permafrost is thin, sporadic or discontinuous, and where thawed ground has an acceptable bearing capacity. On the other hand, if permafrost is well developed, the removal of frozen ground usually proves impracticable, and hence the passive method is employed.

Certain conditions are preferable for the location of buildings, such as a thin active layer with bedrock near the surface, good drainage, non-frost susceptible soil, soil which, when it thaws, has an adequate bearing capacity, and areas not liable to solifluction (Harris, 1986). Foundations frequently are taken through the active layer into the permafrost beneath. Hence, piles often are used as foundations (Andersland and Ladanyi, 1994). The refreezing time for piles in permafrost depends upon the time of emplacement, the ground temperature and the soil moisture. For example, it may take 2–3 months for piles placed in early spring to refreeze and therefore to be loaded, whereas those sunk in autumn may take 6 months. Adfreezing of the ground in the active layer against the piles must be prevented. This can be achieved by insulating, collaring or lubricating the piles in the active layer. In this way, the upward-acting force during freezing does not affect the piles. The piles usually extend to a depth of at least twice the active layer into the permafrost. Floors commonly are raised above the ground to allow the circulation of air.

Frost heave can take place in any climatic regime that experiences freezing weather (see Chapter 5). Where there is a likelihood of frost heave occurring it is necessary to estimate the depth of frost penetration. Once this has been done, provision can be made for the installation of adequate insulation or drainage within the soil and to determine the amount by which the water table may need to be lowered so that it is not affected by frost penetration. The base of footings should be placed below the estimated depth of frost penetration, as should water supply lines and other services. Frost susceptible soils may be replaced by gravels. The addition of certain chemicals to soil can reduce its capacity for water absorption and so can influence frost susceptibility. For example, Croney and Jacobs (1967) noted that the addition of calcium lignosulphate and sodium tripolyphosphate to silty soils were both effective in reducing frost heave. The freezing point of the soil may be lowered by mixing with solutions of calcium chloride or sodium chloride, in concentrations of 0.5–3.0 per cent by weight of the soil mixture. The heave of coarse-grained soils containing appreciable quantities of fines can be reduced or prevented by the addition of cement or bituminous binders. Addition of cement both reduces the permeability of a soil mass and gives it sufficient tensile strength to prevent the formation of small ice lenses as the freezing isotherm passes through.

4.8.9 Glacial deposits

Glacial deposits overlie much of those lowland regions that were covered by ice during the Pleistocene epoch. Two kinds of glacial deposits are distinguished, namely, unstratified drift or till and stratified drift. However, one type commonly grades into the other. Till usually is regarded as being synonymous with boulder clay and was deposited directly by ice while stratified drift was deposited by meltwaters issuing from ice. Generally, these deposits do not present serious problems in geotechnical engineering but at times they can change rapidly and this may present problems, as for example, when clayey tills contain lenses of sand.

The nature of a till deposit depends on the lithology of the material from which it was derived, on the position in which it was transported in the glacier, and on the mode of deposition (Bell, 2000). Till sheets can comprise one or more layers of different material, not all of which are likely to be found at any one locality. Shrinking and reconstituting of an ice sheet can complicate the sequence.

Deposits of till commonly consist of a variable assortment of rock debris ranging from fine rock flour to boulders (Bell, 2002). They are characteristically unsorted. Some tills may consist essentially of sand and gravel with very little binder. Lenses and pockets of sand, gravel and highly plastic slickensided clay frequently are encountered in tills. Some of the masses of sand and gravel are interconnected, due to the action of meltwater, but many are isolated. The compactness of a till varies according to the degree of consolidation undergone, the amount of cementation and size of the grains. Tills that contain less than 10 per cent clay fraction usually are friable while those with over 10 per cent clay tend to be massive and compact.

Distinction has been made between tills derived from rock debris carried along at the base of a glacier and those deposits that were transported within or at the terminus of the ice. The former is referred to as lodgement till while the latter is termed ablation till. Lodgement till is commonly stiff, dense and relatively incompressible. Fissures frequently are present in lodgement till, especially if it is clay matrix dominated. Sub-horizontal fissures have been developed as a result of incremental loading and periodic unloading while sub-vertical fissures owe their formation to the overriding effects of ice and stress relief. Ablation till accumulates on the surface of the ice when englacial debris melts. It therefore is normally consolidated and non-fissile. It is characterized by abundant large rock fragments. The proportion of sand and gravel is high and clay is present only in small amounts (usually less than 10 per cent). The loose packing means that ablation tills often have a low *in situ* density.

Stratified deposits can be subdivided into ice-contact deposits and pro-glacial deposits. Outwash deposits, kames and eskers are examples of the former type of deposits. They usually consist of sands and gravels. The most familiar pro-glacial deposits are varved clays. These sediments accumulated on the floors of glacial lakes and are characteristically composed of alternating laminae of finer and coarser grain size. Each such couplet has been termed a varve. The thickness of the individual varve frequently is less than 2 mm, although much thicker couplets can occur. Generally, the coarser layer is of silt size and the finer of clay size. The plasticity of varved clays ranges from low to very high plasticity, the wide spread reflecting the different proportions of clay and silt present. Metcalf and Townsend (1961) maintained that the two discrete layers of a varve may invalidate normal soil mechanics analyses, based on homogeneous soils and that, in particular, the assessment of the liquid and plastic limits of a bulk sample may not yield a representative result. They assumed that the maximum and minimum values recorded for any one deposit approximate to the properties of the individual layers. However, Bell and Coulthard (1997) suggested that the clay fraction of varved clay has a greater influence on the consistency limits than the silty fraction and that the results obtained from bulk testing do not necessarily present a problem. Varved clays tend to be medium sensitive to sensitive and so may undergo notable reductions in strength when remoulded. They usually are normally consolidated or lightly overconsolidated, and may possess a potential for swelling. Furthermore, the varves are responsible for the anisotropic strength and consolidation behaviour of these clays.

4.8.10 Peat

Peat represents an accumulation of partially decomposed and disintegrated plant remains that has been preserved under conditions of incomplete aeration and high water content (Hobbs, 1986). Surface deposits of peat or blanket bogs found on cool wet uplands where slopes are not excessive and drainage is impeded, have accumulated since the last ice age whilst some buried peats may have been developed during inter-glacial periods. Peats also have accumulated in post-glacial lakes and marshes where they are interbedded with silts and muds. Similarly, they may be associated with salt marshes. Fen deposits are thought to have developed in relation to the eustatic changes in sea level that occurred after the retreat of the last ice sheets. These are areas where layers of peat interdigitate with wedges of estuarine silt and clay.

The void ratio of peat ranges between 9, for dense amorphous granular peat, up to 25, for fibrous types with high contents of sphagnum moss. It usually tends to decrease with depth within a peat deposit. Such high void ratios give rise to phenomenally high water content. The water content of peats varies from a few hundreds per cent dry weight (e.g. 500 per cent in some amorphous granular peats) to over 3000 per cent in some coarse fibrous varieties. Consequently, peat undergoes significant shrinkage on drying out, it ranging between 10 and 75 per cent of the original volume. The pH value of fen peat frequently is in excess of 5, whilst that of bog peat is usually less than 4.5 and may be less than 3. The bulk density of peat is low and variable, being related to the organic content, mineral content, water content and degree of saturation. Gas is formed in peat as plant material decays and has a significant influence on initial consolidation, rate of consolidation, pore pressure under load and permeability.

Differential and excessive settlements are the principal problems confronting the engineer working on peat soils. When a load is applied to peat, settlement occurs because of the low lateral resistance offered by the adjacent unloaded peat. Serious shearing stresses are induced even by moderate loads. Worse still, should the loads exceed a given minimum, then settlement may be accompanied by creep, lateral spread, or in extreme cases by rotational slip and upheaval of adjacent ground. At any given time the total settlement in peat due to loading involves settlement with and without volume change. Settlement without volume change is the more serious for it can give rise to the types of failure mentioned. What is more, it does not enhance the strength of peat.

4.8.11 Fills

A wide variety of materials is used for fills including domestic refuse, ashes, slag, clinker, building waste, chemical waste, quarry waste and all types of soils. The extent to which an existing fill will be suitable as a foundation depends largely on its composition and uniformity. Of particular importance is the time required for a fill to reach a sufficient degree of natural consolidation to make it suitable for a foundation. This depends on the nature and thickness of the fill, the method of placing, and the nature of the underlying ground, especially the groundwater conditions. The best materials in this respect are obviously well graded, hard and granular. Furthermore, properly compacted fills on a sound foundation can be as good as, or better than, virgin soil. Fills containing a large proportion of fine material, by contrast, may take a long while to settle. Similarly, old fills and those placed over low-lying areas of compressible or weak strata should be considered unsuitable unless tests demonstrate otherwise or the structure can be designed for low-bearing capacity and irregular settlement. Frequently, poorly compacted old fills continue to settle for years due to secondary consolidation. Charles and Skinner (2001) maintained that the susceptibility to collapse compression consequent upon inundation often represents a notable hazard when construction is to take place on fill. Most types of partially saturated fill that were placed in a sufficiently loose and/or dry state are susceptible to collapse compression over a wide range of applied loadings when first inundated. Inundation can result from downward percolation of water infiltrated from the surface or can be due to rising groundwater levels, for example, where old quarries or opencast coal workings have been filled (Charles *et al.*, 1993). Such collapse compression can cause serious damage to buildings. Mixed fills that contain materials liable to decay, which may leave voids or involve a risk of spontaneous combustion, afford very variable support and such sites generally should be avoided. Sanitary landfills, in particular, suffer from continuing organic decomposition and physico-chemical breakdown. Methane and hydrogen sulphide are produced in the process and accumulations of these gases in pockets in fills have led to explosions. Settlements are likely to be large and irregular. Some materials such as ashes and industrial wastes may contain sulphate and other products that are potentially injurious as far as concrete is concerned.

Where urban renewal schemes are undertaken it may be necessary to construct buildings on areas covered by fill. In most cases such fills have not been compacted to any appreciable extent and large voids may be present where the rubble has collapsed into old cellars. However, demolition rubble fill is usually comparatively shallow and the most economical method of constructing foundations is either to cut a trench through the fill and backfill it with lean concrete or to clear all the fill beneath the structure and replace it with compacted layers. Deep vibration techniques or dynamic compaction may prove economical in areas where old cellars exist.

One way of avoiding significant settlement on domestic refuse fills is to use piles or to preload the foundation area with embankments of sand or rock. The preloaded fills must remain in position until settlement of the ground surface has ceased or slowed down to an acceptable degree.

4.8.12 The influence of weathering and fissures on clay soils

The greatest variation in the engineering properties of clay soils can be attributed to the degree of weathering that they have undergone. Generally, changes in the clay mineral and quartz content of clay deposits on weathering are slight. However, the degree of illite degradation increases in weathered clay soils and those with the most degraded illite appear to have the lowest strength. In addition, consolidation of a clay deposit gives rise to an anisotropic texture due to the rotation of the platey minerals and diagenetic processes bond particles together. Weathering reverses these processes, altering the anisotropic structure and destroying or weakening interparticle bonds. Therefore, weathering, through the destruction of interparticle bonds, leads to a clay deposit reverting to a normally consolidated sensibly remoulded condition. Higher moisture contents are found in more weathered clay. This progressive degrading and softening also is accompanied by reductions in strength and deformation moduli with a general increase in plasticity. For example, Cripps and Taylor (1981) indicated that the undrained shear strength of the London Clay is reduced by approximately half on weathering. The removal of overburden leads to vertical expansion of a deposit, which facilitates the development of joints and fissures, together with softening. The opening of fissures is accompanied by water entrainment and chemical degradation. Different grades of weathering normally can be recognized in clay deposits. For example, Chandler (1972) recognized four zones of weathering in the Upper Lias Clay in Northamptonshire, England. Subsequently, similar zones of weathering were noted by Russell and Parker (1979) in the Oxford Clay and by Coulthard and Bell (1993) in the Lower Lias Clay.

Fissures play an extremely important role in the failure mechanism of fissured clay soil. Indeed, many clay soils are seriously weakened by the presence of a network of fissures. Terzaghi (1936) provided the first quantitative data relating to the influence of fissures and joints on the strength of clay soils, pointing out that they are characteristic of overconsolidated clays. He maintained that fissures in normally consolidated clays have no significant practical consequences. On the other hand, fissures can have a decisive influence on the engineering performance of an overconsolidated clay, in that the overall strength of such fissured clay can be as low as one-tenth that of the intact clay. In addition to allowing clay soil to soften, fissures and joints allow concentrations of shear stress to exist that locally exceed the peak strength of the soil, thereby giving rise to progressive failure. Under stress, the fissures in clay soil seem to propagate and coalesce in a complex manner. Skempton and La Rochelle (1965) showed that the strength along joints or fissures in clay soil normally is only slightly higher than the residual strength of the intact clay. As intact clay soil has a low tensile strength, there is no resistance to opening of fissures and once open there is no shear resistance along them. Skempton and La Rochelle therefore recommended that, if regular fissure patterns with an unfavourable orientation occur at a site, an attempt should be made to estimate the influence of these fissures on the overall strength of the clay mass.

4.9 Ground treatment

Ground treatment techniques can be used to improve poor ground conditions. Poor ground conditions usually are attributable to a lack of strength, and associated deformability, and/or an excess of groundwater. Such ground conditions could be responsible large total settlements or differential settlements and the latter, in particular, could damage a building if the conditions are not improved. Preloading a site by placing fill is one means of reducing post-construction settlement but can be a time-consuming process. The time period to achieve the required amount of settlement prior to construction can be shorted by using sand drains, sandwicks or band drains beneath the fill. Compaction by preloading and drainage is discussed in Chapter 5.

4.9.1 Grouting

Grouting is widely used in foundation engineering in order to reduce seepage of water or to increase the mechanical performance of the soils or rocks concerned. Grouts may be grouped into two basic types, namely, suspension or particulate grouts, and chemical or non-particulate grouts. Particulate grouts consist of cement-water, clay-water or cement-clay-water mixtures. Other materials may be added such as sand, gravel or pulverized fuel ash to provide bulk when large voids have to be filled. The most common classes of chemical grouts are silicates, lignins, resins, acrylamides and urethanes. The silicates are the most widely used chemical grouts, use of the others being limited by cost and toxicity.

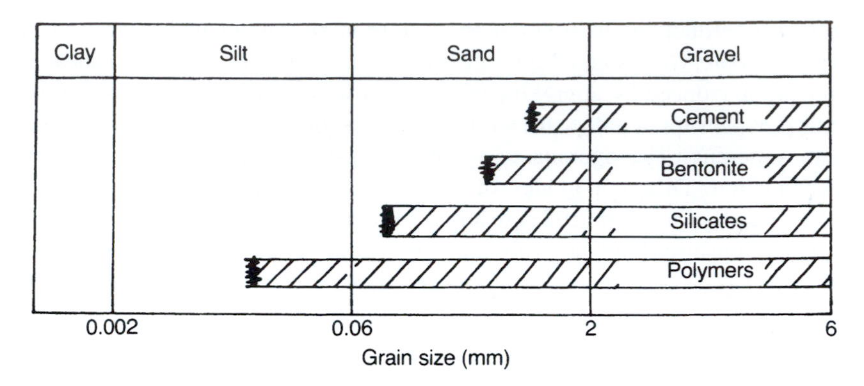

Figure 4.37 Soil particle size limitations on grout permeation.
Source: Mitchell, 1981.

Permeation grouting involves the injection of grout into pores in soils and discontinuities in rock masses. At shallow depths permeation grouting may take place at a single stage from a grout pipe, the grouthole being sunk to full depth and then grouted upwards. Alternatively, grouting may proceed while the hole is drilled. The hole is extended a short distance using a hollow drill rod. It is then withdrawn this distance and grout is injected from the rod. Stage grouting is used when relatively high grouting pressures are required to achieve satisfactory penetration of grout. In this case the holes are drilled to a given depth and then grouted. After the grout has set, the holes are deepened for the next stage of grouting, when the procedure is repeated. Grout also may be used to displace soils as in compaction grouting or to replace soils as in jet grouting.

The groutability of soils and therefore the choice of grout are influenced by their pore size, which is approximately related to particle grading (Fig. 4.37). The limits for suspension grouts generally are regarded as a 10:1 size factor between the D_{15} of the grout and the D_{15} size of the granular system to be injected. Grout penetration also has been assessed in terms of the following groutability ratio:

$$N = \frac{D_{10}\text{soil}}{D_{85}\text{grout}} \tag{4.52}$$

This ratio should not exceed 25, or be less than 11, if grout is to penetrate soil successfully. Generally, particulate grouts are limited to soils with pore dimensions greater than 0.2 mm. Ordinary cement grout cannot enter a fissure of less than 0.1 mm width. In fissured rocks the D_{85} of the grout must be smaller than one-third the width of the discontinuity. There is an upper limit to this ratio, as large quantities of grout have been lost from sites via open fissures. Cavities in rock masses may be grouted in a similar fashion to filling voids due to past mining operations. In order that a grout can achieve effective adhesion, the sides of the fissure or voids must be clean. If they are coated with clay, then they need to be washed prior to grouting. Soils containing less than 10 per cent fines usually can be permeated by chemical grouts. If the fines content exceeds 15 per cent, then effective chemical grouting may prove difficult and it is not possible where the fines content exceeds 20 per cent.

4.9.2 Vibrocompaction

Vibroflotation is a form of vibrocompaction that is used to improve sandy soils by densification. A poker vibrator or vibroflot is used to penetrate the soil and is suspended from a crawler crane.

The technique is best suited to very loose sands below the water table, although maximum densities can be achieved in clean free-draining sands either under dry (water can be jetted from the vibroflot) or saturated conditions. In the latter case, a quick condition is developed that reduces the shearing resistance of the sand, thereby facilitating the movement of particles into a denser state of packing. Liquefaction occurs within the immediate vicinity of the vibroflot (i.e. up to a radius of about 500 mm). The presence of clay layers, excessive amounts of fines, grain cementation or organic matter can cause difficulties for vibroflotation. Silt, clay and organic material damp the vibrations created by the vibroflot and so mitigate against densification. The treatment may reduce settlement in sands by more than 50 per cent and the shearing strength of treated sands is increased substantially. Granular soils of low density also can be compacted by driving and extracting a large open-ended pipe, the terra-probe, using a vibratory pile driver.

A more common type of vibrocompaction technique, which can be applied to most types of soil, involves forming columns of coarse backfill in the soil (Fig. 4.38). The vibroflot is used for compacting these columns, which brings about a reduction in settlement. Since the granular backfill replaces the soil this process is sometimes referred to as vibroreplacement. Vibroreplacement is used in soft normally consolidated compressible clays, saturated silts, alluvial and estuarine soils, and fills. Stone columns have been formed successfully in soils with undrained cohesive strengths as low as 7 kPa. Concrete columns, normally 450 mm in diameter and compacted by a vibrator, can be used to support structures on weak clay, organic or loose granular soils and fills as an alternative to vibroreplacement. Vibrodisplacement involves the vibroflot penetrating the ground by shearing and displacing the ground around it. It accordingly is restricted to strengthening insensitive clay soils that have sufficient cohesion to maintain a stable borehole, that is, to those with over 20 kPa undrained strength. These soils require treatment primarily to boost their bearing capacity, the displacement method inducing some measurable increase in the strength of the clay soil between the columns.

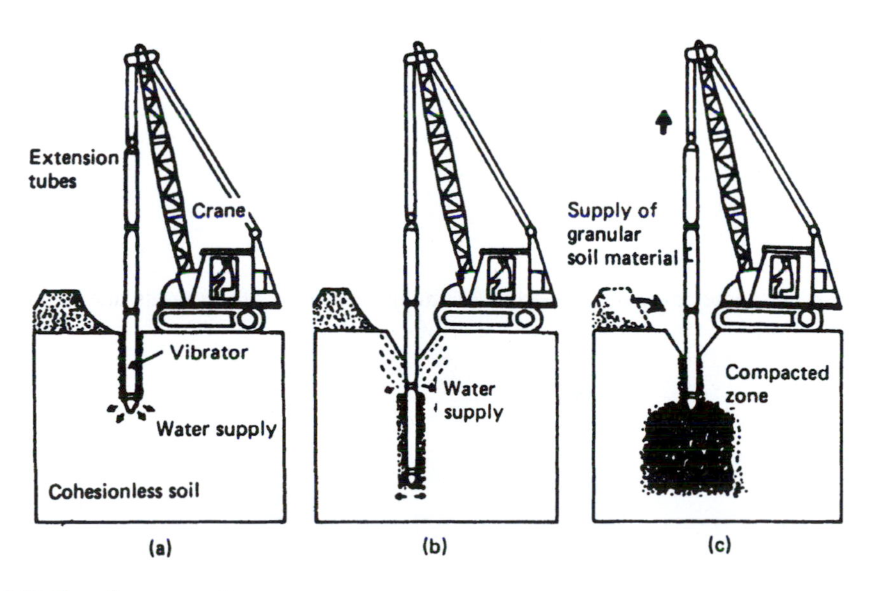

Figure 4.38 The vibrocompaction process.

4.9.3 Compaction piles

Soil can be compacted by driving piles into it, compaction being brought about by displacement. When compaction piles are installed in granular soils, compaction is achieved by both displacement and the vibratory effect of driving the piles. Sand piles generally are emplaced when the sole purpose is to improve the density of granular soils. They are installed by driving a hollow steel mandrel with a false bottom to the required depth, filling the mandrel with sand, applying air pressure to the top of the sand column, and withdrawing the mandrel. A well-compacted layer of soil usually is placed on top of the sand piles to obtain a better load concentration on them. The vibro-composer method is similar and involves driving a casing to the required depth by means of a vibrator. Then a quantity of sand is placed in the casing, which is partially withdrawn as compressed air is blown down the casing. Next, the casing is vibrated to compact the sand. The process is repeated until the ground surface is reached. The formation of high-displacement piles involves boring a hole, 0.5 m in diameter, again to the required depth, by bailing from within a casing. Sand then is rammed from the bottom of the casing by a drop-hammer. The casing is withdrawn in short lifts and a charge of sand is rammed from it at each lift. The Franki technique of placing gravel piles is similar (Bell, 1993).

4.9.4 Dynamic compaction

Dynamic compaction brings about an improvement in the mechanical properties of a soil by the repeated application of very high intensity impacts to the surface. This is achieved by dropping a large weight, typically 10–20 tonnes, from crawler cranes, from heights of 15 to 40 m, at regular intervals across the surface (Fig. 4.39). Repeated passes are made over a site, although several tampings may be made at each imprint during a pass. The first pass at widely spaced centres improves the bottom layer of the treatment zone and subsequent passes then consolidate the upper layers. In finer materials the increased pore water pressures must be allowed to dissipate between passes, which may take several weeks.

Care must be taken in establishing the treatment pattern, tamping energies and the number of passes for a particular site. This should be accompanied by *in situ* testing as the work

Figure 4.39 Dynamic compaction.

proceeds. Coarse granular fill requires more energy to overcome the possibility of bridging action, for similar depths, than finer material. Before subjecting sites that have previously been built over to dynamic compaction underground services, cellars, etc should be located. Old foundations should be demolished to about a metre depth below the proposed new foundation level prior to compaction. Dynamic compaction has proved a particularly satisfactory technique for improving the engineering performance of a wide range of soils from organic and silty clays to loosely packed granular soils and fills, including coarse mine discard.

4.9.5 Lime and cement columns

Lime or cement columns can be used to enhance the carrying capacity and reduce the settlement of sensitive soils. Indeed, the lime column method often can be used economically when the maximum bearing capacity of conventional piles cannot be fully mobilized (Broms, 1991). In such instances, they can be used to support a thin floor slab that carries lightweight buildings. Lime or cement columns are installed by a tool reminiscent of a giant eggbeater, the tool being screwed into the ground to the required depth, then the rotation is reversed and lime or cement slurry is forced into the soil by compressed air from openings just above the blades of the tool. When the depth to a firm base is less than 15 m, then the columns should penetrate the total thickness to ensure that settlement is minimized. The bearing capacity of a column 0.5 m in diameter usually varies between 50 and 500 kN, depending on the type and amount of lime or cement used. The strength and rate of increase in strength are influenced by the curing conditions. Lime or cement columns also facilitate drainage of soil. In fact, Broms and Boman (1977) maintained that the drainage effect of one column was equivalent to that of two or three 100 mm wide band drains or three sand drains 150 mm in diameter.

4.10 Types of foundation structure

The design of foundations embodies three essential operations, namely, calculating the loads to be transmitted by the foundation structure to the soils or rocks supporting it, determining the engineering performance of these soils or rocks, and then designing a suitable foundation structure.

4.10.1 Footings

Footings distribute the load of a structure to the subsoil over an area sufficient to suit the properties of the ground (Hanna, 1987). Their size therefore is governed by the strength of the foundation materials. If a footing supports a single column it is known as a spread or pad footing whereas a footing beneath a wall is referred to as a strip or continuous footing. Spread footings usually provide the most economical type of foundation structure but the allowable bearing pressures must be chosen to provide an adequate factor of safety against shear failure in a soil and to ensure that settlements are not excessive. Settlement for any given pressure increases with the width of footing in almost direct proportion on clay soils and to a lesser degree on sands. There is a tendency for uniformly loaded buildings to settle more at the centre than at the edges. In order to reduce differential settlements, it is advisable to use rather larger pressure under the smaller footings than under the larger ones and, if practicable, to use a larger pressure under the edge footings than under those in the centre of a building.

4.10.2 Rafts

A raft permits the construction of a satisfactory foundation in materials whose strength is too low for the use of spread footings (Tomlinson and Boorman, 1996). The chief function of a raft is to spread the building load over as great an area of ground as possible and thus reduce the bearing pressure to a minimum. In addition, a raft provides a degree of rigidity that reduces differential movements in the superstructure. The settlement of a raft foundation does not depend on the weight of the building it supports. Rather settlement depends on the difference between this weight and the weight of the soil that is removed prior to the construction of the raft, provided the heave produced by the excavation is inconsequential. A raft can be built at a sufficient depth so that the weight of soil removed equals the weight of the building (Fig. 4.40). Such rafts are referred to as buoyancy, compensated, floating or semi-floating foundations (Zeevaert, 1987). The success of this type of foundation structure in overcoming difficult soil conditions has led to the use of deep raft and rigid frame basements for high buildings on clay soils.

4.10.3 Piles

When the soil immediately beneath a proposed structure is too weak or too compressible to provide adequate support, the structural load can be transferred to more suitable material at greater depth by means of piles (Fig. 4.41). Such end-bearing piles must be capable of sustaining the load with an adequate factor of safety, without allowing settlement detrimental to the

Figure 4.40 A buoyancy raft under construction.

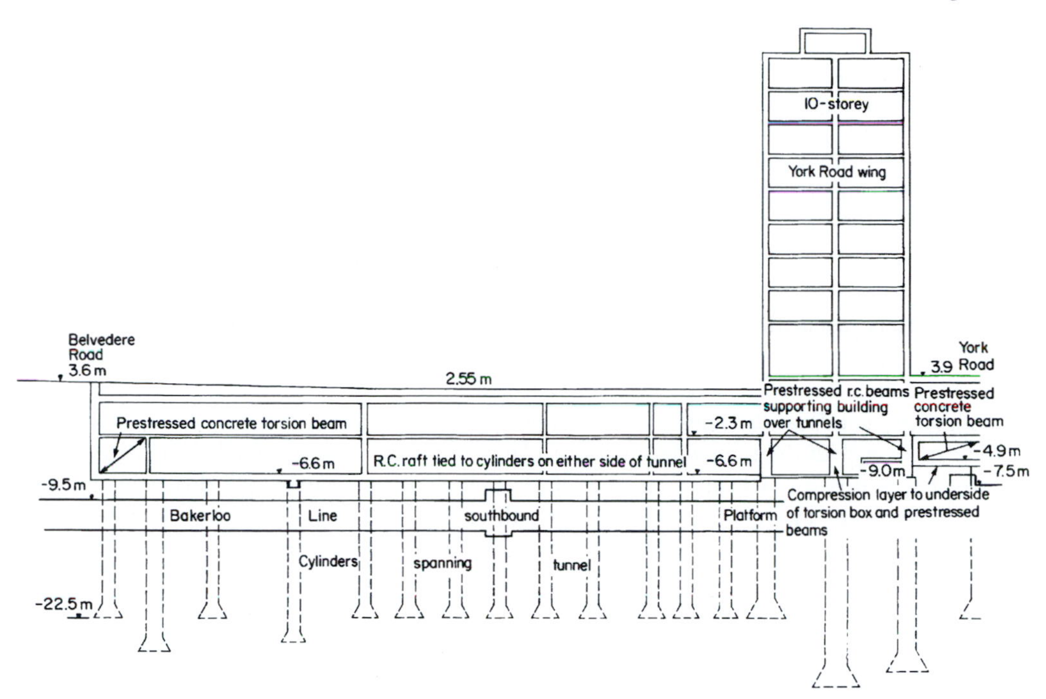

Figure 4.41 Cross-section of the shell centre, London, showing the position of the Bakerloo line underground tunnels in relation to the building.
Source: Measor and Williams, 1962. Reproduced by kind permission of Thomas Telfard.

structure to occur. Piles, however, do not derive their carrying capacity only from end-bearing at their bases, as friction along the pile shaft also contributes towards this end (Fellenius, 1991). Indeed, friction is likely to be predominant when piles are founded in clays and silts whilst end-bearing provides the most of the carrying capacity for piles terminating in or on gravel, hard clay or rock. Accordingly, the ultimate bearing capacity of a pile is equal to the sum of the base resistance and the shaft resistance (Poulos, 1987). The former is the product of the ultimate bearing capacity at the base of the pile (q_b) and the area of the base (A_b). The shaft resistance is the product of the perimeter area (A_s) and the skin friction (f_s) between the pile and the soil. Therefore, the ultimate load (Q) that can be applied to the top of a pile is:

$$Q = A_b q_b + A_s f_s \tag{4.53}$$

An appropriate factor of safety is applied to the ultimate load to obtain the allowable load on the pile. However, the bearing capacity of piles frequently is derived from the results of the standard penetration test, the cone penetrometer test or, more specifically, pile loading tests. Meyerhof (1976) proposed a number of relationships that use the results of the two former tests to determine the load carrying capacity of piles. These are summarized in most texts on soil mechanics such as Craig (1987) and on foundations such as Tomlinson and Boorman (1996). Pile-loading tests may be carried to failure, this giving the ultimate bearing capacity of the particular pile that has been tested. For practical purposes, however, the ultimate bearing capacity may be taken as the load that causes the head of the pile to settle 10 per cent of the pile diameter. The ratio between the settlement of a pile foundation and that of a simple pile acted upon by the design load can have almost any value (Meyerhof, 1976). This is due to the fact that the settlement of an individual pile depends only on the nature of the soil in direct contact with the pile, whereas the settlement of a pile foundation also depends on the number of piles and on the compressibility of the strata located between the level of the ends of the piles and the surface of a relatively incompressible formation.

Piles may be divided into three main types, according to the effects of their installation, namely, displacement piles, small-displacement piles and non-displacement piles (Anon., 1986). Displacement piles are installed by driving and so their volume has to be accommodated below ground by vertical and lateral displacements of soil that may give rise to heave or compaction, which may have detrimental effects upon neighbouring structures. Driving also may cause piles that are already installed to lift. Furthermore, driving piles into clay may affect its consistency (Meyerhof, 1976). In other words, the penetration of the pile, combined with the vibrations set up by the falling hammer, can destroy the fabric of the clay and inaugurate a new process of consolidation that drags the piles in a downward direction. Indeed, they may settle on account of their contact with the remoulded mass of clay even if they are not loaded. Sensitive clays are affected in this way whilst insensitive clays such as tills are not. Small displacement piles include some piles that may be used in soft alluvial ground of considerable depth but they are not suitable in stiff clays or gravels. They also may be used to withstand uplift forces. Non-displacement piles are formed by boring, the hole being filled with concrete. The hole may be lined with casing that is or is not left in place.

Piled rafts are used as a means of supporting tall buildings on various types of soil. It would appear that the basement has a marked influence on the load displacement within a piled raft foundation. During the initial stages of construction, uplift forces resulting from the removal of soil can induce initial pressures on the base of a raft, together with tensile forces in the piles. Subsequent downward loading imposed by the structure slowly increases contact pressures and gives rise to a comparatively rapid build-up in compressive pile loads. The load distribution between the piles and the raft at any stage of construction depends on the ratio of uplift force to vertical structural load. The long-term effect of consolidation is to increase the load carried by the piles and to decrease the raft contact pressures.

4.10.4 Piers

Pier foundations may be constructed within a suitable open excavation and formed to their required dimensions or the excavation may consist of a hole corresponding to the required dimensions of the pier (Haley, 1987). The hole may be sheeted or cased for protection during construction. Where the ground surface is below water the excavation is made within a caisson. The bottoms of piers frequently are belled out to improve their performance, often to about twice the shaft diameter. Piers have been sunk to depths in excess of 60 m in the United States but they have not been used to a great extent in Europe.

4.10.5 Rock-socketed foundations

Foundations in rock masses must satisfy the same criteria as other foundations, that is, provide adequate stability and tolerable deformation. In rock-socketed foundations concrete is cast against a rock mass, frequently to increase the resistance to vertical and lateral loading. Generally in practice, the design of a foundation in a rock mass is governed by displacement rather than by stability criteria, especially when structures impose large loads. Rock-socketed foundations or piers can be designed to carry end-bearing loads, compressive loads in the side-shear wall or a combination of both. The most important factors that have to be taken account of in the design of such foundations include the strength of the rock mass, the fracture index, Young's modulus of the rock mass, the condition of the base and walls of the socket, and the geometry of the socket. A review of design procedures for rock-socketed foundations has been provided by Kulhawy and Carter (1992b), and more recently by Wyllie (1999).

References

Aas, G. 1965. A study of the vane shape and rate of strain on the measured values of *in situ* shear strength of soils. *Proceedings Sixth International Conference Soil Mechanics and Foundation Engineering*, Montreal, **1**, 141–145.

Ambraseys, N.N. 1988. Engineering seismology. *International Journal Earthquake Engineering and Structural Dynamics*, **17**, 1–106.

Andersland, O.B. and Anderson, D.M. (eds) 1978. *Geotechnical Engineering for Cold Regions*. McGraw-Hill, New York.

Andersland, O.B. and Ladanyi, B. 1994. *An Introduction to Frozen Ground Engineering*. Chapman & Hall, New York.

Anon. 1975. *Subsidence Engineer's Handbook*. National Coal Board, London.

Anon. 1976. *Reclamation of Derelict Land: Procedure for Locating Abandoned Mine Shafts*. Department of the Environment, London.

Anon. 1981. *Assessment of Damage in Low-Rise Buildings with Particular Reference to Progressive Foundation Movement*. Building Research Establishment, Digest 251, Her Majesty's Stationery Office, London.

Anon. 1982. *The Treatment of Disused Mine Shafts and Adits*. National Coal Board, London.

Anon. 1986. *Code of Practice for Foundations, BS8004*. British Standards Institution, London.

Anon. 1988. Engineering geophysics. Engineering Group Working Party Report. *Quarterly Journal Engineering Geology*, **21**, 207–273.

Anon. 1990a. *Methods of Test for Soils for Civil Engineering Purposes, BS1377*. British Standards Institution, London.

Anon. 1990b. Tropical residual soils, Engineering Group Working Party Report. *Quarterly Journal Engineering Geology*, **23**, 1–101.

Audric, T. and Bouquier, L. 1976. Collapsing behaviour of some loess soils from Normandy. *Quarterly Journal Engineering Geology*, **9**, 265–278.

Bell, F.G. 1986. Location of abandoned workings in coal seams. *Bulletin International Association Engineering Geology*, **30**, 123–132.

Bell, F.G. 1987. The influence of subsidence due to present day coal mining on surface development. In: *Planning and Engineering Geology*, Engineering Geology Special Publication No. 4, Culshaw, M.G., Bell, F.G., Cripps, J.C. and O'Hara, M. (eds), The Geological Society, London, 359–368.

Bell, F.G. 1988a. The history and techniques of coal mining and the associated effects and influence on construction. *Bulletin Association Engineering Geologists*, **24**, 471–504.

Bell, F.G. 1988b. Subsidence associated with the abstraction of fluids. In: *Engineering Geology of Underground Movements*, Engineering Geology Special Publication No 5. Bell, F.G., Culshaw, M.G., Cripps, J.C. and Lovell, M.A. (eds), The Geological Society, London, 363–376.

Bell, F.G. 1992. Salt mining and associated subsidence in mid-Cheshire, England, and its influence on planning. *Bulletin Association Engineering Geologists*, **22**, 371–386.

Bell, F.G. 1993. *Engineering Treatment of Soils*. E & FN Spon, London.

Bell, F.G. 2000. *Engineering Properties of Soils and Rocks*. Fourth Edition, Blackwell Scientific Publishers, Oxford.

Bell, F.G. 2002. The geotechnical properties of some till deposits occurring along the coastal areas of eastern England. *Engineering Geology*, **63**, 49–68.

Bell, F.G. and Coulthard, J.M. 1997. A survey of some geotechnical properties of the Tees Laminated Clay of central Middlesbrough, north east England. *Engineering Geology*, **48**, 117–133.

Bell, F.G. and Fox, R.M. 1991. The effects of mining subsidence on discontinuous rock masses and the influence on foundations: the British experience. *The Civil Engineer in South Africa*, **33**, 201–210.

Bell, F.G. and Maud, R.R. 1994. Dispersive soils: a review with a South African perspective. *Quarterly Journal Engineering Geology*, **27**, 195–210.

Bell, F.G. and Maud, R.R. 1995. Expansive clays and construction, especially of low-rise structures: a viewpoint from Natal, South Africa. *Environmental and Engineering Geoscience*, **1**, 41–59.

Bell, F.G., Northmore, K.J. and Culshaw, M.G. 2004. A review of collapsible soils with particular emphasis on brickearth. In: *Silt and Siltation, Problems and Engineering Solutions*, Jefferson, I., Rosenbaum, M. and Smalley, I. (eds), Springer-Verlag, Berlin (in press).

Bhattacharya, S. and Singh, M.M. 1985. *Development of Subsidence Damage Criteria*. Office of Surface Mining, Department of the Interior, Contract No. J5120129, Engineering International Inc. Washington, DC.

Bishop, I., Styles, P., Emsley, S.J. and Ferguson, N.S. 1997. The detection of cavities using the microgravity technique: case histories from mining and karstic environments. In: *Modern Geophysics in Engineering Geology*, Engineering Geology Special Publication No. 12, McCann, D.M., Eddleston, M., Fenning, P.J. and Reeves, G.M. (eds), The Geological Society, London, 153–166.

Bjerrum, L. 1963. Discussion, Session IV. *Proceedings European Conference on Soil Mechanics and Foundation Engineering*, Wiesbaden, **2**, 135–137.

Bjerrum, L. 1966. Secondary settlements of structures subjected to large variations in live load. *Proceedings Symposium International Union Theoretical Applied Mechanics and Rheology of Soil Mechanics*, Grenoble, 460–471.

Bjerrum, L., Moun, J. and Eide, O. 1967. Application of electro-osmosis to a foundation problem in a Norwegian quick clay. *Geotechnique*, **17**, 214–235.

Braithewaite, P.A. and Cole, K.W. 1986. Subsurface investigations of abandoned workings in the West Midlands of England by use of remote sensors. *Transactions Institution Mining and Metallurgy*, **95**, Section A, Mining Industry, A181–A190.

Brauner, G. 1973a. *Subsidence due to Underground Mining: Part II, Ground Movements and Mining Damage*. Bureau of Mines, Department of the Interior, US Government Printing Office, Washington, DC.

Brauner, G. 1973b. *Subsidence due to Underground Mining, Part I, Theory and Practice in Predicting Surface Deformation*. Bureau of Mines, Department of the Interior, US Government Printing Office, Washington, DC.

Brinch Hansen, J. 1970. *A Revised and Extended Formula for Bearing Capacity*. Danish Geotechnical Institute, Bulletin No. 28, Copenhagen.

Broms, B.B. 1991. Stabilization of soil with lime columns. In: *Foundation Engineering Handbook*. Second Edition, Fang, H.-Y. (ed.), Chapman & Hall, New York, 833–855.

Broms, B.B. and Boman, P. 1977. Lime columns – a new type of vertical drain. *Proceedings Ninth International Conference Soil Mechanics and Foundation Engineering*, Tokyo, **1**, 427–432.

Brook, C.A. and Alison, T.L. 1986. Fracture mapping and ground subsidence susceptibility modelling in covered karst terrain: the example of Dougherty County, Georgia. *Proceedings Third International Symposium on Land Subsidence*, Venice. International Association of Hydrological Sciences, Publication No. 151, 595–606.

Bryhn, O.R. and Loken, T. 1987. Stabilization of sensitive clays (quick clays) using $Al(OH)_{2.5}Cl_{0.5}$. *Proceedings International Clay Conference*, Denver, Schultz, L.G., van Olphen, H. and Mumpton, F.A. (eds), Clay Minerals Society, Bloomington, Indiana, 427–435.

Burland, J.B. and Burbridge, M.C. 1985. Settlement of foundations on sand and gravel. *Proceedings Institution Civil Engineers*, **78**, 1325–1381.

Burland, J.B. and Wroth, C.P. 1975. Settlement of buildings and associated damage. In: *Settlement of Structures*, British Geotechnical Society, Pentech Press, London, 611–654.

Burland, J.B., Kee, R. and Burford, D. 1974. Short-term settlement of a five storey building on soft chalk. In: *Settlement of Structures*, British Geotechnical Society, Pentech Press, London, 259–265.

Burland, J.B., Broms, B.B. and De Mello, V.F.P. 1977. Behaviour of foundations and structures. *Proceedings Ninth International Conference on Soil Mechanics and Foundation Engineering*, Tokyo, **2**, 495–547.

Cabrera, J.G. and Smalley, I.J. 1973. Quick clays as products of glacial action: a new approach to their nature, geology, distribution and geotechnical properties. *Engineering Geology*, **7**, 115–133.

Carter, P.G. and Mallard, D.J. 1974. A study of the strength, compressibility and density trends within the Chalk of southeast England. *Quarterly Journal of Engineering Geology*, **7**, 43–56.

Casagrande, A. 1932. Discussion on frost heaving. *Proceedings Highway Research Board*, Bulletin No. 12, Washington, DC. 169.

Chandler, R.J. 1972. Lias Clay: weathering processes and their effect on shear strength. *Geotechnique*, **22**, 403–431.

Chandler, R.J., Crilly, M.S. and Montgomery-Smith, G. 1992. A low cost method of assessing clay desiccation for low-rise buildings. *Proceedings Institution Civil Engineers*, **92**, 82–89.

Chaney, R.C. 1987. Cyclic response of soils of varying saturation. *Proceedings International Symposium on Environmental Technology*, Fang, H.-Y. (ed.), Envo Publishing Co., Bethlehem, PA, 2, 181–205.

Charles, J.A. and Skinner, H.D. 2001. Compressibility of foundation fills. *Proceedings Institution Civil Engineers, Geotechnical Engineering*, **149**, 145–157.

Charles, J.A., Burford, D. and Hughes, D.B. 1993. Settlement of opencast mine backfill at Horsley 1973–1992. In: *Engineered Fills*, Clarke, B.G., Jones, C.J.F.P. and Moffat, A.L.B. (eds), Thomas Telford Press, London, 329–440.

Chen, F.H. 1988. *Foundations on Expansive Soils*. Elsevier, Amsterdam.

Cherry, S. 1974. Design input for seismic analysis. In: *Engineering Seismology and Earthquake Engineering*, Solnes, J. (ed.), NATO Advanced Study Institutes Series, Applied Sciences, No 3, 151–162.

Clayton, C.R.I. 1990. The mechanical properties of the Chalk. *Proceedings International Chalk Symposium*, Brighton, Thomas Telford Press, London, 213–232.

Clemence, S.P. and Finbarr, A.O. 1981. Design considerations for collapsible soils. *Proceedings American Society Civil Engineers, Journal Geotechnical Engineering Division*, **85**, 151–180.

Clevenger, M.A. 1958. Experience with loess as a foundation material. *Transactions American Society of Civil Engineers*, **123**, 151–80.

Coulthard, J.M. and Bell, F.G. 1993. The engineering geology of the Lower Lias Clay at Blockley, Gloucestershire, U.K. *Geotechnical and Geological Engineering*, **11**, 185–201.

Craig, R.F. 1987. *Soil Mechanics*. Fourth Edition. Van Nostrand Reinhold, London.

Cripps, J.C. and Taylor, R.K. 1981. The engineering properties of mudrocks. *Quarterly Journal Engineering Geology*, **14**, 325–346.

Croney, D. and Jacobs, J.C. 1967. *The Frost Susceptibility of Soils and Road Materials*. Transport Road Research Laboratory, Report LR90, Crowthorne.

De Bruyn, I.A. and Bell, F.G. 2001. The occurrence of sinkholes and subsidence depressions in the Far West Rand and Gauteng Province, South Africa. *Environmental and Engineering Geoscience*, **6**, 281–295.

Deere, D.U. 1961. Subsidence due to mining – a case history from the Gulf region of Texas. *Proceedings Fourth Symposium on Rock Mechanics*. Bulletin Mining Industries Experimental Station, Engineering Series, Hartman, H.L. (ed.), Pennsylvania State University, 59–64.

De Graft-Johnson, J.W.S. and Bhatia, H. S. 1969. Engineering characteritics of lateritic soils. *Proceedings Speciality Session on Engineering Properties of Lateritic Soils, 7th International Conference Soil Mechanics Foundation Engineering*, Mexico City, **2**, 13–43.

Derbyshire, E. and Mellors, T.W. 1988. Geological and geotechnical characteristics of some loess and loessic soils from China and Britain. *Engineering Geology*, **25**, 135–175.

Denisov, H.Y. 1963. About the nature and sensitivity of quick clays. *Osnov Fudamic Mekhanic Grant*, **5**, 5–8.

Donnelly, L.J. and McCann, D.M. 2000. Location of abandoned mine workings using thermal techniques. *Engineering Geology*, **57**, 39–52.

Driscoll, R. 1983. The influence of vegetation on the swelling and shrinkage of clay soils in Britain. *Geotechnique*, **33**, 93–105.

Emsley, S.J. and Bishop, I. 1997. Application of the microgravity technique to cavity location in investigations for major civil engineering works. In: *Modern Geophysics in Engineering Geology*, Engineering Geology Special Publication No. 12, McCann, D.M., Eddleston, M., Fenning, P.J. and Reeves, G.M. (eds), The Geological Society, London, 183–192.

Evstatiev, D. 1988. Loess improvement methods. *Engineering Geology*, **25**, 341–366.

Fadum, R.E. 1948. Influence values for estimating stresses in elastic foundations. *Proceedings Second International Conference Soil Mechanics and Foundation Engineering*, Rotterdam, **1**, 77–84.

Feda, J. 1966. Structural stability of subsident loess from Praha-Dejvice. *Engineering Geology*, **1**, 201–219.

Feda, J. 1988. Collapse of loess on wetting. *Engineering Geology*, **25**, 263–269.

Fellenius, B.B. 1991. Pile foundations. In: *Foundation Engineering Handbook*. Second Edition, Fang, H.-Y. (ed.), Chapman & Hall, New York, 511–535.

Fookes, P.G. (ed.). 1997. *Tropical Residual Soils: Engineering Group Working Party Revised Report*. The Geological Society, London.

Fookes, P.G. and Best, R. 1969. Consolidation characteristics of some late Pleistocene periglacial metastable soils of east Kent. *Quarterly Journal Engineering Geology*, **2**, 103–128.

Foss, I. 1969. Secondary settlements of buildings in Drammen, Norway. *Proceedings Seventh International Conference Soil Mechanics and Foundation Engineering*, Mexico City, **2**, 168–178.

Garrard, G.E.G. and Taylor, R.K. 1988. Collapse mechanisms of shallow coal mine workings from field measurements. In: *Engineering Geology of Underground Movements*, Engineering Geology Special Publication No. 5, Bell, F.G., Culshaw, M.G., Cripps, J.C. and Lovell, M.A. (eds), The Geological Society, London, 181–192.

Gibbs, H.H. and Bara, J.P. 1962. Predicting surface subsidence from basic soil tests. *American Society for Testing Materials*, Special Technical Publication No. 322, Philadelphia, 231–246.

Gibbs, H.H. and Bara, J.P. 1967. Stability problems of collapsing soil. *Proceedings American Society Civil Engineering, Journal Soil Mechanics Foundations Division*, **93**, 572–594.

Gidigasu, M.D. 1988. The use of non-traditional tropical and residual materials for pavement construction. *Proceedings Second International Conference on Geomechanics in Tropical Soils*, Singapore, **1**, 397–404.

Gillott, J.E. 1979. Fabric, composition and properties of sensitive soils from Canada, Alaska and Norway. *Engineering Geology*, **14**, 149–172.

Gilluly, J. and Grant, U.S. 1949. Subsidence in the Long Beach area, California. *Bulletin Geological Society America*, **60**, 461–560.

Goodman, R.E., Korbay, S. and Buchignani, A. 1980. Evaluation of collapse potential over abandoned room and pillar mines. *Bulletin Association Engineering Geologists*, **17**, 27–37.

Gourley, C.S., Newill, D. and Schreiner, H.D. 1994. Expansive soils: TRL's research strategy. In: *Engineering Characteristics of Arid Soils*, Fookes, P.G. and Parry, R.H.G. (eds), A.A. Balkema, Rotterdam, 247–260.

Grabowska-Olszewska, B. 1988. Engineering geological problems of loess in Poland. *Engineering Geology*, **25**, 177–199.

Grant, R., Christian, J.T. and Vanmarke, E.H. 1974. Differential settlement of buildings. *Proceedings American Society Civil Engineers, Journal Geotechnical Engineering Division*, **100**, 973–991.

Grim, R.E. 1962. *Applied Clay Mineralogy*, McGraw-Hill, New York.

Haley, S.C. 1987. Piers. In: *Ground Engineer's Reference Book*, Bell, F.G. (ed.), Butterworths, London, 53/1–53/12.

Handy, R.L. 1973. Collapsible loess in Iowa. *Proceedings American Society Soil Science*, **37**, 281–284.

Hanna, A.M. 1987. Footings. In: *Ground Engineer's Reference Book*, Bell, F.G. (ed.), Butterworths, London, 49/1–49/16.

Harris, S.A. 1986. *The Periglacial Environment*. Croom Helm, London.

Haskins, D.R., Bell, F.G. and Schall, A. 1998. Weathering and fabric characteristics of a granite saprolite from Mpumalanga Province, South Africa. *Proceedings Eighth Congress International Association Engineering Geology*, Vancouver, Moore, D.P. and Hungr, O. (eds), **5**, 3035–3041.

Head, K.H. 1994. *Manual of Soil Laboratory Testing, Volume 2: Permeability, Shear Strength and Compressibility Tests*. Second Edition. Pentech Press, London.

Healy, P.R. and Head, J.M. 1984. *Construction over Abandoned Mine Workings*. Construction Industry Research and Information Association, Special Publication 32, London.

Hellewell, F.G. 1988. The influence of faulting on ground movement due to coal mining. The UK and European experience. *Mining Engineer*, **147**, 334–337.

Hobbs, N.B. 1986. Mire morphology and the properties and behaviour of some British and foreign peats. *Quarterly Journal Engineering Geology*, **19**, 7–80.

Hoek, E. 1994. Strength of rock and rock masses. *News Journal*, International Society Rock Mechanics, **2**, 4–16.

Hoek, E. and Brown, E.T. 1980. *Underground Excavations in Rock*. Institution of Mining and Metallurgy, London.

Hoek, E. and Brown, E.T. 1997. Practical estimates of rock mass strength. *International Journal Rock Mechanics Mining Science*, **34**, 1165–1186.

Holtz, R.D. 1991. Stress distribution and settlement of shallow foundations. In: *Foundation Engineering Handbook*. Second Edition, Fang, H.-Y. (ed.), Chapman & Hall, New York, 166–223.

Holzer, T.L. 1980. Faulting caused by groundwater level declines, San Joaquin Valley, California. *Water Resources Research,* **16**, 1065–1070.

Hossain, D. and Ali, K.M. 1988. Shear strength and consolidation characteristics of Ob'hor sabkha, Saudi Arabia. *Quarterly Journal Engineering Geology*, **21**, 347–359.

Housner, G.W. 1970. Design spectrum. In: *Earthquake Engineering*, Weigel, R.L. (ed.), Prentice-Hall, Englewood Cliffs, New Jersey, 93–106.

James, A.N. and Little, A.L. 1994. Geotechnical aspects of sabkha at Jubail, Saudi Arabia. *Quarterly Journal Engineering Geology*, **27,** 83–121.

Jammal, S.E. 1986. The Winter Park sinkhole and Central Florida sinkhole type subsidence. *Proceedings Third International Symposium on Land Subsidence*, Venice. International Association of Hydrological Sciences, Publication No. 151, 585–594.

Janbu, N., Bjerrum, L. and Kjaernsli, B. 1956. *Veiledning ved Losning av Fundamenteringsoppgaver*. Norwegian Geotechnical Institute, Publication No. 16, Oslo.

Jennings, J.E. and Knight, K. 1975. A guide to construction on or with materials exhibiting additional settlement due to collapse of grain structure. *Proceedings 6th African Conference Soil Mechanics Foundation Engineering*, Durban, 99–105.

Karnik, V. and Algermissen, S.T. 1979. Seismic zoning. In: *The Assessment and Mitigation of Earthquake Risk*. UNESCO Press, Paris, 11–47.

Kulhawy, F. and Carter, J.P. 1992a. Settlement and bearing capacity of foundations on rock masses. In: *Engineering in Rock Masses*, Bell, F.G. (ed.), Butterworth-Heinemann, Oxford, 231–245.

Kulhawy, F. and Carter, J.P. 1992b. Socketed foundations in rock masses. In: *Engineering in Rock Masses*, Bell, F.G. (ed.), Butterworth-Heinemann, Oxford, 509–529.

Kulhawy, F. and Goodman, R.E. 1980. Design of foundations on discontinuous rock. *Proceedings International Conference on Structural Foundations on Rock*, Sydney, **1**, 209–220.

Larson, M.K. 1986. Potential for fissuring in the Phoenix area, Arizona, USA. *Proceedings Third International Symposium on Land Subsidence*, Venice. International Association of Hydrological Sciences, Publication No. 151, 291–300.

Lin, Z.G. and Wang, S.J. 1988. Collapsibility and deformation characteristics of deep-seated loess in China. *Engineering Geology*, **25**, 271–282.

Litvinov, I.M. 1973. Deep compaction of soils with the aim of considerably increasing their bearing capacity. *Proceedings Eighth International Conference Soil Mechanics and Foundation Engineering*, Moscow, **3**, 392–394.

Lo, K.Y., Inculet, I.I. and Ho, K.S. 1991. Electro-osmostic strengthening of soft sensitive clay. *Canadian Geotechnical Journal*, **28**, 62–73.

Locat, J. Tremblay, H. and Leroueil, S. 1996. Mechanical and hydraulic behaviour of a soft inorganic clay treated with lime. *Canadian Geotechnical Journal*, **33**, 654–669.

Lofgren, B.E. 1968. Analysis of stress causing land subsidence. *United States Geological Survey*, Professional Paper No 600-B, B219–225.

Luttenegger, A.J. and Hallberg, G.R. 1988. Stability of loess. *Engineering Geology*, **25**, 247–261.

Marino, G. and Gamble, W. 1986. Mine subsidence damage from room and pillar mining in Illinois. *International Journal Mining and Geological Engineering*, **4**, 129–150.

McFarlane, M.J. 1976. *Laterite and Landscape*. Academic Press, London.

McQueen, I.S. and Miller, R.F. 1968. Calibration and evaluation of wide ring gravimetric methods for measuring moisture stress. *Soil Science*, **106**, 225–231.

Measor, E.O. and Williams, G.M.J. 1962. Features in the design and construction of the Shell Centre, London. *Proceedings Institution Civil Engineers*, **21**, 475–502.

Meigh, A.C. and Early, K.R. 1957. Some physical and engineering properties of chalk. *Proceedings Fourth International Conference on Soil Mechanics and Foundation Engineering*, London, **1**, 68–73.

Metcalf, J.B. and Townsend, D.L. 1961. A preliminary study of the geotechnical properties of varved clays as reported in Canadian case records. *Proceedings Fourteenth Canadian Conference on Soil Mechanics*, Section 13, 203–225.

Meyerhof, G.G. 1963. Some recent research on the bearing capacity of foundations. *Canadian Geotechnical Journal*, **1**, 16–26.

Meyerhof, G.G. 1976. Bearing capacity and settlement of pile foundations. *Proceedings American Society Civil Engineers, Journal Geotechnical Engineering Division*, **102**, 195–228.

Mitchell, J.K. 1981. Soil improvement – State-of-the-art report. *Proceedings Tenth International Conference on Soil Mechanics and Foundation Engineering*, Stockholm, **3**, 509-563.

Moum, J., Sopp, O.I. and Loken, T. 1968. Stabilization of undisturbed quick clay by salt wells. *Vag och Vattenbtggaren*, **14**, No. 8, 23–29.

Newmark, N.M. 1935. *Simplified Computation of Vertical Pressures in Elastic Foundations*. Engineering Experimental Station, Circular 24, University of Illinois, Urbana.

Norris, G., Gahir, Z. and Siddharthan, R. 1998. An effective stress understanding of liquefaction behavior. *Environmental and Engineering Geoscience*, **4**, 93–101.

Northmore, K.J., Entwisle, D.C., Hobbs, P.R.N., Culshaw, M.G. and Jones, L.D. 1992. *Engineering Geology of Tropical Red Soils, Geotechnical Characterization: Index Properties and Testing Procedures*. Technical Report WN/93/12, British Geological Survey, Keyworth, Nottingham.

Northmore. K.L., Bell, F.G. and Culshaw, M.G. 1996. The engineering properties and behaviour of the brickearth of south Essex. *Quarterly Journal Engineering Geology*, **29**, 147–161.

Ola, S.A. 1980. Mineralogical properties of some Nigerian residual soils in relation with building problems. *Engineering Geology*, **15**, 1–13.

O'Neill, M.W. and Poormoayed, A.M. 1980. Methodology for foundations on expansive clays. *Proceedings American Society Civil Engineers, Journal Geotechnical Engineering Division*, **106**, 1345–1367.

Peck, R.B. 1979. Liquefaction potential: science versus practice. *Proceedings American Society Civil Engineers, Geotechnical Engineering Division*, **105**, 393–398.

Peck, R.B., Hanson, W.E. and Thornburn, T.M. 1974. *Foundation Engineering*. Wiley, New York.

Phien-Wej, N., Pientong, T. and Balasubramanian, A.S. 1992. Collapse and strength characteristics of loess in Thailand. *Engineering Geology*, **32**, 59–72.

Piggott, R.J. and Eynon, P. 1978. Ground movements arising from the presence of shallow abandoned mine workings. *Proceedings First International Conference on Large Ground Movements and Structures*, Cardiff. Geddes, J.D. (ed.). Pentech Press, London, 749–780.

Polshin, D.E. and Tokar, R.A. 1957. Maximum allowable non-uniform settlement of structures. *Proceedings Fourth International Conference Soil Mechanics and Foundation Engineering*, London, **1**, 402–406.

Popescu, M.E. 1986. A comparison between the behaviour of swelling and collapsing soils. *Engineering Geology*, **23**, 145–163.

Poulos, H.G. 1987. Piles and piling. In: *Ground Engineer's Reference Book*, Bell, F.G. (ed.), Butterworths, London, 52/1–52/31.

Price, D.G., Malkin, A.B. and Knill, J.L. 1969. Foundations of multi-storey blocks on Coal Measures with special reference to old mine workings. *Quarterly Journal Engineering Geology*, **1**, 271–322.

Priebe, H.J. 1998. Vibroreplacement to prevent earthquake induced liquefaction. *Ground Engineering*, **31**, No. 9, 30–33.

Russell, D.J. and Parker, A. 1979. Geotechnical, mineralogical and chemical inter-relationships in weathering profiles of an overconsolidated clay. *Quarterly Journal Engineering Geology*, **12**, 197–216.

Sabtan, A., Ali-Saify, M. and Kazi, A. 1995. Moisture retention characteristics of coastal sabkhas. *Quarterly Journal Engineering Geology*, **28**, 37–46.

Serafim, J.L. 1968. Influence of interstitial water on rock masses. In: *Rock Mechanics in Engineering Practice*, Stagg, K.G. and Zienkiewicz, O.C. (eds), Wiley, London, 55–77.

Shadbolt, C.H. 1978. Mining subsidence. *Proceedings First International Conference on Large Ground Movements and Structures*, Cardiff, Geddes, J.D. (ed.), Pentech Press, London, 705–748.

Sieffert, J.G. and Bay-Gress, Ch. 2000. Comparison of European bearing capacity calculation methods for shallow foundations. *Proceedings Institution Civil Engineers, Geotechnical Engineering*, **143**, 65–74.

Simons, N.E. 1975. Normally consolidated and lightly consolidated cohesive materials. In: *Settlement of Structures*, British Geotechnical Society, Pentech Press, London, 500–530.

Simons, N.E. and Menzies, B.K. 2000. *A Short Course in Foundation Engineering*. Second Edition. Thomas Telford Press, London.

Skempton, A.W. 1944. Notes on the compressibility of clays. *Quarterly Journal Geological Society*, **100**, 119–135.

Skempton, A.W. 1951. The bearing capacity of clays. *Building Research Congress*, Watford, Hertfordshire, Division 1, 180–189.

Skempton, A.W. and Bjerrum, L. 1957. A contribution to settlement analysis of foundations on clays. *Geotechnique*, **7**, 168–178.

Skempton, A.W. and La Rochelle, P. 1965. The Bradwell slip: a short term failure in London Clay. *Geotechnique*, **15**, 221–241.

Skempton, A.W. and MacDonald, D.H. 1956. Allowable settlement of buildings. *Proceedings Institution Civil Engineers*, **5**, Part III, 727–768.

Stacey, T.R. and Bell, F.G. 1999. The influence of subsidence on planning and development in Johannesburg, South Africa. *Environmental and Engineering Geoscience*, **5**, 373–388.

Szechy, K. 1966. *The Art of Tunnelling*. Akademia Kiado, Budapest.

Terzaghi, K. 1936. Stability of slopes of natural clay. *Proceedings First International Conference Soil Mechanics Foundation Engineering*, Cambridge, Mass., **1**, 161–165.

Terzaghi, K. 1943. *Soil Mechanics*. Wiley, New York.

Terzaghi, K. 1946. Introduction to tunnel geology. In: *Rock Tunnelling with Steel Supports*, Proctor, R.V. and White, T. (eds), Commercial Shearing and Stamping Company, Youngstown, Ohio, 17–99.

Tincelin, E. 1958. *Pression et Deformations de Terrain dans les Mines de Fer de Lorraine*. Jouve Editeurs, Paris.

Tomlinson, M.J. and Boorman, R. 1996. *Foundation Design and Construction*. Addison Wesley Longman, Harlow, Essex.

Van der Merwe, D.H. 1964, The prediction of heave from the plasticity index and the percentage clay fraction, *The Civil Engineer in South Africa*, **6**, No. 6, 103–107.

Vargas, M. 1985. The concept of tropical soils. *Proceedings First International Conference Geomechanics in Tropical Lateritic and Saprolitic Soils*, Brasilia, **3**, 101–134.

Vargas, M. 1990. Collapsible and expansive soils in Brazil. *Proceedings Second International Conference Geomechanics in Tropical Soils*, Singapore, A.A. Balkema, Rotterdam, **2**, 489–492.

Vaughan, P.R., Maccarini, M. and Mokhtar, S.M. 1988. Indexing the properties of residual soil. *Quarterly Journal Engineering Geology*, **21**, 69–84.

Vesic, A. 1973. Analysis of ultimate loads of shallow foundations. *Proceedings American Society Civil Engineers, Journal Soil Mechanics and Foundations Division*, **99**, 45–73.

Walton, G. and Cobb, A.E. 1984. Mining subsidence. In: *Ground Movements and their effects on Structures*, Attewell, P.B. and Taylor, R.K. (eds), Surrey University Press, London, 216–242.

Ward, W.H., Burland, J.B. and Gallois, R.W. 1968. Geotechnical assessment of a site at Mundford, Norfolk, for a large proton accelerator. *Geotechnique*, **18**, 399–431.

Wardell, K. and Eynon, P. 1968. Structural concept of strata control and mine design. *Transactions Institution Mining and Metallurgy*, 77, Section A, Mining Industry, A125–A150.

Wassmann, T.H. 1980. Mining subsidence in Twente, east Netherlands. *Geologie en Mijnbouw*, **59**, 225–231.

White, H. 1992. Accurate delineation of shallow subsurface structure using ground penetrating radar. *Proceedings Symposium Construction over Mined Areas*, Pretoria, South African Institution Civil Engineers, Yeoville, 23–26.

Whittacker, B.N. and Reddish, D.J. 1989. *Subsidence: Occurrence, Prediction and Control*. Elsevier, Amsterdam.

Williams, A.A.B. and Pidgeon, J.T. 1983. Evapotranspiration and heaving clays in South Africa. *Geotechnique*, **33**, 141–150.

Wilson, G. and Grace, H. 1942. The settlement of London due to underdrainage of the London Clay. *Journal Institution Civil Engineers*, **19**, 107–122.

Wyllie, D.C. 1999. *Foundations on Rock*. Second Edition. E & FN Spon, London.

Yerkes, R.F. and Castle, R.O. 1970. Surface deformation associated with oil and gas field operation in the United States. *Proceedings First International Symposium on Land Subsidence*, Tokyo, International Association of Hydrological Sciences, UNESCO Publication No 88, **1**, 55–66.

Zeevaert, L. 1987. Design of compensated foundations. In: *Ground Engineer's Reference Book*, Bell, F.G. (ed.), Butterworths, London, 51/1–51/20.

5 Routeways

5.1 Introduction

Routeways play a vital and increasingly important role in society. In fact, they have played a major part in the development of civilization. Nonetheless, it perhaps can be argued that the modern era of routeway construction began during the canal age that commenced in the eighteenth century, although canals had been built since ancient times. This was followed by the railway age in the nineteenth century – a century that also saw significant improvements in road construction. The Romans constructed a network of fine roads to link their empire but after the fall of Rome the art of good road construction was lost. Similarly, the Incas built fine roads in South America. It was not until the industrial era that road construction assumed a new importance with, for example, Metcalfe, Telford and McAdam introducing new methods of road construction in Britain that formed the basis of modern road building.

The location of routeways is influenced in the first instance by topography. In the case of canals, the construction of locks, aqueducts and tunnels can help to overcome topography whilst bridges, embankments and tunnels can be constructed to carry railroads and roads with acceptable gradients through areas of more difficult relief. Obviously, the construction of such structures increases the difficulty, time and cost of building routeways. Nonetheless, the distance between the centres that routeways link has to be considered. Although geological conditions often do not determine the exact location of routeways, they can have a highly significant influence on their construction.

Newbury and Subramaniam (1977) referred to four stages in route location, namely, the inception stage when alternative routes are established; next these alternative routes are compared and two or more chosen for further evaluation; then the route location is finalized; and lastly follows the design stage when a detailed study is made of the adopted route. The choice of alternative routes is made primarily in terms of economic and environmental conditions. At this stage the degree of geological input involves a general assessment in order to identify potential problems. More detailed geological data are required in the following stage of comparative assessment of routes in order to evaluate items such as the geological conditions in cuttings and beneath embankments. The information needed for these two stages usually can be obtained from a desk study. Geological conditions, as well as topography, are important because construction costs are dependent to a large extent on earthworks. Hence, site investigation is needed to provide sufficient data for the latter two stages. As routeways are linear structures, they frequently traverse a wide variety of ground conditions along their length (Perry, 1995). In addition, the construction of routeways necessitates the excavation of soils and rocks and the provision of stable foundations as well as construction materials.

Topographic and geological maps, remote sensing imagery and aerial photographs are used in routeway location (Beaumont, 1977). These allow the preliminary plans and profiles of routeways to be prepared. Once a route corridor has been selected, then the geomorphological and ground conditions have to be investigated. Brunsden *et al.* (1975) emphasized the importance of geomorphological mapping during site investigations for routeways. They recognized a number of objectives of such surveys, in particular, geomorphological mapping helps to identify the general characteristics of an area in which a route is to be located and provides information on land forming processes and geohazards that can affect routeway construction, on the character of natural slopes and on the location of construction materials, as well as providing a basis on which to plan the subsequent site exploration (Fig. 5.1). Furthermore, they pointed out that usually there are a number of possible route alignments within a selected route corridor and so there is a need for rapid assessment of these possible alignments so that a decision can be made as to which is the most suitable on the basis of landform and ground conditions. Subsequently, geomorphological mapping can help reduce the amount of excavation, help the preliminary design of cut and fill slopes and land drainage, and help determine the approximate land-take requirements of a routeway. As noted by Hearn (2002), geomorphological mapping has proved especially useful in relation to road construction in mountainous areas. He illustrated this point with a number of examples from Nepal. Terrain evaluation is at times used in the selection of a routeway corridor, especially in developing countries (Dowling and Beaven, 1969).

The site investigation provides the engineer with information on the ground and groundwater conditions on which a rational and economic design for a routeway can be made (Wakeling, 1972). This information should indicate the suitability of the proposed location, both horizontally and vertically, the quantity of earthworks involved, the subsoil and surface drainage requirements and the availability of suitable construction materials. Other factors that have to be taken account of include the safe gradients for cuttings and embankments, locations of river crossings and possible ground treatment. Lovegrove and Fookes (1972) described a site investigation carried out for a highway in Fiji that considered the problems associated with working in residual soils. A number of boreholes, probably supplemented by a geophysical survey, are sunk along the alignment of the routeway to provide a profile of the ground conditions, and from which samples can be obtained for laboratory testing. Various *in situ* tests can be carried out within boreholes, for example, to determine bearing capacity, settlement, and the position of the water table and piezometric pressure. The associated laboratory programme is likely to include tests to determine the moisture content, specific gravity, density, consistency limits, strength and consolidation characteristics of the soils along the routeway alignment, as well as chemical tests for organic, sulphate and carbonate content, and pH value (Anon., 1990a). A mobile laboratory may be used to test soils on site.

5.2 Roadways

Normally, a road consists of a number of layers, each of which has a particular function. In addition, the type of pavement structure depends on the nature and number of the vehicles it has to carry, their wheel loads and the period of time over which it has to last. The wearing surface of a modern road consists either of 'black-top' (i.e. bituminous bound aggregate) or a concrete slab, although a bituminous surfacing may overlie a concrete base (Brown, 1996). A concrete slab distributes the load that a road has to carry, while in a bituminous road the load primarily is distributed by the base beneath. The base and sub-base below it generally consist of granular material, although in heavy-duty roads the base may be treated with cement. The subgrade refers to the soil immediately beneath the sub-base. However, much of the load

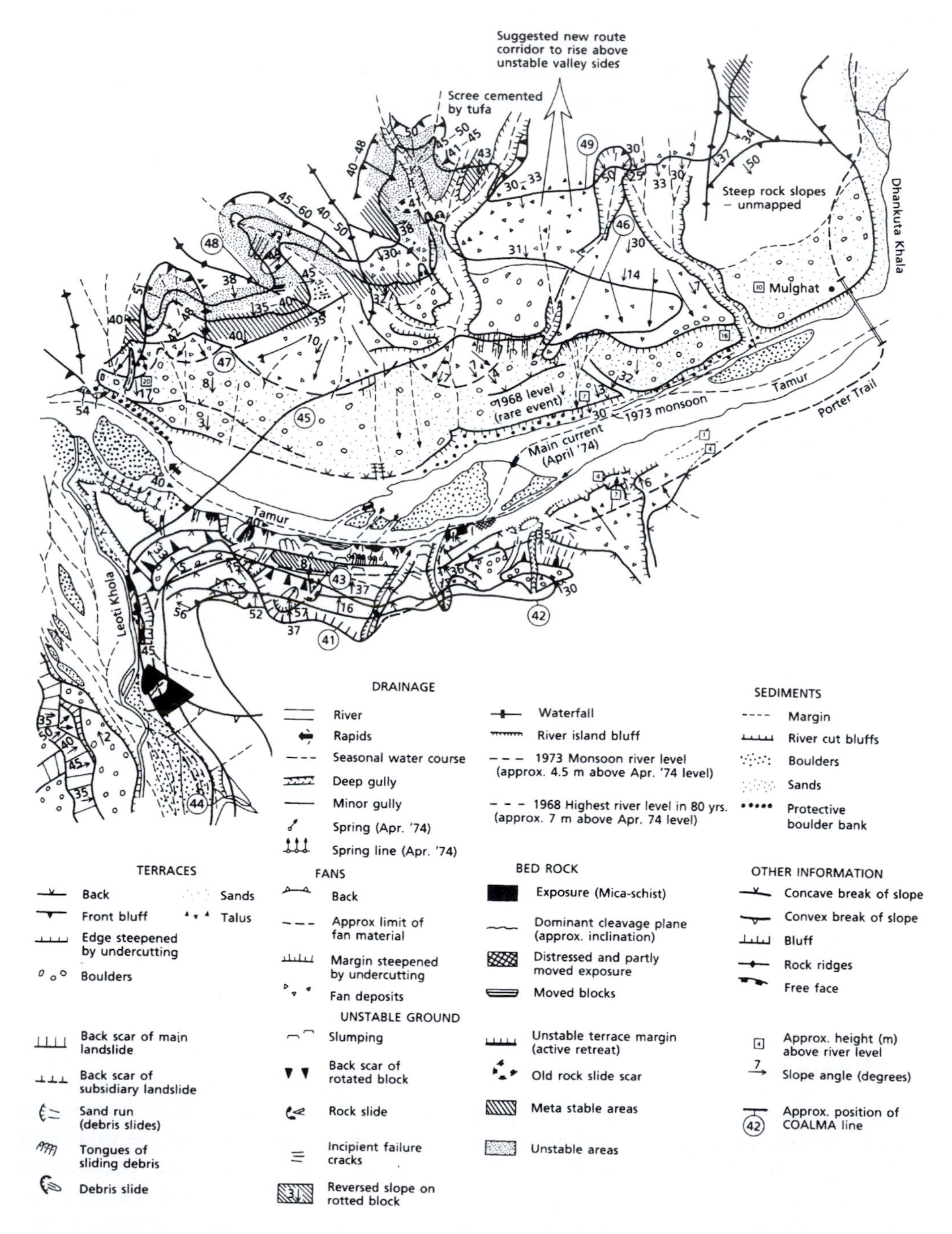

Figure 5.1 Geomorphological map of the site and situation of a proposed bridge crossing on the Tamur River, Nepal.
Source: Brunsden *et al.*, 1975. Reproduced by kind permission of The Geological Society.

carried by a road is distributed by the layers above the subgrade, ultimately the subgrade has to carry the load of the road structure plus that of the traffic. Consequently, the strength of the top of the subgrade may have to be strengthened by compaction or stabilization. The strength of the subgrade, however, does not remain the same throughout its life. Changes in its strength are brought about by changes in its moisture content, by repeated wheel loading, and in some parts of the world by frost action. Although the soil in the subgrade exists above the water table and beneath a sealed surface, this does not stop the ingress of water. As a consequence, partially saturated or saturated conditions can exist in the soil. Also, road pavements are constructed at a level where the subgrade is affected by wetting and drying, which if expansive clay is present may lead to swelling and shrinkage, respectively. Such volume changes are non-uniform and the associated movements may damage the pavement.

A moving wheel load represents a transient stress that is transmitted via the pavement to the soil beneath. Soils being elastoplastic materials, along with the granular layers that form the pavement foundation, possess notably non-linear stress–strain behaviour. In addition, the bituminous layer in flexible pavements has properties that are influenced by the rate of loading and temperature. According to Brown (1996), most pavements respond in a resilient manner when subjected to a single load application, any irrecoverable deformation being small in relation to the resilient component. However, irrecoverable plastic strains may accumulate under repeated loading, depending on its magnitude, leading to fatigue. Hence, the resilient properties of a pavement foundation are important in that they offer resistance to failure. Indeed, the concept of the resilient modulus in terms of pavement behaviour was introduced by Seed *et al.* (1962), who defined it as the repeated deviator stress divided by the recoverable (resilient) strain. In other words, it can be regarded as the resilient Young's modulus. There are two types of traffic-related failure mechanisms that result from the accumulation of vertical permanent strains in bituminous pavements, namely, cracking and rutting. Cracking is a consequence of fatigue and its incidence is related to the support provided by the pavement structure that, in turn, is influenced mainly by the characteristics of the soil. Castell *et al.* (2000) recognized two types of crack growth. One is associated with tensile stresses that develop at the bottom of the surface course and the other is caused by tensile stresses at the top of this course when the load is in front or behind the crack. The growth rate of surface cracks (i.e. those growing downwards) is much slower than the internal crack growth of those growing upwards. Rutting develops as a result of the accumulation of vertical strains beneath the wheel track and can include contributions from all the layers in the pavement.

The thickness of each layer constituting the pavement structure obviously is important in terms of the design of a road. Individual layer thickness affects the resistance of the soil to displacement under wheel loadings, that is, the bearing value. The California Bearing Ratio (CBR) test was developed to assess the bearing value, the CBR being obtained by measuring the relationship between force and penetration when a cylindrical plunger, of cross-sectional area 1935 mm^2, penetrates a soil at a given rate (Anon., 1990a). At any value of penetration, the ratio of the force to a standard force is defined as the CBR. Even though this test was phased out in California, where it was developed, it is still used in road design in many parts of the world. Brown (1996) was critical of the test and pointed out that it has a number of disadvantages, which include the results being influenced markedly by preparation of the sample and reproducible results cannot be obtained with wet clay soils. More importantly, Brown noted that the resilient modulus is not a simple function of the CBR but depends on soil type and the magnitude of loading. Indeed, in the United States the CBR test came to be regarded as an index test for assessment of shear strength. However, pavement behaviour under repeated traffic loading depends to a much greater extent on the elastic moduli of the materials involved than their shear strength.

The ground beneath roads, and more particularly embankments, must have sufficient bearing capacity to prevent foundation failure and also be capable of preventing excess settlements due to the imposed load (Kezdi and Rethati, 1988). Very weak and compressible ground may need to be entirely removed before construction takes place, although this will depend on the quantity of material involved. In some cases, heave that occurs due to the removal of load may cause significant problems. In other cases, improvement of the ground by the use of lime or cement stabilization, compaction, surcharging, the use of drainage, the installation of piles, stone columns or mattresses may be carried out prior to embankment and road building. Usually, the steepest side slopes possible are used when constructing cuttings and embankments as this minimizes the amount of land required for the highway and the quantity of material to be moved. Where potential failure surfaces exist or where adverse groundwater conditions are present, then the slopes probably need to be at a shallower angle. Obviously, attention must be given to the stability of these slopes. Difficulties frequently occur in heterogeneous materials and those with properties that are marginal between soils and rocks. Slight variations in strength, spacing of discontinuities or the grade of weathering can have an effect on the rate of excavation, and may affect the suitability of material that has been removed for use as a construction material. Where the materials excavated are unsuitable for construction, then considerable extra expense is entailed in disposing of waste and importing fill. Geological features such as faults, crush zones and solution cavities, as well as man-made features such as abandoned mine workings, drains and areas of fill, can cause considerable difficulties during construction.

5.2.1 Problem soils and road construction

Unfortunately, many soils can prove problematic in highway engineering, because they expand or shrink, collapse, disperse, undergo excessive settlement, have a distinct lack of strength or are corrosive (Bell, 2000). Such characteristics may be attributable to their composition, the nature of their pore fluids, their mineralogy or their fabric. Soil is formed by the breakdown of rock masses either by weathering or erosion. It may accumulate in a place or undergo a certain amount of transport, either of which influence its character and behaviour. However, the type of breakdown suffered by rock masses is influenced profoundly by the climatic regime in which they are developed, as well as the stage of maturity that they have reached.

Soft alluvial, estuarine and deltaic soils may undergo significant settlement, especially beneath embankments. Indeed, rapid construction of embankments in such conditions can mean that an unacceptably low factor of safety is developed. Consequently, embankment construction may have to be staged, thereby allowing dissipation of excess pore water pressures in both the alluvium and fill, and enhancing stability. Vertical drains may be installed beneath an embankment to increase the rate of settlement and so decrease the length of time of the construction period. Alternatively, lightweight fill such as polystyrene can be used. For example, the use in Germany of expanded polystyrene (EPS) as a lightweight sub-base material for embankments was discussed by Beinbrech and Hillmann (1997). Extruded polystyrene blocks can be used to backfill bridge abutments. Polystyrene, however, is not used in the construction of major highways. Another factor that may have to be considered is the variability of alluvial and estuarine deposits, as pointed out by Cook and Roy (1984). Hawkins (1984), in a consideration of sediments in the Severn estuary, highlighted the significance of their laminated nature. He noted that pore water dissipation and therefore settlement beneath trial embankments for a motorway in southwest England were more rapid, due to the laminated fabric of the sediments, than had been predicted from oedometer tests. In addition, the permeability of laminated sediments is anisotropic, frequently being notably higher parallel to rather than normal to the laminations (Bell, 1998). The pavement conditions of the roads in the Niger Delta, Nigeria, and the geotechnical properties of the soils used in their construction were reviewed by Arumala and Akpokodje (1987). They found that most severe surface deformations, pavement cracking and failures occurred in the seasonally flooded fresh/salt water swamps due to the high water table and poor drainage, as well as very fine silty clays used. They concluded that pavement design should consist of well-compacted subgrade and sub-base courses, a cement stabilized base course and paved shoulders, as well as good drainage in order to ensure good performance.

Some clay soils undergo slow volume changes that occur independently of loading and are attributable to swelling or shrinkage. These volume changes can give rise to ground movements that may damage roads or give rise to shrinkage settlement of embankments that leads to cracking and break-up of the roads they support (Fig. 5.2). Differences in the period and amount of precipitation and evapotranspiration are the principal factors influencing the swell–shrink response of a clay soil. The depth of the active zone in expansive clays (i.e. the zone in which swelling and shrinkage occurs in wet and dry seasons, respectively) varies. For instance, it

Figure 5.2 Break-up of a road in Austin, Texas, due to expansive clay.

may extend to a depth of over 6 m in some semi-arid regions of the world whereas in humid temperate regions, although the potential for significant volume change may exist, in most years any changes are restricted to the upper 1.0–1.5 m in clay soils owing to the damp climate. The potential for volume change of clay soil is discussed in Chapter 4. Cemented and undisturbed expansive clay soils often have a high resistance to deformation and may be able to absorb significant amounts of swelling pressure. Remoulded expansive clays therefore tend to swell more than their undisturbed counterparts. The swell–shrink behaviour of a clay soil under a given state of applied stress in the ground is controlled by changes in soil suction, these changes being brought about by moisture movement through the soil due to evaporation from its surface in dry weather, by transpiration from plants or conversely by recharge consequent upon precipitation. Transpiration from vegetative cover is a major cause of water loss from soils, especially in semi-arid regions. Indeed, the distribution of soil suction in soil is controlled primarily by transpiration. When vegetation is cleared from a site, its desiccating effect also is removed. Hence, the subsequent regain of moisture by clay soils leads to them swelling. For example, swelling movements on expansive clays in South Africa, associated with the removal of vegetation in many areas has amounted to about 150 mm, although movements over 350 mm have been recorded. The amount of heave of expansive soils is reduced significantly when compacted to low densities at high moisture contents. Expansive soils compacted above optimum moisture content undergo negligible swell for any degree of compaction. On the other hand, compaction below optimum results in excessive swell.

Bell and Maud (1995) described a section of the freeway between Durban and Johannesburg, South Africa, at Town Hill near Pietermaritzburg, which developed significant cracks in the pavement surface, as well as along the crests of embankments. At this location, the road crosses over expansive colluvial clay soils. Steep crossfalls on the road had required the construction of fills, and the fill material generally was obtained from the cuttings, or where additional material was required, from borrow pits. Cracking occurred due to desiccation and shrinkage of the colluvial clays and fills during a drought, particularly where thick colluvium occupied palaeovalleys or gullies in the formation below. A survey indicated that 30–40 mm of vertical movement had occurred in this section of the road over a period of 14 months. The situation was further complicated by the fact that creep movements, aggravated by the formation of shrinkage cracks, had occurred on steep crossfalls. It therefore was recommended that the existing road should be realigned, that is, that it should be cut more deeply into the slope and thereby avoid, where possible, location on the fill and colluvium (Fig. 5.3). A monitoring

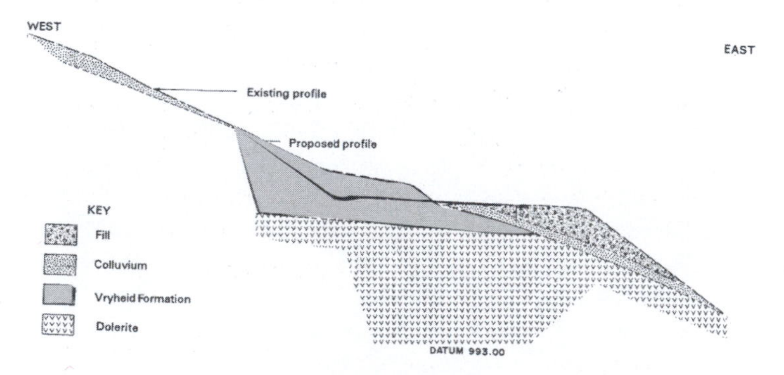

Figure 5.3 Realignment of the freeway at Town Hill, near Pietermaritzburg, South Africa, onto bedrock away from fill and colluvium that are expansive.
Source: Bell and Maud, 1995.

system also was recommended to record any movements that took place during the period prior to the implementation of the remedial works.

Many attempts have been made to reduce the expansiveness of clay soil by chemical stabilization. For example, lime stabilization of expansive soils, prior to construction, can minimize the amount of shrinkage and swelling they undergo. However, significant SO_4 content (i.e. in excess of $5000\,mg\,kg^{-1}$) in clay soils can mean that they react with CaO to form ettringite ($Ca_6Al_2(OH)_12(SO_4)_3.27H_2O$) or thaumasite ($CaSiO_3.CaCO_3.CaSO_4.14.5H_2O$) with resultant expansion and heave, as happened during the construction of a motorway in Oxfordshire, England (Forster *et al.*, 1995). The main sources of sulphate likely to cause heave in lime-stabilized clay soils beneath roads are gypsum and pyrite (Burkart *et al.*, 1999). Nevertheless, Wild *et al.* (1999) claimed that such swelling in lime-stabilized soils can be suppressed by the use of ground granulated blast-furnace slag (GGBS). They recommended the up 60–80 per cent of the lime should be replaced by GGBS in order to minimize or eliminate sulphate expansion and that compaction should be wet of optimum.

Collapsible soils such as loess, brickearth and certain wind-blown silts possess porous textures with high void ratios and relatively low densities. These soils generally consist of 50–90 per cent silt particles and sandy, silty and clayey types have been recognized. At their natural low moisture content, these soils possess high apparent strength but they are susceptible to large reductions in void ratio upon wetting. In other words, the metastable texture collapses as the bonds between the grains break down when the soil is wetted. Collapse on saturation normally only takes a short period of time, although the more clay such a soil contains, the longer the period tends to be. Indeed, significant settlements can take place in collapsible soils after they have been wetted. A fuller outline of collapsible soils and the methods used to treat them is given in Chapter 4.

Dispersion occurs in soils when the repulsive forces between clay particles exceed the attractive forces thus bringing about deflocculation so that in the presence of relatively pure water the particles repel each other (Bell and Maud, 1994). Hence, there is no threshold velocity for erosion in dispersive soil, the colloidal clay particles going into suspension even in quiet water. Therefore, dispersive soils are highly susceptible to erosion and piping. There are no significant differences in the clay fractions of dispersive and non-dispersive soils, except that soils with less than 10 per cent clay particles may not have enough colloids to support dispersive piping. Dispersive soils contain a higher content of dissolved sodium (up to 12 per cent) in their pore water than ordinary soils and their pH value generally ranges between 6 and 8. Dispersive soils occur in semi-arid regions, for example, in South Africa they are found in areas that have less than 850 mm of rain annually. However, the development of dispersive soils in more arid regions generally is inhibited by the presence of free salts.

Dispersive soils can present problems in earthworks, such as those required for roads on both the fill and cut slopes relating thereto. In the case of embankments, dispersive soil can be used provided it is covered by an adequate depth of better class material. Care has to be exercised in the placement and compaction of the fill layers so that no layer is left exposed during construction for such a period that it can shrink and crack, and thus weaken the fill. The dispersive soil material should be placed and compacted at 2 per cent above its optimum moisture content to inhibit shrinkage and cracking. Where seepage areas or springs are located along the alignment of a road embankment that has to be constructed of dispersive material, special care has to be exercised in the provision of adequate subsoil drainage for such areas, otherwise the long term stability of the embankment could be jeopardized by the development of piping. In some instances, to reduce erosion, it is prudent to stabilize the outer 0.3 m or so of dispersive soil material in an embankment with lime, the soil material and lime being mixed in bulk prior to placement rather than being mixed *in situ*. To minimize potential settlement and erosion problems when dispersive soil fill is placed against a structure, such as a bridge abutment and wingwall or a culvert, such structures, where practically possible, should be provided with sloping soil interfaces such that the soil settles on to the interface. In this way the possibility of cracks developing in the soil that could lead to piping erosion is reduced. Unless adequately protected from surface erosion, severe runnel and gulling erosion can develop on cut slopes in dispersive soil. To some extent, this problem can be reduced by providing a steeper than normal slope, for instance, 0.5 vertical in 1.5 horizontal, as the dispersive soil usually possesses adequate cohesion to stand in a stable condition up to a height of about 3 m. The steeper than normal slope means that comparatively less rainfall falls on the slope so that the amount of slope erosion is reduced. Where cuts are located in more than 3 m of dispersive soil, flatter slopes of about 1 vertical in 2 horizontal have to be employed. Such slopes have to be adequately vegetated immediately on completion to limit erosion. However, dispersive soils have low natural fertility. Therefore, it can be difficult to establish and maintain suitable vegetation on cut slopes in, or fill constructed of, dispersive soils in order to inhibit surface erosion. Apart from adequate artificial fertilization, it usually is necessary to place topsoil on such slopes to ensure satisfactory vegetative growth. The steepest slope that can be satisfactorily topsoiled is about 1 vertical in 1 horizontal, and even such a slope may have to have artificial anti-erosion measures installed on it to hold the topsoil and vegetation in place. Adequate open channel drainage (1–2 m in width) has to be provided at the toes of cut slopes. Such channel drains should be concrete-lined to limit erosion along their length.

Most soils in arid regions consist of the products of physical weathering of rock material. This breakdown process gives rise to a variety of rock and mineral fragments that may be transported and deposited under the influence of gravity, wind or water. Many arid areas are dominated by the presence of large masses of sand. For the most part, aeolian sands are poorly (uniformly) graded. Depending on the rate of supply of sand, the wind speed, direction, frequency and constancy, and the nature of the ground surface, sand may be transported and/or deposited in mobile or static dunes. Movement of sand can bury obstacles in its path such as roads and railways. Such moving sand necessitates continuous and often costly maintenance activities. Indeed, areas may be abandoned as a result of sand encroachment. Stipho (1992) pointed out that only a few centimetres of sand on a road surface can constitute a major driving hazard. Obviously, the best way to deal with such a hazard is to avoid it, this being both more effective and cheaper than resorting to control measures. Hence, sensible site selection, based on thorough site investigation, is necessary prior to any development. Removal of moving sand can only happen where the quantities involved are small. Even so, this is expensive and the excavated material has to be disposed of. Often removal is only practical when the sand can be used as fill or ballast, however, the difficulty of compacting aeolian sand frequently precludes its use. Accordingly, a means of stabilizing mobile sand must be employed. One of the best ways to bring this about is by the establishment, if possible, of a vegetative cover. Gravel or coarse aggregate can be placed over a sand surface, depending on its size, to prevent its deflation. A minimum particle diameter of around 20 mm is needed for the gravels to remain unaffected by strong winds. In addition, the gravel layer should be at least 50 mm in thickness. Artificial stabilization, which provides a protective coating or bonds grains together, may be necessary on loose sand. In such cases, rapid-curing cutback asphalt and rapid-setting asphalt emulsion have been used. Chemical sprays also have been used to stabilize loose sand surfaces, and the thin protective layer that develops at the surface helps to reduce water loss. Many chemical stabilizers, however, are only temporary, breaking down in a year or so and therefore they tend to be used together with other methods of stabilization, such as the use of vegetation.

Fences and windbreaks can be used to impound or divert moving sand. Stipho mentioned the use of sand traps such as trenches and pits to prevent sand accumulating on highways. These must be wider than the horizontal leap of saltating particles (up to 3–4 m) and sufficiently deep to prevent scouring. Regular removal of accumulated sand from traps is necessary and so they are expensive to maintain. Stipho also suggested that in some situations it may be necessary to elevate a carriage way well above any surrounding dunes in order to allow the wind to accelerate over the road.

According to Watson (1985), deposition of sand can be reduced by either removing areas of low wind energy or by increasing the velocity of the wind over the surface by shaping the land surface. For example, flat slopes in the range of 1:5–1:6 with rounded shoulders are necessary for small to medium embankments, which help to streamline wind flow. Cuttings require flatter slopes of perhaps 1:10 to allow for free sand transport and usually have a wide ditch at the base of a slope to collect blown sand.

A number of silty deposits formed under arid conditions are liable to undergo considerable volume reduction or collapse when wetted. Such metastability arises due to the loss of strength of interparticle bonds resulting from increases in water content. Many of the deposits within alluvial plains and covering hillsides are poorly consolidated. Consequently, they may undergo large settlements, especially if subjected to vibration due to cyclic loading.

The precipitation of salts in the upper horizons of an arid soil due to evaporation of moisture from the surface commonly means that some amount of cementation has occurred, which has been concentrated in layers, and that the pore water is likely to be saline. High rates of evaporation in hot arid areas may lead to ground heave due to the precipitation of minerals within the capillary fringe. In the absence of downward leaching, surface deposits become contaminated with precipitated salts, particularly sulphates and chlorides. Alluvial plain deposits often contain gypsum particles and cement. Occasional wetting and subsequent evaporation frequently are responsible for a patchy development of weak, mainly carbonate and occasionally gypsum cement, in coarse-grained soils. Like silts, these soils therefore may undergo collapse, especially where localized changes in the soil–water regime are brought about by construction activity.

Sabkhas are extensive saline flats that are underlain by sand, silt or clay and often are encrusted with salt. They tend to occur in low-lying coastal zones and inland plains in arid regions with shallow water tables. In coastal sabkhas, the dominant minerals are calcite, dolomite and gypsum with lesser amounts of anhydrite, magnesite, halite and carnalite, together with various other sulphates and chlorides. The minerals precipitated within the soil are much more variable in continental than coastal sabkhas since they depend on the composition of the local groundwater. Nonetheless, gypsum and anhydrite tend to be of common occurrence. Gypsum tends to occur below the water table whereas anhydrite tends to occur above. Some of these sabkhas, however, contain little carbonate. Although halite frequently forms at the ground surface it commonly is dispersed by wind. Highly developed sabkhas tend to retain a significant proportion of soil moisture and the higher the salinity of the groundwater, the greater the amount of moisture retained by a sabkha at a particular drying temperature (Sabtan *et al.*, 1995). The groundwater surface is held at a certain level by capillary soil moisture but capillary rises of up to 3 m can occur in some fine grained deposits. Salts are precipitated at the ground surface when the capillary fringe extends from the water table to the surface. One of the main problems with sabkha is the decrease in density and strength, and increased permeability that occur, particularly in the uppermost layers, after rainfall, flash floods or marine inundation due to the dissolution of soluble salts that act as cementing materials. Changes in the state of hydration of minerals, such as swelling clays and calcium sulphate, also cause significant volume changes in soils. Excessive settlement also can occur due to the removal of soluble salts by flowing groundwater. Movement of groundwater also can lead to the dissolution of minerals to the extent that small caverns, channels and surface holes can be formed. On the other hand heave, resulting from the precipitation and growth of crystals, can elevate the surface of a sabkha in places by as much as 1 m. Sabkha soils frequently are characterized by low strength (see Chapter 4). Various ground improvement techniques therefore have been used in relation to large highway projects, and soil replacement and preloading have been used when highway embankments have been constructed. Al-Amoudi *et al.* (1995) suggested that sabkha soils can be improved by stabilization with cement and with lime. However, a high water-to-lime ratio does not prove satisfactory, and lime also is not satisfactory, for stabilization of sulphate-rich soils.

Wherever possible, roads should avoid areas in which saline groundwater occurs at or near the surface and aggregate sources should be investigated for salt content prior to use, those with high salt content being rejected. As it is not always possible to avoid the construction of roads on saline ground or the use of materials containing salts, a number of precautions should be taken. If the problems associated with aggressive salty ground are to be avoided, then the first object must be to identify the limits of such ground and the spacial variability of aggressiveness within it. Fookes and French (1977) recognized five moisture zones in arid regions and the influence they have on the behaviour of roads (Fig. 5.4; Table 5.1). They maintained that such zones could be mapped, the resulting map indicating those areas that should be avoided as far as road construction is concerned. Usually, pavement damage attributable to the crystallization of soluble salts is confined to thin bituminous surfaces. Consequently, Blight (1976) suggested that this could be avoided by laying a surface layer of at least 30 mm of dense asphalt concrete to prevent evaporation, and thereby migration and crystallization of salt at or near the surface. Even so, salt may accumulate beneath such surfaces as a result of temperature changes leading to the degradation of base and sub-base material that may lead to a reduction of density with the development of rutting and pot-holing in the long term. Obika *et al.* (1989) indicated that although a thick impermeable surface can be effective in preventing short term damage, especially at the bitumen-base interface, complete impermeability is difficult to achieve. In addition, Obika *et al.* reviewed the various recommendations that have been suggested for salt limits for highway materials. They maintained that as far as soluble salt limits were concerned, caution had to be exercised in their application. They further noted that where pavement material contains more than one type of salt, then the interaction between them can mean that the recommended salt limits may be inappropriate. Indeed, the deleterious effect frequently is enhanced significantly due to mixtures of salt. It appears that cutback bitumen and emulsion bitumen primers perform better than tar primers in relation to the reduction of surface degradation. This may be due to emulsion primers resting on the surface rather than penetrating the pavement layer and thereby providing lower permeability. Januszke and Booth (1984) suggested that evaporation and therefore the accumulation of salt at the surface could be avoided by placing a bituminous surfacing immediately after compaction. Alternatively, French *et al.* (1982) have suggested placement of an impermeable membrane in the base course. A granular layer in the base course also may help reduce the capillary rise.

Salt weathering of bituminous-paved roads built over areas where saline groundwater is at or near the surface is likely to result in notable signs of damage such as heaving, cracking, blistering, stripping, pot-holing, doming and disintegration. For example, Blight (1976) mentioned that the surface layer may be damaged in damp soils as a result of upward capillary migration and concentration of calcium, magnesium and sulphate salts due to evaporation from the surface. Such movements probably are encouraged in the immediate vicinity of a road by the

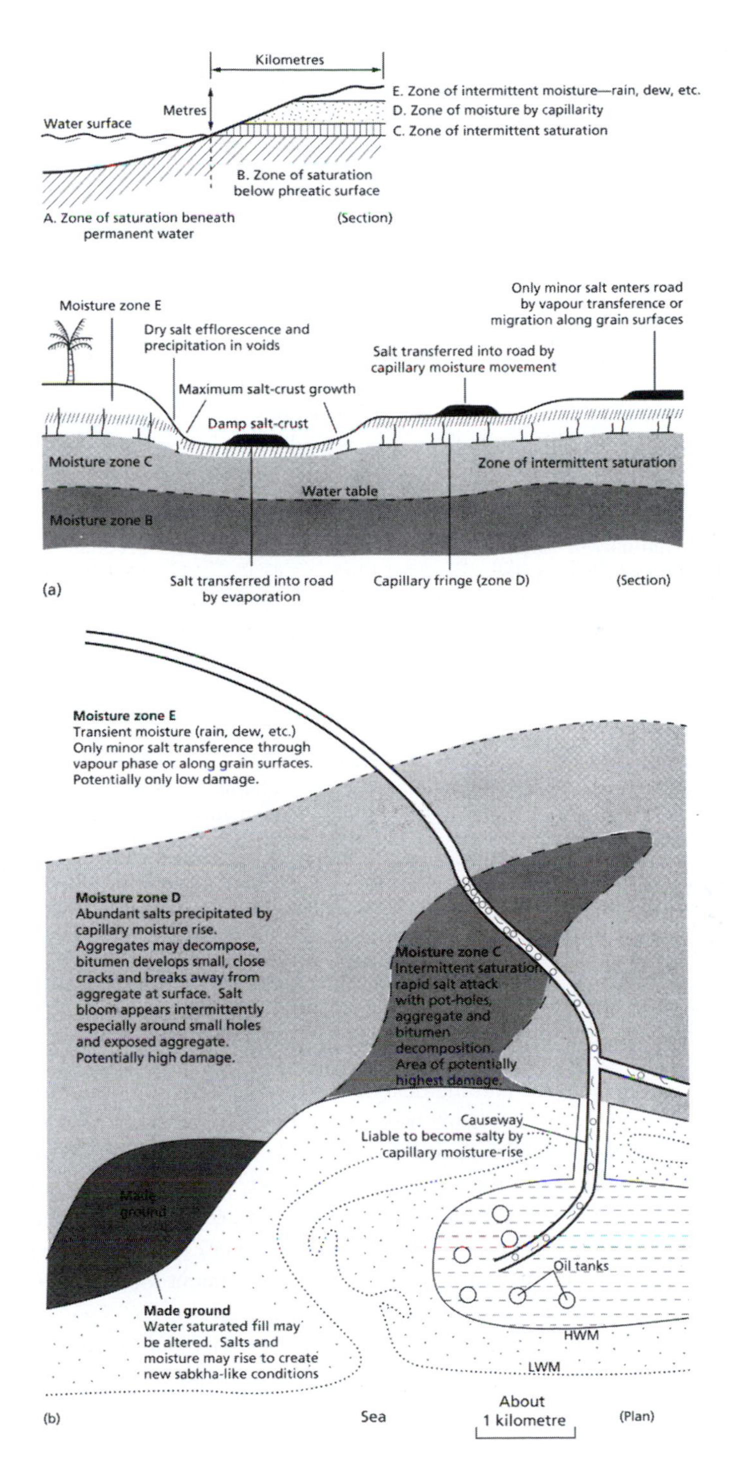

Figure 5.4 (a) Soil moisture zones and road construction in drylands. (b) Hypothetical map to show how identification of soil moisture zones can assist road construction and the determination of maintenance priority areas.
Source: Fookes and French, 1977.

Table 5.1 Summary of the influence of moisture zones A to E on behaviour of roads

Zone	*Moisture conditions*	*Salt conditions*	*Possible damage*
Moisture zone E	Transient water from rain, dew, etc.	Salts may be removed in solution and may accumulate by subsequent evaporation or by vapour transfer, etc.	Damage not serious unless aggregate is rich in salt or road is of thin construction with unsound aggregates, in long term
Moisture zone D	Water present by capillary moisture movement	Salts may be precipitated at all levels of road construction and in large quantities	Aggregate and bitumen may decompose, blisters may develop, small holes and cracks likely. Serious damage only in thin construction
Moisture zone C	Water present by capillary movement or ground may be saturated at times of high water table	Salts precipitated and may be re-dissolved	Large pot-holes develop, aggregates and bitumen decompose rapidly. Irregular surface develops. Maximum damage in thin construction
Moisture zone B	Permanently saturated zone below capillary fringe	Soil and rock properties may be changed in long term	Damage by long term deformation possible
Moisture zone A	Saturated zone below water	May create sabkha conditions in reclaimed ground or embankments	Damage as moisture zones E, D and C depending on elevation of construction

Source: Fookes and French, 1977.

Note
For explanation of moisture zones see Fig. 5.4.

fact that the road has a higher temperature than the surrounding ground. The physical and chemical consequences of the attack on roads mainly depend on the salinity of the groundwater, the type of aggregates used and the design of the road. Once cracks appear, the intensity and extent of salt damage increases with time. As a result of upward migration of capillary moisture, salts may be precipitated beneath the bituminous surface, leading to its degradation and gradually to it being heaved to form ripples in the road, with the ripples containing tension cracks. Aggregates used for base and sub-base courses can be attacked and this can mean that roads undergo settlement and cracking (Fookes and French, 1977). Accordingly, assessment of local groundwater regimes and salt profiles in relation to pavement location is necessary before construction commences in areas where the water table is high.

The precipitation of salts in sediments can lead to the development of various forms of crusts and cretes in arid areas. These may form continuous sheets or patch-like masses at the ground surface when the water table is at or near this level, or at some other position within the ground profile. The most common type of cementing agent is calcite and so calcrete is the commonest form of pedocrete. Pedocretes are mixtures of the parent or host material and cement, and as they develop, the cement content increases until it may form most of the mass. These materials take three forms, namely, indurated pedocretes (e.g. hardpans and nodules), non-indurated pedocretes (soft or powder forms) and mixtures of the two (e.g. nodular pedocretes). The strong crusts may be removed when pedocretes have been subjected to erosion thereby exposing weaker soils that give rise to variable conditions at subgrade level, requiring the design of pavement thickness to be based on the weakest soils. When calcrete is used in highway construction, the excavated material can be highly variable in fragment size. Tomlinson (1978) noted that weakly cemented lumps of calcrete can be broken down by the tracks of earthmoving vehicles or rollers but that if trafficking is excessive, then the materials are broken down to fine white dust that is difficult to compact. Although compaction can be brought about and dust nuisance minimized by water spraying, careful control of added water is required to avoid construction plant becoming bogged down in the calcareous silt. Calcrete used in embankment construction needs to be thoroughly compacted. If voids are left in the fill, then these permit water seepage during rains or sheet flooding, and exposure of lumps of calcrete to flowing water results in their breakdown and consequent settlement of the embankment.

Lateritic soils are residual soils typically found in tropical regions that may be developed above most types of rock masses. Charman (1988) suggested that laterite should be regarded as a highly weathered material that forms as a result of the concentration of hydrated oxides of iron and aluminium in such a way that the character of the deposit in which they occur is affected. These oxides may be present in an unhardened soil, as a hardened layer, as concretionary nodules in a soil matrix or in a cemented matrix enclosing other materials. In fact, Charman proposed a cycle of concretionary development from an original plinthite layer (a layer in which concentration of oxides occurs but with no concretionary development) to a mature hardpan laterite. He recognized intermediate duricrust forms such as nodular laterite and honeycomb laterite. Drying brings about changes in the properties of residual soils not only during sampling and testing, but also *in situ*. The latter occurs as a result of local climatic conditions, drainage and position within the soil profile. Drying initiates two important effects, namely, cementation by the sesquioxides and aggregate formation on the one hand, and loss of water from hydrated clay minerals on the other. In this way, drying can cause almost total aggregation of clay size particles into silt and sand size ranges, and a reduction or loss of plasticity. Unit weight, shrinkage, compressibility and shear strength also can be affected (Northmore *et al.*, 1992). Hence, classification tests should be applied to the soil with as little drying as possible, at least until it can be established from comparative tests that drying has no effect on the results. Lovegrove and Fookes (1972) found that the dry density of residual soil in Fiji that had been air-dried was higher than the dry density of the same soil that had been dried to the moisture content at which it was tested. Because this could affect the compaction characteristics of the residual soil, a trial embankment was constructed, and compaction trials with several types of compaction plant were undertaken in order to prepare a realistic specification for compaction. Although it is frequently assumed that lateritic soils are relatively incompressible due to the presence of cementitious material, cements may weaken on saturation and this eventually may lead to the collapse of the soil structure. In addition, leaching can break down the aggregates, which gives rise to increases in plasticity and compressibility and reductions in strength. Where a hardened crust is present at the surface in lateritic soil, then the strength of the soil frequently decreases with increasing depth.

Peat accumulates in areas where there is an excess of rainfall and the ground is poorly drained, irrespective of latitude or altitude. Nonetheless, peat deposits tend to be most common in those regions with a comparatively cold wet climate (Hobbs, 1986). Physico-chemical and biochemical processes cause this organic material to remain in a state of preservation over a long period of time. The void ratio of peat ranges between 9, for dense amorphous granular peat, and 25, for fibrous types, although it tends to decrease with depth within a peat deposit. Such high void ratios give rise to water contents that may range from 75 to 98 per cent by volume of peat. The pH value of bog peat generally is less than 4.5 and may be less than 3. The principal problem confronting the highway engineer working on peaty soil is settlement and if loads exceed a certain minimum, then settlement may be accompanied by creep, lateral spread, or in extreme cases by rotational slip and upheaval of adjacent ground. The use of precompression, involving surcharge loading in the construction of embankments across peatlands, involves the removal of the surcharge after a certain period of time. This gives rise to some swelling in the compressed peat. Uplift or rebound can be quite significant depending on the actual settlement and surcharge ratio (i.e. the mass of the surcharge in relation to the mass of the fill once the surcharge has been removed). It undergoes a notable increase when surcharge ratios are greater than about 3. Rebound also is influenced by the amount of secondary compression induced prior to unloading. Normally, rebound is between 2 and 4 per cent of the thickness of the compressed layer of peat before surcharge has been removed. Hence, the compressibility of preconsolidated peat is reduced significantly. With few exceptions, improved drainage has no beneficial effect on the rate of consolidation. This is because efficient drainage only accelerates the completion of primary consolidation, which anyhow is completed rapidly. If peat is less than 3–4 m thick and is underlain by a soil with a satisfactory bearing capacity such as gravel or dense sand, then the peat can be removed prior to the construction of an embankment (Perry *et al.*, 2000). Because the water table usually is high in deposits of peat, dewatering by pumping from wells has to take place before peat is removed. Deposits of peat, at times, have been removed during road construction by blasting. There are three methods of bog blasting. Trench shooting is used where the surface deposits are less than 6 m in depth. One or more rows of charges are placed at the base of the peat and fired to form a trench in which fill is placed. Toe shooting is carried out in peat that is soft and liable to slip. The charges are placed close to the fill that has been placed, that is, the peat is removed ahead of the fill. Lastly, in the underfill method, one or more rows of charges are placed beneath fill, the fill then settling into place as the peat is blown aside. Jelisic and Leppanen (1998) referred to the stabilization of peat for road or railway construction by a large mixing tool that formed columns, the stabilizing agents consisting of either pulverized fly ash, ground blast furnace slag or by-products from the paper industry. Barry *et al.* (1995) described the successful construction of a road using a piled geogrid reinforced raft over 11 m of peat in east Sumatra. The piles consisted of local timber and were sunk through the peat into underlying clay. Some settlement was expected as the pile length above the water table rotted. Such a road is designed to carry light loads and the pavement consists of oil-bound stone.

As noted in Chapter 4, frozen ground and frozen ground phenomena occur in regions that experience a tundra climate, that is, in those regions where the winter temperatures rarely rise above freezing point and the summer temperatures are only warm enough to cause thawing in the upper metre or so of the soil (Harris, 1986). The upper surface of perennially frozen ground is termed the permafrost level or table. Continuous permafrost generally has a mean annual temperature of less than −5 °C at a depth of 10–15 m, with the permafrost table being some 0.6 m below the ground surface. In granular material, this may be up to 1.8 m in depth. Discontinuous permafrost is interrupted by thawed areas and is thinner with a lower permafrost table. Because of the permafrost layer, summer meltwater cannot seep into the ground, the active zone then becomes waterlogged and the soils on gentle slopes are liable to flow. Generally, the depth of thaw is less, the higher the latitude. It is at a minimum in peat or highly organic sediments and increases in clay, silt and sand to a maximum in gravel. Layers or lenses of unfrozen ground, termed taliks, may occur, often temporarily, in the permafrost. Occasionally, taliks may be formed of viscous liquid and as such can flow. As far as the construction of roads and runways is concerned, the object is to provide a stable surface across frequently unstable ground conditions. This entails either the use of insulating layers or the replacement of surface material with non-frost susceptible material. Water must be prevented from accumulating beneath a road or runway by the provision of drainage ditches.

5.2.2 *Frost heave and road construction*

Frost heave can cause serious damage to roads, leading to their break-up. A number of factors are necessary in order for frost heave to occur, namely, capillary saturation at the beginning and during freezing of the soil, a plentiful supply of subsoil water and a soil possessing fairly high capillarity together with moderate permeability. Furthermore, the ground surface experiences an increasingly larger amount of heave, the higher the initial water table. Indeed, it has been suggested that frost heave could be prevented by lowering the water table (Andersland and Anderson, 1978). Grain size is another important factor influencing frost heave. For example, gravels, sands and clays are not particularly susceptible to heave whilst silts definitely are. The reason for this is that silty soils are associated with high capillary rises, which allows water to be drawn upwards from the water table, but at the same time their voids are large enough to allow moisture to move quickly enough for them to become saturated relatively rapidly. If ice lenses are present in clean gravels or sands, then they simply represent small pockets of moisture that have been frozen. In fact, Casagrande (1932) suggested that the particle size critical to heave formation was 0.02 mm. If the quantity of such particles in a soil is less than 1 per cent, no heave is to be expected, but considerable heaving may take place if this amount is over 3 per cent in non-uniform soils and over 10 per cent in very uniform soils. This criterion has been used by the United States Army Corps of Engineers (Anon., 1965), together with data from frost heave tests to develop a frost susceptibility system (Fig. 5.5).

Maximum heaving does not necessarily occur at the time of maximum depth of penetration of the 0 °C line, there being a lag between the minimum air temperature prevailing and the maximum penetration of the freeze front. In fact, soil freezes at temperatures slightly lower than 0 °C. Before freezing, soil particles develop films of moisture about them due to capillary action. This moisture is drawn from the water table. As an ice lens grows, the suction pressure it develops exceeds that of the capillary attraction of moisture by the soil particles. Consequently, moisture moves from the soil to the ice lens but the capillary force continues to draw moisture from the water table and so the process continues. This explains why heaves amounting to 30 per cent of the thickness of the frozen layer frequently have been recorded.

Croney and Jacobs (1967) suggested that under the climatic conditions experienced in Britain, well-drained cohesive soils with a plasticity index exceeding 15 per cent could be looked upon as non-frost susceptible. They suggested that where the drainage is poor and the water table is within 0.6 m of formation level the limiting value of plasticity index should be increased to 20 per cent. In addition, in experiments with sand they noted that as the amount of silt added was increased up to 55 per cent or the clay fraction up to 33 per cent, increase in permeability in the freezing front was the overriding factor and heave tended to increase. Beyond these values, the decreasing permeability below the freezing zone became dominant and progressively reduced the heave. This indicates that the permeability below the frozen zone principally is responsible for controlling heave. Croney and Jacobs also suggested that the permeability of soft chalk is sufficiently high to permit very serious frost heave but in the harder varieties the lower permeabilities minimize or prevent heaving.

A frost heave test, developed by the Transport and Road Research Laboratory, has been used to predict frost heave (Jacobs, 1965). A critical review of this test has been provided by Jones (1980), who claimed that it gives poor reproducibility of results. Furthermore, this type

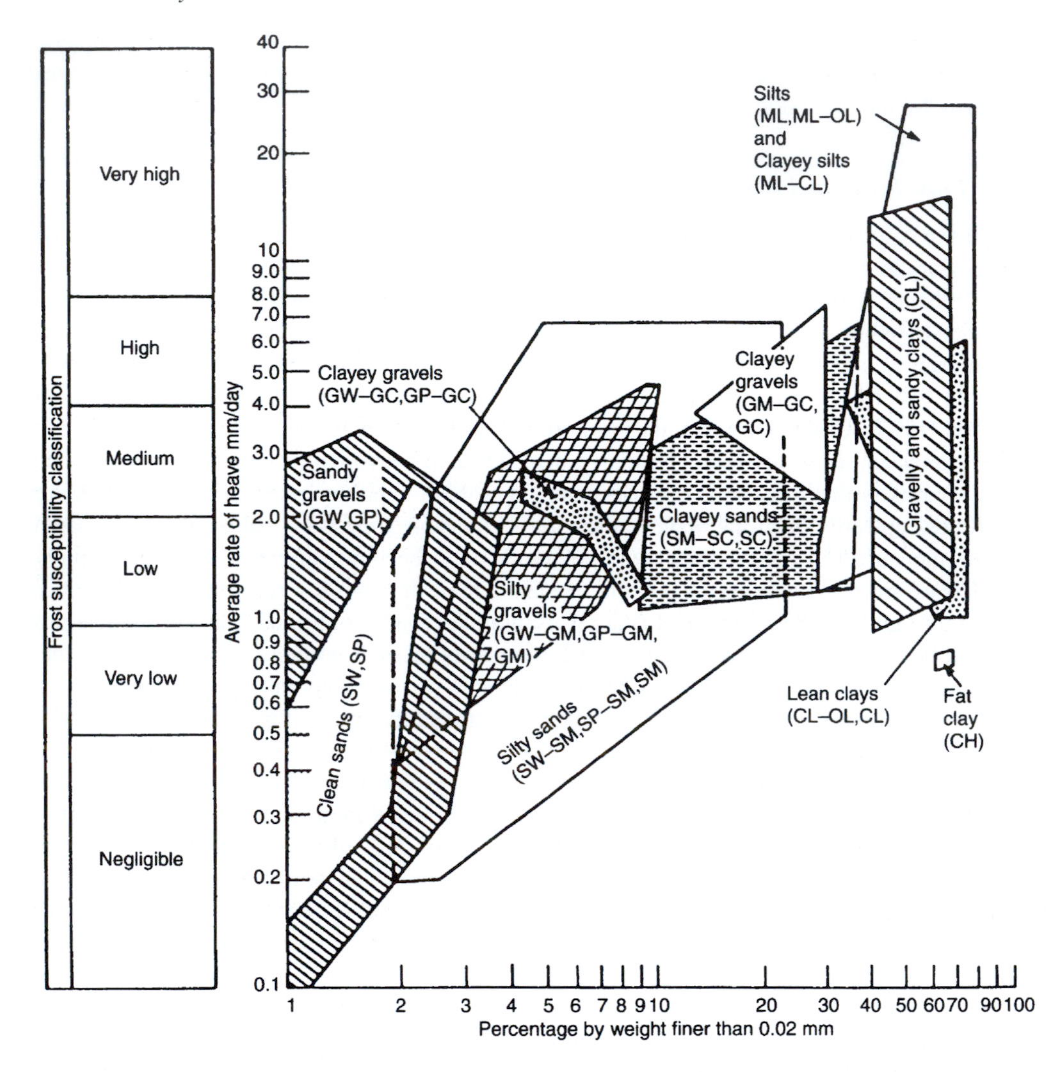

Figure 5.5 Range in the degree of frost susceptibility of soils according to the US Corps of Engineers. *Source*: Anon., 1965.

of test is unfortunately time-consuming, and so a rapid freeze test was developed by Kaplar (1971). Approximate predictions of frost heave also have been based on grain size distribution. However, Reed *et al.* (1979) noted that such predictions failed to take account of the fact that soils can exist at different states of density and therefore porosity, yet have the same grain size distribution. What is more, pore size distribution controls the migration of water in the soil and hence, to a large degree, the mechanism of frost heave. They therefore proposed the following expression, based on pore diameters, to predict the amount of frost heave (H):

$$H = 1.694(D_{40}D_{80}) - 0.3805 \tag{5.1}$$

where D_{40} and D_{80} are the pore diameters whereby 40 and 80 per cent of the pores are larger, respectively. Jones (1980) suggested that if heaving is unrestrained, then the heave can be estimated as follows:

$$H = 1.09kit \tag{5.2}$$

where k is permeability, i the suction gradient (this is difficult to derive) and t time.

Because frozen ground is more or less impermeable this increases the problems due to thaw by impeding the removal of surface water. What is more, when the thaw occurs the amount of water liberated may greatly exceed that originally present in the melted-out layer of the soil. As the soil thaws downwards the upper layers become saturated, and since water cannot drain through the frozen soil beneath, they may suffer a complete loss of strength, excess pore water pressures developing when the rate of ice melt is greater than the discharge capacity of the soil. Loss of bearing capacity and settlement therefore may be associated with thawing of frozen ground (Andersland and Ladanyi, 1994). Repeated cycles of freezing and thawing change the structure of the soil, again reducing its bearing capacity. Rigid concrete pavements are more able to resist frost action than flexible bituminous pavements. Methods of predicting the amount and rate of settlement due to thawing of frozen ground have been discussed by Nixon and Ladanyi (1978). Methods of dealing with frost heave have been mentioned in Chapter 4.

5.2.3 Ground movements and road construction

Notable ground movements can result from mining subsidence, the type of movements and the time of their occurrence being influence by the method of mining used (see Chapter 4). Malkin and Wood (1972) indicated that it was necessary to consider the anticipated effects of subsidence during and after construction of a highway at the design stage and to make final assessment during construction. In areas where mining has ceased, then records of mine plans may exist but even if located, they might not be accurate. Aerial photographs may be of value in determining areas, especially in rural districts, in which mining has gone on in the past. Site investigations in areas of abandoned mines need to determine the existence and pattern of mine workings, and to assess their potential instability in relation to the alignment of a highway and, more specifically, the design of cuttings, embankments and bridges. It probably will be necessary to fill voids with grout, particularly beneath bridge locations. In areas of current mining, the mine operators will be able to provide plans of their workings and data concerning future workings. Predictions may be able to be made from the data received regarding the area affected or to be affected, the amount of subsidence likely to occur and the resulting ground strains. Hence, an assessment of the compatibility of the mining proposals, in particular with the design specification for a route, should be able to be made. In some instances it may be necessary to obtain agreement for a modification or restriction of mining in order to avoid damaging ground movements. Also, close liaison with mine operators may indicate that re-phasing mine working or changes in mine layout can minimize or eliminate subsidence damage. Faults and dykes in mining areas can concentrate the effects mining subsidence giving rise to surface cracking or the development of steps, which can lead to severe surface disruption of a highway. For example, Bell (1987) referred to up to 2 m of subsidence occurring along a fault, consequent on longwall working of coal, that affected a motorway in Lancashire, England, the differential movement across the fault zone being around 1 m, and the strain being in excess of 50 mm m^{-1}. Stacey and Bell (1999) recorded that 150–200 mm of subsidence had occurred along a dyke over a period of a year, and had affected both a road and a railway in Johannesburg. The subsidence profile

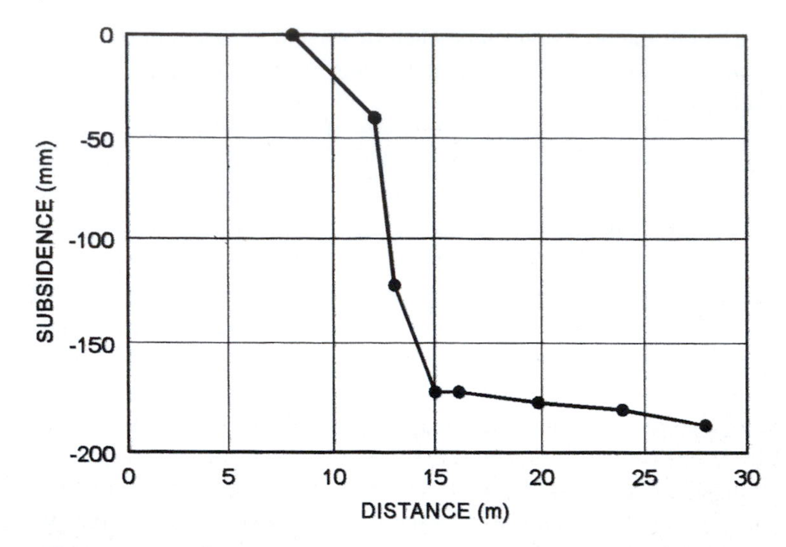

Figure 5.6 Subsidence profile across a dyke affecting a road and railway in Johannesburg, South Africa. *Source*: Stacey and Bell, 1999.

showed the abrupt nature of the movement (Fig. 5.6). Such movement entailed local resurfacing of the motorway and reballasting of the railway track to correct levels. Treatment of faults and dykes involves locating their outcrops. Some idea of their locations, at times, may be obtained from geological maps or mine plans. Their actual location during a site investigation frequently involves trenching or boring large diameter auger holes. The surface effects of movements along fault or dyke planes may be reduced by excavating along the zones of potential movement and replacing with cushions of granular material. Block-jointed rock masses can give rise to irregular movements when affected by mining subsidence with joints gaping by anything up to a metre or so. Because such movements are unpredictable, Malkin and Wood suggested that the rock masses should be broken down in order to reduce the amount of ground movement.

Natural voids and cavities in rock masses also can represent potential subsidence problems in routeway construction. Voids and cavities most commonly are found in carbonate and evaporitic rocks where they occur as a result of dissolution by flowing groundwater. However, the dissolution process may be accelerated by human activity, such as road construction, whereby excessive run-off is channelled into the ground and can lead to the collapse of a road (Fig. 5.7). The collapse of a sinkhole beneath a road obviously can be responsible for disastrous consequences, for example, Boyer (1997) referred to vehicles falling into newly opened sinkholes in Maryland and the death or serious injury of the occupants. De Bruyn and Bell (2001) provided an account of the development of sinkholes and subsidence depressions in the widely distributed Transvaal Dolomite in South Africa, together with methods of dealing with sinkhole problems. Similarly, Lamont-Black *et al.* (2002) gave an account of the mechanisms responsible for the development of karst features in areas underlain by gypsum in England and the problems in construction resulting therefrom. Some of these problems were described by Cooper and Saunders (2002) in relation to the Ripon bypass, North Yorkshire, England. In this area, complex cave systems have developed in gypsum and collapses up to 30 m across and 20 m deep have occurred, with one subsidence event occurring on average once a year. Once sinkholes have been located, for example, by geophysical methods such as ground probing radar and exposed by drilling, they then are filled with bulk grout. For instance, Gregory *et al.* (1988) referred to

Figure 5.7 Collapse of road over a sinkhole, near Pretoria, South Africa.

the use of grouting to protect a major road from the effects of sinkhole development in the Far West Rand, South Africa. Areas of potential high risk were identified and grouting commenced in these areas where the dolomite or cavities occurred at greatest depth. Generally, primary holes were grouted at 20 m centres, this aiding the identification of zones that required further treatment in terms of secondary, tertiary or quaternary grouting, the latter being carried out from a hole spacing of 5 m. Monitoring was undertaken during the grouting programme. If this revealed that any subsidence had occurred because of erosion of wad (a residual clayey material derived from dolomite) by grout, then grouting was carried out in the centre of the area that had subsided. Geogrids are now used in road construction in areas where subsidence, due to either natural causes or mining, could pose a future threat. In the case of cavity collapse, they hopefully prevent the road falling into the cavity before repairs are carried out. For example, Cooper and Saunders referred to the use of geogrids in the construction of the Ripon bypass so that the road could be supported for at least a 24-hour period after being affected by subsidence. Although indicating the location of the subsidence, the resulting depression in the road obviously prohibits its use until repairs have been made. Grykes opened by dissolution in limestone also may represent problems in terms of carriageway foundations. One method of treatment is to cut back the upper parts of the walls of grykes and then infill with mass concrete. Grykes with large gapes can be grouted. Other fissures can be dealt with in similar fashion. Rogers (1972) mentioned the use of a graded rock-blanket formed beneath subgrade level throughout an area where grykes occurred so that the loss of sub-base material was prevented and the effects of strains in the surfacing materials were minimized.

Landslides on either natural or man-made slopes adversely affect roadways. For example, Early and Skempton (1972) referred to a landslide at Walton's Wood, Staffordshire, England, that had been initiated by erosion of an ice-marginal drainage channel during the retreat stage of the last glaciation. The slope had been covered by a thick layer of colluvium that was still potentially unstable when construction of a motorway embankment began. Failure occurred before the embankment was completed. The objectives of the field and laboratory investigations that were carried out after the failure were to define the nature and extent of the landslide, to identify the slip surfaces, to measure the strength of the clay along these surfaces and to determine the pore water pressures within the landslide mass. This was the first occasion when the residual strength of clay, and its significance in relation to slope stability, was determined. Rock falls and debris avalanches from steep hillslopes in northern Vermont that affect highways were described by Lee *et al.* (1997). The rock masses concerned are weakened by freeze-thaw action and tend to move after periods of heavy rainfall. Slides that affected a deep cutting for a road made in a folded sequence of mudstones, siltstones and sandstones in southwest England were described by Sherrell (1971). He concluded that where rocks of differing lithology and shear strength occur in a complex structural arrangement, then not only do the rocks and structures need investigation but the nature of the groundwater regime also needs investigation. This allows assessment of the possible changes that may occur in the regime due to excavation to be made, after which the surface and subsurface drainage requirements, to aid future slope stability, can be designed. Al-Homoud and Tubeileh (1997) also gave an account of damaging landslides in cut slopes along major highways in Jordan, which resulted from interbedded layers of steeply dipping weak mudrock in stronger formations and associated relatively high piezometric conditions. They developed a classification system based on those factors influencing the occurrence of landslides and this was used to grade the stability of slopes. Slope stabilization measures have been dealt with in Chapter 2.

Not only can flooding disrupt road traffic but it can cause the destruction of roads. For example, Bell (1994) reported that during the floods of September 1987, which affected the low-lying and coastal areas of Natal, South Africa, river bank erosion was responsible for undermining some roads, usually on the outside bends of rivers where flow velocities were greatest (Fig. 5.8).

Although earthquake damage to routeways is less directly related to loss of life or property than damage to structures, because urban centres rely upon routeways, their disruption causes disorder to urban living. However, damage to a particular zone of a routeway can affect an area extending beyond the zone and may entail notable diversions of traffic. Geological conditions, especially soil properties, potential relative ground displacement and potential horizontal and vertical strain distribution therefore must be taken into account when designing routeways in seismically active regions. Transverse cracks and vertical displacements may occur across a road pavement due to differential movement at abrupt cut-and-fill contacts. If such a contact lies in the longitudinal direction, the filled section may slide due to the differential response between it and the ground to seismic shocks. Such cut-and-fill sections of railroads also are liable to sliding. Embankments carrying roads or railways can fail due to the sliding of the surface of the slope or the slope failure can involve both the embankment and the ground beneath. Failure also can take the form of the development of vertical cracks near the top of embankment slopes or subsidence of an embankment. Slopes of 1:1.5 and well-graded fill should help to minimize seismic damage to embankments. Kubo and Katayama (1978) indicated that a unit increase in magnitude of an earthquake can give rise to a fourfold increase in the radius of damage and that, for example, damage to embankments has occurred at locations 250 km or more from epicentres of earthquakes that had magnitudes in excess of 7.5. As an illustration, the Kobe earthquake in 1995 in Japan ($M = 7.2$) badly damaged both the elevated expressway and high-speed railway. Esper and Tachibana (1998) recorded that in the former case, large hammerhead-reinforced concrete piers sheared at their bases causing a 500 m long section to collapse. The piers were lacking sufficient transverse reinforcement, having been constructed

Figure 5.8 Erosion and removal of part of the approach road to Mvoti Bridge, Natal, South Africa.

in 1968–1969 under older seismic regulations. In addition, the original strength of the piers had been reduced by over 50 per cent as a result of alkali aggregate reaction in the concrete, even though resin had been injected into cracks that had formed as a result of the reaction. In addition, the main city highway disintegrated in many locations. Railway tracks broke in many locations and trains fell on their sides. Sections of the railway embankment failed and certain overpasses collapsed. An elevated viaduct that carries the railway was severely damaged, when some of the longer spans collapsed. Again, the viaduct was built in the 1960s. By contrast, many modern bridges in the area sustained little or no damage as these were constructed according to the latest codes.

5.3 Embankments

The engineering properties of soils used for embankments, such as their shear strength and compressibility, are influenced by the amount of compaction they undergo. Accordingly, the desired amount of compaction is established in relation to the engineering properties required for the embankment to perform its design function. A specification for compaction needs to indicate the type of compaction equipment to be used, its mass, speed and travel, and any other factors influencing performance such as frequency of vibration, thickness of layers to be compacted and number of passes of the compactor. Embankments are mechanically compacted by laying and rolling soil in thin layers.

Because a high proportion of the cost earthworks is attributable to transportation, attempts generally are made to use locally available material. If the material available locally is of poor quality, then with careful design it still may be possible to use it in embankments. However, poor soil conditions increase the difficulty of earthmoving plant to operate. The moisture condition test determines the strength of a soil in terms of a moisture condition value (MCV). The MCV can be related to the relevant parameters in the design of embankments such as the undrained shear strength and compressibility in that an upper limit of moisture content can be related to an acceptable soil condition. Basically, the moisture condition test consists of determining the compactive effort, in terms of the number of blows of a rammer, required to fully compact a sample of soil and has been described by Parsons (1981). The penetration of the rammer at a given number of blows is compared with the penetration for four times as many blows and the difference in penetration is plotted against the lower number of blows. The MCV is defined as ten times the number of blows corresponding to a change in penetration of 5 mm. One of the factors that has to be considered when choosing the limits of suitability of earthwork material is the ability of the earthmoving plant to operate efficiently on the material compacted in an embankment. Accordingly, Parsons related the MCV of soil to the effective operation of various types of earthmoving plant (Table 5.2).

The most critical period during the construction of an embankment is just before it is brought to grade or shortly thereafter. At this time pore water pressures, due to consolidation in the embankment and foundation, are at a maximum. The magnitude and distribution of pore water pressures developed during construction depend primarily on the construction water content, the properties of the soil, the height of the embankment and the rate at which dissipation by drainage can occur. Water contents above optimum can cause high construction pore water pressures, which increase the danger of rotational slips in embankments. Well-graded clayey sands and sand–gravel–clay mixtures develop the highest construction pore water pressures whereas uniform silts and fine silty sands are the least susceptible.

Geogrids or geomats can be used in the construction of embankments over poor ground without the need to excavate the ground and substitute granular fill. They can allow acceleration of fill placement, often in conjunction with vertical band drains. Layers of geogrid or geowebs can be used at the base of an embankment to intersect potential deep failure surfaces or to help construct an embankment over peat deposits. Geogrids also can be used to encapsulate a drainage layer of granular material at the base of the embankment. The use of band drains or a drainage layer helps reduce and regulate differential settlement. A geocell mattress also can be constructed at the base of an embankment that is to be constructed on soft soil (Fig. 5.9). The cells are generally about 1 m high and are filled with granular material. A geocell mattress also acts as a drainage layer. The mattress intersects potential failure planes and its rigidity forces them deeper into firmer soil. The rough interface at the base of the mattress ensures mobilization

Table 5.2 Minimum moisture condition values of soil for the effective operation of various types of earthmoving plant

Type of plant	*Minimum MCV**
Twin-engined scraper	6–9
Single-engined scraper	8–11
Dump truck – 3-axle, rigid chassis, struck capacity less than 15 m^3	8.5–9.5
Dump truck – 2-axle, rigid chassis, struck capacity 15–25 m^3	10–12
Dump truck – 3-axle, articulated chassis, struck capacity less than 15 m^3	5–7
Towed scraper	Not determined

Source: Parsons, 1981. Reproduced by kind permission of The Geological Society.

Note

* Factors affecting values within the range given are wheel load, wheel diameter, tyre width, number of driven wheels.

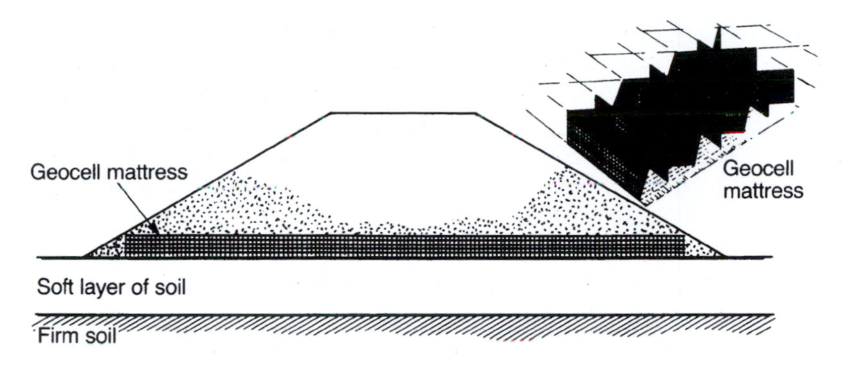

Figure 5.9 A geocell mattress beneath an embankment.

of the maximum shear capacity of the foundation soil and significantly increases stability. Differential settlement and lateral spread are minimized. DuBois *et al.* (1982) described the use of geotextile to reinforce embankment fill and thereby increase the angle of the side slopes.

Cooper and Rose (1999) described the use of stone columns to support an embankment founded on deep alluvial soils at Bristol, England. These soils consisted of very soft clays with interbedded peat, which without treatment would have meant that the embankment would have undergone settlement of some 500–650 mm, with significant differential settlement occurring over lenses of peat. Accordingly, a combination of stone columns and vibrated concrete columns was chosen to facilitate rapid construction, the columns providing the necessary short-term stability and the stone columns, which act as vertical drains, accelerated the rate of settlement. Cooper and Rose emphasized the importance of the correct estimation of stress levels regarding the combined overall shear strength characteristics for the column/soil system.

5.3.1 Soil compaction

Shallow mechanical compaction is used to compact fills and embankments by laying and rolling the soil in thin layers. In other words, it refers to the process by which soil particles are packed more closely due to a reduction in the volume of the void space, resulting from the momentary application of loads such as rolling, tamping or vibration. It involves the expulsion of air from the voids without the moisture content being changed significantly. Hence, the degree of saturation is increased. However, all the air cannot be expelled from a soil by compaction so that complete saturation is not achievable. Nevertheless, compaction does lead to a reduced tendency for changes in the moisture content of a soil to occur. The method of compaction used depends upon the soil type, including its grading and moisture content at the time of compaction, the total quantity of material, layer thickness and rate at which it is to be compacted, and the geometry of the proposed earthworks (Parsons, 1987).

Increasing compaction of a soil with a given moisture content results in closer packing of the soil particles and therefore in increased dry density. This continues until the amount of air remaining in the soil is so reduced that further compaction produces no significant change in volume. Soil is stiff when the moisture content is low and therefore is more difficult to compact. As the moisture content increases it enhances the interparticle repulsive forces, thus separating the particles causing soil to soften and become more workable. This gives rise to higher dry densities and lower air contents. As saturation is approached, however, pore water pressure effects counteract the effectiveness of the compactive effort. Each soil therefore has

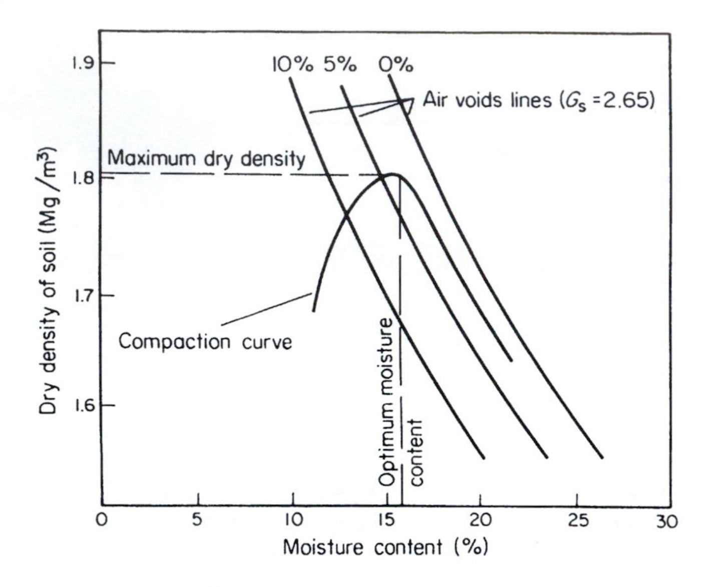

Figure 5.10 Compaction curve showing relationship between dry density and moisture content.

an optimum moisture content at which the soil has a maximum dry density (Fig. 5.10). This optimum moisture content can be determined by the Proctor or standard compaction test (Anon., 1990a). Unfortunately, however, the relationship between maximum dry density and optimum moisture content varies with comparative effort and so test results only can indicate how easily a soil can be compacted. Field tests are needed to assess the actual density achievable by compaction on site.

Most specifications for the compaction of fine-grained soils require that the dry density that is achieved represents a certain percentage of some laboratory standard. For example, the dry density required in the field can be specified in terms of relative compaction or final air void percentage achieved. The ratio between the maximum dry densities obtained *in situ* and those derived from the standard compaction test is referred to as relative compaction. If the dry density is given in terms of the final air percentage, a value of 5–10 per cent usually is stipulated, depending upon the maximum dry density determined from the standard compaction test. In most cases of compaction of clay soils, 5 per cent variation in the value specified by either method is allowable, provided that the average value attained is equal to or greater than that specified.

Specifications for the control of compaction of coarse-grained soils either require a stated relative density to be achieved or stipulate type of equipment, thickness of layer and number of passes required. Field compaction trials should be carried out in order to ascertain which method of compaction should be used, including the type of equipment. Such tests should take account of the variability in grading of the material used and its moisture content, the thickness of the individual layers to be compacted and the number of passes per layer.

Sheepsfoot rollers are best suited to compacting fine-grained soils in which the moisture content should be approximately the same as the optimum moisture content given by the modified AASHO compaction test. At least 24 passes are required in order to achieve reasonably adequate compaction with a sheepsfoot roller and the layer of soil to be compacted should not be more than 50 mm thicker than the length of the feet. The resultant air voids content of soil compacted by sheepsfoot rollers is rather greater than that of soil compacted by smooth-wheel or rubber-tyred rollers.

Rubber-tyred rollers range in weight up to 100 tonnes. The large rollers usually are towed by heavy crawler tractors. Heavy rubber-tyred rollers are not recommended for initial rolling of heavy clay soils but are effective and economical for a wide range of soils from clean sand to silty clay. A rubber-tyred roller produces a smooth compacted surface that unfortunately does not provide significant bonding and blending between successive layers. What is more, the relatively smooth surface of a rubber tyre can neither aerate a wet soil nor mix water into a dry soil. Rubber-tyred rollers are most suited to compacting uniformly graded sands. When used to compact fine-grained soils, rubber-tyred rollers give the best performance when the soil is about 2–4 per cent below the plastic limit. Maximum dry density can be obtained with rubber-tyred rollers by compacting layers of soil, 125–300 mm in thickness, with from 4 to 8 passes. Compaction by rubber-tyred roller is sensitive to the moisture content of a soil. For instance, if the moisture content is on the wet side of the laboratory-determined optimum, then this necessitates an increase in the number of passes to obtain a given density. The optimum moisture content required for compaction of soil by a rubber-tyred roller occurs at a higher degree of saturation than that for a sheepsfoot roller. This may be detrimental in embankments where high construction pore water pressures cannot be tolerated. On the other hand, the construction of earth embankments in regions where rainfall frequently occurs is expedited by the use of rubber-tyred rollers since they help seal the surface of the compacted soil and thereby reduce infiltration.

Smooth-wheel rollers are most suitable for compacting gravels and sands. In coarse-grained soils, the control of moisture content is important in that it should be adjusted to the optimum moisture content. The depth of the layer to be compacted is governed by the nature of the work, as well as the weight of the roller. Generally, however, individual layers vary in thickness, from 50 mm for subgrades to 450 mm for the base of embankments.

Vibratory rollers may be equipped with rubber tyres, smooth wheel drums or tamping feet. Vibration must provide sufficient force (dead weight plus dynamic force) acting through the required distance (amplitude) and give sufficient time for movement of grains (frequency) to take place. The thickness of each compacted layer is governed by the vibration frequency and weight of a vibratory roller. The speed and number of passes of a vibratory roller are critical, inasmuch as they govern the number of dynamic load applications developed at each point of the compacted fill. Vibratory rollers have been used successfully for compacting sand, gravel and some fine-grained soils. Heavyweight rollers with low-frequency vibrations are used to compact gravel; light to medium-weight rollers with high-frequency vibrations are used for sands; and heavyweight rollers with low-frequency vibrations are used for clays. The best compaction is obtained with vibratory rollers when the soil is at or slightly wetter than optimum moisture content.

The engineering properties of soils used in fills, such as their shear strength, consolidation characteristics and permeability, are influenced by the amount of compaction they have undergone (Table 5.3). The desired amount of compaction therefore is established in relation to the engineering properties required for the fill to perform its design function.

The compaction characteristics of clay soil are largely governed by its moisture content (Hilf, 1990). For instance, a greater compactive effort is needed as the moisture content is lowered. It may be necessary to use thinner layers and more passes by heavier compaction plant than required for coarse-grained materials. The properties of clay fills also depend to a much greater extent on the placement conditions than do those of a coarse-grained fill. The shear strength of a given compacted clay soil depends on the density and the moisture content at the time

Table 5.3 Engineering use chart

Typical names of soil groups	*Group symbols*	*Important properties*				*Relative desirability for various uses (graded from 1 (highest) to 14 (lowest))*									
						Rolled earth dams			*Canal sections*		*Foundations*		*Roadways*		
													Fills		
		Permeability when compacted	*Shearing strength when compacted and saturated*	*Compressibility when compacted and saturated*	*Workability as a construction material*	*Homogeneous embankment*	*Core*	*Shell*	*Erosion resistance*	*Compacted earth lining*	*Seepage important*	*Seepage not important*	*Frost heave not possible*	*Frost heave possible*	*Surfacing*
Well-graded gravels, gravel–sand mixtures, little or no fines	*GW*	Pervious	Excellent	Negligible	Excellent	–	–	1	1	–	–	1	1	1	1
Poorly graded gravels, gravel–sand mixtures, little or no fines	*GP*	Very pervious	Good	Negligible	Good	–	–	2	2	–	–	3	3	3	–
Silty gravels, poorly graded gravel–sand–silt mixtures	*GM*	Semi-pervious to impervious	Good	Negligible	Good	2	4	–	4	4	1	4	4	9	5
Clayey gravels, poorly graded gravel–sand–clay mixtures	*GC*	Impervious	Good to fair	Very low	Good	1	1	–	3	1	2	6	5	5	1
Well-graded sands, gravelly sands, little or no fines	*SW*	Pervious	Excellent	Negligible	Excellent	–	–	3 if gravelly	6	–	–	2	2	2	4

Poorly graded sands, gravelly sands, little or no fines	*SP*	Pervious	Good	Very low	Fair	–	–	4 if gravelly	7 if gravelly	–	–	5	6	4	–
Silty sands, poorly graded sand–silt mixtures	*SM*	Semi-pervious to impervious	Good	Low	Fair	4	5	–	8 if gravelly	5 erosion critical	3	7	8	10	6
Clayey sands, poorly graded sand–clay mixtures	*SC*	Impervious	Good to fair	Low	Good	3	2	–	5	2	4	8	7	6	2
Inorganic silts and very fine sands, rock flour, silty or clayey fine sands with slight plasticity	*ML*	Semi-pervious to impervious	Fair	Medium	Fair	6	6	–	–	6 erosion critical	6	9	10	11	–
Inorganic clays of low to medium plasticity, gravelly clays, sandy clays, silty clays, lean clays	*CL*	Impervious	Fair	Medium	Good to fair	5	3	–	9	3	5	10	9	7	7
Organic silts and organic silt–clays of low plasticity	*OL*	Semi-pervious to impervious	Poor	Medium	Fair	8	8	–	–	7 erosion critical	7	11	11	12	–

Table 5.3 (Continued)

Typical names of soil groups	*Group symbols*	*Important properties*				*Relative desirability for various uses (graded from 1 (highest) to 14 (lowest))*									
		Permeability when compacted	*Shearing strength when compacted and saturated*	*Compressibility when compacted and saturated*	*Workability as a construction material*	*Rolled earth dams*			*Canal sections*		*Foundations*		*Roadways*		
						Homogeneous embankment	*Core*	*Shell*	*Erosion resistance*	*Compacted earth lining*	*Seepage important*	*Seepage not important*	*Fills*		
													Frost heave not possible	*Frost heave possible*	*Surfacing*
Inorganic silts, micaceous or diatomaceous fine sandy or silty soils, elastic silts	*MH*	Semi-pervious to impervious	Fair to poor	High	Poor	9	9	–	–	–	8	12	12	13	–
Inorganic clays of high plasticity, fat clays	*CH*	Impervious	Poor	High	Poor	7	7	–	10	8 volume change critical	9	13	13	8	–
Organic clays of medium to high plasticity	*OH*	Impervious	Poor	High	Poor	10	10	–	–	–	10	14	14	14	–
Peat and other highly organic soils	*Pt*	–	–	–	–	–	–	–	–	–	–	–	–	–	–

Source: Wagner, 1957.

of shear. The pore water pressures developed while the soil is being subjected to shear are also of great importance in determining the strength of such soils. For example, if a clay soil is significantly drier than optimum moisture content, then it has a high strength due to high negative pore water pressures developed as a consequence of capillary action. The strength declines as the optimum moisture content is approached and continues to decrease on the wet side of optimum. What is more, there is a rapid increase in pore water pressures as the moisture content approaches optimum. Nevertheless, compaction of clayey soil with moisture contents that are slightly less than optimum frequently gives rise to an increase in strength. This is because the increase in the value of friction more than compensates for the change in pore water pressure. The compressibility of a compacted clay soil also depends on its density and moisture content at the time of loading. However, its placement moisture content tends to affect compressibility more than does its dry density. If a clay soil is compacted significantly dry of optimum and then, if it is saturated, extra settlement occurs on loading. This does not occur when such soils are compacted at optimum moisture content or on the wet side of optimum. A sample of clay soil, especially if it is expansive, compacted on the dry side of optimum moisture content swells more, at the same confining pressure, when given access to moisture than a sample compacted on the wet side. Conversely, a sample compacted wet of optimum shrinks more on drying than a soil, at the same density, which has been compacted dry of optimum. A minimum permeability occurs in a clay soil that is at or slightly above optimum moisture content, after which a slight increase in permeability occurs.

The moisture content has a great influence on both the strength and compaction characteristics of silty soils. For example, an increase in moisture content of 1 or 2 per cent, together with the disturbance due to spreading and compaction, can give rise to very considerable reductions in shear strength, making the material impossible to compact.

Because coarse-grained soils are relatively permeable, even when compacted, they are not affected significantly by their moisture content during the compaction process. In other words, for a given compactive effort, the dry density obtained is high when the soil is dry and high when the soil is saturated, with somewhat lower densities occurring when the soil has intermediate amounts of water. Moisture can be forced from the pores of coarse-grained soils by compaction equipment and so a high standard of compaction can be obtained even if the material initially has a high moisture content. Normally, coarse-grained soils are easy to compact. However, if granular soils are uniformly graded, then a high degree of compaction near to the surface of the fill may prove difficult to obtain particularly when vibrating rollers are used. This problem usually is resolved when the succeeding layer is compacted in that the loose surface of the lower layer also is compacted. According to Anon. (1981), improved compaction of uniformly graded coarse grained material can be brought about by maintaining as high a moisture content as possible by intensive watering and by making the final passes at a higher speed using a non-vibratory smooth wheel roller or grid roller. When compacted, coarse-grained soils have a high load-bearing capacity and are not compressible, and usually are not susceptible to frost action unless they contain a high proportion of fines. If the material contains a significant amount of fines, then high pore water pressures can develop during compaction if the moisture content of the soil is high. It is important to provide an adequate relative density in coarse-grained soil that may become saturated and subjected to static or, more particularly, to dynamic shear stresses. For example, a quick condition may develop in sandy soils with a relative density of less than 50 per cent during ground accelerations of approximately 0.1 g. On the other hand, if the relative density is greater than 75 per cent, liquefaction is unlikely to occur for most earthquake loadings. Consequently, in order to reduce the risk of liquefaction, sandy soils should be densified to a minimum relative density of 85 per cent in foundation areas and to at least 70 per cent in the zone of influence of foundations. Whetton and Weaver (1991) demonstrated that intensive surface compaction with a heavy vibratory roller can be used to achieve the required density in sandy fill to avoid liquefaction.

5.3.2 Precompression

If the soil beneath a proposed embankment is likely to undergo appreciable settlement, then the soil can be treated. One of the commonest forms of treatment is precompression (Johnson, 1970). Precompression has proved an effective means of enhancing the support afforded to embankment foundations and is used to control the magnitude of post-construction settlement. It is well suited for use with soils that undergo large decreases in volume and increases in strength under sustained static loads when there is insufficient time for the required compression to occur. Hence, those soils that are best suited to improvement by precompression include compressible silts, saturated soft clays and organic clays that are either normally consolidated or slightly overconsolidated. Precompression also is used to treat peat soils (Sasaki, 1985).

Precompression normally is brought about by preloading, which involves the placement and removal of a dead load. This compresses the foundation soils thereby inducing settlement prior to construction. If the load intensity from the dead weight exceeds the pressure imposed by the final load, then this is referred to as surcharging. In other words, the surcharge is the excess load additional to the final load and is used to accelerate the process. The ratio of the surcharge load to the final load is termed the surcharge ratio. The surcharge load is removed after a certain amount of settlement has taken place. Application of a surcharge helps reduce the magnitude of any subsequent secondary settlement, especially in soft soils (Alonso *et al.*, 2000). Soil undergoes considerably more compression during the first phase of loading than during any subsequent reloading. If rebound following surcharge removal is to be kept to a minimum, then the amount of permanent load preferably should not be less than approximately one-third that of the surcharge. Normally, however, the amount of expansion following unloading is not significant. The presence of thin layers of sand or silt in compressible soil may mean that rapid consolidation takes place beneath a preloaded area. Unfortunately, this may be accompanied by the development of abnormally high pore water pressures in those layers beyond the edge of the precompression load. This lowers their shearing resistance ultimately to less than that of the surrounding weak soil. Such excess pore water pressures may have to be relieved by vertical drains. Secondary compression may cause significant amounts of settlement in some soils (mainly organic clays and peat). Hence, the effect of surcharge also must take account of secondary compression. The rate of secondary compression appears to decrease with time in a logarithmic manner and its amount is directly proportional to the thickness of a compressible layer at the start of secondary compression.

Accordingly, the geological history of a site and details regarding the types of subsoil, its stratification, strength and compressibility characteristics are of importance as regards the successful use of precompression. Of particular importance is the determination of the amount and rate of consolidation of the soil mass concerned. Hence, sufficient samples must be recovered to locate even thin layers of silt or sand in clay deposits in order to determine their continuity.

The installation of vertical drains (e.g. sand drains, sandwicks or band drains) beneath a precompression load helps shorten the time required to bring about primary consolidation. For example, Robinson and Eivemark (1985) described the use of wick drains and preloading on soft clayey silt. The required settlements under the preload fill were achieved one month after the construction of the preload. Without wick drains a similar amount of settlement would have required up to two years of preloading. The water from the drains flows into a drainage blanket placed at the surface or to highly permeable layers deeper in the soil.

Precompression also can be brought about by vacuum preloading by pumping from beneath an airtight impervious membrane placed over the ground surface and sealed along its edges. In order to ensure the distribution of low pressure, a sand layer is placed on the ground beforehand. The 'negative' pressure, created by the pumps, causes the water in the pores of the soil to move towards the surface because of the hydraulic gradient set up. The degree of vacuum that can be obtained depends on the pump capacity and airtightness of the seals. Values of 60–70 per cent vacuum generally are attained. However, the low pressure in the voids of the soil increases the size of air bubbles that, in turn, may reduce the permeability of the soil. The vacuum method is especially suited to very soft soils where a surcharge may cause instability. Tang and Shang (2000) described the use of vacuum preloading to consolidate alternating layers of silts and clay prior to the construction of a runway at Yaoqiang Airport, China. Some 200–300 mm of settlement was achieved in 80–90 days. A possible advantage of the vacuum method is the use of cyclic preloading of the ground, which can lead to an increased rate of settlement. The vacuum method can produce surcharge loads up to 80 kPa. As in preloading, the method can be improved by use of vertical drains (Woo *et al.*, 1989).

5.4 Reinforced earth

Reinforced earth is a composite material consisting of soil in which occur reinforcing elements that generally consist of strips of galvanized steel or plastic geogrids (Mitchell and Villet, 1987). It also is necessary to provide some form of barrier to contain the soil at the edge of a reinforced earth structure (Fig. 5.11). This facing can be either flexible or stiff but it must be strong enough to retain the soil and to allow the reinforcement to be fixed to it. As reinforced earth is flexible and the structural components are built at the same time as backfill is placed, it is particularly suited for use over compressible foundations where differential settlements may occur during or soon after construction. In addition, as a reinforced earth wall uses small prefabricated components and requires no formwork, it can be readily adapted to required variations in height or shape. Reinforced earth has been used in the construction of embankments, retaining walls and bridge abutments for highways.

Soil, especially granular soil, is weak in tension but if strips of material providing reinforcement are placed within it, the tensile forces can be transmitted from the soil to the strips. The composite material then possesses tensile strength in the direction in which the reinforcement

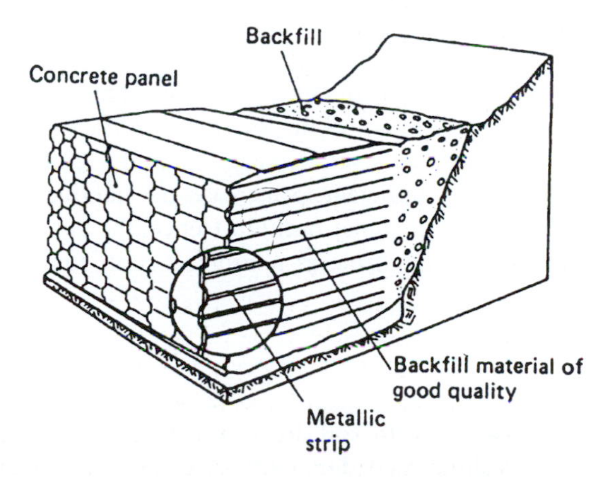

Figure 5.11 Reinforced earth system.

runs. The effectiveness of the reinforcement is governed by its tensile strength and the bond it develops with the surrounding soil. Reinforcing elements can be made from any material possessing the necessary tensile strength, and be of any size and shape that affords the necessary friction surface to prevent slippage and failure by pulling out. They also should resist corrosion or deterioration (Elias, 1990). Steel or aluminium alloy strips, wire mesh, steel cables, glass fibre-reinforced plastic or polymeric geosynthetic materials have been used as reinforcing elements. Grid reinforcement systems consist of polymer or metallic elements arranged in rectangular grids, metallic bar mats and wire mesh. The two-dimensional grid–soil interaction involves both friction along longitudinal members and passive bearing resistance developed on the cross members. Hence, grids are more resistant to pull-out than strips, however, full passive resistance is only developed with large displacements of the order of 50–100 mm (Schlosser, 1990). The most frequently used polymeric materials in earth retaining structures are high-density polyethylene and polypropylene grids. Various attempts have been made to reinforce soil by including fibre within it. The fibres have included geotextile threads, metallic threads and natural fibres such as bamboo. However, there are difficulties associated with the efficient mixing of fibres with soil. In addition, randomly distributed polymeric mesh elements have been included in soils, the meshes interlocking with the soil particles to strengthen the soil. For example, Santoni *et al.* (2001) showed that the unconfined compressive strength of sand could be increased significantly by the inclusion of around 1 per cent by weight of randomly orientated discrete fibres approximately 51 mm in length.

Granular soils compacted to densities that result in volumetric expansion during shear are ideally suited for use in reinforced earth structures (Vidal, 1969). In addition, granular fill usually is free draining and non-frost susceptible, as well as being virtually non-corrosive as far as the reinforcing elements are concerned. If these soils are well drained, then effective normal stress transfer between the strips and soil backfill occurs instantaneously as each lift of backfill is placed. What is more, the increase in shear strength does not lag behind vertical loading. Granular soils behave as elastic materials within the range of loading normally associated with reinforced earth structures. Therefore, no post-construction movements associated with internal yielding or readjustments should be anticipated for structures designed at working stress level.

By contrast, fine-grained materials are not especially suitable for most reinforced earth structures. The adhesion between fill and reinforcement is poor and may be reduced by an increase in pore water pressure. Such soils normally are poorly drained and effective stress transfer is not immediate. Hence, a much slower construction schedule or a low factor of safety during the construction phase is necessary. Moreover, fine-grained materials often exhibit elastoplastic or plastic behaviour, thereby increasing the possibility of post-construction movements. Nonetheless, such fill can be used successfully in structures like embankments. Furthermore, Sridharan *et al.* (1991) showed that fine-grained soil with low internal friction angles can be used as backfill material when layers of sand are used in contact with the reinforcing strips. Generally, a sand layer 15 mm in thickness is sufficient to increase the interfacial friction angle to one almost equal to that of a completely granular backfill. Hence, the pull-out resistance of such a sandwiched system is more or less the same as reinforced earth consisting of coarse-grained soil.

5.5 Soil stabilization

The objectives of mixing additives with soil are to improve volume stability, strength and stress–strain properties, permeability and durability. The development of high strength and stiffness is achieved by reduction of void space, by bonding particles and aggregates together, by maintenance of flocculent structures and by prevention of swelling. The permeability is altered by modification of pore size and distribution. Good mixing of stabilizers with soil is the most important factor affecting the quality of results. The two most commonly used stabilizers are cement and lime.

In order to achieve a uniform material with minimum cement content, good standards of mixing are required. Two basic methods of mixing are employed. In the pre-mix method, all the soil is obtained from a borrow pit. Then it is batched into a mixer. After mixing, the material is transported to site, where it may be spread by hand, grader, stone spreader, concrete spreader or bituminous paver. Compaction is usually by roller. Vibratory rollers are suitable for granular materials, and dead weight rollers for fine-grained soils. Rubber-tyred rollers appear to operate efficiently on a wide range of materials.

In the *mix-in-situ* method, mobile mixers are employed to mix the materials in place on site. If the soil on site is used, savings accrue from a reduction in the volume of earthworks required. However, in some cases it is more economic to treat imported materials that are spread, compacted and levelled, prior to cement or lime spreading and mixing. The machines used for mixing vary from agricultural rotary tillers to large purpose-built units (Fig. 5.12). The smaller machines are limited to processing depths of 150–200 mm, while some of the larger machines can process depths in excess of 400 mm. Standards of mixing are improved, especially in clayey materials, if rotary tillers are used to pulverize the soil into small fragments prior to adding cement or lime. It is often necessary to add water at this stage. After mixing,

Figure 5.12 An autograder producing mix-in-place stabilized soil.

with mechanical spreaders and with any additional water needed to reach the optimum moisture content, compaction and grading to final level is carried out. Finally, the processed layer is covered with a waterproof membrane, commonly bitumen emulsion, to prevent drying out and to ensure cement hydration (Fig. 5.13).

5.5.1 Cement stabilization

The addition of small amounts of cement, that is, up to 2 per cent, modifies the properties of a soil, while larger quantities cause radical changes in the properties. The amount of cement needed to stabilize soil has been related to the durability requirement; put another way, a minimum unconfined compressive strength of 2.8 MPa, after curing at a constant temperature (25°C) and moisture content for seven days, has been widely used. In fact, cement contents may range from 3 to 16 per cent by dry weight of soil, depending on the type of soil and properties required (Table 5.4). Generally, as the clay content of a soil increases, so does the quantity of cement required.

Any type of cement may be used for soil stabilization but ordinary Portland cement is most widely used. The two principal factors that determine the suitability of a soil for stabilization with ordinary Portland cement are, first, whether the soil and cement can be mixed satisfactorily and, second, whether, after mixing and compacting, the soil-cement will harden adequately. Rapid hardening cement with extra calcium is used in organic soils, and a retarded cement will tolerate construction delays. Sulphate-resisting cements are rarely suitable.

Any type of soil, with the exception of highly organic soils or some highly plastic clays, may be stabilized with cement. Although particles larger than 20 mm diameter have been incorporated in soil-cement, a maximum size of 20 mm is preferable since this allows a good

Figure 5.13 Pre-mixed soil–cement being laid as a sub-base for a road in Norfolk, England. Note the cover of bitumen in the centre ground that acts as a waterproof membrane.

Table 5.4 Typical cement requirements for various soil types

Unified soil classification	*Typical range of cement requirement,* (% by wt)*	*Typical cement content for moisture-density test (ASTM D 558),† (% by wt)*	*Typical cement contents for durability tests (ASTM D 559 and D 506),‡ (% by wt)*
GW, GP, GM, SW, SP, SM	3–5	5	3–5–7
GM, GP, SM, SP	5–8	6	4–6–8
GM, GC, SM, SC	5–9	7	5–7–9
SP	7–11	9	7–9–11
CL, ML	7–12	10	8–10–12
ML, MH, CH	8–13	10	8–10–12
CL, CH	9–15	12	10–12–14
MH, CH	10–16	13	11–13–15

Source: Anon., 1990b.

Notes

* Does not include organic or poorly reacting soils. Also, additional cement may be required for severe exposure conditions such as slope protection.

† *ASTM D 558 (1992) Standard Test Method for Moisture-Density Relations of Soil–Cement Mixtures*, American Society for Testing Materials, Philadelphia.

‡ *ASTM D 559 (1982) Standard Methods for Wetting and Drying Tests of Compacted Soil–Cement Mixtures*, American Society for Testing Materials, Philadelphia. ASTM D 506 (1982) *Standard Methods for Freezing and Thawing Tests of Compacted Soil–Cement Mixtures*, American Society for Testing Materials, Philadelphia.

surface finish. At the other extreme, not more than about 50 per cent of the soil should be finer than 0.018 mm. Coarse-grained soils are preferred since they pulverize and mix more easily than fine-grained soils and so result in more economical soil-cement as they require less cement. Typically soils containing between 5 and 35 per cent fines yield the most economical soil-cement. As the grain size of coarse-grained soils is larger than that of cement, the individual grains are coated with cement paste and bonded at their points of contact.

The particles in fine-grained soils are much smaller than cement grains and consequently it is impossible to coat them with cement. In practice, these soils are broken into small fragments that are coated with cement and then compacted. The hydration products formed after short periods of ageing are largely gelatinous and amorphous that, with time, harden due to gradual desiccation. With further curing, poorly ordered varieties of hydrated calcium silicate and hydrated calcium aluminate develop. Ultimately, the hydrated cement forms a skeletal structure, the strength of which depends on the size of the fragments and amount of cement used. A secondary change occurs in clay soils due to the free lime in the cement, which reacts with the clay particles, making the soil less cohesive. However, clay balls may form when the plasticity index exceeds 8 per cent. Where the soil-cement is exposed to weathering, the clay balls break down, which weakens the soil-cement. It is difficult to mix dry cement into heavy clays, and high amounts of cement have to be added to bring about appreciable changes in their properties. Indeed, clay soils with liquid limits exceeding 45 per cent and plasticity indices above 18 per cent are not usually subjected to cement stabilization (Croft, 1968). Heavy clays can, however, be pre-treated with 2–3 per cent cement or, more frequently, with hydrated lime. This reduces the plasticity, thereby rendering the clay more workable. After curing for 1–3 days, the pre-treated clay is stabilized with cement.

Furthermore, the suitability of a clay soil for cement stabilization is controlled by its texture, and chemical and mineralogical composition. Both kaolinite and well-crystallized illite have little or no effect on the hydration and hardening process of cement stabilization. By contrast, the expansive clay minerals, depending upon their relative activities, may have a profound influence on the hardening of cement (Bell, 1995). For instance, the affinity of montmorillonite for lime reduces the pH value of the aqueous phase, and because of their deficiency in lime, the cementitious products developed during curing are inferior to those of non-expansive clays. This means that the strengths developed are lower, and unless enough cement is added to supply the free lime requirement to promote hardening, the properties of the clay are not enhanced. Up to 15 per cent of cement has to be added to montmorillonitic clays to modify them significantly.

The effects of weathered minerals such as degraded illites, chlorites and vermiculites can be similar to that of montmorillonite. Gibbsite, with its high response to lime, may retard stabilization of certain lateritic soils. In general, lime will be more suitable than cement for soils containing these components.

Organic matter and excess salt content, especially sulphates, can retard or prevent hydration of cement in soil–cement mixtures. In fact, soils containing more than 2 per cent organic material usually are considered unacceptable (Anon., 1990b) and soils with pH values of less than 5 are unsuitable for economic stabilization. Organic matter retards the hydration of cement because it preferentially absorbs calcium ions. However, the addition of calcium chloride or hydrated lime can provide a source of calcium and consequently may enable some of these soils to be treated.

The disintegration of cement- (or lime-) stabilized soils, due to sulphate attack, only occurs when the soil has an appreciable clay fraction and when there is an increase in moisture content above that at the time of compaction. There is a risk of deterioration of clay–cement mixtures when the content of SO_3 and SO_4 in the soil is 0.2 and 0.5 per cent or more respectively, or if the SO_3 content in groundwater exceeds $300\,mg\,l^{-1}$. Sulphate resistant cement is no better than ordinary Portland cement for stabilizing clay soils containing sulphates.

The properties developed by compacted cement-stabilized soils are governed by the amount of cement added on the one hand and compaction on the other (Table 5.5). The density achieved is largely a function of compactive effort, soil texture and, in the case of clay soils, the type of clay minerals present, which determine the soil–moisture response (small increases occur in the compacted densities of both kaolinitic and illitic clay soils, but not those containing montmorillonite). Adequate compaction is essential for successful stabilization but prolonged delays between mixing and compaction reduce the maximum density attainable. Soil-cement undergoes shrinkage during drying and soil-cement made with different types of soil shows different crack patterns (crack widths are smaller and more closely spaced in clay soil-cement than in soil-cement made with sand). The development of shrinkage cracks can be reduced by keeping the surface of soil-cement moist beyond the normal period of curing and placing it at slightly below optimum moisture content. The permeability of most soils is reduced by the addition of cement. In multiple-lift construction, the permeability along horizontal surfaces is generally greater than along vertical.

The strength of soil-cement tends to increase in a linear manner with increasing cement content, but in different soils it increases at different rates (Fig. 5.14). Increased pulverization increases the strength of soil-cement. A lengthy period of mixing brings about partial hydration of the cement with a resultant loss of strength at constant density. If compaction is delayed, the cement begins to hydrate and therefore the soil-cement begins to harden. As a result the mixture becomes more difficult to compact. Compaction should be completed within two hours of mixing. The strength of soil-cement gradually increases as the time taken in curing increases. Also, the higher the temperature, the more rapid is the gain of strength. Soil-cement will harden in cold weather providing the temperature does not fall below 0 °C. Excessive drying increases strength but tends to crack the soil-cement. By contrast, strength is reduced by soaking. This is

Table 5.5 Typical average properties of soil–cement and soil–lime mixtures

(a) Typical mean* properties of soil–cement†

Soil type (unified) classification)	*Compressive strength* (MN/m^2)	*Young's Modulus, E* (MN/m^2)	*CBR*	*Permeability* (m/s)	*Shrinkage*	*Comments*
GW, GP, GM, GC, SW	6.5	2×10^4	>600	Decreases ($\approx 2 \times 10^{-7}$)	Negligible	Too strong; liable to wide spaced cracks‡
SM, SC	2.5	1×10^4	600	Decreases	Small	Good material
SP, ML, CL	1.2	5×10^3	200	Decreases ($\approx 1 \times 10^{-9}$)	Low	Fair material
ML, CL, MH, VH	0.6	2.5×10^3	<100	Increases	Moderate	Poor material
CH, OL, OH, Pt	<0.6	1×10^3	<50	Increases ($\approx 1 \times 20^{-11}$)	High	Difficult to mix; needs excessive cement

Source: Ingles and Metcalf, 1972.

Notes
* Variations of approximately 50 per cent around the mean may be expected.
† Values shown are at 10 per cent cement content.
‡ Good material if less cement is used.

(b) Typical mean* properties of soil–lime†

Soil type (unified classification)	*Compressive strength* (MN/m^2)	*Young's Modulus, E* (MN/m^2)	*CBR*	*Permeability* (m/s)	*Linear shrinkage* (%)	*Plasticity index*	*Comments*
GW, GP, GM, GC, SW, SP	≤0.3	–	75	Increases ($\geq 10^{-7}$)	Nil	Non-plastic	Suitable only for plasticity reduction
SM, SC	1.1	$< 1 \times 10^2$	50	Increases	Very low	<5	Poor to fair material
ML, CL, MH, VH	2.5	2×10^4	30	Increases	5	10	Good material
CH	3.5	1×10^3	25	Increases ($\leq 10^{-10}$)	10	20	Good effect, fair to good material
OL, OH, Pt	≤1.0	1×10^2	≤10	–	–	15	Not suitable *per se*†

Source: Ingles and Metcalf, 1972.

Notes
* Variations of approximately 50 per cent around the mean may be expected.
† Values shown are at the additive level optimum for the respective soil types.
‡ Results may be improved by admixture of the lime with gypsum.

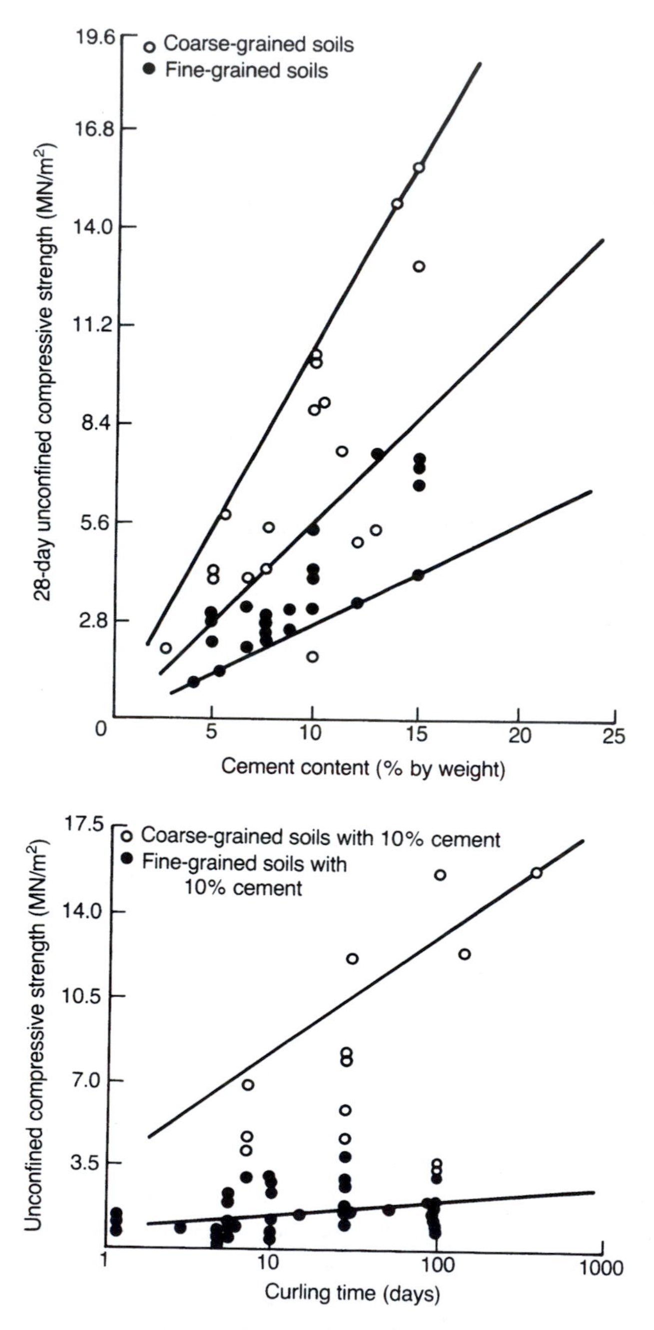

Figure 5.14 (a) Relationship between cement content and unconfined compressive strength for soil–cement mixtures. (b) Effect of curing time on unconfined compressive strength of some soil–cement mixtures. *Source*: Anon., 1990b.

particularly the case with clayey soils. Typical ranges of 7- and 28-day unconfined compressive strengths for soaked soil-cement specimens are given in Table 5.6. Values of Young's modulus vary between 140 and 20 000 MPa depending on soil type and increase with increasing cement content. Plastic deformation occurs on cyclic loading, and under such conditions failure may

Table 5.6 Ranges of unconfined compressive strengths of soil-cement

Soil type	*Soaked compressive strength** (MN/m^2)	
	7-day	*28-day*
Sandy and gravelly soils: Unified groups GW, GC, GP, GM, SW, SC, SP, SM	2.07–4.14	2.76–6.90
Silty soils: Unified groups ML and CL	1.72–3.45	2.07–6.21
Clayey soils: Unified groups MH and CH	1.38–2.76	1.72–4.14

Source: Anon., 1990b.

Note

* Specimens moist-cured 7 or 28 days, then soaked in water prior to strength testing.

occur at 60–70 per cent of the ultimate failure strength. When subjected to constant loading, soil-cement undergoes creep.

The principal use of soil-cement is as a base material underlying pavements. One of the reasons soil-cement is used as a base is to prevent pumping of fine-grained subgrade soils into the pavement above. The thickness of the soil-cement base depends upon subgrade strength, pavement design period, traffic and loading conditions, and thickness of the wearing surface. Frequently, however, soil-cement bases are around 150–200 mm in thickness. Soil-cement has been used to provide slope protection for embankments.

5.5.2 Lime stabilization

Lime stabilization refers to the stabilization of soil by the addition of burned limestone products, either calcium oxide (i.e. quicklime, CaO) or calcium hydroxide, $Ca(OH)_2$. On the whole, quicklime appears as a more effective stabilizer of soil than hydrated lime. Moreover, when quicklime is added in slurry form, it produces a higher strength than when it is added in powder form. The process is similar to cement stabilization except that lime stabilization is applicable to much heavier clayey soils and clayey gravels. The addition of lime has little effect on soils that contain either a small clay content or none at all. It also has little effect on highly organic soils.

Lime usually reacts with most soils with a plasticity index ranging from 10 to 50 per cent. Those soils with a plasticity index of less than 10 per cent require a pozzolan for the necessary reaction with lime to take place, pulverized fly ash commonly being used. Other pozzolans used for the enhancement of lime stabilization include granulated blast-furnace slag and expanded shale. Lime stabilization of heavy clays gives the soil a more friable structure, which is easier to work and compact, although a lower maximum density is obtained. The reaction of lime with montmorillonitic clays is quicker than with kaolinitic clays, in fact, the difference may amount to a few weeks (Bell, 1996). A silica surface, however, should not be considered 'available' if it is bound to a similar surface by ions that are not readily exchangeable. Accordingly, illite and chlorite, although attacked, are much less reactive than montmorillonite.

When lime is used to stabilize clay soil it forms a calcium silicate gel that coats and binds lumps of clay together and occupies the pores in the soil. Reaction proceeds only while water is present and able to carry calcium and hydroxyl ions to the surfaces of the clay minerals (i.e. while the pH value is high). Consequently, reaction ceases on drying and very dry soils do not react with lime. The quantity of lime added should ideally be related to the clay mineral content as the latter is needed for reaction. Addition of up to 3 per cent of lime will modify silty clays, heavy clays and very heavy clays, while 4 per cent is required for the stabilization of silty clay, and 8 per cent for stabilization of heavy and very heavy clays. A useful guide is to allow 1 per cent of lime (by weight of dry soil) for each 10 per cent of clay material in the soil. More exact prescriptions usually can be made after tests at the guide value and at slightly each side of the guide value. The amount of water used in lime stabilization is dictated by the requirements of compaction. However, if quicklime is used, then extra water may be necessary in soils with less than 50 per cent moisture content to allow for the very rapid hydration process. Furthermore, because the lime–soil reaction involves exsolution, it is inhibited if the water content of the soil is too low. Hence, moist curing always is desirable. Mixing is important, and if mixing is delayed after the lime has been exposed to air, then carbonation of the lime will reduce its effectiveness. Therefore, it is desirable that mixing be effected as soon as possible and certainly within 24 hours of exposure to air.

In most cases, the effect of lime on the plasticity of clay soils is more or less instantaneous. In other words, the plasticity is reduced (this is brought about by an increase in the plastic limit and reduction in the liquid limit of the soil), as is the potential for volume change (Fig. 5.15). In kaolinitic clay soils, however, lime treatment at times increases the plasticity index. The addition of lime to clayey soils increases the optimum moisture content and reduces the maximum dry density for the same compactive effort (Fig. 5.16). The significance of these changes depends upon the amount of lime added and the amount of clay minerals present. As lime treatment flattens the compaction curve, a given percentage of the prescribed density can be achieved over a much wider range of moisture contents so that relaxed moisture-control specifications are possible. The permeability of a clayey soil is increased when treated with lime; the higher the clay fraction, the more the permeability of the soil–lime mixture increases. Soils compacted on

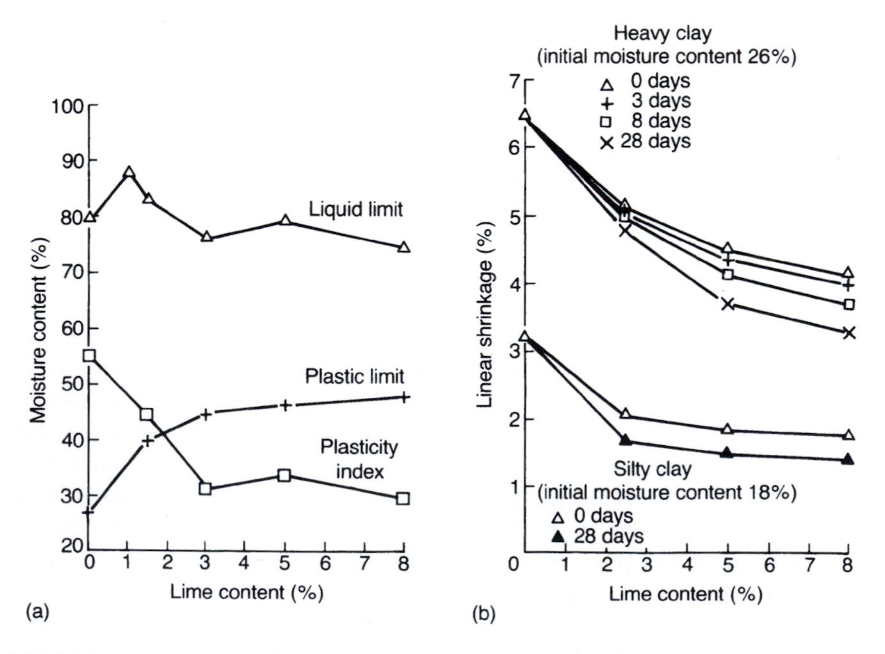

Figure 5.15 (a) Influence of the addition of lime on the plastic liquid limits, and plasticity of a clay soil of high plasticity. (b) Influence of lime on the linear shrinkage of heavy and silty clay soil.
Source: Bell, 1993.

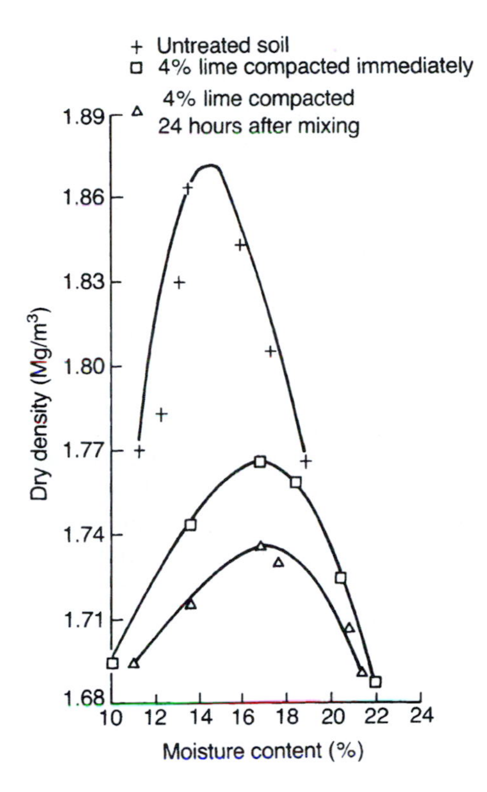

Figure 5.16 Influence of the addition of lime on the compaction curves of clay soil.
Source: Bell, 1993.

the dry side of optimum moisture content develop a higher permeability than those compacted wet of optimum. However, with increasing age the permeability declines.

The strength of soil–lime mixtures depends on several factors such as soil type, and the type and amount of lime added. For example, montmorillonitic clays give lower strengths with dolomitic limes than with high calcium or semi-hydraulic limes (Bell, 1996). Kaolinitic clays, on the other hand, yield the highest strengths when mixed with semi-hydraulic limes, and the lowest strengths are obtained with high-calcium limes. Soil mixed with low lime content attains a maximum strength in less time than that to which a higher content of lime has been added. Strength does not increase linearly with lime content and in fact excessive addition of lime reduces strength. This decrease is because lime itself has neither appreciable friction nor cohesion. The optimum lime content tends to range from 4.5 to 8 per cent, the higher values being required for soils with higher clay fractions. Curing time is another factor influencing strength of lime-stabilized soil, a steady gain in strength over months being characteristic. Higher temperatures accelerate curing, and this gives rise to higher strengths. The soil–lime reaction is retarded or may cease once temperatures fall below 4 °C. Hence, strength development more or less ceases with the onset of cold weather, and strength loss because of cyclic freezing and thawing is cumulative throughout the winter. Residual strength at the end of the freeze-thaw period consequently must be sufficient to guarantee the integrity and stability of the soil–lime layer. Yet another factor influencing the strength of lime-stabilized soils is their natural moisture content, the strength decreasing with increasing moisture content. Even 3 months after stabilization, the shear strengths of soils with high saturated moisture contents remain low. On the other hand, soil–lime mixtures compacted at moisture contents above optimum, after brief periods of curing, attain higher strengths than those compacted with moisture contents less than optimum. This is probably because the lime is more uniformly diffused and occurs in a more homogeneous curing environment. The strength of soils treated at or below optimum moisture content usually can be improved by spraying each lift with water after compaction.

The principal use of the addition of lime to soil is for subgrade and sub-base stabilization and as a construction expedient on wet sites where lime is used to dry out the soil. Lime stabilization also has been used in embankment construction for roads and railways to enhance the shear strength of the soil. As far as lime stabilization for roadways is concerned, stabilization is brought about by the addition of between 3 and 6 per cent lime (by dry weight of soil). Subgrade stabilization involves stabilizing the soil in place or stabilizing borrow materials that are used for sub-bases. After the soil, which is to be stabilized, has been brought to grade, the roadway should be scarified to full depth and width and then partly pulverized. A rooter, grader-scarifier and/or disc harrow for initial scarification, followed by a rotary mixer for pulverization, is employed. An idea of the amount of lime that is required can be obtained from Fig. 5.17. After sprinkling dry lime with water, preliminary mixing is required to distribute the lime thoroughly throughout the soil to the proper depth and width and to pulverize the soil to a depth of some 50 mm. During mixing, water is added to bring the soil slightly above the optimum moisture content. Alternatively, lime slurries of varying concentrations, depending on the percentage of lime required and the optimum moisture content, are applied to the soil. A typical mix consists of 1 tonne of lime to 2500 litres of water. Since lime in the form of slurry is much less concentrated than dry lime applications, usually two or more passes are required to provide the specified amount. Although the slurry method promotes more uniform distribution of lime,

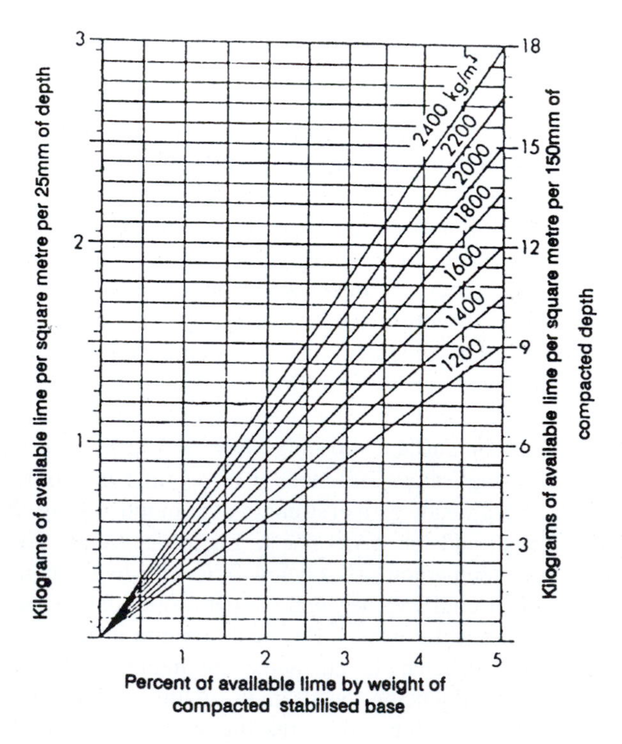

Figure 5.17 Rate of application of lime.
Source: Anon., 1982.

it proves disadvantageous on wet soils. Generally, its use is restricted to projects that require smaller amounts of lime (4 per cent or under) since, at higher percentages, so much water is required to carry the lime in solution that the soil would exceed optimum moisture content most of the time.

After initial mixing, the lime-treated soil should be lightly compacted to reduce carbonation and evaporation loss. Initial curing takes 24–48 hours, water being added to maintain as near optimum mixture conditions as possible. Where necessary the soil–lime should be remixed so that it is friable enough for the required degree of pulverization to be achieved. Soil–lime mixtures should be compacted to high density in order to develop maximum strength and stability. This necessitates compacting at or near the optimum moisture content. Generally, a 5–7 day curing period is required. Two types of curing are employed. First, the soil is kept damp by sprinkling, with light rollers being used to keep the surface knitted together. Second, in membrane curing, the stabilized soil is either sealed with one application of asphalt emulsion within one day of final rolling or primed with increments of asphalt emulsion applied several times during the curing period.

Heavy costs can be incurred when construction equipment and transport become bogged down on site due to heavy rainfall turning clayey ground into mud. In these circumstances, an economical method of drying out the top layers of soil is essential. This can be achieved by the use of quicklime in granular form, it combining with soil moisture to produce hydrated lime. In the process, heat is generated which helps the soil dry out.

The lime slurry pressure injection method involves pumping hydrated lime slurry under pressure into expansive clay soils (Fig. 5.18). The method has been used to improve bearing capacity and reduce differential settlement, to stabilize failed embankment slopes and to minimize subgrade pumping beneath railways and roads. Usually, a depth of 2.1 m is sufficient to emplace lime slurry below the critical zone of changes in moisture content, although depths

Figure 5.18 Lime slurry injection.

of up to 40 m can be stabilized. Injection takes place until refusal. If the points of injection are closely spaced (e.g. at 1.5 m), then the lime slurry forms a network of horizontal sheets interconnected by vertical veins. After injection, the network of soil–lime thickens as a result of calcium ion exchange. This means that there is a gradual improvement in the swell–shrink behaviour of the soil.

5.5.3 Other materials used for soil stabilization

Numerous other materials have been used for stabilizing soil. For example, pulverized fly ash (PFA) can be used by itself to improve the physical properties of a soil or in conjunction with lime or cement to form a binder. Pulverized fly ash is a pozzolan, that is, it reacts with CaO and water to form cementitious material. Because there is a slower gain in strength when PFA is mixed with lime and because PFA has a greater sensitivity to low temperatures, cement usually has been preferred as the mixing agent. It is advantageous to use rapid-hardening cement. Pulverized fly ash–cement or PFA–lime stabilization is best suited to sands and gravels with low clay contents. Well-graded gravel, for example, can be stabilized by the addition of 10 per cent PFA (based on dry weight) and 5 per cent ordinary Portland cement. The compressive strength of the stabilized material is determined by the characteristics of the PFA, the cement, the degree of compaction and the efficiency of mixing.

Bituminous material has been used to stabilize granular soils, providing cohesion and thereby enhancing their strength. In the case of fine-grained soils, the addition of bitumen waterproofs the soil and in this way the loss of strength associated with increasing moisture content is reduced. The quantity of bitumen required for successful waterproofing of fine-grained soils normally increases with increasing clay fraction. Even so, soils with similar clay contents may require different amounts of bitumen depending on their affinity for water. Stabilization of organic soils with bitumen is not successful. Bitumen may be applied hot or as cutbacks or emulsions. Cutbacks, in which the bitumen is thinned by the addition of a volatile oil (e.g. paraffin or naphtha), are more often used than hot bitumen as the latter sometimes proves difficult to mix with soil. As the volatile evaporates from cutback bitumen, the bitumen itself is deposited in the soil. Soil-asphalt is mixed either at a central mixing plant or in place and is compacted by smooth-wheel, pneumatic-tyred or sheepsfoot rollers. Usually, the compacted soil-asphalt is left for two weeks or more so that the volatiles are lost before a seal coat or bituminous treatment is applied to the surface. In temperate climates, cutback bitumen can be used to stabilize sands since excess water can be removed by compaction. In the wet-sand mix, 4–10 per cent of cutback bitumen is mixed with the wet sand to which 1–2 per cent of hydrated lime has been added previously to assist coating. In Britain, it has been found that once the silt fraction exceeds 5 per cent, the mixture becomes difficult to compact. In warmer climates, 10 per cent or more silt material can be tolerated, and in uniformly graded sands the figure may rise to 25 per cent. Al-Homoud *et al.* (1995) investigated the use of cutback asphalt to stabilize expansive and collapsible soils in Jordan. They found that cutback asphalt was more effective than cement in reducing the amount of expansion but less successful than lime, although its use proved more expensive than either cement or lime. Cutback asphalt also reduced the collapse potential of the soils tested.

Bituminous emulsions are used cold and consist of a fine suspension of particles of bitumen in water. The bitumen is deposited as the suspension coagulates or breaks. Bituminous emulsions are generally only suitable for soil stabilization in climates where rapid drying conditions occur. Nevertheless, emulsions have been used in Britain, for example, to treat fine grained soils. In such instances, cement or lime is added to the mix after treatment with bitumen. This causes the emulsion to break, it also absorbs some of the excess moisture as a result of hydration, and affords extra strength to the compacted soil. About 5–7.5 per cent of emulsion and 3–5 per cent cement or lime are added to soils treated in this way.

Construction methods used for bituminous stabilization are similar to those used for soil–cement mixtures. However, the optimum moisture content usually is below that necessary for efficient mixing. Accordingly, except for sands, it is often necessary to allow a period of time for the mix to dry before compaction takes place. In the mix-in-place method, bitumen commonly is added in several passes, each layer being partially mixed before the next pass in order to avoid saturating the surface of the soil.

Some other soil stabilizers have a useful application in certain special circumstances, especially where temporary solutions are acceptable. For example, lateritic soils can be stabilized with the addition of sodium hydroxide, which is applied with the mix water and facilitates compaction. Sodium hydroxide reacts with aluminium-bearing minerals, notably kaolinite, and a substantial increase in strength occurs after an initial curing period.

5.6 Use of geotextiles in road construction

The improvement in the performance of a pavement attributable to the inclusion of geotextiles comes mainly from their separation and reinforcing functions. This can be assessed in terms of either an improved system performance (e.g. reduction in deformation or increase in traffic passes before failure) or reduced aggregate thickness requirements (where reductions of the order 25–50 per cent are feasible for low-strength subgrade conditions with suitable geotextiles).

The most frequent role of geotextiles in road construction is as a separator between the sub-base and subgrade. This prevents the subgrade material from intruding into the sub-base due to repeated traffic loading and so increases the bearing capacity of the system. The savings in sub-base materials, which would otherwise be lost due to mixing with the subgrade, can sometimes cover the cost of the geotextile. The range of gradings or materials that can be used as sub-bases with geotextiles normally is greater than when they are not used. Nevertheless, the sub-base materials preferably should be angular, compactible and sufficiently well graded to provide a good riding surface.

If a geotextile is to increase the bearing capacity of a subsoil or pavement significantly, then large deformations of the soil-geotextile system generally must be accepted as the geotextile has no bending stiffness, is relatively extensible, usually is laid horizontally and restrained from extending laterally. Thus, considerable vertical movement is required to provide the necessary stretching to induce the tension that affords vertical load-carrying capacity to the geotextile. Therefore, geotextiles are likely to be of most use when included within low density sands and very soft clays. Although large deformations may be acceptable for access and haul roads they are not acceptable for most permanent pavements. In this case, the geotextile at the sub-base, subgrade interface should not be subjected to mechanical stress or abrasion. When geotextiles are used in temporary or permanent road construction they help redistribute the load above any local soft spots that occur in the subgrade. In other words, the geotextile deforms locally and progressively redistributes load to the surrounding areas thereby limiting local deflections. As a result, the extent of local pavement failure and differential settlement are reduced. Geotextiles are used in asphaltic paving systems, the geotextiles being impregnated with asphalt. Hence, they function as moisture barriers against the infiltration of water from the road surface, as well as acting as stress absorbing interlayers between pavement layers (Marienfeld and Guram, 1999).

Under wheel loading, the use of geogrid reinforcement in road construction helps restrain lateral expansive movements at the base of the aggregate. This gives rise to improved load redistribution in the sub-base that, in turn, means a reduced pressure on the subgrade. In addition, the cyclic strains in the pavement are reduced. The stiff load bearing platform created by the interlocking of granular fill with geogrids is utilized effectively in the construction of roads over weak soil. Reduction in the required aggregate thickness of 30–50 per cent may be achieved. Geogrids can be used within a granular capping layer when constructing roads over variable subgrades. They also have been used to construct access roads across peat bogs, the geogrid enabling the roads to be 'floated' over the surface. Geogrids are used at times in areas where the potential for subsidence to occur exists.

In arid regions an impermeable geomembrane can be used as a capillary break to stop the upward movement of salts where they would destroy the road surface. Geomembranes also can be used to prevent the formation of ice lenses in permafrost and other frost prone regions. The geomembrane must be located below the frost line and above the water table. Henry and Holtz (2001) found that geocomposites can reduce the amount of frost heave when the soil water suction head above them is high.

Where there is a likelihood of uplift pressure disturbing a road constructed below the piezometric level, it is important to install a horizontal drainage blanket. This intercepts the rising water and conveys it laterally to drains at the side of the road. A geocomposite can be used for effective horizontal drainage. Problems can arise when sub-bases are used that are sensitive to moisture changes, that is, they swell, shrink or degrade. In such instances, it is best to envelop the sub-base, or excavate, replace and compact the upper layers of sub-base in an envelope of impermeable geomembrane.

5.7 Drainage

Drainage systems can be used to control high-pressure gradients or high pore water pressures (Cedergren, 1986). Such systems include aggregate filters and synthetic filter materials, as well as pipes and other conduits. For instance, Perry *et al.* (2000) described the use of filter drains to intercept groundwater below subgrade in cuttings along a bypass. Drains must be capable of removing all water that flows into them without allowing an excessive build-up of head. They also should be designed to prevent the loss of fines from adjacent soil and thereby avoid becoming clogged. Accordingly, when pipes are used in drainage systems, the slots in the pipes should be small enough to retain soil particles thereby preventing their movement into the pipes. Nonetheless, there should be enough slots to allow free movement of water into the pipes. Obviously, the filter materials around the pipes must have gradations that are compatible with the sizes of the slots or holes.

5.7.1 Drainage of slopes (see also Chapter 2)

Drainage improves slope stability by reducing excess pore water pressures. The distribution of groundwater within a slope must be investigated, as must the most likely zone of failure so that the extent of the groundmass that requires drainage treatment can be defined.

Unrestrained flow of surface run-off over a slope should be prevented, for instance, by the installation of a lined drainage ditch at the top of an excavated slope in order to collect the water draining from above. Water from the surface of a slope may be drained by a system of herringbone ditch drainage that leads into an interceptor drain at the foot of the slope (Fig. 2.38). The latter can convey water to watercourses or drainage ponds. Where persistent seepage occurs on a slope, for instance, from a perched water table, it can be intercepted by a slope drain, lined with a geotextile to reduce the ingress of fines and backfilled (Perry *et al.*, 2000).

Support and drainage may be afforded by counterfort-drains, where an excavation is made in sidelong ground, likely to undergo shallow parallel slides (Fig. 5.19). Deep trenches are cut

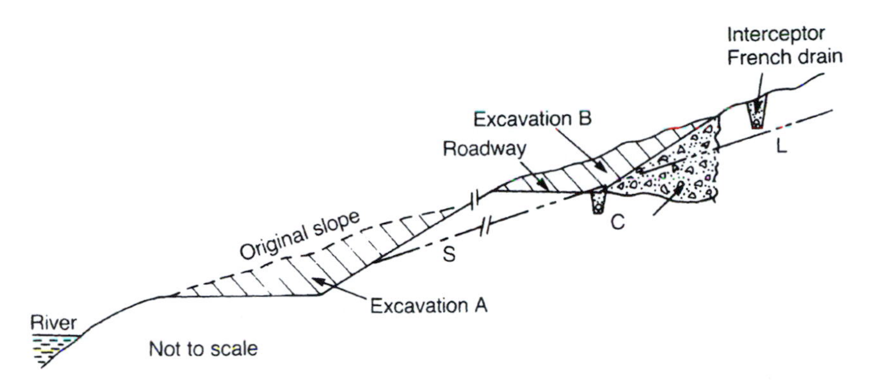

Figure 5.19 Excavation A was made to provide a level recreational area for a school. The ground, being near the bank of a large river in its lower reaches, consisted chiefly of alluvium. Movement started on the slope line SL but went unnoticed. Later, road construction began at excavation B to provide access to the school but was impossible to continue until the hillside was stabilized. Counterfort drains, C, were constructed every 3 m over the length of moving section. The depth of the slip occurred at about 2.5 m and the drains had to be installed below the slip surface. It eventually became possible to construct the roadway.

into the slope, lined with filter fabrics and filled with granular filter material. The granular fill in each trench acts as a supporting buttress or counterfort, as well as providing drainage. However, counterfort-drains must extend beneath the potential failure zone, otherwise they merely add unwelcome weight to the potential slipping mass.

Horizontal drains have been used to stabilize clay slopes. They remove the water by gravity flow. The effectiveness of the drains is governed by their diameter and spacing (the larger the diameter or the lower the spacing, the greater the increase in slope stability), as well as their location in relation to a potential critical slip zone. In other words, the amount of improvement depends on how closely the drains are positioned in relation to the critical zone but there is no additional benefit in extending the length of the drains beyond where this zone intersects the top of the slope.

Subsurface drainage of slopes can be brought about by the construction of drainage galleries. Galleries are indispensable in the case of large slipped masses (Fig. 2.39); in some instances drainage has been carried out over lengths of 200 m or more. Drillholes with perforated pipes are much cheaper than galleries and are satisfactory over short lengths but it is more difficult to intercept water-bearing layers with them. When individual benches are drained by horizontal holes, the latter should lead into a properly graded interceptor trench, which is lined with impermeable material.

5.7.2 Sand drains, sandwicks and band drains

The time taken to reach a certain degree of primary consolidation of a clayey soil is inversely proportional to its permeability. Obviously, if the distance water has to travel in soil is reduced, then consolidation will occur more quickly. This can be achieved by employing drains. A regular pattern of vertical drains permits a radial, as well as a vertical flow of water from the soil. The water from the drains is conveyed into a drainage blanket placed at the surface or to highly permeable layers deeper in the soil. With water escaping radially to the drains, in addition to escaping in the vertical direction, the amount of time taken for consolidation can be reduced to a fraction of that for vertical flow only, the time being governed mainly by the spacing of the drains. Vertical drains also bring about acceleration in the rate of gain in shear strength and reduction in the lateral transmission of excess pressure. They frequently are used beneath road embankments that have to cross over compressible soils in order to allow construction to take place more quickly.

Design values for the coefficient of consolidation for radial flow to drains must take into consideration the influence of variable permeability within a formation, especially the influence of any continuous thin layers of sand or silt that can dominate the permeability of a clay soil. The effect of such layers can be estimated from the results of field permeability tests. Usually, the ratio of horizontal to vertical permeability for such soils ranges between 5 and 10.

The spacing and pattern of vertical drains are now fairly well standardized. At the majority of sites, triangular or square patterns with spacings between 1 and 4 m are used. For example, the diameter of the dewatered cylinder varies from 1.05 times the spacing when sand drains are positioned on a triangular grid to 1.13 times the drain spacing when they are placed on a square grid (Barron, 1948). Spacings of sand drains 1.5–2.5 m are the most commonly adopted. The depth to which drains are installed is normally equal to the thickness of the soil concerned. For depths up to 20 m, vertical drains often prove an economic solution. Beyond 20 m depth, however, the cost of placing drains increases sharply due to the extra effort involved in their installation. For instance, even normally consolidated clays are firm to stiff at depths of 30–40 m. In fact, McGown and Hughes (1981) questioned the value of treating such depths since most settlement and likelihood of shear failure occurs at shallow depths.

Generally, a drainage blanket of granular material is laid over the area where vertical drains are installed. A considerable quantity of water may be discharged into the drainage blanket, especially during the early stages of construction, if a large volume of soil is being drained. The amount of water that can be discharged into the drainage blanket from a given layout of drains can be derived for any particular degree of consolidation. In this way the effectiveness of the drainage blanket can be assessed and, if necessary, its design can be adjusted accordingly.

Sand drains have proved particularly efficient in stratified soils because of the higher permeability of such soils parallel to the bedding. Generally, they are constructed by inserting casing into the ground and then placing sand in the hole under air pressure as the casing is withdrawn. The casing may be driven or jet-placed and the soil displaced by using a closed-end mandrel, jetted from the hole or removed by augers or rotary-drilling methods. The effects of soil disturbance when a drain is installed can give rise to a smear zone being developed around the drain. The formation of a smear zone depends on how the drain is emplaced. For example, in the case of sand drains, mandrels are more likely to give rise to smear zones than augering.

Nowadays wick or band drains are more frequently used than sand drains as they are generally more economic. The sandwick was introduced because it proved difficult to construct sand drains with diameters smaller than 250 mm. Sandwicks are formed by pneumatically filling stockings made of woven-bonded polypropylene with graded sand. The most convenient diameter for sandwicks is around 65 mm. They are installed in holes of small diameter and have proved effective when used to consolidate suspect alluvial deposits. Sandwicks can be emplaced by a variety of methods depending on the type of soil involved (Robinson and Eivemark, 1985). These include solid or hollow stem flight auger rotary drilling, rotary wash boring, jetting, and driven and vibrated casing. They can be placed to greater depths than sand drains and the fabric stocking ensures continuity at the time of placement. The fabric stocking means that a sandwick is continuous, and flexible, and so can adapt to vertical settlement or lateral deformation. It also acts as a filter so reducing the possibility of clogging. The likelihood of intercepting lenses of dubious material in heterogeneous soils is far greater if sandwicks with smaller spacings are installed rather than more widely spaced sand drains.

Since the efficiency of a vertical drain largely depends on its cross-sectional circumference rather than on its cross-sectional area, band-shaped drains are more effective than drains having a circular cross-section. Several types of flat band drains are available, most consisting of a flat plastic core containing drainage channels surrounded by a thin filter layer. The discharge capacity of band-shaped drains varies considerably, depending upon the type of drain used. It also is a function of the effective lateral earth pressure against the drain sleeve. In most cases, the filter is partially squeezed into the channel system of the core due to the pressure of the surrounding soil. Accordingly, the discharge is lowered as a result of the reduction in cross-sectional area. Furthermore, the discharge capacity is reduced by 'ageing' once a band drain is installed. Other factors that may reduce the effectiveness of band drains include fines entering the channel system and buckling of the drain under large vertical strains. According to Bergado *et al.* (1991), as with other types of vertical drains, smear effects due to installation may significantly reduce the effectiveness of band drains. However, the effects of smear caused by installing band drains by displacement methods can be reduced by the correct choice of drain filter fabric and the size of the installation lance. A filter fabric initially allows the finer particles of soil, notably those of clay size, to pass through. Hence, piping occurs, removing fines from the smear zone, which leads to the formation of a natural graded filter in the soil. Band drains generally are installed by displacement methods, usually by a lance (Fig. 5.20). Spacings at 1–2.5 m centres and depths extending to between 10 and 20 m give the highest production rates for most methods of installation.

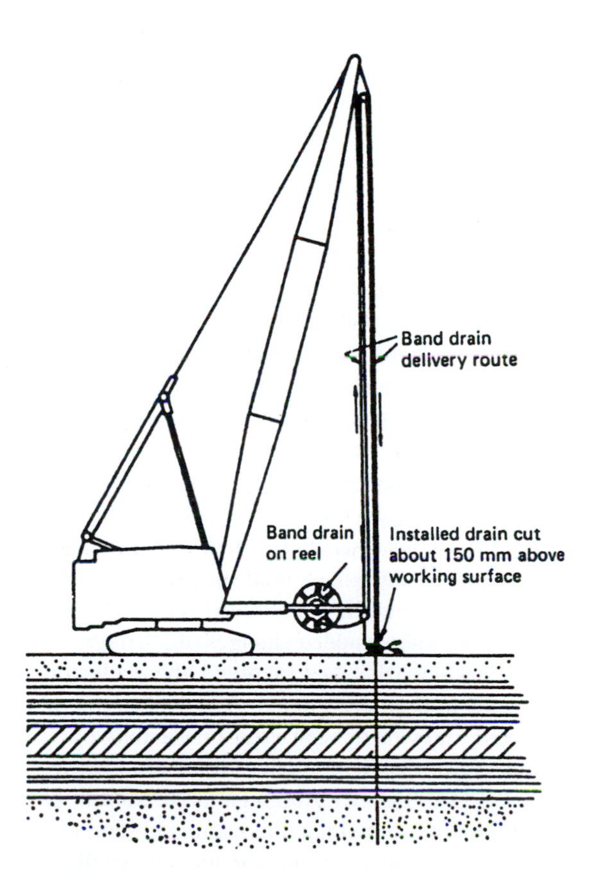

Figure 5.20 Placement of band drains.

Cylindrical columns can be formed in clay soils by mixing the clay with unslaked lime (see Chapter 4). This increases the permeability of the columns to between 100 and 1000 times greater than that of the surrounding clay (Broms and Boman, 1979). Hence, the lime columns act as vertical drains, as well as reinforcing the soil. One lime column of 500 mm diameter has the same drainage capacity as three 100 mm wide band drains. One of the advantages of lime columns is that their installation creates little disturbance in the surrounding soil, which enhances their performance as a drain.

5.8 Railroads

Railroads have and continue to play an important role in national transportation systems, although the construction of new railroads on a large scale is something that belongs to the past. Nonetheless, railroads continue to be built such as those associated with high speed networks like the 109 km long Channel Tunnel Rail Link (CTRL) in southern England. In this case the route was designed to follow existing transport corridors, that is, alongside major roads or railways, in order to limit adverse environmental effects. Consequently, this meant that the influence of geological conditions would be highly significant in that there was little opportunity to benefit from more advantageous ground conditions when selecting the route alignment. A vital part of the work concerned with a high speed railroad, with trains travelling at speeds of up to 300 km h^{-1}, is the trackbed support. In other words, the dynamic behaviour of foundations and earthworks involves a detailed understanding of the soil–structure interaction. This distinguishes a modern high-speed railway from other railways or highways. Obviously, the grades and curvature of railroads impose stricter limits than do those associated with highways. Furthermore, underground systems are being and will continue to be constructed beneath many large cities in order to convey large numbers of people from one place to another quickly and efficiently, and thereby provide some relief to congested surface traffic. Most of what has been said above concerning the location and subsequent investigation of highway alignments can be applied to railroads.

Topography and geology are as important in railroad construction as in highway construction. This can be illustrated by the CTRL that crosses the various formations that occur in the Upper Cretaceous and Tertiary systems in Kent, England, as well as superficial deposits and made ground. These comprise clays, mudstones, limestones, chalk, sandstones, sands, gravels and alluvium. Construction involves the excavation of tunnels beneath the River Thames and part of London, and cut and cover tunnels in soft alluvium, weathered Gault Clay, and the Atherfield and Weald Clays, embankments on soft clays and peats, and cuttings in chalk and other rock types, as well as bridges, notably over the River Medway. Obviously, a very stable trackbed and earthworks are necessary for a high-speed railroad. Accordingly, drainage is an important aspect of trackbed design so that surface water is efficiently removed and groundwater level is maintained below the subgrade (Fig. 5.21). Another important factor is trackbed stiffness. Consequently, where sudden changes of stiffness are likely to occur, notably between embankments and bridges, this has involved the inclusion of service blocks and layers of cement-stabilized and well-graded granular material to provide a gradual transition in stiffness. No overconsolidated clays have been used in the railway embankments in order to achieve the minimum maintenance requirement, the fill material being obtained either from the Upper Chalk or the Folkestone Beds (sands). Although the behaviour of freshly placed chalk fill varies appreciably, depending on its strength and moisture content, both chalk and sand proved good fill. In fact, Parsons (1981) devised a classification of chalk for use in embankments, together with suggested measures that would avoid or minimize instability. However, the sands

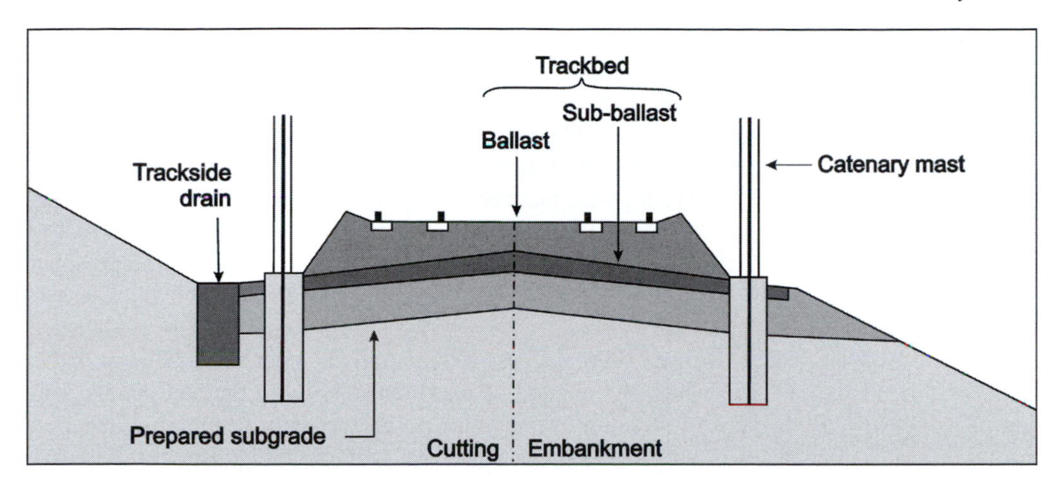

Figure 5.21 Trackside drainage.

of the Folkestone Beds are easily erodible by rain in both embankments and cuttings before vegetation becomes established on them, washouts therefore have been a problem. Solution features in some areas of the Chalk outcrop have called for special drainage measures, and soil nailing has been used to stabilize cuttings through this type of ground. Deep dry-soil mixing has been used where embankments have been constructed over soft clays and peats.

Railroads also may be affected by geohazards. For instance, the effects of earthquakes have been referred to above. In rugged terrain in particular, trains may be interrupted for a time by rockfalls, landslides or mudflows. Areas prone to such hazards along a railroad need to be identified and, where possible, stabilization or protective measures carried out (see Chapter 2). In some areas, because of the nature of the terrain, it may be impossible to stabilize entire slopes. In such cases, it may be possible to install warning devices that will be triggered by such movements and so cause trains to be halted. Be that as it may, lengths of track that are subjected to mass movements should be inspected regularly. A notable example of a landslide-prone area in England that has affected the railroad running through it is Folkestone Warren. Folkestone Warren is a naturally unstable stretch of coastline, 3.2 km long, between Folkestone and Dover in Kent. A succession of slides affected the railway, located at the foot of the cliff, after it was built in the mid-nineteenth century but the most notable occurred in 1915 when a series of rotational slides and chalk falls from the cliffs blocked the line. Fortunately, a train was stopped by a signalman before it entered the section of the line affected by the slide but subsequent movements derailed the train. It took five years to rectify the damage. According to Hutchinson *et al.* (1980), the main causes of the movements were intense marine erosion along the toe of previously slid material and the development of high pore water pressures along potential slip surfaces. Bell and Maud (1996) described a number of slides that had displaced railway lines north of Durban, South Africa. The remedial works involved at the various sites included some combination of subsurface drainage (including sand blanket drains), installation of anchor cables or unloading the slope by bulk excavation. Railway services also may be interrupted seriously by flooding.

Railway track formations normally consist of a layer of coarse aggregate, the ballast, in which the sleepers are embedded. The ballast may rest directly on the subgrade or, depending on the bearing capacity, on a layer of blacketing sand. The function of the ballast is to provide a free-draining base that is stable enough to maintain the track alignment with the minimum of maintenance. The blanketing sand provides a filter that prevents the migration of fines from the subgrade into the ballast due to pumping. The ballast must be thick enough to retain the track in position and to prevent intermittent loading from deforming the subgrade whilst the aggregate beneath the sleepers must be able to resist abrasion and attrition. Hence, strong, good wearing, angular aggregates are required such as provided by many dense siliceous rocks. However, Smith and Collis (1993) suggested that because many limestones offer low resistance to abrasion, they tend not to be used. The thickness of the ballast can vary from as low as 150 mm for lightly trafficked railroads up to 500 mm on railroads that carry high-speed trains or heavy traffic. The blanketing layer of sand normally has a minimum thickness of 150 mm.

In terms of soil mechanics, Brown (1996) suggested that a railway tract could be regarded as a special type of pavement in which the method of load transmission to the soil differs from that of a highway. Both types of structure are in intimate contact with the ground and both consist of one or more layers of unbound granular materials. As far as the design of rail track is concerned, this must take account of vertical plastic strain and vertical permanent deformation at formation level caused by repeated loading (Li and Selig, 1998). The former is necessary to avoid plastic flow that is responsible for progressive shear failure in the upper subgrade whilst the latter involves the overall deformation of the subgrade. The transient deviator stress level in the subgrade is influenced by the thickness of the granular layers and therefore the plastic strain. The design of an adequate granular layer thickness is intended to prevent these two common types of railroad subgrade failures. Under repeated loading, differential permanent strains develop in the ballast of a rail track that bring about a change in the rail line and level. If the voids in the ballast are allowed to fill with fine-grained material, then Brown maintained that a failure condition can develop with high plastic strains. The fines may be derived by pumping from the sub-ballast or subgrade if they become saturated. The failure condition is aided in part by the decrease in permeability of the ballast caused by the fines that, in turn, causes drainage to be impeded. Accordingly, railway track ballast requires regular attention to maintain line and level. Anderson and Key (2000) referred to pneumatic ballast injection (PBI) that involves lifting the sleepers by blowing smaller size aggregate between the ballast and the base of the sleeper, thereby realigning and relevelling the track. This apparently represents a better type of track maintenance than the commonly used ballast tamping method. The size of the aggregate and the thickness of the injected layer are critical in the terms of post-maintenance behaviour. Brown pointed out that permanent deformation of the track is only influenced by the subgrade in the long term except when the track is newly constructed. Unless high moisture contents are developed, normal transient stress levels on clay subgrades tend to give rise to a stable situation after initial plastic strain has occurred. Shin *et al.* (2002) described the use of geogrid reinforcement to strengthen the subgrade beneath the ballast and to reduce settlement due to cyclic loading brought about by passing high-speed trains.

5.9 Bridges

Like tunnels, the location of bridges may be predetermined by the location of the routeways of which they form part. Consequently, this means that the ground conditions beneath bridge locations must be adequately investigated. This is especially the case when a bridge has to cross a river. The geology beneath a river should be correlated to the geology on both banks and drilling beneath the river should go deep enough to determine in place solid rock. The data obtained should enable the bridge, piers and abutments to be designed satisfactorily. The geological conditions may be complicated by the presence of a buried channel beneath a river.

Buried channels generally originated during the Pleistocene epoch when valleys were deepened by glacial action and sea levels were at lower positions. Subsequently, much of these valleys were occupied by various types of sediment, which may include peat.

The ground beneath bridge piers crossing rivers has to support not only the dead load of the bridge but also the live load of the traffic that the bridge will carry, as well as accommodating the horizontal thrust of the river water. The choice of foundations usually is influenced by a number of factors. For example, in the case of the second motorway bridge across the River Severn, linking England and Wales, the two main piers of the central cable-stayed bridge and the 49 supports of the approach viaducts are mostly founded on exposed sandstone and mudrock (Kitchener and Ellison, 1997). The existence of sound rock near the surface meant that spread foundations could be used without the need for widespread piling. However, piled foundations were adopted where alluvial deposits overlie bedrock. Because of the strong currents and the high tidal range in the estuary, 37 of the foundations used precast concrete open-bottom caissons weighing up to 2000 tonnes. These were floated out from the casting yard and put in place by specially adapted barges. The caissons had the advantage of offering permanent formwork shells for the concrete infill. All the caisson foundations were designed as spread foundations except for the backspan caissons of the cable-stayed bridge. Positive loading on the main span of the cable-stayed structure results in uplift forces on the backspan piers so that the total vertical load on these foundations is much less than on similar caissons supporting the viaduct. This meant that resistance to sliding would have to be provided by some means in addition to the self-weight. Hence, the size of the caissons was increased from 10 to 13 m wide, which provided enough weight to prevent overturning, and shear keys in the form of tubular steel piles, 2 m in diameter, were cast into holes excavated 13 m in depth below general foundation level. Barton pier of the suspension bridge across the River Humber, England, is located in the river about 500 m from the south bank and is founded in Kimmeridge Clay (Jurassic). The pier therefore was designed as a cellular structure, 42 m long by 11 m wide and 15 m high, supported on twin hollow cylindrical caissons approximately 24 m in diameter (Simm, 1984). The caissons were founded 36 m below river bed level and were sunk by dredging from an artificial island of sand.

The anchorages for suspension bridges have to resist very high pull-out loads. For example, the Hessle anchorage, on the north side of the river, for the Humber Bridge has to resist a horizontal pull of 38 000 tonnes (Simm, 1984). It is founded at a depth of around 21 m below ground level in chalk. The Barton anchorage on the south bank extends 35 m below ground level into Kimmeridge Clay and has cellular foundation, some 72 m long and 40 m wide, constructed within a framework of diaphragm walls. The anchorage also is designed to resist a horizontal pull from the main cables of 38 000 tonnes. Resistance is derived from friction at the clay/concrete interface at the base of the anchorage, from the passive resistance at the front and from wedge action at the sides.

When a bridge is constructed across a river, its effective cross-sectional area is reduced by the piers that, in turn, leads to an increase in its velocity of flow. This and the occurrence of eddies around the piers enhances scouring action. McManus (1971) noted that the cross-sectional width of the water along the line of the railway bridge across the River Tay, Scotland, was reduced by 10 per cent by the piers. He found that far less scouring took place where the river bed was formed of cobbles and gravel than where it was sand since the critical erosion velocity needed to move the larger material was not attained by scouring action. In addition, McManus maintained that scour had removed more than 6 m of sediment from beneath a wide area within the central zone of the bridge and that subsequently much of the area had been refilled with sediment. Scouring of river bed materials around bridge piers has caused the failure of some bridges.

As illustrated, during the September 1987 floods in Natal, South Africa, 28 bridge structures and 130 bridge approaches were damaged (Bell, 1994). In many cases, damage to bridges was temporary such as that due to clogging with debris or minor erosion. On the other hand, 12 bridges suffered structural failure, one due to the deck floating off, three due to changes in the course of the river and eight due to foundation failures. The damage was caused by very high peak flows, the highest intensity of rain occurring when the rivers were already full. Bridges were damaged, often as a result of failures due to inadequate foundations, exacerbated by build-up of debris at the bridge and excessive scour around supporting caissons. The bridge that suffered the most damage during the flooding was the John Ross Bridge spanning the Tugela River, which is situated on the main road between Richards Bay and Durban. Foundation failure of a pier and caisson caused the collapse of the entire structure, requiring the bridge to be totally rebuilt before the road network could be re-established (Fig. 5.22). The abutments of the John Ross Bridge were on spread footings and the six piers had caisson foundations keyed into bedrock (tillite or diamictite of the Dwyka Group) sunk through river bed sediment. The depth to bedrock varied between 10 and 17 m. The collapse of the structure was initiated

Figure 5.22 The collapsed John Ross Bridge over the Tugela River, Natal, South Africa.

at pier 5. Abnormal scour occurred around the caisson and debris that collected around the pier probably combined with the force of the water to exert pressure on the pier itself. This resulted in movement of the caisson and corresponding failure of the pier it supported. The problem of scouring is accentuated in estuaries, especially where the flow patterns of the ebb and flood tides are different. Furthermore, the positions of the banks and channel of an estuary in particular are liable to change.

Bridges obviously are affected by ground movements such as subsidence. Subsidence movements can cause relative displacements in all directions and so subject a bridge to tensile and compressive stresses. Although a bridge can have a rigid design to resist such ground movements, it usually is more economic to articulate it thereby reducing the effects of subsidence (Fig. 5.23). Bearings and expansion joints must be designed to accommodate the movements. In the case of multi-span bridges, the piers should be hinged at the top and bottom to allow for tilting or change in length, tilting or rocker bearings being incorporated at each pier. Jacking pockets can be used to maintain the level of the deck. As far as shallow abandoned room and pillar workings are concerned, it frequently is necessary to fill voids beneath a bridge with grout. For instance, a major highway is carried on a viaduct through part of Gateshead, England, this concrete structure being carried by piers, the load on each being some 2000 tonnes. The area is underlain by several coal seams, all believed to have been worked, the uppermost being at an average depth of 16.8 m. After the site investigation it was decided, in order to avoid possible subsidence damage to a relatively high risk structure, that the workings in the uppermost seam should be pressure grouted. The construction of the new bridge across the River Ure, which forms part of the Ripon bypass referred to above, can be used to illustrate the case of potential cavity collapse of natural voids. Cooper and Saunders (2002) indicated that the bridge has a strengthened heavy-duty steel girder construction designed with sacrificial supporting piers so that it will withstand the loss of any one pier without collapsing. Each pier has a larger than normal foundation pad to span a future small subsidence event, and monitoring devices are incorporated to detect separation from the deck. Piling was undertaken for one of the piers. Stone columns were used to improve the alluvial ground along the flood plain.

Seismic forces in earthquake-prone regions can cause damage to bridges and so must be considered in bridge design. According to Kubo and Katayama (1978), most seismic damage to low bridges has been caused by failures of substructures resulting from large ground deformation or liquefaction. Indeed, it appears that the worst damage is sustained by bridges located on soft ground, especially that capable of liquefaction. Failure or subsidence of backfill in a bridge approach, leading to an abrupt change in profile, can prevent traffic from using the approach even if the bridge is undamaged. Such failure frequently exerts large enough forces on abutments to cause damage to substructures. On the other hand, Kubo and Katayama maintained that seismic damage to superstructures due purely to the effects of vibrations is rare. Nonetheless, as a result

Figure 5.23 An articulated bridge to reduce the effects of longwall mining subsidence.

of substructure failure, damage can occur within bearing supports and hinges, which combined with excessive movement of substructures can bring about collapse of a superstructure. By contrast, damage to freeway structures during the San Fernando earthquake, California, in 1971, showed that the effects of vibrations can be responsible for catastrophic failures of high bridges that possess relatively little overall stiffness. Arch type bridges are the strongest whereas simple or cantilever beam type bridges are the most vulnerable to seismic effects. Furthermore, the greater the height of substructures and the number of spans, the more likely is a bridge to collapse. The Rion Antirion Bridge near Patras, Greece, is a four-span cable-stayed bridge, located on normally consolidated silty clays extending to depths in excess of 80 m in a region noted for its seismicity. According to Pecker and Teyssandier (1998), the seismic design motion has a peak ground acceleration of 0.48 g at mudline level. As the soil beneath each of the four piers has to support not only the load carried by the piers but also the large seismic forces and the hydrodynamic forces of the water, ground improvement measures were necessary. These consisted of driving 400 hollow steel piles, 20 m in length and 2 m in diameter at 5 m spacings, at each pier location. The seismic design of the piers was performed in a multi-step approach in which the forces applied to the foundation were computed from a dynamic analysis of the soil conditions in order to determine the bearing capacity using limit equilibrium methods and has been described by Pecker and Teyssandier.

Existing bridges may be damaged by seismic forces and even if they are not, because of their age, they also may need to be retrofitted. After the Loma Prieta earthquake in California in 1989 ($M = 7.1$), it was decided to undertake a seismic vulnerability study of the Golden Gate Bridge, San Francisco, which was completed in the 1930s when earthquake engineering was in its early stages. The bridge is located some 80 km to the north of the epicentre of the Loma Prieta earthquake. This study included analyses of the geology and topography adjacent to the bridge; seismic risk investigations examining earthquake probability, strength and proximity; and computer analysis of bridge behaviour under varying earthquake conditions. It showed that although the bridge has performed well in previous earthquakes, it was vulnerable to damage if affected by an earthquake of magnitude 7 or above with an epicentre near to the bridge. Consequently, it was decided that a retrofit should be carried out that would allow the bridge to withstand an earthquake with a magnitude of 8.3 on the nearby San Andreas Fault. Basically the retrofit involves minimizing the impact of seismic energy by giving the Bridge structures more flexibility to move and so help dissipate the seismic forces whilst at the same time strengthening certain structures to accommodate these forces (Fig. 5.24).

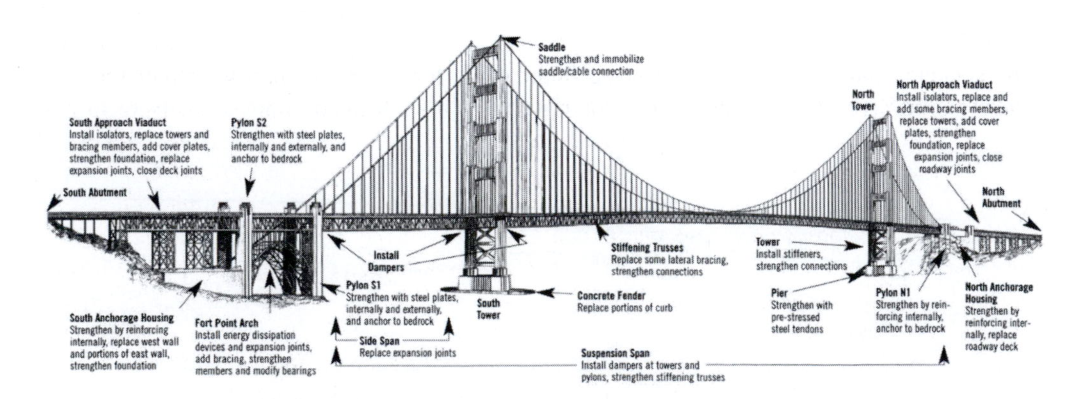

Figure 5.24 Retrofitting the Golden Gate Bridge, San Francisco.

Figure 5.25 Collapse of a section of the San Francisco–Oakland Bridge, California, caused by the Loma Prieta earthquake of October 1989.

During the same earthquake the San Francisco-Oakland Bridge, which is located between the San Andreas Fault and the Hayward Fault, sustained damage, including the partial collapse of one span (Fig. 5.25). It therefore was decided that the eastern part of the bridge should be replaced by a new section constructed to the north of the existing bridge. The bridge type selected was an asymmetrical, single tower, self-anchored suspension bridge to be located in the deep water near Yerba Buena Island and a haunched concrete skyway structure to be located in shallower water to the east. This necessitated a site investigation, the first phase of which consisted of land and marine geophysical surveys, offshore drilling with geophysical logging of drillholes, cone penetrometer testing and extensive sampling for the laboratory-testing programme. The second phase of investigation involved obtaining pier specific data for foundation design. The foundations for the new section will include a main span west pier, main pylon, main span east pier and 14 paired piers for the skyway. Large diameter drilled shafts socketed into bedrock will form the foundations for the main span west pier that will be located on Yerba Buena Island. The main pylon will be located approximately 65 m east of Yerba Buena Island in 12–17 m of water and the foundations will consist of eight bored piles or drilled shafts, 2.5 m in diameter. In this case the foundation footprint consists of a thickening wedge of unconsolidated sediment overlying dipping greywackes interbedded with thin siltstones and claystones. The main span east pier will be situated some 450 m east of Yerba Buena Island overlying the edge of a palaeochannel occupied by Holocene Bay Mud, which generally is soft to firm, normally to slightly overconsolidated clay of high plasticity. Sixteen steel open-ended pipe piles, 2.5 m in diameter, will be driven into dense sand beneath the mud to form the foundation for this pier. The foundations for the skyway similarly will be steel open-ended pipe piles, 2.5 m in diameter, driven some 100 m in depth into dense granular sediments of Pleistocene age.

References

Al-Amoudi, O.S.B., Asi, I.M. and El-Nagger, Z.R. 1995. Stabilization of arid, saline sabkha using additives. *Quarterly Journal Engineering Geology*, **28**, 369–379.

Al-Homoud, A.S. and Tubeileh, T.K. 1997. Damaging landslides of cut slopes along major highways in Jordan: causes and remedies. *Environmental and Engineering Geology*, **3**, 183–204.

Al-Homoud, A.S., Khedaywi, T. and Al-Ajlouni, A.M. 1995. Engineering and environmental aspects of cutback asphalt (MC-76) stabilization of swelling and collapsible soils. *Environmental and Engineering Geology*, **1**, 497–506.

Alonso, E.E., Gens, A. and Lloret, A. 2000. Precompression design for secondary settlement. *Geotechnique*, **50**, 645–656.

Andersland, O.B. and Anderson, D.M. (eds) 1978. *Geotechnical Engineering for Cold Regions*. McGraw-Hill, New York.

Andersland, O.B. and Ladanyi, B. 1994. *An Introduction to Frozen Ground Engineering*. Chapman & Hall, New York.

Anderson, W.F. and Key, A.J. 2000. Model testing of two-layer railway track ballast. *Proceedings American Society Civil Engineers, Journal Geotechnical and Geoenvironmental Engineering*, **126**, 317–323.

Anon. 1965. *Soils and Geology – Pavement Design for Frost Conditions*. Technical Manual TM 5-818-2, United States Corps of Engineers Department of Army, Washington, DC.

Anon. 1981. *Code of Practice on Earthworks, BS 6031*. British Standards Institution, London.

Anon. 1982. *Lime Stabilization Construction Manual*. Seventh Edition. National Lime Association, Washington, DC.

Anon. 1990a. *Methods of Test for Soils for Civil Engineering Purposes, BS 1377:1990*. British Standards Institution, London.

Anon. 1990b. State-of-the-art report on soil-cement. *American Concrete Institute Materials Journal*, **87**, No. 4, 395–417.

Arumala, J.O. and Akpokodje, E.G. 1987. Soil properties and pavement performance in the Niger Delta. *Quarterly Journal Engineering Geology*, **20**, 287–296.

Barron, R.A. 1948. Consolidation of fine grained soils by drains wells. *Transactions American Society Civil Engineers*, **133**, 718–754.

Barry, A.J., Trigunarsyah, B., Symes, T. and Younger, J.S. 1995. Geogrid reinforced piled road over peat. In: *Engineering Geology of Construction*, Engineering Geology Special Publication No. 10, Eddleston, M., Walthall, S., Cripps, J.C. and Culshaw, M.G. (eds), The Geological Society, London, 205–210.

Beaumont, T.E. 1977. *Techniques for the Interpretation of Remote Sensing Imagery for Highway Engineering Purposes*. Transportation and Road Research Laboratory, Report LR753, Department of Environment, Crowthorne, Berkshire.

Beinbrech, G. and Hillmann, R. 1997. EPS in road construction-current situation in Germany. *Geotextiles and Geomembranes*, **15**, 39–57.

Bell, F.G. 1987. The influence of subsidence due to present day coal mining on surface development. In: *Planning and Engineering Geology*, Engineering Geology Special Publication No. 4, Culshaw, M.G., Bell, F.G., Cripps, J.C. and O'Hara, M. (eds), The Geological Society, London, 359–367.

Bell, F.G. 1993. *Engineering Treatment of Soils*. E & FN Spon, London.

Bell, F.G. 1994. Floods and landslides in Natal and notably the greater Durban region, September 1987: a retrospective view. *Bulletin Association Engineering Geologists*, **31**, 59–74.

Bell, F.G. 1995. Cement stabilization of clay soils with examples. *Environmental and Engineering Geoscience*, **1**, 139–151.

Bell, F.G. 1996. Lime stabilization of clay minerals and soils. *Engineering Geology*, **42**, 223–237.

Bell, F.G. 1998. The geotechnical properties and behaviour of a post-glacial lake clay, and its cementitious stabilization. *Geotechnical and Geological Engineering*, **16**, 167–169.

Bell, F.G. 2000. *Engineering Properties of Soils and Rocks*. Fourth Edition, Butterworth-Heinemann, Oxford.

Bell, F.G. and Maud, R.R. 1994. Dispersive soils: a review from a South African perspective. *Quarterly Journal Engineering Geology*, **27**, 195–210.

Bell, F.G. and Maud, R.R. 1995. Expansive clays and construction, especially of low-rise structures: a viewpoint from Natal, South Africa. *Environmental and Engineering Geoscience*, **1**, 41–59.

Bell, F.G. and Maud, R.R. 1996. Landslides associated with the Pietermaritzburg Formation in the greater Durban area, South Africa: some case histories. *Environmental and Engineering Geoscience*, **2**, 557–573.

Bergado, D.T., Asakami, H. and Alfaro, M.C. 1991. Smear effects of vertical drains on soft Bangkok Clay. *Proceedings American Society Civil Engineers, Journal Geotechnical Engineering Division*, **117**, 1509–1530.

Blight, G.E. 1976. Migration of subgrade salts damages thin pavements. *Proceedings American Society Civil Engineers, Transportation Engineering Journal*, **102**, 779–791.

Boyer, B.W. 1997. Sinkholes, soils, fractures and drainage: Interstate 70 near Frederick, Maryland. *Environmental and Engineering Geoscience*, **3**, 469–485.

Broms, B.B. and Boman, P. 1979. Lime columns. A new foundation method. *Proceedings American Society Civil Engineers, Journal Geotechnical Engineering Division*, **105**, 539–556.

Brown, S.F. 1996. Soil mechanics in pavement engineering. *Geotechnique*, **46**, 383–426.

Brunsden, D., Doornkamp, J.C., Fookes, P.G., Jones, D.K.C. and Kelly, J.H.M. 1975. Large-scale geomorphological mapping and highway engineering design. *Quarterly Journal Engineering Geology*, **8**, 227–253.

Burkart, B., Goss, G.C. and Kern, J.P. 1999. The role of gypsum in production of sulfate-induced deformation in lime-stabilized soils. *Environmental and Engineering Geoscience*, **5**, 173–187.

Casagrande, A. 1932. Discussion on frost heaving. *Proceedings Highway Research Board*, Bulletin No. 12, Washington DC, 169.

Castell, M.A., Ingraffea, A.R. and Irwin, L.H. 2000. Fatigue crack growth in pavements. *Proceedings American Society Civil Engineers, Journal Transportation Engineering*, **126**, 283–290.

Cedergren, H.R. 1986. *Drainage of Highway and Airfield Pavements*. Wiley, New York.

Charman, J.M. 1988. *Laterite in Road Pavements*. Special Publication 47, Construction Industry Research and Information Association (CIRIA), London.

Cook, D.A. and Roy, M.R. 1984. A review of the geotechnical properties of Somerset alluvium using data from the M5 motorway and other sources. *Quarterly Journal Engineering Geology*, **17**, 235–242.

Cooper, A.H. and Saunders, J.M. 2002. Road and bridge construction across gypsum karst in England. *Engineering Geology*, **65**, 217–223.

Cooper, M.R. and Rose, A.N. 1999. Stone column support for an embankment on deep alluvial soils. *Proceedings Institution Civil Engineers, Geotechnical Engineering*, **137**, 15–25.

Croft, J.B. 1968. The problem of predicting the suitability of soils for cementitious stabilization. *Engineering Geology*, **2**, 397–424.

Croney, D. and Jacobs, J.C. 1967. *The Frost Susceptibility of Soils and Road Materials.* Transport Road Research Laboratory, Report LR90, Crowthorne.

De Bruyn, I.A. and Bell, F.G. 2001. The occurrence of sinkholes and subsidence depressions in the Far West Rand and Gauteng Province, South Africa, and their engineering implications. *Environmental and Engineering Geoscience*, **7**, 281–295.

Dowling, J.W.P. and Beaven, P.J. 1969. Terrain evaluation for road engineers in developing countries. *Journal Institution Highway Engineers*, **16**, 5–15.

DuBois, D.D., Bell, A.L. and Snaith, M.S. 1982. A fabric reinforced trial embankment. *Quarterly Journal Engineering Geology*, **15**, 217–225.

Early, K.R. and Skempton, A.W. 1972. Investigations of the landslide at Walton's Wood, Staffordshire. *Quarterly Journal Engineering Geology*, **5**, 19–41.

Elias, V. 1990. *Durability/Corrosion of Soil Reinforced Structures*. Federal Highway Administration, Publication No. FHWA-RD-89-186, Department of Transportation, McLean, Virginia.

Esper, P. and Tachibana, E. 1998. Lessons from the Kobe earthquake. In: *Geohazards and Engineering Geology*, Engineering Geology Special Publication No. 15, Maund, J.M. and Eddleston, M. (eds), The Geological Society, London, 110–116.

Fookes, P.G. and French, W.J. 1977. Soluble salt damage to surfaced roads in the Middle East. *Journal Institution Highway Engineers*, **24**, No. 12, 10–20.

Forster, A., Culshaw, M.G. and Bell, F.G. 1995. The regional distribution of sulphate in rocks and soils of Britain. In: *Engineering Geology and Construction*, Engineering Geology Special Publication No. 10, Eddleston, M., Walthall, S., Cripps, J.C. and Culshaw, M.G. (eds), The Geological Society, London, 95–104.

French, W.J., Poole, A.B., Ravenscroft, P. and Khiabani, M. 1982. Results from preliminary experiments on the influence of fabrics on the migration of groundwater and water soluble minerals in the capillary fringe. *Quarterly Journal Engineering Geology*, **15**, 187–199.

Gregory, B.J., Venter, I.S. and Kruger, L.J. 1988. Grouting induced ground movements. In: *Engineering Geology of Underground Movements*, Engineering Geology Special Publication No. 5, Bell, F.G., Culshaw, M.G., Cripps, J.C. and Lovell, M.A. (eds), The Geological Society, London, 153–157.

Harris, S.A. 1986. *The Periglacial Environment.* Croom Helm, London.

Hawkins, A.B. 1984. Depositional characteristics of estuarine alluvium: some engineering implications. *Quarterly Journal Engineering Geology*, **17**, 219–234.

Hearn, G.J. 2002. Engineering geomorphology for road design in unstable mountainous areas: lessons learnt after 25 years in Nepal. *Quarterly Journal Engineering Geology and Hydrogeology*, **35**, 143–154.

Henry, K.S. and Holtz, R.D. 2001. Geocomposite capillary barriers to reduce frost heave in soils. *Canadian Geotechnical Journal*, **38**, 678–694.

Hilf, J.W. 1990. Compacted fill. In: *Foundation Engineering Handbook*, Second Edition, Fang, H.Y. (ed), Chapman & Hall, New York, 249–316.

Hobbs, N.B. 1986. Mire morphology and the properties and behaviour of some British and foreign peats. *Quarterly Journal Engineering Geology*, **19**, 7–80.

Hutchinson, J.N., Bromhead, E.N. and Lupini, J.F. 1980. Additional observations on the Folkestone Warren landslides. *Quarterly Journal Engineering Geology*, **13**, 1–31.

Ingles, O.C. and Metcalf, J.B. 1972. *Soil Stabilization*. Butterworths, Sydney.

Jacobs, J.C. 1965. *The Road Research Laboratory Frost Heave Test*. Transport Road Research Laboratory, Laboratory note LN/766/JCJ, Crowthorne.

Januszke, R.M. and Booth, E.H.S. 1984. Soluble salt damage to sprayed seals on the Stuart highway. *Proceedings Twelfth Australian Road Research Board Conference*, Hobart, **3**, 18–30.

Jelisic, N. and Leppanen, M. 1998. Mass stabilization of peat in road and railway construction. *Proceedings Eighth International Congress International Association Engineering Geology*, Vancouver, Moore, D.P. and Hungr, O. (eds), A.A. Balkema, Rotterdam, **5**, 3449–3454.

Johnson, S.J. 1970. Precompression for improving foundation soils. *Proceedings American Society Civil Engineers, Journal Soil Mechanics and Foundations*, **96**, 145–175.

Jones, R.H. 1980. Frost heave of roads. *Quarterly Journal Engineering Geology*, **13**, 77–86.

Kaplar, C.W. 1971. *Experiments to Simplify Frost Susceptibility Testing of Soils*. Technical Report 223. United States Army Corps Engineers, Cold Regions Research and Engineering Laboratory, Hanover, New Hampshire.

Kezdi, A. and Rethati, L. 1988. *Handbook of Soil Mechanics, Volume 3: Soil Mechanics of Earthworks, Foundations and Highway Engineering*. Elsevier, Amsterdam.

Kitchener, J.N. and Ellison, S.J. 1997. Second Severn crossing – design and construction of the foundations. *Proceedings Institution Civil Engineers, Civil Engineering Supplement*, Special Issue 2, **120**, 22–34.

Kubo, K. and Katayama, T. 1978. Earthquake resistant properties and design of public utilities. In: *The Assessment and Mitigation of Earthquake Risk*, UNESCO, Paris, 171–184.

Lamont-Black, J., Younger, P.L., Forth, R.A., Cooper, A.H. and Bonniface, J.P. 2002. A decision-logic framework for investigating subsidence problems potentially attributable to gypsum karstification. *Engineering Geology*, **65**, 205–215.

Lee, F.T., Odum, J.K. and Lee, J.D. 1997. Slope failures in northern Vermont, USA. *Environmental and Engineering Geoscience*, **3**, 161–182.

Li, D. and Selig, E.T. 1998. Method for railtrack foundation design: I development; II applications. *Proceedings American Society Civil Engineers, Journal Geotechnical and Geoenvironmental Engineering*, **124**, 316–329.

Lovegrove, G.W. and Fookes, P.G. 1972. The planning and implementation of a site investigation for a highway in tropical conditions in Fiji. *Quarterly Journal Engineering Geology*, **5**, 43–68.

Malkin, A.B. and Wood, J.C. 1972. Subsidence problems in route design and construction. *Quarterly Journal Engineering Geology*, **5**, 179–194.

Marienfeld, M.L. and Guram, S.K. 1999. Overview of field installations procedures for paving fabrics in North America. *Geotextiles and Geomembranes*, **17**, 105–120.

McGown, A. and Hughes, F.H. 1981. Practical aspects of the design and installation of deep vertical drains. *Geotechnique*, **31**, 3–18.

McManus, J. 1971. The geological setting of the bridges of the Lower Tay Estuary with particular reference to the fill of the buried channel. *Quarterly Journal Engineering Geology*, **3**, 197–205.

Mitchell, J.K. and Villet, W.C.B. 1987. *Reinforcement of Earth Slopes and Embankments*. National Cooperative Highway Research Program, Report 290, Transportation Research Board, National Research Council, Washington, DC.

Newbury, J. and Subramaniam, A.S. 1977. Geotechnical aspects of route location studies for the M4 north of Cardiff. *Quarterly Journal Engineering Geology*, **10**, 423–441.

Nixon, J.F. and Ladanyi, B. 1978. Thaw consolidation. In: *Geotechnical Engineering for Cold Regions*, Andersland, O.B. and Anderson, D.M. (eds), McGraw-Hill, New York, 164–215.

Northmore, K.J., Entwisle, D.C., Hobbs, P.R.N., Culshaw, M.G. and Jones, L.D. 1992. *Engineering Geology of Tropical Red Soils, Geotechnical Characterization: Index Properties and Testing Procedures*. Technical Report WN/93/12, British Geological Survey, Keyworth, Nottingham.

Obika, B., Freer-Hewish, R.J. and Fookes, P.G. 1989. Soluble salt damage to thin bituminous road and runway surfaces. *Quarterly Journal Engineering Geology*, **22**, 39–73.

Parsons, A.W. 1981. The assessment of soils and soft rocks for embankment construction. *Quarterly Journal Engineering Geology*, **14**, 219–230.

Parsons, A.W. 1987. Shallow compaction. In: *Ground Engineer's Reference Book*, Bell, F.G. (ed.), Butterworths, London, 37/3–37/17.

Pecker, A. and Teyssandier, J.-P. 1998. Seismic design for the foundations of the Rion Anitrion Bridge. *Proceedings Institution Civil Engineers, Geotechnical Engineering*, **131**, 4–11.

Perry, J., Field, M., Davidson, W. and Thompson, D. 2000. The benefits from geotechnics in construction of the A34 Newbury Bypass. *Proceedings Institution Civil Engineers, Geotechnical Engineering*, **143**, 83–92.

Perry, P. 1995. Engineering geology of soils in highway construction: a general overview. In: *Engineering Geology of Construction*, Engineering Geology Special Publication No. 10, Eddleston, M., Walthall, S., Cripps, J.C. and Culshaw, M.G. (eds), The Geological Society, London, 189–203.

Reed, M.A., Lovell, C.W., Altschaeffl, A.G. and Wood, L.E. 1979. Frost heaving rate predicted from pore size distribution. *Canadian Geotechnical Journal*, **16**, 463–472.

Robinson, K.E. and Eivemark, M.M. 1985. Soil improvement using wick drains and preloading. *Proceedings Eleventh International Conference Soil Mechanics and Foundation Engineering*, San Francisco, **3**, 1739–1744.

Rogers, S.H. 1972. Foundation problems of motorway construction. *Quarterly Journal Engineering Geology*, **5**, 145–158.

Sabtan, A., Ali-Saify, M. and Kazi, A. 1995. Moisture retention characteristics of coastal sabkhas. *Quarterly Journal Engineering Geology*, **28**, 37–46.

Santoni, R.L., Tingle, J.S. and Webster, S.L. 2001. Engineering properties of sand-fiber mixtures for road construction. *Proceedings American Society Civil Engineers, Journal Geotechnical and Geoenvironmental Engineering*, **127**, 258–268.

Sasaki, H. 1985. Effectiveness and applicability of the methods of foundation improvement for embankments over peat deposits. In: *Recent Developments in Ground Improvement Techniques*, Balasubramanian, A.S., Chandra, S., Bergado, D.T., Younger, J.S. and Prinzl, F. (eds), A.A. Balkema, Rotterdam, 543–562.

Schlosser, F. 1990. Mechanically stabilized earth retaining structures in Europe. In: *Design and Performance of Earth Retaining Structures*, Lambe, P.C. and Hansen, L.A. (eds), American Society Civil Engineers, Geotechnical Special Publication No. 25, 347–378.

Seed, H.B., Chan, C.K. and Lee, C.E. 1962. Resilience characteristics of subgrade soils and their relation to fatigue failures. *Proceedings International Conference Structural Design of Asphalt Pavements*, Ann Arbor, Michigan, 611–636.

Sherrell, F.W. 1971. The Nag's Head landslips, Cullompton bypass, Devon. *Quarterly Journal Engineering Geology*, **4**, 37–73.

Shin, E.C., Kim, D.H. and Das, B.M. 2002. Geogrid-reinforced railroad bed settlement due to cyclic load. *Geotechnical and Geological Engineering*, **20**, 261–271.

Simm, K.F. 1984. Engineering solutions to geological problems in the design and construction of Humber Bridge. *Quarterly Journal Engineering Geology*, **17**, 301–306.

Smith, M.R. and Collis, L. (eds) 1993. *Aggregates: Sand, Gravel and Crushed Rock Aggregates for Construction Purposes*. Second Edition. Engineering Geology Special Publication No. 9, The Geological Society, London.

Sridharan, A., Srinivisa Murthy, B.R., Bindumadhava, R. and Revanasiddappa, K. 1991. Technique for using fine grained soil in reinforced earth. *Proceedings American Society Civil Engineers, Geotechnical Engineering Division*, **117**, 1174–1190.

Stacey, T.R. and Bell, F.G. 1999. The influence of subsidence on planning and development in Johannesburg, South Africa. *Environmental and Engineering Geoscience*, **5**, 373–388.

Stipho, A.S. 1992. Aeolian sand hazards and engineering design for desert regions. *Quarterly Journal Engineering Geology*, **25**, 83–92.

Tang, M. and Shang, J.Q. 2000. Vacuum preloading consolidation of Yaoqiang Airport runway. *Geotechnique*, **50**, 613–623.

Tomlinson, M.J. 1978. Middle East – highway and airfield construction. *Quarterly Journal Engineering Geology*, **11**, 65–73.

Vidal, H. 1969. *The Principle of Reinforced Earth*. Highway Research Record 282, Washington DC.

Wagner, A.A. 1957. The use of the Unified Soil Classification System for the Bureau of Reclamation. *Proceedings Fourth International Conference Soil Mechanics Foundation Engineering*, London, **1**, 125–134.

Wakeling, T.R.M. 1972. The planning of site investigations for highways. *Quarterly Journal Engineering Geology*, **5**, 7–14.

Watson, A. 1985. The control of wind blown sand and moving dunes: a review of methods of sand control in deserts with observations from Saudi Arabia. *Quarterly Journal Engineering Geology*, **18**, 237–252.

Whetton, M.L. and Weaver, J.W. 1991. Densification of gravelly sand fill using intensive surface compaction. *Proceedings American Society Civil Engineers, Journal Geotechnical Engineering Division*, **117**, 1089–1094.

Wild, S., Kinuthia, J.M., Jones, G.I. and Higgins, D.D. 1999. Suppression of swelling associated with ettringite formation in lime stabilized sulphate bearing clays soils by partial substitution of lime with ground granulated blastfurnace slag. *Engineering Geology*, **51**, 257–277.

Woo, S.M., Van Weele, A.F., Chotivittayathanin, R. and Trangkarahart, T. 1989. Preconsolidation of soft Bangkok Clay by vacuum loading combined with non-displacement sand drains. *Proceedings Twelfth International Conference Soil Mechanics and Foundation Engineering*, Rio de Janiero, **2**, 1431–1434.

6 Reservoirs and dam sites

6.1 Introduction

Although most reservoirs today serve a multipurpose, their principal function, no matter what their size, is to stabilize the flow of water, first, to satisfy a varying demand from consumers or, second, to regulate water supplied to a river course. In other words, water is stored, at times of excess flow to conserve it for later release at times of low flow, or to reduce flood damage downstream.

There is a range of factors that influence the feasibility and economics of a proposed reservoir site. The most important of these is generally the location of the dam. After that consideration must be given to the run-off characteristics of the catchment area, the watertightness of the proposed reservoir basin, the stability of the valley sides, the likely rate of sedimentation in the new reservoir, the quality of the water and, if it is to be a very large reservoir, the possibility of associated seismic activity. Once these factors have been assessed, then they must be weighed against the present land use and social factors. The purposes that the reservoir will serve also must be taken into account in such a survey.

Since the principal function of a reservoir is to provide storage, its most important physical characteristic is its storage capacity. Probably the most important aspect of storage in reservoir design is the relationship between capacity and yield. The yield is the quantity of water that a reservoir can supply at any given time. Obviously, this depends on the size of a reservoir and more particularly upon flow of water into a reservoir, which varies with time. The maximum possible yield equals the mean inflow, less evaporation and seepage loss. In any consideration of yield the maximum quantity of water that can be supplied during a critical dry period (i.e. during the lowest natural flow on record) is of prime importance and is defined as the safe yield. To put it another way, the storage capacity should be sufficient to prevent the reservoir being drawn down to the lowest design level at the intended rate of abstraction during a drought of the greatest severity allowed for. It is, however, impossible to provide sufficient storage to cater for low-flow hydrological risks of great rarity. Hence, a reserve storage allowance is added to the design for a critical risk. Extraordinary droughts have to be met by reducing draft rates, the draft rate being the rate of withdrawal from a reservoir.

Reservoirs may be divided into three categories, direct supply, regulating and pumped storage. The division is solely one of convenience since the first two types are distinguished according to use whilst the third is made on a basis of the method of filling. Impounded direct supply reservoirs include most of those used for water supply. These are filled by natural inflow from their catchments. Water is drawn-off at a more or less constant rate and this, together with the compensation water released to the river below the dam, should not exceed the safe yield if there is no emergency source for drought periods. The yield of impounded regulating reservoirs is assessed on quite a different basis from that used for direct supply reservoirs. In

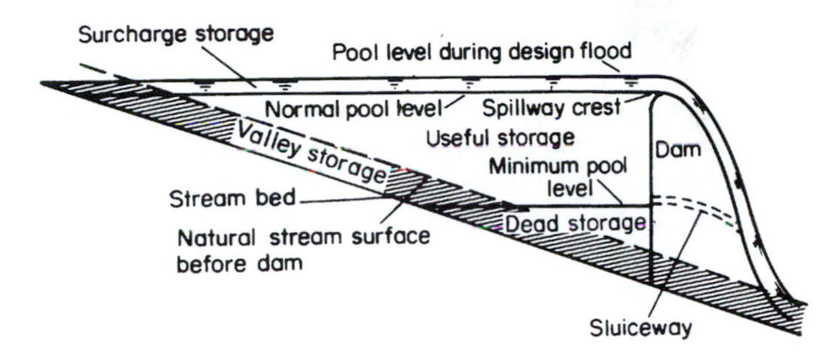

Figure 6.1 Zones of storage in a reservoir.

the case of regulating reservoirs the yield is not simply the amount that can be drawn from the impounded catchment but is the quantity that can be abstracted from a point downstream following regulation of the river by the reservoir. Thus, the yield represents a combination of the run-off from the unreservoired catchment to the abstraction point and that of the regulating reservoir. This, in turn, means that releases from the reservoir need occur mainly during periods of low flow in the river, at other times abstraction is sustained by natural run-off from the unreservoired catchment. The object is to keep the flow in excess of a prescribed minimum, the amount in excess then may be abstracted. One of the problems in operating a regulating reservoir is the difficulty of forecasting stream flows and rainfall conditions between the reservoir and the point of abstraction.

The maximum elevation to which the water in a reservoir basin will rise during ordinary operating conditions is referred to as the top water or normal pool level (Fig. 6.1). For most reservoirs this is fixed by the top of the spillway. Conversely, minimum pool level is the lowest elevation to which the water is drawn under normal conditions, this being determined by the lowest outlet. Between these two levels the storage volume is termed the useful storage, whilst the water below the minimum pool level, because it cannot be drawn upon, is the dead storage. During floods the water level may rise above top water level but this surcharge storage cannot be retained since it is above the elevation of the spillway. Bank storage refers to water that is stored in the rock and soil masses about the perimeter of a reservoir, which when the water level falls, supplements the supply. The amount of bank storage depends upon geological conditions and may account for a significant proportion of the reservoir volume.

6.2 Investigation of reservoir sites

In an investigation of a potential reservoir site, consideration must be given to the amount of rainfall, run-off, infiltration and evapotranspiration that occurs in the catchment area. In addition, vegetative cover, and topographical and geological conditions are important. Accordingly, adequate hydrological records, as well as topographical and geological maps are required. Records of stream flow are required for determination of the amount of water likely to supply a reservoir. Such records also should contain data relating to flood peaks and volumes that can be used to determine the amount of storage needed to control floods, and to design spillways and other outlets. When records of stream flow are not available at a proposed reservoir site, records from a station elsewhere on the river flowing into the site or from a nearby river may be adjusted to the site. Even when records are available, they are often of too short a time span to include a really critical drought period. In such instances, they may be extended by comparison with long duration records of stream flow in the vicinity, by hydrological simulation or other methods. Records of rainfall are used to supplement streamflow records or as a basis for computing stream flow where there are no flow records obtainable. Storm-rainfall records are particularly useful in computing hypothetical maximum floods used for spillway design.

Losses due to evaporation and seepage also must be taken into account. The maximum required reservoir storage is determined from a hydrograph of stream flow or mass curve (Fig. 6.2). A mass curve is a cumulative plot of net reservoir inflow over a given period of time, the slope of the curve at any time being a measure of the rate of inflow at that time. Mass curves may be used to determine the yield that may be expected from a given reservoir capacity (Linsley and Franzini, 1972). The calculation of amount of rainfall lost due to evapotranspiration and infiltration/percolation is an important part of any hydrological survey connected with water supply. Some of the rainfall that infiltrates into the ground and then percolates to the water table eventually reappears at the surface in the form of springs, thereby supplementing run-off. This is especially important during dry weather.

Since evaporation from a free water surface is nearly always in excess of evapotranspiration from a land surface of similar area, there is a net loss of water as a result of reservoir construction. In an arid region the loss may be so great as to defeat the purpose of the reservoir. For instance, annual evaporation losses at the Aswan Reservoir in Egypt are about 7 per cent of gross storage, averaging 7.4 mm per day (Ahmed, 1960). This compares with an average annual evaporation rate from the reservoir at Kempton Park, near London, during the period 1956–1962, of 663 mm (Lapworth, 1965). In the latter case around 80 per cent of the evaporation occurred during the months April–September. Evaporation losses can be substantial in shallow reservoirs or in reservoirs storing more than one year of flow. Conversely, it is a minor consideration in deep reservoirs storing less than a year of flow. Seasonal variations in evaporation are important since maximum losses generally correspond with maximum draw-off draft rates from reservoirs.

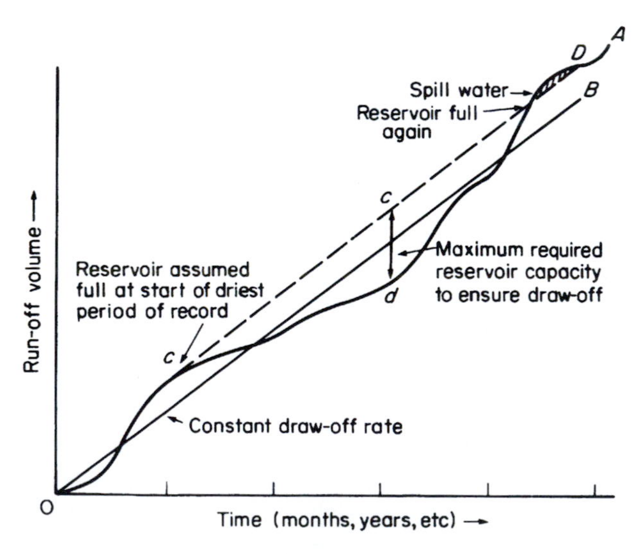

Figure 6.2 The use of mass curves in reservoir design. Mass curves are extremely useful in reservoir-design studies as they provide a ready means of determining the required storage capacity for particular average rates of run-off and draw-off. Suppose that the mass curve OA represents the run-off from a catchment that is to be used for base-load hydroelectric development. If the required constant draw-off is plotted on the same diagram, as line OB, then the required storage capacity to ensure this rate may be found by drawing the line CD parallel to OB from point C at the beginning of the driest period recorded. The storage capacity necessary is denoted by the maximum ordinate cd.

The location of a large impounding direct supply reservoir is very much influenced by topography since this governs the storage capacity of a reservoir. Initial estimates of storage capacity can be made from topographic maps or aerial photographs, more accurate information being obtained, when necessary, from subsequent surveying. Catchment areas and drainage densities also can be determined from maps and aerial photographs. The volume of a reservoir can be estimated, first, by planimetering areas upstream of the dam site for successive contours up to proposed top water level. Second, the mean of two successive contour areas is multiplied by the contour interval to give the interval volume, the summation of the interval volumes providing an approximation of the total volume of the reservoir site.

The field reconnaissance provides indications of the areas where detailed geological mapping may be required and where to locate any drillholes, such as in low, narrow saddles or other seemingly critical areas in the reservoir rim. Drillholes on the flanks of reservoirs should be sunk at least to floor level. Groundwater information should be obtained each day from standing water levels in drillholes before drilling commences. In rock masses with permeabilities less than $10^{-6}\,\mathrm{m\,s^{-1}}$ or in deep drillholes water may take longer than 12 h to stabilize. Changes in permeability with depth may be assessed by packer testing (see Chapter 7). Packer tests usually are carried out on the flanks of reservoirs in order to estimate potential water loss. Piezometers can be installed at various points in drillholes in order to determine water pressures.

Problems may develop both up and downstream in any adjustment of a river regime to the new conditions imposed by a reservoir. If these are anticipated, then they can be considered at the design stage. Deposition around the head of a reservoir may cause serious aggradation upstream resulting in reduced capacity of stream channels to contain flow. Hence, flooding becomes more frequent and the water table rises. The extent of upstream aggradation depends upon stream gradient, the particle size distribution of the load and fluctuations in the water level of the reservoir. Removal of sediment from the outflow of a reservoir can lead to erosion in the river regime downstream of the dam, with consequent acceleration of headward erosion in tributaries and lowering of the water table. For example, approximately 120 000 000 $\mathrm{m^3}$ of material was removed from the channel, for a distance of about 150 km, downstream of Hoover Dam between 1935 and 1951. At Willow Beach, 19.3 km downstream of the dam, the channel was lowered nearly 4.3 m in 14 years.

Waves in reservoirs may damage shoreline structures or earth dams. Consequently, an estimate of wave generation must be made for design purposes. The height of waves and their velocity is dependent upon wind speed and duration on the one hand and fetch on the other. The depth of the water also influences wave height since waves feel bottom and suffer distortion in water that is less than half the wave length in depth. Smaller, shorter waves tend to be produced in deep water. Waves are most critical when a reservoir is near its maximum water level so that maximum wind speeds during that season should be the ones considered.

6.3 Leakage from reservoirs

In most cases the site of a reservoir will be in a river valley, the river water being impounded by a dam. Indeed, the most attractive site for a large impounding reservoir is a valley constricted by a gorge at its outfall with steep banks upstream so that a small dam can impound a large volume of water with a minimum extent of water spread. However, two other factors have to be taken into consideration, namely, the watertightness of the basin and bank stability. Whether or not significant water loss will occur is determined primarily by the groundwater conditions, more specifically by the hydraulic gradient. Accordingly, an assessment of watertightness can be made once the groundwater conditions have been investigated and then any necessary groundwater control measures can be planned.

Leakage from a reservoir may take the form of sudden increases in stream flow downstream of the dam site, with possible boils in the river, and springs may appear on the valley sides. Such leakage may be associated with the presence of major defects in the geological conditions, as for example, solution channels, fault zones or buried channels, through which large and essentially localized flows can occur. Seepage is a more discreet flow, which takes place over a larger area but may be no less in total amount. Generally, seepage is more difficult to trace especially in connection with the investigation of solution channels in limestone. Dyes and radioactive isotopes have been used for tracing subsurface water flow (see Chapter 7).

The economics of reservoir leakage vary from one project to another, for instance, considerable controlled leakage may be tolerated where copius supplies of water are available. Although a highly leaky reservoir may be acceptable in an area where run-off is evenly distributed throughout the year, a reservoir basin with the same rate of water loss may be of little value in an area where run-off is seasonally deficient. A river-regulating scheme can operate satisfactorily despite some leakage from a reservoir, and reservoirs used solely for flood control may be effective even if they are very leaky. By contrast, leakage from a pumped-storage reservoir must be assessed against pumping costs. Unfortunately, serious water loss occasionally has led to the abandonment of a reservoir, examples being the Jerome Reservoir in Idaho, the Cedar Reservoir in Washington, the Hales Bar Reservoir in Tennessee, the Hondo Reservoir in New York and the Monte Jacques Reservoir in Spain.

In some cases, leakage from reservoirs can represent a potential source of trouble. When a dam impounds a body of water behind it, the rock masses forming the floor of the reservoir are subjected to considerable hydraulic pressure. Water will always seek the line of least resistance and, in particular, will flow through any conducting discontinuities or porous rocks in the abutments, floor and ridges about a reservoir, to emerge at lower levels. Water under pressure can give rise to uplift forces and in certain situations may cause piping. Saturation of material on the flanks of a reservoir due to seepage spreads means that the material is weakened and, together with uplift forces, may give rise to unstable slopes.

6.3.1 Piezometric conditions

Apart from the conditions in the immediate vicinity of the dam, the two factors that determine the retention of water in a reservoir basin are the piezometric conditions in, and the natural permeability of, the floor and flanks of the basin. Knill (1971) maintained that four groundwater conditions existed on the flanks of a reservoir (Fig. 6.3), namely:

1 The groundwater divide and piezometric level are at a higher elevation than that of the proposed top water level. In this situation no significant water loss takes place. Knill therefore argued that seepage or leakage through a reservoir margin is determined not by permeability but by pre-existing groundwater behaviour. However, storage at depth in the ground is possible, although it usually is assumed that permeability decreases with depth and that at extreme depth the flow rate is negligible. Roberts (1968) illustrated this type of situation by quoting the Norfolk Dam in Arkansas where no significant seepage has developed, the reservoir being filled in 1944, even though a major fault zone in limestone and thin shales runs through the right abutment. Another example is provided by the Cow Green Reservoir in Teesdale, England, where the water table in the critical col area was at a minimum level of 495 m, some 5.8 m above the proposed top water level. Consequently, it was concluded by Kennard and Knill (1969) that water leakage was not likely to be a problem at that locality. When a reservoir basin is filled, the water table rises in the valley sides and the groundwater divides move towards

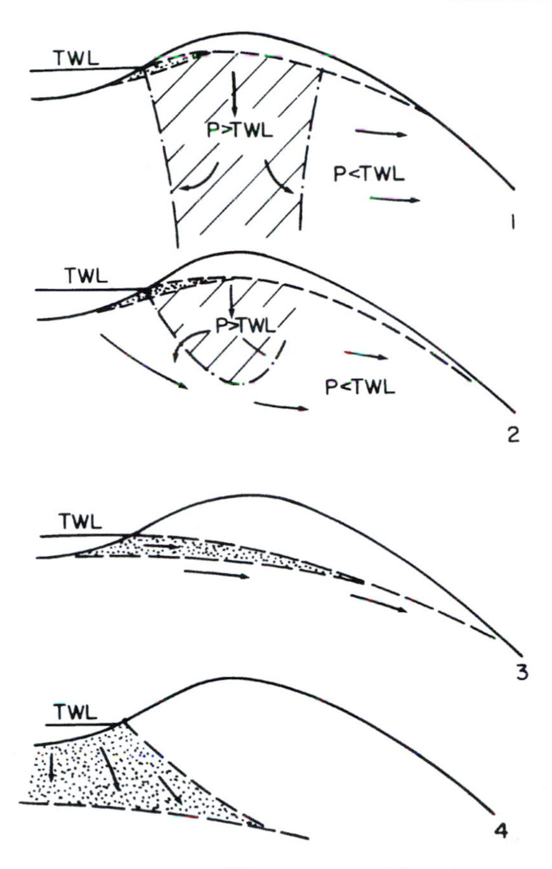

Figure 6.3 Reservoir conditions in relation to differing water table and piezometric situations. TWL top water level: P groundwater pressure; stippled area flooded by reservoir; shaded area groundwater pressures in excess of top water level. Note that the diagrams apply to two-dimensional uniform conditions and are not rigorous.
Source: Knill, 1971. Reproduced by kind permission of The Geological Society.

the reservoir. The flow path to the adjacent valleys is therefore lengthened. However, there generally is an increase in the amount of groundwater discharging into adjoining valleys but at the same time groundwater drains into a reservoir thereby supplementing run-off.

2 The groundwater divide, but not the piezometric level, is higher than the top water level of a reservoir. In these circumstances seepage can take place through a separating ridge into an adjoining valley. Deep seepage can take place, but the rate of flow is determined by the *in situ* permeability.

3 Both the groundwater divide and piezometric level are at a lower elevation than the top water level but higher than that of the reservoir floor. In this case, the increase in groundwater head is low and the flow from the reservoir area is not increased significantly. Deep seepage from the bed of the reservoir may be initiated under conditions of low piezometric pressure in the reservoir flanks. The resulting internal erosion and high seepage pressures can give rise to collapses in the reservoir floor, particularly in limestone terrains. The water table may be low because the rock masses in the reservoir basin are highly permeable, because the amount of groundwater recharge is limited or simply because the precipitation is low and infrequent.

4 The water table is depressed below the base of the reservoir floor. This indicates deep drainage of the rock masses or very limited recharge. A depressed water table does not necessarily mean

that reservoir construction is out of the question but groundwater recharge will take place on filling, which will give rise to a changed hydrogeological environment as the water table rises. In such situations the impermeability of the reservoir floor is important.

Ahmed (1960) showed that seepage from the Aswan Reservoir gradually dropped after filling from 120 to about 9 per cent of the reservoir capacity. In such cases large quantities of seepage water are consumed in charging permeable beds thereby raising the water table. When these beds are more or less saturated, particularly when they have no outlet, seepage is decreased appreciably. At the same time the accumulation of silt on the floor of a reservoir tends to reduce seepage. If, however, any permeable beds have large pore spaces or fissures and they drain from a reservoir, then seepage continues.

6.3.2 Leakage through some soil masses

Alluvial sands and gravels may have porosities that range from around 20 per cent in coarse-grained poorly sorted deposits to 40 per cent or more in clean uniformly sorted material. More importantly, the permeabilites of water-bearing zones within alluvial deposits can vary between 1×10^{-4} and $1 \times 10^{-3}\,\mathrm{m\,s^{-1}}$, although values exceeding $5 \times 10^{-3}\,\mathrm{m\,s^{-1}}$ are not rare. Many fluvio-glacial deposits such as eskers, kames and outwash fans are composed of sands and gravels, and although they usually contain fewer fines than alluvial deposits, their hydrogeological characteristics often are similar. However, kames and eskers frequently are of limited extent. Wind-blown sands that are uniformly sorted may possess porosities of around 35–40 per cent with permeabilities ranging from 5×10^{-5} to $5 \times 10^{-4}\,\mathrm{m\,s^{-1}}$. Loess is characterized by vertical joints and rootholes, which mean that vertical permeability is greater than horizontal permeability. Many deposits of loess possess an open structure that can mean that some deposits have porosities in excess of 40 per cent. However, the porosity of some loess soils suffers a reduction on wetting, due to collapse of their metastable structure. In addition, the range of permeability of loess is much wider than that of wind-blown sands, it varying from as low as 1×10^{-9} to $1 \times 10^{-5}\,\mathrm{m\,s^{-1}}$. After a survey of reservoirs in Kansas and Nebraska that are located on loess and dune sands, Gardner (1968) concluded that even though these deposits are moderately permeable they had not posed major problems as far as leakage was concerned. In other words, the use of cut-offs and impermeable blankets located in the dam areas had been able to control water losses within acceptable amounts.

Buried channels may be filled with coarse granular-stream deposits or deposits of glacial origin and if they occur near the perimeter of a reservoir they almost invariably pose a leakage problem. Indeed, leakage through buried channels, via the perimeter of a reservoir, may be more significant than through the main valley. For example, Coombes (1968) investigated buried channels as a cause of reservoir leakage in the Puget Sound area of Washington. There, glacial processes have disrupted drainage and given rise to many buried channels filled with glacial debris. New river channels have cut through the glacial deposits, often to bedrock, and provide topographically attractive sites for reservoirs. In particular, new valleys are often entrenched through the superficial deposits into bedrock, thus forming a narrow gorge. Upstream the old valley is wide and appears to provide an excellent reservoir site. However, the presence of buried channels occupied by outwash deposits of sand and gravel that have high permeabilities generally means that such sites have to be rejected. Conversely, where the buried channels are occupied by deposits of till, they usually form a seal against leakage. Coombes quoted the Cedar Reservoir as an example of failure due to the presence of a buried channel (such a failure would be unlikely to happen today because of the progress made in geotechnical engineering).

The Cedar River, near Seattle, was dammed in 1914 and the reservoir began to be filled in 1915. However, the northeast abutment of the dam was part of a buried channel consisting of open-textured gravel. Leakage began as filling started, reaching $1.56\,m^3\,s^{-1}$. Despite this, filling was continued slowly but leakage increased. Attempts were made to seal the reservoir and the water level was raised again in 1918. Leakage again occurred and increased rapidly. On 23 December, because of the build-up of hydrostatic pressures due to leakage, movement occurred in the glacial deposits some 1830 m downstream from the dam. Well over $765\,000\,m^3$ of material was washed away by the rush of water that developed, the initial discharge when the failure occurred being estimated at between 850 and $3600\,m^3\,s^{-1}$. The resulting flood was highly destructive, wrecking the small town of Edgewick and other property.

A thin blanket of relatively impermeable material does not necessarily provide an adequate seal against seepage. A controlling factor in such a situation is the groundwater pressure immediately below the blanket. Where artesian conditions exist, springs may emerge through the thinner parts of the superficial cover. If the water table below a blanket is depressed, then there is a risk that the weight of water in the reservoir may puncture it. What is more, on filling there is a possibility that the superficial deposits on the floor of the reservoir may be ruptured or partially removed to expose the underlying rocks. This happened at the Monte Jacques Reservoir in northern Spain where alluvial deposits covered cavernous limestone. The alluvium was washed away to expose a large sinkhole down which the water escaped. Subsequent remedial measures failed to make the reservoir watertight leaving the 73 m high dam redundant.

The Hawthorn Reservoir, Nevada, failed abruptly when it was only one-third full. The reservoir site covered an alluvial fan built up largely by a series of mudflows deposited under arid conditions. These promptly collapsed on being saturated, differential settlement occurring, which at a maximum was approximately 1 m.

6.3.3 Leakage through rock masses

Because of the occurrence of permeable contacts, close jointing, pipes and vesicles, and the possible presence of tunnels and cavities, recent accumulations of basaltic lava flows can rank among the most leaky of rock masses. The permeability of a flow with columnar jointing is highly anisotropic. Usually, vertical permeability, which is attributable primarily to cooling joints, is less than horizontal permeability. Flow in the latter case takes place mainly along the voids between individual lava flows. Furthermore, the upper and lower parts of aa (blocky) lava flows, in particular, may be brecciated and relatively permeable. On the other hand, the centre of a thick lava flow may not exhibit columnar jointing or the joints may be very tight. Lava flows are frequently interbedded, often in an irregular fashion, with pyroclastic deposits. Deposits of ash and cinders tend to be highly permeable. Such mixed sequences, which also may include mud flows, give rise to anisotropic conditions that govern groundwater movement.

According to Moneymaker (1968) reservoir sites in limestone terrains vary considerably in their suitability. Massive horizontally bedded limestones, relatively free from solution features, form excellent sites. On the other hand, well-jointed, cavernous and deformed limestones are likely to present problems in terms of stability and watertightness. Serious leakage usually has taken place as a result of cavernous conditions that were not fully revealed or appreciated at the site investigation stage. Indeed, sites are best abandoned where large numerous solution cavities extend to considerable depths. Where the problem is not so severe, solution cavities can be cleaned and grouted. Under certain circumstances the removal of limestone in solution by groundwater can lead to the progressive opening of discontinuities and solution features within a relatively short period of time, thereby increasing mass permeability. For example,

Moneymaker referred to the leakage abatement programmes that were carried out at the Great Falls and Hales Bar projects in Tennessee. In the former case success was achieved but in the latter, although leakage was initially reduced, it was not. The Great Falls scheme came into operation in 1913 and the dam was heightened by 11.5 m in 1925. Excessive leakage developed immediately the reservoir level was raised, taking place through the left rim where it gave rise to crater falls and cascades. Near the dam the leakage issued just above river level. Leakage accelerated as clay fillings were washed out of cavities in the limestone. The abatement programme was started in 1940. First, the reservoir was drawn down to expose the leakage exits and attempts were made to determine the leakage paths by using dyes. Measurements of leakage were made between July 1940 and June 1943, when it increased from 9.78 to 12.3 $m^3 s^{-1}$. The grouting programme proper followed in 1945 in which the exits were first blocked with asphalt and then sealed with cement grout. Hales Bar Dam is referred to in Section 6.10.2.

Gypsum is more readily soluble than limestone; for example, 2100 $mg l^{-1}$ can be dissolved in non-saline waters as compared with 400 $mg l^{-1}$. Sinkholes and caverns therefore can develop in thick beds of gypsum more rapidly than they can in limestone. Indeed, Eck and Redfield (1965) noted that in certain areas of the United States where beds of gypsum are located below reservoirs, sinkholes have been known to form within a few years. Furthermore, Brune (1965) reported extensive surface cracking and subsidence in reservoir areas in Oklahoma and New Mexico due to the collapse of cavernous gypsum. Uplift, in addition to dissolution, is a problem associated with anhydrite. This occurs when anhydrite is hydrated to form gypsum; in doing so there is a volume increase of between 30 and 58 per cent that exerts pressures that have been variously estimated between 2 and 69 MPa; however, it generally is less than 8 MPa. It would appear that no great length of time is required to bring about such hydration. When it occurs at shallow depths it causes expansion but the process is gradual and usually is accompanied by the removal of gypsum in solution. At greater depths anhydrite is effectively confined during the process. This results in a gradual build-up of pressure and the stress finally is liberated as an explosive force. Such uplifts in the United States have taken place beneath reservoirs, the water providing a constant supply for the hydration process, percolation having taken place via cracks and fissures. Examples are known of ground being elevated by about 6 m. The rapid explosive movement causes strata to fold, buckle and shear, which further facilitates access of water into the ground.

6.3.4 Leakage via faults

Leakage along faults generally is not a serious problem as far as reservoirs are concerned since the length of the flow path usually is too long. However, fault zones occupied by permeable fault breccia running beneath a dam must be given special consideration. When the reservoir basin is filled, the hydrostatic pressure may cause removal of loose material from such fault zones and thereby accentuate leakage. Permeable fault zones can be grouted, or if a metre or so wide, excavated and filled with rolled clay.

The classic case of reservoir failure associated with faults is provided by the Baldwin Hills Reservoir, which was a four-sided basin carved out of a hill top approximately 14 km southeast of Los Angeles (Fig. 6.4). It came into service in 1951 and failed suddenly in 1963, killing five people and causing 15 million dollars worth of damage (Kresse, 1966). The site consisted of poorly consolidated sands, silts and clays with occasional thin limestones, dipping to the southwest into a fault zone that was thought to be active. Consequently, the reservoir was lined with a clay blanket. Not only was the reservoir located in an area that is still active but it was one from which oil was being extracted. In fact, approximately 3 m of subsidence had

Figure 6.4 Balwin Hills Reservoir, a water storage facility of the Los Angeles Department of Water and Power, after rupture, view south. On 14 December 1963, the reservoir failed because of land subsidence that caused displacement in a fault zone beneath the northwest corner of the reservoir (Courtesy of the California Department of Water Resources).

taken place in the immediate neighbourhood of the reservoir due to the extraction of oil during the 50 years prior to 1951. Thus, the reservoir was subjected to movement that led to cracks developing in the floor. After failure these cracks were shown to correspond with minor faults. Analysis of the failure showed that these faults had opened as a result of tension, and large open voids were found in the fault zones. Gradual movements along the faults probably began some time after the basin was filled, but the sudden failure was largely due to rapid movement along the faults that, in turn, was the result of collapse along the voids. Large blocks of the clay lining then sunk into the voids and the full head of the reservoir was exerted on the foundation, rapid failure following.

6.3.5 Leakage and old mine workings

Fractures or zones of high permeability can link abandoned collapsed mine workings with the surface. These, together with shafts, act as conduits conveying surface water to the old workings, thereby reducing the watertightness of potential reservoir sites underlain by abandoned mines. Moreover, old mines often were drained by adits leading into rivers. Knill (1970a) quoted several examples of potential reservoir sites that were quite satisfactory in terms of storage capacity but would probably have suffered serious leakage because they were underlain by old mine workings. The sites therefore were abandoned. However, the Cow Green Reservoir in Teesdale, England, proved an exception, it being constructed above unused lead and barytes workings. After extensive investigations, Kennard and Knill (1969) demonstrated that these old mines would not give rise to serious water loss, this being prevented primarily because the groundwater divides were higher than the proposed top water level of the reservoir.

6.3.6 Control of leakage

Leakage from reservoirs may be controlled by barriers or drainage techniques. Cut-off trenches, carried into bedrock, may be constructed across cols occupied by permeable deposits. Grouting has been used to attempt to control leakage, especially from reservoirs located on limestone terrains. Marinos *et al.* (1997), for instance, referred to the appearance of sinkholes during the construction of two reservoirs in southwest Crete. A drilling programme and geophysical investigation were undertaken in an attempt to locate the presence of subsurface cavities. The remedial works involved filling the cavities with cement grout, and the placement of a shallow grout blanket at one of the reservoirs. However, grouting has not always proved successful in preventing water loss from reservoirs on karstic terrains. Grouting may be effective where localized fissuring is the cause of leakage. Impervious linings consume large amounts of head thereby reducing hydraulic gradients and increasing resistance to seepage loss. Clay blankets or layers of silt have been used to seal exits from reservoirs. For example, Iwao and Gunatilake (1999) referred to the construction of a clay blanket to control leakage from the Samanalawewa Reservoir in Sri Lanka. The first signs of leakage, in the form of springs some 320 m downstream of the toe of the dam, appeared just 9 days after impounding the reservoir basin. Subsequently, water burst from one of the valley sides, some 30 m above the river at more or less the same distance downstream of the dam. Drains allow the free escape of seepage so that it is necessary, for design purposes, to evaluate the probable quantities that must be removed.

6.4 Stability of the sides of reservoirs

The formation of a reservoir upsets the groundwater regime and represents an obstruction to water flowing downhill. The greatest change that occurs is the rise in the water table. Consequently, some rock and soil masses become saturated, which adversely affects their mechanical behaviour so that they may become unstable and fail. This can lead to slumping and sliding on the flanks of a reservoir, and to the reservoir being invaded. Some of the most notable changes have occurred when loess has been inundated. For example, Qian (1982) reported that bank failures in loess around Sanmenxia Reservoir, China, caused the shoreline to retreat some 200 m on initial impounding. Although such action may not give rise to dramatic effects, it means that the land use in the zone affected is altered. It also can destroy infrastructure and recreational facilities located around a reservoir.

Landslides, especially those of appreciable volume, which occur after a reservoir is filled reduce its capacity. They also cause the area immediately surrounding a reservoir to be flooded. Such floods are likely to differ from normal floods in that they may be almost instantaneous, thereby allowing little or no time for warning to be given to those occupying the area that will be affected. Furthermore, sediments derived from floods due to landslides may be too coarse to be flushed from a reservoir and may cause damage to the spillway and outlet structures. An impulse wave generated by the rapid movement of a large landslide into a reservoir may cause damage to any roads and buildings located along the shoreline on which the wave breaks. Worse still, a large impulse wave may overtop the dam, as for example, happened at Pontesei Reservoir, Italy, or at Zhaxi Reservoir, China (Riemer, 1995). The most serious overtopping event on record occurred at the Vajont Dam.

Fortunately, landslide catastrophies associated with reservoirs are comparatively rare events. Nonetheless, Riemer (1995) indicated that at least 1 per cent of all reservoirs have been affected by mass movements. In glaciated valleys, in particular, morainic material generally rests on rock slopes smoothed by glacial erosion, which accentuates the problem of slip. Also, ancient landslipped areas that occur on the rims of a reservoir may be reactivated and may present a leakage hazard. Accordingly, an investigation of a potential reservoir site should include a critical assessment of the slopes around the reservoir basin with regard to their stability and the risk of landslide occurrence (Koukis and Rozos, 1998). When the risk of landslide occurrence has been identified at a reservoir site, then a decision has to be made as to what action should be taken, if any. Three routine options are available, namely, avoid the problem area, increase the resisting forces of the landslide or reduce the motivating forces. In the first case, the potential reservoir site should be investigated to see if it can be relocated away from the hazard zone or alternatively if the storage level can be adjusted. Check dams have been constructed in certain instances to retain debris flows. Conventional methods of landslide (i.e. slope movement) treatment can be used to increase the resisting forces or reduce the motivating forces. Treatment methods to help stabilize slopes include various forms of drainage, prevention of infiltration of water, reafforestation and restraining measures (see Chapter 2).

The worst disaster associated with a landslide moving into and displacing water from a reservoir took place in 1963 in northern Italy. There the River Vajont flows in a steep gorge through the mountains and seemed to provide an ideal location for a hydroelectric scheme. Consequently, a concrete arch dam was constructed in 1961 and the basin upstream was flooded to form a reservoir (Fig. 6.5a). The valley is carved out of a syncline formed of limestones, in which solution features occur, inter-bedded with thin layers of clay and some calcareous mudstones. During glacial times ice scoured the valley along the axis of the syncline and subsequent unloading led to the development of stress relief features parallel to the valley (Fig. 6.5b). For example, joints and bedding planes were opened. As the ice disappeared, river action began to excavate a gorge that eventually attained a depth of some 195–300 m. Landslides periodically occurred and at one time may have dammed the valley. In 1960, sliding movements took place along the reservoir basin with the consequent development of a large tension crack (Kiersch, 1964). The slide contained about $200 \times 10^6\,m^3$ of material and was moving along slip surfaces located some 198 m below ground level. The front of the slide was moving at a rate of up to 100 mm a day and other parts at 30–50 mm per day. To complicate matters, the eastern

Figure 6.5a Vajont valley looking towards the reservoir site during construction.

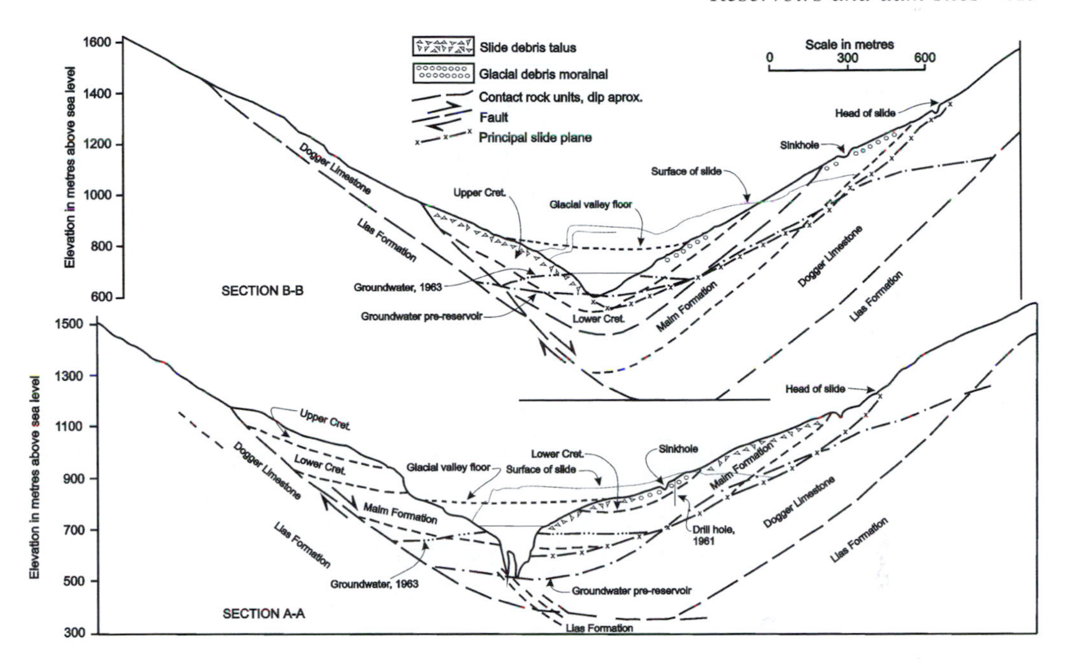

Figure 6.5b Geological cross-sections of the Vajont slide and reservoir basin, running from north to south, showing the principal features of the slide plane, rock units and water levels. See Fig. 6.5c for location of sections.

half of the slide was moving more slowly than the western, which suggested that progressive failure and creep were occurring. The volume of the moving mass meant that the only remedial measures that would reduce the pore water pressures in the slide involved drainage. As would be expected, movement continued after impounding so it was decided to lower the level of water in the reservoir slowly. Since slope stability was related to the water level, the slide seemed to stabilize. Filling the reservoir therefore recommenced in 1962, but in 1963 it had to be lowered again. It appeared to those responsible that the greatest movement occurred when rock was flooded for the first time, and so they concluded that if water levels were raised in stages, then the sliding mass perhaps would eventually reach equilibrium. Further raising and lowering of the water level occurred but on 9 October 1963, there was a violent failure that lasted about a minute. The whole of the disturbed mountain side slid downhill with such momentum that it crossed the gorge, 99 m wide, and rode 135 m up the other side. More than $300 \times 10^6\,m^3$ of rock material had moved, filling the reservoir for a distance of 2 km with slide material, which in places reached heights of 175 m. The slide moved at an estimated speed of about $24\,m\,s^{-1}$ and created strong earth tremors. Recently, Tika and Hutchinson (1999) have suggested that the speed of the slide could be explained by a dramatic drop in shear strength along the slip surface due to the high rate at which the material involved was sheared. The slide displaced water in the reservoir, thereby generating a huge wave that overtopped both abutments to a height of some 100 m above the crest of the dam. The wave was over 70 m high at the confluence of the Vajont valley with the Piave valley, 1.6 km away. Everything in the path of the flood for kilometres downstream was destroyed (Fig. 6.5c). Fortunately, the dam did not fail. Even so the wave destroyed five villages and killed almost 3000 people. One of the most recent reviews of the events that led to the Vajont disaster has been provided by Semenza and Ghirotti (2000).

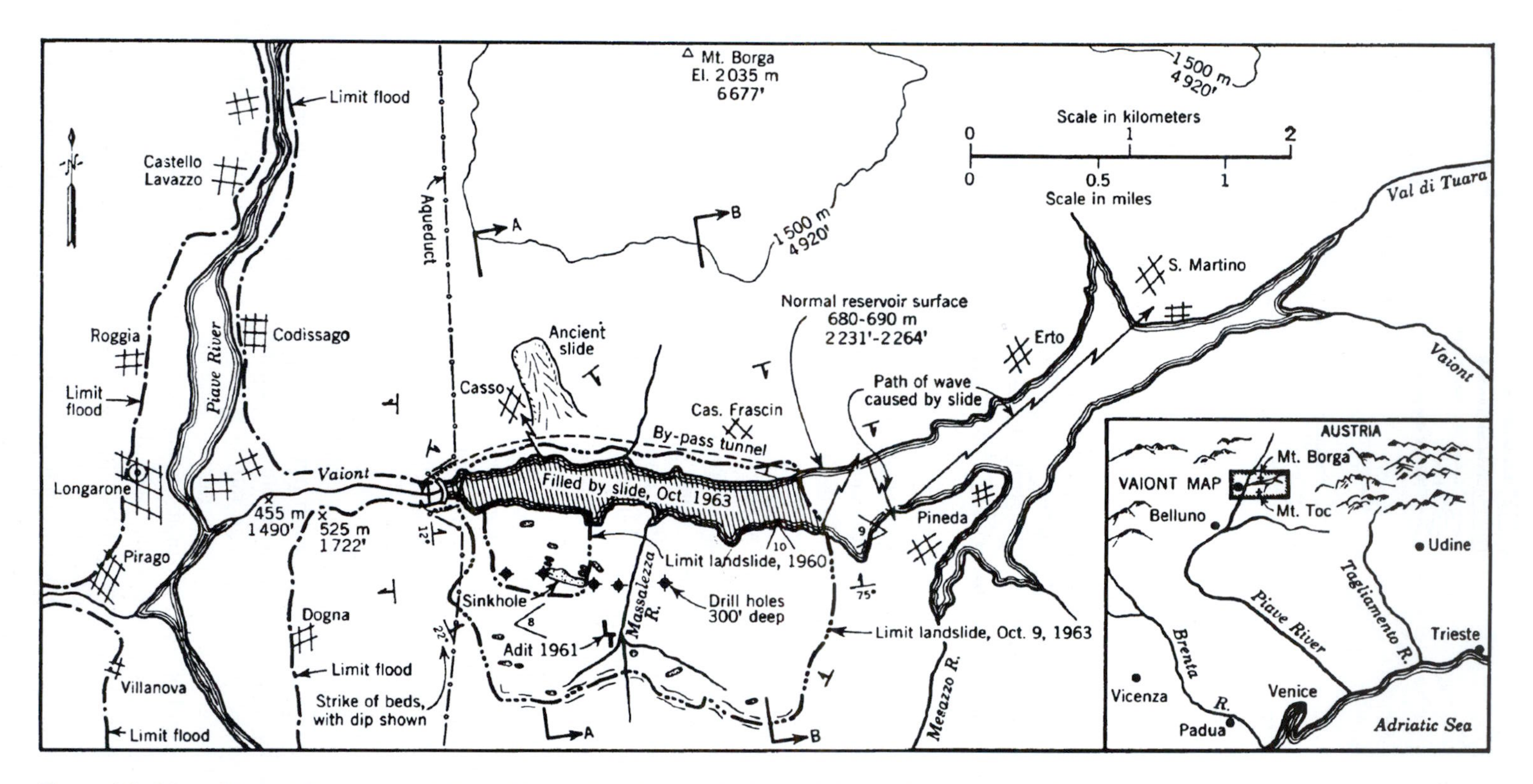

Figure 6.5c Map of Vajont Dam area and Piave River valley showing limits of slide and the destructive flood waves.

6.5 Sedimentation in reservoirs

Although it is seldom a decisive factor in determining location, sedimentation in reservoirs is an important problem in some countries. For instance, in the United States, because of sedimentation, the usefulness of most reservoirs is less than 200 years. Some of the worst sedimentation problems are associated with reservoirs that have been constructed on the Yellow River in China. There, according to Greeman (1998), power generating dams have failed because of the heavy accumulation of silt that has occurred even before the completion of reservoir construction. At peak periods the run-off carries up to $90\,g\,l^{-1}$ and each year an average of about 100 mm of silt accumulates over the river bed.

As can be inferred from above, sedimentation in a reservoir may lead to one or more of its major functions being seriously curtailed or even to it becoming inoperative. For example, Tate and Farquharson (2000) noted that the useful life of Tabela Reservoir on the River Indus, Pakistan, is threatened by a sediment delta that is approaching the intake tunnels of the dam. They suggested that the most effective remedial measure would be an underwater dam to protect the intakes and low-level flushing facilities. Furthermore, reduction of the quantity of sediment downstream of a dam can lead to problems. One such example has been provided by Bailard (2001). Bailard indicated that the Matilija Dam in California impounds around 5 million m^3 of sediment, which has reduced its capacity by almost 90 per cent. In addition, the dam is responsible for the decreased rate of sediment supply to the coast, some 16 km distant, by the Ventura River. The major consequence of this is accelerated beach erosion near the mouth of the Ventura. Consideration is being given to increase the supply of sediment to the coast by pipeline, by transporting it in lorries or by notching the dam to increase the rate of sediment release. Abam (1999) also referred to the effects of reservoirs on the Niger Delta where reduced sediment supply means that instead of accretion taking place in many coastal areas, they are now subject to erosion. In addition, the reduced water levels in the distributaries have led to tidal influences moving further upstream. The latter situation could adversely affect local aquifers if they are used for water supply.

The rates of sedimentation must be estimated accurately in those areas where streams carry heavy sediment loads in order that the useful life of any proposed reservoir may be determined. Such information may be of fundamental importance in evaluating the economic feasibility of a project. Important water losses may occur as a result of deposition in reservoirs, particularly in arid and semi-arid regions. For example, evaporation increases because of the relative increase in exposed water surface for the same volume of water storage. Transpiration from vegetation growing on deposits accumulating about the perimeters of reservoirs also consumes large quantities of water. The size of a drainage basin is the most important consideration as far as sediment yield is concerned, the rock types, drainage density and gradient of slope also being important. The sediment yield also is influenced by the amount and seasonal distribution of precipitation, and the vegetative cover. Poor agricultural practices, overgrazing, improper disposal of mine waste and other human activities may accelerate erosion or contribute directly to stream loads. De Sousa *et al.* (1998) referred to the use of a geographical information system to estimate the rate of sedimentation in the Manso River Reservoir in Brazil. They found that the annual rate of reservoir sedimentation had been underestimated by a factor of almost 9 meaning that the capacity of the reservoir, which began operation in 1991, would be exhausted by 2005. The principal reason for such rapid siltation is that there are extensive mining activities in the catchment area of the reservoir, the tailings therefrom representing the major contribution to the sediment load.

The ability of a reservoir to trap and retain sediment is known as its trap efficiency and is expressed as the percentage of incoming sediment that is retained. Trap efficiency depends on total inflow, rate of flow, sediment characteristics and the size of the reservoir. It decreases with age as the capacity of the reservoir is reduced by sedimentation. The volume of sediment carried varies with stream flow, but usually the peak sediment load occurs prior to the peak stream flow discharge. Frequent sampling accordingly must be made to ascertain changes in sediment transport. Figure 6.6 may be used to estimate the amount of sediment that a reservoir will trap if the average annual sediment load carried by streams is known. The volume occupied annually by this sediment can be calculated approximately by dividing the trapped load by its density. The useful life of a reservoir can be estimated by determining the time required to fill the critical storage volume. These estimates can be refined by taking into account the amount of consolidation that the sediment is likely to undergo. Volumetric measurements of sediment in reservoirs are made by soundings taken to develop the configuration of the sides and bottom of a reservoir below the water surface.

The distribution of deposits in a reservoir depends on the character of the sediment, the inflow–outflow relations, the shape of the reservoir basin and reservoir operation. Where a stream enters a reservoir, its velocity is checked and coarse particles are deposited to build a delta, whereas fine sediments are transported into deeper water. During floods, streams are highly charged with suspended sediment and have higher densities than the reservoir water, which means that they sink to form turbidity currents. The sediment propels the water, the difference in density supplying the driving force to maintain the underflow. Where turbulence and currents are slight, density differences of only a few hundredths of 1 per cent are sufficient to maintain the separate identities of the different water masses. In quiet waters, the interfaces of the flows remain remarkably abrupt. At Lake Mead, the lake created by Hoover Dam, turbidity flows travel the full length of the reservoir, more than 160 km, and are checked by the dam. Bell (1942) estimated that more than 235 million tonnes of sediment were deposited in Lake Mead between 1936 and 1941, deposition occurring at the rate of approximately 877 000 tonnes per week. He suggested that, with proper outlet facilities, from 75 to 90 per cent of this sediment could have been carried beyond the dam by the use of stratified flow. However, at present there is no economical method of desilting a large reservoir. Once capacity is reduced to less than the necessary minimum, new sites must be sought.

Although sedimentation in a reservoir is inevitable it can be retarded. The obvious way to do this is to select a site where the sediment inflow is low and the storage capacity large enough for a useful life. The most common way of dealing with the sediment problem is to

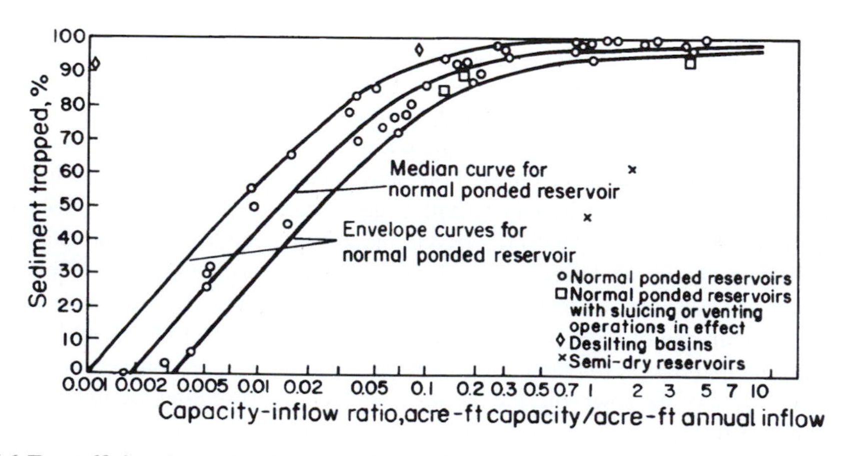

Figure 6.6 Trap efficiency as related to capacity-inflow ratio.
Source: Gosschalk, 1964. Reproduced by kind permission of McGraw-Hill Companies.

designate a part of the reservoir capacity to sediment storage. This is a negative approach that postpones the date when sediment accumulation becomes serious. In other words, the sediment pool is equivalent to the volume of sediment expected to collect during the design life of the reservoir. A good vegetative cover over the catchment area is one of the best ways of reducing sedimentation. However, the success of such catchment protection depends on climate and land management.

Although it is usually cheaper to enlarge a reservoir than to construct sediment traps, in some instances they may be justified. Site selection for sediment traps involves an evaluation of the principal sources of sediment supply in the watershed. Sedimentation basins or traps then are built across the principal streams contributing sediment. A sediment trap is designed principally to catch coarse sediments, and deposits may be removed from it. In a watershed where the primary source of sediment production is in channel erosion, drop inlets, chutes and stream-bank revetments help reduce erosion. Diversion of sediment-laden water around a reservoir, where possible, is another method that has been used to diminish reservoir sedimentation. Sediment accumulation may be reduced to some extent by discharging water through sluices in the dam, for example, trap efficiency may be reduced by up to 10 per cent if turbidity currents can be vented through sluiceways. However, sluicing only removes deposits in the immediate vicinity of the dam. Dredging has on occasions been resorted to for removing sediments from small reservoirs. There then is the problem of disposal. In the case of the Yellow River, the Xiaolangdi Dam has been constructed to control siltation but only has a design life of 20 years as far as siltation is concerned (Greeman, 1998).

6.6 Pumped storage reservoirs

Pumped storage reservoirs for direct supply are commonly sited adjacent to the lower reaches of rivers from which the water is abstracted, thereby taking advantage of the greatest catchment area and maximum available run-off. Yield, however, is influenced by the capacity of the pumping plant to abstract water from the river. For example a series of pumped storage reservoirs occur alongside the River Thames and represent the main supply of water for London. Two of these reservoirs are at Wraysbury and Datchet (Fig. 6.7). Both sites consist of surface gravel underlain by London Clay. The reservoirs are formed by continuous embankments constructed from the local gravels excavated from the reservoir basin, with rolled clay cores that extend 3 m into the London Clay.

Pumped storage projects for hydroelectric power schemes operate with maximum efficiency with heads of about 300 m and relatively short hydraulic systems. There is no need for a natural catchment at the upper reservoir since water is pumped up to it but leakage must be minimized. Some of the best locations for hydroelectric pumped storage schemes are in glaciated uplands in metamorphic and plutonic terrains, which ideally are free from notable faults. A corrie frequently is used for the upper reservoir and therefore it should be suitable in terms of topography and geology for the construction of a dam. The more or less fixed location of tunnels and the underground chambers for the power station means that the geology on site must be thoroughly investigated. Leakage at the lower reservoir site again must be minimal unless there is considerable natural inflow. Generally, large lakes are used as the lower reservoir such as Lake Llanberis for the Dinorwic pumped storage scheme in North Wales. According to Johnstone and Crichton (1966) the Cruachan pumped storage scheme, Scotland, ideally fulfilled the basic economic requirements. For instance, the ratio of length of aqueduct to vertical head is 3.9:1 and the upper reservoir is located in a corrie in the side of Ben Cruachan (Fig. 6.8). The lower reservoir, Loch Awe, is large enough to cater for any fluctuating discharges. The generating station and ancillary works were located within an intrusion of granite.

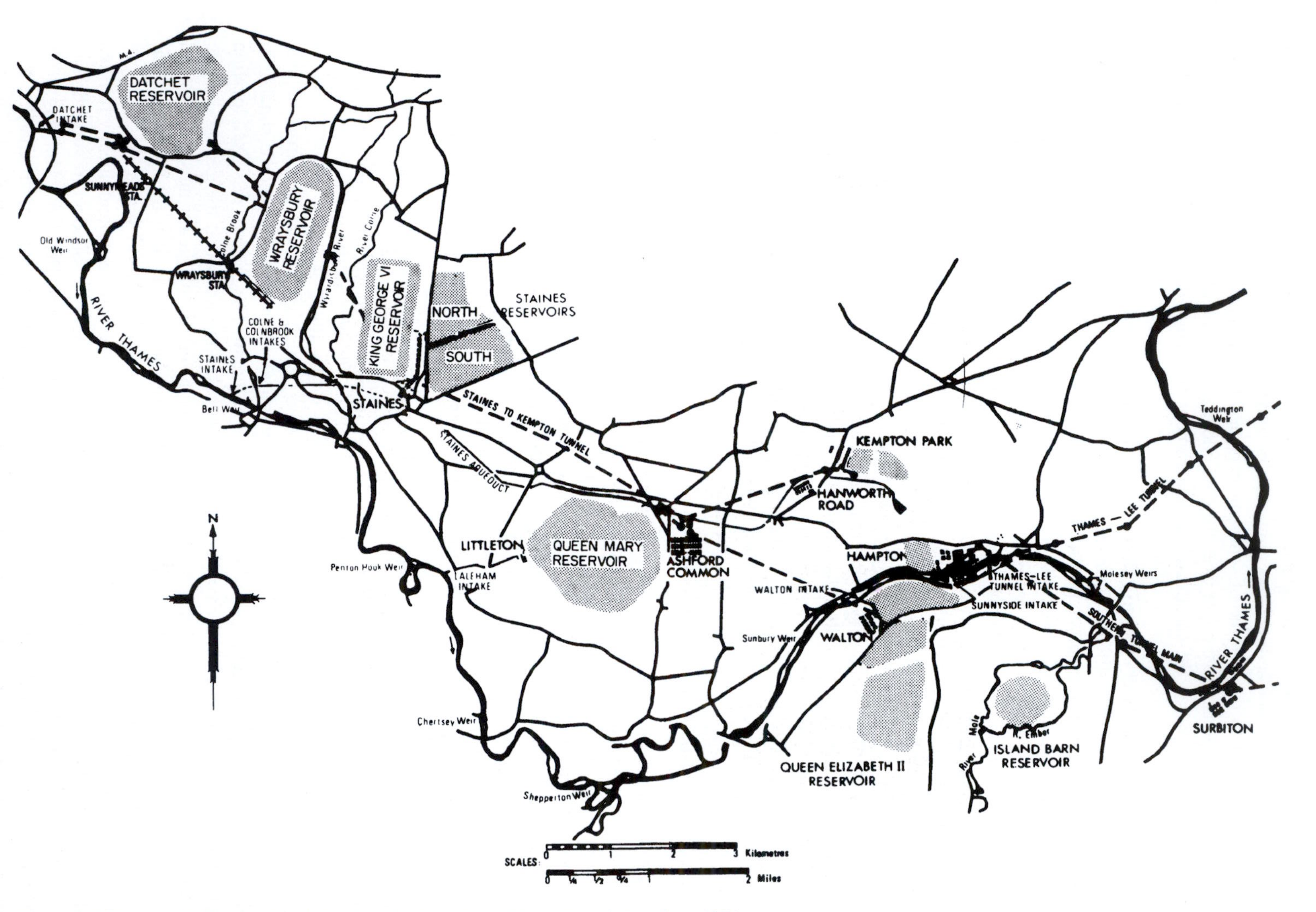

Figure 6.7 Thames Valley works of the then Metropolitan Water Board, London, 1971.

Figure 6.8 Cruachan Dam and intake/outfall works of the power station, Scotland.

6.7 Reservoirs and induced seismicity

A number of earthquakes have been recorded with their epicentres located below or near large reservoirs. Indeed, according to Willmore (1981) some 7 per cent of dams exceeding 100 m in height that impound reservoirs with volumes of 10^9 m^3 or more experience reservoir-induced seismicity. Table 6.1 shows some of the largest reservoirs and dams in the world, at some of which seismicity has been recorded. In addition, the number of earthquakes greater than magnitude 2 that were recorded in the first 10 years of life of Kariba Reservoir, between Zimbabwe and Zambia, is shown in Fig. 6.9. Seismicity associated with reservoir loading was noticed first in the mid-1930s at Lake Mead, Colorado (Carder, 1945). Indeed, the longest range of data has been provided by Lake Mead, where over 10 000 small earthquakes have been recorded. A weak correlation between seismic activity and level of impounding has been reported at Lake Mead, and the epicentres are thought to have been located along faults. The concept of induced seismicity due to reservoir loading, however, is not universally accepted. This is not surprising since so very few reservoirs are instrumented to record local seismic events. Nonetheless, as dams are built higher and reservoirs impound larger volumes of water this cause-and-effect relationship needs to be resolved since induced earthquakes may cause serious damage, which needs to be avoided.

Seismic activity at reservoir sites may be attributable to water permeating the underlying strata thereby increasing pore water pressure and decreasing the effective normal stresses so that shear strength along local faults is reduced (Simpson, 1976). In addition, increased saturation may reduce the strength of rock masses sufficiently to facilitate the release of crustal strains. Such

Table 6.1 Some large reservoirs in relation to seismic activity

Dam	*Country*	*Type of dam*	*Height of dam* (m)	*Volume of dam* ($m^3 \times 10^3$)	*Capacity of reservoir* ($m^3 \times 10^6$)	*Seismicity after construction*
Bhakra	India	Gravity	225	4 130	9 868	No record
Contra	Switzerland	Arch	230	660	86	Slight
Daniel Johnson (Manicopagan 5)	Canada	Multiple arch	214	2 255	141 975	No record
Glen Canyon	USA	Arch	216	3 747	33 304	Nil
Grancarevo	Yugoslavia	Arch	123	376	1 277	Noticeable ($M = 4$)
Grande Dixence	Switzerland	Gravity	284	5 957	400	Nil
Grandval	France	Multiple arch	88	180	292	Noticeable
Hoover	USA	Arch	221	3 364	38 296	Noticeable ($M = 5$)
Kariba	Southern Rhodesia*	Arch	128	1 032	160 368	Strong ($M = 6$)
Koyna	India	Gravity	104	1 300	2 780	Strong ($M = 6.5$)
Kremasta (Roi Paul)	Greece	Earth	160	7 800	4 750	Strong ($M = 6.5$)
Kurobegawa No.4	Japan	Arch	186	1 360	199	No record
Mangla	Pakistan	Earth	115	65 651	6 358	Slight ($M = 3.6$)
Mauvoisin	Switzerland	Arch	237	2 030	180	Slight
Monteynard	France	Arch	155	455	240	Noticeable ($M = 4.9$)
Oroville	USA	Earth	236	59 639	4 298	Slight ($M = 1.5$)
Warragamba	Australia	Gravity	137	1 233	2 052	Noticeable

Source: Lane, 1972. Reproduced by kind permission of Thomas Telford.

Note
*Now Zimbabwe.

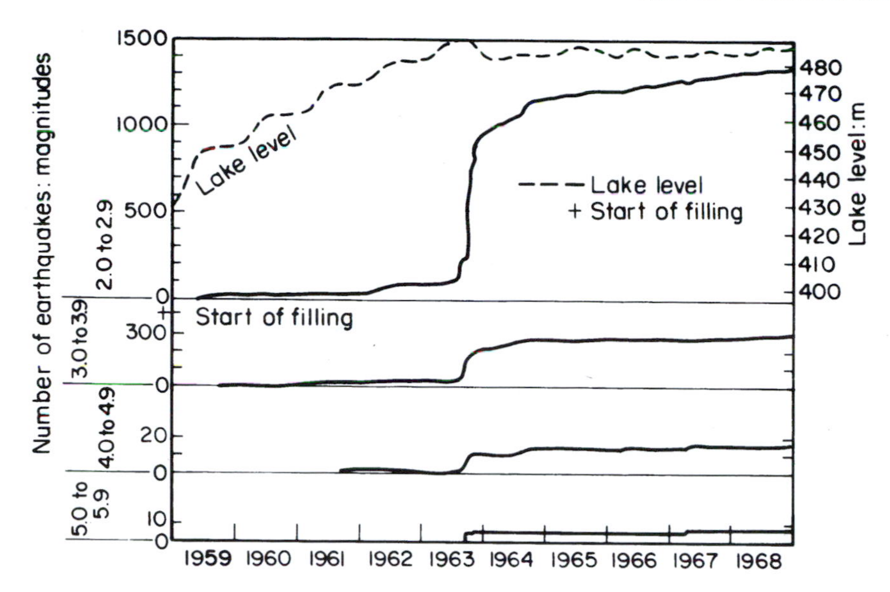

Figure 6.9 Kariba Reservoir, Zambia/Zimbabwe, distribution of induced earthquakes in relation to time and reservoir level.
Source: Lane, 1972. Reproduced by kind permission of Thomas Telford.

activity would not appear to be related to the weight of water stored, for some large impounded reservoirs have not been associated with noticeable seismic activity. Although, seismic activity may be initiated almost as soon as impounding begins, the time of maximum activity sometimes shows an appreciable delay compared with the time when the reservoir reaches its maximum top water level. Earthquakes associated with reservoir impounding tend to occur in swarms, and have shallow foci and modest magnitudes, although they sometimes may cause destruction. They are characterized by an increased number of foreshocks and a slow decay of aftershocks. Reservoir impounding does not increase the maximum values of magnitudes. In fact, reservoirs probably transform and induce seismic phenomena for which the stage is already set by existing tectonic features.

As far as monitoring of induced earthquakes from reservoir loading is concerned, a network of seismometers should be in operation prior to impounding to record the occurrence of small local earthquakes. The seismicity of the area has to be established before filling in order to distinguish whether local earthquakes are either a consequence of reservoir loading, or are part of the general seismic pattern. Such information also is required to evaluate the probable size and location of future shocks.

The most notable earthquake associated with reservoir impounding occurred at Koyna Reservoir, which is located not far from Bombay, India. Impounding began in 1963 and small shocks were recorded a few months later. These continued, gaining in intensity, until in December 1967, a shock with a magnitude of about 6.5 was recorded. This caused significant damage and loss of life in the nearby village of Coynanagar. The 103 m high rubble concrete dam was fissured in several places since the epicentre of the earthquake was near the dam and leakage through dam increased. The measures taken to increase the strength of and reduce the leakage through the dam to ensure its safety have been described by Pendse *et al.* (1999). The focus of the earthquake concerned was located at a depth somewhere between 10 and 20 km. Like Lake Mead, there was a correlation between fluctuating water level and seismicity. Two further notable earthquakes followed in November 1973 and September 1980, and earthquakes of

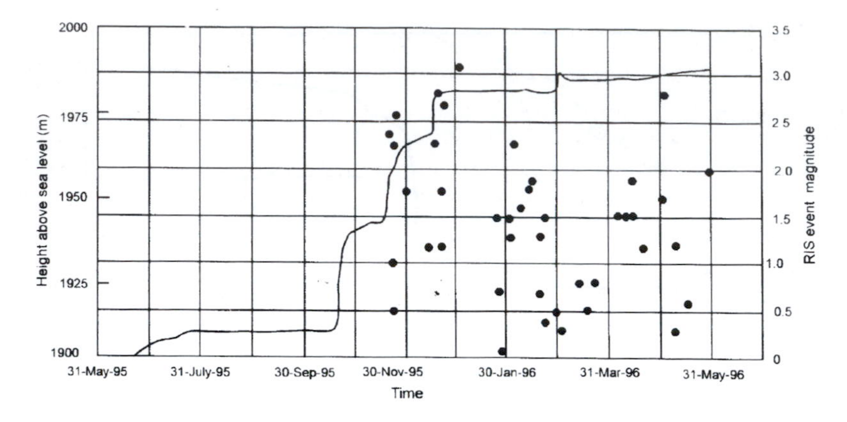

Figure 6.10 Impounding of Katse Reservoir, Lesotho and induced seismic events for the period 31 May 1995, until 31 May 1996.
Source: Bell and Haskins, 1997.

magnitude 4 frequently occurred in the Koyna area (Gupta, 1985). The source of these earthquakes appeared to be located along a north–south trending fault through the reservoir area.

Bell and Haskins (1997) indicated that induced seismicity was recorded at Katse Reservoir, Lesotho, approximately 1 month after impoundment began. Hence, the reservoir-induced seismicity was of rapid response type (Simpson *et al.*, 1988). A network of three seismological stations had been installed in 1991, one of the stations being near the dam wall, while the other two were installed at 8 and 20 km upstream respectively. Filling began on the 20 October 1995, rapidly rising to a level of 1939 m ASL on the 30 October 1995. By December 1995, the water level had risen some 81.5 m above the usual level of the river, to a height of 1981.5 m ASL. On 21 November 1995, the first of a series of low magnitude (1–2.5) earthquakes began (Fig. 6.10). Then on 9 November 1995, an event triggered a series of surface cracks along a linear fracture zone, 1.5 km long, extending away from the reservoir inland through the centre of Mapeleng village. Extensional movement resulted in an average width of crack opening of 10 to 30 mm along the zone. The available data showed a steady increase in seismic activity from November 1995 through to late December 1995, with a decrease of seismic activity in January 1996. This moderation may be attributed to the levelling-off of the water level at 1981 m ASL on the 20 December 1995. A sudden increase of seismic events occurred in February 1996 with a moderation of activity thereafter, until May 1996 (date up to which data extends). The majority of the higher magnitude events tended to occur near the beginning of reservoir impoundment, and the pulses in seismic events suggested that energy was dissipated during the event and took time to build up to the next event.

6.8 Types of dam

The type and size of dam constructed depends upon the need for and the amount of water available, the topography and geology of the site, and the construction materials that are readily obtainable. Dams can be divided into two major categories according to the type of material with which they are constructed, namely, concrete dams and embankment dams. The former category can be subdivided into gravity, arch and buttress dams whilst rolled fill and rockfill embankments comprise the latter type.

A gravity dam is a rigid monolithic structure that is usually straight in plan, although sometimes it may be slightly curved (Fig. 6.11). In cross section a gravity is roughly trapezoidal. Generally, gravity dams can tolerate only the smallest differential movements and require large amounts of concrete since their resistance to dislocation by the hydrostatic pressure of the reservoir water is due to their own weight. Nonetheless, gravity dams have been built on badly fractured variable rock and even on river fill, the ground requiring adequate treatment before their construction. Properly constructed gravity dams with adequate foundations are probably among the safest of all dams. A favourable site is usually one in a constricted area of a valley where sound bedrock is reasonably close to the surface, both in the floor and abutments. An important consideration in some areas of the world is the availability within a reasonable hauling distance of adequate deposits of suitable aggregate for concrete.

An arch dam consists of a high–strength concrete wall, curved in plan, with its convex face pointing upstream (Fig. 6.12). Arch dams are relatively thin walled and lighter in weight than gravity dams. They will stand up to large deflections in the foundation rock provided that the deflections are uniformly distributed. Arch dams transmit most of the horizontal thrust from the reservoir water to the abutments by arch action and this, together with their relative thinness, means that they impose high stresses on narrow zones at the base of the dam, as well as the abutments. Therefore, the strength of the rock mass in the abutments and immediately below and down-valley of the dam must be satisfactory, along with the modulus of elasticity being high enough to ensure that deformation under thrust from the dam does not induce excessive stresses in the dam. Ideal locations for arch dams are provided by narrow gorges where the walls are capable of withstanding the thrust produced by the arch action. The arch itself must

Figure 6.11 Nagle Dam, Natal, South Africa. This is a concrete gravity dam. Note the second dam on the far left-hand side. This helps to control the flow into the reservoir, as well as trapping silt. Excess flow and silt can be released through a sluice to the right-hand side of this dam, the water and silt flowing through a cutting into the river below Nagle Dam.

Figure 6.12 Hoover Dam, Colorado, still one of the largest and most impressive arch dams in the world.

be well keyed into the abutments. If the load on the abutments and foundation are about equal, then the dam is called a gravity-arch dam.

In locations where concrete aggregate is in limited supply and the foundation rocks are competent, buttress dams provide an alternative to gravity dams. Buttress dams also involve more limited excavation of foundations. A buttress dam consists principally of a slab of reinforced concrete that slopes upstream and is supported by a number of buttresses, the axes of which are normal to the slab (Fig. 6.13). The buttresses support the slab and transmit the load of reservoir water to the foundation. They are rather narrow and act as heavily loaded walls that can exert tremendous unit pressures on the foundation. Consequently, buttresses founded in weak rock masses may punch into the ground causing upheaval of material between them. The problem of uplift pressure is practically eliminated in buttress dams.

Embankment dams are constructed of earth or rock with an impermeable core to control seepage (Fig. 6.14). The core usually consists of clayey material or if sufficient quantities are not available, then concrete or asphaltic concrete membranes can be used. The core normally is extended as a cut-off below ground level when seepage beneath the dam has to be controlled. These cut-offs may be very deep and in some cases cut-offs have been extended into the abutments as wing trenches. Drains of sand and/or gravel installed within and beneath an embankment dam also afford seepage control. Because of their broad base embankment dams impose much lower stresses on their foundations than concrete dams. Furthermore, they can accommodate deformation such as that due to settlement more readily. As a consequence, embankment dams have been constructed on a great variety of foundations ranging from weak unconsolidated stream to glacial deposits to high strength rock masses. An embankment dam may be zoned or homogeneous, the former type being more common. A zoned dam is a rolled fill dam composed of several zones that increase in permeability from the core towards the outer slopes (Fig. 6.15). The number of zones depends on the availability and type of borrow material locally. Stability of a zoned dam is due mostly to the weight of the heavy outer zones. If there is only one type of borrow material readily available, then a homogeneous embankment is constructed. In other words, homogeneous dams are constructed entirely or almost entirely of one type of material. The latter usually is fine grained, although sand and sand–gravel mixtures have been used. Rockfill dams usually consist of a loose rockfill dump, an impermeable facing on the upstream side or an impermeable core, and rubble masonry between.

Figure 6.13 Errochty Dam, Scotland, an example of a buttress dam.

Figure 6.14 Harddap Dam, near Marietal, Namibia, and example of an embankment dam.

Some sites that are geologically unsuitable for a specific type of dam design may support one of composite design (Fig. 6.16). For example, a broad valley that has strong rocks on one side and weaker ones on the other can possibly be spanned by a combined gravity and embankment dam.

A spillway of some type is required to allow flood waters to by-pass a dam with safety (Fig. 6.14). There are two design decisions necessary once the discharge capacity of a spillway has been determined. These decisions are very much affected by geology and, according to Woodward (1992), influence the extent to which it will be necessary to provide a concrete lining for, and energy dissipating structures in a spillway, and the extent to which the material

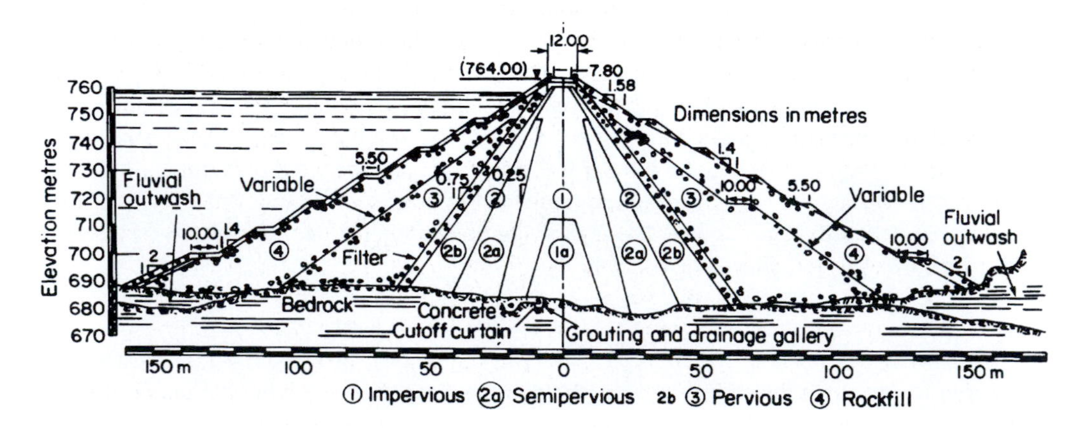

Figure 6.15 Las Pirquitas Dam, Argentina, a zoned embankment dam.

Figure 6.16 Inanda Dam, Natal, South Africa, a composite dam.

excavated from a spillway can be used in dam construction. Good quality rock will be needed if a spillway is to be unlined and for use in dam construction.

6.9 Forces on a dam

The construction of a dam, together with impounding a reservoir behind it, impose loadings on the sides and floor of the valley creating new stress conditions. Indeed, loading on a dam foundation usually is high compared with that of most other structures. A concrete dam behaves as a rigid monolithic structure, the stress acting on the foundation being a function of the weight of the dam as distributed over the total area of the foundation. By contrast, earthfill dams exhibit semi-plastic behaviour and the pressure on the foundation at any point depends on the thickness of the dam above that point. The general requirements for the design of rock foundations for concrete dams are stability against sliding and overturning, acceptable levels of differential deformation, and control of seepage and erosion (Wyllie, 1999). Vertical static forces act downward and include both the mass of the dam and the water, although a large part of the dam is submerged and therefore the buoyancy effect reduces the influence of the load. The most important dynamic forces acting on a dam are wave action, fluctuations in reservoir level, overflow of water, shocks and seismicity.

Horizontal forces are exerted on a dam by the lateral pressure of water behind it. These, if excessive, may cause concrete dams to slide (Deere, 1976). The limit equilibrium method of analysis with regard to a sliding type of failure consists of determining the resisting and motivating forces along the sliding surface, the ratio between the two giving the factor of safety of the foundation (Wyllie, 1999). The tendency towards sliding at the base of concrete dams is of particular significance in fissile rocks such as shales, slates and phyllites. Weak zones, such as interbedded ashes in a sequence of basalt lava flows, can prove troublesome. The presence of flat-lying joints may destroy much of the inherent shear strength of a rock mass and reduce the problem of resistance of a foundation to horizontal forces to one of sliding friction so that the roughness of the joint surfaces become a critical factor. Rock masses that contain clay material also are suspect since this limits the value of the coefficient of friction between the concrete of the dam and the rock of the foundation. The rock surface should be roughened to prevent sliding, and keying the dam some distance into the foundation is advisable. This can be done by building a key wall or by providing a cut-off wall at the heel of the dam. Other methods of reducing sliding include providing a downward slope to the base of the dam in the upstream direction of the valley, by the installation of tensioned anchors and by additional excavation and concreting. Bell and Haskins (1997) mentioned a horizontal shear zone, about 5 m in width and along which movements up to 150 mm had taken place, within one of the thick lava flows at Katse Dam site. This zone of weakness necessitated thickening the base of the dam wall, incorporating a preformed joint in the heel of the dam and including shear keys in the foundation.

The load on the foundation of a gravity dam in inclined downstream, which induces an overturning moment and a non-uniform distribution of stress in the rock masses concerned. Stability analysis against overturning involves calculating the resultant forces acting against the foundation and ensuring that this resultant acts within the middle third of the base (Underwood and Dixon, 1976). The resultant acts through the centre of gravity of the dam and is the vector sum of the total vertical and horizontal forces. The effect of an earthquake on overturning can be determined by adding a pseudostatic force to the resultant force vector. However, overturning is unlikely to happen since others types of failures, such as crushing of toe material and cracking of upstream material, will occur prior to overturning.

A dam settles under its own weight and impounding the reservoir basin causes additional settlement, the amount of settlement depending upon the strength of the rock masses in the foundation. If a dam is constructed on rocks that swell on exposure, then it may undergo more than usual settlement. When load is removed from a rock mass on excavation it moves slightly upward, that is, it is subject to rebound. Rebound may be significant if, during construction, a thick layer of unreliable rock material is removed from the dam site. The amount of rebound depends on the modulus of elasticity of the rocks concerned, the larger the modulus of elasticity, the smaller the rebound. The situation is complicated if the foundation consists of more than one rock type with differing physical properties. This can lead to differential rebound. The rebound process in rocks generally takes a considerable time to achieve completion and will continue after a dam has been constructed if the rebound pressure or heave developed by the foundation material exceeds the effective weight of the dam. Hence, if heave is to be counteracted a dam should impose a load on the foundation equal to or slightly in excess of the load removed.

All foundation and abutment rocks yield elastically to some degree. In particular, the modulus of elasticity of the rock masses is of primary importance as far as the distribution of stresses at the base of a concrete dam is concerned. What is more, tensile stresses may develop in concrete dams when the foundations undergo significant deformation. The modulus of elasticity is used in the design of gravity dams for comparing the different types of foundation rocks with each other and with the concrete of the dam. In the design of arch dams, if Young's modulus of the foundation has a lower value than that of the concrete or varies widely in the rocks against which the dam abuts, dangerous stress conditions may develop in the dam. The elastic properties of a rock and existing strain conditions assume importance in proportion to the height of a dam since they influence the magnitude of the stresses imparted to the foundation and abutments. The influence of geological structures in lowering Young's modulus must be accounted for by the provision of adequate safety factors. It should also be borne in mind that blasting during excavation of a foundation may open fissures and joints, which leads to greater deformability of the rock mass. The deformability of the rock mass, any possible settlements and the amount of increase of deformation with time can be taken into consideration by assuming lower moduli of elasticity in the foundation or by making provisions for pre-stressing.

Rocha (1974) used model testing as a means of investigating how the deformability of a foundation influences the state of stress in a concrete dam. He showed that one of the most important parameters is the E_r/E_c ratio (the ratio of Young's modulus for the rock mass to that of the concrete). If E_r/E_c is less than 1:16, then the behaviour of a dam is governed by the deformability of the foundation, whereas when E_r/E_c exceeds 1:4 the influence of the foundation is very slight. However, if the foundation conditions are more or less homogeneous, then the state of stress in these structures is not much influenced even by values of E_r/E_c as low as 1:16.

The pore water pressure within foundation materials is a variable force that acts in all directions and exerts an important influence on their engineering performance. Estimation of pore water pressure therefore is a fundamental factor in dam design and in the study of the stability of the adjacent slopes (Serafim, 1968). Pore water pressures reduce the strength of rock masses and increase the amount of deformation they undergo. If stratified rock masses occur in a dam foundation, then pore water reduces the coefficient of friction between the individual beds, and between the foundation and the dam. Increasing pore water pressure may lift beds, thereby decreasing the shearing strength and resistance to sliding within the rock mass.

Percolation of water through the foundations of concrete dams, even when the rock masses concerned are of good quality and of minimum permeability, is always a decisive factor in the safety and performance of dams. Such percolation can remove filler material that may be occupying joints, which in turn can lead to differential settlement of the foundation. It also may open joints and so decreases the strength of the rock mass. In highly permeable rocks excessive seepage beneath a dam may damage the foundation. Seepage rates can be lowered by reducing the hydraulic gradient beneath a dam by incorporating a cut-off into the design. A cut-off lengthens the flow path thereby reducing the hydraulic gradient. It extends to an impermeable horizon or some specified depth and usually is located below the upstream face of the dam. The rate of seepage also can be reduced by placing an impervious earth fill against the lower part of the upstream face of a dam.

Uplift pressure acts against the base of a dam and is caused by water seeping beneath it that is under hydrostatic head from the reservoir. The uplift pressure on the heel of a dam is equal to the depth of the foundation below water level multiplied by the unit weight of the water. In the simplest case, it is assumed that the difference in hydraulic heads between the heel and the toe of the dam is dissipated uniformly between them. The uplift pressure can be reduced by allowing water to be conducted downstream by drains incorporated into the foundation and base of the dam.

Kaloustian (1984) reported that over 60 per cent of the 110 failures and deteriorations of 4489 concrete dams between 1900 and 1978 were attributable to seepage and uplift problems, and that 81 per cent occurred when the reservoir was being impounded. The latter indicates that impounding represents a critical time as far as a dam is concerned since both the dam and its foundation experience rapid changes in gravity and seepage stress. Kaloustian concluded from his survey that deterioration due to loss of foundation strength took place within the first two years of operation of a dam, deterioration because of seepage and uplift occurring within the first five years. Failures generally took place within the first four years.

6.10 Geology and dam sites

Of the various natural factors that directly influence the design of dams none is more important than the geological, not only do they control the character of the foundation but they also govern the type of materials available for construction. Every site has some geological peculiarity but the major questions that need answering include the depth at which an adequate foundation exists, the strengths of the rock masses involved, the likelihood of water loss and any special features that have a bearing on excavation. The character of the foundation upon which a dam is built and its reaction to the new conditions of stress and strain, and of hydrostatic pressure must be ascertained so that the proper factor of safety may be adopted to ensure against subsequent failure. What is more, as far as the foundation for a concrete dam is concerned, it is also necessary to investigate how the properties of the rock mass(es) concerned influence the behaviour of the dam, since the dam and foundation should be regarded as a structural unit. As a consequence, projects should not be embarked upon until all reasonable doubts relating to the geological feasibility have been removed.

6.10.1 Investigations for dam sites

Surface investigations include a general study of the topography, hydrology and geology of the area concerned. Subsurface exploration at dam sites should aid the production of detailed geological maps and sections showing the succession, geological structures, depth of weathering, position of the water table and information on the physical properties of the foundation rocks. Most of the detailed geological information is obtained from drilling. Initially, one hole may be put down in the middle of each abutment, and one or more in the river section to determine the depth of river fill. Additional holes then can be located from the results of these holes. Drillholes always should be taken into bedrock unless the weathered zone or superficial material is extremely deep. The presence of boulders above the rock formation at a dam site may be misleading if they are of the same composition as the local solid rock, and they usually are. Therefore, when rock is met with on drilling, it usually is recommended that drilling should continue for at least another 6 m. Cores should be examined closely for weathered surfaces or linear structures that may help determine whether the stick comes from sound rock or a boulder.

Adits provide an effective means of exploring dam abutments, especially if the valley walls are steep. They are preferable to all other methods for exploring steeply dipping joints, faults, shear zones, creep zones and cavernous structures in valley walls. Shafts provide the most reliable means of exploring and sampling overburden. Trenches can be used to explore weathered zones and for exposing rock formations under shallow overburden.

Seismic refraction frequently has been used in investigations of dam sites. It has proved most useful in the detection of buried channels and in the approximate location of bedrock. However, the results should not be used to fix the location of structures or establish grade lines. Seismic refraction also can be used to determine the *in situ* value of Young's modulus. The resistivity method has been used at prospective dam sites with varying degrees of success in locating buried channels, in determining the depths to bedrock and in detecting permeable beds in valley alluvium. However, it is much less dependable than seismic refraction.

Rocha (1974) reviewed the various *in situ* testing methods that are available for obtaining data relating to foundation rock masses and more importantly how to interpret such information in relation to the design of concrete dams. It is advisable to test every foundation zone that, according to the data available, can be distinguished with respect to deformability. Some *in situ* tests such as plate load tests, shear tests and flat-jack tests may be carried out in adits, whilst others like the dilatometer test are carried out down drillholes (see Chapter 1). Hydrostatic pressure chambers occasionally have been used to measure the reaction of a rock mass to stress over large areas, providing values of Young's modulus, elastic recovery, inelastic deformation and creep. The results are used to evaluate the behaviour of a dam foundation and related strain distribution in the structure, and to help estimate the behaviour of pressure tunnel linings. Such chambers may exceed 5 m in length and 2 m in diameter and therefore are costly, hence their infrequent use. They are lined with reinforced concrete, which is divided into a number of independent sections, and reinforced concrete plugs close both ends. Water loss is prevented by a rubber lining and in sound rock testing pressures may reach 5 MPa. Deformations are measured by special extensometers. The first attempt to perform *in situ* triaxial tests using hydraulic jacks and flat jacks on a rock cube was made at the Kurobe IV Dam project in Japan. A later testing programme at Ghiona in Greece was described by Voort and Lotgers (1974). The use of the Geological Strength Index, introduced by Hoek (1994), to estimate the *in situ* strength parameters and Young's modulus of rock masses has been referred to in Chapter 3.

Percolation of water along joints, fissures, fault zones, and altered and crushed zones must be assessed in order to design grout curtains and drainage systems, otherwise dangerous uplift forces may develop in the foundations.

6.10.2 Rock masses and dam sites

In their unaltered state, plutonic rocks are essentially sound and durable with adequate strength for any engineering requirement. Sites with sound rock exposed at the surface are often found in regions where glaciers have removed the weathered mantle. In some instances, however, intrusives may be highly altered by weathering or hydrothermal attack. In tropical humid regions, in particular, valleys carved in granite may be covered with residual soils that extend to depths often in excess of 30 m. Fresh rocks may be exposed only in valley bottoms that have actively degrading streams. At such sites it is necessary to determine the extent of weathering and the engineering properties of the weathered products. Generally, the weathered product of plutonic rocks has a large clay content, although that of granite rocks is sometimes porous with a permeability comparable to that of medium-grained sand, so that it requires some type of cut-off or special treatment of the upstream surface at a dam site. The effect of weathering on construction materials is to put hard rock quarries at a premium while residual soil suitable for earth fill is abundant. Three main design alternatives exist in such ground conditions. First, to excavate down to sound rock in order to provide a foundation for a concrete dam. Second, to construct an earth fill dam in which stripping is largely confined to the removal of top-soil. Cut-off problems in the residual soil could involve concrete walls, chemical grouting or deep core trenches. If foundations are inadequate for chute spillways, resort may be made to bellmouth spillways. Third, a composite dam may be built, for example, a buttress section including the spillway in the valley bottom with fill shoulders.

Thick massive basalt lava flows often make satisfactory dam sites. However, discontinuities and shear zones apart, dam foundations may be affected adversely by weathering, hydrothermal alteration and autobrecciation of basalt. For example, the degree of weathering and autobrecciation were responsible for reducing the strength in the foundation area at Katse Dam, Lesotho, which meant that an extra 240 000 m^3 had to be excavated (Bell and Haskins, 1997). Many basalts of comparatively young geological age, however, are highly permeable, transmitting water via their open joints, pipes, cavities, tunnels and contact zones. Foundation problems in young volcanic sequences are two-fold. First, weak beds of ash and tuff may occur between the basalt flows that give rise to problems of differential settlement or sliding. Second, weathering during periods of volcanic inactivity may have produced fossil soils, these being of much lower strength. Concrete dams may be used where thick beds of sound basalt adequately confine such weak zones, otherwise foundation conditions dictate the adoption of an earth fill embankment. Rhyolites, and frequently andesites, do not present the same severe leakage problems as some basalts. They frequently offer good foundations for concrete dams although at some sites chemical weathering may mean that embankment designs have to be adopted.

Pyroclastics usually give rise to extremely variable foundation conditions due to wide variations in strength, durability and permeability. Their behaviour very much depends upon their degree of induration; for example, many agglomerates have high enough strength to support a concrete dam and also have low permeability. By contrast, ashes are invariably weak, often undergo hydrocompaction when wetted and generally are highly permeable. Clay-cement grouting at high pressures, however, may turn ash into a satisfactory foundation. Ashes and tuffs frequently are prone to sliding. Dam sites that contain young ashes and tuffs dipping toward the valley in either abutment are especially questionable. Montmorillonite is not an uncommon constituent in basic ashes and tuffs when they are weathered, and its presence should be given special attention.

The metamorphic rocks vary considerably in their suitability for dam sites. Fresh thermally metamorphosed rocks such as quartzite and hornfels are very strong and afford excellent dam sites. Marble has the same advantages and disadvantages as other carbonate rocks. Cleavage, schistosity and foliation in regional metamorphic rocks may affect their strength adversely and make them more susceptible to decay. Moreover, areas of regional metamorphism usually have suffered extensive folding so that rocks may be fractured and deformed. Some schists, slates and phyllites are variable in quality, some being excellent for dam sites; others, regardless of the degree of their deformation or weathering, are so poor as to be wholly undesirable in foundations and abutments. For instance, chlorite and sericite schists are weak rocks that contain closely spaced planes of schistosity. Large-scale field tests may be required to measure shear strength and sliding potential, the latter being especially critical if the rock masses dip downstream. Some schists become slippery on weathering and therefore fail under a moderately light load. On the other hand, slates and phyllites tend to be durable. Particular care is required in blasting slates, phyllites and schists, otherwise considerable overbreak or shattering may result. It may be advantageous to use pre-splitting for final trimming purposes. Generally, gneiss has proved a good foundation rock for dams, although a notable exception was at the Malpasset Dam site. The rupture of the Malpasset Dam, near Frejus, France, occurred on 2 December 1959. Over 400 people lost their lives and part of Frejus was destroyed. This arch dam was founded in gneiss in which there are magmatic intrusions. It would appear that fissures opened in the rock under the heel of the dam, which then was subjected to tensile stresses. Consequently, this zone became very pervious, allowing the slow build-up of water pressure in the gneiss that, in turn, led to increased fissuration and further weakening, so allowing the dam to slide. The dam underwent a double rotation movement. These displacements may have caused a fissure, 10–20 mm wide, to open up on the upstream side of the dam, some 6 months before rupture occurred. At this junction it is very likely that the foundation had been weakened all along the periphery of the dam. The displacement of the dam foot increased and an active arch was formed within the dam. Because the dam was more or less loose from its foundation a tremendous thrust was transferred to the left abutment. A blowout therefore occurred in the rock mass on the left bank and the left concrete abutment slid causing the dam to collapse. Failure occurred when the reservoir reached its top water level (Jaeger, 1963).

Sandstones have a wide range of strength depending largely upon the amount and type of cement-matrix material occupying their voids. With the exception of shaley sandstone, sandstone is not subject to rapid surface deterioration on exposure. As a foundation rock, even poorly cemented sandstone is not susceptible to plastic deformation. However, friable sandstones introduce problems of scour at a dam foundation. Moreover, sandstones are highly vulnerable to the scouring and plucking action at the overflow from dams and have to be adequately protected by suitable hydraulic structures. A major problem of dam sites located in sandstones results from the fact that they generally are transected by joints, which reduce resistance to sliding. Generally, however, sandstones have high coefficients of internal friction that give them high shearing strengths, when restrained under load. Sandstones frequently are interbedded with shale, which may constitute potential sliding surfaces. Sometimes such interbedding accentuates the undesirable properties of the shale by permitting access of water to the shale–sandstone contacts. Contact seepage may weaken shale surfaces and cause slides in formations that dip away from abutments and spillway cuts. Severe uplift pressures also may develop beneath beds of shale in a dam foundation and appreciably reduce its resistance to sliding. Foundations and abutments composed of interbedded sandstones and shales also present problems of settlement and rebound, the magnitude of these factors depending upon the character of the shales. The permeability of sandstone depends upon the amount of cement in the voids and, more particularly, on the incidence of discontinuities. The porosity of sandstones generally does not introduce leakage problems of consequence. The sandstones in a valley floor may contain many open joints that wedge out with depth, these often being caused by rebound of interbedded shales. Conditions of this kind in the abutments and foundations of dams greatly increase the construction costs for several reasons. They have a marked influence on the depth of stripping, especially in the abutments. They must be cut-off by pressure grouting and drainage to prevent excessive leakage, as well as to reduce the undesirable uplift effects of hydrostatic pressure of reservoir water on the base of the dam or on the base of some bedding contact within the dam foundation. Where beds of sandstone in a hillside dip downstream, wing trenches can be constructed upstream from the main cut-off in order to prevent the impounded water from gaining access to the hillside. If the dip is upstream, the wing trench can be carried in a downstream direction. Many sandstones found in valleys have been fractured by valley bulging and cambering. Spectacular valley bulges were recorded in the foundations of the Howden, Derwent and Ladybower dams in South Yorkshire, England (Fig. 6.17). In the latter, the fold was present to a depth of almost 60 m. Cambering produces fissures that run along the valley sides that may gape up to 250 mm wide at the surface. Extensive grouting programmes may be necessary in such situations.

Thickly bedded horizontally lying limestones relatively free from solution features afford excellent dam sites. On the other hand, thinly bedded highly folded or cavernous limestones are likely to present serious foundation or abutment problems involving bearing strength, sliding and watertightness. Furthermore, beds separated by layers of clay or shale, especially those inclined downstream may serve as sliding planes. Some solution features are always present in limestone. The size, form, abundance and downward extent of these features depend upon the geological structure and the presence of interbedded impervious layers. Individual cavities may be open, partially or completely filled with sediment, or they may be occupied by water. Solution cavities present numerous problems in the construction of large dams, among which bearing strength and watertightness are paramount. Sufficient bearing strength generally may be obtained in cavernous rock by deeper excavation than otherwise would be necessary. Watertightness may be attained in most instances by removing the material from cavities and refilling with concrete. Small filled cavities may be sealed by washing out and then by grouting with cement. Nonetheless, Bozovic *et al.* (1981) referred to large cavities in limestone at the Keban Dam in Turkey that exceeded 100 000 m^3 in volume. Even though the cavities were filled and an extensive grouting programme carried out, leakage on reservoir impoundment amounted

Figure 6.17 Valley bulging in interbedded shales and thin sandstones of Namurian age, exposed during the construction of Howden Dam in 1933, South Yorkshire, England (Courtesy of the Severn-Trent Water Authority).

to some $26\,m^3\,s^{-1}$. Indeed, Foyo *et al.* (1997) maintained that grouting programmes in karstic limestones may be difficult because on the one hand, normal joints may be tight or have small apertures and so not be groutable, whilst on the other, open joints, galleries and cavities allow grout to flow over large distances involving huge grout takes. The establishment of a watertight cut-off through cavernous limestone presents difficulties in proportion to the size and extent of the solution openings. A classic case of leakage is associated with the Hales Bar Dam, Tennessee, which was founded on the Bangor Limestone and located at the downstream limit of the gorge where the River Tennessee emerges from the mountains (Moneymaker, 1968). The limestone was dissolved in places to depths of more than 30 m below the original bed of the river, dissolution being controlled predominantly by the minor faults. Leakage in the worst areas meant that copious quantities of cement grout were pumped into the limestone. Leakage continued after completion of the dam in 1917 and so numerous attempts were made to seal the limestone by pumping molten asphalt via heated pipes or by placing concrete via tremie pipes into the cavities but all were unsuccessful. By 1939 leakage amounted to about $48\,m^3\,s^{-1}$ and the stability of certain parts of the dam was in doubt. A subriver cut-off wall along the upstream toe of the dam was constructed, which extended from the foundation to depths that ranged from 7.6 to 32 m below the open cavernous rock. However, by the late 1950s leakage had increased to more than $54\,m^3\,s^{-1}$. A further grouting programme was undertaken in 1960 but without success. Consequently, Hales Bar Dam was demolished in 1968. In fact, potential dam sites should be abandoned where cavities are large and numerous, and extend to considerable depths.

The adverse affects of dissolution of gypsum or hydration of anhydrite associated with some reservoirs in the United States has been referred to above. However, anhydrite and gypsum generally have proved sound where they are interbedded with mudstone at sites in arid areas.

According to Burwell (1950) well-cemented shales, under structurally sound conditions, present few problems at dam sites, though the strength limitations and elastic properties of such shales may be factors of importance in the design of concrete dams of appreciable height. However, these shales have lower elastic moduli and lower shear strength than concrete and therefore are unsatisfactory foundation materials for arch dams. Moreover, if the lamination is horizontal and well developed, then a foundation may offer little shear resistance to the horizontal forces exerted by a dam. A structure keying the dam into such a foundation then is required. On the other hand, severe settlements may take place in low grade compaction shales. As a consequence, embankment dams usually are constructed at such sites but associated concrete structures such as spillways are likely to undergo settlement. Rebound in deep spillway cuts may cause buckling of spillway linings and differential rebound movements in a foundation may require special design provisions. The stability of slopes in excavations in shale may present problems both during and after construction. For instance, major slides into a spillway channel must be avoided since blockage might cause overtopping and possible failure of the dam. Burwell suggested that two expedients should be resorted to in building concrete dams against shale abutments. First, high steps should be avoided and, second, the abutment monoliths of the dam should be plug-poured in regular succession from lower to higher elevations against the final grade surfaces as soon after exposure as possible. The opening of joints and the development of shear planes in shales for considerable distances behind the normal zones of creep on valley sides result from a combination of elastic rebound, oversteepening of slopes and super-incumbent loading. These deep-seated disturbances may give rise to dangerous hydrostatic pressures on the abutment rock masses downstream from the dam, leakage around the ends of the dam and reduced resistance of the rock mass(es) to the horizontal forces.

6.10.3 Joints, faults, shear zones and dams

Joints, shear zones and faults are responsible for most of the unsound rock encountered at dam sites on plutonic and metamorphic rocks. Unless they are sealed they may allow leakage through foundations and abutments. Slight opening of joints on excavation leads to imperceptible rotations and sliding of rock blocks, which nonetheless are large enough to reduce the strength and stiffness of the rock mass. Sheet or flat-lying joints tend to lie approximately parallel to the topographic surface and introduce a dangerous element of weakness into valley slopes. Indeed, Terzaghi (1962) observed that the most objectionable feature in terms of the foundation at Mammoth Pool Dam, California, which is in granodiorite, was the sheet joints orientated parallel to the rock surface. Their width varied and if they had remained untreated large quantities of water could have escaped through them from the reservoir. Moreover, the joints could have transmitted hydrostatic pressures to the rock mass downstream from the abutments. If joint openings are very wide and located close to the surface of a rock mass, then they may close up under the weight of the dam so causing differential settlement. Sharma *et al.* (1999) noted the presence of a sheared contact between dolertite and slate below the foundation for the Lakhwar Dam, India, which dipped at 40–50° upstream. In order to avoid settlement of the dam along the sheared zone, it was proposed that three shear keys should be constructed at 20, 30 and 40 m below the foundation of the dam, and that pressure grouting should be carried out along the length of the shear keys.

Intraformational shears are zones of shearing parallel to bedding that are associated with faulting. Such shear zones tend to die out when traced away from the faults concerned and are probably formed as a result of flexuring of strata adjacent to faults. A shear zone may consist of a single polished or slickensided shear plane, whereas a more complex shear zone may be up to 300 mm in thickness. Intraformational shear zones often occur in clays, mudstones and shales, but are not restricted to argillaceous rocks. Their presence means that the strength of the rock along the shear zone has been reduced to its residual value. Xiao *et al.* (2000) described two types of shear zones at Gaobazhou Dam site, China, one type being less well developed than the other. The less well-developed type is associated with incompetent rock whilst the well-developed type is associated with interbed folds, as well as the rocks in which they occur. However, it is only the latter type that has a significant influence on engineering.

If the movement along a fault zone has been severe, then the rocks involved may have been crushed, sheared or pulverized. Where shales or clays have been faulted, the fault zone may be occupied by clay gouge. Fault breccias and conglomerates, which respectively consist of jumbled masses of angular or rounded fragments containing high proportions of voids, occur when more competent rocks are faulted. Notable fault movements in granitic masses may have reduced the material occupying the zones to fine gravel or sand size, which may or may not contain significant amounts of clayey material. If not treated, such zones can be responsible for differential settlement or leakage beneath a dam. Treatment may involve removal of material from the fault zone and replacement by cement or the material may be grouted with cement if the particle size distribution allows. Movements of greater intensity are responsible for the occurrence of mylonite along a fault zone. Fault zones occupied by shattered or crushed material, because of their weakness, may give rise to landslides on excavation for a dam foundation. The occurrence of faults in a river is not unusual and this generally means that the material along the fault zone is highly altered. In such a situation a deep cut-off is necessary. For example, a fault zone 6.1 m wide was found on excavating the river bed for the Rodrigues Dam on the Tijuana in California. This entailed the construction of a trench along the fault zone to 91.4 m below the river bed. Major cavitation in the limestones beneath the Hales Bar Dam on the River Tennessee was associated with fault zones, and as pointed out above, this gave rise to considerable leakage.

6.10.4 Soils and dam sites

The major problems associated with dam foundations on alluvial deposits generally result from the fact that alluvial silts and clays are poorly consolidated. Silts and clays are subject to plastic deformation or shear failure under relatively light loads and undergo consolidation for long periods of time when subjected to appreciable loads. Many large embankment dams have been built upon such materials but this demands a thorough exploration and testing programme in order to design safe structures. The slopes of an embankment dam may be flattened in order to mobilize greater foundation shear strength, or berms may be introduced. Soft alluvial clays at ground level generally have been removed if economically feasible. Where soft alluvial clays are not more than 2.3 m thick they should consolidate during construction if covered with a drainage blanket, especially if resting on sand and gravel. With thicker deposits it may be necessary to incorporate vertical drains within the clays, the spacing of which will depend on the horizontal permeability of the deposit (Bell, 1993). By contrast, coarser sands and gravels in alluvial deposits undergo comparatively little consolidation under load and therefore afford good foundations in terms of support for embankment dams. The primary problems associated with such soils result from their permeability. For example, Malkawi and Al-Sheriadeh (2000) referred to serious leakage problems (i.e. $400\,l\,s^{-1}$) through alluvial soils at Kafrein Dam, Jordan. The alluvial deposits are some 60 m in depth and consist of poorly sorted sandy gravel with occasional lenses of boulders, silty clays and silty sands that may be partially cemented. Alluvial sands and gravels form natural drainage blankets under the higher parts of an earth or rock fill dam, so that seepage through them beneath the dam must be controlled. Seepage through pervious strata may be dealt with by a grout curtain. Alternatively, seepage may be checked by the construction of an impervious upstream blanket to lengthen the path of percolation and the installation on the downstream side of suitable drainage facilities to collect the seepage.

Embankment dams usually are constructed on clay soils as they lack the load bearing capacity necessary to support concrete dams. Clay soils beneath valley floors frequently are contorted, fractured and softened due to valley creep so that the load of an embankment dam may have to spread over a wider area than is the case with shales and mudstones. Settlement beneath an embankment dam constructed on soft clay soils can present problems and may lead to the development of excess pore water pressures in the foundation soils (Olson, 1998). As noted previously, the rate of consolidation of soft clay soils can be enhanced by the use of vertical drains that lead into an overlying drainage blanket (Almeida *et al.*, 2000). Al-Homoud and Tanash (2001) described how the Karameh Dam, Jordan, was constructed in stages on clay soils to allow settlement to be monitored and controlled. Rigid ancillary structures necessitate spread footings or raft foundations. Deep cuts involve problems of rebound if the weight of removed material exceeds that of the structure. Slope stability problems also arise, with rotational slides a hazard. It is essential to carry out a thorough site investigation in order to determine the consolidation characteristics of the foundation clay soils and their behaviour as embankment materials. This involves obtaining continuous undisturbed samples from the foundation and from prospective borrow pit areas.

Among the many manifestations of glaciation are the presence of buried channels, disrupted drainage systems, deeply filled valleys, sand-gravel terraces, narrow overflow channels connecting open valleys, and extensive deposits of lacustrine silts and clays, tills, and outwash sands and gravels. Deposits of peat and solifluction debris may be interbedded with some glacial deposits. Consequently, glacial deposits may be variable in composition, both laterally and vertically. As a result, dam sites in glaciated areas are among the most difficult to appraise on the basis of surface evidence. Knowledge of the pre-glacial, glacial and post-glacial history of a locality is of vital importance in the search for the most practicable sites. A primary consideration in glacial terrains is the discovery of sites where rock foundations are available for spillway, outlet and, if part of the project, powerhouse structures. Generally, embankment dams are constructed in areas of glacial deposits. Concrete dams, however, are feasible in post-glacial, rock-cut valleys, or composite dams are practicable in valleys containing rock benches.

The glacial deposits in the buried channel at the Derwent Dam, England, were extremely complex and included an upper and lower aquifer beneath laminated silty clays (Fig. 6.18a). The aquifers contained water under artesian pressure (Ruffles, 1965). The maximum depth to bedrock was some 60 m. Because of the threat of potential seepage beneath the dam it was initially proposed to build a concrete cut-off trench about 65 m deep. However, subsequent

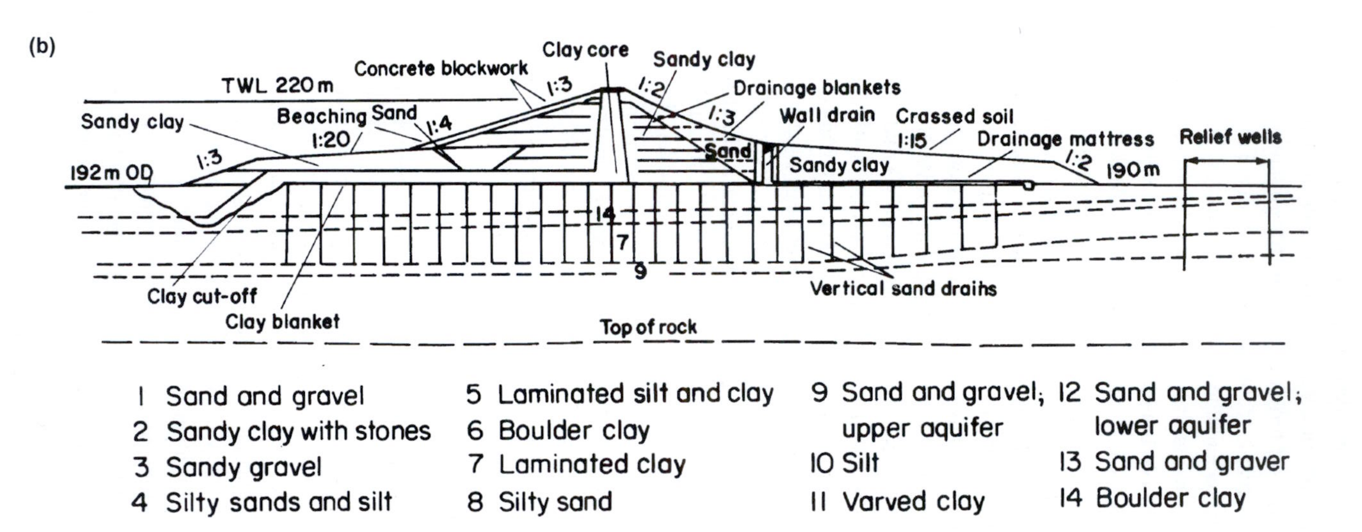

Figure 6.18 (a) Derwent Dam showing the complexity of glacial deposits under the deepest part of the centre-line section. (b) Section through the dam and foundation, showing the horizontal clay blanket linking the clay core with the clay cut-off, the vertical sand drains that hastened consolidation of the laminated clay and the relief wells into the upper aquifer.
Source: Ruffles, 1965.

pumping tests indicated that the total seepage losses would be within acceptable limits, even supposing that the impounded water could gain access to the aquifers. There was therefore no necessity for a deep cut-off, and a relatively shallow upstream cut-off was formed in open cut. It was linked by a clay blanket to the rolled clay core to obviate seepage through the superficial alluvial sands and gravels (Fig. 6.18b). A number of relief wells were sunk under the downstream toe to reduce uplift and to recover the seepage losses. Over 4000 vertical sand drains were installed in the laminated clays to accelerate consolidation during the construction of the dam, the shoulders of which were formed of the local till and incorporated horizontal drainage layers at 5 m centres. The upstream slope varied from 1 in 3 to 1 in 4 and the downstream slope from 1 in 2 to 1 in 3. Weight blocks were placed at each toe to counter possible deep-seated slips through the laminated clay.

Dispersive soils occur in semi-arid regions; for example, in South Africa they tend to be found in areas that have less than 850 mm of rain annually but not in arid regions. Dispersion occurs in soils when the repulsive forces between clay particles exceed the attractive forces so that in the presence of relatively pure water, the particles repel each other, deflocculating to form a suspension (Anon., 1990). The clay particles go into suspension even in quiet water and therefore are highly susceptible to erosion and piping. Although dispersive soils normally contain moderate to high contents of clay, there are no significant differences in the clay fractions of dispersive and non-dispersive soils, except that soils with less than 10 per cent clay particles may not have enough colloids to support dispersive piping. Dispersive soils contain a higher content of dissolved sodium (up to 12 per cent) in their pore water than ordinary soils. The presence of exchangeable sodium is the main chemical factor contributing towards dispersive behaviour in soil. This is expressed in terms of the exchangeable sodium percentage (ESP):

$$\mathrm{ESP} = \frac{\text{exchangeable sodium}}{\text{exchange capacity}} \times 100 \tag{6.1}$$

where the units are given in meq/100 g of dry clay. A threshold value of ESP of 10 per cent has been suggested by Elges (1985), above which soils that have their free salts leached by seepage of relatively pure water are prone to dispersion. Soils with ESP values above 15 per cent, are highly dispersive (Bell and Walker, 2000). Those with low cation exchange values (15 meq/100 g of clay) have been found to be completely non-dispersive at ESP values of 6 per cent or below. Similarly, soils with high cation exchange capacity values and a plasticity index greater than 35 per cent swell to such an extent that dispersion is not significant. Dispersion is usually a problem when the eroding water has an electrical conductivity lower than that of the pore water in the soil. The sodium adsorption ratio (SAR) is used to quantify the role of sodium where free salts are present in the pore water and is defined as:

$$\mathrm{SAR} = \frac{\mathrm{Na}}{\sqrt{0.5(\mathrm{Ca}+\mathrm{Mg})}} \tag{6.2}$$

with units expressed in meq litre^{-1} of the saturated extract. There is a relationship between the electrolyte concentration of the pore water and the exchangeable ions in the adsorbed layers of clay particles. This relationship is dependent upon pH value and also may be influenced by the type of clay minerals present. Hence, it is not necessarily constant. Gerber and Harmse (1987) considered a SAR value greater than 10 indicative of dispersive soils, between 6 and 10 as intermediate, and less than 6 as non-dispersive. However, Aitchison and Wood (1965) regarded soils in which the SAR exceeded 2 as dispersive. Another property that has been claimed to govern the susceptibility of soils to dispersion is the total content of dissolved salts (TDS, in this case Ca + Mg + Na + K) in the pore water (Sherard *et al.*, 1976). The lower this is, the greater the susceptibility of sodium-saturated clays to dispersion. Furthermore, there is a threshold value for total dissolved salts in the pore water (for a given ESP) above which the soil remains flocculated.

Serious piping damage and failures have occurred when dispersive soils have been used for the construction of earth dams. Early indications of piping take the form of small leakages of muddy water from an earth embankment after initial filling of the reservoir. Dispersive erosion may be caused by initial seepage through the more pervious areas in an earth dam. This is especially true in areas where compaction may not be so effective such as at the contacts with conduits, against concrete structures and the foundation interface, through desiccation cracks, or through cracks formed by differential settlement or hydraulic fracturing. Another mechanism involved in the failure of earth dams could be water that initially flowed along small tunnels made by termites (Hall *et al.*, 1993). In fact, most earth dams that have failed in South Africa did so on first wetting because that is when the fill is most vulnerable to hydraulic fracturing (Bell and Maud, 1994). Fractures represent paths along which piping can develop. The pipes can enlarge rapidly and this can lead to failure of a dam. Far more failures have occurred in small homogeneous earth dams, which generally are more poorly engineered and seldom have filters, than in major dams (Fig. 6.19).

Brink and Wagener (1990) outlined some measures that can be taken to avoid leakage at the interface with concrete structures. For example, the backfill surrounding a conduit in the key area of the Tweedraai Dam, South Africa, consisted of a sand–bentonite mixture with a 10:1 ratio by mass and extended at least 10 m on either side of the cut-off key centre line. The backfill was placed at optimum moisture content and compacted to 98 per cent of Proctor maximum dry

Figure 6.19 Failure of a small earth dam constructed of dispersive soil that was subjected to piping (note pipe exits), Natal, South Africa.

density. Bentonite expands as it absorbs water, thereby sealing the contact between the backfill and the surface of the conduit. The concrete-surrounds to the pipes were cambered to facilitate compaction at the interface of the concrete and backfill. Concrete collars were constructed to lengthen the interface between the conduit and earth dam.

Compaction and density control can help reduce the potential for crack initiation and thereby avoid the development of piping. As far as compaction is concerned, maximum resistance to piping failure occurs at slightly wet of optimum moisture content. Accordingly, lifts should be placed at 1.5–2 per cent above optimum moisture content so that the fill is more flexible. Compaction should be continuous with minimum exposure of any layer in order to minimize drying and subsequent shrinkage, and formation of fine cracks. Based on Australian experience, Tadanier and Ingles (1985) proposed that soils used in earth dams should have an air voids percentage of less than 6 to avoid unnecessarily dangerous conditions. In addition, they recommended that the clay content of soil used in earth dams should exceed 20 per cent and the linear shrinkage should be less than 7 per cent.

A placement density of 94 per cent of Proctor maximum dry density usually is recommended as sufficient to ensure uniformity and minimize differential settlement in an earth dam thereby reducing the likelihood of crack development. However, Brink and Wagener (1990) considered it necessary to specify a higher density for Tweedraai Dam to prevent the occurrence of areas of higher soil mass permeability that could occur if the placement moisture content was somewhat dry and so would help in the development of cracks. Therefore, a minimum density of 98 per cent of Proctor was specified for the clay core and cut-off key. An earth dam should possess a sufficiently low permeability if the soil is compacted slightly on the wet side of optimum moisture content. If a permeability of $10^{-7}\,\mathrm{m\,s^{-1}}$ or less is achieved, then Tadanier and Ingles (1985) maintained that this should prevent piping in dispersive soils. However, Sherard *et al.* (1977) pointed out that no matter how good the compaction (or for that matter the low permeability), if the foundation conditions permit excessive settlement, then cracking develops in an earth dam, which if constructed of dispersive soil, increases the potential for piping. Proper construction obviously includes a thorough assessment of foundations conditions so that, where necessary, they can be treated successfully to avoid excessive settlement.

The formation of shrinkage cracks during dry weather at the crest of a dam constructed of dispersive soil can be prevented by placing a capillary barrier of cohesionless sand and gravel over the top of the dam or if it has a clay core, over the top of the core and down part of the upstream side of the core. This cohesionless material tends to be washed into any cracks that develop in the upper part of a dam, thereby sealing the cracks.

Experience, however, indicates that if an earth dam is built with careful construction control and incorporates filters, then it should be safe even if it is constructed with dispersive soil. Indeed, Sherard *et al.* (1984b), and Sherard and Dunnigan (1989) showed that the incorporation of filters into earth dams represents the principal form of defence against failure by dispersive erosion and piping. Sand filters will seal and safely control leaks in dispersive soils. For example, Sherard *et al.* suggested that sand filters with a D_{15} of 0.5 mm or smaller will control and seal concentrated leaks through most dispersive soils with D_{85} larger than about 0.03 mm. Sand filters with a D_{15} of 0.2 mm or less should be used for dispersive clays. If geotextiles are used as filters, the soil–geotextile interaction must be tested for the proposed application (Watermeyer *et al.*, 1991). The Soil Conservation Service in the United States maintained that a chimney filter of sand is the most effective design measure that can be incorporated into an earth dam to prevent internal erosion (Anon., 1991). If no supply of clean pervious sand of the type used for a chimney filter is available, then a zone of silty sand or sandy silt can be considered as a line of defence against piping in dispersive soil used for a homogeneous earth dam. Such a chimney may not be pervious enough to act as a drain but it may inhibit piping.

Alternatively hydrated lime, aluminium sulphate and gypsum have been used to treat dispersive clays used in earth dams. The type of stabilization undertaken depends on the properties of the soil, especially the ESP and the SAR. The percentage of lime used should be that which raises the shrinkage limit to a value near saturation moisture content based on the compaction density to be achieved in the embankment. McDaniel and Decker (1979) found that the addition of 4 per cent, by weight, of hydrated lime converted dispersive soil to non-dispersive soil. Usually, lime treatment is applied to the outer 0.3 m of the surface of the embankment, the purpose being to prevent surface erosion. However, it may be impossible to achieve homogeneous mixing of small quantities of lime. Furthermore, mixing with lime, besides introducing brittleness and thereby increasing the potential for crack development, disrupts work that can lead to shrinkage cracks developing in a dam. Both granular and liquid aluminium sulphate or alum also have been used to treat dispersive clays. Soils can be compacted immediately after mixing, with no curing period, since alum does not affect the plasticity or brittleness of soils appreciably (Anon., 1991). Due to its relative solubility in water, gypsum in a very finely divided powder form, is another stabilizing material that can be used for dispersive soils. The gypsum is mixed with the soil during construction, the quantity added being equivalent to the excess sodium that has to be replaced in order to bring ESP values within the desired limits. Alternatively, the water in a reservoir can be dosed with gypsum thereby increasing the electrical conductivity of water moving into the earth dam so that if seepage occurs, deflocculation in the soil is prevented.

Talus or scree may clothe the lower slopes in mountainous areas. Because of its high permeability and unstable nature, scree should be avoided when locating a dam site, unless it is sufficiently shallow to be economically removed from the footprint of the dam.

Landslips are a common feature of valleys in mountainous areas, and large slips often cause narrowing of a valley so that it looks topographically suitable for a dam. Unless they are shallow seated and can be removed or effectively drained, it is prudent to avoid landslipped areas in a dam location, because their unstable nature may result in movement during construction or subsequently on filling or drawdown.

6.10.5 Earthquakes and dam sites

Fault breaks not only occur in association with large and infrequent earthquakes but they also occur in association with small shocks and continuous slippage known as fault creep. Earthquakes resulting from displacement and energy release on one fault can sometimes trigger small displacements on other unrelated faults many kilometres distant. Breaks on subsidiary faults have occurred at distances as great as 25 km from the main fault, but obviously with increasing distance from the main fault, the amount of displacement decreases. For instance, displacements on branch and subsidiary faults located more than 3 km from the main fault break are generally less than 20 per cent of the main fault displacement.

In most known instances of historic fault breaks the fracturing has occurred along a pre-existing fault. However, whilst it seems probable that a given fault would break again at the same location as the last break, this cannot be concluded with certainty. On the other hand, the likelihood of a new fault interfering with a dam is remote. There is little information on the frequency of breaking along active faults, all that can be said is that some master faults have suffered repeated movements, in some cases recurring in less than 100 years. On the other hand, much longer intervals, totalling many thousands of years, have occurred between successive breaks. Therefore, because movement has not been recorded in association with a particular fault in an active area it cannot be concluded that the fault is inactive. Individual fault breaks during simple earthquakes have ranged in length from less than a kilometre to several hundred kilometres. However, the length of the fault break during a particular earthquake is generally only a fraction of the true length of the fault. The longer fault breaks have greater displacements and generate larger earthquakes (Vallejo and Shettima, 1996). Also, the smaller the fault displacement, the greater is the number of observed fault breaks. The maximum displacement is less than 6 m for the great majority of fault breaks and the average displacement along the length of a fault is less than 50 per cent of the maximum. These figures suggest that zoned embankment dams can be built with safety at sites with active faults. Offset displacements are generally less than 3 m and at a maximum are 8 m. The critical zones of an embankment dam therefore should be made much larger than fault offsets.

As far as dam design is concerned, Sherard *et al.* (1974) suggested that all major faults located in regions where strong earthquakes have occurred should be regarded as potentially active unless convincing evidence exists to the contrary. In stable areas of the world little evidence exists of fault displacements in the recent past. Nevertheless, these authors suggested that an investigation should be carried out to confirm the absence of active faults at and near any proposed major dam in any part of the world. Where there is little or no evidence of activity it generally is considered reasonable to proceed with dam construction on the assumption that it is highly unlikely that a fault will break during the lifetime of the dam and that if it should, then the amount of movement probably will not be great enough to cause serious damage.

Dams have to be designed to be safe under the normal and the design earthquake loadings. Two different design earthquakes have been recommended by the International Commission on Large Dams (Anon., 1983), namely, the design basis earthquake (DBE) and the maximum credible earthquake (MCE). The DBE is the largest earthquake that is likely to occur during the life of a dam whilst the MCE is the maximum earthquake event that can be conceived to affect a dam. Okamota (1978) noted that it usually is assumed that the design earthquake load is reduced by half when reservoirs are surcharged or empty, and that a strong earthquake and an extraordinary flood are unlikely to occur at the same time. There are two methods of analysis used to ascertain the safety of dams in relation to earthquake shocks, namely, the pseudostatic method and the dynamic method. In the former method the inertial force and the seismic water pressure on the dam are regarded as a static force. The magnitude of the inertial force is determined by the mass of the dam multiplied by a seismic coefficient. Seismic coefficients vary between 0.05 and 0.25 in the horizontal direction, depending on the type of dam and the seismicity of the dam site. The seismic coefficient in the vertical direction is taken as between 0 and 0.5 the horizontal coefficient.

However, pseudostatic analysis of the Lower San Fernando Dam in 1966 indicated that it would not fail in the event of an earthquake. Unfortunately, on 9 February 1971, it was subjected to an earthquake that caused about 12 s of strong shaking, with a peak acceleration of about 0.5 g in the rock mass beneath the dam site. The upstream section, including 9.2 m of the crest, slid over 21 m into the reservoir (Fig. 6.20). Fortunately the dam held, but only just. Consequently, Seed *et al.* (1975) suggested that the pseudostatic method appeared unsatisfactory for evaluating the seismic stability of earth dams. Accordingly, dynamic analysis provides a more satisfactory basis for assessing the stability and deformation of embankment dams during earthquakes. In the dynamic method, the ground motion during earthquakes is specified initially, and then the vibration of the dam and of the reservoir water in response to ground motion are calculated. The inertial force and the dynamic water pressure applied to the dam are then determined.

Dams in regions subjected to severe earthquake should be flexible and self-healing such as earth and earth-rock embankments. A concrete dam should not be considered where active faults cross the foundation, for fault movement can break the contact between foundation and

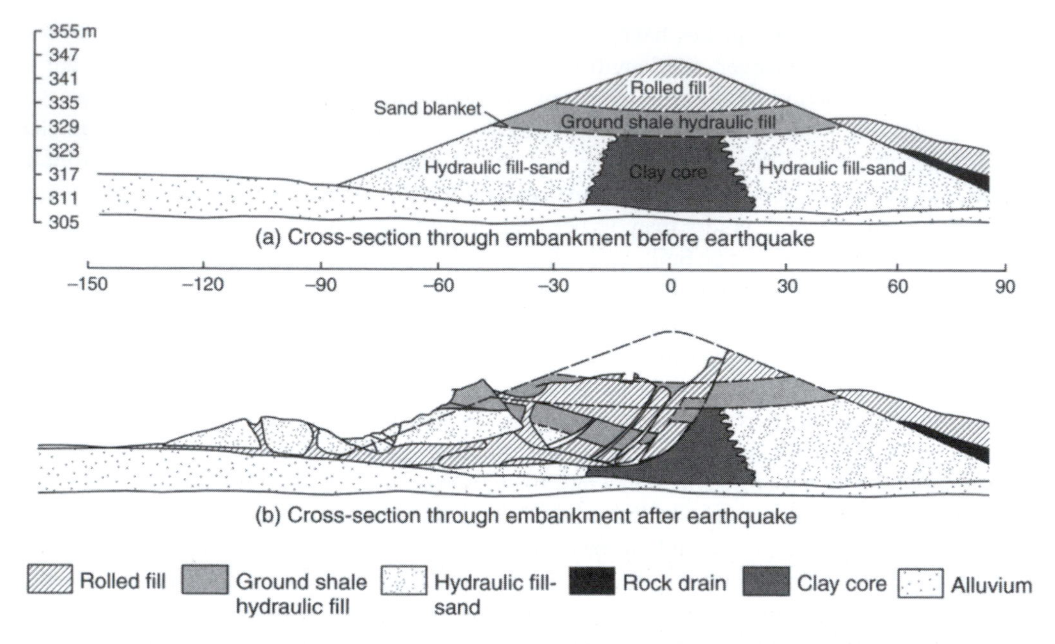

Figure 6.20 Cross-section through the Lower San Fernando embankment dam before and after sliding took place due to the earthquake of 9 February 1971. There was no evidence of failure in the foundation soils, the slide occurring as a result of failure in a zone of soil about 6 m thick in the hydraulic fill near the base of the dam. The failure was accompanied by some liquefaction of the hydraulic sand fill.
Source: Seed *et al.*, 1975. Reproduced by kind permission of Thomas Telford.

dam. This allows full uplift pressure to act from beneath, thereby reducing shearing resistance along the base of the dam and causing failure by sliding on the foundation. Arch dams, although very resistant to earthquake shocks are too brittle and slender to survive large displacements in the foundation or abutments. For example, an abrupt displacement of two opposing sectors of an arch dam, of the order of 0.25–0.5 m in almost any direction, may cause complete sudden failure. In such an instance the concrete may be crushed or one end of the dam may be lifted off its abutment.

Earthquake resisting earth dams are designed with a high freeboard and wider crest than those in stable regions. A higher camber and flatter slopes provide against slumping and sliding. The core of an earth dam must be larger and in dams of lesser height more plastic material makes self-healing more likely. If the maximum effective stress near the centre of the core exceeds 400 kPa in dams over 40 m in height, then lean saturated clays will tend to flow on dynamic loading, thereby closing cracks and fissures. Large filter zones of sand and gravel, with suitable transition characteristics, are arranged on the upstream and downstream side of the core. Filters also are provided wherever different materials with different percolation characteristics are brought into contact with each other. This ensures control of leakage through transverse cracks in the core. Large outlets in such embankments are desirable to provide a means of lowering the reservoir quickly in an emergency. For example, Cedar Springs Dam was completed in 1971 in an area of southern California where strong earthquakes are expected (Sherard *et al.*, 1974). Indeed, it is only 8 km from the San Andreas Fault and recently active faults traverse its foundation. Accordingly, the dam consists of a zoned embankment with thick exterior shells of quarried rock and thick transitions of well-graded coarse sand–gravel mixture, with a silty sand zone separating them from the rolled clay core (Fig. 6.21). The crest of the dam is 19 m wide,

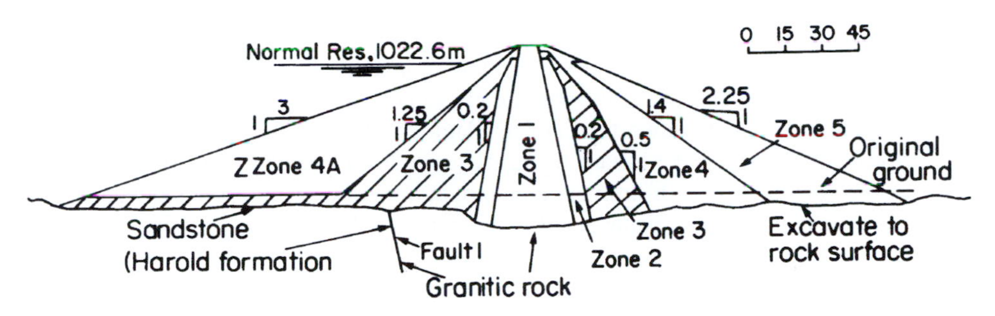

Description of zones

Zone 1	Clay core from lake bed deposit
Zone 2	Silty sand from Harold formation
Zone 3	Processed sand gravel transition from river alluvium or crushed rock
Zone 4 & 4a	Rolled, processed rockfill (75 mm minimum and 750 mm maximum)
Zone 5	Dumped rockfill, processed (450 mm minimum)

Figure 6.21 Cross-section of Cedar Springs Dam, California. Zone 1 = clay core from lake bed deposit; zone 2 = silty sand from Harold Formation; zone 3 = processed sand-gravel transition from river alluvium or crushed rock; zones 4 and 4a = rolled processed rockfill (75 mm minimum, 750 mm maximum); zone 5 = dumped processed rockfill (450 mm minimum).
Source: Sherard *et al.*, 1974. Reproduced by kind permission of Thomas Telford.

which is twice the usual width. The sand–gravel zone is too clean and cohesionless to allow an open crack to exist and its permeability, irrespective of the clay core, means that the quantity of water that could pass through would be safely controlled by the zones of rockfill. It represents an internal cohesionless transition zone for safe leakage control. Since the permeability of the transition zone will always be at least an order of magnitude less than that of the rockfill zone, the amount of leakage water will always fall well below the quantity that would tax the hydraulic capacity of the latter zone. The gradation of the transition zone is such that material cannot be washed from it into the voids of the rockfill zone. In such instances, the maximum leakage will emerge safely through the toe of the downstream rockfill zone. The clay of the core is highly non-dispersive and resistant to erosion.

One of the problems that may develop in certain soils, notably loosely packed saturated sands and silts, when subjected to dynamic loading by an earthquake is liquefaction. If such soils occur in the foundation of an embankment dam, then liquefaction during an earthquake event could give rise to excessive deformation and settlement of the dam, to slide movements on the sides of the dam, and to fissuring of the dam and associated movements of blocks of dam material. Seed and De Alba (1986) described how the results of the standard penetration test and the cone penetration test could be used to evaluate the liquefaction potential of soils. Some of the methods that may be used to improve the resistance of such soils to liquefaction prior to dam construction include deep compaction, *in situ* stabilization and grouting. Deep compaction methods include vibroflotation, dynamic compaction and compaction piles (see Chapter 4). In addition, explosives have been used to compact loose sandy soils. According to Carpentier *et al.* (1985) a saturated condition gives a more uniform propagation of shock waves and the relative densities of loose sands can be improved by 15–30 per cent. Lime or cement may be mixed into suspect soils by large augers to increase their stability. Grouts improve the strength and stiffness of soils but they are likely to reduce the ability of soils to dissipate pore water pressures in the event of an earthquake. In addition, berms may be incorporated into the design of a dam to improve its stability and drainage layers used to aid the dissipation of pore water pressures. Pressure relief wells, exiting upward into a drainage blanket, can be installed in the soil beneath the downstream side of the dam to relieve pore water pressure, and relief well also can be installed immediately downstream of the dam.

The seriousness of earthquake hazard for many large dams, together with the importance of the project, dictate that seismic-instrumentation programmes are set up. Bolt and Hudson (1975) discussed the manner in which dams could be instrumented so that the severity of earthquake ground motions and the dam response could be evaluated. They recommended that not less than four strong motion accelerographs should be installed at each site, two of which should be located so as to record earthquake motions in the foundation, and two used to measure the response of the dam. The latter two are usually situated near the top or on the crest of a dam. La Villita Dam is a zoned earth and rockfill dam located in southwest Mexico, which is a seismically active region. The dam is 420 m in length and 60 m high, and is founded on alluvial soils up to 70 m in depth. Between 1975 and 1985 the dam experienced five major earthquake events that caused significant vertical and horizontal displacements. These displacements were recorded at reference points along the surface of the dam and within the dam by inclinometers. In addition, bedrock accelerations at the right-bank and crest accelerations near the central section of the dam were recorded. After an analysis of this date, Succarieh *et al.* (1993) concluded that future strong earthquakes will mean that the dam will undergo larger deformations than previously. Their investigation demonstrated the significance of localized stick-slip deformation as a failure mechanism in embankment dams. Seismically induced deformations are highly dependent on the presence, locations and residual strength of such interfaces.

The Karameh Dam in Jordan is a zoned earthfill dam with a clay core that crosses a wrench fault zone and associated faults in the Jordan Valley Rift Zone. According to Al-Homoud (2000) evidence suggests that some 8–15 m of displacement has been associated with this fault over a rupture length of approximately 130 km and earthquakes with magnitudes as high as 7.8 have occurred in the area. What is more, layers of sand that occur in the dam foundation could be liquefied by a peak ground acceleration of 0.5 g. Al-Homoud maintained that the dam is under-designed as far as an earthquake with a magnitude of 7.8 is concerned and so represents a serious risk to safety. He indicated that if the dam is to resist a large earthquake event safely, then its freeboard should be widened by 2 m; the foundation should be stabilized against liquefaction by densification of the sand layers; relief measures should be taken downstream to avoid the build-up of excessive pore water pressures in the foundation, and the chimney filter and drainage zones should be widened.

6.11 Embankment dams

6.11.1 Types of embankment dams

An embankment dam is basically trapezoidal in cross-section and has to be impervious enough to prevent excessive loss of water from the reservoir it impounds. The design has to ensure stable slopes, in particular, the upstream slope of the dam must be protected from the destructive action of waves. Post-construction settlement of the crest of the dam must be limited so that adequate free-board is maintained. Embankments, of course, can be raised. Seepage and excessive hydrostatic uplift must be controlled by proper drainage.

Embankment dams are constructed where material is available readily and/or the rock foundations are suspect. Indeed, they are often more economical to construct than concrete dams. Moreover, the broad crest of an embankment dam can accommodate a highway where it is necessary to route a road across a valley. An embankment dam may be zoned or homogeneous,

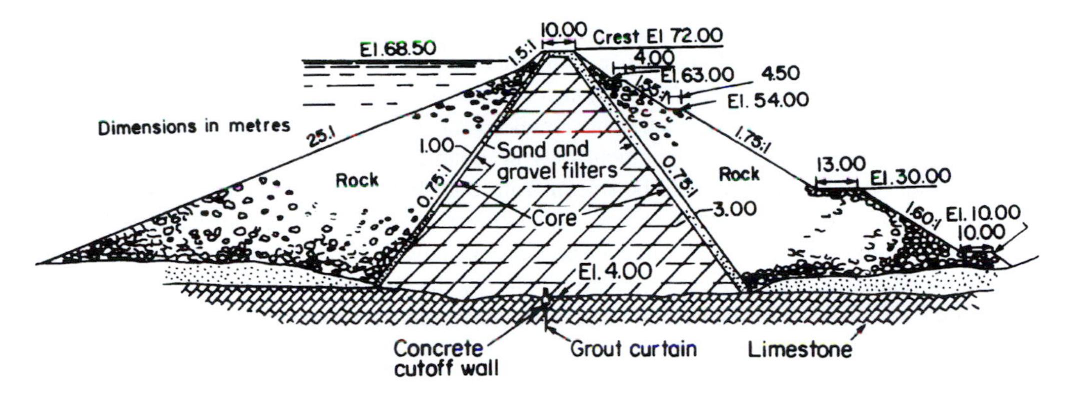

Figure 6.22 President Aleman Dam, Mexico, an example of a rockfill dam.

the former type being more common. A zoned dam is a rolled fill dam composed of several zones that increase in permeability from the core towards the outer slopes (Fig. 6.15). The number of zones depends on the availability and type of borrow material (Fell *et al.*, 1992). Stability of a zoned dam is due mainly to the weight of the heavy outer zones.

If there is only one type of borrow material readily available, a homogeneous embankment is constructed. In other words, homogeneous dams are constructed entirely or almost entirely of one type of material. The latter is usually fine grained, although sand and sand–gravel mixtures have been used. Zones of lower permeability can be formed in homogeneous embankments by using either more compaction or a higher water content during construction. To compensate for the absence of zonal loading in a homogeneous fill, the slopes of the dam are flattened, which also contributes to seepage control by decreasing the velocity of percolating water. These dams are often of low to moderate height, indeed very low dams are almost always homogeneous, otherwise their construction would be unduly complicated.

Rockfill dams generally consist of three basic elements, a loose rockfill dump, which constitutes the bulk of the dam and resists the thrust of the reservoir water, an impermeable facing on the upstream slope or an impermeable core, and rubble masonry between to act as a cushion for the membrane and to resist destructive deflections (Fig. 6.22). The disadvantage of an artificial impervious facing, such as concrete facing, is its relative inflexibility. Consolidation of the main rock body may tend to leave the face unsupported with the result that cracks are formed through which seepage takes place. Flexible asphalt membranes overcome this problem. One advantage of impervious faced rockfills is their ability to withstand overtopping by floods. A flexible rolled sloping impervious earth core has been used in some dams. Rockfill dams may prove less expensive in areas where concrete is expensive, where foundations are not favourable for concrete dams, where there are insufficient adequate earth materials for a rolled fill dam, where proper quality rock is readily available or where earthquakes are likely.

6.11.2 River diversion

Wherever dams are built there are problems concerned with keeping the associated river under control. These have a greater influence on the design of an embankment than a concrete dam. In narrow steep-sided valleys the river is diverted through a tunnel or conduit before any foundation treatment is completed over the floor of the river. However, the abutment sections of an embankment can be constructed in a wider valley prior to river diversion. In such instances, suitable borrow materials must be set aside for the closure section, as this often has to be constructed rapidly. However, rapid placement of the closure section can give rise to differential settlement and associated cracking. Hence, extra filter drains may be required to control leakage through such cracks. Sherard *et al.* (1967) suggested that compaction of the closure section at a higher average water content means that it can adjust more easily to differential settlement without cracking. Earthmoving equipment may be unable to cross a large river until closure is effected and so materials have to be drawn from both banks. This may mean that different design sections have to be adopted for the embankment on opposite sides of the river.

The construction programme at Tarbela Dam in Pakistan was divided into three main stages, each being related to the location of the river. During the first stage the river was allowed to flow in its own channel. In the second stage it was diverted into a specially excavated channel and in the third stage it flowed through four tunnels in the right abutment (Cartmel, 1971). The diversion channel varied in width from 200 m for the upstream section to 210 m for the downstream section, being 4633 m in length and averaging 13.7 m in depth. It was designed to pass a flow of 750 000 $m^3 s^{-1}$, which was well in excess of the maximum flood discharge. The main embankment and the silt blanket were constructed between the river bed and the diversion channel during stage 1. This part of the embankment initially was constructed with processed materials from the diversion channel and tunnels. When these were exhausted construction continued with material from the borrow area. The embankment was completed during the two following stages.

6.11.3 Construction materials

Wherever possible construction materials for an embankment dam should be obtained from within the future reservoir basin. Accordingly, the investigation of a dam site and the surrounding area should determine the availability of impervious and pervious materials for the embankment, sand and gravels for drains and filter blankets, and stone for rip-rap. In some cases only one type of soil is obtainable easily for an embankment dam. If this is impervious, then a homogeneous embankment is constructed, which incorporates a small amount of permeable material in order to control internal seepage. On the other hand, where sand and gravel are in plentiful supply, a very thin earth core may be built into the dam if enough impervious soil is available, otherwise an impervious membrane may be constructed of concrete or interlocking steel sheet piles. However, since concrete can withstand very little settlement such core walls must be located on sound foundations. Sites that provide a variety of soils lend themselves to the construction of zoned dams. The finer more impervious materials are used to construct the core whilst the coarser materials provide strength and drainage in the upstream and downstream zones. When the material that is most readily available at a site is so variable that it cannot be relied upon to have the requisite properties for an impervious core or other embankment zones, then its most economical use may be in random zones. Where two types of soil occur in two different layers in the same borrow pit they either can be excavated separately and placed in different zones in the dam or excavated together and blended into a single material with intermediate properties. To some extent the properties of the blended material can be controlled by varying the excavation procedure to obtain different proportions of the two soils. Materials also can be blended from different borrow pits, although this often proves uneconomic.

Materials generally are used without processing, although the larger cobbles and boulders normally are removed from embankment material to facilitate compaction. Boulders were crushed and the material sorted into five grades for use in the embankment at Tarbela Dam (Cartmel, 1971). Because of gap-grading problems with the gravels, crushing also was undertaken to meet the required specification. Cobbles and boulders can be used for pervious, semi-pervious or random zones, and the coarsest material can be used for rip-rap providing it does not slake.

According to Sherard *et al.* (1967), the volumes of rock excavated at many major dam sites for cut-off trenches, spillway(s), outlet works and other appurtenant structures have exceeded the volumes of the embankments. In such cases, it generally is cheaper to dispose of the waste in the embankment rather than in spoil heaps. Consequently, such material should be used whenever possible, even though it may have less desirable properties, and be more difficult to place than soil from borrow pits. Such rockfill can be used to form berms at both the upstream and downstream toes of the embankment when the foundation consists of soft ground and therefore requires enhancement of stability. It also can be used for free-draining rockfill zones, or material with suspect properties can be placed in random zones.

6.11.4 Compaction

Embankment soils need to develop high shear strength, low permeability and low water absorption, and undergo minimal settlement. This can be achieved by compaction. The relative compaction achieved on site, if possible, generally should come within 90–95 per cent of the maximum dry density obtained in a compaction test. Full-scale tests probably will be necessary for a large earth dam, a test section being compacted with the plant that is going to be employed. In this way it is possible to determine the number of passes a machine needs to make in order to obtain the desired dry density. Compaction of clay soils should be carried out when the moisture content of the soil is not more than 2 per cent above the plastic limit. If it exceeds this figure, then the soil must be allowed to dry. As far as granular material is concerned, it can be compacted at its natural moisture content. Overcompaction of soil on site, that is, compacting the soil beyond the optimum moisture content, should be avoided since this means that the soil becomes softer. The bulk density and moisture content of the compacted material should be assessed regularly so that proper control can be maintained.

As far as the shear strength of compacted soil is concerned, the greatest shear strength for a given degree of compaction is achieved when the moisture content is somewhat lower than the optimum. Decreasing permeability accompanies an increase in moisture content on the dry side of the optimum moisture content, the minimum permeability occurring at or slightly above optimum. At moisture contents higher than optimum there is a slight tendency for settlement to occur under steady and repeated loading.

6.11.5 Slopes of embankment dams

The permissible gradients of the slopes of an embankment dam depend upon the strengths of the foundation and embankment materials, and the internal zoning. Generally, slopes range between 1 in 2 and 1 in 4. Slopes may exceed 1 in 2 if the foundation is strong and the dam is designed with large rockfill zones. For example, the downstream slope of a dam consisting of excavated rock or pervious granular material, with a central earth core, and founded on rock, commonly varies between 1 in 1.6 and 1 in 1.8. By contrast, where the foundation is weak it may be necessary to construct much flatter slopes. Slopes also may be influenced by the rate of construction and the width of the valley.

The higher homogeneous dams formed of fine-grained material are, the flatter their slopes should be. On the other hand, the permissible slopes for thin-core dams, which consist mostly of pervious granular material, are independent of height, except for the extent that core strength contributes to embankment stability. Zoned dams usually have steeper slopes because, first, the stronger materials can be positioned where they provide most resistance to shearing stresses and, second, the internal drainage systems control pore water pressures. For any given factor of safety against shear failure an embankment with minimum volume usually is obtained when the slopes are steeper at the upper elevations and flatter near the bottom. Variable slopes should be considered for all embankment dams higher than 30 m. Dams constructed in gorges can have somewhat steeper slopes than otherwise because of the added stability provided by the confining effect of the steep sides.

In order to keep a uniform vertical pressure at all points of an embankment dam, the slopes are flattened gradually from the top towards the base. Average slopes for the upstream face built of soil materials are 1 in 2.5 or 1 in 3 below the top water level and 1 in 2 above. The downstream face is generally 1 in 2 or flatter. Slopes may incorporate berms. Usually, these are spaced vertically about 36 m apart and are provided with proper drainage for surface water. When material is available, a rockfill or gravel toe is placed on the downstream side of a dam. Sometimes a toe may be placed on the upstream face. These toes tend to increase the stability of a dam and afford some control over seepage.

It is not practical to construct slopes steeper than 1 in 1.5 for angular gravels and about 1 in 2 for smooth rounded gravels. The finer the particle sizes used in the fill, the flatter are the slopes. Therefore, an embankment composed entirely of homogeneous silts may have slopes as flat as 1 in 4 below the water line. Dams consisting mainly of clay soils sometimes have been built with slopes of 1 in 10 near their base. As mentioned above, the gradient of the slope also is dependent upon the competency of the underlying foundation, the less competent the foundation, the flatter the slope. In this way the load is spread more widely over the foundation materials thus reducing settlement and the risk of sliding. In the analysis of both upstream and downstream embankment slopes for stability during construction, a minimum safety factor of 1.5 normally is specified. In order to achieve stability of the upstream slope with reservoir empty and of the downstream slope with steady seepage from full reservoir head, a minimum factor of safety varying from 1.5 for clean granular materials to about 2.0 for highly cohesive clays, is needed.

One of the critical aspects of embankment dam design is the analysis of stability and safety of the dam under operating conditions. Any analysis of stability should consider the potential mechanism and geometry of sliding, the shear strength of the materials involved, especially of the zones of lower strength, and the pore water pressures. Generally, in the case of embankment dams there are three potential failure mechanisms that have to be assessed. According to Fell *et al.* (1992), these are the downstream slope for steady state seepage, the upstream slope in terms of drawdown, and the downstream and upstream slopes for the construction condition. Fell *et al.* maintained that analysis should be carried out in terms of effective stress except when an embankment dam is constructed on soft clay. They pointed out that in the latter case the short-term undrained strength condition is critical and so total stress analysis is appropriate. Traditionally, the stability of embankment dams is carried out in terms of limit equilibrium methods (e.g. Bishop, 1955; Morgenstern and Price, 1963; Sarma, 1973). However, finite element methods have been used for the analysis of slopes. For example, finite element techniques were used by Potts *et al.* (1990) in an analysis of progressive failure at Carsington Dam, Derbyshire, England, and more recently by Day *et al.* (1998) at Thika Dam, Kenya. In the latter case high pore water pressures developed in the fill during the early stages of construction giving rise to concern about the subsequent short-term stability. Construction was halted and a reassessment of stability was undertaken using finite element analyses. These highlighted zones of high shear stress and shear strain, and the formation of failure surfaces. The analyses also indicated the level of drainage required to bring about stability, to control movement and to reduce the risk of the development residual shear surfaces. Liang *et al.* (1999) contended that because of the uncertainties involved with obtaining certain parameters used in conventional stability analysis that a reliability-based approach should be taken for the evaluation of the stability of embankment dams, especially zoned dams. They used this approach in their stability assessment of the King Talal Dam in Jordan.

6.11.6 Protection of embankment slopes

Waves generated by wind blowing across a reservoir, or exceptionally by earthquakes or massive slides along the sides of a reservoir basin mean that the upstream face of an embankment dam requires protection. Such protection is afforded by either a concrete pavement or a layer of rip-rap. Concrete pavements may be articulated or monolithic. Gravel is placed beneath rip-rap to prevent the soils of the embankment from being removed by water action. It also helps stop the rip-rap sinking into the dam. On the other hand, if the stones in the rip-rap are not heavy enough to resist wave action, then they are removed. Beaching results when a few stones are moved out of place, a wave cut notch being formed to expose the compacted embankment material. The downstream slope of an earth dam can be protected by covering it with graded gravel or crushed rock, or by seeding it with protective grasses.

6.11.7 Pore water pressures and cracking

The most critical period in the construction of the dam is just before it is brought to grade or shortly thereafter. At this time, pore water pressures, due to consolidation in the embankment and foundation, are at a maximum. Piezometers should be installed in critical areas if there is the possibility of adverse pore water pressures developing during construction. The data thereby obtained allows the design assumptions to be checked and if necessary the design can be modified during construction. On the other hand, if suspect pore water pressures do not develop, then there usually is no question concerning the stability of an embankment dam, since most soils possess adequate strength when fully consolidated.

The magnitude and distribution of pore water pressures that develop during construction depend primarily on the construction water content, the properties of the soil, the height of the dam and the rate at which dissipation by drainage can occur. Casagrande (1950) maintained that a low placement water content produces a brittle fill, which can give rise to cracking in the core of an embankment dam, if differential settlement occurs. He added that any reduction of water content below Proctor optimum should be related to the plasticity index of the fill material, for example, 2 per cent below optimum would produce a fairly plastic clay but a brittle silt. Casagrande's views were supported subsequently by De Mello (1977). Water contents above the Proctor optimum can cause high construction pore water pressures that increase the danger of rotational slips in an earth dam.

The critical stability condition for the upstream slope of an earth dam occurs when the water in the reservoir is lowered, after the reservoir has been full for some length of time. Removal of the supporting reservoir load, together with the slow dissipation of pore water pressures in fine-grained material, means that slope stability is reduced. It therefore has been common practice to make the upstream slope flat enough to be stable under the maximum possible pore water pressures. Except for embankments consisting of very fine-grained silts and compressible clays this is probably a conservative procedure. Filter drains provide a reliable method of controlling pore water pressures developed on reservoir drawdown. If, however, the material comprising the upstream section of the dam is free draining, crushed rock or sand and gravel, then water will flow out of the pores as rapidly as the reservoir is lowered, and the problem of excess pore water pressure does not arise. Rapid drawdowns are continually involved in the operation of pumped storage schemes, and have produced failures in the upstream slopes of some of the embankment dams concerned.

It has been found that much higher vertical pressures can develop in the shoulders of an earth dam than in the core. Indeed, the measured vertical pressure in the latter can be as little as half the nominal overburden pressure. This may be due to an arching effect in the core, as during construction, the core tends to compress more under the weight of the overlying fill than do the shoulders. Therefore, part of the weight of the core is transferred to the shoulders by shearing stresses and arching. Unfortunately, hydraulic fracturing can occur in the core when the upper part is supported by the arching effects of the adjoining shoulders while the lower part of the core settles. In such a situation the total stress in the core can be reduced below the value of the pressure from the reservoir water. Such a development is detrimental to watertightness and the performance of the core. However, Penman (1977) observed that if the pore water pressures throughout the core at the end of the construction period exceed those imposed by the reservoir water, then the latter cannot give rise to hydraulic fracturing. Penman (1975) had suggested previously that arching action was reduced if a core was soft enough.

Cracking in embankment dams is influenced by soil properties and construction methods. For example, inorganic clay soils with plasticity indices of less than 15 are more susceptible to cracking when compacted drier than optimum moisture content than either finer or coarser materials. By contrast, clay soils with plasticity indices exceeding 20 can undergo larger deformations without cracking. There is a high likelihood of cracking occurring in embankments formed of residual soils that contain coarse particles of soft rock that break down when compacted. It is difficult to mix sufficient water into such soils so that they are often compacted at a lower water content than intended. This produces a brittle material.

6.11.8 Piping

Internal erosion of a foundation or embankment caused by seepage is referred to as piping. Usually, erosion begins at the downstream toe and works backwards towards the reservoir, forming channels or pipes in or under the dam (Fig. 6.19). These channels develop along paths of maximum permeability and sometimes may not begin to form until many years after a dam has been completed. By contrast, piping associated with dispersive soils, as noted above, may occur on first filling of a reservoir. Water may emerge downstream as a small spring, which gradually increases in size, and when muddy water appears at the toe a failure may occur within hours. Sometimes a boiling sand-water suspension occurs at the toe of the dam. When an embankment pipes, a progressive backward sloughing or ravelling of the saturated downstream slope occurs.

Resistance of an embankment or foundation to piping depends on the plasticity of the soil, its gradation and degree of compactness. As referred to above, the most susceptible soils to piping are dispersive soils. Settlement cracks, even in resistant materials, also may produce piping. Piping can be avoided by the incorporation of appropriate filters into a dam (see Section 6.10.4) or by lengthening the path of percolation of water within a dam and its foundation. The latter decreases the hydraulic gradient of the water flow and hence its velocity. It can be accomplished by installing impervious blankets from the upstream face or by widening the base of the dam, particularly where blanketing materials are scarce. The hydrostatic pressure head can be decreased by means of relief wells.

The Teton Dam, Idaho, was 90 m in height and failed on 6 June 1976, before the reservoir reached top water level. Eleven people lost their lives and some $400 million worth of damage was done. The dam was founded in a gorge carved out of interbedded volcanic and alluvial deposits. Because the rock at the sides of the gorge was too fractured for effective grouting, it was decided to replace the upper part of the grout curtain with a key trench, 21 m deep, filled with core material. Compaction in the key trench proved difficult, although great care was taken to ensure that the material was placed 0.6 per cent dry of optimum. The grout curtain extended 300 m into the right abutment and 150 m into the left abutment. It was formed by a single row of holes at 3 m spacing, located between two outer rows, 3 m away, in which the groutholes occurred at 6 m centres. Local deposits of aeolian silt were used as the fill for this more or less homogeneous dam. The silt was compacted at a moisture content of 0.5–1.5 per cent below Proctor optimum in layers of 150 mm, over which a sheepsfoot roller made 12 passes. This produced a strong brittle fill. Unfortunately, because the silt when placed had a low dry density, about 1.6 $\mathrm{Mg\,m^{-3}}$, it tended to crack when subjected to differential settlements and was eroded easily. Leakage was noticed downstream of the dam two days before the failure but this was not considered serious. However, on the morning of the failure two large springs, in which the water was turbid, and with a combined flow of over 400 $\mathrm{l\,s^{-1}}$, were issuing from the right abutment. By mid-morning a tunnel, approximately 2 m in diameter and 10 m long, had formed in the fill where the water was emerging. The tunnel gradually extended into the embankment and a whirlpool developed in the reservoir, opposite the tunnel. By now the flow discharging from the downstream slope of the dam had increased to approximately 3000 $\mathrm{l\,s^{-1}}$. After a further hour a large section of the dam moved into the whirlpool and part of the crest dropped into the tunnel. Flow increased rapidly, forming a deep channel through the dam (Fig. 6.23a). By 18.00 hours the 27 km long reservoir was essentially empty. Failure was initiated by a pipe eroding through the silt in the key trench, the silt being washed into fissures in the downstream wall of the trench (Penman, 1977). Because the fissures in the bedrock were inadequately grouted, the velocity of flow of water along the contact between core and foundation had not been reduced to a safe value. In addition, low-placement water content meant that the fill was strong enough to allow the formation of substantial pipes. It also increased the risk of hydraulic fracture. Piping and poor construction were responsible for the failure of the Zoeknog Dam, Mpumalanga Province, South Africa (Fig. 6.23b). The grout curtain had been omitted and fissures in the foundation rock adjacent to the conduit had not been sealed, which led to hydraulic fracturing and piping in the poorly compacted fill around the conduit. The situation was further compounded by the blanket drain being installed above rather than at foundation level so that large sections of central embankment below the drain were unprotected and prone to piping.

6.11.9 Internal drainage of embankments

According to Sherard *et al.* (1967), the design of the internal drainage system is governed mainly by the height of the dam, the cost and availability of pervious material, and the permeability of the foundation. A simple toe drain, to depress the piezometric level, can be used in a low homogeneous embankment dam (Fig. 6.24a). However, De Mello (1977) pointed out that failures have occurred in dams with toe drains. For example, under extreme conditions piping can develop above the toe. A horizontal drainage blanket often is constructed at the base of an embankment, on the downstream side, when the depth of the reservoir exceeds 15 m (Fig. 6.24b). One of the disadvantages of such a blanket is attributable to stratification in the embankment, which means that it is more permeable in the horizontal than vertical direction. As a consequence, seepage flows through the more pervious layers, discharging on the downstream slope where it can cause slumping. Vertical drains can be installed that lead into a horizontal blanket. Chimney drains have been used to avoid trouble due to stratification

Figure 6.23a Failure of the Teton Dam, Idaho (Courtesy of the US Bureau of Reclamation).

Figure 6.23b Failed Zoeknog Dam, Mpumalanga Province, South Africa.

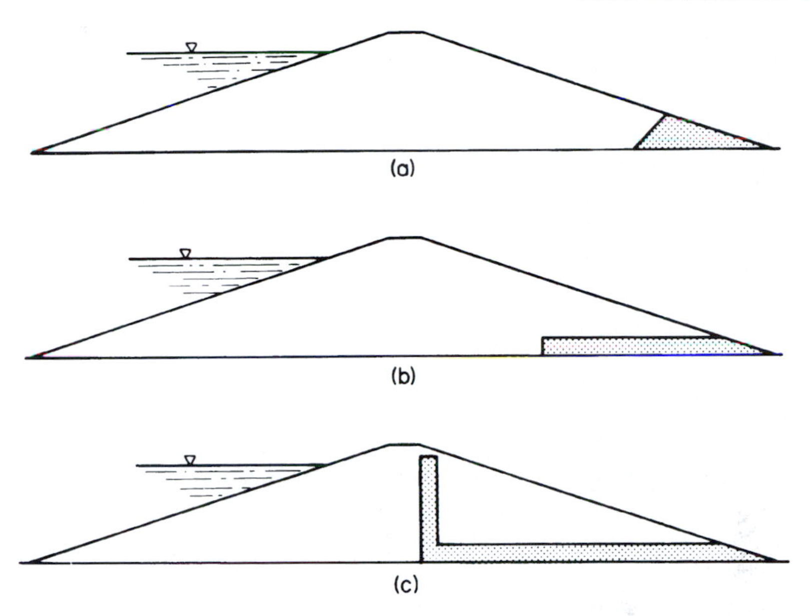

Figure 6.24 Drains used in dams: (a) toe drains; (b) horizontal blanket drain; (c) chimney drain.

and to intercept water before it reaches the downstream slope (Fig. 6.24c). However, De Mello contended that the use of vertical chimney drains should be avoided and that filters inclined upstream were more effective in terms of stability. Although this may double or treble the seepage losses he argued that this was perfectly acceptable. In addition to controlling seepage and producing a stable downstream slope when the reservoir is full, such an interceptor drain linked to a horizontal drainage blanket, is effective in reducing pore water pressures both during construction and following rapid reservoir drawdown.

In zoned dams the relative grading of adjacent zones should meet established filter criteria so that appreciable migration of soil particles is avoided. If this cannot be achieved between fine and coarse zones, then zones of intermediate gradation must be provided. Progressive zoning, without graded filter zones, may provide the necessary transition from fine to coarse material that prevents piping in a zoned dam with an internal impervious core. Filters offer protection if the core of an earth dam develops cracks as a result of movement. The filter material may either enter or bridge the cracks and accordingly reduces the likelihood of piping.

Aggregate filters and synthetic filter materials are used to control high pressure gradients or high pore water pressures and to collect the water and conduct it to outlets. An effective filter should satisfy three principal requirements. First, it should have a coefficient of permeability at least 10–100 times greater than that of the average embankment material it drains. Second, it should be large enough to cope with anticipated flow, with an adequate margin of safety for unexpected leakage. Third, it must be graded so as to prevent soil particles being washed from the embankment into its voids and so causing clogging. It is desirable to reduce the loss of head due to flow through a filter to the lowest value compatible with the grain size requirements. Hence, effective aggregate filters almost invariably require one layer of graded fine aggregate for filtering and a coarse layer of relatively high permeability to remove groundwater.

The following criterion has been used for the design of filters:

$$\frac{D_{15}(\text{filter})}{D_{85}(\text{soil})} < 4 \text{ to } 5 > \frac{D_{15}(\text{filter})}{D_{15}(\text{soil})}$$

The ratio of D_{15} of the filter to D_{85} of the soil is termed the piping ratio. In other words, the piping criterion dictates that the D_{15} of the filter material must not be more than 4 or 5 times the D_{85} of the surrounding soil. If the piping criterion is satisfied it is more or less impossible for piping to occur, even under extremely large hydraulic gradients. According to the permeability criterion (on the right-hand side), the D_{15} of the filter must be at least 4 or 5 times the D_{15} of the soil. Although this means that filter layers are several times more permeable than the surrounding soil it does not always guarantee hydraulic conductivity in filters. The above filter criterion was proposed by Bertram (1940) and since then there have been many studies of filter criteria. For example, the United States Army Corps of Engineers (Anon., 1955) limited the piping ratio to 5 and used the following criterion:

$$\frac{D_{50}(\text{filter})}{D_{50}(\text{soil})} \leq 25$$

If filters are placed in plastic clay, then they allow higher piping ratios, that is, for such soils the D_{15} size of the filter may be as high as 0.4 mm and the above mentioned criterion is disregarded. This relaxation in criteria for protecting plastic clays allows the use of a one-stage filter material. However, the filter must be well graded to ensure its non-segregation and therefore a coefficient of uniformity (D_{60}/D_{10}) not exceeding 20 is necessary.

Sherard *et al.* (1984a) showed that in well-graded soils (e.g. sandy silts and clays, and well-graded pervious sands), a filter sufficiently fine to catch the D_{85} (soil) size also will catch the finer particles. They also showed that the Bertram filter criterion was conservative. It can be considered to have a safety factor of about 2. Hence, Sherard *et al.* (1984a) concluded that it is still appropriate to use:

$$\frac{D_{15}(\text{filter})}{D_{85}(\text{soil})} \leq 5$$

as the principal filter acceptance criterion. This conclusion applies generally to filters with D_{15} larger than about 1.0 mm. For certain gap-graded and broadly graded coarse soils, usually graded from clay sizes to gravels, with D_{85} larger than 2 mm, the soil fines may be able to enter the voids in the filter even if the coarser particles cannot. The recommendations made by Sherard *et al.* (1984b) for filters in silts and clays are summarized in Table 6.2. It frequently has been suggested that a particle size distribution curve of the filter should be approximately the same shape as that of the soil. Sherard *et al.* (1984a), however, maintained that generally this

Table 6.2 Sand or gravelly sand filters and fine-grained soils

Soil type	*D_{85} particle size of soil* (mm)	*D_{15} particle size of filter* (mm)	*Comments on D_{15}(filter) $\leq$ 5 D_{85} (soil) criteria*
Sandy silts and clays	0.1–0.5	Around 0.5	Satisfactory
Fine-grained silts	0.03–0.1	Less than 0.3	Satisfactory
Clays with some sand content	Greater than 0.1	Around 0.5	Satisfactory
Fine-grained clays	0.03–0.1	Less than 0.5	Reasonable
Very fine-grained clays and silts	Less than 0.02	0.2 or smaller	Satisfactory

Source: Sherard *et al.*, 1984b. Reproduced by kind permission of American Society of Civil Engineers.

is neither necessary nor desirable. They also considered that the average particle size of a sand or gravel filter, D_{50}, does not provide a satisfactory measure of the minimum pore sizes and that filter criteria using D_{50} therefore were unsatisfactory. They further maintained that filter criteria using D_{15} (soil), such as:

$$\frac{D_{15}(\text{filter})}{D_{15}(\text{soil})} < 40$$

were even less satisfactory. Hence, they recommended that the use of such filter criteria should be abandoned.

Honjo and Veneziano (1989) undertook a survey of filter criteria of granular soils and showed that a D_{95}/D_{75} of 2 or less provided an indication of the capability of the soil to form self-healing layers (i.e. to prevent grains washing through). This they referred to as the self-healing index. They therefore proposed a new criterion as follows:

$$\frac{D_{15}(\text{filter})}{D_{85}(\text{soil})} \leq 5.5 - 0.5 \frac{D_{95}(\text{soil})}{D_{75}(\text{soil})} \quad \text{for} \quad \frac{D_{95}}{D_{75}} \leq 7$$

They maintained that this modified criterion provided a uniform safety (i.e. a probability of filter failure of about 0.1) over a wide range of soils. However, according to Talbot (1991), their single criterion for all soil types would result in filters that are too coarse for certain broadly graded soils and soils with significant sand and gravel fractions. Hence, the filter criteria advanced by Sherard *et al.* (1984a,b) are more reliable. Foster and Fell (2001) recently described a method by which the particle size distribution of a filter can be compared with the surrounding soil that it has to protect and so determine whether the filter is sufficiently fine to resist erosion. Their method is based on the analysis of the results of laboratory tests and the characteristics of dams that have experienced piping incidents.

Vaughan and Soares (1982) showed that a relationship exists between the size of particles retained by a filter and its permeability. Hence, they suggested that the permeability of a granular filter represented a better way of quantifying particle retention than its grading.

Geotextiles have been used for filters in embankment dams. For example, they often are used to enclose aggregate filters, in which case one of their purposes is to prevent clogging and thereby reduce the need for grading the aggregate. The performance of a filter drain incorporating geotextiles is governed, on the one hand, by the properties of the soils of the earth dam (particle size distribution, shape and packing) and the hydraulic flow conditions (undirectional steady state), and, on the other, by the properties of the fabric (pore size distribution, permeability, thickness, and variations of these with time and structural loading). Geotextiles also can be used to encapsulate drainage blankets beneath earth dams. The permeability of a geotextile, however, does reduce with time. A fuller discussion of filters and embankment dams has been provided by Fell *et al.* (1992).

6.12 Grouting

Grouting has proved effective in reducing percolation of water through dam foundations so that sites previously considered unsuitable because of adverse geological conditions can now be utilized. The design of a successful grouting programme requires the selection of a suitable grout material, and the correct drilling equipment, procedures and grouthole patterns. It is more important to ensure that the full design volume is permeated with grout when the objective is water cut-off than when the objective is to improve the mechanical properties of the ground.

The grout pattern includes the layout of the holes, the sequence in which each hole is placed and grouted, and the vertical thickness and sequence of grouting the stages for each hole (Kennedy, 2001). The layout of holes may follow some geometrical pattern or this may be modified in relation to the ground conditions. Hole spacings of about 1.3–2.5 m are typical.

6.12.1 Groutability

The rate at which grout can be injected into the ground generally increases with an increase in the grouting pressure, but this is limited since excessive pressures cause the ground to fracture and lift. The safe maximum pressure depends on the weight of overburden, the strength of the ground, the *in situ* stresses, the pore water pressures and the permissible amount of ground surface movement, if any. However, there is no simple relationship between these factors, and so a common rule-of-thumb for safe maximum grouting pressure is to relate the pressures to the weight of overburden. For example, the pressures used, as measured at the top of the hole, may start at 70 kPa for the first 3.1 m stage and increase by 70 kPa in each successive 3.1 m stage, while not exceeding 350 kPa for the fifth and lower stages.

The penetrability of grouts in terms of groutability ratios has been referred to in Chapter 4. As far as cement grouts are concerned, which tend to be the most commonly used in relation to dam sites, Nonveiller (1989) quoted the limits of penetration for ordinary Portland cement (OPC), high early strength cement (HESC), colloidal fine cement (CFC) and ultra-fine cement (UFC) into granular soils given in Table 6.3. Alternatively, permeability may be used to assess groutability. For instance, Littlejohn (1985) suggested that cement grouts cannot be used to treat soils with permeabilities of less than $5 \times 10^{-3}\,\mathrm{m\,s^{-1}}$ because cement particles are filtered out during injection. As far as clay grouts are concerned, the small sizes of the particles allow them to penetrate voids in soil with a permeability of $10^{-5}\,\mathrm{m\,s^{-1}}$. Initial estimates of the groutability of ground commonly are based upon the results of pumping-in tests, such as the Lefranc test in soils or the Lugeon test in rocks (a form of packer test), in which water is pumped into the ground via a borehole or drillhole (Cambefort, 1987). In the case of particulate grouts, penetration also depends on their shear strength in that their initial shear strength must be overcome before the grout begins to flow. The penetrability of chemical grouts is governed by their viscosity, injection pressure and period of injection, as well as permeability in the case of soils (Bodocsi and Bowers, 1991). Chemical grouts with viscosities of less than $2 \times 10^{-3}\,\mathrm{N\,s\,m^{-2}}$ (2 cP), such as acrylamide-based grouts, usually can be injected without difficulty into soils with permeabilities as low as $10^{-6}\,\mathrm{m\,s^{-1}}$. Those with viscosities around $5 \times 10^{-3}\,\mathrm{N\,s\,m^{-2}}$ (5 cP), for example, chrome-lignin grouts, are limited to soils that possess permeabilities in excess of $10^{-5}\,\mathrm{m\,s^{-1}}$. At viscosities about $1 \times 10^{-2}\,\mathrm{N\,s\,m^{-2}}$ (10 cP), typical of silicate based formulations, such grouts may not penetrate soils with permeabilities lower than $10^{-4}\,\mathrm{m\,s^{-1}}$.

Table 6.3 Limits of penetration of cement grout into granular soils

Type of cement	*Specific surface of cement* ($\mathrm{cm^2/g}$)	*Permeability, k* (m/s)	D_{85} *of cement* (mm)	D_{15} *of soil* (mm)
OPC	3.17	2.3×10^{-3}	0.047	0.87
HESC	4.32	1.3×10^{-3}	0.333	0.67
CFC	6.27	3.2×10^{-4}	0.019	0.38
UFC	8.15	3.5×10^{-5}	0.006	0.12

The results of the Lugeon test are described in terms of lugeon units, one lugeon being equal to a flow of one litre per metre per minute at a pressure of 1 MPa. A lugeon unit is approximately equal to a coefficient of permeability of 10^{-7} m s^{-1}. A rock absorbing less than one lugeon unit can be considered watertight (Lugeon, 1933). An impermeable rock formation that possesses discontinuties 0.2 mm wide at regular one metre intervals has a directional permeability of approximately 50 lugeon units, a high hydraulic conductivity, but the minimum groutable opening for ordinary cement grout is about 0.2 mm. Hoek and Londe (1974) therefore maintained that this meant that consolidation grouting with cement probably was useless in rock zones where water tests indicated less than 50 lugeon units. In such cases chemical grouts have to be used.

Houlsby (1992) proposed that grout curtains should achieve an approximate degree of water-tightness, which can be specified in terms of standards related to permeability. He described a modified form of lugeon testing that provided an assessment of the need for foundation grouting at dam sites (Fig. 6.25). Five consecutive tests are performed, for 10 min each at pressures A, B, C, B, A. The interpretation that Houlsby placed upon these five lugeon values is summarized in Fig. 6.26. Houslby pointed out that the lugeon scale decreases in sensitivity as values increase, the low values, 1–5, being the most sensitive and important. At 50 an accuracy better than ±10 units is not warranted and at 100 units ±30 is adequate. Beyond 100 units values become meaningless. Houlsby suggested that grouting foundations with permeabilities up to 3 lugeon units usually is unnecessary (this is in the range of laminar flow). Nonetheless, there are two principal reasons for warranting grout curtains tighter than 3 lugeon units, namely, the value

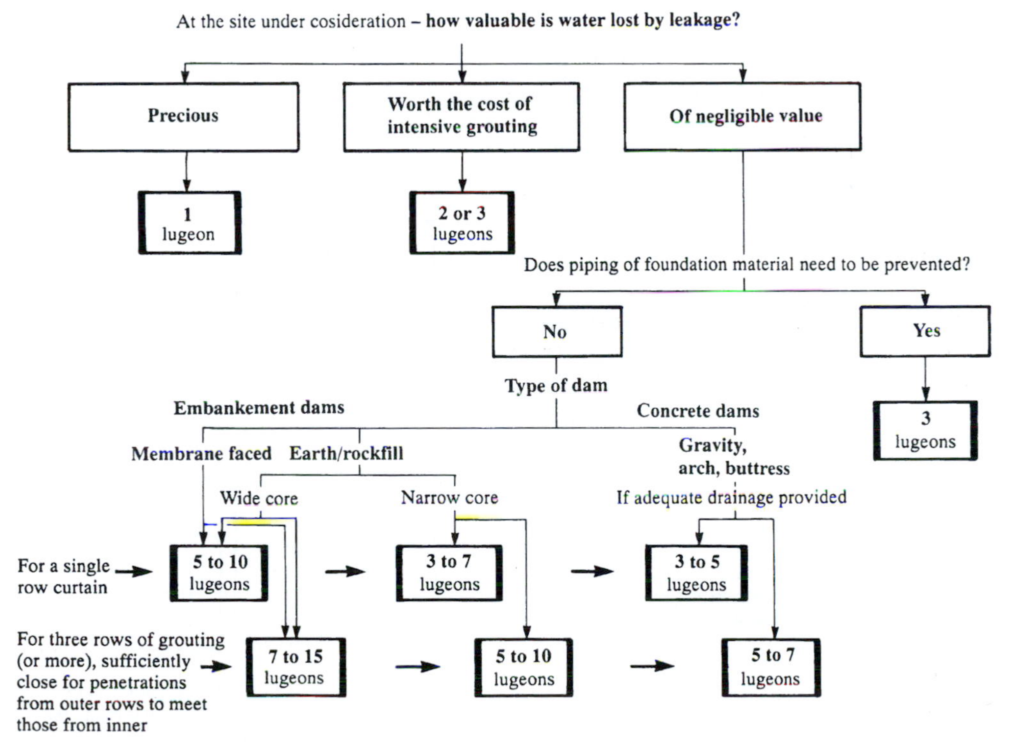

Figure 6.25 Criteria for deciding whether a grout curtain is necessary, and if so, to what standard of tightness the grouting should be taken.
Source: Houlsby, 1992.

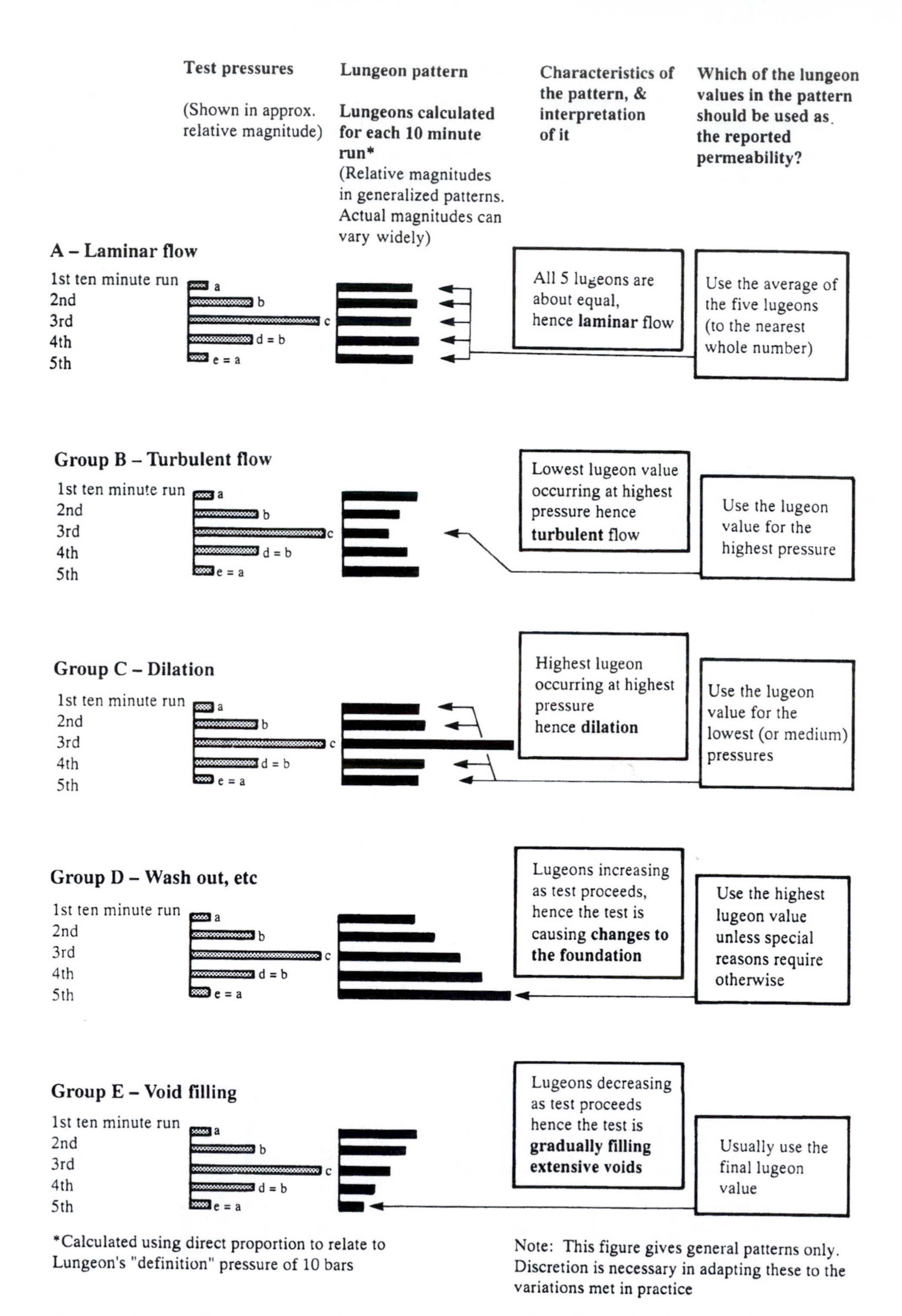

Figure 6.26 Permeability testing for exploratory purposes. The use of five pressure sequences for determining the type of flow and hence the lugeon value to report for the test.
Source: Houlsby, 1992.

of water lost by leakage and the stabilization of material susceptible to removal from joints by seepage. If these requirements do not apply, then the type of dam and the thickness of curtain become relevant to the desirable standard of grouting. For instance, requirements in earth dams with wide cores usually are less rigid than for those with narrow cores, and for concrete dams. When the lugeon value is 3 or over, the flow is turbulent and grouting generally is employed.

6.12.2 Grouting methods

Permeation grouting at shallow depths may take place at a single stage from a grout pipe, the grouthole being sunk to full depth and then grouted upwards. Alternatively, grouting may proceed while the hole being drilled, that is, the hole is extended a short distance using a hollow drill rod; it then is withdrawn this distance and grout is injected from the rod. The cycle then is repeated. Stage grouting is used when relatively high grouting pressures have to be employed to achieve satisfactory penetration of grout in deep holes or tighter sections of holes. In stage grouting the hole is drilled to a given depth and then grouted. After the grout has set, the hole is deepened for the next stage of grouting when the procedure is repeated (Fig. 6.27). Stage grouting allows increasing grout pressures to be used for increasing depth of grouthole and reduces the loss of grout due to leakage at the surface.

The *tube-a-manchette* commonly is used to grout alluvial soils. It consists of a steel tube perforated with rings of small holes at intervals of approximately 0.3 m. Each ring of holes is enclosed by a tightly fitting rubber sleeve that acts as a one-way valve (Fig. 6.28). A drillhole is sunk, with the aid of casing, to the full depth to be treated and the *tube-a-manchette* placed in it. The casing is withdrawn and grout, termed *sleeve grout* (clay-cement or bentonite), is poured into the annular space left behind. Grouting then is carried out through the *tube-a-manchette* by lowering into it a small-diameter injection pipe perforated at its lower end and fitted with two U-packers. The packers can be centred over any one of the rings of injection holes. When injection starts the pressure in the grout pipe rises until the grout lifts the rubber sleeve, rupturing the sleeve grout and escaping through the small holes into the soil. The rubber sleeves stop any return of the grout into the *tube-a-manchette* and the sleeve grout prevents any leakage of grout at the surface. Use of the *tube-a-manchette* offers great flexibility since the same hole can be grouted more than once and different grouts can be used. In this way coarser soils can be treated first and finer ones later.

Claquage or fracture grouting is a technique that frequently is used to treat alluvial soils and makes use of the *tube-a-manchette*. The aim is to develop a network of grouted fractures

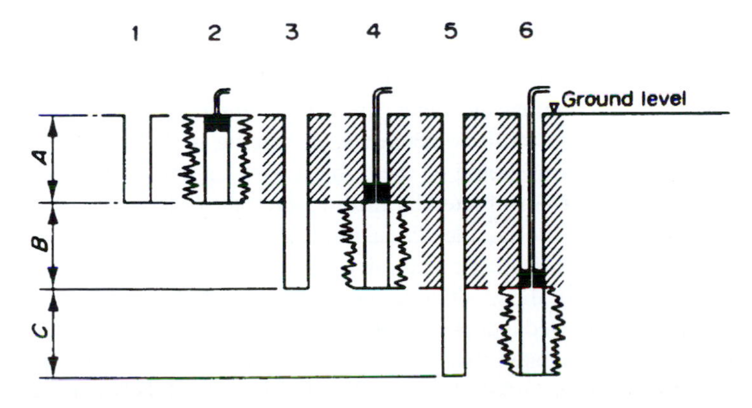

Figure 6.27 Stage and packer grouting.

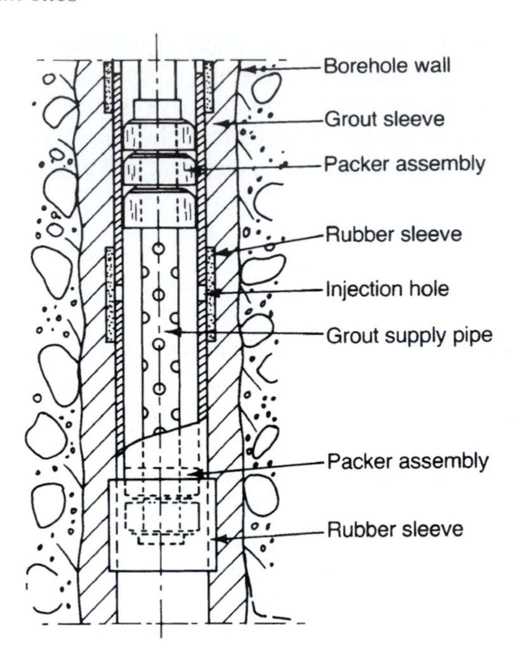

Figure 6.28 Detail of the *tube-a-manchette*.

as silts, and clays are not amenable to permeation grouting (Ischy and Glossop, 1962). As injection proceeds claquages spread rapidly, forming an intermeshing network of grout-filled fractures and in this way reduce the permeability, as well as compacting and improving the mechanical properties of the soil. Claquages appear when the pressure in the grouthole exceeds a certain value depending on the characteristics of the soil and depth of overburden. Considerable pressure may be required to start the flow, but once the grout begins to move the pressure can be reduced. Claquages may form in any direction. According to Cambefort (1977), vertical fractures tend to develop before horizontal fractures. The grout occupying horizontal fractures, in particular, may cause the ground to heave. If damaging heave is to be avoided, ground movement has to be monitored.

A grout curtain is constructed to form a low-permeability barrier beneath a dam (Fig. 6.29). Holes are sunk and grouted, from the base of the cut-off, or heel trench downwards. Once the standard of permeability has been decided, for the whole or a section of a grout curtain, it is achieved by split spacing or closure methods in which primary, secondary, tertiary, etc. sequences of grouting are carried out until water tests in the groutholes approach the required standard. In multiple row curtains, the outer rows are completed first thereby allowing the innermost row to effect closure on the outer rows. A spacing of 1.5 m between rows usually is satisfactory. The upstream row should be the tightest row, tightness decreasing downstream. Single row curtains usually are constructed by drilling alternate holes first and then completing the treatment by intermediate holes. Ideally, a grout curtain is taken to a depth where the requisite degree of tightness is available naturally. This is determined either by investigatory holes sunk prior to the design of the grout curtain or by primary holes sunk during grouting. The search usually does not go beyond a depth equal to the height of the storage head above ground surface. Cambefort (1977) indicated that average permeability in completed grout curtains in sands and gravels could not be reduced below $5 \times 10^{-6}\,\mathrm{m\,s^{-1}}$, whilst in fissured rock it was possible to achieve $5 \times 10^{-7}\,\mathrm{m\,s^{-1}}$.

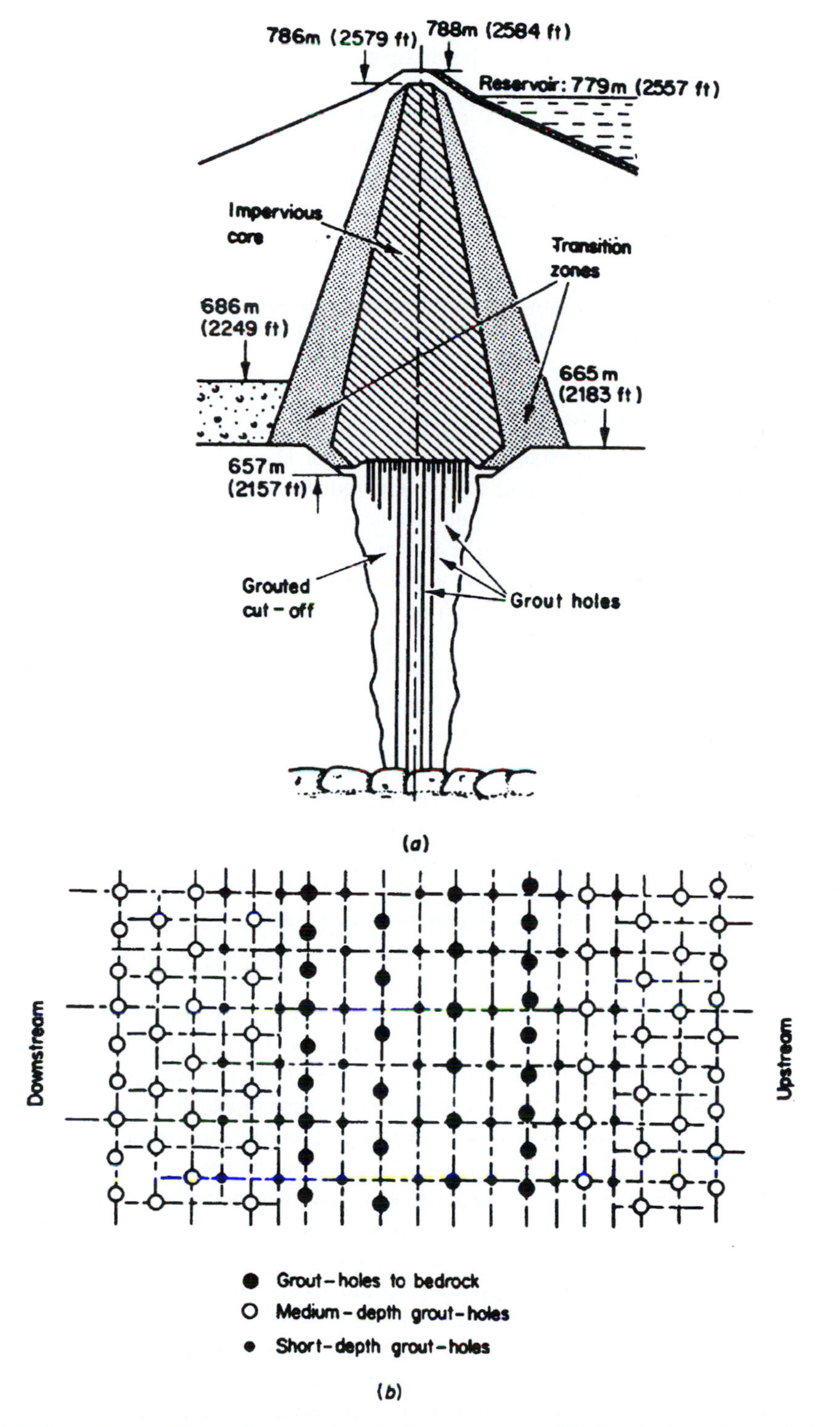

Figure 6.29 Serre Poncon Dam, France: (a) typical cross-section; (b) distribution of grout holes in the central part of the grouted cut-off or curtain.

The foundation of an earth dam should allow seepage to pass at a slightly lower pore pressure than that in the core since this allows the dissipation of pore water pressure from the core. However, when a tight grout curtain occurs at the centre of the core excess pore water pressures are developed in the foundation. On the other hand, if a tight curtain is constructed near the upstream edge of the core it induces very little excess of foundation pore water pressure over that in the core, but curtains are rarely positioned that far upstream. Hence, the tighter the grout curtain, the further upstream it should be located. Houlsby (1977) suggested that high pore water pressures could be reduced by leaky grouting. Even so, the best course of action would be to make the surface stage of grouting as tight as necessary whilst deeper stages are left sufficiently permeable to allow regulated seepage through them. Grouting of abutments frequently is necessary. For example, grouting was carried out in both abutments from adits at Bakoyianni Dam, an embankment dam, in Greece (Kalkani, 1997).

Jet grouting can be used to form an impermeable barrier in all types of soils. For example, Sembenelli and Sembenelli (1999) described the construction of cut-offs beneath two high cofferdams (the upstream one exceeding 60 m) at Ertan Dam site in the United States. An injection pipe is inserted into the soil to the required depth, then the soil is subjected to a horizontally rotating jet of water and simultaneously mixed with grout (cement or cement-bentonite) to form plastic soil-cement. The injection pipe is raised gradually. In replacement jet, grouting soil is removed from the zone to be treated by a high-energy jet of water and air. The water and air are jetted under very high pressure from closely spaced nozzles at the base of a triple fluid phase drill pipe, which also places the grout at the same time. The operation proceeds from the base of a borehole upwards, the soil that is removed being brought to the surface by air-lift pressure. In granular soils some of the coarsest particles are not removed and are incorporated in the grout. Treatment has been completed successfully to depths in excess of 60 m. There are three methods of replacement jet grouting, namely, column, panel or wing grouting (Fig. 2.28). In column grouting, the columns are formed by rotating the jet pipe as it is raised (Coomber, 1986). A barrier is produced by overlapping the columns. In panel grouting, jetting takes place from the pipe in a single vertical plane as it is raised. Interconnecting panels form cut-offs of low permeability. Wing jetting uses two nozzles, without rotation, to form a fan-shaped grouted mass. The use of pre-bored guideholes, usually 150 mm in diameter, facilitates discharge of soil, helps to maintain verticality and provides a visual check on the continuity of adjacent grouted areas.

Consolidation grouting is usually shallow, the holes seldom extending more than 10 m. It is intended to improve the ground by increasing its strength and reducing its permeability. In the case of rock masses, it also improves the contact between concrete and rock, and makes good any slight loosening of the rock surface due to blasting operations. In addition, consolidation grouting increases the stiffness and affords a degree of homogeneity to the foundation, which is desirable if differential settlement and unbalanced stresses are to be avoided. Holes usually are drilled normal to the foundation surface, but in certain instances they may be orientated to intersect specific features. They are set out on a grid pattern at 3 to 14 m centres, depending on the nature of the rock.

In a consideration of the efficiency of grouting operations at dam sites Ferguson and Lancaster-Jones (1964) introduced the concept of reduction ratio, which refers to the reduction in acceptance of grout as treatment progresses. They suggested that a reduction ratio between 0.2 and 0.8 may be regarded as satisfactory. A ratio approaching or exceeding unity is indicative of groutholes too widely spaced so that gaps may exist in a grout curtain, whereas a very low ratio shows that the initial hole spacing was too close. If the spacing of primary holes is correctly estimated, then grout acceptance is appreciably less in the secondary holes. However, Ferguson and Lancaster-Jones pointed out that a satisfactory ratio on its own does not imply that treatment is complete.

The leakage potential of a rock mass is commonly investigated by water pressure tests (Foyo *et al.*, 1997). These are carried out either in exploratory holes drilled before grouting operations begin, thereby allowing determination of the initial permeability, or in injection holes prior to an injection stage. This is a pre-injection test. A post-injection test is carried out in the injection hole after injection, the hole having been re-drilled. Test holes can be specially drilled to determine the effect of treatment at any particular time. An assessment of the grout take also can be made from an examination of core sticks from drillholes in the grouted zone.

Knill (1970b) showed that seismic refraction could be used to predict the grout take. What is more, he suggested that the relationship between grout take and longitudinal seismic velocity could be used as a means of quantitative control of grouting, whilst velocity measurements before and after grouting could demonstrate the effectiveness of the grouting process.

6.13 Drainage systems

Casagrande (1961) cast doubts on the need for grout curtains, maintaining that a single row grout curtain constructed prior to reservoir filling frequently is inadequate. What is more, he maintained that expensive grouting was useless as far as reducing pore water pressures was concerned and that drainage systems were the only efficient method of controlling the piezometric level and therefore uplift forces along a dam foundation. He further maintained that drainage is the only efficient treatment available for rock masses of low hydraulic conductivity, that is, rock with fine fissures. Drainage can control the hydraulic potential on the downstream side of a dam thus achieving what is required of a grout curtain except, of course, that drainage does not reduce the amount of leakage. However, leakage is not of consequence in most rock masses where the hydraulic conductivity is low. In other words, Casagrande contended that for fissured rocks of low permeability (less than 5 lugeon units), drainage generally is essential whereas grouting constitutes a wasted effort. Conversely, if the permeability is high (in excess of 50 lugeon units), grouting is necessary to control groundwater leakage beneath a dam. Drainage and grouting may be carried out in rock masses of medium permeability, the decision on whether or not to construct a grout curtain usually depending on economics. Casagrande's views on grout curtains have not been universally accepted but it must be remembered that he was referring, in particular, to the reduction of uplift pressures beneath dams. The purpose of grouting is not to reduce uplift pressures but to check water seepage beneath and around a dam. As far as the reduction of water loss is concerned, it is accepted that grout curtains have proved successful.

6.14 Impervious blankets

Where the thickness of alluvial deposits are such that the construction of a grout curtain to bedrock is not feasible, an impervious blanket may be placed upstream of the dam. This increases the length of the seepage path to the extent that the seepage gradient and therefore seepage velocity are reduced significantly, thereby decreasing the quantity of seepage. An example is provided by Tabela Dam where the alluvium contained lenses of open-work gravel (Haq, 1996). Because of the depth of alluvium an upstream impervious blanket, over 2000 m in length, and in continuation with the core of the dam was used to control seepage beneath the dam. The blanket was formed of silty sand and silt. Unfortunately, on first filling of the reservoir in 1974 the fill material was eroded and some 200 relief wells downstream of the dam discharged some 0.028 $m^3 s^{-1}$ each. The reservoir therefore was emptied exposing 362 holes and 140 cracks in the blanket. The holes and cracks were filled with blanket material and the blanket was thickened and extended upstream. Additional relief wells were installed downstream to facilitate drainage and reduce uplift pressure. The reservoir was refilled in 1975 and side-scan sonar was used to monitor any holes that formed in the blanket. Holes that did develop were treated by dumping from a barge. This method proved effective, no further holes appearing by 1978.

References

Abam, T.K.S. 1999. Impact of dams on the hydrology of the Niger Delta. *Bulletin Engineering Geology and the Environment*, **57**, 239–251.

Ahmed, A.A. 1960. An analytical study of the storage losses in the Nile Basin, with special reference to Aswan Dam Reservoir and the High Aswan Dam. *Proceedings Institution Civil Engineers*, **35**, 181–200.

Aitchison, G.D. and Wood, C.C. 1965. Some interactions of compaction, permeability and post-construction deflocculation affecting the probability of piping failures in small dams. *Proceedings Sixth International Conference on Soil Mechanics and Foundation Engineering, Montreal*, **2**, 442–446.

Al-Homoud, A.S. 2000. Geologic hazards at an embankment dam constructed across a major active plate boundary fault. *Environmental and Engineering Geoscience*, **6**, 353–382.

Al-Homoud, A.S. and Tanash, N. 2001. Monitoring and analysis of settlement and stability of an embankment dam constructed in stages on soft ground. *Bulletin Engineering Geology and the Environment*, **59**, 259–284.

Almeida, M.S.S., Santa Maria, P.E.L., Martins, I.S.M., Spotti, A.P. and Coelho, L.B.M. 2000. Consolidation of a very soft clay with vertical drains. *Geotechnique*, **50**, 633–643.

Anon. 1955. Drainage and erosion control. Subsurface drainage for airfields (Chapter 2). *Engineering Manual, Military Construction*, United States Army Corps of Engineers, Washington, DC.

Anon. 1983. *Seismicity and Dam Design*. Bulletin 46, International Commission on Large Dams, Paris.

Anon. 1990. *Dispersive Soils in Earth Dams, Review*. Bulletin 77, International Commission on Large Dams, Paris.

Anon. 1991. *Dispersive Clays*. Soil Mechanics Note No. 13, United States Department of Agriculture, Soil Conservation Service, Engineering Division, Washington, DC.

Bailard, J.A. 2001. Sedimental journey. *Civil Engineering*, American Society Civil Engineers, **71**, No. 3, 66–69.

Bell, F.G. 1993. *Engineering Treatment of Soils*. E & FN Spon, London.

Bell, F.G. and Haskins, D.R. 1997. A geotechnical overview of Katse Dam and Transfer Tunnel, Lesotho, with a note on basalt durability. *Engineering Geology*, **46**, 175–198.

Bell, F.G. and Maud, R.R. 1994. Dispersive soils and earth dams: a South African view. *Bulletin Association Engineering Geologists*, **31**, 433–446.

Bell, F.G. and Walker, D.J.H. 2000. A further examination of the nature of dispersive soils in Natal, South Africa. *Quarterly Journal Engineering Geology and Hydrogeology*, **33**, 187–199.

Bell, H.S. 1942. Density currents as agents for transporting sediments. *Journal Geology*, **50**, 512–547.

Bertram, G.E. 1940. An experimental investigation of protective filters. Harvard University, Publication No. 267.

Bishop, A.W. 1955. The use of the slip circle in the stability of slopes. *Geotechnique*, **5**, 7–17.

Bodocsi, A. and Bowers, M.T. 1991. Permeability of acrylate, urethane and silicate grouted sands with chemicals. *Proceedings American Society Civil Engineers, Journal Geotechnical Engineering Division*, **117**, 1227–1244.

Bolt, B.A. and Hudson, D.E. 1975. Seismic instrumentation of dams. *Proceedings American Society Civil Engineers, Journal Geotechnical Engineering Division*, **101**, 1095–1104.

Bozovic, A., Budanur, H., Nonveiller, E. and Pavlin, B. 1981. The Keban Dam foundation on karstified limestone – a case history. *Bulletin International Association Engineering Geology*, **24**, 45–51.

Brink, A.B.A. and Wagener, F. 1990. Tweedraai Dam: design and construction with dispersive clay. *Proceedings Symposium on Dam Safety: Four Years On*. South African National Committee on Large Dams, Pretoria, 1–13.

Brune, G. 1965. Anhydrite and gypsum problems in engineering geology. *Bulletin Association Engineering Geologists*, **3**, 26–38.

Burwell, E.B. 1950. Geology in dam construction, Part 1. In: *Application of Geology to Engineering Practice*, Berkey Volume, Paige, S. (ed.), Geological Society America, 11–33.

Cambefort, N. 1977. The principles and applications of grouting. *Quarterly Journal Engineering Geology*, **10**, 57–96.

Cambefort, N. 1987. Grouts and grouting. In: *Ground Engineer's Reference Book*, Bell, F.G. (ed.), Butterworths, London, 32/1–32/30.

Carder, D.S. 1945. Seismic investigations in the Boulder Dam area, 1940–44, and the influence of reservoir loading on earthquake activity. *Bulletin Seismological Society America*, **35**, 175–192.

Carpentier, R., De Wolf, P., Van Damme, L., De Rouck, J. and Bernard, A. 1985. Compaction blasting in offshore harbour construction. *Proceedings Eleventh International Conference on Soil Mechanics and Foundation Engineering*, San Francisco, **3**, 1687–1692.

Cartmel, R.M. 1971. Construction work at Tarbela Dam enters the second stage. *Water Power*, **23**, 197–206.

Casagrande, A. 1950. Notes on the design of earth dams. *Journal Boston Society Civil Engineers*, **37**, 405–429.

Casagrande, A. 1961. Control of seepage through foundations and abutments of dams. *Geotechnique*, **11**, 161–181.

Coomber, D.B. 1986. Groundwater control by jet grouting. In: *Groundwater in Engineering Geology*, Engineering Geology Special Publication No. 3, Cripps, J.C., Bell, F.G. and Culshaw, M.G. (eds), The Geological Society, London, 445–454.

Coombes, H.A. 1968. Leakage through buried channels. *Bulletin Association Engineering Geologists*, **6**, 45–60.

Day, R.A., Hight, D.W. and Potts, D.M. 1998. Finite element analysis of construction stability of Thika Dam. *Computers and Geotechnics*, **23**, 205–219.

Deere, D.U. 1976. Dams on rock foundations – some design questions. *Proceedings Speciality Conference on Rock Engineering for Foundations and Slopes*, Boulder, Colorado, American Society Civil Engineers, Geotechnical Engineering Division, New York, **2**, 55–86.

De Mello, V.F.B. 1977. Reflections on the design decisions of practical significance to embankment dams.*Geotechnique*, **27**, 279–355.

De Sousa, S.P., Calijuri, M.L. and Meira, A.D. 1998. Reservoir silting up: the Manso System case history. *Proceedings Eighth International Congress International Association of Engineering Geology and the Environment*, Vancouver, Moore, D. and Hungr, O. (eds), A.A. Balkema, Rotterdam, **5**, 3291–3295.

Eck, W. and Redfield, R.C. 1965. Engineering geology problems at Sanford Dam, Borger, Texas. *Bulletin Association Engineering Geologists*, **3**, 15–25.

Elges, H.F.W.K. 1985. Dispersive soils. *The Civil Engineer in South Africa*, **27**, 347–355.

Fell, R., MacGregor, P. and Stapleton, D. 1992. *Geotechnical Engineering of Embankment Dams*. A.A. Balkema, Rotterdam.

Ferguson, F.F. and Lancaster-Jones, P.F.F. 1964. Testing the efficiency of grouting operations at dam sites. *Proceedings Eighth International Congress on Large Dams*, Edinburgh, **1**, 121–139.

Foster, M. and Fell, R. 2001. Assessing embankment dam filters that do not satisfy design criteria. *Proceedings American Society Civil Engineers, Journal Geotechnical and Geoenvironmental Engineering*, **127**, 398–407.

Foyo, A., Tomillo, C., Maycotte, J.I. and Willis, P. 1997. Geological features, permeability and groutability characteristics of Zimapan Dam foundation, Hidalgo State, Mexico. *Engineering Geology*, **46**, 157–174.

Gardner, W.I. 1968. Dams and reservoirs in Pleistocene eolian deposit terrain of Nebraska and Kansas. *Bulletin Association Engineering Geologists*, **6**, 31–44.

Gerber, A. and Harmse, H.J. von M. 1987. Proposed procedure for identification of dispersive soils by chemical testing. *The Civil Engineer in South Africa*, 29, 397–399.

Gosschalk, L.C. 1964. Reservoir sedimentation. In: *Handbook of Applied Hydrology*, Chow, V.T. (ed.), McGraw Hill, New York, 17/1–17/33.

Greeman, A. 1998. Stopping the silt. *Ground Engineering*, **31**, No. 2, 16–17.

Gupta, M.K. 1985. The present status of reservoir induced seismicity investigations with special emphasis on Koyna earthquakes. *Tectonophysics*, **118**, 257–279.

Hall, B.E., Watermeyer, C.F. and Frame, J.A. 1993. Improving dam safety using a fin drain. *Proceedings Conference on Geotechnical Practice in Dam Rehabilitation, Raleigh, North Carolina*, Geotechnical Special Publication No. 35, American Society Civil Engineers, New York, 521–535.

Haq, U. 1996. Tarbela Dam: resolution of seepage. *Proceeding Institution Civil Engineers, Geotechnical Engineering*, **199**, 49–56.

Hoek, E. 1994. Strength of rock and rock masses. *News Journal* (ISRM), **2**, 4–16.

Hoek, E. and Londe, P. 1974. Surface workings in rock. *Proceedings Third International Congress Rock Mechanics* (ISRM), Denver, **1**, 613–654.

Honjo, Y. and Veneziano, D. 1989. Improved filter criterion for cohesionless soils. *Proceedings American Society Civil Engineers, Journal Geotechnical Engineering Division*, **115**, 75–96.

Houlsby, A.C. 1977. Engineering of grout curtains to standards. *Proceedings American Society Civil Engineers, Geotechical Engineering Division*, **103**, 953–970.

Houlsby, A.C. 1992. Grouting in rock masses. In: *Engineering in Rock Masses*, Bell, F.G. (ed.), Butterworth-Heinemann, London, 334–350.

Ischy, E. and Glossop, R. 1962. An introduction to alluvial grouting. *Geotechnique*, **22**, 449–474.

Iwao, Y. and Gunatilake, J. 1999. A geotechnical overview of the reservoir leakage problem of the Samanalawewa dam, Sri Lanka. *Proceedings Thirty Fourth Japanese National Conference on Geotechnical Engineering*, Tokyo, Japanese Geotechnical Society, **2**, 1275–1276.

Jaeger, C. 1963. The Malpasset Report. *Water Power*, **15**, 55–61.

Johnstone, G.S. and Crichton, J.R. 1966. Geological and civil engineering aspects of hydroelectric developments in the Scottish Highlands. *Engineering Geology*, **1**, 311–342.

Kalkani, E.C. 1997. Geological conditions, seepage grouting, and evaluation of piezometer measurements in the abutments of an earth dam. *Engineering Geology*, **46**, 93–104.

Kaloustian, E.S. 1984. Statistical analysis of distribution of concrete dam foundation failures. *Proceedings Conference on Safety of Dams*, Coimbra, Portugal, A.A. Balkema, 311–319.

Kennard, M.F. and Knill, J.L. 1969. Reservoirs on limestone, with particular reference to the Cow Green scheme. *Journal Institution Water Engineers*, **23**, 87–136.

Kennedy, A. 2001. Drilling and grouting for dam construction in southern Turkey. *Geodrilling International*, **9**, No. 3, 1–16.

Kiersch, G.A. 1964. Vajont Reservoir disaster. *Civil Engineering*, American Society Civil Engineers, **34**, 32–39.

Knill, J.L. 1970a. The engineering geology of old mine workings. *Midland Society Soil Mechanics and Foundation Engineering*, **4**, 1–25.

Knill, J.L. 1970b. The application of seismic methods to the prediction of grout take. In: *In Situ Testing in Soils and Rock*, British Geotechnical Society, London, 93–99.

Knill, J.L. 1971. Assessment of reservoir feasibility. *Quarterly Journal Engineering Geology*, **4**, 355–372.

Koukis, G. and Rozos, D. 1998. Slope stability problems in Polyphyton dam site, Kozani, Greece. *Proceedings Eighth International Congress International Association of Engineering Geology and the Environment*, Vancouver, Moore, D. and Hungr, O. (eds), A.A. Balkema, Rotterdam, **5**, 3067–3071.

Kresse, L.C. 1966. Baldwin Hills Reservoir failure of 1963. In: *Engineering Geology in Southern California*, Special Publication Association Engineering Geologists, Lung, R. and Proctor, R. (eds), Glendale, California, 93–103.

Lane, R.G.T. 1972. Seismic activity at mad-made reservoirs. *Proceedings Institution Civil Engineers*, **50**, 15–24.

Lapworth, C.F. 1965. Evaporation from a reservoir near London. *Journal Institution Water Engineers*, **16**, 163–181.

Liang, R.Y., Nusier, O.K. and Malkawi, A.H. 1999. A reliability based approach for evaluating the slope stability of embankment dams. *Engineering Geology*, **54**, 271–285.

Linsley, R.K. and Franzini, J.B. 1972. *Water Resources Engineering*. McGraw-Hill, New York.

Littlejohn, G.S. 1985. Chemical grouting. *Ground Engineering*, **18**; Part 1, No. 2, 13–18; Part 2, No. 3, 23–28; Part 3, No. 4, 29–34.

Lugeon, M. 1933. *Barrage et Geologie*. Dunod, Paris.

Malkawi, A.H. and Al-Sheriadeh, M. 2000. Evaluation and rehabilitation of seepage problems. A case study: Kafrein dam. *Engineering Geology*, **56**, 335–345.

Marinos, P.G., Cavounidis, S., Benissi, M. and Kaplanides, A. 1997. Development of sinkholes during reservoir construction. *Proceedings International Symposium on Engineering Geology and the Environment*, Athens, Marinos, P.G., Koukis, G.C., Tsiambaos, G.C. and Stournaras, G.C. (eds), A.A. Balkema, Rotterdam, **2**, 2769–2776.

McDaniel, T.N. and Decker, R.S. 1979. Dispersive soil problem at Los Esteros Dam. *Proceedings American Society Civil Engineers, Journal Geotechnical Engineering Division*, **105**, 1017–1030.

Moneymaker, B.C. 1968. Reservoir leakage in limestone terrains. *Bulletin Association Engineering Geologists*, **6**, 3–30.

Morgenstern, N.R. and Price, V.E. 1963. The analysis of the stability of general slip surfaces. *Geotechnique*, **13**, 121–131.

Nonveiller, E. 1989. *Grouting Theory and Practice*. Elsevier, Amsterdam.

Okamota, S. 1978. Present trend of earthquake resistant design of large dams. In: *The Assessment and Mitigation of Earthquake Risk*, UNESCO, Paris, 185–197.

Olson, R.E. 1998. Settlements of embankments on soft clays. *Proceedings American Society Civil Engineers, Journal Geotechnical and Geoenvironmental Engineering*, **124**, 278–288.

Pendse, M.D., Huddar, S.N. and Kulkarni, S.Y. 1999. Rehabilitation of Koyna Dam – a case study. *Proceedings Symposium on Rehabilitation of Dams*, New Delhi, Varma, C.V.J., Visvanathan, N. and Rao, A.R.G. (eds), 127–138.

Penman, A.D.M. 1975. Earth pressures measured with hydraulic piezometers. *Ground Engineering*, **9**, No. 3, 17–23.

Penman, A.D.M. 1977. The failure of Teton Dam. *Ground Engineering*, **10**, No. 6, 18–27.

Potts, D.M., Dounias, G.T. and Vaughan, P.R. 1990. Finite element analysis of progressive failure of Carsington embankment. *Geotechnique*, **40**, 79–101.

Qian, N. 1982. General report. *Proceedings Fourteenth International Congress on Large Dams*, **Q.54**, 639–690.

Riemer, W. 1995. Landslides and reservoirs. *Proceedings Sixth International Symposium on Landslides*, Christchurch, Bell, D.H. (ed.), A.A. Balkema, Rotterdam, **3**, 1973–2004.

Roberts, D.G. 1968. Predictions of reservoir leakage. *Bulletin Association Engineering Geologists*, **6**, 70–82.

Rocha, M. 1974. Present possibilities of studying foundations of concrete dams. *Proceedings Third International Congress Rock Mechanics* (ISRM), Denver, **1**, 879–897.

Ruffles, N.J. 1965. Derwent Reservoir. *Journal Institution Water Engineers*, **19**, 361–370.

Sarma, S.K. 1973. Analysis of embankments and slopes. *Geotechnique*, **23**, 423–433.

Seed, H.B. and De Alba, P. 1986. Use of the SPT and CPT tests for evaluating the liquefaction resistance of sands. *Proceedings Speciality Conference on Use of In Situ Tests in Geotechnical Engineering*. Geotechnical Special Publication No. 6, Clemence, S.P. (ed.), American Society Civil Engineers, New York.

Seed, H.B., Lee, K.L., Idriss, M.L. and Madisi, F.I. 1975. The slides in the San Fernando Dams during the earthquake of February 9th, 1971. *Proceedings American Society Civil Engineers, Journal Geotechnical Engineering Division*, **101**, 651–688.

Sembenelli, P.G. and Sembenelli, G. 1999. Deep jet grouted cut-offs in riverine alluvia for Ertan cofferdams. *Proceedings American Society Civil Engineers, Journal Geotechnical and Geoenvironmental Engineering*, **125**, 142–153.

Semenza, E. and Ghirotti, M. 2000. History of the 1963 Vaiont slide: the importance of geological factors. *Bulletin Engineering Geology and the Environment*, **59**, 87–97.

Serafim, J.L. 1968. Influence of interstitial water on the behaviour of rock masses. In: *Rock Mechanics in Engineering Practice*, Stagg, K.G. and Zienkiewicz, O.C. (eds), Wiley, London, 55–97.

Sharma, S., Raghuvanshi, T. and Sahai, A. 1999. An engineering geological appraisal of the Lakhwar Dam, Grahwal Himalaya, India. *Engineering Geology*, **53**, 381–398.

Sherard, J.L. and Dunnigan, L.P. 1989. Critical filters for impervious soils. *Proceedings American Society Civil Engineers, Journal Geotechnical Engineering Division*, **115**, 927–947.

Sherard, J.L., Woodward, R.L., Gizienski, S.F. and Clevenger, W.A. 1967. *Earth and Earth-Rock Dams*. Wiley, New York.

Sherard, J.L., Cluff, L.S. and Allen, L.R. 1974. Potentially active faults in dam foundations. *Geotechnique*, **24**, 367–429.

Sherard, J.L., Dunnigan, L.P. and Decker, R.S. 1976. Identification and nature of dispersive soils. *Proceedings American Society Civil Engineers, Journal Geotechnical Engineering Division*, **102**, 287–301.

Sherard, J.L., Dunnigan, L.P. and Decker, R.S. 1977. Some engineering problems with dispersive clays. *Proceedings Symposium on Dispersive Clays, Related Piping and Erosion in Geotechnical Projects*, Sherard, J.L. and Decker, R.S. (eds), American Society Testing Materials Special Publication 623, Philadelphia, 3–12.

Sherard, J.L., Dunnigan, L.P. and Talbot, J.R. 1984a. Basic properties of sand and gravel filters. *Proceedings American Society Civil Engineers, Journal Geotechnical Engineering Division*, **110**, 684–700.

Sherard, J.L., Dunnigan, L.P. and Talbot, J.R. 1984b. Filters for silts and clays. *Proceedings American Society Civil Engineers, Journal Geotechnical Engineering Division*, **110**, 701–718.

Simpson, D.W. 1976. Seismicity changes associated with reservoir loading. *Engineering Geology*, **10**, 123–150.

Simpson, D.W., Leith, W.S. and Scholtz, C.H. 1988. Two types of reservoir induced seismicity. *Bulletin Seismological Society America*, **78**, 2025–2040.

Succarieh, M.F., Elgamal, A.-W. and Yan, L. 1993. Observed and predicted earthquake response of La Villita Dam. *Engineering Geology*, **34**, 11–26.

Tadanier, R. and Ingles, O.G. 1985. Soil security test for water retaining structures. *Proceedings American Society Civil Engineers, Journal Geotechnical Engineering Division*, **111**, 289–301.

Talbot, J.R. 1991. Discussion: Improved filter criterion for cohesionless soils. *Proceedings American Society Civil Engineers, Journal Geotechnical Engineering Division*, **117**, 1633–1634.

Tate, E.L. and Farquharson, F.A.K. 2000. Simulating reservoir management under the threat of sedimentation: the case of the Tarbela Dam on the River Indus. *Water Resources Management*, **14**, 191–208.

Terzaghi, K. 1962. Dam foundations on sheeted granite. *Geotechnique*, **12**, 199–208.

Tika, Th.E. and Hutchinson, J.N. 1999. Ring shear tests on soil from the Vaiont landslide slip surface. *Geotechnique*, **49**, 59–74.

Underwood, L.B. and Dixon, N.A. 1976. Dams on rock foundations. *Proceedings Speciality Conference on Rock Engineering for Foundations and Slopes*, Boulder, Colorado, American Society Civil Engineers, Geotechnical Engineering Division, New York, **2**, 125–146.

Vallejo, L.E. and Shettima, M. 1996. Fault movement and its impact on ground deformation and engineering structures. *Engineering Geology*, **43**, 119–133.

Vaughan, P.A. and Soares, H.H. 1982. Design of filters for clay core dams. *Proceedings American Society Civil Engineers, Journal Geotechnical Engineering Division*, **108**, 17–31.

Voort, H. and Lotgers, G. 1974. *In situ* determination of the deformation behaviour of a cubical rock mass sample under triaxial load. *Rock Mechanics*, **6**, 65–79.

Watermeyer, C.F., Botha, G.R. and Hall, R.E. 1991. Countering potential piping at an earth dam on dispersive soils. In: *Geotechnics in the African Environment*, Blight, G.E., Fourie, A.B., Luker, I., Mouton, D.J. and Scheurenburg, R.J. (eds), A.A. Balkema, Rotterdam, 321–328.

Willmore, P.L. 1981. Hazards of natural and induced seismicity in the vicinity of large dams. *Proceedings Conference on Dams and Earthquakes*, Thomas Telford Press, London, 261–266.

Woodward, R.C. 1992. The geology of dam spillways. *Engineering Geology*, **32**, 243–254.

Wyllie, D.C. 1999. *Foundations on Rock*. E & FN Spon, London.

Xiao, Y.J., Lee, C.F. and Wang, S.X. 2000. Spacial distribution of inter-layer shear zones at Gaobazhou dam site, Qingjiang River, China. *Engineering Geology*, **55**, 227–239.

7 Hydrogeology

7.1 Introduction

The hydrological cycle involves the movement of water in all its forms over, on and through the earth (Fig. 7.1). As such, the cycle can be visualized as beginning with the evaporation of water from the oceans. This is followed by the transport of the resultant water vapour by winds, with some water vapour condensing over land and falling to the surface of the earth as precipitation. To complete the cycle this precipitation then makes its way back to the oceans via rivers and underground flow. However, some precipitation may be evapotranspired and describe several subcycles before completing its journey. Groundwater forms an integral component of the hydrological cycle. Most groundwater recharge is brought about by the infiltration and percolation of precipitation into the ground. Less significant, though important locally, is the direct contribution from rivers and lakes. In fact, the hydrological cycle can be regarded as a series of storage components, with water moving slowly from one to another until one circuit has been completed. Table 7.1 shows the estimated amount of water available within the various storage components and the total quantity that would be available if it could all be released from storage. Only 0.5 per cent of the total water resources of the world is in the form of groundwater. Not all this is available for exploitation since about half is below 800 m, and therefore is too deep for economic utilization. However, the capacity of the underground resource should not be underestimated in that about 98 per cent of the usable freshwater of the earth is stored underground.

The principal source of groundwater is meteoric water, that is, precipitation (rain, sleet, snow and hail). However, two other sources are very occasionally of some consequence, that is, juvenile water and connate water. The former is derived from magmatic sources whilst connate water was trapped in the pore spaces of sediments as they were formed.

As can be inferred from above, precipitation is dispersed as run-off, infiltration/percolation and evapotranspiration. Run-off is made up of two basic components, surface water run-off and groundwater discharge. The former is usually the more important and is responsible for the major variations in river flow. Run-off generally increases in magnitude as the time from the beginning of precipitation increases.

Infiltration refers to the process whereby water penetrates the ground surface and starts moving down through the zone of aeration. The subsequent gravitational movement of the water down to the zone of saturation is termed percolation, although there is no clearly defined point where infiltration becomes percolation. The amount of water that infiltrates into the ground depends upon how precipitation is dispersed, that is, on what proportions are assigned to immediate run-off and to evapotranspiration, the remainder constituting the proportion allotted to infiltration/percolation. The infiltration capacity is influenced by the rate at which rainfall occurs, the vegetation cover, the porosity of the soils and rocks concerned, their initial moisture

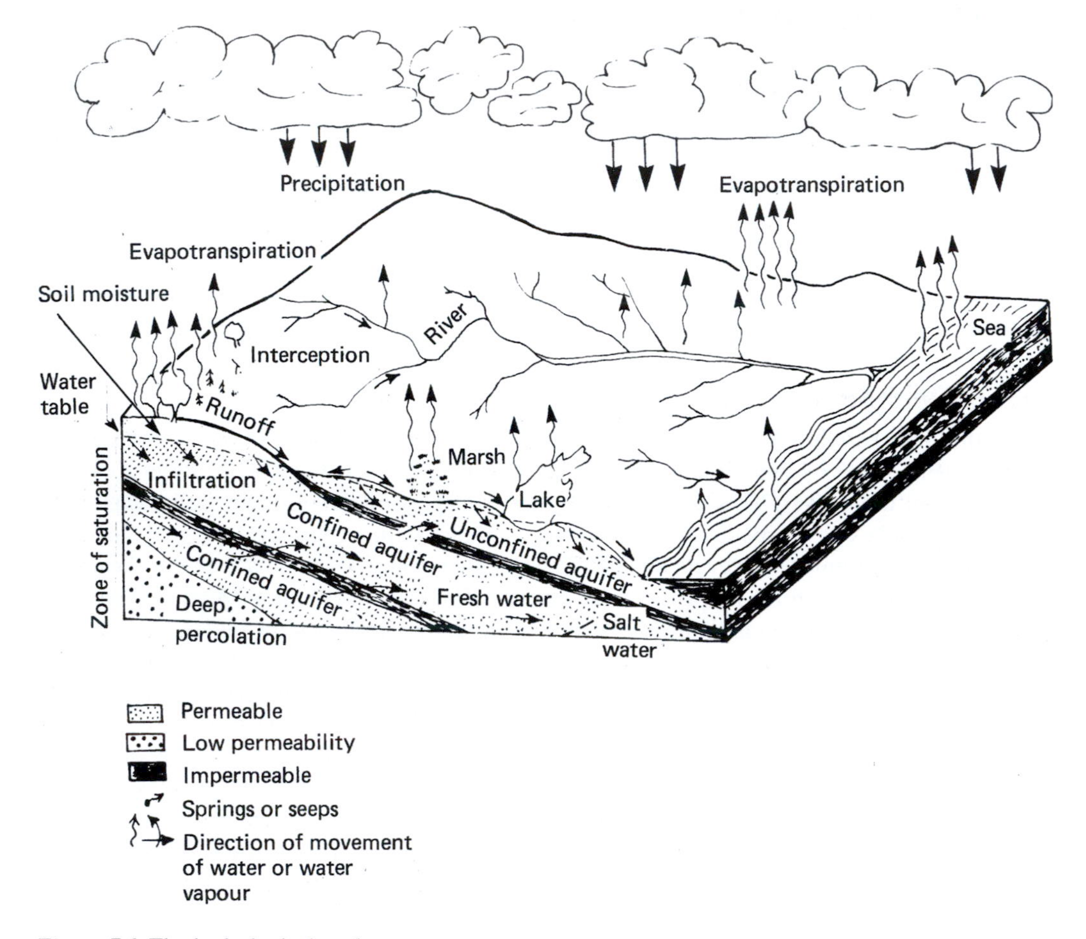

Figure 7.1 The hydrological cycle.

Table 7.1 The water inventory of the earth

Storage component	*Volume of water* (10^{12} m^3)	*Total water* (%)
Oceans	1 350 400	97.6
Saline lakes and inland seas	105	0.008
Ice caps and glaciers	26 000	1.9
Soil moisture	150	0.01
Groundwater	7 000	0.5
Freshwater lakes	125	0.009 usable fresh
Rivers	2	0.0001 water = 51%
Atmosphere	13	0.009
Total	1 384 000	100

Source: Nace, 1969.

Note
All figures are approximate estimates and rounded.

content and the position of the zone of saturation. Accordingly, the rate at which groundwater is replenished is dependent basically upon the quantity of precipitation falling on the recharge area of an aquifer, although rainfall intensity also is very important. Frequent rainfall of moderate intensity is more effective in recharging groundwater resources than short concentrated periods of high intensity. This is because the rate at which the ground can absorb water is limited, any surplus water tending to become run-off. As noted, some precipitation is lost through evaporation and transpiration. The rate at which water can be lost from the ground surface through evapotranspiration is dependent to some extent upon the amount of water that is present in the soil.

If lower strata are less permeable than the surface layer, the infiltration capacity is reduced so that some of the water that has penetrated the surface moves parallel to the water table and is called interflow. The water that becomes interflow normally is discharged to a river at some point and forms part of the baseflow component of the river. The remaining water may continue down through the zone of aeration until it reaches the water table and becomes groundwater recharge. This can be a slow process (typically about 1 m year^{-1}), since the percolating water may become temporarily suspended in the zone of aeration as a result of the various dynamic forces that operate in the zone.

Although infiltration may be high in a dry soil, the fact that the soil is dry means that water is more likely to be held in the surface layers of the soil and either evaporated or transpired by plants, and therefore less likely to reach the water table. Consequently, most groundwater recharge occurs when the ground is comparatively wet and evapotranspiration is relatively insignificant. When evapotranspiration exceeds precipitation and vegetation has to draw on reserves of water in the soil to satisfy transpiration requirements, then soil moisture deficits occur. The soil moisture deficit (SMD) at any time is the difference between the moisture remaining and the field capacity of the soil, which is the amount of water retained in the soil by capillary forces after excess water has been drained from it. Before recharge can happen, any soil moisture deficits must be made up so that the soil is returned to its field capacity. Once this has been achieved, then any additional or excess water becomes groundwater recharge. Plants extract water from the soil until, as the soil dries out, it becomes impossible for them to continue doing so. At that point plants wilt and die if the soil is not rewetted. The point at which permanent wilting starts is referred to as the permanent wilting point. The field capacity and the permanent wilting point can be defined in terms of soil suction, their respective pF values being 2 and 4.2.

The retention of water in a soil depends upon the capillary force and the molecular attraction of the particles. As the pores in a soil become thoroughly wetted, the capillary force declines so that gravity becomes more effective. In this way downward percolation can continue after infiltration has ceased but as the soil dries, so capillarity increases in importance. No further percolation occurs after the capillary and gravity forces are balanced. Thus, water percolates into the zone of saturation when the retention capacity is satisfied. This means that the rains that occur after the deficiency of soil moisture has been catered for are those that count as far as supplementing groundwater is concerned.

The pores within the zone of saturation are filled with water, generally referred to as phreatic water. The upper surface of this zone therefore is known as the phreatic surface but more commonly is termed the water table. A perched water table is one that forms above a discontinuous impermeable layer such as a lens of clay in a formation of sand, the clay impounding a water mound. The zone of aeration, in which both air and water occupy the pores, occurs above the zone of saturation. The water in the zone of aeration is referred to as vadose water. Meinzer (1942) divided this zone into three belts, those of soil water, the intermediate belt

Zones and sub zones

Aeration	Soil water	Vadose water	Hygroscopic	Percolation	Discontinuous capillary saturation
	Intermediate		Pellicular		Semi-continuous capillary saturation
	Capillary fringe		Capillary		Continuous capillary saturation
Saturation	Phreatic zone	Phreatic water	Ground-water	Seepage	Unconfined groundwater

Water table

Figure 7.2 Zones and subzones of groundwater.

and the capillary fringe (Fig. 7.2). The uppermost or soil water belt discharges water into the atmosphere in perceptible quantities by evapotranspiration. In the capillary fringe, which occurs immediately above the water table, water is held in the pores by capillary action. An intermediate belt occurs when the water table is far enough below the surface for the soil water belt not to extend down to the capillary fringe.

The geological factors that influence percolation not only vary from one soil or rock outcrop to another but may do so within the same one. This, together with the fact that rain does not fall evenly over a given area, means that the contribution to the zone of saturation is variable, which influences the position of the water table, as do the points of discharge. A rise in the water table as a response to percolation is controlled partly by the rate at which water can drain from the area of recharge. Mounds and ridges form in the water table under the areas of greatest recharge. Superimpose upon this the influence of water that may drain from lakes and streams, and it can be appreciated that a water table is continually adjusting towards equilibrium. Because of the low flow rates in most rock masses this equilibrium is rarely, if ever, attained before another disturbance occurs.

As noted, the water table fluctuates in position, particularly in those climates where there are marked seasonal changes in rainfall. Thus, permanent and intermittent water tables can be distinguished, the former marking the level beneath which the water table does not sink, whilst the latter is an expression of the fluctuation. Usually, water tables fluctuate within the lower and upper limits rather than between them, especially in humid regions, since the periods between successive recharges are small. The position at which the water table intersects the surface is termed the spring line. Similarly, intermittent and permanent springs can be distinguished.

Dykes often act as barriers to groundwater flow so that the water table on one side may be higher than on the other. Fault planes occupied by clay gouge may have a similar effect. Conversely, faults may act as conduits where the fault planes are not sealed. The movement of water across a permeable boundary that separates aquifers of different permeabilities leads to deflection of flow, the bigger the difference the larger the deflection. When groundwater meets an impermeable boundary it flows along it and, as noted previously, in some situations, such as the occurrence of a dyke, may be impounded. The nature of a rock mass also influences whether flow is steady or unsteady. Generally, it is unsteady since it usually is due to discharge from storage.

7.2 Capillary movement in soil and soil suction

Capillary movement in a soil refers to the movement of moisture through the minute pores between the soil particles that act as capillaries. It takes place as a consequence of surface tension, therefore moisture can rise from the water table. In fact, this movement can occur in any direction, not just vertically upwards. It occurs whenever evaporation takes place from the surface of the soil, thereby exerting a 'surface tension pull' on the moisture, the forces of surface tension increasing as evaporation proceeds. Hence, capillary moisture is in hydraulic continuity with the water table and is raised against the force of gravity, the degree of saturation decreasing from the water table upwards. Equilibrium is attained when the forces of gravity and surface tension are balanced.

The boundary separating capillary moisture from the gravitational water in the zone of saturation is ill-defined and cannot be determined accurately. That zone immediately above the water table that is saturated with capillary moisture is referred to as the closed capillary fringe, whilst above this air and capillary moisture exist together in the pores of the open capillary fringe. The depth or the thickness of the capillary fringe is dependent largely upon the particle size distribution and density of a soil mass that, in turn, influence pore size. In other words, the smaller the pore size, the greater is the depth. For example, capillary moisture can rise to great heights in clay soils (Table 7.2a) but the movement is very slow. The height of the capillary fringe in soils that are poorly graded generally varies whereas in uniformly textured soils it attains roughly the same height. Where the water table is at shallow depth and the maximum capillary rise is large, moisture is attracted continually from the water table, due to evaporation from the surface, so that the uppermost soil is near saturation. For instance, under normal conditions peat deposits may be assumed to be within the zone of capillary saturation. This means that the height to which water can rise in peat by capillary action is greater than the depth below ground to which the water table can be reduced by drainage. The coarse fibrous type of peat, containing appreciable sphagnum, may be an exception.

Drainage of capillary moisture cannot be effected by the installation of a drainage system within the capillary fringe as only that moisture in excess of that retained by surface tension can be removed, but the capillary fringe can be lowered by lowering the water table. Furthermore, the capillary ascent can be interrupted by the installation of impermeable membranes or layers of coarse aggregate. These two methods can be used in the construction of embankments to control the capillary rise, or the height of the fill can be raised.

Table 7.2 Capillary rise, capillary pressures and suction pressure in soil

(a) Capillary rise and pressures

Soil	*Capillary rise* (mm)	*Capillary pressure* (kPa)
Fine gravel	Up to 100	Up to 1.0
Coarse sand	100–150	1.0–1.5
Medium sand	150–300	1.5–3.0
Fine sand	300–1000	3.0–10.0
Silt	1000–10 000	10.0–100.0
Clay	Over 10 000	Over 100.0

Table 7.2 (b) Suction pressure and pF values

pF value	*Equivalent suction*	
	(mm water)	(kPa)
0	10	0.1
1	100	1.0
2	1 000	10.0
3	10 000	100.0
4	100 000	1000.0
5	1 000 000	10 000.0

Below the water table the water contained in the pores is under normal hydrostatic load, the pressure increasing with depth. Because these pressures exceed atmospheric pressure they are designated positive pressures. On the other hand, the pressures existing in the capillary zone are less than atmospheric and so are termed negative pressures. Thus, the water table is regarded as a datum of zero pressure between the positive pore water pressure below and the negative above.

At each point where moisture menisci are in contact with soil particles the forces of surface tension are responsible for the development of capillary or suction pressure (Table 7.2a). The air and water interfaces move into the smaller pores. In so doing the radii of curvature of the interfaces decrease and the soil suction increases. Hence, the drier the soil, the higher is the soil suction. Soil suction is a negative pressure and indicates the height to which a column of water could rise due to such suction. Since this height or pressure may be very large, a logarithmic scale has been adopted to express the relationship between soil suction and moisture content, the relationship being given in terms of the pF value (Table 7.2b). Soil suction tends to force soil particles together and these compressive stresses contribute towards the strength and stability of the soil. As Table 7.2 indicates, there is a particular suction pressure for a particular moisture content in a given soil. For instance, as a clay soil dries out the soil suction may increase to several megapascals. However, the strength of a soil attributable to soil suction is only temporary and is destroyed upon saturation. At that point soil suction is zero.

7.3 Aquifers, aquicludes and aquitards

An aquifer is the term given to a rock or soil mass that not only contains water but from which water can be abstracted readily in significant quantities. The ability of an aquifer to transmit water is governed by its permeability. Indeed, the permeability of an aquifer usually is in excess of $10^{-5}\,m\,s^{-1}$. By contrast, a formation with a permeability of less than $10^{-9}\,m\,s^{-1}$ is one that, in engineering terms, is regarded as impermeable and is referred to as an aquiclude. For example, clays and shales are aquicludes. Even when such rocks are saturated they tend to impede the flow of water through stratal sequences. An aquitard is a formation that transmits water at a very slow rate but which, over a large area of contact, may permit the passage of large amounts of water between adjacent aquifers that it separates.

7.3.1 Unconfined and confined aquifers

An aquifer is described as unconfined when the water table is open to the atmosphere, that is, the aquifer is not overlain by material of lower permeability (Fig. 7.3a). Conversely, a

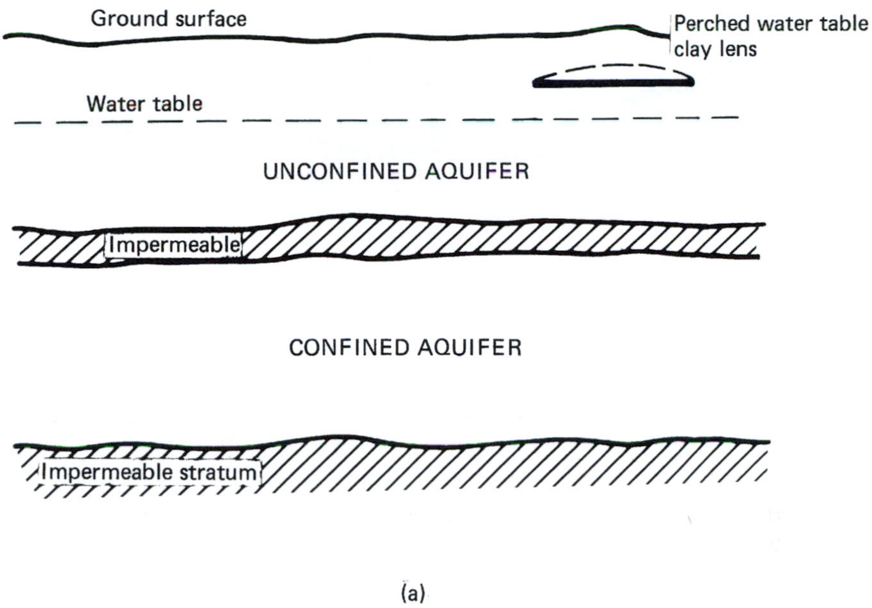

Figure 7.3 (a) Diagram illustrating unconfined and confined aquifers, with a perched water table in the vadose zone. (b) Diagram illustrating a leaky aquifer.

confined aquifer is one that is overlain by impermeable rocks. Confined aquifers may have relatively small recharge areas as compared with unconfined aquifers and therefore may yield less water. An aquifer that is overlain and/or underlain by aquitard(s) is described as a leaky aquifer (Fig. 7.3b). Even though the semi-pervious enclosing formations offer relatively high resistance to the flow of water through them, large amounts of water may flow from aquitard to aquifer and vice versa as a consequence of the extensive contact area between them. The

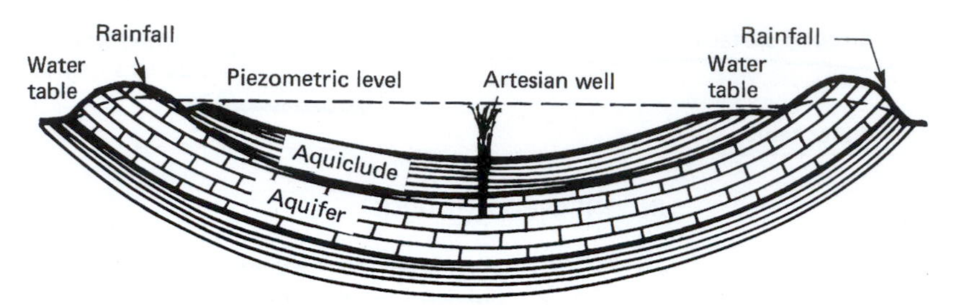

Figure 7.4 An artesian basin.

direction and quantity of leakage in either case depends on the difference in piezometric head that exists across the semi-pervious aquitard.

Very often, the water in a confined aquifer is under piezometric pressure, that is, there is an excess of pressure sufficient to raise the water above the base of the overlying bed when the aquifer is penetrated by a well. Piezometric pressures are developed when the buried upper surface of a confined aquifer is lower than the water table in the aquifer at its recharge area. Where the piezometric surface is above ground level, then water overflows from a well. Such wells are described as artesian. A synclinal structure is the commonest cause of artesian conditions (Fig. 7.4). The term subartesian is used to describe those conditions in which water is not under sufficient piezometric pressure to rise to the ground surface.

7.3.2 Basement aquifers

Basement aquifers are developed within the weathered overburden and fractured bedrock of rocks of intrusive igneous or metamorphic origin. Because basement aquifers generally provide a low yield, development usually is from point sources, utilizing handpumps or bucket and windlass systems. Viable aquifers that occur completely within fractured bedrock, according to Clark (1985), are rare due to the low storativity of fracture systems. Hence, to be effective, the bedrock component should be in continuity with storage in the weathered regolith, or other suitable formations such as alluvium. Furthermore, because of their low storativity, basement aquifers may be depleted significantly during periods of drought. The fact that the groundwater in such aquifers commonly is contained within fissures in rock at shallow depth means that it is susceptible to surface pollution. There is a close relationship between groundwater occurrence on the one hand and relief, soil, surface water hydrology and vegetation cover on the other. Because of the localized nature of many basement aquifers, there frequently is a high failure rate associated with groundwater supply boreholes, it being in the range of 10–40 per cent, and there is a wide range of yields (Wright, 1992).

The thickness of a weathered horizon or regolith depends on the nature of the bedrock, notably rock type and structure, the age of the land surface, its relief and the climate. The degree of weathering increases towards the surface with an increasing proportion of minerals being broken down, often to form clay minerals. As weathering is most effective in the vadose zone and the zone within which the water table fluctuates, the upper part of a saprolite usually contains more clay material and so has a low permeability. The boundary with the underlying saprock, the weathered bedrock, may be sharp (e.g. against coarse-grained massive rocks) or transitional (e.g. against finer-grained rocks). Regolith permeability values quoted by Wright (1992) for certain areas in Malawi and Zimbabwe generally did not exceed 0.5 m per day. However, the weathered residium of some granites and gneisses, especially in semi-arid regions, may be quite sandy and so may provide limited groundwater supply for small rural communities. In addition, springs may emerge from the contact between the weathered regolith and the bedrock beneath, which often provide a source of water for local communities (Bell and Maud, 2000).

The regolith may contain throughflow channels such as basal breccias, conglomerates and gravel beds, old termite tunnels, tree roots and residual quartz veins. Residual fractures can be orientated more or less in any direction from vertical to horizontal and frequently are open below the water table.

Fracture or discontinuity systems are related to the dissipation of residual stress as overburden is removed by erosion or to tectonic forces. In the former type of system, the discontinuities tend to be subhorizontal and decline in number with depth. Discontinuities developed by tectonic forces normally are subvertical and often are in zonal concentrations. Fissure permeability depends upon the nature of the discontinuities, in particular, on their frequency and whether they are open or filled (if open, the amount of gape is important). Clay illuviation, especially in the weathered zone, often is responsible for sealing discontinuities. Whether tectonic discontinuities close with depth is debatable, however, Black (1987) observed no relationship between depth and decreasing permeability to a depth of 700 m, and groundwater flow via discontinuities at times is a problem in deep mines.

Development of basement aquifers, particularly in much of Africa, is principally for rural water supply with some usage for irrigation or urban supply. Recharge usually exceeds the maximum requirement of rural water supply appreciably. Hence, some of the excess could be made available for other uses. Boreholes, dug wells and, to a much lesser extent, collector wells are used to abstract water from basement aquifers. Substantially larger yields can be obtained from collector wells than boreholes with the added advantage of low drawdowns. Radial collectors can be drilled at any angle from large diameter (2–3 m) wells.

7.4 Springs

Springs develop at the points where underground conduits discharge water at the surface. The location of a spring is dependent upon a number of factors, climate and geology being two of the most important. Intermittent springs are very much influenced by climate. After heavy rainfall the water table may rise to intersect the ground surface and so produce a spring. In dry weather the water table sinks and the stream disappears. Water rarely moves uniformly throughout an entire rock mass and most springs therefore issue as concentrated flows. Springs that percolate from many small openings have been termed seepage springs and they may discharge so little water that they are barely noticeable.

The commonest geological setting for a spring is at the contact between two beds of differing permeability. This setting often gives rise to a spring line and such springs are referred to as stratum springs. There are two types of stratum springs (Fig. 7.5a). If the downward percolation of water in a permeable horizon is impeded by an underlying impermeable layer, then the spring that issues is termed a contact spring. Conversely, an overflow spring may be formed where a permeable bed dips beneath an impermeable one. Unconformities may give rise to springs along their outcrop. When a permeable formation is thrown against an impermeable formation by faulting, water may issue at the surface along the fault plane (Fig. 7.5b). Water table or valley springs emerge where a valley is carved beneath the water table in thick permeable formations (Fig. 7.5c). The discharge from such springs usually is small.

The most common type of thermal spring is the hot spring, however, the temperatures of thermal springs may range from lukewarm to near boiling point. Hot springs are found in all volcanic districts, even some of those where the volcanoes are extinct. They presumably

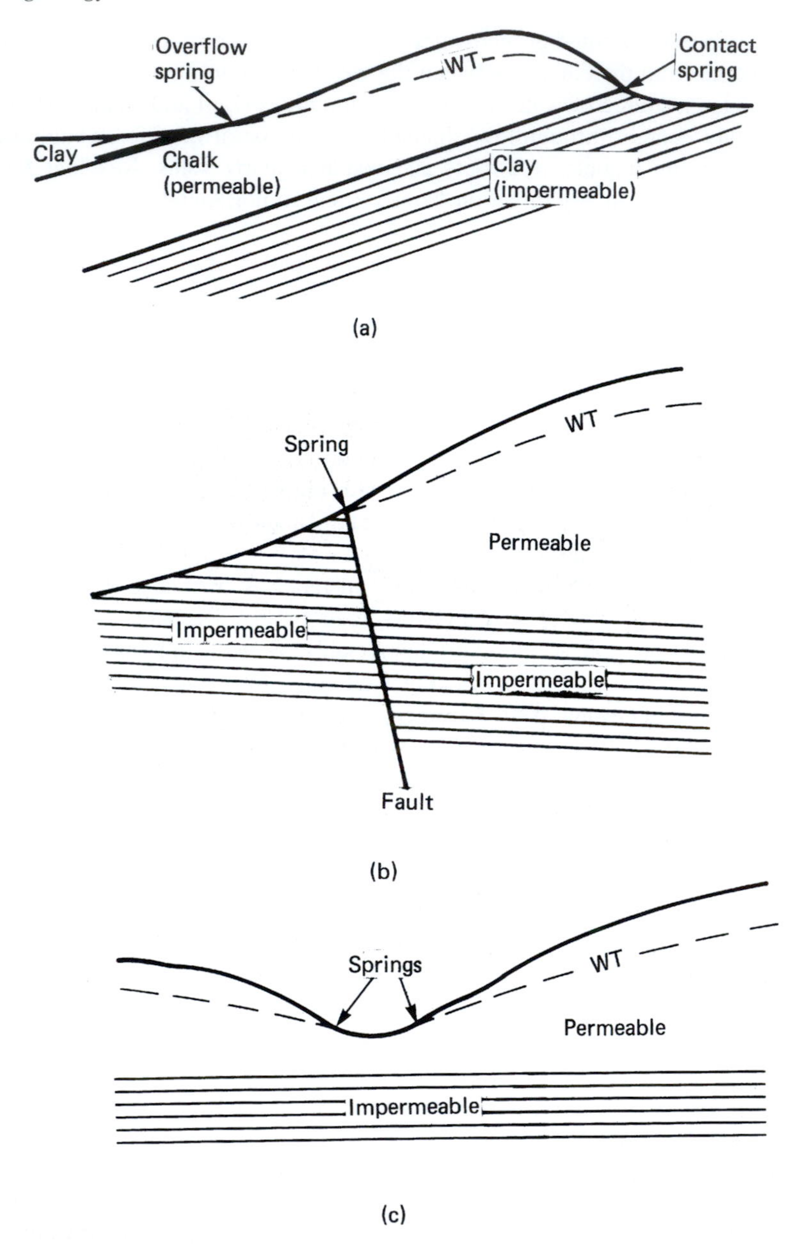

Figure 7.5 Types of spring: (a) stratum spring; (b) fault spring; (c) valley spring. WT, water table.

originate from steam given off by a magmatic source. On its passage to the surface the steam commonly encounters groundwater, which consequently is heated and contributes to the hot springs. A geyser is a hot spring that periodically discharges a column of hot water (Fig. 7.6). Geothermal energy may be captured by removing the heat from hot groundwaters that are pumped to the surface via wells. Ideally, geothermal reservoirs of practical interest should have temperatures exceeding 180 °C, an adequate reservoir volume and sufficient reservoir permeability to guarantee continuous delivery of hot water to the wells, together with location

Figure 7.6 Old Faithful Geyser, Yellowstone National Park, Wyoming.

at shallow depth for more economical exploitation. However, Barker *et al.* (2000) suggested that the sandstones in several deep sedimentary basins of Permo-Triassic age in England, with low temperature (40–100 °C) hot water resources, might have the potential for geothermal development. They estimated that these resources possess an estimated 69.1×10^{18} J, which is the equivalent to 2576 million tonnes of coal. In several places in the world hot springs are utilized for the generation of electric power, for example, at Larderello in Italy, at Wairakei in New Zealand and in Somona County, California. In Iceland, hot springs are used for domestic heating. However, hot springs generally contain dissolved gases and minerals such as carbon dioxide, hydrogen sulphide, calcium carbonate and silica. These may cause serious problems of corrosion and precipitation in a plant that harnesses such springs.

7.5 Water budget studies

Water budget studies attempt to quantify the average annual flow through an aquifer by considering the disposition of the rain that falls on the recharge area. Basically, the rain must either run off the surface and become streamflow, be evapotranspired or infiltrate the surface and thereby add to groundwater storage. The potential infiltration rate and subsequently the change in groundwater storage can be estimated from the following expression:

$$\mathrm{d}S_g = P - \mathrm{AET} - R_o \tag{7.1}$$

where $\mathrm{d}S_g$ is the change in groundwater storage (m year^{-1}), P the precipitation (m year^{-1}), AET the actual evapotranspiration (m year^{-1}) and R_o the total run-off, including groundwater discharge and interflow (m^3 m^{-1} of catchment/year).

If the area of the recharge zone of an aquifer is known, the annual volume of recharge to the aquifer is given by:

$$Q = (P - \mathrm{AET} - R_o)A \tag{7.2}$$

where Q is the annual volume of groundwater recharge (m^3 year^{-1}) and A the area of the recharge zone (m^2). If potential evapotranspiration (PET) is used in Eq. 7.2 instead of AET, then a more conservative estimate of groundwater recharge is obtained since PET always exceeds AET. Alternatively, if the infiltration rate over the recharge area has been determined using a percolation gauge or by analysing groundwater discharge then:

$$Q = f_e A \tag{7.3}$$

where f_e is the effective infiltration rate (m year^{-1}).

The reliability of water budget methods depends to some extent upon how accurately the values of the variables are known or can be estimated. For good results, quite intensive instrumentation of the recharge area is necessary. Another limitation of this technique is the assumption that water infiltrating the surface of the recharge area will eventually reach any well located down gradient. Some potential groundwater recharge may become interflow and be discharged to a surface watercourse without even reaching the water table. In addition, some proportion of the water that has succeeded in percolating down to the zone of saturation may be lost as groundwater discharge to rivers and springs. The perennial yield to wells will only be a fraction of the total recharge volume calculated. Conversely, recharge may take place as a result of seepage from rivers or other bodies of surface water. In such circumstances, the use of river gauging or tracer techniques may be necessary to identify and quantify the leakage.

The water budget technique attempts to quantify groundwater recharge by considering the balance between precipitation, actual evapotranspiration and surface run-off in a given area. However, in areas that have highly permeable surface strata, such as limestone areas, there may be little or no surface run-off. Consequently, any increase in groundwater storage is equal to the difference between precipitation (P) and actual evapotranspiration (AET) over the recharge area (A), so that the annual volume of recharge (Q) is given by:

$$Q = (P - \mathrm{AET})A \text{ m}^3 \text{ year}^{-1} \tag{7.4}$$

Unfortunately, some water is likely to be lost as interflow or groundwater discharge to neighbouring catchments, and this adds some element of uncertainty to the use of Eq. 7.4.

If a region has very permeable surface deposits, but also has some sizeable streams and rivers, then it is quite possible (or even probable) that these water courses are maintained by groundwater discharge. In an area where a very high proportion of the available groundwater resource discharges naturally to surface streams, it may be possible to quantify the surplus capacity of an aquifer from an analysis of river discharge hydrographs. The assumption inherent in this method is that the aquifer is overflowing into the surface water courses and that this water could be diverted to wells instead. Thus, the amount of water that is available, assuming natural overflow is stopped, is equal to the total groundwater component of river discharge.

Most aquifers discharge either directly or indirectly to rivers and seas by way of seepage and springs. The most common form of groundwater discharge is that to a river. The groundwater discharge that becomes the baseflow of a river is the outflow from unconfined or artesian aquifers bordering the river, which go on discharging more and more slowly with time as the differential head falls. Bank storage is the water temporarily held in store in the ground adjacent to a river between the low and high water levels. This water is released as the river level falls.

The groundwater contribution to streamflow is most important during a dry season, when surface run-off is reduced as a result of soil moisture deficits. During the dry months, a very high proportion of streamflow may be derived from groundwater sources. In fact, it may be the groundwater contribution to flow during this period that prevents streams from drying up.

7.6 Hydrogeological properties

Porosity and permeability are the two most important factors governing the accumulation, migration and distribution of groundwater. However, both porosity and permeability may change within a rock or soil mass in the course of its geological evolution. Furthermore, it is not uncommon to find variations in both per metre of depth beneath the ground surface.

7.6.1 Porosity

The porosity of a rock can be defined as the percentage pore space within a given volume. Total or absolute porosity is a measure of the total void volume and is the excess of bulk volume over grain volume per unit of bulk volume. It usually is determined as the excess of grain density (i.e. specific gravity) over dry density, per unit of grain density, and can be obtained from the following expression:

$$\text{Absolute density} = \left(1 - \frac{\text{dry density}}{\text{grain density}}\right) \times 100 \qquad (7.5)$$

The effective, apparent or net porosity is a measure of the effective void volume of a porous medium and is determined as the excess of bulk volume over grain volume and occluded pore volume. It may be regarded as the pore space from which water can be removed.

The factors affecting the porosity of soils or clastic sedimentary rocks include particle size distribution, sorting, grain shape, fabric, degree of compaction and cementation, solution effects and lastly mineralogical composition, particularly the presence of clay particles. In experiments with packing arrangements, Frazer (1935) found that for a given mode of packing of equal-sized spheres porosity was independent of size. Rhombohedral packing (Fig. 7.7) was the tightest form and produced a porosity of 26 per cent whilst the loosest type of packing gave rise to a porosity of 87.5 per cent.

However, in natural assemblages as grain sizes decrease so friction, adhesion and bridging become more important because of the higher ratio of surface area to volume. Therefore, as the grain size decreases the porosity increases. For example, in coarse sands it ranges from

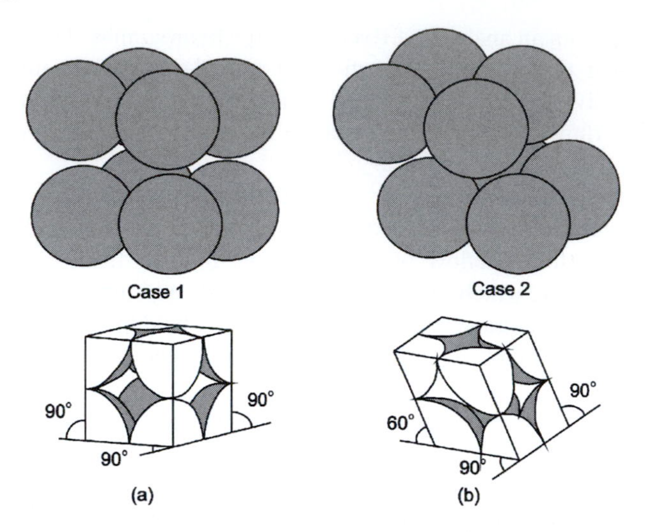

Figure 7.7 Packing of spherical grains: (a) cubic packing; (b) rhombohedral packing.

39–41 per cent, medium sands 41–48 per cent and fine sands 44–49 per cent. Whether the grain size is uniform or non-uniform is of fundamental importance with respect to porosity. The highest porosity commonly is attained when all the grains are the same size. The addition of grains of different size to such an assemblage lowers its porosity and this is, within certain limits, directly proportional to the amount added. As would be expected, the skewness of the size distribution influences porosity. For example, sands with a negative skewness, that is, an excess of coarse particles in relation to fines, tend to have higher porosities.

Irregularities in grain shape result in a larger possible range of porosity, as irregular forms may theoretically be packed either more tightly or more loosely than spheres. Similarly, angular grains may either cause an increase or a decrease in porosity, although the only type of angularity that has been found experimentally to produce a decrease is that in which the grains are mildly and uniformly disc-shaped.

After a sediment has been buried and indurated several additional factors help determine its porosity. The chief amongst these are closer spacing of grains, deformation and granulation of grains, recrystallization, secondary growth of minerals, cementation and, in some cases, dissolution. For instance, when chemical cements are present in sandstones in large amounts their influence on porosity is dominant and masks the control of other factors. Thus, two types of porosity may be distinguished, original and secondary. Original porosity is an inherent characteristic of a soil or rock in that it was determined at the time they were formed. The process by which a given sediment has accumulated affects its porosity in two ways. First, the nature and variety of the materials deposited affects the entire deposit by controlling the range and uniformity of the sizes present, as well as their degree of rounding. Second, is the manner in which the material is packed. Hence, the original porosity results from the physical impossibility of packing grains in such a way as to exclude interstitial voids of a conjugate nature. On the other hand, secondary porosity results from later changes undergone by a rock that may either increase or decrease its original porosity.

The porosity of a deposit does not necessarily provide an indication of the amount of water that can be obtained therefrom, for example, clay soils possess high porosities but are regarded as impermeable. Nevertheless, the water content of a soil or rock mass is related to its porosity.

The natural moisture content (m) of a porous material usually is expressed as the percentage mass of the moisture content (W_w) in relation to the mass of the solid material (W_s), that is:

$$m = \left(\frac{W_w}{W_s}\right) \times 100 \tag{7.6}$$

where W_w is the weight of the water. The degree of saturation (S_r) refers to the relative volume of water (V_w) in the voids (V_v) and is expressed as a percentage:

$$S_r = \left(\frac{V_w}{V_v}\right) \times 100 \tag{7.7a}$$

or

$$S_r = \frac{mG_s}{e} \tag{7.7b}$$

where G_s is the specific gravity and e the void ratio ($e = V_v/V_s$, i.e. the volume of the voids divided by the volume of the solids; or $e = mG_s/S_r$). In the case of a fully saturated soil, the void ratio is obtained from the moisture content and specific gravity ($e = mG_s$). The void ratio for clayey soils can be obtained indirectly from the oedometer test. The porosity (n) of soil usually is derived from the void ratio [$n = e/(1 + e)$].

The porosity of a rock generally is determined experimentally by using the standard saturation method (Brown, 1981). Alternatively, an air porosimeter can be used (Ramana and Venkatanarayana, 1971). Both tests give an effective value of porosity, although that obtained by the air porosimeter may be somewhat higher because air can penetrate pores more easily than water. The actual value of porosity obtained does not provide an indication of the way in which the pore space is distributed within a rock specimen or whether it consists of many fine pores or a smaller number of coarse pores. Two tests have been used to determine the microporosity of a rock specimen, namely, the suction plate test and the mercury porosimeter test. The microporosity tends to be defined as the percentage of pores with an effective diameter of less than 5 μm.

7.6.2 *Specific retention and specific yield*

The capacity of a material to yield water is of greater importance than is its capacity to hold water, as far as supply is concerned. Even though a rock or soil mass may be saturated, only a certain proportion of water can be removed by drainage under gravity or pumping, the remainder being held in place by capillary or molecular forces. The ratio of the volume of water retained, V_{wr} to that of the total volume of rock or soil, V, expressed as a percentage, is referred to as the specific retention, S_{re}:

$$S_{re} = \frac{V_{wr}}{V} \times 100 \tag{7.8}$$

The amount of water retained varies directly in accordance with the surface area of the pores and indirectly with regard to the pore space. The specific surface of a particle is governed by its size and shape. For example, particles of clay have far larger specific surfaces than do those of sand. As an illustration, a grain of sand, 1 mm in diameter, has a specific surface of about $0.002\,m^2\,g^{-1}$, compared with kaolinite that varies from approximately 10 to $20\,m^2\,g^{-1}$. Hence, clays have a much higher specific retention than sands (Fig. 7.8).

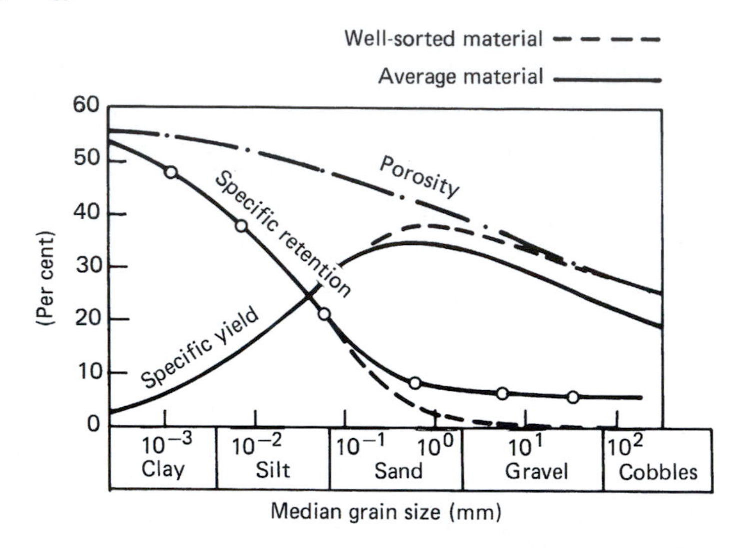

Figure 7.8 Relationship between grain size, porosity, specific retention and specific yield.

The specific yield, S_y, of a rock or soil mass refers to its water yielding capacity attributable to gravity drainage, as occurs when the water table declines. It was defined by Meinzer (1942) as the ratio of the volume of water, after saturation, that can be drained by gravity, V_{wd}, to the total volume of the aquifer, expressed as a percentage, hence:

$$S_y = \frac{V_{wd}}{V} \times 100 \tag{7.9}$$

The specific yield plus the specific retention is equal to the porosity of the material when all the pores are interconnected. The relationship between the specific yield and particle size distribution is shown in Fig. 7.8. Examples of the specific yield of certain common types of soil and rock are given in Table 7.3 (it must be appreciated that individual values of specific yield can vary considerably from those quoted).

The storage coefficient or storativity (S) of an aquifer is defined as the volume of water released from or taken into storage per unit surface area of the aquifer, per unit change in

Table 7.3 Some examples of specific yield

Materials	*Specific yield* (%)
Gravel	15–30
Sand	10–30
Dune sand	25–35
Sand and gravel	15–25
Loess	15–20
Silt	5–10
Clay	1–5
Till (silty)	4–7
Till (sandy)	12–18
Sandstone	5–25
Limestone	0.5–10
Shale	0.5–5

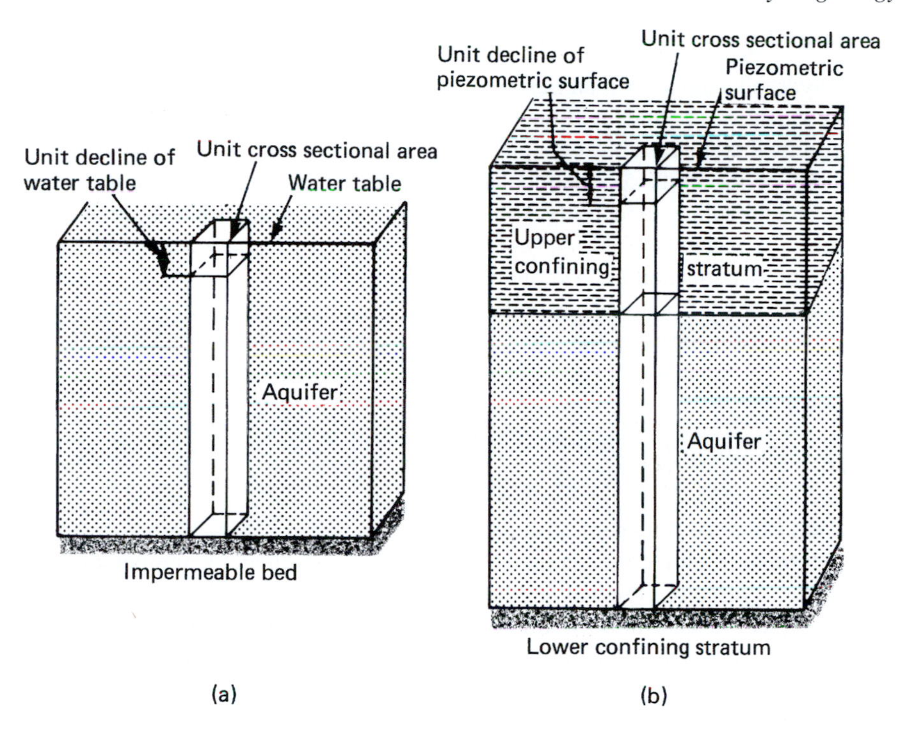

Figure 7.9 Diagram illustrating the storage coefficient of: (a) an unconfined aquifer; (b) a confined aquifer.

head normal to that surface (Fig. 7.9). It is a dimensionless quantity. Changes in storage in an unconfined aquifer represent the product of the volume of the aquifer between the water table before and after a given period of time and the specific yield. The storage coefficient of an unconfined aquifer virtually corresponds to the specific yield as more or less all the water is released from storage by gravity drainage, and only an extremely small part results from compression of the aquifer and the expansion of water. In confined aquifers, water is not yielded simply by gravity drainage from the pore space because there is no falling water table and the material remains saturated. Hence, other factors are involved regarding yield such as consolidation of the aquifer and expansion of water consequent upon lowering of the piezometric surface. Therefore, much less water is yielded by confined than unconfined aquifers. According to Lohman (1972), the values of the storage coefficient in most confined aquifers fall within the range 10^{-5}–10^{-3}, which indicates that significant changes in pressure are required over extensive areas in order to produce substantial yields of water. He suggested a rule-of-thumb means for estimating the coefficient of confined aquifers, which is:

$$S = 3 \times 10^{-6} H_s \qquad (7.10)$$

where H_s is the saturated thickness of the aquifer.

Both the specific yield and storage coefficient can be determined by pumping tests or by using a neutron moisture probe. In the latter case, the probe is used to determine the moisture content of the material saturated and when drained (Meyer, 1962).

7.6.3 Permeability

In ordinary hydraulic usage, a material is termed permeable when it permits the passage of a measurable quantity of fluid in a finite period of time and impermeable when the rate at which it transmits that fluid is slow enough to be negligible under existing temperature–pressure conditions. The permeability of a particular soil or rock type is defined by its coefficient of permeability or hydraulic conductivity, it being the flow in a given time through a unit cross-sectional area of a material (Table 7.4). The permeability of a particular material is defined by its coefficient of permeability. Alternatively, the transmissivity or flow in cubic metres per day through a section of aquifer 1 m wide under a hydraulic gradient of unity is sometimes used as a convenient quantity in the calculation of groundwater flow instead of the coefficient of permeability. The transmissivity (T) and coefficient of permeability (k) are related to each other as follows:

$$T = kH_s \tag{7.11}$$

where H_s is the saturated thickness of the aquifer.

The flow through a unit cross-section of material is modified by temperature, hydraulic gradient and the hydraulic conductivity. The latter is affected by the uniformity and range of grain size, shape of the grains, stratification, the amount of consolidation and cementation undergone, and the presence and nature of discontinuities. Temperature changes affect the flow rate of a fluid by changing its viscosity.

Permeability and porosity are not necessarily as closely related as would be expected, for instance, very fine textured sandstones frequently have a higher porosity than coarser ones, though the latter are more permeable. More particularly, the permeability of a clastic material is affected by the interconnections between the pore spaces. If these are highly tortuous, then the permeability is accordingly reduced. Consequently, tortuosity influences the extent and rate of free water saturation. It can be defined as the ratio of the total path covered by a current flowing in the pore channels between two given points to the straight line distance between them.

Stratification in a formation varies within limits both vertically and horizontally. It frequently is difficult to predict what effect stratification has on the permeability of the beds. Nevertheless, in the great majority of cases where a directional difference in permeability exists, the greater permeability is parallel to, rather than normal to, the bedding. For example, the Permo-Triassic sandstones of the Mersey and Weaver Basins, in England, are notably anisotropic as far as permeability is concerned, the flow parallel to the bedding being higher than across it. Ratios of 5:1 are not uncommon and occasionally values of 100:1 have been recorded where fine marl partings occur.

As far as rock masses are concerned, the frictional resistance to flow through discontinuities is frequently much lower than that offered by their porosity; hence appreciable quantities of water can be transmitted by discontinuities. Indeed, the permeability of intact rock (primary permeability) is usually several orders less than the *in situ* permeability (secondary permeability). In other words, permeability in the field generally is governed by fissure flow. Consequently, as far as the assessment of flow through rock masses is concerned, field tests provide more reliable results than can be obtained from testing intact samples in the laboratory. An example of the difference between primary and secondary permeability is provided by the massive limestones of Lower Carboniferous age of the Pennine area, England, the permeability of an intact rock being much lower than that obtained by field tests (10^{-16}–10^{-11} m s^{-1} and 10^{-5}–10^{-1} m s^{-1}, respectively). The significantly higher permeability found in the field is attributable to the joint

Table 7.4 Relative values of permeabilities

Permeability range (m s^{-1})												
		Porosity	*10^{-0}*	*10^{-2}*	*10^{-4}*	*10^{-6}*	*10^{-8}*	*10^{-10}*	*Well yields*			
Rock types	*Primary (grain)* %	*Secondary (fracture)**	*Very high*	*High*	*Medium*	*Low*	*Very low*	*Impermeable*	*High*	*Medium*	*Low*	*Type of water-bearing unit*
Sediments, unconsolidated												
Gravel	30–40		—	—					—			Aquifer
Coarse sand	30–40		—	—	—				—	—		Aquifer
Medium to fine sand	25–35			—	—				—	—		Aquifer
Silt	40–50	Occasional				—	—			—	—	Aquiclude
Clay, till	45–55	Often fissured				—	—	—			—	Aquiclude
Sediments, consolidated												
Limestone, dolostone	1–50	Solution joints, bedding planes	—	—	—	—	—	—	—	—	—	Aquifer or aquiclude
Coarse, medium sandstone	<20	Joints and bedding planes	—	—	—	—			—	—		Aquifer or aquiclude
Fine sandstone	<10	Joints and bedding planes			—	—	—			—	—	Aquifer or aquiclude
Shale, siltstone	–	Joints and bedding planes				—	—	—			—	Aquifer or aquiclude
Volcanic rocks, e.g. basalt	–	Joints and bedding planes	—	—	—	—	—	—	—	—	—	Aquifer or aquiclude
Plutonic and metamorphic rocks		Weathering and joints decreasing as depth increases				—	—	—	—	—	—	Aquiclude or Aquifer

Note
*Rarely exceeds 10 per cent.

Table 7.5 Assessment of seepage from discontinuities

	A. Open discontinuities	*B. Filled discontinuities*
Seepage rating	*Description*	*Description*
1	The discontinuity is very tight and dry, water flow along it does not appear possible	The filling material is heavily consolidated and dry, significant flow appears unlikely due to very low permeability
2	The discontinuity is dry with no evidence of water flow	The filling materials are damp but no free water is present
3	The discontinuity is dry but shows evidence of water flow, i.e. rust staining, etc.	The filling materials are wet, occasional drops of water
4	The discontinuity is damp but no free water is present	The filling materials show signs of outwash, continuous flow of water (estimate $l\,min^{-1}$)
5	The discontinuity shows seepage, occasional drops of water but no continuous flow	The filling materials are washed out locally, considerable water flow along outwash channels (estimate $l\,min^{-1}$ and describe pressure, i.e. low, medium, high)
6	The discontinuity shows a continuous flow of water (estimate $l\,min^{-1}$ and describe pressure, i.e. low, medium, high)	The filling materials are washed out completely, very high water pressures are experienced, especially on first exposure (estimate $l\,min^{-1}$ and describe pressure)

Source: Barton, 1978. Reproduced by kind permission of Elsevier.

systems and bedding planes that have been opened by dissolution. The mass permeability of sandstones also is very much influenced by discontinuities. For instance, the average laboratory permeability for the Fell Sandstone Group from Shirlawhope Well near Longframlington, Northumberland, England, was found to be $17.4 \times 10^{-7}\,m\,s^{-1}$ by Bell (1978). This is compared with an estimated value of $2.4 \times 10^{-3}\,m\,s^{-1}$ obtained from field tests. Although the secondary permeability is affected by the frequency, continuity and openness, and amount of infilling, of discontinuities, a rough estimate of the permeability can be obtained from their frequency (Table 7.5). Admittedly, such estimates must be treated with some caution.

7.7 Flow through soils and rocks

Water possesses three forms of energy, namely, potential energy attributable to its height, pressure energy owing to its pressure and kinetic energy due to its velocity. The latter usually can be discounted in any assessment of flow through soils. Energy in water usually is expressed in terms of head. The head possessed by water in soil or rock masses is manifested by the height to which water will rise in a standpipe above a given datum. This height is referred to as the piezometric level and provides a measure of the total energy of the water. If at two different points within a continuous area of groundwater there are different amounts of energy, then there will be a flow towards the point of lesser energy and the difference in head is expended in maintaining that flow. Other things being equal, the velocity of flow between two points is directly proportional to the difference in head between them. The hydraulic gradient, i, refers to the loss of head or energy of water flowing through the ground. This loss of energy by the water is due to the friction resistance of the ground material and in soils is greater in fine- than coarse-grained types. Thus, there is no guarantee that the rate of flow will be uniform, indeed this is exceptional. However, if it is assumed that the resistance to flow is constant, then for a given difference in head the flow velocity is directly proportional to the flow path.

7.7.1 Darcy's law

Before any mathematical treatment of groundwater flow can be attempted, certain simplifying assumptions have to be made, namely: that the material is isotropic and homogeneous; that there is no capillary action; and that a steady state of flow exists. Since rocks and soils are anisotropic and heterogeneous, as they may be subject to capillary action and as flow through them is characteristically unsteady, any mathematical assessment of flow must be treated with caution.

The basic law concerned with flow in porous media is that enunciated by Darcy (1856) that states that the rate of flow (v) is proportional to the gradient of the potential head (i), measured in the direction of flow:

$$v = ki \tag{7.12}$$

where k is the coefficient of permeability. For a particular rock or soil or part of it, of area, A:

$$Q = vA = Aki \tag{7.13}$$

where Q is the quantity in a given time. The ratio of the cross-sectional area of the pore spaces in a soil to that of the whole soil is given by $e/(1+e)$, where e is the void ratio. Hence, the seepage velocity (v_s) is:

$$v_s = \left(\frac{1+e}{e}\right) ki \tag{7.14}$$

Darcy's law is valid as long as laminar flow exists. Departures from Darcy's law therefore occur when the flow is turbulent, such as when the velocity of flow is high. Such conditions exist in very permeable media normally when the Reynolds number can attain values above four. Reynolds number (N_R) is commonly used to distinguish between laminar and turbulent flow, and is expressed as follows:

$$N_R = \rho \frac{vR}{\mu} \tag{7.15}$$

where ρ is density, v mean velocity, R hydraulic radius and μ dynamic viscosity. Flow is laminar for small values of Reynolds number. Accordingly, it usually is accepted that Darcy's law can be applied to those soils that have finer textures than gravels. Furthermore, Darcy's law probably does not accurately represent the flow of water through a porous medium of extremely low permeability because of the influence of surface and ionic phenomena, and the presence of gases.

Apart from an increase in mean velocity, the other factors that cause deviations from the linear laws of flow include, first, the non-uniformity of pore spaces, since differing porosity gives rise to differences in the seepage rates through pore channels. A second factor is an absence of a running-in section where the velocity profile can establish a steady-state parabolic distribution. Lastly, such deviations may be developed by perturbations due to jet separation from wall irregularities.

Darcy (1856) omitted to recognize that permeability also depends upon the density (ρ) and dynamic viscosity of the fluid (μ) involved, and the average size (D_n) and shape of the pores in a porous medium. In fact, permeability is directly proportional to the unit weight of the fluid concerned and is inversely proportional to its viscosity. The latter is influenced very much by temperature. The following expression attempts to take these factors into account:

$$k = \frac{CD_n^2 \rho}{\mu} \tag{7.16}$$

where C is a dimensionless constant or shape factor that takes note of the effects of stratification, packing, particle size distribution and porosity. It is assumed in this expression that both the porous medium and the water are mechanically and physically stable, but this may never be true. For example, ionic exchange on clay and colloid surfaces may bring about changes in mineral volume that, in turn, affect the shape and size of the pores. Moderate to high groundwater velocities tend to move colloids and clay particles. Solution and deposition may result from the pore fluids. Small changes in temperature and/or pressure may cause gas to come out of solution, which may block pore spaces.

It has been argued that a more rational concept of permeability would be to express it in terms that are independent of fluid properties. Thus, the intrinsic permeability, k_i, characteristic of the medium alone has been defined as:

$$k_i = CD_n^2 \tag{7.17}$$

However, it has proved impossible to relate C to the properties of the medium. Even in uniform spheres, it is difficult to account for the variations in packing arrangement. In this context, a widely accepted relationship for laminar flow through a permeable medium is that given by Fair and Hatch (1935):

$$k = \frac{1}{m\left[\frac{(1-n)^2}{n^3}\left(\frac{\theta}{100}\sum\frac{P}{D_m}\right)^2\right]} \tag{7.18}$$

where n is the porosity, m the packing factor found by experiment to have a value of 5, θ the particle shape factor varying from 6.0 for spherical to 7.7 for angular grains, P the percentage of particles by weight held between each pair of adjacent sieves and D_m is the geometric mean opening $(D_1 D_2)^{1/2}$ of the pair.

The Kozeny-Carmen equation for deriving the coefficient of permeability also takes the porosity into account, as well as the specific surface area of the porous medium, S_a, which is defined per unit volume of solid as:

$$k = C_o \frac{n^3}{(1-n)^2 S_a^2} \tag{7.19}$$

where C_o is a coefficient, the suggested value of which is 0.2.

7.7.2 General equation of flow

If an element of saturated porous material is taken, with the dimensions dx, dy and dz (Fig. 7.10) and flow is taking place in the x–y plane, then the generalized form of Darcy's Law is:

$$v_x = k_x i_x \tag{7.20}$$

$$= k_x \frac{\delta h}{\delta x} \tag{7.21}$$

and

$$v_y = k_y i_y \tag{7.22}$$

$$= k_y \frac{\delta h}{\delta y} \tag{7.23}$$

where h is the total head under steady state conditions and k_x, i_x and k_y, i_y are the coefficients of permeability and the hydraulic gradients in the x- and y-directions, respectively. Assuming that the fabric of the medium does not change and that water is incompressible, then the volume of water entering the element is the same as that leaving in any given time, hence:

$$v_x \mathrm{d}y\mathrm{d}z + v_y \mathrm{d}x\mathrm{d}z = \left(v_x + \frac{\delta v_x}{\delta x}\mathrm{d}x\right)\mathrm{d}v\mathrm{d}z + \left(v_y + \frac{\delta v_y}{\delta y}\mathrm{d}y\right)\mathrm{d}x\mathrm{d}z \tag{7.24}$$

In such a situation, the difference in volume between the water entering and leaving the element is zero, therefore:

$$\frac{\delta v_x}{\delta x} + \frac{\delta v_y}{\delta y} = 0 \tag{7.25}$$

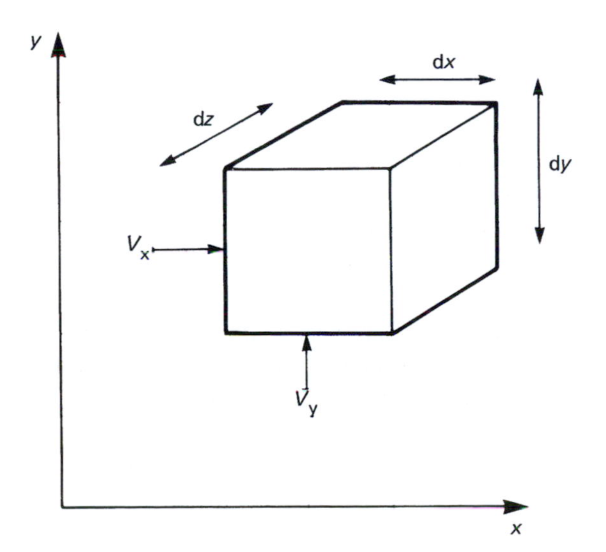

Figure 7.10 Seepage through an element of soil.

Equation 7.25 is referred to as the flow continuity equation. If Eqs 7.21 and 7.23 are substituted in the continuity equation, then:

$$k_x \frac{\delta^2 h}{\delta x^2} + k_y \frac{\delta^2 h}{\delta y^2} = 0 \tag{7.26}$$

If there is a recharge or discharge to the aquifer (–w and +w, respectively), then this term must be added to the right-hand side of the above equation.

If it is assumed that the hydraulic conductivity is isotropic throughout the media so that $k_x = k_y$, then Eq. 7.26 becomes:

$$\frac{\delta^2 h}{\delta x^2} + \frac{\delta^2 h}{\delta y^2} = 0 \tag{7.27}$$

This is the two-dimensional Laplace equation for steady-state flow in an isotropic porous medium.

7.7.3 *Flow through stratified deposits*

In a stratified sequence of deposits, the individual beds, no doubt, have different permeabilities, so that vertical permeability differs from horizontal permeability. Consequently, in such situations, it may be necessary to determine the average values of the coefficient of permeability normal to (k_v) and parallel to (k_h) the bedding. If the total thickness of the sequence is H_T and the thickness of the individual layers are $H_1, H_2, H_3, \ldots, H_n$, with corresponding values of the coefficient of permeability $k_1, k_2, k_3, \ldots, k_n$, then k_v and k_h can be obtained as follows:

$$k_v = \frac{H_T}{H_1/k_1 + H_2/k_2 + H_3/k_3 + \cdots + H_n/k_n} \tag{7.28}$$

and

$$k_h = \frac{H_1 k_1 + H_2 k_2 + H_3 k_3 + \cdots + H_n k_n}{H_T} \tag{7.29}$$

7.7.4 *Fissure flow*

Generally, it is the interconnected systems of discontinuities that determine the permeability of a particular rock mass. Indeed, as mentioned above, the permeability of a jointed rock mass usually is several orders higher than that of intact rock. According to Serafim (1968), the following expression can be used to derive the filtration through a rock mass intersected by a system of parallel-sided joints with a given opening, e, separated by a given distance, d:

$$k = \frac{e^3 \gamma_w}{12 d \mu} \tag{7.30}$$

where γ_w is the unit weight of water and μ its viscosity. The velocity of flow (v) through a single joint of constant gape (e) is expressed by

$$v = \left(\frac{e^2 \gamma_w}{12\mu}\right) i \tag{7.31}$$

where i is the hydraulic gradient.

Wittke (1973) suggested that where the spacing between discontinuities is small in comparison with the dimensions of the rock mass, it is often admissible to replace the fissured rock with regard to its permeability, by a continuous anisotropic medium, the permeability of which can be described by means of Darcy's law. He also provided a resumé of procedures by which three-dimensional problems of flow through rocks under complex boundary conditions could be solved.

Lovelock *et al.* (1975) suggested that it can be shown that the contribution of the fissures, T_f, to the transmissivity of an idealized aquifer can be approximated from the following expression:

$$T_f = \frac{g}{12\mu_k} \sum_{x=1}^{n} b_x^3 \tag{7.32}$$

where b_x is the effective aperture of the xth of n horizontal parallel sided smooth walled openings; g the acceleration due to gravity, μ_k is the kinematic viscosity of the fluid, and flow is laminar. The third power relationship means that a small variation in effective aperture (b) gives rise to a large variation in the fissure contribution (T_f).

7.8 Groundwater exploration

A groundwater investigation requires a thorough appreciation of the hydrology and geology of the area concerned, and a groundwater inventory needs to determine possible gains and losses affecting a subsurface reservoir. Of particular interest is the information concerning the lithology, stratigraphical sequence and geological structure, as well as the hydrogeological characteristics of the subsurface materials. Also of importance are the positions of the water table and piezometric level, and their fluctuations.

In major groundwater investigations, records of precipitation, temperatures, wind movement, evaporation and humidity may provide essential or useful supplementary information. Similarly, data relating to stream flow may be of value in helping to solve the groundwater equation since seepage into or from streams constitutes a major factor in the discharge or recharge of groundwater. The chemical and bacterial qualities of groundwater obviously require investigation.

Essentially, an assessment of groundwater resources involves the location of potential aquifers within economic drilling depths. Whether or not an aquifer will be able to supply the required amount of water depends on its thickness and spatial distribution, its porosity and permeability, whether it is fully or partially saturated, and whether or not the quality of the water is acceptable. Pumping lift and the effect of drawdown upon an aquifer also have to be considered.

7.8.1 Desk study and mapping

The desk study involves a consideration of the hydrological, geological, hydrogeological and geophysical data available concerning the area in question. Particular attention should be given to assessing the lateral and vertical extent of any potential aquifers, to their continuity and structure, to any possible variations in formation characteristics and to possible areas of recharge and discharge. Additional information relating to groundwater chemistry, the outflow of springs and surface run-off, data from pumping tests, from any mine workings, from waterworks, or meteorological data should be considered. Information on vegetative cover, land utilization, topography and drainage pattern can prove of value at times.

Geological maps will provide an indication of those rock masses that should be investigated as potential sources of groundwater. Some idea of the dimensions, continuity and geological structure of potential aquifers, the depth at which they occur, and the nature of the cover rocks of confined aquifers can be obtained from maps and sections. Perhaps some indication of the quality of the groundwater can be gleaned from the type of rock forming the aquifer, for instance, carbonate rocks are likely to yield hard water.

Aerial photographs may aid recognition of broad rock and soil types, and thereby help locate potential aquifers. The combination of topographical and geological data may help identify areas of likely groundwater recharge and discharge. In particular, the nature and extent of superficial deposits may provide some indication of the distribution of areas of recharge and discharge. Aerial photographs allow the occurrence of springs to be recorded.

Variations in water content in soil and rock masses that may not be readily apparent on black and white photographs are often depicted by false colour. The specific heat of water is usually two to ten times greater than that of most rocks, which therefore facilitates its detection in the ground. Indeed, the specific heat of water can cause an aquifer to act as a heat sink that, in turn, influences near-surface temperatures.

Furthermore, because the occurrence of groundwater is much influenced by the nature of the ground surface, aerial photographs can yield useful information. Also, the vegetative cover may be identifiable from aerial photographs and as such may provide some clue as to the occurrence of groundwater. In arid and semi-arid regions, the presence of phreatophytes, that is, plants that have a high transpiration capacity and derive water directly from the water table, indicates that the water table is near the surface. By contrast, xerophytes can exist at low moisture contents in soil and their presence suggests that the water table is at an appreciable depth. Thus, groundwater prediction maps can sometimes be made from aerial photographs (Howe *et al.*, 1956). These can be used to help locate sites for test wells.

The advantages of using geographical information system (GIS) in groundwater investigations are that they provide the ability to integrate multiple layers of data, to derive additional information, to visualize spatial data and to model results. For example, Lachassagne *et al.* (2001) described a methodology based on a GIS to delineate favourable sites for prospecting for groundwater. The method also can be used to assess the spatial distribution and thickness of an aquifer. In addition, Hiscock *et al.* (1995) used a GIS for mapping areas of groundwater vulnerability in the Midlands and southeast England, recognizing four classes, namely, extreme, high, moderate and low vulnerability. Although the resulting maps are of value, they must be interpreted with caution since the concept of general vulnerability to some universal pollutant does not have much meaning. It therefore is better to assess vulnerability to pollution in terms of a contaminant.

Geological mapping frequently forms the initial phase of exploration and should identify potential aquifers such as sandstones and limestones, and distinguish them from aquicludes. Argillaceous rocks represent the most common aquicludes and represent hydrogeological barriers in sedimentary sequences. The situation is further complicated where facies changes occur in a horizontal direction. Superficial deposits may perform a confining function in relation to a major aquifer that they overlie, or because of their lithology they may play an important role in controlling recharge to an aquifer. Moreover, geological mapping should locate igneous intrusions, weathered horizons and major faults, and it is important during the mapping programme to establish the geological structure. Large intrusions can have a notable influence on the pattern of groundwater movement. At one extreme, fault zones may be highly permeable whilst at the other they may act as barriers to groundwater flow, depending on the type of material occupying a fault zone. Furthermore, faulting can either completely or partially reduce the hydraulic conductivity between a confined aquifer and its outcrop, that is, the recharge area.

Geophysical methods of exploration and drilling are dealt with in Chapter 1.

7.8.2 Maps

Isopachyte maps can be drawn to show the thickness of a particular aquifer and the depth below the surface of a particular bed. They can be used to estimate the positions and depths of drillholes. They also provide an indication of the distribution of potential aquifers.

Maps showing groundwater contours are compiled when there is a sufficient number of observation wells to determine the configuration of the water table (Fig. 7.11). Data on surface water levels in reservoirs and streams that have free connection with the water table also should be used in the production of such maps. These maps are usually compiled for the periods of maximum, minimum and mean annual positions of the water table. A water table contour map is most useful for studies of unconfined groundwater.

As groundwater moves from areas of higher potential towards areas of lower potential and as the contours on groundwater contour maps represent lines of equal potential, the direction of groundwater flow moves from highs to lows at right angles to the contours or equipotential lines. Analysis of conditions revealed by groundwater contours is made in accordance with Darcy's law. Accordingly, spacing of contours is dependent on the flow rate. If continuity of flow rate is assumed, then the spacing depends upon aquifer thickness and permeability. Hence, areal changes in contour spacing may be indicative of changes in aquifer conditions. However, because of the heterogeneity of most aquifers, changes in gradient must be carefully interpreted in relation to all factors. The shape of the contours portraying the position of the water table helps to indicate where areas of recharge and discharge of groundwater occur. Groundwater mounds can result from the downward seepage of surface water. In an ideal situation, the gradient from the centre of such a recharge area will decrease radially and at a declining rate. An impermeable boundary or change in transmissivity will affect this pattern.

Depth-to-water table maps show the depth to water from the ground surface. They are prepared by overlaying a water table contour map on a topographical map of the same area and scale, and recording the differences in values at the points where the two types of contours intersect. Depth-to-water contours then are interpolated in relation to these points. A map indicating the depth to the water table also can provide an indication of areas of recharge and discharge. Both are most likely to occur where the water table approaches the surface.

Water level change maps are constructed by plotting the change in the position of the water table recorded at wells during a given interval of time (Fig. 7.12). The effect of local recharge or discharge often shows as distinct anomalies on water level change maps, for example, it may indicate that the groundwater levels beneath a river have remained constant while falling everywhere else. This would suggest an influent relationship between the river and the aquifer. Hence, such maps can help identify the locations where there are interconnections between surface water and groundwater. These maps also permit an estimation to be made of the change in groundwater storage that has occurred during the lapse in time involved.

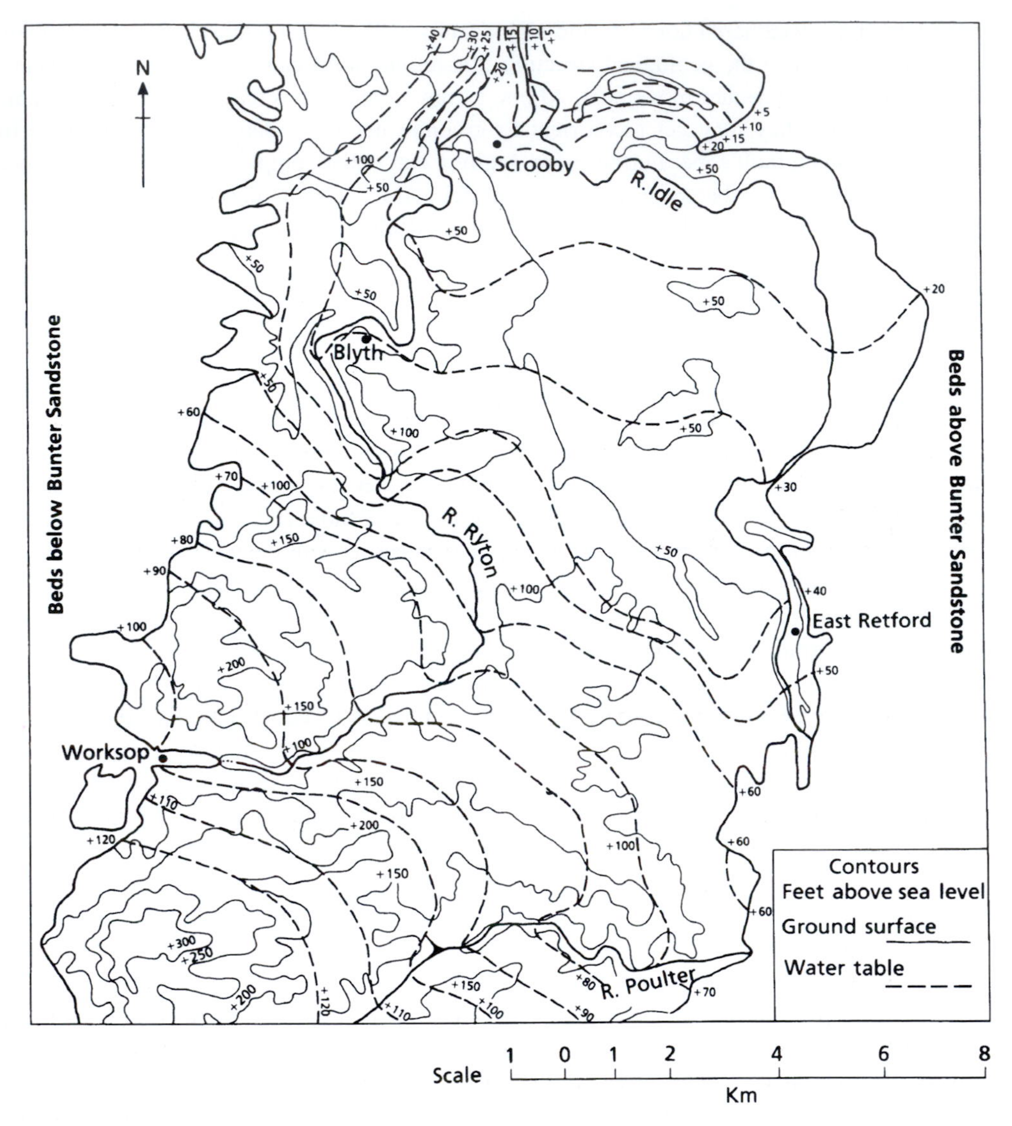

Figure 7.11 Sketch map of north Nottinghamshire, England, showing the water table in the Sherwood (formerly Bunter) Sandstone.

7.9 Assessment of permeability and flow

7.9.1 Assessment of permeability in the laboratory

Permeability is assessed in the laboratory by using either a constant head or a falling head permeameter (Anon., 1990). A constant head permeameter is used to measure the permeability of granular materials such as gravels and sands (Fig. 7.13a). A sample is placed in a cylinder of known cross-sectional area (A) and water is allowed to move through it under a constant head. The amount of water discharged (Q) in a given period of time (t) together with the difference in head (h) measured by means of manometer tubes, over a given length of sample (l) is obtained.

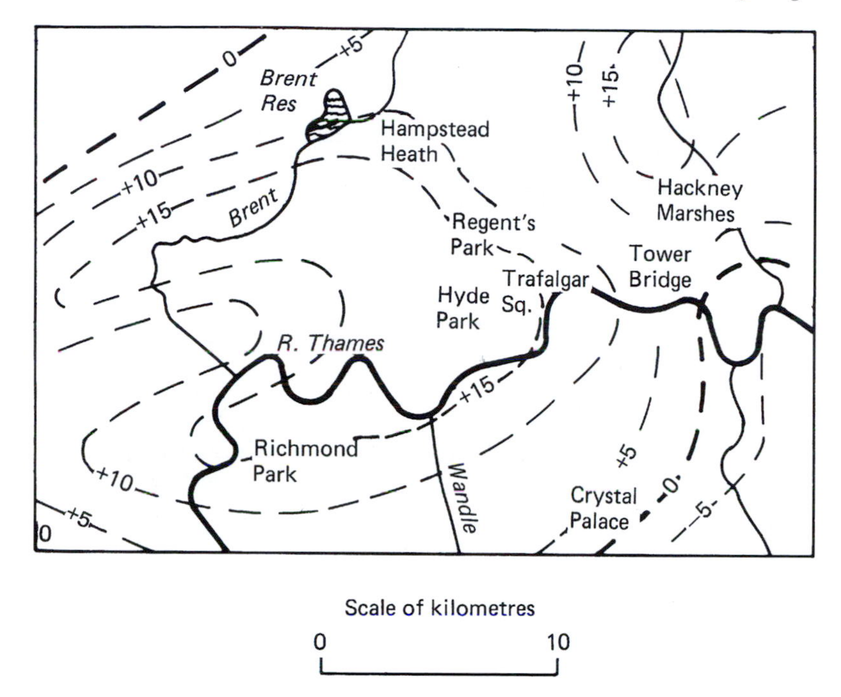

Figure 7.12 Changes in groundwater levels in the Chalk below London, 1965–1980 (contours in metres). *Source*: Marsh and Davies, 1983. Reproduced by kind permission of Thomas Telford.

The results are substituted in the Darcy expression and the coefficient of permeability (k) is derived from:

$$\frac{Q}{t} = \frac{(Ak)h}{l} = Aki \tag{7.33}$$

where i is the hydraulic gradient.

Determination of the permeability of fine sands and silts, as well as many rock types, is made by using a falling head permeameter (Fig. 7.13b). The sample is placed in the apparatus that then is filled with water to a certain height (h_1) in the standpipe. Then, the stopcock is opened and the water infiltrates through the sample, the height of the water in the standpipe falling to h_2. The times at the beginning (t_1) and end (t_2) of the test are recorded. These, together with the cross-sectional area (A) and length of sample (l), are then substituted in the following expression, which is derived from Darcy's law, to obtain the coefficient of permeability:

$$k = \frac{2.303al}{A(t_2 - t_1)} \times \log_{10}\left(\frac{h_1}{h_2}\right) \tag{7.34}$$

where A is the cross-sectional area of the standpipe. The permeability of clay cannot be measured using a permeameter; it must be determined indirectly, for example, from the consolidation test.

Bernaix (1969) examined the variations of permeability in rocks under stress by using a radial percolation test. A cylindrical specimen with an axial hole drilled in it is placed in a cell in which water can be pressurized. The flow is radial over almost the whole height of the sample and is convergent when the water pressure is applied to the outer faces of the specimen, and

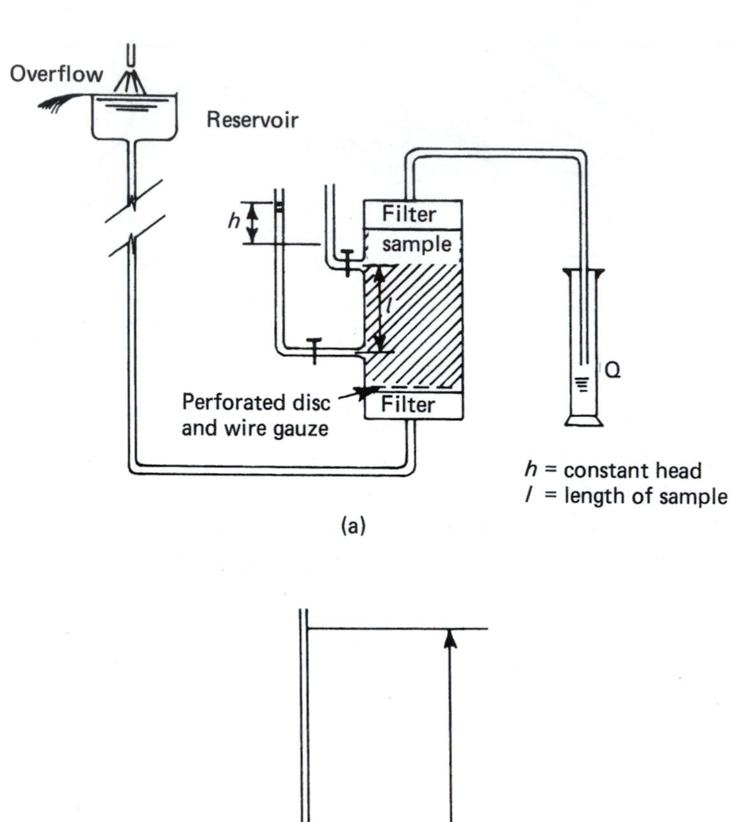

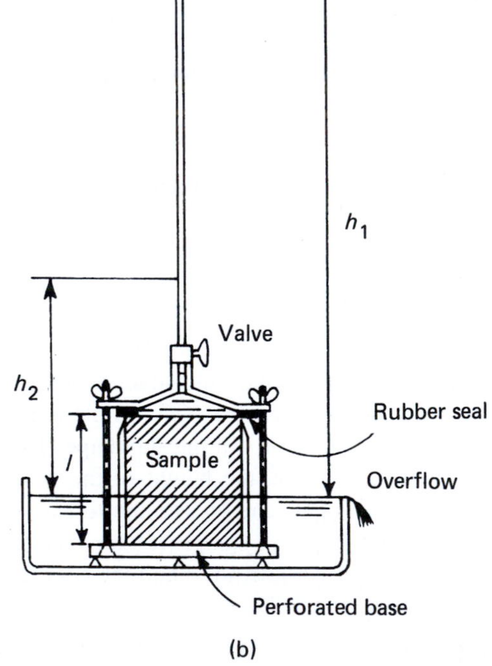

Figure 7.13 (a) The constant head permeameter. (b) The falling head permeameter.

divergent when the water is under pressure inside the specimen. For radial flow from a cylinder of unfractured rock with interconnected pores, the permeability (k) is given by:

$$k = \frac{Q\gamma_w}{2\pi l \delta p} \ln\left(\frac{R_2}{R_1}\right) \quad (7.35)$$

in which R_1 and R_2 are the radii of the inner and outer surfaces, respectively, Q the flow discharge, γ_w the unit weight of water, l the length of cylinder over which flow is occurring and δp is the difference between external and internal water pressures and is positive for convergent flow. Bernaix found that porous rocks remain more or less unaffected by pressure changes whereas fissured rocks exhibit far greater permeability in divergent flow than in convergent flow. Moreover, fissured rocks tend to exhibit a continuous increase in permeability as the pressure attributable to divergent flow is increased. Indeed, some amount of hydraulic fracturing may occur when testing in divergent flow, for example, some of the laminated sandstones tested in this way by Bell and Jermy (2002) underwent hydraulic fracturing. Goodman and Sundaram (1980) noted the same type of behaviour when they tested tuff, schist, sandstone and limestone.

7.9.2 *Assessment of permeability in the field*

An initial assessment of the magnitude and variability of *in situ* permeability can be obtained from tests carried out in boreholes as the hole is advanced. By artificially raising the level of water in a borehole (falling head test) above that in the surrounding ground, the flow rate from the borehole can be measured. However, in very permeable soils it may not be possible to raise the level of water in the borehole. Conversely, the water level in a borehole can be artificially depressed (rising head test) so allowing the rate of water flow into the borehole to be assessed. Wherever possible, a rising and a falling head test should be carried out at each required level and the results averaged.

In a rising or falling head test in which the piezometric head varies with time, the permeability is determined from the expression:

$$k = \frac{A}{F(t_2 - t_1)} \times \ln\left(\frac{h_1}{h_2}\right) \quad (7.36)$$

where h_1 and h_2 are the piezometric heads at times t_1 and t_2, respectively, A the inner cross-sectional area of the casing in the borehole and F an intake or shape factor (Anon., 1999). Where a borehole of diameter (D) is open at the base and cased throughout its depth, F = 2.75D. If the casing extends throughout the permeable bed to an impermeable contact, then F = 2D. The test procedure involves observing the water level in the casing at given times and recording the results, from which a graph of water level against time is constructed (Fig. 7.14).

The constant head method of *in situ* permeability testing is used when the rise or fall in the level of the water is too rapid for accurate recording (i.e. occurs in less than 5 min). This test normally is conducted as an inflow test in which the flow of water into the ground is kept under a sensibly constant head (e.g. by adjusting the rate of flow into the borehole so as to maintain the water level at a position on the inside of the casing near the top (Sutcliffe and Mostyn, 1983). The method is only applicable to permeable ground such as gravels, sands and broken rock, when there is a negligible or zero time for equalization. The rate of flow (Q) is measured

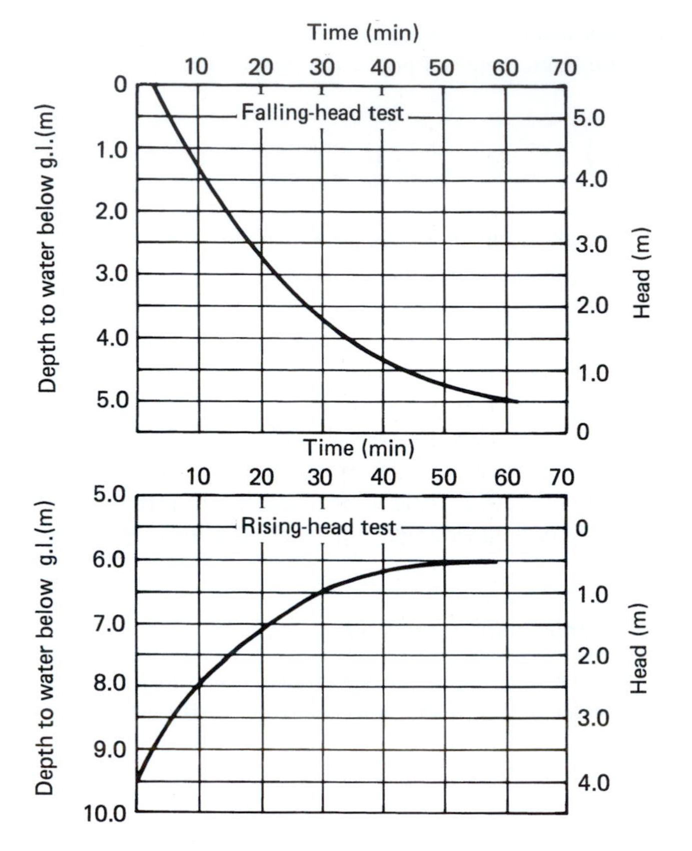

Figure 7.14 Rising and falling head permeability test curves.

once a steady flow into and out of the borehole has been attained over a period of some 10 min. The permeability (k) is derived form the following expression:

$$k = \frac{Q}{Fh_c} \tag{7.37}$$

where F is the intake factor and h_c the applied constant head.

The permeability of an individual bed of rock can be determined by a water-injection or packer test carried out in a drillhole (Bliss and Rushton, 1984; Brassington and Walthall, 1985). This is done by sealing off a length of uncased hole with packers and injecting water under pressure into the test section (Fig. 7.15). Usually, because it is more convenient, packer tests are carried out after the entire length of a hole has been drilled. Two packers are used to seal off selected test lengths and the tests are performed from the base of the hole upwards. The hole must be flushed to remove sediment prior to a test being performed. Sutcliffe and Mostyn (1983) found that using wireline pneumatic packers was much more effective and considerably quicker, especially in poor ground conditions, than using mechanical or hydraulic packers. With double packer testing the variation in permeability throughout the test hole can be determined. The rate of flow of water over the test length is measured under a range of constant pressures and recorded. The permeability is calculated from a flow-pressure curve (Fig. 7.16). Water generally is pumped into the test section at steady pressures for periods of 15 min. The test

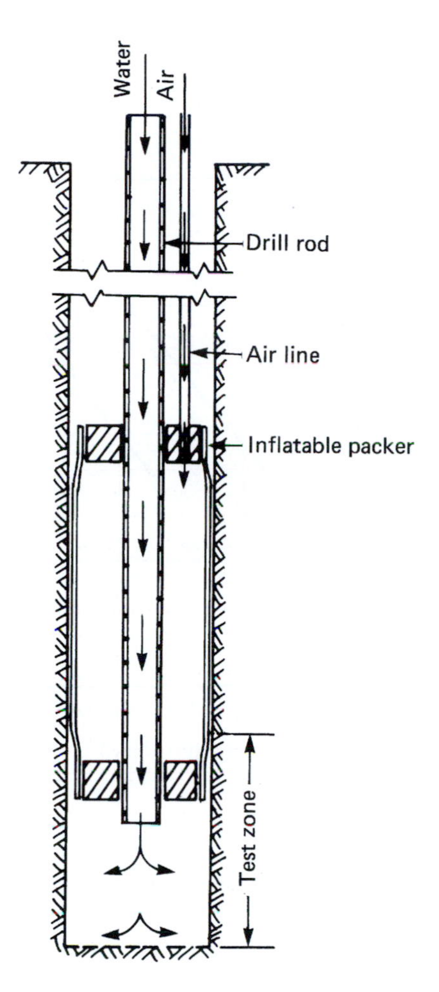

Figure 7.15 Packer test equipment.

usually consists of five cycles at successive pressures of 6, 12, 18, 12 and 6 kPa for every metre depth of the packer below the surface. The evaluation of the permeability from packer tests normally is based upon methods using a relationship of the form:

$$k = \frac{Q}{C_s rh} \tag{7.38}$$

where Q is the steady flow rate under an effective applied head h (corrected for friction losses), r the radius of the drillhole and C_s a constant depending upon the length and diameter of the test section.

The slug test involves injecting into or abstracting a known volume of water from a well and has been used to assess the transmissivity or storativity of confined aquifers. Immediately after injection or abstraction of a slug of water, the water level in the well has a certain elevation (h_o) above or below the initial elevation. As the level of the water rises or falls in the well, the difference (h) in elevation between that at the original head and that at a given time (t) is measured. The ratio of the measured head to that of the subsequent head (h/h_o) is plotted

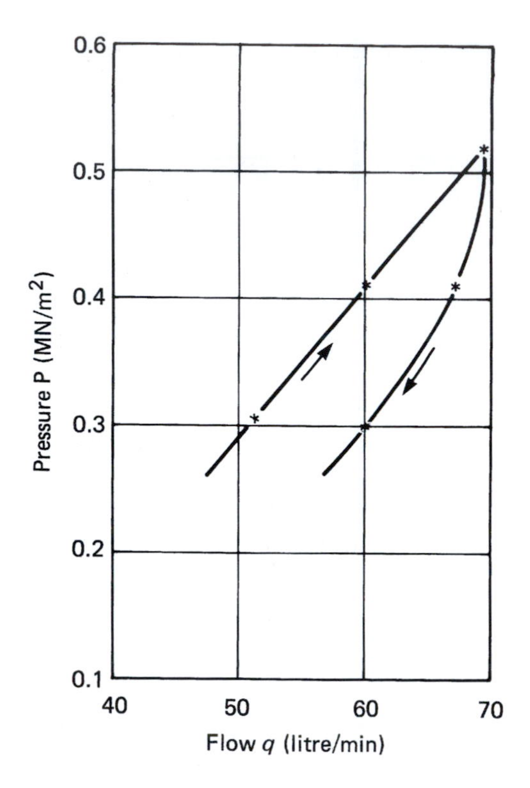

Figure 7.16 Flow pressure curve.

against time on semi-logarithmic paper, the latter being plotted on the logarithmic scale. This curve is drawn at the same scale as that of a series of standard curves, above which it is overlain and fitted to a type curve that has the same curvature (Fig. 7.17). The vertical time axis for $Tt/r_c^2 = 1.0$ is selected and the transmissivity (T) is obtained from:

$$T = \frac{1.0r_c^2}{t_1} \tag{7.39}$$

where r_c is the radius of the well casing. The value of the storage coefficient or storativity (S) is found from:

$$S = \frac{r_c^2 \mu}{r_s^2} \tag{7.40}$$

where μ is the value of the μ curve for the field data and r_s the radius of the well screen.

Field pumping tests allow the determination of the coefficients of permeability and storage, as well as the transmissivity, of a larger mass of ground than the aforementioned tests. A pumping test involves abstracting water from a well at known discharge rate(s) and observing the resulting water levels as drawdown occurs (Lovelock *et al.*, 1975; Anon., 1983). At the same time, the behaviour of the water table in the aquifer can be recorded in observation wells radially arranged about the abstraction well. There are two types of pumping test: the constant pumping rate aquifer test and the step-performance test. In the former test, the rate of discharge is constant whereas in a step-performance test there are a number of stages, each of equal length

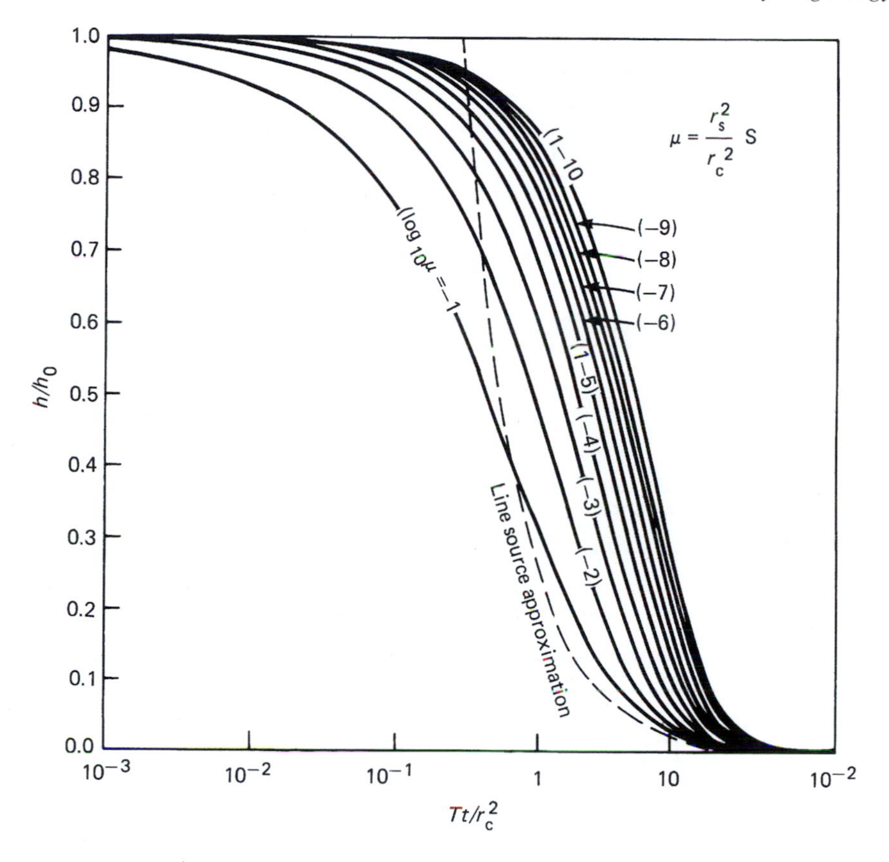

Figure 7.17 Type curves for the slug test in a well of finite diameter.

of time, but at different rates of discharge (Clark, 1977). The step-performance test usually is carried out before the constant-pumping-rate aquifer test. Yield drawdown graphs are plotted from the information obtained (Fig. 7.18). The hydraulic efficiency of the well is indicated by the nature of the curve(s); the more vertical and straighter they are, the more efficient the well.

7.9.3 Assessment of flow

A flowmeter log provides a record of the direction and velocity of groundwater movement in a drillhole. Flowmeter logging requires the use of a velocity-sensitive instrument, a system for lowering the instrument into the hole, a depth measuring device to determine the position of the flowmeter and a recorder located at the surface. The direction of flow of water is determined by slowly lowering and raising the flowmeter through a section of hole, 6–9 m in length and recording velocity measurements during both traverses. If the velocity measured is greater during the downward than the upward traverse, then the direction of flow is upward and vice versa. A flowmeter log made while a drillhole is being pumped at a moderate rate or by allowing water to flow if there is sufficient artesian head, permits identification of the zones contributing to the discharge. It also provides information on the thickness of these zones and the relative yield at that discharge rate. Because the yield varies approximately directly with the drawdown of water level in a well, flowmeter logs made by pumping, should be

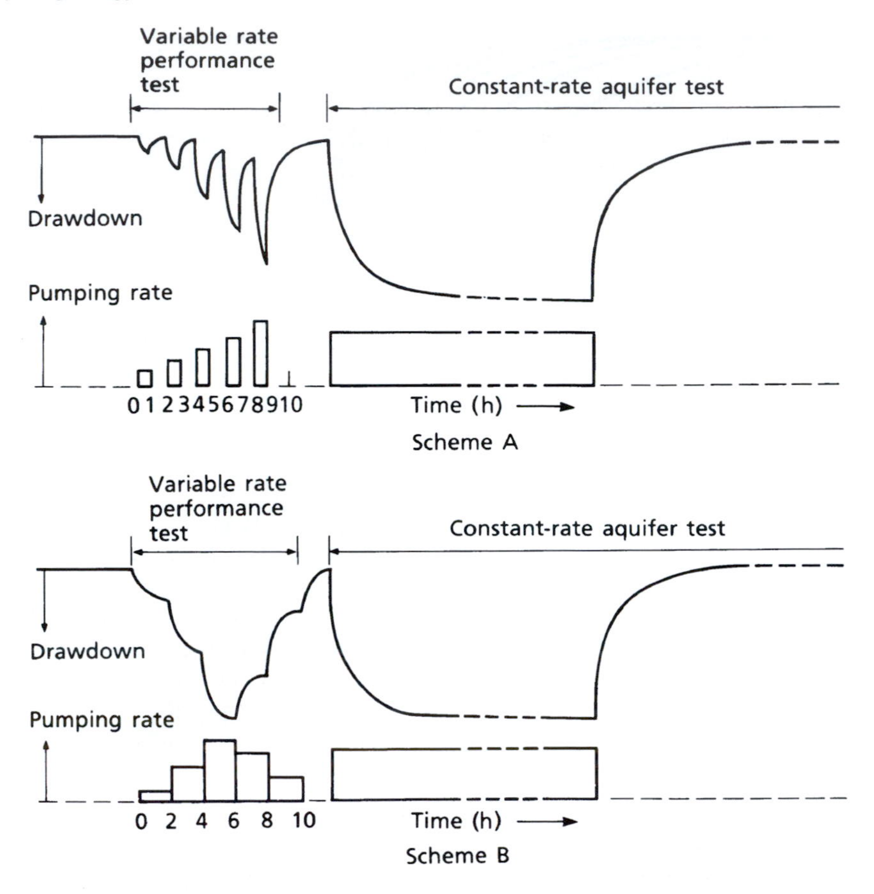

Figure 7.18 Pumping test procedures.

pumped at least at three different rates. The drawdown of water level should be recorded for each rate (Patten and Bennett, 1962). A low velocity flowmeter can be used to measure the vertical flow in a drillhole. For example, Bruzzi *et al.* (1983) used a flowmeter, which was almost entirely insensitive to horizontal flow effects and had a minimum measurable velocity of about 0.005 $m s^{-1}$, for measuring the flows in boreholes with slotted casing in alluvial ground. Readings were taken at every half metre of depth, which enabled the velocity curves to be checked with the stratigraphical log.

Use of an axial viewing head on television equipment enables the effects of horizontal or inclined flow to be observed directly in a drillhole. The flow movement is indicated by streamers suspended from beneath the television camera. Not only are they observed but they can be recorded on film or videotape. Also, visible tracers allow assessment of the flow direction to be made.

A number of different types of tracers have been used to investigate the movement of groundwater, and the interconnection between surface and groundwater resources. The ideal tracer should be easy to detect quantitatively in minute concentrations; it should not change the hydraulic characteristics of, or be adsorbed by the media through which it is flowing; it should be more or less absent from and should not react with the groundwater concerned; and it should have a low toxicity. The type of tracers in use include water soluble dyes that can be detected by colorimetry; sodium chloride or sulphate salts that can be detected chemically; and strong electrolytes that can be detected by electrical conductivity. Radioactive tracers also are used and one of their advantages is that they can be detected in minute quantities in water (Mather *et al.*, 1973). Such tracers should have a useful half-life and should present the minimum of hazard. For example, tritium is not the best of tracers because of its relatively long half-life. In addition, because it is introduced as tritiated water it is preferentially adsorbed by montmorillonite. Some of the more frequently used types of tracer are as follows:

Chemical	*Colorimetric*	*Nuclear*	*Stable isotopes*
Copper sulphate	Sodium fluorescein	Bromide 82	Deuterium
Sodium iodide	Methylene blue	Chromium 51	Helium 4
Dextrose		Cobalt 60	Oxygen 18
		Iodine 131	
		Phosphorus 32	
		Rubidium 86	
		Tritium	

When a tracer is injected via a drillhole into groundwater it is subject to diffusion, dispersion, dilution and adsorption. Dispersion is a result of very small variations in the velocity of laminar flow through porous media. Molecular diffusion is probably negligible unless the velocity of flow is unusually low. Even if these processes are not significant, flow through an aquifer may be stratified or concentrated along discontinuities. Therefore, a tracer may remain undetected unless the observation drillholes intersect these discontinuities.

Turkmen *et al.* (2002) used two dye tracers and salt in an attempt to trace seepage losses through karstic limestone at Kalecik Dam, Turkey. Their investigation revealed the seepage paths beneath the dam and spillway, and that some fill material occupying voids and enlarged discontinuities was being removed. It was recommended that a new grout curtain should be constructed beneath the spillway in order to control seepage loss.

The vertical velocity of water movement in a drillhole can be assessed by using tracers. A tracer is injected at the required depth and the direction and rate of movement is monitored by a probe (Tate *et al.*, 1970). For instance, Ineson and Gray (1963) determined the velocity of groundwater flow in a drillhole by recording the rate of movement of the peak of an injected saline slug with an electrical conductivity probe. Changes in the form of the electrical conductivity profile indicate variations in the pattern of groundwater flow, due to inflow or outflow from surrounding rocks. From a study of the differences between the original and superimposed profiles, quantitative assessment of loss from, or inflow of groundwater into a well can sometimes be derived. However, Patten and Bennett (1962) maintained that brine tracing is a more troublesome means of monitoring flow than investigating it with a flowmeter. Furthermore, the technique is not exact enough to take measurements at precise depths within a drillhole, cannot produce satisfactory results when used where water is entering the drillhole and does not allow a number of measurements to be taken in rapid succession. Nonetheless, Patten and Bennett noted that a relatively high degree of accuracy can be achieved when a significant length of drillhole is used for measurement and that brine tracing can indicate the direction of flow at velocities that are too low to be recorded by a flowmeter.

Determination of permeability can be done by measuring the time it takes for a tracer to move between two test holes. Like pumping tests, this tracer technique is based on the assumption that the aquifer is homogeneous and that observations taken radially at the same distance from the well are comparable (Keeley and Scalf, 1969). This method of assessing permeability requires that injection and observation wells are close together thereby avoiding excessive travel time and that the direction of flow is known so that observation holes are correctly sited. Several observation holes improve the chances of the tracer flowing into one of them but increase costs. Since the tracer flows through the aquifer with an average interstitial velocity (v_i), then:

$$v_i = \frac{kh}{nL} \tag{7.41}$$

where k is the coefficient of permeability, n the porosity, h the difference in height between water levels in the test holes and L the distance between them. However, v_i can also be obtained from:

$$v_i = \frac{L}{t} \tag{7.42}$$

where t is the time of travel over the distance involved, hence:

$$k = \frac{nL^2}{ht} \tag{7.43}$$

In the point dilution method, a tracer is introduced into the test hole and thoroughly mixed with the water therein. As water flows into and out of the test hole, measurements are taken of tracer concentration. Analysis of the resulting dilution curve provides an indication of the groundwater velocity. The velocity of groundwater also can be recorded by measuring the rate of dilution of a tracer in observation wells (Malkki and Vihuri, 1983).

Flow nets provide a graphical representation of the flow of water through the ground and indicate the loss of head involved (Fig. 7.19). They also provide data relating to the changes in head velocity and effective pressure that occur, for example, in a foundation subjected to flowing groundwater conditions. In other words, where the flow lines of a flow net move closer together this indicates that the flow increases, although their principal function is to indicate the direction of flow. The equipotential lines indicate equal losses in head or energy as the water flows through the ground, so that the closer they are, the more rapid is the loss in head. Hence, a flow net can provide quantitative data related to a flow problem, for instance, seepage

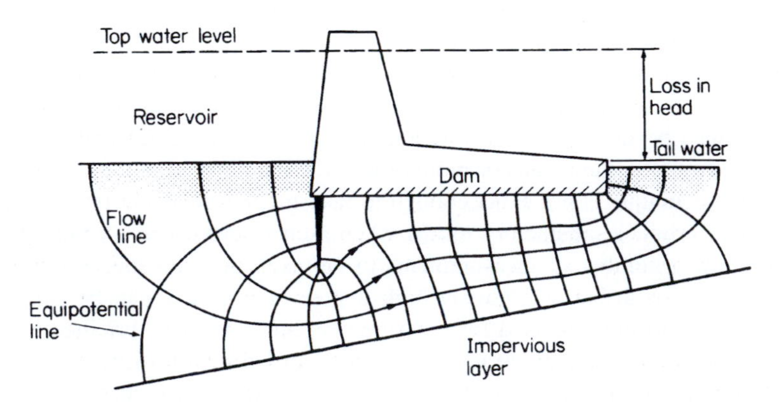

Figure 7.19 Flow net beneath a concrete gravity dam with a cut-off at the heel, showing 17 equipotential drops and four flow channels.

pressures can be determined at individual points within the net. It is possible to estimate the amount of water flowing through a soil from a flow net (Cedergren, 1986). If the total loss of head and the permeability of the soil are known, then the quantity of water involved can be calculated by using Darcy's law. However, it is not really as simple as that, for the area through which the water flows usually varies, as does the hydraulic gradient, since the flow paths vary in length. By using the total number of flow paths (f) the total number of equipotential drops (d) and the total loss of head (i_t) together with the permeability (k) in the following expression:

$$Q = \frac{k i_t f}{d} \tag{7.44}$$

the quantity of water flow (Q) can be estimated.

7.10 Water quality and uses

In any evaluation of water resources, the quality of the water is of almost equal importance to the quantity available. In other words, the physical, chemical and biological characteristics of water are of major importance in determining whether or not it is suitable for domestic, industrial or agricultural use. However, the number of major dissolved constituents in groundwater is quite limited and natural variations are not as large as might be expected. Lastly, the quality of water in the zone of saturation reflects that of the water that has percolated to the water table and the subsequent reactions between water and rock that occur. The factors that influence the solute content include the original chemical quality of water entering the zone of saturation, the distribution, solubility, exchange capacity and exchange selectivity of the minerals involved in reaction, the porosity and permeability of the soil and rock masses involved, and the flow path of the water. Of critical importance in this context is the residence time of the water since this determines whether there is sufficient time for dissolution of minerals to proceed to the point where the solution is in equilibrium with the reaction. Residence time depends on the rate of groundwater movement and this usually is very slow beneath the water table.

The quality of groundwater often compares favourably with that from surface sources and the groundwater, especially from deep aquifers, can be remarkably pure. However, this does not mean that the quality can be relied upon. In fact, groundwater can undergo cyclic changes in quality, and natural variations in groundwater quality can occur with depth, and soil and rock type. Thus, the purity of groundwater should not be taken for granted and untreated water should not be put directly into supply.

The quality of water in the zone of saturation is affected by the fluctuation of the water table. In particular, if the water table occurs at shallow depth, then losses by transpiration, and possibly evaporation, will increase when it rises. This means that the salt content increases. Conversely, when the water table is lowered this may cause lateral inflow from surrounding areas with a consequent change in salinity.

The uppermost layers of soil and rock act as purifying agents. In the soil, organisms such as fungi and bacteria attack pathogenic bacteria, as well as reacting with certain other harmful substances. The other important factor in purifying groundwater is the filtering action of soil and rock. This depends on the size of the pores, the proportion of argillaceous and organic matter present, and the distance travelled by the water involved. Unconfined water at shallow depth is highly susceptible to pollution but at greater depth recharge is by water that has been partially or wholly purified.

There is a frequent tendency for the salt content of groundwater to increase with depth. The reasons for this are, first, the greater the depth at which the groundwater occurs, the slower its movement and so it is less likely to be replaced by other water, especially that infiltrating from the surface. Second, the longer residence time of the water provides more time for reaction with the host rock and so more material goes into solution until an equilibrium condition is attained. Also, connate or fossil water may occur at greater depth. As the character of groundwater in an aquifer frequently changes with depth, it is possible at times to recognize zones of different quality of groundwater. For example, Elliot *et al.* (2001) recognized three major hydrogeochemical zones in the Chalk of Yorkshire, England. With increasing depth, cation exchange reaction increases in importance, and there is a gradual replacement of calcium and magnesium in the water by sodium. Any nitrates present near the surface of an aquifer invariably decrease with depth. On the other hand, sulphates tend to increase with depth. However, at appreciable depth, sulphates are reduced, which produces a low sulphate–high bicarbonate water. The chloride content also tends to increase with depth. In fact, with increasing depth groundwater may become non-potable, due to its high chloride content. Most highly saline, chlorine groundwater (not associated with evaporites) occurring at depth, where groundwater circulation is restricted, is connate or fossil water.

7.10.1 Physical characteristics of water

Temperature, colour, turbidity, odour and taste are the most important physical properties of water in relation to water supply. Groundwater only undergoes appreciable fluctuations in temperature at shallow depth, beneath which temperatures remain relatively constant. In fact, the depth at which temperatures are more or less uniform occurs at about 10 m in the tropics, increasing to about 20 m in polar regions, although rock type, elevation, precipitation and wind can produce significant local deviations. Below the zone of surface influence, groundwater temperatures increase by approximately 1 °C for every 30 m of depth, that is, in accordance with the geothermal gradient. The colour of groundwater may be attributable to organic or mineral matter carried in solution. For example, light to dark brown discolorations can occur in groundwater that has been in contact with peat or other organic deposits. Brownish discoloration also can result if groundwater that contains dissolved ferrous iron is exposed to the atmosphere. This leads to insoluble ferric hydroxides being formed. Turbidity of groundwater is caused mainly by the presence of clay and silt particles derived from the aquifer. Oxidation of dissolved ferrous iron to form insoluble ferric hydroxides also contributes towards turbidity. The natural filtration that occurs when groundwater flows through unconsolidated deposits largely removes such material from groundwater. Tastes and odours may be derived from the presence of mineral matter, organic matter, bacteria or dissolved gases.

7.10.2 Biological characteristics of water

Although groundwater usually is of good quality, contamination can occur for a number of reasons such as inadequate sanitary completion of wells, siting wells too near to on-site sanitation, leaking sewers and disposal of sewage sludge on land. For example, Craun (1985) reported that untreated or inadequately treated groundwater was responsible for 51 per cent of all outbreaks of waterborne disease and 40 per cent of all waterborne illness in the United States between 1971 and 1982. Contaminated, untreated or inadequately disinfected groundwater caused 65 per cent of the waterborne outbreaks and 60 per cent of waterborne illness in non-community and individual water systems as compared with 32 per cent of the outbreaks and 31 per cent of the illness in community water systems. In fact, the control of microbiological contamination is of paramount importance as the transmission of waterborne diseases through contaminated drinking water is a major source of morbidity and mortality worldwide, particularly in developing countries. According to Pedley and Howard (1997), around 80 per cent of the population in rural and semi-urban communities in developing countries use groundwater as a source of drinking water. They maintained that although the threat of cholera and typhoid has been reduced significantly in developing countries, they have been supplemented by protozoal infections such as cryotosporidiosis and viral infections. On the other hand, micro-organisms in suitable conditions may prove useful in the biodegradation of some contaminants in groundwater (West and Chilton, 1997).

Standard tests used to determine the safety of water for drinking purposes involve identifying whether or not bacteria belonging to the coliform group are present. One of the reasons for this is that this group of bacteria (*Escherichia coli* in particular) is relatively easy to recognize. Since the whole coliform group is foreign to water, a positive *E. coli* test indicates the possibility of bacteriological contamination. If *E. coli* are present, then it is possible that the less numerous or harmful pathogenic bacteria, which are much more difficult to detect, also are present. On the other hand, if there are no *E. coli* in a sample of water, then the chances of faecal contamination and of pathogens being present generally are regarded as negligible. The results of coliform tests are reported in terms of the most probable number (MPN) of coliform group organisms present in a given volume of water. Viruses in groundwater are more critical than bacteria in that they tend to survive longer and some viruses are more resistant to disinfectant. In addition, one virus unit (one plague-forming unit in a cell culture) may cause an infection when ingested. By contrast, ingestion of thousands of pathogenic bacteria may be required before clinical symptoms are developed. Fortunately, groundwater generally is free from pathogenic bacteria and viruses, except perhaps from very shallow aquifers. Micro-organisms can be carried by groundwater but tend to attach themselves by adsorption to the surfaces of clay particles. In fine-grained soils, bacteria generally move less than a few metres, but they can migrate much larger distances in coarse-grained soils or discontinuous rocks. The maximum rate of travel of bacterial pollution appears to be about two-thirds that at which the groundwater is moving. Viruses that retain all their characteristics for more than 50 days may migrate 250 m or more in soils where organic matter is present to supply a food source. The recommended safe distances between domestic wells and sources of pollution in non-karstic terrains are indicated in Table 7.6. Nonetheless, biological pollution of groundwater generally does not occur because the soil represents a fairly effective filter between a source of pollution and the water table. Pathogenic bacteria, viruses and other micro-organisms not native to the subsurface environment generally do not multiply underground and eventually die.

Table 7.6 Recommended safe distance between domestic water wells and sources of pollution

Source of pollution	*Distance*(m)
Septic tank	15
Cess pit	45
Sewage farms	30
Infiltration ditches	30
Percolation zones	30
Pipes with watertight joints	3
Other pipes	15
Dry wells	15

Source: Hamill and Bell, 1986.

Most cases of contamination result form poor well construction, from over-abstraction, or are associated with aquifers that possess large pores such as gravel deposits, or open discontinuities such as some limestones. Both afford connection between surface water, which may be polluted, and groundwater. In cavernous or fissured limestone the distances travelled may be several kilometres.

7.10.3 Chemical characteristics of water

The chemical elements present in groundwater are derived from precipitation that infiltrates into the ground, from organic processes that go on in the soil, and from the breakdown of minerals in the rocks through which the groundwater flows (Table 7.7). The solution of carbonates, especially calcium and magnesium carbonate, is due principally to the formation of weak carbonic acid in the soil horizons where CO_2 is dissolved by soil water. Calcium in sedimentary rocks is derived from calcite, aragonite, dolomite, anhydrite and gypsum. In igneous and metamorphic rocks, calcium is supplied by the feldspars, pyroxenes and amphiboles, and the less common minerals such as apatite and wollastonite.

Because of its abundance, calcium is one of the most common ions in groundwater. When calcium carbonate is attacked by carbonic acid, bicarbonate is formed. Calcium carbonate and bicarbonate are the dominant constituents found in the zone of active circulation and for some distance under the cover of younger strata. The normal concentration of calcium in groundwater ranges from 10 to $100\,mg\,l^{-1}$. Such concentrations have no effect on health and it has been suggested that as much as $1000\,mg\,l^{-1}$ may be harmless.

Magnesium, sodium and potassium are less common cations, and sulphate and chloride and, to some extent, nitrate are less common anions, although in some groundwaters the latter may be present in significant concentrations. Dolomite is the common source of magnesium in sedimentary rocks. The rarer evaporite minerals such as epsomite, kierserite, kainite and carnallite are not significant contributors. Olivine, biotite, hornblende and augite are among those minerals that make significant contributions in the igneous rocks, and serpentine, talc, diopside and tremolite are amongst the metamorphic contributors. Despite the higher solubilities of most of its compounds (magnesium sulphate and magnesium chloride are both very soluble), magnesium usually occurs in lesser concentrations in groundwaters than calcium. Common concentrations of magnesium range from about 1 to $40\,mg\,l^{-1}$, and concentrations above $100\,mg\,l^{-1}$ are rarely encountered.

Sodium does not occur as an essential constituent of many of the principal rock-forming minerals, plagioclase feldspar being the exception. Plagioclase is the primary source of most sodium in groundwaters; in areas of evaporitic deposits halite is important. Sodium salts are highly soluble and will not precipitate unless concentrations of several thousand parts per million are reached. The only common mechanism for removal of large amounts of sodium ions from water is through ion exchange, which operates if the sodium ions are in great abundance. The conversion of calcium bicarbonate to sodium bicarbonate accounts for the removal of some sodium ions from sea water that has invaded freshwater aquifers. This process is reversible. All groundwaters contain measurable amounts of sodium, up to $20\,mg\,l^{-1}$ being the most common concentrations.

Common sources of potassium are the feldspars and micas of the igneous and metamorphic rocks. Potash minerals such as sylvite occur in certain evaporitic sequences but their contribution is not important. Although the abundance of potassium in the earth's crust is similar to that of sodium, its concentration in groundwaters is usually less than one-tenth that of sodium. Most groundwaters contain less than $10\,mg\,l^{-1}$. Like sodium, potassium is highly soluble and therefore is not easily removed from water except by ion exchange.

Table 7.7 Examples of groundwater quality from different types of rock masses (in $mg\,l^{-1}$)

Rock type	*TDS*	*Ca*	*Mg*	HCO_3	*Na*	*K*	*Cl*	SO_4	*Fe*	SiO_4	*Location*
Granite	223	27	6.2	93	9.5	1.4	5.2	32	1.6	39	Maryland
Rhyolite	148	12	2.2	80	6.8	0.6	2.0	0.1	1.1	39	N. Carolina
Gabbro	359	32	16	203	25	1.1	13	10	0.06	56	N. Carolina
Basalt	505	62	28	294	24	–	37	30	–	30	Hyderabad, India
Diorite	346	72	4.1	114	10	2.8	6.5	115	0.04	22	N. Carolina
Syenite	80	9.5	2.3	38	2.8	0.6	2.1	2.8	0.14	19	New York
Andesite	70	12	0.5	38	1.8	2.6	–	6.3	–	8.9	Idaho
Quartzite	52	1.6	5.8	18	2.8	–	9.9	2.0	–	8	South Africa
Marble	236	39	10	162	2.7	0.3	3.8	2.4	0.03	9.9	Alabama
Schist	221	27	5.7	138	16	0.7	2.5	9.6	0.11	21	Georgia
Gneiss	135	19	5.1	39	4.4	3.2	5.8	30	0.09	13	Connecticut
Sandstone		60	60				22	38			Northumberland
Sandstone	210	40	12	67	7.6	0.4	19	26	–	12	Worcestershire
Sandstone	439	65	38	326	44	–	63	79	–	14	France
Arkose	101	9.6	1.9	38	5.1	–	1.8	7.4	0.2	35	Colorado
Greywacke	553	74	20	381	34	1.2	2.7	26	0.62	12	New York
Limestone	247	48	5.8	168	4	0.7	0.1	4.8	0.05	8.9	Florida
Limestone	720	124	28	460	14	3	18	57	0.22	9.2	Tennessee
Chalk	384	115	5	152	10.2	1.2	20	39	–	1.1	Hertfordshire
Chalk	491	125	9	–	18	4	24	150	–		Lincolnshire
Dolostone	546	67	39	390	7.6	0.4	–	17	–	24	South Africa
Gypsum	2480	636	43	143	16.1	0.9	24	1570	–	29	New Mexico
Lignite	2580	74	53	702	624	5.4	25	1080	0.9	11	North Dakota
Shale	260	29	16	126	12	1.1	12	22	0.02	16	New Jersey
Shale	1100	123	70	539	61	2.2	3.5	283	1.3	19	Ohio
Alluvium	371	45	20	207	16	2.6	17	35	0.05	25	Nevada
Glacial deposits	548	86	27	33	5.1	3	6	60	–	24	Minnesota

Sedimentary rocks such as shales and clays may contain pyrite or marcasite from which sulphur can be derived. Most sulphate ions probably are derived from the solution of calcium and magnesium sulphate minerals found in evaporitic sequences, gypsum and anhydrite being the most common. The concentration of sulphate ions in water can be affected by sulphate-reducing bacteria, the products of which are hydrogen sulphide and carbon dioxide. Hence, a decline in sulphate ions frequently is associated with an increase in bicarbonate ions. Concentration of sulphate in groundwaters usually is less than $100\,mg\,l^{-1}$ and may be less than $1\,mg\,l^{-1}$ if sulphate-reducing bacteria are active.

The chloride content of groundwaters may be due to the presence of soluble chlorides from rocks, saline intrusion, connate and juvenile waters, or contamination by industrial effluent or domestic sewage. In the zone of circulation, the chloride ion concentration normally is relatively small and it is a minor constituent in the earth's crust, sodalite and apatite being the only igneous and metamorphic minerals containing chlorite as an essential constituent. Halite is one of the principal mineral sources. As with sulphate ions, the atmosphere probably makes a significant contribution to the chloride content of surface waters, these, in turn, contributing to the groundwaters. Usually, the concentration of chloride in groundwater is less than $30\,mg\,l^{-1}$ but concentrations of $1000\,mg\,l^{-1}$ or more are common in arid regions.

Nitrate ions generally are derived from the oxidation of organic matter with a high protein content. Their presence may indicate a pollution source and their occurrence usually is associated with shallow groundwater sources. Concentrations in fresh water generally do not exceed $5\,mg\,l^{-1}$, although in rural areas where nitrate fertilizer is liberally applied, concentrations may exceed $600\,mg\,l^{-1}$.

Although silicon is the second most abundant element in the earth's crust and is present in almost all the principal rock-forming minerals, its low solubility means that it is not one of the most abundant constituents of groundwater. Groundwater generally contains between 5 and $40\,mg\,l^{-1}$, although high values may be recorded in water from volcanic rocks.

Iron forms approximately 5 per cent of the earth's crust and is contained in a great many minerals in rocks, as well as occurring as ore bodies. Most iron in solution is ionized. Normally, iron occurs in groundwater in the form of Fe^{2+}, $Fe(OH)_3$ or $FeOH^+$. When iron occurs in concentrations of $1\,mg\,l^{-1}$ or above, it does so in the ferrous state. If such groundwater is exposed to air, the iron is oxidized to the ferric condition and precipitated as ferric hydroxide. Concentrations of ferrous iron in groundwater are typically in the range $1–10\,mg\,l^{-1}$.

Ion exchange affects the chemical nature of groundwater. The most common natural cation exchangers are clay minerals, humic acids and zeolites. The replacement of Ca^{2+} and Mg^{2+} by Na^+ may occur when groundwater moves beneath argillaceous rocks into a zone of more restricted circulation. This produces soft water. Changes in temperature–pressure conditions may result in precipitation (e.g. a decrease in pressure may liberate CO_2 causing the precipitation of calcium carbonate).

Certain dissolved gases such as oxygen and carbon dioxide affect groundwater chemistry. Others affect the use of water, for example, hydrogen sulphide in concentrations more than $1\,mg\,l^{-1}$ renders water unfit for consumption because of the objectionable odour. Methane coming out of solution may accumulate and present a fire or explosion hazard. The minimum concentration of methane in water sufficient to produce an explosive methane–air mixture above the water from which it has escaped depends on the volume of air into which the gas evolves. Theoretically water containing as little as $1–2\,mg\,l^{-1}$ of methane can give rise to an explosion in a poorly ventilated air space.

Several minor elements are a matter of concern because of their toxic effects. For instance, arsenic should not exceed $0.01–0.1\,mg\,l^{-1}$ in drinking water, neither barium nor copper should exceed 1 mg l^{-1}, neither chromium nor lead 0.05 mg l^{-1}, cadmium 0.01 mg l^{-1} or mercury 0.002 mg l^{-1} (Bell, 1998). Although boron is essential to healthy plant growth, it is injurious if present in groundwater in significant quantities. However, the sensitivities of plants to boron vary widely. For example, citric trees may be damaged by as little as 0.5 mg l^{-1} if soil drainage is good. The normal amount contained by groundwater varies from 0.01 to 1.0 mg l^{-1}. On the other hand, deficiencies of certain trace elements can be injurious to health. For instance, iodine deficiency can lead to the development of goitre in humans. Groundwater commonly is richer in iodine than surface water and high baseline concentrations may occur in carbonate aquifers. Edmunds and Smedley (1996) found that the median concentration in the Chalk aquifer of the London Basin was 32 μg l^{-1}, which is about four times higher than the non-carbonate aquifers in England. Dental caries may result from low concentrations of fluorine, that is, less than about 0.5 mg l^{-1} whereas a chronic exposure to higher concentrations can result in dental fluorosis (around 1.5 mg l^{-1}) and skeletal fluorosis (around 4 mg l^{-1}). However, Warnakulasuriya *et al.* (1990) suggested that in hot dry climates, where water consumption is higher, then dental fluorosis could develop when the fluorine content in groundwater is less than 0.3 mg l^{-1}. The occurrence of high amounts of aluminium in drinking water has been linked with the development of Alzheimer's disease. The World Health Organization (Anon., 1993) recommended that the maximum concentration of Al in drinking water should be 0.2 mg l^{-1}, however, they noted that this should not be regarded as a health-based guideline.

The total dissolved solids (TDS) in a sample of water includes all solid material in solution, whether ionized or not. Water for most domestic and industrial uses should contain less than 1000 mg l^{-1} and the TDS content of water for most agricultural purposes should not be above 3000 mg l^{-1}. Groundwater has been classified by Hem (1985) according to its TDS content as follows:

Fresh	Less than 1000 mg l^{-1}
Slightly saline	1000–3000 mg l^{-1}
Moderately saline	3000–10 000 mg l^{-1}
Very saline	10 000–35 000 mg l^{-1}
Briny	Over 35 000 mg l^{-1}

The hardness of water relates to its reaction with soap and to the scale and encrustations that form in boilers and pipes where water is heated and transported. It is attributable to the presence of divalent metallic ions, calcium and magnesium being the most abundant in groundwater. Waters derived from limestone or dolostone aquifers containing gypsum or anhydrite may have 200–300 mg l^{-1} hardness or more. Water for domestic use should not contain more than 80 mg l^{-1} total hardness. Hardness (H_T) generally is expressed in terms of the equivalent of calcium carbonate, hence:

$$H_T = \text{Ca} \times \left(\frac{\text{CaCO}_3}{\text{Ca}}\right) + \text{Mg} \times \left(\frac{\text{CaCO}_3}{\text{Mg}}\right) \tag{7.45}$$

where H_T, Ca and Mg are measured in mg l^{-1} and the ratios in equivalent weights. This equation reduces to:

$$H_T = 2.5\text{Ca} + 4.1\text{Mg} \tag{7.46}$$

The degree of hardness in water has been described as follows:

Description	*Hardness* ($mg\,l^{-1}$ as $CaCO_3$)	
	(after Sawyer and McCarthy, 1967)	*(after Hem, 1985)*
Soft	Below 75	Below 60
Moderately hard	75–150	61–120
Hard	150–300	121–180
Very hard	Over 300	Over 180

A relationship between water hardness and cardiovascular disease has long been suspected, and numerous investigations have been undertaken (e.g. Masironi, 1979; Piispanen, 1993). The British Committee on Medical Aspects of Food Policy (Anon., 1994) did find a relationship between water hardness and cardiovascular disease mortality but noted that the size of the effect was small and most clearly seen at levels of water hardness less than $170\,mg\,l^{-1}$ (as $CaCO_3$). Several suggestions have been advanced to explain the connection with water. These include the potential for Ca and/or Mg to protect against some forms of cardiovascular disease; that some trace elements that are more prevalent in hard water, may be beneficial; and that many metals are more soluble in soft water and hence may promote cardiovascular disease.

7.10.4 Water use

Water for human consumption must be free from organisms and chemical substances in concentrations large enough to affect health adversely (Table 7.8). In addition, drinking water should be aesthetically acceptable in that it should not possess unpleasant or objectionable taste, odour, colour or turbidity. For example, the maximum concentration of chloride in drinking water is $250\,mg\,l^{-1}$, primarily for reasons of taste. Again for reasons of taste, and also to avoid staining, the recommended maximum concentration of iron in drinking water is $0.3\,mg\,l^{-1}$. Oxidation of manganese in groundwater can produce black stains and therefore the maximum permitted concentration of manganese for domestic water is $0.05\,mg\,l^{-1}$. The pH value of drinking water should be close to 7 but treatment can cope with a range of 5–9. Beyond this range treatment to adjust the pH to 7 becomes less economical.

The bacterial quality of drinking water that has been established by the United States Environmental Protection Agency requires that tests reveal no more than one total coliform organism per 100 ml as the arithmetic mean of all water samples examined per month, with no more than 4 per 100 ml in more than any sample if the number of samples is less than 20 per month, or no more than 4 per 100 ml in 5 per cent of the samples if the number of samples is more than 20 per month. Samples should be taken at frequent intervals of time. These vary according to the population supplied, for example, one sample per month if less than 1000 individuals are supplied to 300 per month if 1 000 000 are supplied. In addition, drinking water should contain less than 1 virus unit per 400 to 4000 *l*. Viruses can be eliminated by effective chlorination. The water should be free from suspended solids in which viruses can harbour and thereby be protected against disinfectant.

The quality of water required in different industrial processes varies appreciably, indeed it can differ within the same industry. Nonetheless, salinity, hardness and silica content are three parameters that usually are important in terms of industrial water. On the one hand, water used in the textile industries should contain a low amount of iron, manganese and other heavy metals likely to cause staining. Hardness, total dissolved solids, colour and turbidity also must be low. On the other hand, the quality of water required by the chemical industry varies widely depending on the processes involved. Similarly, water required in the pulp and paper industry is governed by the type of products manufactured. Groundwater generally is preferable to surface water since it shows less variation in chemical and physical quality.

Table 7.8 Standards for drinking water

	European standards (1971)	*International standards* (1972)
*Biology**		
Coliform bacteria	Nil	Nil
Escherichia coli	Nil	Nil
Streptococcus faecalis	Nil	Nil
Clostridium perfringens	Nil	
Virus	Nil	
	Less than 1 plaque-forming unit	
	per litre per examination in 10	
Microscopic organisms	1	
	of water	
	Nil	
Radioactivity		
Overall α radioactivity	<3 pCi l^{-1}	<pCi l^{-1}
Overall β radioactivity	<30 pCi l^{-1}	
Chemical elements	(mg l^{-1})	(mg l^{-1})
Pb	<0.1	<0.1
As	<0.05	<0.05
Se	<0.01	<0.01
Hexavalent Cr	<0.05	<0.05
Cd	<0.01	<0.01
Cyanides (in CN)	<0.05	<0.05
Ba	<1.00	<1.00
Cyclic aromatic hydrocarbon	<0.20	
Total Hg	<0.01	<0.01
Phenol compounds (in phenol)	<0.001	<0.001–0.002
NO_3 recommended	<50	
acceptable	50–100	
not recommended	>100	
Cu	<0.05	0.05–1.5
Total Fe	<0.1	0.10–1.0
Mn	<0.05	0.10–0.5
Zn	<5	5.00–15
Mg if SO_4 > 250 mg l^{-1}	<30	<30
if SO_4 < 250 mg l^{-1}	<125	<125
SO_4	<250	250–400
H_2S	0.05	
Cl recommended	<200	
acceptable	<600	
NH_4	<0.05	
Total hardness	2–10 meq/l	2–100 meq/l
Ca	75 200 mg l^{-1}	75–200 mg l^{-1}

Table 7.8 (Continued)

F	*In the case of fluorine the limits depend upon air temperature*			
Mean annual maximum day time temperature (°C)	*Lower limit* ($mg\,l^{-1}$)	*Optimum* ($mg\,l^{-1}$)	*Upper limit* ($mg\,l^{-1}$)	*Unsuitable* ($mg\,l^{-1}$)
10–12	0.9	1.2	1.7	2.4
12.1–14.6	0.8	1/1	1.5	2.2
14.7–17.6	0.8	1.0	1.3	2.0
17.7–21.4	0.7	0.9	1.2	1.8
21.5–26.2	0.7	0.8	1.0	1.6
26.3–32.6	0.6	0.7	0.8	1.4

Source: WHO, 1993; CEC, 1980; Brown *et al.*, 1972.

Note
* No 100 ml sample to contain *E. coli* or more than 10 coliform.

7.11 Wells

The commonest way of recovering groundwater is to sink a well and lift water from it (Fig. 7.20). The most efficient well is developed so as to yield the greatest quantity of water with the least drawdown and the lowest velocity in the vicinity of the well. The specific capacity of a well is expressed in litres of yield per metre of drawdown when the well is being pumped. It is indicative of the relative permeability of the aquifer. Location of a well obviously is important if an optimum supply is to be obtained and a well site should always be selected after a careful study of the geological setting.

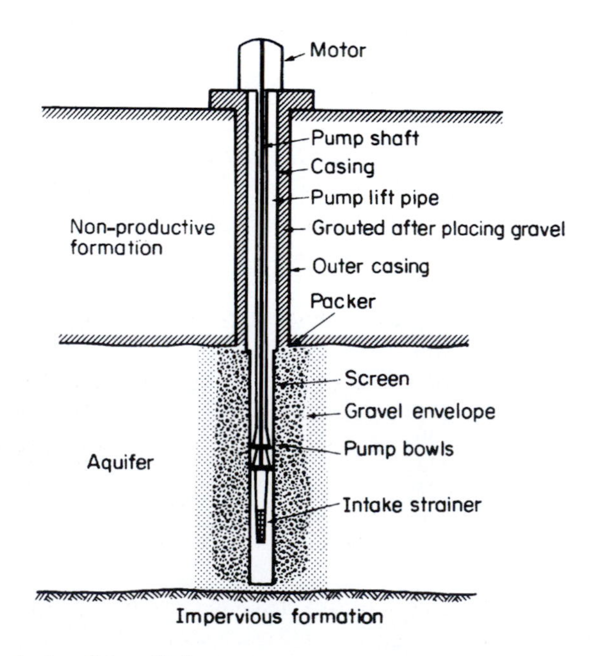

Figure 7.20 Gravel-packed well installation.

Completion of a well in an unconsolidated formation requires that it be cased so that the surrounding deposits are supported. Sections of the casing must be perforated to allow the penetration of water from the aquifer into the well, or screens can be used. The casing should be as permeable as, or more permeable than, the deposits it confines.

Wells that supply drinking water should be properly sealed. However, an important advantage of groundwater is its comparative freedom from bacterial pollution. Abandoned wells should be sealed to prevent aquifers being contaminated.

The yield from a well in granular material can be increased by surging that removes the finer particles from the zone about the well. Water supply from wells in rock masses can be increased by driving galleries or adits from the bottom of deep wells. Yields from rock formations also can be increased by fracturing the rocks with explosives, or with fluid pumped into the well under high pressure, or in the case of carbonate rocks, such as chalk, by using acid to enlarge the discontinuities. The use of explosives in sandstones has led to increases in yield of up to 40 per cent whilst acidification of wells in carbonate rocks has increased yields by over 100 per cent.

From the hydrological point of view, the long-term yield of a well, according to Hamill and Bell (1986), depends upon the following factors:

1 The annual rate of groundwater recharge. This determines the rate of flow in the aquifer and therefore the amount of water available for abstraction.
2 The location of the well within an aquifer. There may be some advantage in siting a well near to a recharge area, so that a surface water resource is diverted underground to augment aquifer flow by induced infiltration. This could increase well yields. Alternatively, a well could be sited near the discharge area of an aquifer with the objective of diverting as much of the natural discharge as possible to the well. Neither of these options should be undertaken without a careful evaluation of the possible consequences (see Section 7.14.2). An alternative strategy may be to exploit the storage of an aquifer without interfering with the flow of groundwater through the aquifer. In this case, the wells should be sited some distance from the areas of natural recharge and discharge, so that it takes some significant time for the pumping effects to reach them. An indication of the delays to be expected is given by the response time, which is SL^2/T, where S is the storage coefficient, T the transmissivity and L the length of the flow path from the well to the zone of affected natural discharge.
3 The permeability of an aquifer in the area surrounding a well. The higher the permeability the easier it is for water to flow to a well during periods of abstraction.
4 The thickness of an aquifer at a well site. A well should be located where the saturated thickness is greatest. If the aquifer material has only a small variation in permeability, then transmissivity increases with increasing saturated thickness. This again facilitates flow to a well.
5 The location and orientation of any faults or notable discontinuities. These may act as preferred flow channels and greatly increase the flow to a well. Many wells in relatively impermeable material have been successful as a result of flow through secondary permeability features. However, a fault may also act as a barrier to flow if it is filled with impermeable gouge material or if the throw of the fault places the aquifer against an impermeable stratum. Faults or other potential boundaries should be evaluated using pumping test analysis techniques.
6 The location of wells with respect to any features that may jeopardize the quality and quantity of discharge or the groundwater resource as a whole. It is important that a well should be able to operate at its design discharge and drawdown without the quantity and

quality of abstracted water being adversely affected, and without the abstraction having an adverse effect upon ecological or environmental features or resulting in the derogation of existing groundwater sources. For example, induced infiltration can be used to augment a groundwater supply, but if the surface source is polluted, it also can mean that the pumping rates of wells in the area have to be severely limited. In coastal aquifers saline intrusion may cause a progressive decrease in the water abstracted from wells and if the discharge becomes unacceptably saline the wells may have to be abandoned.

The development of wells for groundwater supplies in rural areas, especially in developing countries, frequently is of major importance. In such areas, fractured zones and weathered horizons, the latter may be extensive and of appreciable depth, of granitic or gneissic masses may provide sufficient water for small communities. In addition, the fractured and weathered contact zones of thick dykes and sills may yield similar quantities. For example, Bell and Maud (2000) referred to the four categories of well yield recognized by the South African Department of Water Affairs. These are high well yields (over $3.0\,l\,s^{-1}$) that are suitable for the supply of medium- to large-scale water schemes supporting small towns and/or small- to medium-scale irrigation schemes. Moderate well yields (0.5–$3.0\,l\,s^{-1}$) are suitable for reticulation schemes for villages, clinics and schools. Low well yields (0.1–$0.5\,l\,s^{-1}$) can be used to supply a hand pump for a non-reticulating water supply for a small community and stock watering purposes. Lastly, very low well yields (less than $0.1\,l\,s^{-1}$) only provide marginal supplies.

7.12 Safe yield

The yield of a surface water resource generally is related to a return interval, being defined as the steady supply that can be maintained through a drought of specified severity. For example, the yield that could be maintained through a 1 in 50-year (2 per cent) drought is greater than that which could be maintained through a 1 in 100-year (1 per cent) drought. Thus, yield is not an absolute quantity, but a variable that depends upon the specified frequency of occurrence of the limiting drought conditions. Yield also may be defined as the steady supply that could be maintained through the worst drought on record. In this case, the severity of the limiting drought conditions depends upon the rainfall recorded in the years preceding the drought period, the length of the record available for analysis, and chance that a short record may contain a particularly severe drought, whilst a much longer record elsewhere may not. For this reason, the first definition is to be preferred.

The abstraction of water from the ground at a greater rate than it is being recharged leads to the water table being lowered and upsets the equilibrium between discharge and recharge. The concept of safe yield has been used for many years to express the quantity of water that can be withdrawn from the ground without impairing the aquifer as a water source. Draft in excess of safe yield is overdraft. Estimation to the safe yield is a complex problem that must take into account the climatic, geological and hydrogeological conditions. As such, the safe yield is likely to vary appreciably with time. Nonetheless, the recharge–discharge equation, the transmissivity of the aquifer, the potential sources of contamination and the number of wells in operation must all be given consideration if an answer is to be found. The safe yield (G) often is expressed as follows:

$$G = P - Q_s - E_T + Q_g - \Delta S_g - \Delta S_s \qquad (7.47)$$

where P is the precipitation on the area supplying the aquifer, Q_s the surface stream flow over the same area, E_T the evapotranspiration, Q_g the net groundwater inflow to the area, ΔS_g the change in groundwater storage and ΔS_s the change of surface storage. With the exception of precipitation, all the terms of this expression can be subjected to artificial change. The equation cannot be considered an equilibrium equation or solved in terms of mean annual values. It can be solved correctly only on the basis of specified assumptions for a stated period of years.

The transmissivity of an aquifer may place a limit on the safe yield even though this equation may indicate a potentially large draft. This can only be realized if the aquifer is capable of transmitting water from the source area to wells at a rate high enough to sustain the draft. Where contamination of the groundwater is possible, then the location of wells, their type and the rate of abstraction must be planned in such a way that conditions permitting contamination cannot be developed.

Once an aquifer is developed as a source of water supply, then effective management becomes increasingly necessary if it is not to suffer deterioration. Management should not merely be concerned with the abstraction of water but also should consider its utilization, since different qualities of water can be put to different uses. Pollution of water supply is most likely to occur when the level of the water table has been so lowered that all the water that goes underground within a catchment area drains quickly and directly to the wells. Such lowering of the water table may cause reversals in drainage so that water drains from rivers into the groundwater system (induced infiltration) rather than the other way around. This river water may be contaminated.

A progressive decline in groundwater level is frequently an indication of future management problems, since this is often the consequence of exceeding the safe or perennial yield of an aquifer. The result is likely to be an unacceptable pumping lift, a reduced yield due to the restricted drawdown available and possibly a deterioration in water quality. The latter often occurs as a result of old highly mineralized water being drawn from deep within an aquifer into the wells (upconing), or through induced infiltration or saline intrusion. These problems can necessitate a reduction in the output of a wellfield, or even its abandonment. Falling groundwater levels also may result in the loss of natural marshes and wetlands, with potentially serious agricultural and ecological implications.

7.13 Artificial recharge

Artificial recharge may be defined as an augmentation of the natural replenishment of groundwater storage by artificial means. Its main purpose is water conservation, often with improved quality as a second aim, for example, soft river water may be used to reduce the hardness of groundwater. Artificial recharge therefore is used for reducing overdraft, for conserving and improving surface run-off and for increasing available groundwater supplies (Brown and Signor, 1973).

The suitability of a particular aquifer for artificial recharge must be investigated. For instance, it must have adequate storage and the bulk of the water recharged should not be lost rapidly by discharge into a nearby river. The hydrogeological and groundwater conditions must be amenable to artificial replenishment. An adequate and suitable source of water for recharge must be available. The source of water for artificial recharge may be storm run-off, river or lake water, water used for cooling purposes, industrial waste water, or sewage water. Many of these sources require some kind of pre-treatment.

Interaction between artificial recharge and groundwater may lead to precipitation, for example, of calcium carbonate, and iron and magnesium salts, resulting in lower permeability. Nitrification or denitrification, and possibly even sulphate reduction, may occur during the early stages of infiltration. Bacterial action may lead to the development of sludges, which reduce the rate of infiltration.

Artificial recharge may be accomplished by various surface-spreading methods utilizing basins, ditches or flooded areas; by spray irrigation; or by pumping water into the ground via vertical shafts, horizontal collector wells, pits or trenches. The most widely practised methods are those of water spreading, which allow increased infiltration to occur over a wide area when the aquifer outcrops at or near the surface. Therefore these methods require that the ground has a high infiltration capacity. In the basin method water is contained in a series of basins formed by a network of dykes, constructed to take maximum advantage of local topography. Care has to be taken in spray irrigation not to flush salts and nutrients from the soil into the groundwater. In regions with hot dry climates, there may be a danger of excessive evaporation of recharge water leading to salinization. Recharge wells are employed most frequently when the aquifer to be recharged is deep or confined, or there is insufficient space for recharge basins. Some methods by which the amount of recharge can be estimated have been reviewed by Simmers (1998). Credible estimates of the amount of recharge are essential for the assessment of sustainable water resources in semi-arid regions that rely on groundwater supplies for domestic use and for irrigation.

Zoetemann *et al.* (1975) referred to an artificial recharge scheme developed in the dunes of the Veluwe area southeast of the Zuydee Zee in the Netherlands. Recharge is by means of wells that store river water in the ground during periods of peak flow, the river water being treated before it is used for recharge.

7.14 Groundwater pollution

Pollution can be regarded as impairment of water quality by the addition of chemical, physical or biological substances, or of heat, to an extent that does not necessarily create an actual public health hazard, but that it does adversely affect such waters for domestic, agricultural or industrial use. In fact, most of man's activities have a direct, and usually adverse effect upon water quality. In 1964, it was estimated that more than 900 million people were without a public water service of any kind, while in the developing countries some 500 million people per year were affected by water-borne or water-related diseases (Pickford, 1979). By 1980, it was estimated that 1320 million people (57 per cent) of the developing world (excluding China) were without a clean water supply, while 1730 million (75 per cent) were without adequate sanitation. At least 30 000 people per day die in the developing world because they have inadequate water and sanitation facilities. Hence, the control of water pollution in these countries is of vital importance, and some disastrous environmental results have occurred through lack of attention to the problem.

The greatest danger of groundwater pollution is from surface sources such as excessive application of fertilizers, leaking sewers, polluted streams, mining and mineral wastes, domestic and industrial waste, and so on. Areas, with thin soil cover or where an aquifer is exposed as in the recharge area, are the most critical from the point of view of pollution potential. Any possible source of pollution in these areas should be carefully evaluated, both before and after well construction, and the viability of groundwater protection measures considered (Raucher, 1983). Changes in land use may pose new threats to water quality. Obvious precautions against pollution are to locate the wells as far from any potential source of contamination as possible and to fence off the tops of wells so that animals cannot defecate adjacent to the well. Good well design and construction also is important. However, it should be appreciated that the slow rate of travel of pollutants in underground strata means that a case of pollution may go undetected for a number of years. During this period, a large part of the aquifer may become polluted and cease to have any potential as a source of water (Selby and Skinner, 1979).

Attenuation of a pollutant occurs as it moves through the ground as a result of four major processes. First, the soil has an enormous purifying power due to the communities of bacteria and fungi that live in the soil. These organisms are capable of attacking pathogenic bacteria and also can react with certain other harmful substances. Second, as water passes through fine-grained porous media suspended impurities are removed by filtration. Third, some substances react with minerals in the soil/rock, and some are oxidized and precipitated from solution. Adsorption also may occur in argillaceous or organic material. Fourth, dilution and dispersion of a pollutant may lead to its concentration being reduced until it eventually becomes negligible at some distance from the source.

The form of a pollutant is an important factor with regard to its susceptibility to the various purifying processes. For instance, pollutants that are soluble, such as fertilizers and some industrial wastes, are not affected by filtration. Metal solutions may not be susceptible to biological action. Solids, on the other hand, are amenable to filtration provided that the transmission media is not coarse-grained, fractured or cavernous. Non-soluble liquids such as hydrocarbons generally are transmitted through porous media, although some fraction may be retained in the host material. Usually, however, the most dangerous forms of groundwater pollution are those that are miscible with the water in an aquifer.

Generally, the concentration of a pollutant decreases as the distance it has travelled through the ground increases. Thus, the greatest pollution potential exists for wells tapping shallow aquifers that intersect or lie near ground level. An aquifer that is exposed or overlain by a relatively thin formation in the recharge area is also at risk, particularly when the overlying material consists of sand or gravel. Conversely, deeply buried aquifers overlain by relatively impermeable shale or clay beds usually can be considered to have low pollution potential and are less prone to severe contamination. When assessing pollution potential, an additional consideration is the way in which the pollutant enters the ground. If it is evenly distributed over a large area, then its effect probably will be less than the same amount of pollutant concentrated at one point. Concentrated sources are most undesirable because the self-cleansing ability of the soil/rock in the area concerned is likely to be exceeded. As a result, a raw pollutant may be able to enter an aquifer and travel some considerable distance from the source before being reduced to a negligible concentration.

A much greater hazard exists when the pollutant is introduced into an aquifer beneath the soil such as can happen with faulty well design, and from poorly maintained domestic septic tanks and soakaways. Leaking pipelines and underground storage tanks have resulted in the abandonment of several million domestic and other water wells in the United States. This situation is most critical when the pollutant is added directly to the zone of saturation, because the horizontal component of permeability is usually much greater than the vertical component. For instance, intergranular seepage in the unsaturated zone may have a typical velocity of less than $1\,\mathrm{m\,year^{-1}}$, whereas lateral flow beneath the water table may be as much as $2\,\mathrm{m\,day^{-1}}$ under favourable conditions. In karst or weathered limestone areas, in particular, pollutants may be able to travel quickly over large distances. For instance, Hagerty and Pavoni (1973) observed the spread of polluted groundwater over a distance of 30 km through limestone in approximately 3 months. They also noted that the degradation, dilution and dispersion of harmful constituents were less effective than in surface waters. Consequently, a pollutant beneath the water table can travel a much greater distance before significant attenuation occurs.

Biological pollution in the form of micro-organisms, viruses and pathogens is quite common. Not all bacteria are harmful, on the contrary, many are beneficial and perform valuable functions, such as attacking and biodegrading pollutants as they migrate through the soil.

The bacteria that normally inhabit the soil thrive at temperatures of around 20 °C. Bacteria that are of animal origin prefer temperatures of around 37 °C and so generally die quite quickly outside the host body. Consequently, it is sometimes erroneously assumed that all pathogenic bacteria cannot survive long underground. However, Brown *et al.* (1972) pointed out that some bacteria in groundwater can have a life span of up to four years. It generally is assumed that bacteria move at a maximum rate of about two-thirds the groundwater velocity. Since most groundwaters only move a few metres per year, the distances travelled are usually quite small and, in general, it is unusual for bacteria to spread more than 33 m from the source of the pollution. Of course, in openwork gravel, cavernous limestone or fissured rock, the bacteria may spread over distances of many kilometres.

Viruses are parasites that require a host organism before they can reproduce but, as noted, some are capable of retaining all their characteristics for 50 days or more in other environments. Because of their very small size, viruses are not greatly affected by filtration but are prone to adsorption, particularly when the pH is around 7. Although viruses are capable of spreading over distances 250 m or more, the more usual distance is 20–30 m.

7.14.1 Faulty well design and construction

Perhaps the greatest risk of groundwater contamination arises when pollutants can be transferred from the ground surface to an aquifer. In this case, the purifying processes that take place within the soil are bypassed and attenuation of the pollutant is reduced. Therefore, it is not surprising that one of the most common causes of groundwater pollution is poor well design, construction and maintenance. Indeed, a faulty well can ruin a high-quality groundwater resource.

During the construction of a well there is an open hole that affords a direct route from the surface to the aquifer. Apart from the possibility of surface run-off entering the hole during periods of rainfall, various unsanitary materials such as non-potable quality water, drilling fluids, chemicals, casings, screens, and so on, are deliberately placed in the hole. This provides an ideal opportunity for chemical and bacteriological pollution to occur, but lasting damage can be avoided if the well is completed, disinfected and pumped within a short space of time. Under these circumstances, most of the potentially harmful substances are discharged from the well. However, if there is a lengthy delay in completing the well, the possibility of pollution increases. Similarly, if a well is constructed in a cavernous or highly permeable formation, the chances of recovering all the harmful materials introduced during construction are decreased, while the possibility of pollution at nearby wells is increased.

As far as the well structure itself is concerned, Campbell and Lehr (1973) noted the following ways by which pollution can take place:

1. Via an opening in the surface cap or seal, through seams or welds in the casing, or between the casing and the base of a surface mounted pump.
2. As a result of reverse flow through the discharge system.
3. Through the disturbed zone immediately surrounding the casing.
4. Via an improperly constructed and sealed gravel pack.
5. As a result of settlement due to the inability of the basal formation to support the weight of the well structure, sand pumping or seepage resulting from reduced effectiveness of the surface cap or seal.
6. As a result of the grout or cementing material forming the seals failing through cracking, shrinking, etc.
7. Through breaks or leaks in the discharge pipes leading to scour and failure of the cement-grout seals.

When a drillhole penetrates only one aquifer, the principal concern is the transfer of pollution from the ground surface. However, with multiple aquifers there is the additional possibility of inter-aquifer flow. In such cases, each aquifer (or potential source of pollution) must be isolated using cement-grout seals in the intervening strata. Failure to do this, particularly when the well incorporates a gravel pack, provides a conduit through which water can be transferred from the ground surface and from one aquifer to another. This could be potentially disastrous where a shallow contaminated aquifer overlies a deeper unpolluted aquifer. In such a situation, pollution also could occur as a result of leakage under the bottom of the surface casing, or through the casing itself if it has seams, welded joints or is severely corroded.

Even after a well has been abandoned, it still provides a means of entry to an aquifer for pollutants and may even be a greater hazard than it was during its operational life. Abandoned wells make suitable receptacles for all kinds of wastes and refuges for vermin. In addition, as a well becomes progressively older, the strength and effectiveness of the casing and sanitary seals deteriorate, thereby increasing the possibility of potential for structural failure and pollution. If a well has been abandoned, any pollution that does happen might go unnoticed until detected at a nearby well. Accordingly, abandoned wells should be sealed to prevent the access of pollutants.

7.14.2 Induced infiltration

Induced infiltration occurs where a stream is connected hydraulically to an aquifer and lies within the area of influence of a well. When water is abstracted from a well, the water table in the immediate vicinity is lowered and assumes the shape of an inverted cone that is referred to as a cone of depression (Fig. 7.21). The steepness of the cone is governed by the soil/rock type(s), it being flatter in highly permeable materials. The size of the cone depends on the rate of pumping, equilibrium being achieved when the rate of abstraction balances the rate of recharge. However, if abstraction exceeds recharge, then the cone of depression increases in size. As the cone of depression spreads, water is withdrawn from storage over a progressively increasing area of influence and the groundwater levels about the well continue to be lowered. Some of the larger wells in the Chalk aquifer in England have cones with radii up to 1.5 km, and this figure may be exceeded by some of those in the Sherwood Sandstone aquifer. Eventually, such a situation leads to an aquifer being recharged by influent seepage of surface water, so that some proportion of the water abstracted from the well is obtained from a surface source. In effect, the surface source (e.g. a stream or streams) constitutes a recharge boundary and the situation is similar to that shown in Fig. 7.21. Assuming relatively uniform conditions, the flow in the cone generally conforms to Darcy's law, so that the flow towards the well is greatest on the side nearest the source where the gradients are steepest. As pumping continues, the proportion of water entering the cone of depression that is derived from the stream increases progressively.

If influent seepage of surface water is less than the amount required to balance the discharge from a well, the cone of depression spreads up and downstream until the drawdown and the area of the stream bed intercepted are sufficient to achieve the required rate of infiltration. If the stream bed has a high permeability, the cone of depression may extend over only part of the width of the stream. Conversely, if it has a low permeability the cone may expand across and beyond the stream. If pumping is continued over a prolonged period, a new condition of equilibrium is established with essentially steady flows. Most of the abstracted water then is derived from the surface source. Since stream bed infiltration occurs as a result of groundwater abstraction and would not occur otherwise, this is termed induced infiltration.

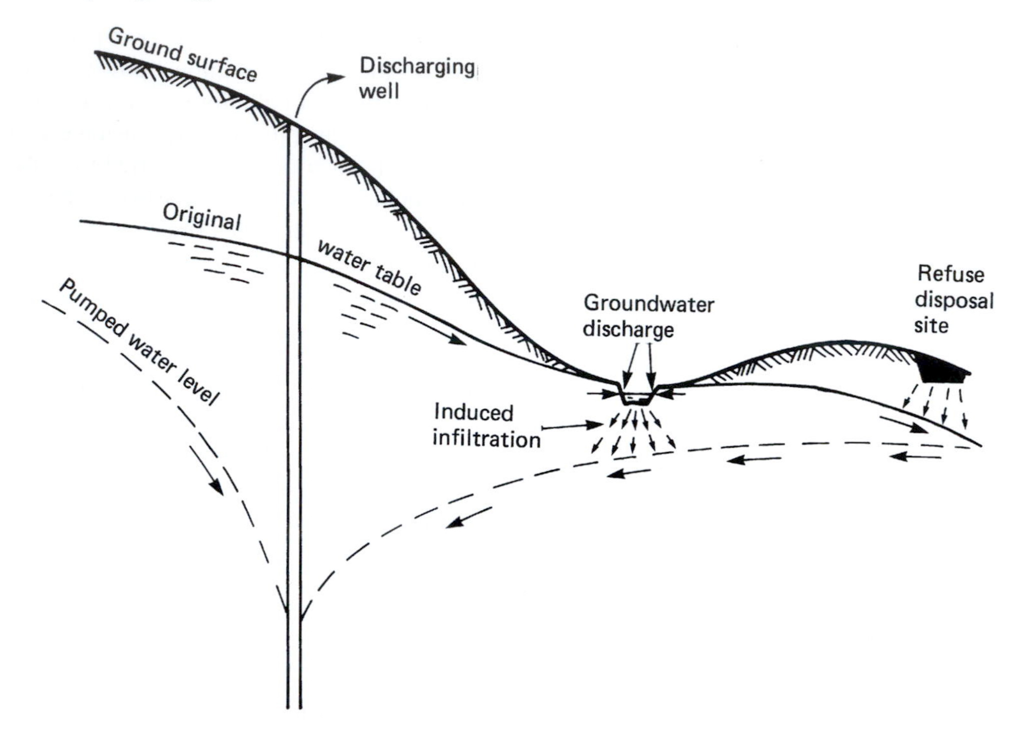

Figure 7.21 An example of induced infiltration caused by overpumping. The original hydraulic gradient over much of the area has been reversed so that pollutants can travel in the opposite direction, that is, towards the well. Additionally, the aquifer has become influent (i.e. water drains from the river into the aquifer) instead of effluent, as it was originally.

Induced infiltration is significant from the point of view of groundwater pollution in two respects. First, the new condition of equilibrium that is established may involve the reversal of the non-pumping hydraulic gradients, particularly when groundwater levels have been significantly lowered by groundwater abstraction (Fig. 7.21). This may result in any pollutants travelling in the opposite direction to that which they normally do. Second, surface water resources are often less pure than the underlying groundwater so induced infiltration introduces the danger of pollution. However, whether or not induced infiltration gives rise to pollution depends upon the quality of the surface water source, the nature of the aquifer, the quantity of infiltration involved and the intended use of the abstracted groundwater. Conversely, induced infiltration often represents a method of augmenting groundwater supplies. Accordingly, induced infiltration at one extreme can have potentially disastrous consequences or at the other can provide an addition to the groundwater resources of an area.

As far as estimating the proportion of well discharge that originates from a surface source is concerned, there are basically two approaches that can be adopted. The first uses field data to form a correlation between groundwater level and stream leakage, determined from the analysis of gauging station records. In most situations, this approach will not be feasible, so recourse must be made to one of the many theoretical techniques that are available. One of the first useful methods was outlined by Theis (1941), who considered a stream as an idealized straight line of infinite extent. By using an image well technique and assuming that the groundwater level under the stream did not change, Theis was able to show that:

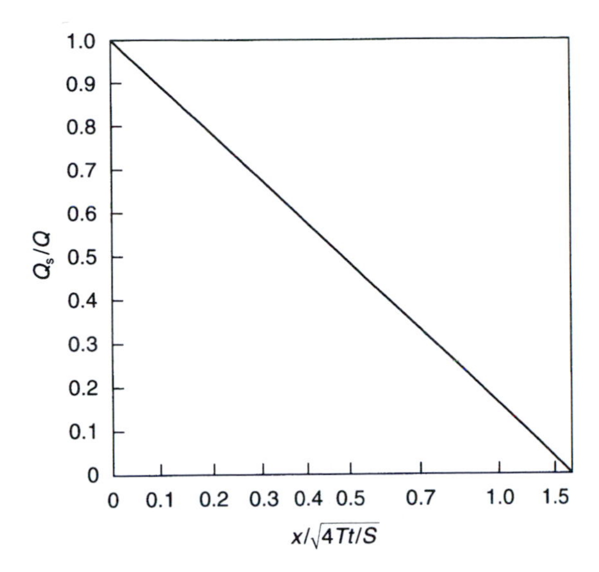

Figure 7.22 Graph for determining the proportion of well discharge derived from a nearby stream. x, the perpendicular distance from the stream to the well; T, the coefficient of transmissivity; S, the storage coefficient; t, the duration of pumping; and Q_S/Q, – the proportion derived from the stream.
Source: Glover and Balmer, 1954. Reproduced by kind permission of American Geophysical Union.

1 Over half the well discharge derived directly from a stream (or derived indirectly by preventing or diverting groundwater flow to a stream) always originates between points at a distance, x, upstream and downstream of the well, where x is the distance between the well and the stream. The proportion originating from beyond these two points diminishes rapidly with distance.
2 If there is no significant increase in recharge or decrease in evapotranspiration (which would significantly reduce the effect of the well on the stream) the proportion of water taken from streamflow varies widely and depends mainly upon the coefficient of transmissivity of the aquifer and the distance of the well from the stream.

Subsequently, Glover and Balmer (1954) produced a graph for determining the proportion of well discharge derived from a nearby stream (Fig. 7.22). Once the quantity and quality of the water derived from the surface source has been established, it is possible to assess whether or not this constitutes a significant threat to groundwater quality.

7.14.3 Landfill and groundwater pollution

A problem of increasing concern at the present day arises from the disposal of wastes in landfill sites (see Chapter 9). The decomposition of such waste produces liquids and gases that may present a health hazard by contamination of groundwater supply. They also can threaten the water resource potential of a region and therefore can represent an economic problem of considerable magnitude. Widespread deterioration of groundwater quality means increases in the cost of treatment prior to its use. When treatment costs become prohibitive, the volume of unusable water within the system increases and occupies storage space at the expense of potable supplies.

Leachate is formed when rainfall infiltrates a landfill and dissolves the soluble fraction of the waste, along with those soluble products formed as a result of the chemical and biochemical processes occurring within the decaying waste. If the rate of percolation is very slow, then degradation by natural processes may render these liquids innocuous. Leachates from landfill sites are prone to variations in viscosity that, in turn, influence their rate of flow. Domestic refuse is a heterogenous collection of material, much of which is capable of being extracted by water to give a liquid rich in organic matter, mineral salts and bacteria. The organic carbon content is especially important since this influences the growth potential of pathogenic organisms. Domestic refuse therefore has a pollution potential.

Generally, the conditions within a landfill are anaerobic, so leachates often contain high concentrations of dissolved organic substances resulting from the decomposition of organic material such as vegetable matter and paper. Recently emplaced wastes may have a chemical oxygen demand (COD) of around 11 600 $mg\,l^{-1}$ and a biochemical oxygen demand (BOD) in the region of 7250 $mg\,l^{-1}$. The concentration of chemically reduced inorganic substances like ammonia, iron and manganese varies according to the hydrology of the site and the chemical and physical conditions within the site. Barber (1982) listed the concentration of ammoniacal nitrogen in a recently emplaced waste as 340 $mg\,l^{-1}$, chloride as 2100 $mg\,l^{-1}$, sulphate as 460 $mg\,l^{-1}$, sodium as 2500 $mg\,l^{-1}$, magnesium as 390 $mg\,l^{-1}$, iron as 160 $mg\,l^{-1}$ and calcium as 1150 mg l^{-1}. Previously, Brown *et al.* (1972) had calculated that 1000 m^3 of waste can yield 1.25 tonnes of potassium and sodium, 0.8 tonnes of calcium and magnesium, 0.7 tonnes of chloride, 0.19 tonnes of sulphate and 3.2 tonnes of bicarbonate. Furthermore, Barber estimated that a small landfill site with an area of 1 hectare located in southern England could produce up to 8 m^3 of leachate per day, mainly between November and April, from a rainfall of 900 mm $year^{-1}$, assuming that evaporation is close to the average for the region and run-off is minimal. A site with an area ten times as large would produce a volume of effluent with approximately the same BOD load per year as that received by a small rural sewage treatment works. Accordingly, it can be appreciated that the disposal of domestic wastes in landfill sites can produce large volumes of effluent with a high pollution potential. For this reason, the location and management of these sites must be carefully controlled.

Gray *et al.* (1974) considered that the major criterion for the assessment of a landfill site as a serious risk was the presence of toxic or oily liquid waste, although sites on impermeable substrata often merited a lower assessment of risk, depending on local conditions. In this instance, serious risk meant that there was a serious possibility of an aquifer being polluted and not necessarily that there was a danger to life. As such, it may involve restricting the type of materials that can be disposed off at the site concerned. The range of toxic wastes varies from industrial effluents on the one hand to chemical and biological wastes from farms on the other. Unfortunately, the effects of depositing several types of waste together are unknown. Thus, site selection for waste disposal must first of all take into account the character of the material that is likely to be disposed of. Will this cause groundwater pollution that, because of its toxic nature, will give rise to a health hazard; although not toxic, will the material increase the concentrations of certain organic and inorganic substances to such an extent that the groundwater becomes unusable; will the wastes involved be inert and therefore give no risk of pollution?

Selection of a landfill site for a particular waste or a mixture of wastes involves a consideration of economic and social factors, as well as the hydrogeological conditions (Naylor *et al.*, 1978). As far as the latter are concerned, then argillaceous sedimentary, massive igneous and metamorphic rocks have low permeabilities, and therefore afford the most protection to water supply. By contrast, the least protection is offered by rocks intersected by open discontinuities or in which solution features are developed. In this respect, limestones and some sequences of volcanic rocks may be suspect. Granular material may act as a filter. The position of the water table is important as it determines whether wet or dry tipping can take place, as is the thickness of unsaturated material underlying a potential site (see Chapter 9).

There are two ways in which pollution by leachate can be tackled, first, by concentrating and containing or, second, by diluting and dispersing. Infiltration through granular ground of liquids from a landfill may lead to their decontamination and dilution. Hence, sites for disposal of domestic refuse can be chosen where decontamination has the maximum chance of reaching completion and where groundwater sources are located far enough away to enable dilution to be effective. Consequently, disposal can occur, according to Gray *et al.* (1974), at dry sites on granular material, which has a thickness of at least 15 m. Water supply sources should be located at least 0.8 km away from the landfill site. Sanitary landfills should not be located on discontinuous rocks unless overlain by 15 m of superficial deposits. The fill should be completely covered with clay at the end of each day.

As far as potentially toxic waste is concerned it is perhaps best to contain it. Gray *et al.* (1974) recommended that such sites should be underlain by at least 15 m of impermeable strata and that any source abstracting groundwater for domestic use and confined by such impermeable strata, should be at least 2 km away. Furthermore, the topography of the site should be such that run-off can be diverted from the landfill so that it can be disposed of without causing pollution of surface waters. Containment can be achieved if an artificial impermeable lining is provided over the bottom of the site. However, there is no guarantee that clay, soil-cement, asphalt or plastic linings will remain impermeable, for example, they may be ruptured by settlement. Therefore, migration of leachate from a landfill into the substrata will occur eventually, only the length of time before this happens being in doubt. In some instances, this will be sufficiently long for the problem of pollution to be greatly diminished.

In order to reduce the amount of leachate emanating from a landfill, it is advisable to construct cut-off drains around the site to prevent the flow of surface water into the landfill area. The leachate that originates from direct precipitation, or other sources, should be collected in a sump and either pumped to a sewer, transported away by tanker, or treated on site (Robinson, 1984; Bell *et al.*, 1996). Under no circumstances should untreated leachate be allowed to enter a surface watercourse, or be allowed to percolate to the water table if this is likely to pose a significant threat to quality.

As can be concluded from above, waste disposal sites can represent potential sources of groundwater pollution, the risk increasing when toxic chemicals are involved. Certain precautions should be undertaken. These include, first, the location of water wells up the hydraulic gradient, if possible, from any waste disposal site, or at least ensuring that wells are not directly down gradient of the source of pollution. A flow net may provide a suitable means of study initially, followed by tracer surveys if a serious risk is involved. Second, a well should be located so that the cone of depression or area of diversion to the well does not reach the source of pollution. In fact, the greater the distance is between a well and the source of pollution, the better. Third, a check should be maintained on land use, or changes in land use, in critical areas such as the recharge zone of an aquifer and in the vicinity of shallow wells. This may detect illegal dumping or other harmful practices. An estimate of the probable rainfall at the landfill site may give some idea of the potential volume of leachate that could be produced (Holmes, 1984). The dilution with water within an aquifer then can be calculated. Flow lines can be used to estimate the proportion of leachate that will arrive at supply wells and hence the quality of the abstracted water. While it is difficult to assess the effects of attenuation, dilution and dispersion on a pollutant, nevertheless such calculations are better than nothing. Lastly, an emergency plan should be available that can be implemented quickly if polluted water appears in monitoring or supply wells.

Groundwater pollution can be detected by a suitable groundwater monitoring scheme (see below). Monitoring of groundwater quality should be undertaken at points around the periphery of a disposal site, and between the site and any water supply wells. Over 20 drillholes may be required to identify the hydrogeology and migration plume at a particular site (Cherry, 1983). Typical indicators of deteriorating groundwater quality include increases in water hardness and in the concentration of sulphates and chlorides. Concentrations of free CO_2 greater than $20\,mg\,l^{-1}$ create corrosive conditions; nitrate in excess of $50\,mg\,l^{-1}$ is considered dangerous, while chloride levels of $200\,mg\,l^{-1}$ may cause taste problems in potable water. By the judicious use of routine water quality surveys, it may be possible to detect groundwater deterioration at an early stage. This should ensure that any water wells in the affected area are closed down before the supply becomes contaminated.

7.14.4 Saline intrusion

Although saline groundwater, originating as connate water or from evaporitic deposits, may be encountered, the problem of saline intrusion is specific to coastal aquifers. An interface exists near a coast between the overlying fresh groundwater and the underlying salt groundwater (Fig. 7.23). Excessive lowering of the water table along a coast leads to saline intrusion, the salt water entering an aquifer via submarine outcrops thereby displacing fresh water. However, the fresh water still overlies the saline water and continues to flow from the aquifer to the sea. In the past, the two groundwater bodies usually have been regarded as immiscible from which it was assumed that a sharp interface exists between them. In fact, there is a transition zone that may vary from 0.5 or so to over 100 m in width, although the latter figure may be atypically high. Herbert and Lloyd (2000) have provided a method of modelling saline intrusion in terms of the assessment of groundwater resources for small islands.

The shape of the interface is governed by the hydrodynamic relationship between the flowing fresh and saline groundwater. However, if it is assumed that hydrostatic equilibrium exists between the fresh and salt water, then the depth of the interface can be approximated by the Ghyben–Herzberg formula:

$$Z = \frac{\rho_w}{\rho_{sw} - \rho_w} \times h_w \qquad (7.48)$$

where ρ_{sw} is the density of sea water, ρ_w the density of fresh water and h_w the head of fresh water above sea level at the point on the interface (Fig. 7.23a). If ρ_{sw} is taken as $1025\,kg\,m^{-3}$ and ρ_w as $1000\,kg\,m^{-3}$, then $Z = 40/h_w$. Thus, at any point in an unconfined aquifer there is approximately 40 times as much fresh water below mean sea level as there is above it. However, the above expression implies that there is no flow but groundwater is invariably moving in coastal areas. Where flow is moving upward, near the coast, the relationship gives too small a depth to salt water but further inland where the flow lines are nearly horizontal the error is negligible. This relationship also can be applied to confined aquifers if the height of the water table is replaced by the elevation of the piezometric surface above mean sea level. If the aquifer overlies an impermeable stratum, this formation will intercept the interface and prevent any further saline intrusion (Fig. 7.23b).

The Ghyben–Herzberg relationship applies quite accurately to two-dimensional flow at right angles to a shoreline. When flow is three-dimensional, such as when a well is discharging near the coast, most formulae for the position of the interface and the radius of influence of the well are inaccurate. However, the problem is amenable to mathematical modelling (Volker and Rushton, 1982).

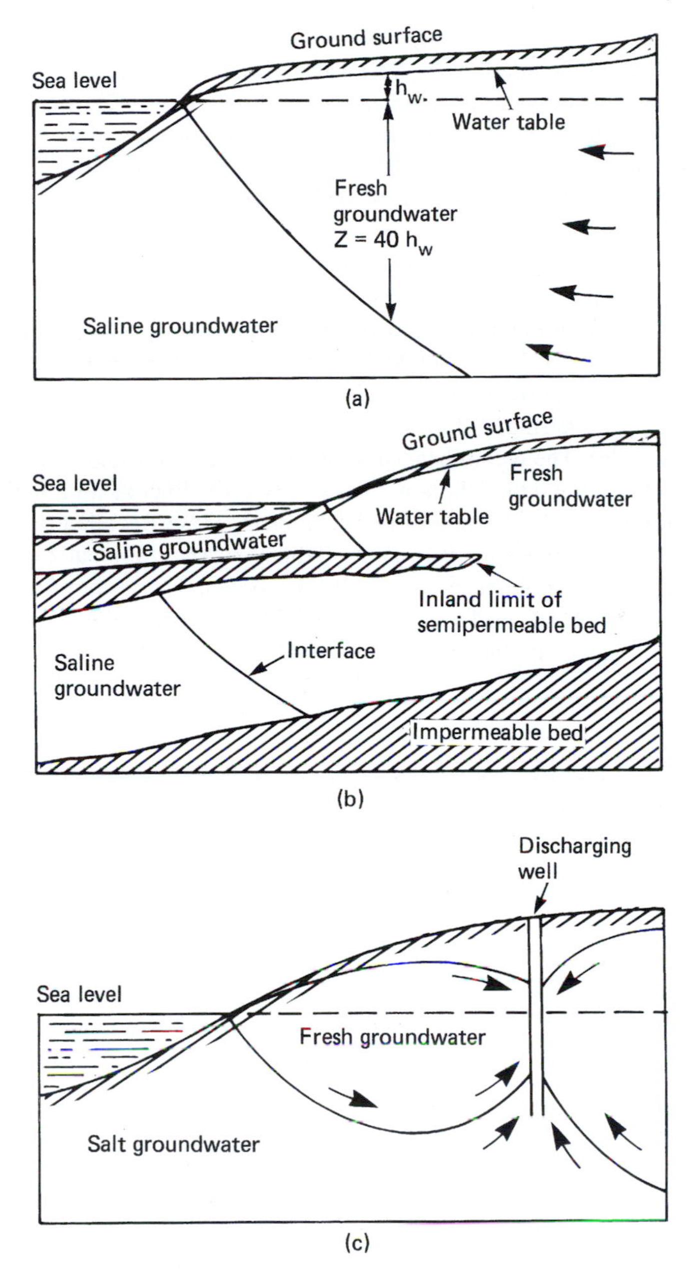

Figure 7.23 Ghyben-Herzberg hydrostatic relationship in: (a) a homogeneous coastal aquifer; (b) a layered coastal aquifer; (c) a pumped coastal aquifer.

The problem of saline intrusion usually starts with the abstraction of groundwater from a coastal aquifer, which leads to the disruption of the Ghyben–Herzberg equilibrium condition. Generally, saline water is drawn up towards the well and is sometimes termed upconing (Fig. 7.23c). This is a dangerous condition that can occur even if the aquifer is not overpumped and a significant proportion of the freshwater flow still reaches the sea. A well may be ruined by an increase in salt content even before the actual 'cone' reaches the bottom of the well. This is due to leaching of the interface by freshwater. Again, this situation is best studied using modelling techniques (Nutbrown, 1976), although Linsley *et al.* (1982) offered a rule of thumb that a drawdown at a well of 1 m will result in a salt water rise of approximately 40 m. Upconing is a particular problem on small oceanic islands where the freshwater lens is shallow and everywhere rests on salt water. In such cases, shallow wells or horizontal infiltration galleries may be adopted.

The encroachment of salt water may extend for several kilometres inland leading to the abandonment of wells. The first sign of saline intrusion is likely to be a progressively upward trend in the chloride concentration of water obtained from the affected wells. Chloride levels may increase from a normal value of around 25 mg l^{-1} to something approaching 19 000 mg l^{-1}, which is the concentration in sea water. In the Gravesend area of Kent, England, chloride ion concentrations of up to 5000 mg l^{-1} have been recorded, which are consistent with sea water encroachment (Fig. 7.24). The recommended limit for chloride concentration in drinking water in Europe is 200 mg l^{-1}, after which the water has a salty taste. Encroaching sea water, however, may be difficult to recognize as a result of chemical changes, so Revelle (1941) recommended the use of the chloride–bicarbonate ratio as an indicator. An additional complication is that there are likely to be frequent fluctuations in chloride content as a result of tides, varying rates of freshwater flow through the aquifer, meteorological phenomena, and so on. This also means that the saline–fresh water interface moves seaward or landward according to the prevailing conditions.

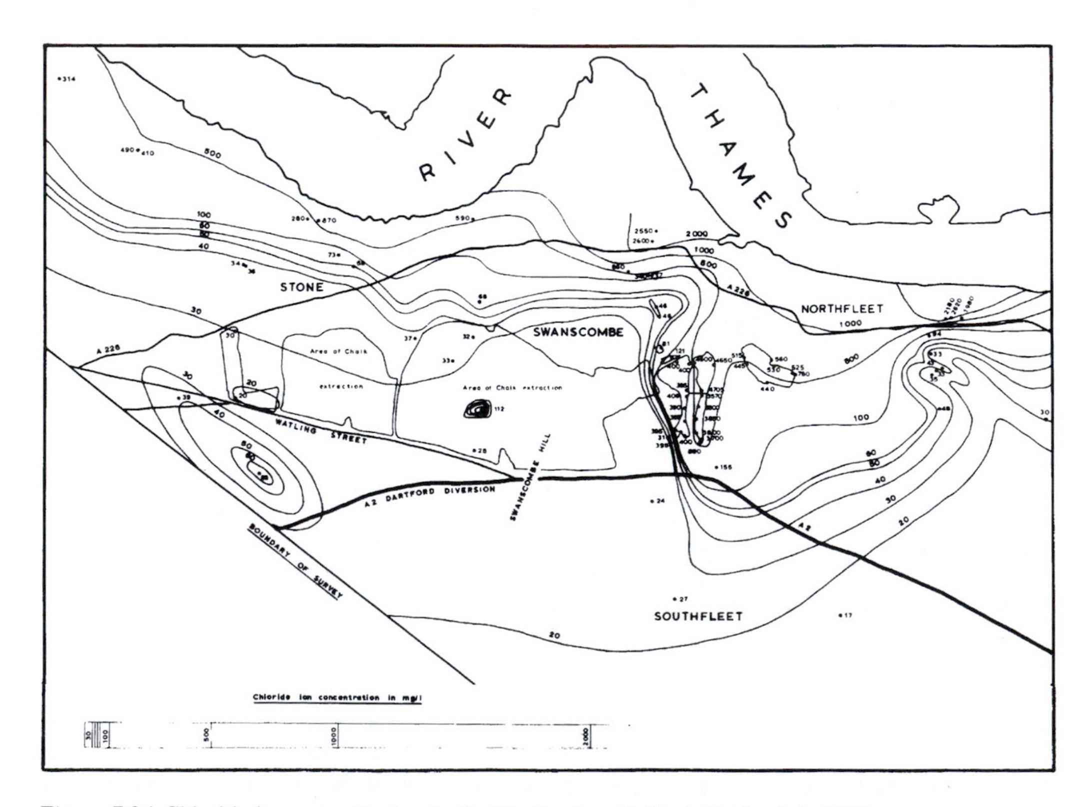

Figure 7.24 Chloride ion concentration in the Chalk of north Kent, England, in 1972.

Resistivity surveys often have been used to investigate areas affected by saline intrusion because of the contrast between salt and fresh water. For example, Oteri (1983) used a resistivity survey to delineate the extent of saline intrusion in a shingle aquifer at Dungeness, Kent, England. However, if the depth to water is too great, then the thickness of the overlying unsaturated sediments can mask any anomaly between the two types of groundwater. McDonald *et al.* (1998) described the use of resistivity tomography and ground conductivity surveys to delineate saline intrusion in a tidal coastal wetland in the south of England. They found that the fluctuations of saline intrusion were out of phase with the tidal cycles in areas of low lateral permeability and in phase in areas where buried channels and inlets existed.

Once saline intrusion develops in a coastal aquifer, it is not easy to control. The slow rates of groundwater flow, the density differences between fresh and salt waters and the flushing required, usually mean that contamination, once established, may take years to remove under natural conditions. Overpumping is not the only cause of salt water encroachment, continuous pumping or inappropriate location and design of wells also may be contributory factors. In other words, the saline–fresh water interface is the result of a hydrodynamic balance, hence if the natural flow of fresh water to the sea is interrupted or significantly reduced by abstraction, then saline intrusion is almost certain to occur. Once salt water encroachment is detected, pumping should cease, whereupon the denser saline water returns to lower levels of the aquifer.

Although it is difficult to control saline intrusion and to effect its reversal, the encroachment of salt water can be checked by maintaining a freshwater hydraulic gradient towards the sea, which means that there is a flow of fresh water towards the sea. This gradient can be maintained naturally, or by some artificial means such as artificial recharge to form a groundwater mound between the coast and the area where abstraction is taking place. However, the technique requires an additional supply of clean water. Alternatively, an extraction barrier, that is, a line of wells parallel to the coast, which abstracts encroaching salt water before it reaches the protected inland wellfield, can be employed. The abstracted water will be brackish and therefore generally is pumped back into the sea. There will probably be a progressive increase in salinity at the extraction wells, which must be pumped continuously if an effective barrier is to be created.

Neither of these two methods of artificially controlling saline intrusion offers an inexpensive or foolproof solution. Consequently, when there is a possibility of saline intrusion, the best policy is probably to locate wells as far from the coast as possible, select the design discharge with care and use an intermittent seasonal pumping regime (Ineson, 1970). Ideally, at the first sign of progressively increasing salinity, pumping should be stopped. Ineson described a system of selective pumping from the Chalk aquifer in Sussex that represented an attempt to control groundwater deterioration due to saline intrusion. In this area of the south coast of England, groundwater development after 1945 had been limited due to increasing salinity. Ineson recorded that the chloride ion concentration at one point reached $1000\,mg\,l^{-1}$, as compared with a normal value not exceeding $40\,mg\,l^{-1}$. The scheme adopted involved pumping from the wells near the coast during the winter months, whereby output was reduced from the inland stations. This led to the interception of fresh groundwater at the coastal boundary that would otherwise have flowed into the sea. As a result, the storage of groundwater in the landward part of the aquifer increased. During the summer months, the pumping scheme was reversed. This method of operation not only controlled saline intrusion but also achieved an overall increase in abstraction.

7.14.5 Nitrate pollution

There are at least two ways in which nitrate pollution of groundwater is known or suspected to be a threat to health. First, the build-up of stable nitrate compounds in the bloodstream reduces its oxygen-carrying capacity. Infants under one year are most at risk and excessive amounts of nitrate can cause methaemoglobinaemia, commonly called 'blue-baby'. Consequently, if the limit of 50 mg l^{-1} of NO_3 recommended by the World Health Organisation (Anon., 1993) for European countries is exceeded frequently, or if the concentration lies within the minimum acceptable range of 50–100 mg l^{-1}, bottled low-nitrate water should be provided for infants. Second, there is a possibility that a combination of nitrates and amines through the action of bacteria in the digestive tract results in the formation of nitrosamines, which are potentially carcinogenic. For example, Jenson (1982) reported that the water supply of Aarlborg in Denmark had a relatively high nitrate content of approximately 30 mg l^{-1}, and there was a slightly greater frequency of stomach cancer in Aarlborg than in other towns during the period 1943–1972. This appeared to be linked to the nitrate content of the drinking water.

Nitrate pollution is basically the result of intensive cultivation. The major source of nitrate is the large quantity of synthetic nitrogenous fertilizer that has been used in arable farming, although over-manuring can have the same result. Foster and Crease (1974) estimated that in the 11 years between 1956 and 1967, the application of nitrogen fertilizer to all cropped land in certain areas of east Yorkshire, England, increased by about a factor of four (from around 20 to 80 kgN ha^{-1} year^{-1}). Regardless of the form in which nitrogen fertilizer is applied, within a few weeks it will have been transformed to NO_3^-. This ion is neither adsorbed nor precipitated in the soil and therefore is easily leached by heavy rainfall and infiltrating water. However, the nitrate does not have an immediate affect on groundwater quality, possibly because most of the leachate that percolates through the unsaturated zone as intergranular seepage has a typical velocity of about 1 m year^{-1} (Smith *et al.*, 1970). Thus, there may be a considerable delay between the application of fertilizer and the subsequent increase in the concentration of nitrate in the groundwater. So although the use of nitrogenous fertilizer increased sharply in east Yorkshire after 1959, a corresponding increase in groundwater nitrate was not apparent until after 1970, when the level rose from around 14 mg l^{-1} NO_3 (about 3 mg l^{-1} NO_3—N) to between 26 and 52 mg l^{-1} NO_3 (6–11.5 mg l^{-1} NO_3—N). This just exceeded the EEC's and WHO's recommended limit of 50 mg l^{-1} NO_3 (11.3 mg l^{-1} NO_3—N).

The effect of the time lag, which is frequently of the order of 10 years or more, is to make it very difficult to correlate fertilizer application with nitrate concentration in groundwater. In this respect, one of the greatest concerns is that if nitrate levels are unacceptably high now, they may be even worse in the future because the quantity of nitrogenous fertilizer used has continued to increase steadily. Foster and Young (1980) studied various sites in England and found that in the unsaturated zone the nitrate concentration of the interstitial water was closely related to the history of agricultural practice on the overlying land. Sites subjected to long-term, essentially continuous, arable farming were found to yield the highest concentrations, while lower figures were generally associated with natural grassland. In many of the worst locations, nitrate concentrations exceeded 100 mg l^{-1} NO_3 (22.6 mg l^{-1} NO_3—N) while individual wells with values in excess of 440 mg l^{-1} NO_3 (100 mg l^{-1} NO_3—N) were encountered.

Lawrence *et al.* (1983) reported on nitrate pollution of groundwater in the Chalk of east Yorkshire a decade after the original study. It appeared that there was a rising trend of nitrate concentration in most of the public groundwater supplies obtained by abstraction from the Chalk aquifer. There also was a marked seasonal variation in groundwater quality that showed a strong correlation with groundwater level. This was attributed tentatively to either rapid recharge directly from agricultural soils, or elusion of pore water solutes from within the zone of groundwater level fluctuation. Since nitrate concentrations in the pore water above the water table beneath arable land were found to be two to four times higher than in groundwater supplies, the long-term trend must be towards substantially higher levels of nitrate in groundwater sources.

Measures that can be taken to alleviate nitrate pollution include better management of land use, mixing of water from various sources or the treatment of high-nitrate water before it is put into use (Cook, 1991). According to Foster (2000), nitrate leaching in England varies appreciably according to management practice, especially in relation to the timing of the application of the fertilizer and the method of soil cultivation. He noted that most nitrate is leached in late autumn when the period of active nitrification is followed by one of excess rainfall. Hence, he concluded that leaching could be reduced by avoiding the application of fertilizer at this period and by ensuring the presence of a growing crop. In general, the ion exchange process has been recommended as the preferred means of treating groundwaters, although this may not be considered cost-effective at all sources. The *in situ* denitrification capacity of groundwater results in the removal of nitrate and is most likely to occur in aquifers where a notable content of natural organic carbon and/or pyritic minerals are present. Robertson *et al.* (2000) carried out an investigation to assess the value of reactive porous barriers as an *in situ* method of treatment of nitrate in groundwater. The barriers consisted of waste cellulose solids (wood mulch, sawdust and leaf compost), which provided the source of carbon for heterotrophic denitrification. The trials were successful in reducing the amount of nitrate significantly and showed that such barriers could remain active for a decade or more without carbon replenishment. Robertson *et al.* reported that reactive barriers now were being used to treat nitrate pollution from a number of sources including agricultural run-off, leakage from septic tank systems, leachate from landfill and leakage from some industrial operations. Tompkins *et al.* (2001) pointed out that biological denitrification is a naturally occurring process in the saturated zone, reducing nitrate to nitrogen gases and nitrous oxide. These processes can be autotrophic or heterotrophic and the latter, which probably is the more notable in groundwater, can be enhanced by supplying an appropriate source of organic carbon.

7.14.6 The problem of acid mine drainage

The term acid mine drainage (AMD) is used to describe drainage resulting from the natural oxidation of sulphide minerals that occurs in mine rock or waste that is exposed to air and water. This is a consequence of the oxidation of sulphur in the mineral to a higher oxidation state and, if aqueous iron is present and unstable, the precipitation of ferric iron with hydroxide occurs. It can be associated with underground workings, with open pit workings, with spoil heaps, with tailings ponds or with mineral stockpiles (Brodie *et al.*, 1989).

Acid mine drainage is responsible for problems of water pollution in major coal and metal mining areas around the world. The character and rate of release of acid mine drainage is influenced by various chemical and biological reactions at the source of acid generation. If acid mine drainage is not controlled, it can pose a serious threat to the environment since acid generation can lead to elevated levels of heavy metals and sulphate in water that obviously has a detrimental effect on its quality (Table 7.9). The development of AMD is time-dependent and at some mines may evolve over a period of years.

A major source of AMD may result from the closure of a mine (Bell *et al.*, 2002). When a mine is abandoned and dewatering by pumping ceases, the water level rebounds. However, the workings often act as drainage systems so that the water does not rise to its former level. Consequently, a residual dewatered zone remains that is subject to continuing oxidation. The mine water quality is determined by the hydrogeological system and the geochemistry of the upper mine levels. The large areas of fractured rock exposed in opencast mines or open pits can give rise to large volumes of acid mine drainage. Spoil heaps may consist of waste produced by mining or waste produced by any associated smelting or beneficiation. Consequently, the

Table 7.9 Composition of acid mine water from a South African coalfield

Determinand (mg l^{-1})	*Sample 1*	*Sample 2*	*Sample 3*	*Sample 4*	*Sample 5*	*Sample 6*
TDS	4844	2968	3202	2490	3364	3604
Suspended solids	33	10.4	12	10.0	7.6	
EC (mS m^{-1})	471	430	443	377	404	340
pH value	1.9	2.4	2.95	2.9	2.3	2.8
Langelier S1	−8.2	−7.5	−6.9	−6.9	−7.7	
Turbidity as NTU	5.5	0.6	2.0	0.9	1.7	
Nitrate NO_3 as N	0.1	0.1	0.1	0.1	0.1	0.1
Chlorides as CI	310	431	406	324	353	611
Fluoride as F	0.6	0.5	0.33	0.6	0.6	0.84
Sulphate as SO_4	3250	1610	1730	1256	2124	1440
Total hardness		484	411	576	585	377
Calcium hardness as $CaCO_3$		285	310	327	282	
Magnesium hardness as $CaCO_3$		199	101	249	305	
Calcium as Ca		114.0	124	131	113	84
Magnesium as Mg	173.8	48.4	49.5	60.5	49.3	31
Sodium as Na	89.4	326.0	311	278	267	399
Potassium as K	247.0	9.4	8.9	6.4	3.8	
Iron as Fe	7.3	128	140	87	89.9	193
Manganese as Mn	248.3	15	9.9	13.4	13.9	9.3
Aluminium as Al	17.9	124		112	204	84

Source: Bell *et al.*, 2002.

Note

Samples 1 to 5 are from surface water courses fed by springs from the mine; sample 6 is from coal seam level. NTU, nephelometric turbidity unit.

sulphide content of the waste can vary significantly. Acid generation tends to occur in the surface layers of spoil heaps where air and water have access to sulphide minerals (Bullock and Bell, 1995). Tailings deposits that have a high content of sulphide represent another potential source of acid generation. However, the low permeability of many tailings deposits, together with the fact that they commonly are flooded, means that the rate of acid generation and release is limited. Consequently, the generation of AMD can continue to take place long after a tailings deposit has been abandoned. Mineral stockpiles may represent a concentrated source of AMD. Major acid flushes commonly occur during periods of heavy rainfall after long periods of dry weather.

The ability of a particular mine rock or waste to generate net acidity depends on the relative content of acid-generating minerals and acid-consuming or neutralizing minerals. Acid waters produced by sulphide oxidation of mine rock or waste may be neutralized by contact with acid-consuming minerals. As a result, water draining from the parent material may have a neutral pH value and negligible acidity despite ongoing sulphate oxidation. If the acid-consuming minerals are dissolved, washed out or surrounded by other minerals, then acid generation continues. Where neutralizing carbonate minerals are present, metal hydroxide sludges, such as iron hydroxides and oxyhydroxides are formed. Sulphate concentration generally is not affected by neutralization unless mineral saturation with respect to gypsum is attained. Hence, sulphate sometimes may be used as an overall indicator of the extent of acid generation after neutralization by acid-consuming minerals.

The primary chemical factors that determine the rate of acid generation include pH value; temperature; oxygen content of the gas phase if saturation is less than 100 per cent; concentration of oxygen in the water phase; degree of saturation with water; chemical activity of Fe^{3+}; surface area of exposed metal sulphide; and chemical activation energy required to initiate acid generation. In addition, the biotic micro-organism, *Thiobacillus ferrooxidans,* may accelerate reaction by its enhancement of the rate of ferrous iron oxidation. It also may accelerate reaction through its enhancement of the rate of reduced sulphur oxidation. *Thiobacillus ferrooxidans* is most active in waters with a pH value around 3.2 and biotic oxidation of pyrite is four times faster than the abiotic reaction at pH 3.0. The formation of sulphuric acid in the initial oxidation reaction and concomitant decrease in the pH make conditions more favourable for the biotic oxidation of the pyrite by *Thiobacillus ferrooxidans*. The presence of *Thiobacillus ferrooxidans* also may accelerate the oxidation of sulphides of antimony, arsenic, cadmium, cobalt, copper, gallium, lead, molybdenum, nickel and zinc. In order to generate severe AMD ($pH<3$) sulphide minerals must create an optimum micro-environment for rapid oxidation and must continue to oxidize long enough to exhaust the neutralization potential of the rock.

Accurate prediction of AMD is required in order to determine how to bring it under control. The objective of AMD control is to satisfy environmental requirements using the most cost-effective techniques. The options available for the control of polluted drainage are greater at proposed rather than existing operations as control measures at working mines are limited by site-specific and waste disposal conditions. For instance, the control of AMD that develops as a consequence of mine dewatering is helped by the approach taken towards the site water balance. In other words, the water resource management strategy developed during mine planning will enable mine water discharge to be controlled and treated prior to release or to be reused. The length of time that the control measures require to be effective is a factor that needs to be determined prior to the design of a system to control AMD.

Prediction of the potential for acid generation involves the collection of available data and carrying out static and kinetic tests. A static test determines the balance between potentially acid-generating and acid-neutralizing minerals in representative samples. One of the frequently used static tests is acid–base accounting. Acid–base accounting allows determination of the proportions of acid-generating and neutralizing minerals present. However, static tests cannot be used to predict the quality of drainage waters and when acid generation will occur. If potential problems are indicated, the more complex kinetic tests should be used to obtain a better insight of the rate of acid generation. Kinetic tests involve weathering of samples under laboratory or on-site conditions in order to confirm the potential to generate net acidity, to determine the rates of acid formation, sulphide oxidation, neutralization and metal dissolution, and to test control and treatment techniques. The static and kinetic tests provide data that may be used in various models to predict the effect of acid generation and control processes beyond the time frame of kinetic tests.

There are three key strategies in AMD management, namely, control of the acid-generation process, control of acid migration, and collection and treatment of AMD (Connelly *et al.*, 1995). Control of AMD may require different approaches, depending on the severity of potential acid generation, the longevity of the source of exposure and the sensitivity of the receiving waters. Mine water treatment systems installed during operation may be adequate to cope with both operational and long-term post-closure treatment with little maintenance. On the other hand, in many mineral operations, especially those associated with abandoned workings, the long-term method of treatment may be different from that used while a mine was operational. Hence, there may have to be two stages involved with the design of a system for treatment of AMD, one during mine operation and another after closure.

Obviously, the best solution is to control acid generation, if possible. Source control of AMD involves measures to prevent or inhibit oxidation, acid generation or contaminant leaching. If acid generation is prevented, then there is no risk of contaminants entering the environment. Such control methods involve the removal or isolation of sulphide material or the exclusion of water or air. The latter is much more practical and can be achieved by the placement of a cover over acid-generating material such as waste or air-sealing adits in mines.

Migration control is considered when acid generation is occurring and cannot be inhibited. Since water is the transport medium, control relies on the prevention of water entry to the source of AMD. Water entry may be controlled by diversion of surface water flowing towards the source by drainage ditches; by prevention of groundwater flow into the source by interception and isolation of groundwater (this is very difficult to maintain over the long term); and by prevention of infiltration of precipitation into the source by the placement of cover materials but again their long-term integrity is difficulty to ensure.

Release control is based on measures to collect and treat AMD. In some cases, especially at working mines, this is the only practical option available. Collection requires the collection of both ground and surface water polluted by AMD, and involves the installation of drainage ditches and collection trenches and wells. Treatment processes have concentrated on neutralization to raise the pH and precipitate metals. Lime or limestone commonly is used, although offering only a partial solution to the problem. Other alkalis such as sodium hydroxide may be used, but are more expensive although they do not produce such dense sludges as does lime. The sludges recovered from alkali neutralization, followed by sedimentation and consolidation, are of relatively low density (2–5 per cent dry solids), but drain fairly well.

The oxidation of pyrite can be eliminated by flooding a mine with water. However, it obviously is not possible to flood an active mine. It also is not possible to flood abandoned mine workings that are above the water table and that are drained by gravity. More sophisticated processes (active treatment methods) involve osmosis (waste removal through membranes), electrodialysis (selective ion removal through membranes), ion exchange (ion removal using resin), electrolysis (metal recovery with electrodes) and solvent extraction (removal of specific ions with solvents).

Cambridge (1995) pointed out that conventional active treatment of mine water requires the installation of a treatment plant, with continuous operation and maintenance. Hence, the capital and operational costs of active treatment are high. Acid mine water treated with active systems tends to produce a solid residue that has to be disposed of in tailings lagoons. This sludge contains metal hydroxides. However, according to Cambridge, the long-term disposal of sludge in tailings lagoons is not appropriate. Alternatively, sludges could be placed in hazardous waste landfills but such sites are limited.

Alternatively, passive systems try to minimize input of energy, materials and manpower, and so reduce operational costs. Passive treatment involves engineering a combination of low maintenance biochemical systems (e.g. anoxic limestone drains, aerobic and anaerobic wetlands, and rock filters). Such treatment does not produce large volumes of sludge and the metals are precipitated as oxides or sulphides in the substrate materials.

Due to the impact on the environment of AMD, regular monitoring is required. The major objectives of a monitoring programme developed for AMD are, first, to detect the onset of acid generation before AMD develops to the stage where environmental impact occurs. If required, control measures should be put in place as quickly as possible. Second, it is necessary to monitor the effectiveness of the prevention-control-treatment techniques and to detect whether the techniques are unsuccessful at the earliest possible time.

7.14.7 Waste waters and effluents from coal mines

Generally, the major pollutants in water associated with coal mining are suspended solids, dissolved salts (especially chlorides), acidity and iron compounds (Bell and Kerr, 1993). However, the character of drainage from coal mines varies from area to area and from coal seam to coal seam. Hence, mine drainage waters are liable to vary in both quality and also in quantity, sometimes unpredictably, as the mine workings develop. Nonetheless, drainage water, for example, from coal mines in Britain can be classified as hard, alkaline, moderately saline, highly saline, alkaline and ferruginous, and acidic and ferruginous (Best and Aikman, 1983;). Colliery discharges have little oxygen demand, the BOD normally being very low. Elevated levels of suspended matter are associated with most coal mining effluents, with occasionally high values being recorded. Although not all mine waters are highly mineralized, a high level of mineralization is typical of many coal mining discharges and is reflected in the high values of electrical conductivity. Highly mineralized mine waters usually contain high concentrations of sodium and potassium salts, and mine waters that do not contain sulphate may contain high levels of strontium and barium. Similarly, not all mine waters are ferruginous, and in fact some are of the highest quality and can be used for potable supply. Nonetheless, they are commonly high in iron and sulphates. The low pH values of many mine waters are commonly associated with highly ferruginous discharges.

The high level of dissolved salts that is often present in mine waters represents the most intractable water pollution problem connected with coal mining. This is because dissolved salts are not readily susceptible to treatment or removal. The range of dissolved salts encountered in mine water is variable, with electrical conductivity values up to 335 000 $\mu S\,cm^{-1}$ and chloride levels of 60 000 $mg\,l^{-1}$ being recorded. The principal groups of salts in mine discharge waters are chlorides and sulphates. These salts are released into the workings by mining operations. In general, the salinity increases with depth below the surface and with distance from the outcrop or incrop. The more saline waters contain significant concentrations of barium, strontium, ammonium and manganese ions.

Movement of pollutants through strata is often very slow and is difficult to detect. Hence, effective remedial action is often either impractical or prohibitively expensive. Because of this there are few successful recoveries of polluted aquifers.

In the old coalfields of the east Midlands and south Staffordshire, England, colliery spoil was often tipped on top of the Sherwood Sandstone, which is the second most important aquifer in Britain. Although modern tipping techniques may render spoil impervious, surface water run-off can leach out soluble salts, especially chloride. This may result in the loss of up to 1 tonne of chloride per hectare of exposed spoil heap per annum under average rainfall conditions (spoil heaps may extend to many hundreds of hectares in extent). The run-off from these spoil heaps may discharge directly into drainage ditches or to land around the periphery of the heap and infiltrate into the aquifer. Figure 7.25 shows the isopleths for chloride ion concentration in the groundwater of the Sherwood Sandstone in the concealed part of the Nottinghamshire coalfield in the 1970s. It also indicates the locations of the collieries and demonstrates the relationship between elevated chloride ion level in the groundwater and mining activity. Such pollution of an aquifer can be alleviated by lining the beds of influent streams that flow across the aquifer or by providing pipelines to convey mine discharge to less-sensitive water courses that do not flow across the aquifer.

7.14.8 Volatile organic chemicals (non-aqueous phase liquids)

According to Mackay (1998), volatile organic chemicals (VOCs) are the most frequently detected organic contaminants in water supply wells. Of the VOCs, by far the most commonly

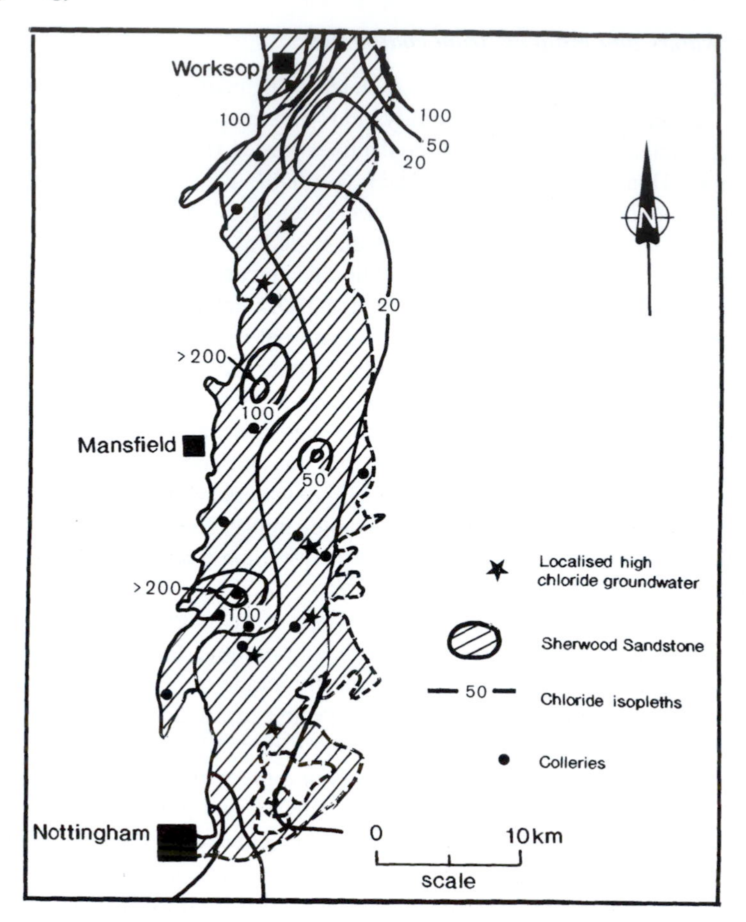

Figure 7.25 Chloride ion isopleths (in mg l^{-1}) in groundwater in the Sherwood Sandstone aquifer of the concealed coalfield, Nottinghamshire, England, in the late 1970s.
Source: Bell and Kerr, 1993.

found in the United States are chlorinated hydrocarbon compounds. Conversely, petroleum hydrocarbons rarely are present in supply wells. This may be due to their *in situ* biodegradation. Mackay further mentioned that the most frequently detected chlorinated hydrocarbons were perchloroethane (PCE), trichloroethane (TCE) and 1,1,1-tricholoethane (TCA), and carbon tetrachloride (TC), which are used as industrial solvents. Several other types of chlorinated hydrocarbons have been noted in groundwater, some of which may have been produced by biotic or abiotic transformations of PCE, TCE or TCA.

Many of the VOCs are liquids and usually are referred to as non-aqueous phase liquids (NAPLs), which are sparingly soluble in water. Those that are lighter than water, such as the petroleum hydrocarbons are termed LNAPLs whereas those that are denser than water, that is, the chlorinated solvents are called DNAPLs. Of the VOCs, the DNAPLs are the least amenable to remediation. If they penetrate the ground in a large enough quantity, depending on the hydrogeological conditions, DNAPLs may percolate downwards into the saturated zone. This can occur in granular soils or discontinuous rock masses. Plumes of dissolved VOCs develop from the source of pollution. Although dissolved VOCs migrate more slowly than the average velocity of groundwater, Mackay and Cherry (1989) indicated that there are many examples of chlorinated VOC plumes several kilometres in length in the United States occurring in sand and gravel aquifers. Such plumes contain billions of litres of contaminated water. Because VOCs are sparingly soluble in water, the time taken for complete dissolution, especially of DNAPLs, by groundwater flow in granular soils is estimated to be decades or even centuries. Bishop *et al.* (1998) described two extensive investigations of point source pollution of groundwater by chlorinated solvents. The investigations involved soil gas sampling, this being a rapid method of detecting such pollution, as well as sampling of pore water obtained from drillholes. They pointed out that careful consideration should be given to the problem of cross-contamination before the drilling programme commences. Groundwater pollution almost always is stratified and consequently drillholes afford the opportunity for vertical migration of contaminants. If such cross-contamination represents a threat, then drilling should be carried out layer by layer. Acworth (2001) also described an investigation of groundwater contamination by DNAPL in Sydney, Australia. He noted that a prominent thin band of high total dissolved solids was associated with the top of the zone containing DNAPL, which formed a sharp electrical conductivity anomaly. Analysis of the pore fluids obtained from boreholes sunk in the contaminated sands indicated a low pH (3–4) and sodium to chloride ratios of less than 0.2. The influence of subsurface geology on the migration of DNAPL (i.e. PCE) at Sussex, New Brunswick, was noted by Broster and Pupek (2001). Two sand and gravel aquifers were affected by surface spillage, the upper had PCE concentrations ranging up to $28\,\mu g\,l^{-1}$ whilst concentrations of $1.6\,\mu g\,l^{-1}$ were noted in the lower. Subsurface migration through the upper aquifer was influenced significantly by a buried river channel cut into the aquitard separating the two aquifers. Scouring in the base of the channel had eroded windows through which the PCE had gained access to the lower aquifer.

The most frequently used method of remediation or control of contaminated groundwater is the pump and treat method by which the affected groundwater is abstracted and then treated at the surface. However, the pump and treat method has not always proved successful, especially where contaminant characteristics and hydrogeological conditions are not simple, and large volumes of water have been affected. Usually, significant mass removal is achieved in the initial stages of a pump and treat operation but this declines dramatically because of the development of residual NAPL saturation, which does not flow out of the pore space. In addition, the time taken for a clean-up operation may be several years (Muldoon *et al.*, 1998). Nonetheless, the pump and treat method does appear to be effective in terms of limiting the migration of contaminant plumes. Clark and Sims (1998) gave an account of the investigation and subsequent remediation of leakage of kerosene from a fuel pipeline at Heathrow Airport, London. The kerosene, which was almost a metre in depth in places, floated on top of a shallow water table (about 2.5 m below the surface) in gravel deposits. Remediation was brought about by the construction of concrete-lined wells, 1.5 m in diameter, sunk to 2 m below the water table. The recovery of the kerosene took approximately four years of continuous effort. Attempts have been made to contain contaminated groundwater by installing an impermeable barrier but such a measure is not always feasible. Chapple *et al.* (1998) described the use of interceptor trenches to recover LNAPLs. They indicated that such a trench may be open or filled, for example, with gravel, and is excavated in an unconfined aquifer to collect contaminated groundwater. As the LNAPL floats on the water, it can be removed by skimming. An impermeable membrane commonly is installed on the down-gradient side of the trench, suspended below the water table, to prevent the leakage of LNAPL. Various other techniques that have been used, or are being developed, to remediate NAPL contamination of groundwater include bioremediation, solvent flushing, steam flushing and soil vapour extraction. For example, Londergan *et al.* (2001) described the use of surfactants to remove residual DNAPL from an alluvial aquifer in Utah. They claimed that the total recovery of DNAPL amounted to about 98.5 per cent. Permeable reactive barriers (PRBs) are a cost-effective method used for *in situ* clean up of contaminated groundwater (Scherer *et al.*, 2000). A wide variety of materials can be used in PRBs such as zero-valent metals (e.g. metal iron), oxides, humic materials, surfactant-modified zeolites (SMZs), and oxygen- and nitrate-releasing compounds. Permeable reactive barrier materials remove dissolved groundwater contaminants by immobilization within the barrier or transformation to less harmful products. The principal removal processes include chemical reactions, sorption and precipitation, and biologically mediated reactions. For instance, Soudain (1997) reported the employment of a reactive gate and barrier system to treat groundwater contaminated by TCE in Belfast, Northern Ireland. The system used cut-off walls to channel the contaminated groundwater through a reactor vessel filled with iron filings that break down the contaminants to carbon dioxide and water. Finkel *et al.* (1998) also referred to a funnel and gate system to treat groundwater contaminated with polycyclic aromatic hydrocarbons (PAHs). The treatment walls consisted of granular-activated carbon.

Methyl tertiary butyl ether (MTBE) is an additive to petroleum that can represent up to 10 per cent by volume of unleaded petrol. According to Burgess *et al.* (1998), although it has a low toxicity, it has a low odour and taste threshold as far as drinking water is concerned and so constitutes a problem at concentration levels exceeding 2–3 $\mu g\,l^{-1}$. It is ten times more soluble than the BTEX (benzene, toluene, ethylbenzene, xylene) components, which have led to concern about its behaviour in aquifers. Moreover, it would appear that MTBE is not retarded by sorption, neither is it biodegradable. Fortunately, it can be detected at concentrations as low as 0.1 $\mu g\,l^{-1}$ in groundwater by high-pressure liquid chromatography. Burgess *et al.* noted that skimmer pumps, pump and treat, and volatization have been used as remedial methods for MTBE contamination of groundwater, their effectiveness depending on the pattern of contamination and the hydrogeological conditions.

7.14.9 Other causes of groundwater pollution

The list of potential groundwater pollutants is almost endless. For example, sewage sludge arises from the separation and concentration of most of the waste materials found in sewage (Davis, 1980). Since the sludge contains nitrogen and phosphorous, it has a value as a fertilizer. As a result, about 50 per cent of the 1.24 million dry tonnes per year of sludge produced in the United Kingdom is used on agricultural land. While this does not necessarily lead to groundwater pollution, the presence in the sludge of contaminants such as metals, nitrates, persistent organic compounds and pathogens, does mean that the practice must be carefully controlled (Andrews *et al.*, 1998). In particular, cadmium may give cause for concern. Too much cadmium can cause kidney damage in humans and this metal was implicated in *itai-itai* disease in Japan. Food is the usual source of the cadmium found in humans, although small amounts also are present in water. The European standard for drinking water recommends a cadmium concentration of less than 0.01 $mg\,l^{-1}$.

Disposal of animal wastes on agricultural land can pollute groundwater if not properly undertaken. This is mainly because of the high content of nitrate and phosphate in animal wastes. Garnier *et al.* (1998) indicated that such disposal was a matter of concern in the Chiana valley in Italy where there is a high concentration of pig breeding farms centred on a shallow unconfined aquifer. Consequently, Garnier *et al.* undertook an investigation of the maximum rate at which animal waste could be disposed of without exceeding the maximum limit of nitrate concentration in the groundwater. They found that even though the application of manure was carried out in accordance with regulations some pollution may arise beneath sensitive soils, that is, the amount of nitrate leaching is halved on fine-grained soils.

Agricultural activity can be responsible for the generation of microbial and pathogen loading in the soil and streams that, in turn, can affect groundwater. As a consequence, the more vulnerable aquifers at times can be infected by faecal coliforms. Foster (2000) indicated that the pathogens that may be present in contaminated groundwater included some types of *Escherichia coli*, the cysts of the protozoa *Cryptosporidium parvum* and *Giardia lamblia*, as well as certain enteric viruses. *Cryptosporidium parvum* is of most concern at present. It affects humans, livestock and various wild animals, and those infected excrete large numbers of oocysts. These are about 4–6 μm in diameter, are able to survive dormant for long periods in water and to resist chlorine disinfectant. Hence, they pose a serious potential hazard. A methodology for the assessment of risk of *Cryptosporidium* contamination of groundwater has been developed (Boak and Packman, 2001). The pathways by which such microbial contamination can reach an aquifer are of paramount importance in any risk assessment.

The widespread use of chemical and organic pesticides and herbicides is another possible source of groundwater pollution. However, Chilton *et al.* (1998) pointed out that it would be prohibitively expensive to monitor for all the large number of pesticides in regular use. They therefore suggested that attention should be focused on high-risk pesticides such as those that are mobile, toxic, persistent and widely used on the most vulnerable aquifers. They also noted that it appears that pesticides are more persistent in groundwater than in soils and that degradation rates of pesticides within aquifers can vary significantly. Chilton *et al.* concluded that the greatest risk to groundwater supplies from pesticides is likely to occur where aquifers are overlain by permeable soils and travel times to the water table are short. Other factors that have to be taken into consideration include the porosity of the aquifer since if this is low, then dilution is less effective, and the relative stability of the pesticide. Sweeney *et al.* (1998) described the pollution of a limestone aquifer in north Cambridgeshire, England, by herbicide. The European Community Drinking Water Directive sets a maximum admissible concentration for an individual pesticide or herbicide of $0.1\,\mu g\,l^{-1}$. In their investigations both Chilton *et al.* and Sweeney *et al.* found much higher concentrations than the permitted limits referred to.

Irrigation water may pose a pollution hazard to groundwater, especially in arid and semi-arid regions where soluble salts may be present in soil. These salts can be leached from the soil and so become concentrated in irrigation water, the situation being worsened if poor quality water is used for irrigation purposes. Indeed, salinization of soils is a problem in many parts of the world. In such instances, shallow groundwater may be recharged by irrigation water. Walton *et al.* (1999), and Hibbs and Boghici (1999) referred to two areas along the Rio Grande in Texas where shallow aquifers have been adversely affected by the intensive use of irrigation water. They noted that there is a tendency for the salinity of the groundwater in the aquifer to increase downstream, the TDS increasing from between 1000 and $3500\,mg\,l^{-1}$ to between 3000 and $6000\,mg\,l^{-1}$. Smith and Guitjens (1998) also reported a problem of increasing salinity in the groundwater at Henderson, Nevada, due to irrigation. There the TDS concentrations ranged from 3000 to $12\,000\,mg\,l^{-1}$ in the uppermost groundwater.

The potential of run-off from roads to cause pollution often is overlooked. This water can contain chemicals from many sources, including those that have been dropped, spilled or deliberately spread on the road. For instance, hydrocarbons from petroleum products, and urea and chlorides from de-icing agents are all potential pollutants and have caused groundwater contamination. There also is the possibility of accidents involving vehicles carrying large quantities of chemicals. The run-off from roads can cause bacteriological contamination.

Cemeteries and graveyards form another possible health hazard. The minimum distance between a potable water well and a cemetery required by law in England is 91.4 m (100 yards). However, a distance of around 2500 m is better because the purifying processes in the soil can sometimes break down. Decomposing bodies produce fluids that can leak to the water table if a non-leakproof coffin is used. Typically, the leachate produced from a single grave is of the order of 0.4 m^3 $year^{-1}$ and this may constitute a threat for about 10 years. It is recommended that the water table in cemeteries should be at least 2.5 m deep and that an unsaturated depth of 0.7 m should exist below the bottom of a grave.

7.15 Groundwater monitoring and groundwater protection zones

In almost all situations where groundwater monitoring is undertaken, it is important that adequate background samples are obtained before a groundwater abstraction scheme is inaugurated. Without this background data, it is impossible to assess the effects of new development. It also should be remembered that groundwater can undergo cyclic changes in quality, so any apparent changes must be interpreted with caution. Hence, routine all-year-round monitoring is essential.

Pfannkuch (1982) pointed out that the first important step in designing an efficient groundwater-monitoring system is the proper understanding of the mechanics and dynamics of contaminant propagation (e.g. soluble or multiphase flow), the nature of the controlling flow mechanism (e.g. vadose or saturated flow) and the aquifer characteristics (e.g. permeability, porosity). He also itemized the objectives of a monitoring programme. Initially, the extent, nature and degree of contamination should be determined, as should the propagation mechanism and hydrological parameters so that the appropriate counter measures can be initiated. Next, a detection system should be established that can provide warning of any movement into critical areas. An assessment of the effectiveness of the immediate counter measures undertaken to offset the effects of contamination should be undertaken, and data recorded for long-term evaluation and compliance with standards. These objectives may have to be changed to suit the physical, political or other conditions prevailing at a site of a particular pollution incident.

The design of a water quality monitoring well and its method of construction must be related to the geology of the site. The depth and diameter of the well should be as small as possible so as to reduce the cost, but not so small that the well becomes difficult to use or ineffective. However, an additional requirement when water quality is concerned is that the well structure should not react with the groundwater. Clearly, if the groundwater does react with the well casing or screen, any subsequent analyses of samples taken from the well will be affected. Thus, monitoring wells frequently are constructed using plastic casings and screens. Plastic, however, does react with some types of pollutant, so that well materials must be selected to suit the anticipated conditions. The well screen should be provided with a gravel or sand pack to prevent the migration of fine material into the well. The selection of a suitable screen slot width and particle size distribution for the pack is accomplished using the same procedures as for a water supply well, although the design entrance velocity should be lower for a monitoring well than a water well. Generally, a sample of the aquifer material is obtained, its grain size distribution determined and the appropriate particle size for the pack decided. The pack should extend at least 0.3 m above and below the screened zone.

When deciding the diameter of a monitoring well, some consideration must be given to the method that is to be used to obtain a water sample. If a bailer or some form of sampler is to be inserted down the casing, then the well must be of a sufficiently large diameter to permit this. There are, however, several alternatives. For instance, water samples can be obtained using a specially designed submersible pump that is small enough to fit into a 50-mm diameter well.

After a well has been completed it should be developed, by pumping or bailing, until the water becomes clear. Background samples then should be collected over a lengthy period prior to the commencement of whatever it is that the well has been constructed to monitor.

When designing a monitoring network to detect pollution, the problem is to ensure that there are sufficient wells to allow the extent, configuration and concentration of the pollution plume to be determined, without incurring the unnecessary expense of constructing more wells than are actually required. The network of monitoring wells must be designed to suit a particular location and modified, as necessary, as new information is obtained or in response to changing conditions.

Diefendorf and Ausburn (1977) recommended that at least three monitoring wells should be used to observe the effects of a new development such as a landfill site. However, as a result of aquifer heterogeneity, non-uniform flow, variations in quality with depth and so on, it was suggested that three well clusters (rather than individual wells) may be required. Each well cluster should contain two or more wells located at different depths within the aquifer, or within different aquifers in the case of a multiple aquifer system (Fig. 7.26). One cluster should be located close to the source of the pollution, for early warning purposes, with another installation some suitable distance down gradient to assess the propagation of the plume. A third installation should be located up gradient of the monitored site to detect changes in background quality attributable to other causes.

Under favourable conditions, resistivity surveying can be used to determine the boundaries of a plume of polluted groundwater. Vertical electrical soundings are made in areas of known pollution in an attempt to define the top and bottom of the plume. Drillhole logs are used to establish geological control. Next, resistivity profiling is carried out to determine the lateral extent of the polluted groundwater. In this way, a quantitative assessment can be made of groundwater pollution. The method is based on the fact that formation resistivity depends on the conductivity of the pore fluid, as well as the properties of the porous medium. Generally, the resistivity of a rock mass is proportional to the water it is saturated with. The resistivity of the groundwater decreases if its salinity increases and hence the resistivity of the rock mass concerned decreases. Obviously, there must be a contrast in resistivity between the contaminant and groundwater in order to obtain useful results. Contrasts in resistivity may be attributed

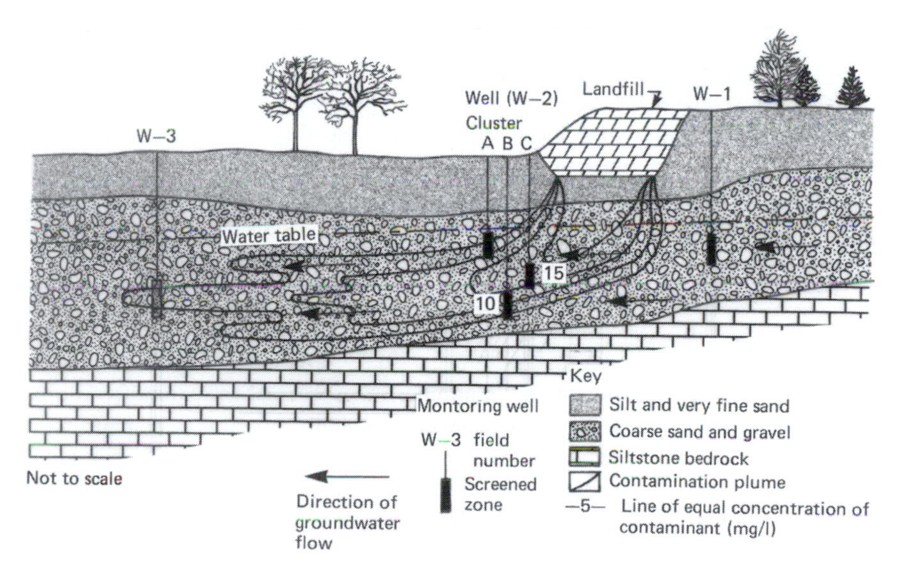

Figure 7.26 Hypothetical example of a well field to monitor a landfill contamination plume. *Source*: Diefendorf and Ausburn, 1977.

to mineralized groundwater with a higher than normal specific conductance due to pollution. However, the resistivity of a saturated porous sandstone or limestone aquifer depends not only on the salinity of the saturated groundwater but also on its porosity and the amount of conductive minerals, notably clay, in the rock matrix. The accurate determination of water quality only can take place if the effects of porosity and clay content are insignificant, or at least understood (Anon., 1988). Fortunately, it is often the case, especially in coastal aquifers, that extreme variations in groundwater salinity preclude the necessity of considering porosity and matrix conduction effects.

Ground conductivity profiling can be used to detect plumes of polluted water in simple situations. For example, a ground conductivity survey can be carried out if saline groundwater is near the surface. If the depth to water is too great, then the thickness of overlying unsaturated sediments can mask any contrasts between polluted and natural groundwater. In addition, the geology of the area has to be relatively uniform so that the conductivity values and profiles can be compared with others. Reported uses of this technique include tracing polluted groundwater from landfills, septic tanks, oil field brine disposal pits, AMD, sewage treatment effluent, industrial process waters and spent sulphur liquor.

The ionic content of the groundwater in a drillhole can be monitored by measuring the resistivity of the fluid, at a short electrode spacing. The quality of groundwater also can be estimated from the spontaneous potential deflection on an electric log, provided the specific conductance of the drilling mud is less than the specific conductance of the pore water. The method tends to give better results as the salinity of the pore water increases.

The designation of protection zones around sources of groundwater is now a major factor in groundwater management. According to Anon. (1998), the Environmental Agency in England and Wales recognizes three protection zones. First, there is an inner zone located immediately about the source defined by a 50-day travel time from any point below the water table to the source, which is based on bacteriological decay criteria, and in addition has a minimum 50-m radius from the source. Second, an outer zone is defined by a 400-day travel time or 25 per cent of the catchment area of the source, whichever is larger. The travel time of the latter zone is based on the minimum time needed to provide delay, dilution and attenuation of slowly degrading pollutants. However, the zone generally is not delineated for confined aquifers. Third, the source capture zone is defined as the area required to support the protected yield by long-term groundwater recharge. The source capture zone, in particular, needs to take account of hydrogeological conditions. Keating *et al.* also discussed the methods that can be used to determine protection zones around groundwater sources. The position in Scotland and Ireland regarding the delineation of protection zones is somewhat different and partly reflects the different hydrogeological conditions. The situation in Scotland has been discussed by Fox (2000) and in Ireland by Misstear and Daly (2000).

7.16 Rising water tables

Rising water tables can be responsible for a number of problems, for example, in urban areas, they can cause reductions in pile-bearing capacity beneath structures, damage to basement floors, leakages into tunnels and deep basements, and disruption to utility conduits. One of the basic reasons for rising water tables in England is decreased demand for groundwater due to changing industrial and land use patterns. Marsh and Davies (1983) reported that since the mid-1960s, the rate of abstraction of groundwater from the Chalk aquifer beneath London had decreased significantly so that groundwater levels increased by as much as 1 m year^{-1} in some areas. By 1997, the water table had risen by as much as 39 m in some places and was beginning to affect the clay above the Chalk. This has given rise to concern since it could give rise to differential settlement in building foundations. A scheme has been proposed to pump 70 million litres per day from the aquifer to relieve the clay. About half of this water would be used for domestic purposes. Rising water tables also occurred at the same time in the sandstone aquifers beneath parts of Birmingham and Liverpool. The control of groundwater levels therefore has an importance that extends beyond water supply considerations. Ideally, if structures are built during a period when the water table is at a particular level, then care should be taken to ensure that changing water levels do not diminish the integrity of these structures. This requires skilful long-term management of the groundwater resource.

A rather special case of a rising water table in Cardiff, Wales, was mentioned by Thomas (1997). The impoundment of the Rivers Taff and Ely by the Cardiff Bay Barrage will cause a rise in groundwater level in the south of Cardiff that will affect an old gasworks site. Potential contaminants at such sites include ammoniacal liquors, coal tars and other hydrocarbons, heavy metals, cyanides, and sulphur and sulphates. Groundwater modelling suggested that the water table in the gravel aquifer beneath the site would rise by 1.7–2.5 m and that the rise in the perched water table in the made ground above would amount to 0.1 m. The potential existed for dispersion of pollutants into the surrounding made ground through the perched water table and via local connections into the gravel aquifer below. Thomas claimed that the predicted rise in groundwater level at the site will reduce the hydraulic gradient between the two aquifers, which will result in a reduced potential for vertical pollutant migration where the aquifers are interconnected.

Subsurface mining commonly entails dewatering to avoid mines flooding. Consequently, when mines are abandoned groundwater rebound occurs. Groundwater rebound frequently results in a deterioration of groundwater quality within the rock masses in which mining took place so that after rebound is complete overlying aquifers or even surface waters can be polluted. For example, polluted groundwater may drain to the surface from old drainage adits, river bank mouths, faults, springs and shafts that intersect strata that is under artesian pressure. The problem of AMD has been dealt with in Section 7.14.6, and Burke and Younger (2000) and Dumpleton *et al.* (2001) described modelling methods used to assess such problems in two coalfield areas in England. A number of other consequences have been noted in areas in Britain where coal mines have been abandoned and groundwater levels have risen as a result. These include mine gas being driven ahead of the rising groundwater and at times being emitted at the surface, fault reactivation and associated ground movement, and accelerated deterioration of pillars in old shallow mines leading to their collapse and resultant ground subsidence. Fault reactivation is now a matter of concern in some areas, for instance, in parts of the old Durham coalfield in northeast England. Fault reactivation probably is the result of rising groundwater causing increases in pore water pressures within the fault zones so affected (Donnelly *et al.*, 1998). Increasing pore water pressures reduce the normal stress acting across a fault thereby reducing the shear strength along the fault.

References

Acworth, R.I. 2001. Physical and chemical properties of a DNAPL contaminated zone in a sand aquifer. *Quarterly Journal Engineering Geology and Hydrogeology*, **34**, 85–98.

Andrews, R.J., Lloyd, J.W. and Lerner, D.N. 1998. Sewage sludge disposal to agricultural land and other options in the UK. In: *Groundwater Contaminants and Their Migration*, Special Publication No. 128, Mather, J., Banks, D., Dumpleton, S. and Fermor, M. (eds), The Geological Society, London, 63–74.

Anon. 1983. *Code of Practice for Test Pumping Water Wells*, *BS 6316*, British Standards Institution, London.

Anon. 1988. Engineering geophysics. Engineering Group Working Party Report. *Quarterly Journal Engineering Geology*, **27**, 207–273.

Anon. 1990. *Methods of Test for Soils for Civil Engineering Purposes, BS 1377*. British Standards Institution, London.

Anon. 1993. *Guidelines for Drinking Water Quality*. World Health Organization (WHO), Geneva.

Anon. 1994. *Nutritional Aspects of Cardiovascular Disease*. No. 46, Committee on Medical Aspects of Food Policy, Her Majesty's Stationery Office, London.

Anon. 1998. *Policy and Practice for the Protection of Groundwater*. Environmental Agency, Her Majesty's Stationery Office, London.

Anon. 1999. *Code of Practice on Site Investigations, BS 5930*. British Standards Institution, London.

Barber, C. 1982. Domestic Waste and Leachate, Notes on Water Research No. 31, *Water Research Centre*, Medmenham, England.

Barker, J.A., Downing, R.A., Gray, D.A., Findlay, J., Kellaway, G.A., Parker, R.H. and Rollin, K.E. 2000. Hydrogeothermal studies in the United Kingdom. *Quarterly Journal Engineering Geology and Hydrogeology*, **33**, 41–58.

Barton, N.R. 1978. Suggested methods for the quantitative description of discontinuities in rock masses. ISRM Commission on Standardization of Laboratory and Field Tests. *International Journal Rock Mechanics Mining Science and Geomechanical Abstracts*, **15**, 319–368.

Bell, F.G. 1978. Some petrographic factors relating to porosity and permeability in the Fell Sandstones of Northumberland. *Quarterly Journal Engineering Geology*, **11**, 113–126.

Bell, F.G. 1998. *Environmental Geology*. Blackwell Scientific Publishers, Oxford.

Bell, F.G. and Kerr, A. 1993. Coal mining and water quality with illustrations from Britain. In: *Environmental Management, Geowater and Engineering Aspects*, Chowdhury, R.N. and Sivakumar, S. (eds), A.A. Balkema, Rotterdam, 607–614.

Bell, F.G. and Maud, R.R. 2000. A groundwater survey of the greater Durban area and environs, Natal, South Africa. *Environmental Geology*, **39**, 925–936.

Bell, F.G. and Jermy, C.A. 2002. Permeability and fluid flow in stressed strata and mine stability, Eastern Transvaal Coalfield, South Africa. *Quarterly Journal Engineering Geology and Hydrogeology*, **35**, (in press).

Bell, F.G., Sillito, A.J. and Jermy, C.A. 1996. Landfills and associated leachate in the greater Durban area, two case histories. In: *Engineering Geology of Waste Disposal*, Engineering Geology Special Publication No. 11, Bentley, S.P. (ed.), The Geological Society, London, 15–35.

Bell, F.G., Halbich, T.F.J. and Bullock, S.E.T. 2002. The effects of acid mine drainage from an old mine in the Witbank Coalfield, South Africa. *Quarterly Journal Engineering Geology and Hydrogeology*, **35**, 265–278.

Bernaix, J. 1969. New laboratory methods for studying the mechanical properties of rocks. *International Journal Rock Mechanics Mining Sciences*, **6**, 43–90.

Best, G.T. and Aikman, D.T. 1983. The treatment of ferruginous groundwater from an abandoned colliery. *Water Pollution Control*, **82**, 557–566.

Bishop, P.K., Lerner, D. and Stuart, M. 1998. Investigation of point source pollution by chlorinated solvents in two different geologies: a multi-layered Carboniferous sandstone-mudstone sequence and the Chalk. In: *Groundwater Contaminants and Their Migration*, Special Publication No. 128, Mather, J., Banks, D., Dumpleton, S. and Fermor, M. (eds), The Geological Society, London, 229–252.

Black, J.H. 1987. Flow and flow mechanisms in crystalline rock. In: *Fluid Flow in Sedimentary Basins and Aquifers*, Goff, J.C. and Williams, B.P.J. (eds), Special Publication No. 34, The Geological Society, London, 185–200.

Bliss, J.C. and Rushton, K.R. 1984. The reliability of packer tests for estimating the hydraulic conductivity of aquifer. *Quarterly Journal Engineering Geology*, **17**, 81–91.

Boak, R.A. and Packman, M.J. 2001. A methodology for the assessment of risk of *Cryptosporidium* contamination of groundwater. *Quarterly Journal Engineering Geology and Hydrogeology*, **34**, 187–194.

Brassington, F.C. and Walthall, S. 1985. Field techniques using borehole packers in hydrogeological investigations. *Quarterly Journal Engineering Geology*, **18**, 181–194.

Brodie, M.J., Broughton, L.M. and Robertson, A. 1989. A conceptional rock classification system for waste management and a laboratory method for ARD prediction from rock piles. In: *British Columbia Acid Mine Drainage Task Force*, Draft Technical Guide, **1**, 130–135.

Broster, B.E. and Pupek, D.A. 2001. The significance of buried landscape in subsurface migration of dense non-aqueous phase liquids: the case of perchloroethylene in the Sussex aquifer, New Brunswick. *Environmental and Engineering Geoscience*, **7**, 17–29.

Brown, E.T. (ed.) 1981. *Rock Characterization Testing and Monitoring*. Pergamon Press, Oxford.

Brown, R.F. and Signor, D.C. 1973. Artificial recharge – state of the art. *Ground Water*, **12**, No. 3, 152–160.

Brown, R.H., Konoplyantsev, A.A., Ineson, J. and Kovalevsky, V.S. (eds) 1972. *Groundwater Studies*. An International Guide for Research and Practice, Studies and Reports in Hydrology 7, UNESCO, Paris.

Bruzzi, D., Cabrobbi, L. and Nobile, M. 1983. Use of a down-hole flowmeter for the study of an aquifer. *Bulletin International Association Engineering Geology*, **26–27**, 367–375.

Bullock, S.E.T. and Bell, F.G. 1995. An investigation of surface and groundwater quality at a mine in the north west Transvaal, South Africa. *Transactions Institution Mining and Metallurgy*, **104**, Section A Mining Industry, A125–A133.

Burgess, W.G., Dottridge, J. and Symington, R.M. 1998. Methyl tertiary butyl ether (MTBE): a groundwater contaminant of growing concern. In: *Groundwater Contaminants and Their Migration*, Special Publication No. 128, Mather, J., Banks, D., Dumpleton, S. and Fermor, M. (eds), The Geological Society, London, 29–34.

Burke, S.P. and Younger, P.L. 2000. Groundwater rebound in the South Yorkshire coalfield: a first approximation using the GRAM model. *Quarterly Journal Engineering Geology and Hydrogeology*, **33**, 149–160.

Cambridge, M. 1995. Use of passive systems for treatment of mine outflows and seepages. *Minerals Industry International Bulletin, Institution Mining and Metallurgy*, **1024**, 35–42.

Campbell, M.D. and Lehr, J.H. 1973. *Water Well Technology*. McGraw-Hill, New York.

CEC. 1980. *Directive Relating to Quality of Water for Human Consumption*, 80/778/EEC. Commission of the European Community, Brussels.

Cedergren, H. 1986. *Seepage, Drainage and Flow Nets*. Second Edition. Wiley, New York.

Chapple, M.C., Vittorio, L.F., Tucker, W.A. and Richey, M.G. 1998. Capillary influence on the operation and effectiveness of LAPL interceptor trenches. In: *Contaminated Land and Groundwater: Future Directions*. Engineering Geology Special Publication No. 14, Lerner, D.N. and Walton, N.R.G. (eds), The Geological Society, London, 13–18.

Cherry, J.A. (ed.) 1983. Migration of contaminants in groundwater at a landfill: a case study. *Journal Hydrology*, Special issue, 1–398.

Chilton, P.J., Lawrence, A.R. and Stuart, M.E. 1998. Pesticides in groundwater: some preliminary results from recent research in temperate and tropical environments. In: *Groundwater Contaminants and Their Migration*, Special Publication No. 128, Mather, J., Banks, D., Dumpleton, S. and Fermor, M. (eds), The Geological Society, London, 333–345.

Clark, L. 1977. The analysis and planning of step-drawdown tests. *Quarterly Journal Engineering Geology*, **10**, 125–143.

Clark, L. 1985. Groundwater abstraction from basement complex areas in Africa. *Quarterly Journal Engineering Geology*, **18**, 25–34.

Clark, L. and Sims, P.A. 1998. Investigation and clean-up of jet-fuel contaminated groundwater at Heathrow International Airport, UK. In: *Groundwater Contaminants and Their Migration*, Special Publication No. 128, Mather, J., Banks, D., Dumpleton, S. and Fermor, M. (eds), The Geological Society, London, 147–157.

Connelly, R.J., Harcourt, K.J., Chapman, J. and Williams, D. 1995. Approach of remediation of ferruginous discharge in the South Wales coalfield and its application to closure planning. *Minerals Industry International Bulletin, Institution Mining and Metallurgy*, **1024**, 43–48.

Cook, H.F. 1991. Nitrate protection zones: targeting and land-use over an aquifer. *Land Use Policy*, **8**, 16–28.

Craun, G.F. 1985. A summary of waterborne illness through contaminated groundwater. *Journal Environmental Health*, **48**, 122–127.

Darcy, H. 1856. *Les Fontaines Publiques de la Ville de Dijon*. Dalmont, Paris.

Davis, R.D. 1980. *Control of Contamination Problems in the Treatment and Disposal of Sewage Sludge*, Technical Report 156, Water Research Centre, Medmenham, England.

Diefendorf, A.F. and Ausburn, R. 1977. Groundwater monitoring wells. *Public Works*, **108**, No. 7, 48–50.

Donnelly, L.J., Young, B. and Dumpleton, S. 1998. *Whittle Colliery, Northumberland, and the Environmental and Geotechnical Implications of Mine Water Recovery*. British Geological Survey, Technical Report No. WN/98/15C, Keyworth, Nottinghamshire.

Dumpleton, S., Robins, N.S., Walker, J.A. and Merrin, P.D. 2001. Mine water rebound in south Nottinghamshire: risk evaluation using 3-D visualization and predictive modelling. *Quarterly Journal Engineering Geology and Hydrogeology*, **34**, 307–319.

Edmunds, W.M. and Smedley, P.L. 1996. Groundwater geochemistry and health. In: *Environmental Geochemistry and Health with Special Reference to Developing Countries*, Appleton, J.D., Fuge, R. and McCall, G.J.H. (eds), Special Publication No. 113, The Geological Society, London, 91–105.

Elliot, T., Chadha, D.S. and Younger, P.L. 2001. Water quality impacts and paleohydrogeology in the Yorkshire chalk aquifer, UK. *Quarterly Journal of Engineering Geology and Hydrogeology*, **34**, 385–398.

Fair, G.M. and Hatch, L.P. 1935. Fundamental factors governing the streamline flow of water through sand. *Journal American Water Works Association*, **25,** 1151–1165.

Finkel, M., Leidl, R. and Teutsch, G. 1998. A modelling study of the efficiency of groundwater treatment walls in heterogeneous aquifers. *International Association Hydrological Sciences*, Publication 250, 467–474.

Foster, S.S.D. 2000. Assessing and controlling the impacts of agriculture on groundwater – from barley barons to beef bans. *Quarterly Journal Engineering Geology and Hydrogeology*, **33**, 263–280.

Foster, S.S.D. and Crease, R.I. 1974. Nitrate pollution of Chalk groundwater in east Yorkshire – a hydrological appraisal. *Journal Institution Water Engineers Scientists*, **28**, 178–194.

Foster, S.S.D. and Young, C.P. 1980. Groundwater contamination due to agricultural land-use practices in the United Kingdom. In: *Aquifer Contamination and Protection*, Jackson, R.E. (ed.), Chapter 11-3, Project 8.3 of the International Hydrological Programme, Studies and Reports in Hydrology, **30**, UNESCO, Paris.

Fox, I.A. 2000. Groundwater protection in Scotland. In: *Groundwater in the Celtic Regions: Studies in Hard Rock and Quaternary Hydrogeology*, Special Publication No. 182, Robins, N.S. and Misstear, B.D. (eds) The Geological Society, London, 67–70.

Frazer, H.J. 1935. Experimental study of porosity and permeability of clastic sediments. *Journal Geology*, **43**, 910–1010.

Garnier, M., Leone, A., Uricchio, V. and Marini, R. 1998. Application of the GLEAMS model to assess groundwater pollution risk caused by animal waste land disposal. In: *Contaminated Land and Groundwater – Future Directions*, Engineering Geology Special Publication No. 13, Lerner, D.N. and Walton, N. (eds), The Geological Society, London, 93–99.

Glover, R.E. and Balmer, C.E. 1954. River depletion resulting from a well pumping near a river. *Transactions American Geophysical Union*, **35**, 468–470.

Goodman, R.E. and Sundaram, P.N. 1980. Permeability and piping in fractured rocks. *Proceedings American Society Civil Engineers, Journal Geotechnical Engineering Division*, **103**, 485–498.

Gray, D.A., Mather, J.D. and Harrison, I.B. 1974. Review of groundwater pollution for waste sites in England and Wales with provisional guidelines for future site selection. *Quarterly Journal Engineering Geology*, **7**, 181–196.

Hagerty, D.J. and Pavoni, J.L. 1973. Geologic aspects of landfill refuse disposal. *Engineering Geology*, **7**, 219–230.

Hamill, L. and Bell, F.G. 1986. *Groundwater Resource Development*. Butterworths, London.

Hem, J.D. 1985. Study and interpretation of the chemical characteristics of natural water. *United States Geological Survey*, Water Supply Paper 2254.

Herbert, A.W. and Lloyd, J.W. 2000. Approaches to modelling saline intrusion for assessment of small island water resources. *Quarterly Journal Engineering Geology and Hydrogeology*, **33**, 77–86.

Hibbs, B.J. and Boghici, R. 1999. On the Rio Grande aquifer: flow relationships, salinization, and environmental problems from El Paso to Fort Quitman, Texas. *Environmental and Engineering Geoscience*, **5**, 51–59.

Hiscock, K.M., Lovett, A.A., Brainard, J.S. and Parfitt, J.P. 1995. Groundwater vulnerability assessment: two case studies using GIS methodology. *Quarterly Journal Engineering Geology*, **28**, 179–194.

Holmes, R. 1984. Comparison of different methods of estimating infiltration at a landfill site in south Essex with implications for leachate management and control. *Quarterly Journal Engineering Geology*, **17**, 9–18.

Howe, R.H., Wilke, H.R. and Bloodgood, D.E. 1956. Application of air photo interpretation in the location of groundwater. *Journal American Water Works Association*, **48**, 1380–1390.

Ineson, J. 1970. Development of groundwater resources in England and Wales. *Journal Institution Water Engineers*, **24**, 155–177.

Ineson, J. and Gray, D.A. 1963. Electrical investigations of borehole fluids. *Journal Hydrology*, **1**, 204–218.

Jensen, O.M. 1982. Nitrate in drinking water and cancer in northern Jutland, Denmark, with special reference to stomach cancer. *Ecotoxicology and Environmental Safety*, **6**, 258–267.

Keeley, J.W. and Scalf, M.R. 1969. Aquifer storage determination by radio tracer techniques. *Ground Water*, **7**, 17–22.

Lachassagne, P., Wyns, R. and Berard, P. 2001. Exploitation of high yields in hard rock aquifers: downscaling methodology combining GIS and multicriteria analysis to delineate field prospecting zones. *Ground Water*, **39**, 568–581.

Lawrence, A.R., Foster, S.S.D. and Izzard, P.W. 1983. Nitrate pollution of chalk groundwater in east Yorkshire – a decade on. *Journal Institution Water Engineers and Scientists*, **37**, 410–420.

Linsley, R.K., Kohler, M.A. and Paulhus, J.L.H. 1982. *Hydrology for Engineers*. Third Edition, McGraw-Hill, New York.

Lohman, S.W. 1972. *Groundwater Hydraulics*. United States Geological Survey, Professional Paper 708, Washington, DC.

Londergan, J.T., Meinardus, H.W. and Mariner, P.E. 2001. DNAPL removal from a heterogeneous alluvial aquifer by surfactant-enhanced aquifer remediation. *Environmental Science and Technology*, **21**, No. 4, 57–67.

Lovelock, P.E.R., Price, M. and Tate, T.K. 1975. Groundwater conditions in the Penrith Sandstone at Cliburn, Westmoreland. *Journal Institution Water Engineers*, **29**, 157–174.

Mackay, D.M. 1998. Is clean-up of VOC-contaminated groundwater feasible? In: *Contaminated Land and Groundwater – Future Directions*, Engineering Geology Special Publication No. 13, Lerner, D.N. and Walton, N. (eds), The Geological Society, London, 3–11.

Mackay, D.M. and Cherry, J.A. 1989. Groundwater contamination: limits of pump-and-treat remediation. *Environmental Science and Technology*, **23**, 630–636.

Malkki, E. and Vihuri, H. 1983. Measurement of hydraulic conductivity by tracer dilution method. *Bulletin International Association Engineering Geology*, **26–27**, 473–476.

Marsh, T.J. and Davies, P.A. 1983. The decline and partial recovery of groundwater levels below London. *Proceedings Institution Civil Engineers*, Part 1, **74**, 263–276.

Masironi, R. 1979. Geochemistry and cardiovascular diseases. *Philosophical Transactions Royal Society London*, **B288**, 193–203.

Mather, J.D., Gray, D.A., Allen, R.A. and Smith, D.B. 1973. Groundwater recharge in the Lower Greensand of the London Basin – results from tritium and carbon-14 determinations. *Quarterly Journal Engineering Geology*, **6**, 141–142.

McDonald, R.J., Russill, N.R.W., Miliorizosis, M. and Thomas, J.W. 1998. A geophysical investigation of saline intrusion and geological structure beneath areas of tidal coastal wetland at Langstone Harbour, Hampshire, UK. In: *Groundwater Pollution, Aquifer Recharge and Vulnerability*, Special Publication No. 130, Robins, N.S. (ed.), The Geological Society, London, 77–94.

Meinzer, O. 1942. Occurrence, origin and discharge of groundwater. In: *Hydrology*, Meinzer, O. (ed.), Dover, New York, 385–443.

Meyer, W.R. 1962. Use of a retention moisture probe to determine the storage coefficient of an unconfined aquifer. *Research in Experimental Hydrology*, United States Geological Survey, Professional Paper 450-E, E174–E176, Washington, DC.

Misstear, B.D. and Daly, D. 2000. Groundwater protection in a Celtic region: the Irish example. In: *Groundwater in the Celtic Regions: Studies in Hard Rock and Quaternary Hydrogeology*, Special Publication No. 182, Robins, N.S. and Misstear, B.D. (eds), The Geological Society, London, 53–65.

Muldoon, D.G., Connelly, P.J., Makovitch, A.W., Holden, J.M.W. and Tunstall-Pedoe, N. 1998. Groundwater remediation of chlorinated hydrocarbons at an electronics facility in northeastern USA. In: *Groundwater Contaminants and Their Migration*, Special Publication No. 128, Mather, J., Banks, D., Dumpleton, S. and Fermor, M. (eds), The Geological Society, London, 183–200.

Nace, R.L. 1969. World water inventory and control. In: *Water, Earth and Man*, Chorley, R.J. (ed.), Methuen, London, 31–42.

Naylor, J.A., Rowland, C.D., Young, C.P. and Barber, C. 1978. *The Investigation of Landfill Sites*. Technical Report 91, Water Research Centre, Medmenham, England.

Nutbrown, D.A. 1976. Optimal pumping regimes in an unconfined coastal aquifer. *Journal Hydrology*, **31**, 271–280.

Oteri, A.U.E. 1983. Delineation of saline intrusion in the Dungeness shingle aquifer using surface geophysics. *Quarterly Journal Engineering Geology*, **16**, 43–52.

Patten, E.P. and Bennett, G.D. 1962. Methods of flow measurement in well bores: general groundwater techniques. *United States Geological Survey*, Water Supply Paper 1544-C, Washington DC.

Pedley, S. and Howard, G. 1997. The public health implications of microbiological contamination of groundwater. *Quarterly Journal Engineering Geology*, **30**, 179–188.

Pfannkuch, H.A. 1982. Problems of monitoring network design to detect unanticipated contamination, *Ground Water Monitoring Review*, **2**, No. 1, 67–76.

Pickford, J.A. 1979. Control of pollution and disease in developing countries. *Water Pollution Control*, **78**, 239–253.

Piispanen, R. 1993. Water hardness and cardiovascular mortality in Finland. *Environmental Geochemistry and Health*, **15**, 201–208.

Ramana, Y.V. and Venkatanarayana, R. 1971. An air porosimeter for the porosity of rocks. *International Journal Rock Mechanics Mining Science*, **8**, 29–53.

Raucher, R.L. 1983. A conceptual framework for measuring the benefits of groundwater protection. *Water Resources Research*, **19**, 320–326.

Revelle, R. 1941. Criteria for recognition of sea water in ground-waters. *Transactions American Geophysical Union*, **22**, 593–597.

Robertson, W.D., Blowes, D.W., Ptacek, C.J. and Cherry, J.A. 2000. Long-term performance of in situ reactive barriers for nitrate remediation. *Ground Water*, **38**, 689–695.

Robinson, H.D. 1984. On site treatment of leachate using aerobic biological techniques. *Quarterly Journal Engineering Geology*, **17**, 31–37.

Sawyer, C.N. and McCarthy, P.L. 1967. *Chemistry for Sanitary Engineers*. Second Edition, McGraw-Hill, New York.

Scherer, M.M., Richter, S., Valentine, R.L. and Alvares, P.J.J. 2000. Chemistry and microbiology of permeable reactive barriers for *in situ* groundwater clean up. *Critical Reviews in Environmental Science and Technology*, **30**, 363–411.

Selby, K.H. and Skinner, A.C. 1979. Aquifer protection in the Severn-Trent region: policy and practice. *Water Pollution Control*, **78**, 320–326.

Serafim, J.L. 1968. The influence of interstitial water on rock masses. In: *Rock Mechanics in Engineering Practice*, Stagg, K.G. and Zienkiewicz, O.C. (eds), Wiley, London.

Simmers, I. 1998. Groundwater recharge: an overview of estimation 'problems' and recent developments. In: *Groundwater Pollution, Aquifer Recharge and Vulnerability*, Special Publication No. 130, Robins, N.S. (ed.), The Geological Society, London, 107–115.

Smith, D. and Guitjens, J.C. 1998. Characterization of urban surfacing ground water in northwest Henderson, Clark County, Nevada. *Environmental and Engineering Geoscience*, **4**, 455–477.

Smith, D.B., Wearn, P.L., Richards, H.J. and Rowe, P.C. 1970. Water movement in the unsaturated zone of high and low permeability strata by measuring natural tritium, *Proceeding Symposium Isotope Hydrology*, International Atomic Energy Association, Vienna, 73–87.

Soudain, M. 1997. Iron constitution. *Ground Engineering*, **30**, No. 6, 20–21.

Sutcliffe, G. and Mostyn, G. 1983. Permeability testing for the O K Tedi (Papua New Guinea). *Bulletin International Association Engineering Geology*. **26–27**, 501–508.

Sweeney, J., Hart, P.A. and McConvey, P.J. 1998. Investigation and management of pesticide pollution in the Lincolnshire Limestone aquifer in eastern England. In: *Groundwater Contaminants and Their Migration*, Special Publication No. 128, Mather, J., Banks, D., Dumpleton, S. and Fermor, M. (eds), The Geological Society, London, 347–360.

Tate, T.K., Robertson, A.S. and Gray, D.A. 1970. The hydrogeological investigation of fissure flow by borehole logging techniques. *Quarterly Journal Engineering Geology*, **2**, 195–215.

Theis, C.V. 1941. The effect of a well on the flow of a near stream. *Transactions American Geophysical Union*, **22**, 734–738.

Thomas, B.R. 1997. Possible effects of rising groundwater levels on a gasworks site: a case study from Cardiff Bay, UK. *Quarterly Journal Engineering Geology*, **30**, 79–93.

Tompkins, J.A., Smith, S.R., Cartmell, E. and Whealer, H.S. 2001. *In situ* bioremediation is a viable option for denitrification of Chalk groundwaters. *Quarterly Journal Engineering Geology and Hydrogeology*, **34**, 111–125.

Turkmen, S., Ozguler, E., Taga, H. and Karaogullarindan, T. 2002. Seepage problems in the karstic limestone foundation of Kalecik Dam (south Turkey). *Engineering Geology*, **63**, 247–257.

Volker, R.E. and Rushton, K.R. 1982. An assessment of the importance of some parameters for seawater intrusion in aquifers and a comparison of dispersive and sharp-interface modelling approaches. *Journal Hydrology*, **56**, 239–250.

Walton, J., Ohlmacher, G., Utz, D. and Kutianawala, M. 1999. Response of the Rio Grande and shallow ground water in the Mesilla Bolson to irrigation, climate stress and pumping. *Environmental and Engineering Geoscience*, **5**, 41–50.

Warnakulasuriya, K.A.A.S., Balasuriya, S. and Perera, P.A.J. 1990. Prevalence of dental fluorosis in four selected schools from different areas of Sri Lanka. *Ceylon Medical Journal*, **35**, 125–128.

West, J.M. and Chilton, P.J. 1997. Aquifers as environments for microbiological activity. *Quarterly Journal Engineering Geology*, **30**, 147–154.

WHO. 1992. Guidelines for Drinking Water Quality. World Health Organisation, Geneva.

Wittke, W. 1973. Percolation through fissured rock. *Bulletin International Association Engineering Geology*, **7**, 3–28.

Wright, E.P. 1992. The hydrogeology of crystalline basement aquifers in Africa. In: *The Hydrogeology of Crystalline Basement Aquifers in Africa*, Wright, E.P. and Burgess, W.G. (eds), Special Publication No. 66, The Geological Society, London, 1–28.

Zoetemann, B.C., Hribec, J. and Brinkmann, F.J. 1975. The Veluwe artificial recharge plan, water quality aspects. *Journal Institution Water Engineers*, **30**, 123–137.

8 River and coastal engineering

All rivers form part of a drainage system, the form of which is influenced by rock type and structure, the nature of the vegetation cover and the climate. For instance, Schumm and Spitz (1996) provided an account of the influence of geology on the lower Mississippi River and its alluvial valley in which they indicated that the river had reacted to uplift, faults, plutonic intrusions, clay plugs and Pleistocene gravels. Indeed, they maintained that the directions of streams are similar to those of the major structural trends of the area, the positions of streams reflecting movements in the rock masses in which the alluvial valley has been formed. An understanding of the processes that underlie river development forms the basis of correct river management.

Rivers also form part of the hydrological cycle in that they carry precipitation run-off. This run-off is the surface water that remains after evapotranspiration and infiltration into the ground have taken place. Some precipitation may be frozen, only to contribute to run-off at some other time, while any precipitation that has infiltrated into the ground may reappear as springs where the water table meets the ground surface. Although, due to heavy rainfall or in areas with few channels, the run-off may occur as a sheet, usually it becomes concentrated into channels that become eroded by the flow of water and so eventually form valleys.

In addition, rivers play a part in the coastal environment in that they bring water and sediment to the coastal zone. They, in turn, are influenced by coastal processes, notably tides. If changes occur within a river system that affect the sediment supply to the coast, in particular, if this is reduced, then erosion may be accelerated in those areas so affected. Conversely, beaches may be expanded where extra sediment is introduced into the coastal zone by changes in river regime.

Johnson (1919) distinguished three elements in a shoreline, the coast, the shore and the offshore. The coast was defined as the land immediately behind the cliffs whilst the shore was regarded as that area between the base of the cliffs and low-water mark; the area that extended seawards from the low-water mark was termed the offshore. The shore was further divided into foreshore and backshore, the former embracing the intertidal zone whilst the latter extended from the foreshore to the cliffs. Those deposits that cover the shore constitute the beach.

8.1 Fluvial processes

Fluvial processes comprise the full range of movement and action of surface water on sloping terrain. Unchannelled flow is termed sheet wash or overland flow. Rills and gullies are features associated with small-scale incisement. Rills possess negligible drainage areas whilst gullies have steep banks with entrenched channels with steep headwalls. Streams and rivers represent larger channelled flow. They develop a characteristic pattern and obey quantitative and geometrical laws in their development. The shape and size of a river basin have an important influence on water flow characteristics, and the amount of erosion, transportation and deposition that occurs (Schumm, 1969).

Carlston (1968) proposed that the mean annual discharge (Q_m) of a stream was related to channel slope (S) in the following manner:

$$S \propto Q_m^b \tag{8.1}$$

He found that the value of the constant, b, varied significantly with the character of the channel. Where the channel was cut into an alluvial bed, and on sections that were in a steady state and b could be defined, the correlation between Q_m and S was good. In ungraded streams, however, there was no significant correlation between these two parameters. As the discharge necessary for the formation of channels is that which approaches bankfull stage, the mean annual flood should be closely related to channel morphology. Subsequently, expressions were developed by Schumm (1977) to relate channel hydrology and other morphological characteristics to channel width (an index of discharge) and channel width/depth ratio (an index of the type of sediment load). For example, he showed that channel width (W) and depth (Z) were related to mean annual discharge (Q_m) and mass (M) as follows:

$$W = 2.3\frac{Q_m^{0.38}}{M^{0.39}} \tag{8.2a}$$

and

$$Z = 0.6M^{0.34} \times Q_m^{0.29} \tag{8.2b}$$

Water discharge influences channel width (W), depth (Z), mean wavelength (λ) and gradient (S) of a river as follows:

$$Q = \frac{WZ\lambda}{S} \tag{8.3}$$

In stable rivers with sand-beds the relation between bedload (L_b) and channel morphology is given by:

$$L_b \cong \frac{W\lambda S}{Zp} \tag{8.4}$$

where p is sinuosity (the ratio of channel length to valley length). Also, as bedload and sediment size (d) increase, either water discharge (Q) or slope (S) or both increase to compensate (i.e. $L_b d \cong QS$). Streams in rugged terrain tend to carry coarse bedload and generally are less sinuous. Their width/depth cross-sections are large in relative terms to other streams because a wide shallow channel is more efficient for transportation of coarse bedload since higher velocities occur nearer the channel floor. As a stream moves farther from its source it increases in sinuosity, and bedload becomes finer. Sinuous streams minimize their total work expenditure by adjusting their curvature to their slopes. By increasing their sinuosity, energy is dissipated more uniformly through each unit distance along the channel.

Leopold and Wolman (1957) maintained that there are three different flow patterns in river development, namely straight, meandering and braided types, in which slope, sediment load, and discharge are the controlling factors. Straight streams possess essentially straight banks, having a sinuosity of less than 1.5, but flow between the banks is not necessarily straight.

Turbulent flow does not move water in a straight line because secondary currents develop with transverse flow producing various bedforms, pools and riffles. A pool is a deep reach whereas a riffle is shallower with elongated sediment bars. Braided stream patterns are characteristic of a main channel that is divided into a network of anastomozing and branching smaller channels, within which are small lateral and horizontal bars and islands. The network has a sinuosity that usually exceeds 1.5. Braided streams are common in glacial, proglacial, arid and semi-arid environments, and regions where weathering produces debris that is impoverished in terms of fine-grained sediment (Fig. 8.1). Rapid variations in discharge capable of creating alternating erosion and deposition sequences also are important in the formation of braided streams. Usually, stream banks are low and have poor cohesiveness so that they are erodible. The gradient in the bifurcated reaches increases as a result of divided flow and the necessity to maintain discharge. Tributary streams supply coarser material than the main channel can transport during normal flow regimes, so bars and various bedforms develop in the lower flow regime. Braids become more elaborate in sandbed streams as particle size is reduced. Meandering streams are characterized by sinuosity greater than 1.5, the channel being singular and free-swinging over a plain (Fig. 8.2). Empirical relationships have been developed concerning the geometry of alluvial channels, for example, meander wavelength (λ) and bankfull width (W) are related by:

$$\lambda = aW^b \tag{8.5}$$

where a varies between 7 and 10, and b is approximately 1.0. The relationship between the mean radius of curvature (R) and wavelength (λ) is expressed as:

$$\lambda = eR^f \tag{8.6}$$

Figure 8.1 The Rakaia River, South Island, New Zealand. This braided river flows over a sequence of thick gravel deposits.

Figure 8.2 Meanders and cut-offs along part of the Mudjalik River, Saskatchewan, Canada.

in which e and f are approximately 5 and 1.0, respectively. The mean radius of curvature is approximately twice the bankfull width. Such relationships can only define general tendencies and so may not apply to particular situations. A more complex expression used to determine meander wavelength was proposed by Schumm (1967), that is:

$$\lambda = 1890 \frac{Q_{\mathrm{m}}^{0.34}}{M^{0.74}} \tag{8.7}$$

in which Q_{m} is the mean annual discharge and M is the percentage of silt and clay in the river load.

8.1.1 River flow

It generally is considered that, except at times of flood, a stream has a steady uniform flow, that is, one in which, at any given point, the depth does not vary with time. If the flow is uniform, the depth is constant over the length of the stream concerned (Fig. 8.3). Although these two assumptions are not strictly true, they make possible simple and satisfactory solutions for river engineering problems. Assuming a steady uniform flow, the rate at which water passes through successive cross-sections of a stream is constant. River flow is measured as discharge of volume of water passing a given point per unit time. The height or stage of water in a channel depends on discharge, as well as the shape and capacity of a river channel (i.e. the total amount of sediment carried). Bankfull discharge refers to the maximum volume of water at a certain velocity of flow that a river can sustain without overbank spillage. Hydraulic geometry and bankfull discharge alter if the channel is changed by erosion.

If water flows along a smooth, straight channel at very low velocities, it moves in laminar flow, with parallel layers of water shearing one over the other. In laminar flow the layer of

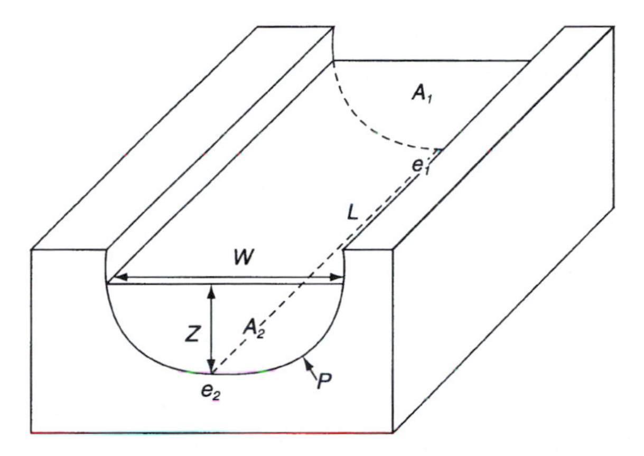

Figure 8.3 Stream channel morphometry. Stream width, W, is the actual width of water in the channel. The wetted perimeter, P, is the outline of the edge where water and channel meet. Cross-section, A, is the area of a traverse section of the river. Depth, Z, is approximately the same as the hydraulic radius, R, which is the cross-section divided by the wetted perimeter ($R = AP$). Stream gradient, S, is the drop in elevation ($e_1 - e_2$) between two points on the bottom of a channel divided by the projected horizontal distance between them, L.

maximum velocity lies below the water surface whilst around the wetted perimeter it is least. Laminar flow cannot support particles in suspension and is not found in natural streams except near the bed and banks. When the velocity of flow exceeds a critical value it becomes turbulent. Fluid components in turbulent flow follow a complex pattern of movement, components mixing with each other and secondary eddies are superimposed on the main forward flow. The Reynolds number (N_R) is the starting point for calculation of erosion and transportation within a channel system and commonly is used to distinguish between laminar and turbulent flow, it being expressed as:

$$N_R = \rho \left(\frac{vR}{\mu} \right) \tag{8.8}$$

where ρ is the fluid density, v is the mean velocity, R is the hydraulic radius (the ratio between the cross-sectional area of a river channel and the length of its wetted perimeter) and μ is the viscosity. Flow is laminar for small values of Reynolds number and turbulent for higher ones. The Reynolds number for flow in streams is frequently over 500, varying from 300 to 600.

There are two kinds of turbulent flow, namely streaming and shooting flows. Streaming flow refers to the ordinary turbulence found in most streams whereas shooting flow occurs at higher velocities such as found in rapids. Whether turbulent flow is shooting or streaming it is determined by the Froude number (F_n):

$$F_n = \frac{v}{\sqrt{gZ}} \tag{8.9}$$

where v is the mean velocity, g is the force of gravity and Z is the depth of water. If the Froude number is less than 1, then a stream is in the streaming flow regime, whilst if greater than 1 it is in the shooting flow regime.

Generally, the highest velocity of a stream is at its centre, below or extending below the surface. The exact location of maximum velocity depends on channel shape, roughness and sinuosity but it usually lies between 0.05 and 0.25 of the depth. In a symmetrical river channel, the maximum water velocity is below the surface and centred. Regions of moderate velocity but high turbulence occur outward from the centre, being greatest near the bottom. Near the wetted perimeter velocities and turbulence are low. On the other hand, in an asymmetrical channel the zone of maximum velocity shifts away from the centre toward the deeper side. In such instances, the zone of maximum turbulence is raised on the shallow side and lowered on the deeper side. Consequently, channel morphology has a significant influence on erosion.

Turbulence and velocity are very closely related to erosion, transportation and deposition. The work done by a stream is a function of the energy it possesses. Potential energy is converted by downflow to kinetic energy, which is mostly dissipated in friction. Stream energy therefore is lost owing to friction from turbulent mixing and, as such, frictional losses are dependent upon channel roughness and shape. Total energy is influenced mostly by velocity, which is a function of the stream gradient, volume and viscosity of water in flow, and the characteristics of the channel cross-section and bed. This relationship has been embodied in the Chezy formula, which expresses velocity as a function of hydraulic radius (R) and slope (S):

$$v = C\sqrt{RS} \tag{8.10}$$

where v is the mean velocity and C is a constant that depends upon gravity and other factors contributing to the friction force. The minimum bed erosion takes place when the gradient of a channel is low and the wetted perimeter is large compared with the cross-sectional area of the channel.

The Manning formula represents an attempt to refine the Chezy equation in terms of the constant C

$$v = \frac{1.49}{n} R^{2/3} S^{1/2} \tag{8.11}$$

where the terms are the same as the Chezy equation and n is a roughness factor. The velocity of flow increases as roughness decreases for a channel of particular gradient and dimensions. The roughness factor has to be determined empirically and varies not only for different streams but also for the same stream under different conditions and at different times. In natural channels the value of n is 0.01 for smooth beds, and about 0.02 for sand and 0.03 for gravel beds. The roughness coefficients of some natural streams are given in Table 8.1. Anything that affects the roughness of a channel changes n, including the size and shape of grains on the bed, sinuosity, and obstructions in the channel section. Variation in discharge also affects the roughness factor since depth of water and volume influence the roughness.

The shear stress (τ) exerted on a fixed boundary by a turbulent fluid is a function of fluid density and shear velocity of the fluid. The mean boundary shear stress equation for open channels is:

$$\tau = \rho g R S \tag{8.12}$$

where ρ is the density, g is the acceleration due to gravity, R is the hydraulic radius and S is the slope of the channel. Since shear stress is related to the velocity gradient, as well as the energy gradient and the hydraulic radius (R), then frictional velocity (U^*) is given by:

$$U^* = gRS \tag{8.13}$$

Table 8.1 Values of roughness coefficient n for natural streams

Description of stream	*Normal n*
On a plain	
Clean straight channel, full stage, no riffs or deep pools	0.030
Same as above but with more stones and weeds	0.035
Clean winding channel, some pools and shoals	0.040
Sluggish reaches, weedy, deep pools	0.070
Mountain streams	
No vegetation, steep banks, bottom of gravel, cobbles and a few boulders	0.040
No vegetation, steep banks, bottom of cobbles and large boulders	0.050
Flood plains	
Pasture, no brush, short grass	0.030
Pasture, no brush, high grass	0.035
Brush, scattered to dense	0.050–0.10
Trees, dense to cleared, with stumps	0.150–0.04

Source: Chow, 1964. Reproduced by kind permission of The McGraw-Hill Companies.

The hydraulics of sediment transport involves dissolved load that offers no resistance and suspended load that serves to dampen turbulence thereby increasing stream efficiency. In order to determine which set of flow conditions cause entrainment of a particle it is important to calculate factors such as the critical flow velocity, the critical boundary shear stress and the critical lifting force. These factors are contained in the relationship provided by the entrainment function (F_s):

$$F_s = \frac{\tau}{(\rho_g - \rho_w)gd} \tag{8.14}$$

where τ is the boundary shear stress, ρ_g is the particle density, ρ_w is the density of the water, g is the acceleration due to gravity and d is the diameter of the particle. The energy and forces of a stream are derived from its gravitational component as reflected by velocity and discharge. Thus, each stream contains potential energy, which is a function of the weight and head of water. This energy is converted to kinetic energy (E_k) during downhill travel. Most kinetic energy is lost to friction but that which remains is available for erosion and transportation processes. This relationship is embodied in the kinetic equation:

$$E_k = \frac{M}{2}v^2 \tag{8.15}$$

where M is the mass and v is the velocity, so that velocity determines energy and varies according to stream gradient and channel characteristics. When the channel cross-sectional area increases, the stream velocity decreases because water is not compressible.

8.1.2 River erosion

The work undertaken by a river is threefold, it erodes soils and rocks, transports the products thereof, which it eventually deposits. Erosion occurs when the force provided by river flow exceeds the resistance of the material over which it flows. Hence, the erosion velocity is appreciably higher than that required to maintain stream movement. The amount of erosion accomplished by a river in a given time depends upon the quantity of energy it possesses, which is influenced by its volume and velocity of flow, and the character and size of its load. The soils, rock types and geological structures over which a river flows, the infiltration capacity of the area it drains and the vegetation cover influence the rate of erosion.

Channel erosion essentially consists of the detachment of grains or aggregates from the bed and banks of a river followed by fluvial entrainment (Thorne *et al.*, 1997). An assessment of the rate of erosion, according to Allen *et al.* (1999), can be made by comparing measured or modelled channel velocities or shear stress with empirically derived threshold values, by measurement of actual erosion rates by use of erosion pins, repeated planimetric and cross-section surveys, terestrial photogrammetry or optoelectronic devices (Lawler, 1993), or by modelling erosion by means of suitable erosion coefficients together with data on channel hydraulic conditions. The latter method assumes that rates of bank erosion or detachment from the channel perimeter are principally governed by shear stress due to flow movement. The shear stress or stream power increases as the rate of flow increases until the critical shear stress needed to detach material is exceeded.

Once bank erosion starts, and the channel is widened locally, the process is self-sustaining. Bank recession allows the current to wash more directly against the downstream side of the eroded area, and so erosion and bank recession continue (Abam, 1997). Sediment accumulates on the opposite side of the channel, or the inside of a bend, so that the channel gradually shifts in the direction of bank attack. In such a way, meandering is initiated and this process is responsible for most channel instability in that the concave banks become oversteepened, leading to bank caving and slumping. Stream meanders usually occur in series and there is normally a downstream progression of meander loops. Meander growth often is stopped by the development of shorter chute channels across bars formed on the inside of the bed. Chutes may develop because the resistance to flow around a lengthening bend becomes greater than that across a bar, or because changes in alignment caused by channel shifting upstream tend to direct flow across a bar inside the bend. Meander loops may be abandoned because cut-offs develop from adjacent bends, the loops either migrating into each other, or channel avulsions form across the necks between adjacent bends during periods of overbank flooding (Fig. 8.4). Meander cut-offs or shortening by chute developments reduce channel lengths and increase slopes, and hence generally are beneficial for reducing flood heights or improving drainage. However, they may cause much local damage by channel shifting and bank erosion, and the resulting unstable bed conditions may interfere with navigation.

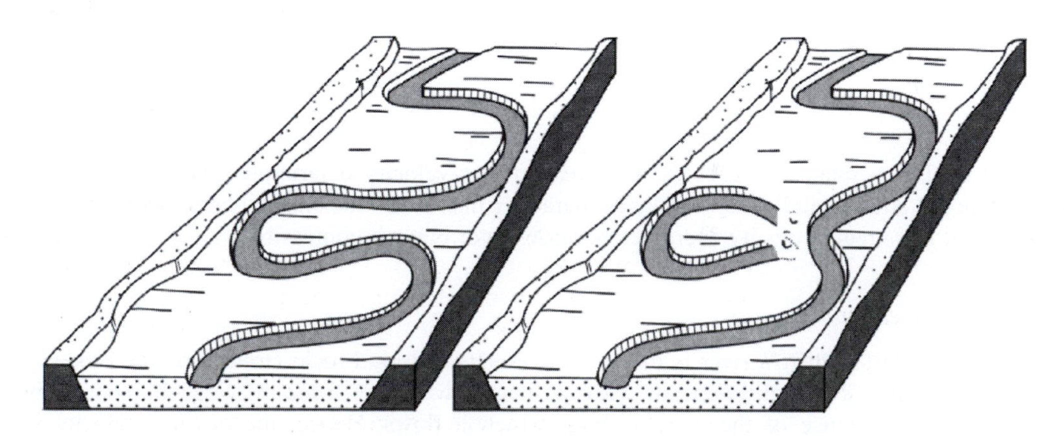

Figure 8.4 Formation of an oxbow lake.

During flood the volume of a river is increased significantly, which leads to an increase in its velocity. The principal effect of flooding in the upper reaches of a river is to accelerate the rate of erosion; much of the material so produced, then is transported downstream and deposited over the flood plain (Lewin, 1989). The vast increase in erosive strength during maximum flood is well illustrated by the devastating floods that occurred on Exmoor, England, in August 1952. It was estimated that the River Lyn moved 153 000 m^3 of rock debris into Lynmouth, some of the boulders weighing up to 10 tonnes. Scour and fill are characteristic of flooding. Often a river channel is filled during the early stages of flooding but as discharge increases scour takes over. For example, streams flooding on alluvial beds normally develop an alternating series of deep and relatively narrow pools, typically formed along the concave sides of bends, together with shallow wider reaches between bends where the main current crosses the channel diagonally from the lower end of one pool to the upper end of the next. During high flows, the pools or bends tend to be scoured more deeply, while the crossing bars are built higher by sediment deposition, although deposition does not equal rise in stage, and hence water depth increases on the bars. When the stage falls, erosion takes place from the top of crossing bars leading to some filling in the pools. As low stage activity is less effective, the general shape of the bed usually reflects the influence of the flood stage.

8.1.3 River transport

The load that a river carries is transported by traction, saltation, suspension or solution. The competence of a river to transport its load is demonstrated by the largest boulder it is capable of moving. This varies according to its velocity and volume, being at a maximum during flood. Generally, the competence of a river varies as the sixth power of its velocity. As mentioned above, the capacity of a river refers to the total amount of sediment that it carries, and varies according to the size of the particles that form the load on the one hand and its velocity on the other. When the load consists of fine particles the capacity is greater than when it is comprised of coarse material. Usually, the capacity of a river varies as the third power of its velocity. Both the competence and capacity of a river are influenced by changes in the weather, and the lithology and structure of the rocks over which it flows.

The sediment discharge of a river is defined as the mass rate of transport through a given cross-section measured as mass per second per metre width, and can be divided into the bedload and suspended load. The force necessary to entrain a given particle is referred to as the critical tractive force and the velocity at which this force operates on a given slope is the erosion velocity. The critical erosion velocity is the lowest velocity at which loose grains of a given size on the bed of a channel will move. The value of the erosion velocity varies according to the characteristics and depth of the water, the size, shape and density of particles being moved, and the slope and roughness of the floor. The Hjulstrom (1935) curves provide an indication of the threshold boundaries for erosion, transportation and deposition (Fig. 8.5a). Modified versions of these curves have appeared subsequently (Fig. 8.5b). It can be seen from Fig. 8.5a that fine- to medium-grained sands are more easily eroded than clays, silts or gravels. The fine particles are resistant because of the strong cohesive forces that bind them and because fine particles on the channel floor give the bed a smoother surface. There are accordingly few protruding grains to aid entrainment by giving rise to local eddies or turbulence. However, once silts and clays are entrained they can be transported at much lower velocities. For example, particles 0.01 mm in diameter are entrained at a critical velocity of about 600 mm s^{-1} but remain in motion until the velocity drops below 1 mm s^{-1}. Gravel is hard to entrain simply because of the size and weight of the particles involved. Particles in the bedload move slowly and intermittently. They

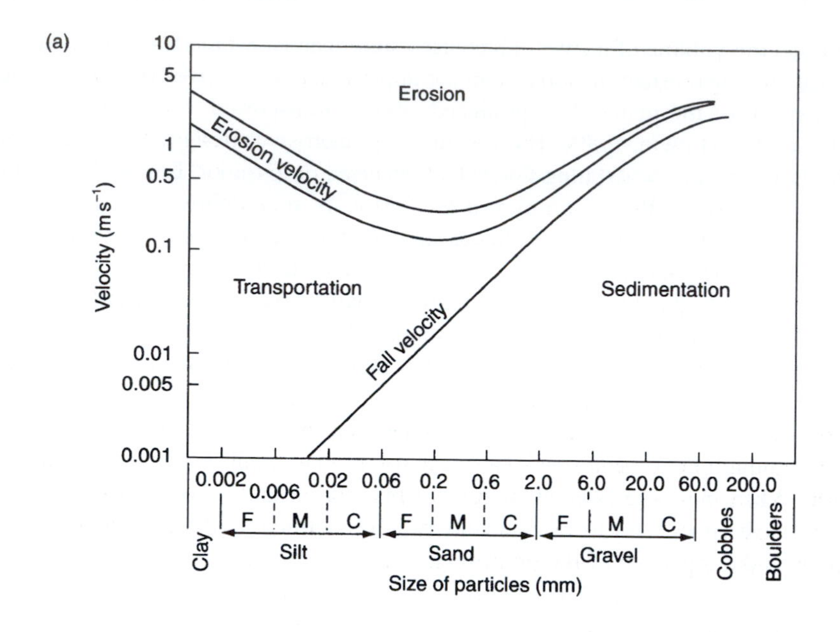

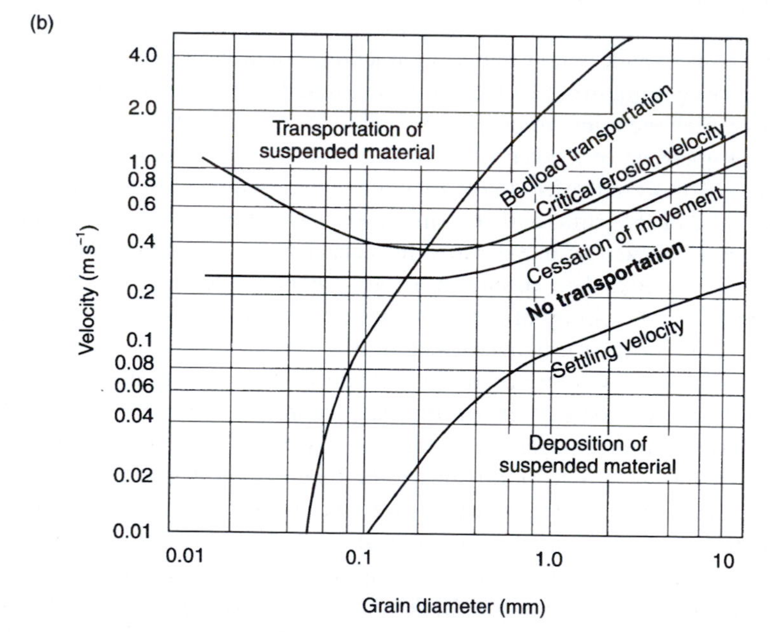

Figure 8.5 (a) Curves for erosion, transportation and deposition of uniform sediment. F, fine; M, medium; C, coarse grained. Note that fine sand is the sediment eroded most easily.
Source: Hjulstrom, 1935. Reproduced by kind permission of Dagmar Fense-Kroy. (b) Relationship between flow velocity, grain size, entrainment and deposition for uniform grains with a specific gravity of 2.65. The velocities are those 1 m above the bottom of the body of water. The curves are not valid for high sediment concentrations which increase the fluid viscosity.
Source: Sundberg, 1956.

generally move by rolling or sliding, or by saltation if the instantaneous hydrodynamic lift is greater than the weight of the particle. Deposition takes place wherever the local flow conditions do not re-entrain the particles.

Bedload transport in sand bed channels depends upon the regime of flow, that is, on streaming or shooting flow. When the Froude number is much smaller than 1, flow is tranquil, velocity is low, the water surface is placid and the channel bottom is rippled. Resistance to flow is great in this streaming regime and sediment transport is small with only single grains moving along the bottom. As the Froude number increases, but remains within the streaming flow regime, the form of the bed changes to dunes or large scale ripples (Fig. 8.6a). Turbulence is now generated at the water surface and eddies form in the lee of the dunes (Fig. 8.6b). Movement of grains takes place up the leeside of the dunes to cascade down the steep front, causing the dunes to move downstream. When the Froude number exceeds 1, flow is rapid, velocity is high, resistance to flow is small and bedload transport is great. At the transition to the upper flow regime, planar beds are formed. As the Froude number increases further, standing waves form and then antidunes are developed. The particles in the suspended load have settling velocities that are less than the buoyant velocity of the turbulence and vortices. Once particles are entrained and are part of the suspended sediment load, little energy is required to transport them. Indeed, as mentioned above, they can be carried by a current with a velocity less than the critical erosion velocity needed for their entrainment. Moreover the suspended load decreases turbulence that, in turn, reduces frictional losses of energy and makes the stream more efficient.

The quantity of suspended load increases rapidly with depth below the surface of a stream, the highest concentration generally occurring near the bed. However, there is a variation in suspended sediment concentrations at various depths of a stream for grains of different sizes. Most of the sand grains are carried in suspension near the bottom, whereas there is not very much change in silt concentration with depth.

Suspended load commonly is calculated from a sample obtained by a depth integrating sampler that is moved up and down along a vertical level in a stream. The weight of sediment in a sampler is determined and referred to the weight of water carrying it. This is the concentration of the suspended load and is expressed in terms of parts per million. The suspended load concentration usually increases with an increase in stage of a stream. In large streams the peak sediment concentration generally is close to the peak discharge. In fact, during flood the amount of suspended sediment load generally increases more quickly than discharge and reaches a peak concentration, may be several hours before the floodwater peak. In such cases, the suspended load carried during the highest water flow is considerably less than capacity.

Because the discharge of a river varies, sediments are not transported continuously, for instance, boulders may be moved only a few metres during a single flood. In other words, there is a threshold discharge below which no movement of bedload occurs so that there is a direct relationship between bedload movement and flood discharge. Once the rate of flow falls beneath the threshold value the bedload remains stationary until the next flood of equal or higher magnitude occurs. Nonetheless, the contribution of bedload to total streamload frequently is high. Often the amount of bedload moved downstream during times of flood exceeds that of suspended load by several times.

As remarked, the total amount of sediment carried by a river increases significantly during flood. Part of the increase in load is derived from within the channel and part from outside the channel. Consequently, channel form changes during flood, the depth and width of the channel adapting to scour and deposition (Gupta, 1988). Scour at one point is accompanied by deposition at another. Initially, during a flood, bedload may be moved into small depressions in the floor of a river and it is not removed easily. Major floods can bring about notable changes

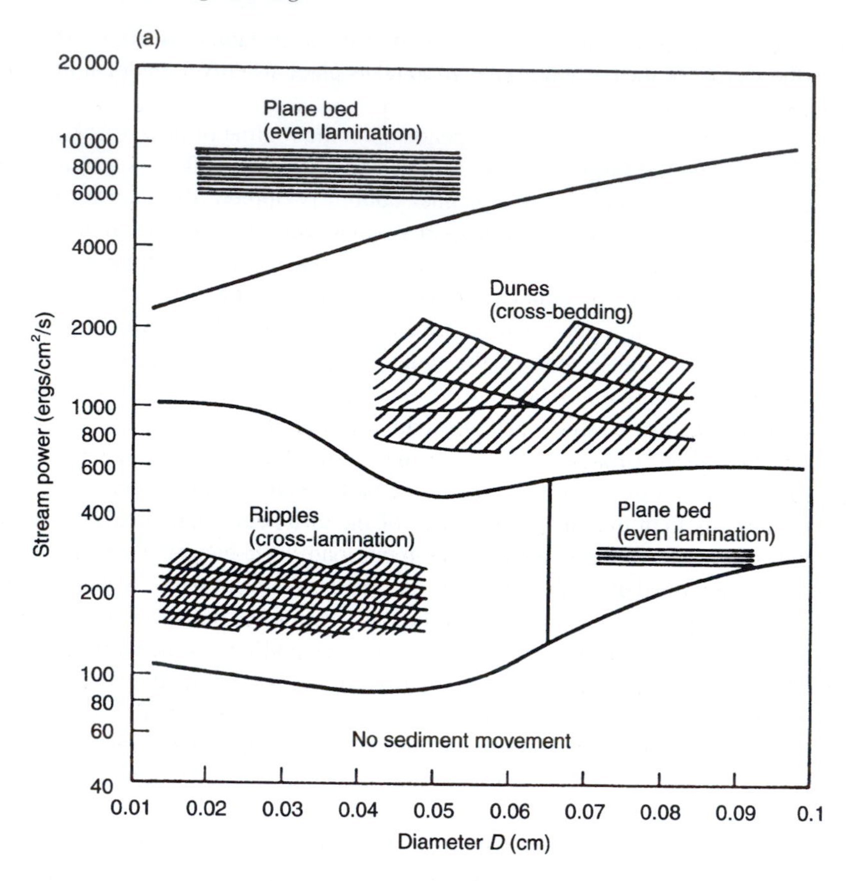

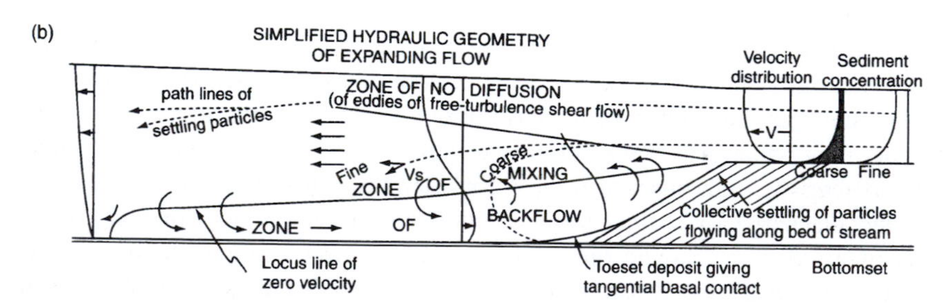

Figure 8.6 (a) Bedforms in relation to stream power and size of bedload material.
Source: Simons and Richardson, 1961. Reproduced by kind permission of American Society of Civil Engineers. (b) Hydraulic conditions at the lip of a small delta produced in a flume. The flow structure developed over the slip slope is the same as that produced over a slip slope on a dune or ripple.
Source: Jopling, 1963. Reproduced by kind permission of Blackwell.

in channel form by bank scour and slumping. However, after a flood, deposition may make good some of the areas that were removed.

Some alluvial deposits such as channel bars are transitory, existing for a matter of days or even minutes. Hence, the channels of most streams are excavated mainly in their own sedimentary deposits, which streams continually rework by eroding the banks in some places and redepositing the sediment farther downstream. Indeed, sediments that are deposited over a flood plain may be regarded as being stored there temporarily. Consequently, any sediment management strategy must recognize that sediment variation is both location and time dependent.

8.1.4 Deposition of sediments

Deposition occurs where turbulence is at a minimum or where the region of turbulence is near the surface of a river. For example, lateral accretion occurs with deposition of a point bar on the inside of a meander bend. A point bar grows as the meander moves downstream or new ones are built as the river changes course during or after floods. Old meander scars can often be seen on flood plains (Fig. 8.7a). The combination of point bar and filled slough results in what is called ridge and swale topography. The ridges are composed of sandbars and the swales are the depressions that subsequently were filled with silt and clay. A detailed description of the fluvial geomorphological features found along the lower Mississippi River has been provided by Smith (1996).

An alluvial floodplain is the most common depositional feature of a river. Ward (1978) suggested that a floodplain can be regarded as a store of sediment across which channel flow takes place and that, in the long term, is comparatively unchanging in amount. However, dramatic changes may occur in the short term (Rahn, 1994). In fact, the morphology of a floodplain often displays an apparent adjustment to flood discharge in that different floodplain or terrace levels may be related to different frequencies of flood discharge, the higher levels being inundated by the largest floods.

As the peak concentration of suspended load usually occurs prior to the discharge peak, much of the load of a stream is contained within its channel. The water that overflows onto the floodplain accordingly possesses less suspended sediment (Macklin *et al.*, 1992). As a consequence, the processes within the channel responsible for lateral accretion usually are more important in the formation of a floodplain than overbank flow, which results in vertical accretion. In fact, lateral accretion may account for between 60 and 80 per cent of the sediment that is deposited (Leopold *et al.*, 1964).

The alluvium of flood plains is made up of many kinds of deposits, laid down both in the channel and outside it (Marsland, 1986). Vertical accretion on a floodplain is accomplished by in-channel filling and the growth of overbank deposits during and immediately after floods. Gravels and coarse sands are moved chiefly at flood stages and deposited in the deeper parts of a river. As the river overtops its banks, its ability to transport material is lessened so that coarser particles are deposited near the banks to form levees (Smith, 1996). Levees therefore slope away from the channels into the flood basins, which are the lowest part of a floodplain. At high stages, low sections and breaks in levees may mean that there is a concentrated outflow of water from the channel into the floodplain. This outflow rapidly erodes a crevasse, leading to the deposition in the flood basin of a crevasse splay. Finer material is carried farther and laid down as backswamp deposits (Fig. 8.7b). At this stage a river sometimes aggrades its bed, eventually raising it above the level of the surrounding plain. In such a situation, when the levees are breached by flood water, hundreds of square kilometres may be inundated.

8.2 Floods

Floods are due to excessive surface run-off and represent the commonest type of geological hazard (Fig. 8.8). They probably affect more individuals and their property than all the other hazards put together. However, the likelihood of flooding is more predictable than some other

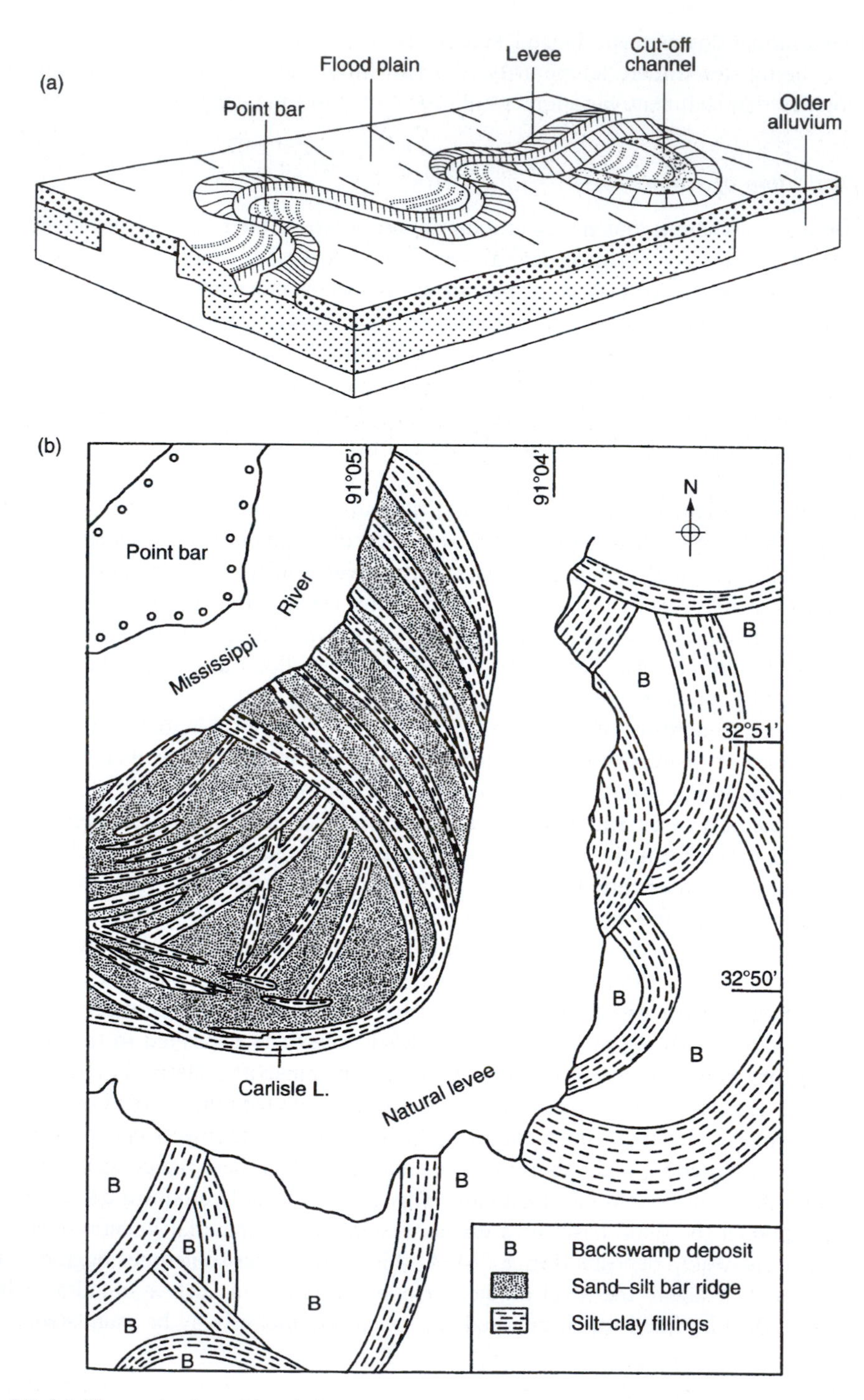

Figure 8.7 (a) The main depositional features of a meandering channel. (b) Map of a portion of the Mississippi River flood plain, showing various kinds of deposit.
Source: Fisk, 1944.

Figure 8.8 Flooding in Ladysmith, South Africa, February 1994.

types of hazards. Most floods are the result of excessive rainfall or snowmelt and in most regions they occur more frequently in certain seasons than others. Meltwater floods are seasonal and are characterized by a substantial increase in river discharge so that there is a single flood wave (Church, 1988). The latter may have several peaks. The two most important factors that govern the severity of floods due to snowmelt are the depth of snow and the rapidity with which it melts. Sinha (1998) noted that in addition to inundation, the two most notable hazards due to flooding of the rivers in the north Bihar plain, India, were rapid lateral migration and extensive bank erosion.

Identical flood-generating mechanisms, especially those associated with climate, can generate different floods within different catchments, or within the same catchment at different times. Such differences, according to Newson (1994), are attributable to the effects of flood intensifying factors such as certain basin, network and channel characteristics that may increase the rate of flow through the system. For instance, flood peaks may be low and attenuated in a long narrow basin with a high bifurcation ratio whereas in a basin with a rounded shape and low bifurcation ratio the flood peaks are higher. Generally, drainage patterns that give rise to merging flood flows from major tributaries in the lower part of a river basin are associated with high magnitude flood peaks in that part of the system. In addition, the capacity of the ground to allow water to infiltrate and the amount that can be stored influences the size and timing of a flood. Frozen ground, by inhibiting infiltration, is an important flood intensifying factor and it may extend the source area over the whole of a catchment. On the other hand, a catchment area with highly permeable ground conditions may have such a high infiltration capacity that it is rarely subjected to floods. Flood discharges are at a maximum per unit area in small drainage basins. This is because storms tend to be local in occurrence. High drainage density watersheds produce simultaneous flooding in tributaries, which then overload junction capacity to transmit flow. In low drainage density basins the procession of flood waves have longer delay periods allowing the main channel to absorb incoming flows one at a time. Hence, it usually takes some time to accumulate enough run-off to cause a major disaster.

Flash floods prove the exception (Ristic and Malosevic, 1997). Flash floods are short-lived extreme events. They usually occur under slowly moving or stationary thunderstorms that last for less than 24 hours. The resulting rainfall intensity exceeds infiltration capacity so that run-off takes place very rapidly. Flash floods frequently are very destructive as the high energy flow can carry much sedimentary material (Clarke, 1991).

On the other hand, long-rain floods are associated with several days or even weeks of rainfall, which may be of various intensities (Bell, 1994). They are the most common cause of major flooding. Single event floods have a single main peak but are of notably longer duration than flash floods. Nevertheless, some of the most frequently troublesome floods are associated with a series of flood peaks that follow closely on each other. Such multiple event floods are caused by more complex weather conditions than those associated with single event floods. The effects of multiple event floods usually are severe because of the duration over which they extend. This could be several weeks, or even months in the case of certain monsoon or equatorial regions where seasonal floods frequently are of extended duration.

As the volume of water in a river is greatly increased during times of flood, its erosive power increases accordingly. Thus, a river carries a much higher sediment load. Deposition of the latter where it is not wanted also represents a serious problem.

The influence of human activity can bring about changes in drainage basin characteristics, for example, removal of forest from parts of a river basin can lead to higher peak discharges that generate increased flood hazard. A most notable increase in flood hazard can arise as a result of urbanization, the impervious surfaces created mean that infiltration is reduced, and together with stormwater drains, they give rise to increased run-off. Not only does this produce higher discharges but lag-times also are reduced. The problem of flooding is particularly acute where rapid expansion has led to the development of urban sprawl without proper planning, or worse where informal settlements have sprung up. Heavy rainfall can prove disastrous as far as informal settlements are concerned (Fig. 8.9). Lo and Diop (2000) discussed the problems associated with recurrent flooding in Dakar, Senegal, which include not only degradation of roads, damage to buildings and clogged drains and sewers but the proliferation of mosquitoes and other insects with attendant health risks.

A flood can be defined in terms of height or stage of water above some given point such as the banks of a river channel. However, rivers generally are considered in flood when their level has risen to an extent that damage occurs. The discharge rate provides the basis for most methods of predicting the magnitude of flooding, the most important factor being the peak discharge that is responsible for maximum inundation. Not only does the size of a flood determine the depth and area of inundation but it primarily determines the duration of a flood. These three parameters, in turn, influence the velocity of flow of flood waters, all four being responsible for the damage potential of a flood. As noted above, the physical characteristics of a river basin together with those of the stream channel affect the rate at which discharge downstream occurs. The average time between a rainstorm event and the consequent increase in river flow is referred to as the lag-time. Lag-time can be measured from the commencement of rainfall to the peak discharge or from the time when actual flood conditions have been attained (e.g. bankfull discharge) to the peak discharge. This lag-time is an important parameter in flood forecasting. Calculation of the lag-time, however, is a complicated matter. Nonetheless, once enough data on rainfall and run-off versus time have been obtained and analysed, an estimate of where and when flooding will occur along a river system can be made.

An estimate of future flood conditions is required for either forecasting or design purposes (Malamud *et al.*, 1996). In terms of flood forecasting more immediate information is needed

Figure 8.9 Damage to informal settlements at Ntuzuma, north of Durban, South Africa, September 1987.

regarding the magnitude and timing of a flood so that appropriate evasive action can be taken. In terms of design, planners and engineers require data on magnitude and frequency of floods. Hence, there is a difference between flood forecasting for warning purposes and flood prediction for design purposes. A detailed understanding of the run-off processes involved in a catchment and stream channels is required for the development of flood forecasting but is less necessary for long-term prediction. The ability to provide enough advance warning of a flood means that it may be possible to reduce the resulting damage by allowing people to evacuate the area likely to be affected, along with some of their possessions. The most reliable forecasts are based on data from rainfall or melt events that have just taken place or are still occurring. Hence, advance warning generally is measured only in hours or sometimes in a few days. The longer-term forecasts commonly are associated with snowmelt. Basically, flood forecasting involves the determination of the amount of precipitation and resultant run-off within a given catchment area. The volume of run-off then is converted into a time-distributed hydrograph and flood routing procedures are used, where appropriate, to estimate the changes in the shape of the hydrograph as the flood moves downstream. In other words, flood routing involves determination of the height and time of arrival of the flood wave at successive locations along a river. As far as prediction for design purposes is concerned, the design flood, namely, the maximum flood against which protection is being designed, is the most important factor to determine (see Section 8.6). Methods of flood forecasting and prediction have been reviewed by Ward (1978).

8.3 Factors affecting run-off

The total run-off from a catchment area generally consists of four component parts, namely, direct precipitation on the stream channels, surface run-off, interflow and baseflow. Unless the catchment area contains a large number of lakes or swamps, direct precipitation onto water surfaces and into stream channels normally represents only a small fraction of the total volume of water flowing in streams. Even where the area of lakes is large, evaporation from them may equal the amount of precipitation they receive. Consequently, this component is usually ignored in run-off calculations. However, where lakes and swamps occur in the drainage basin, they tend to absorb high peaks of surface run-off, which is particularly beneficial in catchments with rather low infiltration capacities.

Surface run-off comprises the water that travels over the surface as sheet or channel flow. It is the first major component of flood and peak discharges during a rainstorm. Some proportion of rainfall infiltrates into the ground where it may meet a relatively impermeable layer that causes it to flow laterally, just below the surface, towards streams. This is referred to as interflow and, as would be expected, it moves more slowly than surface run-off. The interflow contribution to total run-off depends mainly on the soil characteristics of the catchment and the depth of the water table. In some areas interflow may account for up to 85 per cent of the total run-off.

The infiltration capacity is the rate at which water is absorbed by a soil. Absorption starts with an initial value, decreases rapidly, then reaches a steady value that is taken as the infiltration capacity. Rainfall occurring after a steady rate of infiltration is reached, is rainfall excess and flows off as surface run-off. Conversely, rain that is not capable of satisfying the infiltration capacity produces no rainfall excess and thus no run-off. During periods of extended rainfall the infiltration capacity is reduced so that the amount of run-off increases. The amount of surface moisture in the soil obviously influences the infiltration capacity. For instance, if the soil is saturated, most rainfall will go towards run-off and flooding may occur. The infiltration capacity of a particular soil is influenced by soil texture, that is, the size and arrangement of grains and their state of aggregation, since they influence porosity and permeability, by vegetative cover, by biological structures such as root and worm holes, by antecedent soil moisture, that is, the moisture remaining from a previous rain, and by the condition of the soil surface, for example, whether it is hardened due to drying or compacted. Rain itself can reduce infiltration capacity by packing the soil, breaking down the structure of aggregates, washing down fine grains to fill pores, and causing colloids and clay particles to swell as a result of wetting.

Often a significant proportion of total run-off is stored in river banks, which therefore is referred to as bank storage. Bank storage takes place above the normal phreatic surface. As stream levels fall, water from bank storage is released into them. Most of the rainfall that percolates to the water table eventually reaches the mainstream channels as baseflow or effluent seepage. Since water moves very slowly through the ground, the outflow of groundwater into stream channels not only lags behind the occurrence of rainfall by several days, weeks or even years, but also is very regular. Baseflow therefore normally represents the major long-term component of total run-off and may be particularly important during long dry spells when low water flow may be entirely derived from groundwater supplies.

The flow of any stream is governed by climatic and physiographic factors. As far as the climatic factors are concerned, the type, intensity, duration and distribution of precipitation contribute towards streamflow, whilst evapotranspiration has the opposite effect and is influenced by temperature, wind velocity and relative humidity. Indeed, the most obvious and probably the most effective influence on the total volume of run-off is the long-term balance between the amount of water gained by a catchment area in the form of precipitation and the amount of water lost in the form of evapotranspiration.

Most rivers show a seasonal variation in flow that, although influenced by many factors, is largely a reflection of climatic variations. The pattern of seasonal variations, which tends to be repeated year after year, frequently is referred to as the river regime. Obviously, the study of river regimes plays an important part in the understanding of problems associated with flood prevention and sediment transport. The type of precipitation is important, for example, the contribution to run-off of rainfall is almost immediate, providing that its intensity and magnitude are great enough. Gentle rain that falls over a period of several hours or even days gives rise to a relatively modest flood peak with a comparatively long time base, but when the same quantity of rainfall is concentrated in a much shorter time period, then significant flooding may occur. In cold climates, in particular, a large proportion of steam-flow may be derived from melting snow and ice. Where melting occurs gradually the contribution resembles that of baseflow. In other words, the snow or ice blanket acts as a store of water supply and makes a stable contribution to run-off. On the other hand, if melting occurs suddenly as a result of a rapid thaw, a large volume of water enters streams during a short period of time, giving a peak run-off. Nevertheless, the effects of accumulation and snowmelt are of long-term significance only in high latitude and high altitude regions.

Surface run-off does not usually become a significant feature, except in the case of intense storms, until most of the soil moisture deficit has been replenished. However, once this has happened run-off increases quite rapidly in amount, representing an increasing proportion of the rainfall during the rest of the fall. However, the increase in stream flow does not occur at the same rate as the increase in rainfall excess because of the lag effect resulting from storage. If rainfall occurs over a frozen surface, infiltration cannot take place, so that once the initial interception and depression storage have been satisfied, the remaining rainfall contributes towards run-off.

There is a critical period for an individual drainage basin for which all storms of a particular duration, irrespective of intensity, produce a period of surface run-off that is essentially the same whilst for rains of longer duration, the period of surface run-off is increased. The effectiveness of rainfall duration varies with the size and relief of the drainage basin. In a small catchment with steep slopes maximum potential run-off is likely to be attained by a rainfall of shorter duration than in a large catchment with gentle slopes.

Storms that produce floods in large drainage basins very rarely are distributed uniformly. The high peak flows in large basins usually are produced by storms that occur over large areas whilst high peak flows in small drainage basins commonly are the result of intense thunderstorms that extend over limited areas. The amount of run-off resulting from any rainfall depends to a large extent on how the rainfall is distributed; if it is concentrated in a particular area of a basin, then the run-off is greater than if it is uniformly distributed throughout the basin. This is because in the former instance the infiltration capacity is exceeded quickly. The distribution coefficient provides an assessment of the run-off that results from a particular distribution of rainfall. It is expressed as:

$$\text{Distribution coefficient} = \frac{\text{Maximum rainfall at any point}}{\text{Mean of the basin}} \tag{8.16}$$

The peak run-off increases as the distribution coefficient increases.

Every drainage basin or catchment area is defined by a topographic divide or watershed that bounds the area from which the surface run-off is derived. Similarly, the groundwater contribution to a given catchment is bounded by a phreatic divide. These two divides are not necessarily coincident and intershed leakage accordingly can occur. The location of the phreatic divide tends to move with fluctuations in the water table, and the higher the water table the more nearly do the two divides coincide.

The area of a basin affects the size of floods likely to occur, as well as influencing minimum flow levels. For instance, the larger the drainage basin, the longer it takes for the total flood flow to pass a given location. What is more, the peak flow decreases relatively as the area of the basin increases since storms become less effective and infiltration increases. Because local rains contribute to the discharge of the main stream, larger basins are likely to provide a more sustained flow than smaller ones.

One of the principal factors that governs the rate at which run-off is supplied to the main river is the shape of the drainage basin. The outlines of large drainage basins are generally fixed, at least in part, by major geological structures whilst erosional features usually form the limits of small drainage basins. The effect of shape can be demonstrated best by considering three differently shaped catchments with the same area, subjected to rainfall of the same intensity (Fig. 8.10a). If each catchment is divided into concentric segments, which may be assumed to have all points within an equal distance along the stream channels from the control point, it may be seen that shape A requires 10 time units to pass before every point on the catchment is contributing to the discharge. Similarly, B requires 5 and C 8.5 time units. The shape factor also affects the run-off when a rainstorm does not cover the whole catchment at once but moves over it from one end to the other. The direction in which it moves in relation to the direction of flow can have a decided influence upon the resulting peak flow and also upon the duration of surface run-off. For example, consider catchment A to be slowly covered by a storm moving upstream that just covers the catchment after 5 time units. The flood contribution of the last segment will not arrive at the control until 15 time units from commencement. Alternatively, if the storm were moving at the same rate downstream, the flood contribution of time segment 10 would arrive at the control point simultaneously with that of all the others, so that an extremely rapid flood rise would occur. The effect of changing the direction of storm movement on the other catchments is less marked but still appreciable. The effect of storm location within a catchment is illustrated in Figure 8.10b, the steeper of the two flood hydrographs (see Section 8.4) being associated with the storm located closer to the outlet of the basin.

The index related to the shape of a drainage basin is termed the compactness coefficient. This is the ratio of the watershed perimeter to the circumference of a circle whose area is equal to that of the drainage basin. The less compact a basin is, the less likely it is to have intense rainfall simultaneously over its entire extent. The lower the value of the coefficient, the more rapidly is water likely to be discharged from the catchment area via the main streams.

The variation in and mean elevation of a drainage basin obviously influence temperatures and precipitation that, in turn, influence the amount of run-off. Generally, precipitation increases with altitude but the effect of reduced evaporation is more important, as is the temporary storage of precipitation in snow and ice. This affects the distribution of the mean monthly run-off, reducing it to a minimum in winter in cold climates.

Surface run-off and infiltration are related to the gradient of a drainage basin. Indeed, the slope of a drainage basin is one of the major factors controlling the time of overland flow and concentration of rainfall in stream channels, and is of special importance as far as the magnitude of floods is concerned. Obviously, with steep slopes there is a greater chance that the water will move off the surface before it has time to infiltrate, so that surface run-off is large.

The efficiency of a drainage system is dependent upon the stream pattern. For instance, if a basin is well-drained the length of overland flow is short, the surface run-off concentrates quickly, the flood peaks are high and in all probability the minimum flow is correspondingly low. The drainage pattern also generally reflects the geological conditions found within the drainage basin. A drainage network can be described in terms of stream order, length of tributaries, stream density and drainage density, and length of overland flow. First-order streams are unbranched, and when two such streams become confluent they form a second-order stream. When two of the latter types join, they form a third-order stream, and so on. It is only when

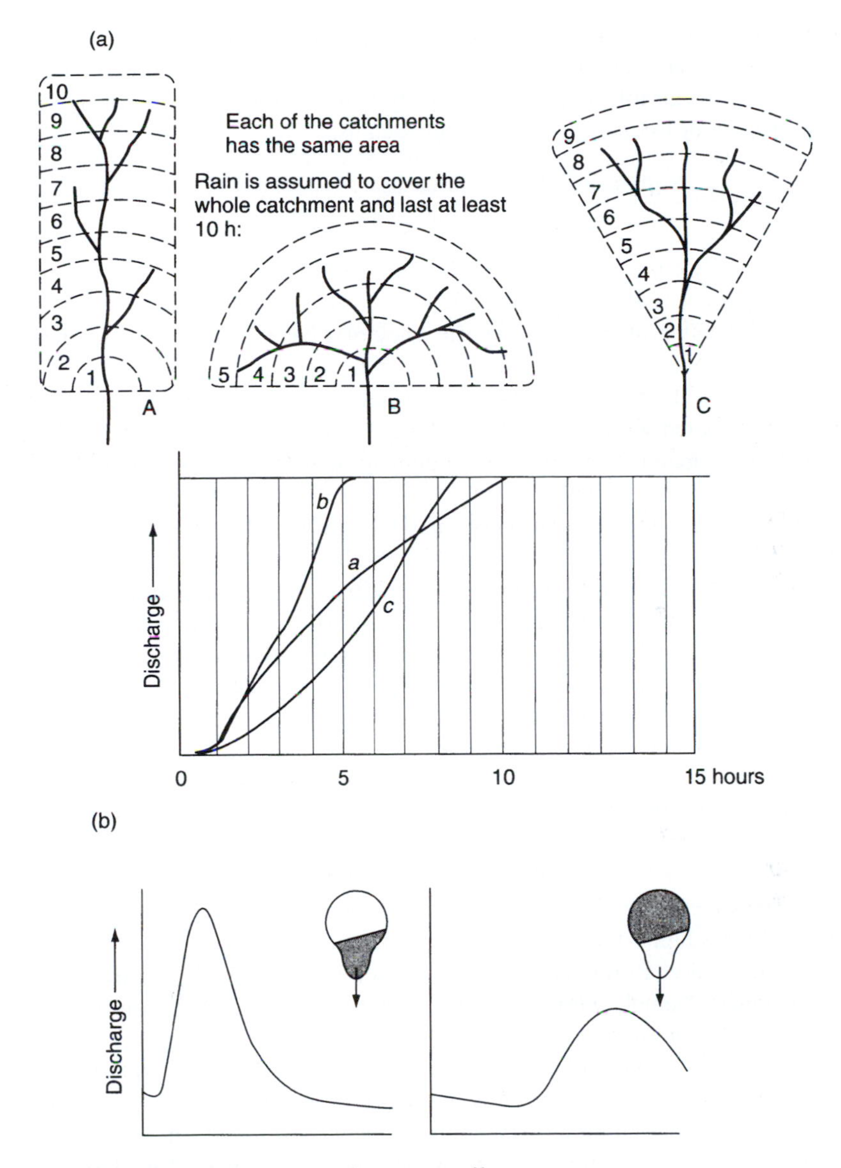

Figure 8.10 (a) The effect of shape on catchment run-off.
Source: Wilson, 1983. (b) Identical storms located in different parts of a drainage basin giving different run-offs.

streams of the same order meet, that they produce one of higher rank. The frequency with which streams of a certain order flow into those of the next order above is referred to as the bifurcation ratio and is derived by dividing the number of any given order of streams by that of the next highest order. Dendritic river systems generally have bifurcation ratios in the range 3.5–4.0 whereas those associated with trellis patterns are much higher. The order of the main stream gives an indication of the size and extent of the drainage pattern. Such an identification procedure provides a technique for comparing streams and basins in a numerical manner.

The length of tributaries is an indication of the steepness of the drainage basin, as well as the degree of drainage. Steep well-drained areas usually have numerous small tributaries whereas on plains with deep permeable soils only relatively long tributaries are generally perennial.

The stream density or frequency may be expressed as:

$$\text{Stream density} = \frac{\text{Number of streams in basin}}{\text{Total area of basin}} \tag{8.17}$$

Stream density does not provide a true measure of the drainage efficiency. Drainage density is the stream length per unit of area and varies inversely as the length of overland flow thus providing some indication of the drainage efficiency of a basin. It reflects the concentration or stream frequency of an area and is inversely related to the rate of infiltration. In other words, a river basin in which the floor consists of permeable rocks tends to have fewer streams per unit area than one floored with impermeable rocks. The drainage density also is influenced by climate. The shape of drainage basins and their drainage density influence streamflow conditions (see Section 8.2).

The circularity index is the calculation of the degree of circularity of the basin. It is a ratio that expresses the area of a circle to the area of the basin when the same circumference or perimeter length is used. Dendritic basins have circularity indices that are about 0.6–0.7.

The geometry of a stream channel influences run-off. A wide dish-shaped channel, for example, gives a rapid rate of increase in width with increasing discharge, whilst a rectangular channel has a rapid rate of increase of depth with increasing discharge.

A permeable soil or rock mass allows water to percolate to the zone of saturation from where it can be slowly discharged as springs. Open-textured sandy soils have much higher infiltration capacities than clay soils and therefore give rise to much less surface run-off. A more dramatic example is provided by the virtual disappearance of surface drainage in some areas where massive limestones are exposed. By contrast, basins on impermeable rock produce high volumes of direct run-off and very little baseflow. In flat low-lying areas, the soil type also influences the position of the water table. The water table in clay soils rises after rainfall, perhaps causing waterlogging, whereas rapid drainage through gravels tends to allow the water table to remain below the ground surface. In the latter case, the baseflow contribution to streamflow is likely to reach the drainage channels with little delay. Although direct surface run-off is reduced by percolation, the final total amount is not.

The geological structure may influence the movement of groundwater towards streams and it generally explains the lack of correlation between the topographical and hydrological divides of adjacent catchments. The long-term relationship between groundwater and surface run-off determines the main characteristics of a stream and provides a basis for classifying streams into ephemeral, intermittent or perennial types. Ephemeral streams are those that contain only surface run-off and therefore only flow during and immediately after rainfall or snow melt. Normally, there are no permanent or well-defined channels and the water table is always below the bed of the stream. Intermittent streams flow during a wet season, drying up during drought. Streamflow consists mainly of surface run-off but baseflow makes some contribution during a wet season. Perennial streams flow throughout the year because the water table is always above the bed of the stream, making a continuous and significant contribution to total run-off. It is seldom possible to classify the entire length of a stream in this way.

The most important effect of the vegetation cover is to slow down the movement of water over the surface after rainfall and therefore to allow more time for infiltration to take place. In this way the timing of run-off after rainfall may be considerably modified and peak stream flows may be much lower, although more prolonged.

Human factors such as agricultural practices and land-use also affect run-off. In urban areas, sewers, drains and paving increase run-off efficiency and reduce lag-time and ground storage of precipitation, producing accelerated and high run-off peaks. Thus, stream channel networks become more efficient in collecting water quickly and may give rise to flash flooding.

8.4 Assessment of run-off

In dealing with run-off the hydrologist has to try to provide answers relating to the occurrence, size and duration of floods and droughts. Of special concern is the magnitude and duration of run-off from a particular catchment with respect to time. This can be resolved by producing graphs of the frequency and duration of individual discharges from observations over a long period of time, though if such observations are not available, estimations may be made at various probabilities. Even if measurements of rainfall and evapotranspiration were completely reliable, there would still be a need for direct measurement of streamflow. However, streamflow is perhaps the most difficult, and is certainly the most costly, of the hydrological parameters to measure accurately. Being a widely variable quantity there is no direct way of continuously monitoring flows in a river. Basically, however, there are three related operations in the measurement of run-off. The first of these involves the determination of the height or stage of the river, the second involves the determination of the mean velocity of the water flowing in the river channel, and the third involves the derivation of a known relationship between stage and total volume of discharge.

8.4.1 Measurement of river stage

The measurement of the stage of a river may be made periodically or continuously, depending upon the degree of accuracy required and the hydrological characteristics of the stream. Generally, the larger the catchment area and the more permeable the ground, the less important it is for river stage to be continuously monitored. Periodic observations of river stage normally are made by reference to a staff gauge. An alternative method is to use a surface contact gauge or wire weight gauge. It may be necessary during floods to know the height of the flood peak on a large number of tributary streams in order to ascertain which area of a drainage basin is contributing the greatest proportion of surface run-off. A peak gauge can be used in such an instance. It may consist of a hollow tube set vertically in the water, peak levels being indicated inside the tube by a non-returning float. At most, major gauging stations water levels are reproduced autographically by means of continuous recorders.

8.4.2 Measurement of current velocity

The measurement of current velocity in streams usually is carried out by means of a current meter of which there are two principal types. In the first of these, the water flows against a number of cups attached to a vertical spindle, whilst in the second type, the flowing water acts directly on the upstream surface of a propeller that is attached to a horizontal spindle. The speed at which the cups or propeller are turned provides an indication of the current velocity. Due to friction between the flowing water and the wetted perimeter of the stream channel, the velocity profile of a stream is not constant. Consequently, mean velocity has to be established from the average of a number of current meter observations located in such a way as to detect the differences in velocity. Generally, the intervals between adjacent measuring points should not exceed one-fifteenth of the stream width where the bed profile is regular and one-twentieth the width where it is irregular. Due care must be given to the selection of the reach of stream used for current meter measurements, the ideal being a reach where the velocity profile is both regular and symmetrical. The length of channel chosen preferably should be straight for a distance of approximately three times the bankfull width of the river, the bed should be smooth so as to reduce turbulent flow to a minimum and the direction of flow should be normal to the section of measurement.

The total volume of water or discharge flowing past a given point in a given time is the product of the cross-sectional area of the stream and its velocity. If the stream bed and banks have been accurately surveyed at the place of measurement, then only data on the stage of the stream is needed to enable the cross-sectional area of the water in the stream channel to be calculated. The volume of discharge is calculated once the mean velocity of the current is known. Discharge may be accurately measured by means of a weir or flume on streams where a physical obstruction in the channel is permissible. If a continuous record of discharge is required, then it must be correlated with river stage. When discharge is plotted against corresponding stages, the curve drawn through plotted points is referred to as a rating curve (Fig. 8.11). In other words, a rating curve is a graph relating the stage of a river channel at a certain cross-section to the corresponding discharge at that section. Hence, it can be used to estimate the quantity of water passing a particular location at a given time. When this ratio is greater than 1.0, flood conditions are imminent. A rating curve can be used to predict the severity of flooding as measured by the height of the flood wave that overtops bankfull height. This information then can be added to the curve developed from calculation of the frequency of flooding as related to discharge. This establishes the recurrence interval for floods of different magnitude. These data allow determination of which parts of the floodplain will be affected by a storm of a given intensity and duration.

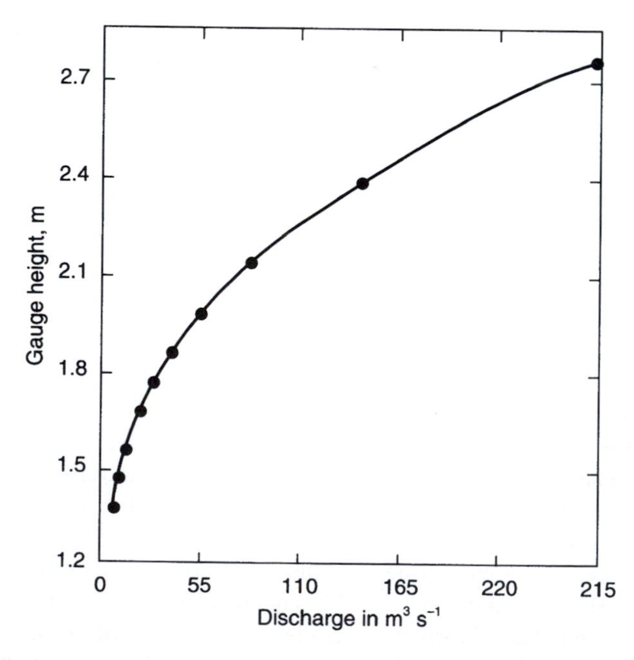

Figure 8.11 Stage discharge rating curve showing the relationship between height of water at gauge and amount of discharge.

8.4.3 Peak flow

There are two specific problems related to run-off predictions. First, there is the need to forecast peak flows associated with sudden increases in surface run-off and, second, there is the prediction of minimum flow that involves decreasing volume of baseflow. The accuracy of run-off predictions tends to improve as the time interval is increased. Estimates of annual or even seasonal run-off totals for given catchment areas may, in some cases, be made from annual or seasonal rainfall totals, using a simple straight line regression between the two variables. However, correlations between rainfall and run-off generally may be expected to yield forecasts of only token accuracy for they take no account of the contributions made by interflow and baseflow. Increases in surface run-off after rainfall or snowmelt tend to be rapid, leading to a short-lived peak, which is followed by a rather longer period of declining run-off. Since it is the peak flow that is likely to cause problems, then the principal object of prediction concerns the magnitude and timing of this peak, and the frequency with which it is likely to occur.

In order to carry out a flood frequency analysis either the maximum discharge for each year or all discharges greater than a given discharge are recorded from gauging stations according to magnitude (Dalrymple, 1960; Benson, 1968). Then the recurrence interval, that is, the period of years within which a flood of given magnitude or greater occurs, is determined from:

$$T = \frac{n+1}{m} \tag{8.18}$$

where T is the recurrence interval, n is the number of years of record and m is the rank of the magnitude of the flood, with $m = 1$ as the highest discharge on record. Each flood discharge is plotted against its recurrence interval and the points are joined to form the frequency curve (Fig. 8.12). The flood frequency curve can be used to determine the probability of the size of the discharge that could be expected during any given time interval. The larger the recurrence interval, the longer is the return period and the greater the magnitude of flood flow (Malamud *et al.*, 1996). The probability that a given magnitude of flow will occur or be exceeded in

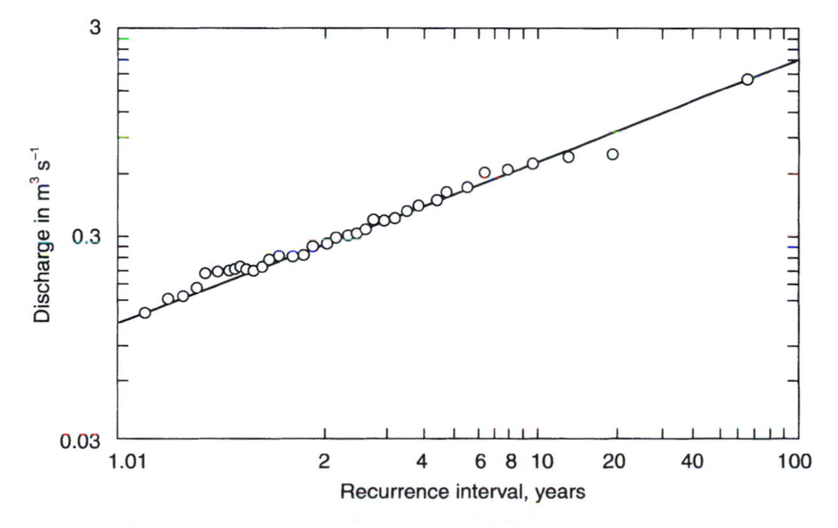

Figure 8.12 Flood frequency curve of the Locking River, Tobaso, Ohio.
Source: Dalrymple, 1960. Reproduced by kind permission of US Geological Survey.

a given time is the reciprocal of the recurrence interval. The probability (P) of a flood of recurrence interval T years being equalled or exceeded in x years is given by:

$$P = 1 - \left(1 - \frac{1}{T}\right)^x \tag{8.19}$$

Estimates of the recurrence intervals of floods of different sizes can be improved by using alluvial stratigraphy to date the sediments they deposited, the more widespread thicker deposits being formed by larger floods (Jarret, 1990). The technique can be used to reconstruct past changes in climate and so used to determine whether floods were more or less frequent in the past at a particular location (Costa, 1978). Clarke (1996) described how to use boulder size to estimate the probable maximum flood that has occurred in the past.

A flow duration curve shows the percentage of time a specified discharge is equalled or exceeded. In order to prepare such a curve, all flows during a given period are listed according to their magnitude. The percentage of time each one equalled or exceeded a given discharge is calculated and plotted (Fig. 8.13). The shape of the curve affords some insight into the characteristics of the drainage basin concerned. For instance, if the curve has an overall steep slope, this means that there is a large amount of direct run-off. On the other hand, if the curve is relatively flat, there is substantial storage, either on the surface or as groundwater. This tends to stabilize streamflow.

8.4.4 Hydrograph analysis

Hydrograph analysis generally is used in run-off prediction. The hydrograph of a river is a graph that shows how the stream flow varies with time. As such, it reflects those characteristics

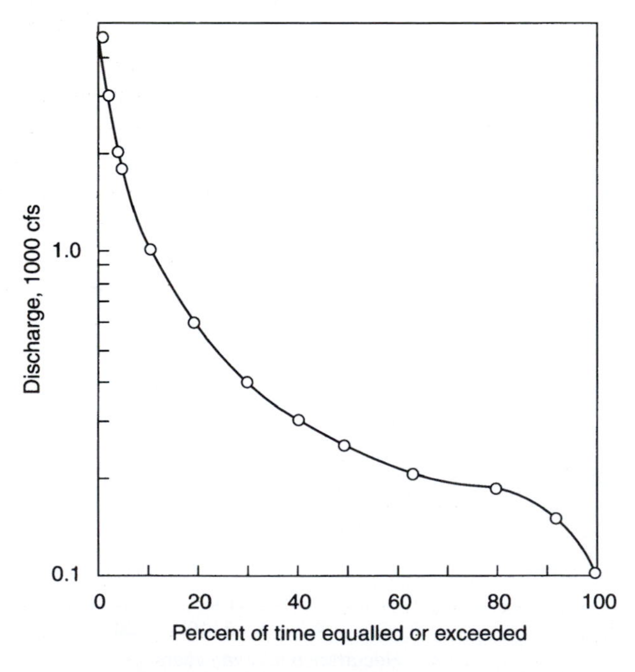

Figure 8.13 Flow duration curve (daily flow) for the Bowie Creek near Hattiesburg, Mississippi, for the period 1939–1948.
Source: Searcy, 1959. Reproduced by kind permission of US Geological Survey.

of the watershed that influence run-off. Hydrographs are sensitive to the size of river basin in that smaller basins show quick responses to fluctuations in precipitation. It may show yearly, monthly, daily or instantaneous discharges. Accordingly, the total flow, baseflow and periods of high and low flows can be determined from hydrographs. Storm hydrographs can be used to predict the passage of flood events. The rising limb of the curve generally is concave upward and reflects the infiltration capacity of a watershed (Fig. 8.14). The time before the steep climb represents the time before infiltration capacity is reached. A sudden, steeply rising limb reflects large, immediate surface run-off. The peak of the curve marks the maximum run-off. Some basins may have two or more peaks for a single storm, depending upon the time distribution of the rain and the basin characteristics. The recession limb represents the outflow from basin storage after inflow has ceased. Its slope therefore is dependent upon the physical characteristics that determine storage. Meanwhile infiltration and percolation result in an elevated water table that therefore contributes more at the end of the storm flow than at the beginning, but thereafter declines along its depletion curve. The dividing line between run-off and baseflow on a hydrograph is indeterminate and can vary widely. If a second precipitation event follows closely on the one preceding, then the latter storm hydrograph may rise on the recession limb of the former. This complicates hydrograph analysis.

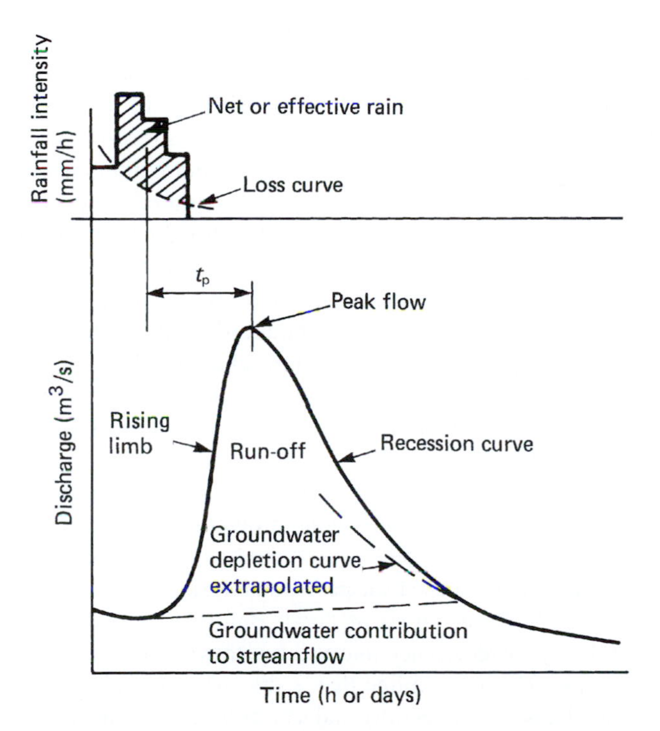

Figure 8.14 Component parts of a hydrograph. There is an initial period of interception and infiltration when rainfall commences before any measurable run-off reaches the stream channels. During the period of rainfall, these losses continue in a reduced form, so the rainfall graph must be adjusted to show effective rain. When the initial losses are met, surface run-off begins and continues to a peak value, which occurs at time t_p, measured from the centre of gravity of the effective rain on the graph. Thereafter, surface run-off declines along the recession limb until it disappears. Baseflow represents the groundwater contribution along the banks of the river.

The unit hydrograph method is one of the most dependable and most frequently used techniques for predicting stream flow. The basis of the method depends on the fact that a stream hydrograph reflects many of the physical characteristics of a drainage basin, so that similar hydrographs can be produced by similar rainfalls. Accordingly, once a typical or unit hydrograph has been derived for certain defined conditions, it is possible to estimate run-off from a rainfall of any duration or intensity. The unit hydrograph is the hydrograph of 25 mm (now normally taken as 10 mm) of run-off from the entire catchment area resulting from a short, uniform unit rainfall. A unit storm is defined as a rain of such duration that the period of surface run-off is not appreciably less for any rain of shorter duration. Its duration is equal to or less than the period of rise of a unit hydrograph, that is, the time from the beginning of surface run-off to the peak. For all unit storms, regardless of their intensity, the period of surface run-off is approximately the same.

It sometimes is necessary to determine unit hydrographs for catchments with few, if any, run-off records. In such instances, close correlations between the physical characteristics of the catchment area and the resulting hydrographs are required. Snyder (1938) was one of the earlier workers to derive synthetic unit hydrographs, and he found that the shape of the catchment and the time from the centre of the mass of rainfall to the hydrograph peak were the main influencing characteristics.

8.5 Hazard zoning, warning systems and adjustments

The lower stage of a river, in particular, can be divided into a series of hazard zones based on flood stages and risk (Fig. 8.15). A number of factors have to be taken into account when evaluating flood hazard such as the loss of life and property, erosion and structural damage, disruption of socio-economic activity including transport and communications, contamination of water, food and other materials, and damage of agricultural land and loss of livestock. Hazard zones are based on historical evidence related to flooding that includes the magnitude of each flood and the elevation it reached, as well as the recurrence intervals, the amount of damage involved, the effects of urbanization and any further development, and an engineering assessment of flood potential. Maps then are produced from such investigations that, for example, show the zones of most frequent flooding and the elevation of the flood waters. Flood hazard maps provide a basis for flood management schemes. For example, Kenny (1990) recognized four geomorphological flood hazard map units in central Arizona, which formed the basis of a flood management plan (Table 8.2).

Floodplain zones can be designated for specific types of land-use, that is, in the channel zone water should be allowed to flow freely without obstruction, for example, bridges should allow sufficient waterway capacity. Another zone could consist of that area with a recurrence interval of 1–20 years that could be used for parks, and agricultural and recreational purposes. Such areas act as washlands during times of flood. Buildings would be allowed in the zone encompassing the 20–100-year recurrence interval, but they would have to have some form of protection against flooding. However, a line that is drawn on a map to demarcate a floodplain zone may encourage a false sense of security and as a consequence development in the upslope area may be greater than it otherwise would. At some point in time this line is likely to be transgressed that will cause more damage than would have been the case without a floodplain boundary being so defined. In addition, urbanization, as seen, is a flood intensifying land-use.

Land-use regulation seeks to obtain the beneficial use of floodplains with minimum flood damage and minimum expenditure on flood protection. In other words, the purpose of land-use regulation is to maintain an adequate floodway (i.e. the channel and those adjacent areas of

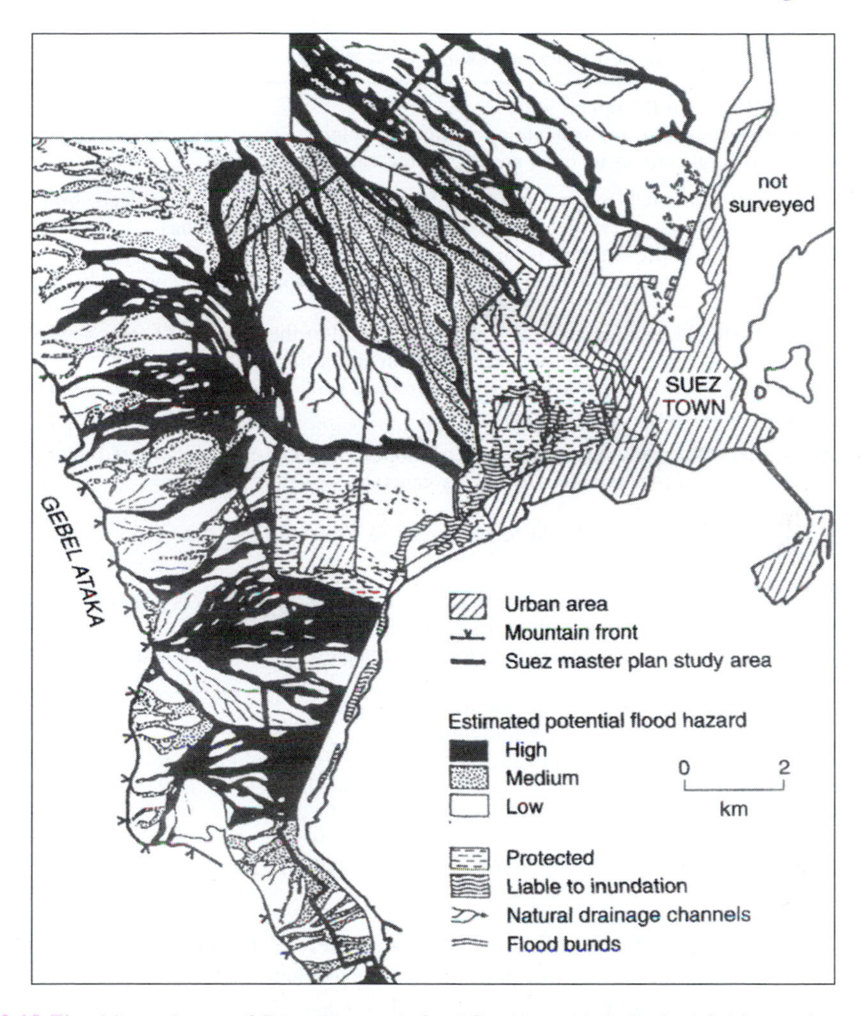

Figure 8.15 Flood hazard map of Suez, Egypt, derived from geomorphological field mapping. *Source*: Cooke *et al*., 1982. Reproduced by kind permission of The Geological Society.

the floodplain that are necessary for the passage of given flood discharge) and to regulate land-use development alongside it (i.e. in the floodway fringe, which in the United States is the land between the floodway and the maximum elevation subject to flooding by the 100-year flood). Land-use in the floodway in the United States now is severely restricted to, for example, agriculture and recreation. The floodway fringe also is a restricted zone in which land-use is limited by zoning criteria based upon the degree of flood risk. However, land-use control involves the cooperation of the local population and authorities, and often central government. Any relocation of settled areas that are of high risk involves costly subsidization. Such cost must be balanced against the cost of alternative measures and the reluctance of some to move. Bell and Mason (1998) outlined such a problem at Ladysmith, South Africa. In fact, change of land-use in intensely developed areas usually is so difficult and so costly that the inconvenience of occasional flooding is preferred. The purchase of land by government agencies to reduce flood damage is rare.

Table 8.2 Generalized flood hazard zones and management strategies

Flood hazard zone I (Active floodplain area) Prohibit development (business and residential) within floodplain Maintain area in a natural state as an open area or for recreational uses only
Flood hazard zone II (Alluvial fans and plains with channels less than a metre deep, bifurcating, and intricately interconnected systems subject to inundation from overbank flooding) Flood proofing to reduce or prevent loss to structures is highly recommended Residential development densities should be relatively low; development in obvious drainage channels should be prohibited Dry stream channels should be maintained in a natural state and/or the density of native vegetation should be increased to facilitate superior water drainage retention and infiltration capabilities Installation of upstream stormwater retention basins to reduce peak water discharges. Construction should be at the highest local elevation site where possible
Flood hazard zone III (Dissected upland and lowland slopes; drainage channels where both erosional and depositional processes are operative along gradients generally less than 5%) Similar to flood hazard zone II Roadways which traverse channels should be reinforced to withstand the erosive power of a channelled stream flow
Flood hazard zone IV (Steep gradient drainages consisting of incised channels adjacent to outcrops and mountain fronts characterized by relatively coarse bedload material) Bridges, roads, and culverts should be designed to allow unrestricted flow of boulders and debris up to a metre or more in diameter Abandon roadways which presently occupy the wash flood plains Restrict residential dwelling to relatively level building sites Provisions for subsurface and surface drainage on residential sites should be required Stormwater retention basins in relatively confined upstream channels to mitigate against high peak discharges

Source: Kenny, 1990. Reproduced by kind permission of The Geological Society.

In some situations, properties of high value that are likely to be threatened by floods can be flood proofed. For instance, an industrial plant can be protected by a flood wall, or buildings may have no windows below high water level and possess watertight doors, with valves and cut-offs on drains and sewers to prevent backing-up. Other structures may be raised above the most frequent flood levels. Most flood-proofing measures can be more readily and economically incorporated into buildings at the time of their construction rather than subsequently. Hence, the adoption and implementation of suitable design standards should be incorporated into building codes and planning regulations to ensure the integrity of buildings during flood events.

Reafforestation of slopes denuded of woodland tends to reduce run-off and thereby lowers the intensity of flooding (Sotir, 1998). As a consequence, forests are commonly used as a watershed management technique. They are most effective in relation to small floods where the possibility exists of reducing flood volumes and delaying flood response. Nonetheless, if the soil is saturated, differences in interception and soil moisture-storage capacity due to forest cover will be ineffective in terms of flood response. Agricultural practices such as contour ploughing and strip cropping are designed to reduce soil erosion by reducing the rate of run-off. Accordingly, they can influence flood response.

Emergency action involves erection of temporary flood defences and possible evacuation. The success of such measures depends on the ability to predict floods and the effectiveness of the warning systems. A flood warning system, broadcast by radio or television, or from helicopters or vehicles, can be used to alert a community to the danger of flooding. Warning systems may use rain gauges and stream sensors equipped with self-activating radio transmitters to convey data to a central computer for analysis (Gruntfest and Huber, 1989). Alternatively, a radar rainfall scan can be combined with a computer model of a flood hydrograph to produce real time forecasts as a flood develops (Collinge and Kirby, 1987). However, widespread use of flood warning usually is only available in highly sensitive areas. The success of a flood warning system depends largely on the hydrological characteristics of the river. Warning systems often work well in large catchment areas that allow enough time between the rainfall or snowmelt event and the resultant flood peak to allow evacuation, and any other measures to be put into effect. By contrast, in small tributary areas, especially those with steep slopes or appreciable urban development, the lag-time may be so short that, although prompt action may save lives, it is seldom possible to remove or protect property. In the United States, flash flood warning systems have been installed in certain areas, at which sensors at an upstream station detect critical water levels and relay them to an alarm station in the community about to be affected. The flood warning is issued from the station.

Financial assistance in the form of government relief or insurance payouts do nothing to reduce flood hazard. Indeed, by attempting to reduce the economic and social impact of a flood, they encourage repair and rebuilding of damaged property that may lead to the next flood of similar size giving rise to more damage. What is more, the expectation that financial aid will be made available in such an emergency may result in further development of flood prone areas. Hence, it would be more realistic to adjust insurance premiums for flooding in relation to the degree of risk. Be that as it may, the basic principle of insurance needs to be modified in the case of floods so that premiums paid over many years cover the large losses that will be encountered in a few years. In the United States, the Federal Flood Insurance Program is administered by the Federal Emergency Management Agency (FEMA) from which interested parties can purchase subsidized flood insurance. The programme provides incentives to local governments to plan and regulate land-use in flood hazard areas. The boundary of the 100-year flood defines the area that is subject to flood damage compensation and if a community wishes to qualify for federal aid it must join the Federal Flood Insurance Program. All property owners within the 100-year flood boundary then must purchase flood insurance. The aim was to achieve flood damage abatement and efficient use of the floodplain.

8.6 River control and flood regulation

River control refers to projects designed to hasten the run-off of flood waters or confine them within restricted limits, to improve drainage of adjacent lands, to check stream bank erosion or to provide deeper water for navigation (Brammer, 1990). What has to be borne in mind, however, is that a river in an alluvial channel is continually changing its position due to hydraulic forces acting on its banks and bed. As a consequence, any major modifications to a river system that are imposed without consideration of the channel will give rise to a prolonged and costly struggle to maintain the change.

8.6.1 River training

River training works have evolved from practices centuries old. Today many of these methods still are largely unsophisticated and vary from area to area depending on the availability of suitable material. They have two objectives, namely, to prevent erosion of river banks and to improve the discharge capacity of a river channel. A careful study of the current pattern in a river or estuary should be made before deciding the location of training works as they may cause changes in the behaviour of river flow that may be undesirable (Dutnell, 1999). In estuaries, it may be desirable to check the proposed line of training works with the aid of hydraulic models.

Willow piling is one of the most simple methods of controlling erosion, although revetments composed of fascines of willows (faggots) have been more widely used in Britain (Fig. 8.16). Fascines are used especially to protect bank toes or shallow river banks and should not be used on steep banks because they can dry and be washed away during a flood. Nicholls and Leiser (1998) referred to the use of rows of cylinders of coconut fibre to stabilize the base of the banks of the Petaluma River, California, the upper slopes being stabilized by bundles of live willow cuttings staked out in rows. The areas between the bundles and the fibre cylinders were secured with coconut fibre matting, 50.1 mm in thickness. Gerstgraser (1999) carried out a series of tests in a flume and found that brush mattresses of willows can resist large hydrodynamic forces and so can protect steep river banks. Vegetated geotextiles and geogrids also can be used to protect shallow river banks but only above water level. When stone is readily available it is used in preference to faggotting as it is more durable. Stone frequently has been used as pitching to counteract erosion and for training walls to stabilize river channels. Pre-cast concrete blocks are used more extensively as revetment than stone. Blockwork revetment is more inflexible than stone and requires a rigid toe beam to maintain its alignment. Gabions consist of steel-mesh rectangular boxes filled with stone (Fig. 2.38). Along rivers where there is an ample supply of boulders gabions have some advantages. In urban areas channels may be lined with concrete.

Channel regulation can be brought about by training dykes, jetties or wing dams, which are used to deflect channels into more desirable alignments or confine them to lesser widths. Dykes and dams can be used to close secondary channels and thus divert or concentrate a river into a preferred course. Permeable pile dykes are the principal means of training and contracting the lower Mississippi River, restricting the low-water channel to about its normal average width,

Figure 8.16 Placing fascines along the west bank of the River Ouse, Lincolnshire, England.

thereby eliminating local sections of excessive width where shoaling is most troublesome. Sand-fill dams and dykes, built by hydraulic dredging, are used in many places to direct the current or close off secondary channels. The sand-fill dams are not expected to be permanent. In some cases ground sills or weirs need to be constructed to prevent undesirable deepening of the bed by erosion. Bank revetment by pavement, rip-rap or protective mattresses to retard erosion, is usually carried out along with channel regulation (Pilarczyk, 1998).

Morris and Moses (1999) described the design and construction of a rehabilitation project for a small suburban river in Moscow, Idaho. This essentially involved the re-establishment of the meander channel pattern that had been straightened and dredged, thereby forming a new flood plain. A 50-year flood six months after the project had been completed has no significant impact on the newly constructed channel and revetments.

8.6.2 Dredging

River channels may be improved by dredging. When a river is dredged, its floor should not be lowered to such a degree that the water level is appreciably lowered. In addition, the nature of the materials occupying the floor should be investigated. First, this gives an indication of which plant may be suitably employed. Indeed, removal of unconsolidated material usually revolves around the selection of suitable floating equipment for dredging work. Of special importance is the possible presence of boulders. Suction dredgers should always be fitted with a simple trap device in the suction line for catching boulders and rock fragments. Underwater rock excavation can be carried out by underwater drilling and blasting or from a floating rock breaker. Second, it provides information relating to the stability of the slopes of the channel. The rate at which sedimentation takes place provides some indication of the regularity with which dredging should be carried out. Dredging of river mouths and deposition of the sediment at sea, however, may lead to erosion of neighbouring beaches subjected to reduced sediment supply.

Rock dredging is expensive and so good prediction of dredgeability is important. In an assessment of the dredgeability of carbonate rocks in Qatar, Vervoort and De Wit (1997) developed a rating system based on the unconfined compressive strength, the fissure index and the thickness of the individual layers of rock. The values of each parameter for each layer of rock provided the rating value. The values for each layer encountered in a drillhole then were summed to give a total rating. Rating values were related to the volume excavation rate and the rate of wear of picks. Previously, Verhoef (1993) had investigated wear as it affects tool consumption in relation to the abrasivity of sandstone.

8.6.3 The design flood and flood control

No structure of any importance, either in or adjacent to a river, should ever be planned or built without due consideration being given to the damage it may cause by its influence on flood waters or the damage to which it may be subjected by those same waters (Oostinga and Daemen, 1997). To avoid disaster, bridges must have the required waterway opening, flood walls and embankments must be high enough for overtopping not to occur, reservoirs must have sufficient capacity and dams must have sufficient spillway capacity as well as adequate protection against scour at the toe.

The maximum flood that any such structure can safely pass is called the design flood. If a flood of a given magnitude occurs on average once in 100 years there is a 1 per cent chance that such a flood will occur during any one year. The important factor to be determined for any design flood is not simply its magnitude but the probability of its occurrence. In other words, is the structure safe against the 2, 1 or the 0.1 per cent chance flood or against the maximum flood that may ever be anticipated? Once this has been answered the magnitude of the flood that may be expected to occur with that particular average frequency has to be determined. The design of any flood protection works also must consider:

1 The extent to which human life will be endangered. Any structure whose failure would seriously endanger human lives ideally should be designed to pass the greatest flood, that will probably ever occur at that point, with safety.
2 The value of any property that would be destroyed by any particular flood. This can be weighed against the estimated cost of the necessary flood protection works.
3 The inconvenience resulting from failure of a structure. A flood control system has a psychological effect in that it provides a sense of security against floods.

In the United States the Flood Disaster and Protection Act (1973) specified the 100-year flood as the limit of the flood plain for flood insurance purposes and this became widely accepted as the standard of risk. However, flood mitigation projects should not be constrained by the design standard being linked to a particular flood event. When assessing the future effectiveness of a specific flood mitigation project, its likely reduction of damage and safety afforded to the community over the entire spectrum of possible floods should be evaluated. The design flood therefore should be chosen in relation to all relevant economic, social and environmental factors.

A relatively simple solution to the problem of flooding is to build flood defences consisting of either earth embankments or masonry or concrete walls around the area to be protected (Thampapillai and Musgrave, 1985). Levees have been used extensively in the United States to protect flood plains from overflow (Fig. 8.17a). Their use along the Lower Mississippi River is given in an account of river engineering activities, along this most engineered of rivers, by Smith and Winkley (1996). Levees are earth embankments, the slopes of which should be protected against erosion by planting trees and shrubs, by paving or with rip-rap. Without protection against bank erosion, a river will probably begin to meander again. The rate at which river channels revert to their former condition depends upon many factors. Nevertheless, this means that there is a need for maintenance. Levees reduce the storage of flood water by eliminating the natural overflow basins of a river on the flood plain. Furthermore, they contract the channel and so increase flood stages within, above and below the leveed reach. Accordingly, wherever possible levees should be located away from river channels and ideally outside the meander belt (Fig. 8.17b). In rural areas, this normally poses no great problems but it is impractical in urban areas. On the other hand, because levees confine a river to its channel this means that its efficiency is increased, hence they expedite run-off. This means, however, that consideration must be given to the fact that more rapid movement of water through a section of river that has been hydraulically improved can enhance flood peaks further downstream and be responsible for accelerated erosion. Levees often encourage new development at lower levels where previously no one was inclined to build. Consequently, when an exceptional flood occurs and overtops the levees, the hazard may be reduced by building fuseplugs into the levees, that is, making certain sections deliberately weaker than the standard levee section, thereby determining that if breaks occur, they do so at locations where they cause minimum damage.

Levees or embankments for flood waters along a river inevitably come in contact with tributaries. In such instances, the protective structures must be continued along the tributary until high ground is reached. Alternatively, the tributary channel can be blocked off during times

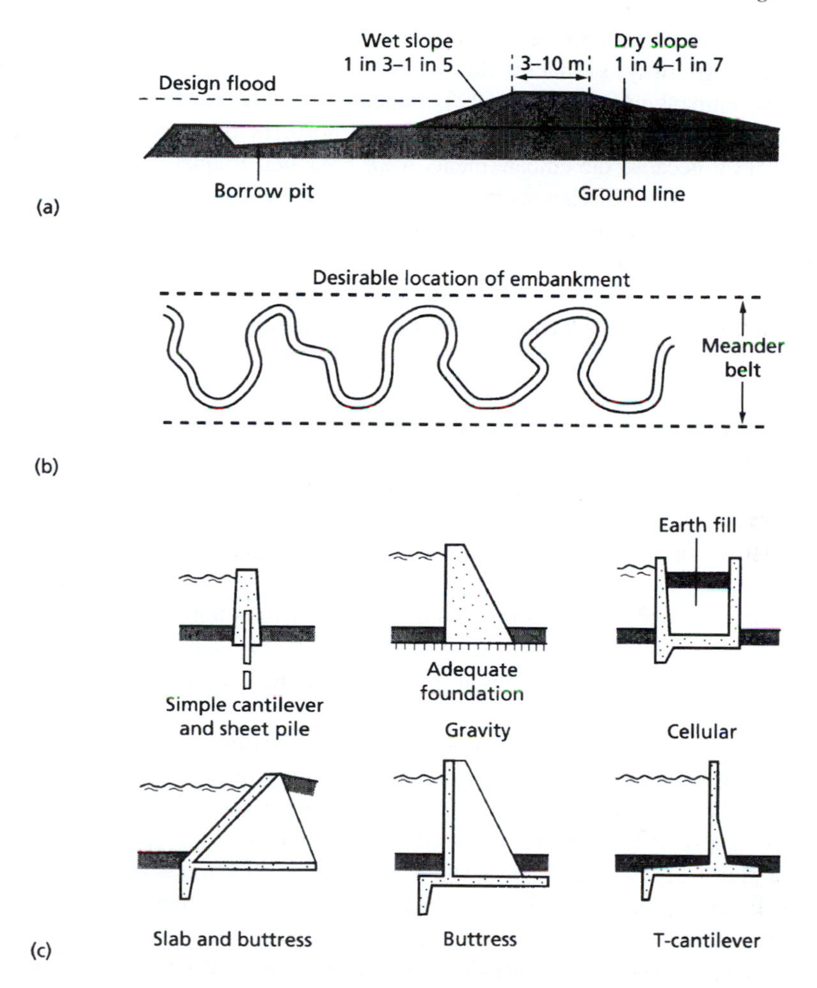

Figure 8.17 (a) Cross-section of a typical flood embankment (levee). (b) Preferred location of embankments along a meandering channel. (c) Cross-sections of typical flood walls.

of flood, for example, by sluice gates, but this causes a problem of interior drainage (i.e. water collecting behind the embankment). This problem may be solved by collecting the water during a flood and pumping it over the embankment, by conveying the water in an open channel on the landward side of the embankment to some point downstream where it can be discharged, or by collecting water in a storage basin for subsequent disposal. Embankment stability also must be considered. Excessive seepage of flood water into an embankment may bring about instability or induce piping. The latter can cause an embankment to collapse. If breached, an embankment can accentuate the flood problem in the protected area by inhibiting drainage of floodwater downstream of the breach back into the channel. Gilvear *et al.* (1994) recorded 116 breaches in the embankments along the Rivers Tay and Earn in Scotland after the 100-year flood in 1993, mostly as a result of overtopping. Extensive flood plain damage was associated with a number of the breaches. It appeared that embankments located on the outside of bends and overlying old river courses were highly vulnerable to failure. Embankments bordering tributaries also seemed vulnerable to failure because of being perpendicular to the direction of flow across the flood plain during inundation. Gilvear *et al.* concluded that when breaches are repaired, they should incorporate spillways so that the future flood waters can move into the flood plain before overtopping the embankments. Sinha (1998) indicated that the construction of embankments along the major sections of the rivers in the plains of northern Bihar, India, had proved only a short-term solution because the embankments were breached frequently. This is not only due to the extremely high river discharges but also because these rivers carry high sediment loads that give rise to rapid deposition and aggradation of the beds of the rivers and the water levels they contain. In certain situations underseepage beneath levees can lead to problems such as sand boils or piping that ultimately could result in failure of the levees. Control measures need to be put in place to prevent such an occurrence and commonly include the installation of relief wells. For example, Mansur *et al.* (2000) indicated that the levees alongside the Mississippi River in the St Louis district were raised in the 1940s and that an investigation was initiated in 1950 to determine the amount of seepage that took place beneath the levees. The results led to 2480 relief wells being installed along about 470 km of mainstream and tributary levees. The flood of 1993 along this section of the river equalled or exceeded the highest level to which the levees had ever been subjected. Mansur *et al.* found that the relief wells had functioned successfully during this flood and had prevented the development of any significant sand boils or piping.

Flood walls may be constructed in urban areas where there is not enough space for embankments. They should be designed to withstand the hydrostatic pressure (including uplift pressure) exerted by the water when at design flood level. If the wall is backed by an earth-fill, then it also must act as a retaining wall.

Diversion is another method used to control flooding. This involves opening a new exit for part of the river water. Diversion schemes may be temporary or permanent. In the former case, the river channel is supplemented or duplicated by a flood relief channel, a bypass channel or a floodway. These operate during times of flood. Flood water is diverted into a diversion channel via a sluice or fixed crest spillway. A permanent diversion acts as an intercepting or cut-off channel that replaces the existing river channel diverting either all or a substantial part of the flow away from a flood-prone reach of river. Such schemes normally are used to protect intensively developed areas. Temporary diversion channels are most effective in reducing flood water levels when they do not return further downstream into the river concerned, that is, they exit into the sea, a lake or another river system. Old river channels can be used as diversions to relieve the main channel of part of its flood water, the Atchafalaya floodway on the lower Mississippi providing an example (Fig. 8.18). Usually, however, it is necessary to return the diverted water into the river at a point downstream. Therefore, the diversion channel should be long enough to minimize backwater effects in the stretch of river being protected. Any diversion must be designed in such a way that it does not cause excessive deposition to occur in the main channel otherwise it defeats its purpose.

The flood hazard often can be lessened by stage reduction by improving the hydraulic capacity of the channel without affecting the rate of discharge. This may be accomplished by straightening, widening and deepening a river channel. Inasmuch as the quantity of discharge through any given cross-section of a river during a given time depends upon its velocity, the stage can be reduced by increasing the velocity (Brookes, 1988). However, the extent of the benefits that can be obtained by straightening depends upon the initial conditions of the channel. For example, even though a river course is extremely sinuous, if the fall is slight, then the amount of stage reduction that can be accomplished by straightening usually is quite limited. Morris *et al.* (1996) mentioned that during the twentieth century many streams in western Iowa had been channelized to reduce flooding and to convert adjacent marshland for

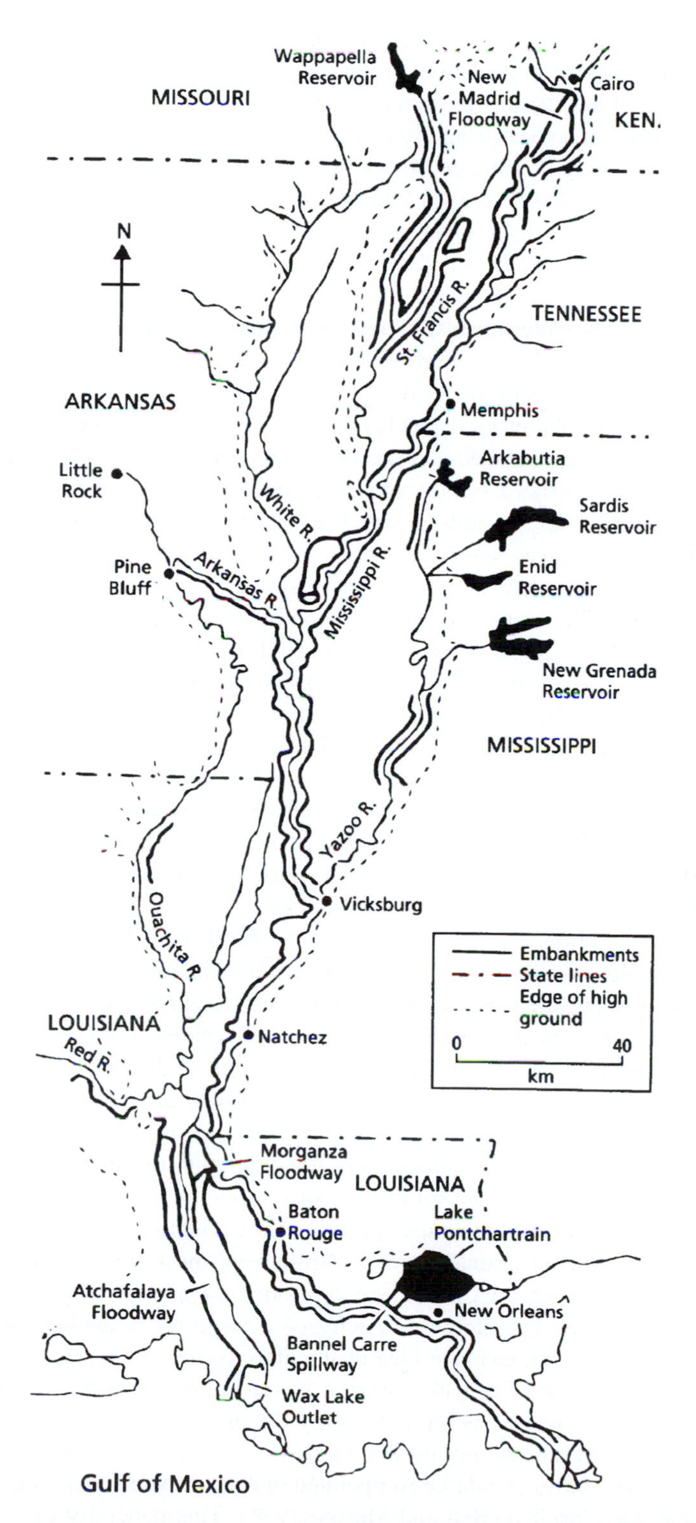

Figure 8.18 Major embankments in the Lower Mississippi valley and Atchafalaya floodway.

cultivation. Unfortunately, the enhanced stream velocities that channelization had given rise to also resulted in increased degradation within the streams. This, in turn, led to loss of land and adversely affected the streams as communication systems. Nevertheless, the velocity and therefore the efficiency of a river generally is increased by cutting through constricted meander loops. River channels may be enlarged to carry the maximum flood discharges within their banks without overspill. However, over-widened channels eventually revert to their natural sizes unless continuously dredged. Over-large channels also mean that during periods of low flow the depths of water are shallow and that riparian land may become over drained.

Flood routing is a procedure by which the variation of discharge with time at a point on a stream channel may be determined by consideration of similar data for a point upstream (Wilson, 1983). In other words, it is a process that indicates how a flood wave may be reduced in magnitude and lengthened in time by the use of storage in a reach of the river between the two points. Flood routing therefore depends on a knowledge of storage in a particular reach. This can be evaluated by either making a detailed topographical and hydrographical survey of the river reach and the riparian land, thereby determining the storage capacity of the channel at different levels, or by using records of past levels of flood waves at the limits of the reach and hence deducing its storage capacity.

Peak discharges can be reduced by temporarily storing a part of the surface run-off until after the crest of the flood has passed. This is done by inundating areas where flood damage is not important, such as water meadows or waste land. If, however, storage areas are located near or in towns and cities, they sterilize large areas of land that, if usable for other purposes, could be extremely valuable. On the other hand, it may be feasible to develop these storage areas as recreational centres. This method is seldom sufficient in itself and should be used to complement other measures.

Reservoirs help to regulate run-off, thereby helping to control floods and improve the utility of a river (Hager and Sinniger, 1985). There are two types of storage, regardless of the size of the reservoir, controlled and uncontrolled. In controlled storage, gates in the impounding structure may regulate the outflow. Only in unusual cases does such a reservoir have sufficient capacity to completely eliminate the peak of a major flood. As a result the regulation of the outflow must be planned carefully. This necessitates estimation of how much of the early portion of a flood may be safely impounded that, in turn, requires an assessment of the danger that can arise if the reservoir is filled before the peak of a flood is reached. Where reservoirs exist on several tributaries, the additional problem of timing of the release of the stored waters becomes a matter of great importance, since to release these waters in such a way that the peak flows combine at a downstream point can bring disaster. In uncontrolled storage there is no regulation of the outflow capacity and the only flood benefits result from the modifying and delaying effects of storage above the spillway crest.

A significant, although largely qualitative criterion for evaluating a flood-mitigation reservoir is the percentage of the total drainage area controlled by the reservoir. Generally, one-third or more of the total drainage area should be controlled by a reservoir if flood control is to be effective (Ward, 1978). The storage capacity of a reservoir is another factor that has to be taken into account. An approximate idea of the effectiveness of a reservoir can be derived by comparing its storage capacity with potential storm rainfall over the catchment area. The maximum capacity that is needed is represented by the difference in volume between the safe release from the reservoir and the design flood inflow.

Reservoirs for flood control should be so operated that the capacity required for storing flood water is available when needed (Bell and Mason, 1998). This generally can be accomplished by lowering the water level of the reservoir as soon as practicable after a flood passes. On the other hand, the greatest effectiveness of reservoirs for increasing the value of a river for utilization, is realized by keeping them as near full as possible. Hence, there must be some compromise in operation between these two purposes. Furthermore, after a flood has occurred, a part of the storage capacity will be occupied by the floodwater and will not be available until this water is released. However, another part of the storage capacity should be reserved just in case there is a second flood that follows before floodwater release is effected. In other words, the full capacity of a reservoir cannot be assumed available for a single flood event. Accordingly, the effectiveness of a reservoir in controlling the regimen of a river increases as the reservoir capacity is augmented and is measured by the ratio of capacity to total run-off. A reservoir generally must be designed and situated so that the quantity of inflow that has to be stored ordinarily does not exceed reservoir capacity. The effect of a reservoir on the regimen of flow in any part of a river varies inversely with the distance from the reservoir because of the time involved in transit, natural losses, fluctuations in flow of the intervening tributaries and decreasing relative effect as river flow increases with increase in drainage area.

The economic aspect of controlling rivers by means of reservoirs is affected by the availability of possible reservoirs sites. Reservoir sites are located principally in the middle reaches of a river course (Fig. 8.19). Thus, there are limitations placed upon the use of reservoirs for river control. The value of small reservoirs in the headwaters of a river as a method of flood control is questionable. This is because such reservoirs are likely to be full or partly full at the time of a flood producing rain. Moreover, there is little point in inundating a large area of land to protect other areas if they are only slightly more valuable. Reservoirs reduce the flow velocity of streams, as they enter it, thereby causing siltation. In time, the growth of the

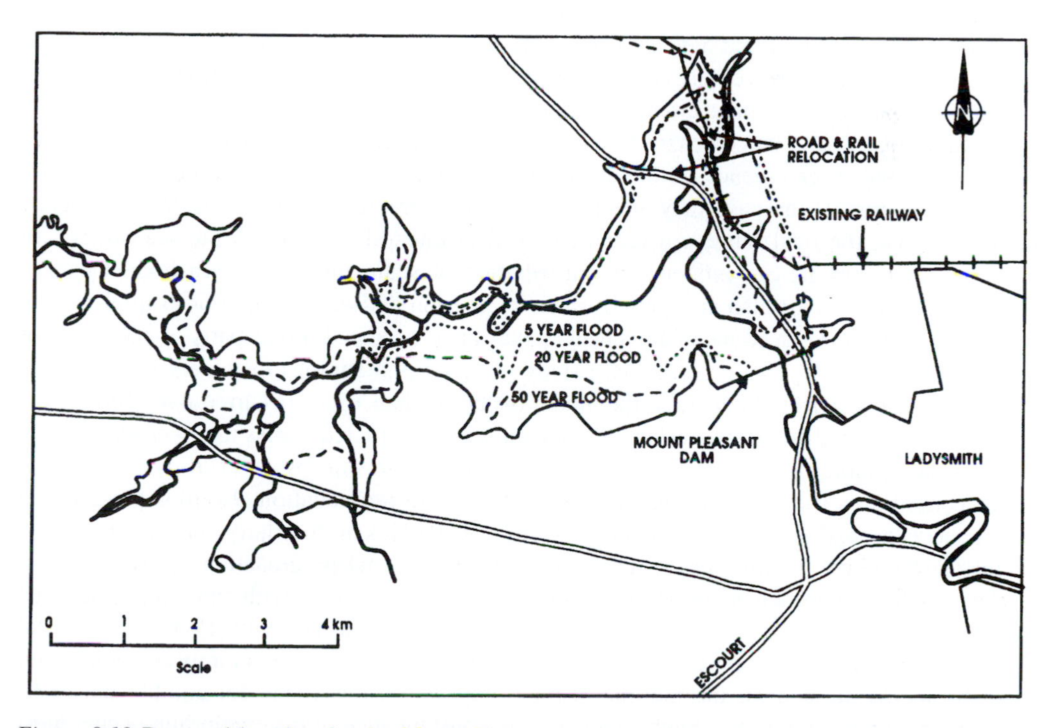

Figure 8.19 Proposed location for the Mount Pleasant Dam and Reservoir to protect Ladysmith, South Africa showing the 5, 20 and 50 year flood boundaries.
Source: Beli and Mason, 1998.

resulting delta may reach upstream and can affect the economy and installations at upstream localities. Water discharged from the reservoir below a dam has renewed ability to erode and entrain sediment from the channel immediately below the dam. Such channel lowering causes lowering of tributary channels. The sediment eroded from positions near the dam is transported further downstream where it is deposited at slack water sites. These new deposits upset the normal channel system and can lead to renewed flooding.

Retarding basins are much less common than reservoirs. They are provided with fixed ungated outlets that regulate outflow in relation to the volume of water in storage. An ungated sluiceway acting as an orifice tends to be preferable to a spillway functioning as a weir. The discharge from the outlet at full reservoir capacity ideally should be equal to the maximum flow that the downstream channel can accept without causing serious flood damage. One of the advantages of retarding basins is that only a small area of land is permanently removed from use after their construction. Although land will be inundated at times of flood, this will occur infrequently and so the land can be used for farming but obviously not for permanent habitation. As with reservoirs, the planning of a system of retarding basins must avoid making a flood event worse by synchronizing the increased flow during drawdown with flood peaks from tributaries. Consequently, retarding basins are better suited to small rather than large catchment areas.

8.7 The coastal zone

Marine activity in coastal environments acts within a narrow and often varying vertical zone. Significant changes in sea level, associated with the Pleistocene glaciation, have taken place in the recent geological past that have influenced the character of present day coastlines. When ice sheets expanded, the sea level fell, rising when ice melted. In addition, as ice sheets retreated, the land beneath began to rise in an attempt to regain isostatic equilibrium, thereby complicating the situation. Hence, coasts can be characterized by features developed in response to submergence on the one hand or emergence on the other. Such changes in the relative level of the sea are not only a feature of the past, they are still occurring at present. Tides also have an influence upon the coast, the tidal range governing the vertical interval over which the sea can act. In addition, tidal currents can influence the distribution of sediments on the sea floor. Changes in the outline of a coast also are brought about by erosion and deposition. For example, Cetin *et al.* (1999) gave an account of the changes that have occurred along parts of the coast of Turkey in the last 50 years or so. After a study of aerial photographs they were able to show that progradation across the delta at the mouth of the Seyhan River occurred over 198 125 m^2 between 1947 and 1954. However, in 1954 a dam was constructed across the river that greatly reduced the amount of sediment carried to the coast. As a result, by 1995 over 1 000 000 m^2 of the coastal area had been lost due to erosion. Extensive progradation also took place in the coastal areas around the mouths of the Rivers Ceyhan and Goksu. Similarly, a dam constructed on the Ceyhan River in 1984 has reduced the amount of coastal progradation notably.

Waves acting on beach material are a varying force. They vary with time and place due to changes in wind force and direction over a wide area of sea, and with changes in coastal configuration and offshore relief. This variability means that the beach is rarely in equilibrium with the waves, in spite of the fact that it may only take a few hours for equilibrium to be attained under new conditions. Such a more or less constant state of disequilibrium occurs most frequently where the tidal range is considerable, as waves are continually acting at a different level on the beach.

8.7.1 Waves

When wind blows across a surface of deep water it causes an orbital motion in those water particles in the plane normal to the wind direction (Fig. 8.20). The motion decreases in significance with increasing depth, dying out at a depth equal to that of the wavelength. Because adjacent particles are at different stages in their circular course a wave is produced. However, there is no progressive forward motion of the water particles in such a wave, although the form of the wave profile moves rapidly in the direction in which the wind is blowing. Such waves are described as oscillatory waves.

Forced waves are those formed by the wind in the generating area, which are usually irregular. On moving out of the area of generation, these waves become long and regular. They then are referred to as free waves. As these waves approach a shoreline, they feel bottom, which disrupts their pattern of motion, changing them from oscillation to translation waves. Where the depth is approximately half the wavelength, the water-particle orbits become ellipses with their major axes horizontal.

The forward movement of the water particles, as a whole, is not entirely compensated by the backward movement. As a result there is a general movement of the water in the direction in which the waves are travelling. This is known as mass transport. The time required for any one particle to complete its orbital revolution is the same as the period of the waveform. The orbital velocity (u) is equal to the length of orbital travel divided by the wave period (T). Hence:

$$u = \frac{2\pi r}{T} \tag{8.20}$$

Similarly, the wave velocity or celerity (c) is:

$$c = \frac{2\pi R}{T} \tag{8.21}$$

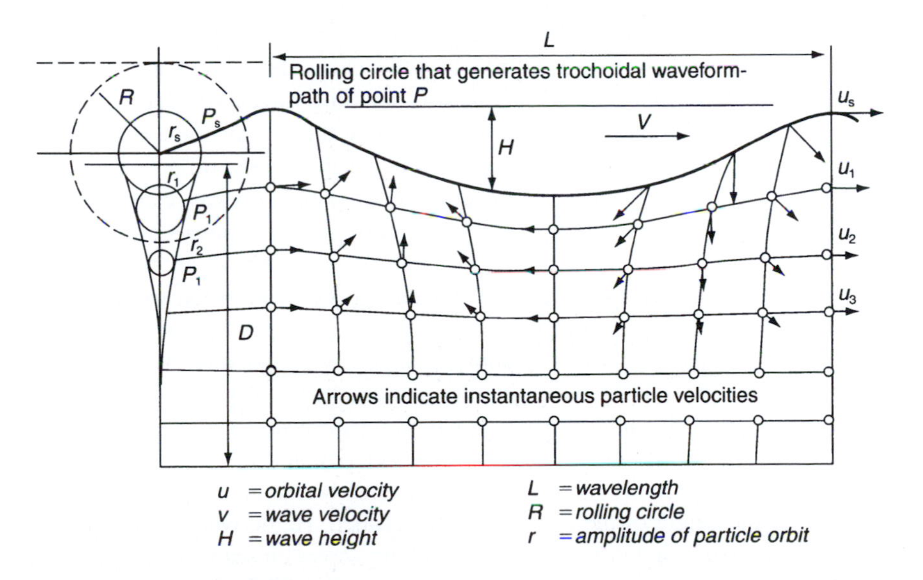

Figure 8.20 Trochoidal water waves.

where R is the rolling circle required to generate a trochoidal waveform and r in Eq. 8.20 is the amplitude of particle orbit. The celerity of waves also may be derived from the following expression:

$$c = \left[\frac{gL}{2\pi}\tanh\left(\frac{2\pi Z}{L}\right)\right]^{1/2} \tag{8.22}$$

where g is the acceleration due to gravity, Z is the still water depth and L is the wavelength. If Z exceeds L, then $\tanh(2Z/L)$ becomes equal to unity, so that in deep water:

$$\begin{aligned} c &= \frac{gL}{2\pi} \\ &= 2.26\sqrt{L} \end{aligned} \tag{8.23}$$

When the depth is less than one-tenth of the wavelength, $\tanh(2\pi Z/L)$ approaches $2\pi Z/L$ and c equals $\sqrt{gZ}$. As waves move into shallow water their velocity is reduced and so their wavelength decreases.

8.7.2 *Force and height of waves*

The force exerted by waves includes jet impulse, viscous drag and hydrostatic pressure. The source of dynamic wave action lies in the inertia of the moving particles. Each particle may be considered as having a tangential velocity due to rotation about the centre of its orbit, and a velocity of translation corresponding to mass transport. The vectorial sum of these two components is the actual velocity of the particle at any instant; when the particle is at a wave crest this resultant velocity is horizontal.

The effectiveness of wave impact on a shoreline or marine structure depends on the depth of water and the size of the wave, it drops sharply with increasing depth. If deep water occurs alongside cliffs or sea walls, then waves may be reflected without breaking and by doing so they may interfere with incoming waves. In this way standing waves, which do not migrate, are formed, in which the water surges back and forth between the obstruction and a distance equal to half the wavelength away. Their crests are much higher than in the original wave. This form of standing wave is known as clapotis. The oscillation of standing waves causes an alternating increase and decrease of pressure along any discontinuities in rock masses or cracks in marine structures that occur below the water line. It is assumed that such action gradually dislodges blocks of material. It has been estimated that translation waves reflected from a vertical face exert six times as much pressure on that face as oscillation waves of equal dimensions. When waves break, jets of water are thrown at approximately twice the wave velocity, which also causes increases in the pressure in discontinuities and cracks thereby causing damage.

It has been shown experimentally that the maximum pressure exerted by breaking waves on a sea wall is $1.3(c+u_c)^2$ where c is the wave celerity and u_c is the orbital velocity at the crest. Alternatively, Little (1975) gave the velocity of waves breaking in shallow water as $5.8H_c$ where H_c is the height of the crest above the seabed. He derived the simple expression $p = 10.7H$ for the pressure (in kPa), exerted by breaking waves on sea works. This is applicable to most short waves, and beach and wall slopes exceeding 1 in 50. In the case of long waves he suggested $p = 13.5H$, H being the wave height out at sea. These expressions overestimate the pressure but in design it is always better to overestimate rather than to underestimate. Bascom (1964) estimated that a storm that has wave heights averaging 3 m can develop a maximum force equal 145 kPa, that a 6.1 m swell could develop 115 kPa, and that strong gales could develop waves with forces in excess of 290 kPa.

Fetch is the distance that a wind blows over a body of water and is the most important factor determining wave size and efficiency of transport. For instance, winds of moderate force that blow over a wide stretch of water generate larger waves than do strong winds that blow across a short reach. For a given fetch, the stronger the wind, the higher are the waves (Fig. 8.21a). Their period also increases with increasing fetch. The maximum wave height (H_{max}) for a given fetch can be derived from the expression:

$$H_{max} = \frac{0.3 v_w}{g} \tag{8.24}$$

where v_w is the velocity of a uniform wind in cm s^{-1} and g is the acceleration due to gravity. Wind duration also influences waves, the longer it blows, the faster the waves move, whilst their lengths and periods increase (Fig. 8.21b). Usually, wavelengths in the open sea are less than 100 m and the speed of propagation is approximately 48 km h^{-1}. They do not normally exceed 8.5 m in height. Those waves that are developed in storm centres in the middle of an ocean may journey outwards to the surrounding landmasses. This explains why large waves may occur along a coast during fine weather. Waves frequently approach a coastline from different areas of generation; if they are in opposition, then their height is decreased, whilst their height is increased if they are in phase. Moreover, when the wind shifts in direction or intensity the new generation of waves differ from the older waves, which still persist in their original course. The more recent waves may overtake or, if the storm centre has migrated meanwhile, may intercept the earlier waves, thereby complicating the wave pattern.

8.7.3 Wave refraction

Wave refraction is the process whereby the direction of wave travel changes because of changes in the topography of the nearshore sea floor (Fig. 8.22). When waves approach a straight beach at an angle, they tend to swing parallel to the shore due to the retarding effect of the shallowing water. At the break point, such waves seldom approach the coast at an angle exceeding 20° irrespective of the offshore angle to the beach. As waves approach an irregular shoreline refraction, causes them to turn into shallower water so that the wave crests run roughly parallel to the depth contours. Along an indented coast shallower water is first met with off-headlands. This results in wave convergence and an increase in wave height with wave crests becoming concave towards headlands. Conversely, where waves move towards a depression in the sea floor, they diverge, are decreased in height and become convex towards the shoreline. In both cases the wave period remains constant. The concentration of erosion on headlands leads to the coast being gradually smoothed.

Refraction diagrams often form part of a shoreline study, graphic or computer methods being used to prepare them from hydrographic charts or aerial photographs. These diagrams indicate the direction in which waves flow and the spacing between the orthogonal lines is inversely proportional to the energy delivered per unit length of shore. Because of refraction, it is possible by constructing one or more raised areas on the sea bed to deviate waves passing over them so that calm water occurs at harbour entrances.

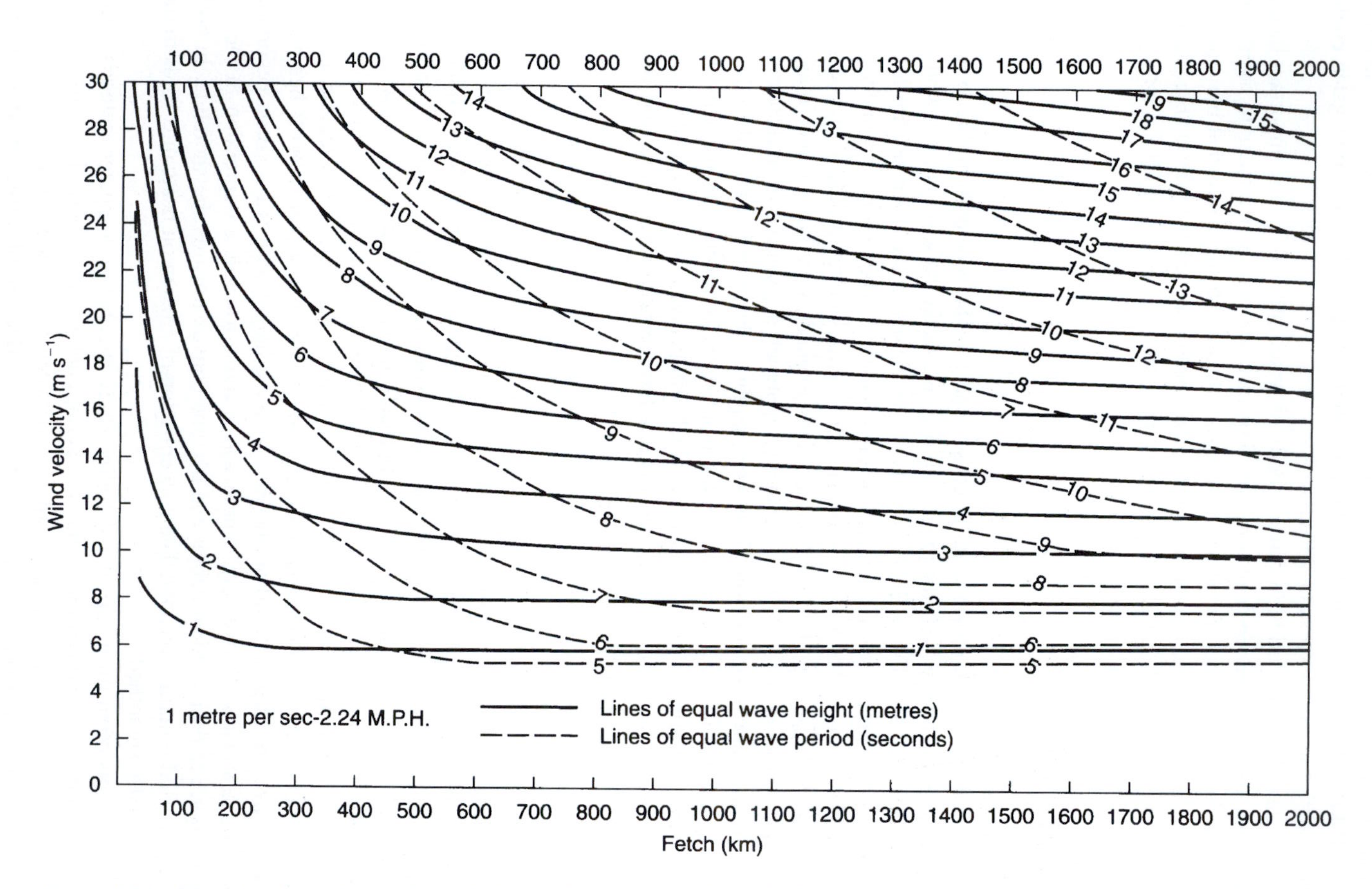

Figure 8.21a Wave height and period in relation to fetch and wind speed. Caution must be exercised using these curves since the quantities are rarely known accurately, so the results read from the curves will be correspondingly open to doubt.

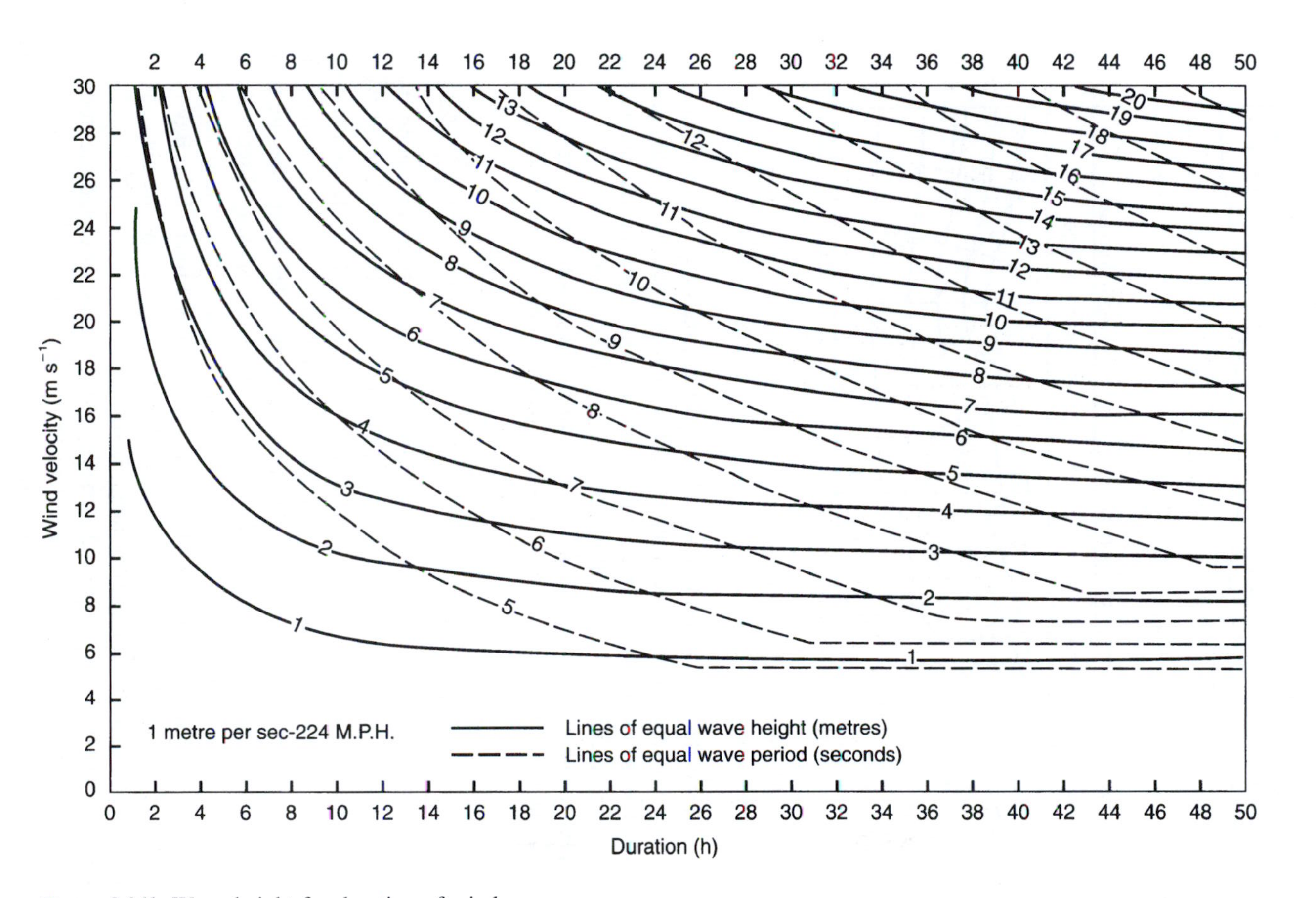

Figure 8.21b Wave height for duration of wind.
Source: Sverdrup and Munk, 1946. Reproduced by kind permission of American Geophysical Union.

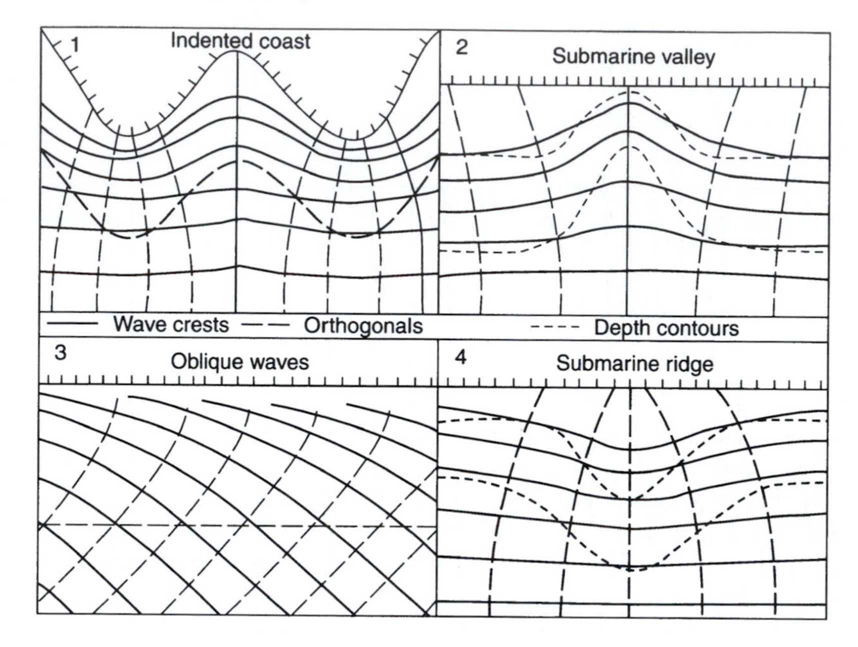

Figure 8.22 Wave refraction patterns.

Waves are smaller in the lee of a promontory or off a partial breakwater. The nature of the wave pattern within this sheltered area is determined by diffraction as well as refraction phenomena. That portion of the advancing wave crest that is not intercepted by the barrier immediately spreads out into the sheltered area and the wave height shrinks correspondingly. This lateral dissipation of wave energy is termed diffraction. Diffraction results from interference with the horizontal components of wave motion. The depth of water is not a relevant factor.

8.7.4 Tides

Tides are the regular periodic rise and fall of the surface of the sea, observable along shorelines. They are formed under the combined attraction of the moon and the sun, chiefly the moon. Normal tides have a dual cycle of about 24 hours and 50 min. The tide height is at a maximum or minimum when sun, moon and earth are in line, so that the attractive force of both celestial bodies upon the ocean waters is combined and thus creates spring tides. Neap tides are generated when the attractive forces of the sun and moon are aligned in opposing directions. In this instance the range between high and low tide is lowest. The tidal range also is affected by the configuration of a coastline. For example, in the funnel-shaped inlet of the Bay of Fundy, Canada, a range of 16 m is not uncommon. By contrast, along relatively straight coastlines or in enclosed seas like the Mediterranean the range is small.

The tidal current that flows into bays and estuaries along a coast is called the flood current whereas the current that returns to the sea is named the ebb current. These are of unequal duration as the higher flood tide travels more quickly than the lower ebb tide. The inshore current all along a coast need not necessarily be in the same direction as the offshore tidal current, for example, the effect of a headland may be to cause an eddy giving an inshore current locally in the opposite direction from the tidal current. Local conditions affecting tidal estuaries often are so variable that a detailed study of each case usually is required.

8.7.5 Beach zones

Four dynamic zones have been recognized within the nearshore current system of the beach environment. They are the breaker zone, the surf zone, the transition zone and the swash zone (Fig. 8.23). The breaker zone is that in which waves break. The surf zone refers to that region between the breaker zone and the effective seaward limit of backwash. The presence and width of a surf zone is primarily a function of the beach slope and tidal phase. Those beaches that have gentle foreshore slopes often are characterized by wide surf zones during all tidal phases whereas this zone may not be represented on steep beaches. The transition zone includes that region where backwash interferes with the water at the leading edge of the surf zone and it is characterized by high turbulence. The region where water moves up and down the beach is termed the swash zone.

The breaking of a wave is influenced by its steepness, the slope of the sea floor and the presence of an opposing or supplementary wind. When waves enter water equal in depth to approximately half their wavelength, they begin to feel bottom, their velocity and length decrease whilst their height, after an initial decrease, increases. The wave period remains constant. As a result the wave grows steeper until it eventually breaks.

Three types of breaking waves can be distinguished, namely, plunging breakers, spilling breakers and surging breakers (Fig. 8.24). The beach slope influences the type of breaking waves that develop, for example, spilling waves are commonest on gentle beaches whereas surging breakers are associated with steeper beaches. In fact, Galvin (1968) showed that the type of breaker that occurs could be predicted from $H_w/L_w\beta$, where H_w is the wave height in deep water, L_w is the wavelength in deep water and β is the angle of the beach slope. Spilling breakers would appear to have values in excess of 4.8, plunging breakers values between 0.09 and 4.8, and those of surging waves are below 0.09.

Plunging breakers collapse suddenly when their wave height is approximately equal to the depth of the water. At the plunge point, wave energy is transformed into energy of turbulence and kinetic energy, and a tongue of water rushes up the beach. The greater part of the energy released by the waves is used in overcoming frictional forces or in generating turbulence, and relatively lesser amounts are utilized in shifting bottom materials or in developing longshore currents. The plunge point is defined as the final breaking point of a wave just before water rushes up the beach. For any given wave there is a plunge line along which it breaks, but the variations in positions where waves strike the shore usually result in a plunge zone of limited width. The distribution of sand at the plunge line is almost uniform from the bottom to the water surface during the breaking of a wave, but seaward of the plunge line, the sand content of the surface water rapidly decreases. Accordingly, the maximum disturbance of the sand due to turbulence occurs in the vicinity of the plunge line, it tending to migrate from the plunge line towards less turbulent water on both sides. This may result in a trough along the plunge line. Seaward of the plunge line, shingle is not generally moved by waves in depths much greater than the wave height but sand is moved for a considerable distance offshore. Spilling breakers begin to break when the wave height is just over one-half the water depth. They do so gradually over some distance. Surging breakers or swash rush up the beach and usually are encountered on beaches with a steep profile. The height to which the swash rises determines the height of shingle crests and whether or not sea walls or embankments are overtopped. The more impermeable and steep the slope, the higher is the swash height. The term backwash is used to describe the water that subsequently descends the beach slope.

Swash tends to pile water against the shore and thereby gives rise to currents that move along it, which are termed longshore currents. After flowing parallel to the beach the water runs back to the sea in narrow flows called rip currents (Fig. 8.25). In the neighbourhood of

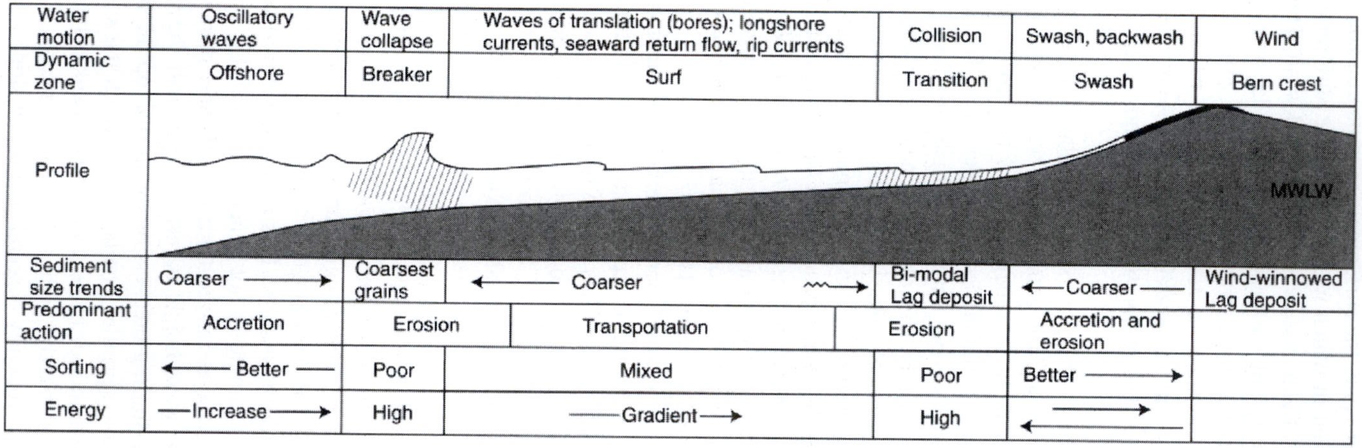

Water motion	Oscillatory waves	Wave collapse	Waves of translation (bores); longshore currents, seaward return flow, rip currents	Collision	Swash, backwash	Wind
Dynamic zone	Offshore	Breaker	Surf	Transition	Swash	Bern crest
Profile						MWLW
Sediment size trends	Coarser →	Coarsest grains	← Coarser →	Bi-modal Lag deposit	← Coarser —	Wind-winnowed Lag deposit
Predominant action	Accretion	Erosion	Transportation	Erosion	Accretion and erosion	
Sorting	← Better —	Poor	Mixed	Poor	Better →	
Energy	—Increase→	High	—Gradient→	High	→ ←	

Figure 8.23 Effect of the four main dynamic zones in the beach environment. Hatched areas represent zones of high concentrations of suspended grains. The surf zone is bounded by two high energy zones, namely, the breaker zone and the transition zone. MWLW, mean water low water.
Source: Ingle, 1966. Reproduced by kind permission of Professor James Ingle.

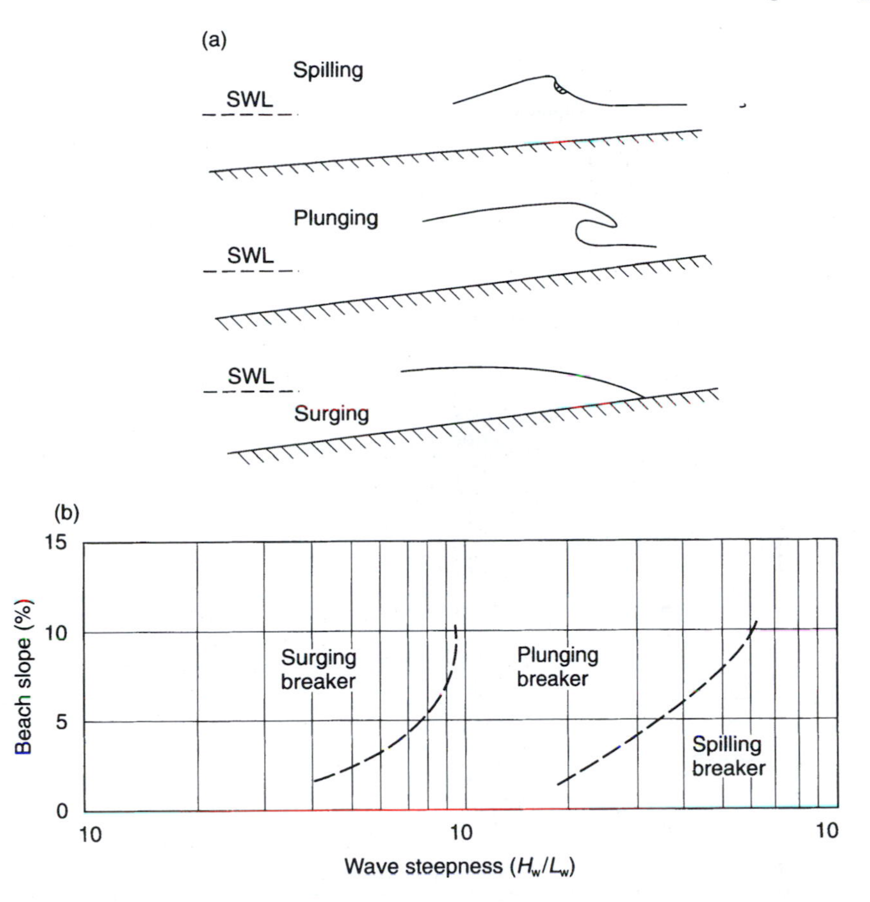

Figure 8.24 (a) Cross-sections of the three breaker types (SWL, still water line). (b) Type of breaking wave in relation to steepness and beach gradient.

the breaker zone, rip currents extend from the surface to the floor whilst in the deeper reaches they override the bottom water that still maintains an overall onshore motion. The positions of rip currents are governed by submarine topography, coastal configuration, and the height and period of waves. They frequently occur on the up-current sides of points and on either side of convergence where the water moves away from the centre of the convergence and turns seawards.

Material is sorted according to its size, shape and density, and the variations in the energy of the transporting medium. During the swash, material of various sizes may be swept along in traction or suspension whereas during the backwash, the lower degree of turbulence results in a lessened lifting effect so that most of the movement of grains is by rolling along the bottom. This can mean that the maximum size of particles thrown on a beach is larger than the maximum size washed back to the surf zone. However, grains of larger diameter may roll down-slope farther than smaller particles. The continued operation of waves on a beach, accompanied by the winnowing action of wind on dry sand, tends to develop patterns of variation in average particle size, sorting, firmness, porosity, permeability, moisture content, mineral composition and other attributes of the beach.

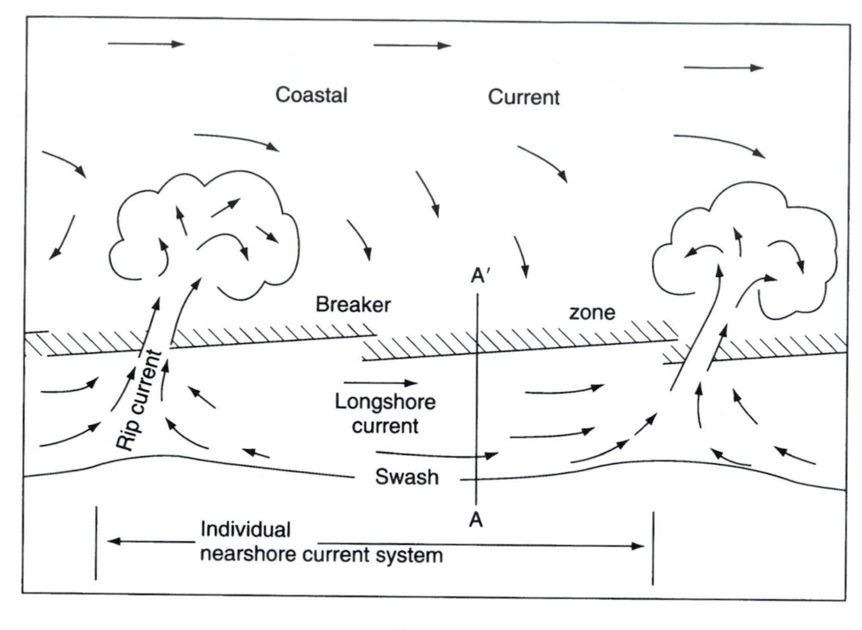

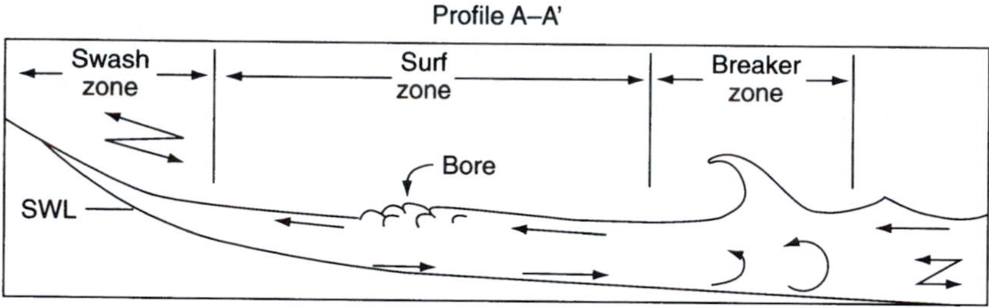

Figure 8.25 Terminology of near-shore current systems. Each individual system begins and ends with a seaward flowing rip-current. Arrows indicate the direction of water movement in plan and profile. Seaward return flow along the foreshore–inshore bottom (profile A–A′). Also, seaward bottom flow often occurs at the same time that surface flow is shorewards. The surf zone is defined as the area between the seaward edge of the swash zone and the breaker zone. SWL, still water line.
Source: Ingle, 1966. Reproduced by kind permission of Professor James Ingle.

8.8 Coastal erosion

Coastal erosion is not necessarily a disastrous phenomenon, however, problems arise when erosion and human activity come into conflict. Increases in the scale and density of human development along coastlines is accelerating rapidly and has led to increased vulnerability that, in turn, means that better coastal planning is an absolute necessity (Kay and Alder, 1999).

Those waves with a period of approximately 4 s usually are destructive whilst those with a lower frequency, that is, a period of about 7 s or over are constructive. When high-frequency waves collapse they form plunging breakers and the mass of water accordingly is directed downwards at the beach. In such instances, swash action is weak and, because of the high frequency of the waves, is impeded by backwash. As a consequence, material is removed from the top of the beach. The motion within waves that have a lower frequency is more elliptical and produces a strong swash, which drives material up the beach. In this case, the backwash is reduced in strength because water percolates into the beach deposits and therefore little material is returned down the beach. Although large waves may throw material above the high water level and thus act as constructive agents, they nevertheless have an overall tendency to erode a beach, whilst small waves are constructive. For example, when steep storm waves attack a sand beach they usually are destructive and the coarser the sand, the greater the quantity that is removed to form a submarine bar offshore. In some instances, a vertical scarp is left on a beach at the high tide limit. It is by no means a rarity for the whole beach to be removed by storm waves, witness the disappearance of sand from some of the beaches along the Lincolnshire coast of England after the storm flood of January–February 1953, which exposed their clay base. Storm waves breaking on shingle throw some shingle to the backshore to form a storm beach ridge that may extend far above high tide level. Storm waves also may bring shingle down the foreshore so that a step is developed at their break point.

The rate at which coastal erosion proceeds is influenced by the nature of the coast itself. Marine erosion is most rapid where the sea attacks soft unconsolidated sediments (Fig. 8.26). Some such stretches of the coast may disappear at a rate of a metre or more annually. In addition, beaches that are starved of sediment supply due to, for example, natural or artificial barriers inhibiting movement of longshore drift are exposed to erosion.

Steep cliffs made of unconsolidated materials are prone to landsliding (Hutchinson *et al.*, 1991). For example, clay may be weakened by wetting due to waves or spray, or the toe of the cliffs may be removed by wave action. Thus, rotational slips and mud flows may develop (Fig. 8.27) and they may carry away protective structures. For erosion to continue, the debris produced must be removed by the sea. This usually is accomplished by longshore currents. On the other hand, if material is deposited to form extensive beaches and the detritus is reduced in size, then the submarine slope, because of its small angle of rest, is very wide. Wave energy

Figure 8.26 Marine erosion of till along the Holderness coast near Skipsea, England, causing destruction of holiday cottages.

Figure 8.27 Slides along the Palisades coast in southern California, northwest of Santa Monica.

therefore is dissipated as water moves over the beach and any cliff erosion ceases. Boggett *et al.* (2000) described a geomorphological survey that they undertook to identify active slips along the South Shore Cliffs, Whitehaven, England. The data gathered during the mapping exercise was used for a risk assessment of landslip along this section of cliffs and to produce a risk zonation plan. The major slip along these cliffs required emergency works to protect the public and an existing sewage-outfall penstock chamber on the beach below. Coastal erosion, together with landslides at Blackgang, Isle of Wight, England, resulted in a dramatic retreat of the coastal cliffs in 1994. Two cottages and an access road were destroyed, and 12 homes had to be evacuated. Consequently, an investigation was undertaken to determine the causes and extent of coastal instability and cliff recession. Moore *et al.* (1998) showed that the cliffs had been oversteepened significantly and appeared to be sensitive to rainfall. Because cliff recession is complex, involving processes not just operating on or within the cliff but also on the foreshore, Moore *et al.* recognized and carried out their investigation in terms of individual cliff behaviour units (CBUs). The identification and characterization of CBUs allowed Moore *et al.* to suggest cliff management strategies (e.g. managed retreat and the installation of a continuous monitoring and early warning system) that sought to reduce the level of risk whilst maintaining the environmental and amenity value of the cliffs, coastal protection works being economically non-viable.

Marine erosion obviously is concentrated in areas along a coast where the rocks offer less resistance. In most cases, a retreating coast is characterized by steep cliffs, at the base of which a beach is excavated by wave action. Erosive forms of local relief include such features as wave-cut notches, caves, blow-holes, marine arches and stacks (Fig. 8.28). Debris from the cliff is washed seaward to form a terrace, which marks the outer limit of the beach. The degree to which rocks are traversed by discontinuities affects the rate at which they are removed. In particular, the attitude of joints and bedding planes is important. Where the bedding planes are

Figure 8.28a Coastal features formed by erosion. Stacks, the Apostles, south coast of Victoria, Australia.

Figure 8.28b Coastal features formed by erosion. Marine arch, coast of southern California.

Figure 8.28c Coastal features formed by erosion. Blowhole, pancake rocks, west coast of South Island, New Zealand.

vertical or dip inland, the cliff recedes vertically under marine attack. Conversely, if beds dip seawards, blocks of rock are dislodged more readily since the removal of material from the cliff means that the rock above lacks support and tends to slide onto the beach where it is further broken down by marine action. Marine erosion also occurs along fault planes. In addition, the height of a cliff influences the rate at which erosion takes place. The higher the cliff, the more material falls when its base is undermined by wave attack. This, in turn, means that a greater amount of debris has to be broken down and removed before the cliff is once more attacked with the same vigour.

8.9 Beaches and longshore drift

Sandy beaches are supplied with sand that usually is derived from the adjacent sea floor, although in some areas, a larger proportion is produced by cliff erosion. During periods of low waves, the differential velocity between onshore and offshore motion is sufficient to move sand onshore except where rip currents are operational. Onshore movement is notable particularly when long-period waves approach a coast whereas sand is removed from the foreshore by high waves of short period. Of course, all beaches are continually changing. Some develop during periods of small waves and disappear during periods of high waves whereas other beaches change in height and width during stormy seasons. If seasonal changes bring about changes in the wave approach, then sand is shifted along a beach, beaches tending to form at right angles to the direction of the wave approach. If sand forms dunes and these migrate landwards, then sand is lost to the beach.

The beach slope is produced by the interaction of swash and backwash. Beaches undergoing erosion tend to have steeper slopes than prograding beaches. The beach slope also is related to the grain size, in general, the finer the material, the gentler the slope and the permeability of the beach. For example, the loss of swash due to percolation into beaches composed of grains of 4 mm in median diameter is ten times greater than those where the grains average 1 mm. As a result there is almost as much water in the backwash on fine beaches as there is in the swash. So the beach profile is gentle and the sand is hard packed. The grain size of beach material, however, is continually changing because of breakdown and removal or addition of material.

Storm waves produce the most conspicuous constructional features on a shingle beach, but they remove material from a sandy beach. A small foreshore ridge develops on a shingle beach at the limit of the swash when constructional waves are operative. Similar ridges or berms may form on a beach composed of coarse sand, these being built by long low swells. Berms represent a marked change in slope and usually occur at a small distance above high water mark. They are not conspicuous features on beaches composed of fine sand. A fill on the upper part of a beach may be balanced by a cut lower down the foreshore.

Dunes are formed by onshore winds carrying sand-sized material landward from the beach along low-lying stretches of coast where there is an abundance of sand on the foreshore. Leathermann (1979) maintained that dunes act as barriers, energy dissipators, and sand reservoirs during storm conditions, for example, the broad sandy beaches and high dunes along the coast of the Netherlands present a natural defence against inundation during storm surges (Fig. 8.29). Because dunes provide a natural defence against erosion, once they are breached, the ensuing coastal changes may be long lasting (Gares, 1990). On the other hand, along parts of the coast of North Carolina where dune protection is limited, washovers associated with storm tides have been responsible for high rates of erosion (Cleary and Hosier, 1987). In spite of the fact that dunes inhibit erosion, Leathermann (1979) stressed that without beach nourishment (see below) they cannot be relied upon to provide protection, in the long term, along rapidly eroding shorelines. Beach nourishment widens the beach and maintains the proper functioning of the beach-dune system during normal and storm conditions.

Figure 8.29 Planting marram grass to protect the base of coastal dunes in the Netherlands.

When waves approach a coast at an angle, material is moved up the beach by the swash, in a direction normal to that of wave approach and then it is rolled down the beach slope by the backwash. In this manner, material is moved in a zig-zag path along the beach, the phenomenon being referred to as longshore or littoral drift. Currents, as opposed to waves, seem incapable of moving material coarser than sand grade. Since the angle of the waves to the shore affects the rate of drift, there is a tendency for erosion or accretion to occur where the shoreline changes direction. The determining factors involved are the rate of arrival and the rate of departure of beach material over the length of foreshore concerned. Moreover, the projection of any solid structure below mean tide level results in the build-up of drift material on the up-drift side of the structure and perhaps in erosion of material from the other side. If the drift is of any magnitude, the effects can be serious, especially in the case of works protecting harbour entrances. On the other hand, sand can be trapped by deliberately stopping longshore drift in order to build up a good beach section. Under normal conditions the bulk of the sand in longshore drift is moved within a relatively shallow zone with perhaps 80 per cent of shore drift being moved within depths of 2 m or less.

The amount of longshore drift that occurs along a coast is influenced by coastal outline and wavelength. Short waves can approach the shore at a considerable angle and generate consistent down-drift currents. This is particularly the case on straight or gently curving beaches and can result in serious erosion where the supply of beach material reaching the coast from up-drift is inadequate. Conversely, long waves suffer appreciable refraction before they reach the coast.

An indication of the direction of longshore drift is provided by the orientation of spits along a coast. Spits are deposits that grow from the coast and are chiefly supplied by longshore drift. Their growth is spasmodic and alternates with episodes of retreat. The distal end of a spit frequently is curved (Fig. 8.30). Ciavola (1997) found that only 6 per cent of the longshore drift produced by cliff erosion along the Holderness coast, England, was transported along Spurn Head Spit, mainly by waves from the north and northeast. Much of the drift is trapped by coastal defences whilst the rest is stored in an area of sand and gravel banks at the tip of the spit. Medium-sized sand is carried around the tip by south-easterly winds and is spread over a tidal flat behind the recurved spits. A spit that extends from the mainland to link up with an island is referred to as a tombola. A bay-bar is constructed across the entrance to a bay by the growth of a spit being continued from one headland to the other. A cuspate bar arises where a change in the direction of spit growth takes place so that it eventually joins the mainland again, or where two spits coalesce. If progradation occurs, then cuspate bars give rise to cuspate

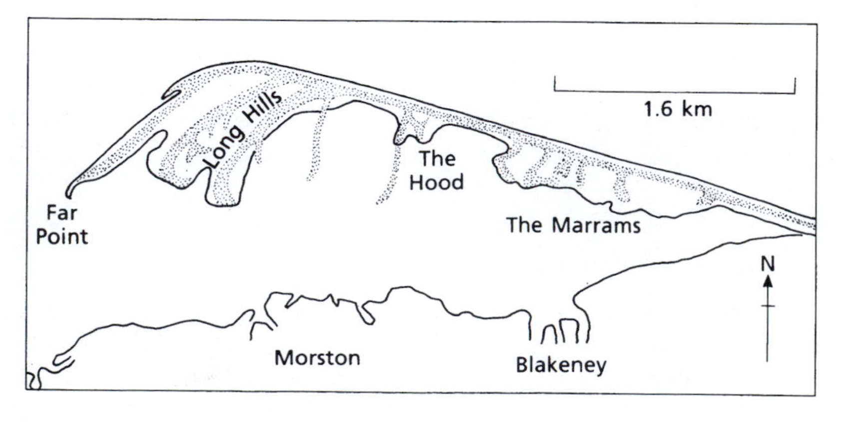

Figure 8.30 A spit with recurved laterals, Blackeney Point, north coast of Norfolk, England.

Figure 8.31 Cuspate foreland, Dungeness, Kent, England.

forelands (Fig. 8.31). Bay-head beaches are one of the commonest types of coastal deposits and tend to straighten a coastline. Wave refraction causes longshore drift to move from headlands to bays, sweeping sediments with it. Since the waves are not so steep in the innermost reaches of bays, they tend to be constructive. Marine deposition also helps straighten coastlines by building beach plains and prograding spits may block the mouths of estuaries (Lindsay and Bell, 1998).

Offshore bars consist of ridges of shingle or sand that may extend for several kilometres and may be located up to a few kilometres offshore. They usually project above sea level and may cut off a lagoon on their landward side. The distance from the shore at which a bar forms is a function of the deep water wave height, the water depth and the deep water wave steepness. If water depth and wave steepness remain constant, and wave height is reduced, the bar moves shoreward. If wave height and wavelength stay the same, and the depth of water increases, again the bar moves shorewards. If water depth and wave height remain unchanged but steepness is reduced, then the bar moves seaward. During storms, sand is shifted outward to form an offshore bar, and during the ensuing quieter conditions, this sand is wholly or in large part moved back to the beach.

8.10 Shoreline investigation

Protective works or littoral barriers have to be planned and built to prevent the destruction of a shoreline by marine action. Before any project or beach planning and management scheme can be started, a complete study of the beach must be made. The preliminary investigation of the area concerned should therefore first consider the landforms and rock formations along the beach and adjacent rivers, paying particular attention to their durability and stability. In addition, consideration must be given to the width, slope, composition, and state of accretion or erosion of the beach, the presence of bluffs, dunes, marshy areas, or vegetation in the backshore area, and the presence of beach structures such as groynes.

The stage of development of the beach area in terms of the cycle of shore erosion, the sources of supply of beach material, probable rates of growth and probable future history merit considerable discussion. Estimates of the rates of erosion and the proportion and size of eroded material contributed to a beach must be made, as well as whether they are influenced by seasonal effects (Martinez *et al*., 1990). Not only must the character of the rocks be investigated, but the presence of joints and fractures also must be studied. The latter afford some estimate of the average size of fragments resulting from erosion. The rapidity of breakdown of rock fragments into individual particles should be noted. The behaviour, when weathered, of any unconsolidated materials that form cliffs, together with their slope stability and likelihood of sliding has to be taken into account. Samples of the beach and underwater material have to be collected and analysed for such factors as their particle size distribution and mineral content. Mechanical analysis may prove useful in helping to determine the amount of material that is likely to remain on the beach, for beach sand is seldom finer than 0.1 mm in diameter. The amount of material moving along the beach must be investigated as much of the effectiveness of the structures erected may depend upon the quantity of drift available.

A characteristic of sand and shingle foreshores is their permeability, which is an important factor in foreshore stability. Any introduction of an impermeable structure along a beach tends to alter stability and hence it often is advisable to use permeable defences. In some instances, erosion problems may arise as a result of artificial structures along the shore robbing down-drift areas of sediment supply.

8.10.1 Recording devices

Valuable data may be obtained from a wave gauge that can be a graduated pole erected on the foreshore, from which the height of waves may be estimated. The wave period can be determined by timing the rise and fall of the surface of the sea at the wave gauge. The tide level is derived from gauge board readings in a tidal basin or from recorders on nearby maritime structures. A number of different types of wave-measuring instruments have been developed. A step-resistance wave gauge consists of a series of graduated resistors connected to a common source of electric potential. The circuit is arranged so that the rising water surface furnishes an electrical path that short-circuits successive resistors and causes an increase in the current. Pressure gauges also may be used to measure wave height and, like electrical resistance gauges, generally are open-framed structures. Wave data also can be obtained by using wave recording buoys or by ship-borne wave recorders. Analysis of wave records is accomplished by measuring a number of wave heights and periods. Aerial photography may be used to study refraction patterns of long swell.

Surface floats are used to obtain information on the pattern of tidal streams over an appreciable area. In a float test, the tidal stream at any particular locality has to be estimated throughout the tidal cycle. Velocity profiles determined by current meter usually are required to translate the results of float tests into overall transport. The direct current reading meter is used extensively for measuring currents at sea, the measurement being obtained from the speed of rotation of a propeller when the meter is suspended in the sea. The depth of the instrument usually is recorded automatically and current profiles are obtained by raising and lowering the instrument in stages, and taking readings at pre-determined levels. Recording current meters measure the speed and direction of currents in the same way, but the information is recorded on computer.

The depth of water may be recorded by an echo-ranging device, the technique being referred to as transit sonar or oblique asdic. A narrow transverse beam of sound energy pulses is transmitted to one side of a vessel following a steady course. Echoes are recorded continuously and the surface topography may be deduced from their interpretation, augmented as required by visual, radiowave or laser instrument sights. Seismic refraction also can be used to measure the depth of water to the sea floor, the energy pulses being generated by an air-gun or sparker. A survey vessel traverses the area concerned, releasing the pulses and recording the refracted shock waves. In addition, it may be possible to estimate the depth of sediment on the sea floor by using this method.

The most reliable method for measuring suspended load at some distance away from the sea bed is to use oceanographers sampling bottles, suspended at given depth intervals. Sampling is done by using an automatic recorder that counts the sediment particles. The Delft bottle may be used to measure the rate of transport of material in suspension or saltation near the sea bed. The velocity of the water entering the orifice of this bottle is reduced and thus sediment down to about 0.1 mm in size falls from suspension and is retained in the bottle. There is no satisfactory method of direct measurement of bed load transport at sea, although a trap like the bottom transport meter may be used. However, this is unsuitable for measuring bed transport where the sediments are soft. There are a number of techniques for obtaining samples from the sea bed, each being appropriate to a fairly narrow range of sediments. The drop sampler falls freely into sediments whilst the piston sampler or explosive sampler is forced into the sea bed.

Where physiographical or structural features provide a complete barrier, it is only necessary to measure the rate at which material is accumulating on the up-drift side or diminishing on the down-drift side. If sand is involved, the survey has to extend some distance offshore and inshore. On the other hand, shingle drift usually is confined to the littoral zone, the exceptions occurring in those localities where rapid currents transport shingle into deep water. Submarine investigation at shallow depth is sometimes of special importance. In such cases, divers may be employed.

Tracer techniques may be used to assess the movement of sedimentary particles. For example, radioactive tracers may be attached to the particles concerned, or alternatively irradiated artificial material of similar size may be introduced into the sedimentary material under study. The movement and distribution of the tracer is surveyed subsequently at intervals by means of a Geiger counter or scintillation counter lowered by cable to the sea bed. While radioactive tracers have proved of the greatest value offshore, fluorescent tracers have been adopted most widely for determination of longshore drift, with particular success on shingle beaches. The usual procedure is to deposit the tracer (this, on a shingle beach, is often crushed concrete) at a steady rate at pre-determined positions on the foreshore and to detect its presence, at regular intervals of time and distance along the foreshore. Calculations of longshore drift are based on recording the count of visible particles and assuming uniform mixing of the tracer with the natural beach.

8.10.2 Topographic and hydrographic surveys

Topographic and hydrographic surveys of an area allow the compilation of maps and charts from which a study of the changes along a coast may be made. Topographic maps extending back to the earliest editions, remote sensing imagery and aerial photographs taken at successive intervals of time prove of value in evaluating changes in coastlines over the recent past. Historical evidence also has been used to determine coastal changes, the study of the retreat of the Holderness coast in England by Steers (1948) providing a classic example. Bathymetric charts provide information on variations in offshore topography.

Observations should be taken of winds, waves and currents, and information gathered on streams that enter the sea in or near the area concerned. Differentiation should be made between periods of light and strong winds. It is the distribution and direction of the latter that concerns the coastal engineer.

Where there is an appreciable tidal range, it is necessary to consider the most severe combination of water depth and wave height, since a breaking or surging wave gives rise to a very much greater force and water velocity than does a wave that strikes a structure without breaking. The maximum wave is determined by a statistical method similar to that used to find the exceptional flood. It may likewise be defined in terms of a 100-year period. Where structural failure could lead to loss of life or other unacceptable risk, the maximum wave must be considered. The selection of the design wave in deep water is, according to the nature and importance of the particular structure, normally contained between the limits of the significant wave height (H_s) and the maximum wave height (H_{max}), for the period corresponding to the aggregate length of storms of such intensity as anticipated during the life of the structure. For example, for a revetment that may be readily repaired following storm damage without risk of major resultant costs, it may be economic to design the structure to withstand waves of height, say $H_{1/10}$ (i.e. the average height of the highest 10 per cent of all waves).

Local currents affect the layout and siting of marine works and, after determination by hydrographic surveying, they usually must be integrated into design. The contributions made by large streams may vary, for example, material brought down by large floods may cause a temporary, but nevertheless appreciable, increase in the beach width around the mouths of rivers. Inlets across a beach need particular evaluation. During normal times there may be relatively little longshore drift but if up-beach break throughs occur in any bars off the inlet mouth, then sand is moved down-beach and is subjected to longshore drift.

Selection of the measures to be taken necessitates consideration of whether the beach conditions represent a long-term trend or whether they are cyclical phenomena that may recur at some time in the future. The recent history of a beach and the marine processes operating on it may have to be evaluated. Observations must be extended over at least one complete storm cycle because beach slopes and other features may change rapidly during such times. The effects of any likely changes in sea level also should be taken into consideration (Clayton, 1990). For example, as the width of a beach declines due to accelerated erosion associated with a rise in sea level, areas behind the beach are likely to become increasingly exposed to wave action during storms (Nicholls *et al.*, 1995). According to Nicholls (1998) around 70 per cent of the sandy beaches around the world have been eroded within the last few decades, and therefore he maintained that long-term beach recession is a major hazard. Nicholls went on to propose a means of assessment of possible beach erosion due to sea level rise. Previously, Leatherman *et al.* (1995) had used aerial videotape as part of a vulnerability assessment of the impacts of sea level rise.

8.10.3 Modelling

Scale models of particular stretches of coastline can be constructed in wave tanks in laboratories in an attempt to simulate coastal processes and so evaluate the results that individual projects may have if developed along a coastline. Such small-scale modelling allows variation of various factors in the coastal regime and represents a three-dimensional reproduction of the situation. Computers also can be used to model changes in coastlines. Again the consequences of varying parameters can be evaluated, as can the influence that various protection works will have on a coastline.

8.11 Protective barriers

8.11.1 Sea walls

Training walls are designed to protect inlets and harbours, as well as to prevent serious wave action in the area they protect. They also impound any longshore drift material up-beach of an inlet and thereby prevent sanding of the channel. However, as noted previously, this may cause or accelerate erosion down-drift of the walls. Training walls usually are built at right angles to the shore, although their outer segments may be set at an angle. Two parallel training walls may extend from each side of a river mouth for some distance out to sea and because of this confinement, the velocity of river flow is increased, which lessens the amount of deposition that takes place between the walls. Training walls may be built on rubble mounds.

Sea walls may rise vertically or they may be curved or stepped in cross section (Fig. 8.32). They are designed to prevent wave erosion and overtopping by high seas by dissipating or absorbing destructive wave energy. Sea walls are expensive so that they tend to be used where property requires protection. They must be stable and this is usually synonymous with weight. They also should be impermeable and resist marine abrasion. Sea walls can be divided into two main classes, that is, those from which waves are reflected and those on which waves break. The second category can be subdivided into those types where the depth of the water in front of the wall is such that waves break on the structure and those types where the bigger waves break on the foreshore in front of the wall.

It generally is agreed that a wall that combines reflection and breaking sets up very severe erosive action immediately in front of it. Consequently, in the layout of a sea wall, particular attention should be given to the wave reflection and the possibility of the crests of the waves increasing in height as they travel along the wall due to their high angle of obliquity (Pilarczyk,

Figure 8.32 Seawall near Runswick Bay, near Whitby, England.

1990). The toe of the wall should not become exposed. If base erosion becomes serious, a row of sheet piling may be driven along the seaward edge of the foot of the wall. Unless the eroding foreshore in front of a sea wall is stabilized, the apron of the wall must be repeatedly extended seawards as the foreshore falls. Generally speaking, on a deteriorating foreshore, shingle drops much more rapidly than sand and the estimated rate of fall determines the extent to which an apron should be extended. Shingle fill must extend slightly above swash height to achieve stability and the seaward profile must have a stable slope.

If a sea wall can be positioned near the top of the swash, it has a negligible effect on the beach profile. Also, if the wave energy in front of a sea wall can be dampened, the deleterious effects are reduced. This can be accomplished by means of a permeable revetment or armouring. However, a stepped profile absorbs only a small fraction of the energy of storm waves, since the steps are usually too small by comparison with the wave height. A curved profile helps to avoid a violent reflection, and a bullnosed coping at the top of the wall tends to deflect the plume from a breaking wave towards the sea.

A simple sloping wall is not subject to the full force of the waves. As they travel up the wall their break is accelerated, if they have not already broken, and much of their energy is expended in their run-up the slope, the impact is a glancing rather than a direct blow. The effect of roughness is greater on a gentle slope where velocities are greater and the total distance travelled by the swash also is greater. An inclined berm in a swash wall provides an effective means of reducing wave run-up. Providing the width of berm represents a significant part, say 20 per cent of the wavelength, then its effect is approximately the same as if its slope were continued to the crest of the wall.

A sea wall is subjected to earth pressure on its landward side. The position of the water table therefore is important and is complicated by the probable existence of appreciable fluctuations due to the rise and fall of the tides. For example, water can saturate the backfill of a sea wall during the flood tide and descends during the ebb tide. As the water level behind the wall does not fall as quickly as the ebb water, the resultant lag leads to an increase of pore water pressure against the wall. In the case of high tides, this lag should be estimated in relation to the permeability of the material.

8.11.2 Breakwaters

Breakwaters disperse the waves of heavy seas and provide shelter for harbours. As such, they commonly run parallel to the shore or at slight angles to it and are attached to the coast at one end, their orientation being chosen with respect to the direction that storm waves approach the coast. The detached offshore breakwater is constructed as a barrier parallel to the shoreline and is not connected to the coast. Attached breakwaters cause sand or shingle of longshore drift, on shorelines so affected, to accumulate against them and so rob down-drift beaches of sediment. Consequently, severe erosion may take place in the down-drift area after the construction of a breakwater. Furthermore, accumulation of sand can eventually move around the breakwater to be deposited in the protected area. Komar (1976) described this as happening at Santa Barbara, California, after a breakwater had been built to protect the harbour. As a result, the harbour entrance has to be dredged and the sand is deposited on the down-drift shoreline. Although long offshore breakwaters shelter their leeside, they cause wave refraction and may generate currents in opposite directions along the shore towards the sheltered area with resultant impounding of sand.

The character of the sea bed at the site of a proposed breakwater must be carefully investigated. Not only must the sea bed be investigated fully with respect to its stability, bearing capacity and ease of removal if dredging is involved, but the adjacent stretches of coast must be studied in order to determine local currents, longshore drift and any features that may in any way be affected by the proposed structure. Moreover, breakwaters may be subject to scour. Sumer and Fredsoe (2000) carried out an investigation to determine the effects of scour on a model rubble mound breakwater. They found that areas of scour and deposition developed parallel to the breakwater, and that scour depth was greater in the case of regular waves than irregular waves. The effects of wave reflection and transmission, as well as wave loads on a caisson founded on a rubble mound were examined by Sulisz (1997). He found that the height of the base of the rubble mound had a significant effect on wave reflection, transmission and wave loads whereas the width of the bench had little effect. In addition, thick, less permeable layers of armour increase the effect of the rubble mound.

The oldest form of breakwater is a pile of dumped rock, in other words, a rubble mound. Rubble mound breakwaters dissipate storm waves by turbulence, as sea water penetrates the voids between the blocks of rock. Most of the attenuation occurs near the surface of the water, with oscillatory motions of water at depth being reduced somewhat. Rubble mound breakwaters are adaptable to any depth of water, are suitable for nearly all types of foundations from solid rock to soft mud and can be repaired readily. The stability of such a structure is not determined by the integrity of the entire mass in the presence of wave action, but upon the ability of the individual stones to resist displacement. The quality and durability of the stone also is important (Latham, 1991). The structural design of a rubble mound breakwater therefore is mainly concerned with specifications for rock size, density and durability, and the selection of an appropriate slope angle. In general, the wave force acting upon an individual stone is proportional to the exposed area of the stone, while the resistance of the stone is proportional to its volume. Hence, the larger the stone the greater is its stability. Tetrapods are four-legged concrete blocks that weigh up to 40 tonnes (Fig. 8.33). They have been used mainly in harbour

Figure 8.33 Tetrapods used to construct a breakwater near Port Elizabeth, South Africa.

works. Their design enables them to interlock and so give relatively steep slopes whilst their roughness reduces swash height. Vertical wall breakwaters include masonry walls, timber cribs, caissons, or sheet piling, alone or in combination. This type of breakwater is not very much used, but a combined type consisting of a wall placed on top of a rubble mound structure has many applications in practice.

A pneumatic breakwater provides protection against relatively short waves by releasing bubbles from a compressed air pipe (Silvester, 1974). Pipelines supplying air for pneumatic breakwaters can soon become silted up if placed too close to the sea bed. The release of air bubbles into a wave system reduces wave height and therefore wave energy. However, the pneumatic breakwater is inefficient as far as protection against long waves is concerned.

Sea walls and breakwaters often are made up of interconnected earth-filled cylinders of steel sheet piling. Each constituent cell is a stable unit, deriving its stability from the composite strength of both the earth-fill and the steel shell. It is undesirable to carry clay back-fill above the water line because poor drainage allows the development of excess pore water pressures, hence sand-fill should be used above the low water level. The sand-fill should be adequately drained so that groundwater levels do not lag much behind water levels. With a weak foundation the back-fill should be all sand.

8.11.3 Embankments

Embankments have been used for centuries as a means of coastal protection. The weight of an embankment must be sufficient to withstand the pressure of the water. The material composing an embankment should be impervious or at least it should be provided with a clay core to ensure that seepage does not occur. It requires a protective stone apron to withstand the destructive power of the waves. Clay is the usual material for embankment construction.

The top of a clay embankment preferably should be at least 1 m above the highest known tide level so that the risk of water flowing through surface fissures is reduced to a minimum. This fissure zone is caused by the surface drying out. If the nature of the soil and the space available is such that the desirable level for the top of an embankment cannot be attained without the possibility of rotational slip occurring, a common cause of embankment failure, then two alternatives are possible. The first is to construct a crest wall and the second is to provide adequate protection to the landward side of the embankment so that if it is overtopped during severe storms it does not suffer any significant damage. Usually, if overtopping occurs, then it is of short duration and the quantity of water involved is often small, and so can be dealt with easily by a system of dykes, whereby it can be discharged back into the sea.

The general effect of flattening the slope of an embankment is to lessen the swash height, lessen foreshore erosion and reduce the wave pressure on the upper part of the apron. The slope therefore may vary from 1 in 2 for an estuary embankment well protected by a wide mud flat at a high level, to 1 in 5 or even flatter for one in deep water that permits large waves to approach and break on it. A berm generally is constructed on the seaward side of an embankment at about the level of high water to reduce swash height.

Clay embankments may fail due to direct frontal erosion by wave action, flow through the fissured zone causing a shallow slump to occur on the landward face of the wall, scour of the back of the wall by overtopping, or rotational slip. If potential scouring of the back of the wall is likely to be too severe for turf to provide sufficient protection, then generally it can only be combated by constructing a protective revetment. Delayed slips sometimes occur up to nine months or more after construction, and usually are due to a redistribution of pore water pressure, excessive pore water pressures resulting from changes in tidal level. If a permeable stratum lies beneath a clay embankment, then uplift pressures could develop under it at high tides. These can reduce stability by lessening the shear strength of the soil and even by causing the landward toe to heave. Suitable remedies are to provide more weight on the landward slope and to provide relief filter drains near the landward toe to reduce the uplift pressures. Even where failure does not occur, slow settlement of embankments is very common particularly where the underlying material is very soft.

8.11.4 Revetments

A revetment affords an embankment protection against wave erosion. Stone is used most frequently for revetment work, although concrete also is employed. The chief factor involved in the design of a stone revetment is the selection of stone size, it being important to guard against erosion between stones. Consequently, coarse rip-rap must be isolated from the earth embankment by one or more courses of filter stone (Fig. 8.34). Stone pitching is an ancient form of embankment protection, consisting of stone properly placed on the clay face of the wall and keyed firmly into position. Revetments of this type are flexible and have proved to be very satisfactory in the past. Flexibility is an important requirement of revetment since slow settlement is likely to occur in a clay embankment. Alternatively, to keying, stones may be grouted or asphalt jointed. Individual blockwork units may be used so that in the event of a wash-out the spread of damage is limited. Blockwork formed of igneous rock is one of the finest forms of revetment for resisting severe abrasion. On walls where wave action is not severe, interlocking concrete blocks provide a very satisfactory form of revetment. The effectiveness of a revetment lies more in its water tightness than in the weight of the stones.

Other forms of coastal protection include steel sheet-piled revetments, gabions, steel reinforced concrete contiguous piles, which may have a capping beam and be secured by ground anchors, and boulder splash aprons. Maharaj (1998) referred to the use of such coastal defences along the east coast of Trinidad. However, he also mentioned that the construction of a coastal road that necessitated the removal of mangrove areas and the construction of a bridge across a local river led to increased erosion. This, in turn, meant that some of the coastal defence works performed badly or even failed. Maharaj placed the blame for the poor performance and failure of some of the coastal defences on oversights in planning and design, and claimed that a more systematic and integrated approach should have been taken in the analysis of such coastal engineering problems.

8.11.5 Bulkheads

Bulkheads are vertical walls either of timber or of steel sheet piling that support an earth bank. A bulkhead may be cantilevered out from the foundation soil or in addition it may be restrained by tie rods. Except for installations of moderate height and stable foundation, a cantilever bulkhead will be rejected in favour of a design having one or two anchorages. The tie rod on an anchored bulkhead must be carried to a secure anchorage. Generally, this is provided by a deadman, the placement of which must be such that the potential slip surface behind it does not invade the zone influencing bulkhead pressures. Foundation conditions for bulkheads must be given careful attention and due consideration must be given to the likelihood of scour occurring at the foot of the wall and to changes in beach conditions. Cut-off walls of steel sheet piling below reinforced concrete super-structures provide an effective method of construction, ensuring protection against scouring. Providing the wall structure does not encroach far below high tide mark, the existence of this wall should not affect the stability of the beach. If a

Figure 8.34 Rip-rap used as stone revetment for embankment protection, Freemantle, Western Australia.

bulkhead wall has to be constructed either within the tidal range on a beach or below the low water mark, then study must be made of the possible consequences of construction.

8.12 Stabilization of longshore drift

Before any scheme for beach stabilization is put into operation it is necessary to determine the prevailing direction of longshore drift, its magnitude and whether there is any seasonal variation. Also, it is necessary to know whether the foreshore is undergoing a net gain or loss of beach material and what is the annual rate of change.

8.12.1 Groynes

A groyne is the structure used most frequently to stabilize or increase the width of a beach by arresting longshore drift (Fig. 8.35). Groynes are used to limit the movement of beach material, and to stabilize the foreshore by encouraging accretion. However, groynes do not usually halt all drift. Groynes should be constructed transverse to the mean direction of the breaking crest of storm waves, which usually means that they should be approximately at right angles to the coastline. Standard types usually slope at about the same angle as the beach. With abundant longshore drift and relatively mild storm conditions almost any type of groyne appears satisfactory, whilst when the longshore drift is lean, the choice is much more difficult.

By arresting longshore drift, groynes cause the beach line to be re-orientated between successive groynes so that the prevailing waves arrive more nearly parallel to the beach. If the angle of wave incidence is greater than that for maximum rate of drift, groynes serve no useful purpose, since accumulation of beach material against groynes orientates the beach immediately up-drift

Figure 8.35 Groynes used to impede longshore drift at Ventor, Isle of Wight, England. Note that the beach is starved of sand beyond the far groyne.

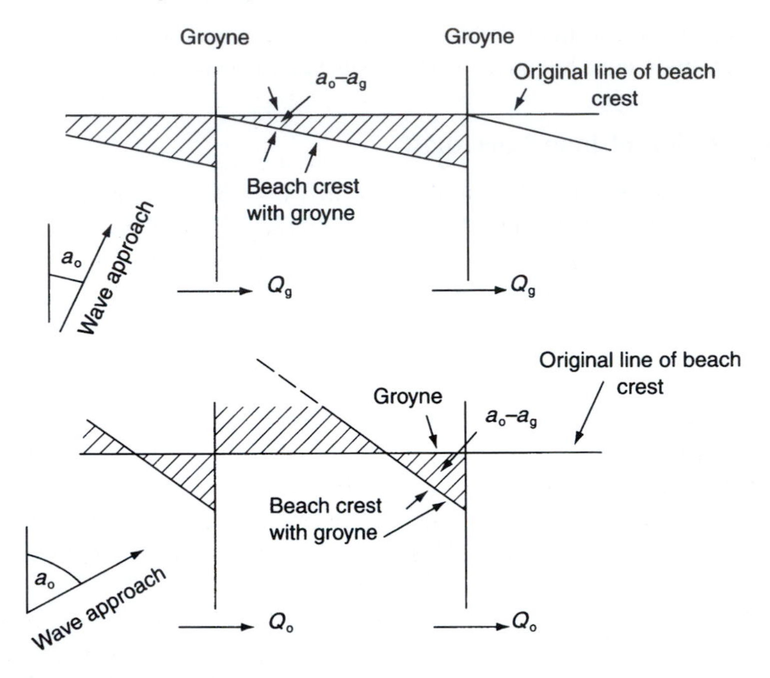

Figure 8.36 Diagram of a groyned shore.
Source: Muir Wood and Fleming, 1983.

of them in such a way as to lead initially to increased longshore drift (Fig. 8.36). The beach crest rapidly swings around until the original rate of drift is restored, resulting in loss of beach material (Muir Wood and Flemming, 1983).

The height of a groyne determines the maximum beach profile up-drift of it. In general, there is little advantage in building groynes that are higher than the highest level normally reached by the sea. Low groynes are more simple to construct. The stability of a groyne must take account of the storm profile and local scour. On a sandy foreshore the height of groyne should not exceed 0.5–1.0 m above the beach if it is to minimize scour from wave and tidal currents, particularly at the seaward end. It therefore is advisable to provide low groynes at relatively close spacing to achieve the desired value of Q_g/Q_o. Closely spaced groynes on a sand beach have the additional advantage that they tend to control rip currents so that these occur at correspondingly more frequent intervals and consequently, the longshore drift in the surf zone is reduced. Where long groynes are too close, a proportion of the beach material is not captured by each groyne bay. The effect is more marked for sand than shingle, because of beach gradient and the mode of transport.

The length of groynes often is determined by the tidal range and beach slope. Limiting groynes to the beach above the level of low spring tides is utterly defenceless in terms of efficiency since the total rate of sand movement in the coastwise direction is greater below the level of low spring tides than above it. Length also is related to the desired effectiveness of the groyne system. In other words, as the ratio of the quantity of longshore drift with groynes (Q_g) to the quantity without (Q_o) becomes smaller, the necessary length of the groyne becomes greater until when $Q_g/Q_0 = 0$, the groyne must extend to the limit of longshore drift. Intermediate groynes sometimes have been used to form littoral cells capable of retaining virtually all the shingle on the foreshore. Groynes should extend landwards to the cliff or protective structure or on shingle shores to somewhat higher than the highest swash height at high water. This is to protect them from being outflanked by storm waves.

The maximum groyne spacing frequently is governed by the resulting variation in beach level on either side (Flemming, 1990). For instance, in bays where the direction of wave attack is confined, groynes may be more widely spaced than on exposed promontories. Indeed, it may be necessary to provide a more substantial groyne where a change occurs in the direction of a coastline, to counter the accentuated wave attack and change in the rate of drift. The common spacing rule for groynes is to arrange them at intervals of one to three groyne lengths. More specifically when groyne spacing is being considered, estimation of the desired value of Q_g/Q_o, the beach profile, the direction and strength of the prevailing or storm waves, the amount and direction of longshore drift, the relative exposure of the shore, and the angle of the beach crest between groynes, should be taken into account.

Permeable groynes have openings that increase in size seawards and thereby allow some drift material to pass through them. The use of permeable groynes for a sand foreshore may help to reduce local scour but may only be considered where Q_g/Q_o is to remain fairly high, since they cause no more than a slight reduction in the rate of longshore drift. On a shingle foreshore, groyne-induced scour need not be a serious problem and a permeable groyne is not an economic expedient.

Groynes reduce the amount of material passing down-drift and therefore can prove detrimental to those areas of the coastline (Silvester, 1974). Their effect on the coastal system therefore should be considered before installation. Beach nourishment may be used to prevent accelerated erosion in those areas down-drift of a groyned beach.

8.12.2 Beach replenishment

Artificial replenishment or nourishment of the shore by building beach-fills is used either if it is economically preferable or if artificial barriers fail to defend the shore adequately from erosion (Whitcombe, 1996). In fact, beach nourishment represents the only form of coastal protection that does not adversely affect other sectors of the coast. Unfortunately, it often is difficult to predict how frequently a beach should be renourished. Ideally, the beach-fill used for nourishment should have a similar particle size distribution to that of the natural beach material.

The material for beach replenishment may be obtained from a borrow pit away from the coastline, in which case it is necessary to consider the effects of using a material of different grain size and sorting on the stability of the beach. Alternatively, it may be obtained from points of accretion on the foreshore. A common method of beach replenishment is to take material from the down-drift end of a beach and return it at the up-drift end. This is usually the case when beach nourishment is combined with groynes. However, investigation of the foreshore may indicate that drift increases in the down-drift direction and therefore that recharge is required at intermediate points to make good such a deficiency. Special attention should be given to areas where there is a change in the direction of coastline. The immediate advantage from beach nourishment is experienced up-drift rather than down-drift of the supply point.

Sediment bypassing involves mechanical or hydraulic transfer of sediment across a shoreline structure, for instance, across a harbour entrance to the down-drift zone where there is the potential for sediment starvation. The purpose of bypasses is to prevent the detention of too much sediment by a structure or the diversion of sediment into deeper water out of reach of the beach regime. Sediment bypassing systems, however, are expensive to build and maintain.

Sand dunes to protect a coast may be built up by the construction of permeable screens. As the screens become buried more screens are constructed above them to allow sand to accumulate continuously. Marram grass generally is planted to stabilize the dunes.

8.12.3 Coastal land-use management

Because of the large cost of engineering works and the expense of rehabilitation, coastal management represents an alternative way of tackling the problem of coastal erosion and involves prohibiting or even removing developments that are likely to destabilize the coastal regime (Gares and Sherman, 1985; Ricketts, 1986). The location, type and intensity of any development along a coast must be planned and controlled. Public safety is the primary consideration. Generally, the level at which risk becomes unacceptable is governed by the socio-economic cost. Conservation areas and open space zones may need to be established. However, in areas that have been developed the existing properties along a coastline may be expensive, so that public purchase may be out of the question. What is more, owners of private property may resist acquisition attempts.

Environmental impact assessment (EIA) is a methodology that is commonly used in coastal management. The aim of any environmental impact process is to improve the effectiveness of planning by providing objective information relating to the effects on the environment of a project or course of action. It therefore involves data collection relevant to environmental impact prediction. Solutions should be proposed to any problems that may arise as a result of interactions between the environment and the project. Environmental evaluation must involve comparative analyses of reasonable alternatives, and plans must be analysed so that the benefits of a project are appreciated fully. The negative effects as well as the positive effects must be conveyed to the decision-makers. Kay and Alder (1999) provided a useful summary of the EIA process in relation to coastal planning and management.

In recent years attitudes relating to coastal defence, planning and management have tended to change, moving away from traditional engineering schemes towards those that incorporate natural processes and pay greater attention to the environment. As noted above, some types of engineered coastal defence systems interfere with the movement of sediment in the shore zone and so may have severe impacts on other locations along a coast, as well as conflicting with the environment. Consequently, a more holistic approach to coastal defence and planning is evolving, involving the concept of sustainable coastal defence that is unlikely to result in disruptions to natural processes. Such an approach is based on an understanding of coastal processes and geomorphological change. For instance, Bray and Hooke (1998) advocated that each coastal plan should develop a conceptual model of the behaviour of the coastal system and its component landforms, explaining the evolution of the system in terms of its geomorphological history. Nonetheless, in developed coastal areas where there is substantial economic investment, there will be a continuing demand for engineered protection works. Obviously, such coastal defences need to be maintained. For example, Ciavola (1997) indicated that because the coastal defences along Spurn Head Spit, England, had not been maintained, the spit probably would be breached by any future storm surge.

8.13 Storm surges and marine inundation

Except where caused by failure of protection works, marine inundation almost always is attributable to severe meteorological conditions giving rise to abnormally high sea levels, referred to as storm surges. For example, Murty (1987) described storm surges as oscillations of coastal waters in the period range from a few minutes to a few days that were brought about by weather systems such as hurricanes, cyclones and deep depressions. Low pressure and driving winds during a storm may lead to marine inundation of low-lying coastal areas, particularly if this coincides with high spring tides. This is especially the case when the coast is unprotected. Floods may be frequent, as well as extensive where flood plains are wide and the coastal area is flat. Coastal areas that have been reclaimed from the sea and are below high tide level are particularly suspect if coastal defences are breached (Fig. 8.37). A storm surge can be regarded as the magnitude of sea level along the shoreline that is above the normal seasonally adjusted high tide level. Storm surge risk often is associated with a particular season. The height and location of storm damage along a coast over a period of time, when analysed, provide some idea of the maximum likely elevation of surge effects. The seriousness of the damage caused by storm surge tends to be related to its height and velocity of water movement.

Factors that influence storm surges include the intensity in the fall in atmospheric pressure, the length of water over which the wind blows, the storm motion and offshore topography. Obviously, the principal factor influencing storm surge is the intensity of the causative storm, especially the speed of the wind piling up the sea against the coastline. For instance, threshold wind speeds of approximately $120\,\mathrm{km\,h^{-1}}$ tend to be associated with central pressure drops of around 34 mbar. Normally, the level of the sea rises with reductions in atmospheric pressure associated with intense low pressure systems. In addition, the severity of a surge is influenced by the size and track of a storm and, particularly in the case of open coastline surges, by the nearness of the storm track to the coastline. The wind direction and the length of fetch also are important, both determining the size and energy of waves. Because of the influence of the topography of the sea floor, wide shallow areas on the continental shelf are more susceptible to damaging surges than those where the shelf slopes steeply. Surges are intensified by converging coastlines that exert a funnel effect as the sea moves into such inlets.

Figure 8.37 Marine inundation flooding polderland near Philipsand, the Netherlands, January 1953.

Ward (1978) distinguished two types of major storm surge. First, there are storm surges on open coastlines that travel as running waves over large areas of the sea. Such surges are associated with tropical cyclones, typhoons and hurricanes. Second, surges occur in enclosed or partially enclosed seas where the area of sea is small compared with the size of the atmospheric disturbance. Hence, the surges more or less affect the whole sea at any one time. This type of surge may be frequent, as well as damaging.

Most deaths have occurred as a result of storm surges generated by hurricanes. The height of a storm surge associated with hurricanes is influenced by the distance from the centre of the storm to the point at which maximum wind speeds occur, the barometric pressure in the eye of a hurricane, the rate of forward movement of the storm, the angle at which the centre of the storm crosses the coastline, and the depth of sea water off the shore. In addition to the storm surge, enclosed bays may experience seiching. Seiches involve the oscillatory movement of waves generated by a hurricane. Seiching also may occur in open bays if the storm moves forward very quickly and in low-lying areas this can be highly destructive.

The east coast of India and Bangladesh experience serious inundation due to storm surges once in a decade or so (Sharma and Murty, 1987). Several major rivers flow into the Bay of Bengal and excessive siltation in their estuaries, together with the formation of sand bars, provide ideal conditions for the propagation of surges. The deltaic area of the River Ganges is especially prone to storm surges. The width of the continental shelf, in places up to 300 km, also favours storm surge development. Furthermore, the orientation of the track of the cyclone in relation to the coast is of importance where the width of the continental shelf is significant. For example, a surge induced at an angle of incidence of 90° may be almost twice one that is produced by a cyclone of the same intensity when the angle of incidence is 135°. The winds associated with tropical cyclones can blow at speeds between 50 and 120 km h^{-1}. Storm surges of over 5 m in height have been recorded. The duration of marine inundation may be related to the duration of the surge, or in relatively slow draining areas such as tidal marshes to the time it takes for drainage to occur.

Obviously, prediction of the magnitude of a storm surge in advance of its arrival can help to save lives. Storm tide warning services have been developed in various parts of the world. Warnings usually are based upon comparisons between predicted and observed levels at a network of tidal gauges such as those along the coast of the North Sea where coastal surges tend to move progressively along the coast. To a large extent the adequacy of a coastal flood forecasting and warning system depends on the accuracy with which the path of the storm responsible can be determined. Storms can be tracked by satellite, and both satellite and ground data are used for forecasting. Numerous attempts have been made to produce models that predict storm surges (Jelesmanski, 1978).

Because of the increased urbanization of many coastal areas, the damaging effects of storm surges has increased in the last 60 years. For instance, this is the case with several large coastal cities in Japan, Osaka being a notable example (Tsuchiya and Kawata, 1987). Typhoons frequently move northeastward before striking the Japanese coast and unfortunately Osaka Bay is elliptical in shape with its major axis running northeast to southwest. This means that water can pile up at the northeast part of the bay during storm surges. During high spring tides, the tidal range is approximately 1.85 m. Hence, a meteorological tide due to a storm surge combined with such an astronomical tide can give rise to an exceptional increase in sea level. The characteristics of three storm surges and associated typhoons are given in Table 8.3. After the Muroto typhoon of 1934, embankments were constructed in low-lying areas. Dykes of wooden piles, concrete or soil, together with locks and pumping stations, were constructed along rivers and canals, and in harbour areas. Tens of kilometres of breakwaters were constructed.

Table 8.3 Characteristics of the Muroto, Jane and Daini Muroto Typhoons and their accompanying storm surges

Items	*Muroto Typhoon*	*Typhoon Jane*	*Daini-Muroto Typhoon*
Date	9 September 1934	3 September 1950	16 September 1961
Lowest atmospheric pressure (mbar)	954.3	970.3	937.3
Moving velocity of typhoon (km h^{-1})	60	58	50
Maximum mean wind velocity for 10 min (m s^{-1})	42.0	28.1	33.3
Instantaneous maximum wind velocity (m s^{-1})	More than 60	44.7	50.6
Highest tidal level above OP (m)	4.2	3.85	4.12
Total precipitation (mm)	22.3	62.2	44.2

Source: Tsuchiya and Kawata, 1987.

Note
OP, Osaka Port.

Also, an early warning system was introduced so that people could be evacuated from areas that could be affected by storm surges. Historical data relating to storm surges is available for over 1000 years for this area and suggests a mean recurrence interval of some 150 years for severe storm surges, that is, a surge having a height of 3.5 m or more above astronomical tide level.

Protective measures, such as noted above, are not likely to be totally effective against every storm surge. Nevertheless, flood protection structures are used extensively and often prove substantially effective. Sand dunes and high beaches offer natural protection against floods and can be maintained by artificial nourishment with sand. Dunes also can be constructed by tipping and bulldozing sand or by aiding the deposition of sand by the use of screens and fences. Coastal embankments usually are constructed of local material. Earth embankments are susceptible to erosion by wave action if not protected and if overtopped can be scoured on the landward side. Slip failures also may occur. Sea walls may be constructed of concrete or sometimes of steel sheet piling. Permeable sea walls may be formed of rip-rap or concrete tripods. The roughness of their surfaces reduces swash height and the scouring effect of backwash on the foreshore.

Barrages may be used to shorten the coastline along highly indented coastlines and at major estuaries. Although expensive, barrages can serve more than one function, not just flood protection. They may produce hydro-electricity, carry communications, and freshwater lakes can be created behind them. The Delta Scheme in the Netherlands provides one of the best examples (Fig. 8.38). The scheme was developed after the disastrous flooding that resulted from the storm surge of February 1953. This inundated some 150 000 ha of polders and damaged over 400 000 buildings. The death toll amounted to 1835 and 72 000 people were evacuated. Some 47 000 cattle were killed. The scheme involved the construction of four massive dams to close-off sea inlets and a series of secondary dams, with a flood barrier northwest of Rotterdam. The length of the coastline was reduced by some 700 km and coastal defences were improved substantially.

Estuaries that have major ports require protection against sea flooding. The use of embankments may be out of the question in urbanized areas as space is not available without expensive

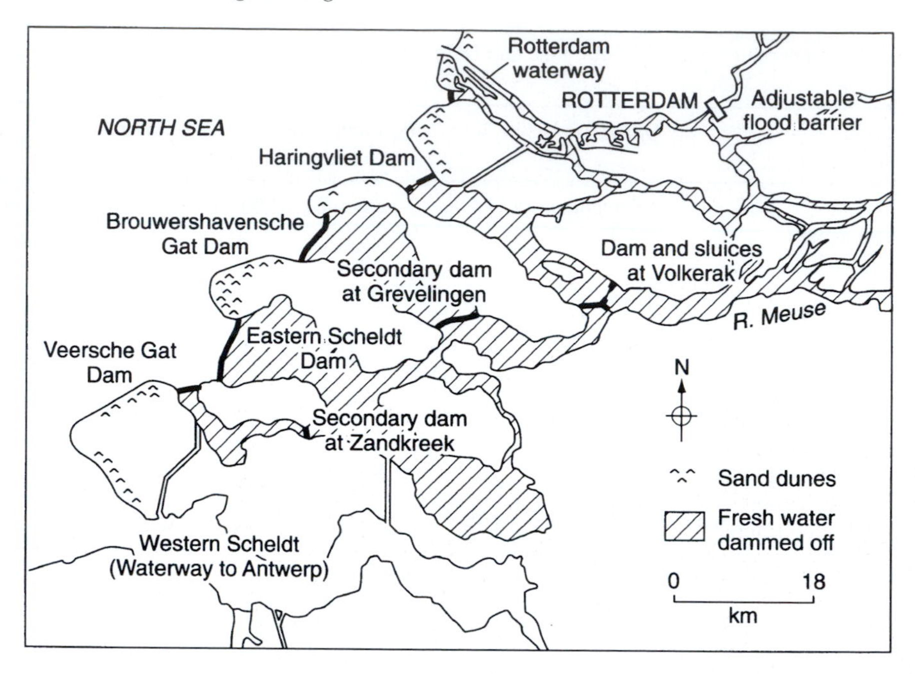

Figure 8.38 The Delta project, the Netherlands.

compulsory purchase of waterfront land and barrages with locks would impede sea-going traffic. In such situations, movable flood barriers may be used that can be emplaced when storm surges threaten. The Thames Barrier in London offers an example (Fig. 8.39).

Marine inundation also may be brought about by co-seismic subsidence. For example, the Alaskan earthquake of 1964 meant that some areas of the coast were flooded, crustal subsidence of up to 2 m being recorded. Peterson *et al.* (2000) carried out an investigation of great prehistoric earthquakes in the central Cascadia margin of the west coast of the United States and found that coastal subsidence had resulted in serious lowland flooding and catastrophic beach erosion. They concluded that any similar seismic event would mean the 525 km of bay shorelines in Washington and Oregon would be threatened by chronic flooding and 250 km to disastrous beach erosion.

8.14 Tsunamis

One of the most terrifying phenomena that occur along coastal regions is inundation by large masses of water called tsunamis (Fig. 8.40). Most tsunamis originate as a result of fault movement generating earthquakes on the sea floor, though they also can be generated by submarine landslides or volcanic activity. However, even the effects of large earthquakes are relatively localized compared with the impact of tsunamis. Seismic tsunamis are most common in the Pacific Ocean and usually are formed when submarine faults have a significant vertical movement. Faults of this type occur along the coasts of South America, the Aleutian Islands and Japan. Resulting displacement of the water surface generates a series of tsunami waves, which travel in all directions across the ocean. As with other forms of waves, it is the energy of tsunamis that is transported not the mass. Oscillatory waves are developed with periods of 10–60 min, which affect the whole column of water from the bottom of the ocean to the

Figure 8.39 The Thames Barrier, London.

Figure 8.40 The northern end of Resurrection Bay at Seward, Alaska, after it had been affected by a tsunamis in 1964. The epicentre of the earthquake responsible was some 75 km distant.

surface. The magnitude of an earthquake, its depth of focus and the amount of vertical crustal displacement determine the size, orientation and destructiveness of tsunamis. Horizontal fault movements such as occur along the Californian coast, do not result in tsunamis.

In the open ocean, tsunamis normally have a very long wavelength and their amplitude is hardly noticeable. Their wavelength (L) is equal to:

$$L = vT \tag{8.25}$$

in which v is the velocity and T is the period. Successive waves may be from five minutes to an hour apart. They travel rapidly across the ocean at speeds of around $650\,\mathrm{km\,h^{-1}}$. In fact, the velocity (v) of propagation of tsunamis in deep water is given by:

$$v = \sqrt{gZ} \tag{8.26}$$

where g is the acceleration due to gravity and Z is the depth of water. This allows prediction to be made of the arrival times of tsunamis along coasts. Unfortunately, however, the vertical distance between the maximum height reached by the water along a shoreline and mean sea level, that is, the tsunami run-up, is not possible to predict. Because their speed is proportional to the depth of water, this means that the wave fronts always move towards shallower water. It also means that in coastal areas, the waves slow down and increase in height, rushing onshore as highly destructive breakers. Waves have been recorded up to nearly 20 m in height above normal sea level. Tsunamis, like other waves, are refracted by offshore topography and by the differences in the configuration of a coastline.

Due to the long period of tsunamis, the waves are of great length (e.g. 200–700 km in the open ocean, 50–150 km on the continental shelf). It therefore is almost impossible to detect tsunamis in the open ocean because their amplitudes are extremely small (0.1–1.0 m) in relation to their length. They can only be detected near the shore. The size of a wave as it arrives at the shore depends upon the magnitude of the original displacement of water at the source and the distance it has travelled, as well as underwater topography and coastal configuration. Soloviev (1978) devised a classification of tsunami intensity that is given in Table 8.4.

Free oscillations develop when tsunamis reach the continental shelf that modify their form. Usually, the largest oscillation level is not the first but one of the subsequent oscillations. However, to an observer on the coast, a tsunami appears not as a sequence of waves but as a quick succession of floods and ebbs (i.e. a rise and fall of the ocean as a whole) because of the great wavelength involved. Shallow water permits tsunamis to increase in amplitude without significant reduction in velocity and energy. On the other hand, where the water off a shoreline is relatively deep, the growth in the size of wave is restricted. Large waves, several metres in height, are most likely when tsunamis move into narrowing inlets. Such waves cause terrible devastation and can wreck settlements. For example, if the wave is breaking as it crosses the shore, it can destroy any beach front houses merely by the weight of water. In fact, water flow exerts a force on obstacles that are in its path that is proportional to the product of the depth of water and the square of its velocity. Wiegel (1970) suggested that the maximum pressure (P_{max}) on an obstacle can be derived from:

$$P_{\mathrm{max}} = 0.5 C_{\mathrm{D}} \rho v_{\mathrm{c}}^2 F_{\mathrm{n}}^2 \tag{8.27}$$

where C_{D} is a factor including the shape of the obstacle, ρ is the density of water, $v_{\mathrm{c}} = 2\sqrt{gZ_{\mathrm{c}}}$, Z_{c} is the depth of the inundation zone and F_{n} is the Froude number. The subsequent backwash

Table 8.4 Scale of tsunami intensity

Intensity	*Run-up height* (m)	*Description of tsunami*	*Frequency in Pacific Ocean*
I	0.5	*Very slight.* Wave so weak as to be perceptible only on tide gauge records	One per hour
II	1	*Slight.* Waves noticed by people living along the shore and familiar with the sea. On very flat shores waves generally noticed	One per month
III	1	*Rather large.* Generally noticed. Flooding of gently sloping coasts. Light sailing vessels carried away on shore. Slight damage to light structures situated near the coast. In estuaries, reversal of river flow for some distance upstream	One per eight months
IV	4	*Large.* Flooding of the shore to some depth. Light scouring on made ground. Embankments and dykes damaged. Light structures near the coast damaged. Solid structures on the coast lightly damaged. Large sailing vessels and small ships swept inland or carried out to sea. Coasts littered with floating debris	One per year
V	8	*Very large.* General flooding of the shore to some depth. Quays and other heavy structures near the sea damaged. Light structures destroyed. Severe scouring of cultivated land and littering of the coast with floating objects, fish and other sea animals. With the exception of large ships, all vessels carried inland or out to sea. Large bores in estuaries. Harbour works damaged. People drowned, waves accompanied by a strong roar	Once in three years
≥VI	16	*Disastrous.* Partial or complete destruction of man-made structures for some distance from the shore. Flooding of coasts to great depths. Large ships severely damaged. Trees uprooted or broken by the waves. Many casualties	Once in ten years

Source: Soloviev, 1978.

may carry many partially destroyed buildings out to sea and may remove several metres depth of sand from dune coasts. Damage also is caused when the resultant debris smashes into buildings and structures.

Usually, the first wave is like a very rapid change in tide, for example, the sea level may change 7 or 8 m in 10 min. A bore occurs where there is a concentration of wave energy by funnelling, as in bays, or by convergence, as on points. A steep front rolls over relatively quiet water. Behind the front the crest of such a wave is broad and flat, the wave velocity being around 30 km h^{-1}. Along rocky coasts large blocks of material may be dislodged and moved shoreward.

Because of development in coastal areas within the last 60 or more years, the damaging effects of future tsunamis will probably be much more severe than in the past. Hence, it is increasingly important that the hazard is evaluated accurately and that the potential threat is estimated correctly. This involves an analysis of the risk associated with tsunamis for the purpose of planning and putting into place mitigation measures. However, the tsunami hazard is not frequent and when it does affect a coastal area its destructiveness varies with both location and time. Accordingly, analysis of the historical record of tsunamis is required in any risk assessment. This involves a study of the seismicity of a region to establish the potential threat from earthquakes of local origin. In addition, tsunamis generated by distant earthquakes must be evaluated. The data gathered may highlight spatial differences in the distribution of the destructiveness of tsunamis that may form the basis for zonation of the hazard. If the historic record provides sufficient data, then it may be possible to establish the frequency of recurrence of tsunami events, together with the area that would be inundated by a 50-, 100- or even 500-year tsunami event. On the other hand, if insufficient information is available, then tsunamis modelling may be resorted to either using physical or computer models. The latter are now much more commonly used and provide reasonably accurate predictions of potential tsunami inundation that can be used in the management of tsunamis hazard. Such models permit the extent of damage to be estimated and the limits for evacuation to be established. The ultimate aim is to produce maps that indicate the degree of tsunami risk that, in turn, aids the planning process thereby allowing high risk areas to be avoided or used for low-intensity development (Pararas-Carayannis, 1987). Models also facilitate the design of defence works.

Various instruments are used to detect and monitor the passage of tsunamis. These include sensitive seismographs that can record waves with long-period oscillations, pressure recorders placed on the sea floor in shallow water, and buoys anchored to the sea floor and used to measure changes in the level of the sea surface. The locations of places along a coast affected by tsunamis can be hazard mapped that, for example, show the predicted heights of tsunami at a certain location for given return intervals (e.g. 25, 50 or 100 years). Homes and other buildings can be removed to higher ground and new construction prohibited in the areas of highest risk. Resettlement of coastal communities and prohibition of development in high risk areas has occurred at Hilo, Hawaii. However, the resettlement of all coastal populations away from possible danger zones is not a feasible economic proposition. Hence, there are occasions when evacuation will be necessary. This depends on estimating just how destructive any tsunami will be when it arrives on a particular coast. Furthermore, evacuation requires that the warning system is effective and that there is an adequate transport system to convey the public to safe areas.

Breakwaters, coastal embankments, and groves of trees tend to weaken a tsunami wave, reducing its height and the width of the inundation zone. Sea walls may offer protection against some tsunamis. Buildings that need to be located at the coast can be constructed with reinforced concrete frames and elevated on reinforced concrete piles with open spaces at ground level (e.g. for car parks). Consequently, the tsunami may flow through the ground floor without adversely affecting the building. Buildings usually are orientated at right angles to the direction of approach of the waves, that is, perpendicular to the shore. It is, however, more or less impossible to protect a coastline fully from the destructive effects of tsunamis.

Ninety per cent of destructive tsunamis occur within the Pacific ocean, averaging more than two each year. For example, in the past 200 years the Hawaiian Islands have been subjected to over 150 tsunamis. Hence, a long record of historical data provide the basis for prediction of tsunamis in Hawaii and allow an estimation of the events following a tsunamigenic earthquake to be made.

The Pacific Tsunami Warning System (PTWS) is a communications network covering the countries bordering the Pacific Ocean and is designed to give advance warning of dangerous tsunamis (Dohler, 1988). The system uses 69 seismic stations and 65 tide stations in major harbours about the Pacific Ocean. Earthquakes with a magnitude of 6.5 or over cause alarms to sound and those over 7.5 give rise to around-the-clock tsunami watch. Nevertheless, it is difficult to predict the size of waves that will be generated and to avoid false alarms. Clearly, the PTWS cannot provide a warning of an impending tsunami to those areas that are very close to the earthquake epicentre that is responsible for the generation of the tsunami. In fact, 99 per cent of the deaths and much of the damage due to tsunamis occur within 400 km of the area where it was generated. Considering the speed of travel of a tsunami wave, the warning afforded in such instances is less than 30 min. Recently, however, a system has been developed whereby an accelerometer transmits a signal, via a satellite over the eastern Pacific Ocean, to computers when an earthquake of magnitude 7 or more occurs within 100 km of the coast. The computers decode the signal, cause water-level sensors to start monitoring and transmit messages to those responsible for carrying out evacuation plans. On the other hand, waves generated off the coast of Japan take 10 hours to reach Hawaii. In such instances the PTWS can provide a few hours for evacuation to take place if it appears to be necessary (Bernard, 1991).

References

Abam, T.K.S. 1997. On the development and failure of overhangs in alluvial channel banks. *Journal Mining and Geology*, **33**, 1–6.

Allen, P.M., Arnold, J. and Jakubowski, E. 1999. Prediction of stream channel erosion potential. *Environmental and Engineering Geoscience*, **5**, 339–351.

Bascom, W. 1964. *Waves and Beaches.* Doubleday and Co., New York.

Bell, F.G. 1994. Floods and landslides in Natal and notably the greater Durban area, September 1987, a retrospective view. *Bulletin Association Engineering Geologists*, **31**, 59–74.

Bell, F.G. and Mason, T.R. 1998. The problems of flooding in Ladysmith, Natal, South Africa. In: *Geohazards and Engineering Geology*, Engineering Geology Special Publication No. 14, Maund, J.G. and Eddleston, M. (eds), The Geological Society, London, 3–10.

Benson, M.A. 1968. Uniform flood frequency estimating methods for federal agencies. *Water Resources Research*, **4**, 891–908.

Bernard, E.N. 1991. Assessment of Project THRUST: past, present and future. *Natural Hazard*, **4**, 285–292.

Boggett, A.D., Mapplebeck, N.J. and Cullen, R.J. 2000. South Shore Cliffs, Whitehaven – geomorphological survey and emergency cliff stabilization works. *Quarterly Journal Engineering Geology and Hydrogeology*, **33**, 213–226.

Brammer, H. 1990. Floods in Bangladesh: flood mitigation and environmental aspects. *Geographical Journal*, **156**, 158–165.

Bray, M. and Hooke, J. 1998. Spacial perspective in coastal defence and conservation strategies. In: *Coastal Defence and Earth Science Conservation*, Hooke, J. (ed.), The Geological Society, London, 115–132.

Brookes, A. 1988. *Channelized Rivers; Perspectives for Environmental Management.* Wiley, Chichester.

Carlston, C.W. 1968. Slope discharge relations for eight rivers in the United States. *United States Geological Survey*, Professional Paper, 600-D, 45–47, Washington, DC.

Cetin, H., Bal, Y. and Demirkol, C. 1999. Engineering and environmental effects of coastline changes in Turkey, northeastern Mediterranean. *Environmental and Engineering Geoscience*, **5**, 315–330.

Chow, C.T. (ed.). 1964. *Handbook of Applied Hydrology.* McGraw-Hill, New York.

Church, M. 1988. Floods in cold climates. In: *Flood Geomorphology*, Baker, V.R., Kochel, R.C. and Patton, P.C. (eds), Wiley, New York, 205–229.

Ciavola, P. 1997. Coastal dynamics and impact of coastal protection works on the Spurn Head spit (UK). *Catena*, **30**, 369–389.

Clarke, A.O. 1991. A boulder approach to estimating flash flood peaks. *Bulletin Association Engineering Geologists*, **28**, 45–54.

Clarke, A.O. 1996. Estimating probable maximum floods in the upper Santa Ana Basin, southern California, from stream boulder size. *Environmental and Engineering Geoscience*, **2**, 165–182.

Clayton, K.M. 1990. Sea level rise and coastal defences in the United Kingdom. *Quarterly Journal Engineering Geology*, **23**, 283–288.

Cleary, W.J. and Hosier, P.E. 1987. North Carolina coastal geologic hazards, an overview. *Bulletin Association Engineering Geologists*, **24**, 469–488.

Collinge, V. and Kirby, C. (eds) 1987. *Weather, Radar and Flood Forecasting.* Wiley, Chichester.

Cooke, R.U., Brunsden, D., Doornkamp, J.C. and Jones, D.K.C. 1982. *Urban Geomorphology in Drylands*. Oxford University Press, Oxford.

Costa, J.E. 1978. Holocene stratigraphy and flood frequency analysis. *Water Resources Research*, **14**, 626–632.

Dalrymple, T. 1960. Flood frequency analysis. *United States Geological Survey*, Water Supply Paper, 1543A, 1–80, Washington, DC.

Dohler, G.C. 1988. A general outline of the ITSU Master Plan for the tsunami warning system in the Pacific. *Natural Hazards*, **1**, 295-302.

Dutnell, R.C. 1999. Applying fluvial geomorphology for natural channel design and stream bank stabilization. *Proceeding Thirtieth Conference International Erosion Control Association on Investing in the Protection of Our Environment*, Nashville, 331–339.

Fisk, H.N. 1944. *Geological Investigation of the Alluvial Valley of the Lower Mississippi River*. United States Army Corps of Engineers, Mississippi River Commission, Vicksburg, Mississippi.

Flemming, C.A. 1990. Principles and effectiveness of groynes. In: *Coastal Protection*, Pilarczyk, K.W. (ed.), A.A. Balkema, Rotterdam, 121–156.

Galvin, C.J. 1968. Breaker type classification on three laboratory beaches. *Journal Geophysical Research*, **73**, 3651–3659.

Gares, P.A. 1990. Predicting flooding probability for beach-dunes systems. *Environmental Management*, **14**, 115–123.

Gares, P.A. and Sherman, D.J. 1985. Protecting an eroding shoreline: the evolution of management response. *Applied Geography*, **5**, 55–69.

Gerstgraser, C. 1999. The effect and resistance of soil bioengineering methods for streambank protection. *Proceeding Thirtieth Conference International Erosion Control Association on Investing in the Protection of Our Environment*, Nashville, 381–391.

Gilvear, D.J., Davies, J.R. and Winterbottom, S.J. 1994. Mechanisms of floodbank failure during large flood events on the rivers Tay and Earn, Scotland. *Quarterly Journal Engineering Geology*, **27**, 319–332.

Gruntfest, E.C. and Huber, C. 1989. Status report on flood warning systems in the United States. *Environmental Management*, **13**, 357–368.

Gupta, A. 1988. Large floods as geomorphic events in the humid tropics. In: *Flood Geomorphology*, Baker, V.R., Kochel, R.C. and Patton, P.C. (eds), Wiley, New York, 301–315.

Hager, W.H. and Sinniger, R. 1985. Flood storage in reservoirs. *Proceedings American Society Civil Engineers, Journal Irrigation and Drainage Engineering*, **111**, 76–85.

Hjulstrom, F. 1935. Studies of the morphological activity of rivers, as illustrated by the river Fynis, *Uppsala University Geological Institute*, Bulletin 25, Uppsala.

Hutchinson, J.N., Bromhead, E.N. and Chandler, M.P. 1991. Investigation of the coastal landslides at St Catherins'e Point, Isle of Wight. *Proceedings Conference on Slope Stability Engineering – Applications and Developments*, Thomas Telford Press, London, 151–161.

Ingle, J.G. 1966. *The Movement of Beach Sand*. Elsevier, Amsterdam.

Jarret, R.D. 1990. Palaeohydrologic techniques used to define the special occurrence of floods. *Geomorphology*, **3**, 181–195.

Jelesmanski, C.P. 1978. Storm surges. In: *Geophysical Predictions*. National Academy of Sciences, Washington, DC, 185–192.

Johnson, D.W. 1919. *Shoreline Processes and Shoreline Development*. Wiley, New York.

Jopling, A.V. 1963. Hydraulic studies on the origin of bedding. *Sedimentology*, **2**, 115–121.

Kay, R.C. and Alder, J. 1999. *Coastal Planning and Management*. E & FN Spon, London.

Kenny, R. 1990. Hydrogeomorphic flood hazard evaluation for semi-arid environments. *Quarterly Journal Engineering Geology*, **23**, 333–336.

Komar, P.D. 1976. *Beach Processes and Sedimentation*. Prentice-Hall, Englewood Cliffs, New Jersey.

Latham, J.-P. 1991. Degradation model for rock armour in coastal engineering. *Quarterly Journal Engineering Geology*, **24**, 101–118.

Lawler, D.M. 1993. The measurement of channel erosion and lateral channel changes: a review. *Earth Surface Processes and Landforms*, **18**, 777–821.

Leathermann, S.P. 1979. Beach and dune interactions during storm conditions. *Quarterly Journal Engineering Geology*, **12**, 281–290.

Leatherman, S.P., Nicholls, R.J. and Dennis, K.C. 1995. Aerial videotape-assisted vulnerability analysis: a cost-effective approach to assess sea level rise impacts. *Journal Coastal Research*, Special Issue, **14**, 15–25.

Leopold, L.B. and Wolman, M.G. 1957. River channel patterns, braided, meandering and straight. *United States Geological Survey*, Professional Paper, 282-B, Washington, DC.

Leopold, L.B., Wolman, M.G. and Miller, J.P. 1964. *Fluvial Processes in Geomorphology*. Freeman and Co., San Francisco.

Lewin, J. 1989. Floods in fluvial geomorphology. In: *Floods: Hydrological, Sedimentological and Geomorphological Implications*, Beven, K. and Carling, F. (eds), Wiley, Chichester, 265–284.

Lindsay, P. and Bell, F.G. 1998. Integrated natural and anthropogenic response in two South African estuaries. *Proceedings Eighth Congress International Association Engineering Geology*, Vancouver, Moore, D.P. and Hungr, O. (eds), A.A. Balkema, Rotterdam, **4**, 2733–2740.

Little, D.H. 1975. Harbours and docks. In: *Civil Engineer's Reference Book*, Blake, L.S. (ed.), Newnes-Butterworths, London, Section 24, 2–40.

Lo, P.G. and Diop, M.D. 2000. Problems associated with flooding in Dakar, western Senegal: influence of geological setting and town management. *Bulletin Engineering Geology and the Environment*, **58**, 145–149.

Macklin, M.G., Rumsby, M.T. and Newson, M.D. 1992. Historic overbank floods and floodplain sedimentation in the lower Tyne valley, northeast England. In: *Gravel Bed Rivers*, Hey, R.D. (ed.), Wiley, Chichester.

Maharaj, R.J. 1998. The performance of some coastal engineering structures for shoreline stabilization and coastal defence in Trinidad, West Indies. In: *Geohazards in Engineering Geology*, Engineering Geology Special Publication No. 15, Maund, J.G. and Eddleston, M. (eds), The Geological Society, London, 61–69.

Malamud, B.D., Turcotte, D.L. and Barton, C.C. 1996. The 1993 Mississippi River flood: a one hundred or a one thousand year event? *Environmental and Engineering Geoscience*, **4**, 479–486.

Mansur, C.I., Postol, G. and Salley, J.R. 2000. Performance of relief well systems along Mississippi River levees. *Proceedings American Society Civil Engineers, Journal Geotechnical and Geoenvironmental Engineering*, **126**, 727–738.

Marsland, A. 1986. The flood plain deposits of the Lower Thames. *Quarterly Journal Engineering Geology*, **19**, 223–247.

Martinez, M.J., Espejo, R.A., Bilbao, I.A. and Cabrera, M.D. del R. 1990. Analysis of sedimentary processes on the Las Canteras beach (Las Palmas, Spain) for its planning and management. *Engineering Geology*, **29**, 377–386.

Moore, R., Clark, A.R. and Lee, E.M. 1998. Coastal cliff behaviour and management: Blackgang, Isle of Wight. In: *Geohazards in Engineering Geology*, Engineering Geology Special Publication No. 15, Maund, J.G. and Eddleston, M. (eds), The Geological Society, London, 49–59.

Morris, S. and Moses, T. 1999. Urban stream rehabilitation: a design and construction case study. *Environmental Management*, **23**, 165–177.

Morris, L.L., McVey, M.J., Lohnes, R.A. and Baumel, C.P. 1996. Estimates of future impacts of degrading streams in the deep loess soil region of western Iowa on private and public infrastructure costs. *Engineering Geology*, **43**, 255–264.

Muir Wood, A.M. and Flemming, C.A. 1983. *Coastal Hydraulics*, Second Edition, Macmillan, London.

Murty, T.S. 1987. Mathematical modelling of global storm surge problems. In: *Natural and Man-Made Hazards*, El-Sabh, M.I. and Murty, T.S. (eds), Riedel Publishing Company, Dordrecht, 183–192.

Newson, M.D. 1994. *Hydrology and the River Environment*. Clarendon Press, Oxford.

Nicholls, R.J. 1998. Assessing erosion of sandy beaches due to sea-level rise. In: *Geohazards in Engineering Geology*, Engineering Geology Special Publication No. 15, Maund, J.G. and Eddleston, M. (eds), The Geological Society, London, 71–76.

Nicholls, R.J. and Leiser, A.T. 1998. Biotechnical bank stabilization on the Petaluma River, California. *Proceedings Twenty-Ninth Conference International Erosion Control Association on Winning Solutions for Risky Problems*, Reno, 283–288.

Nicholls, R.J., Leatherman, S.P., Dennis, K.C. and Volonte, C.R. 1995. Impacts and responses of sea level rise: qualitative and quantitative assessments. *Journal Coastal Research*, Special Issue, **14**, 26–43.

Oostinga, H. and Daemen, I. 1997. Construction of river training works for the Jamuna Bridge Project in Bangladesh. *Terra et Aqua*, **69**, 3–13.

Pararas-Carayannis, G. 1987. Risk assessment of the tsunami hazard. In: *Natural and Man-Made Hazards*, El-Sabh, M.I. and Murty, T.S. (eds), Riedel Publishing Company, Dordrecht, 183–192.

Peterson, C.D., Doyle, D.L. and Barnett, E.T. 2000. Coastal flooding and beach retreat from coseismic subsidence in the central Cascadia margin, USA. *Environmental and Engineering Geoscience*, **6**, 255–269.

Pilarczyk, K.W. 1990. Design of sea wall and dikes – including an overview of revetments. In: *Coastal Protection*, Pilarczyk, K.W. (ed.), A.A. Balkema, Rotterdam, 197–288.

Pilarczyk, K.W. (ed.). 1998. *Dikes and Revetments: Design, Maintenance and Safety Assessment*. A.A. Balkema, Rotterdam.

Rahn, P.H. 1994. Flood plains. *Bulletin Association Engineering Geology*, **31**, 171–183.

Ricketts, P.J. 1986. National policy and management responses to the hazard of coastal erosion in Britain and the United States. *Applied Geography*, **6**, 197–221.

Ristic, R. and Malosevic, D. 1997. Torrential floods – natural and man-made hazard. *Proceedings International Symposium Engineering Geology and the Environment*, Athens, Marinos, P.G., Koukis, G.C., Tsiambaos, G.C. and Stournaras, G.C. (eds), A.A. Balkema, Rotterdam, **1**, 993–995.

Schumm, S.A. 1967. Meander wavelength of alluvial rivers. *Science*, **157**, 1549–1550.

Schumm, S.A. 1969. River metamorphosis. *Proceedings American Society Civil Engineers, Journal Hydraulics Division*, **95**, 255–273.

Schumm, S.A. 1977. Applied fluvial geomorphology. In: *Applied Geomorphology*, Hails, J.R. (ed.), Elsevier, Amsterdam, 119–156.

Schumm, S.A. and Spitz, W.J. 1996. Geological influences on the Lower Mississippi River and its alluvial valley. *Engineering Geology*, **45**, 245–261.

Searcy, J.M. 1959. Flow-duration curves. *United States Geological Survey*, Water Supply Paper 1542A, Washington, DC.

Sharma, G.S. and Murty, A.J. 1987. Storm surges along the east coast of India. In: *Natural and Man-Made Hazards*, El-Sabh, M.I. and Murty, T.S. (eds), Riedel Publishing Company, Dordrecht, 257–278.

Silvester, R. 1974. *Coastal Engineering*, Volumes 1 and 2. Elsevier, Amsterdam.

Simons, D.B. and Richardson, E.V. 1961. Forms of bed roughness in alluvial channels. *Proceedings American Society Civil Engineer, Journal Hydraulics Division*, **87**, 87–105.

Sinha, R. 1998. On the controls of fluvial hazards in the north Bihar plains, eastern India. In: *Geohazards in Engineering Geology*, Engineering Geology Special Publication No. 15, Maund, J.G. and Eddleston, M. (eds), The Geological Society, London, 35–40.

Smith, L.M. 1996. Fluvial geomorphic features of the Lower Mississippi River alluvial valley. *Engineering Geology*, **45**, 139–165.

Smith, L.M. and Winkley, B.R. 1996. The response of the Lower Mississippi River to river engineering. *Engineering Geology*, **45**, 433–455.

Snyder, F.F. 1938. Synthetic unit graphs, *Transactions American Geophysical Union*, **19**, 447–463.

Soloviev, S.L. 1978. Tsunamis. In: *The Assessment and Mitigation of Earthquake Risk*, UNESCO Press, Paris, 91–143.

Sotir, R.B. 1998. Watershed management for streambank protection and riverine restoration. *Proceedings Twenty-Ninth Conference International Erosion Control Association on Winning Solutions for Risky Problems*, Reno, 453–462.

Steers, J.A. 1948. *The Coastline of England and Wales.* Cambridge University Press, Cambridge.

Sulisz, W. 1997. Wave loads on caisson founded on multilayered rubble base. *Proceedings American Society Civil Engineers, Journal Waterway, Port, Coastal and Ocean Engineering*, **123**, 91–101.

Sumer, B.M. and Fredsoe, J. 2000. Experimental study of 2D scour and its protection at a rubble mound breakwater. *Coastal Engineering*, **40**, 59–87.

Sundberg, A. 1956. The river Klarelren, a study of fluvial processes. *Geografiska Annaler*, **38**, 127–316.

Sverdrup, H.U. and Munk, W.H. 1946. Empirical and theoretical relations. In: *Forecasting Breakers and Surf*, American Geophysical Union, **27**, 823–827.

Thampapillai, D.J. and Musgrave, W.F. 1985. Flood damage and mitigation: a review of structural and non-structural measures and alternative decision frameworks. *Water Resources Research*, **21**, 411–424.

Thorne, C.R., Hey, R.D. and Newson, M.D. 1997. *Applied Fluvial Geomorphology for River Engineering and Management.* Wiley, Chichester.

Tsuchiya, Y. and Kawata, K. 1987. Historical changes of storm-surge disasters in Osaka. In: *Natural and Man-Made Hazards*, El-Sabh, M.I. and Murty, T.S. (eds), Riedel Publishing Company, Dordrecht, 279–304.

Verhoef, P.N.W. 1993. Abrasivity of the Hawkesbury Sandstone (Sydney, Australia) in relation to rock dredging. *Quarterly Journal Engineering Geology*, **26**, 5–17.

Vervoort, A. and De Wit, K. 1997. Correlation between dredgeability and mechanical properties of rock. *Engineering Geology*, **47**, 259–267.

Ward, R.C. 1978. *Floods: A Geographical Perspective.* Macmillan, London.

Whitcombe, L.-J. 1996. Behaviour of an artificially replenished shingle beach at Hayling Island, UK. *Quarterly Journal Engineering Geology*, **29**, 265–272.

Wiegel, R.L. 1970. Tsunamis. In: *Earthquake Engineering*, Wiegel, R.L. (ed.), Prentice-Hall, Englewood Cliffs, New Jersey.

Wilson, E.M. 1983. *Engineering Hydrology.* Macmillan, London.

9 Waste and its disposal

9.1 Introduction

With increasing industrialization, technical development and economic growth, the quantity of waste that is produced has increased immensely. In addition, in developed countries the nature and composition of waste has evolved over the decades, reflecting changes in industrial and domestic practices. For example, in Britain domestic waste has changed significantly since the 1950s, from largely ashes and little putrescible content of relatively high density to low-density highly putrescible waste. Many types of waste material are produced by society of which domestic waste, commercial waste, industrial waste, mining waste and radioactive waste are probably the most notable. Over and above this, waste can be regarded as non-hazardous and hazardous. Waste may take the form of solids, sludges, liquids and gases or any combination thereof. Depending on the source of generation, some of these wastes may degrade into harmless products whereas others may be non-degradable and/or hazardous, thus posing health risks and environmental problems if not managed properly. A further problem is the fact that deposited waste can undergo changes through chemical reactions, resulting in dangerous substances being developed. Indeed, waste disposal is one of the most expensive environmental problems to deal with. Dealing with the waste problem is one of the fundamental tasks of environmental protection.

As waste products differ considerably from one another, the storage facilities they require also differ. Despite increased efforts at recycling wastes and avoiding their production, many different kinds of special wastes are produced that must be disposed of in special ways. Wastes that do not decompose within a reasonable time, mainly organic and hazardous wastes, and liquids that cannot be otherwise disposed of, ideally should be burnt. All the organic materials are removed during burning, converting them to less hazardous forms and leaving an inorganic residue. Solid unreactive immobile inorganic wastes can be disposed of at above-ground disposal sites. It is sometimes necessary to treat these wastes prior to disposal. In order to provide long-term isolation from the environment, highly toxic non-degradable wastes should be disposed of underground if they cannot be burnt.

The best method of disposal is determined on the basis of the type and amount of waste on the one hand, and the geological conditions of the waste disposal site on the other. In terms of locating a site, a desk study is undertaken initially. The primary task of the site exploration that follows is to determine the geological and hydrogeological conditions. Their evaluation provides the basis of models used to test the reaction of the system to engineering activities. Chemical analysis of groundwater, and at times mineralogical analysis of soils and rocks, may help yield information regarding the future development of the site. At the same time, the leaching capacity of the water is determined, which allows prediction of reactions between water, wastes, soil or rock. If groundwater must be protected, or highly mobile toxic and/or very slowly degradable substances are present in wastes, then impermeable liners may be used to inhibit infiltration of leachate into the surrounding ground.

In terms of waste disposal by landfill, a landfill is environmentally acceptable if it is correctly engineered. Unfortunately, if it is not constructed to sufficiently high standards, a landfill may have an adverse impact on the environment. Surface water or groundwater pollution may result. Consequently, a physical separation between waste on the one hand and ground and surface water on the other, as well as an effective surface water diversion drainage system, is fundamental to design.

The objectives of a landfill classification are to consider waste disposal situations in terms of the type of waste, the amount of waste and the potential for water pollution. The different classes of landfill then can be used as a basis for the establishment of minimum requirements for site selection. Although no waste is completely non-hazardous, non-hazardous waste includes builders, garden, domestic, commercial and general dry industrial waste, which can be disposed of in a landfill. Nevertheless, some such landfills may produce significant amounts of leachate and so must incorporate leachate control systems. The size of a non-hazardous landfill operation depends on the daily rate of deposition of waste that, in turn, depends upon the size of population served. Hazardous waste has the potential, even in low concentrations to have a detrimental effect on public health and/or the environment because it may be, for instance, ignitable, corrosive or carcinogenic.

In order to carry out a detailed assessment of an existing or potential waste disposal site, a large volume of technical information and site-specific data are required (Sara, 1994). The technical information includes:

1 The interaction of wastes in the landfill, their degradation and leaching characteristics and how these change with time.
2 The composition of the leachate, including the solubilities and speciation of toxic elements, organics and major ion pollutants.
3 Changes in the composition and rate of leachate production with time.
4 The generation and behaviour of gases at the site, and their migration through surrounding permeable strata.
5 Groundwater and mixed fluid phase movement away from the site, and how these will change with the geomorphological evolution of the area during the period considered in the assessment.
6 The physical, chemical and biological interactions of the leachate with the rocks and soils of the site, including sorption (desorption reactions, matrix diffusion, dispersion and other diluting/retarding mechanisms).
7 The long-term geotechnical behaviour of the site with respect to engineering structures (including geochemical reactions of leachates and groundwaters with such structures).

One of the most important factors is the requirement for a thorough geological and hydrogeological survey of the site to a standard that will allow predictive leachate transport and gas migration modelling to be carried out. The results of such modelling, then can be used, via a sensitivity analysis that takes account of the wastes to be disposed, to design an engineered landfill that meets specific performance criteria. Until a performance assessment is available, it is not possible, based solely on a survey of the geology, to judge what the engineering requirements or range of acceptable waste types and volumes might be. The form of engineering envisaged for a hazardous waste site depends on whether a safety assessment indicates that containment or predictable dispersion is more appropriate for the wastes concerned.

9.2 Domestic refuse and sanitary landfills

At the present time, increasing concern is being expressed over the disposal of domestic waste products. Although domestic waste is disposed of in a number of ways, quantitatively the most important method is placement in a landfill (Fig. 9.1). For example, in Britain about 76 per cent of approximately 19 million tonnes of domestic solid waste produced each year is disposed of in landfills.

Domestic refuse is a heterogeneous collection of almost anything (i.e. waste food, garden rubbish, paper, plastic, glass, rubber, cloth, ashes, building waste, metals, etc.; Fang, 1995), much of which is capable of reacting with water to give a liquid rich in organic matter, mineral salts and bacteria, namely, leachate. Leachate is formed when rainfall infiltrates a landfill and dissolves the soluble fraction of the waste, as well as from the soluble products formed as a result of the chemical and biochemical processes occurring within decaying wastes. The organic carbon content of waste is especially important since this influences the growth potential of pathogenic organisms.

Matter exists in the gaseous, liquid and solid states in landfills, and all landfills comprise a delicate and shifting balance between the three states. Any assessment of the state of a landfill and its environment must take into consideration the substances present in a landfill, their mobility now and in the future, the potential pathways along which pollutants can travel and the targets potentially at risk from the substances involved.

Waste materials disposed of in sanitary landfills have dry densities varying from 158 to 685 kg m^{-3}, moisture contents of 10–35 per cent and low bearing capacities of 19.2–33.5 kPa. Baling or shredding the waste materials before deposition improves their *in situ* properties but at greatly increased cost.

As far as the location of landfills in concerned, decisions have to be made on site selection, project extent, finance, construction materials and site rehabilitation. The major requirement

Figure 9.1 A landfill site in Durban, South Africa.

in planning a landfill site is to establish exactly, by survey and analysis, the types, nature and quantities of waste involved. A waste survey is undertaken, after which future trends are forecast. These forecasts form the basis of the decision on which a potential site for a landfill is made. Initially, the potential life of a site is estimated and its distance from the proposed waste catchment area assessed. The design of a landfill is influenced by the character of the material that it has to accommodate.

Modern landfill disposal facilities require detailed investigations to ensure that appropriate design and safety precautions are undertaken. Furthermore, legislation generally requires those responsible for waste disposal facilities to guarantee that sites are suitably contained so as to prevent harming the environment. This may require that investigation continues during or after the construction of a landfill. However, sinking boreholes through the base of a leachate-filled landfill to check what is happening is unacceptable without precautions being taken to avoid leakage. Geophysical methods are being used increasingly for this purpose.

Selection of a landfill site for a particular waste or a mixture of wastes involves a consideration of economic and social factors, as well as geological and hydrogeological conditions. Basagaoglu *et al.* (1997) described the use of a geographical information system to help select sites for more detailed investigation for landfills in the Golbasi region of Turkey. Some of the factors that were taken account of included groundwater, surface water, soil types, erosion susceptibility, ecological features, road network and built-up areas. Langer (1995) reviewed the engineering geological factors that should be considered during a site investigation for waste disposal purposes. The ideal landfill site should be hydrogeologically acceptable, posing no potential threat to surface water or groundwater quality when used for waste disposal; be free from running or static water; and have a sufficient store of material suitable for covering each individual layer of waste. It also should be situated at least 200 m away from any residential development. As far as the hydrogeological conditions are concerned, most argillaceous sedimentary, massive igneous and metamorphic rock formations have low intrinsic permeability and therefore are likely to afford the most protection to water supply. By contrast, the least protection is provided by rocks intersected by open discontinuities or in which solution features are developed. Sandy materials may act as filters leading to dilution and decontamination. Hence, sites for disposal of domestic refuse can be chosen where decontamination has the maximum chance of reaching completion and where groundwater sources are located far enough to enable dilution to be effective. The position of the water table is important as it determines whether wet or dry tipping is involved. Generally, unless waste is inert, wet tipping should be avoided. Aquifers that contain potable supplies of water must be protected. If near a proposed landfill, a thorough hydrogeological investigation is necessary to ensure that site operations will not pollute the aquifer. If pollution is a possibility, the site must be designed to provide some form of artificial protection, otherwise the proposal should be abandoned.

Barber (1982) identified three classes of landfill site based upon hydrogeological criteria (Table 9.1). When assessing the suitability of a site, two of the principal considerations are the ease with which a pollutant can be transmitted through the substrata and the distance it is likely to spread from the site. Consequently, the primary and secondary permeability of the formations underlying a potential landfill area is of major importance. It is unlikely that the first type of site mentioned in Table 9.1 would be considered suitable. There also would be grounds for an objection to a landfill site falling within the second category of Table 9.1 if the site was located within the area of diversion to a water supply well. Generally, the third category, in which the leachate is contained within the landfill area, is preferable. In many situations, total containment only can be achieved if an artificial impermeable lining is provided over the bottom of the site. However, there is no guarantee that clay, soil-cement, asphalt or plastic

Table 9.1 Classification of landfill sites based upon their hydrogeology

Designation	*Description*	*Hydrogeology*
Fissured site or site with rapid subsurface liquid flow	Material with well-developed secondary permeability features	Rapid movement of leachate via fissures, joints, or through coarse sediments. Possibility of little dispersion in the groundwater, or attenuation of pollutants
Natural dilution, dispersion and attenuation of leachate	Permeable materials with little or no significant secondary permeability	Slow movement of leachate into the ground through an unsaturated zone. Dispersion of leachate in the groundwater, attenuation of pollutants (sorption, biodegradation, etc.) probable
Containment of leachate	Impermeable deposits such as clays or shales, or sites lined with impermeable materials or membranes	Little vertical movement of leachate. Saturated conditions exist within the base of the landfill

Source: Barber, 1982.

linings will remain impermeable permanently. Thus, the migration of materials from a landfill site into the substrata will occur eventually, although the length of time before this happens may be subject to uncertainty. Ideally, the delay should be sufficiently long for the pollution potential of the leachate to be greatly diminished. As mentioned, one of the methods of tackling the problem of pollution associated with landfills is by dilution and dispersal of the leachate. Otherwise leachate can be collected by internal drains within the landfill and conveyed away for treatment.

One of the difficulties in predicting the effect of leachate on ground or surface water is the continual change in the characteristics of leachate as a landfill ages. Leachate may be diluted where it gains access to run-off or groundwater, but this depends on the quantity and chemical characteristics of the leachate, as well as the quantity and quality of the receiving water.

9.2.1 Design considerations for a landfill

The design of landfill sites is influenced by the physical and the biochemical properties of the wastes and the need to control leachate production. The latter is dependent on the extent to which possible pollution problems may arise at a landfill site. Site selection therefore has an important influence on the need for leachate control and so a thorough investigation must be undertaken to determine site suitability. The hydraulic properties of landfills, which may differ with depth and age, influence the development of leachate and control its flow. They also affect the performance of any leachate-extraction wells (Powrie and Beaven, 1999). The use of leachate control and/or treatment methods may permit unsuitable sites to be used for the disposal of solid wastes. Leachate control should be planned before landfill development rather than after the landfill has been constructed, especially if control techniques such as drainage are to be installed beneath the waste.

The frequency and duration of rainfall for the catchment area of a landfill has to be considered since this influences the amount of water that can penetrate into a landfill. Data on evapotranspiration and infiltration rates for landfills also are important since this affects leachate generation. Furthermore, it is necessary to determine how much leachate can be absorbed by the area before run-off occurs. Surface run-off should be considered when the design of a landfill is undertaken. The rate of run-off is affected by the topography of the catchment area, as well as the nature and type of strata surrounding and below the proposed site.

Various techniques have been developed using water budget methods to estimate the amount of free water in a landfill. These methods consider a mass balance between precipitation, evapotranspiration, surface run-off and waste moisture storage. In other words, the quantity of leachate produced is influenced by the amount of groundwater in a landfill (leachate percolation = net percolation − water absorbed by waste + liquid disposal into landfill). Some fraction of precipitation will infiltrate the waste, after passing through the cover material. A field capacity or moisture-holding capacity ranging from 50 to 60 per cent by dry weight for compacted refuse has been suggested. However, the quantity of water absorbed by waste depends on the age of placement of the waste. Initially, the water-absorbing capacity of the waste exceeds net percolation and the leachate flow is zero. At the other extreme, if the waste is totally saturated, then the leachate flow equals net percolation plus groundwater or subsurface water flow into a landfill plus liquid disposal into landfill. As an example, Bell *et al.* (1996) used the area of a landfill, the percentage percolation, the quantity of liquid co-disposed and the quantity of liquid retained in storage, which is related to the absorptive capacity of the waste, in a simple water balance calculation to help determine the amount of leachate generated in a landfill. Water balance calculations can be used to determine the size of a working cell within a site needed to control the amount of free leachate that will be produced from rainfall or other water inputs. This excess free-draining leachate then may lead to the development of a zone of saturated refuse. The rate of build-up of leachate levels within the site and the amount of leachate held within the saturated zone is directly related to the effective porosity or storativity of the refuse. The ability to manage leachate within a site through leachate drainage schemes and vertical pumping wells is related to both the storativity and permeability of the waste.

The most common means of controlling leachate is to minimize the amount of water infiltrating the site by encapsulating the waste in impermeable material. Hence, well-designed landfills usually possess a cellular structure, as well as a lining and a cover, that is, the waste is contained within a series of cells formed of clay (Fig. 9.2). The cells are covered at the end of each working day with a layer of soil and compacted. The dimensions of a cell depend on the volume of waste received and the availability of cover material but they tend to range from 2.5 to 9.0 m. The thickness of the daily cover is 150–300 mm, its purpose being to control flying paper, minimize gas and moisture percolation, and to provide rodent and fire control, as well as improved appearance. Panagiotakopoulos and Dokas (2001) reviewed the dimensions of cells in relation to the soil to refuse ratio (S/R), final refuse density, length of working face, lift height and cover thickness. They suggested that the sensitivity of minimum S/R to variations in cell dimensions decreases with cell size, and that the length of a working face and lift height affect the S/R ratio significantly.

The ability of water to move within a landfill depends on a number of factors. Cell cover may act as an aquiclude. However, uneven settlement of a landfill can cause fractures within the cover that allow water to percolate downwards into the fill. In addition, successful landfill design involves the layers of waste being capable of absorbing precipitation during layer formation. Waste consolidation either on site or before tipping also helps reduce the quantity of leachate produced.

At the present time, there is no standard rule as to how waste should be dumped or compacted. Nonetheless, compaction is important since it reduces settlement and hydraulic conductivity, whilst increasing shear strength and bearing capacity. Furthermore, the lesser the quantity of air trapped within landfill waste, the lower is the potential for spontaneous combustion. Fang (1995) emphasized that compaction requires planning during the waste disposal process.

Figure 9.2 Cellular construction of a landfill in Kansas City, United States.

Wherever possible, waste should be uniformly distributed in thin layers prior to compaction. If non-uniform spreading cannot be achieved, then Fang recommended that heavier items should be placed in the centre of a landfill to help control its stability. Locally available soil can be mixed with the waste. As far as landfill stability is concerned, the potential for slope failure in a landfill is related to compaction control during disposal and the heavier the roller that is used for compaction the better. Even so, conventional compaction techniques do not always achieve effective results, especially with highly non-uniform waste. In such instances, dynamic compaction has been used with good results (Van Impe and Bouazza, 1996).

The settlement of wastes is attributable to physical mechanisms and decomposition. Prediction of settlement at landfill sites has become a vital element of their effective design, particularly in relation to post-closure settlement site redevelopment. However, determination of the amount and rate of settlement of a landfill is not a simple task. The heterogeneous nature of the materials involved, the fact that different materials may have been disposed of in different places and different times, means that a traditional soil mechanics approach for settlement prediction generally is unsatisfactory. Not only would settlement, especially large differential settlement, adversely affect any future construction on such a site but it may damage the cover system or lining, for example, geomembrane linings may be torn. Accordingly, Ling *et al.* (1998) reviewed the various empirical methods of estimating settlement of landfills and then proposed a further method for evaluating the amount and rate of settlement. Watts and Charles (1999) also discussed the settlement characteristics of landfills and suggested various ground improvement techniques that could be used, notably the use of surcharge loading, to reduce post-construction settlement.

Stabilization of soft landfills can be brought about by mixing with soil, fly ash, incinerator residue, lime or cement. When a minimum unconfined compressive strength of 24 kPa occurs due, for example, to irregularities in mixing at the site, geosynthetic material (e.g. a geoweb) should be used to span such areas and should be securely anchored in a trench at the perimeter of the site. A geomembrane should be placed over the geoweb to provide a secondary seal protecting against the infiltration of surface liquids. Clay soil above the geomembrane acts as a seal but it must be as thin as possible so as not to overload the geoweb-geomembrane.

Another factor that should be taken into account in the design of landfills is their stability. Any stability analyses must consider the stability of the soil or rock beneath a landfill, as well as the stability of the landfill material. Numerous assessments of the shear strength of landfill material have been made as this is required in any stability analysis (Jones *et al.*, 1997). Stability analyses are even more important if landfills are located on sloping sites. Eid *et al.* (2000) described the largest failure of a landfill that has taken place in the United States, it occurring in the Cincinnati area. It involved 1.2 million m^3 of waste material, and lateral and vertical displacements of 275 and 61 m, respectively. The failure surface developed through the weak underlying colluvial/residual soils. Eid *et al.* also described the laboratory- and field-testing programmes undertaken to assess the shear strength of the waste and the weak soils. Bromhead *et al.* (1996) described the stabilization of a landfill near Ancona, Italy, which had been constructed on a slope of 8–10° above sandy–silty colluvial clays. Active slip surfaces had been noted in an inspection shaft, and subsequently slow mass movements were confirmed by the installation of inclinometers. Stabilization was brought about by the construction of subhorizontal drains drilled from pits located near the toe of the landfill.

9.2.2 Landfill liners and drainage systems

Mitchell (1986) maintained that a properly designed and constructed liner and cover offer long-term protection for ground and surface water. Landfill liners are constructed from a wide variety of materials (Fig. 9.3). A review of the different methods of liner construction for landfills has been provided by Bouazza and Van Impe (1998). Adequate site preparation is necessary if a lining system is to perform satisfactorily. Nonetheless, no liner system, even if perfectly designed and constructed, will prevent all seepage losses. For instance, no liner, no matter how rigid or highly reinforced, can withstand large differential settlement without eventually leaking or possibly failing completely. Jessberger *et al.* (1995) briefly outlined how to assess the amount of deformation a liner is likely to undergo and whether unacceptable cracking will develop as a result. If extreme concern is warranted, an underdrainage system can be placed beneath the primary liner to collect any leakage passing through, which can either be treated or it can be recirculated back to the containment area. A secondary liner can be placed beneath the underdrainage system. However, Blight and Fourie (1999) maintained that there is justification for relaxation of the rigorous requirements for landfill liners and leachate collection systems in arid and semi-arid areas where low rainfall means that very little leachate is produced.

Clay, bentonite, geomembrane, soil-cement or bitumen cement can be placed beneath a landfill to inhibit movement of leachate into the soil. Clay liners are suitable for the containment of many wastes because of their low permeability (generally 10^{-9} m s^{-1} or less) and their ability to adsorb some wastes (Daniel, 1993). However, clay liners may shrink and so crack under unsaturated conditions that, in turn, may give rise to increased leakage rates. The permeability of clay liners also may be increased as a result of clay–pollutant reaction. Clay liners are constructed by compaction in lifts of about 150 mm thickness. Care must be taken during construction to ensure that the clay is placed at the specified moisture content and density, and to avoid cracking due to drying out after construction (Quigley *et al.*, 1988). The factors that govern the behaviour of a clay liner include the compaction moisture content, method of compaction, quality control during compaction and potential increase in the permeability

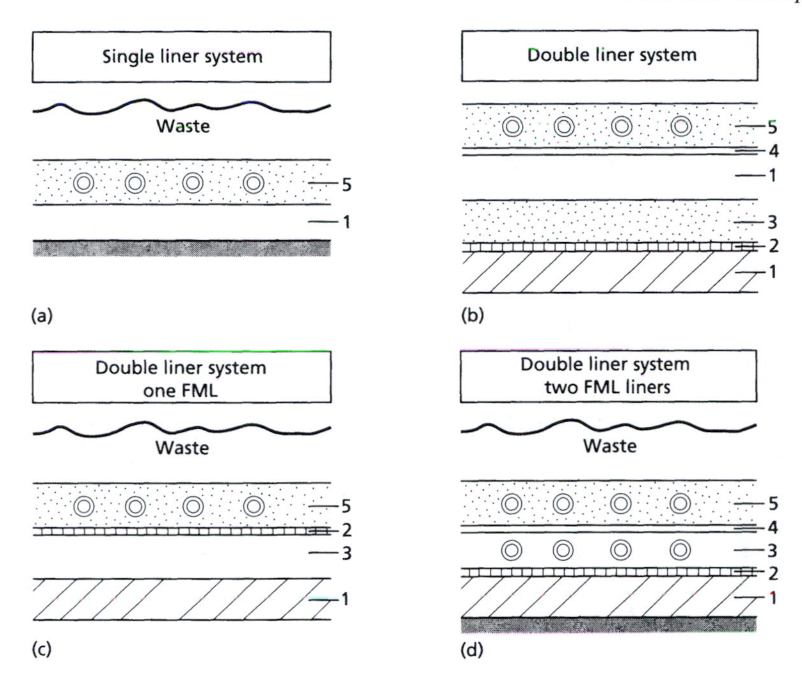

Figure 9.3 Landfill liner systems. 1, compacted low permeability clay; 2, flexible membrane liner (FML); 3, leachate collection/detection system; 4, FML; 5, primary leachate collection system; O, collection pipes.

due to interaction with waste liquids. Clark and Davis (1996) maintained that a tamping roller should be used for compaction rather than a smooth drum roller as the feet of a tamping roller provide a remoulding action, thereby breaking down lumps of clay and producing a more homogeneous material. Clay liners compacted at wet of optimum moisture content are less permeable compared with those compacted dry of optimum.

Another function of a clay liner for a landfill is to attenuate contaminants in the transport of leachates in and through it. According to Yong *et al.* (1999), if the clay material possesses a good chemical buffering capacity, low permeability and a high sorption capacity, then it should provide a competent contaminant attenuation barrier system against leachate transport. Data relating to the properties and behaviour of clay soil, along with evaluation procedures, represent the basis for assessment of the competency of clay soil to operate as an engineered barrier system for a landfill. Yong *et al.* itemized a number of physico-chemical properties that should be included in an evaluation of the suitability of clay soils as barrier systems. These included the natural moisture content, the specific gravity, the liquid limit, the plasticity index, the maximum dry density, the permeability, the cation exchange capacity, the specific surface area and the chemistry of the pore water. They also used a soil column leaching test to determine the attenuation characteristics of compacted clay soil, together with the distribution of contaminants in the soil in relation to the amount of fluid moving it. It has been suggested that the liquid limit of suitable clay soils should be less than 90 per cent and that the plasticity index should be less than 65 per cent but greater than 12 per cent (Murray *et al.*, 1996).

Certain leachates have proved especially troublesome in the case of clay liners. Those consisting of organic solvents and those containing high levels of dissolved salts, acids or alkalis can give rise to cracking and the development of pipes in clay. In such situations, some leachate could seep into a clay liner and could migrate to the groundwater at some future time. Clays, however, can be treated with polymers that reduce their sensitivity to potential contaminants. Also, cation exchange can take place between the ions on the surfaces of clay particles and those present in the leachate. Nevertheless, the effectiveness of clay liners in containing many hazardous wastes has been questioned.

Colliery spoils have been used in the construction of liners for landfills, the important factors affecting their permeability being the moulding moisture content, the degree of compaction and the nature of the material. Normally, most of the materials comprising colliery spoils are derived from mudrock, and their particle size distributions, liquid limits and plasticity indices fall within the acceptable range for clay liners. Smith *et al.* (1999) carried out a series of laboratory tests to assess the value of colliery spoil for landfill liners and found that the most critical factor affecting permeability was the moulding moisture content. Preferably, the latter should be about 0.5–1.5 per cent wet of optimum moisture content. Furthermore, they showed that a maximum particle size of up to 50 mm has little effect on the permeability of the material, which they attributed partly to its grading and partly to good mixing. Smith *et al.* concluded that with good construction control procedures, colliery spoil should prove acceptable for use in the construction of landfill liners.

Fly ash is a waste product that is produced in large quantities by coal-fired power stations. The use of pozzolanic fly ash as a material for landfill liners has been discussed by Prashanth *et al.* (2001). Those pozzolanic fly ashes that contain sufficient reactive silica (over 4 per cent) and free lime (over 3 per cent) develop good strength and exhibit little shrinkage and so do not crack. Pozzolanic fly ashes that contain sufficient free lime possess low permeability on curing because of the formation of gelatinous compounds that tend to occupy the pore spaces. Moreover, under the highly alkaline conditions that prevail in pozzolanic fly ashes, most toxic elements that are present in leachates are precipitated and so their migration is limited. Those pozzolanic fly ashes that contain sufficient reactive silica but insufficient free lime can be improved by the addition of lime. Accordingly, Prashanth *et al.* suggested that pozzolanic fly ashes could be used for the construction of liners to contain alkaline leachate. Mollamahmutoglu and Yilmaz (2001) suggested that a mixture of fly ash with bentonite (about 20 per cent by weight), compacted at optimum moisture content, could be used for both landfill liners and covers.

Bentonite is used widely in the construction of liners because of its low permeability and sorption capacity. However, Spooner and Giusti (1999) pointed out that sodium bentonite may be adversely affected by long-term exposure to particular inorganic and organic substances in landfills, for example, their structure may collapse, leading to shrinkage. In addition, because bentonite swells on wetting, it also shrinks on drying, which can lead to the development of desiccation cracks thereby increasing the permeability of the material. Sand can be added to bentonite in order to reduce the amount of shrinkage on drying. Indeed, bentonite-enhanced sand mixtures (BES) are being used increasingly as landfill liners. Between 3 and 6 per cent, by weight, of bentonite may be mixed with soil obtained from borrow pits and this is used as a liner. Sands with 20–30 per cent fines usually are preferable. Compatibility with the anticipated leachate should be determined because sodium bentonite is vulnerable to some organic chemicals. The thickness of a bentonite–soil liner is usually 300 mm or more. The material is mixed on site and applied wet of optimum moisture content in two lifts and compacted.

A series of tests were carried out by Tay *et al.* (2001) to evaluate the shrinkage, desiccation and permeability of compacted bentonite–sand mixtures. The mixtures contained 10 and 20 per cent bentonite by weight and were compacted at moisture contents varying between 8 and 32 per cent. All the mixtures exhibited some amount of shrinkage on air drying, it increasing with increasing moisture content of specimens. However, the top surface of BES when compacted at 15 per cent moisture content showed no visible desiccation cracking when air dried and only minor cracking when compacted at 20 per cent. Cracking only appeared when BES underwent more than 4 per cent volumetric shrinkage on air drying. If desiccation does not occur, then the permeability of BES is not affected. Tay *et al.* concluded that BES mixtures similar to those that they tested were unlikely to undergo significant damage due to desiccation cracking if placed by heavy compaction unless the moisture content was allowed to increase after a liner was constructed.

Horseman *et al.* (1999) investigated gas migration in relation to bentonite clay barriers. They showed that passage of gas through water-saturated clay is possible only if the gas pressure marginally exceeds the sum of the pore water pressure and the swelling pressure. If bentonite is compacted to a high dry density, then it possesses a large swelling pressure and a correspondingly high gas entry pressure. In addition, Horseman *et al.* suggested that the capillary threshold for the entry of gas into saturated clay is so large that normal two-phase flow, together with the displacement of pore water, is impossible. It would appear that clay ruptures at a gas pressure that is less than the gas entry pressure, irrespective of the increase in gas pressure. This is due to the smallness of the pores and throat areas in bentonite, the strong attraction between water molecules and clay particle surfaces and the extremely large meniscus curvatures associated with the penetration of gas into the pore spaces. They concluded that if gas is to move through a bentonite barrier, then it can only do so via an interconnected network of cracks produced by the tensile rupture of the clay under high gas pressure, otherwise saturated bentonite is impermeable to gas. However, their testing programme indicated that if the flow of gas is re-established in a test specimen, then the threshold pressure for gas breakthrough is less than it was initially. This shows that voids filled with residual gas occur along the pathways of gas migration.

Soil–cement liners make use of local soils. The soil ideally should have less than 20 per cent silt/clay fraction and when mixed with 3–12 per cent cement by weight, should perform adequately as a liner. Lining slopes with soil-cement, however, can present difficulties. In addition, soil-cement can be degraded in acidic conditions that tend to attack the cement. Furthermore, the likelihood of soil-cement shrinking and cracking leads to an increase in its permeability. When used, the minimum design-compacted thickness is usually around 300 mm. The mixing is carried out *in situ* in a central plant, and the soil-cement is laid by a mechanical spreader in 150–225 mm lifts.

Concrete liners can be reinforced with wire mesh or reinforcement bars, depending on the conditions of the subgrade. Unfortunately, construction and expansion joints must be incorporated in the pour, both of which are subject to leaks. Waterstops have to be used but they are expensive and might not be chemically compatible with the leachate. The paving of side slopes with angles greater than 30° to the horizontal presents major construction problems.

Shotcrete or gunite may be applied to the bottom and sides of a waste pit. Wire mesh may be used with shotcrete or gunite to reinforce it and to it prevent cracking. The technique often is used when a landfill is constructed on rock.

Bitumen cement and bituminous concrete have been used for lining landfills. In the latter case, bitumen, cement and high-quality mineral aggregates are hot mixed. Both are compacted in place and the compacted thickness ranges from 38 to 150 mm. Usually, the material is placed in more than one layer. However, a major consideration is the chemical compactibility between the contents of the landfill and the bituminous material.

When geomembranes are used as liners the chemical compatibility between waste material and geomembrane should be assessed as not all geomembranes are chemically compatible with certain types of liquid waste products and solvents (Tissinger *et al.*, 1991). High-density polyethylene (HDPE) appears to be one of the least affected materials and has a permeability of around 10^{-15} m s^{-1}. Hence, it is commonly used as a basal liner for landfill and waste storage facilities. A geomembrane must possess sufficient thickness and strength to avoid failure due to physical stresses (Koerner, 1993). The foundation beneath the liner must be able to support the liner and to resist pressure gradients. If the support system settles, compresses or lifts, the liner may rupture. Installation of flexible membrane liners (FMLs) involves seaming the sheets together. Mollard *et al.* (1996) pointed out that the layout of geomembrane panels should be orientated parallel to the maximum slope of the base and that the number of seams should be minimized in corners and other geometrically complex situations in order to avoid complicated welding layouts and stress concentrations. They also described the two principal techniques used for seaming geomembranes, namely, fusion welding and extrusion welding. The quality of the seams is checked by visual examination, vacuum testing or ultrasonic testing. Underliners and covers may be used to protect the geomembranes from puncture, tearing, abrasion and ultraviolet rays. For example, a bedding course of sand may be placed below the liner and another layer of sand may be placed on top. Composite liners incorporate both clay blankets and geomembranes. Other measures used to protect the liner include a minimum soil cover, elimination of folds, no seaming in cold weather, removal of sharp objects in the subgrade and prohibition of equipment on top of the liner. Table 9.2 outlines the general advantages and disadvantages of some geosynthetic materials.

Old quarries continue to be used for the disposal of domestic refuse. This practice makes economic use of the quarries, is a natural progression from quarrying activity at a site and restores a site that would otherwise represent a scar of the landscape. The exception to this is when a quarry represents a site of scientific interest and therefore needs to be preserved. However, unless the rock mass in which an old quarry occurs is more or less impermeable, then the quarry needs to be lined to prevent the migration of leachate and landfill gas. Lining the slopes of a quarry can prove difficult. High-density polyethylene (HDPE) geomembranes, 1.5–2.5 mm thick, frequently are used in such situations. Unfortunately, the faces of old quarries are by no means smooth so without protection a geomembrane could be punctured or torn.

Table 9.2 Broad advantages and disadvantages of some polymers

	Advantages	*Disadvantages*
Polyvinyl chloride	Resistant to inorganics, good tensile strength, elongation, puncture and abrasion resistance, easy to seam	Low resistance to organic chemicals, including hydrocarbons, solvents and oil, poor resistance to exposure
High-density polyethylene	Good, resistance to oils, chemicals and high temperature, available in thick sheets (20–150 mm)	Require more field seams, subject to stress cracking, punctures at low thickness, poor tear propagation
Ethylene propylene rubber	Resistant to dilute concentrations of acids, alkalis, silicates, phosphates and brine, tolerates extreme temperatures, flexible at low temperatures, excellent resistance to exposure	Not recommended for petroleum solvents or halogenated solvents, difficult to seam or repair, low seam strength
Chlorosulphonated polyethylene	Good resistance to ozone, heat, acids and alkalis, easy to seam	Poor resistance to oil, good tensile resistance if reinforced

Accordingly, quarry faces should be cleaned of loose material, after which they may need to be stabilized by rock bolts, wire mesh and shotcrete. Jones (1996) suggested that where the profile of a quarry face is particularly difficult, then perhaps an artificial face could be constructed by using gabions or reinforced earth. He further suggested that a heavy non-woven geotextile, along with a geonet to provide strength, could be used between the treated face and the geomembrane, and so offer it protection. Gallagher *et al.* (2000) referred to a number of other non-mineral liners that have been used in former quarries. These included a vertical barrier system, reinforced earth with polystyrene former and revetment. The vertical barrier system consisted of a triple row of HDPE pipes, the central row being filled with sand–bentonite–fly ash slurry that had a permeability of approximately $7.9 \times 10^{-10}\,m\,s^{-1}$, whilst the outer rows are filled with gravel. The tubes were fabricated in groups of five and groups were welded together on site. In the case of the reinforced earth with polystyrene formers, geogrid-reinforced earth is constructed in 3–3.5 m lifts, at 2.5–3.0 m widths. Polystyrene formers are inclined at a similar slope angle to that of the quarry (the system tends to be used in quarries with slide slopes of up to 70°) and form a facing on which a geomembrane is laid. The geomembrane, in turn, is overlain by a geocomposite that acts as a leachate drainage and gas venting layer. Lastly, a layer of inert material, 500 mm in thickness, is placed against the geocomposite. The system relies on the waste providing lateral support. The revetment system essentially consists of a geomembrane sandwiched between two gabions, with a protection layer each side of the geomembrane.

The double liner system is intended to prevent leakage. The primary liner of such a system is the upper geomembrane. Leachate should be properly collected above the primary liner at the bottom of the landfill, adequately treated and disposed of in accordance with accepted environmental principles. A perforated pipe collector system is located below the primary liner and is bedded within a crushed stone or sand drainage blanket. A secondary geomembrane liner occurs beneath the drainage blanket, which only need function if leakage from the primary liner is found in the underdrainage system. If leakage from the primary liner does occur, a downstream well monitoring system must be deployed to check for possible leakage from the secondary liner.

Leachate drainage systems should be designed to collect the anticipated volume of leachate likely to be produced by the landfill (Jessberger *et al.*, 1995). This will vary during the life of a site and can be estimated by a water balance calculation. One of their functions is to prevent the level of leachate rising to an extent that it can drain from a landfill and pollute nearby water courses. The system should be robust enough to withstand the load of waste imposed upon it, as well as that of equipment used in the construction of the landfill and should be able to accommodate any settlement. It should be capable of resisting chemical attack in the corrosive environment of a landfill. Ideally, a leachate drainage collection system should occur over the whole base of a landfill and, if below ground level, it should extend up the sides. The drainage blanket, consisting of granular material, should be at least 300 mm in thickness, with a minimum permeability of $1\text{–}10\,m\,s^{-1}$. The perforated collection pipes within the drainage blanket convey leachate to collection points or wells from which leachate can be removed by pumping.

If lining is considered too expensive, then drainage is a relatively inexpensive alternative. For instance, a drainage ditch can be combined with a layer of free-draining granular material overlying a low-permeability base. The granular material is graded down towards the perimeter of the landfill so that leachate can flow into the drainage ditches. The leachate can be pumped or flow away from the ditches. Because the drainage layer facilitates the flow of leachate to the drains, it avoids the development of a leachate mound. The permeability of the drainage layer usually is at least an order of magnitude higher than that of the waste (typically $10\,m\,s^{-1}$) and should have a minimum thickness of 300 mm. Synthetic drainage layers are available and vary from 4 to over 40 mm in thickness. Several layers may be used to achieve the drainage capacity. A filter medium may be used between the waste and the drainage layer to allow leachate to pass through but to prevent the migration of fines and so avoid clogging. A minimum filter thickness of 150 mm is recommended. A case of a clogged drainage layer, formed of coarse-crushed limestone, was reported by Fleming *et al.* (1999). The drainage layer had been exposed to leachate for 14 years and was clogged with mineral precipitates, fine particles and biofilm. Geotextiles often are used as filter fabrics at the base of a landfill but they may not ensure trouble-free operation of the system since they may be prone to clogging. The capability of pipes, filters, drainage and other materials that come in contact with leachate should be assessed by testing.

Any leachate that infiltrates into the ground beneath a landfill may be subject to attenuation as it flows through the surrounding soil and rocks. The slower the flow of leachate, the more effective attenuation is likely to be. The existence of an unsaturated zone of soil beneath a landfill can be particularly beneficial as the presence of liquid and gas phases delays the movement of leachate towards the water table thereby promoting gaseous exchange, oxidation, adsorption by clay minerals and biodegradation.

If a liner beneath a landfill fails, then a leachate plume may develop within the ground beneath. The size of a plume at any particular time depends upon the rate at which the leachate can spread, which is governed by the permeability of the soil or rock and the hydraulic gradient.

9.2.3 Cover systems

The principal function of a cover is to minimize infiltration of precipitation into a landfill; other functions include gas control, future site use and aesthetics. Traditionally, this has been achieved by using a clay layer with a low permeability overlain by a protection layer that can be used for planting when the site is restored (Cherrill and Phillips, 1997). A drainage layer frequently is included in the cover (Smith and Staff, 1997). A properly designed and maintained cover can prevent the entry of liquids into a landfill and hence minimize the formation of leachate. The nature of the soil used for covering waste materials is very important. At many sites, however, the quality of borrow material is less than the ideal and so some blending with imported soil may be required. The relative suitability of soils for various functions is provided in Table 9.3. Because clays generally will last longer than synthetic materials, a clay cover usually should be chosen.

Nonetheless, in cool or cold regions freeze-thaw action can cause a landfill clay cover to crack with associated reduction in its strength and increase in its permeability. Hence, a minimum thickness of cover is needed to accommodate any freeze-thaw activity. This thickness should allow for some degradation of the upper part of a cover at the same time as ensuring that the part of the cover beneath is unaffected. Unfortunately, according to Miller and Lee (1999), data relating to the depth of frost penetration in landfill covers are very limited, although they did quote depths of 320 and 450 mm of frost penetration in landfill covers in the mid-western United States. Miller and Lee maintained that frost heave is likely to occur in clay covers formed of clay–silt mixtures of low plasticity but that the amount of heave will vary, possibly reflecting the difficulty in achieving uniform placement conditions throughout a clay cover. Furthermore, with increasing time and increasing freeze-thaw cycles, a cover may suffer increasing frost penetration and heave, and associated adverse effects.

According to Daniel (1995), most cover systems in the United States consist of a number of components. However, not all the components shown in Fig. 9.4 need be present in a cover

Table 9.3 Ranking of soil types according to performance of some cover function

Soil type USCS symbol	*Impedence of water percolation*	*Hydraulic conductivity (approx.)* (cm s^{-1})	*Support vegetation*	*Impedence of gas migration*	*Resistance to water erosion*	*Frost resistance*	*Crack resistance*
GW	X	10^{-2}	X	X	I	I	I
GP	XII	10^{-1}	X	IX	I	I	I
GM	VII	5×10^{-4}	VI	VII	IV	IV	III
GC	V	10^{-4}	V	IV	III	VII	V
SW	IX	10^{-3}	IX	VIII	II	II	I
SP	XI	5×10^{-2}	IX	VII	II	II	I
SM	VIII	10^{-3}	II	VI	IV	V	II
SC	VI	2×10^{-4}	I	V	VI	VI	IV
ML	IV	10^{-5}	III	III	VII	X	VI
CL	II	3×10^{-8}	VII	II	VIII	VIII	VIII
OL			IV		VII	VIII	VII
MH	III	10^{-7}	IV		IX	IX	IX
CH	I	10^{-9}	VIII	I	X	III	X
OH			VIII				IX
PT			III				

Note
Ranking: I (best) to XIII (poorest).

and some layers may be combined, for example, the surface layer and protection layer may be combined into a single soil layer. The surface soil layer offers protection to the layers beneath and allows for vegetative growth, thereby helping to stabilize the cover and reducing erosion. Vegetation also helps to reduce infiltration by the process of transpiration. A protection layer not only protects the layers in the cover beneath from excessive wetting and drying, and from freezing that can cause cracking, but helps separate the waste from burrowing animals and plant roots. A sand–gravel drainage layer or bulky needle-punched geotextile can be used to divert surface water from the clay seal to drains along the boundary of the landfill that convey water away. Sand–gravel drainage layers must be adequately graded. Where infiltration is substantial, a composite geosynthetic may be required. The drainage layer reduces pore water pressures in the cover soils, as well as the head on the barrier layer beneath. The barrier layer usually consists of clay and provides a relatively impermeable barrier, helping prevent surface water entering a landfill. The clay must be compatible with the material in the landfill. It is compacted to a high strength to help reduce the potential for cover cracking. A gas collection layer may be incorporated into a cover and is composed of coarse to medium gravel. Its purpose is to collect gas for processing or discharge via vents. A filter of sand layer is placed between the clay seal and the underlying gas collection layer. Alternatively, a geotextile filter can be used or a thick geotextile can be used as both drain and filter. A single clay layer cover may be placed on a landfill that is reasonably stable (i.e. it has an unconfined compressive strength of 70 kPa or above). In some circumstances, a two-layer cover, consisting of a geomembrane overlain by a clay layer, may help minimize infiltration of precipitation.

Geosynthetic clay liners (GCLs) are thin layers of dry bentonite (approximately 5 mm thick) attached to one or more geosynthetic materials. In other words, the bentonite is sandwiched between an upper and a lower sheet of geotextile or the bentonite is mixed with adhesive and glued to a geomembrane. The primary purpose of the geosynthetic component(s) is to hold the bentonite together in a uniform layer. When the bentonite is wetted, it swells and provides a

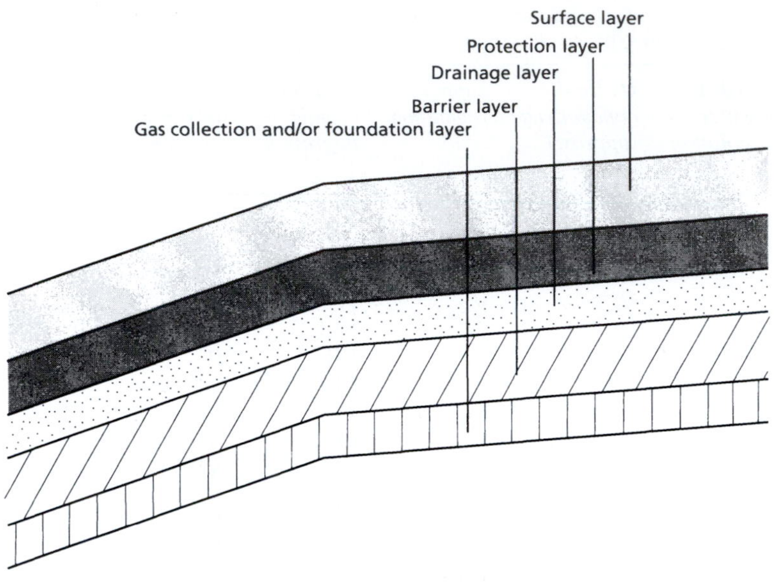

Layer	Description of layer	Typical materials	Typical thickness (m)
1	Surface layer	Topsoil; geosynthetic erosion control layer; cobbles; paving material	0.15
2	Protection layer	Soil; recycled or reused waste material; cobbles	0.3–1
3	Drainage layer	Sand or gravel; geonet or geocomposite	0.3
4	Barrier layer	Compacted clay; geomembrane; geosynthetic clay liner; waste material; asphalt	0.3–1
5	Gas collection layer and/or foundation layer	Sand or gravel; soil; geonet or geotextile; recycled or reused waste material	0.3

Figure 9.4 Basic components of a landfill cap.

cover with a low permeability. A GCL can be used in a number of ways in a cover system but should always be overlain by a layer of protective soil (Fig. 9.5). Daniel (1995) compared the performance of GCLs with compacted clay liners (Table 9.4).

The two key geotechnical factors in cover design are its stability and its resistance to cracking. Cracking of a clay cover may be brought about by desiccation or by the build-up of gas pressure beneath, if a venting system is not functioning correctly or not installed. It also may be difficult to maintain the integrity of the cover if large differential settlements occur in the landfill. A permeability of $10^{-9}\,m\,s^{-1}$ usually is specified for clay covers but perhaps never attained because of cracking. Reinforcement with high-strength geotextile on the top and bottom of a clay liner may help reduce the likelihood of cracking. Differential settlement can be reduced

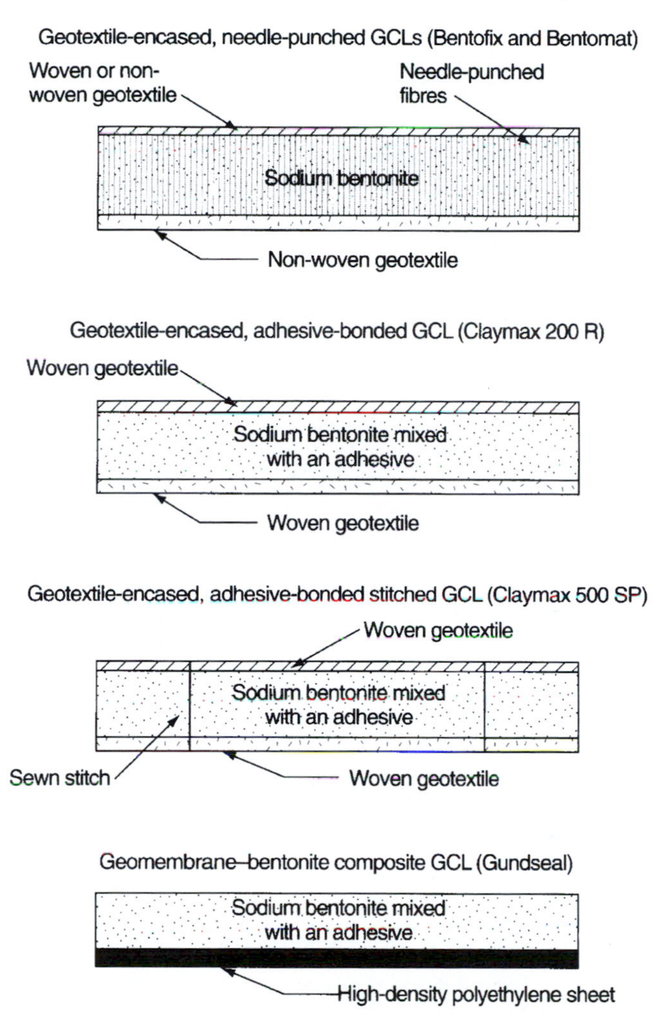

Figure 9.5 Four types of geosynthetic clay liner.

by dynamic compaction of refuse. Controlled percolation may not be critical if treatment of leachate is regarded as economically more feasible than an expensive cover.

The slope of the surface of the landfill influences infiltration. Water tends to collect on a flat surface and subsequently infiltrates, whereas water tends to run off steeper slopes. However, when the surface slope exceeds about 8 per cent, then the possibility of surface run-off eroding the cover exists. Surface water should be collected in ditches and routed from the site.

As mentioned, a landfill ideally should not be located where wastes are likely to come in contact with the groundwater, otherwise a leachate collection system has to be installed. For instance, leachate collection pipes can be installed above the underliner to remove the leachate to a sump from which it is conveyed to a treatment works (Fig. 9.6). Similarly, a landfill should not be located in an area drained by large quantities of surface water. However, streams can be diverted away from a landfill, or can be piped or culverted through a landfill.

Table 9.4 Differences between GCLs and compacted clay liners

Characteristic	*Geosynthetic clay liner*	*Compacted clay liner*
Materials	Bentonite, adhesives, geotextiles and geomembranes	Native soils or blend of soil and bentonite
Thickness	Approximately 12 mm, consumes very little landfill volume	Typically 300–600 mm, consumes more landfill volume
Hydraulic conductivity	$\leq 1-5 \times 10^{-11}\,\mathrm{m\,s^{-1}}$	$\leq 1 \times 10^{-9}\,\mathrm{m\,s^{-1}}$
Speed and ease of construction	Rapid, simple installation	Slow, complicated construction
Ease of quality assurance (QA)	Relatively simple, straightforward, common-sense procedures	Complex QA procedures requiring highly skilled and knowledgeable people
Vulnerability to damage during construction from desiccation and freeze thaw	GCLs are essentially dry; GCLs cannot desiccate during construction; not particularly vulnerable to damage from freeze thaw	Compacted clay liners are nearly saturated; can desiccate during construction; vulnerable to damage from freeze thaw
Vulnerability to damage from puncture	Thin GCL is vulnerable to puncture	Thick compacted clay liner cannot be punctured
Vulnerability to damage from differential settlement	Can withstand much greater differential settlement than compacted clay liner	Cannot withstand much differential settlement without cracking
Availability of materials	Materials easily shipped to any site	Suitable materials not available at all sites
Cost	Reasonably low, highly predictable cost that does not vary much from project to project	Highly variable – depends greatly on characteristics of locally available soils
Ease of repair	Easy to repair with patch placed over problem area	Very difficult to repair; must mobilize heavy earth-moving equipment if large area requires repair
Experience	Limited due to newness	Has been used for many years
Regulatory approval	Not explicitly allowed in most regulations – owner must gain approval on the basis of equivalence in meeting performance objectives	Compacted clay liners are usually required by regulatory agencies

Source: Daniel, 1995.

9.2.4 Barrier systems

The major function of all barrier systems is to provide isolation of wastes from the surrounding environment, so offering protection to soil and groundwater from contamination. Hence, the hydraulic and gas conductivities of any cut-off systems are of paramount importance and generally the hydraulic conductivity of an earth or earth-treated cut-off should be less than $10^{-9}\,\mathrm{m\,s^{-1}}$. However, the overall effectiveness of a cut-off system depends on its thickness as well as its permeability. Mitchell (1994) maintained that the permeability divided by the thickness, i.e. the permittivity, provided a basis for the comparison of the relative effectiveness of such cut-off systems.

Figure 9.6 Leachate sump for a landfill in Durban, South Africa.

If leachate flows from an existing landfill, then the construction of a seepage cut-off system provides a solution. Where an impermeable horizon exists beneath the site, a cut-off wall should be keyed into it. A very deep cut-off is required where no impermeable stratum exists. Steel sheet piles can be used to form a continuous cut-off wall, the integrity of the wall depending on the interlock between individual piles. Sealable joints have been developed to reduce the possibility of leakage. Geosynthetic sheet piles with sealable joints are now available. Watertight cut-off walls also can be formed by using secant piles. In this system, alternating bentonite–cement piles are constructed first, then concrete piles are constructed between them, parts of the bentonite–cement piles being bored out in order to form the concrete piles and to achieve interlock (see Chapter 2). A grout curtain may be used as a cut-off wall. A grout curtain can be formed by 2 or 3 rows of groutholes approximately 1.5–2.5 m apart. Particular concerns with grout curtains are the difficulty of ensuring sufficient overlap of the grout columns and the assessment of the overall integrity of the curtain. The latter cannot be determined until the post-construction monitoring data are available. However, cut-off walls in soils are more likely to be constructed by jet grouting than by injection (see Chapter 2).

Compacted clay barriers can be constructed in soils above the water table in trenches and may be suitable for shallow depths. However, most seepage cut-off systems consist of slurry trenches that may be 1 m thick. Once completed, the trench generally is backfilled with soil–bentonite. Usually, the addition of 4–5 per cent, by weight, of bentonite to soil is sufficient to reduce the permeability of the soil to less than $10^{-9}\,m\,s^{-1}$. Cement–bentonite cut-off walls are formed by mixing cement into bentonite slurry so that the slurry sets in place to form a barrier (Fig. 9.7). However, the permeability of such trenches is usually somewhat higher than that of soil–bentonite trenches. Plastic concrete cut-off walls are similar to cement–bentonite

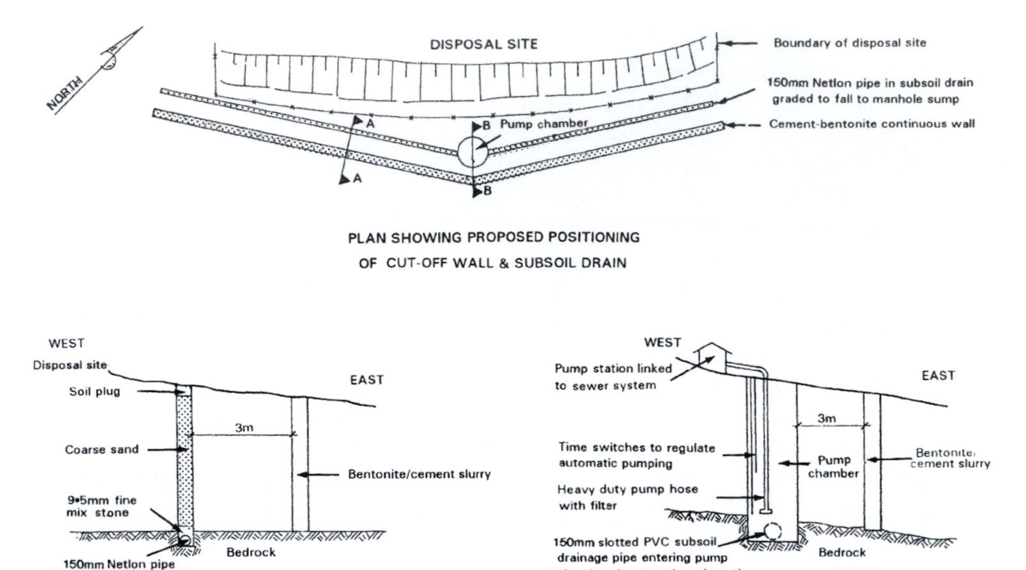

Figure 9.7 Remedial works, including a cut-off barrier, sub-drain and pump chamber, for a landfill leaking leachate.
Source: Bell *et al*., 1996.

cut-offs except that they contain aggregate. Moreover, they generally are constructed in panels rather than continuously, as is done with soil–bentonite and cement–bentonite walls. Mitchell (1994) pointed out that the greater the strength and stiffness of plastic concrete cut-off walls, then the better they are suited to situations where ground stability and movement control are important. Concrete diaphragm cut-off walls also can be used where high structural strength is required. They also are constructed in panels that are excavated under bentonite slurry. Once excavation is complete, the concrete is tremied into the panel from the bottom up (Bell and Mitchell, 1986).

A geomembrane sheeting-enclosed cut-off wall consists of a geomembrane that is fabricated to form a U-shaped envelope that fits the dimensions of a trench. Ballast is placed within the geomembrane in order to sink it into the slurry trench. After initial submergence into the trench, the envelope is filled with wet sand. A system of wells and piezometers can be installed in the sand for monitoring water quality and pore water pressure, respectively. If the sheeting is damaged, then the system will detect any infiltration of leachate. This can be abstracted by the wells in the sand (Ressi and Cavalli, 1984).

A cut-off wall also can be constructed by forming a narrow cavity in the soil by repeatedly driving or vibrating an I-beam within the soil. As the I-beam is extracted, a mixture of bentonite and cement is pumped into the cavity. Successive penetrations of the I-beam are overlapped to develop a continuous membrane. In the driven pile system, a clutch of H-piles is forced into the soil by a vibrator hammer along the line of advance, the end pile being withdrawn and redriven at the head end. Grouting, by a tube fixed to the web, is carried out during withdrawal. Walls are limited to a maximum of 20 m depth with these systems due to the danger of pile deviation during driving.

Leakage from a landfill where the water table is high and there is no impermeable layer beneath, can be treated by jet grouting. The grout pipe is inserted successively into a series of holes at predetermined centres at the base of the landfill and rotated through 360°. This allows the formation of interlocking discs of grouted soil to be formed. Alternatively, inclined drilling can be used for the injection of grout or jet grouting, provided the overall width of the site is not too great. Inclined drilling from opposite sides of a landfill forms an interlocking V-shaped barrier, which can be used along with vertical cut-off walls.

Adverse interactions of wastes on the materials of which liners, covers and cut-offs are constructed can give rise to increases in their permeability. Of particular concern are those landfills where there are concentrated free-phase organics, very acid or basic solutions and/or high concentrations of dissolved salts. High plasticity clays at high water content such as soil–bentonite mixtures are more susceptible to shrinkage and cracking than densely compacted lower water content soils when exposed to aggressive fluids. Advection and diffusion are involved in the transfer of organics and salts through clay barriers. According to Mitchell and Madsen (1987), diffusion coefficients tend to fall within the range 2×10^{-10} to $2 \times^{-9}$ $m^2 s^{-1}$ for diffusion through soils and around $3 \times 10^{-14} m^2 s^{-1}$ for diffusion through geomembranes. Nonetheless, total chemical transfer through soil or geomembrane barriers by diffusion is low when the permeability of the barrier system is low and the difference in head across the barrier also is low.

9.2.5 *Degradation of waste in landfills*

As remarked above, a major problem associated with landfills is the production of leachate. The movement of leachate into the surrounding soil, groundwater or surface water can cause pollution problems. Leachate is formed by the action of liquids, primarily water, within a landfill. The generation of leachate occurs once the absorbent characteristics of the refuse are exceeded.

The waste in a landfill site generally has a variety of origins. Many of the organic components are biodegradable. Initially, decomposition of waste is aerobic. Bacteria flourish in moist conditions, and waste contains varying amounts of liquid that may be increased by infiltration of precipitation. Once decomposition starts the oxygen in the waste rapidly becomes exhausted, and so the waste becomes anaerobic.

There are basically two processes by which anaerobic decomposition of organic waste takes place. Initially, complex organic materials are broken down into simpler organic substances, which are typified by various acids and alcohols. The nitrogen present in the original organic material tends to be converted into ammonium ions, which are readily soluble and may give rise to significant quantities of ammonia in the leachate. The reducing environment converts oxidized ions such as those in ferric salts to the ferrous state. Ferrous salts are more soluble and therefore iron is leached from a landfill. The sulphate in a landfill may be reduced biochemically to sulphides. Although this may lead to the production of small quantities of hydrogen sulphide, the sulphide tends to remain in a landfill as highly insoluble metal sulphides. In a young landfill, the dissolved salt content may exceed $10\,000 \, mg \, l^{-1}$, with relatively high concentrations of sodium, calcium, chloride, sulphate and iron, whereas as a landfill ages the concentration of inorganic materials usually decreases. Suspended particles may be present in leachate due to washout of fine material from the landfill.

The second stage of anaerobic decomposition involves the formation of methane. In other words, methanogenic bacteria use the end products from the first stage of anaerobic decomposition to produce methane and carbon dioxide. Methanogenic bacteria prefer neutral conditions so that if acid formation in the first stage is excessive, their activity can be inhibited.

The characteristics of leachate depend on the nature of the waste and the stage of waste decomposition (Table 9.5). Its composition varies due to the material of which it consists,

Table 9.5 Leachate composition

(a) Typical composition of leachates from recent and old domestic wastes at various stages of decomposition

Determinand	*Leachate from recent wastes*	*Leachate from old wastes*
pH	6.2	7.5
Chemical oxygen ($mg\,l^{-1}$)	23 800	1160
Biochemical oxygen ($mg\,l^{-1}$)	11 900	260
Total organic carbon ($mg\,l^{-1}$)	8000	465
Fatty acids ($mg\,l^{-1}$)	5688	5
Ammoniacal-N ($mg\,l^{-1}$)	790	370
Oxidized ($mg\,l^{-1}$)	3	1
o-Phosphate ($mg\,l^{-1}$)	0.73	1.4
Chloride ($mg\,l^{-1}$)	1315	2080
Sodium (Na) ($mg\,l^{-1}$)	960	1300
Magnesium (Mg) ($mg\,l^{-1}$)	252	185
Potassium (K) ($mg\,l^{-1}$)	780	590
Calcium (Ca) ($mg\,l^{-1}$)	1820	250
Manganese (Mn) ($mg\,l^{-1}$)	27	2.1
Iron (Fe) ($mg\,l^{-1}$)	540	23
Nickel (Ni) ($mg\,l^{-1}$)	0.6	0.1
Copper (Cu) ($mg\,l^{-1}$)	0.12	0.3
Zinc (Zn) ($mg\,l^{-1}$)	21.5	0.4
Lead (Pb) ($mg\,l^{-1}$)	8.4	0.14

(b) Chemical analyses of leachate from sump at a landfill in Durban, South Africa, taken over a five-year period from 1986 to 1991

	pH	*Suspended solids* ($mg\,l^{-1}$)	*Total dissolved solids* ($mg\,l^{-1}$)	*Conductivity* ($ms\,m^{-1}$)	*Oxygen adsorption* ($mg\,l^{-1}$)	*Chemical oxygen demand* ($mg\,l^{-1}$)
Max	8.9	6044	45 041	8080	6200	70 900
Min	6.6	200	900	450	145	11.2
Mean	7.7	965	17 029	1174	649	13 546
MPL	5.5–9.5		500	300	10	75
	Sodium (l^{-1})	*Potassium* ($mg\,l^{-1}$)	*Calcium* ($mg\,l^{-1}$)	*Magnesium* ($mg\,l^{-1}$)	*Sulphate* ($mg\,l^{-1}$)	*Ammonium* ($mg\,l^{-1}$)
Max	8249	2646	1236	759	2237	3530
Min	80	355	60	70	6.5	92
Mean	1393	613	146	109	837	1093
MPL	400	400	200	100	600	2

Source: Bell *et al*., 1996.

Note

MPL, South African maximum permissible limit for domestic water insignificant risk (Anon., 1993).

because of interactions between the waste, the age of the fill, the hydrogeology of the site, the climate, the season and moisture routing through the fill. The height of each cell, the overall depth of the fill, top cover and refuse compaction also are important.

As a landfill ages, because much of the readily biodegradable material has been broken down, the organic content of the leachate decreases. The acids associated with older landfills are not readily biodegradable. Consequently, the biological oxygen demand (BOD) in the leachate changes with time. The BOD rises to a peak as microbial activity increases. This peak is reached between 6 months and 2.5 years after tipping. Thereafter, the BOD concentration decreases until the landfill is stabilized after 6–15 years.

The microbial activity in a landfill generates heat. Hence, the temperature in a fill rises during biodegradation to between 24 and 45 °C, although temperatures up to 70 °C have been recorded.

9.2.6 Attenuation of leachate

Physical and chemical processes are involved in the attenuation of leachate in soils. These include precipitation, ion exchange, adsorption and filtration. In the reduction zone, where organic pollution is greatest, insoluble heavy metal sulphides and soluble iron sulphide are formed. In the area between reducing and oxidizing zones, ferric and inorganic hydroxides are precipitated. Other compounds may be precipitated, especially with ferric hydroxide. The reduction of nitrate may yield nitrite, nitrogen gas or possibly ammonia, although ammonia is usually present due to the biodegradation of nitrogen-bearing organic material. The ion exchange and adsorption properties of soil or rock are attributable primarily to the presence of clay minerals. Consequently, a soil with a high clay content has a high ion exchange capacity. The humic material in soil also has a high ion exchange capacity. Adsorption is susceptible to changes in pH value as the pH affects the surface charge on a colloid particle or molecule. Hence, at low pH values the removal rates due to adsorption can be reduced significantly. However, the removal of pollutants by ion exchange and adsorption can be reversible. For example, after high-level pollution from a landfill subsides and more dilute leachate is produced, the soil or rock may release pollutants back into the leachate. However, some ions will be irreversibly adsorbed or precipitated.

The degree of filtration brought about by a soil or rock depends on the size of the pores. In many porous soils or rocks, filtration of suspended matter occurs within a short distance. On the other hand, a rock mass containing open discontinuities may transmit leachate for several kilometres (Hagerty and Pavori, 1973). Pathogenic micro-organisms, which may be found in a landfill, do not usually travel far in the soil because of the changed environmental conditions. In fact, pathogenic bacteria normally are not present within a few tens of metres of a landfill.

9.2.7 Surface and groundwater pollution

Leachates contain many contaminants that may have a deleterious effect on surface water. If, for example, leachate enters a river, then oxygen is removed from the river by bacteria as they break down the organic compounds in the leachate. In cases of severe organic pollution, the river may be completely denuded of oxygen with disastrous effects on aquatic life. The net effect of oxygen depletion in a river is that its ecology changes and at dissolved oxygen levels below $2\,mg\,l^{-1}$ most fish cannot survive. The oxygen balance is affected by several factors. For instance, the rate of chemical and biochemical reactions increases with temperature, whereas maximum dissolved oxygen concentration decreases as temperature rises. At high river flows, bottom mud may be re-suspended and exert an extra oxygen demand, but the extra turbulence also increases aeration. At low flows, organic solids may settle out and reduce the oxygen demand of the river.

The principal inorganic pollutants that may cause problems due to leachate are ammonia, iron, heavy metals and, to a lesser extent, chloride, sulphate, phosphates and calcium. Ammonia can be present in landfill leachates up to several hundred milligrams per litre, whereas unpolluted rivers have a very low content of ammonia. Discharge of leachate high in ammonia into a river exerts an oxygen demand on the receiving water; ammonia is toxic to fish (lethal concentrations range from 2.5 to $25\,mg\,l^{-1}$), and as ammonia is a fertilizer, it may alter the ecology of the river. Leachate that contains ferrous iron is particularly objectionable in a river since ochreous deposits are formed by chemical or biochemical oxidation of the ferrous compounds to ferric compounds. The turbidity caused by the oxidation of ferrous iron can lower the amount of light and so reduce the number of flora and fauna. Heavy metals can be toxic to fish at relatively low concentrations. They also can affect the organisms on which fish feed.

Physically leachate affects river quality in terms of suspended solids, colour, turbidity and temperature. Suspended solids, colour and turbidity reduce light intensity in a river. As noted, this can affect the food chain and the lack of photosynthetic activity by plants reduces oxygen replacement in a river. The suspended solids may settle on the bed of a river in significant qualities. This can destroy plant and animal life.

If a leachate enters the phreatic zone, it mixes and moves with the groundwater (Bell *et al*., 1996). The organic carbon content in the leachate leads to an increase in the BOD in the groundwater, which may increase the potential for reproduction of pathogenic organisms. Organic matter is stabilized slowly because the oxygen demand may deoxygenate the water rapidly and usually no replacement oxygen is available. If anaerobic conditions develop, then metals such as iron and manganese, may dissolve in the water causing further problems. If the groundwater has a high buffering capacity, then the effects of mixing of acidic or alkaline leachate with the groundwater are reduced. The worst situation occurs in discontinuous rock masses where groundwater movement is dominated by fissure flow, or where groundwater movement is slow in a shallow water table so that little dilution occurs. The velocity of groundwater flow is important in that a high velocity gives rise to more dispersion in the direction of flow and relatively less laterally so that the leachate plume forms a narrow cone in the direction of groundwater flow. Low groundwater velocity leads to the formation of a wider plume.

Ideally, a groundwater monitoring programme should be established prior to tipping and continue for anything up to 20 years after completion of the site. The number, location and depth of monitoring wells depend on the particular site (see Chapter 7). Monitoring should include sampling from the wells and analysis for pH, chlorides, dissolved solids, organic carbon and the concentration of any particular hazardous waste that has been tipped in the landfill.

The most serious effect that a leachate may have on groundwater is mineralization. Much of the organic carbon and organic nitrogen is biodegraded as it moves through the soil. Heavy metals may be attenuated due to ion exchange onto clay minerals. Mineralization is brought about by inorganic ions, such as chloride. When the concentrations are high, they may only be reduced by dilution. Continuing input of inorganic ions into potable groundwater eventually means that groundwater becomes undrinkable.

9.2.8 Landfill and gas formation

The biochemical decomposition of domestic and other putrescible refuse in a landfill produces gas consisting primarily of methane (CH_4), with smaller amounts of carbon dioxide and volatile organic acids. In the initial weeks or months after placement, the landfill is aerobic and gas production is mainly CO_2, but also contains O_2 and N_2. As the landfill becomes anaerobic, the evolution of O_2 declines to almost zero and N_2 to less than 1 per cent. The principal gases produced during the anaerobic stage are CO_2 and CH_4, with CH_4 production increasing slowly as methanogenic bacteria establish themselves. The factors that influence the rate at which gas is produced include the character of the waste, the moisture content, temperature and pH of the landfill. Concentration of salts, such as sulphate and nitrate, also may be important.

If the refuse is pulverized, then microbial activity in the landfill is higher, which may give rise to a higher production rate of gases. However, the period of time over which gas is produced may be reduced. Compaction or baling of refuse may decrease the rate of water infiltration into a landfill, retarding bacterial degradation of the waste, with gas being produced at a lower rate over a longer period. If toxic chemicals are present in a landfill, then bacterial activity, and especially methanogenesis, may be inhibited. A moisture content of 40 per cent or higher is desirable for optimized gas production. Generally, the rate of gas production increases with temperature. The pH value of a landfill should be around 7.0 for optimum production of gas, methanogenesis tending to cease below a pH of 6.2. The amount of gas produced by domestic refuse varies appreciably and a site investigation is required to determine the amount if such information is required. Nonetheless, it has been suggested that between 2.2 and 250 litres per kilogram dry weight may be produced (Oweis and Khera, 1990).

Methane production can constitute a dangerous hazard because methane is combustible, and in certain concentrations explosive (5–15 per cent by volume in air), as well as asphyxiating. Appropriate safety precautions must be taken during site operation. In many instances, landfill gas is able to disperse safely to the atmosphere from the surface of a landfill. However, when a landfill is completely covered with a soil capping of low permeability in order to minimize leachate generation, the potential for gas to migrate along unknown pathways increases and there are cases of hazard arising from methane migration. Furthermore, there are unfortunately numerous cases on record of explosions occurring in buildings due to the ignition of accumulated methane derived from landfills near to or on which the houses were built (Williams and Aitkenhead, 1991). The source of the gas should be identified so that remedial action can be taken. Identification of the source can involve a drilling programme and analysis of the gas recovered. For instance, in a case referred to by Raybould and Anderson (1987), distinction had to be made between household gas and gas from sewers, old coal mines and a landfill before remedial action could be taken to eliminate the gas hazard affecting several houses. Generally, the connection between a source of methane and the location where it is detected can be verified by detecting a component of the gas that is specific to the source or by establishing the existence of a migration pathway from the source to the location where the gas is detected. An analysis of the gas obviously helps identify the source. For example, methane from landfill gas contains a larger proportion of carbon dioxide (16–57 per cent) than does coal gas (up to 6 per cent). Analysis also may involve trace components or isotopic characterization using stable isotope ratios ($^{13}C/^{12}C; ^{2}H/^{1}H$) or the radio-isotope ^{14}C. This again allows distinction between landfill gas and coal gas as in the latter all the radiocarbon has decayed.

The movement of methane gas may be upwards and out through a landfill cover. The slow escape of gas from the top of a landfill can present a problem, particularly if it accumulates in pockets on the site. Artificial channels, such as drains, soakaways, culverts or shafts can act as pathways for gas movement. Hence, proper closure of a landfill site can require gas management to control methane gas by passive venting, power-operated venting or the use of an impermeable barrier. In addition, the identification of possible migration paths in the assessment of a landfill site is highly important, especially where old mine workings may be present or where there is residential property nearby. For instance, Raybould and Anderson (1987) described grouting old mine workings that had acted as a conduit for the migration of methane from a landfill to residential properties. Although methane is not toxic to plant life, the generation of significant quantities can displace oxygen from the root zone and so suffocate plant roots. Large concentrations of carbon dioxide or hydrogen sulphide can produce the same result.

Monitoring of landfill gas is an important aspect of safety. Instruments usually monitor methane as this is the most important component of landfill gas. Gas may be sampled from monitoring wells in the fill or from areas into which the gas has migrated. Leachate-monitoring wells can act as gas collectors. Due caution must be taken when sampling.

Measures to prevent or control the migration of the gas include impermeable barriers (clay, bentonite, geomembranes or cement) and gas venting. An impermeable barrier should extend to the base of the fill or the water table, whichever is the higher. A problem, however, is ensuring that the integrity of the barrier is maintained during its installation and subsequent operation.

Venting either wastes the gas into the atmosphere or facilitates its collection for utilization. Passive venting involves venting gases to locations where it can be released to the atmosphere or burned. The vents are placed at locations where gas concentrations and/or pressures are high (Fig. 9.8). This usually occurs towards the lower sections of a landfill. A vent that penetrates to the leachate is most effective. Atmospheric vents may be spaced at 30–45 m intervals. Care must be taken to ensure that vents do not become blocked or clogged. In addition, the vent must remain relatively dry in order to maintain its permeability to gases. Therefore, either natural or pumped drainage must be present to allow the escape of water percolating into the vent. Such vents are not effective in controlling the migration of gas laterally. Perimeter wells are used where there are no geological constraints preventing lateral migration of gas. Alternatively, gas may be intercepted by a trench filled with coarse aggregate. Wilson (2002) described a new

Figure 9.8 Gas vent at a landfill site in Kansas City, United States.

system of passive-venting barrier that uses geocomposite vent nodes, which are driven into the ground and connected to a collection/dilution duct for safe venting to the atmosphere. The spacing of the vent nodes has to be assessed to ensure that the gas is vented to the atmosphere in acceptable concentrations. Where atmospheric venting is insufficient to control the discharge of gas, an active or forced venting system is used. Active ventilation involves connecting a vacuum pump to the discharge end of the vent. The radius of influence depends primarily on the flow rate and the depth of the well screen (e.g. radii of influence of 6 m at 0.85 $m^3\,m^{-1}$ flow rate to 22.5 m at a flow rate of 1.4 $m^3\,m^{-1}$).

If an impermeable cover is placed over an entire landfill, methane eventually will move laterally into any permeable soil and escape. Hence, in such situations methane gas should be properly vented. Each cell cover should be shaped so that a vent is located at the uppermost slope of the bottom of the cover.

A sand–gravel drainage layer located above the liner of a landfill can be used to collect carbon dioxide. Alternatively, a geocomposite of adequate transmissivity, along with a perforated pipe collection system, can be used for gathering the gas.

9.3 Hazardous wastes

A hazardous waste can be regarded as any waste or combination of wastes of inorganic or organic origin that, because of its quantity, concentration, physical, chemical, toxicological or persistency properties, may give rise to acute or chronic impacts on human health and/or the environment, when improperly treated, stored, transported or disposed of. Such waste can be generated from a wide range of commercial, industrial, agricultural and domestic materials and can take the form of liquid, sludge or solid. The characteristics of the waste not only influence its degree of hazard but also are of great importance in the choice of a safe and environmentally acceptable method of disposal. Hazardous wastes may involve one or more risks such as explosion, fire, infection, chemical instability or corrosion, or acute toxicity (in particular, is the waste carcinogenic, mutagenic or teratogenic?). An assessment of the risk posed to health and/or the environment by hazardous waste must take into consideration its biodegradability, persistency, bioaccumulation, concentration, volume production, dispersion and potential for leakage into the environment. Hazardous wastes therefore require special treatment and cannot be released into the environment, added to sewage or be stored in a situation that is either open to the air or from which aqueous leachate could emanate.

Some of the primary criteria for identification of hazardous wastes include the type of hazard involved (flammability, corrosivity, toxicity and reactivity), the origin of the products, including industrial origins (e.g. medicines, pesticides, solvents, electro-plating, oil refining), and the presence of specific substances or groups of substances (e.g. dioxin, lead compounds, PCBs). These criteria and others are used alone or in combination but in different ways in different countries. The compositional characteristics of the waste may or may not be quantified and where levels of substance concentration are set, they again vary from country to country.

A number of attempts have been made to produce a qualitative system of comparison of wastes in terms of their toxicity or the relative hazards they present. For example, Smith *et al.* (1980) defined a geotoxicity hazard index (GHI) for buried underground toxic materials as:

$$\mathrm{GHI} = \mathrm{TI} \times P \times A \times C \tag{9.1}$$

where TI is the toxicity index and is equivalent to the volume of water required to dilute the substance to acceptable drinking water levels, P the persistence factor that is a function of the decay half-life of the material, *C* the build-up correction factor that takes care of any more toxic daughter products that may be produced during decay and *A* the availability factor. The latter represents an attempt to relate the availability of the substance to man to the amount naturally available in a reference volume of soil. It takes account of the leach rate of a substance, its chemical behaviour and interaction with the soil. It is at this stage where the hazard index concept becomes impractical, especially for substances with no simple natural analogue (e.g. many organic contaminants found in groundwater).

Assessment of the suitability of a site for hazardous waste disposal is a complex matter that involves the use of models to predict the chemical behaviour of the waste in the ground and the potential for mobilization and migration in groundwater. The form and rates of release of the waste to the environment, together with the time and place at which releases occur, can be produced with reasonable degrees of confidence. To translate these results into risk assessment requires a parallel prediction of the consequences of a release of waste. Risk associated with the disposal of wastes involves the probability of an event or process that leads to release occurring within a given time period, multiplied by the consequences of that release (or alternatively, the probability that an individual or group will be exposed to a pollutant, multiplied by the probability that this exposure will give rise to a serious health effect). Given adequate epidemiological data on the health effects of toxic substances, the risk can be calculated.

The classification of hazardous wastes and hazard rating facilitates the choice of a cost-effective method of waste management. Obviously, there is no truly non-hazardous waste, in that some risk always exists when dealing with waste. The quantity and nature of waste involved, and the susceptibility of man or other living things to a certain waste can be used to determine its degree of hazard. The first step in classifying waste is to identify the hazardous substances present. It then can be placed into a particular group of a classification system, perhaps according to the most dangerous constituent recognized. Hazard ratings can be categorized as extreme, high, moderate or low. On the one hand, waste that contains significant concentrations of extremely hazardous material, including certain carcinogens, teratogens and infectious substances, is of primary concern. On the other, the low category of hazardous waste contains potentially hazardous constituents but in concentrations that represent only a limited threat to health or the environment. If the hazard rating is less than the low category, then the waste can be regarded as non-hazardous and disposed of as general waste.

The minimum requirements for the treatment and disposal of hazardous waste involve ensuring that certain classes of waste are not disposed of without pretreatment. The objective of treating a waste is to reduce or destroy the toxicity of the harmful components in order to minimize the impact on the environment. In addition, waste treatment can be used to recover materials during waste-minimization programmes. The method of treatment chosen is influenced by the physical and chemical characteristics of the waste, that is, is it gaseous, liquid, in solution, sludge or solid; is it inorganic or organic; and what is the concentration of hazardous and non-hazardous components? Physical treatment methods are used to remove, separate and concentrate hazardous and toxic materials. Chemical treatment is used in the application of physical treatment methods and to lower the toxicity of a hazardous waste by changing its chemical nature. This may produce essentially non-hazardous materials. In biological treatments, microbial activity is used to reduce or destroy the toxicity of a waste. For example, microbial action can be used to degrade organic substances or reduce inorganic compounds. The principal objective of such processes as immobilization, solidification or encapsulation is to convert hazardous waste into an inert mass with very low leachability. Macro-encapsulation involves the containment of waste in drums or other approved containers within a reinforced concrete cell that is stored within a landfill. Incineration can be regarded as a means both of treatment and of disposal.

Safe disposal of hazardous waste is the ultimate objective of waste management, disposal being in a landfill, by burial, by incineration or marine disposal. When landfill is chosen as the disposal option, the capacity of the site to accept certain substances without exceeding a specified level of risk has to be considered. The capacity of a site to accept waste is influenced by the geological and hydrogeological conditions, the degree of hazard presented by the waste, the leachability of the waste and the design of the landfill (Fig. 9.9). Certain hazardous wastes may be prohibited from disposal in a landfill, such as explosive wastes or flammable gases. Obviously, medium- to high-level radioactive waste cannot be disposed of in a landfill.

Protection of groundwater from the disposal of toxic waste in landfills can be brought about by containment. A number of containment systems have been developed that isolate wastes and include compacted clay barriers, slurry trench cut-off walls, geomembrane walls, sheet piling, grout curtains and hydraulic barriers (Mitchell, 1986). A compacted clay barrier consists of a trench that has been backfilled with clay compacted to give a low hydraulic conductivity. Slurry trench cut-off walls are narrow trenches filled with soil–bentonite mixtures that, where possible, extend downwards into an impermeable layer. Again they have a low hydraulic conductivity.

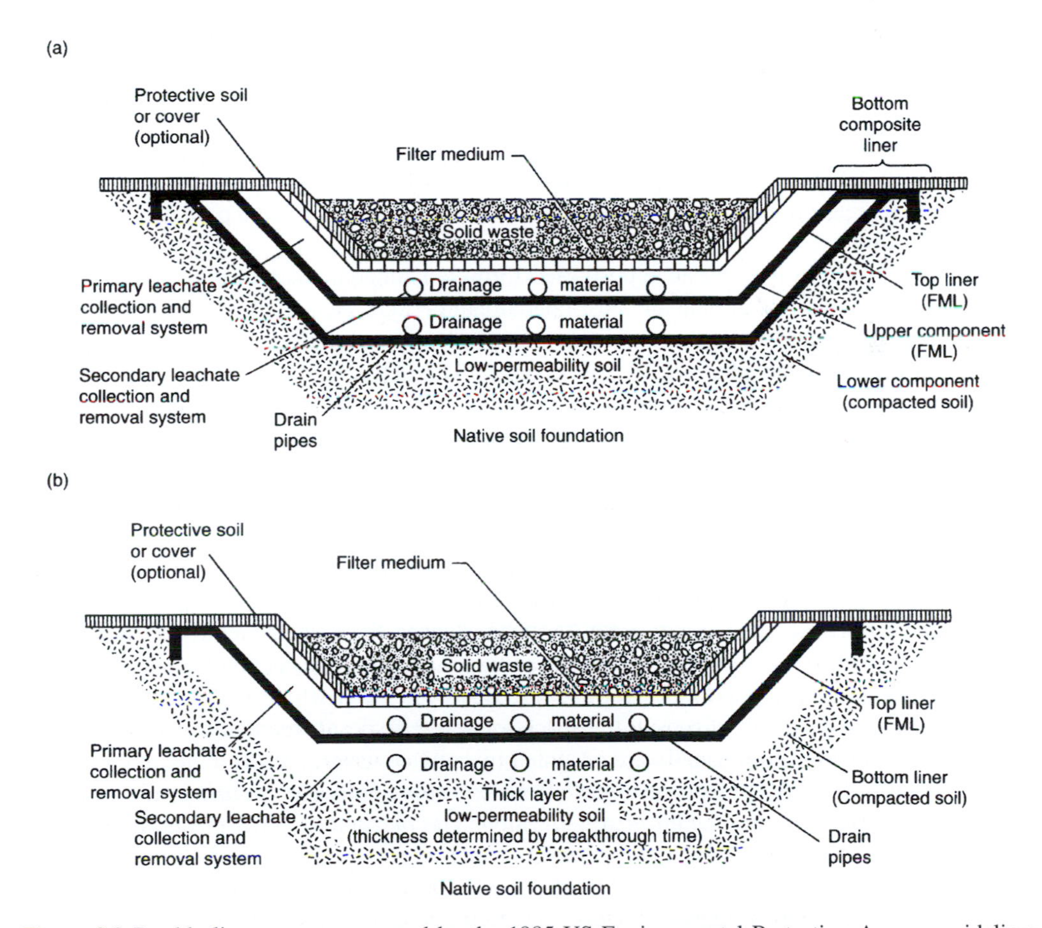

Figure 9.9 Double liner system proposed by the 1985 US Environmental Protection Agency guidelines (FML, flexible membrane liner). The leachate collection layer also is considered to function as the geomembrane protection layer. (a) FML/composite double layer. (b) FML/compacted soil double liner. *Source*: Mitchell, 1986. Reproduced by kind permission of The Geological Society.

Diaphragm walls are an expensive form of containment. Their use therefore is restricted to situations where high structural stability is required. Grout curtains may be used in certain situations. Extraction wells can be used to form hydraulic barriers and are located so that the contaminant plume flows towards them.

Low-level hazardous waste can be co-disposed, that is, mixed with very much larger quantities of non-hazardous domestic waste and buried in disused quarries in rock masses of low permeability, clay pits and other acceptable holes in the ground (Chapman and Williams, 1987). The objectives of co-disposal are to absorb, dilute and neutralize liquids, and to provide a source of biodegradable materials to encourage microbial activity. In other words, co-disposal makes use of the attenuation processes in a landfill to minimize the impact of hazardous waste on the environment. The technique helps bring about changes in the waste so that it is not retained in its original form with the potential to give rise to pollution.

Christiansson and Jerias (1996) discussed the possibilities of storing hazardous waste in crystalline rock above the water table. The waste materials concerned included batteries containing mercury, coal ash (pressurized fluidized bed combustion product, PFBC), flue gas cleaning by-product (FGCB) and metal hydroxide sludge (MeOH). Heavy metals such as As, Cd, Co, Cr, Cu, Hg, Mo, Ni, Pb and Zn were present in these wastes, some in large amounts, especially the metal hydroxide sludge. Christiansson and Jerias suggested that these materials could be stored in either rock caverns or silos located at shallow depth. In each case, a drainage tunnel system would be constructed around and beneath the repository from which drainage holes would be drilled in order to ensure that the water table was kept permanently below the base of the repository. Any fractures in the rock at the ground surface would be sealed and the surface smoothed to increase run-off, it being calculated that sealing would reduce percolation to about 1–2 mm annually.

Monitoring of hazardous waste repositories forms an inherent part of the safety requirements governing their operational and post-operational periods. Hence, the repository operators are required to conduct monitoring programmes to detect any failure in the waste containment systems so that remedial action can be taken. The nature and duration of a monitoring programme needed to ensure continuing safe isolation of waste depends upon a number of parameters including the physical condition, composition and nature of the host formation. In addition, the regulatory authorities impose conditions and limitations on the disposal of hazardous wastes, in terms of the type and quantity of waste, and the level of activity for each particular repository.

Disposal of liquid hazardous waste also has been undertaken by injection into deep wells located in rock below fresh water aquifers, thereby ensuring that contamination or pollution of underground water supplies does not occur. In such instances, the waste generally is injected into a permeable bed of rock several hundreds or even thousands of metres below the surface, which is confined by relatively impervious formations. However, even where geological conditions are favourable for deep well disposal, the space for waste disposal frequently is restricted and the potential injection zones usually are occupied by connate water. Accordingly, any potential formation into which waste can be injected must possess sufficient porosity, permeability, volume and confinement to guarantee safe injection. The piezometric pressure in the injection zone influences the rate at which the reservoir can accept liquid waste. A further point to consider is that induced seismic activity has been associated with the disposal of fluids in deep wells (Evans, 1966). Two important factors relating to the cost of construction of a well are its depth and the ease with which it can be drilled.

Monitoring is especially important in deep well disposal that involves toxic or hazardous materials. A system of observation wells sunk into the subsurface reservoir concerned, in the vicinity of the disposal well, allows the movement of liquid waste to be monitored. In addition, shallow wells sunk into freshwater aquifers permit monitoring of water quality so that any upward migration of the waste can be noted readily. Effective monitoring requires that the geological and hydrogeological conditions are accurately evaluated and mapped before the disposal programme is started.

9.4 Radioactive waste

Radioactive waste may be of low, intermediate or high level. Low-level waste contains small amounts of radioactivity and so does not present a significant environmental hazard if properly dealt with. Intermediate-level waste comes from nuclear plant operations and consists of items such as filters used to purify reactor water, discarded tools and replaced parts. This waste has to be stored for approximately 100 years. When reactors are closed down the decommissioning waste has to be disposed of safely.

Chapman and Williams (1987) maintained that low level radioactive waste can be disposed of safely by burying in carefully controlled and monitored sites where the hydrogeological and geological conditions severely limit the migration of radioactive material. According to Rogers (1994), the most recent trend in low-level radioactive waste disposal in the United States is to provide engineered (concrete) barriers in the disposal facility that prevent the short-term release of contamination. Such disposal facilities include below-ground vaults, above-ground vaults, modular concrete canisters, earth-mounded concrete bunkers, augered holes and mined cavities (Fig. 9.10). A below-ground vault involves placing waste containers in a structure consisting of a reinforced concrete floor, walls and roof, and covering with soil. In this way, infiltration of water into the waste is restricted and gamma radiation at the surface reduced. An above-ground vault is a structure consisting of a reinforced concrete floor, walls and roof that remains uncovered after closure. The spaces between the waste containers placed in the structure are filled with earth. Thc modular concrete canisters concept involves placing waste containers in reinforced concrete canisters in trenches in the ground, with the spaces between the containers being filled with grout. When filled, the trench is covered by a layer of soil. The earth-mounded concrete bunker includes a below-ground disposal of waste containers in a concrete bunker and an above-ground disposal of containers with a soil cover, which usually is less than 5 m in thickness. The bunker comprises a number of cells with reinforced concrete floors, walls and roofs. Void spaces between the containers within a cell are filled with concrete. All waste placed above ground is solidified and placed in steel drums or compacted and grouted into concrete canisters. These are stacked on a concrete floor that separates this part of the structure from the concrete bunker and the voids between the containers are filled with soil. More active low-level radioactive waste has been disposed of in augered holes that are either lined or unlined. The linings may be of fibreglass, concrete or steel. A cover (about 3 m thick) is placed over the augered hole. A mined cavity can be either a disused mine or a purpose-built mine and can be used for the disposal of all types of low-level radioactive waste.

High-level radioactive waste unfortunately cannot be made non-radioactive and so disposal has to take account of the continuing emission of radiation. Furthermore, as radioactive decay occurs the resulting daughter product is chemically different from the parent product. The daughter product also may be radioactive but its decay mechanism may differ from that of the parent. This is of particular importance in the storage and disposal of radioactive material (Krauskopf, 1988). The half-life of a radioactive substance determines the time that a hazardous waste must be stored for its activity to be reduced by half. However, as noted previously, if a radioactive material decays to form another unstable isotope and if the half-life of the

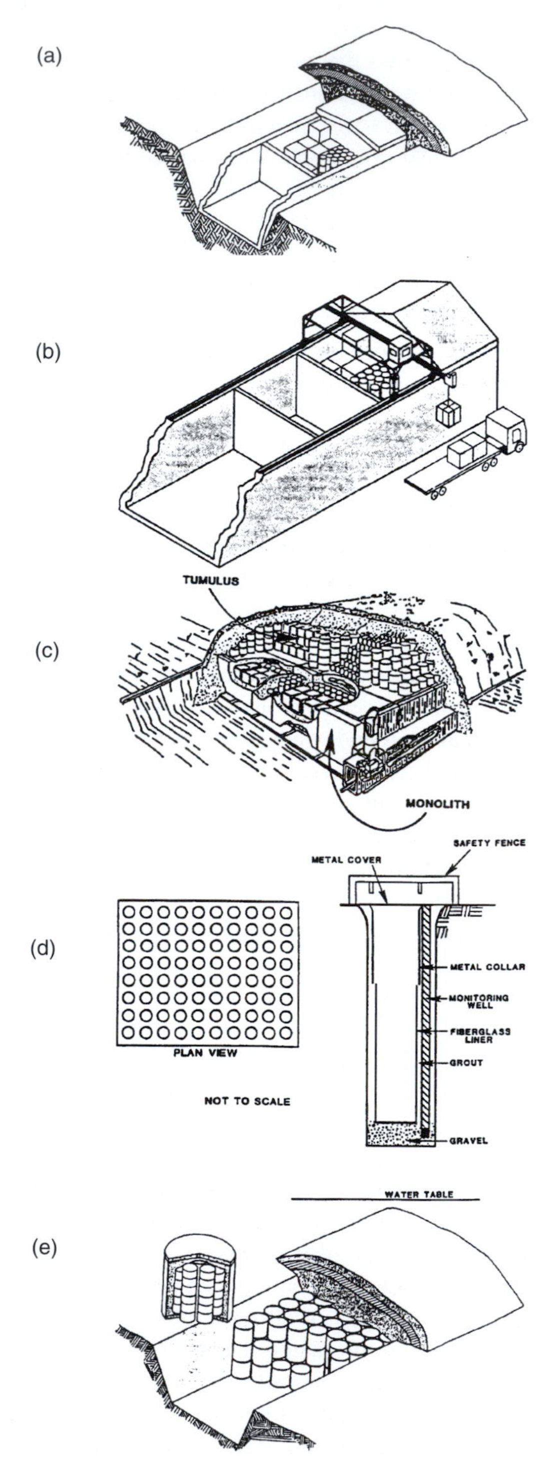

Figure 9.10 Disposal of low-level radioactive waste: (a) shallow land burial; (b) above-ground vault burial; (c) earth mounded bunker disposal; (d) augered hole disposal; (e) modular concrete, canister disposal, showing cutaway of a canister.
Source: Rogers, 1994. Reproduced by kind permission of University of Alberta.

daughter product is long, as compared with that of the parent, then although the activity of the parent will decline with time, that of the daughter will increase since it is decaying more slowly. Consequently, the storage time of radioactive wastes must consider the half-lives of the products that result from decay. The most long-lived radioactive elements need to be stored safely for hundreds of thousands of years.

In general, two types of high-level radioactive waste are being produced, namely, spent fuel rods from nuclear reactors and reprocessed waste. At present, both kinds of waste are isolated from the environment in container systems. However, because much radioactive waste remains hazardous for hundreds or thousands of years it should be disposed of far from the surface environment and where it requires no monitoring. Baillieul (1987) reviewed various suggestions that have been advanced to meet this end, such as disposal in ice sheets, disposal in the ocean depths, disposal on remote islands and even disposal in space.

9.4.1 High-level radioactive waste disposal

Ice sheets have been suggested as a repository for isolating high-level radioactive waste. The presumed advantages are disposal in a cold remote area and in a material that would entomb the wastes for many thousands of years. The high cost, adverse climate and uncertainties of ice dynamics are factors that do not favour such a means of disposal.

Disposal on the deep sea bed involves emplacement in sedimentary deposits at the bottom of the sea (i.e. thousands of metres beneath the surface). Such deposits have sorptive capacity for many radionuclides that might leach from breached waste packages. In addition, if any radionuclides escaped they would be diluted by dispersal. Currently, however, disposal of radioactive waste beneath the sea floor is prohibited by international convention.

Disposal of radioactive waste on a remote island involves the placement of wastes within deep stable geological formations. It also relies on the unique hydrological system associated with island geology. The remoteness, of course, is an advantage in terms of isolation.

The rock melt concept involves the direct placement of liquids or slurries of high-level wastes or dissolved spent fuel in underground cavities. After evaporation of the slurry water, the heat from radioactive decay would melt the surrounding rock. In about 1000 years, the waste rock mixture would resolidify, trapping the radioactive material in a relatively insoluble matrix deep underground. The rock melt concept, however, is suitable only for certain types of wastes. Moreover, since solidification takes about 1000 years, the waste is most mobile during the period of greatest fission product hazard. Gibb (1999) discussed the use of deep drillholes for the disposal of vitrified high-level radioactive waste in granitic host rocks. The method would utilize the heat from the waste to partially melt the host rock, which after a period of time would recrystallize. Gibb maintained that recent advances in the knowledge of continental crystalline rocks and fluids at depths of several kilometres suggested that deep disposal might offer a safer and environmentally more acceptable solution to the high-level radioactive waste problem.

Placement of encapsulated nuclear waste in drillholes as deep as 1000 m in stable rock formations cannot dispose of high volumes of waste. Similarly, injection of liquid waste into porous or fractured strata, at depths from 1000 to 5000 m, can only accommodate limited quantities of waste. Such waste is isolated by relatively impermeable overlying strata and relies on the dispersal and diffusion of the liquid waste through the host rock. The limits of diffusion need to be well defined. Alternatively, thick beds of shale, at depths between 300 and 500 m, can be fractured by high pressure injection and then waste, mixed with cement or clay grout, can be injected into the fractured shale and allowed to solidify in place. The fractures need to be produced parallel to the bedding planes. This requirement limits the depth of injection. The concept is applicable only to reprocessed wastes or to spent fuel that has been processed in liquid or slurry form. Again this type of disposal can only dispose of limited amounts of waste.

Morfeldt (1989) described a tunnel system in Sweden where spent nuclear fuel, contained in copper canisters, has been placed in drillholes in the floor of a tunnel and the holes backfilled with highly compacted bentonite. The copper cannisters will provide sufficient protection against radiation over a long period of time. The surrounding bentonite swells and seals any fissures in the drillholes. Also, the tunnels will be backfilled with a mixture of sand and bentonite. Such a design means that the rock mass in the floor of the tunnel can be investigated to locate the best position of the holes for the canisters. If the tunnel should cross an unexpected fracture zone in the rock mass, it would be avoided by the holes for the cannisters.

The most favoured method, since it probably will give rise to the least problems subsequently, is disposal in chambers excavated deep within the earth's crust in geologically acceptable conditions (Morfeldt, 1989). Deep disposal of high-level radioactive waste involves the multiple barrier concept, which is based upon the principle that uncertainties in performance can be minimized by conservation in design (Horseman and Volckaert, 1996). In other words, a number of barriers, both natural and man-made, exist between the waste and the surface environment. These can include encapsulation of the waste, waste containers, engineered barriers such as backfills and the geological host rocks that are of low permeability.

A deep disposal repository will consist of a large underground system located at least 200 m, and preferably 300 m or more, beneath the ground surface, in which there is a complex of horizontally connected tunnels for transportation, ventilation and emplacement of high-level radioactive waste (Eriksson, 1989). It also will require a series of inclined tunnels and vertical shafts to connect the repository with the surface. Modern blasting techniques can produce maintenance-free storage chambers and tunnels in rock masses, as can tunnel-boring machines. Ideally, the waste should be so well entombed that none will reappear at the surface or if it does so, in amounts minute enough to be acceptable (Brotzen, 1995).

If not already in solid form (e.g. spent fuel rods), then waste can be treated to convert it into solid, ideally non-leachable material. A variety of different solidification process materials have been proposed including cement, concrete, plaster, glass and polymers. Currently, borosilicate glass is the most popular agent as it can incorporate wastes of varying composition and has a low solubility in water. The glass would be placed within a metal canister and surrounded by cement or clay. The purpose of the container system is to provide a shield against radiation and so it must be corrosion resistant. Nonetheless, any container system has to remain intact for a very long time (as mentioned, storage of high-level radioactive waste will have to be for thousands of years) and behaviour over such a period of time cannot be predicted.

These metal canisters would be stored in deep underground caverns excavated in relatively impermeable rock types in geologically stable areas, that is, areas that do not experience volcanic activity, in which there is a minimum risk of seismic disturbance and that are not likely to undergo significant erosion (Fig. 9.11). Although earthquakes may represent a potential risk factor for rock chambers and tunnels, experience in mines in earthquake-prone regions has shown that vibrations in rock decrease with depth, for example, even at a depth of 30 m vibrations are only one-seventh those measured at the surface. Deep structural basins are considered as possible locations (Baillieul, 1987; Eriksson, 1989).

A monitored interim retrievable storage facility can be integrated into a waste management system. Such a monitored retrievable storage facility would act as a central receiving point for spent fuel and other waste forms. The waste material would be packaged into standardized disposal canisters at the facility and temporarily stored pending transfer to the main repository. It could store waste for 30–40 years before final disposal.

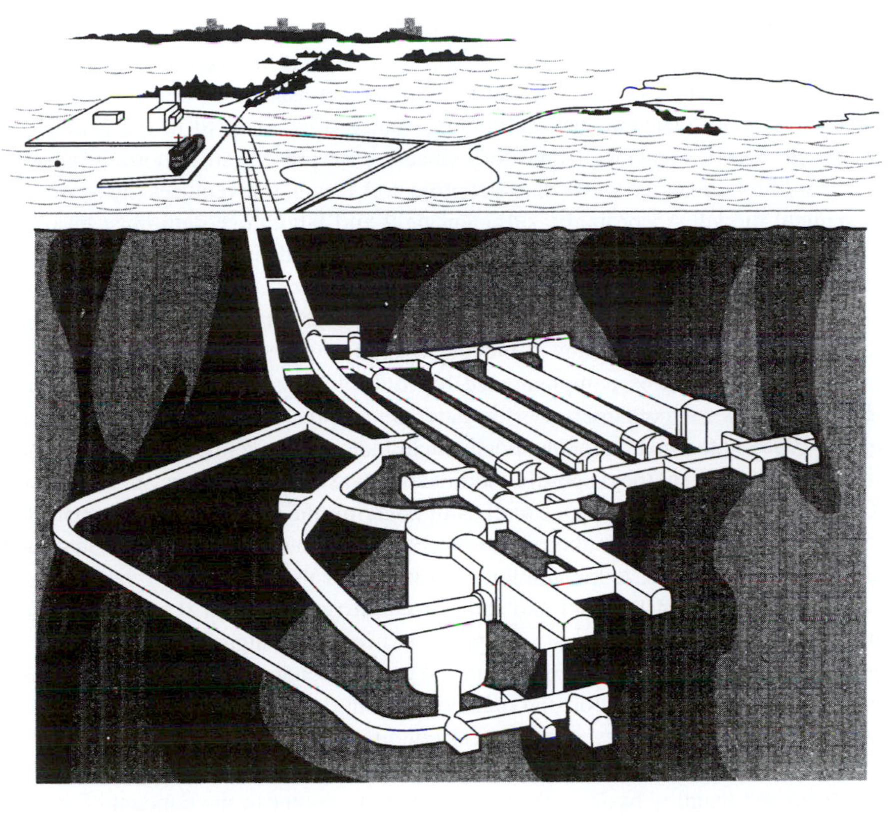

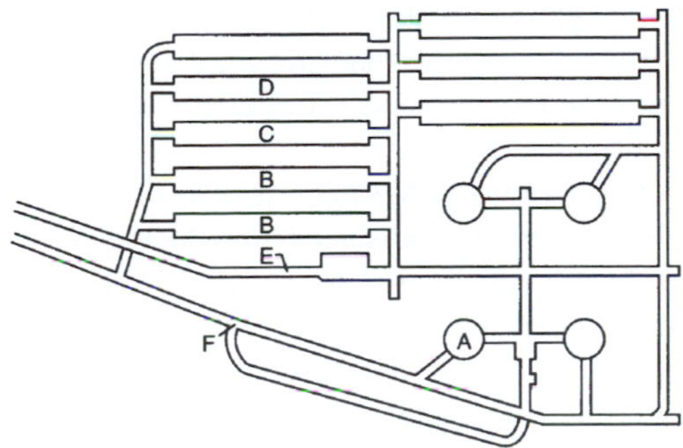

Figure 9.11 Possible final repository for reactor waste.
Source: Morfeldt, 1989. Reproduced by kind permission of Springer-Verlag.

9.4.2 Geological conditions and disposal

The necessary safety of a permanent repository for radioactive wastes has to be demonstrated by a site analysis that takes account of the site geology, the type of waste and their interrelationship (Langer, 1989). The site analysis must assess the thermomechanical load capacity of the host rock so that disposal strategies can be determined. It must determine the safe dimensions of an underground chamber and evaluate the barrier systems to be used. According to the multibarrier concept, the geological setting for a waste repository must be able to make an appreciable contribution to the isolation of the waste over a long period of time. Hence, the geological and tectonic stability (e.g. mass movement or earthquakes), the load-bearing capacity (e.g. settlement or cavern stability), geochemical and hydrogeological development (e.g. groundwater movement and potential for dissolution of rock) are important aspects of safety. Subsequently, Langer and Heusermann (2001) emphasized the importance of an analysis of the geotechanical stability and integrity of a permanent repository since they represent an important part of any safety assessment of a radioactive waste disposal project. As such, this requires the development of a geomechanical model. Accordingly, sufficient field and laboratory data must be obtained to develop the model, which should simulate the geological and hydrogeological conditions, including the stress conditions and the behaviour of the rock masses concerned. The best disposal conditions would be in a geological environment with little or no groundwater circulation as groundwater is the most probable means of moving waste from the repository to the biosphere. The geological setting would be complemented by multiple engineered barriers such as the form of the waste, the waste containers, buffer materials and the backfill.

Savage *et al.* (1999) noted that although geochemical factors have tended to play a secondary role in the selection of sites for the disposal of radioactive waste, these factors are of importance in terms of engineered barrier systems, as well as in delaying the transport of radionuclides. In particular, they indicated that the rock-groundwater system should be capable of maintaining suitable pH (between 8 and 10), redox conditions (Eh greater than $-200\,\text{mV}$), partial pressure of carbon dioxide (less than 10^{-4} bars) and ionic strength (greater than 0.5 M). They further indicated the colloid (<0.01 ppm), organic (<5 ppm) and inorganic ligand ($Cl^{-1} > HCO_3^{-1}$, low abundances of phosphate) content of groundwater. These geochemical conditions should help sorb, ion exchange, filter, precipitate or trap radionuclides in dead-end pore space. Savage *et al.* concluded that the most satisfactory geochemical conditions for high-level radioactive waste disposal are found in basic and intermediate igneous rock masses in tectonically stable low-lying regions away from sedimentary basins, which is somewhat counter to the concept of storage in thick deposits of rock salt or shales. However, although they realized that geochemical criteria cannot be used on their own in the selection of radioactive waste repositories, they maintained that suitable geochemical conditions may help relax the requirements for optimum performance indicated by geological and hydrogeological conditions.

Because of the length of time that high-level radioactive waste has to be stored, Talbot (1999) argued that future ice ages pose a threat to cavities located at depths between 400 and 800 m in regions likely to be so affected. He maintained that the threat comes not just from glacial erosion but also from over-pressurized meltwater and gases in the earth's crust, a crust that would have been flexed by major ice sheets. For instance, groundwater in subhorizontal fractures dilated by glacial unloading may reach pressures capable of lifting huge blocks of bedrock thereby increasing both their permeability and susceptibility to erosion. Talbot also mentioned other factors that were associated with glaciations that could adversely affect waste repositories. These included deepening and lengthening of fiords; radical changes in the patterns of glacial valleys in highland regions; the formation of river gorges as sea level falls; the formation of large lakes in lowland regions; and the reactivation of major faults by earthquakes associated with rebound at the end of glaciation. Obviously, not all glaciations and deglaciations would pose equal threats. Morner (2001) went further and contended that long-term prediction in terms of storage of nuclear waste was absurd. He illustrated his assertion by noting that although the Fennoscandian Shield is a geologically stable area at present, some 10 000 years ago during the last deglacial phase, Sweden was a region of very high seismic activity. Accordingly, Morner maintained that the only safe way to store radioactive waste is to do so in a dry rock repository under constant control and monitoring, accessible for maintenance and possible future methods of rendering the waste harmless or even removal.

Stress redistribution due to subsurface excavation and possible thermally induced stresses should not endanger the state of equilibrium in the rock mass and should not give rise to any inadmissible convergence or support damage during the operative period. The long-term integrity of the rock formations must be assured. Therefore, it is necessary to determine the distribution of stress and deformation in the host rock of the repository. This may involve consideration of the temperature-dependent rheological properties of the rock mass in order to compare them with its load-bearing capacity. Obviously, substantial strength is necessary for engineering design of subsurface repository facilities, especially in maintaining the integrity of underground openings.

9.4.3 Rock types and disposal

Completely impermeable rock masses are unlikely to exist although many rock types may be regarded as practically impermeable, such as large igneous rock massifs, thick sedimentary sequences, metamorphic rocks and rock salt. The permeability of a rock mass depends mainly on the discontinuities present, their surfaces, width, amount of infill and their intersections. Stringent requirements apply to the storage facilities of high-level, long-lived radioactive waste. In particular, the repository needs to be watertight to prevent the transport of radionuclides by groundwater to the surface. Control and test pumping of the groundwater system and sealing by injection techniques may be necessary.

Rock types such as thick deposits of salt or shale, or granites or basalts at depths of 300–500 m are regarded as the most feasible in which to excavate caverns for disposal of high-level radioactive waste. Once a repository is fully loaded, it can be backfilled, with the shafts being sealed to prevent the intrusion of water. Once sealed, the system can be regarded as isolated from the human environment.

As far as the disposal of high-level radioactive waste in caverns is concerned, thick deposits of salt have certain advantages. Salt has a high thermal conductivity and so will rapidly dissipate any heat associated with high-level nuclear waste; it possesses gamma-ray protection similar to concrete; it undergoes only minor changes when subjected to radioactivity; and it tends not to provide paths of escape for fluids or gases (Langer and Wallmer, 1988). Moreover, salt is 'plastic' at proposed repository depths so that any fractures that may develop due to construction operations can self-heal. However, as Schultz *et al.* (2001) pointed out plastic deformation of rock salt without the formation and propagation of dilation cracks only takes place as long as the state of stress remains within the non-dilatant stress domain. They showed that loading in the dilatant domain gives rise to cracking and to increasing damage with increasing strain. In addition, they found that pore fluid pressures only affected the mechanical properties of rock salt that had undergone dilation. The attractive feature of deep salt deposits is their lack of water and the inability of water from an external source to move through them. These advantages may be compromised if the salt contains numerous interbedded clay or mudstone horizons, open cavities containing brine or faults cutting the salt beds so providing conduits for external water. The solubility of salt requires that unsaturated waters are totally isolated from underground openings in beds of salt by watertight linings, isolation seals or cut-offs and/or by collection systems. If suitable precautions are not taken in more soluble horizons in salt, any dissolution that occurs can lead to the irregular development of a cavity being excavated. Any water that does accumulate in salt will be a concentrated brine that, no doubt, will be corrosive to metal canisters. The potential for heavy groundwater inflows during shaft sinking requires the use of grouting or ground freezing. Rock salt is a material that exhibits short- and long-term creep (Hunsche and Hampel, 1999). Hence, caverns in rock salt are subject to convergence as a result of plastic deformation of the salt. The rate of convergence increases with increasing temperature and stress in the surrounding rock mass. Since temperature and stress increase with depth, convergence also increases with depth. If creep deformation is not restrained or not compensated for by other means, excessive rock pressures can develop on a lining system that may approach full overburden pressure.

Not all shales are suitable for the excavation of underground caverns in that soft compacted shales would present difficulties in terms of wall and roof stability. Caverns also may be subject to floor heave. Caverns could be excavated in competent cemented shales. Not only do these possess low permeability, they also could adsorb ions that move through it. A possible disadvantage is that if temperatures in a cavern exceeded 100 °C, then clay minerals could lose water and therefore shrink. This could lead to the development of fractures. In addition, the adsorption capacity of the shale would be reduced. Bonin (1998) described the investigation that took place at the Tournemire test site in France to assess the value of shales as host media for the disposal of radioactive waste. The programme examined the physical and physico-chemical properties of the shales; rock and water chemistry; long-term behaviour; and diffusion and adventive transport. Although it was concluded that fluid circulation in shales is very slow, it nevertheless would be enhanced by the presence of fractures. The 100-year-old tunnel at the site exhibits a damaged excavation zone with large fractures that would allow water to drain through them rather quickly. Obviously, the presence of any such fractures around the perimeter of a waste repository would have to be sealed.

Granite is less easy to excavate than rock salt or shale but is less likely to suffer problems of cavern support. It provides a more than adequate shield against radiation and will disperse any heat produced by radioactive waste. The quantity of groundwater in granite masses is small and its composition is generally non-corrosive. However, fissure and shear zones do occur within granites along which copious quantities of groundwater can flow. Discontinuities tend to close with depth and faults may be sealed. For location and design of a repository for spent nuclear fuel, the objective is to find solid blocks that are large enough to host the tunnels and caverns of a repository. The Pre-Cambrian shields represent stable granite–gneiss regions at present.

The large thicknesses of lava flows in basalt plateaux mean that such successions also could be considered for disposal sites. Like granite, basalt also can act as a shield against radiation and can disperse heat. Frequently the contact between flows is tight and little pyroclastic material is present. Joints may not be well developed at depth, and strengthwise basalt should support a cavern. However, the durability of basalts on exposure may be suspect, which could give rise to spalling from the perimeter of a cavern (Haskins and Bell, 1995). Furthermore, such basalt formations can be interrupted by feeder dykes and sills that may be associated with groundwater. Groundwater associated with basalts generally is mildly alkaline with a low redox potential.

In the United States extensive geological studies have been conducted in six states to identify sites for detailed site characterization studies and ultimately the construction of that country's first high-level nuclear waste disposal facility. The host rock types that were considered included basalts in Washington state, tuffs in Nevada, evaporites (salt) in Utah and Texas, and diapiric salt in Mississippi and Louisiana. The Nuclear Waste Policy Act of 1982 mandated that extensive site-characterization activities had to be performed to evaluate the suitability of potential sites for the construction of a high-level nuclear waste repository. The site characterization activities included comprehensive surface and subsurface investigations. Characterization of the subsurface involved the use of geophysical techniques, exploratory drillholes with soil, rock and groundwater sampling, and the construction of an exploratory shaft and underground facilities for *in situ* testing of the target repository geological horizon. The exploratory shaft facility provided a means of verifying the repository design parameters. In 1988, one candidate geological repository site was chosen, namely, the Yucca Mountain site in Nevada.

9.5 Waste materials from mining

Mine wastes result from the extraction of metals and non-metals. In the case of metalliferous mining, high volumes of waste are produced because of the low or very low concentrations of metal in the ore. In fact, mine wastes represent the highest proportion of waste produced by industrial activity, billions of tonnes being produced annually. Waste from mines has been and is deposited on the surface in spoil heaps or tailings lagoons. The chemical characteristics of mine waste and waters arising therefrom depend upon the type of mineral being mined, as well as the chemicals that are used in the extraction or beneficiation processes. Because of its high volume, mine waste historically has been disposed of at the lowest cost, often without regard for safety and often with considerable environmental impacts. Catastrophic failures of spoil heaps and tailings dams, although uncommon, have led to the loss of lives.

Mining waste may be inert or contain hazardous constituents but generally is of low toxicity. Nonetheless, in some areas where metals were mined in the past, because little regard was given to the disposal of waste, relatively high concentrations of heavy metals can represent an environmental problem. For example, Smith and Williams (1996) referred to the United States Superfund Site near Kellogg, Idaho, where over 20 lead, zinc and silver mines had been operational in the district since the late 1880s, and mine and mill waste were discarded primarily into the South Fork of the Coeur d'Alene River. One area within the Superfund Site, namely, Smelterville Flats required urgent attention since the wastes, which contained various heavy metals, were a source of airborne particulates. Risk assessments undertaken by the USEPA had suggested that a major threat to human health was posed by the ingestion of airborne tailings and smelter wastes, and elevated levels of lead had been found in the blood of children who lived in the neighbourhood. Furthermore, the wastes also had degraded the quality of the groundwater. According to Smith and Williams, reclamation of Smelterville Flats was imperative if the objectives of the Comprehensive Environmental Response, Compensation and Liability Act (CERCLA) were to be met. They suggested that typical reclamation procedures for such wastes included removing them to an engineered disposal facility; stabilizing them *in situ* either chemically or biochemically; or selective *in situ* leaching or remining and reprocessing them. After sampling the wastes and carrying out a statistical analysis to determine metal content, Smith and Williams suggested that selective treatment was a viable option. Selective treatment of wastes can reduce clean-up costs by avoiding handling or processing wastes that possess lower levels of contamination. In addition, recycling mine wastes to recover metals can be worthwhile economically since metal-processing procedures in the past were not as efficient as they are today.

The character of waste rock from metalliferous mines reflects that of the rock hosting the metal, as well as the rock surrounding the ore body. The type of waste rock disposal facility depends on the topography and drainage of the site, and the volume of waste. Van Zyl (1993) referred to the disposal of coarse mine waste in valley fills, side-hill dumps and open piles. Valley fills normally commence at the upstream end of a valley and progress downstream, increasing in thickness. Side-hill dumps are constructed by the placement of waste along hillsides or valley slopes, avoiding natural drainage courses. Open piles tend to be constructed in relatively flat-lying areas. Obviously, an important factor in the construction of a spoil heap is its slope stability, which includes its long-term stability. Acid mine drainage from spoil heaps is another environmental concern (see Chapter 7).

9.5.1 Basic properties of coarse discard associated with colliery spoil heaps

Spoil heaps associated with coal mines represent ugly blemishes on the landscape and have a blighting effect on the environment. They consist of coarse discard, that is, run-of-mine material that reflects the various rock types that are extracted during mining operations. As such, coarse discard contains varying amounts of coal that has not been separated by the preparation process. Obviously, the characteristics of coarse colliery discard differ according to the nature of the spoil. The method of tipping also influences the character of coarse discard. In addition, some spoil heaps, particularly those with relatively high coal contents, may be burnt, or still be burning, and this affects their mineralogical composition.

In Britain illites and mixed-layer clays are the principal components of unburnt spoil in English and Welsh tips (Taylor, 1975). Although kaolinite is a common constituent in Northumberland and Durham, it averages only 10.5 per cent in the discard of other areas. Quartz exceeds the organic carbon or coal content, but the latter is significant in that it acts as the major diluent, that is, it behaves in an antipathetic manner towards the clay mineral content. Sulphates, feldspars, calcite, siderite, ankerite, chlorite, pyrite, rutile and phosphates average less than 2 per cent.

The chemical composition of spoil material reflects that of the mineralogical composition. Free silica may be present in concentrations up to 80 per cent and above, and combined silica in the form of clay minerals may range up to 60 per cent. Concentrations of aluminium oxide may be between a few per cent and 40 per cent or so. Calcium, magnesium, iron, sodium, potassium and titanium oxides may be present in concentrations of a few per cent. Lower amounts of manganese and phosphorus also may be present, with copper, nickel, lead and zinc in trace amounts. The sulphur content of fresh spoils often is less than 1 per cent and occurs as organic sulphur in coal, and in pyrite.

Pyrite is a relatively common iron sulphide in some of the coals and associated argillaceous rock. It also is an unstable mineral, breaking down quickly under the influence of weathering. The primary oxidation products of pyrite are ferrous and ferric sulphates, and sulphuric acid. Oxidation of pyrite within spoil heap waste is governed by the access of air that, in turn, depends upon the particle size distribution, amount of water saturation and the degree of compaction. However, any highly acidic oxidation products that may form may be neutralized by alkaline materials in the waste material.

The moisture content of spoil increases with increasing content of fines. It also is influenced by the permeability of the material, the topography and climatic conditions. Generally, it falls within the range 5–15 per cent (Table 9.6).

The range of specific gravity depends on the relative proportions of coal, shale, mudstone and sandstone in the waste, and tends to vary between 1.7 and 2.7. The proportion of coal is of particular importance, the higher the coal content, the lower the specific gravity. The bulk density of material in spoil heaps shows a wide variation, most material falling within the range 1.5–2.5 $Mg\,m^{-3}$. Low densities are mainly a function of low specific gravity. Bulk density tends to increase with increasing clay content.

The argillaceous content influences the grading of spoil, although most spoil material is essentially granular. In fact, as far as the particle size distribution of coarse discard is concerned there is a wide variation, often most material may fall within the sand range but significant proportions of gravel and cobble range also may be present. Indeed, at placement coarse discard very often consists mainly of gravel-cobble size but subsequent breakdown on weathering reduces the particle size. Once buried within a spoil heap, coarse discard undergoes little further reduction in size. Hence, surface samples of spoil contain a higher proportion of fines than those obtained from depth.

Table 9.6 Examples of soil properties of coarse discard

	Yorkshire Main	*Brancepath*	*Wharncliffe*
Moisture content, %	8.0–13.6	5.3–11.9	6–13 (7.14)**
Bulk density, $Mg\,m^{-3}$	1.67–2.19	1.27–1.88	1.58–2.21
Dry density, $Mg\,m^{-3}$	1.51–1.94	1.06–1.68	1.39–1.91
Specific gravity	2.04–2.63	1.81–2.54	2.16–2.61
Plastic limit, %	16–25	Non-plastic–35	14–21
Liquid limit, %	23–44	23–42	25–46
Permeability, ms^{-1}	$1.42–9.78 \times 10^{-6}$		
Size, <0.002 mm, %	0.0–17.0	Most material of	2.0–20
Size, >2.0 mm, %	30.0–57.0	sand size range	38–67
Shear strength, ϕ'	31.5–35.0°	27.5–39.5°	29–37°
Shear strength, c'	19.44–21.41 kPa	3.65–39.03 kPa	16–40 kPa

Source: Bell, 1996.

Note
**Average value.

The liquid and plastic limits can provide a rough guide to the engineering characteristics of a soil. In the case of coarse discard, however, they are only representative of that fraction passing the 425 μm BS sieve, which frequently is less than 40 per cent of the sample concerned. Nevertheless, the results of consistency tests indicate a low to medium plasticity, whilst in certain instances spoil has proved to be virtually non-plastic (Fig. 9.12). Plasticity increases with increasing clay content.

The shear strength parameters of coarse discard do not exhibit any systematic variation with depth in a spoil heap and so are not related to age, that is, time dependent. This suggests that coarse discard is not seriously affected by weathering. As far as the effective shear strength of coarse discard is concerned, the angle of shearing resistance usually varies from 25 to 45°. The angle of shearing resistance and therefore the strength increases in spoil that has been burnt.

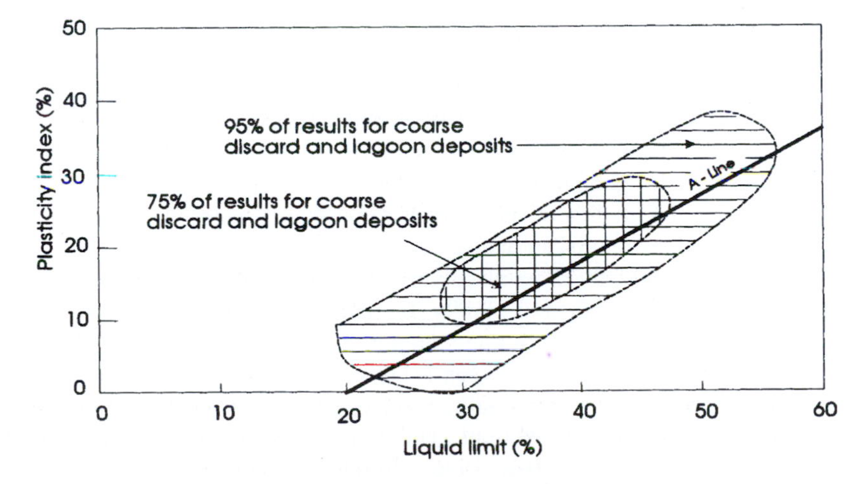

Figure 9.12 Plasticity chart of coarse colliery discard and tailings.

With increasing content of fine coal, the angle of shearing resistance is reduced. Also, as the clay mineral content in spoil increases, so its shear strength decreases.

The shear strength of discard within a spoil heap, and therefore its stability, is dependent upon the pore water pressures developed within it. Pore water pressures in spoil heaps may be developed as a result of the increasing weight of material added during construction or by seepage though the heap of natural drainage. High pore water pressures usually are associated with fine-grained materials that have a low permeability and high moisture content. Thus, the relationship between permeability and the build-up of pore water pressures is crucial. In fact, in soils with a coefficient of permeability of less than $5 \times 10^{-9}\,\mathrm{m\,s^{-1}}$ there is no dissipation of pore water pressures, whilst above $5 \times 10^{-7}\,\mathrm{m\,s^{-1}}$ they are completely dissipated. The permeability of colliery discard depends primarily upon its grading and its degree of compaction. It tends to vary between 1×10^{-4} and $5 \times 10^{-8}\,\mathrm{m\,s^{-1}}$, depending upon the amount of degradation in size that has occurred.

The most significant change in the character of coarse colliery discard brought about by weathering is the reduction of particle size. The extent to which breakdown occurs depends upon the type of parent material involved and the effects of air, water and handling between mining and deposition on the spoil heap. After a few months of weathering, the debris resulting from sandstones and siltstones usually is greater than cobble size. After that, the degradation to component grains takes place at a very slow rate. Mudstones, shales and seatearth exhibit rapid disintegration to gravel size. Although coarse discard may reach its level of degradation within a matter of months, with the degradation of many mudstones and shales taking place within days, once it is buried within a spoil heap it suffers little change. When spoil material is burnt, it becomes much more stable as far as weathering is concerned.

9.5.2 Spoil heaps and spontaneous combustion

Spontaneous combustion of carbonaceous material, frequently aggravated by the oxidation of pyrite, is the most common cause of burning spoil. It can be regarded as an atmospheric oxidation (exothermic) process in which self-heating occurs. Coal and carbonaceous materials may be oxidized in the presence of air at ordinary temperatures, below their ignition point. Generally, the lower rank coals are more reactive and accordingly more susceptible to self-heating than coals of higher rank.

Oxidation of pyrite at ambient temperature in moist air leads, as mentioned above, to the formation of ferric and ferrous sulphate, and sulphuric acid. This reaction also is exothermic. When present in sufficient amounts, and especially when in finely divided form, pyrite associated with coaly material increases the likelihood of spontaneous combustion. When heated, the oxidation of pyrite and organic sulphur in coal gives rise to the generation of sulphur dioxide. If there is not enough air for complete oxidation, then hydrogen sulphide is formed.

The moisture content and grading of spoil are also important factors in spontaneous combustion. At relatively low temperatures an increase in free moisture increases the rate of spontaneous heating. Oxidation generally takes place very slowly at ambient temperatures but as the temperature rises, so oxidation increases rapidly. In material of large size, the movement of air can cause heat to be dissipated whilst in fine material the air remains stagnant, and this means that burning ceases when the supply of oxygen is consumed. Accordingly, ideal conditions for spontaneous combustion exist when the grading is intermediate between these two extremes and hot spots may develop under such conditions. These hot spots may have temperatures around 600 °C or occasionally up to 900 °C (Bell, 1996). Furthermore, the rate of oxidation generally increases as the specific surface of particles increases.

Spontaneous combustion may give rise to subsurface cavities in spoil heaps, the roofs of which may be incapable of supporting a person. Burnt ashes also may cover zones that are red-hot to appreciable depths. Anon. (1973) recommended that during restoration of a spoil heap, a probe or crane with a drop-weight could be used to prove areas of doubtful safety that are burning. Badly fissured areas should be avoided and workmen should wear lifelines if they walk over areas not proved safe. Any area that is suspected of having cavities should be excavated by drag line or drag scraper rather than allow plant to move over suspect ground.

When steam comes in contact with red-hot carbonaceous material, watergas is formed, and when the latter is mixed with air, over a wide range of concentrations, it becomes potentially explosive. If a cloud of coal dust is formed near burning spoil when reworking a heap, then this also can ignite and explode. Damping with a spray may prove useful in the latter case.

Noxious gases are emitted from burning spoil. These include carbon monoxide, carbon dioxide, sulphur dioxide and, less frequently, hydrogen sulphide. Each may be dangerous if breathed in certain concentrations, which may be present at fires on spoil heaps (Table 3.1). The rate of evolution of these gases may be accelerated by disturbing burning spoil by excavating into or reshaping it. Carbon monoxide is the most dangerous since it cannot be detected by taste, smell or irritation and may be present in potentially lethal concentrations. By contrast, sulphur gases are readily detectable in the aforementioned ways and usually are not present in high concentrations. Even so, when diluted they still may cause distress to persons with respiratory ailments. Nonetheless, the sulphur gases are mainly a nuisance rather than a threat to life. In certain situations, a gas-monitoring programme may have to be carried out. Where danger areas are identified, personnel should wear breathing apparatus.

Burning spoil material may represent a notable problem when reclaiming old tips (Fig. 9.13). Spontaneous combustion of coal in colliery spoil can be averted if the coal occurs in an

Figure 9.13 Restoration of a colliery spoil heap near Barnsley, England. The earth-moving equipment is moving over a hot spot.

oxygen-deficient atmosphere that is humid enough with excess moisture to dissipate any heating that develops. Cook (1990), for example, described shrouding a burning spoil heap with a cover of compacted discard to smother the existing burning and prevent further spontaneous combustion. Anon. (1973) also recommended blanketing and compaction, as well as digging out, trenching, injection with non-combustible material and water, and water spraying as methods by which spontaneous combustion in spoil material may be controlled. Bell (1996) described the injection of a curtain wall of pulverized fuel ash, extending to original ground level, as a means of dealing with hot spots.

9.5.3 Restoration of spoil heaps

The configuration of a spoil heap depends upon the type of equipment used in its construction and the sequence of tipping the waste. The shape, aspect and height of a spoil heap affects the intensity of exposure, the amount of surface erosion that occurs, the moisture content in its surface layers and its stability. Although the mineralogical composition of coarse discard from different mines varies, pyrite frequently occurs in the shales and coaly material present in spoil heaps. When pyrite weathers it gives rise to the formation of sulphuric acid, along with ferrous and ferric sulphates and ferric hydroxide, which promote acidic conditions in the weathered material. Such conditions do not aid the growth of vegetation. Indeed, some spoils may contain elements that are toxic to plant life. Furthermore, the uppermost slopes of a spoil heap frequently are devoid of near-surface moisture. However, in order to support vegetation a spoil heap should have a stable surface in which roots can become established, must be non-toxic and contain an adequate and available supply of nutrients. Hence, spoil heaps are often barren of vegetation.

Restoration of a spoil heap represents an exercise in large-scale earthmoving. Since it invariably involves spreading the waste over a larger area, this may mean that additional land beyond the site boundaries has to be purchased. Where a spoil heap is very close to the disused colliery, spoil may be spread over the latter area. This involves the burial or removal of derelict colliery buildings, and sometimes old mine shafts or shallow old workings may have to be treated (Johnson and James, 1990). Water courses may have to be diverted, as may services, notably roads.

Landscaping of spoil heaps frequently is to allow them to be used for agriculture or forestry. This type of restoration generally is less critical than when structures are to be erected on the site since bearing capacities are not so important, and steeper surface gradients are acceptable. Most spoil heaps offer no special handling problems other than the cost of regrading and possibly the provision of adjacent land, so that the gradients on the existing site can be reduced by the transfer of spoil to the adjacent land. However, as noted, some spoil heaps present problems due to spontaneous combustion of coaly material that they contain.

Surface treatments of spoil heaps vary according to the chemical and physical nature of the spoil, and the climatic conditions. Preparation of the surface of a spoil heap also depends upon the use to which it is to be put subsequently. Where it is intended to sow grass or plant trees, the surface layer should not be compacted. Drainage plays an important part in the restoration of a spoil heap. If the spoil is acid, then it can be neutralized by liming. The chemical composition influences the choice of fertilizer used and in some instances spoil can be seeded without the addition of top soil, if suitably fertilized. More commonly, top soil is added prior to seeding or planting.

9.5.4 Spoil heaps and stability

The stability of a spoil heap is influenced by the material of which it is composed, its height, the gradients and lengths of its slopes, and the nature and rate of erosion to which it is exposed that, in turn, is influenced by the climatic regime. Occasionally, landslips may be associated with spoil heaps. In particular, old colliery spoil heaps were formed by tipping the waste without any compaction, which means that their stability is adversely affected if the material becomes saturated. Flow slides can develop under such conditions. Both flow slides and rotational slides can damage property, with the former being more likely to claim lives. The worst disaster in Britain due to the failure of a spoil heap occurred at Aberfan, in South Wales (Fig. 9.14). There, spoil heaps from the local colliery had been built on the steep valley slopes above the village. On 21 October 1966, about 107 000 m^3 of spoil, representing one-third of a heap, flowed into the village. The Pennant Sandstone in which the valley is carved is well jointed and the joints had been affected by mining subsidence. This allowed rain falling on the upper slopes to be discharged as springs on the lower slopes. Accordingly, water issued beneath the toe of the spoil heap-eroding material. This had given rise to intermittent slips previously. Heavy rains

Figure 9.14 Flow slide/mudflow from a spoil heap at Aberfan, South Wales, October 1966.

two days prior to the fatal slide meant that spoil became saturated and, with its shear strength very much reduced, the spoil became liable to flow, which it subsequently did with disastrous consequences (Bishop, 1973). The resultant mudflow invaded a school and several houses, killing 116 children and 28 adults. Over two years later, work began on removing the spoil to a new site. The spoil was placed in a number of terraces and extensive drainage works were undertaken.

9.5.5 Waste disposal in tailings dams

Tailings are fine-grained residues that result from crushing rock that contains ore or are produced by the washeries at collieries. Mineral extraction usually takes place by wet processes in which the ore is separated from the parent rock by flotation, dissolution by cyanide, washing, etc. Consequently, most tailings originate as slurries, in which form they are transported for hydraulic deposition. The water in tailings may contain certain chemicals associated with the metal recovery process such as cyanide in tailings from gold mines, and heavy metals in tailings from copper–lead–zinc mines. Tailings also may contain sulphide minerals like pyrite that can give rise to acid mine drainage. The particle size distribution, permeability and resistance to weathering of tailings affect the process of acid generation (Bell and Bullock, 1996). Acid drainage also may contain elevated levels of dissolved heavy metals. Accordingly, contaminants carried in the tailings may represent a source of pollution for both ground and surface water, as well as soil.

Tailings are deposited as slurry generally in specially constructed tailings dams (Fig. 9.15). The embankment dams can be constructed to their full height before the tailings are discharged. They are constructed in a similar manner to earth fill dams used to impound reservoirs. They may be zoned with a clay core and have filter drains. Such dams are best suited to tailings impoundments with high water storage requirements. However, tailings dams usually consist of

Figure 9.15 A tailings lagoon for waste from the processing of tar sands, Fort McMurdo, Alberta.

raised embankments, that is, the construction of the dam is staged over the life of the impoundment. Raised embankments consist initially of a starter dyke that normally is constructed of earth fill from a borrow pit. This dyke may be large enough to accommodate the first two or three years of tailings production. A variety of materials can be used subsequently to complete such embankments including earth fill, mine waste or tailings themselves. Tailings dams consisting of tailings can be constructed using the upstream, centreline or downstream method of construction (Fig. 9.16). In the upstream method of construction, tailings are discharged from spigots or small pipes to form the impoundment (Fig. 9.17). This allows separation of particles according to size, with the coarsest particles accumulating in the centre of the embankment beneath the spigots and the finer particles being transported down the beach.

Alternatively, cycloning may be undertaken to remove coarser particles from tailings so that they can be used in embankment construction. Centreline and downstream construction of embankments uses coarse particles separated by cyclones for the dam. For equivalent embankment heights, water retention embankment dams and downstream impoundments require approximately three times more fill than an upstream embankment. A centreline embankment requires about twice as much fill as an upstream embankment of similar height (Vick, 1983). The design of tailings dams must pay due attention to their stability both in terms of static and dynamic loading.

Erosion of the outer slopes of tailings dams should be minimized by erosion control measures such as berms, surface drainage, rip-rap or geofabrics, and where appropriate the rapid establishment of vegetation. Blight and Amponsah-Da Costa (1999) pointed out that climate has an influence on erosion control measures in that water and wind erosion differ in effectiveness between humid, semi-arid and arid regions. Furthermore, vegetation is less easy to establish on slopes in arid regions so that a cover of rip-rap over a slope may be more appropriate.

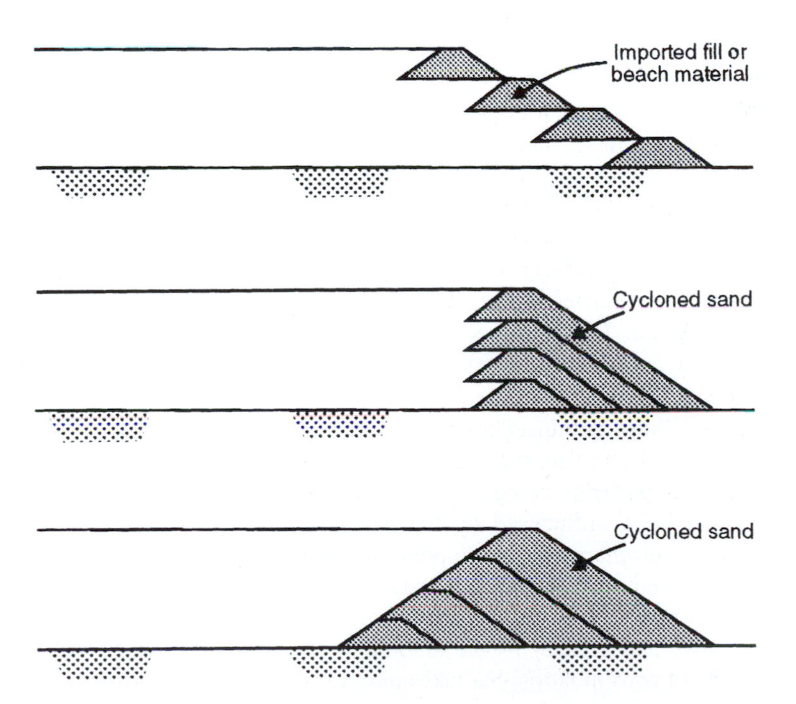

Figure 9.16 Upstream, centreline and downstream methods of embankment construction for tailings dams. *Source*: Vick, 1983.

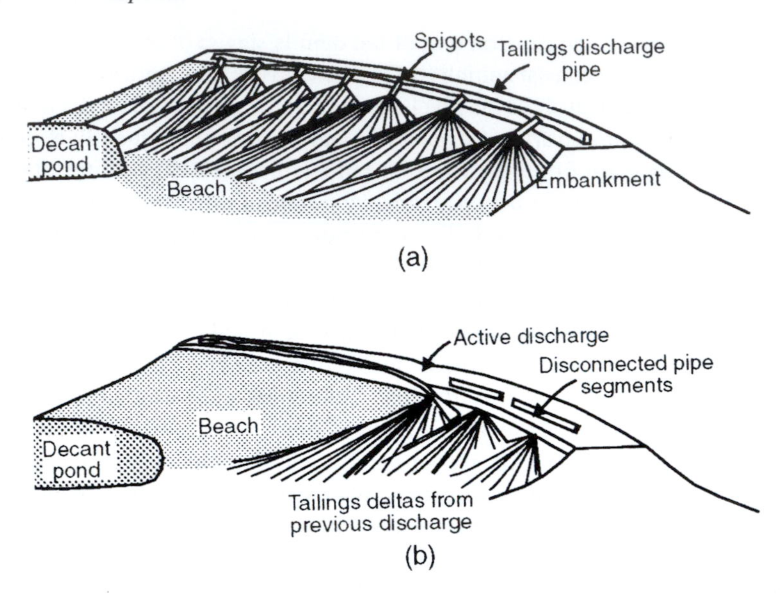

Figure 9.17 Tailings discharge by spigotting.

They undertook an evaluation of the cost effectiveness of various slope protection measures and found that a cover of 300 mm of fine rock was most effective.

Failure of a dam can lead to catastrophic consequences. For example, failure of the tailings dam at Buffalo Creek in West Virginia after heavy rain in February 1972, destroyed over 1500 houses and cost 118 lives. The failure of the Merriespruit gold tailings dam in South Africa in 1994 resulted in 17 deaths. In the latter case, there may have been insufficient freeboard provision and poor pool control. Fourie *et al.* (2001) suggested that high void ratios in some parts of the dam could have meant that these zones were in a metastable condition, overtopping and erosion of the dam wall exposing such zones. This resulted in liquefaction and consequent flow failure.

The tailings slurry can be discharged under water in the impoundment or it may be disposed of by subaerial deposition. In subaerial disposal, the slurry is discharged from one or more points around the perimeter of the impoundment with the slurry spreading over the floor to form beaches or deltas. The free water drains from the slurry to form a pond over the lowest area of the impoundment. The discharge points can be relocated around the dam to allow the exposed solids to dry and thereby increase in density. Alternatively, discharge can take place from one fixed point.

The deposition of tailings within a dam may lead to the formation of a beach or mudflat above the water level. When discharged, the coarser particles in tailings settle closer to the discharge point(s) with the finer particles being deposited further away (Blight, 1994). The amount of sorting that takes place is influenced by the way in which the tailings are discharged, for instance, high volume discharge from one point produces little sorting of tailings. On the other hand, discharge from multiple points at moderate rates gives rise to good sorting. After being deposited, the geotechnical properties of tailings such as moisture content, density, strength and permeability are governed initially by the amount of sorting that occurs, and subsequently by the amount and rate of consolidation that takes place. For example, if sorting has led to material becoming progressively finer as the pool is approached, then the permeability of the tailings is likely to decrease in the same direction. The moisture content of deposited tailings can vary from 20 to over 60 per cent and dry densities from 1 to 1.3 $Mg\,m^{-3}$. By the end of deposition, the density of the tailings generally has increased with depth due to the increasing self-weight on the lower material. The coarser particles that settle out first also drain more quickly than the finer material, which accumulates further down the beach, and so develop shear strength more quickly than the latter. These variations in sorting also affect the permeability of the material deposited.

The quantity of tailings that can be stored in a dam enclosing a given volume is dependent upon the density that can be achieved. The latter is influenced by the type of tailings, the method by which they are deposited, whether they are deposited in water or subaerially, the drainage conditions within the dam and whether or not they are subjected to desiccation. Ideally, the pool should be kept as small as possible and the penstock inlet, if present, should be located in the centre of the pool. This depresses the phreatic surface, thus helping to consolidate the tailings above the latter. Blight and Steffen (1979) referred to the semi-dry or subaerial method of tailings deposition whereby in semi-arid or arid climates a layer of tailings is deposited in a dam and allowed to dry before the next layer is placed. Since this action reduces the volume of the tailings, it allows more storage to take place within the dam. However, drying out can give rise to the formation of desiccation cracks in the fine discard that, in turn, can represent locations where piping can be initiated. In fact, Blight (1988) pointed out that tailings dams have failed as a result of desiccation cracks and horizontal layering of fine-grained particles leading to piping failure. The most dangerous situation occurs when the ponded water on the discard increases in size and thereby erodes the cracks in the dam to form pipes that may emerge on the outer slopes of the dam.

The permeability and rheological properties of tailings are, according to East and Morgan (1999), the principal factors influencing the design of effective drainage systems at a tailings impoundment. As referred to, tailings commonly are transported as slurries via a pipeline to the impoundment where the solid particles settle out of the liquid fraction, so that a pond or lagoon often occurs within a tailings dam. For example, slurries produced by gold processing usually contain between 25 and 50 per cent solid particles by weight. However, slurry is added to the saturated tailings waste within the dam faster than the excess pore water pressures can be dissipated. This means that undrained tailings are difficult to reclaim once the life of an impoundment has come to an end because working surfaces for light earth moving equipment are costly to establish until appreciable consolidation has taken place, which may take several years. In addition, ponding and saturated tailings can adversely affect the stability of the embankments and therefore are important factors to be considered in the design of tailings dams. Consequently, removal of pond water and reduction of pore water pressures, together with internal drainage of the embankment should be incorporated into the design of a tailings dam. Removal of pond water can be achieved by decanting via penstock towers, by pumping or by sloping filter drains. Underdrainage can help reduce pore water pressures in the waste and can consist of a simple system of finger drains on the one hand or a drainage blanket of sand or gravel with internal collection pipework on the other. Alternatively, geotextile and geonet sandwiches can be used for underdrainage. Underdrainage conveys water to collection centres. However, whether such measures are put in place depends upon the size of the impoundment, which influences their cost.

The rate at which seepage occurs from a tailings dam is governed by the permeability of the tailings and the ground beneath the impoundment. Climate and the way in which the tailings dam is managed also will have some influence on seepage losses. In many instances, because of the relatively low permeability of the tailings compared with the ground beneath, a partially saturated flow condition will occur in the foundation. Nonetheless, the permeability of tailings can vary significantly within an impoundment depending on the nearness to discharge points, the degree of sorting, the amount of consolidation that has taken place, the stratification of coarser and finer layers that has developed, and the amount of desiccation that the discard has undergone.

According to Fell *et al.* (1993), one of the most cost-effective methods of controlling seepage loss from a tailings dam is to cover the whole floor of the impoundment with tailings from the start of the operation (Fig. 9.18). This cover of tailings, provided it is of low permeability, will form a liner. Tailings normally have permeabilities between 10^{-7} and $10^{-9}\,m\,s^{-1}$ or less. Nevertheless, Fell *et al.* mentioned that a problem could arise when using tailings to line an impoundment, namely, if a sandy zone develops near the point(s) of discharge, then localized higher seepage rates will occur if water covers this zone. This can be avoided by moving the points of discharge or by placing fines. Alternatively, a seepage collection system can be placed beneath the sandy zone prior to its development. Clay liners also represent an effective method of reducing seepage from a tailings dam. The permeability of properly compacted clay soils usually varies between 10^{-8} and $10^{-9}ms^{-1}$. However, clay liners may have to be protected from drying out, with attendant development of cracks, by placement of a sand layer on top. The sand on slopes may need to be kept in place by using geotextiles. Geomembranes have tended not to be used for lining tailings dams, primarily because of the cost. As mentioned, filter drains may be placed at the base of tailings, with or without a clay liner. They convey water to collection ponds. A toe drain may be incorporated into the embankment. This will intercept seepage that emerges at that location. Where a tailings dam has to be constructed on sand or sand and gravel, a slurry trench may be used to intercept seepage water but slurry trenches are expensive.

Seepage losses from tailings dams that contain toxic materials can have an adverse effect on the environment. For example, Sharma and Al-Busaidi (2001) described a serious pollution problem associated with a tailings dam in Oman. Between 1982 and 1994, 11 million tonnes of sulphide-rich tailings from a copper mine and 5 million m^3 of sea water (used in the processing of the copper ore) were disposed of within an unlined tailings dam. The resultant pollution plume extended some 14 km downstream of the tailings dam and the TDS of the groundwater

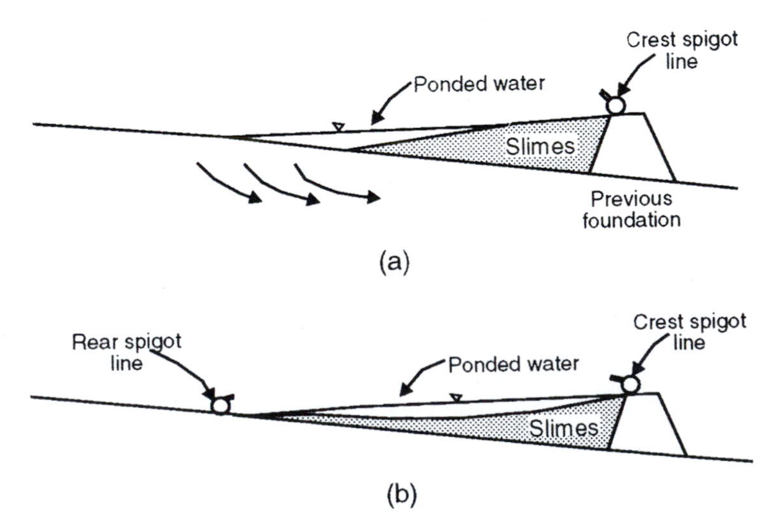

Figure 9.18 Control of seepage by spigotting procedures: (a) major seepage at water-foundation contact; (b) foundation sealing by near spigotting of tailings.
Source: Vick, 1983.

in some monitoring wells increased from around $1000\,mg\,l^{-1}$ in 1984 to some $55\,000\,mg\,l^{-1}$ in 1996. In fact, it usually is necessary to operate a dam as a closed or controlled system from which water can be released at specific times. It therefore is necessary to carry out a water balance determination for a dam, and any associated catchments such as a return water reservoir, to assess whether and/or when it may be necessary to release water from or make up water within the system. Because of variations in precipitation and evaporation, water balance calculations should be undertaken frequently. If provision has been made for draining the pool of the dam, then after tailings deposition has ceased it should be possible to maintain a permanent water deficit that will reduce seepage of polluted water from the dam. Bowell *et al.* (1999) reviewed various chemical methods of dealing with mine waste, especially in relation to water quality (see Chapter 7).

The objectives of rehabilitation of tailings impoundments include their long-term mass stability, long-term stability against erosion, prevention of environmental contamination and return of the area to productive use. Normally, when the discharge of tailings comes to an end the level of the phreatic surface in the embankment falls as water replenishment ceases. This results in an enhancement of the stability of embankment slopes. However, where tailings impoundments are located on slopes, excess run-off into the impoundment may reduce embankment stability, or overtopping may lead to failure by erosion of the downstream slope. The minimization of inflow due to run-off by judicious siting is called for when locating an impoundment that may be so affected. Diversion ditches can cater for some run-off but have to be maintained, as do abandonment spillways and culverts. Accumulation of water may be prevented by capping the impoundment, the capping sloping towards the boundaries. Erosion by water or wind can be impeded by placing rip-rap on slopes and by the establishment of vegetation on the waste. The latter also will help to return the impoundment to some form of productive use (Troncoso and Troncoso, 1999). However, tailings may contain high concentrations of heavy metals and be very low in plant nutrients, with low pH values. Accordingly, the application of lime and/or fertilizer may be necessary to establish successful plant growth. Even then, the number of species that can be grown initially without their growth being stunted may be limited. Where long-term potential for environmental contamination exists, particular precautions need to be taken. For example, as the water level in an impoundment declines, the rate of oxidation of any pyrite present in the tailings increases, reducing the pH and increasing the potential for heavy metal contamination. In the case of tailings from uranium mining, radioactive decay of radium gives rise to radon gas. Diffusion of radon gas does not occur in saturated tailings but after abandonment radon reduction measures may be necessary. In both these cases, a clay cover can be placed over the tailings impoundment to prevent leaching of contaminants or to reduce emission of radon gas.

Once the discharge of tailings ceases, the surface of the impoundment is allowed to dry. Drying of the decant pond may take place by evaporation and/or by drainage to an effluent plant. Desiccation and consolidation of the slimes may take a considerable time. Stabilization can begin once the surface is firm enough to support equipment. As mentioned, this normally will involve the establishment of a vegetative cover.

References

Anon. 1973. *Spoil Heaps and Lagoons: Technical Handbook*. National Coal Board, London.

Anon. 1993. *South African Water Quality Guidelines, Volume 1: Domestic Use*. Department of Water Affairs and Forestry, Pretoria.

Baillieul, T.A. 1987. Disposal of high level nuclear waste in America. *Bulletin Association Engineering Geologists*, **24**, 207–216.

Barber, C. 1982. *Domestic Waste and Leachate*. Notes on Water Research No. 31, Water Research Centre, Medmenham.

Basagaoglu, H., Celenk, E., Marini, M.A. and Usul, N. 1997. Selection of waste disposal sites using GIS. *Journal American Water Resources Association*, **33**, 455–464.

Bell, F.G. 1996. Dereliction: colliery spoil heaps and their rehabilitation. *Environmental and Engineering Geoscience*, **2**, 85–96.

Bell, F.G. and Bullock, S.E.T. 1996. The problem of acid mine drainage, with an illustrative case history. *Environmental and Engineering Geoscience*, **2**, 369–392.

Bell, F.G. and Mitchell, J.K. 1986. Control of groundwater by exclusion. In: *Groundwater in Engineering Geology*, Engineering Geology Special Publication No. 3, Cripps, J.C., Bell, F.G. and Culshaw, M.G. (eds), The Geological Society, London, 429–443.

Bell, F.G., Sillito, A.J. and Jermy, C.A. 1996. Landfills and associated leachate in the greater Durban area: two case histories. In: *Engineering Geology of Waste Disposal*, Engineering Geology Special Publication No. 11, Bentley, S.F. (ed.), The Geological Society, London, 15–35.

Bishop, A.W. 1973. The stability of tips and spoil heaps. *Quarterly Journal Engineering Geology*, **6**, 335–376.

Blight, G.E. 1988. Some less familiar aspects of hydraulic fill structures. In: *Hydraulic Fill Structures*, Van Zyl, D.J.A. and Vick, S.G. (eds), American Society Civil Engineers, Geotechnical Special Publication No. 21, New York, 1000–1064.

Blight, G.E. 1994. Environmentally acceptable tailings dams. *Proceeding First International Congress on Environmental Geotechnics*, Edmonton, Carrier, W.D. (ed.), BiTech Publishers Ltd, Richmond, BC, 417–426.

Blight, G.E. and Amponsah-Da Costa, F. 1999. Improving the erosional stability of tailings dam slopes. *Proceedings Sixth International Conference on Tailings and Mine Waste '99*, Fort Collins, Colorado, A.A. Balkema, Rotterdam, 197–206.

Blight, G.E. and Fourie, A.B. 1999. Leachate generation in landfills in semi-arid climates. *Proceedings Institution Civil Engineers, Geotechnical Engineering*, **137**, 181–187.

Blight, G.E. and Steffen, O.K.H. 1979. Geotechnics of gold mining waste disposal. In: *Current Geotechnical Practice in Mine Waste Disposal*. American Society Civil Engineers, New York, 1–52.

Bonin, B. 1998. Deep geological disposal in argillaceous formations: studies at the Tournemire test site. *Journal Contaminant Hydrology*, **35**, 315–330.

Bouazza, A. and Van Impe, W.F. 1998. Liner design for waste disposal sites. *Environmental Geology*, **35**, 41–54.

Bowell, R.J., Williams, K.P., Connelly, R.J., Sadler, P.J.K. and Dodds, J.E. 1999. Chemical containment of mine waste. In: *Chemical Containment of Waste in the Geosphere*. Special Publication No. 157, Metcalfe, R. and Rochelle, C.A. (eds), The Geological Society, London, 213–240.

Bromhead, E.N., Coppola, L. and Rendell, H.M. 1996. Stabilization of an urban refuse dump and its planned extension near Ancona, Marche, Italy. In: *Engineering Geology of Waste Disposal*, Engineering Geology Special Publication No. 11, Bentley, S.F. (ed.), The Geological Society, London, 87–92.

Brotzen, O. 1995. Public acceptance and real testing of a nuclear repository. *Waste Management*, **15**, 559–566.

Chapman, N.A. and Williams, G.M. 1987. Hazardous and radioactive waste management: a case of dual standards? In: *Planning and Engineering Geology*, Engineering Geology Special Publication No. 4, Culshaw, M.G., Bell, F.G., Cripps, J.C. and O'Hara, M. (eds), The Geological Society, London, 489–493.

Cherrill, H.E. and Phillips, A. 1997. The capping of landfill – towards the re-use of material which would otherwise be waste. In: *Engineering Geology of Waste Disposal*, Engineering Geology Special Publication No. 11, Bentley, S.F. (ed.), The Geological Society, London, 403–408.

Christiansson, R. and Jernias, R. 1996. Storage of hazardous waste at shallow depths. In: *Engineering Geology of Waste Disposal*, Engineering Geology Special Publication No. 11, Bentley, S.F. (ed.), The Geological Society, London, 237–244.

Clark, R.G. and Davis, G. 1996. The construction of clay liners for landfills. In: *Engineering Geology of Waste Disposal*, Engineering Geology Special Publication No. 11, Bentley, S.P. (ed.), The Geological Society, London, 171–176.

Cook, B.J. 1990. Coal discard-rehabilitation of a burning heap. In: *Reclamation, Treatment and Utilization of Coal Mining Wastes*, Rainbow, A.K.M. (ed.), A.A. Balkema, Rotterdam, 223–230.

Daniel, D.E. 1993. Clay cores. In: *Geotechnical Practice for Waste Disposal*, Daniel, D.E. (ed.), Chapman & Hall, London, 137–162.

Daniel, D.E. 1995. Pollution prevention in landfills using engineered final covers. *Proceedings Symposium on Waste Disposal, Green '93 – Geotechnics Related to the Environment*, Bolton, Sarsby, R.W. (ed.), A.A. Balkema, Rotterdam, 73–92.

East, D.E. and Morgan, D.J.T. 1999. The use of geomembranes in the design of mineral waste storage facilities. *Proceedings Sixth International Conference on Tailings and Mine Waste '99*, Fort Collins, Colorado, A.A. Balkema, Rotterdam, 371–376.

Eid, H.T., Stark, T.D., Evans, W.D. and Sherry, P.E. 2000. Municipal solid waste slope failure. 1: Waste and foundation soil properties; 2: Stability analyses. *Proceedings American Society Civil Engineers, Journal Geotechnical and Geoenvironmental Engineering*, **126**, 378–387, 408–419.

Eriksson, L.G. 1989. Underground disposal of high level nuclear waste in the United States of America. *Bulletin International Association Engineering Geology*, **39**, 35–52.

Evans, D.M. 1966. Man-made earthquakes in Denver. *Geotimes*, **10**, 11–18.

Fang, H.Y. 1995. Engineering behaviour of urban refuse, compaction control and slope stability analysis of landfill. *Proceedings Symposium on Waste Disposal by Landfill, Green '93 – Geotechnics Related to the Environment*, Bolton, Sarsby, R.W. (ed.), A.A. Balkema, Rotterdam, 47–72.

Fell, R., Miller, S. and de Ambrosio, L. 1993. Seepage and contamination from mine waste. *Proceedings Conference on Geotechnical Management of Waste and Contamination*, Sydney, Fell, R., Phillips, A. and Gerrard, C. (eds), A.A. Balkema, Rotterdam, 253–311.

Fleming, I.R., Rowe, R.K. and Cullimore, D.R. 1999. Field observations of clogging in a landfill leachate collection system. *Canadian Geotechnical Journal*, **36**, 685–707.

Fourie, A.B., Blight, G.E. and Papageorgiou, G. 2001. Static liquefaction as a possible explanation for the Merriespruit tailings dam failure. *Canadian Geotechnical Journal*, **38**, 707–719.

Gallagher, E.M., Needham, A.D. and Smith, D.M. 2000. Non-mineral steepwall liner systems for landfills. *Ground Engineering*, **33**, No. 10, 32–36.

Gibb, F.G.F. 1999. A new scheme for the very deep geological disposal of high-level radioactive waste. *Journal Geological Society*, **157**, 27–36.

Hagerty, D.J. and Pavori, J.L. 1973. Geologic aspects of landfill refuse disposal. *Engineering Geology*, **7**, 219–230.

Haskins, D.R. and Bell, F.G. 1995. Drakensberg basalts: their alteration, breakdown and durability. *Quarterly Journal Engineering Geology*, **28**, 287–302.

Horseman, S.T. and Volckaert, G. 1996. Disposal of radioactive wastes in argillaceous formations. In: *Engineering Geology of Waste Disposal*, Engineering Geology Special Publication No. 11, Bentley, S.P. (ed.), The Geological Society, London, 179–191.

Horseman, S.T., Harrington, J.F. and Sellin, P. 1999. Gas migration in clay barriers. *Engineering Geology*, **54**, 139–149.

Hunsche, U. and Hampel, A. 1999. Rock salt – the mechanical properties of the host material for a radioactive waste repository. *Engineering Geology*, **52**, 271–291.

Jessberger, H.L., Manassero, M., Sayez, B. and Street, A. 1995. Engineering waste disposal (Geotechnics of landfill design and remedial works). *Proceedings Symposium on Waste Disposal by Landfill, Green '93 – Geotechnics Related to the Environment*, Bolton, Sarsby, R.W. (ed.), A.A. Balkema, Rotterdam, 21–33.

Johnson, A.C. and James, E.J. 1990. Granville colliery land reclamation/coal recovery scheme. In: *Reclamation and Treatment of Coal Mining Wastes*, Rainbow, A.K.M. (ed.), A.A. Balkema, Rotterdam, 193–202.

Jones, D.R.V. 1996. Waste disposal in steep-sided quarries: geomembrane based barrier systems. In: *Engineering Geology of Waste Disposal*, Engineering Geology Special Publication No. 11, Bentley, S.F. (ed.), The Geological Society, London, 127–131.

Jones, R., Taylor, D. and Dixon, N. 1997. Shear strength of waste and its use in landfill stability analysis. In: *Geoenvironmental Engineering. Contaminated Ground: Fate of Pollutants and Remediation*, Yong, R.N. and Thomas H.R. (eds), Thomas Telford Press, London, 343–350.

Koerner, R.M. 1993. Geomembrane liners. In: *Geotechnical Practice for Waste Disposal*, Daniel, D.E. (ed.), Chapman & Hall, London, 164–186.

Krauskopf, K.B. 1988. *Radioactive Waste Disposal and Geology*. Chapman & Hall, London.

Langer, M. 1989. Waste disposal in the Federal Republic of Germany: concepts, criteria, scientific investigations. *Bulletin International Association Engineering Geology*, **39**, 53–58.

Langer, M. 1995. Engineering geology and waste disposal: scientific report and recommendations of the IAEG Commission No.14. *Bulletin International Association Engineering Geology*, **51**, 5–29.

Langer, M. and Heusermann, S. 2001. Geomechanical stability and integrity of waste disposal in salt structures. *Engineering Geology*, **61**, 155–161.

Langer, M. and Wallmer, M. 1988. Solution-mined salt caverns for the disposal of hazardous chemical wastes. *Bulletin International Association Engineering Geology*, **37**, 61–70.

Ling, H.I., Leshchinsky, D., Mohri, Y. and Kawabata, T. 1998. Estimation of municipal solid waste settlement. *Proceedings American Society Civil Engineers, Journal Geotechnical and Geoenvironmental Engineering*, **124**, 21–28.

Miller, C.J. and Lee, J.Y. 1999. Response of landfill clay liners to extended periods of freezing. *Engineering Geology*, **51**, 291–302.

Mitchell, J.K. 1986. Hazardous waste containment. In: *Groundwater in Engineering Geology*, Engineering Geology Special Publication No. 3, Cripps, J.C., Bell, F.G. and Culshaw, M.G. (eds), The Geological Society, London, 145–157.

Mitchell, J.K. 1994. Physical barriers for waste containment. *Proceedings First International Congress on Environmental Geotechnics*, Edmonton, Carrier, W.D. (ed.), BiTech Publishers Ltd, Richmond, BC, 951–962.

Mitchell, J.K. and Madsen, F.T. 1987. Chemical effects on clay hydraulic conductivity. In: *Geotechnical Practice for Waste Disposal '87*, American Society Civil Engineers, Geotechnical Special Publication No. 13, 87–116.

Mollamahmutoglu, M. and Yilmaz, Y. 2001. Potential use of fly ash and bentonite mixture as liner or cover at waste disposal areas. *Environmental Geology*, **40**, 1316–1324.

Mollard, S.J., Jefford, C.E., Staff, M.G. and Browning, G.R.J. 1996. Geomembrane landfill liners in the real world. In: *Engineering Geology of Waste Disposal*, Engineering Geology Special Publication No. 11, Bentley, S.F. (ed.), The Geological Society, London, 165–170.

Morfeldt, C.O. 1989. Different subsurface facilities for the geological disposal of radioactive waste (storage cycle) in Sweden. *Bulletin International Association Engineering Geology*, **39**, 25–34.

Morner, N.-A. 2001. In absurdum: long-term predictions and nuclear waste handling. *Engineering Geology*, **61**, 75–82.

Murray, E.J., Rix, D.W. and Humphrey, R.D. 1996. Evaluation of clays as linings to landfill. In: *Engineering Geology of Waste Disposal*, Engineering Geology Special Publication No. 11, Bentley, S.F. (ed.), The Geological Society, London, 251–258.

Oweis, I.S. and Khera, R.P. 1990. *Geotechnology of Waste Management*. Butterworths, London.

Panagiotakopoulos, D. and Dokas, I. 2001. Design of landfill daily cells. *Waste Management and Research*, **19**, 332–341.

Powrie, W. and Beaven, R.P. 1999. Hydraulic properties of household waste and implications for landfills. *Proceedings Institution Civil Engineers, Geotechnical Engineering*, **137**, 235–247.

Prashanth, J.P., Sivapullaiah, P.V. and Sridharan, A. 2001. Pozzolanic fly ash as a hydraulic barrier in landfills. *Engineering Geology*, **60**, 245–252.

Quigley, R.M., Fernandez, F. and Crooks, V.E. 1988. Engineered clay liners: a short review. *Proceedings Symposium on Environmental Geotechnics and Problematic Soils and Rocks*, Bangkok, Balasubramanian, A.S., Chandra, S., Bergado, D.T. and Natalaya, P. (eds), A.A. Balkema, Rotterdam, 63–74.

Raybould, J.G. and Anderson, J.G. 1987. Migration of landfill gas and its control by grouting – a case history. *Quarterly Journal Engineering Geology*, **20**, 78–83.

Ressi, A. and Cavalli, N. 1984. Bentonite slurry trenches. *Engineering Geology*, **21**, 333–339.

Rogers, V. 1994. Present trends in nuclear waste disposal. *Proceedings First International Congress on Environmental Geotechnics*, Edmonton, Carrier, W.D. (ed.), BiTech Publishers Ltd, Richmond, BC, 837–845.

Sara, M.N. 1994. *Standard Handbook for Solid and Hazardous Waste Facilities Assessment*. Lewis Publishers, Boca Raton, Florida.

Savage, D., Arthur, R.C. and Saito, S. 1999. Geochemical factors in the selection and assessment of sites for deep disposal of radioactive wastes. In: *Chemical Containment of Waste in the Geosphere*. Special Publication No. 157, Metcalfe, R. and Rochelle, C.A. (eds), The Geological Society, London, 27–45.

Schultz, O., Popp, T. and Kern, H. 2001. Development of damage and permeability in deforming rock salt. *Engineering Geology*, **61**, 163–180.

Sharma, R.S. and Al-Busaidi, T.S. 2001. Groundwater pollution due to a tailings dam. *Engineering Geology*, **60**, 235–244.

Smith, C.C., Cripps, J.C. and Wymer, M.J. 1999. Permeability of compacted colliery spoil – a parametric study. *Engineering Geology*, **53**, 187–193.

Smith, C.F., Cohen, J.J. and McKone, T.E. 1980. *Hazard Index for Underground Toxic Material*. Lawrence Livermore Laboratory, Report No UCRL-52889.

Smith, G.J. and Staff, M.G. 1997. Design of restoration capping for landfills. In: *Engineering Geology of Waste Disposal*, Engineering Geology Special Publication No. 11, Bentley, S.F. (ed.), The Geological Society, London, 377–382.

Smith, M.L. and Williams, R.E. 1996. Examination of methods for evaluating remining a mine waste site. Part 1: Geostatistical characterization methodology. Part 2: Indicator kridging for selective remediation. *Engineering Geology*, **43**, 11–21, 23–30.

Spooner, A.J. and Giusti, L. 1999. Geochemical interactions between landfill leachate and sodium bentonite. In: *Chemical Containment of Waste in the Geosphere*, Special Publication No. 157, Metcalfe, R. and Rochelle, C.A., (eds), The Geological Society, London, 131–142.

Talbot, C.J. 1999. Ice ages and nuclear waste isolation. *Engineering Geology*, **52**, 177–192.

Tay, Y.Y., Stewart, D.I. and Cousens, T.W. 2001. Shrinkage and desiccation cracking in bentonite-sand landfill liners. *Engineering Geology*, **60**, 263–274.

Taylor, R.K. 1975. English and Welsh colliery spoil heaps – mineralogical and mechanical relationships. *Engineering Geology*, **7**, 39–52.

Tissinger, L.G., Peggs, I.D. and Haxo, H.E. 1991. Chemical compatibility testing of geomembranes. In: *Geomembranes Indentification and Performance Testing*. Chapman & Hall, London, 96–111.

Troncoso, J.H. and Troncoso, D.A. 1999. Rehabilitation of abandoned deposits of mineral residues. *Proceedings Sixth International Conference on Tailings and Mine Waste '99*, Fort Collins, Colorado, A.A. Balkema, Rotterdam, 693–700.

Van Impe, W.F. and Bouazza, A. 1996. Densification of domestic waste fills by dynamic compaction. *Canadian Geotechnical Journal*, **33**, 879–887.

Van Zyl, D.J.A. 1993. Mine waste disposal. In: *Geotechnical Practice for Waste Disposal*, Daniel, D.E. (ed.), Chapman & Hall, London, 269–287.

Vick, S.G. 1983. *Planning, Design and Analysis of Tailings Dams*. Wiley, New York.

Watts, K.S. and Charles, J.A. 1999. Settlement characteristics of landfill wastes. *Proceedings Institution Civil Engineers, Geotechnical Engineering*, **137**, 225–233.

Williams, G.M. and Aitkenhead, N. 1991. Lessons from Loscoe: the uncontrolled migration of landfill gas. *Quarterly Journal Engineering Geology*, **24**, 191–208.

Wilson, S.A. 2002. Design and performance of a passive dilution gas migration barrier. *Ground Engineering*, **35**, No. 1, 34–37.

Yong, R.N., Tan, B.K., Bentley, S.P. and Thomas, H.R. 1999. Competency assessment of two clay soils from South Wales for landfill liner contaminant attenuation. *Quarterly Journal Engineering Geology*, **32**, 261–270.

10 Derelict and contaminated land

10.1 Derelict land and urban areas

Derelict land can be regarded as land that has been damaged by industrial use or other means of exploitation to the extent that it has to undergo some form of remedial treatment before it can be of beneficial use. Such land usually has been abandoned in an unsightly condition and often is located in urban areas where land for development is scarce. Consequently, not only is derelict land a wasted resource but it also has a blighting effect on the surrounding area and can deter new development. Its rehabilitation therefore is highly desirable, not only by improving the appearance of an area but also by making a significant contribution to its economy by bringing derelict land back into worthwhile use. Accordingly, there is both economic and environmental advantage in the regeneration of derelict land. Land recycling in urban areas also can be advantageous since the infrastructure generally is still in place. Moreover, its regeneration should help reduce the exploitation of greenfield sites and encourage that of brownfield sites. The term 'brownfield' normally is applied in a broad sense to land that has been developed previously and as such includes vacant land, with or without buildings, together with derelict and contaminated land. The regeneration of brownfield sites is linked with the process of sustainable development. Hence, the potential for redevelopment of a brownfield site needs to be assessed in terms of economic, environmental and social factors that contribute to the overall concept of sustainable development, as well as site-based factors (Anon., 1999a). The environmental geotechnics of a brownfield site influence its subsequent re-use as well as affecting the costs of bringing the site into a state suitable for redevelopment.

The existence of abandoned derelict sites runs counter to the concept of sustainable development. Sustainable land use is essential if present-day development needs are to be met without compromising the ability of future generations to meet their needs. A sustainable land use therefore requires derelict sites to be re-used, thereby recovering such sites as a land resource. It has been estimated that two-thirds of the population of the earth will live in cities by 2015, which emphatically highlights the need for more efficient use of urban space, especially in those cities that are growing at a rapid rate. In the European Community there are 52 000 ha of derelict land associated with the mining industry alone, 80 per cent of which occurs in Belgium, Britain, France, Germany and Spain (Ferber, 1995). Mabey (1991) reviewed the position of derelict land in Britain, noting that in 1988 there were some 40 500 ha of derelict land but that the amount was decreasing. Of the land attributable to industrial dereliction, 94 per cent was considered to justify remediation. Fortunately, attitudes towards derelict land have changed and legislation has been enacted in many countries in the form of planning acts to limit its development and to facilitate its restoration.

The use to which derelict land is put should suit the needs of the surrounding area and be compatible with other forms of land use. Restoring a site to a condition that is well integrated into its surroundings also upgrades the character of the environment beyond the confines of the site. In the past derelict land in rural areas usually has been reclaimed for agricultural purposes. However, in the European Community in particular, where more efficient food production has meant that areas of farm land are now surplus to requirements, this type of goal has been re-assessed. Furthermore, there is a growing awareness that the quality of life is influenced by cultural heritage, protection of biodiversity and nature conservation. Therefore, derelict land in rural areas can offer opportunities for habitat restoration. In addition, restoration for conservation may prove less demanding financially than restoration to agriculture. Nonetheless, any type of restoration must take account of safety (e.g. in terms of derelict buildings, mine shafts, etc.) and potential problems of pollution off-site.

Any project involving the restoration of derelict land requires a feasibility study. This needs to consider accessibility to the site; land use and market value; land ownership and legal issues; topography and geological conditions; site history and contamination potential; and the local environment and existing infrastructure. The results of the feasibility study allow an initial assessment to be made of possible ways to develop a site and the costs involved. This is followed by a site investigation. Aerial photographs may prove of value, especially for large sites. The investigation provides essential input data for the design of remedial measures and indicates whether or not demolition debris can be used in the rehabilitation scheme.

Derelict land may present hazards, for example, disposal of industrial wastes may have contaminated land, in some cases so badly that earth has to be removed, or the ground may be severely disturbed by the presence of massive old foundations and subsurface structures such as tanks, pits and conduits for services (Fig. 10.1). Contaminated land may emit gases or may represent a fire hazard. Details relating to such hazards should be determined during the site investigation (Pearson, 1992). Site hazards result in constraints on the freedom of action, necessitate following stringent safety requirements, may involve time-consuming and costly working procedures, and affect the type of development. For instance, Leach and Goodyear (1991) noted that constraints may mean that the development plan has to be changed so that the more sensitive land uses are located in areas of reduced hazard. Alternatively, where notable hazards exist, then a change to a less sensitive end use may be advisable. Settlement is another problem that frequently has to be faced when derelict industrial land, which consists of a substantial thickness of rubble fill, is to be built on. If this is not contaminated, then dynamic compaction or vibro-compaction can be used to minimize the amount of settlement. Where the soils are soft, then they can be preloaded, with or without vertical drains, or stabilized by deep mixing whereby cement, lime or cementitious fly ash is mixed into the ground by auger rigs (Thom and Hausmann, 1997). Nonetheless, Leach and Goodyear maintained that there are few derelict sites that cannot be brought back into beneficial use.

Derelict sites may require varying amounts of filling, levelling and regrading. As far as fill is concerned, this should be obtained from on site if possible, otherwise from an area nearby. Reid and Buchanan (1987), for example, mentioned that the derelict Queen's Dock in Glasgow, Scotland, had been filled with 1.3 million m^3 of demolition rubble from old buildings in the vicinity, degradable material such as timber having been removed. Similarly, Hartley (1991) referred to the use of old foundation material that was crushed and used as fill at a site in Middlesbrough, England. Once regrading has been completed, the actual surface needs restoring. This is not so important if the area is to be built over (e.g. if it is to be used for an industrial estate), as it is if it is to be used for amenity or recreational purposes. In the case where buildings are to be erected, however, the ground must be adequately compacted so that they are not subjected to adverse settlement. On the other hand, where the land is to be used

Figure 10.1 Derelict land in the Ruhr, Germany, showing massive old foundations.

for amenity or recreational purposes, then soil fertility must be restored so that the land can be grassed and trees planted. This involves laying top-soil (where appropriate) or substitute materials, the application of fertilizers and seeding (frequently by hydraulic methods). Adequate subsoil drainage also needs to be installed.

Derelict land caused by some earlier industrial operation may contain structures or machinery of historical interest and worthy of preservation. According to Brown (1998), any assessment of the industrial archaeological value of a site must consider its scientific and engineering interest, its state of preservation and completeness, and its representativeness and rarity. The amenity, recreation and tourism aspects also need to be considered (Fig. 10.2). Brown provided a number of examples of partly or wholly derelict sites in Shropshire and West Yorkshire, England, which had been developed as sites of archaeological interest including the multi-site Ironbridge Gorge Museum in the area where the Industrial Revolution began.

Figure 10.2 The 'Big Hole' near Kimberley, South Africa. This famed old diamond working is a popular tourist attraction.

10.2 Derelict land: restoration of old quarries, pits and mines

Some of the worst dereliction has been associated with past mineral workings and mining activities (Fig. 10.3). Abandoned quarries and pits, where ground conditions are suitable (e.g. old bricks pits located in thick formations of clay), frequently have been used for the disposal of domestic waste and then subsequently reclaimed. Others, where the water table is high, have become flooded and as such can represent a hazard. In the case of abandoned mines, old buildings and spoil heaps represent notable forms of surface dereliction. The presence of old shafts represents a particular problem (the location and treatment of abandoned shafts is dealt with in Chapter 4), another being subsidence, especially where it has been responsible for extensive flooding necessitating reclamation. Abandoned quarries, pits and spoil heaps are particularly conspicuous, and can be difficult to rehabilitate into the landscape. Like other forms of dereliction such features have a blighting effect on both the local environment and economy. The reclamation of colliery spoil heaps has been dealt with in Chapter 9.

According to Bailey and Gunn (1991) reclamation strategies for hard rock quarries have been surprisingly limited. In the past quarries have been screened from view by planting trees

Figure 10.3a Derelict old mining works, Butte, Montana.

around them or, where conditions allow, used for waste disposal. Some large old quarry voids have been used for storage or for the location of industrial units. Such uses usually require that the quarry faces are treated to reduce the potential for rock fall. This involves removal of loose rock from quarry faces and possibly stabilization measures. It also may involve material being placed at the base of faces, and then being covered with soil and planted. Bailey and Gunn suggested that in limestone areas in particular, landform replication represents the best method of quarry reclamation. Landform replication involves the construction of landforms and associated habitats similar to those of the surrounding environment. Restoration blasting along quarry faces can be used to replicate scree slopes and to produce multi-faceted slope sequences. Different grades of crushed limestone can be used to dress some scree slopes to encourage the growth of characteristic limestone flora.

The extraction of sand and gravel deposits in low-lying areas along the flanks of rivers frequently means that the workings eventually extend beneath the water table. On restoration it is not necessary to fill the flooded pits completely. Partial filling and landscaping can convert such sites into recreational areas offering such facilities as sailing, fishing and other water sports. It is necessary to carry out a thorough survey of flooded workings, with soundings being taken so that accurate plans and sections can be prepared. The resultant report then forms the basis for the design of the measures involved in rehabilitation. Bell and Genske (2000) described old sand and gravel workings to the west of Brighouse in West Yorkshire, England, which were in need of rehabilitation. The site was approximately 16.3 ha in extent and had

Figure 10.3b Waste from the old copper smelter at Anaconda, Montana, reclaimed as a golf course. Note the black bunkers.

been an active sand and gravel pit located in the valley of the River Calder. As much of the excavation was below water, two berms had been constructed across the workings to reduce the amount of pumping (Fig. 10.4). It was decided to convert the site into a marina after the workings closed. The western part of the site was substantially higher than the east and so the site had to be regraded, excavation being carried out on a cut-and-fill basis by scrapers, after which layers of subsoil and topsoil were placed. A caravan park and various facilities were provided. In addition to boating and fishing, the lake and islands provide a sanctuary for wild life.

Opencast working of coal involves excavation to depths of up to around 100 m below the surface. Working is advanced on a broad front with face lengths of three to five kilometres not being uncommon. Once a site is no longer operational, then it is restored, which influences the way in which excavation takes place in that the material excavated, minus the coal, is used to fill the void. The topsoil is removed by a scraper and placed in dumps around the site. Similarly, the subsoil is stripped and put into separate dumps about the site. Face shovels are used to excavate the initial box-cut and for subsequent forward reduction of overburden. The material from the box-cut is placed above ground in a suitable position in relation to void filling so that later rehandling is minimized. Draglines excavate the lower seams in a progressive strip cut behind the face shovels. Drilling and blasting is carried out where necessary. Restoration can begin before a site is closed, indeed this usually is the more convenient method. Hence, worked out areas behind the excavation front are filled with rock waste. This means that the final contours can be designed with less spoil movement than if the two operations were undertaken separately. Furthermore, more soil for spreading can be conserved when restoration and coal working are carried out simultaneously. Because of high stripping ratios (often 15:1–25:1), there usually is enough spoil to more or less fill the void. The restored land generally is used for agriculture or forestry but it can be used for country parks, golf courses, etc. The water

Figure 10.4 Berms in old sand and gravel workings, created at the start of reclamation for a marina near Brighouse, West Yorkshire, England.

table at many opencast sites is lowered by pumping in order to provide dry working conditions in the pit. If a site is to be restored for agricultural use or for forestry, then this takes place without compaction control. However, if a site is to be built over, then settlement is likely to be a problem without proper compaction. Significant settlements of opencast backfill can occur when the partially saturated material becomes saturated by rising groundwater after pumping has ceased. For example, Charles *et al.* (1993) referred to an opencast site in Northumberland, England, where the waste was backfilled without any systematic compaction. They recorded that when the water table rose, then some 0.33 m of settlement occurred where the fill was 63 m in depth. Settlement due to wetting collapse is more significant than that due to the self-weight of the backfill, which can give rise to difficulties in settlement prediction (Blanchfield and Anderson, 2000). Therefore, as remarked, if a site is to be built upon after restoration, then the waste should be properly compacted. Alternatively, Kilkenny (1968) recommended that where opencast fills exceed 30 m in depth, then the minimum time that should elapse before development takes place should be 12 years after restoration is complete. This was based on comprehensive observations at an opencast site where the fill varied from 23 to 38 m in depth. Approximately 50 per cent of settlement was complete within two years and 75 per cent within five years. Restoration of other opencast mineral workings can be dealt with in a similar way to those associated with coal.

One of the consequences of subsidence due to mining, especially longwall working of coal in low-lying areas alongside rivers, is flooding. One of the most notable regions where coal mining has taken place is the Ruhr Basin in Germany (Fig. 10.5a), where the maximum subsidence recorded is 24 m. By the end of the nineteenth century subsidence had caused the reversal of natural drainage in extensive areas, and this gave rise to problems with sanitation and associated outbreaks of typhus and cholera. In fact, flooding was characteristic of this area before mining

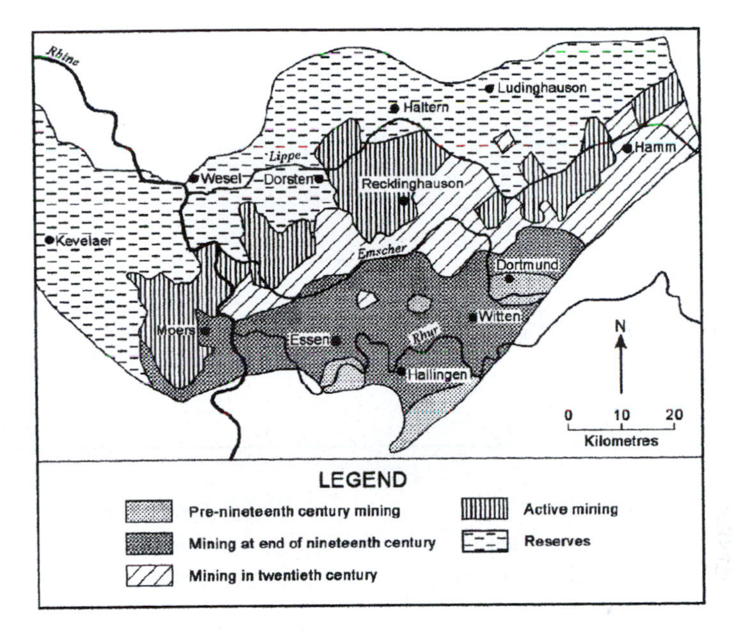

Figure 10.5a Coal mining in the Ruhr district of Germany.
Source: Bell and Genske, 2001.

of coal began and consequently has been exacerbated by subsidence, giving rise to a situation where much of the Emscher area is now a 'polderland' (Bell and Genske, 2001). Accordingly, areas now have to be drained by a large number of pumping stations to protect them from flooding.

Similar conditions exist in the eastern lowlands of the River Lippe so that a belt affected by subsidence also extends from the Rhine along the northern Ruhr district to Hamm. By 1989 the 'polderland' along the River Emscher totalled approximately 340 km^2, while along the River Lippe there were around 243 km^2 of 'polderland'. The development of subsidence troughs below the water table leads to surface areas being inundated, resulting in the formation of ponds and lakes. In densely populated areas these can have amenity value in that they can be developed for recreational purposes and as nature reserves. On the other hand, the development of such lakes may mean the roads and railways have to be re-routed, that existing buildings have to be protected and that agriculture is adversely affected. Some lakes in the above-mentioned areas have been filled and rivers have had to be realigned and channelled. The Dellwig Creek near Dortmund provides one such example. Around 1930 an enlarged concrete lined streambed was constructed for Dellwig Creek that was large enough to deal with the sewage of the surrounding communities. Then, in 1977, the area was declared a development zone for recreation, Dellwig Creek no longer being a sewage channel. The rehabilitation involved removal of the concrete lining of the canalized creek and the construction of a diversion canal with the surrounding area being turned into a parkland (Fig. 10.5b).

Bell *et al.* (2000) described the utilization of methane for a reclamation project at an old coal mine site in North Rhine-Westphalia, Germany. Methane gas is emitted from three old shafts associated with a coal mine. This is the first time in the Ruhr area that methane utilization has happened even though it is estimated that 120 million cubic metres of methane escape annually from decommissioned coal mines in the area, which is equivalent to 100 000 tonnes of fuel oil. Formerly, the gas was burnt without it being utilized, which resulted in the emission of approximately 8 million tonnes of CO_2. It was estimated that about 1 million cubic metres of

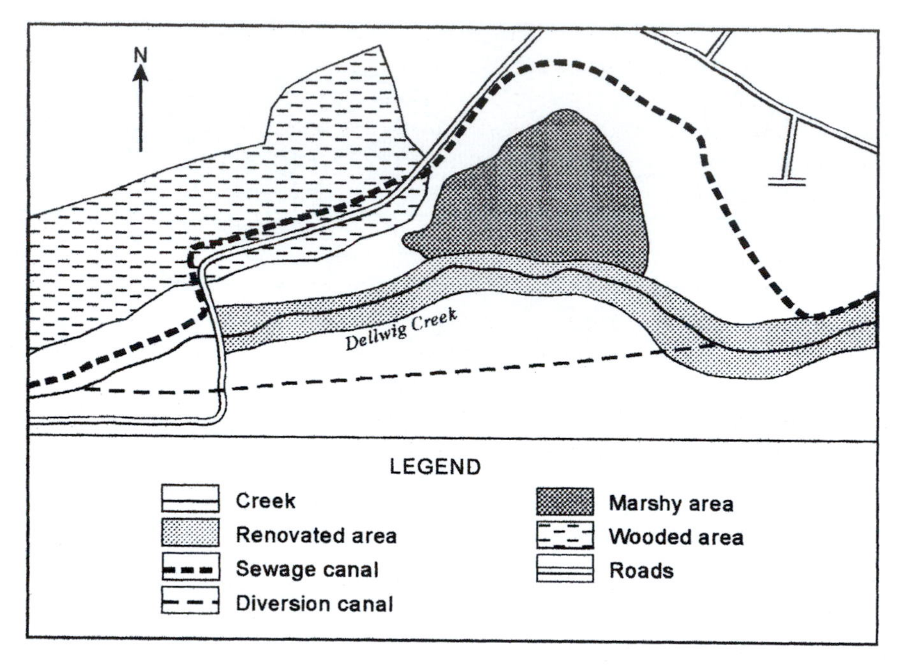

Figure 10.5b Rehabilitation of Dellwig Creek, near Dortmund, Germany.
Source: Bell and Genske, 2001.

gas is available for use from the mine each year, which converts to 2 million kWh of electrical energy and 3 million kWh of heat for new buildings. However, the emission of gas is not constant, fluctuating according to atmospheric pressure. Accordingly, a supplementary natural gas plant of 1800 kWh plus a hot water storage tank was constructed to secure a constant energy and heat supply. To buffer peak consumption the installation also is connected to the municipal supply, the connection also allowing discharge of surplus energy produced from mine gas.

10.3 Contaminated land

Land can contain substances that are undesirable or even hazardous as a result of natural processes, for example, as a result of mineralization. However, most cases of contaminated land are associated with human activity. In particular, in many of the industrialized countries of the world, one of the legacies of the past two centuries is that land has been contaminated. The reason for this is that industry and society tended to dispose of their waste with little regard for future consequences. Hence, when such sites are cleared for redevelopment they can pose problems (Wood, 1994). Contamination can be brought about by atmospheric fallout, by the disposal of liquid wastes, by leakage or spillage of liquids, and by the disposal of solid waste. Parry and Bell (1987) catalogued some of the principal sources of contamination as those attributable to process works, mining, landfill (see Chapter 9), bulk wastes and transportation.

Contamination can take many forms and can be variable in nature across a site, and each site has its own characteristics (Attewell, 1993). In some cases only a single previous use of a site may be identified, which has a characteristic pattern of contamination. For instance, Thorburn and Buchanan (1987) referred to a site in Glasgow, Scotland, on which a church was constructed, which contained some three million tonnes of alkali waste from an old chemical works that consisted primarily of calcium sulphate and calcium carbonate. On the other hand, some sites will have had a number of former uses, especially when industry was established on them over a long period of time. In such cases there may be no particular characteristic pattern of contamination present. The types of contaminants that may be encountered include heavy metals, sulphates, asbestos, various organic compounds, toxic and flammable gases, combustible materials and radioactive materials. A review of the chief types of contaminants has been provided by Haines and Harris (1987), and includes a survey of their sources and principal hazards. Moreover, the increasing scarcity of acceptable land for development in many western countries means that poorer quality sites are considered for re-development. However, the increasing awareness of environmental issues has meant that within the last 20 years or so contaminated land has become a matter of public concern (Harris, 1987). Indeed, there are few geotechnical tasks that are as complex as the remediation of contaminated land. It is not only the spectrum of possible pollutants and their migration behaviour but also the variety of boundary conditions that makes this problem so demanding. Hence, remediation of contaminated land is not a standard task and requires an interdisciplinary effort.

10.3.1 Contaminated land and hazard

Contaminated land is by no means easy to define but it can be regarded as land that contains substances that, when present in sufficient concentrations, are likely to cause harm, directly or indirectly, to man, to the environment or to other targets. The British Standards Institution Draft Document (Anon., 1988a) defined contaminated land as *land, that because of its nature or former uses, may contain substances that could give rise to hazards likely to affect a proposed form of development.* Such a definition poses the question of what is meant by hazard? Hazard implies a degree of risk, but the degree of risk varies according to what is being risked. It depends, for example, upon the mobility of the contaminant(s) within the ground and different types of soils have different degrees of reactivity to compounds that are introduced. It also is influenced by the future use of a site. In fact, no simple definition of hazard can incorporate the variety of circumstances that may arise. Therefore, Beckett (1993) suggested that contamination should be regarded as a concept rather than something that is capable of exact definition. In any given circumstances the hazards that arise from contaminated land will be peculiar to the site and will differ in significance. Consequently, the concept of risk analysis for contaminated land assessment is complex and needs a great deal of investigation and assessment of data before it can be used with effect. One of the draft documents released by the Department of the Environment (Anon., 1995) in Britain indicates that there was no consensus as to what was an acceptable risk. This draft document defined contaminated land as *any land which appears to be in such a condition, by reason of substances in, on or under the land that significant harm is being caused or there is a significant possibility of such harm being caused; or pollution of controlled waters is being, or is likely to be caused.* Harm was defined as *harm to the health of living organisms or other interference with the ecological systems of which they form part, and in the case of man includes harm to his property.* Severe, moderate, mild and minimal degrees of harm were recognized by Anon. (1995), as well as high, medium, low and very low degrees of possibility of harm being caused. A later draft circular on contaminated land was issued by the Department of the Environment, Transport and the Regions in September 1999 (Anon., 1999b). It considers the identification and remediation of contaminated land, as well as providing guidance on statutes and regulations referring to such land.

A fundamental objective of risk assessment and risk management is the need to define whether risks are real or are perceived. The nature of the relationship between the source of the contaminants, the pathway(s) and the receptor(s) determines the degree of risk (Smith, 1998).

As noted, contamination is site specific in that the variation of contaminants and the host media are peculiar to the particular site. Accordingly, a prescriptive solution generally is inappropriate and so adoption of a risk management framework allows proper characterization and evaluation of a site, selection of appropriate remedial strategies, effective reduction or control of defined risks, and thereby effective technical and financial control of a project. Risk assessment requires the acquisition of data by a site investigation so that hazards are identified and evaluated. The assessment should ensure that any unacceptable health or environmental risks are identified and dealt with appropriately.

The presence of potentially harmful substances at a site may not necessarily require remedial action, if it can be demonstrated that they are inaccessible to living things or materials that may be detrimentally affected. However, consideration must always be given to the migration of soluble substances. The migration of soil borne contaminants is associated primarily with groundwater movement, and the effectiveness of groundwater to transport contaminants is dependent mainly upon their solubility. The quality of water can provide an indication of the mobility of contamination and the rate of dispersal. In an alkaline environment, the solubility of heavy metals becomes mainly neutral due to the formation of insoluble hydroxides. Providing groundwater conditions remain substantially unchanged during the development of a site, then the principal agent likely to bring about migration will be percolating surface water. On many sites the risk of migration off site is of a very low order because the compounds have low solubility, and frequently most of their potential for leaching has been exhausted. Liquid and gas contaminants, of course, may be mobile. Obviously, care must be taken on site during working operations to avoid the release of contained contaminants (e.g. liquors in buried tanks) into the soil. Where methane has been produced in significant quantities, it can be oxidized by bacteria as it migrates through the ground with the production of carbon dioxide. However, this process does not necessarily continue.

According to Genske and Thein (1994), the key to effective rehabilitation of contaminated land involves the harmonized management of ground investigation, risk assessment and clean-up strategies. However, the accumulation of large amounts of data from a major site, which are derived from different sources, can represent a problem of processing. The data has to be sorted, simplified and represented on plans, sections and three-dimensional models. Digitization of data for the production of various site plans can be accomplished by a regular computer aided design (CAD) application or a geographical information system (GIS). Although the GIS approach is more sophisticated, CAD systems can be used to generate the necessary site plans for most remediation projects. In addition, data can be represented in three dimensions with the aid of a suitable work station.

10.3.2 Attitudes towards contamination

A number of countries have developed criteria for assessing the risk posed by contamination of ground and of groundwater. Such criteria tend to reflect the environmental and legal conditions that prevail in the countries concerned. The assessment of sites generally is related to clean-up standards and/or to the amount of contamination measured against some critical concentration above which humans, animals and the environment can be caused harm if some type of action is not undertaken. Hence, two types of criteria have been recognized, namely, one for a clean-up standard and the other for a level beyond which intervention takes place, that is, remedial action is required. Regulatory intervention takes place at a higher concentration of contamination than that for a clean-up standard.

In Britain the Government maintains a commitment to the 'suitable for use' approach to the control and treatment of existing contaminated land. This supports sustainable development by reducing damage from past activities and by permitting contaminated land to be kept in, or returned to, beneficial use wherever possible. Such an approach only requires remedial action where the contamination poses unacceptable actual or potential risks to health and/or the environment, and where there are appropriate and cost effective means available to do so. A contaminated site in Britain therefore will only undergo rehabilitation if it is to be redeveloped, and then when the contaminants exceed certain levels. Previously, threshold levels had been proposed by the Interdepartmental Committee on the Redevelopment of Contaminated Land (Anon., 1987; Table 10.1a). These were the most commonly used guidelines and provided practitioners with some guidance relating to the different types of historical contamination and hazards that are present in Britain. The threshold values offered an indirect method of

Table 10.1a Guidance on the assessment and redevelopment of contaminated land

Contaminants	*Use code*	*Reference value trigger concentrations mg kg^{-1} air-dried soil*	
		Threshold	*Action*
Group A: Selected inorganic contaminants that may pose hazards to health			
Arsenic	1	10	NS
	2	40	NS
Cadmium	1	3	NS
	2	15	NS
Chromium total	1	600	NS
	2	1000	NS
Chromium (hexavalant) (1)	1,2	25	NS
Lead	1	500	NS
	2	2000	NS
Mercury	1	1	NS
	2	20	NS
Selenium	1	3	NS
	2	6	NS
Group B: Contaminants that are phytotoxic, but not normally hazards to health			
Boron (water soluble) (2)	4	3	NS
Copper (3) (4)	4	130	NS
Nickel (3) (4)	4	70	NS
Zinc (3) (4)	4	300	NS
Contaminants associated with former coal carbonization sites			
Polyaromatic hydrocarbons (5) (6) (7)	1	50	500
	3, 5, 6	1000	10000
Phenols (5)	1	5	200
	3, 5, 6	5	1000
Free cyanide (5)	1, 3	25	500
	5, 6	25	500
Complex cyanides (5)	1	250	1000
	3	250	5000
	5, 6	250	NL
Thiocyanate (5) (7)	All	50	NL

Table 10.1a (Continued)

Contaminants	*Use code*	*Reference value trigger concentrations mg kg*$^{-1}$ *air-dried soil*	
		Threshold	*Action*
Sulphate (5)	1, 3	2000	10 000
	5	2000 (8)	50 000 (8)
Sulphide	All	250	1000
Sulphur	All	500	20 000
Acidity (pH less than)	1, 3	pH 5	pH 3
	5, 6	NL	NL

Source: ICRCL (Anon., 1987).

Use codes

1 Domestic gardens and allotments
2 Parks, playing fields, open space
3 Landscaped areas
4 Any use where plants are grown (applies to contaminants that are phytotoxic, but not normally hazards to health)
5 Buildings
6 Hard cover.

Conditions

1 Tables are invalid if reproduced without the conditions and footnotes.
2 All values are for concentrations determined on 'spot' samples based on adequate site investigation carried out prior to development. They do not apply to analysis of averaged, bulked or composited samples, nor to sites that have already been developed.
3 Many of these values are preliminary and will require regular updating. For contaminants associated with former coal carbonization sites, the values should not be applied without reference to the current edition of the report (Anon., 1987). *Problems Arising from the Redevelopment of Gas Works and Similar Sites,* Second edition, Department of the Environment, London.
4 If all samples values are below the threshold concentrations, then the site may be regarded as uncontaminated as far as the hazards from these contaminants are concerned, and development may proceed. Above these concentrations remedial action may be needed, especially if the contamination is still continuing. Above the action concentration, remedial action will be required or the form of development changed.

Footnotes

NS Not specified.
NL No limit set as the contaminant does not pose a particular hazard for this use.

1 Soluble hexavalant chromium extracted by 0.1 M HCl at 37°C; solution adjusted to pH 1.0 if alkaline substances present.
2 Determined by standard ADAS method (soluble in hot water).
3 Total concentration (extraction by $HNO_3/HClO_4$).
4 Total phytotoxic effects of copper, nickel and zinc may be additive. The trigger values given here are those applicable to the worst case phytotoxic effects that may occur at these concentrations in acid, sandy soils. In neutral or alkaline soils phytotoxic effects are unlikely at these concentrations.
The soil pH value is assessed to be about 6.5 and should be maintained at this value. If the pH falls, the toxic effects and uptake of these elements will be increased.
Grass is more resistant to phytotoxic effects than most other plants and its growth may be adversely affected at these concentrations.
5 Many of these values are preliminary and will require irregular updating. They should not be applied without reference to the current edition of the report *Problems Arising from the Redevelopment of Gas Works and Similar Sites.*
6 Used here as a marker for coal tar, for analytical reasons. See *Problem Arising from the Redevelopment of Gas Works and Similar Sites, Annex A1.*
7 See *Problem Arising from the Redevelopment of Gas Works and Similar Sites* for details of analytical methods.
8 See also BRE Digest 250: *Concrete in Sulphate-Bearing Soils and Groundwater.* Building Research Establishment, Watford.

assessing the risk from levels of contamination in soil according to land-use. Three possible contamination zones were recognized for each contaminant, namely, areas of acceptable risk and of unacceptable risk, which were separated by a zone for professional judgement. In theory, the threshold and action values that were based on the concentration of the contaminant were supposed to establish the boundaries between these zones. Unfortunately, in practice only the threshold values of many of the common metal contaminants were established. With time their purpose became confused with that of remediation standards. As a result, the ICRCL guidelines (Anon., 1987, Guidance Note 59/83) were withdrawn by the Department for Environment, Food and Rural Affairs (DEFRA) in December, 2002. DEFRA maintained that although the guidelines had proved a useful tool, they are now technically out of date and their approach is not in line with the current statutory regime (Part IIA of the Environmental Protection Act 1990) and associated policy. In particular, DEFRA noted that the guidelines are not suitable for assessing the significant possibility of significant harm to human health that is now required, significant harm being defined as death, disease (e.g. cancer, liver dysfunction, extensive skin ailments, etc.), serious injury, genetic mutation, birth defects or impairment of reproductive functions.

Consequently, DEFRA has published a comprehensive package, referred to as the Contaminated Land Exposure Assessment (CLEA) that refers specifically to the assessment of risk to human health arising from long-term exposure to soil contamination. This package, which forms part of the Contaminated Land Report (Anon., 2002), deals with the direct assessment of risks to human health from soil contamination. It is based on two principal factors. First, on toxicological criteria that establish a level of unacceptable human intake of a contaminant derived from the soil, and, second, on estimates of human exposure to soil contamination based on generic land-use. The latter takes into consideration the characteristics of adults and children, and their activity patterns, as well as the fate and transport of the contaminant in the soil. As mentioned, the ICRCL (Anon., 1987) guidelines related hazards to land-uses. In other words, they recognized lower thresholds for certain uses such as residential developments with gardens than for hard cover areas. This concept was adopted by the CLEA guidelines, which recognize four categories of land-use, namely, residential with and without plant uptake, allotments and commercial/industrial uses.

Obviously, there are difficulties in establishing a particular level of concentration for a contaminant beyond which there is an unacceptable risk to human health as this requires toxicological data on health effects and definition of what is unacceptable risk. Accordingly, the Contaminated Land Report (CLR) consists of a series of documents that have been produced by DEFRA and the Environment Agency to provide interested parties with appropriate and authoritative information and advice on the assessment of risks due to soil being contaminated (Anon., 2002). However, it must be borne in mind that soil is not the only source of contamination. Document CLR7 of the Contaminated Land Report gives an overview of the development and use of Soil Guideline Values (SGVs) and research related to the assessment of risks to human health from contaminated land. It also considers approaches to soil sampling, testing and data analysis in relation to the determination of site contamination values. Obviously, contaminant concentrations vary across sites, and sampling and analysis are likely to introduce some errors into the data obtained. Consequently, the mean concentration determined from sampling will have some degree of uncertainty associated with it. This must be taken account of when comparison of data obtained from a site is made with SGVs referred to in the CLR10 supplementary documents. CLR8, which deals with potential contaminants, identifies those priority contaminants, or families of contaminants, which are likely to be present in concentrations on current or former sites affected by industrial activity or the disposal of waste that may cause harm. The collation of toxicological data on soil contaminants and intake values for humans are considered in CLR9. This sets out the approach to the selection of tolerable daily intakes and index doses for contaminants in order to support the SGVs. It is supplemented by a number of documents on toxicological data for individual contaminants. At the time of writing these include arsenic, benzo(a)pyrene, cadmium, chromium, inorganic cyanide, lead, mercury, nickel, selenium, and dioxins, furans and dioxin-like PCBs. CLR10 considers the CLEA model, describing the conceptual exposure models for each standard land-use that are used to determine soil guideline values, that is, intervention values that can indicate a potentially significant risk to human health. Nonetheless, exceedance of a SGV need not mean that there is an actual risk to health as site specific factors may reduce the risk to an acceptable level. CLR10 also recognizes a number of exposure pathways by which humans can be subjected to contamination, which are included in the model. DEFRA provides software to run the CLEA model into which the characteristics of the site and nature of the receptor (residential with or without plant uptake, allotments, industrial/commercial) are programmed to derive the soil guideline values at a site. In addition, CLR10 is supplemented by a series of documents that outline the derivation of the guideline values for arsenic, cadmium, chromium, inorganic mercury, lead, nickel and selenium (Table 10.1b). As has been noted, these values depend on a number of assumptions, including the type and behaviour of contaminants, exposure pathways, soil conditions, land-use patterns and availability of receptors. Any assessment of a site needs to take account of the extent to which these assumptions are applicable. All the documents referred to can be obtained from the relevant DEFRA website (see Anon., 2002).

In Europe, as a comparison, the Dutch attitude towards contaminated land is more demanding than that of the British. The Dutch recognize the possibility of a regular change in land use and insisted that when contaminated land was redeveloped, the clean-up involved had to return the land to a standard that would allow any future use of the site in question. However, the Dutch have found that it has been impossible to organize, fund and execute all the clean-ups that have been deemed necessary (Cairney, 1993). Indeed, the concept of total clean-up is a standard of excellence that in practice is usually cost prohibitive. Hence, standards of relevance become a necessary prerequisite in order to avoid negative land values that would mean that

Table 10.1b Soil guideline values (in $mg\,kg^{-1}$ dry weight of soil) as a function of land-use

Contaminant	*Standard land-use*			
	Residential with plant uptake	*Residential without plant uptake*	*Allotments*	*Commercial/ industrial*
Arsenic	20	20	20	500
Cadmium	pH 6 – 1, pH 7 – 2, pH 8 – 8	30	pH 6 – 1, pH 7 – 2, pH 8 – 8	1400
Chromium (IV)	130	200	130	5000
Inorganic mercury	8	15	8	480
Lead	450	450	450	750
Nickel	50	75	50	5000
Selenium	35	260	35	8000

Source: Anon., 2002.

Notes

1 Values for residential areas and allotments are set for children; for commercial/industrial sites they are set for adults.
2 Values are based only on the oral exposure pathway and generally are on sandy soils.
3 The availability of cadmium to plants is influenced by pH value.

remedial action would not take place. Dutch standards for contaminant concentrations are given in Table 10.2.

The United States was the first country to establish a national fund for the remediation of contaminated sites. Under the Comprehensive Environmental Response, Compensation and Liability Act (CERCLA) of 1980 a federally controlled fund, known as the Superfund, was established. A national priority listing of contaminated sites based on risk assessment criteria

Table 10.2a Dutch standards for concentration values of contamination

Present in: component/concentration	*Soil* ($\mu g\,l^{-1}$ *dry matter*)			*Groundwater* ($\mu g\,l^{-1}$)		
	A	*B*	*C*	*A*	*B*	*C*
I. Metals						
Cr	*	250	800	20	50	200
Co	20	50	300	*	50	200
Ni	*	100	500	*	50	200
Cu	*	100	500	*	50	200
Zn	*	500	3000	*	200	800
As	*	30	50	*	30	100
Mo	10	40	200	5	20	100
Cd	*	5	20	*	2.5	10
Sn	20	50	300	10	30	150
Ba	200	400	2000	50	100	500
Hg	*	2	10	*	0.5	2
Pb	*	150	600	*	50	200
II. Inorganic compounds						
NH_4 (as N)	–	–	–	*	1000	3000
F (as total)	*	400	2000	*	1200	4000
CN (total, free)	1	10	100	5	30	100
CN (total, combined)	5	50	500	10	50	200
S (total, sulphide)	2	20	200	10	100	300
Br (total)	20	50	300	*	500	2000
PO_4 (as P)	–	–	–	*	200	700
III. Aromatic compounds						
Benzene	0.05	0.5	5	0.2	1	5
Ethylbenzene	0.05	5	50	0.2	20	60
Toluene	0.05	3	30	0.2	15	50
Xylenes	0.05	5	50	0.2	20	60
Phenols	0.05	1	10	0.2	15	50
Aromatics (total)	–	7	70	–	30	100
IV. Polycyclic hydrocarbons						
Napthalene	*	5	50	0.2	7	30
Anthracene	*	10	100	0.005	2	30
Phenanthrene	*	10	100	0.005	2	10
Fluoranthene	*	10	100	0.005	2	5
IV. Polycyclic hydrocarbons (continued)						
Chrycene	*	5	50	0.005	0.2	2
Benzo (*a*) anthracene	*	5	50	0.005	0.5	2
Benzo (*a*) pyrene	*	1	10	0.005	0.2	1
Benzo (*k*) fluoranthene	*	5	50	0.005	0.2	2
Indeno (1, 2, 3*cd*) phyrene	*	5	50	0.005	0.5	2
Benzo (*ghy*) perylene	*	10	100	0.005	1	5
Total polycyclic	1	20	200	0.2	10	40

Table 10.2a (Continued)

Present in: component/concentration	*Soil (μg l⁻¹ dry matter)*			*Groundwater (μg l⁻¹)*		
	A	*B*	*C*	*A*	*B*	*C*
V. Chlorinated hydrocarbons						
Aliphatic (indiv.)	*	5	50	0.01	10	50
Aliphatic (total)	–	7	70	–	15	70
Chlorobenzenes (indiv.)	*	1	10	0.01	0.5	2
Chlorobenzenes (total)	–	2	20	–	1	5
Chlorophenols (indiv.)	*	0.5	5	0.01	0.3	1.5
Chlorophenols (total)	–	1	10	–	0.5	2
Chlor. polycyclic (total)	*	1	10	–	0.2	1
PCBs (total)	*	1	10	0.01	0.2	1
EOCI (total)	0.1	8	80	1	15	70
VI. Pesticides						
Chlor.organics (indiv.)	*	0.5	5	1/0.1	0.2	1
Chlor.organics (total)	–	1	10	–	0.5	2
Non-chlor. (indiv.)	*	1	10	1/0.1	0.5	2
Non-chlor. (total)	–	2	20	–	1	5
VII. Other pollutants						
Tetrahydrothurane	0.1	4	40	0.5	20	60
Pyridine	0.1	2	20	0.5	10	30
Tetrahydrothiopene	0.1	5	50	0.5	20	60
Cyclohexanes	0.1	5	60	0.5	15	50
Styrene	0.1	5	50	0.5	20	60
Phthalates (total)	0.1	50	500	0.5	10	50
Total polycyclic hydrocarbons oxidized	1	200	2000	0.2	100	400
Mineral oil	*	1000	5000	50	200	600

The Dutch use three values of ascending levels of contaminant concentration, namely, A, B, C. These are differentiated according to the nature of the pollution.

- Level A acts as a reference value. This level may be regarded as an indicative level above which there is demonstrable pollution and below which there is no demonstrable pollution.
- Level B is an assessment value. Pollutants above the B level should be investigated more thoroughly. The question asked is: to what extent are the nature, location, and concentration of the pollutant(s) of such a nature that it is possible to speak of a risk of exposure to man or the environment?
- Level C is to be regarded as the assessment value above which the pollutant(s) should generally be treated.

Note
* Reference value (level A) for soil quality (For I and II, see Tables 10.2b and 10.2c).

specified in CERCLA was drawn up. In practice, however, this national listing has had the effect of causing contaminated sites and the surrounding areas to be avoided by developers so blighting such areas. It is not just the contamination that has deterred development but the fear of liability associated with such sites since the Act involves retroactive, joint and several legal liability and application of the polluter pays principle. In other words, an entire chain of owners of a site, along with their advisers and investors, can be held liable for any contamination found on the site. Furthermore, the cost of clean-up in some cases may exceed the potential market value of the land. In 1986, the Superfund Amendment and Reauthorization Act made landowners liable for environmental contamination on their properties.

A contaminated site in Britain will only undergo clean-up if it is to be redeveloped, and then when the contaminants exceed certain threshold levels as, for example, those that were proposed

Table 10.2b Reference values (level A) for heavy metals, arsenic and fluorine

Component	*Soil* (mg kg^{-1} *dry matter*)		*Groundwater* (μg l^{-1})
	Method of calculation	*Standard soil* ($H = 10/L = 25$)	
Cr (chromium)	50 + 2L	100	1
Ni (nickel)	10 + L	35	15
Cu (copper)	15 + 0.6 (L + H)	36	15
Zn (zinc)	50 + 1.5 (2L + H)	140	150
As (arsenic)	15 + 0.4 (L + H)	29	10
Cd (cadmium)	0.4 + 0.007 (L + H)	0.8	1.5
Hg (mercury)	0.2 + 0.0017 (2L + H)	0.3	0.05
Pb (lead)	50 + L + H	85	15
F (fluorine)	175 + 13L	500	–

Notes
Reference values (level A) for heavy metals, arsenic and fluorine in all types of soil can be calculated by means of the formula given for each element. This formula expresses the reference value in terms of the clay content (L) and/or the organic materials content (H). The clay content is taken to be the weight of mineral particles smaller than 2 μm as a percentage of the total dry weight of the soil. The organic materials content is taken to be the loss in weight due to burning as a percentage of the total dry weight of the soil. As examples, the reference values (level A) are given for an assumed standard soil containing 25% clay (L) and 10% organic material (H). For groundwater in the saturated zone, the reference values are considered independently of the type of soil.

Table 10.2c Reference values (level A) for other inorganic compounds

Component	*Groundwater*	*Remarks*
Nitrate	5.6 mg N/l	Lower values may be
Phosphate	0.4 mg P/l sandy soils	specified for the protection of
(Total phosphate)	3.0 mg P/l clay and peaty soils	soils low in nutrients
Sulphate	150 mg l^{-1}	In maritime areas higher
Bromides	0.3 mg l^{-1}	values occur naturally (saline
Chlorides	100 mg l^{-1}	and brackish groundwater)
Fluorides	0.5 mg l^{-1}	
Ammonium compounds	2 mg N^{-1} l^{-1}	
	10 mg N^{-1} l^{-1} clay and peaty soils	

by the Interdepartmental Committee on the Redevelopment of Contaminated Land (Anon., 1987; Table 10.1). These were, although the most commonly used, not the only guideline reference values available. The ICRCL guidelines have recently been replaced by the Contaminated Land Exposure Assessment (CLEA) guidelines that refer specifically to the assessment of risk to human health arising from long-term exposure to soil contamination. The development of these guidelines was outlined by Ferguson and Denner (1998). Therefore, prior to any set of values being used, the original publication(s) should be checked as their legal standing and applicability differ. There were no reference limits for certain organic contaminants in the ICRCL (Anon., 1987) guidelines and it contained no standards on groundwater quality. In the latter case the practitioner had to fall back on drinking water abstraction standards. However, in many situations such water quality standards are restrictive. Accordingly, the lack of quantitative data on reference limits of some potentially hazardous substances found in soils and groundwater in areas likely to be contaminated, at times, presents both practitioners and developers with problems. Nonetheless, after an investigation of a site, if the results are found to be below the threshold levels, then the site may be regarded as uncontaminated. Furthermore, the ICRCL (Anon., 1987) guidelines related hazards to land uses thereby differing from assessment systems based on concentration limits. In other words, they recognized lower thresholds for certain uses such as residential developments with gardens than for hard cover areas. As noted, this concept was adopted by the CLEA guidelines, which recognize four categories of land use, namely, residential with and without plant uptake, allotments and commercial/industrial uses.

10.4 Investigation of contaminated sites

If any investigation of a site that is suspected of being contaminated is going to achieve its purpose, then its objectives must be defined and the level of data required determined, the investigation being designed to meet the specific needs of the project concerned (Anon., 1988a). This is of greater importance when related to potentially contaminated land than an ordinary site. According to Johnson (1994) all investigations of potentially contaminated sites should be approached in a staged manner. This allows for communication between interested parties, and helps minimize costs and delays by facilitating planning and progress of the investigation. After the completion of each stage, an assessment should be made of the degree of uncertainty and of acceptable risk in relation to the proposed new development. Such an assessment should be used to determine the necessity for, and type of further investigation.

The first stage in any investigation of a site suspected of being contaminated is a desk study that provides data for the design of the subsequent ground investigation. The desk study should identify past and present uses of the site, and the surrounding area, and the potential for and likely forms of contamination. The objectives of the desk study are to identify any hazards and the primary targets likely to be at risk; to provide data for health and safety precautions for the site investigation; and to identify any other factors that may act as constraints on development. Hence, the desk study should provide, wherever possible, information on the layout of the site, including structures below ground, its physical features, the geology and hydrogeology of the site, the previous history of the site, the nature and quantities of materials handled, the processing involved, health and safety records, and methods of waste disposal. It should allow a preliminary risk assessment to be made and the need for further investigation to be established. For example, Bell *et al.* (2000) described a desk study of a former mine and coking plant site in the Ruhr district of Germany where a historical analysis was undertaken that included examination of old building permits, plans and documents, as well as old maps and aerial photographs. Previously, Genske *et al.* (1994) had indicated that the interpretation of such historical data could be done with the aid of CAD and GIS programmes. The evaluation of the data obtained by the desk study aids the planning of the site investigation programme (Fig. 10.6). Howland (2001) also referred to the development of a series of special computer database systems to store information gathered for the regeneration of London Docklands and to the use of GIS to process the data. A huge amount of documentary information was available, including that from some 10 000 borehole records, and was used initially to form four separate systems, namely, a geotechnical database, a geochemical database, a historic land use register and a groundwater model. In this way, a regional understanding of the area was developed that allowed a rapid assessment of any site to be made prior to an actual ground investigation. Howland maintained that such an approach is well suited to risk assessments of geoenvironmental hazards that require a source–pathway–target characterization of any contaminants and therefore is not constrained by individual site boundaries.

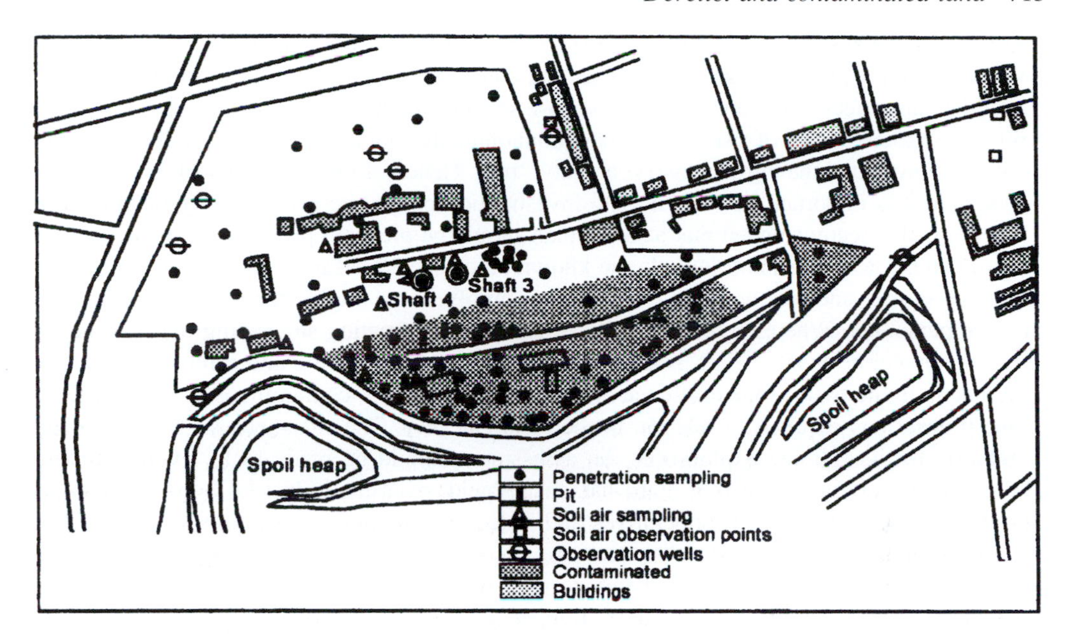

Figure 10.6 Outline of sampling at the former Graf Moltke mine near Essen, Germany.
Source: Bell *et al.*, 2000.

Aerial photographs, especially infra-red colour photographs, can help detect contaminated ground. Vegetation affected by chemical contamination of ground or groundwater, or by methane does not reflect infra-red waves as well as healthy vegetation. Consequently, this can be reflected in contrasts on infra-red photographs, thereby identifying areas for further investigation.

The preliminary reconnaissance, which is often referred to as a land quality appraisal, provides the data for planning the site exploration, which includes the personal and equipment needs, the sampling and analytical requirements, and the health and safety requirements. In other words, it is a fact-finding stage that should confirm the chief hazards and identify any additional ones so that the site exploration can be carried out effectively. It thereby supplements the desk study. The preliminary reconnaissance also should refer to any short-term or emergency measures that are required at the site before the commencement of full-scale operations and should formulate objectives so that the work is cost effective (Hobson, 1993). A number of factors should be noted during the preliminary reconnaissance. These include the state of the site and its topography; the location of any buildings and any evidence of previous buildings; the location of disposed materials, storage tanks, ponds and pits; man-made and natural drainage; unusual colours, fumes and odours; site hazards and signs of contamination; presence or absence of vegetation, its type and condition; and site access and adjacent land use. Simple on-site testing and sampling of soil materials and water can be undertaken. Gases can be tested, for example, by hand-held photo-ionization detectors (Anon., 1992).

Contaminated sites may pose a health hazard to the personnel involved in an investigation and exposure to certain contaminants could have serious consequences. Accordingly, Clark *et al.* (1994) proposed that contaminated sites should be categorized according to the potential level of risk, thereby providing an indication of the safety and protective precautions required throughout the investigation process including the laboratory testing programme. In this way some appreciation of the potential hazards that may exist can forewarn those involved.

Clark *et al.* suggested three categories (Table 10.3). The first was the least hazardous and applied primarily to inert materials (green). The second category, which they referred to as 'yellow', contained substances that were unlikely to cause serious impairment of health but nevertheless represented some degree of risk. Sites where contaminants could endanger health, even leading to death, represented the third category, namely, 'red'. Hence, if the possibility of contaminants exists, then it is important that a historical investigation of the site precedes any on-site work. Clark *et al.* also recommended that a safety plan should be prepared prior to commencement of work on site. The plan should include the known or suspected hazards, provide guidance as to how to recognize and deal with the contaminants involved and outline the effects that they can have on personnel. Where there is initially little or no information suggesting the presence of hazardous material on site, Clark *et al.* proposed that primary probing and sampling should be undertaken on the basis of the existence of potentially dangerous contaminants, that is, in terms of the precautions required in relation to the third category. The site then can be categorized in relation to the full site exploration. An assessment should be made as to whether protective clothing and equipment (e.g. hard hat, face shield, overalls, industrial boots, respiratory equipment) should be worn by operatives and a decontamination unit should be provided on hazardous sites. If unidentified substances are encountered, then the investigation should be suspended until a specialist can visit the site and determine the nature of the substance and whether it poses a hazard. For instance, Bell *et al.* (1996) described an investigation for a relief sewer in Glasgow, Scotland, which met with contaminated ground. Chemicals began to appear at shallow depth in wet sandy silt that contained slag, which together formed a dark bluish green slurry with a slightly oily appearance and a sulphurous odour. Boring operations were suspended and the area was fenced off. Chemical analysis of samples indicated that the material was typical of waste that probably originated from a chemical works that existed

Table 10.3 Site categorization

Category	*Broad description of contaminants*
Green	Subsoil, Topsoil, Hardcore, Bricks, Stone, Concrete, Clay, Excavated Road Materials, Glass, Ceramics, Abrasives, etc. Wood, Paper, Cardboard, Plastics, Metals, Wool, Cork, Ash, Clinker, Cement, etc. Note: There is a possibility that bonded asbestos could be present in otherwise inert areas
Yellow	Waste Food, Vegetable Matter, Floor Sweepings, Household Waste, Animal Carcasses, Sewage Sludge, Trees, Bushes, Garden Waste, Leather, etc. Rubber and Latex, Tyres, Epoxy Resin, Electrical Fittings, Soaps, Cosmetics, Non-Toxic Metal and Organic Compounds, Tar, Pitch, Bitumen, Solidified Wastes, Dye Stuffs, Fuel Ash, Silica Dust, etc.
Red	All substances that could subject persons and animals to risk of death, injury or impairment of health Wide range of Chemicals, Toxic Metal and Organic Compounds, etc. Pharmaceutical and Veterinary Wastes, Phenols, Medical Products, Solvents, Beryllium, Micro-organisms, Asbestos, Thiocyanates, Cyanides, etc. Hydrocarbons, Peroxides, Chlorates, Flammable and Explosive Materials. Materials that are particularly corrosive or carcenogenic, etc.

Source: Clark *et al.*, 1994. Reproduced by kind permission of University of Albert.

in the area in the late nineteenth century. The material had high pH, and chromium levels that could cause skin irritation and give rise to health risks. An assessment of the potential environmental hazards presented to site investigation staff as a requirement of the employer was carried out in accordance with Control of Substances Hazardous to Health Regulations (Anon., 1988b). This included an evaluation of the likely hazards and forms of exposure, and the specification of health and safety measures including safe working practices, supervision, monitoring and protective clothing. In addition to protective clothing, staff were issued with respirators to prevent the inhalation of possible harmful dusts. Washing and shower facilities were provided for the operatives. A supply of clear water was always available at the location of each excavation along with a first aid kit. No eating, drinking or smoking was allowed on site in order to minimize ingestion of contaminants. Lastly, a training seminar was held to inform all staff of the possible hazards associated with the work and to discuss the appropriate safety precautions.

Just as a normal site investigation, one that is involved with the exploration for contamination needs to determine the nature of the ground. In addition, it needs to assess the ability of the ground to transmit any contaminants either laterally, or upward by capillary action. Permeability testing therefore is required. The investigation also must establish the location of perched water tables and aquifers, and any linkages between them, as well as determining the chemistry of the water on site. The exploratory methods used in the site exploration can include manual excavation, trenching and the use of trial pits, light cable percussion boring, power auger drilling, rotary drilling, and water and gas surveys (Bell *et al.*, 1996). Excavation of pits and trenches is among the most widely used techniques for investigating contaminated land (Fig. 10.7). Visually different materials should be placed in different stockpiles as a trench or pit is dug to facilitate sampling. Of the geophysical methods, resistivity and electromagnetic techniques have the more general application (Jewell *et al.*, 1993). They can be used for the detection of cavities and subsurface structures, and for contaminant mapping, in addition to assessment of soil type and distribution (Table 10.4). For example, Bell *et al.* (2000) described the use of an electromagnetic survey to detect subsurface structures, which remained in the ground after the superstructures had been demolished. The resulting map revealed zones of low conductivity that reflected the natural ground conditions together with zones of high conductivity suggesting anomalies associated with old foundations, pipes and tunnels (Fig. 10.8).

If thermal activity is suspected at a site, then temperature surveys should be carried out. Leach and Goodyear (1991) recommended that thermocouples should be set out on a grid pattern, initially at 25–100 m centres, and subsequently at closer centres at suspected or discovered hot spots. Subsurface temperatures can be recorded by lowering thermocouples down boreholes, but movement of air or groundwater may affect results, or by probes driven into the ground. The latter can be left in place for long-term monitoring.

Guidelines on how to conduct a sampling and testing programme are provided by the United States Environmental Protection Agency (Anon., 1988c) and Anon. (1988a). A sampling plan should specify the objectives of the investigation, the history of the site, analyses of any existing data, types of samples to be used, sample locations and frequency, analytical procedures and operational plan. Historical data should be used to ensure that any potential 'hot' spots are sampled satisfactorily. A sampling grid should be used in such areas (Anon., 1988a; Sara, 1994). An account of various sampling techniques has been provided by Leach and Goodyear (1991).

Most sites require careful interpretation of the ground investigation data. Sampling procedures are of particular importance and the value of the data obtained therefrom is related to how representative the samples are. Eccles and Retford (1999) mentioned the use of dynamic

Figure 10.7 Subsurface structures revealed by trenching at a derelict site in Leeds, England.

sampling on contaminated sites, noting that it is a fast flexible method with health and safety advantages relative to trial pitting and light cable percussive boring. The sampler consists of a lightweight sampling unit that uses specially constructed sampling tubes driven into the ground using a high frequency percussion hammer. The method also can be used for the installation of a range of instrumentation including piezometers and gas monitoring wells. Some materials can change as a result of being disturbed when they are obtained or during handling. Hence, sampling procedure should take account of the areas of a site that require sampling; the pattern, depth, types and numbers of samples to be collected; their handling, transport and storage; as well as sample preparation and analytical methods. Bell *et al.* (1996) indicated that on-site chemical analysis in mobile laboratories is becoming increasingly important in relation to the investigation of contaminated sites. Such on-site investigation for characterization of toxic chemicals has many advantages over conventional laboratories since transportation time is more or less eliminated, sample integrity is maintained and in cases where heavy metal occurrences affect agricultural land, preventative measures can be suggested on the spot by fast-screening samples over a wide area. Furthermore, on-site analysis of many samples in a short time allows a directional investigation of 'hot' spots to be made in days.

Table 10.4 Geophysical methods frequently used in environmental site assessment

Method	*Dependent physical property*	*Major applications*
Resistivity	Electrical conductivity	Soil type, stratigraphy, depth to bedrock Degree of saturation Contaminant mapping; cavity detection
Frequency domain electromagnetic (EM) techniques	Electrical conductivity	Soil type, stratigraphy Degree of saturation Landfill investigations Contaminant mapping Cavity detection Location of buried objects
Transient electromagnetic (TEM) techniques	Electrical conductivity	Faults, shear zones Contaminant mapping Soil type, stratigraphy, depth to bedrock
Ground penetrating radar (GPR)	Dielectric permittivity	Location of water table Contaminant mapping Cavity detection Location of buried objects
Seismic refraction	Elastic moduli	Bedrock lithology and profiling Overburden characteristics Degree of saturation Location of landfill boundaries
Vertical seismic profiling (surface-to-downhole)	Elastic moduli	Vertical and lateral changes in bedrock properties Anisotropy due to linear features such as faults and joints
Magnetometry	Magnetic susceptibility	Fault location, bedrock lithology, cavity detection, location of buried objects

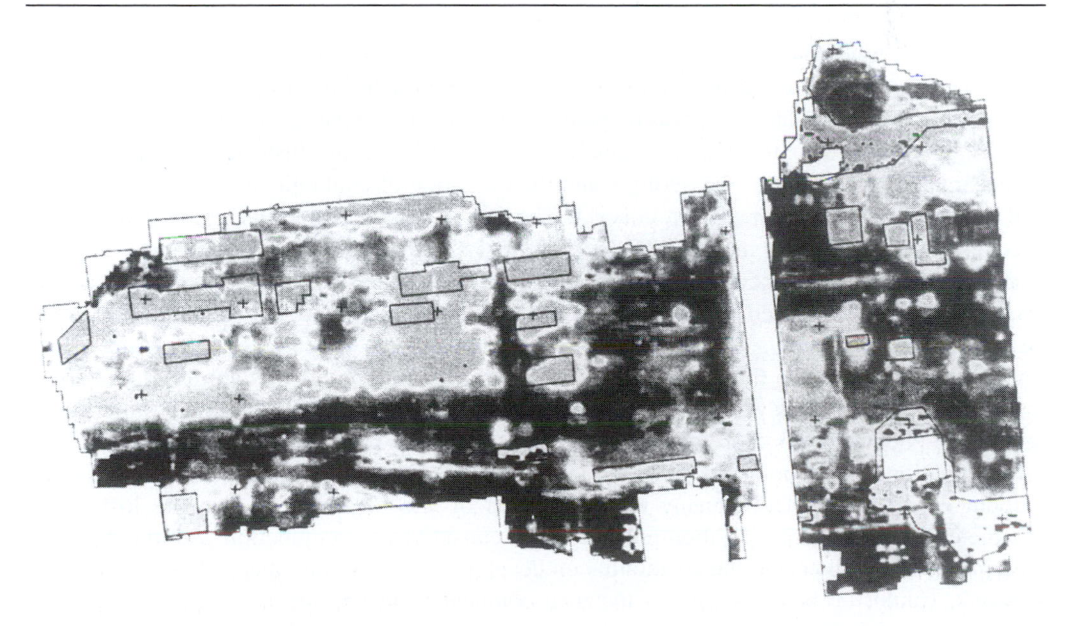

Figure 10.8 Anomalies revealed by an electromagnetic survey of the old Minister Auchenbach mine, Lunen, Germany. Zones of high conductivity suggested the present of buried structures.
Source: Bell and Genske, 2000.

Volatile contaminants or gas-producing material can be determined by sampling the soil atmosphere by using a hollow gas probe, inserted into the ground to the required depth. The probe is connected to a small vacuum pump and a flow of soil gas induced. A sample is recovered by using a syringe. Care must be taken to determine the presence of gas at different horizons by the use of sealed response zones. Analysis usually is conducted on site by portable gas chromatography or photo-ionization. Standpipes can be used to monitor gases during the exploration work.

One of the factors that should be avoided is cross contamination. This is the transfer of materials by the exploratory technique from one depth into a sample taken at a different depth. Consequently, cleaning requirements should be considered, and ideally a high specification of cleaning operation should be carried out on equipment between both sampling and borehole locations.

To ensure that the site investigation is conducted in the manner intended and the correct data recorded, the work should be carefully specified in advance. As the investigation proceeds, it may become apparent that the distribution of material about the site is not as predicted by the desk study or preliminary reconnaissance. Hence, the site investigation strategy then needs to be adjusted. The data obtained during the investigation must be accurately recorded in a manner whereby it can be understood subsequently. The testing programme should identify the types, distribution and concentration (severity) of contaminants, and any significant variations or local anomalies. Comparisons with surrounding uncontaminated areas can be made. The data obtained on the distribution of the contaminants may need to be analysed in order to estimate any possible degree of error. For example, in a survey of a site near Essen, Germany, which had been contaminated by a former coal mine and coking plant, Bell *et al.* (2000) carried out a statistical survey of the distribution of contamination. After the analysis of samples to determine the nature of the contamination on site, the grades of contamination of the sampling points were interpreted as regionalized variables and a block kriging routine was applied. In addition to a map depicting the contaminated zones, the prediction error was quantified on a second map (Fig. 10.9). The site investigation programme was further optimized in relation to the error map, zones of high contamination and of large error indicating where additional investigation was necessary. Bell *et al.* also described how, in order to help visualize the complex task of remediating the site, a video-tape was prepared for viewing by all parties involved in the project. The video consisted of three parts, namely, demonstration of the historical analysis, three-dimensional visualization of the geology and the migration of contaminants into the substrata, and explanation of the remediation concept. The first part, that is, the historical analysis was accomplished by merging the map of the present position of the site with the data obtained from historical building documents and aerial photographs. The second part, three-dimensional visualization, was produced as an animated idealization of the geological situation that could be viewed by means of a virtual camera performing circuits around the three-dimensional image. Subsequently, the migration of contaminants into the ground was animated. As far as the rehabilitation process was concerned, the virtual camera intersected the three-dimensional model and zoomed in on the rehabilitation system.

Investigation of contaminated sites frequently requires the use of a team of specialists and without expert interpretation many of the benefits of site investigation may be lost. Once completed, the site characterization process, when considered in conjunction with the development proposals, will enable the constraints on development to be identified. These constraints, however, cannot be based solely on the data obtained from the site investigation but must take account of financial and legal considerations. If hazard potential and associated risk are regarded as too high, then the development proposals will need to be reviewed. When the

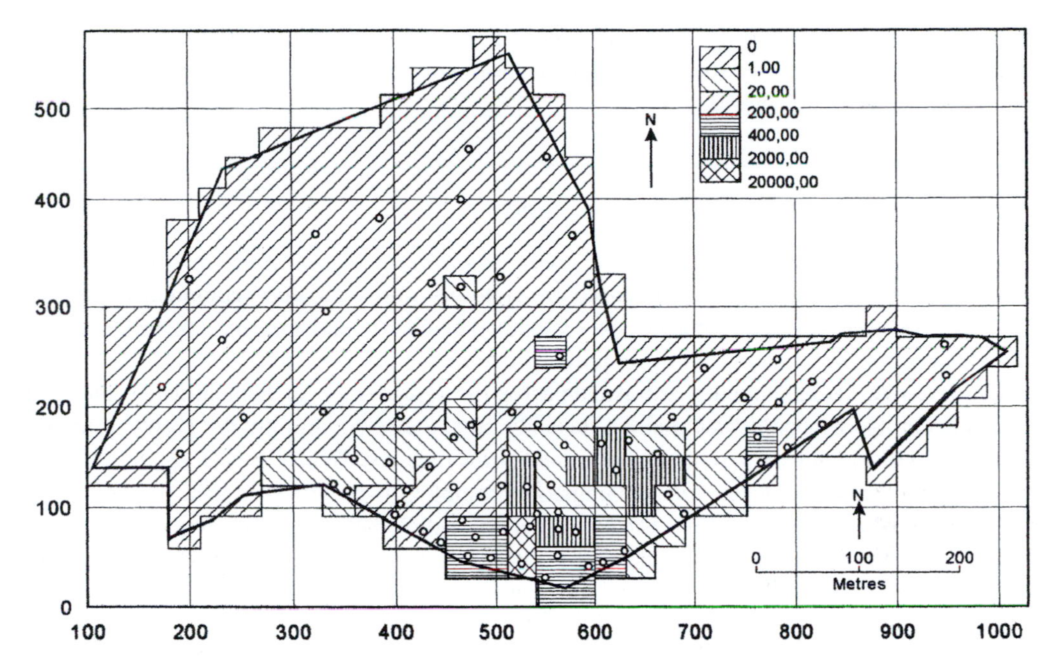

Figure 10.9a Distribution of contamination, as revealed by site investigate, at the abandoned Graf Moltke mine roking plant, near Essen, Germany.

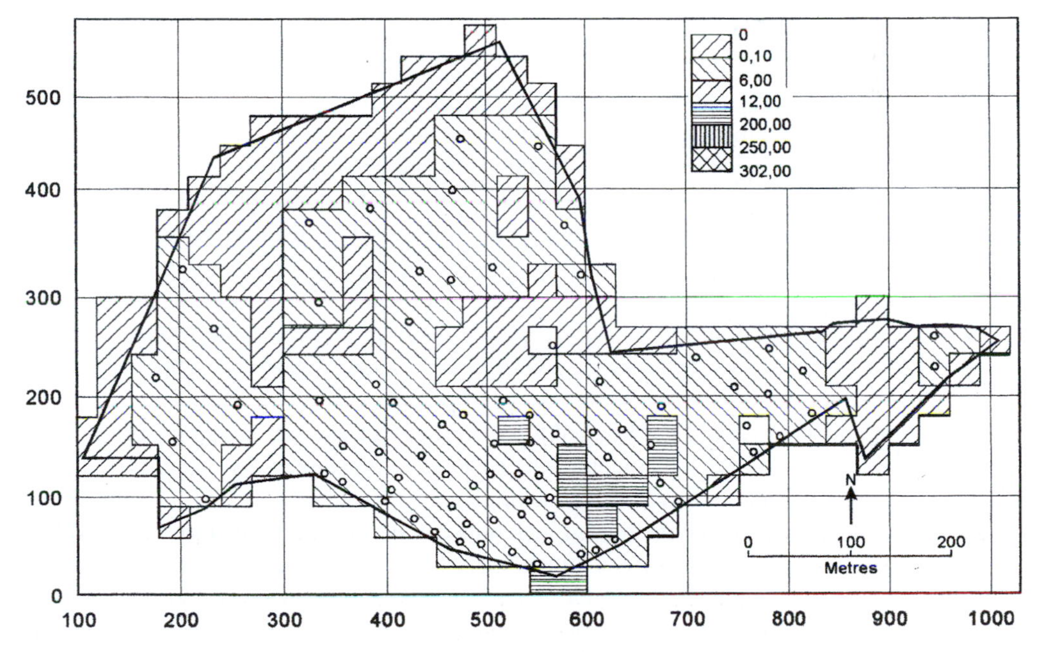

Figure 10.9b Estimation of error in the distribution of contamination (in mg kg^{-1}).
Source: Bell *et al.*, 2000.

physical constraints and hazards have been assessed, then a remediation programme can be designed, which allows the site to be economically and safely developed. It is at this stage when clean-up standards are specified in conjunction with the assessment of the contaminative regime of the surrounding area.

10.5 Remediation of contaminated land

Remedial planning and implementation is a complex process that not only involves geotechnical methodology but may be influenced by statutory and regulatory compliance. Indeed, it is becoming increasingly common for regulatory standards and guidance to provide very prescriptive procedural and technical requirements (Attewell, 1993; Holm, 1993).

A wide range of technologies are available for the remediation of contaminated sites and the applicability of a particular method depends on the site conditions, the type and extent of contamination, and the extent of the remediation required. When the acceptance criteria are demanding and the degree of contamination complex, it is important that the feasibility of the remediation technology is tested in order to ensure that the design objectives can be satisfied (Swane *et al.*, 1993). This can necessitate a thorough laboratory testing programme being undertaken, along with field trials. In some cases the remediation operation requires the employment of more than one method.

It obviously is important that the remedial works do not give rise to unacceptable levels of pollution either on site or in the immediate surroundings. Hence, the design of the remedial works should include measures to control pollution during the operation. The effectiveness of the pollution control measures needs to be monitored throughout the remediation programme.

In order to verify that the remediation operation has complied with the clean-up acceptance criteria for the site, a further sampling and testing programme is required (Anon., 1989). Swane *et al.* (1993) recommended that it is advisable to check that the clean-up standards are being attained as the site is being rehabilitated. In this way any parts of the site that fail to meet the criteria can be dealt with there and then, so improving the construction schedule.

The nature of the remedial action depends upon the nature of the contaminants present on site. In some situations it may be possible to rely upon natural decay or dispersion of the contaminants. Removal of contaminants from a site for disposal in an approved disposal facility frequently has been used. However, the costs involved in removal are increasing and can be extremely high, especially if an approved disposal site is not within a reasonable transport distance. On-site burial can be carried out by the provision of an acceptable surface barrier or an approved depth of clean surface filling over the contaminated material. This may require considerable earthworks. Clean covers are most appropriate for sites with various previous uses and contamination. They should not be used to contain oily or gaseous contamination.

Another method is to isolate the contaminants by *in situ* containment by using, for example, cut-off barriers. Encapsulation involves immobilization of contaminants, for instance, by the injection of grout. Low-grade contaminated materials can be diluted below threshold levels by mixing with clean soil. However, there are possible problems associated with dilution. The need for quality assurance is high (unacceptable materials cannot be reincorporated into the site). The dilution process must not make the contaminants more leachable. Other less used methods include bioremediation of organic material, soil flushing and soil washing, incineration, vacuum extraction and venting of volatile constituents, or removal of harmful substances to be buried in containers at a purpose built facility.

10.5.1 Soil remediation

Most of the remediation technologies differ in their applicability to treat particular contaminants in the soil. Landfill disposal and containment are the exceptions in that they are capable of dealing with most soil contamination problems. However, it frequently is argued that the removal of contaminated soil from a site for disposal in a special landfill facility transfers the problem from one location to another. Hence, the in-place treatment of sites, where feasible, is a better course of action. Containment is used to isolate contaminated sites from the environment by the installation of a barrier system such as a cut-off wall. In addition, containment may include a cover placed over the contaminated zone(s) to reduce infiltration of surface water and to act as a separation layer between land users and the contaminated ground. Buchanan and Thorburn (1987) described a site, 8 ha in size, in Glasgow, Scotland, which was designated for industrial development and that had been occupied formerly by a chemical and tar works. The site investigation revealed that the site was covered by up to 6.5 m of made ground consisting of ash, clay and coarse colliery discard. This made ground contained waste from the former chemical works and zones were impregnated with tar products. It was decided to remove the near-surface material and replace it with 2 m of compacted granular material. Fortunately, a waste disposal site that was approved for dealing with such materials was available within a reasonable distance. Excavation had the advantage of exposing any unsuitable material that might otherwise have remained concealed. Problem materials such as the contents of the tar ponds were removed by specialist contractors. At the site near Essen, referred to above, the authorities required that excavation was kept to a minimum. Hence, a reinforced cover design, using geotextiles, was chosen for the heavily contaminated zones. The surface confinement of the contaminated ground had to meet three goals. First, it had to be waterproof in order to prevent the penetration of precipitation into the affected zones. Second, it had to be gas proof to stop the migration of any toxic gas to the surface. Third, it had to be stiff enough to allow any construction to proceed. The third aspect was particularly important in relation to this site as large massive remnant foundations occurred within loose fill that could lead to differential settlement beneath structures erected on the site. A composite soil-geosynthetic system was designed to take account of the three factors mentioned (Fig. 10.10). Basically, the system consisted of a lower reinforced supporting layer, a draining and sealing layer incorporating geotextiles and geomembranes, and an upper reinforced foundation layer to accommodate the structural loading.

Fixation or solidification processes reduce the availability of contaminants to the environment by chemical reactions with additives (fixation) or by changing the phase (e.g. liquid to solid) of the contaminant by mixing with another medium. Such processes usually are applied to concentrated sources of hazardous wastes. Various cementing materials such as Portland cement and quicklime can be used to immobilize heavy metals (Harris *et al.*, 1995). Cementitious stabilization permits the use of site-specific mixtures and appropriate methods of mixing for a wide range of situations. For example, Al-Tabbaa *et al.* (1998) used *in situ* auger mixing, with seven different soil–grout mixes to treat contaminated made ground, and sand and gravel, at a site in London. The mixes were made up of varying proportions of cement, lime and pulverized fuel ash and were injected to form overlapping stabilized soil columns. The auger consisted of cutting flights and mixing rods, and was designed to produce homogeneous mixing of soil and grout with minimal exposure of contaminated material. Contaminants are held by entrapment in the cementitious compounds rather than by the formation of binding chemical compounds. Accordingly, there is the possibility that contaminants could leach out in the long term if the cementitious compounds begin to break down. In order to investigate the possibility of break down, Reid and Brookes (1999) undertook a test programme to assess the durability of lime stabilized contaminated materials. Leaching tests showed an initial concentration of mobile ions

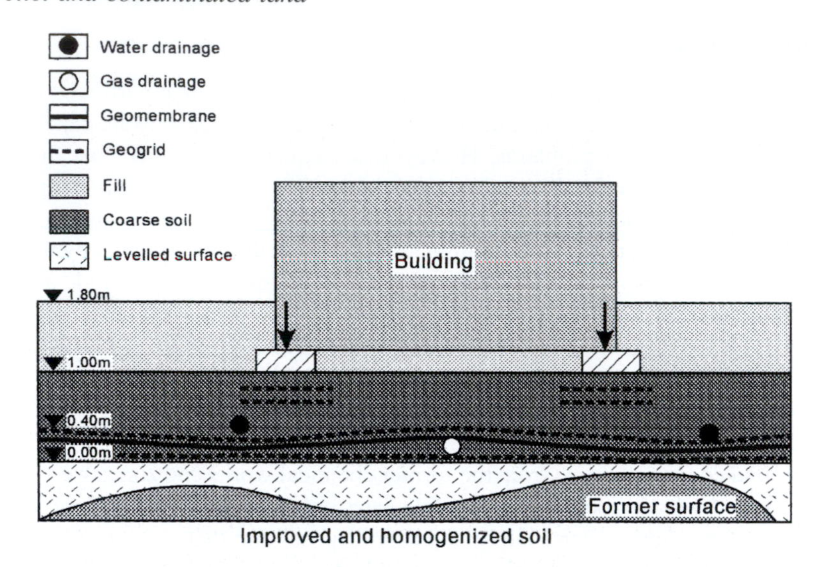

Figure 10.10 A geotextile reinforced cover with a draining layer and geomembranes used for the reclamation of the Graf Moltke site near Essen, Germany.
Source: Bell *et al.*, 2000.

such as Na and Cl, with concentrations of Cu, Ni and phenols exhibiting a similar pattern, that is, high initial concentrations that decreased as testing progressed. Overall, however, the test results showed that the stabilized material maintained its long-term integrity and the concentrations of most metals in the leachate were below detection limits. Reid and Brookes therefore concluded that the mobility of the individual contaminants appears to depend upon their speciation. They supported the use of cementitious stabilization as a method of dealing with contaminated ground but pointed out that it would be necessary to carry out a site-specific risk assessment to show that both people and the environment would be unaffected by such treatment.

Soil washing involves using particle size fractionation; aqueous-based systems employing some type of mechanical and/or chemical process; or counter current decantation with solvents for organic contaminants, and acids/bases or chelating agents for inorganic contaminants, to remove contaminants from excavated soils (Trost, 1993). Particle size fractionation is based on the premise that contaminants generally are more likely to be associated with finer particle sizes. Hence, in particle size fractionation the finer material is subjected to high pressure water washing, which can include the use of additives. Aqueous-based soil wash systems generally make use of the froth flotation process to separate contaminants. Basically, counter current decantation uses a series of thickeners in tanks into which the contaminated soil is introduced and allowed to settle. The contaminated soil is pumped from one tank to the next and the solvent, acid/base or chelating agent is in the last tank. Variation in the type of soil and contaminants at a site means that soil washing is more difficult and that a testing programme needs to be carried out to determine effectiveness. Welsh and Burke (1991) described the use of high pressure soil washing to treat soil contaminated with phenol beneath three old factory buildings in Hamburg, Germany. In this method, a drillhole is advanced to the required depth, then water and air are jetted from closely spaced nozzles at the base of a triple fluid phase pipe, as in jet grouting, to displace the contaminated soil to the surface. At the same time bentonite slurry is introduced to stabilize the ground. The removed soil was cleaned by oxidation in a self-contained decontamination unit, then mixed with cement and returned in place.

Steam injection and stripping can be used to treat soils in the vadose zone contaminated with volatile compounds. The steam provides the heat that volatizes the contaminants, which then are extracted (i.e. stripped). The efficiency of the process depends upon the ease with which steam can be injected and recovered. As steam is used in near-surface soils and since much of it is lost due to gas pressure fracturing the soil, an impermeable cover may be used to impede the escape of steam. One of the disadvantages is that some steam turns to water on cooling that means that some contaminated water remains in the soil.

Solvents can be used to remove contaminants from the soil. For example, soil flushing makes use of water, water-surfactant mixtures, acids, bases, chelating agents, oxidizing agents and reducing agents to extract semi-volatile organics, heavy metals and cyanide salts from the vadose zone of the soil. The concept of soil flushing is based on the premise that a selective liquid extractant that is introduced into the soil will concentrate and remove contaminants as it moves downwards (Dawson and Gilman, 2001). Ultimately, hydraulic capture such as pump and treat is required to recover the contaminated liquid. The technique is used in soils that are sufficiently permeable (ideally not less than $10^{-5}\,m\,s^{-1}$) to allow the solvent to permeate and the more homogeneous the soil is, the better. However, prefabricated vertical drains can be used to facilitate the flushing operation by reducing the drainage path and the travel times between injection and extraction points in less permeable soils. Every effort should be made to prevent the contaminated extractant from invading the groundwater. Other disadvantages include the difficulty in achieving good coverage and mixing, and the incomplete capture of the contaminated liquid. In fact, due to the possibility of soil flushing having an adverse impact on the environment, the USEPA (Anon., 1990) recommended that the technique should be used only when others with lower impacts on the environment are not applicable. Solvents also can be used to treat soils that have been excavated.

Some contaminants can be removed from the soil by heating. For instance, soil can be heated to between 400 and 600 °C to drive off or decompose organic contaminants such as hydrocarbons, solvents, volatile organic compounds and pesticides. Mobile units can be used on site, the soil being removed, treated and then returned as backfill. Norris *et al.* (1999) referred to the use of low temperature thermal desorption (LTTD) to treat soil on part of a site affected by polychlorinated biphenyl (PCB) and chlorinated solvent contamination. Low temperature thermal desorption at around 400 °C is suitable for dealing with soils on site that are contaminated with low to middle distillate organic compounds such as solvents, petrol, diesel fuel and lubricating oils. The contaminated soil is fed continuously through a rotary kiln where it is heated to temperatures sufficient to evaporate or combust the contaminants, so stripping them from the soil. The exhaust gases and any non-combusted vapours pass through dust filters into a thermal oxidizer unit or after burner, where a minimum temperature of 850 °C ensures their destruction. The treated soil passes out of the plant and is available for re-use.

Incineration, whereby wastes are heated to between 1500 and 2000 °C, is used for dealing with hazardous wastes containing halogenated organic compounds such as polychlorinated biphenyl (PCB) and pesticides that are difficult to remove by other techniques. Incineration involves removal of the soil and it then usually is crushed and screened to provide fine material for firing. The ash that remains may require additional treatment since heavy metal contamination may not have been removed by incineration. It then is disposed of in a landfill.

In situ vitrification transforms contaminated soil into a glassy mass. It involves electrodes being inserted around the contaminated area and sufficient electric current being supplied to melt the soil (the required temperatures can vary from 1600 to 2000 °C). According to Dawson and Gilman (2001), a layer of graphite conducting material and glass frit (a fusable ceramic mixture) is added to the surface of the soil in order to facilitate the development of the high temperatures required. The volatile contaminants are either driven off or destroyed, whilst the non-volatile contaminants are encapsulated in the glassy mass when it solidifies. It may be used in soils that contain heavy metals.

Bioremediation involves the use of microbial organisms to bring about the degradation or transformation of contaminants so that they become harmless (Loehr, 1993). The micro-organisms involved in the process either occur naturally or are artificially introduced. In the case of the former, the microbial action is stimulated by optimizing the conditions necessary for growth (Singleton and Burkes, 1994). The principal use of bioremediation is in the degradation and destruction of organic contaminants, although it also has been used to convert some heavy metal compounds into less toxic states. The rate of bioremediation may be slow in some cases because of the physical and/or chemical nature of the contaminants. Physical behaviour may influence the availability of micro-organisms due to low solubility or strong adsorption to soil particles. Biodegradation also may be inhibited if special organisms are required or if different conditions are needed for various stages of the degradation process. Bioremediation can be carried out on ground *in situ* or ground can be removed for treatment. *In situ* bioremediation depends upon the amenability of the organic compounds to biodegradation, the ease with which oxygen and nutrients can reach the contaminated area, the permeability of the soil, its temperature and the pH value. Air circulation brought about by soil vacuum extraction enhances the supply of oxygen to micro-organisms in the vadose zone of the soil. Other systems used for *in situ* bioremediation in the vadoze zone include infiltration galleries or injection wells for the delivery of water carrying oxygen and nutrients. In the phreatic zone water can be extracted from the contaminated area, which then is treated at the surface with oxygen and nutrients, and subsequently injected back into the contaminated ground. This allows groundwater to be treated. However, the *in situ* method of treatment when applied to soil or water alone may prove unsatisfactory due to recontamination by the remaining untreated soil or untreated groundwater. For example, contaminants in the groundwater, notably oils, solvents and creosote that are immiscible, recontaminate soil as groundwater levels fluctuate. On the other hand, contamination in the soil may be washed out continuously over many years so contaminating groundwater. Therefore, as Adams and Holroyd (1992) recommended, it frequently is worthwhile cleaning both soil and groundwater simultaneously. *Ex situ* bioremediation involves the excavation of contaminated ground, placing it in beds where it is treated and then returning it to where it was removed as cleaned backfill. There are a number of different types of *ex situ* bioremediation that include land farming, composting, purpose-built treatment facilities and biological reactors.

10.5.2 Groundwater remediation and soil vapour extraction

Contaminated groundwater either can be treated *in situ* or it can be abstracted and treated (see also Chapter 7). The solubility in water and volatility of contaminants influence the selection of the remedial technique used. Some organic liquids are only slightly soluble in water and are immiscible. These are known as non-aqueous phase liquids (NAPLs), when dense they are referred to as DNAPLs or 'sinkers' and when light as LNAPLs or 'floaters'. Examples of the former include many chlorinated hydrocarbons such as trichloroethylene and trichloroethane, whilst petrol, diesel oil and paraffin provide examples of the latter. The permeability of the ground influences the rate at which contaminated groundwater moves and therefore the ease and rate at which it can be extracted.

The pump and treat method is the most widely used means of remediation of contaminated groundwater (Haley *et al.*, 1991). The latter is abstracted from the aquifer concerned by wells, trenches or pits and treated at the surface. It then is injected back into the aquifer. The pump and treat method proves most successful when the contaminants are highly soluble and are not readily adsorbed by clay minerals in the ground. Methods of treatment that are used to remove contaminants that are dissolved in water, once it has been abstracted, include standard water treatment techniques, air stripping of volatiles, carbon adsorption, microfiltration and bioremediation (Swane *et al.*, 1993).

As far as LNAPLs are concerned, they can be separated from the groundwater either by using a skimming pump in a well or by using oil–water separators at the surface. It usually does not prove possible to remove all light oil in this way so that other techniques may be required to treat the residual hydrocarbons. Oily substances and synthetic organic compounds normally are much more difficult to remove from an aquifer. In fact, the successful removal of DNAPLs is impossible at the present. As such, they can be dealt with by containment.

Active containment refers to the isolation or hydrodynamic control of contaminated groundwater. The process makes use of pumping and recharge systems to develop zones of stagnation or to alter the flow pattern of the groundwater. Cut-off walls are used in passive containment to isolate the contaminated groundwater. Additives capable of capturing and holding or degrading contaminants can be placed in permeable barriers that are located in the flow path of a plume. Such barriers must be wide enough for the residence time of the captured liquid to be sufficient to complete its task.

In situ bioremediation makes use of microbial activity to degrade organic contaminants in the groundwater so that they become non-toxic. Oxygen and nutrients are introduced into an aquifer to stimulate activity in aerobic bioremediation whereas methane and nutrients may be introduced in anaerobic bioremediation.

Electrokinetic remediation involves using electro-osmosis and electromigration to remove contaminants (Taha, 1997). A direct current is sent between a series of electrodes located across the area to be treated, causing charged particles in the pore fluid to move towards the electrodes. This movement, in turn, causes water and associated non-ionic contaminants to flow in the same direction. Sorbents or other media are placed in the path of induced flow between the electrodes to recover the contaminants. Groundwater also is pumped from the zones about the electrodes.

Vacuum extraction involves the removal of the gaseous and liquid phases of contaminants by the use of vacuum extraction wells. It can be applied to volatile organic compounds (VOCs) residing in the unsaturated soil or to volatile light non-aqueous phase liquids (LNAPLs) resting on the water table. The method can be used in most types of soils, although its efficiency declines in heterogeneous and high permeability soils. The VOCs are transferred from the liquid to the vapour stage during extraction and vapour from the exhaust pipes can be treated when necessary. Lindhult *et al.* (1995) referred to the use of vacuum extraction to remediate a petrol station site containing volatile organic compounds (benzene, toluene, ethylbenzene and xylene).

The applicability of vacuum systems can be extended to less volatile constituents and to clay soils where the moisture content inhibits the flow of soil gases during the extraction process, by heat-enhanced soil vapour extraction. Heat may be introduced into the ground by way of electrodes, by injection of hot air or steam, or by the application of radio frequency waves. The heat increases the vapour pressure of the contaminants and evaporates pore water thereby opening additional conduits for vapour movement. Dawson and Gilman (2001) indicated that in some cases enough heat can be added to dewater saturated zones and provide access to DNAPL that might otherwise be inaccessible.

Air sparging is a type of *in situ* air stripping in which air is forced under pressure through an aquifer in order to remove volatile organic contaminants. It also enhances desorption and bioremediation of contaminants in saturated soils. The air is removed from the ground by soil venting systems. The injection points, especially where contamination occurs at shallow depth, are located beneath the area affected. According to Waters (1994), the key to effective air sparging is the amount of contact between the injected air and the contaminated soil and groundwater. Because air permeability is higher than that of water, greater volumes of air have to be used to treat sites. Nevertheless, air sparging can be used to treat sites when pump and treat methods are ineffective. The air that is vented may have to be collected for further treatment as it could be hazardous.

Biosparging is a closed loop system used to remove gases via wells from a contaminated plume, treat the vapour in a bioadsorption tank, enrich the vapour stream with oxygen and heat, and re-inject it into screened wells beneath groundwater level at the source of the plume for complete degradation to occur (Madhav *et al.*, 1997). However, disadvantages include the difficulty in achieving good coverage and mixing, and the incomplete capture of contaminated liquid. Ozone is injected when the contaminants are chlorinated solvents or wood preservatives, the ozone stripping the compounds of chlorine. Extraction wells are located around the perimeter of the plume to induce the migration of contaminants from the more concentrated zone towards the wells. In this way, the concentration is reduced to the point where natural degradation by micro-organisms can take over.

10.6 Contamination, mining and associated industries

Mining represents one of man's earliest activities and as such, together with mineral processing and associated industries, has a notable effect on the environment. Many abandoned sites are heavily contaminated. For instance, crude metal extraction techniques used in the past meant that tailings contained high contents of associated metals. The large amounts of waste associated with mining represent a source of contamination of land, and both ground and surface water. These wastes normally possess chemical and physical characteristics that prevent the re-establishment of plants without some form of prior remedial treatment. Furthermore, as the scale of mining operations increased so did the degree of contamination.

Soil may be contaminated by mining waste from old workings. Davies (1972) quoted an example of lead contamination from the Tamar Valley in southwest England where some garden soils contained as much as $522\,mg\,kg^{-1}$ Pb. However, a number of factors decrease the solubility of Pb in waste (Davis *et al.,* 1994). These include the mineral composition, the degree of encapsulation in pyrite or silicate matrices, the nature of the alteration rinds and the particle size. As such they decrease the bioavailability of Pb in soils derived from or contaminated by mining wastes. A notable example of lead contamination is provided by Parc Mine in North Wales. This mine produced lead and zinc from the beginning of the nineteenth century until the mid-twentieth century. A tailings dam, containing some 250 000 tonnes of fine waste, was located in a valley head on a tributary of the River Conwy. Storms during the winter of 1963–1964 eroded the tailings dam spreading tailings over a distance of approximate 1 km, which effectively destroyed or seriously damaged 11 ha of agricultural land. It was estimated that several hundred tonnes of lead and zinc were deposited in the area affected and subsequently a number of instances of lead poisoning in cattle were detected, together with zinc toxicity in cereals.

Gold mining in South Africa has meant that huge amounts of tailings have been impounded within tailings dams. Poor construction and management, in particular, of some of the older tailings dams resulted in seepage loss that adversely affected both ground and groundwater. Rosner and van Schalkwyk (2000) noted that some tailings dams have been partially or totally reclaimed thereby leaving behind contaminated footprints. They found that the topsoil in such areas has been highly acidified and contains heavy metals. As such, it poses a serious threat to the underlying dolomitic aquifers. Rosner and van Schalkwyk recommended that soil management measures such as liming could be used to prevent the migration of contaminants from the topsoil into the subsoil and groundwater, and would aid the establishment of vegetation. The removal of contaminated soil, because of the cost, can only be undertaken in situations where small volumes are involved.

Acid mine drainage is responsible for contamination of ground, and ground and surface water in coal and in metal mining areas around the world (see Chapter 7). Acid generation gives rise to elevated levels of heavy metals in the drainage water, which pollutes natural waters and can be precipitated on or in sediments (Table 10.5). Bell *et al.* (2002) suggested that the accumulation of Zn, Cu, Ni, Mn, Cr and Pb in material precipitated from a stream polluted by acid mine drainage could be due to adsorption to iron and sulphate minerals, which are characteristically associated with acid mine drainage. For example, jarosite [$KFe_3(OH)_6(SO_4)_2$] acted as an important trace element accumulator with high concentrations of rubidium being present in precipitates containing this mineral. The high concentrations of lead in some of the sediments concerned may be the result of sorption to goethite [FeO(OH)]. High concentrations of heavy metals in surface water and soils can decimate vegetation, and can seriously affect the health of animals.

According to Hatheway (2002) abandoned manufactured gas sites represent the most common and problematic of all contaminated waste sites. Former manufactured gas plants (FMGPs) roasted coal to drive off gas and in the process produced toxic wastes, notably tar. Unfortunately, most tar residuals and gas oils are highly resistant to natural degradation or attenuation in the environment and therefore potential problems associated with tar could persist for centuries. Tar residuals and gas oils are composed of complex mixtures of hundreds of aliphatic and aromatic organic hydrocarbons. The polycyclic aromatic hydrocarbons (PAHs) are of particular concern as they are suspected of being carcinogenic. Hatheway further pointed out that most of the broad advances in dealing with toxic and persistent groundwater contaminants have been focused on dealing with solvents, pesticides and heat dissipation oils. However, solvents are

Table 10.5 XRF analysis of sediments in the Blesbokspruit and catchment

(a) Major oxides

Oxides/elements	*Sample number*										
	S1	*S2*	*S3*	*S4*	*S5*	*S6*	*S7*	*S8*	*S9*	*S10*	*S11*
SiO_2	52	80	83	67	46	84	87	76	87	64	102
TiO_2	0.91	0.53	0.42	0.56	0.49	0.27	0.51	0.59	0.43	0.83	0.59
Al_2O_3	22	7.7	3.8	8.4	7.9	3.9	4.6	8.1	2.0	8.4	7.4
Fe_2O_3	17	5.5	10	15	36	6.1	4.4	4.2	6.8	8.2	3.2
MgO	0.18	0.07	0.03	0.10	0.08	0.05	0.06	0.11	0.02	0.24	0.28
CaO	0.04	0.02	0.02	0.03	0.04	0.08	0.02	0.03	0.02	0.11	0.04
Na_2O	0.08	0.04	0.05	0.13	0.16	0.51	0.04	0.06	0.04	0.05	0.04
K_2O	0.93	0.34	0.26	0.78	0.97	0.16	0.18	0.35	0.12	0.68	0.73
P_2O_4	0.35	0.02	0.03	0.04	0.03	0.05	0.02	0.03	0.03	0.21	0.10
SO_3	1.6	0.4	0.85	1.9	4.0	0.85	0.35	0.83	0.48	1.0	0.83
Cl	0.04	0.01	0.02	0.02	0.03	0.02	0.04	0.01	0.01	0.05	0.02
Organic carbon (%)	nd	0.33	0.48	0.47	0.61	1.76	0.48	0.94	0.68	9.5	3.9

Table 10.5 (b) Trace elements (mg kg^{-1})

Elements	*Sample number*										
	S1	*S2*	*S3*	*S4*	*S5*	*S6*	*S7*	*S8*	*S9*	*S10*	*S11*
Zn	67	13	12	25	38	140	13	16	14	73	43
Cu	39	13	16	21	30	24	26	26	27	56	21
Ni	29	14	20	16	11	41	26	28	35	63	40
V	181	49	47	92	109	56	71	148	68	107	47
Cr	157	118	165	305	225	169	179	186	204	244	138
Mn	141	82	112	108	108	181	147	106	209	243	158
Co	9.7	3.9	4.1	3.8	5.0	14	3.9	5.7	3.9	25	19
Zr	271	617	624	514	383	294	655	442	920	240	345
Y	42	19	15	25	23	14	17	24	13	77	45
Sr	152	14	11	29	30	17	8.5	14	9.4	14	8.0
Rb	26	19	16	38	44	10	15	25	7.2	36	36
Pb	34	7.8	10	23	36	21	21	64	21	26	10
U	6.0	2.6	1.9	4.1	5.3	3.2	1.8	3.9	2.9	11	3.3

Source: Bell *et al.*, 2002.

Notes
S1 sample from seepage point.
S2, S3, S4 and S5 samples from decantation ponds.
S10 sample from wetland.

volatile organic compounds (VOCs) that are very different in character from the predominant semi-volatile organic compounds (SVOCs) associated with gas manufacture. In addition, a gas-manufacturing site produced significant amounts of solid, as well as liquid waste. For instance, Hatheway estimated that three tons of brick were removed from, and replaced, at each generator set per year. Other waste solids included ash, clinker, slag, scurf (hard carbon deposits formed on the interior surfaces of retorts and generators), spent lime, spent wood chips, spent iron spirals (for capturing sulphur) and retort and bench fragments. The solid material, some of which may be contaminated, was disposed of in dumps, which usually were located around the periphery of the plant, along an adjacent stream or occupying topographic hollows. Because dumps had high void ratios, toxic wastes and sludges may have been disposed within them. Adjacent low land often was used as unlined tar ponds. Once in the ground most manufactured gas wastes become immobile. Semi-volatile organic compounds are highly viscous and have a low solubility in water, and so come to rest in the vadose zone. Site exploration proceeds once the past layout of a site has been determined and wastes located, as far as possible, from documentary data.

Prosper III coal mine was established in 1906 and, with associated industries, covered an area of about 290 000 m^2 (Bell and Genske, 2000). Most of the site was occupied by a coking plant and chemical factories. The mine closed in 1986, leaving a derelict site, part of which was identified as highly contaminated. Since this former mine site is located in the centre of Bottrop, in the Ruhr district, Germany, it was decided that it should be rehabilitated in order to allow new industries and residential areas to be developed. Rehabilitation of the site involved excavation of contaminated areas including the massive foundation remnants to depths of around 2 m. The ground that was excavated was replaced by coarse cohesionless material, the soil properties of which were to be good enough to allow the construction of the industrial and residential buildings. The contaminated soil and foundation remnants were stored within a sector of the site that was identified as highly contaminated whereas the less contaminated material was distributed over the remaining sectors. This involved the movement of some 180 000 m^3 of soil and foundation remnants. Such excavation work, however, could lead to the release of contaminated dust into the air. Consequently, special protective measures were taken by the work force so that they were not affected. As the degree of the contamination present in the excavated soil involved a risk, the soil was continuously tested and if a given contamination threshold was exceeded, then the soil affected was removed to a special waste site. Fortunately, this was necessary only occasionally. The excavated material was used to form an undulating topography, part of which is a recreation area. The highly contaminated sectors were covered with a drain and seal system, consisting of geosynthetics and clay cover, to avoid immediate contact with contaminated soil, and to prevent the percolation of meteoric water through the contaminated ground that would cause migration of contaminants into the saturated zone beneath the water table. In addition, a number of observation wells were installed. These are sampled continuously in order to detect any possible contamination. If increased values are detected, then the observation wells will be used to as recovery wells to extract and clean the contaminated groundwater.

10.7 Contamination in estuaries

The ultimate receptacle of much of the effluent of the industrialized world is the sea, effluent entering the sea via estuaries that form the final outlets of rivers draining the land. Eventually discharged substances are deposited on marine sediments. For example, Fukue *et al.* (1999) carried out a survey of marine sediments in several bays in Japan and found that they were seriously polluted. Sediments, in particular the fine and organic fractions, are regarded as carriers and possible sources of contaminants (heavy metal and organic) in aquatic systems. Heavy metals may bio-accumulate in food webs, resulting in possible detrimental effects on biota and humans. Be that as it may, heavy metals are not necessarily fixed permanently to sediment but may be remobilized via various chemical, physical and biological processes. Potentially toxic metals can be detected and monitored by analysing water, sediment or biota. It generally is accepted, however, that the measurement of pollutants in water is not conclusive due to discharge fluctuations and short residence times. An advantage of analysing sediments is that they record the history of contamination and indicate how pollution enters an aquatic system. Geochemical analysis of sediments thus plays a fundamental role within the framework of environmental forensic investigations, such as locating sources of pollution, identifying contaminants that could escape detection in water analysis and ascertaining anthropogenic induced accumulation trends over time. Although sediment analysis alone is insufficient to evaluate the degree of toxicity in an aquatic system, it does give an idea of the degree of contaminant loading and identifies areas prone to environmental impacts from anthropogenic activities.

Estuaries, in particular, often tend to act as receptacles for contaminants as they frequently are the sites of harbours and industrial activity, with surrounding large centres of population. Hence, many estuaries in the industrialized world are major sources and accumulation sites (sinks) of contaminants and so have ecological problems related to contamination. The level of heavy metal contamination within sediments consequently has important environmental and operational management implications for harbours and estuaries. Limits on marine and estuarine heavy metal contamination have been established by various countries. Unfortunately, these limits tend to vary between countries, mainly due to different background values produced by different geological settings. Therefore, for effective assessment of heavy metal contamination in estuaries and harbours the pre-anthropogenic background values should be established for each system and compared with post-industrial levels. Control limits such as special care and prohibition levels, then can be determined for each individual system.

When trace metals are released into the water column they can be transferred rapidly to the sediment phase by adsorption onto suspended particulate matter (SPM) followed by sedimentation. However, some heavy metals have a particular affinity to particulates whereas other heavy metals preferentially stay in solution in the water column. For example, Pb has a strong affinity to particulates, with approximately 80 per cent of Pb released attaching onto particulates. By contrast, Zn has no particular affinity to particulates and prefers to remain in solution. Consequently, if significant amounts of Zn occur in particulate material, then it must be present in the water column in even greater amounts. Intertidal flats may be considered as important trace metal sinks since they accumulate large amounts of SPM. In polluted estuaries the deposition of SPM on intertidal flats therefore may result in contaminated intertidal sediment. An understanding of the distribution and movement of contaminated SPM in estuaries and over intertidal flats is important as far as the role of SPM in determining the transport and fate of particle-bound contaminants is concerned. It also can be relevant to other operations such as dredging and spoil dumping.

When a contaminant becomes associated with SPM or sediment, then particle dynamics are more important than water movement in determining its fate. Large-scale sediment transport patterns in a particular estuary may concentrate contaminants in specific areas remote from their point of introduction (McLaren and Little, 1987). In estuaries the sediment transport may be inland or seawards due to variations in tidal and freshwater flow, and the sediment may spend long periods stored as bed material in intertidal zones where storm waves are the main erosion influence. However, although fine sediment motion and associated contaminant transport have been studied intensively for over 50 years, the understanding of the exchange of fine sediment and contamination within estuaries, and between estuaries and shelf seas is not completely understood due to the complexities of sediment cohesive processes, current distributions, chemical interactions and wave patterns in estuaries.

The sediment and associated contaminant patterns in medium to high tidal range estuaries are complicated by the existence of turbidity maxima. Turbidity maxima is the accumulation of fine-grained material in the upper reaches of estuaries in the region of the limit of saline intrusion. Much, or all, of this material is suspended during each tidal cycle and the entire region undergoes a seasonal variation that appears to depend on fluvial input. Attenuation of sunlight may cause reduced organic activity and dissolved oxygen levels may be depleted when SPM levels are high due to the oxygen demand of decomposing organic material. Any increase in chemical reactivity on particle surfaces may have implications for the dispersal of some pollutants, whereas contaminants that adhere to solid particles may be accumulated in these regions.

In most estuaries that act as harbours and that are surrounded by industrial areas, hydraulic and mineralogical particulate fractionation usually results in increasing heavy metal concentrations with decreasing grain size. Therefore, it is necessary to normalize the effect of grain size distribution and provenance on natural metal variability before any effects of anthropogenic metallic input can be determined (Negrel, 1997). Since contaminants tend to concentrate in the fine fraction of sediments, correlation between total concentration of trace metals and the weight per cent of the fine fractions constitutes a simple, yet often effective method of normalization. Geochemical normalization of trace metal data in relation to elements such as Al and Fe can be used to compensate for natural metal variability caused by both grain size and mineralogy. In such an approach, it is assumed that there is a linear relationship between the normalizing element and the trace metal. In other words, should the concentration of the normalizing element vary because of changing mineralogy or particle size, then the concentration of the trace metal adjusts accordingly. The normalizing element therefore should be an important constituent of one or more of the major fine-grained trace metal carrier(s) and reflect their granular variability in the sediments. As a result, the regional trace metal normalizing element ratio remains constant irrespective of changes in grain size (Loring and Rantala, 1992). For example, Al can be used to reflect the granular variation of the alumino-silicate fraction, particularly the clay fraction. For linear geochemical normalization to be of value there must be a highly significant relationship, at least at the 95 per cent confidence level between the normalizing element and the grain size distribution, and between the normalizing element and the trace metal.

The content of organic matter in estuarine sediments has the ability to concentrate trace metals in sediments. For instance, organic matter appears to have a strong association with zinc, lead, arsenic and copper, and a weaker association with chromium, tin and nickel. In addition, the water in estuaries surrounded by industrial areas is oxygen deficient and under anoxic conditions bacterial sulphate reduction allows dissolved metals in the water column to be deposited as sulphides. As an example, in Durban harbour, South Africa, dissolved oxygen concentrations as low as $1.25\,mg\,l^{-1}$ have been recorded and in the bay-head area, in particular, enriched metal values are related to elevated sulphur content and low dissolved oxygen concentrations. Total sulphur content in bay-head sediment can be up to 5.7 times ($32\,580\,mg\,kg^{-1}$) the average value of sulphur ($5704\,mg\,kg^{-1}$) for harbour sediments. The presence of pyrite in sediment lends support to the view that metal sulphides are being precipitated in anoxic organic-rich areas of the harbour. The physico-chemical properties of water tend to change rapidly in estuaries from freshwater, entering from rivers and canals, to those of sea water. Increasing salinity induces flocculation of fine-grained particles (clays + organics) suspended in water. Flocs settle to form mud layers, which aids metal enrichment in sediments as both clays and organics, as noted, have a high capacity for binding metals.

Lindsay and Bell (1997) examined how contamination had affected the Forth estuary in Britain. The upper part of the Forth estuary has received a variety of industrial and domestic wastes, notably from sewage works, breweries and distilleries whilst some of the more problematic wastes received by the lower section of the estuary included chlorinated thermal effluent from a power station, wastes from a petrochemical refinery and inputs from dockyards. Heavy metal distribution does not change in a regular fashion from river to sea because contaminants enter the estuary from various sources from the land and seaward ends leading to contamination being site specific. However, lower heavy metal concentrations in surface sediment show that in recent years some industrial discharges have been reduced due to pollution control legislation. For example, in the less consolidated surface material in the lower part of the estuary Cu, Pb and Zn levels are 50–300 per cent lower than concentrations in the more consolidated lower material. In other words, recent surface sediments indicate no heavy metal layers coming within the upper end of the special care and action levels whereas older more contaminated sediments do fall within the special care range.

The degree of contamination in sediments varies within Durban harbour, South Africa, with extremely high concentrations of Zn, Co and Pb occurring in the bay-head and central basin areas, and decreasing seawards (Lindsay *et al.*, 1997). For example, in the bay-head area sediments may be 21 times the background value, with copper 18 and lead 12 times this figure. The main sources of these metals appear to be the various canals discharging into the bay, ship repairing in the central basin and inputs from sewers. The highest concentrations of arsenic occur in the bay-head area, where values exceed the background level by up to 11 times and decline seawards. Arsenic also appears to be entering the bay primarily from the canals discharging it and from sewers. Like other heavy metals, elevated concentrations of Sn occur in the bay-head and central basin areas where levels are up to 2.5 times that of the background value. Contamination appears to originate from one of the canals and the dry/floating dock area.

The investigation for the Sydney Airport parallel runway revealed areas of silt contaminated with heavy metals in the old mouth of the Cook River (Herbert and Blumer, 1994). Most of the heavy metals were in the slight to medium levels of contamination but the level of mercury contamination was high with samples exceeding the Australian guideline criteria by 10–80 times. Accordingly, 400 000 m^3 of contaminated silt was removed from the site of the new runway in Botany Bay to an adjacent disposal area where it was sealed by a cap of sand below the existing floor of the bay. A water quality monitoring system was installed and subsequently showed that no detectable heavy metals had been released into the bay.

Backfilling of Western Dock, which was the largest dock in the London Docks, was undertaken as part of the reclamation of the Docklands area and provided 25 per cent of the total area required for housing in this reclamation programme (Lord *et al.*, 1987). Dredging schedules had not been maintained towards the end of the life of the docks and as a result the basins that remained as water areas became heavily silted. This silt was highly contaminated with heavy metals and contained significant amounts of organic matter that gave rise to the generation of methane. Western Dock contained around 160 000 m^3 of contaminated silt. As removal of the silt would have proved extremely expensive, it was decided to fill the dock. The method of filling was constrained by the need to ensure stability of the dock wall and adequate drainage. Consequently, the walls of the dock were stabilized by end-tipping fill against them to form an embankment around the perimeter of the dock before the water level was lowered. The water level was lowered by submersible electric pumps and discharged into the River Thames. In order to avoid blowouts due to artesian pressures in the flood plain gravels below the dock, some water remained in the dock so that fill had to be tipped both into water and above it. The principal material used for backfill was demolition arising, for example, from the concrete jetty, which was crushed on site. Dynamic compaction was used to minimize future settlement, after which the surface level was brought up to grade by rolled rock fill or soil fill. Surface drainage from the reclaimed area is either to an adjacent dock retained as a water area or to a canal. The houses were provided with voids to allow methane ventilation.

References

Adams, D. and Holroyd, M. 1992. *In situ* soil bioremediation. *Proceedings Third International Conference on Construction on Polluted and Marginal Land*, London, Forde, M.C. (ed.), Engineering Technics Press, Edinburgh, 291–294.

Al-Tabbaa, A., Evans, C.W. and Wallace, C.J. 1998. Pilot *in situ* auger mixing treatment of a contaminated site, Part 2: site trial. *Proceedings Institution Civil Engineers, Geotechnical Engineering*, **131**, 89–95.

Anon. 1987. *Guidelines on the Assessment and Redevelopment of Contaminated Land: Guidance Note 59/83.* Interdepartmental Committee on the Redevelopment of Contaminated Land (ICRCL), Second Edition, Department of the Environment, Her Majesty's Stationery Office, London.

Anon. 1988a. *Draft for Development, DD175:1988, Code of Practice for the Identification of Potentially Contaminated Land and its Investigation.* British Standards Institution, London.

Anon. 1988b. *Control of Substances Hazardous to Health Regulation.* Statutory Instrument No. 1657, Her Majesty's Stationery Office, London.

Anon. 1988c. *Guidance for Conducting Remedial Investigations and Feasibility Studies under CERCLA.* United States Environmental Protection Agency, Office of Emergency and Remedial Response, US Government Printing Office, Washington, DC.

Anon. 1989. *Methods for Evaluating the Attainment of Clean-up Standards, Volume 1: Soils and Solid Media, EPA/540/2-90/011.* United States Environmental Protection Agency, Office of Policy, Planning and Evaluation, US Government Printing Office, Washington, DC.

Anon. 1990. *Subsurface Contamination Reference Guide, EPA/540/5-91/008.* United States Environmental Protection Agency, Office of Emergency and Remedial Response, US. Government Printing Office, Washington, DC.

Anon. 1992. *Standard Guide for Soil Gas Monitoring in the Vadose Zone, ASTM: D5314-92*. American Society for Testing Materials, Philadelphia, Pennsylvania.

Anon. 1995. *A Guide to Risk Assessment and Risk Management for Environmental Protection*. Department of Environment, Her Majesty's Stationery Office, London.

Anon. 1999a. *A Better Quality of Life: A Strategy for Sustainable Development in the United Kingdom*. Department of Environment, Transport and Regions, Her Majesty's Stationery Office, London.

Anon. 1999b. *Draft Circular on Contaminated Land*. Department of Environment, Transport and Regions. Her Majesty's Stationery Office, London.

Anon. 2002. *The Contaminated Land Report*. Department for the Environment, Food and Rural Affairs and the Environment Agency, Environment Agency, Bristol. Obtainable from http://www.defra.gov.uk/environment/landliability/pubs.htm#new.

Attewell, P.B. 1993. *Ground Pollution; Environmental Geology, Engineering and Law*. E & FN Spon, London.

Bailey, D. and Gunn, J. 1991. Landform replication as an approach to the reclamation of limestone quarries. In: *Land Reclamation, an End to Dereliction*, Davies, M.C.R. (ed.), Elsevier Applied Science, London, 96–105.

Beckett, M.J. 1993. Land contamination. In: *Contaminated Land, Problems and Solutions*, Cairney, T. (ed.), Blackie, Glasgow, 8–28.

Bell, F.G. and Genske, D.D. 2000. Restoration of derelict mining sites and mineral workings. *Bulletin Engineering Geology and the Environment*, **59**, 173–185.

Bell, F.G. and Genske, D.D. 2001. The influence of subsidence attributable to coal mining on the environment, development and restoration: some examples from western Europe and South Africa. *Environmental and Engineering Geoscience*, **7**, 81–99.

Bell, F.G., Bell, A.W., Duane, M.J. and Hytiris, N. 1996. Contaminated land: the British position and some case histories. *Environmental and Engineering Geoscience*, **2**, 355–368.

Bell, F.G., Genske, D.D. and Bell, A.W. 2000. Rehabilitation of industrial areas: case histories from England and Germany. *Environmental Geology*, **40**, 121–134.

Bell, F.G., Halbich, T.F.J. and Bullock, S.E.T. 2002. The effects of acid mine drainage from an old mine in the Witbank Coalfield, South Africa. *Quarterly Journal Engineering Geology and Hydrogeology*, **35**, 265–278.

Blanchfield, R. and Anderson, W.F. 2000. Wetting collapse in opencast coalmine backfill. *Proceedings Institution Civil Engineers, Geotechnical Engineering*, **143**, 139–149.

Brown, I.J. 1998. Industrial archaeological aspects of land reclamation in the UK. *Proceedings Fourth International Conference International Affiliation of Land Reclamationists, Land Reclamation Achieving Sustainable Benefits*, Nottingham, Fox, H.R., Moore, H.M. and McIntosh, A.D. (eds), A.A. Balkema, Rotterdam, 305–313.

Buchanan, N.W. and Thorburn, S. 1987. Development of land occupied formerly by tar and chemical works. *Proceedings Conference on Building on Marginal and Derelict Land*, Glasgow, Thomas Telford Press, London, 423–433.

Cairney, T. 1993. International responses. In: *Contaminated Land, Problems and Solutions*, Cairney, T. (ed.), Blackie, Glasgow, 1–6.

Charles, J.A., Burford, D. and Hughes, D.B. 1993. Settlement of opencast backfill at Horsley 1973–1992. *Proceedings Conference on Engineered Fills*, Newcastle upon Tyne, Clarke, B.G., Jones, C.J.F.P. and Moffat, A.I.B. (eds), Thomas Telford Press, London, 429–440.

Clark, R.G., Scarrow, J.A. and Skinner, R.W. 1994. Safety considerations specific to the investigation of landfills and contaminated land. *Proceedings First International Congress on Environmental Geotechnics*, Edmonton, Carrier, W.D. (ed.), Bi-Tech Publications Ltd, Richmond, British Columbia, 167–172.

Davies, B.E. 1972. Occurrence and distribution of lead and other metals in two areas of unusual disease incidence in Britain. *Proceedings International Symposium on Environmental Health Aspects of Lead*, Amsterdam, 125–134.

Davis, A., Ruby, M.V. and Bergstrom, P.D. 1994. Factors controlling lead bioavailability in the Butte mining district, Montana, USA. *Environmental Geochemistry and Health*, **16**, 147–157.

Dawson, G.W. and Gilman, J. 2001. Land reclamation technology – expanding the geotechnical engineering envelope. *Proceedings Institution Civil Engineers, Geotechnical Engineering*, **149**, 49–61.

Eccles, C.S. and Redford, R.P. 1999. The use of dynamic (window) sampling in the site investigation of potentially contaminated ground. *Engineering Geology*, **53**, 125–130.

Ferber, U. 1995. Fächenrecycling in Europa – stragegien und empfehlungen. *Brach Flachen Recycling/Recycling Derelict Land*, **1**, 14–18.

Ferguson, C.C. and Denner, J.M. 1998. Human health risk assessment using UK guideline values for contaminants in soils. In: *Contaminated Land and Groundwater: Future Directions*, Engineering Geology Special Publication No. 14, Lerner, D.N. and Walton, N.R.G. (eds), The Geological Society, London, 37–43.

Fukue, M., Nakamura, T., Kato, Y. and Yamasaki, S. 1999. Degree of pollution for marine sediments. *Engineering Geology*, **53**, 131–137.

Genske, D.D. and Thein, J. 1994. Recyclycling derelict land. *Proceedings First International Congress on Environmental Geotechnics*, Edmonton, Carrier, W.D. (ed.), Bi-Tech Publications Ltd, Richmond, British Columbia, 493–498.

Genske, D.D., Kappernagel, T. and Noll, P. 1994. Computer aided remediation of contaminated sites. *Proceedings Seventh International Congress International Association of Engineering Geology*, Lisbon, Oliveira, R., Rodriques, L.F., Coelho, A.G. and Cunha, A. (eds), A.A. Balkema, **4**, 4557–4562.

Haines, R.C. and Harris, M.R. 1987. Main types of contaminants. In: *Reclaiming Contaminated Land*, Cairney, T. (ed.), Blackie, Glasgow, 39–61.

Haley, J.L., Hanson, B., Enfield, C. and Glass, J. 1991. Evaluating the effectiveness of groundwater extraction systems. *Ground Water Monitoring Review*, **12**, 119–124.

Harris, M.R. 1987. Recognition of the problem. In: *Reclaiming Contaminated Land*, Cairney, T. (ed.), Blackie, Glasgow, 1–29.

Harris, M.R., Herbert, S.M. and Smith, M.A. 1995. *Remedial Treatment for Contaminated Land.* Construction Industry Research and Information Association (CIRIA), Special Publications SP101–SP112, London.

Hartley, D. 1991. The use of derelict land – thinking the unthinkable. In: *Land Reclamation, an End to Dereliction*, Davies, M.C.R. (ed.), Elsevier Applied Science, London, 65–74.

Hatheway, A.W. 2002. Geoenvironmental protocol for site and waste characterization of former manufactured gas plants; worldwide remediation challenge in semi-volatile organic wastes. *Engineering Geology*, **64**, 317–338.

Herbert, C.P. and Blumer, D. 1994. Sydney Airport parallel runway project: successful disposal of contaminated silts. *Proceedings Institution Civil Engineers, Geotechnical Engineering*, **107**, 59–64.

Hobson, D.M. 1993. Rational site investigations. In: *Contaminated Land, Problems and Solutions*, Cairney, T. (ed.), Blackie, Glasgow, 29–67.

Holm, L.A. 1993. Strategies for remediation. In: *Geotechnical Practice for Waste Disposal*, Daniel, D.E. (ed.), Chapman & Hall, London, 289–310.

Howland, A.F. 2001. The history of the development of procedures for the rapid assessment of environmental conditions to aid the urban regeneration process at London Docklands. *Engineering Geology*, **60**, 117–125.

Jewell, C.M., Hensley, P.J., Barry, D.A. and Acworth, I. 1993. Site investigation and monitoring techniques for contaminated sites and potential waste disposal sites. *Proceedings Conference on Geotechnical Management of Waste and Contamination*, Sydney, Fell, R., Phillips, A. and Gerrard, C. (eds), A.A. Balkema, Rotterdam, 3–38.

Johnson, A.C. 1994. Site investigation for development on contaminated sites – how, why and when? *Proceedings Third International Conference on Re-use of Contaminated Land and Landfills*, London, Forde, M.C. (ed.), Engineering Technics Press, Edinburgh, 3–7.

Kilkenny, W.M. 1968. *A Study of the Settlement of Restored Opencast Coal Sites and Their Suitability for Development.* Bulletin No. 38, Department of Civil Engineering, University of Newcastle upon Tyne, Newcastle upon Tyne.

Leach, B.A. and Goodyear, H.K. 1991. *Building on Derelict Land.* Special Publication 78, Construction Industry Reseach and Information Association (CIRIA), London.

Lindhult, E.C., Tarsavage, J.M. and Fukaris, K.A. 1995. Remediation of clay using two-phase vacuum extraction. *Proceedings Conference on Innovative Technologies for Site Remediation and Hazardous Waste Management*, American Society Civil Engineers, New York, 13–20.

Lindsay, P. and Bell, F.G. 1997. Contaminated sediment in two United Kingdom estuaries. *Environmental and Engineering Geoscience*, **3**, 375–387.

Lindsay, P., Bell, F.G. and Wright, C.I. 1997. Heavy metal distribution in non-industrialized microtidal southern hemisphere estuaries and an industrialized macrotidal northern hemisphere estuary. *Proceedings International Symposium on Engineering Geology and the Environment*, Athens, Marinos, P.G., Koukis, G.C., Tsiambaos, G.C. and Stournaras, G.C. (eds), A.A. Balkema, Rotterdam, **2**, 1967–1974.

Loehr, R.C. 1993. Bioremediation of soils. In: *Geotechnical Practice for Waste Disposal*, Daniel, D.E. (ed.), Chapman & Hall, London, 520–550.

Lord, J.A., Mudd, I.G. and Williams, G.J. 1987. Western Dock and Hermitage Basin reclamation, London Docks. *Proceedings Conference on Building on Marginal and Derelict Land*, Glasgow, Thomas Telford Press, London, 521–537.

Loring, D.H. and Rantala, R.T.T. 1992. Manual for geochemical analyses of marine sediments and suspended particulate matter. *Earth Science Reviews*, **32**, 235–283.

Mabey, R. 1991. Derelict land – recent developments and current issues. In: *Land Reclamation, an End to Dereliction*, Davies, M.C.R. (ed.), Elsevier Applied Science, London, 3–39.

Madhav, M.R., Bouazza, A. and Van Impe, W.F. 1997. Reclamation of landfills and contaminated ground – a review. In: *Environmental Geotechnics – Geoenvironment 97*, Bouazza, A., Kodikara, J. and Parker, R. (eds), A.A. Balkema, Rotterdam, 505–510.

McLaren, P. and Little, D.I. 1987. The effects of sediment transport on contaminant dispersal: An example from Milford Haven. *Marine Pollution Bulletin*, 18, 586–594.

Negrel, P. 1997. Multi-element chemistry of the Loire Estuary sediments: anthropogenic *vs* natural sources. *Estuarine, Coastal and Shelf Science*, **44**, 395–410.

Norris, G., Al-Dhahir, Z., Birnstingl, J., Plant, S.J., Cui, S. and Mayell, P. 1999. A case study of the management and remediation of soil contaminated with polychlorinated biphenyls. *Engineering Geology*, **53**, 177–185.

Parry, G.D.R. and Bell, R.M. 1987. Types of contaminated land. In: *Reclaiming Contaminated Land*, Cairney, T. (ed.), Blackie, Glasgow, 30–38.

Pearson, C.F.C. 1992. Site investigation of gas contaminated sites. In: *Proceedings Second International Conference on Polluted and Marginal Land*, London, Forde, M.C. (ed.), Engineering Technics Press, Edinburgh, 83–89.

Reid, J.M. and Brookes, A.H. 1999. Investigation of lime stabilised contaminated material. *Engineering Geology*, **53**, 217–231.

Reid, W.M. and Buchanan, N.W. 1987. The Scottish Exhibition Centre. *Proceedings Conference on Building on Marginal and Derelict Land*, Glasgow, Thomas Telford Press, London, 435–448.

Rosner, T. and van Schalkwyk, A. 2000. The environmental impact of gold mine tailings footprints in the Johannesburg region, South Africa. *Bulletin Engineering Geology and the Environment*, **59**, 137–148.

Sara, M.N. 1994. *Standard Handbook for Solid and Hazardous Waste Facilities Assessment.* Lewis Publishers, Boca Raton, Florida.

Singleton, M. and Burke, G.K. 1994. Treatment of contaminated soil through multiple bioremediation technologies and geotechnical engineering. *Proceedings Third International Conference on Reuse of Contaminated Land and Landfills*, London, Forde, M.C. (ed.), Engineering Technics Press, Edinburgh, 97–107.

Smith, S.L. 1998. Risk management strategies. *Proceedings Fourth International Conference International Affiliation of Land Reclamationists, Land Reclamation Achieving Sustainable Benefits*, Nottingham, Fox, H.R., Moore, H.M. and McIntosh, A.D. (eds), A.A. Balkema, Rotterdam, 219–225.

Swane, I.C., Dunbavan, M. and Riddell, P. 1993. Remediation of contaminated sites in Australia. *Proceedings Conference on Geotechnical Management of Waste and Contamination*, Sydney, Fell, R., Phillips, A. and Gerrard, C. (eds), A.A. Balkema, Rotterdam, 127–163.

Taha, M.R. 1997. Some aspects of electrokinetic remediation of soil. In: *Environmental Geotechnics – Geoenvironment 97*, Bouazza, A., Kodikara, J. and Parker, R. (eds), A.A. Balkema, Rotterdam, 511–516.

Thom, M.J. and Hausmann, M.R. 1997. Construction on derelict land. In: *Environmental Geotechnics – Geoenvironment 97*, Bouazza, A., Kodikara, J. and Parker, R. (eds), A.A. Balkema, Rotterdam, 143–160.

Thorburn, S. and Buchanan, N.W. 1987. Building on chemical waste. *Proceedings Conference on Building on Marginal and Derelict Land*, Glasgow, Thomas Telford Press, London, 281–296.

Trost, P.B. 1993. Soil washing. In: *Geotechnical Practice for Waste Disposal*, Daniel, D.E. (ed.), Chapman & Hall, London, 585–603.

Waters, J. 1994. *In Situ* remediation using air sparging and soil venting. *Proceedings Third International Conference on Re-use of Contaminated Land and Landfills,* London, Forde, M.C. (ed.), Engineering Technics Press, Edinburgh, 109–112.

Welsh, J.P. and Burke, G.K. 1991. Jet grouting for soil improvement. In: *Geotechnical Engineering*, Geotechnical Special Publication No. 27, American Society Civil Engineers, New York, 334–345.

Wood, A.A. 1994. Contaminated land – an engineer's viewpoint. *Proceedings Third International Conference on Re-use of Contaminated Land and Landfills,* London, Forde, M.C. (ed.), Engineering Technics Press, Edinburgh, 205–209.

11 Geological materials used in construction

11.1 Building or dimension stone

Stone has been used as a construction material for thousands of years. One of the reasons for this was its local availability. Furthermore, stone requires little energy for extraction and processing. Indeed, stone is used more or less as it is found except for the seasoning, shaping and dressing that is necessary before it is used for building purposes. Yet other factors, as many ancient buildings testify, are the attractiveness, durability and permanence of stone.

11.1.1 Factors that determine whether a rock is worked for building stone

A number of factors determine whether a rock will be worked as a building stone. These include the volume of material that can be quarried, the ease with which it can be quarried, the wastage consequent upon quarrying and the cost of transportation, as well as its appearance and physical properties. As far as volume is concerned, the life of a quarry should be at least 20 years. The amount of overburden that has to be removed also affects the economics of quarrying. Obviously, there comes a point when amount of overburden that has to be removed makes operations uneconomic, although if geological conditions are suitable, then the rock could be mined from this time on. Weathered rock normally represents waste, therefore the ratio of fresh to weathered rock is another factor of economic importance. The ease with which a rock can be quarried depends to a large extent upon geological structures, notably the geometry of discontinuities. Ideally, rock for building stone should be massive; certainly it must be free from closely spaced joints or other discontinuities as these control block size. The stone should be free of fractures and other flaws. In the case of sedimentary rocks, where beds dip steeply, quarrying has to take place along the strike. Steeply dipping rocks also can give rise to problems of slope stability when excavated. On the other hand, if beds of rock dip gently it is advantageous to develop the quarry floor along the bedding planes. The massive nature of igneous rocks such as granite, means that a quarry can be developed in any direction, within the constraints of planning permission.

A uniform appearance generally is desirable in building stone. The appearance of a stone largely depends upon its colour, which in turn is determined by its mineral composition. Texture also affects the appearance of a stone, as does the way in which it weathers. For example, the weathering of some minerals, such as pyrite, may produce stains. Generally speaking, rocks of light colour are used as building stone.

For usual building purposes, an unconfined compressive strength of 35 MPa is satisfactory and the strength of most rocks used for building stone is well in excess of this figure. In certain instances tensile strength is important, for example, tensile stresses may be generated in a stone subjected to ground movements. However, the tensile strength of a rock, or more particularly its resistance to bending, is a fraction of its compressive strength.

Hardness as far as building stone is concerned is a factor of small consequence. The exception to this is if a stone is likely to be subjected to continual wear such as in the case of steps or pavings.

The texture and porosity of a rock affect its ease of dressing and the amount of expansion, freezing and dissolution it may undergo. For example, fine-grained rocks are more easily dressed than coarse varieties. The retentivity of water in a rock with small pores is greater than in one with large pores, and so they are more prone to frost attack.

In some cases, stone may be obtained by splitting along bedding and/or joint surfaces by using a wedge and feathers. Another method of quarrying rock for building stone consists of drilling a series of closely spaced holes (often with as little as 25 mm spacing between them) in line in order to split a large block from the face. Stone also may be cut from the quarry face by using a wire saw (the stone is cut by sand fed between the wire and the rock) or by a diamond impregnated wire saw (Fig. 11.1a). Flame torch cutting has been used primarily for winning granitic rocks (Fig. 11.1b). It is claimed that this technique is the only way of cutting stone in areas of high stress relief.

Figure 11.1a Working marble with diamond-impregnated wire saws, Estremos, Portugal.

Figure 11.1b Quarrying blocks of granite by flame torch, Cornwall, England.

If explosive is used to work building stone, then the blast should only weaken the rock along joint and/or bedding planes, and not fracture the material. The object is to obtain blocks of large dimension that can be sawn to size (Fig. 11.2). Hence, the blasting pattern and amount of charge (black powder) are very important. Every effort should be made to keep rock wastage and hair cracking to a minimum.

When stone is won from a quarry it contains a certain amount of pore water referred to as quarry sap. As this dries out it causes the stone to harden. Consequently, it is wise to shape the material as soon as possible after it has been got from the quarry. Blocks are first sawn to the required size, after which they may be planed or turned, before final finishing. Alternatively, lengths of sawn stone may be split to size by a guillotine. Careless operation of dressing machines or tooling of the stone may produce bruising. Subsequently, scaling may develop at points where the stone was bruised, thereby spoiling its appearance.

11.1.2 Durability of building stone

The durability of stone is a measure of its ability to resist weathering and thereby retain its original size, shape, strength and appearance over an extensive period of time. It is one of the most important factors that determines whether or not a rock will be worked for building stone (Sims, 1991). The amount of weathering undergone by a rock in field exposures or quarries affords some indication of its qualities of resistance. However, there is no guarantee that the durability is the same throughout a rock mass and if it changes, then it is far more difficult to detect than, for example, a change in colour. Another means of assessing how a rock will

Figure 11.2 Cutting a block of sandstone, Birchover, Derbyshire, England.

behave on exposure is to observe how it has performed in existing buildings. Again, however, there is no way of knowing that the quality and durability of a stone used in an existing building is the same as that which is now available from the same quarry. Furthermore, the microclimate experienced by a stone in an existing building may differ from that in a new building elsewhere.

The type and rate of weathering varies from one climatic regime to another. In humid regions, chemical and chemico-biological processes generally are much more significant than those of mechanical disintegration. The degree and rate of weathering in humid regions depend primarily on the temperature and amount of moisture present in the atmosphere. If the temperature is high, then weathering is active; indeed an increase of 10 °C, more than doubles the rate at which chemical reactions occur. In dry air, rocks decay very slowly. The presence of moisture hastens the rate, first, because water is itself an effective agent of weathering and, second, because water holds in solution substances that react with the component minerals of rock, carbon dioxide being especially important in the case of limestone.

The response of a rock to weathering is directly related to its internal surface area and average pore size. Hence, coarse-grained rocks generally weather more rapidly than fine-grained ones. The degree of interlocking between component minerals also is a particularly important textural factor, since the more strongly a rock is bonded together, the greater is its resistance to weathering. The closeness of the interlocking of grains governs the porosity of rock. This, in turn, determines the amount of water it can hold and so the more porous the rock, the more susceptible it is to chemical attack (Ordoñez *et al.*, 1997).

The chemical processes of weathering are aided by mechanical breakdown that leads to the enlargement of mineral surfaces. This provides greater accessibility for oxygen and moisture, further accelerating chemical breakdown. Minerals tend to be attacked by a solvent until saturation is reached, dissolution proceeding more aggressively the lower the saturation of the solution. The rate of solution generally depends on the solubilities and specific solution rate constants of the minerals concerned, the degree of saturation of the solvent, the area presented to the solvent and on the motion of the solvent (which may keep it undersaturated). The form a mineral adopts does not influence its solution rate. Chemical weathering also aids rock disintegration by weakening the fabric and by emphasizing structural weaknesses, however slight. Enlargement of pores enhances pore water movement and brings about an increase in stress within the fabric of the rock, thus reducing its strength and increasing stress corrosion.

The severity of the conditions to which a stone is exposed in a building is influenced by the nature of the environment and the degree to which this is affected by the features of, and the location within, the building itself. Hence, different zones can be recognized in terms of the protection afforded (Fig. 11.3). If 60 years is taken as the design life of domestic buildings, then there are few stones that do not afford satisfactory performance if located in a sheltered position. Durability needs to be taken into account when the stone is intended for longer life, for more exposed positions (e.g. string courses or copings) or for harsher environments (e.g. coastal situations).

Many urban atmospheres were polluted by acidic gases during the second half of the nineteenth and much of the twentieth centuries, and more recently some rural areas have been affected by acid rain. Such polluted air is believed to have accelerated the process of weathering compared with the rate in natural environments. If stone in such affected environments is susceptible to acid attack, as are carbonate rocks, then even plain wall areas suffer damage from acidic pollutants, and the effects can be conspicuous and unsightly.

Exposure of a stone to intense heating causes expansion of its component minerals with subsequent exfoliation at its surface. The most suspect rocks in this respect appear to be those that contain high proportions of quartz and alkali feldspars, such as granites and sandstones. Indeed, quartz is one of the most expansive minerals, expanding by 3.76 per cent between normal temperatures and 570°C. When limestones and marbles are heated to 900°C or dolostones to 800°C, superficial calcination begins to produce surface scars. Generally, finely textured rocks offer a higher degree of heat resistance than do coarse grained varieties.

11.1.3 Moisture and salts in stone

Moisture in stone is of great importance as a disruptive agent and as a means of transporting salts. Most of the moisture present in stone is derived by condensation from the atmosphere, from rain and from rising ground moisture. The latter contains more ions than rain water (Arnold, 1981).

Stone can be damaged by alternate wetting and drying. What is more, water in the pores of a stone of low tensile strength can expand enough when warmed to cause its disruption. For example, when the temperature of water is raised from 0 to 60°C it expands some 1.5 per cent and this can exert a pressure of up to 52 MPa in the pores of a rock. Indeed, water can cause expansion within granite ranging from 0.004 to 0.009 per cent, in marble from 0.001 to 0.0025 per cent and in quartz arenites from 0.01 to 0.044 per cent. The stresses imposed upon masonry by expansion and contraction, brought about by changes in temperature and moisture content, can result in masonry between abutments spalling at the joints, and blocks may even be shattered and fall out of place.

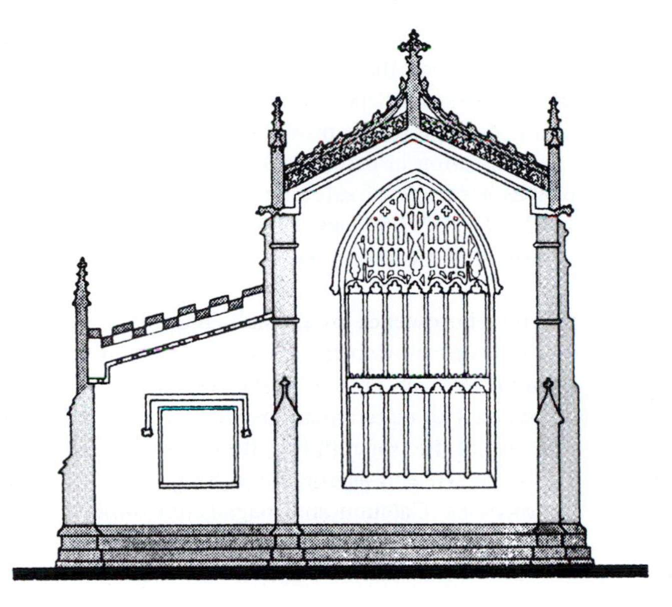

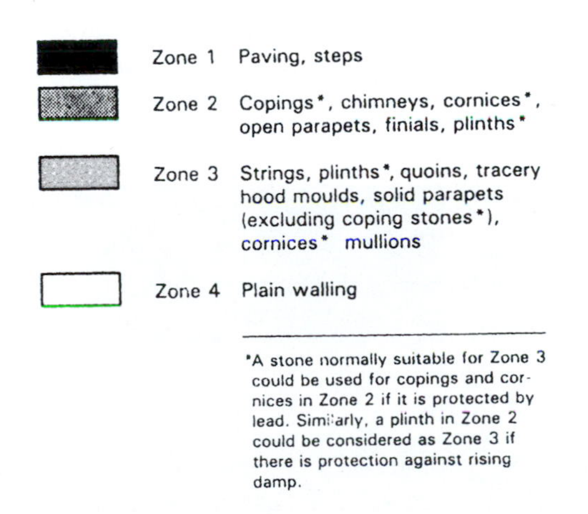

Figure 11.3 Exposure zones of a building in which stone is used.
Source: Anon., 1983a. Reproduced by kind permission of BRE. Crown copyright.

Soluble salts also may occur in the pores of stone and tend to move outwards. The more soluble chlorides and some sulphates remain in solution and move back and forth in a stone with changes in weather and moisture gradient. The less soluble salts crystallize at or near the surface.

There are three ways whereby salts within a stone can cause it to breakdown, namely, by pressure of crystallization, by hydration pressure and by differential thermal expansion. Under certain conditions some salts may crystallize or recrystallize to different hydrates, which occupy a larger space (being less dense) and exert additional pressure, that is, the hydration pressure. The crystallization pressure depends on the temperature and degree of supersaturation of the solution, whereas the hydration pressure depends on the ambient temperature and the relative humidity. Calculated crystallization pressures provide an indication of the potential pressures that may develop during crystallization in narrow closed channels. In particular, salt corrosion or salt fretting, which involves a variety of salt actions, occurs along the upper fringe of the capillary zone (Arnold, 1982). The effectiveness of salt action depends on the kinds of salts present, on the size and shape of the capillary system, on the pore moisture content and on exposure to solar radiation.

Salts in a stone give rise to efflorescence by crystallizing on its surface at the open ends of capillary systems, due to outward movement of moisture. Efflorescences consist of water-soluble salts that initially may have been present in the stone or may have formed as a result of reactions of atmospheric gases with materials in the stone. In subflorescence, crystallization takes place just below the surface of the stone. In fact, the outer skin frequently loses its support and exfoliation occurs. This is referred to as contour scaling (Fig. 11.4). Subflorescence and exfoliation are continuing processes. Calcium and magnesium sulphates are among the most common of the salts involved.

Dolostones have the potential for magnesium sulphate formation. However, the surface skin of a dolostone usually shows the amount of magnesium sulphate to be very much less than calcium sulphate (Hart, 1988). This is because magnesium sulphate is much more soluble than calcium sulphate and is readily washed away by rain in exposed situations.

It is the frequency of wetting and drying, and heating and cooling of surface layers (referred to as fatigue loading) that can make salt weathering so destructive. Salts expand and contract at different rates from the host stone, producing changing internal stresses as temperatures rise and fall. Resistance to crystallization damage is strongly dependent on the internal structure of the stone and decreases as the proportion of fine pores increases. The pressures produced

Figure 11.4 Contour scaling on sandstone used in a building at Tynemouth, England.

on crystallization in small pores are appreciable. For instance, gypsum ($CaSO_4.nH_2O$) exerts a pressure of up to 100 MPa, anhydrite ($CaSO_4$), 120 MPa, kieserite ($MgO_4.nH_2O$), 100 MPa, and halite (NaC1), 200 MPa, and these pressures are sufficient to cause disruption (Winkler, 1978). Hence, rock salt used on icy streets can represent one of the most damaging factors in stone decay, salt fretting and spalling being most noticeable a metre or so above the ground level of buildings where damage can be readily observed along edges and corners.

Crystallization caused by freely soluble salts such as sodium chloride, sodium sulphate or sodium hydroxide normally leads to the surface of the stone crumbling or powdering. Deep cavities may be formed in magnesium limestone (Fig. 11.5). Magnesium and calcium sulphates are formed when sulphur dioxide reacts with the stone and the sulphate skin formed on its surface may be very dense and hinder evaporation. When the skin is broken, soluble salts are drawn to this area, where evaporation can take place more easily. Hence, decay is concentrated preferentially in this area causing the formation of a cavity. Salt action can give rise to honeycomb weathering in porous limestones and sandstones (Pye and Mottershead, 1995). Conversely, surface induration of a stone by the precipitation of salts may give rise to a protective hard crust, that is, case hardening. If the stone is the sole supplier of these salts, then the interior is correspondingly weaker.

Crystallization of salts within the pores of a stone may generate sufficiently large stresses to cause local fragmentation. Calcium sulphite or sulphate may form within limestones or sandstones with calcareous cement when $CaCO_3$ reacts with sulphur dioxide dissolved in rainwater. The solubility of calcium sulphite is more or less the same as that of calcite in pure water but is neither influenced by temperature nor by the CO_2 content of the solvent. Calcium sulphate, both anhydrite and gypsum, are much more soluble than calcite and calcium sulphite. In addition, the formation of the hydrated form of $CaSO_4$, namely, gypsum can give

Figure 11.5 A deep cavity formed in magnesium limestone, parish church, Retford, Nottinghamshire, England.

rise to an eightfold increase in volume over the original sulphide and can exert pressures of up to 500 kPa, which can facilitate disintegration and surface spalling of stone. Hence, those features of buildings that are frequently wetted by rain are subjected to repeated crystallization of calcium sulphate, which dislodges particles of stone. These subsequently are washed off, together with calcium sulphate, during very heavy rainfall. The surface of the stone therefore is eroded gradually. In the more sheltered parts of a building, such as under ledges or in protected areas of decoration, calcium sulphate remains in position to form hard black crusts (Butlin, 1988). Black crusts are a mixture of gypsum and soot particles. They have a dramatic effect on the appearance of buildings (Fig. 11.6). Scanning electron microscope studies have shown that black crusts have an open crystalline structure that permits penetration of moisture. This moisture can carry dissolved salts into the stone with further disruptive consequences.

Disruption in stone may also take place due to the considerable contrasts in thermal expansion of salts in the pores. For instance, halite expands by some 0.5 per cent from 0 to 60°C and this may aid the decay of stone.

Figure 11.6 Black crust developed on a column of limestone, Lincoln Cathedral, Lincoln, England.

11.1.4 Polluted atmospheres and carbonate stone

The effects of acid rain have been the focus of much attention over recent years. However, the term acid rain has now been replaced by acid deposition, which is considered to better reflect the complex processes operating. Acids can be deposited on buildings in a variety of ways, either dry or wet, and arise principally from the combustion products of fossil fuels or from further reactions of the products. Deposition of pollutants on building stones depends not only on their concentration in the atmosphere and turbulence profile of the boundary layer, but also on the strength and direction of the wind and the intensity of rainfall (if any), both in the open and immediately around the place of deposition. Examples of dry deposition include sulphur dioxide (SO_2), oxides of nitrogen (NO_x), chlorides, sulphuric and nitric acids and particles (soot, fly ash, etc.). These particles accumulate during dry periods to be mobilized as acidic solutions when it rains or by other forms of precipitation or simply by humid air. Acid rain or wet deposition is defined as rainfall with a pH value less than 5 (rain containing no pollutants but saturated with carbon dioxide has a pH of 5.6). In addition, acids can be deposited on the surface of stone by mists, fog or dew, this is occult deposition. Although the amount of moisture is not large, the pollution is concentrated and acidity can be 20 times that of acid rain in the same area.

A number of questions arise regarding the effects of pollution on carbonate stone, namely:

1 What are the rates of decay attributable to present day pollution relative to natural rates of decay caused by weathering, in particular, is atmospheric pollution accelerating the rate of weathering?
2 Are current rates of decay related to present pollution levels or to those of the past, more especially, are the rates of decay less than, the same as or greater than those that occurred before the major changes in smoke and sulphur dioxide emissions took place?
3 Can quantitative relationships be established between concentrations and rates of decay?

Obviously, the major sources of pollution emissions that affect carbonate stones need to be determined, as does the principal mechanism of attack (e.g. dry or wet deposition), as well as determining whether the rate of decay is related to the pH value of rain or to its volume. How these factors interact and the timescale on which they operate need investigating in order to assess the effects of acid deposition. Furthermore, the concentration and chemical and physical forms of pollutants have to be measured on and around buildings, so that quantitative relationships between rates of decay and pollutants can be established. Procedures used to characterize weathering of carbonate-building stones include the character of surface crusts, weight loss measurements, microweathering measurements and comparison of current stone dimensions with historic records.

The absence of quantitative studies of deterioration of stone in urban areas due to atmospheric pollution during the past 150 years prevents the period over which damage has occurred, being accurately assessed. This, in turn, leads to confusion over the current rates of stone degradation, both in absolute and relative terms.

One aspect of research into damage of carbonate building stone has been directed at the specific influence of sulphur dioxide and its net long-term effect. The situation, however, is complicated by the fact that in many urban areas major reductions in sulphur dioxide and smoke emissions have taken place in the last 30 or so years. If London is taken as an example, levels of sulphur dioxide fell from around $220\,\mu g\,m^{-3}$ (annual average) in 1960 to about $50\,\mu g\,m^{-3}$ by 1980 and are currently even less. Smoke levels dropped from approximately $100\,\mu g\,m^{-3}$ to about $25\,\mu g\,m^{-3}$ during the same period. This is further emphasized by a report published in 1989 by the Building Effects Review Group (BERG) that noted that over the previous 40 years the average concentrations of smoke and sulphur dioxide had fallen by up to 90 per cent. Similar trends of decreasing air pollution are found in the northeast United States and southeastern Canada. Although levels of sulphur dioxide and smoke are below what they were, nevertheless significant fluctuations occur over short periods of time (Fig. 11.7). By contrast, the contribution to smoke levels made by vehicle exhausts rose, for instance, it accounted for 70 per cent of all smoke in central London in the 1970s. In fact, the emission of nitrogen oxides has more than doubled over the past 50 years. Nitric acid is derived from oxides of nitrogen (NO_x), chiefly given off in the exhaust gases from internal combustion engines. However, there is little evidence at present that supports direct attack of nitrogen oxides on stone surfaces. Nevertheless, it has been suggested that oxides of nitrogen might act as catalysts for the formation of sulphates, that is, they have a synergistic effect (Smith *et al.*, 1988). Admittedly, the evidence of attack on carbonate stone of nitric acid, and hydrochloric acid, is difficult to establish because the reaction products, calcium nitrate and calcium chloride, are soluble and therefore quickly washed out of stonework.

Be that as it may, acid deposition in southeast England and northeastern United States appear to be at high levels. Furthermore, the rate of deterioration of stone in many towns and cities is a matter of grave concern and in some, such as Athens, is thought to be accelerating. Although the consensus of opinion would agree, the evidence in support is by no means overwhelming.

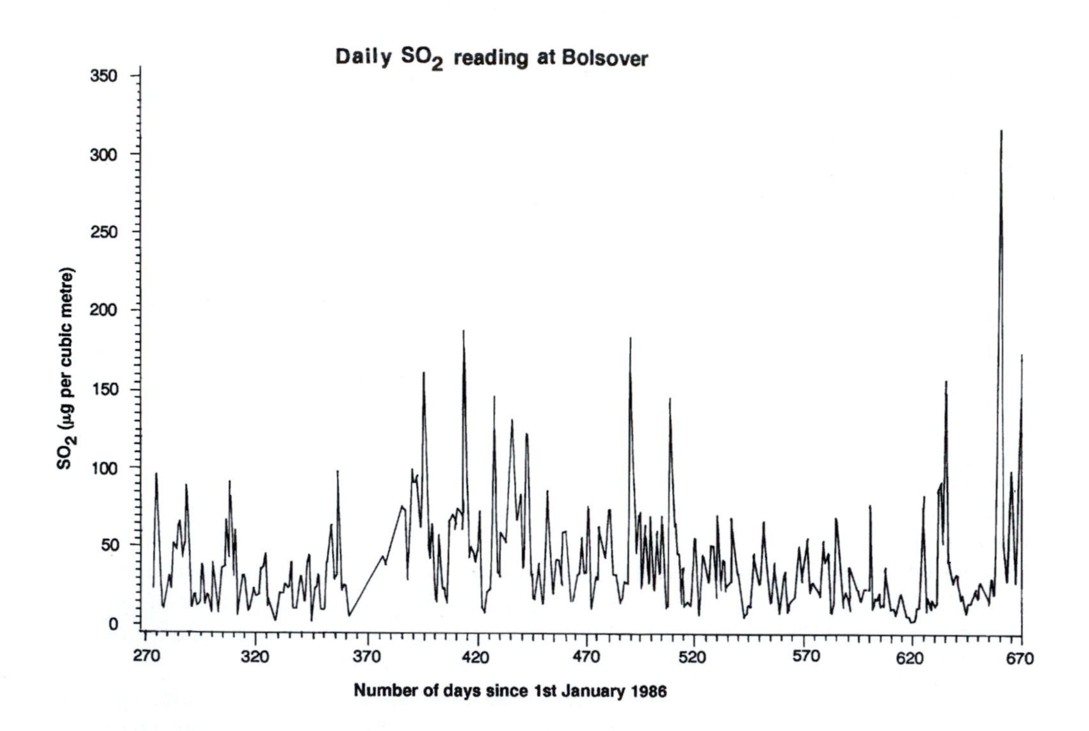

Figure 11.7 Concentration of sulphur dioxide ($\mu g\,m^{-3}$) in air during 1986–1987 at Bolsover Castle, Derbyshire, England.
Source: Yates *et al.*, 1988.

For instance, the influence of sulphur oxides on deterioration of marble in St Louis was found by Haynie (1983) to be insignificant.

Since sulphur dioxide in the atmosphere has been regarded as a factor responsible for damage of carbonate stone, Reddy (1988) devised an index termed the antecedent sulphur dioxide exposure, which is derived by multiplying the average sulphur dioxide concentration by the number of days between precipitation events. In other words, the transport and deposition of sulphur dioxide on a stone surface is related to its concentration in the atmosphere and the length of time the stone is exposed between washing by rainfall.

The uptake of sulphur dioxide by limestone is a complex process and is influenced by the moistness of the stone surface so that the humidity of the atmosphere is assumed to play a significant role. Nonetheless, Reddy's (1988) initial investigations indicated that the influence of sulphur dioxide deposition in terms of damage to limestone was not significant (Table 11.1). Similarly, in Britain no relationship has been found between the level of atmospheric pollution, especially due to sulphur dioxide, and the rate of decay of carbonate stone (Butlin *et al.*, 1985). This is probably because of the complexity of the process of decay and the variation in pollution concentrations and meteorological variables. Table 11.2 indicates some measured rates of decay of Portland Stone. One of the major difficulties of interpreting such data, however, is that the high levels of pollution in the past, especially the late 1800s and early 1900s, may have some lasting effect on building stone that is still contributing to decay mechanisms (the memory effect). If sulphur dioxide is the main cause of decay of limestone, the calcium sulphate formed some time ago may still be leaching out of the stone and contributing to current measurements of decay (Jones *et al.*, 1998).

That there does not seem to be any direct correlation between pollution levels and rates of decay appears to be borne out by the work done by Jaynes and Cooke (1987). They collected data from 25 sites in southeast England at which two types of limestone were exposed, some to wet and dry deposition, others to dry deposition only. They showed that for both exposed and sheltered samples, weight loss was significantly higher in central London than in provincial

Table 11.1 Summary of ranges of stone rainfall run-off experiments for three sites; Research Triangle Park, North Carolina; Chester, New Jersey; and Newcomb, New York

	RTP, NC	*Chester, NJ*	*Newcomb, NY*
Limestone experiment			
Rain events	13	10	16
Rain (mm)	5–55	6–57	4–28
Incident (pH)	4–4.5	3.9–4.7	4.0–4.7
Run-off (pH)	7.4–9.1	5.9–8.1	7.7–8.6
H^+ load (meq m^{-2})	0.2–0.6	0.3–3.6	0.1–1.7
$CaCO_3$ lost (μm)	0.02–0.6	0.04–0.3	0.04–0.17
Antecedent SO_2 (ppb day^{-1})	4–40	10–50	2–16
Marble experiment			
Rain events	18	9	15
Rain (mm)	2–50	6–41	4–28
Incident rain pH	3.8–5.5	3.9–4.9	4.0–4.6
Run-off (pH)	7.0–8.8	6.5–7.8	7.1–8.3
H^+ load (meq m^{-2})	0.03–2.3	0.2–3.7	0.2–1.7
$CaCO_3$ lost (μm)	0.02–0.35	0.05–0.28	0.01–0.16
Antecedent SO_2 (ppb day^{-1})	4–40	20–50	2–16

Source: Reddy, 1988.

Table 11.2 Measured decay rates for stone exposed in the United Kingdom

Location	*Stone type*	*Decay rate*	*Period*
London (St Paul's Cathedral)	Portland limestone	220 μm year^{-1} (run-off analysis)	1980–1981
		139 μm year^{-1} (direct measurement of erosion)	1980–1982
		78 μm year^{-1} (differential erosion)	1727–1982
Garston, Herts	Portland limestone	0.13% year^{-1}	1955–1965
London (Whitehall)	Portland limestone	0.33% year^{-1}	1955–1965
Southeast England	Portland limestone	0.46% year^{-1} (urban)	1981–1983
		0.34% year^{-1} (rural)	1981–1983

Source: Butlin *et al.*, 1985. Reproduced by kind permission of BRE.

centres, and both were higher than in rural areas. However, the weight losses did not correlate with the concentration of sulphur dioxide in the atmosphere. Nonetheless, the results supported the belief that weathering is accelerated as a consequence of atmospheric pollution in urban areas, but they also indicated that urban–rural gradients are not simple.

The reaction of acid rain with a porous surface of carbonate rock is influenced by the amount of infiltration, dissolution and precipitation reactions within the stone, and leaching and subsequent movement of solutes from the interior of the stone to the surface. In addition to the character of the stone, the amount and intensity of rainfall influences run-off characteristics from a porous limestone. If none of the rain runs off the surface of the stone, then no calcium ion is removed. The infiltrating solution subsequently may reprecipitate calcium carbonate within pore spaces.

The volume of rainfall and the amount of rainfall run-off from stone have been used to determine the acid rain loading and damage to stone by Reddy and Werner (1985). They showed that a direct correlation exists between increasing quantity of rainfall and surface recession of carbonate stone. Physical differences in stone surfaces such as grain size and porosity are of secondary importance in determining removal of calcium ions as compared with composition. Nonetheless, calcite dissolution and removal of calcium ions from a stone surface may reach a limiting value at high rainfall intensities. In addition, the pH value of rain is a significant factor in terms of surface recession of carbonate stone (Table 11.1). Since the surface recession of carbonate stone depends on the amount of rainfall and its pH value, the product of these two factors, the hydrogen ion loading, correlates directly and significantly with stone recession (Fig. 11.8).

11.1.5 Frost activity and stone

Frost damage is one of the major factors causing deterioration in a building stone in many temperate climates. Sometimes small fragments are separated from the surface of a stone due to frost action but the major effect is gross fracture. Frost damage is most likely to occur on steps, copings, cills and cornices where rain and snow can collect. Damage to susceptible stone may be reduced if it is placed in a sheltered location. Most igneous rocks and the better quality sandstones and limestones are immune. Obviously, frost damage to stone occurs where temperature contrasts frequently lower the ambient air temperatures below freezing. Winkler (1978) showed that although temperatures are less severe in urban than rural environments, freeze-thaw cycles are more frequent in the former so that frost damage is more likely to occur in urban areas.

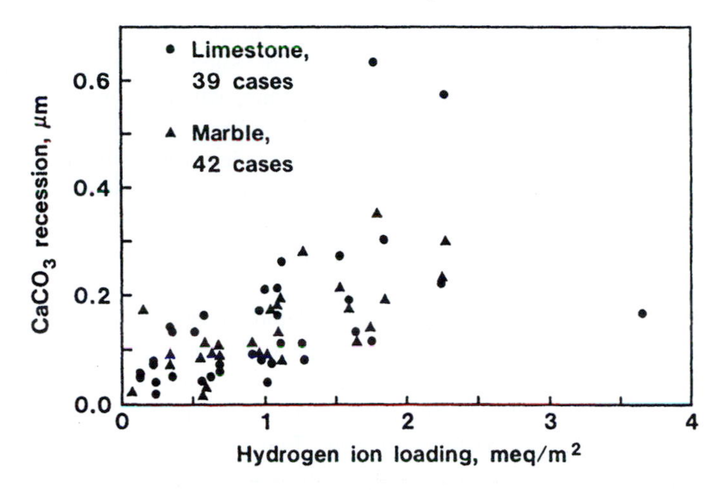

Figure 11.8 Stone surface recession for limestone and marble versus hydrogen ion loading at Newcomb, New York; Chester, New Jersey; and Research Triangle Park, North Carolina, during the summer and autumn of 1984.
Source: Reddy, 1988.

Frost activity depends upon a combination of factors such as the expansion in volume when water moves into the ice phase, the degree of saturation of the water in the pore system, the critical pore size, the amount of pore space and the continuity of the pore system. In particular, the pore structure governs the degree of saturation and the magnitude of stresses that can be generated on freezing.

As water turns to ice it increases in volume by some 9 per cent thus giving rise to an increase in pressure within the pores of a stone. This action is enhanced by the displacement of unfrozen pore water away from the developing ice front that generates a hydraulic pressure. Once ice has formed the ice pressures rapidly increase with decreasing temperature, so that at approximately −22°C ice can exert a pressure of 200 MPa. Coarse-grained and coarse-pored stones usually withstand freezing better than the fine-grained types. Indeed, the critical pore size for freeze-thaw durability appears to be about 0.005 mm (Larsen and Cady, 1969). In other words, larger mean pore diameters allow outward drainage and escape of fluid from the frontal advance of the ice line, and are therefore less frost susceptible. For instance, fine-grained limestones that have over 5 per cent absorbed water often are very susceptible to frost action whilst they are very durable below 1 per cent water sorption. Mamillan (1976), however, maintained that (especially in the case of French limestones), if a large quantity of water occupied the pores, then the stone fractured when the water froze. Hence, he recognized a certain critical moisture content above which porous stones failed on freezing. This critical moisture content tended to vary from 75 to 96 per cent of the total volume of the pores. The rapidity with which a stone attained its critical moisture content was supposedly governed by its initial state of saturation and its ability to absorb water during thawing.

While there are French and Belgian Standards stipulating test requirements for assessing frost susceptibility of stone used as masonry, there is no equivalent British Standard (Honeyborne, 1982). The system used in France requires measurement of porosity, capillarity coefficient and frost resistance in order to assess the suitability of a stone for use in a particular zone in a building. The Belgian system places a greater emphasis upon frost resistance, its assessment being based on the saturation coefficient and the diameter of the pores. The mercury porosimeter is used to obtain a quantitative classification of pore size (Ordoñez *et al.*, 1997). A correlation therefore can be established between stones that have significant numbers of pores of small diameter (D < 2.5 μm) and their natural moisture content. Such small pores facilitate capillary absorption of water and thereby aid saturation of a stone, which reduces its ability to resist freeze-thaw action. Although freeze-thaw tests are used in French and Belgian procedures they were discontinued in Britain because of the difficulty of developing an inexpensive test that would reproduce accurately the behaviour of building stones under natural conditions. Also, during a freeze-thaw test it is vital to determine the degree of saturation of the specimens, since any stone will fail if it is completely saturated with water when frozen and a dry stone remains unaffected.

The Building Research Establishment (BRE) in Britain developed a test that made use of natural frosts (Anon., 1989a). Although this test appeared to give reliable results, its reliance on the occurrence of natural frosts meant that several years were required before significant results could be obtained. Accordingly, the BRE went over to using the crystallization test to provide an indication of the frost resistance of stone. Furthermore, this test also can be used to assess the deterioration of stone due to crystallization of salts within the pores. In other words, the crystallization test measures the ability of a stone to withstand the pressures generated by the repeated crystallization and hydration of soluble salts within its pores. The crystallization test uses either magnesium or sodium sulphate. There are two types of test, namely, the severe test and the mild test. The former test uses a saturated solution that is very aggressive and only is recommended for use when the natural weathering conditions are particularly severe, or the stone is expected to have a particularly long life. The mild test uses a 15 per cent solution. The test itself is carried out on up to six cubic specimens (usually with edges between 40 and 50 mm long). The specimens are oven dried at 105 °C until they reach a constant weight. After cooling the specimens are completely immersed in a 15 per cent solution of sodium sulphate decahydrate (the specific gravity of the solution at 20°C is 1.055) for 2 hours, the temperature of the solution being kept at 20°C. The specimens then are dried in an oven that has a high relative humidity in the early stages of drying and the temperature of the specimens is raised gradually to 105 °C within 10–15 hours. The initial high relative humidity is obtained by placing a tray containing about 300 ml of water in the cold oven and switching it on about a half an hour before the specimens are placed in it. All the specimens remain in the oven for at least 16 hours, after which they are cooled to room temperature, then re-immersed in fresh sodium sulphate solution. The specimens are subjected to 15 cycles of immersion and, after final washing, to remove sulphate, and drying are weighed to determine the weight loss. The results are reported in terms of weight loss, expressed as a percentage of the initial dry weight, or as the number of cycles required to produce failure if a specimen is too fractured to be weighed before the fifteenth cycle has been completed. Conclusions relating to durability are obtained by comparing the results of the tests with the performance of stone of known weathering behaviour. This unfortunately is one of the shortcomings of the test since specific reference stones are seldom available. Nonetheless, the crystallization test enables stone to be graded into six categories ranging from the best available to that which is normally considered suitable mainly for interior use. The exposure zones where these various grades of stones may be used with safety in buildings in various climates is given in Table 11.3. Moh'd *et al.* (1996) provided a rapid means of predicting the BRE durability class for limestone based on porosity and saturation.

11.1.6 Some commonly used building stones

Granite is ideally suited for building, engineering and monumental purposes. Its crushing strength varies between 160 and 240 MPa. As a building stone, it has exceptional weathering

Table 11.3 Effect of change of environment on suitability of limestones for various building zones

Limestone class	*Suitability zones for various limestones in a range of climatic conditions*							
	Inland				*Exposed coastal*			
	Low pollution		*High pollution*		*Low pollution*		*High pollution*	
	No frost	*Frost*	*No frost*	*Frost*	*No frost*	*Frost*	*No frost*	*Frost*
A	Zones 1–4	Zones 1–4	Zones 1–4	Zones 1–4	Zones 1–4	Zones 1–4	Zones 1–4	Zones 1–4
B	Zones 2–4	Zones 2–4	Zones 2–4	Zones 2–4	Zones 2–4	Zones 2–4	Zones 2*–4	Zones 2*–4
C	Zones 2–4	Zones 2–4	Zones 3–4	Zones 3–4	Zones 3*–4	Zone 4		
D	Zones 3–4	Zone 4	Zones 3–4	Zone 4				
E	Zone 4	Zone 4	Zones 4*					
F	Zone 4	Zone 4						

Source: Anon., 1983a. Reproduced by kind permission of BRE. Crown copyright.

Notes
For zones see Fig.11.3.
* Probably limited to 50 years' life.

properties and most granites are virtually indestructible under normal climatic conditions. There are examples of granite polished over 120 years ago on which the polish has not deteriorated to any significant extent. Indeed, it is accepted that the polish on granite is such that it is only after exposure to very heavily polluted atmospheres, for a considerable length of time, that any sign of deterioration is apparent. The maintenance cost of granite compared with other materials therefore is very much less and in most cases there is no maintenance cost at all for a considerable number of years. Some granites may weather to a different colour, for instance, within several weeks some light-grey granites may alter to various shades of pink, red, brown or yellow. This is caused by the hydration of the iron oxides in them.

Although the rate of weathering of igneous rocks usually is slow, there are exceptions and once weathering penetrates a rock, the rate accelerates. For example, some basalts used for monumental purposes in Germany have proved exceptional in that they have deteriorated rapidly, crumbling after about 5 years of exposure. Such basalts have been referred to as sun-burnt basalts. On petrological examination these basalts were found to contain analcite, the development within which of micro-cracks is presumed to have produced the deterioration. Haskins and Bell (1995) commented on the rapid breakdown of basalts on exposure and attributed this to the presence of smectitic clay minerals formed by the deuteric alteration of primary minerals, the clay minerals swelling and shrinking on wetting and drying respectively. This type of basalt has been termed slaking basalt by Higgs (1976).

Carbonate minerals give limestones and dolostones colours varying from white to buff. Texture also affects the appearance of a stone, and those that are fossiliferous are highly attractive when cut and polished. The compressive strength of carbonate rocks normally is satisfactory for building stone (Table 11.4). However, carbonate stone can undergo dissolution by acidified water. This results in dulling of polish, surface discolouration and structural weakening. Carvings and decoration are subdued and eventually may disappear; natural features such as oolites, pisolites, fossils, etc. are emboldened. Some limestones have suffered significant deterioration, especially in urban atmospheres. For instance, weak sulphuric acid reacts with the calcium carbonate of limestones to produce calcium sulphate. The latter often forms just below the surface of a stone and the expansion that takes place upon crystallization causes slight disruption. If this reaction continues, then the outer surface of the limestone begins to flake off. Furthermore, Trudgill and Viles (1998) quoted calculated dissolution rates of calcite of 0.06–0.11 mm $year^{-1}$ at pH 5.5 and 2.18–2.69 mm $year^{-1}$ at pH 4.0. Bellanger *et al.* (1993) investigated the frost resistance of some limestones from Lorraine in northeast France. They found that those limestones that possessed a bimodal porosity (i.e. grain-supported calcarenites that are lacking in fine material) had high values of air trapped in the pores and were more permeable. Hence, their degree of saturation was restricted to 60 per cent and they were more frost resistant. On the other hand, calcilutites with a unimodal well-connected pore network possessed a low volume of trapped air, which gave rise to a high degree of saturation and low frost resistance. Bellanger *et al.* maintained that frost damage is mainly the result of capillary pressures developed on the curved water–ice interface at the interconnection between large spherical pores, where the freezing point is reached, and fine capillaries where water is not frozen.

Sandstone is a common sedimentary rock type and as such is widely used as building stone. Those sandstones that are of use for building purposes are found in most of the geological systems, the exception being those of the Cainozoic era. The sandstones of this age generally are too soft and friable to be of value. The colour of sandstone varies from white to buff, red or brown and is largely influenced by the type of cement that binds together the constituent grains. More importantly, the strength of sandstone also is governed largely by the amount and type

Table 11.4 Some physical properties of British limestones used for building purposes

Stone	*Orton Scar*	*Anstone*	*Doulting*	*Ancaster*	*Bath*	*Portland*	*Purbeck*
Age	*Lr. Carboniferous*	*Magnesian Limestone*	*Inferior Oolite*	*Lincolnshire Limestone*	*Great Oolite*	*Portland*	*Purbeck*
Location	*Orton*	*Kiveton Park*	*Shepton Mallet*	*Ancaster*	*Monks Park*	*Isle of Portland*	*Swanage*
Property							
Specific gravity	2.72	2.83	2.7	2.7	2.71	2.7	2.7
Dry density ($Mg\,m^{-3}$)	2.59	2.51	2.34	2.27	2.3	2.25	2.21
Porosity (%)*	4.4	10.4	12.8	19.3	18.3	22.4	9.6
Microporosity (% saturation)	54	23	30	60	77	43	62
Saturation coefficient	0.68	0.64	0.69	0.84	0.94	0.58	0.62
Unconfined compressive strength (MPa)$^{+}$	96.4	54.6	35.6	28.4	15.6	20.2	24.1
Young's modulus (GPa)	60.9	41.3	24.1	19.5	16.1	17.0	17.4
Velocity of sound ($m\,s^{-1}$)	4800	3600	2900	2900	2800	3000	3700
Crystallization test (% wt loss)	1	5	8	20	52	13	3
Durability classification (see Table 11.3)	A	B	C	D	E	C	B

Source: Bell, 1993.

Notes

Classification of porosity: 1–5%, low; 5–15%, medium; 15–30%, high (after Anon., 1979).

Classification of unconfined compressive strength: 12.5–50 MPa, moderately strong; 50–100 MPa, strong (after Anon., 1977).

Durability: A – highest (passes severe acid immersion and severe crystallization tests); to F – poorest (fails mild acid immersion and mild crystallization tests).

of cement present; the cement content also influences porosity and water absorption. The latter three factors affect the durability of sandstone. The mechanical properties of sandstone are influenced by grain size, density of packing of grains, grain contact and degree of induration. The origin of a sandstone may therefore influence its mechanical properties. For example, sandstone deposited by wind in arid environments usually is less well cemented than sandstone deposited in water and therefore generally is weaker. Representative values of some physical properties of sandstones that have been used for building purposes in Britain are given in Table 11.5. As can be seen from Table 11.5, the Triassic sandstones (wind blown) do not have such high strengths as those of Carboniferous age (waterlain). Again, it can be seen from Table 11.5 that strength has an obvious influence on durability.

The degree of resistance that sandstone offers to weathering depends upon its mineralogical composition, texture, porosity, amount and type of cement/matrix, and the presence of any planes of weakness. Accordingly, the best type of sandstone for external use for building purposes is a quartz arenite that is well bonded with siliceous cement, has a low porosity and is free from visible laminations. The tougher the stone, the more expensive it is to dress. Sandstones are composed chiefly of quartz grains that are highly resistant to weathering but other minerals present in lesser amounts may be suspect, for example, feldspars may be kaolinized. Calcareous cements react with weak acids in urban atmospheres, as do iron oxides that produce rusty surface stains. The reactions caused by acid attack may occasionally lead to the surface of a stone flaking off irregularly or, in extreme cases, to it crumbling. Laminated sandstone usually weathers badly when it is used in the exposed parts of buildings, it decaying in patches. Leary (1986) suggested that an initial assessment of the durability of sandstone can be made by the acid immersion test. This involves immersing specimens for 10 days in sulphuric acid of density 1.145 mg m^{-3}. Sandstones that are unaffected by the test are regarded as being resistant to attack by acidic rain water. Those sandstones that fail are not recommended for external use in polluted environments. A more severe test consists of immersing specimens in sulphuric acid with a density of 1.306 mg m^{-3}. The latter test is of particular value when the design life of a proposed building is exceptionally long. If a sandstone survives the acid immersion test intact, then it is subjected to the crystallization test.

11.1.7 Preservation of stone

Stone preservation involves the use of chemical treatments that prolong the life of a stone, either by preventing or retarding the progress of stone decay or by restoring the physical integrity of the decayed stone. A stone preservative therefore may be defined as a material that, when applied, will avert or compensate for the harmful effects of time and the environment. When applied, the preservative must not change the natural appearance or architectural value of the stone to any appreciable extent. Before a preservative is used on stone, there should be evidence that indicates that it will perform as intended for a reasonable period of time. It should be borne in mind that stone cannot be preserved by a single treatment but that it requires regular maintenance. Unfortunately, although numerous materials and stone preservative techniques have been used, data regarding their long-term effectiveness under natural conditions is lacking. What is more, preservative solutions frequently behave very differently on different varieties of stone and are influenced by location, exposure and the position of the stone in a building.

As the agents of weathering are responsible for the decay of stone, attention should be given to weather protection before stone preservation treatments are considered. For example, in some instances decay of stone is due to the poor condition of protective cornices and canopies or jointing in masonry, or to the inefficient disposal of rainfall. Obviously, such faults must be

Table 11.5 Some physical properties of British sandstones used for building purposes

Stone	*Colour*	*Grain size*	*Specific Gravity*	*Dry density* ($Mg\,m^{-3}$)	*Porosity* (%)	*Unconfined compressive strength* (MPa)	*Young's modulus* (GPa)	*Acid immersion test*	*Crystallization test*	*Saturation coefficient*	*Durability classification*
Hollington Trias Near Uttoxeter	Pink to red, mottled buff	Fine to medium grained	2.71	2.04	23.5	29	13.6	Passed	F9	0.71	D
Red St Bees Trias St Bees	Red	Fine grained	2.68	2.15	19.6	12.5	8.9	Failed	–	0.66	E,F
Lazonby Permian Near Penrith	Dark pink to red	Fine to medium grained	2.68	2.38	9.3	40	21.8	Passed	37	0.47	B,C
Cat Castle Coal Measures Barnard Castle	White to buff	Medium grained	2.68	2.34	11.5	74	38.7	Passed	F11	0.66	D
Ladycross Coal Measures Near Hexham	Light grey to buff	Fine to medium grained	2.69	2.36	11.6	82	41.2	Passed	9	0.62	A
Birchover Namurian Near Matlock	Buff to pink	Medium to coarse grained	2.69	2.34	12.4	48	25.6	Passed	40	0.65	B,C

Table 11.5 (Continued)

Stone	*Colour*	*Grain size*	*Specific gravity*	*Dry density* (Mg m^{-3})	Porosity (%)	*Unconfined compressive strength* (MPa)	*Young's modulus* (GPa)	*Acid immersion test*	*Crystallization test*	*Saturation coefficient*	*Durability classification*
Stancliffe Namurian Near Matlock	Fine to buff	Medium grained	2.67	2.38	11.5	72	41.5	Passed	20	0.63	A,B
Blaxter Lr Carbonif-erous Elsdon	Fine to buff	Medium grained	2.67	2.24	16.6	50	35.4	Passed	56	0.59	B,C
Monmouth Old Red Sandstone	Red to pinkish brown	Fine to medium grained	2.69	2.43	8.8	22	17.4	Failed	–	0.59	E,F

Source: Bell, 1992a.

corrected before any preservation treatment is undertaken. Furthermore, where new stonework is easily accessible it probably is better to maintain it in sound condition rather than use preservatives. Conversely, where access is difficult, the use of preservatives may be desirable.

There are two principal techniques used to preserve stone (Anon., 1975). The first is based on a change in chemistry of the stone when the preservative is introduced, so that the resultant material becomes more able to resist atmospheric attack. The second provides, by means of impregnation, water repellency in the form of a protective coating and a cement between dislodged grains. This consists of introducing monomers and polymers into the pores of a stone and allowing them to polymerize at a given depth within the stone.

Many of the water repellents are based upon silicone resins, siliconates and siliconesters. When water repellent treatments are applied to limestone they retard the solution of the stone by acidic rain water but their influence may last only for a few months, after which the rate of surface decay actually increases (Bell and Coulthard, 1990). The reason for this is that the preservative coating only forms a skin that penetrates the stone to a depth of 2 or 3 mm and this frequently promotes the crystallization of salts behind the skin, which cause the treated surface to blister and flake off.

Consequently, many of the new preservatives being developed are aimed at impregnating stone to much greater depths, for example, to depths of 50 mm or more, with minimum penetration of 25 mm. Not only does such deep penetration consolidate friable material but, hopefully, it immobilizes the salts that bring about decay by making the salts inaccessible to water or by making the stone more resistant to crystallization damage. Moreover, the concentration of soluble salts at depth is much less than it is near the surface. The frequency of damaging cycles of crystallization thereby is reduced or even eliminated by deep penetration of preservative.

Prevention of migration of soluble salts must be achieved without inhibiting the movements of vapour in stone. In this respect, the preservative should develop a lining around the grains, without filling the pores and their interconnections. This not only modifies the pore structure but it also enhances the tensile strength of stone, both of which increase the resistance to damage by salt crystallization.

In order to achieve adequate penetration a preservative must have a very low viscosity, a high surface tension and a low contact angle at the time of the treatment. In addition, the choice of preservative may be restricted by such properties as toxicity, flammability, water miscibility, vapour pressure and elastic modulus on curing. No preservative is ideal in every respect but viscosity, is second to cost as the governing factor. This requirement usually is obtained either by the use of monomers followed by *in situ* polymerization or by dissolving a resin in a solvent of low viscosity. The second method may suffer from the disadvantage of possible migration of the resin back to the surface as the solvent evaporates. Another disadvantage is that larger polymer molecules may be unable to enter the smallest pores of a stone.

Before any programme of stone treatment is carried out, it must be preceded by a survey of the prevailing environmental conditions, the structural condition of the building and the surface condition of the stone (Clarke and Ashurst, 1972). The environmental conditions should include the effects of prevailing winds, exposure to direct sunlight, exposure to rain, water run and drip effects, humidity patterns, local pollution level, proximity to heating outlets and nuisance from roosting birds. The structural survey should consider the presence of soft weathered material, failures due to face or edge bedding, diagnosis of other crack patterns, damage from impacts, identification of position and type of fixing, and type of any stiffening or reinforcing present. As far as the surface condition of the stone is concerned, the type and physical state of the stone(s), the cause and pattern of decay and soiling, the type of existing repairs and fillings (especially those associated with decay and any previous treatments), efflorescence, and any organic growths should be investigated. In addition, an estimate of the probable future course of decay should be made. A testing programme can eliminate the use of preservatives that prove unsatisfactory. When the use of a preservative would appear to be acceptable, it still is wise to carry out trials on site to check that there are no unforeseen effects such as a change in the appearance of the stone.

Some degree of cleaning almost always is necessary before preservation commences, and as such cleaning becomes an integral part of the process. In fact, it can have as great an effect on the final appearance as the method of treatment itself. Generally, preservatives are applied by brush or sprayed onto stonework. All preservatives penetrate a stone more successfully when it is dry.

The polymerization of monomers within a stone appears to offer a promising approach to stone preservation and has centred on three systems, namely, vinyl monomers, epoxies and alkoxysilanes. Methyl methacrylate (MMA) and butyl methacrylate are two vinyl monomers that have been investigated. The former is cheaper and penetrates slightly faster than the latter, which is less volatile but yields a more flexible polymer. They may be polymerized by heating (to above 80 °C) with an initiator, such as benzoyl peroxide. However, heating large masses of masonry is impracticable.

Most of the common epoxy resins available are far too viscous to impregnate the pores of stone, but a few compounds are available that contain two epoxide groups in a relatively small molecule that have low viscosities. These include 1:2, 3:4 diepoxybutane, diglycidyl ether and 1:4 butanediol diglycidyl ether cured with 1:8 diamino-p-methane. In order to lower the viscosity further, the preservative can be diluted with tetraethoxysilane or tetramethoxysilane. Unfortunately, the hardener can react with carbon dioxide to give a white efflorescence.

Some alkoxysilanes have very low viscosities (i.e. comparable with water), and can polymerize at outdoor temperatures within a stone before evaporation of the solvent brings the solution back to the surface. They are also colourless. Alkoxysilanes react with water to give a gel that coats the pores and then eventually hardens. Polymerization is preceded by hydrolysis. The water of hydrolysis may come from the moisture that is already present in the stone or else it may be added before application. In the latter case, the silane must be made miscible with water. Curing is dependent on the loss of alcohol by evaporation and may take several days. On polymerizing, the alkoxysilanes may produce materials that are chemically similar to some cements present in stones of sedimentary origin and therefore may be more stable in the long term. They are resistant to heat and moisture. Alkoxysilanes are intended to penetrate up to 50 mm or more, depending on the porosity of a stone, and its moisture and salt content. The object is to penetrate more deeply than the layer that is subjected to the normal cycle of wetting and drying.

There are several types of silanes available, most using the same resin base but with different catalysts and solvent mixtures for particular design characteristics. None of these treatments should be used in cases where there is severe contamination of stone with soluble salts (e.g. where there is rising damp or where the atmosphere is charged with sulphur dioxide). They must be applied to a dry surface; consequently, cleaning by an abrasive is preferable. However, silanes are expensive materials and large quantities are needed to achieve deep penetration. For instance, 5 litres per square metre is required to fill all the pores to a depth of 25 mm in a stone with a porosity of 20 per cent. The labour costs (e.g., one operative can treat approximately one square metre per day) also are high. Therefore, in terms of maintenance of plain walling, the costs are prohibitive, however, they are not unrealistic for the preservation of valuable architectural detail and statuary.

Brethane is a silane that has been used in Britain (Price, 1981). It is a colourless liquid with a viscosity comparable with water, so that it can penetrate stone to a depth in excess of 25 mm. It then sets as a gel that hardens over a few days to form a permanent protective coating of silicone polymer, both on the exposed surface and on the internal pore walls. As Brethane sets it gives off methanol and as a result vacates spaces within the pores, thereby leaving the stone permeable. The coating, which is inconspicuous, binds the friable decayed stone to sound stone, encases the salts that cause decay thereby rendering them harmless, and in the case of limestones, protects the stone from attack by acidic air pollutants. The stonework must be as dry as possible for deep penetration of Brethane. Other preparatory work includes cleaning, application of biocides, fracture grouting, pinning and mortar filling.

The lime method has been used to treat the limestone sculptures of certain historic churches in England such as on the west fronts of Wells and Exeter Cathedrals (Fig. 11.9). The stone first must be cleaned. Next, a hot quicklime poultice is applied to the stone. After about three weeks, the poultice is removed and during this time it is alleged that there is a superficial transfer of lime to the stone. Any old fillings must be removed and dirt or fine debris flushed away by clean water from hand-operated sprays. The cleaned surfaces and cavities then are coated with about 40 applications of limewater over several days in an attempt to consolidate the more friable zones. The consolidation treatment is followed by placing mortar repairs. All mortar repairs are based on high-calcium non-hydrated lime. The final stage of the lime method involves the application of the shelter coat, made from lime but with the consistency of thin cream. This is a thin surface coating that is applied to all the cleaned and repaired surfaces. It represents a sacrificial layer that slows down the rate of weathering. The lime method is surrounded by controversy in that some authorities doubt its effectiveness. For example, Price and Ross (1984) carried out a series of laboratory experiments that suggested that the use of a hot poultice has no influence on the removal of calcium sulphate contained at depth in stone, and they obtained variable results as far as its removal from the surface skin of stone is concerned. Furthermore,

Figure 11.9 Use of the lime method to treat statuary, notably the Twelve Apostles, on the west front of Wells Cathedral, Wells, Somerset, England.

they found no conclusive evidence to suggest that multiple applications of limewater serve to consolidate friable limestone, in other words it does not reduce its porosity and enhance its strength. Nevertheless, the results achieved by the use of the lime method at the West Front of Wells Cathedral produced a dramatic improvement in its appearance.

11.2 Roofing and facing materials

Rocks used for roofing purposes must possess a sufficient degree of fissility to allow them to split into thin slabs, as well as being durable and impermeable. Consequently, slate is one of the best roofing materials available and has been extensively used. Today, however, more and more tiles are being used for roofing, these being cheaper than stone, which has to be won and cut to size.

Slates are derived from argillaceous rocks that, because they were involved in major earth movements, were metamorphosed (Prentice, 1990). They are characterized by their cleavage, which allows the rock to break into thin slabs. Some slates, however, may possess a grain that runs at an angle to the cleavage planes and may tend to fracture along it. Therefore, the grain should run along its length when slate is used for roofing purposes. In Britain, slates have been worked for roofing purposes primarily in Wales. Welsh slates are differently coloured, they may be grey, blue, purple, red or mottled. Red slates contain more than twice as much ferric as ferrous oxide. A slate may be greenish coloured if the reverse is the case. Manganese is responsible for the purplish colour of some slates. Blue and grey slates contain little ferric oxide. The green coloured slates of the Lake District, England, are obtained from the Borrowdale Volcanic Series and, in fact, are cleaved tuffs. They are somewhat coarser grained than Welsh slates but more attractive. Other fissile rock types have been worked in England for roofing purposes such as the Colleyweston slates and the Stonesfield slates, which are sandy limestones of Jurassic age that were worked in Lincolnshire and the Cotswold Hills respectively (Smith, 1999). The Caithness Flagstones were used in the north of Scotland as roofing stones, their thickness and weight helping to resist the strong winds of that region.

The specific gravity of a slate is about 2.7–2.9 with an approximate density of 2.59 Mg m^{-3}. The maximum permissible water absorption of a slate is 0.37 per cent. Calcium carbonate may be present in some slates of inferior quality, which may result in them flaking and eventually crumbling upon weathering. Accordingly, a sulphuric acid test is used to test their quality. Top-quality slates, which can be used under moderate to severe atmospheric pollution conditions, reveal no signs of flaking, lamination or swelling after the test.

There is a large amount of wastage when explosives are used to quarry slate. Accordingly, they are sometimes quarried by using a wire saw. Waste slate can be used for floor tiles, paving or for stone walls. The slate, once won, is sawn into blocks, and then into slabs about 75 mm thick. These slabs are split into slate tiles by hand. Riven facing stones also are produced in the same way (Fig. 11.10).

An increasingly frequent method of using stone at the present day is as relatively thin slabs, applied as a cladding or facing to a building to enhance its appearance. Facing stone also provides a protective covering. Various thicknesses are used from 20 mm in the case of granite, marble and slate in certain positions at ground floor level, up to 40 mm at first floor level or above. If granite or syenite is used as a facing stone, then it should not be overdried, but should retain some quarry sap, otherwise it becomes too tough and hard to fabricate. As far as limestones and sandstones are concerned, the slabs are somewhat thicker, that is, varying between 50 and 100 mm. Because of their comparative thinness facing stones should not be

Figure 11.10 Splitting 'green slate' by hand, Broughton Moor, Cumbria, England.

too rigidly fixed otherwise differential expansion, due to changing temperatures, can produce cracking (Smith, 1999).

When fissile stones are used as facing stone and are given a riven or honed finish they are extremely attractive. Facing stones usually have a polished finish and are even more attractive, the polished finish enhancing the textural features of the stone. Polishing is accomplished by carborundum-impregnated discs that rotate over the surface of the stone, successively finer discs being used to produce the final finish (Fig. 11.11). The discs are cooled and lubricated by water. A flame-textured finish can be produced by moving a high temperature small flame across the flat surface of a stone, which causes the surface to spall thereby giving a rippled effect. Facing stones are almost self-cleansing.

Rocks used for facing stones have a high tensile strength in order to resist cracking. The high tensile strength also means that thermal expansion is not a great problem when slabs are spread over large faces.

Figure 11.11 Polishing limestone to be used as facing stone, Orton, Cumbria, England.

11.3 Armourstone

Armourstone refers to large blocks of rock that are used to protect civil engineering structures such as earth dams or vulnerable sections of coastlines or rivers. Large blocks of rock, which may be single size or more frequently widely graded (rip-rap), are used to protect the upstream face of dams against wave action, and in the construction of river training schemes, in river bank and bed protection and stabilization, as well as in the prevention of scour around bridge piers. In coastal engineering, armourstone is used in the construction of rubble mound breakwaters, for revetment covering embankments, for the protection of sea walls, and for rubble rock groynes. Indeed, breakwaters and sea defences represent a major use of armourstone.

11.3.1 Armourstone used in coastal engineering

The marine environment is one of the most aggressive in which construction takes place and so rock that is suitable for use in coastal engineering works is suitable when used in a protective or stabilizing role elsewhere. One of the most important factors as far as the use of armourstone in coastal engineering is concerned is its stability against wave action; accordingly block size and density are all important. Shape also is important since this affects how blocks interlock together. In addition, the rock material must be able to withstand rapid and severe changes in hydraulic pressure, alternating wetting and drying, thermal changes, wave and sand/gravel impact and abrasion, as well as salt and solution damage. Consequently, the size, grading, shape, density, water absorption, abrasion resistance, impact resistance, strength and durability of the rock material used for armourstone must be considered during the design stage of a particular project. The thickness of the protective layer and the need for a granular filter or geotextile beneath it depend on the design application, as well as the geotechnical and hydraulic conditions at the site (Thorne *et al.*, 1995).

Primary armourstone usually is specified by weight, a median weight of between 1 and 10 tonnes normally being required. However, blocks up to 20 tonnes may be required for breakwaters that will be subjected to large waves. As far as primary armourstone is concerned, the size of blocks determines the effectiveness of a breakwater, as well as the slope at which it can be maintained. The median weight of secondary armourstone and underlayer rock material may range upwards from 0.1 tonne. In the case of rip-rap used for revetment and river bank protection, the weight of the blocks required usually is less than 1.0 tonne and may grade down to 0.05 tonne. The size of blocks that can be produced at a quarry depends on the incidence of discontinuities and to a lesser extent on the method of extraction. Detailed discontinuity surveys can provide the data required for prediction of *in situ* block size and shape, and a description of a methodology using computer analysis of discontinuity data has been provided by Wang *et al.* (1991).

The armourstone on the seaward side of a breakwater is taken down to a depth of around 1.5 times the design significant wave height, below which the armourstone may be reduced in weight and/or the slope may be steepened. The size of the armourstone that is needed for the leeside and crest of a breakwater is governed to a large extent by the purpose of the breakwater. For instance, if a roadway, together with a wall to reflect waves, are constructed at the seaward crest of a breakwater, then the armourstone on the leeside of a breakwater may be reduced in size. By contrast, if no access is provided along a breakwater and it has been designed to be overtopped by severe waves, the armourstone on the leeside should be the same size as that on the seaward side.

The location of armourstone on a breakwater, according to Dibb *et al.* (1983), is an important factor that should be considered when making an assessment of rock durability. In this regard, Fookes and Poole (1981) recognized four zones, occurring one above the other, of a breakwater (Fig. 11.12). Zone I is the splash zone, which occurs above maximum sea level, and is subjected to some abrasion by sand and shingle carried by waves or wind, to salt action and to weathering. Zone II occurs above high water mark and is affected by wave upwash and abrasion by shingle, alternating wetting and drying, and weathering. Zone III is the intertidal zone and tends to suffer the most damage due to the abrasive action of waves, alternate wetting and drying, and possibly the effects of marine boring organisms. Lastly, zone IV is permanently submerged and usually is the least aggressive environment. The durability of rock involves its ability to resist breakdown during its working life, that is, to resist chemical decay and mechanical disintegration, including reduction in size and change of shape. The intrinsic properties of a rock such as its mineralogy, fabric, grain size, grain interlock, porosity, and in the case of sedimentary rocks, type and amount of cementation, all affect its resistance to breakdown. In addition, the amount of damage that armourstone undergoes is influenced by the action it has to face. For instance, the damage suffered by armourstone used on breakwaters depends on the type of waves (plunging or breaking), their height, period and duration, notably during storm conditions, on the one hand, and the slope and permeability of the structure on the other. Abrasion due to wave action is the principal reason for the reduction in size undergone by blocks of armourstone used in breakwaters, as well as rounding of their shape.

11.3.2 Selection of armourstone

The selection of a suitable source of rock for armourstone involves inspection and evaluation of the quarry or quarries concerned, assessment of the quality of the intact and processed

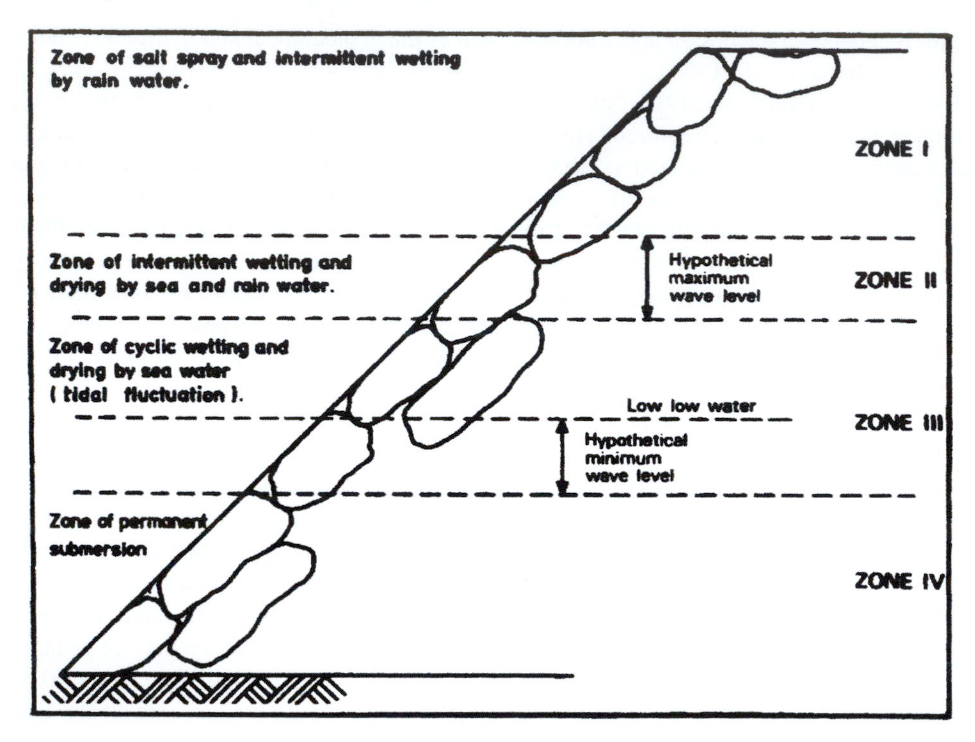

Figure 11.12 Four main weathering zones of the coastal marine environment.
Source: Fookes and Poole, 1981. Reproduced by kind permission of The Geological Society.

stone, which involves testing the processed stone, as well as consideration of the method of transporting and placing the stone. In this regard, Latham (1991) developed a rather complicated rating system for armourstone used in coastal engineering that took into consideration rock fabric strength, size of block, grading, initial shape, incident wave energy, zone of structure, meteorological climate effects, water-borne attrition agents, concentration of wave attack and mobility of armourstone in design concept (Table 11.6). Rock fabric strength refers to the intrinsic properties of the rock material, which in this case includes mineralogy, microtexture, weathering grade, resistance to weathering, abrasion and impact breakage. Latham (1993) suggested that the rock fabric strength could be assessed by a particular abrasion mill test. Unfortunately, this test equipment is of limited availability. Another rating system for assessing the suitability of armourstone was proposed by Lienhart in 1998.

Subsequently, Latham (1998) discussed the assessment and specification of armourstone quality, especially in relation to the CIRIA/CUR manual (Anon., 1991) on rock for coastal and shoreline engineering. The parameters included in this assessment are the rock type, its density, water absorption, degree of weathering and potential for further weathering, the potential for development of planes of weakness and the abrasion resistance. Latham gave a brief outline of the tests used to evaluate these parameters. In the case of resistance to weathering the recommended tests are the methylene blue test, the magnesium sulphate test, the water absorption test, the freeze thaw test and boiling to assess the Sonnenbrand effect (see sun-burnt basalts above). Methylene blue is used to detect the presence of swelling secondary clay minerals in rocks, which can mean that they breakdown rapidly (see slaking basalts above). The magnesian sulphate test is the crystallization test referred to above and is used

Table 11.6 Ratings estimates for factors in degradation model

Parameter	*Rating estimates*							*Parameter influence* X_{max}/X_{min}	*Quality of calibration*
(k_s)	Rock fabric strength								
	Use Mill Abrasion Test to plot W/W_0 versus revs, or select from plots of similar material tested in mill							>20	Very good
X_1	Size effect given by 0.5 $(W_{50})^{1/3}$ for W_{50} in tonnes								
	W_{50}		15	8	1	0.1	0.01		
	Rating		**1.23**	**1.0**	**0.5**	**0.23**	**0.11**	~ 10 (~2 armour)	Good
X_2	Grading								
	$(W_{85}/W_{15})^{1/3}$		1.1–1.4		1.4–2.5		2.5–4.0		
	Rating		**1.2**		**1.0**		**0.5**	~ 2.5	Fair
X_3	Initial shape								
	P_R (Asperity roughness)		>0.013 (irregular, tabular, elongate)		0.013–0.010 (equant)		<0.010 (semi-rounded rounded),		
	Rating		**1.0**		**1.5**		**2.0**	~ 2.0	Fair
X_4	Incident wave energy								
	Wave height								
	H_s (m)		>8		4–8		<4		
	Integrity of blocks	good	poor	good	poor	good	poor		
	Rating	**1**	**0.3**	**2**	**1**	**3**	**2**	~ 10	Poor
X_5	Zone of structure								
		supra-tidal hot	temperate	inter-tidal		submerged			
	Rating	**2.5**	**8**	**1**		**10**		~ 10	Good
X_6	Meteorological climate—effects of specific rock types and water absorption (W_{ab}%)								
	Hot + dry		Hot + humid		Freezing	winters	Temperate		
	$W_{ab}>2$	$W_{ab}<2$	basic	acidic	$W_{ab}>2$	$W_{ab}<2$	all		
	0.2	**0.5**	**0.2**	**0.8**	**0.5**	**1.0**	**1.0**	~ 5	Fair
X_7	Water-borne attrition agents								
		shingle	gravel	sand	silt	none			
	Rating	**0.2**	**0.5**	**1.0**	**1.2**	**1.5**		~ 7.5	Poor
X_8	Concentration of wave attack								
	Tidal range (m)		<2		2–6	>6			
	Seaward slope- (cot α)	<2.5	>3	<2.5	>3	<2.5	>3		
	Rating	**1**	**1.5**	**1.2**	**1.8**	**1.5**	**2.0**	~ 2.0	Fair
X_9	Mobility of armour in design concept								
	$H_s/\Delta D_{n50}$		1–3	4–6	6–20	(20–500)			
	Rating		**2**	**1**	**0.5**	**0.2**		~ 4(~10)	Fair

Source: Latham, 1991. Reproduced by kind permission of The Geological Society.

Note
W_{50} =median weight of block.

to assess salt crystallization, as well as frost damage. The water absorption test provides a means of determining the saturation density of stone. An acceptable value of water absorption for armourstone is less than 2 per cent. If a stone has a water absorption value of less than 0.5 per cent, then it is regarded as frost resistant. If not, then a freeze-thaw test can be carried out. Latham recognized two types of breakage, namely, that due to the presence of planes of weakness or cracks induced during blasting (type 1 breakage), and breakage caused by impact (type 2 breakage). Anon. (1991) recommended a drop test to assess the integrity of blocks of stone, the test subsequently being investigated by Latham and Gauss (1995). The fracture toughness test (Anon., 1988a) and the point load test (Anon., 1985a) were recommended by Anon. (1991) to assess type 2 breakage. Anon. (1991) also suggested that the aggregate impact strength test (Anon., 1975), in which the specimens were saturated, could be used to evaluate type 2 breakage, but Latham maintained that this test was not particularly suitable, especially in the case of coarse-grained rocks. The use of a specially developed mill abrasion test to determine the resistance to abrasion has been referred to above. Lastly, Latham pointed out that as special tests and specifications for armourstone are a relatively new development, existing national standards, which usually are based on aggregate tests, do not provide suitable test methods.

11.4 Crushed rock: concrete aggregate

Crushed rock is produced for a number of purposes, the chief of which is for concrete and road aggregate. As approximately 75 per cent of the volume of concrete consists of aggregate, its properties have a significant influence on the engineering behaviour of concrete. Aggregate is divided into coarse and fine varieties, the former usually consists of rock material that passes the 37 mm sieve and is retained on the 4.8 mm mesh sieve. The latter passes through this sieve and is caught on the 100-mesh sieve. Fines passing through the 200-mesh sieve should not exceed 10 per cent by weight of the aggregate.

11.4.1 Quarrying operations

The amount of overburden that has to be removed is an important factor in quarrying operations for if this increases, and is not useable, then a time comes when quarrying operations become uneconomic. In certain instances, however, rock may be mined. The removal of weak overburden usually is undertaken by scrapers and bulldozers, the material being disposed of in spoil dumps on site. Unfortunately, in the case of weathered overburden, weathered profiles frequently are not a simple function of depth below the surface and can be highly variable. Furthermore, in humid tropical areas in particular, weathered horizons may extend to appreciable depths. Accordingly, assessment of the amount of overburden that has to be removed can be more complicated. Indurated overburden may require drilling and blasting before being removed by dump trucks to the dump area. Spoil normally is used to backfill worked out areas as part of the restoration programme.

High explosives such as gelegnite, dynamite or trimonite are used in drillholes when quarrying crushed rocks. ANFO, a mixture of diesel oil and ammonium nitrate, also is used frequently. The holes are drilled at an angle of about 10–20° from vertical for safety reasons, and are usually located 3–6 m from the working face and a similar distance apart. Usually one, but sometimes two or more, rows of holes are drilled. The explosive does not occupy the whole length of a drillhole; lengths of explosive alternate with zones of stemming, which is commonly quarry dust or sand. Stemming occupies the top six or so metres of a drillhole. A single detonation

Figure 11.13 A large quarry in granite that is worked for aggregate in Hong Kong.

fires a cordex instantaneous fuse, which has been fed into each hole. It is common practice to have millisecond delay intervals between firing individual holes, in this way the explosions are complementary. The object of blasting is to produce stone with an acceptable fragmentation index, that is, of a size that is readily lifted by a front-wheeled loader into a dump truck. If this is not achieved, then large stone must be reduced in size by using a drop-ball or by secondary blasting. The height of the face largely depends upon the stability of the rock mass concerned. When the height of a working face begins to exceed 20–30 m it may be worked in tiers (Fig. 11.13).

After quarrying the rock is fed into a crusher and then screened to separate the broken rock material into different grade sizes (Fig. 11.14). Because crushed rock can be produced on a large scale by comparatively inexpensive operations it is relatively cheap. Fortunately, the raw materials that can be used for crushed rock generally are widespread, and indeed in large construction operations if the rock excavated is suitable, then it can be crushed and used as aggregate.

11.4.2 Aggregate properties

Hammersley (1989) noted that the petrography of a rock mass, involving field inspection, can be of value in any assessment of its potential suitability for use as aggregate. In addition, petrographic examination can indicate the presence of deleterious materials and defects.

The size and shape of aggregate particles are important properties, and are governed by the type of crusher plant on the one hand and the fracture pattern of a rock mass on the other. Rocks like basalts, dolerites, andesites, granites, quartzites and limestones tend to produce angular fragments when crushed. However, argillaceous limestones when crushed produce an excessive amount of fines, which influence the water requirement of concrete. The crushing characteristics of a sandstone depend upon the closeness of its texture, and the amount and type of cement. Angular fragments may produce a mix that is difficult to work, that is, it can be placed less easily and offers less resistance to segregation. Nevertheless, it is claimed that angular particles produce denser concrete. Rounded, smooth fragments produce workable

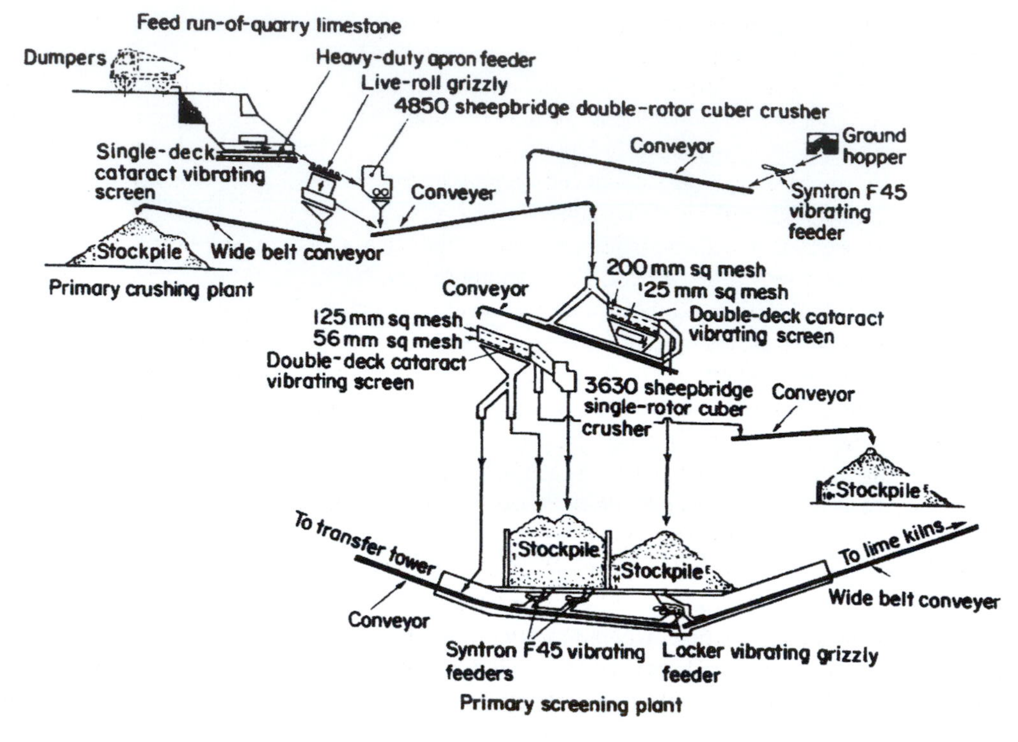

Figure 11.14 Flow sheet of the primary crushing and primary screening plant at a typical aggregate quarry.

mixes. The less workable the mix, the more sand, water and cement must be added to produce a satisfactory concrete. Fissile rocks such as those that are strongly cleaved, schistose, foliated or laminated have a tendency to split and, unless crushed to a fine size, give rise to tabular- or planar-shaped particles. Planar and tabular fragments not only make concrete more difficult to work but they also pack poorly and so reduce its compressive strength and bulk weight. Furthermore, such particles tend to lie horizontally in the cement thereby allowing water to collect beneath them, which inhibits the development of a strong bond on their under surfaces and so tends to reduce the strength of concrete. The requirements of the British Standard, BS 882 (Anon., 1992), are that the flakiness index (see Section 11.6.1.) should not exceed 40.

The surface texture of aggregate particles largely determines the strength of the bond between the cement and themselves (French, 1991). A rough surface creates a good bond whereas a smooth surface does not, thereby affecting strength accordingly.

According to Smith and Collis (1993), the particle density (i.e. specific gravity) of aggregate usually is not subject to limits in contract specifications except when, for design reasons, it is necessary to ensure a minimum concrete density, as for example, in concrete dams. Nonetheless, particle density does affect concrete mix design. Dense rock aggregate can be used for making wear-resistant concrete such as that used for roads. Generally, the bulk density of aggregates varies from 1.2 to 1.8 $Mg\,m^{-3}$. The bulk density commonly is used to enable concrete mixes specified according to volume to be converted into gravimetric proportions, thereby permitting batch mixes to be determined.

The porosity of an aggregate particle influences its specific gravity, water absorption, abrasion resistance, strength, elasticity and durability (Gillott, 1980). Water absorption can affect the strength, shrinkage/expansion and durability of concrete aggregate. Ideally, aggregate particles should possess water absorption values below 3 per cent. However, limits tend not to be specified unless water absorption is likely to adversely affect some undesirable property such as frost susceptibility (see Section 11.1.5) or volume change. For example, Bell and Haskins (1997) reported that the specified limit for water absorption of the coarse aggregate for the concrete for Katse Dam, Lesotho, was 2 per cent. This was to avoid the possibility of volume change due to presence of expansive clay minerals in the basalt material used as aggregate, which could have affected the performance of the concrete adversely.

It usually is assumed that shrinkage in concrete should not exceed 0.045 per cent, this taking place in the cement. However, basalt, gabbro, dolerite, mudstone and greywacke have been shown to be shrinkable, that is, they have relatively large wetting and drying movements of their own, so much so that they affect the total shrinkage of concrete. Clay and shale absorb water and are likely to expand if they are incorporated in concrete, and on drying they shrink, causing injury to the cement. Consequently, the proportion of clay material in a fine aggregate should not exceed 3 per cent. Granite, limestone, quartzite and felsite remain unaffected.

The strength of rock is influenced by many factors, in particular by the mineral composition and texture. In the case of sedimentary rocks, the size of the pores and the amount of pore space, together with the type and amount of cement also are important. The initial stages of weathering reduce the strength of a rock somewhat. Crushed rock aggregate normally has a higher strength than the concrete in which it forms part and so represents a significant factor in terms of the strength of concrete. For example, the unconfined compressive strength of rock used for aggregate generally ranges between 70 and 300 MPa. On drying cement shrinks. If the aggregate is strong, the amount of shrinkage is minimized and the cement-aggregate bond remains good.

Fookes (1980) regarded the soundness of an aggregate as its ability to resist volume changes due to temperature changes, freezing and thawing, expansion of salts in pore spaces, and moisture absorption. Aggregates that are physically unsound lead to the deterioration of concrete, inducing cracking, popping or spalling and therefore should be avoided. An aggregate soundness test (Anon., 1983b, 1989b) is basically a crystallization test, a sample of graded aggregate being subjected to a number of cycles of immersion in a saturated solution of magnesium or sodium sulphate after each of which it is oven dried. The maximum acceptable weight loss, according to Anon. (1983b), is 18 per cent after five cycles in magnesium sulphate solution, and 12 per cent after the same number of cycles in sodium sulphate solution. However, experience has shown that these limits do not always lead to satisfactory performance in aggressive environments such as those in which freeze-thaw activity or salt weathering is common.

11.4.3 Alkali aggregate reaction

As concrete sets, hydration takes place and alkalies (Na_2O and K_2O) are released. These react with siliceous material. Table 11.7 lists some of the reactive rock types. If any of these types of rock are used as aggregate in concrete made with high alkali cement the concrete is liable to expand and crack, thereby losing strength. The expansion can bring about misalignment of structures and can adversely affect their structural integrity. French (1980) discussed the nature of the reaction and provided a number of formulae that relate expansion to the proportion of reactive aggregate. When concrete is wet the alkalies that are released are dissolved by its water content and as the water is used up during hydration the alkalies are concentrated in the remaining liquid. This caustic solution attacks reactive aggregates to produce alkali-silica gels. The osmotic pressures developed by these gels as they absorb more water eventually may rupture the cement around reacting aggregate particles. The gels gradually occupy the cracks

Table 11.7 Rocks which react with high-alkali cements

Reactive rocks	Reactive component
Siliceous rocks	Opal
Opaline cherts	Chalcedony
Chalcedonic cherts	Chalcedony and/or opal
Siliceous limestones	
Volcanic rocks	Glass, devitrified glass and tridymite
Rhyolites and rhyolitic tuffs	
Dacites and dacitic tuffs	
Andesites	
Metamorphic rocks	Hydromica (illite)
Phyllites	
Miscellaneous rocks	
Any rocks containing veinlets, inclusions, coatings, or detrital grains of opal, chalcedony or tridymite. Quartz highly fractured by natural processes	

Source: McConnell *et al.*, 1950.

thereby produced and these eventually extend to the surface of the concrete. If alkali reaction is severe, a polygonal pattern of cracking develops on the surface (Fig. 11.15). According to Smith and Collis (1993), the reactivity of silica minerals depends primarily on the amount of disorder in the crystal structure. For instance, opal has a highly disordered structure and so is the most reactive form of silica whilst well-ordered unstrained quartz normally is unreactive.

Figure 11.15 Alkali aggregate reaction in concrete at a multi-storey car park, Plymouth, England.

The proportion of reactive silica has a major influence on the amount of expansion that occurs, the amount increasing with increasing content of reactive silica up to a maximum, beyond which further increase in reactive silica content causes a reduction in the amount of expansion. The reactive silica content at which maximum expansion takes place is referred to as the pessimum (French, 1980; Anon., 1988b) and appears to be governed by the type of reactive silica present and the character of the combination of aggregate. These troubles can be avoided if a preliminary petrological examination is made of the aggregate (McConnell *et al.*, 1950). In other words, material that contains over 0.25 per cent opal, over 5 per cent chalcedony, or over 3 per cent glass or crypto-crystalline acidic to intermediate volcanic rock, by weight, will be sufficient to produce an alkali reaction in concrete unless low alkali cement is used. The latter contains less than 0.6 per cent of Na_2O and K_2O. Two tests are available to investigate alkali reactivity of aggregate and can be used if a petrological examination of aggregate material indicates the likelihood of its occurrence. The mortar-bar method assesses the long-term expansion of concrete (Anon., 1987a) and the other assesses the chemical reactivity (Anon., 1987b). If aggregate contains reactive material surrounded by or mixed with inert matter a deleterious reaction may be avoided. The deleterious effect of alkali aggregate reaction also can be avoided if a pozzolan is added to the mix, the reaction taking place between it and the alkalis.

Reactivity may be related not just to composition but also to the percentage of strained quartz that a rock contains. For instance, Gogte (1973) maintained that rock aggregates containing 40 per cent or more of strongly undulatory or highly granulated quartz were highly reactive whilst those with between 30 and 35 per cent were moderately reactive. He also showed that basaltic rocks with 5 per cent or more secondary chalcedony or opal, or about 15 per cent palagonite showed deleterious reactions with high-alkali cements. Sandstones and quartzites containing 5 per cent or more chert behaved in a similar manner.

Alkali-silicate reaction refers to reactions that occur between phyllosilcate minerals and the caustic mix water of the concrete (Gillott, 1980). These minerals include mica, chlorite and vermiculite, so that slates, phyllites and schists may be suspect. However, these rocks frequently are unsuitable as aggregate because the material they produce when crushed has an unacceptable flakiness index. This type of alkali aggregate reaction also has occurred when greywacke has been used as aggregate. Again the main cause of expansion probably is swelling of alkali-silica gel formed when the finely divided silica in the suspect aggregate is attacked.

Certain argillaceous dolostones have been found to expand when used as aggregates in high alkali cement, thereby causing failure in concrete. This phenomenon has been referred to as alkali-carbonate rock reaction and an explanation has been provided by Gillott and Swenson (1969). They proposed that the expansion of such argillaceous dolostones in high alkali cements was due to the uptake of moisture by the clay minerals. This was made possible by dedolomitization caused by alkalis that provided access for moisture. Moreover, they noted that expansion only occurred when the dolomite crystals were less than 75 microns. Subsequently, Poole and Sotiropoulos (1980) reached similar conclusions.

11.5 Road aggregate

Aggregate constitutes the basic material for road construction and is quarried in the same way as aggregate for concrete. Because it forms the greater part of a road surface, aggregate has to bear the main stresses imposed by traffic such as slow crushing loads and rapid impact loads, and has to resist wear. The rock material used therefore should be fresh and have a high strength (Fookes, 1991). In addition, the aggregate used in the wearing course should be able to resist the polishing action of traffic. The aggregate in blacktop should possess good adhesion properties with bituminous binders. Accordingly, aggregate used as road metal must, as well as having a high strength, have a high resistance to impact and abrasion, polishing and skidding, and frost action. It also must be impermeable, chemically inert and possess a low coefficient of expansion.

A number of tests have been developed to evaluate the properties of rocks used as aggregates for road construction. The British Standard tests include the aggregate crushing test, the 10 per cent fines test, the aggregate impact test, the aggregate abrasion test and the test for the assessment of the polished stone value. Other tests of consequence are those for water absorption, specific gravity and density, and the aggregate shape tests (Anon., 1975). The aggregate shape tests are referred to in Section 11.6. The aggregate crushing value (ACV) measures the percentage of fines (i.e. the material less than 2.36 mm) produced when a sample of aggregate of specified size (i.e. less than 12.7 mm and greater than 9.5 mm) is subjected to a continuous load of 400 kN for 10 min (Anon., 1990). The 10 per cent fines value measures the load required to produce 10 per cent fines from a sample of aggregate. In both tests, the lower the result, the more resistant the aggregate is to crushing. The aggregate impact value (AIV) is derived by dropping a hammer of standard weight (13.5–14.1 kg) 15 times from a standard height of 381 ± 6.5 mm on to a sample of similarly sized aggregate. It is expressed as the weight of fines produced as a percentage of the original weight of the sample. The aggregate abrasion value (AAV) is the percentage weight loss of a sample of similarly sized aggregate that is fixed with epoxy resin to a small tray and subjected to 1000 revolutions of a disc rotating at approximately 28 rpm. The sample is loaded with a weight of 1.25 kg and standard sand is fed between the sample and the disc. Some typical values of roadstone properties are given in Table 11.8.

A further series of tests have been developed in the United States to assess the value of aggregate as roadstone. The Dorry test is carried out on a cylindrical sample of rock, 25 mm both in length and diameter, which is held against a revolving disc under a pressure of 2.5 MPa, the total load equalling 1.25 kg (Anon., 1981a). Standard crushed quartz, sized between 30 and 40 ASTM mesh screens, is fed onto the revolving disc. The loss in weight obtained by subjecting both ends to a total of 1000 revolutions gives the hardness index. The Los Angeles abrasion test subjects a graded sample to wear due to collision between rock pieces and the impact forces produced by an abrasive charge of steel spheres, the sample and charge being placed in a container and rotated at 30–33 rpm. If the aggregate particles are smaller than 38 mm, then they are subjected to 500 revolutions and to 1000 revolutions if they are larger than 19 mm (Anon., 1981b,c). After the test, the sample is shaken through a No. 12 US sieve

Table 11.8 Some representative values of the roadstone properties of some common aggregates

Rock type	*Water absorption*	*Specific gravity*	*Aggregate crushing value*	*Aggregate impact value*	*Aggregate abrasion value*	*Polished stone value*
Basalt	0.9	2.91	14	13	14	58
Dolerite	0.4	2.95	10	9	6	55
Granite	0.8	2.64	17	20	15	56
Micro-granite	0.5	2.65	12	14	13	57
Hornfels	0.5	2.81	13	11	4	59
Quartzite	1.8	2.63	20	18	15	63
Limestone	0.5	2.69	14	20	16	54
Greywacke	0.5	2.72	10	12	7	62

(approximately 1.7 mm aperture). The amount of wear is given as the loss in weight expressed as a percentage of the original weight. The Deval test is an attrition test in which 50 pieces of rock, weighing about 5 kg, are revolved in a cylinder 10 000 times, the value being given as the weight of fines less than 0.06 mm as a percentage of the original weight of the sample (Anon., 1968). This test is infrequently used today.

The properties of an aggregate are related to the texture and mineralogical composition of the rock from which it was derived. Most igneous and contact metamorphic rocks meet the requirements demanded of good roadstone. On the other hand, many rocks of regional metamorphic origin are either cleaved or schistose and therefore are unsuitable for roadstone. This is because they tend to produce flaky particles when crushed. Such particles do not achieve a good interlock and therefore impair the development of dense mixtures for surface dressing. The amount and type of cement and/or matrix material that bind grains together in a sedimentary rock influence roadstone performance.

The way in which alteration develops can strongly influence roadstone durability. Weathering may reduce the bonding strength between grains to such an extent that they are easily plucked from the aggregate. Chemical alteration, however, is not always detrimental to all the mechanical properties, indeed a small amount of alteration may improve the resistance of a rock to polishing. On the other hand, resistance to abrasion decreases progressively with increasing content of altered minerals, as does the crushing strength. The combined hardness of the minerals in a rock together with the degree to which they are cleaved, as well as the texture of the rock, also influence its rate of abrasion. The crushing strength is related to porosity and grain size, the higher the porosity and the larger the grain size, the lower the crushing strength.

One of the most important parameters of road aggregate is the polished stone value, which influences skid resistance. A skid resistant surface is one that is able to retain a high degree of roughness whilst in service. The rate of polish initially is proportional to the volume of traffic, and straight stretches of road are less subject to polishing than bends. The latter may polish up to seven times more rapidly. Stones are polished when fine detrital powder is introduced between tyre and road surface. Investigations have shown that detrital powder on a road surface tends to be coarser during wet, than dry periods. This suggests that polishing is more significant when the road surface is dry than wet, the coarser detritus more readily roughening the surface of stone chippings. An improvement in skid resistance can be brought about by blending aggregates.

Rocks within the same major petrological group may differ appreciably in their polished stone characteristics. The best resistance to polish occurs in rocks containing a proportion of softer alteration materials. Coarser grain size and the presence of cracks in individual grains also tends to improve resistance to polishing. In the case of sedimentary rocks, the presence of hard grains set in a softer matrix produces a good resistance to polish. Sandstones, greywackes and gritty limestones offer a good resistance to polishing, but unfortunately not all of them possess sufficient resistance to crushing and abrasion to render them useful in the wearing course of a road. Purer limestones show a significant tendency to polish. In igneous and contact metamorphic rocks a good resistance to polish depends on the variation in hardness between the minerals present.

The petrology of an aggregate determines the nature of the surfaces to be coated, the adhesion attainable depending on the affinity between the individual minerals and the binder, as well as the surface texture of the individual aggregate particles. If the adhesion between the aggregate and binder is less than the cohesion of the binder, then stripping may occur. Insufficient drying and the non-removal of dust before coating are, however, the principal causes of stripping. Acid igneous rocks generally do not mix well with bitumen as they have a poor ability to absorb it.

By contrast, basic igneous rocks such as basalts possess a high affinity for bitumen, as does limestone.

Igneous rocks are commonly used for roadstone. Dolerite has been used extensively. It has a high strength and resists abrasion and impact, but its polished stone value usually does not meet motorway specification in Britain, although it is suitable for trunk roads. Felsite, basalt and andesite are also much sought after. The coarse-grained igneous rocks such as granite are not generally so suitable as the fine-grained types, as they crush more easily. On the other hand, the very fine-grained and glassy volcanics are often unsuitable since when crushed they produce chips with sharp edges, and they tend to develop a high polish.

Igneous rocks with a high silica content resist abrasion better than those in which the proportion of ferromagnesian minerals is high; in other words acid rocks like rhyolites are harder than basic rocks such as basalts. Some rocks that are the products of thermal metamorphism such as hornfels and quartzite, because of their high strength and resistance to wear, make good roadstones. By contrast, many rocks of regional metamorphic origin, because of their cleavage and schistosity, are unsuitable. Coarse-grained gneisses give a similar performance to granites. Of the sedimentary rocks, limestone and greywacke are frequently used as roadstone. Greywacke, in particular, has a high strength, resists wear and develops a good skid resistance. Some quartz arenites are used, as are gravels. In fact the use of gravel aggregates is increasing.

11.6 Gravels and sands

11.6.1 Gravel

Gravel deposits usually represent local accumulations, for example, river channel fillings. In such instances they are restricted in width and thickness but may have considerable length. Fan-shaped deposits of gravels or aprons may accumulate at the snouts of ice masses, or blanket deposits may develop on transgressive beaches. The latter type of deposits are usually thin and patchy whilst the former are frequently wedge-shaped.

A gravel deposit consists of a framework of pebbles between which are voids. The voids are rarely empty, being occupied by sand, silt or clay material. River and fluvio-glacial gravels are notably bimodal, the principal mode being in the gravel grade, the secondary in the sand grade. Marine gravels, however, are often unimodal and tend to be more uniformly sorted than fluvial types of similar grade size.

The shape and surface texture of the pebbles in a gravel deposit are influenced by the agent responsible for its transportation and the length of time taken in transport, although shape also is dependent on the initial shape of the fragment, which in turn is controlled by the fracture pattern within the parental rock. The shape of gravel particles is classified in BS 812 (Anon., 1975) as rounded, irregular, angular, flaky and elongated. It also defines a flakiness index, an elongation index and an angularity number. The flakiness index of an aggregate is the percentage of particles, by weight, whose least dimension (thickness) is less than 0.6 times their mean dimension. The elongation index of an aggregate is the percentage, by weight, of particles whose greatest dimension (length) is greater than 1.8 times their mean dimension. The angularity number is a measure of relative angularity based on the percentage of voids in a compacted aggregate. The least angular aggregates are found to have about 33 per cent voids and the angularity number is defined as the amount by which the percentage of voids exceeds 33. The angularity number ranges from 0 to about 12. The same British Standard recognizes the following types of surface texture, namely, glassy, smooth, granular, rough, crystalline and honeycombed.

The composition of a gravel deposit reflects not only the type of rocks in the source area, but it is also influenced by the agent(s) responsible for its formation and the climatic regime in which it was or is being deposited. Furthermore, relief influences the character of a gravel deposit, for example, under low relief, gravel production is small and the pebbles tend to be chemically inert residues such as vein quartz, quartzite, chert and flint. By contrast, high relief and rapid erosion yield coarse immature gravels. Be that as it may, a gravel achieves maturity much more rapidly than does a sand under the same conditions. Gravels that consist of only one type of rock fragment are termed oligomictic. Such deposits are usually thin and well sorted. Polymictic gravels usually consist of a varied assortment of rock fragments and occur as thick poorly sorted deposits.

Gravel particles generally possess surface coatings that may be the result of weathering or may represent mineral precipitates derived from circulating groundwaters. The latter type of coating may be calcareous, ferruginous, siliceous or occasionally gypsiferous. Clay also may form a coating about pebbles. Surface coatings generally reduce the value of gravels for use as concrete aggregate, thick and/or soft and loosely adhering surface coatings are particularly suspect. Clay and gypsum coatings, however, can often be removed by screening and washing. Siliceous coatings tend to react with the alkalies in high alkali cements and are therefore detrimental to the concrete.

In a typical gravel pit, the material is dug from the face by a mechanical excavator. This loads the material into trucks or onto a conveyor that transports it to the primary screening and crushing plant. After crushing, the material is further screened and washed. This sorts the gravel into various grades and separates it from the sand fraction (Fig. 11.16). The latter usually is sorted into coarser and finer grades, the coarser is used for concrete and the finer is preferred

Figure 11.16 Screening sand from gravel at a pit in Derbyshire, England.

for mortar. Because gravel deposits are highly permeable, if the water table is high, then a gravel pit will flood. The gravels then have to be worked by dredging. Sea-dredged aggregates are becoming increasingly important.

11.6.2 Sand

The textural maturity of sand varies appreciably. A high degree of sorting coupled with a high degree of rounding characterizes a mature sand. The shape of sand grains, however, is not greatly influenced by length of transport. Maturity also is reflected in their chemical or mineralogical composition and it has been argued that the ultimate sand is a concentration of pure quartz. This is because the less stable minerals disappear due to mechanical or chemical breakdown during erosion and transportation or even after the sand has been deposited.

Sands are used for building purposes to give bulk to concrete, mortars, plasters and renderings. For example, sand is used in concrete to lessen the void space created by the coarse aggregate. A sand consisting of a range of grade sizes gives a lower proportion of voids than one in which the grains are of uniform size. Indeed, grading is probably the most important property as far as the suitability of sand for concrete is concerned. British Standard 882 (Anon., 1992) recognizes four grades of sand that can produce good-quality concrete. In any concrete mix, consideration should be given to the total specific surface of the coarse and fine aggregates, since this represents the surface that has to be lubricated by the cement paste to produce a workable mix. Poorly graded sands can be improved by adding the missing grade sizes to them, so that a high quality material can be produced with correct blending.

It is alleged that generally sand with rounded particles produces slightly more workable concrete than sand consisting of irregularly shaped particles. Sands used for building purposes are usually siliceous in composition and should be as free from impurities as possible. They should contain no significant quantity of silt or clay (i.e. less than 3 per cent by weight), since these need a high water content to produce a workable concrete mix. This, in turn, leads to shrinkage and cracking on drying. Furthermore, clay and shaley material tend to retard setting and hardening, or they may spoil the finished appearance of concrete. If sand particles are coated with clay they form a poor bond with cement and produce weaker and less durable concrete. The presence of feldspars in sands used in concrete has sometimes given rise to hair cracking, and mica and particles of shale adversely affect the strength of concrete. Organic impurities may adversely affect the setting and hardening properties of cement by retarding hydration, and thereby reduce its strength and durability. Organic and coaly matter also cause popping, pitting and blowing. If iron pyrite occurs in sand, then it gives rise to unsightly rust stains when used in concrete. The salt content of marine sands is unlikely to produce any serious adverse effects in good quality concrete although it probably will give rise to efflorescence. Salt can be removed by washing sand.

High grade quartz sands are used for making silica bricks used for refractory purposes. Glass sands must have a silica content of over 95 per cent (over 96 per cent for plate glass). The amount of iron oxides present in glass sands must be very low, in the case of clear glass under 0.05 per cent. Uniformity of grain size is another important property as this means that the individual grains melt in the furnace at approximately the same temperature.

11.6.3 Gravel and sand deposits

Scree material or talus accumulates along mountain slopes as a result of freeze-thaw action. Talus frequently is composed of one rock type. The rock debris has a wide range of size distribution and the particles are angular. Because scree simply represents broken rock material, then it is suitable for use as aggregate, if the parent rock is suitable. Such scree deposits, if large enough, only need crushing and screening and therefore are generally more economical to work than the parent rock.

The composition of a river gravel deposit reflects the rocks of its drainage basin. Sorting takes place with increasing length of river transportation, the coarsest deposits being deposited first, although during flood periods large fragments can be carried great distances. Thus, river deposits possess some degree of uniformity as far as sorting is concerned. Naturally, differences in gradation occur in different deposits within the same river channel but the gradation requirements for aggregate generally are met with or they can be made satisfactory by a small amount of processing. Moreover, as the length of river transportation increases softer material is progressively eliminated, although in a complicated river system with many tributaries new sediment is being added constantly. Deposits found in river beds usually are characterized by rounded particles. This is particularly true of gravels. River transportation also roughens the surfaces of pebbles.

River terrace deposits are similar in character to those found in river channels. The pebbles of terrace deposits may possess secondary coatings, due to leaching and precipitation. These are frequently of calcium carbonate, which does not impair the value of the deposit but if they are siliceous, then this could react with alkalies in high alkali cement and therefore could be detrimental to concrete. The longer the period of post-depositional weathering to which a terrace deposit is subjected, the greater is the likelihood of its quality being impaired.

Alluvial cones are found along valleys located at the foot of mountains. They are poorly stratified and contain rock debris with a predominantly angular shape and great variety in size.

Gravels and sands of marine origin are increasingly used as natural aggregate. The winnowing action of the sea leads to marine deposits being relatively clean and uniformly sorted. For the latter reason these sands may require some blending. The particles are generally well rounded with roughened surfaces. Gravels and sands that occur on beaches generally contain deleterious salts and therefore require vigorous washing. By contrast, much of the salt may have been leached out of the deposits found on raised beaches.

Wind-blown sands are uniformly sorted. They are composed predominantly of well-rounded quartz grains that have frosted surfaces.

Glacial deposits are poorly graded, commonly containing an admixture of boulders and rock flour. Furthermore, glacial deposits generally contain a wide variety of rock types and the individual rock fragments normally are subangular. The selective action of physical and chemical breakdown processes is retarded when material is entombed in ice and therefore glacial deposits often contain rock material that is unsuitable for use as aggregate. As a consequence, glacial deposits are usually of limited value as far as aggregate is concerned.

Conversely, fluvio-glacial deposits are frequently worked for this purpose. These deposits were laid down by melt waters that issued from bodies of ice. They take the form of eskers, kames and outwash fans (Fig. 11.17). The influence of water on these sediments means that they have undergone a varying degree of sorting. They may be composed of gravels or, more frequently, of sands. The latter are often well sorted and may be sharp, thus forming ideal building material.

11.7 Lime, cement and plaster

Lime is made by heating limestone, including chalk, to a temperature of between 1100 and 1200 °C in a current of air, at which point carbon dioxide is driven off to produce quicklime

Figure 11.17 A kame worked for sand, north of Lillehammer, Norway.

(CaO). Approximately 56 kg of lime can be obtained from 100 kg of pure limestone. Slaking and hydration of quicklime take place when water is added, giving calcium hydroxide. Carbonate rocks vary from place to place both in chemical composition and physical properties so that the lime produced in different districts varies somewhat in its behaviour. For example, dolostones also produce lime, however, the resultant product slakes more slowly than does that derived from limestones.

Portland cement is manufactured by burning pure limestone or chalk with suitable argillaceous material (clay, mud or shale) in the proportions 3 to 1. The raw materials are first crushed and ground to a powder, and then blended. They then are fed into a rotary kiln and heated to a temperature of over 1800 °C (Fig. 11.18). Carbon dioxide and water vapour are driven off and the lime fuses with the aluminium silicate in the clay to form a clinker. This is ground to a fine powder and less than 3 per cent gypsum is added to retard setting. Lime is the principal constituent of Portland cement but too much lime produces a weak cement. Silica constitutes approximately 20 per cent and alumina 5 per cent, both are responsible for the strength of the cement. A high content of the former produces a slow setting cement whilst a high content of the latter gives a quick setting cement. The percentage of iron oxides is low and in white Portland cement it is kept to a minimum. The proportion of magnesia (MgO) should not exceed 4 per cent otherwise the cement is unsound. Similarly, sulphate (SO_4) must not exceed 2.75 per cent. Sulphate resisting cement is made by the addition of a very small quantity of tricalcium aluminate to normal Portland cement.

When gypsum ($CaSO_4.nH_2O$) is heated to a temperature of 170 °C it loses three quarters of its water of crystallization, becoming calcium sulphate hemi-hydrate or plaster of Paris. Anhydrous calcium sulphate forms at higher temperatures. These two substances are the chief materials used in plasters. Gypsum plasters have now more or less replaced lime plasters.

Figure 11.18 A rotary kiln for producing Portland cement from chalk and clay, Gravesend, Kent, England.

11.8 Clay deposits and refractory materials

The principal clay minerals belong to the kandite, illite, smectite, vermiculite and palygorskite families. The kandites, of which kaolinite is the chief member, are the most abundant clay minerals. Deposits of kaolin or china clay are associated with granite masses that have undergone kaolinization. The soft china clay is excavated by strong jets of water under high pressure, the material being washed to the base of the quarry. This process helps separate the lighter kaolin fraction from the quartz. The lighter material is pumped to the surface of the quarry where it is fed into a series of settling tanks. These separate mica, which is itself removed for commercial use, from china clay. Washed china clay has a comparatively coarse size, only approximately 20 per cent of the constituent particles being below 0.01 mm in size, accordingly the material is non-plastic. Kaolin is used for the manufacture of white earthenwares and stonewares, in white Portland cement and for special refractories.

Ball clays are composed almost entirely of kaolinite and as between 70 and 90 per cent of the individual particles are below 0.01 mm in size, these clays have a high plasticity. Their plasticity at times is enhanced by the presence of montmorillonite. Ball clays contain a low percentage of iron oxide and consequently when burnt give a light cream colour. They are used for the manufacture of sanitary ware and refractories.

If a clay or shale can be used to manufacture refractory bricks it is termed a fireclay. Such material should not fuse below 1600 °C and should be capable of taking a glaze. Ball clays and china clays are in fact fireclays, fusing at 1650 and 1750 °C respectively, however, they are too valuable except for making special refractories. Most fireclays are highly plastic and contain kaolinite as their predominant material. Some of the best fireclays are found beneath coal seams, indeed in Britain fireclays are almost entirely restricted to strata of Coal Measures age.

The material in a bed of fireclay that lies immediately beneath a coal seam is often of better quality than that found at the base of the bed. Since fireclays represent fossil soils that have undergone severe leaching, they consist chiefly of silica and alumina, and contain only minor amounts of alkalies, lime and iron compounds. This accounts for their refractoriness (alkalies, lime, magnesia and iron oxides in a clay deposit tend to lower its temperature of fusion and act as fluxes). Very occasionally a deposit contains an excess of alumina and in such cases it possesses a very high refractoriness. After a fireclay has been quarried or mined it usually is left to weather for an appreciable period of time to allow it to breakdown before it is crushed. The crushed fireclay is mixed with water and moulded. Bricks, tiles and sanitary ware are made from fireclay.

Bentonite is formed by the alteration of volcanic ash, the principal clay mineral being either montmorillonite or beidellite. When water is added to bentonite it swells to many times its original volume to produce a soft gel. Bentonite is markedly thixotropic and this, together with its plastic properties, has given the clay a wide range of uses. For example, it is added to poorly plastic clays to make them more workable and to cement mortars for the same purpose. In the construction industry it is used as a material for clay grouting and for drilling mud, as well as in slurry trenches and diaphragm walls.

11.8.1 Evaluation of mudrocks for brickmaking

The suitability of a raw material for brick making is determined by its physical, chemical and mineralogical character and the changes that occur when it is fired. The unfired properties such as plasticity, workability (i.e. the ability of clay to be moulded into shape without fracturing and to maintain its shape when the moulding action ceases), dry strength, dry shrinkage and vitrification range are dependent upon the character of the source material. On the other hand, the fired properties such as colour, strength, total shrinkage on firing, porosity, water absorption, bulk density and tendency to bloat are controlled by the nature of the firing process. The price that can be charged for a brick depends largely upon its attractiveness, that is, its colour and surface appearance. The ideal raw material should possess moderate plasticity, good workability, high dry strength, total shrinkage on firing of less than 10 per cent and a long vitrification range. However, the suitability of a mudrock for brick manufacture can be determined only by running it through a production line or by pilot plant firing tests.

The mineralogy of the raw material influences its behaviour during the brickmaking process and hence the properties of the finished product (Bell, 1992a). Mudrocks consist of clay minerals and non-clay minerals, mainly quartz. The clay mineralogy varies from one deposit to another. Although bricks can be made from most mudrocks, the varying proportions of the different clay minerals have a profound effect on the processing and character of the fired brick. Those clays that contain a predominant clay mineral have a shorter temperature interval between the onset of vitrification and complete fusion, than those consisting of a mixture of clay minerals. This is more true of montmorillonitic and illitic clays than those composed chiefly of kaolinite. Also, those clays that consist of a mixture of clay minerals do not shrink as much when fired as those composed predominantly of one type of clay mineral. Mudrocks containing significant amounts of disordered kaolinite tend to have moderate to high plasticity and therefore are easily workable. They produce lean clays that undergo little shrinkage during brick manufacture. They also possess a long vitrification range and produce a fairly refractory product. However, mudrocks containing appreciable quantities of well-ordered kaolinite are poorly plastic and less workable. Illitic mudrocks are more plastic and less refractory than those in which disordered kaolinite is dominant, and fire at somewhat lower temperatures. Smectites are the most plastic and least refractory of the clay minerals. They show high shrinkage on drying since they require high proportions of added water to make them workable. As far as the unfired properties of the raw materials are concerned, the non-clay minerals present act mainly as a diluent, but they may be of considerable importance in relation to the fired properties. The non-clay material also may enhance the working properties, for instance, colloidal silica improves the workability by increasing plasticity.

The presence of quartz, in significant amounts, gives strength and durability to a brick. This is because during the vitrification period quartz combines with the basic oxides of the fluxes released from the clay minerals on firing to form glass, which improves the strength. However, as the proportion increases, the plasticity of the raw material decreases.

The accessory minerals in mudrocks play a significant role in brickmaking. The presence of carbonates is particularly important and can influence the character of the bricks produced. When heated above 900 °C carbonates break down yielding carbon dioxide and leave behind reactive basic oxides, particularly those of calcium and magnesium. The escape of carbon dioxide can cause lime popping or bursting if large pieces of carbonate, for example, shell fragments, are present, thereby pitting the surface of a brick. To avoid lime popping the material must be finely ground to pass a 20 mesh sieve. The residual lime and magnesia form fluxes that give rise to low viscosity silicate melts. The reaction lowers the temperature of a brick and hence, unless additional heat is supplied, lowers the firing temperature and shortens the range over which vitrification occurs. The reduction in temperature can result in inadequately fired bricks. If excess oxides remain in a brick it will hydrate on exposure to moisture, thereby destroying the brick. The expulsion of significant quantities of carbon dioxide can increase the porosity of bricks, reducing their strength. Engineering bricks must be made from raw material that has a low carbonate content.

Sulphate minerals in mudrocks are detrimental to brickmaking. For instance, calcium sulphate does not decompose within the range of firing temperature of bricks. It is soluble and, if present in trace amounts in a fired brick, causes effluorescence when the brick is exposed to the atmosphere. Soluble sulphates dissolve in the water used to mix the clay. During drying and firing they often form a white scum on the surface of a brick. Barium carbonate may be added to render such salts insoluble and so prevent scumming.

Iron sulphides, such as pyrite and marcasite, frequently occur in mudrocks. When heated in oxidizing conditions, the sulphides decompose to produce ferric oxide and sulphur dioxide. In the presence of organic matter oxidation is incomplete yielding ferrous compounds that combine with silica and basic oxides, if present, to form black glassy spots. This may lead to a black vitreous core being present in some bricks, which can reduce strength significantly. If the vitrified material forms an envelope around the ferrous compounds and heating continues until they decompose, then the gases liberated cannot escape causing bricks to bloat and distort. Under such circumstances the rate of firing should be controlled in order to allow gases to be liberated prior to the onset of vitrification. Too high a percentage of pyrite or other iron-bearing minerals gives rise to rapid melting that can lead to difficulties on firing.

Pyrite, and other iron-bearing minerals such as hematite and limonite, provide the iron that primarily is responsible for the colour of bricks. The presence of other constituents, notably calcium, magnesium or aluminium oxides, tends to reduce the colouring effect of iron oxide, whereas the presence of titanium oxide enhances it. High original carbonate content tends to produce yellow bricks.

Organic matter commonly occurs in mudrock. It may be concentrated in lenses or seams, or be finely disseminated throughout the mudrock. Incomplete oxidation of the carbon upon firing may result in black coring or bloating. In the latter case, even minute amounts of carbonaceous material can give black coring in dense bricks if it is not burned out. Black coring can be prevented by ensuring that all carbonaceous material is burnt out below the vitrification temperature. This means that if a raw material contains much carbonaceous material it may be necessary to admit cool air into the firing chamber to prevent the temperature rising too quickly. On the other hand, the presence of oily material in a clay can be an advantage in that it can reduce the fuel costs involved in brickmaking. For instance, the Lower Oxford Clay in parts of England contains a significant proportion of oil, so that when it is heated above approximately 300 °C it becomes almost self-firing.

Mineralogical and chemical information is essential for determining the brickmaking characteristics of a mudrock. Differential thermal analysis and thermogravimetric analysis can identify clay minerals in mudrocks but provide only very general data on relative abundance. X-ray diffraction methods are used to determine the relative proportions of clay and other minerals present (Fig. 11.19a). The composition of the clay minerals present also can be determined by plotting ignition loss against moisture absorption (Fig. 11.19b). The moisture absorption characterizes the type of clay mineral present while ignition loss provides some indication of the quantity present.

The presence of most of the accessory minerals can be determined by relatively simple analysis. Wet chemical methods can be used to determine total calcium, magnesium, iron and sulphur. The organic matter can be estimated by oxidation or calorimetric methods. Carbon and sulphur are important impurities, which are critical to the brickmaking process, and the quantities present are always determined.

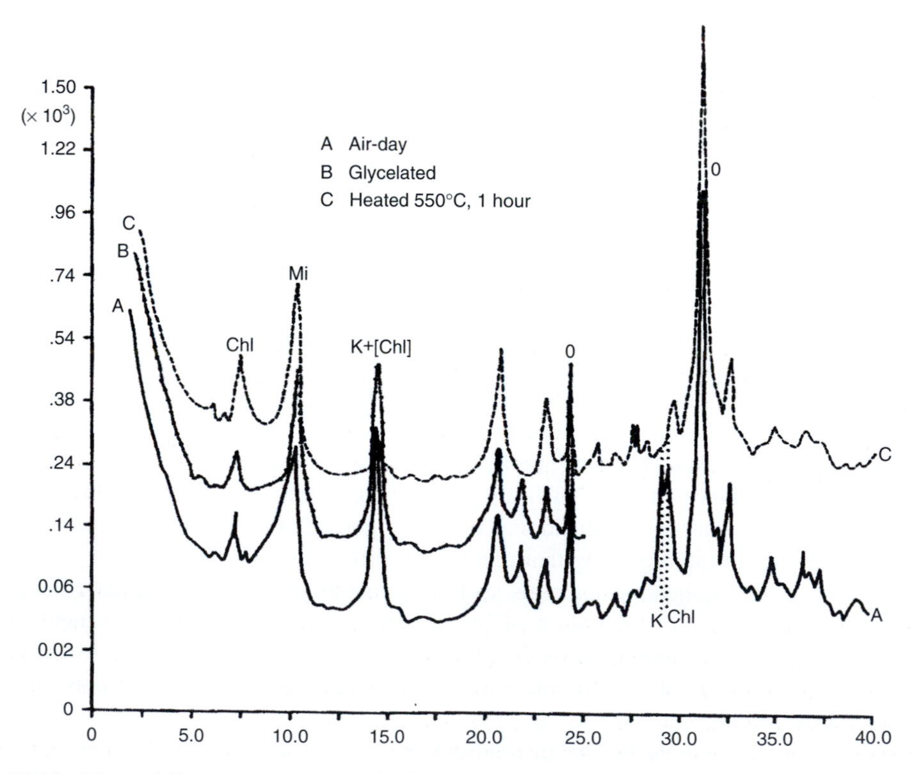

Figure 11.19a X-ray diffraction traces of a sample of mudrock used for brickmaking. Chl = chlorite; Mi, mica; K, kaolinite; Q, quartz.

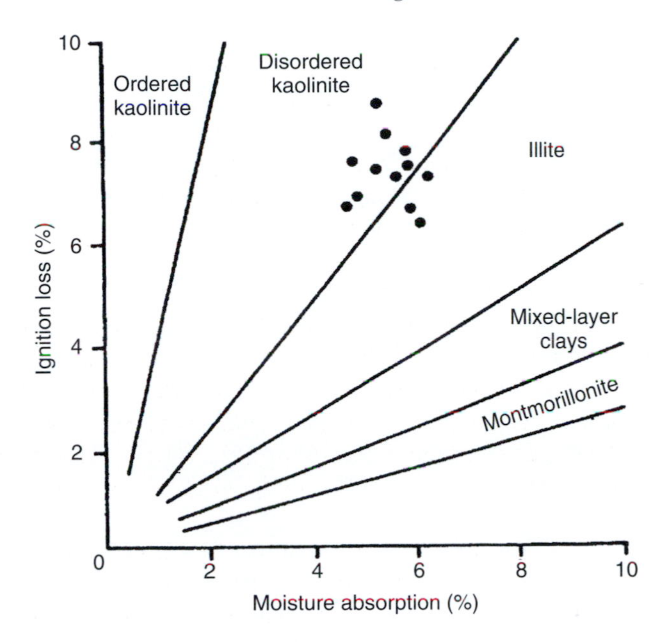

Figure 11.19b Clay mineral determination using Keeling's method.
Source: Bell, 1992b.

Physical tests can provide some indication of how the raw material will behave during brick manufacture. The particle size distribution affects the plasticity of the raw material and hence its workability. Plasticity is related to shrinkage and ignition loss during firing. It can be assessed in terms of the Atterberg limits or, less frequently, by means of the Pfefferkorn test. Shrinkage, loss on ignition and the temperature range for firing are determined from briquettes in a laboratory furnace. Other tests that are carried out in relation to brickmaking such as water absorption, bulk density, strength, liability to effluorescence are described in BS 3921 (Anon., 1985b).

Sufficient quantities of suitable raw material must be available at a site before a brickfield can be developed (Fig. 11.20). The volume of suitable mudrock must be determined, as must the amount of waste, that is, the overburden and unsuitable material within the sequence that is to be extracted (Bell, 1992b). The first stages of the investigation are topographical and geological surveys, followed by a drillhole programme. This leads to a lithostratigraphic and structural evaluation of the site. It also should provide data on the position of the water table and the stability of the slopes that will be produced during excavation of the brick pit.

11.8.2 Bricks

The nature of brick production has changed in the last 50 years. Most of the small units that produced bricks for local needs have closed so that brickmaking is now concentrated in large units that supply bricks over much wider areas and that require large sources of raw material. The technology of brickmaking also has changed in response to economic pressures that favour the concentration of brickmaking in large, highly mechanized units.

There are four main methods of brick production in Britain, namely, the wire-cut process, the semi-dry pressed method, the stiff plastic method and moulding by hand or machine. One of the distinguishing factors between these methods is the moisture content of the raw material

Figure 11.20 Continuous strip working of the Lower Oxford Clay for brickmaking near Peterborough, England.

when the brick is fashioned. This varies from as little as 10 per cent in the case of semi-dry pressed bricks to 25 per cent or more in hand-moulded bricks. Hence, the natural moisture content of a mudrock can have a bearing on the type of brickmaking operation. For example, many mudrocks have a natural moisture content in excess of 15 per cent and are therefore unsuitable for the semi-dry pressed or even the stiff plastic methods of production unless they are dried.

The first process in brick manufacture is to dig the raw material from the pit. It then usually is stored in heaps for a period of time to weather before being crushed, sieved, mixed with water and pressed to shape. Then the green bricks are dried before being placed in the kiln. Three stages can be recognized in brick burning. During the water smoking stage, which takes place up to approximately 600 °C, water is given off. Pore water and the water with which the clay was mixed is driven off at about 110 °C, whilst water of hydration disappears between 400 and 600 °C. The next stage is that of oxidation during which the combustion of carbonaceous matter and the decomposition of iron pyrite takes place, and carbon dioxide, sulphur dioxide and water vapour are given off. The final stage is that of vitrification. Above 800 °C the centre of the brick gradually develops into a highly viscous mass, the fluidity of which increases as the temperature rises. The components now are bonded together by the formation of glass. Bricks are fired at temperatures around 1000–1100 °C for about 3 days. The degree of firing depends on the fluxing oxides, principally H_2O, Na_2O, CaO and Fe_2O. Mica is one of the chief sources of alkalies in clay deposits. Because illites are more intimately associated with micas than kaolinites, illitic clays usually contain a higher proportion of fluxes and so are less refractory than kaolinitic clays.

The strength of a brick depends largely on the degree of vitrification it underwent. Theoretically, the strength of bricks made from mudrocks containing fine-grained clay minerals such as illite should be higher than those containing the coarser-grained kaolinite. Illitic clays, however, vitrify more easily and there is a tendency to underfire, particularly if they contain fine-grained calcite or dolomite. Kaolonitic clays are much more refractory and can stand harder firing, greater vitrification therefore is achieved.

Permeability of a brick also depends on the degree of vitrification. Mudrocks containing a high proportion of clay minerals produce less permeable products than those with a high proportion of quartz, but the former types of mudrocks may give a high drying shrinkage and high moisture absorption.

The colour of the mudrock prior to burning gives no indication of the colour it will have after leaving the kiln. Indeed, a chemical analysis can only offer an approximate guide to the colour of a finished brick. The iron content of clay, however, is important in this respect. For instance, as there is less scope for iron substitution in kaolinite than in illite, this often means that kaolinitic clays give a whitish or pale yellow colour on firing whilst illitic clays generally produce red or brown bricks. More particularly, a clay possessing about 1 per cent of iron oxides when burnt tends to produce a cream or light yellow colour, 2–3 per cent gives buff, and 4–5 per cent red.

Other factors, however, must be taken into account. For instance, a clay containing 4 per cent Fe_2O_3 under oxidizing conditions, burns pink below 800°C, turns brick red at about 1000°C and at 1150°C, as vitrification approaches completion, it adopts a deep red colour. By contrast, under reducing conditions ferrous silicate develops and the clay product has a blackish colour. Reducing conditions are produced if carbonaceous material is present in clay or they may be brought about by the addition of coal or sawdust to the clay before it is burnt. Blue bricks also are produced under reducing conditions. The clay should contain about 5 per cent iron together with lime and alkalies. An appreciable amount of lime in clay tends to lighten the colour of the burnt product, for example, 10 per cent of lime does not affect the colour at 800°C, but at higher temperatures, with the formation of calcium ferrites, a cream colour is developed. This occurs in clays with 4 per cent of Fe_2O_3 or less. The presence of manganese in clay may impart a purplish shade to the burnt product.

References

Anon. 1968. *Test for Abrasion of Rock by Use of the Deval Machine, ASTM D-233*. American Society Testing Materials, Philadelphia.

Anon. 1975. *Methods for Sampling and Testing Mineral Aggregates, Sands and Fillers, BS 812*. British Standards Institution, London.

Anon. 1977. The description of rock masses for engineering purposes. Working Party Report. *Quaterly Journal Engineering Geology*, **10**, 355–388.

Anon. 1979. Classification of rocks and soil for engineering geological mapping. Part 1 – Rock and soil materials. *Bulletin International Association Engineering Geology*, **19**, 364–371.

Anon. 1981a. *Abrasion of Rock by Use of the Dorry Machine, ASTM C-241*. American Society Testing Materials, Philadelphia.

Anon. 1981b. *Test for Resistance to Abrasion of Small Size Coarse Aggregate by Use of the Los Angeles Machine, C-131*. American Society Testing Materials, Philadelphia.

Anon. 1981c. *Test for Resistance to Degradation of Large Size Coarse Aggregate by Abrasion and Impact in the Los Angeles Machine, C-535*. American Society Testing Materials, Philadelphia.

Anon. 1983a. *The Selection of Natural Building Stone*. Digest 260, Building Research Establishment, Her Majesty's Stationery Office, London.

Anon. 1983b. *Test for Soundness of Aggregates by Use of Sodium Sulphate or Magnesium Sulphate, ASTM C88-83*. American Society Testing Materials, Philadelphia.

Anon. 1985a. Suggested method for determining the point load strength (revised version). ISRM Commission on Testing Methods. *International Journal Rock Mechanics, Mining Science and Geomechanical Abstracts*, **22**, 51–60.

Anon. 1985b. *Specification for Clay Bricks, BS 3921*. British Standards Institution, London.

Anon. 1987a. *Test for Potential Reactivity of Cement: Aggregate Combinations (Mortar-Bar Method), ASTM C227-87*. American Society Testing Materials, Philadelphia.

Anon. 1987b. *Test for Potential Reactivity of Aggregates (Chemical Method), ASTM C289-87*. American Society Testing Materials, Philadelphia.

Anon. 1988a. Suggested method for determining the fracture toughness of rock. ISRM Commission on Testing Methods. *International Journal Rock Mechanics, Mining Science and Geomechanical Abstracts*, **25**, 71–96.

Anon. 1988b. *Alkali Aggregate Reactions in Concrete*. Digest 330, Building Research Establishment, Her Majesty's Stationery Office, London.

Anon. 1989a. *The Selection of Natural Building Stone*. Digest 260, Building Research Establishment, Her Majesty's Stationery Office, London.

Anon. 1989b, 1990. *Testing Aggregates, BS 812: Parts 101–124*. British Standards Institution, London.

Anon. 1991. *Manual on the Use of Rock in Coastal and Shoreline Engineering*. Construction Industry Research and Information Association, Publication 83, London/ Centre for Civil Engineering Research, Codes and Specifications, Report 154, Gouda.

Anon. 1992. *Specification for Aggregates from Natural Sources for Concrete, BS 882*. British Standards Institution, London.

Arnold, A. 1981. Nature and reactions of saline minerals in walls. *Proceedings Conference on the Conservation of Stone II*, Bologna, 13–23.

Arnold, A. 1982. Rising damp and saline minerals. *Proceedings Fourth International Congress on the Deterioration and Preservation of Stone*. Louisville, 11–28.

Bell, F.G. 1992a. The durability of sandstone as building stone, especially in urban environments. *Bulletin Association Engineering Geologists*, **24**, 49–60.

Bell, F.G. 1992b. An investigation of a site in Coal Measures for brickmaking materials: an illustration of procedures. *Engineering Geology*, **32**, 39–52.

Bell, F.G. 1993. Durability of carbonate rock as building stone with comments on its preservation. *Environmental Geology*, **21**, 87–200.

Bell, F.G. and Coulthard, J.M. 1990. Stone preservation with illustrative examples from the United Kingdom. *Journal Environmental Geology and Water Science*, **16**, 75–81.

Bell, F.G. and Haskins, D.R. 1997. A geotechnical overview of Katse Dam and Transfer Tunnel, Lesotho, with a note on basalt durability. *Engineering Geology*, **46**, 175–198.

Bellanger, M., Homand, F. and Remy, J.M. 1993. Water behaviour in limestones as a function of pore structure: application to frost resistance of some Lorraine limestones. *Engineering Geology*, **36**, 99–108.

Building Effects Review Group. 1989. *The Effects of Acid Rain Deposition on Buildings and Building Materials in the United Kingdom*. Her Majesty's Stationery Office, London.

Butlin, R.N. 1988. Acid deposition and stone. *Structural Surveyor*, **7**, No. 3, 1–6.

Butlin, R.N., Cooke, R.U., Jaynes, S.M. and Sharp, A.D. 1985. Research on limestone decay in the United Kingdom. *Proceedings Fifth International Congress on Deterioration and Conservation of Stone*, Lausanne, 536–546.

Clarke, B.L. and Ashurst, J. 1972. *Stone Preservation Experiments*. Building Research Establishments, Her Majesty's Stationery Office, Watford.

Dibb, T.E., Hughes, D.W. and Poole, A.B. 1983. Controls on the size and shape of natural armourstone. *Quaterly Journal Engineering Geology*, **16**, 31–42.

Fookes, P.G. 1980. An introduction to the influence of natural aggregates on the performance and durability of concrete. *Quarterly Journal Engineering Geology*, **13**, 207–229.

Fookes, P.G. 1991. Geomaterials. *Quarterly Journal Engineering Geology*, **24**, 3–16.

Fookes, P.G. and Poole, A.B. 1981. Some preliminary considerations on the selection and durability of rock and concrete materials for breakwaters and coastal protection works. *Quarterly Journal Engineering Geology*, **14**, 97–128.

French, W.J. 1980. Reactions between aggregates and cement paste – an interpretation of the pessimum. *Quarterly Journal Engineering Geology*, **13**, 231–247.

French, W.J. 1991. Concrete petrography: a review. *Quarterly Journal Engineering Geology*, **24**, 17–48.

Gillott, J.E. 1980. Properties of aggregates affecting concrete in North America. *Quarterly Journal Engineering Geology*, **13**, 289–303.

Gillott, J.E. and Swenson, E.G. 1969. Mechanism of alkali carbonate reaction. *Quarterly Journal Engineering Geology*, **2**, 7–24.

Gogte, B.S. 1973. An evaluation of some common Indian rocks with special reference to alkali aggregate reactions. *Engineering Geology*, **7**, 135–154.

Hammersley, G.P. 1989. The use of petrography in the evaluation of aggregates. *Concrete*, **23**, No. 10.

Hart, D. 1988. *The Building Magnesian Limestones of the British Isles*. Building Research Establishment, Her Majesty's Stationery Office, Watford.

Haskins, D.R. and Bell, F.G. 1995. Drakensberg basalts: their alteration, breakdown and durability. *Quarterly Journal Engineering Geology*, **28**, 287–302.

Haynie, F.H. 1983. Deterioration of marble. *Durability of Building Materials*, **1**, 241–254.

Higgs, N.B. 1976. Slaking basalts. *Bulletin Association Engineering Geologists*, **13**, 151–162.

Honeyborne, D.S. 1982. *The Building Limestones of France*. Building Research Establishment. Her Majesty's Stationery Office, Watford.

Jaynes, S.M. and Cooke, R.U. 1987. Stone weathering in south east England. *Atmospheric Environment*, **21**, 1601–1622.

Jones, M.S., Horbury, A. and Thompson, G.E. 1998. Characterization of freshly quarried and decayed Doulting limestone. *Quarterly Journal Engineering Geology*, **31**, 325–331.

Larsen, T.D. and Cady, P.D. 1969. *Identification of Frost Susceptible Particles in Concrete Aggregates*, National Cooperative Research Program, Report 66, Highway Research Board, Washington DC.

Latham, J.-P. 1991. Rock degradation model for armourstone in coastal engineering. *Quarterly Journal Engineering Geology*, **24**, 101–118.

Latham, J.-P. 1993. A mill abrasion test for wear resistance of armourstone. In: *Rock for Erosion Control*, McElroy, C.H. and Lienhart, D.A. (eds), American Society Testing Materials, Philadelphia, 46–61.

Latham, J.-P. 1998. Assessment and specification of armourstone quality: from CIRIA/CUR (1991) to CEN (2000). In: *Advances in Aggregates and Armourstone Evaluation*. Engineering Geology Special Publication No. 13. Latham, J.-P. (ed.), Geological Society, London, 65–85.

Latham, J.-P. and Gauss, G.A. 1995. The drop test for armourstone integrity. In: *River, Coastal and Shoreline Protection*, Thorne, C.A.R., Abt, S.R., Barends, F.B.J., Maynard, S.T. and Pilarczky, K.W. (eds), Wiley, Chichester.

Leary, E. 1986. *The Building Sandstones of the British Isles*. Building Research Establishment. Her Majesty's Stationery Office, London.

Lienhart, D.A. 1998. Rock engineering rating system for assessing the suitability of armourstone sources. In: *Advances in Aggregates and Armourstone Evaluation*. Engineering Geology Special Publication No. 13. Latham, J.-P. (ed.), The Geological Society, London, 91–106.

Mamillan, M. 1976. Nouwelles connaissances pour l'utilisation et la protection des pierres de construction. *Annales de l'Institut Technique du Bâtiment et des Travoux Publics*, Serie Materiaux, No. 48, Supplément No. 335, 18–48.

McConnell, D., Mielenz, R.C., Holland, W.Y. and Greene, K.T. 1950. Petrology of concrete affected by cement aggregate reaction. In: *Application of Geology to Engineering Practice*, Paige, S. (ed.), Berkey Volume, Memoir American Geological Society, 222–250.

Moh'd, B.K., Howarth, R.J. and Bland, C.H. 1996. Rapid prediction of building research establishment limestone durability class from porosity and saturation, *Quarterly Journal Engineering Geology*, **29**, 285–297.

Ordoñez, S., Fort, R. and Garcia del Cura, M.A. 1997. Pore size distributon and the durability of a porous limestone. *Quarterly Journal Engineering Geology*, **30**, 221–230.

Poole, A.B. and Sotiropoulos, P. 1980. Reactions between dolomitic aggregate and alkali pore fluids in concrete. *Quarterly Journal Engineering Geology*, **13**, 281–287.

Prentice, J.E. 1990. *Geology of Construction Materials*. Chapman & Hall, London.

Price, C.A. 1981. Brethane stone preservatives. *Current paper CP1/81*, Building Research Establishment, Her Majesty's Stationery Office, Watford.

Price, C.A. and Ross, K.D. 1984. The cleaning and treatment of limestone by the lime method. Part II: A technical appraisal of stone conservation techniques employed at Wells Cathedral. *Monumentum*, **27**, 301–312.

Pye, K. and Mottershead, D.N. 1995. Honeycomb weathering of Carboniferous sandstone in a sea wall at Weston-super-Mare, UK. *Quaterly Journal Engineering Geology*, **28**, 333–347.

Reddy, M.M. 1988. Acid rain damage to carbonate stone: a quantitative assessment based on aqueous geochemistry of rainfall run-off from stone. *Earth Surface Processes and Landforms*, **13**, 335–354.

Reddy, M.M. and Werner, M. 1985. Composition of rainfall run-off from limestone and marble at Research Triangle Park, North Carolina. *United States Geological Survey Open-File Report*, 1–6, 85–630.

Sims, I. 1991. Quality and durability of stone for construction. *Quarterly Journal Engineering Geology*, **24**, 67–74.

Smith, B., Whalley, B. and Fassina, V. 1988. Elusive solution to monumental decay. *New Scientist*, 2nd June, 49–53.

Smith, M.R. (ed.). 1999. *Stone: Building Stone, Rock Fill and Armourstone in Construction*. Engineering Geology Special Publication No. 16, The Geological Society, London.

Smith, M.R. and Collis, L. (eds). 1993. *Aggregates: Sand, Gravel and Crushed Rock Aggregates for Construction Purposes*. Engineering Geology Special Publication No. 9, The Geological Society, London.

Thorne, C.R., Abt, S.R., Barends, F.B.J., Maynord, S.T. and Pilarczyk, K.W. 1995. *Coastal and Shoreline Portection*. Wiley, Chichester.

Trudgill, S.T. and Viles, H.A. 1998. Field and laboratory approaches to limestone weathering. *Quarterly Journal Engineering Geology*, **31**, 333–341.

Wang, H., Latham, J.-P. and Poole, A.B. 1991. Predictions of block size distribution for quarrying. *Quarterly Journal Engineering Geology*, **24**, 91–99.

Winkler, E.M. 1978. Stone decay in urban atmospheres. In: *Decay and Preservation of Stone*. Engineering Geology Case Histories No. 11, Winkler, E.M. (ed.), Geological Society America, New York, 53–58.

Yates, T.J.S., Coote, A.T. and Butler, R.N. 1988. The effect of acid deposition on buildings and building materials. *Construction Building Materials*, **2**, No. 1, 1–7.

Index